Biology of Plants

SIXTH EDITION

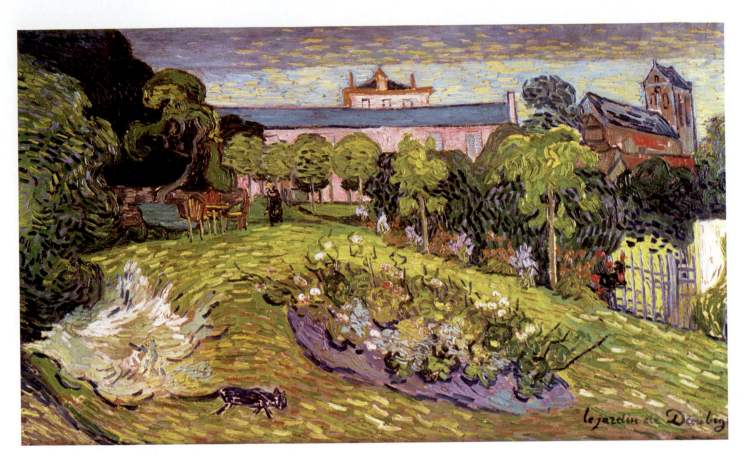

Vincent van Gogh (1853–1890) Dutch
Daubigny's Garden, 1890

Soon after his arrival in May 1890 in Auvers, France, Vincent van Gogh wrote to his brother Theo, "Some pictures are vaguely present in my mind; it will take time to get them clear, but that will come little by little." A month later, he painted a study of the garden of Charles-Francois Daubigny, a Barbizon painter much admired by van Gogh. The study, painted on a 20-inch-square linen towel, was followed by a larger version painted from nature (seen here) and still another version rendered in his studio.

Van Gogh described the final version: "Daubigny's garden, foreground of grass in green and pink. To the left a green and lilac bush and the stem of a plant with whitish leaves. In the middle a border of roses, to the right a wicket, a wall, and above the wall a hazel tree with violet foliage. Then a lilac hedge, a row of rounded yellow lime trees, the house itself in the background, pink, with a roof of bluish tiles. A bench and three chairs, a figure in black with a yellow hat and in the foreground a black cat. Sky pale green."

Behind Daubigny's house, the top floor of the Villa Ida is visible, and to the right is the Church of Auvers; the figure in the garden is the widowed Mme. Daubigny. Van Gogh completed his final painting of the garden in July 1890, only a few weeks before his death.

Biology of Plants

SIXTH EDITION

Peter H. Raven
*Missouri Botanical Garden and
Washington University, St. Louis*

Ray F. Evert
University of Wisconsin, Madison

Susan E. Eichhorn
University of Wisconsin, Madison

**W.H. FREEMAN AND COMPANY
WORTH PUBLISHERS**

Biology of Plants SIXTH EDITION

Manufactured in the United States of America

Library of Congress Catalog Card Number: 98-60171

ISBN: 1-57259-041-6
ISBN: 1-57259-611-2 (comp)

Printing: 1 2 3 4 5 — 02 01 00 99 98

Development Editor: Sally Anderson

Design: Malcolm Grear Designers

Art director: George Touloumes

Production editor: Deena Cloud

Production managers: Bernadine Richey, Sarah Segal

Photographs: John Miller and Townsend P. Dickinson

Line art: Rhonda Nass (section-opening paintings and line art), Kandis Elliot,
Shirley Baty, Carol Watkins, Jane Nass, Ellen Marie Dudley, and Demetrios Zangos

Typographer: New England Typographic Service

Prepress: Creative Graphic Services

Printing and Binding: R. R. Donnelley

Cover: Vincent van Gogh, *Daubigny's Garden* (detail), 1890, oil on canvas,
$21\frac{1}{4}$" $\times$ $39\frac{3}{4}$," Rudolf Staechelin Foundation, on extended loan to the
Öffentliche Kunstsammlung Basel

Illustration credits begin on page 915 and constitute an extension of the
copyright page

Examination copy information: 1-800-446-8923

W.H. Freeman and Company
41 Madison Avenue
New York, New York 10010 U.S.A.
www.whfreeman.com

Preface

As a man tramps the woods to the lake
he knows he will find ponds and lilies,
blue heron and golden shiners,
shadows on the rocks,
and the glint of light on the wavelets,
just as they were in the summer of 1354,
as they will be in 2054 and beyond.
He can stand on a rock by the shore
and be in a past he could not have known,
in a future he will never see.
He can be a part of time that was
and time yet to come.

　　　William Chapman White, *Adirondack Country*, 1954

Only a half-century ago, such was the perception that educated people enjoyed of the permanence of beloved natural settings, in this case, the Adirondack Mountains of northern New York State. Since it was written, we have learned of the extraordinary effects of human activity on climate and species, ponds and lilies included. In an astonishingly short period, climate changes could force plant species to migrate or to perish. Vastly increased human populations are altering nature as we have known it. Knowledge of the natural world, which depends ultimately on the botanical world, is all the more essential if we are to have any hope of a sustainable future.

It was with this in mind that we approached the writing of the sixth edition of *Biology of Plants*. If we are to be the proper stewards of this planet, we must know how and why it functions as it does. Throughout this new edition, therefore, we have strengthened the coverage of environmental issues, and we close the book with an essay on "A New Millennium: The Transition to Sustainability."

Major changes in our understanding of the botanical world and shifts in how we view its components have occurred since the last edition. Reactions to that edition, as well as previous editions, have been gratifying. However, as we scrutinized the existing chapters in light of the many advances that have been made in the plant biological sciences over the past eight years, it became obvious to us that this new edition would require a substantial level of revision. Extensive updating and rewriting has taken place, and virtually all of the chapters have been reorganized to provide an even more logical sequence. Indeed, this edition represents the most thorough revision that *Biology of Plants* has undergone. We believe it provides a solid and exciting botanical foundation for our entry into the twenty-first century.

Major reorganizational changes involved moving the chapter on The Molecular Composition of Plant Cells (now Chapter 2) before the chapter dealing with Introduction to the Plant Cell (now Chapter 3), which provides the chemical foundation for a more complete discussion of the structure and function of plant cells. A section on Secondary Metabolites has been added to Chapter 2 so that, early on in the book, students are made aware of the wide variety of familiar chemical products manufactured by plant cells. More emphasis on cellular functioning and cell-to-cell communication at the molecular level meant a new approach to the plasma membrane, particularly in Chapter 4 (Membrane Structure and Function), and new diagrams clarify chemiosmotic coupling and other features of Respiration (Chapter 6) and Photosynthesis (Chapter 7).

Consideration of cell division has been moved from the plant cell chapter to Section 3 (Genetics and Evolution) as Chapter 8 (The Reproduction of Cells), and the entire section has been expanded to provide a solid introduction to genetic engineering, population genetics, and speciation. Included are many advances based on the application of recombinant DNA techniques and procedures, which are discussed in detail.

The vast changes in taxonomic relationships, based on molecular sequencing, required a complete overhaul of the Diversity section. The introductory chapter to the Diversity section has been substantially enlarged to consider cladistic methods and molecular systematics, and with this edition we have incorporated the taxonomic scheme that recognizes the three domains of *Bacteria, Archaea,* and *Eukarya.* The new methods of genetic analysis have caused major shifts across organismal groups, with the most profound effects seen among the relationships of the protists.

New findings, especially those based on *Arabidopsis* research, have been integrated throughout. The most significant impact has been on portions of the Anatomy and Physiology chapters (Sections 5 and 6), where new discussions, micrographs, and diagrams have been added.

As with past revisions, we found explanations that would benefit from rewriting and artwork that could be improved by adding labels or clarifying captions. For this edition, we also added many new illustrations, especially in the early cellular, energetics, and genetics chapters. We were grateful to be able to adapt a number of well-conceived illustrations drawn by Shirley Baty for *Invitation to Biology*, Fifth Edition, by Helena Curtis and N. Sue Barnes, published by Worth Publishers.

After taking a hard look at the matter of accessibility, we added a number of pedagogical features to this edition. Each chapter now opens with an Overview that previews in general terms the content of the chapter. The Overview is followed by Checkpoints, a list of specific questions that help the student read with greater understanding. Within each chapter, subheadings have been reworded as full sentences, providing a series of guideposts through the material. Similar headings also appear in the end-of-chapter summary. Lists of key terms, with page references, allow students to test their vocabulary, and new end-of-chapter questions place the emphasis on critical thinking rather than memorization of facts. In all, the effect is a clearer, more cohesive presentation of the material in each chapter.

As always, we have appreciated the support and constructive recommendations made by teachers who used the last edition in their classes. We have also received substantial assistance from: Kay Robinson-Beers, who read the previous edition and made detailed recommendations regarding the incorporation of molecular biology; Wayne Becker, of the University of Wisconsin, who carefully read and reread the important and heavily revised chemistry, cell, energetics, and genetics chapters; Peter Crane, Vice President, Academic Affairs and Director of the Field Museum in Chicago, who scrutinized the Diversity chapters through several drafts, introduced a number of cladograms, and integrated important information concerning systematics and paleobotany; Linda Graham, of the University of Wisconsin, who helped significantly with the major revision of the protist and bryophyte chapters undertaken in this edition; Anthony Bleecker, also of the University of Wisconsin, who contributed enormously to the chapters on growth regulation and growth response; and Bob Evans, of Rutgers University, Camden, who skillfully wrote the Overviews, Checkpoints, and subheadings, in addition to reviewing every single chapter in various drafts of manuscript.

We also wish to express our sincere thanks to the following people, who provided critiques of chapters in the last edition or reviewed various drafts of revised manuscript:

Richard Amasino, *University of Wisconsin*
Michael Balick, *New York Botanical Garden*
Bruce Baldwin, *University of California, Berkeley*
Phillip Barak, *University of Wisconsin*
Dana Bergstrom, *University of Queensland*

Peter Bernhardt, *St. Louis University*
David Bilderback, *University of Montana*
Brian Boom, *New York Botanical Garden*
Peter Bretting, USDA-ARS, *Iowa State University*
Rita Calvo, *Cornell University*
Mark Chase, *Royal Botanic Gardens, Kew*
Joby Marie Chesnick, *Lafayette College*
Calvin Clyde, *Portland State University*
Nigel Crawford, *University of California, San Diego*
Joanne Dannenhoffer, *Central Michigan University*
Jerry Davis, *University of Wisconsin, Eau Claire*
James Doyle, *University of California, Davis*
Roland Dute, *Auburn University*
Peter Endress, *University of Zurich*
Donna Fernandez, *University of Wisconsin*
Ned Friedman, *University of Colorado*
Thomas Givnish, *University of Wisconsin*
Jo Handelsman, *University of Wisconsin*
Christopher Haufler, *University of Kansas*
D. L. Hawksworth, *CAB International, Mycological Institute, Surrey, UK*
Kent Holsinger, *University of Connecticut*
Robert Hunter, *University of Wisconsin*
L.C.W. Jensen, *University of Auckland, New Zealand*
William Jordan, *University of Wisconsin*
Arthur Kelman, *North Carolina State University*
Ken Kilborn, *Shasta College*
Wayne Kussow, *University of Wisconsin*
David Lee, *Florida International University*
Donald Les, *University of Connecticut*
O.A.M. Lewis, *University of Capetown*
Paul Ludden, *University of Wisconsin*
Brian McCarthy, *Ohio University*
Elliott Meyerowitz, *California Institute of Technology*
Brent Mishler, *University of California, Berkeley*
Steven Pallardy, *University of Missouri*
James W. Perry, *University of Wisconsin Center, Fox Valley*
Peter Quail, *The Plant Gene Expression Center*
Richard Robinson
Scott Russell, *University of Oklahoma*
Fred Sack, *Ohio State University*
Jozef Schell, *Max Planck Institute, Cologne*
Leslie Sieburth, *McGill University*
Beryl Simpson, *University of Texas, Austin*
Susan Singer, *Carleton College*
Edgar Spalding, *University of Wisconsin*
Michael Sussman, *University of Wisconsin*
Kenneth Sytsma, *University of Wisconsin*
Edith Taylor, *University of Kansas*
Jennifer Thorsch, *University of California, Santa Barbara*
Kenneth Todar, *University of Wisconsin*
Sue Tolin, *Virginia Polytechnic Institute and State University*
Kate VandenBosch, *Texas A & M*
Thomas Volk, *University of Wisconsin, La Crosse*
Warren Wagner, *University of Wisconsin*

Again, we are most grateful to Rhonda Nass for her exquisite paintings that grace each section opener and for the beautifully drawn new artwork, as well as for the illustrations she created for previous editions that we have continued to use, most notably the superb life cycles. Added to this edition are expertly rendered electronic illustrations by Kandis Elliot, who performed with uncommon skill and good humor.

We also wish to thank the following people at the University of Wisconsin: Lee Wilcox for providing the summarizing tables of protists; Mark Wetter and Theodore Cochrane for plant identification; Claudia Lipke for photographic work; and Sharon Pittman and Carri Van Ells for help with manuscript preparation.

As in the past, our work on the book took far more time and effort than we originally envisioned, and we wish to thank Mary Evert for her enthusiastic support and encouragement throughout the difficult and lengthy process, and Henry (Ike) Eichhorn for his patience and steady presence during the evenings and weekends that were devoted to work on the book.

Once again, we have relied on the capable people associated with Worth Publishers. Sally Anderson, our developmental editor, has worked with us on three editions now, and the association has been warm, supportive, and rewarding. We are enormously grateful for her many outstanding contributions at every stage of the process, from the early stages of planning the new edition through all of the manuscript stages to the finished book. Others who have made important contributions are most notably Deena Cloud, our dedicated project editor, as well as Sarah Segal, Bernadine Richey, George Touloumes, Michael Weinstein, Demetrios Zangos, Jennie Nichols, Lee Mahler, John Miller, and Yuna Lee. We extend our sincerest appreciation to Linda Strange, who was instrumental in improving the text and art throughout the copyediting and proofreading stages, and to Laura Evert for her skilled proofreading of galleys. We also wish to thank Susan Driscoll, President of Worth Publishers, for her enthusiasm and support from the beginning of her tenure until the present. And, with this edition, we have been grateful for the marketing efforts of the following people at W. H. Freeman: John Britch, Sara Tenney, Todd Elder, and Nicole Folchetti. The combined talents and hard work of an enormous number of people, only some of whom are mentioned here, have contributed in important and essential ways to the book you now hold in your hand, and we extend to them our sincerest appreciation.

Peter H. Raven
Ray F. Evert
Susan E. Eichhorn

Supplements

Student's *Raven PLUS+* CD-ROM and Web Site
The *Raven PLUS+* CD-ROM is organized by text chapter and features: (a) Plants and People: Sixteen essays linked to a key concept in selected chapters focus on topics of general interest, including: "Cellulase: A Bright Future in Faded Jeans," "The World's Largest Organism," "The Coffee Connection," "Kudzu: Way Too Much of a Good Thing," "Sweetness and Light: The Making of Maple Syrup," and "Murder Preserved: Tales of the Bogmen." (b) Video and Animations: Approximately thirty minutes of newly created three-dimensional animation illustrating the life cycle of the moss and pine, as well as well-conceived animations of photosynthesis, are supplemented with original live-action footage. (c) Drill and Practice: Numerous study aids, including multiple-choice quizzes, chapter outlines, drag-and-drop interactive exercises, and flashcards. (d) Annotated Web links: A collection of selected Web sites that illustrate concepts from each chapter. (e) Ongoing updates: The student version of *Raven PLUS+*, like the Instructor's version, can be updated via the Web.

Instructor's *Raven PLUS+* CD-ROM and Web Site
Raven PLUS+ is a unique multimedia tool that combines the audio-visual quality, interactive precision, and delivery speed of CD-ROM with the currency and spontaneity of the World Wide Web. It is also an electronic integrator, offering many existing supplements and new interactive materials in an easy-to-navigate, browserlike environment, all carefully coordinated around the book's contents and customizable around the instructor's electronic syllabus.

Transparencies
A set of 150 four-color acetate transparencies including all life-cycle drawings. (ISBN 1-57259-608-2)

Test Bank (printed)
Robert Evans, Rutgers University (ISBN 1-57259-607-4)

Computerized Test Bank
The Test Bank includes multiple-choice, true-false, and short-answer questions that can be used to test student understanding of all major topics in the textbook. The Computerized Test-Generation System helps instructors to use this database of questions to customize exams quickly and easily. (Windows ISBN: 1-57259-609-0; Macintosh ISBN: 1-57259-610-4)

Laboratory Topics in Botany
Ray F. Evert and Susan E. Eichhorn, University of Wisconsin, Madison
The lab manual offers several exercises within each topic that can be selected for coverage that suits individual course needs. Questions and problems follow each topic. This edition includes new topics and new exercises, with refinements and updating throughout. (ISBN 1-57259-605-8)

Preparation Guide for Laboratory Topics in Botany
Susan E. Eichhorn, James W. Perry, and Ray F. Evert, University of Wisconsin, Madison
The Preparation Guide offers helpful suggestions on sources of material, ordering schedules, laboratory set-up, and propagation and culturing methods. (ISBN 1-57259-606-6)

Contents in Brief

Chapter 1   Botany: An Introduction   1

SECTION 1   *The Biology of the Plant Cell*   16

Chapter 2   The Molecular Composition of Plant Cells   17

Chapter 3   Introduction to the Plant Cell   40

Chapter 4   Membrane Structure and Function   73

SECTION 2   *Energetics*   92

Chapter 5   The Flow of Energy   93

Chapter 6   Respiration   108

Chapter 7   Photosynthesis, Light, and Life   126

SECTION 3   *Genetics and Evolution*   154

Chapter 8   The Reproduction of Cells   155

Chapter 9   Meiosis and Sexual Reproduction   169

Chapter 10   Genetics and Heredity   183

Chapter 11   Gene Expression   207

Chapter 12   The Process of Evolution   235

SECTION 4   *Diversity*   260

Chapter 13   Systematics: The Science of Biological Diversity   261

Chapter 14   Prokaryotes and Viruses   281

Chapter 15   Fungi   306

Chapter 16   *Protista* I: Euglenoids, Slime Molds,
              Cryptomonads, Red Algae,
              Dinoflagellates, and
              Haptophytes   347

Chapter 17   *Protista* II: Heterokonts and
              Green Algae   370

Chapter 18   Bryophytes   400

Chapter 19   Seedless Vascular Plants   424

Chapter 20   Gymnosperms   466

Chapter 21   Introduction to the
              Angiosperms   495

Chapter 22   Evolution of the
              Angiosperms   517

SECTION 5   **The Angiosperm Plant Body: Structure
              and Development**   554

Chapter 23   Early Development of the
              Plant Body   555

Chapter 24   Cells and Tissues of the
              Plant Body   570

Chapter 25   The Root: Structure and
              Development   589

Chapter 26   The Shoot: Primary Structure
              and Development   610

Chapter 27   Secondary Growth in Stems   647

SECTION 6   **Physiology of Seed Plants**   672

Chapter 28   Regulating Growth and
              Development: The Plant
              Hormones   673

Chapter 29   External Factors and
              Plant Growth   702

Chapter 30   Plant Nutrition and Soils   726

Chapter 31   The Movement of Water and
              Solutes in Plants   750

SECTION 7   **Ecology**   772

Chapter 32   The Dynamics of Communities
              and Ecosystems   773

Chapter 33   Global Ecology   796

Chapter 34   The Human Prospect   823

Suggested Readings   851

Appendix A: Fundamentals of Chemistry   863

Appendix B: The Hardy–Weinberg Equation   877

Appendix C: Metric Table and Temperature
                Conversion Scale   879

Appendix D: Classification of Organisms   881

Glossary   889

Illustration Credits   915

Index   920

Geologic Eras   inside front cover

Contents

Chapter 1   Botany: An Introduction   1

Evolution of Plants   2
Evolution of Communities   9
Appearance of Human Beings   11
Summary   14

SECTION 1   The Biology of the Plant Cell   16

Chapter 2   The Molecular Composition of
Plant Cells   17

Organic Molecules   18
Carbohydrates   18
Essay: Representations of Molecules   19
Lipids   22
Essay: Amino Acids and Nitrogen   26
Proteins   26
Nucleic Acids   30
Secondary Metabolites   32
Summary   37

Chapter 3   Introduction to the Plant Cell   40

Development of the Cell Theory   41
Prokaryotic Cells and Eukaryotic Cells   41
Essay: Viewing the Cellular World   42
The Plant Cell: An Overview   45
Plasma Membrane   46
Essay: Cytoplasmic Streaming in Giant Algal Cells   47
Nucleus   47

Chloroplasts and Other Plastids   48
Mitochondria   52
Peroxisomes   53
Vacuoles   54
Oil Bodies   55
Ribosomes   55
Endoplasmic Reticulum   56
Golgi Complex   57
Cytoskeleton   58
Flagella and Cilia   60
Cell Wall   61
Essay: Cell Theory versus Organismal Theory   65
Plasmodesmata   66
Summary   68

Chapter 4   Membrane Structure
and Function   73

Structure of Cellular Membranes   74
Movement of Water and Solutes   76
Cells and Diffusion   78
Essay: Imbibition   80
Osmosis and Living Organisms   80
Transport of Solutes across Membranes   82
Essay: Patch-Clamp Recording in the Study of
Ion Channels   83
Vesicle-Mediated Transport   85
Cell-to-Cell Communication   86
Summary   89

SECTION 2   Energetics   92

Chapter **5**   The Flow of Energy   93

The Laws of Thermodynamics   95
Oxidation-Reduction   98
Enzymes   99
Cofactors in Enzyme Action   102
Metabolic Pathways   103
Regulation of Enzyme Activity   103
The Energy Factor: ATP Revisited   105
Summary   106

Chapter **6**   Respiration   108

An Overview of Glucose Oxidation   109
Glycolysis   109
The Aerobic Pathway   113
Essay: Bioluminescence   122
Other Substrates for Respiration   122
Anaerobic Pathways   122
The Strategy of Energy Metabolism   123
Summary   124

Chapter **7**   Photosynthesis, Light, and Life   126

Photosynthesis: A Historical Perspective   127
The Nature of Light   128
Essay: The Fitness of Light   130
The Role of Pigments   130
The Reactions of Photosynthesis   133
Essay: Chemiosmotic Coupling in Chloroplasts and
      Mitochondria   137
The Carbon-Fixation Reactions   139
Essay: The Carbon Cycle   150
Summary   152

SECTION 3   *Genetics and Evolution*   154

Chapter **8**   The Reproduction of Cells   155

Cell Division in Prokaryotes   156
Cell Division in Eukaryotes   156
The Cell Cycle   157
Interphase   158
Cell Division in Plants   158
Essay: Immunofluorescence Microscopy   159
Cell Division and the Reproduction of the Organism   166
Summary   166

Chapter **9**   Meiosis and Sexual Reproduction   169

Haploid and Diploid   170
Meiosis, the Life Cycle, and Diploidy   171
The Process of Meiosis   172
The Phases of Meiosis   173
Asexual Reproduction: An Alternative Strategy   179
Essay: Vegetative Reproduction:
      Some Ways and Means   180
Advantages of Sexual Reproduction   180
Summary   181

Chapter **10**   Genetics and Heredity   183

The Concept of the Gene   184
The Principle of Segregation   185
Essay: Mendel and the Laws of Probability   188
The Principle of Independent Assortment   188
Discovery of the Chromosomal
      Basis of Mendel's Laws   189
Linkage   190
Mutations   192
Broadening the Concept of the Gene   194
The Chemical Basis of Heredity   197
The Chemistry of the Gene: DNA versus Protein   197
The Structure of DNA   197
DNA Replication   201
The Problem of the Ends of Linear DNA   203
The Energetics of DNA Replication   203
DNA as a Carrier of Information   204
Summary   204

Chapter **11**   Gene Expression   207

From DNA to Protein: The Role of RNA   208
The Genetic Code   209
Protein Synthesis   210
Regulating Gene Expression   214
The Prokaryotic Chromosome   216
The Eukaryotic Chromosome   218
Regulation of Gene Expression in Eukaryotes   220
The DNA of the Eukaryotic Chromosome   221
Transcription and Processing of mRNA in Eukaryotes   223

Recombinant DNA Technology   224
Essay: Arabidopsis thaliana: The Model Plant   228
Summary   231

Chapter 12   The Process of Evolution   235

Darwin's Theory   237
The Concept of the Gene Pool   239
The Behavior of Genes in Populations:
    The Hardy-Weinberg Law   239
The Agents of Change   240
Preservation and Promotion of Variability   242
Responses to Selection   243
The Result of Natural Selection: Adaptation   245
The Origin of Species   248
How Does Speciation Occur?   249
Essay: Adaptive Radiation in Hawaiian Tarweeds   250
Maintaining Reproductive Isolation   256
The Origin of Major Groups of Organisms   256
Summary   258

SECTION 4   *Diversity*   260

Chapter 13   Systematics: The Science of
                  Biological Diversity   261

Taxonomy and Hierarchical Classification   262
Classification and Phylogeny   264
Essay: Convergent Evolution   266
Methods of Classification   267
Molecular Systematics   268
The Major Groups of Organisms: *Bacteria, Archaea,*
    and *Eukarya*   271
Origin of the Eukaryotes   272
The Eukaryotic Kingdoms   275
Summary   279

Chapter 14   Prokaryotes and Viruses   281

Characteristics of the Prokaryotic Cell   282
Diversity of Form   284
Reproduction and Gene Exchange   285
Endospores   286
Metabolic Diversity   286
Bacteria   287
Archaea   294
Viruses   296
Viroids: Other Infectious Particles   302
The Origin of Viruses   303
Summary   303

Chapter 15   Fungi   306

The Importance of Fungi   307
Biology and Characteristics of Fungi   310
Evolution of the Fungi   312
Chytrids: Phylum *Chytridiomycota*   312
Phylum *Zygomycota*   313
Essay: Phototropism in a Fungus   315
Phylum *Ascomycota*   317
Phylum *Basidiomycota*   320
Yeasts   330
Deuteromycetes   332
Essay: Predaceous Fungi   333
Symbiotic Relationships of Fungi   334
Essay: From Pathogen to Symbiont:
    Fungal Endophytes   335
Summary   344

Chapter 16   *Protista* I: Euglenoids, Slime Molds,
                  Cryptomonads, Red Algae,
                  Dinoflagellates, and
                  Haptophytes   347

Ecology of the Algae   348
Euglenoids: Phylum *Euglenophyta*   350
Plasmodial Slime Molds: Phylum *Myxomycota*   352
Cellular Slime Molds: Phylum *Dictyosteliomycota*   354
Cryptomonads: Phylum *Cryptophyta*   356
Red Algae: Phylum *Rhodophyta*   357
Dinoflagellates: Phylum *Dinophyta*   361
Essay: Red Tides/Toxic Blooms   364
Haptophytes: Phylum *Haptophyta*   366
Summary   368

Chapter 17   *Protista* II: Heterokonts
                  and Green Algae   370

The Heterokonts   371
Oomycetes: Phylum *Oomycota*   371
Diatoms: Phylum *Bacillariophyta*   375
Chrysophytes: Phylum *Chrysophyta*   378

Brown Algae: Phylum *Phaeophyta*   379
Essay: Algae and Human Affairs   380
Green Algae: Phylum *Chlorophyta*   383
Summary   398

Chapter **18** Bryophytes   400

The Relationships of Bryophytes to Other Groups   401
Comparative Structure and Reproduction
    of Bryophytes   403
Liverworts: Phylum *Hepatophyta*   407
Hornworts: Phylum *Anthocerophyta*   412
Mosses: Phylum *Bryophyta*   412
Summary   423

Chapter **19** Seedless Vascular Plants   424

Evolution of Vascular Plants   425
Organization of the Vascular Plant Body   425
Reproductive Systems   430
The Phyla of Seedless Vascular Plants   431
Phylum *Rhyniophyta*   432
Phylum *Zosterophyllophyta*   434
Phylum *Lycophyta*   435
Phylum *Trimerophytophyta*   443
Phylum *Psilotophyta*   443
Phylum *Sphenophyta*   445
Phylum *Pterophyta*   449
Essay: Coal Age Plants   456
Summary   463

Chapter **20** Gymnosperms   466

Evolution of the Seed   468
Progymnosperms   470
Extinct Gymnosperms   472
Living Gymnosperms   472
Phylum *Coniferophyta*   474
The Other Living Gymnosperm Phyla:
    Cycadophyta, Ginkgophyta, and *Gnetophyta*   486
Essay: Wollemia nobilis: A Newly Discovered
    Living Fossil   488
Summary   493

Chapter **21** Introduction to the Angiosperms   495

Diversity in the Phylum *Anthophyta*   496
The Flower   498
The Angiosperm Life Cycle   503
Essay: Hay Fever   508
Summary   514

Chapter **22** Evolution of the Angiosperms   517

Relationships of the Angiosperms   518
Origin and Diversification of the Angiosperms   519
Evolution of the Flower   524
Essay: An Ambiguous Aquatic Plant   526
Evolution of Fruits   543
Biochemical Coevolution   549
Summary   552

SECTION 5   *The Angiosperm Plant Body:*
Structure and Development   554

Chapter **23** Early Development of
    the Plant Body   555

Formation of the Embryo   556
The Mature Embryo and Seed   562
Essay: Wheat: Bread and Bran   563
Requirements for Seed Germination   564
From Embryo to Adult Plant   566
Summary   568

Chapter **24** Cells and Tissues of
    the Plant Body   570

Apical Meristems and Their Derivatives   571
Growth, Morphogenesis, and Differentiation   571
Internal Organization of the Plant Body   572
Ground Tissues   573
Vascular Tissues   576
Dermal Tissues   583
Summary   588

Chapter 25   The Root: Structure and
Development   589

Root Systems   590
Origin and Growth of Primary Tissues   591
Primary Structure   595
Effect of Secondary Growth on the Primary Body
    of the Root   600
Origin of Lateral Roots   603
Aerial Roots and Air Roots   603
Essay: Getting to the Root of Organ Development   604
Adaptations for Food Storage: Fleshy Roots   606
Summary   608

Chapter 26   The Shoot: Primary
Structure and Development   610

Origin and Growth of the Primary Tissues
    of the Stem   611
Primary Structure of the Stem   614
Essay: Plants, Air Pollution, and Acid Rain   616
Relation between the Vascular Tissues of the
    Stem and the Leaf   622
Morphology of the Leaf   624
Essay: Leaf Dimorphism in Aquatic Plants   626
Structure of the Leaf   626
Grass Leaves   632
Development of the Leaf   632
Sun and Shade Leaves   636
Leaf Abscission   636
Transition between Vascular Systems of the Root
    and the Shoot   636
Development of the Flower   639
Stem and Leaf Modifications   641
Summary   644

Chapter 27   Secondary Growth in Stems   647

Annuals, Biennials, and Perennials   648
The Vascular Cambium   648
Effect of Secondary Growth on the
    Primary Body of the Stem   650
The Wood: Secondary Xylem   659
Essay: The Truth about Knots   668
Summary   670

SECTION 6   *Physiology of Seed Plants*   672

Chapter 28   Regulating Growth and
Development: The Plant
Hormones   673

Auxins   675
Cytokinins   679
Ethylene   682
Abscisic Acid   683
Gibberellins   684
The Molecular Basis of Hormone Action   687
Plant Biotechnology   693
Essay: Totipotency   695
Summary   700

Chapter 29   External Factors and Plant Growth   702

The Tropisms   703
Circadian Rhythms   706
Photoperiodism   709
Chemical Basis of Photoperiodism   711
Hormonal Control of Flowering   715
Genetic Control of Flowering   716
Dormancy   717
Cold and the Flowering Response   719
Nastic Movements   719
Generalized Effects of Mechanical Stimuli on Plant Growth
    and Development: Thigmomorphogenesis   721
Solar Tracking   722
Summary   723

Chapter 30   Plant Nutrition and Soils   726

Essential Elements   727
Functions of Essential Elements   729
The Soil   731
Essay: The Water Cycle   734
Nutrient Cycles   735

Nitrogen and the Nitrogen Cycle   736
Essay: Carnivorous Plants   737
Assimilation of Nitrogen   742
The Phosphorus Cycle   742
Human Impact on Nutrient Cycles and
    Effects of Pollution   743
Essay: Halophytes: A Future Resource?   744
Soils and Agriculture   744
Plant Nutrition Research   745
Essay: Compost   746
Summary   747

**Chapter 31   The Movement of Water and
            Solutes in Plants**   750

Movement of Water and Inorganic Nutrients
    through the Plant Body   751
Assimilate Transport: Movement of Substances
    through the Phloem   764
Summary   769

SECTION 7   *Ecology*   772

**Chapter 32   The Dynamics of Communities and
            Ecosystems**   773

Interactions between Organisms   774
Essay: Competing for Light   778
Essay: Pesticides and Ecosystems   781
Nutrient Cycling   782
Trophic Levels   783
Development of Communities and Ecosystems   786
Essay: The Great Yellowstone Fire   790
Summary   793

Chapter 33   Global Ecology   796

Life on the Land   797
Essay: Alexander von Humboldt   802
Rainforests   803
Savannas and Deciduous Tropical Forests   805
Deserts   807
Essay: How Does a Cactus Function?   809
Grasslands   810
Temperate Deciduous Forests   812
Essay: Jobs versus Owls   815
Temperate Mixed and Coniferous Forests   815
Mediterranean Scrub   816
Taiga   818
Arctic Tundra   819
Summary   821

Chapter 34   The Human Prospect   823

The Agricultural Revolution   824
Essay: The Origin of Maize   829
The Growth of Human Populations   835
Agriculture in the Future   837
Essay: A New Millennium: The Transition to
    Sustainability   844
Summary   847

Suggested Readings   851

Appendix A: Fundamentals of Chemistry   863

Appendix B: The Hardy–Weinberg Equation 877

**Appendix C: Metric Table and Temperature
                Conversion Scale**   879

Appendix D: Classification of Organisms 881

Glossary   889

Illustration Credits   915

Index   920

Geologic Eras   inside front cover

Life on Earth depends on the ability of plants to capture the sun's energy and use it to produce the molecules needed to maintain living organisms. The hummingbird seen here is about to consume the energy-rich nectar contained deep within the columbine flower.

Botany: An Introduction

OVERVIEW

When you begin your study of botany, you start on a journey that leads both to the future and to the past. To understand—and appreciate—the structure, function, and diversity of plants, it is first necessary to mentally travel back billions of years to a time soon after the Earth was formed and follow the hypothesized sequence of events that gave rise first to the chemical building blocks of life in the early oceans and then to cells—the smallest units of life itself. As you will see in this chapter, a key event occurred when some of these cells began to photosynthesize—that is, to use the sun's energy to manufacture their own food from simple building blocks present in the marine environment. Such photosynthetic cells changed the composition of the early atmosphere and influenced the evolution of plants and animals alike. Eventually, plants from the sea colonized the land, and many of the structures seen in present-day terrestrial plants—such as roots, stems, and leaves—can be considered evolutionary adaptations for surviving in this relatively dry environment.

Your study of botany leads to the future because it provides the background you will need to understand and perhaps solve the many challenges facing us in the years to come. Problems such as pollution, food shortages, global warming, and the destruction of the ozone layer, as well as such solutions as the development of new crops using genetic engineering, require a knowledge of plant biology.

CHECKPOINTS

By the time you finish reading this chapter, you should be able to answer the following questions:

1. What are the main factors thought to be responsible for the origin of life on Earth, and what evidence supports the hypothesis that life arose in the oceans?

2. What is the principal difference between a heterotroph and an autotroph, and what role did each play on the early Earth?

3. Why is the evolution of photosynthesis thought to be such an important event in the evolution of life in general?

4. What were some of the problems encountered by plants as they made the transition from the sea to the land, and what structures in terrestrial plants apparently solve those problems?

5. What are biomes, and what are the principal roles of plants in an ecosystem?

"What drives life is . . . a little current, kept up by the sunshine," wrote Nobel laureate Albert Szent-Györgyi. With this simple sentence, he summed up one of the greatest marvels of evolution—photosynthesis. During the photosynthetic process, radiant energy from the sun is captured and used to form the sugar on which all life, including our own, depends. Oxygen, also essential to our existence, is released as a by-product. The process begins when a particle of light strikes a molecule of the green pigment chlorophyll, boosting one of the electrons in the chlorophyll to a higher energy level. The "excited" electron, in turn, initiates a flow of electrons that ultimately converts the radiant energy from the sun to the chemical energy of sugar molecules. Sunlight striking a leaf of the columbine shown on the preceding pages, for example, is the first step in the process leading to production of the sugary nectar that will provide nourishment for the hummingbird.

Only a few types of organisms—plants, algae, and some bacteria—possess chlorophyll, which is essential for a living cell to carry out photosynthesis. Once light energy is trapped in chemical form, it becomes available as an energy source to all other organisms, including human beings. We are totally dependent upon photosynthesis, a process for which plants are exquisitely adapted.

The word "botany" comes from the Greek *botanē*, meaning "plant," derived from the verb *boskein*, "to feed." Plants, however, enter our lives in innumerable ways other than as sources of food. They provide us with fiber for clothing; wood for furniture, shelter, and fuel; paper for books (such as the page you are reading at this moment); spices for flavor; drugs for medicines; and the oxygen we breathe. We are utterly dependent on plants. Plants also have enormous sensory appeal, and our lives are enhanced by the gardens, parks, and wilderness areas available to us. The study of plants has provided us with great insight into the nature of all life and will continue to do so in the years ahead. With modern technology, including continued development of molecular and computer techniques, we have just entered the most exciting period in the history of botany.

Evolution of Plants

Life Originated Early in Earth's Geologic History

Like all other living organisms, plants have had a long history during which they have **evolved,** or changed, over time. The planet Earth itself—an accretion of dust and gases swirling in orbit around the star that is our sun—is some 4.5 billion years old (Figure 1–1). The earliest known fossils are found in rocks of Western Australia about 3.5 billion years old and consist of several kinds of small, relatively simple cells resembling

bacteria (Figure 1–2). Evidence obtained by analysis of carbon particles embedded in the oldest rocks on Earth—from Akilia Island in southern West Greenland—indicate, however, that life already existed on Earth 3.85 billion years ago.

It is believed that Earth sustained a lethal meteor bombardment that ended 3.8 billion years ago, marking the end of Earth's earliest geologic period. Vast chunks of rubble slammed into the planet, helping to keep it hot. As the molten Earth began to cool, violent storms raged, accompanied by lightning and the release of electrical energy. Radioactive substances in the Earth emitted large quantities of energy, and widespread volcanism spewed molten rock and boiling water from beneath Earth's surface. Evidence of the presence of life on Earth as early as 3.85 billion years ago might mean that life was eliminated and reemerged, or that it originated elsewhere and reached Earth through space in the form of spores—resistant reproductive cells—or by some

1–1

Of the nine planets in our solar system, only one, as far as we know, has life on it. This planet, Earth, is visibly different from the others. From a distance, it appears blue and green, and it shines a little. The blue is water, the green is chlorophyll, and the shine is sunlight reflected off the layer of gases surrounding the planet's surface. Life, at least as we know it, depends on these visible features of Earth.

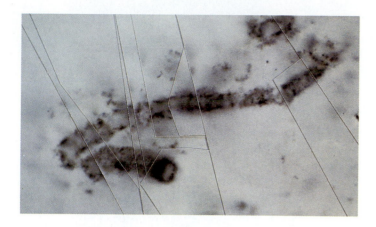

1–2

The earliest known fossils from ancient rocks in northwestern Western Australia, dated at 3.5 billion years of age. They are about a billion years younger than the Earth itself, but there are few suitable older rocks in which to look for earlier evidence of life. More complex organisms—those with eukaryotic cellular organization—did not evolve until about 1.5 billion years ago. For at least 2 billion years, therefore, prokaryotes were the only forms of life on Earth. These so-called "microfossils" have been magnified 260 times.

these compounds accumulated in the oceans where, with the driving forces of the energy of lightning and solar radiation, they gave rise to the first forms of life. Oparin's proposal seemed to have been borne out in 1953 with experiments carried out by Stanley L. Miller, then a graduate student working with Dr. Harold Urey at the University of Chicago. Using a similar gas mixture over an "ocean" of heated water and electric sparks to simulate lightning, Miller obtained a variety of complex organic molecules similar to those that form the fundamental building blocks of all life (Figure 1–3).

One problem with the Miller–Urey experiments is that the gas mixture included methane and ammonia, but these compounds may not have been present in Earth's first atmosphere. In the absence of an ozone layer, these gases would have been destroyed by ultraviolet radiation. The major atmospheric gases at the time probably were carbon dioxide and nitrogen emitted by volcanoes, in addition to water vapor. These three molecules contain the chemical elements carbon, oxygen, nitrogen, and hydrogen, which make up about 98 percent of the material found in living organisms today.

other means. Life may have formed on Mars, for example, whose early history apparently paralleled that of Earth. This possibility has been raised by the conclusion of NASA scientists that a Martian meteorite discovered in Antarctica in 1984 contains remains of bacteria-like life forms about 3.6 billion years old. In addition, evidence provided by the Galileo spacecraft in 1996 suggested that one of Jupiter's moons, Europa, might have liquid water beneath its frozen surface—raising the possibility of environments beyond Earth that could support life. We will continue to assume, however, that life on Earth originated on Earth.

The Chemical Building Blocks of Life Accumulated in the Early Oceans

In 1871, Charles Darwin speculated that life started in "a warm little pond," and this concept of the origin of life from some primordial soup has persisted. This view was first elaborated upon in the 1930s by the Russian scientist A. I. Oparin, who proposed that vast quantities of carbon- and hydrogen-containing compounds were formed in the early atmosphere from volcanic gases composed of methane, ammonia, water vapor, and hydrogen. Washed out of the atmosphere by driving rains,

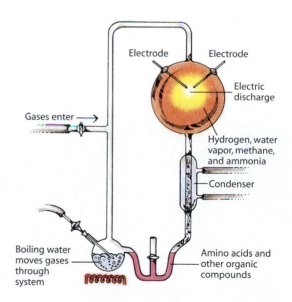

1–3

Stanley Miller, while a graduate student at the University of Chicago in the 1950s, used apparatus such as that shown here to simulate conditions he believed existed on the primitive Earth. Hydrogen, methane, and ammonia were circulated continuously between a lower "ocean," which was heated, and an upper "atmosphere," through which an electric discharge was transmitted. At the end of 24 hours, about half of the carbon originally present in the methane gas had been converted to amino acids and other organic molecules. This was the first test of Oparin's hypothesis.

Nevertheless, subsequent experiments using different mixtures of these gases have yielded a variety of organic compounds following electrical discharge. Among the compounds synthesized in laboratories are the nucleotides of which the nucleic acids RNA and DNA are formed.

Another theory for the origin of the chemical precursors essential to life on Earth points to comets as their source. In addition to being bombarded by meteors, the early Earth is believed to have been subjected to bombardment by vast numbers of comets, rich in large amounts of simple molecules, and possibly some complex ones as well, that gave rise to life. Even now, some 300 tons of complex organic matter in the form of comet and asteroid dust filters through the Earth's atmosphere every year.

Support for this theory was bolstered by the comet Hale-Bopp, which was discovered in July 1995 and thrilled scientists and nonscientists alike during the spring of 1997 (Figure 1–4). As Hale-Bopp moved through the solar system it spewed forth tons of water, methyl alcohol, formaldehyde, carbon monoxide, hydrogen cyanide, hydrogen sulfide, and many compounds rich in carbon. Never before had such a rich array of important molecules, potentially containing the basic building blocks for the origin of life, been identified with a single comet.

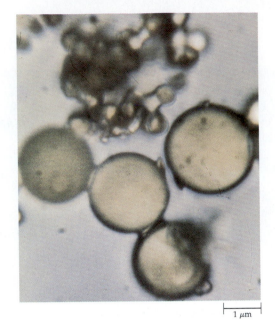

1 μm

1–5
When dry mixtures of amino acids are heated at moderate temperatures, polymers known as thermal proteinoids are formed. Each of these polymers may contain as many as 200 amino acid subunits. When the polymers are placed in water solution and maintained under suitable conditions, they spontaneously form proteinoid microspheres, as shown here. The microspheres are separated from the surrounding solution by a membrane that appears to be two-layered.

The short, straight line at the bottom of this micrograph and those that follow provide a reference for size; a micrometer, abbreviated μm, is 1/10,000 of a centimeter. The same system is used to indicate distances on a road map.

1–4
The comet Hale-Bopp, seen here above Smith Rocks State Park, Oregon, in the spring of 1997, is made up of dirty ice that contains many of the chemical precursors of life. Some scientists believe that comets falling on the early Earth in vast numbers provided the chemical "seeds" that eventually gave rise to the Earth's rich diversity of living organisms.

Most Likely, the Forerunners of the First Cells Were Simple Aggregations of Molecules Some organic molecules have a tendency to aggregate in groups. In the primitive oceans these groups probably took the form of droplets, similar to the droplets formed by oil in water. Such droplets of organic molecules appear to have been the forerunners of primitive cells, the first forms of life. Sidney W. Fox and his coworkers at the University of Miami have produced proteins that aggregate into cell-like bodies in water. Called proteinoid microspheres, these bodies grow slowly by the accumulation of additional proteinoid material and eventually bud off smaller microspheres (Figure 1–5). Although Fox likens this process to a type of reproduction, the microspheres are not living cells. It has also been suggested that clay particles or even bubbles may have played a role in life's origin on Earth by collecting chemicals and concentrating them for synthesis into complex molecules.

According to current theories, these organic molecules also served as the source of energy for the earliest forms of life. The primitive cells or cell-like structures were able to use these abundant compounds to satisfy their energy requirements. As they evolved and became more complex, these cells were increasingly able to control their own destinies. With this increasing complexity, they acquired the *ability to grow, to reproduce,* and *to pass on their characteristics to subsequent generations (heredity).* Together with *cellular organization,* these three properties characterize all living things on Earth.

Autotrophic Organisms Make Their Own Food, but Heterotrophic Organisms Must Obtain Their Food from External Sources Cells that satisfy their energy requirements by consuming the organic compounds produced by external sources are known as **heterotrophs** (Gk. *heteros,* "other," and *trophos,* "feeder"). A heterotrophic organism is dependent on an outside source of organic molecules for its energy. Animals, fungi (Figure 1–6a), and many of the one-celled organisms, such as certain bacteria and protists, are heterotrophs.

As the primitive heterotrophs increased in number, they began to use up the complex molecules on which their existence depended—and which had taken millions of years to accumulate. Organic molecules in free solution (that is, not inside a cell) became more and more scarce, and competition began. Under the pressure of this competition, cells that could make efficient use of the limited energy sources now available were more likely to survive than cells that could not. In the course of time, by the long, slow process of elimination of the most poorly adapted, cells evolved that were able to make their own energy-rich molecules out of simple inorganic materials. Such organisms are called **autotrophs,** "self-feeders." Without the evolution of these early autotrophs, life on Earth would soon have come to an end.

The most successful of the autotrophs were those that evolved a system for making direct use of the sun's energy—that is, the process of photosynthesis (Figure 1–6b). The earliest photosynthetic organisms, although simple in comparison with plants, were much more complex than the primitive heterotrophs. Use of the sun's energy required a complex pigment system to capture the light energy and, linked to this system, a way to store the energy in an organic molecule.

Evidence of the activities of photosynthetic organisms has been found in rocks 3.4 billion years old, about 100 million years after the first fossil evidence of life on Earth. We can be almost certain, however, that both life and photosynthetic organisms evolved considerably earlier than the evidence suggests. In addition, there seems to be no doubt that heterotrophs evolved before autotrophs. With the arrival of autotrophs, the flow of energy in the **biosphere** (that is, the living world and its environment) came to assume its modern form: radiant energy from the sun channeled through the photosynthetic autotrophs to all other forms of life.

Photosynthesis Altered Earth's Atmosphere, Which in Turn Influenced the Evolution of Life

As photosynthetic organisms increased in number, they changed the face of the planet. This biological revolution came about because one of the most efficient strategies of photosynthesis—the one employed by nearly all living autotrophs—involves splitting the water molecule (H_2O) and releasing its oxygen as free oxygen molecules (O_2). Thus, as a result of photosynthesis, the amount of oxygen gas in the atmosphere increased. This increase in oxygen level had two important consequences.

1–6

A modern heterotroph and a photosynthetic autotroph. (a) A fungus, Coprinus atramentaris, *growing on a forest floor in California.* Coprinus, *like other fungi, absorbs its food (often from other organisms). (b) Large-flowered trillium (*Trillium grandiflorum*), one of the first plants to flower in spring in the deciduous woods of eastern and midwestern North America. Like most vascular plants, trilliums are rooted in the soil; photosynthesis occurs chiefly in the leaves. Flowers are produced in well-lighted conditions before leaves appear on surrounding trees. The underground portions (rhizomes) of the plant live for many years and spread to produce new plants vegetatively under the thick cover of decaying material on the forest floor. Trilliums also reproduce by producing seeds, which are dispersed by ants.*

(a)　　　　　*(b)*

First, some of the oxygen molecules in the outer layer of the atmosphere were converted to ozone (O_3) molecules. When there is a sufficient quantity of ozone in the atmosphere, it absorbs the ultraviolet rays—rays highly destructive to living organisms—from the sunlight that reaches the Earth. By about 450 million years ago, organisms, protected by the ozone layer, could survive in the surface layers of water and on the land.

Second, the increase in free oxygen opened the way to a much more efficient utilization of the energy-rich carbon-containing molecules formed by photosynthesis. It enabled organisms to break down those molecules by the oxygen-utilizing process known as respiration. As is discussed in Chapter 6, respiration yields far more energy than can be extracted by any **anaerobic,** or oxygenless, process.

Before the atmosphere accumulated oxygen and became **aerobic,** the only cells that existed were **prokaryotic**—simple cells that lacked a nuclear envelope and did not have their genetic material organized into complex chromosomes. It is quite likely that the first prokaryotes were heat-loving organisms called archaea (meaning "ancient ones"), the descendants of which are now known to be widespread, with many thriving at extremely high temperatures hostile to life. Bacteria also are prokaryotes. Some archaea and bacteria are heterotrophic, and others are autotrophic. According to the fossil record, the increase of relatively abundant free oxygen was accompanied by the first appearance of **eukaryotic** cells—cells with nuclear envelopes, complex chromosomes, and organelles, such as mitochondria (sites of respiration) and chloroplasts (sites of photosynthesis), surrounded by membranes. Eukaryotic organisms, in which the individual cells are usually much larger than those of the bacteria, appeared about 1.5 billion years ago and were well established and diverse by 1 billion years ago. Except for archaea and bacteria, all organisms—from amoebas to dandelions to oak trees to human beings—are composed of one or more eukaryotic cells.

The Seashore Environment Was Important in the Evolution of Photosynthetic Organisms

Early in evolutionary history, the principal photosynthetic organisms were microscopic cells floating below the surface of the sunlit waters. Energy abounded, as did carbon, hydrogen, and oxygen, but as the cellular colonies multiplied, they quickly depleted the mineral resources of the open ocean. (It is this shortage of essential minerals that is the limiting factor in any modern plans to harvest the seas.) As a consequence, life began to develop more abundantly toward the shores, where the waters were rich in nitrates and minerals carried down from the mountains by rivers and streams and scraped from the coasts by the ceaseless waves.

1–7

A fossil of Cooksonia, *one of the earliest and simplest plants known, from the Late Silurian period (414–408 million years ago).* Cooksonia *consisted of little more than a branched stem with terminal sporangia, or spore-producing structures.*

The rocky coast presented a much more complicated environment than the open sea, and, in response to these evolutionary pressures, living organisms became increasingly complex in structure and more diversified. Not less than 650 million years ago, organisms evolved in which many cells were linked together to form an integrated, multicellular body (Figure 1–7). In these primitive organisms we see the early stages in the evolution of plants, fungi, and animals. Fossils of multicellular organisms are much easier to detect than those of simpler ones. The history of life on Earth, therefore, is much better documented from the time of their first appearance.

On the turbulent shore, multicellular photosynthetic organisms were better able to maintain their position against the action of the waves, and, in meeting the challenge of the rocky coast, new forms developed. Typically, these new forms evolved relatively strong cell walls for support, as well as specialized structures to anchor their bodies to the rocky surfaces (Figure 1–8). As these organisms increased in size, they were confronted with the problem of how to supply food to the dimly lit, more deeply submerged portions of their bodies, where

1–8

Multicellular photosynthetic organisms anchored themselves to rocky shores early in the course of their evolution. These kelp, seen at low tide on the rocks along the coast of Tatoosh Island, Washington, are brown algae (Phaeophyta), a group in which multicellularity evolved independently of other groups of organisms.

Land animals, generally speaking, are mobile and able to seek out water just as they seek out food. Fungi, though immobile, remain largely below the surface of the soil or within whatever damp organic material they feed upon. Plants utilize an alternative evolutionary strategy. **Roots** anchor the plant in the ground and collect the water required for maintenance of the plant body and for photosynthesis, while the **stems** provide support for the principal photosynthetic organs, the **leaves.** A continuous stream of water moves upward through the roots and stems, and then out through the leaves. The outermost layer of cells, the **epidermis,** of all of the aboveground portions of the plant that are ultimately involved in photosynthesis is covered with a waxy **cuticle,** which retards water loss. However, the cuticle also tends to prevent the exchange of gases between the plant and the surrounding air that is necessary for both photosynthesis and respiration. The solution to this dilemma is found in pairs of specialized epidermal cells (the guard cells), which regulate small openings between them called **stomata** (singular: stoma). The stomata open and close in response to environmental and physiological signals, thus helping the plant maintain a balance between its water losses and its oxygen and carbon dioxide requirements (Figure 1–9).

In younger plants and in those with a life span of one year (annuals), the stem is also a photosynthetic organ. In longer-lived plants (perennials), the stem may become thickened and woody and covered with **cork,** which, like the cuticle-covered epidermis, retards water loss. In both annuals and perennials, the stem serves to conduct, via the **vascular system** (conducting system), a variety of

photosynthesis was not taking place. Eventually, specialized food-conducting tissues evolved that extended the length of their bodies and connected the upper, photosynthesizing parts with the lower, nonphotosynthesizing structures.

Colonization of the Land Was Associated with the Evolution of Structures to Obtain Water and Minimize Water Loss

The body of a plant can best be understood in terms of its long history and, in particular, in terms of the evolutionary pressures involved in the transition to land. The requirements of a photosynthetic organism are relatively simple: light, water, carbon dioxide for photosynthesis, oxygen for respiration, and a few minerals. On land, light is abundant, as are oxygen and carbon dioxide—both of which circulate more freely in air than in water—and the soil is generally rich in minerals. The critical factor, then, for the transition to land—or as one investigator prefers to say, "to the air"—is water.

20 μm

1–9

Open stomata on the surface of a tobacco (Nicotiana tabacum) leaf. The stomata, or small openings, in the aerial parts of the plant are regulated by two guard cells that border each stoma.

substances between the photosynthetic and nonphotosynthetic parts of the plant body. The vascular system has two major components: the **xylem,** through which water passes upward through the plant body, and the **phloem,** through which food manufactured in the leaves and other photosynthetic parts of the plant is transported throughout the plant body. It is this efficient conducting system that has given the main group of plants—the **vascular plants**—their name (Figure 1–10).

Plants, unlike animals, continue to grow throughout their lives. All plant growth originates in **meristems**—localized regions of perpetually young tissues capable of adding cells indefinitely to the plant body. Meristems located at the tips of all roots and shoots—**apical meristems**—are involved with the extension of the plant body. Thus the roots are continuously reaching new sources of water and minerals, and the photosynthetic regions are continuously extending toward the light. The type of growth that originates from apical meristems is known as **primary growth. Secondary growth,** which results in a thickening of stems and roots, originates from two **lateral meristems**—the vascular cambium and the cork cambium.

Plants also underwent further adaptations that made it possible for them to reproduce on land. The first of these adaptations was the production of drought-resistant spores. This was followed by the evolution of complex, multicellular structures in which the gametes, or reproductive cells, were held and protected from drying out by a layer of sterile cells. In the **seed plants,** which

1–10

Diagram of a young broad bean (Vicia faba) plant, showing the principal organs and tissues of the modern vascular plant body. The organs—root, stem, and leaf—are composed of tissues, which are groups of cells with distinct structures and functions. Collectively, the roots make up the root system, and the stems and leaves together make up the shoot system of the plant. Unlike roots, stems are divided into nodes and internodes. The node is the part of the stem at which one or more leaves are attached, and the internode is the part of the stem between two successive nodes. In the broad bean, the first few foliage leaves are divided into two leaflets each. Buds, or embryonic shoots, commonly arise in the axils—the upper angle between leaf and stem—of the leaves. Lateral, or branch, roots arise from the inner tissues of the roots. The vascular tissues—xylem and phloem—occur together and form a continuous vascular system throughout the plant body. They lie just inside the cortex in root and stem. The mesophyll tissue of leaves is specialized for photosynthesis. In this diagram, a cotyledon, or seed leaf, can be seen through a tear in the seed coat.

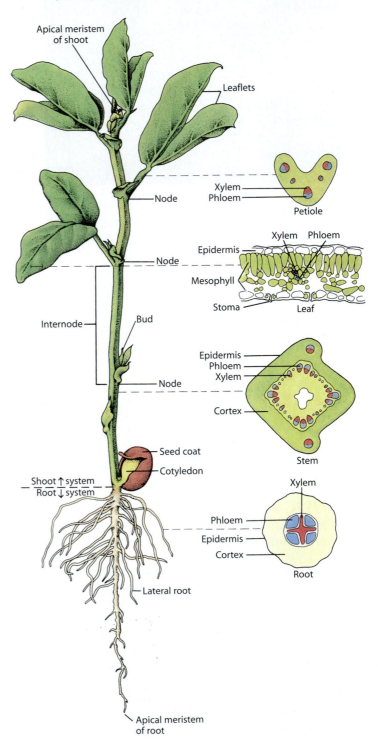

include almost all familiar plants except the ferns, mosses, and liverworts, the young plant, or embryo, is enclosed within a specialized covering (seed coat) provided by the parent. There the embryo is protected from both drought and predators and provided with a supply of stored food. The embryo, the supply of stored food, and the seed coat are the components of the **seed.**

Thus, in summary, the vascular plant (Figure 1–10) is characterized by a root system that serves to anchor the plant in the ground and to collect water and minerals from the soil; a stem that raises the photosynthetic parts of the plant body toward its energy source, the sun; and leaves, highly specialized photosynthetic organs. Roots, stems, and leaves are interconnected by a complicated and efficient vascular system for the transport of food and water. The reproductive cells of plants are enclosed within multicellular protective structures, and in seed plants the embryos are protected by resistant coverings. All of these characteristics are adaptations to a photosynthetic existence on land.

Evolution of Communities

The invasion of the land by plants changed the face of the continents. Looking down from an airplane on one of Earth's great expanses of desert or on one of its mountain ranges, one can begin to imagine what the world looked like before the appearance of plants. Yet even in these regions, the traveler who goes by land will find an astonishing variety of plants punctuating the expanses of rock and sand. In those parts of the world where the climate is more temperate and the rains are more frequent, communities of plants dominate the land and determine its character. In fact, to a large extent, they *are* the land. Rainforest, savanna, woods, desert, tundra— each of these words brings to mind a portrait of a landscape (Figure 1–11). The main features of each landscape are its plants, enclosing us in a dark green cathedral in our imaginary rainforest, carpeting the ground beneath our feet with wildflowers in a meadow, moving in great golden waves as far as the eye can see across our imaginary prairie. Only when we have sketched these **biomes**—natural communities of wide extent, characterized by distinctive, climatically controlled groups of plants and animals—in terms of trees and shrubs and grasses can we fill in other features, such as deer, antelope, rabbits, or wolves.

How do vast plant communities, such as those seen on a continental scale, come into being? To some extent we can trace the evolution of the different kinds of plants and animals that populate these communities. Even with accumulating knowledge, however, we have only begun to glimpse the far more complex pattern of development, through time, of the whole system of organisms that make up these various communities.

Ecosystems Are Relatively Stable, Integrated Units That Are Dependent upon Photosynthetic Organisms

Such communities, along with the nonliving environment of which they are a part, are known as ecological systems, or **ecosystems.** Ecosystems are discussed later in Chapter 32. For now, it is sufficient to regard an ecosystem as a kind of corporate entity made up of transient individuals. Some of these individuals, the larger trees, live as long as several thousand years; others, the microorganisms, live only a few hours or even minutes. Yet the ecosystem as a whole tends to be remarkably stable (although not static); once in balance, it does not change for centuries. Our grandchildren may someday walk along a woodland path once followed by our great-grandparents, and where they saw a pine tree, a mulberry bush, a meadow mouse, wild blueberries, or a towhee, these children, if this woodland still exists, will see roughly the same kinds of plants and animals in the same numbers.

An ecosystem functions as an integrated unit, although many of the organisms in the system compete for resources. Virtually every living thing, even the smallest bacterial cell or fungal spore, provides a food source for some other living organism. In this way, the energy captured by green plants is transferred in a highly regulated way through a number of different types of organisms before it is dissipated. Moreover, interactions among the organisms themselves, and between the organisms and the nonliving environment, produce an orderly cycling of elements such as nitrogen and phosphorus. Energy must be added to the ecosystem constantly, but the elements are cycled through the organisms, returned to the soil, decomposed by soil bacteria and fungi, and recycled. These transfers of energy and the cycling of elements involve complicated sequences of events, and in these sequences each group of organisms has a highly specific role. As a consequence, it is impossible to change a single component of an ecosystem without the risk of destroying the balance upon which the stability of the ecosystem depends.

At the base of productivity in virtually all ecosystems are the plants, algae, and photosynthetic bacteria. These organisms alone have the ability to capture energy from the sun and to manufacture organic molecules that they and all other kinds of organisms may require for life. There are roughly half a million kinds of organisms capable of photosynthesis, and at least eight or ten times that many heterotrophic organisms, which are completely dependent upon the photosynthesizers. For animals, including human beings, many kinds of molecules—including essential amino acids, vitamins, and minerals—can be obtained only through plants or other photosynthetic organisms. Furthermore, the oxygen that is released into the atmosphere by photosynthetic organ-

(a)

(b)

1–11

Some examples of the enormous diversity of biological communities on Earth. (a) The temperate deciduous forest, which covers most of the eastern United States and southeastern Canada, is dominated by trees that lose their leaves in the cold winters. Here are paper birches and a red maple photographed in early autumn in the Adirondack Mountains of New York State. (b) Underlain with permafrost, Arctic tundra is a treeless biome characterized by a short growing season. Shown here are tundra plants in full autumn color, photographed in Tombstone Valley, Yukon, Canada. (c) In Africa, savannas are inhabited by huge herds of grazing mammals, such as these zebras and wildebeests. The tree in the foreground is an acacia. (d) The tropical rainforest, shown here in Costa Rica, is the richest, most diverse biome on Earth, with at least two-thirds of all species of organisms on Earth found there. (e) Deserts typically receive less than 25 centimeters of rain per year. Here in the Sonoran desert in Arizona, the dominant plant is the giant saguaro cactus. Adapted for life in a dry climate, saguaro cactuses have shallow, widespreading roots and thick stems for storing water. (f) Mediterranean climates are rare on a world scale. Cool, moist winters, during which the plants grow, are followed by hot, dry summers, during which the plants become dormant. Shown here is a pine-oak chapparal photographed on Mount Diablo in California.

(c)

(d)

(e)

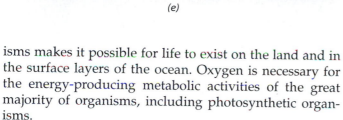

(f)

isms makes it possible for life to exist on the land and in the surface layers of the ocean. Oxygen is necessary for the energy-producing metabolic activities of the great majority of organisms, including photosynthetic organisms.

Appearance of Human Beings

Human beings are relative newcomers to the world of living organisms (Figure 1–12). If the entire history of the Earth were measured on a 24-hour time scale starting at midnight, cells would appear in the warm seas before dawn. The first multicellular organisms would not be present until well after sundown, and the earliest appearance of humans (about 2 million years ago) would be about half a minute before the day's end. Yet humans more than any other animal—and almost as much as the plants that invaded the land—have changed the surface of the planet, shaping the biosphere according to their own needs, ambitions, or follies.

The development of agriculture, starting at least 11,000 years ago, in time made it possible to maintain large populations of people in towns and cities. This development (reviewed in detail in Chapter 34) allowed specialization and the diversification of human culture. One characteristic of this culture is that it examines itself and the nature of other living things, including plants. Eventually, the science of biology developed within the human communities that had been made possible through the domestication of plants. That part of biology that deals with plants and, by tradition, with prokaryotes, fungi, and algae is called botany, or plant biology.

Plant Biology Includes Many Different Areas of Study

The study of plants has been pursued for thousands of years, but like all branches of science, it has become diverse and specialized only during the past three centuries. Until little more than a century ago, botany was a branch of medicine, pursued chiefly by physicians who used plants for medicinal purposes and who were interested in determining the similarities and differences between plants and animals. Today, however, it is an important scientific discipline that has many subdivisions: **plant physiology,** which is the study of how plants function, that is, how they capture and transform energy and how they grow and develop; **plant morphology,** the study of the form of plants; **plant anatomy,** the study of their internal structure; **plant taxonomy** and **systematics,** involving the naming and classifying of plants and studying the relationships among them; **cytology,** the study of cell structure, function, and life histories; **genetics,** the study of heredity and variation; **molecular biology,** the study of the structure and function of biological molecules; **economic botany,** the study of past, present, and future uses of plants by people; **ecology,** the study of the relationships between organisms and their environment; and **paleobotany,** the study of the biology and evolution of fossil plants.

Included in this book are all organisms that have traditionally been studied by botanists: plants as well as prokaryotes, viruses, fungi, and autotrophic protists (algae). Only the animals have traditionally been the province of zoologists. Although we do not regard algae, fungi, prokaryotes, or viruses as plants, and shall not refer to them as plants in this book, they are included

1–12

The clockface of biological time, which shows when important events in the Earth's past would have occurred if the Earth's 4.5-billion year history were condensed into one day. Life first appears relatively early, sometime before 6:00 A.M. on a 24-hour time scale. The first multicellular organisms do not appear until the early evening of that 24-hour day, and Homo, *the genus to which humans belong, is a late arrival—at about 30 seconds to midnight.*

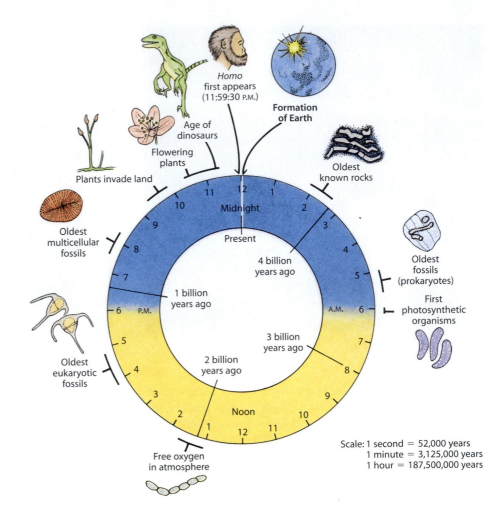

A Knowledge of Botany Is Important for Dealing with Today's—and Tomorrow's—Problems

here because of tradition and because they are normally considered as part of the botanical portion of the curriculum, just as botany itself used to be considered a part of medicine. Virology, bacteriology, phycology (the study of algae), and mycology (the study of fungi) are well-established fields in their own right, but they still fall loosely under the umbrella of botany.

A Knowledge of Botany Is Important for Dealing with Today's—and Tomorrow's—Problems

In this chapter, we have ranged from the beginnings of life on this planet to the evolution of plants and ecosystems to the development of agriculture and civilization. These broad topics are of interest to many people other than botanists, or plant biologists. The urgent efforts of botanists and agricultural scientists will be needed to feed the world's rapidly growing human population (Figure 1–13), as is discussed in Chapter 34. Modern plants, algae, and bacteria offer the best hope of providing a renewable source of energy for human activities, just as extinct plants, algae, and bacteria have been responsible for the massive accumulations of gas, oil, and coal on which our modern industrial civilization de-

pends. In an even more fundamental sense, the role of plants, along with that of algae and photosynthetic bacteria, commands our attention. As the producers of energy-containing compounds in the global ecosystem, these organisms are the route by which all other living things, including ourselves, obtain energy, oxygen, and many other materials necessary for our continued existence. As a student of botany, you will be in a better position to assess the important ecological and environmental issues of the day and, by understanding, help to build a healthier world.

As we approach the beginning of the twenty-first century, it has become apparent that human beings, who will number well over 6 billion by the year 2000, are managing the Earth with an intensity that would have been unimaginable a few decades ago. Every hour, manufactured chemicals fall on every square centimeter of the planet's surface. The protective stratospheric ozone layer formed 450 million years ago has been seriously depleted by the use of chlorofluorocarbons (CFCs), and damaging ultraviolet rays penetrating the depleted layer have increased the incidence of skin cancer in people all over the world. Moreover, it has been estimated that, by the middle of the next century, the average temperature will have increased between 1° and 6°C due to the greenhouse effect. This phenomenon—the trapping of heat radiating from the Earth's surface out into space—is intensified through the increased amounts of carbon dioxide, nitrogen oxides, CFCs, and methane in the atmosphere, resulting from human activities. Most serious of all, a large portion of the total number of species of plants, animals, fungi, and microorganisms is disappearing during our lifetime— the victims of human exploitation of the Earth. All of these trends are alarming, and they demand our utmost attention.

Marvelous new possibilities have been developed during the past few years for the better utilization of plants by people. We discuss these developments throughout this book. It is now possible to stimulate the growth of plants, to deter their pests, to control weeds in crops, and to form hybrids between plants with much more precision and far wider possibilities than ever before. The potential of these new discoveries is growing

with every passing year, as additional discoveries are made and new applications are developed. For example, the methods of genetic engineering discussed in Chapters 11 and 28 now make it possible to transfer either natural or synthetic genes from one plant or animal to another in order to produce certain characteristics. These methods, first applied in 1973, have already been the basis of billions of dollars in investments and increased hope for the future. Discoveries still to be made undoubtedly will far exceed our wildest dreams and go far beyond the facts that are available to us now.

As we turn to Chapters 2 and 3, in which our attention narrows to a cell so small it cannot be seen by the unaided eye, it is important to keep these broader concerns in mind. A basic knowledge of plant biology is useful in its own right and is essential in many fields of endeavor. It is also increasingly relevant to some of society's most crucial problems and to the difficult decisions that will face us in choosing among the proposals for diminishing them. Our own future, the future of the

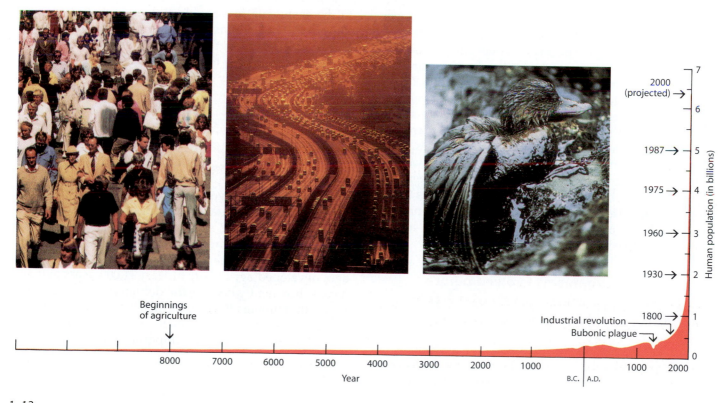

1–13

Over the last 25,000 years, the human population has grown from several million to approximately 6 billion. A significant increase in the rate of population growth occurred as a result of the agricultural revolution, and an even more dramatic increase began with the advent of the industrial revolution, which began in the mid-eighteenth century and has continued to the present time.

The consequences of the rapid growth of the human population are many and varied. In the United States and other parts of the developed world, they include not only the sheer numbers of people but also heavy consumption of nonrenewable fossil fuels and the resulting pollution—both as the fuels are burned and as a result of accidents, such as oil spills during their transport. In less developed parts of the world, the consequences include malnutrition and, all too often, starvation, coupled with a continuing vulnerability to infectious diseases. The consequences for other organisms include not only the direct effects of pollution but also—and most important—the loss of habitat.

world, and the future of all kinds of plants—as individual species and as components of the life-support systems into which we all have evolved—depend upon our knowledge. Thus, this book is dedicated not only to the botanists of the future, whether teachers or researchers, but also to the informed citizens, scientists and laypeople alike, in whose hands such decisions lie.

Summary

Only a few kinds of organisms—plants, algae, and some bacteria—have the capacity to capture energy from the sun and to fix it in organic molecules by the process of photosynthesis. Virtually all life on Earth depends, directly or indirectly, on the products of this process.

The Chemical Building Blocks of Life Accumulated in the Early Oceans

The planet Earth is some 4.5 billion years old. The oldest known fossils date back 3.5 billion years, but evidence indicates that life already existed on Earth as early as 3.85 billion years ago. The major atmospheric gases at the time probably were carbon dioxide and nitrogen emitted by volcanoes, in addition to water vapor. These three molecules contain the chemical elements carbon, hydrogen, nitrogen, and oxygen, which make up about 98 percent of the material found in all living organisms today. In the turbulent early atmosphere, the gases were recombined into new, larger molecules. Oxygen was essentially absent until photosynthetic organisms began to produce it in quantity. As a result, ultraviolet rays (now largely blocked from the surface of the Earth by ozone) bombarded the surface of the Earth and assisted in the synthesis of molecules.

Heterotrophic Organisms Evolved before Autotrophic Organisms, Prokaryotes before Eukaryotes, and Unicellular Organisms before Multicellular Ones

Heterotrophs, organisms that feed on organic molecules or other organisms, were the first to evolve. Autotrophic

organisms, those that could produce their own food by photosynthesis, evolved no less than 3.4 billion years ago. Until about 1.5 billion years ago, the prokaryotes—archaea and bacteria—were the only organisms that existed. Eukaryotes, with larger, much more complex cells, evolved at that time. Multicellular eukaryotes began to evolve at least 650 million years ago, and they began to invade the land about 450 million years ago.

The Colonization of the Land Was Associated with the Evolution of Structures to Obtain Water and Minimize Water Loss

Plants, which are basically a terrestrial group, have achieved a number of specialized characteristics that suit them for life on land. These characteristics are best developed among the members of the dominant group known as the vascular plants. Among them are a waxy cuticle, penetrated by specialized openings known as stomata through which gas exchange takes place, and an efficient conducting system. This system consists of xylem, in which water and minerals pass from the roots to the stems and leaves, and phloem, which transports the products of photosynthesis to all parts of the plant. Plants increase in length by primary growth and in girth by secondary growth through the activity of meristems, which are perpetually young tissues capable of adding cells indefinitely to the plant body.

Ecosystems Are Relatively Stable, Integrated Units That Are Dependent upon Photosynthetic Organisms

As plants have evolved, they have come to constitute biomes, great terrestrial assemblages of plants and animals. The interacting systems made up of biomes and their nonliving environments are called ecosystems. Human beings, which appeared about 2 million years ago, developed agriculture at least 11,000 years ago and have subsequently become the dominant ecological force on Earth. Humans have used their knowledge of plants to foster their own development and will continue to do so with increasingly greater importance in the future.

Selected Key Terms

aerobic   p. 6

autotrophs   p. 5

biomes   p. 9

biosphere   p. 5

cuticle   p. 7

cytology   p. 11

ecology   p. 11

economic botany   p. 11

ecosystems   p. 9

epidermis   p. 7

eukaryotic cells   p. 6

genetics   p. 11

heterotrophs   p. 5

leaves   p. 7

meristems   p. 8

molecular biology   p. 11

paleobotany   p. 11

phloem   p. 8

plant anatomy   p. 11

plant morphology   p. 11

plant physiology   p. 11

plant taxonomy   p. 11

primary growth   p. 8

prokaryotic cells   p. 6

roots   p. 7

secondary growth   p. 8

seeds   p. 9

stomata   p. 7

stems   p. 7

vascular plants   p. 8

vascular system   p. 7

xylem   p. 8

Questions

1. Some scientists believe that other planets in our galaxy may well contain some form of life. If you were seeking such a planet, what features would you look for?

2. What criteria would you use to determine whether an entity is a form of life?

3. What role did oxygen play in the evolution of life on Earth?

4. What advantages do terrestrial plants have over their aquatic ancestors? Can you think of any disadvantages to being a terrestrial plant?

5. Plants enter our lives in innumerable ways other than as sources of food. How many ways can you list? Have you thanked a green plant today?

6. A knowledge of botany—of plants, fungi, algae, and bacteria—is key to our understanding of how the world works. How is that knowledge important for dealing with today's and tomorrow's problems?

The Biology of the Plant Cell

SECTION 1

Plants capture the sun's energy and use it to form the organic molecules essential to life. This process—photosynthesis—requires the green pigment chlorophyll, which is present in the leaves of this honeysuckle. The organic molecules formed during photosynthesis provide not only the energy that powers living systems but also the larger structural molecules that make up living organisms.

Chapter 2

The Molecular Composition of Plant Cells

OVERVIEW

As you probably know, the growth and development—the very survival—of plants depends on the proper functioning of their component cells. The survival of these cells, in turn, depends on the interactions of a tremendous variety of chemical constituents. Some, such as potassium ions (K^+) and water (H_2O), are small even on an atomic scale. Others are relatively large and are composed of carbon atoms linked together in long chains. This chapter focuses on these carbon-containing molecules, primarily carbohydrates (including sugars and starches), lipids (including fats and oils), proteins (including enzymes), and nucleic acids (including DNA). Although these molecules appear complex, most are composed of similar subunits joined together like the links of a chain. Learning the characteristic features of the subunits is thus a key step in understanding the structure of the whole molecule. This chapter also presents information about other carbon-containing molecules, such as caffeine, nicotine, and morphine, many of which have commercial or medical uses but whose functions in the plant are only beginning to be understood.

CHECKPOINTS

By the time you finish reading this chapter, you should be able to answer the following questions:

1. What are the four main types of organic molecules found in plant cells, and what are the basic structural subunits of each?

2. What are the principal functions of each of the four main types of organic molecules in plant cells?

3. By what process are all four types of organic molecules split into their subunits, and by what process can these subunits be joined together?

4. How do energy-storage polysaccharides and structural polysaccharides differ from one another? What are some examples of each?

5. What is an enzyme, and why are enzymes important in cells?

6. How is ATP different from ADP, and why is ATP important in cells?

7. What is the difference between a primary metabolite and a secondary metabolite?

8. What are the main types of secondary metabolites, and what are some examples of each?

Everything on Earth—including everything you can see at the moment, as well as the air surrounding it—is made up of chemical elements in various combinations. Elements are substances that cannot be broken down into other substances by ordinary means. Carbon is an element, as are hydrogen and oxygen. Of the 92 elements that occur naturally on Earth, only six of them were selected in the course of evolution to form the complex, highly organized material of living organisms. These six elements—carbon, hydrogen, nitrogen, oxygen, phosphorus, and sulfur—make up 99 percent of the weight of all living matter. The unique properties of each element depend on the structure of its atoms and on the way those atoms can interact and bond with other atoms to form molecules. (You may find it helpful to read Appendix A, Fundamentals of Chemistry, before proceeding with this chapter.)

Water, a molecule consisting of two hydrogens and one oxygen (H_2O), makes up more than half of all living matter and more than 90 percent of the weight of most plant tissues. (Appendix A also covers some important properties of water.) By contrast, electrically charged ions, such as potassium (K^+), magnesium (Mg^{2+}), and calcium (Ca^{2+}), account for about 1 percent. Almost all the rest of a living organism, chemically speaking, is composed of **organic molecules**—that is, of molecules that contain carbon.

In this chapter, we present some of the types of organic molecules that are found in living things. The molecular drama is a grand spectacular with, literally, a cast of thousands. A single bacterial cell contains some 5000 different kinds of organic molecules, and an animal or plant cell has at least twice that many. As we noted, however, these thousands of molecules are composed of relatively few elements. Similarly, relatively few types of molecules play the major roles in living systems. Consider this chapter an introduction to the principal players in the drama. The plot begins to unfold in the next chapter.

Organic Molecules

The special bonding properties of carbon permit the formation of a great variety of organic molecules. Of the thousands of different organic molecules found in cells, just four different types make up most of the dry weight of living organisms. These four are **carbohydrates** (composed of sugars), **lipids** (most of which contain fatty acids), **proteins** (composed of amino acids), and **nucleic acids** (DNA and RNA, which are made up of complex molecules known as nucleotides). All of these molecules—carbohydrates, lipids, proteins, and nucleic acids—consist mainly of carbon and hydrogen, and most of them contain oxygen as well. In addition, proteins contain nitrogen and sulfur. Nucleic acids, as well as some lipids, contain nitrogen and phosphorus.

Carbohydrates

Carbohydrates are the most abundant organic molecules in nature and are the primary energy-storage molecules in most living organisms. In addition, they form a variety of structural components of living cells. The walls of young plant cells, for example, are made up of the carbohydrate cellulose embedded in a matrix of other carbohydrates and proteins.

Carbohydrates are formed from small molecules known as **sugars.** There are three principal kinds of carbohydrates, classified according to the number of sugar subunits they contain. **Monosaccharides** ("single, or simple, sugars"), such as ribose, glucose, and fructose, consist of only one sugar molecule. **Disaccharides** ("two sugars") contain two sugar subunits linked covalently. Familiar examples are sucrose (table sugar), maltose (malt sugar), and lactose (milk sugar). Cellulose and starch are **polysaccharides** ("many sugars"), which contain many sugar subunits linked together.

Macromolecules (large molecules), such as polysaccharides, that are made up of similar or identical small subunits are known as **polymers** ("many parts"). The individual subunits of polymers are called **monomers** ("single parts"); **polymerization** is the stepwise linking of monomers into polymers.

Monosaccharides Function As Building Blocks and Sources of Energy

Monosaccharides are the simplest carbohydrates. They are made up of linked carbon atoms to which hydrogen atoms and oxygen atoms are attached in the proportion of one carbon atom to two hydrogen atoms to one oxygen atom. Monosaccharides can be described by the formula $(CH_2O)_n$, where n can be as small as 3, as in $C_3H_6O_3$, or as large as 7, as in $C_7H_{14}O_7$. These proportions gave rise to the term *carbohydrate* (meaning "carbon with water added") for sugars and for the larger molecules formed from sugar subunits. Examples of several common monosaccharides are shown in Figure 2–1. Notice that each monosaccharide has a carbon chain (the carbon "skeleton") with a hydroxyl group (—OH) attached to every carbon atom except one. The remaining carbon atom is in the form of a carbonyl group (—C=O). Both these groups are **hydrophilic** ("water-loving"), and thus monosaccharides, as well as many other carbohydrates, readily dissolve in water. The five-carbon sugars (pentoses) and six-carbon sugars (hexoses) are the most common monosaccharides in nature. They occur in both a chain form and a ring form, and, in fact, they are normally found in the ring form when dissolved in water (Figure 2–2). As the ring is formed, the carbonyl group is converted to a hydroxyl group. Thus, the carbonyl group is a distinguishing feature of monosaccharides in the chain form but not in the ring form.

Representations of Molecules

When representing complex molecules, chemists usually use molecular formulas or structural formulas. A **molecular formula** indicates the number of atoms of each kind within the molecule, whereas a **structural formula** shows how the atoms are bonded to one another.

Glucose, for example, has 6 carbon atoms, 12 hydrogen atoms, and 6 oxygen atoms. Its molecular formula is $C_6H_{12}O_6$. However, fructose also contains 6 carbons, 12 hydrogens, and 6 oxygens and therefore has the same molecular formula as glucose. Moreover, the two molecules have similar structures—a chain of carbon atoms to which hydrogen and oxygen atoms are attached. The differences between glucose and fructose are determined by the way in which the oxygen and hydrogen atoms are attached to the carbon atoms. The molecules can therefore be distinguished by their structural formulas.

In these formulas, the symbol — represents a single covalent bond, and the symbol = represents a double covalent bond.

Although it is not necessary to number the carbon atoms, doing so often makes it easier to interpret the structural formulas. The terminal carbon atom at the more oxidized end of the molecule is considered number 1. Notice that glucose and fructose differ only in the location of the —C=O (carbonyl) group.

Glucose (chain form)

Fructose (chain form)

When glucose and fructose are in solution, however, they form ring structures and so are more accurately represented by these structural formulas:

Glucose (ring form)

Fructose (ring form)

The lower edges of the rings are drawn thicker to hint at a three-dimensional structure. The ring is perpendicular to the page, with the thick edges projecting toward you; the thin edges project behind the page. By convention, the carbon atoms at the intersections of the links in an organic ring structure are understood to be present and are not labeled.

Structural formulas provide convenient tools for examining the molecules involved in the structures and processes of living systems.

(a) Glyceraldehyde ($C_3H_6O_3$)

(b) Ribose ($C_5H_{10}O_5$)

(c) Glucose ($C_6H_{12}O_6$)

2–1

Examples of biologically important monosaccharides. (a) Glyceraldehyde, a three-carbon sugar, is an important energy source and provides the basic carbon skeleton for numerous organic molecules. (b) Ribose, a five-carbon sugar, is found in the nucleic acids DNA and RNA and in the energy-carrier molecule ATP. (c) The six-carbon sugar glucose serves important structural and transport roles in the cell. The terminal carbon nearest the double bond is designated carbon atom 1.

alpha-Glucose (ring form)

Glucose (chain form)

beta-Glucose (ring form)

2–2

In aqueous solution, the six-carbon sugar glucose exists in two different ring structures, alpha and beta, which are in equilibrium with each other. The molecules pass through the chain form to get from one structure to the other. The sole difference in the two ring structures is the position of the hydroxyl (—OH) group attached to carbon atom 1; in the alpha form it is below the plane of the ring, and in the beta form it is above the plane.

Monosaccharides are the building blocks—the monomers—from which living cells construct disaccharides, polysaccharides, and other essential carbohydrates. Moreover, the monosaccharide **glucose** is the form in which sugar is transported in the circulatory system of humans and other vertebrate animals. As we will see in Chapter 6, glucose and other monosaccharides are primary sources of chemical energy for plants as well as for animals.

The Disaccharide Sucrose Is a Transport Form of Sugar in Plants

Although glucose is the common transport sugar for many animals, sugars are often transported in plants and other organisms as disaccharides. **Sucrose,** a disaccharide composed of glucose and fructose, is the form in which sugar is transported in most plants from the photosynthetic cells (primarily in the leaves), where it is produced, to other parts of the plant body. The sucrose we consume as table sugar is commercially harvested from sugarbeets (enlarged roots) and sugarcane (stems), where it accumulates as it is transported from the photosynthetic parts of the plant.

In the synthesis of a disaccharide from two monosaccharide molecules, a molecule of water is removed and a new bond is formed between the two monosaccharides. This type of chemical reaction, which occurs when sucrose is formed from glucose and fructose, is known as **dehydration synthesis,** or a condensation reaction (Figure 2–3). In fact, the formation of most organic polymers from their subunits occurs by dehydration synthesis.

When the reverse reaction occurs—for example, when a disaccharide is split into its monosaccharide subunits—a molecule of water is added. This splitting, which occurs when a disaccharide is used as an energy source, is known as **hydrolysis,** from *hydro,* meaning "water," and *lysis,* "breaking apart." Hydrolysis reactions are energy-yielding processes that are important in energy transfers in cells. Conversely, dehydration synthesis reactions—the reverse of hydrolysis reactions—require an input of energy.

Polysaccharides Function As Storage Forms of Energy or As Structural Materials

Polysaccharides are polymers made up of monosaccharides linked together in long chains. Some polysaccharides function as storage forms of sugar, and others serve a structural role.

Starch, the primary storage polysaccharide in plants, consists of chains of glucose molecules (also known, in polymers, as glucose residues). There are two forms of starch: amylose, which is an unbranched molecule, and amylopectin, which is branched (Figure 2–4). Amylose and amylopectin are stored as starch grains within plant

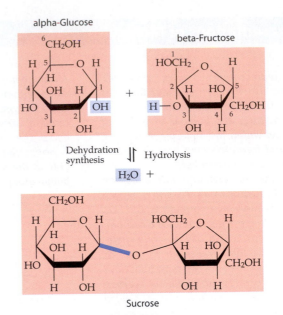

2–3

Sugar is generally transported in plants as the disaccharide sucrose. Sucrose is made up of two monosaccharide subunits, one alpha-glucose and one beta-fructose, bonded in a 1,2 linkage (the carbon 1 of glucose is linked to the carbon 2 of fructose). The formation of sucrose involves the removal of a molecule of water (dehydration synthesis). The new chemical bond formed in the course of this reaction is shown in blue. In cells, the formation of this bond, called a glycosidic bond, always involves an activated monomer (such as uridine diphosphate glucose, or UDP-glucose) and is therefore a more complicated, multistep process than is implied here. The reverse reaction—splitting sucrose into its constituent monosaccharides—requires the addition of a water molecule (hydrolysis). Formation of sucrose from glucose and fructose requires an energy input of 5.5 kcal per mole (see Appendix A) by the cell. Hydrolysis releases the same amount of energy.

cells. **Glycogen,** the common storage polysaccharide in prokaryotes, fungi, and animals, is also made up of chains of glucose molecules. It resembles amylopectin but is more highly branched. In some plants—most notably grains, such as wheat, rye, and barley—the principal storage polysaccharides in leaves and stems are polymers of fructose called **fructans.** These polymers are water-soluble and can be stored in much higher concentrations than starch.

Polysaccharides must be hydrolyzed to monosaccharides and disaccharides before they can be used as energy sources or transported through living systems. The plant breaks down its starch reserves when monosaccharides and disaccharides are needed for growth and development. We hydrolyze these polysaccharides when our digestive systems break down the starch stored by plants in such foods as maize (a grain) and potatoes (tubers), making glucose available as a nutrient for our cells.

(a) Amylose—linear chain of repeated alpha-glucose monomers

(b) Amylopectin—branched chain of repeated alpha-glucose monomers

Branch point

Branch point

2–4

In most plants, accumulated sugars are stored in the form of starch. Starch occurs in two forms: unbranched (amylose) and branched (amylopectin). (a) A single molecule of amylose may contain 1000 or more alpha-glucose monomers with carbon 1 of one glucose ring linked to carbon 4 of the next (known as a 1,4 linkage) in a long, unbranched chain that winds to form a uniform coil. (b) A molecule of amylopectin may contain 1000 to 6000 or more alpha-glucose monomers; short chains of about 8 to 12 alpha-glucose monomers branch off the main chain at intervals of about every 12 to 25 alpha-glucose monomers. (c) Perhaps because of their coiled nature, starch molecules tend to cluster into grains. In this scanning electron micrograph of a single storage cell of potato (Solanum tuberosum), the spherical structures are starch grains.

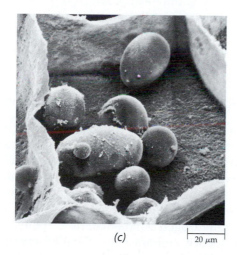

(c)

20 μm

Polysaccharides are also important structural compounds. In plants, the principal component of the cell wall is the polysaccharide **cellulose** (Figure 2–5). In fact, half of all the organic carbon in the biosphere is contained in cellulose, making it the most abundant organic compound known. Wood is about 50 percent cellulose, and cotton fibers are nearly pure cellulose.

Cellulose is a polymer composed of monomers of glucose, as are starch and glycogen, but there are important differences. Starch and glycogen can be readily used as fuels by almost all kinds of living systems, but only some microorganisms—certain prokaryotes, protozoa, and fungi—and a very few animals—such as silverfish—can hydrolyze cellulose. Cows, termites, and cockroaches can use cellulose for energy only because it is broken down by microorganisms that live in their digestive tracts.

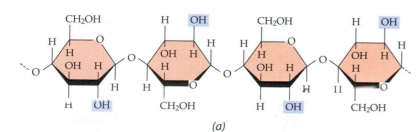

(a)

(b)

2–5

(a) Cellulose resembles starch in that it consists of glucose monomers in 1,4 linkages. Cellulose, however, consists of beta-glucose monomers, whereas starch is made up of alpha-glucose monomers. (b) Cellulose molecules, bundled together into microfibrils, are important structural components of plant cell walls. The —OH groups (blue), which project from both sides of the cellulose chain, form hydrogen bonds (dashed lines) with —OH groups on neighboring chains, resulting in microfibrils made up of cross-linked parallel cellulose molecules. Compare the structure of cellulose with that of starch in Figure 2–4.

To understand the differences between structural polysaccharides, such as cellulose, and energy-storage polysaccharides, such as starch or glycogen, we must look again at the glucose molecule. You will recall that the molecule is basically a chain of six carbon atoms and that when it is in solution, as it is in the cell, it assumes a ring form. The ring may close in either of two ways (Figure 2–2): one ring form is known as alpha-glucose, the other as beta-glucose. The alpha and beta forms are in equilibrium, with a certain number of molecules changing from one form to the other all the time, with the chain form as the intermediate. Starch and glycogen are both made up entirely of alpha-glucose subunits (Figure 2–4), whereas cellulose consists entirely of beta-glucose subunits (Figure 2–5). This seemingly slight difference has a profound effect on the three-dimensional structure of the cellulose molecules, which are long and unbranched. As a result, cellulose is impervious to the enzymes that easily break down the storage polysaccharides. Once glucose molecules are incorporated into the plant cell wall in the form of cellulose, they are no longer available to the plant as an energy source.

Cellulose molecules form the fibrous part of the plant cell wall. The long, rigid cellulose molecules combine to form microfibrils, each consisting of hundreds of cellulose chains. In plant cell walls, the cellulose microfibrils are embedded in a matrix containing two other complex, branched polysaccharides, namely, pectins (Figure 2–6) and hemicelluloses (Figure 2–7). Plant cell walls have been compared to reinforced concrete, in which embedded steel rods—the cellulose microfibrils—are used to strengthen the concrete—the matrix.

Chitin is another important structural polysaccharide. It is the principal component of fungal cell walls and also of the relatively hard outer coverings, or exoskeletons, of insects and crustaceans, such as crabs and lobsters. The monomer of chitin is the six-carbon sugar glucose to which a nitrogen-containing group has been added.

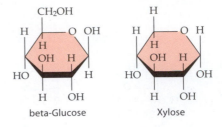

beta-Glucose   **Xylose**

2–7

Building-block sugars of the hemicellulose xyloglucan. The backbone of xyloglucan consists of a chain of beta-glucose subunits to which side chains of xylose are attached. The xyloglucans play an important structural role in stabilizing the primary cell wall by hydrogen bonding to the cellulose microfibrils.

Until the late 1960s, carbohydrates were thought of solely as energy sources and structural materials. Since then, evidence has accumulated that certain cell wall polysaccharides, dubbed "oligosaccharins," may function as hormones, regulating plant growth and development.

Lipids

Lipids are fats and fatlike substances. They are generally **hydrophobic** ("water-fearing") and thus are insoluble in water. Typically, lipids serve as energy-storage molecules—usually in the form of fats or oils—and for structural purposes, as in the case of phospholipids and waxes. Although some lipid molecules are very large, they are not, strictly speaking, macromolecules because they are not formed by the polymerization of monomers.

Fats and Oils Are Triglycerides That Store Energy

Plants, such as the potato, ordinarily store carbohydrates as starch. Some plants, however, also store food energy as oils (Figure 2–8), especially in seeds and fruits. Animals, which have a limited capacity for storing carbohydrates as glycogen, readily convert excess sugar to fat. Fats and oils contain a higher proportion of energy-rich carbon-hydrogen bonds than do carbohydrates and, as a consequence, contain more chemical energy. On the average, fats yield about 9.1 kilocalories (kcal) per gram, compared with 3.8 kcal per gram of carbohydrate or 3.1 kcal per gram of protein.

Fats and **oils** have similar chemical structures (Figure 2–9). Each consists of three fatty acid molecules bonded to one glycerol molecule. As with the formation of disaccharides and polysaccharides from their subunits, each of these bonds is formed by dehydration synthesis, which involves the removal of a molecule of water. Fat and oil molecules, also known as **triglycerides** (or triacylglycerols), contain no polar (hydrophilic) groups. Nonpolar molecules tend to cluster together in water, just as droplets of fats tend to coalesce, for example, on the surface of chicken soup. Nonpolar molecules are therefore hydrophobic, or insoluble in water.

(a) alpha-Galacturonic acid

(b) Pectic acid

2–6

Pectins are built up of monomers of alpha-galacturonic acid (a), which is a derivative of glucose. Polymers of this sugar derivative are known as pectic acid (b). Calcium and magnesium salts of pectic acid make up most of the middle lamella, a layer of intercellular material that cements together the walls of adjacent plant cells.

2–8

Two cells from the fleshy, underground stem (corm) of the quillwort Isoetes muricata. During winter these cells contain a large quantity of oil stored as oil bodies. In addition, carbohydrate in the form of starch grains is stored within amyloplasts, cellular structures in which the starch grains are formed. Several vacuoles (liquid-filled cavities) can be seen in each of these cells.

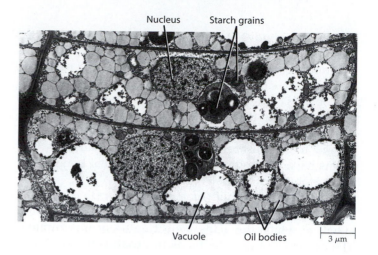

2–9

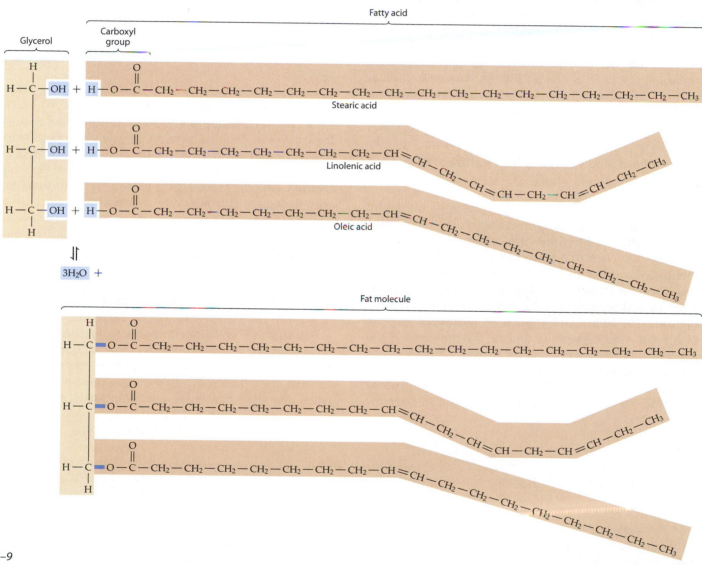

A fat molecule consists of three fatty acid molecules joined to a glycerol molecule (hence the term "triglyceride"). The long hydrocarbon chain of each fatty acid terminates in a carboxyl (—COOH) group, which becomes covalently bonded to the glycerol molecule. Each bond is formed when a molecule of water (blue) is removed (dehydration synthesis). The physical properties of a fat—such as its melting point—are determined by the lengths of its fatty acid chains and by whether the chains are saturated or unsaturated. Three different fatty acids are shown here. Stearic acid is saturated, and linolenic and oleic acids are unsaturated, as you can see by the double bonds in the hydrocarbon chains.

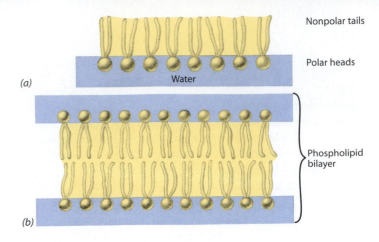

You have undoubtedly heard a lot about "saturated" and "unsaturated" fats. A fatty acid in which there are no double bonds between carbon atoms is said to be **saturated.** Each carbon atom in the chain has formed covalent bonds to four other atoms, and its bonding possibilities are therefore complete. By contrast, a fatty acid that contains carbon atoms joined by double bonds is said to be **unsaturated.** The double-bonded carbon atoms have the potential to form additional bonds with other atoms.

The physical nature of a fat is determined by the length of the carbon chains in the fatty acids and by the extent to which the acids are saturated or unsaturated. The presence of double bonds in unsaturated fats leads to kinks in the hydrocarbon chains, which prevents close packing among molecules. This tends to lower the melting point of the fat, and therefore unsaturated fats tend to be liquid (oily) at room temperature. Examples found in plants are safflower oil, peanut oil, and corn oil, which are obtained from oil-rich seeds. A notable exception is coconut oil, which is almost totally saturated. Animal fats and their derivatives, such as butter and lard, contain highly saturated fatty acids and are usually solid at room temperature.

Phospholipids Are Modified Triglycerides That Are Components of Cellular Membranes

Lipids, especially phospholipids, play very important structural roles, particularly in cellular membranes. Like triglycerides, **phospholipids** are composed of fatty acid molecules attached to a glycerol backbone. In the phospholipids, however, the third carbon of the glycerol molecule is occupied not by a fatty acid but by a phosphate group to which another polar group is usually attached (Figure 2–10). Phosphate groups are negatively charged. As a result, the phosphate end of the molecule is hydrophilic and therefore soluble in water, whereas the fatty acid end is hydrophobic and insoluble. If phospholipids are added to water, they tend to form a film along its surface, with their hydrophilic "heads" under the

2–11

(a) Because phospholipids have polar heads and nonpolar tails, they tend to form a thin film on a water surface. The hydrophilic (water-loving) heads are in the water, and the hydrophobic (water-fearing) tails extend above the water. **(b)** Surrounded by water, phospholipids spontaneously arrange themselves in two layers, with their hydrophilic heads extending outward (into the water) and their hydrophobic tails inward (away from the water). This arrangement—a phospholipid bilayer— forms the structural basis of cellular membranes.

water and their hydrophobic "tails" protruding above the surface. If phospholipids are surrounded by water, as in the watery interior of the cell, they tend to align themselves in double layers, with their phosphate heads directed outward and their fatty acid tails oriented toward one another (Figure 2–11). As we shall discuss further in Chapter 4, such configurations are important not only to the structure of cellular membranes but also to their functions.

Cutin, Suberin, and Waxes Are Lipids That Form Barriers to Water Loss

Cutin and **suberin** are unique lipids that are important structural components of many plant cell walls. The major function of these lipids is to form a matrix in which **waxes**—long-chain lipid compounds—are embedded. The waxes, in combination with cutin or suberin, form barrier layers that help prevent the loss of water and other molecules from plant surfaces.

2–10

A phospholipid molecule consists of two fatty acid molecules linked to a glycerol molecule, as in a triglyceride, but the third carbon of glycerol is linked to the phosphate group of a phosphate-containing molecule. The letter "R" denotes the atom or group of atoms that makes up the "rest of the molecule." The phospholipid tail is nonpolar and uncharged and is therefore hydrophobic (insoluble in water); the polar head containing the phosphate and R groups is hydrophilic (soluble in water).

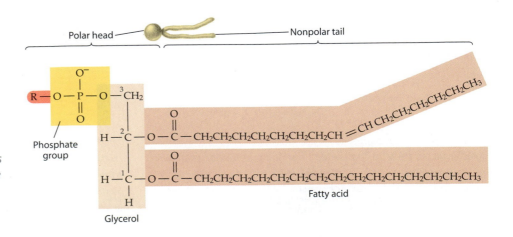

2–12

Scanning electron micrograph of the upper surface of a leaf of the eucalyptus tree (Eucalyptus cloeziana) showing deposits of epicuticular wax. Beneath these deposits is the cuticle, a wax-containing layer or layers covering the outer walls of the epidermal cells. Waxes help protect exposed plant surfaces from water loss.

A protective **cuticle,** which is characteristic of plant surfaces exposed to air, covers the outer walls of epidermal cells (the outermost cells) of leaves and stems. Composed of wax embedded in cutin **(cuticular wax),** the cuticle is frequently covered by a layer of **epicuticular wax** (Figure 2–12). When you polish a freshly picked apple on your sleeve, you are polishing this layer of epicuticular wax.

Suberin is a major component of the walls of cork cells, the cells that form the outermost layer of bark. As seen with the electron microscope, suberin-containing, or suberized, cell walls have a lamellar (layered) appearance, with alternating light and dark bands (Figure 2–13). The light bands are believed to be composed of waxes and the dark bands of suberin.

Waxes are the most water-repellent of the lipids. Carnauba wax used for car and floor polishes is harvested from the leaves of the carnauba wax palm *(Copernicia cerifera)* from Brazil.

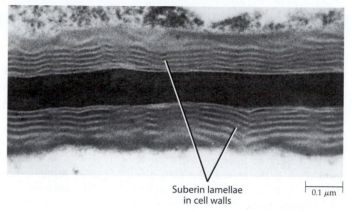

Suberin lamellae
in cell walls

2–13

Electron micrograph showing suberin lamellae in the walls between two cork cells of a potato tuber. Note the alternating bands, which are interpreted as consisting of wax (light bands) and suberin (dark bands). Cork cells form the outermost layer of a protective covering in plant parts such as potato tubers and woody stems and roots.

Steroids Have Linked Hydrocarbon Rings and Play a Variety of Roles in Plants

Steroids can be easily distinguished from the other classes of lipids by the presence of four interconnected hydrocarbon rings. In living organisms, hydrocarbon chains of various lengths as well as hydroxyl and/or carbonyl groups may be attached to this skeleton, making possible a large variety of molecules. When a hydroxyl group is attached at the carbon-3 position, the steroid is called a **sterol** (Figure 2–14a). Sitosterol (Figure 2–14b) is the most abundant sterol in green algae and plants, and ergosterol (Figure 2–14c) is found frequently in fungi. Cholesterol (Figure 2–14d), so common in animal cells, is present in only trace amounts in plants. In all organisms except prokaryotes, sterols are important components of membranes, where they stabilize the phospholipid tails.

2–14

Sterols. (a) General structure of a sterol. (b) β-Sitosterol, the most abundant sterol in green algae and plants. (c) Ergosterol, found frequently in fungi. (d) Cholesterol, common in animals.

(a) General structure of a sterol

(b) β-Sitosterol
(most abundant sterol in green
algae and plants)

(c) Ergosterol
(found frequently in fungi)

(d) Cholesterol
(common in animals)

Amino Acids and Nitrogen

Like fats, amino acids are formed within living cells using sugars as starting materials. Whereas fats contain only carbon, hydrogen, and oxygen atoms—all available in the sugar and water of the cell—amino acids also contain nitrogen. Most of the Earth's supply of nitrogen exists in the form of gas in the atmosphere. Only a few organisms, all of which are microorganisms, are able to incorporate nitrogen from the air into compounds—ammonia, nitrites, and nitrates—that can be used by living systems. Hence, only a small proportion of the Earth's nitrogen supply is available to the living world.

Plants incorporate the nitrogen from ammonia, nitrites, and nitrates into carbon-hydrogen compounds to form amino acids. Animals are able to synthesize some of their own amino acids, using ammonia derived from their diet as a nitrogen source. The amino acids they cannot synthesize, the so-called **essential amino acids**, *must be obtained in the diet, either from plants or from the meat of other animals that have eaten plants. For adult human beings, the essential amino acids are lysine, tryptophan, threonine, methionine, histidine, phenylalanine, leucine, valine, and isoleucine.*

People who eat meat usually get enough protein and the correct balance of amino acids. People who are vegetarians, whether for philosophical, aesthetic, or economic reasons, have to be careful that they get enough protein and, in particular, all of the essential amino acids.

For many years, agricultural scientists concerned with the world's hungry people concentrated on developing plants with a high caloric yield. More recently, increasing recognition of the role of plants as a major source of amino acids for human populations has led to emphasis on the development of high-protein strains of food plants. Of particular importance has been the development of plants, such as "high-lysine" maize, with increased levels of one or more of the essential amino acids.

Another approach to obtaining the right balance of amino acids from plant sources is to combine certain foods. Beans, for instance, are likely to be deficient in tryptophan and in the sulfur-containing amino acids cysteine and methionine, but they are a good-to-excellent source of isoleucine and lysine. Rice is deficient in isoleucine and lysine but provides an adequate amount of the other essential amino acids. Thus rice and beans in combination make just about as perfect a protein menu as eggs or steak, as some vegetarians seem to have known for quite a long time.

Steroids may also function as hormones. For example, the sterol antheridiol serves as a sex-attractant in the aquatic fungus *Achlya bisexualis*, and a group of steroid derivatives called brassins promote the growth of certain stems. There is also evidence that plants produce estrogen, one of the mammalian sex hormones, but its role in the plant is unknown.

Proteins

Proteins are among the most abundant organic molecules. In the majority of living organisms, proteins make up 50 percent or more of the dry weight. Only plants, with their high cellulose content, are less than half protein. Proteins perform an incredible diversity of functions in living systems. In their structure, however, proteins all follow the same simple blueprint: they are all polymers of nitrogen-containing molecules known as **amino acids,** arranged in a linear sequence. Some 20 different kinds of amino acids are used by living systems to form proteins. (See "Amino Acids and Nitrogen" above.)

Protein molecules are large and complex, often containing several hundred or more amino acid monomers. Thus, the possible number of different amino acid sequences, and therefore the possible variety of protein molecules, is enormous—about as enormous as the number of different sentences that can be written with our own 26-letter alphabet. Organisms, however, synthesize only a very small fraction of the proteins that are theoretically possible. A single cell of the bacterium *Escherichia coli*, for example, contains 600 to 800 different kinds of proteins at any one time, and a plant or animal cell has several times that number. A complex organism has at least several thousand different kinds of proteins, each with a special function and each, by its unique chemical nature, specifically fitted for that function.

In plants, the largest concentration of proteins is found in certain seeds, in which as much as 40 percent of the dry weight may be protein. These specialized proteins function as storage forms of amino acids to be used by the embryo when it resumes growth upon germination of the seed.

Amino Acids Are the Building Blocks of Proteins

Each specific protein is made up of a precise arrangement of amino acids. Amino acids have the same basic structure, consisting of an amino group ($-NH_2$), a carboxyl group ($-COOH$), and a hydrogen atom all bonded to a central carbon atom. The differences arise from the fact that every amino acid has an "R" group—an atom or a group of atoms bonded to the central carbon (Figure 2–15). It is the R group (R can be thought of as the rest of the molecule) that determines the identity of each amino acid.

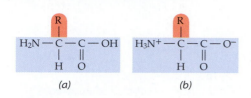

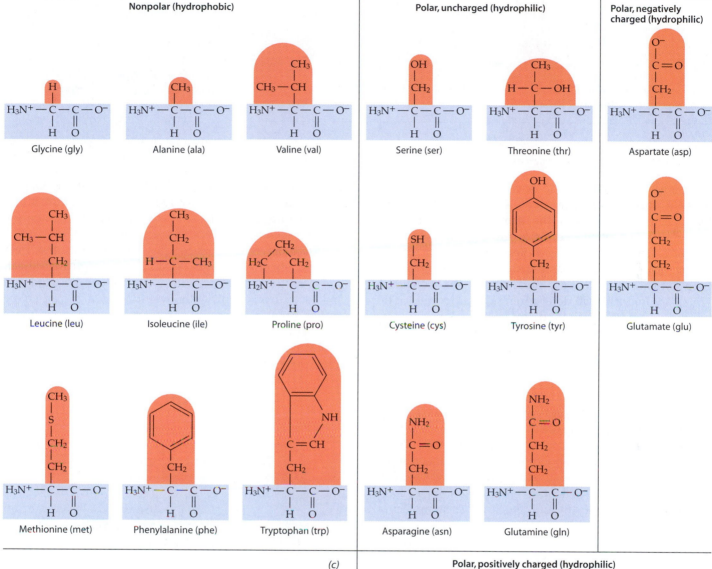

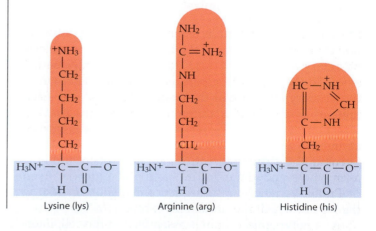

(c)

2–15

(a) The general formula of an amino acid. Every amino acid contains an amino group (—NH$_2$) and a carboxyl group (—COOH) bonded to a central carbon atom. A hydrogen atom and a side group (R) are also bonded to the same carbon atom. This basic structure is the same in all amino acids, but the R side group is different in each kind of amino acid. (b) At pH 7, both the amino and the carboxyl groups are ionized. (c) The 20 amino acids present in proteins. As you can see, the essential structure is the same in all 20 molecules, but the R side groups differ. Amino acids with nonpolar R groups are hydrophobic and, as proteins fold into their three-dimensional form, these amino acids tend to aggregate on the inside of proteins. Amino acids with polar, uncharged R groups are relatively hydrophilic and are usually on the surface of proteins. Those amino acids with acidic (negatively charged) and basic (positively charged) R groups are very polar, and therefore hydrophilic; they are almost always found on the surface of protein molecules. All of the amino acids are shown in the ionized state predominating at pH 7. The letters in parentheses following the name of each amino acid form the standard abbreviation for that amino acid.

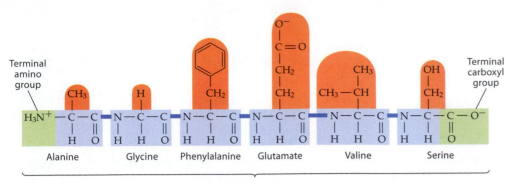

2–16

Polypeptides are polymers of amino acids linked together, one after another, by peptide bonds. Consequently, the basic structure of a polypeptide is a long, unbranched molecule. The short polypeptide chain shown here contains six different amino acids, but proteins consist of polypeptides with several hundred or even as many as 1000 linked amino acid monomers. This linear arrangement of the amino acids is known as the primary structure of the protein.

A large variety of amino acids is theoretically possible, but only 20 different kinds are used to build proteins. And it is always the same 20, whether in a bacterial cell, a plant cell, or a cell in your own body. Figure 2–15 shows the complete structure of the 20 amino acids found in proteins. The amino acids are grouped according to their polarity and electric charge, which are important in determining the properties of the individual amino acids and, more importantly, of the proteins formed from them.

In yet another example of a dehydration synthesis, the amino group of one amino acid links to the carboxyl group of the adjacent amino acid by the removal of a molecule of water. Again, this is an energy-requiring process. The covalent bond formed is known as a **peptide bond,** and the molecule that results from the linking of many amino acids is known as a **polypeptide** (Figure 2–16). Proteins are large polypeptides; these macromolecules have molecular weights ranging from 10^4 (10,000) to more than 10^6 (1,000,000). In comparison, water has a molecular weight of 18, and glucose has a molecular weight of 180.

A Protein's Structure Can Be Described in Terms of Levels of Organization

In a living cell, a protein is assembled in a long polypeptide chain, one amino acid at a time. The linear sequence of amino acids, which is dictated by the information stored in the cell for that particular protein, is known as the **primary structure** of the protein (Figure 2–16). Each kind of protein has a different primary structure—a unique protein "word" consisting of a unique sequence of amino acid "letters." The sequence of amino acids determines the structural features of the protein molecule as a whole and therefore its biological function. Even one small variation in the sequence may alter or destroy the way in which the protein functions.

As a polypeptide chain is assembled in the cell, interactions among the various amino acids along the chain cause it to fold into a pattern known as its **secondary structure.** Because peptide bonds are rigid, a chain is

limited in the number of shapes it can assume. One of the two most common secondary structures is the **alpha helix** (Figure 2–17). The shape of the helix is maintained by hydrogen bonds.

Another common secondary structure is the **beta pleated sheet** (Figure 2–18). In the beta pleated sheet, polypeptide chains are lined up in parallel and are linked by hydrogen bonds, resulting in a zigzag shape rather than a helix. In some proteins, two or more polypeptide chains are aligned with one another to form a pleated sheet. In other proteins, a single polypeptide chain loops back and forth in such a way that adjacent portions of the chain form a beta pleated sheet.

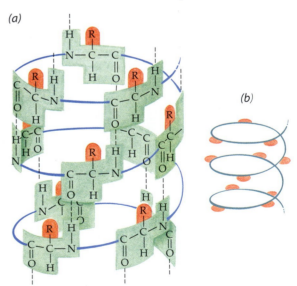

2–17

In many different types of proteins, the polypeptide chain forms a secondary structure known as an alpha helix. (a) The helix is held in shape by hydrogen bonds, indicated by the dashed lines. The hydrogen bonds form between the double-bonded oxygen atom in one amino acid and the hydrogen atom of the amino group in another amino acid situated four amino acids further along the chain. The R groups, which appear flattened in this diagram, actually extend out from the helix, as shown in (b). In some proteins, virtually all of the molecule is in the form of an alpha helix. In other proteins, only certain regions of the molecule have this secondary structure.

2–18

*Another common secondary structure of proteins is the beta pleated sheet. (**a**) The pleats result from the alignment of the zigzag pattern of the atoms that form the backbone of polypeptide chains. The sheet is held together by hydrogen bonding between adjacent chains. The R groups extend above and below the pleats, as shown in (**b**). In some proteins, two or more polypeptide chains are aligned with one another to form a pleated sheet. In other proteins, a single polypeptide chain loops back and forth in such a way that adjacent portions of the chain form a pleated sheet.*

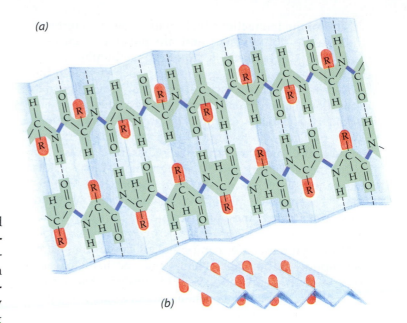

(a)

(b)

Proteins that exist for most of their length in a helical or pleated-sheet secondary structure are known as **fibrous proteins.** Fibrous proteins play a variety of important structural roles, providing support and shape in organisms. In other proteins, known as **globular proteins,** the secondary structure folds to form a **tertiary structure** (Figure 2–19). For some proteins, the folding occurs spontaneously, that is, by a self-assembly process. For others, certain proteins known as **molecular chaperones** facilitate the process by inhibiting incorrect folding. Globular proteins tend to be structurally complex, often having more than one type of secondary structure. Most biologically active proteins, such as enzymes, membrane proteins, and transport proteins, are globular, as are some important structural proteins. The microtubules that occur within the cell, for example, are composed of a large number of spherical subunits, each of which is a globular protein (see Figure 3–25).

The tertiary structure forms as a result of complex interactions among the R groups in the individual amino acids. These interactions include attractions and repulsions among amino acids with polar R groups and repulsions between nonpolar R groups and the surrounding water molecules. In addition, the sulfur-containing R groups of the amino acid cysteine can form covalent bonds with each other. These bonds, known as **disulfide bridges,** lock portions of the molecule into a particular position.

2–19

*The four levels of protein organization. (**a**) The primary structure of a protein consists of a linear sequence of amino acids linked together by peptide bonds. (**b**) The polypeptide chain may coil into an alpha helix, one form of secondary structure. (**c**) The alpha helix may fold to form a three-dimensional, globular structure, the tertiary structure. (**d**) The combination of several polypeptide chains into a single functional molecule is the quaternary structure. The polypeptides may or may not be identical.*

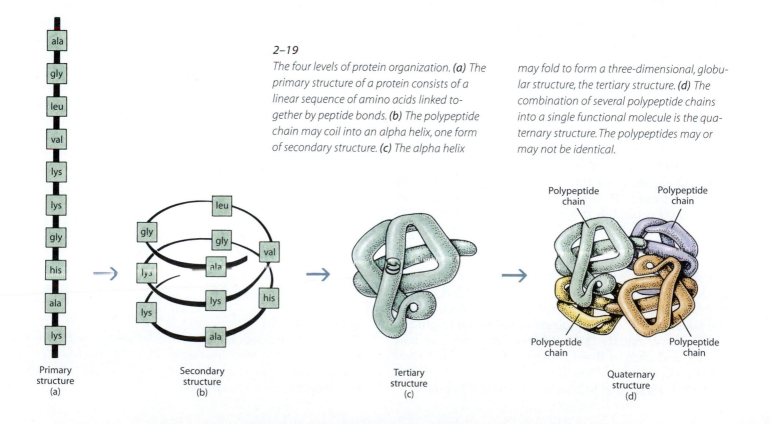

Primary structure (a)

Secondary structure (b)

Tertiary structure (c)

Quaternary structure (d)

Polypeptide chain

Polypeptide chain

Polypeptide chain

Polypeptide chain

Most of the interactions that give a protein its tertiary structure are not covalent and are therefore relatively weak. They can be disrupted quite easily by physical or chemical changes in the environment, such as heat or increased acidity. This structural breakdown is called **denaturation.** The coagulation of egg white when an egg is cooked is a familiar example of protein denaturation. When proteins are denatured, the polypeptide chains unfold and the tertiary structure is disrupted, causing a loss of the biological activity of the protein. Most organisms cannot live at extremely high temperatures or outside a specific pH range because their enzymes and other proteins become unstable and nonfunctional due to denaturation.

Many proteins are composed of more than one polypeptide chain. These chains may be held to each other by hydrogen bonds, disulfide bridges, hydrophobic forces, attractions between positive and negative charges, or, most often, by a combination of these types of interactions. This level of organization of proteins—the interaction of two or more polypeptides—is called a **quaternary structure** (Figure 2–19d).

Enzymes Are Proteins That Catalyze Chemical Reactions in Cells

Enzymes are large, complex globular proteins that act as catalysts. By definition, **catalysts** are substances that accelerate the rate of a chemical reaction by lowering the energy of activation (see Figure 5–6), but remain unchanged in the process. Because they remain unaltered, catalyst molecules can be used over and over again and so are typically effective at very low concentrations.

In the laboratory, the rates of chemical reactions are usually accelerated (up to a point) by the application of heat, which increases the force and frequency of collisions among molecules. In nature, however, hundreds of different reactions are going on in the cell at the same time, and heat would speed up all these reactions indiscriminately. Moreover, heat would melt the lipids, denature the proteins, and have other generally destructive effects on the cell. Because of enzymes, cells are able to carry out chemical reactions at great speeds at relatively low temperatures. If enzymes were not present, the reactions would occur, but at a rate so slow that their effects would be negligible, and life at "room temperature" would not be possible.

Enzymes are often named by adding the ending *-ase* to the root of the name of the substrate (the reacting molecule or molecules). Thus, amylase catalyzes the hydrolysis of amylose (starch) into glucose molecules, and sucrase catalyzes the hydrolysis of sucrose into glucose and fructose. Nearly 2000 different enzymes are now known, and each of them is capable of catalyzing some specific chemical reaction. The behavior of enzymes in biological reactions is further explained in Chapter 5.

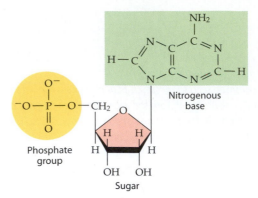

2–20

A nucleotide is made up of three different subunits: a phosphate group, a five-carbon sugar, and a nitrogenous base. The nitrogenous base in this nucleotide is adenine, and the sugar is ribose. Because there is a single phosphate group, this nucleotide is called adenosine monophosphate, abbreviated AMP.

Nucleic Acids

The information dictating the structures of the enormous variety of proteins found in living organisms is encoded in and translated by molecules known as nucleic acids. Just as proteins consist of long chains of amino acids, nucleic acids consist of long chains of molecules known as **nucleotides.** A nucleotide, however, is a more complex molecule than an amino acid.

As shown in Figure 2–20, a nucleotide consists of three subunits: a phosphate group, a five-carbon sugar, and a nitrogenous base—a molecule that has the properties of a base and contains nitrogen. The phosphate group (PO_4^{3-}) is an ion of phosphoric acid (H_3PO_4); it is the source of "acid" in the term "nucleic acid." The sugar subunit of a nucleotide may be either **ribose** or **deoxyribose,** which contains one less oxygen atom than ribose (Figure 2–21). Five different nitrogenous bases occur in the nucleotides that are the building blocks of nucleic acids. In the nucleotide in Figure 2–20, the nitrogenous base is adenine. In adenine and other nitrogenous bases, each of the nitrogen atoms in the molecule has an unshared pair of electrons in the outer energy level. These electrons exert a weak attraction for hydrogen ions (H^+). Thus the molecule is a base, capable of combining with H^+ ions and thereby increasing the rela-

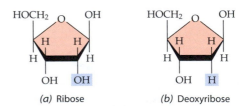

(a) Ribose   *(b)* Deoxyribose

2–21

The sugar subunit of a nucleotide may be either (a) ribose or (b) deoxyribose. The structural difference between the two sugars is highlighted in blue. RNA is formed from nucleotides that contain ribose, and DNA from nucleotides that contain deoxyribose.

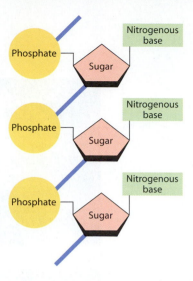

2–22

Nucleic acid molecules are long chains of nucleotides in which the sugar subunit of one nucleotide is linked to the phosphate group of the next nucleotide. The covalent bond linking one nucleotide to the next—shown here in blue—is formed by a dehydration synthesis reaction. This reaction is reversible; with the addition of a molecule of water in hydrolysis, the bond can be broken.

RNA molecules consist of a single chain of nucleotides, as shown here. DNA molecules, by contrast, consist of two chains of nucleotides, coiled around each other in a double helix.

tive number of OH⁻ (hydroxide ions) in a solution (see Appendix A).

Two types of nucleic acids are found in living organisms. In **ribonucleic acid (RNA),** the sugar subunit in the nucleotides is ribose. In **deoxyribonucleic acid (DNA),** it is deoxyribose. Like polysaccharides, lipids, and proteins, RNA and DNA are formed from their subunits in dehydration synthesis reactions. The result is a linear macromolecule consisting of one nucleotide after another (Figure 2–22). DNA molecules, in particular, are exceedingly long and, in fact, are the largest macromolecules in cells.

Although their chemical components are very similar, DNA and RNA generally play different biological roles. DNA is the carrier of the genetic message. It contains the information, organized in units known as **genes,** that we and other organisms inherit from our parents. RNA molecules are involved in the synthesis of proteins based on the genetic information provided by

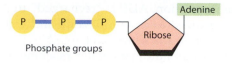

Adenosine triphosphate (ATP)

2–23

A schematic diagram of the ATP (adenosine triphosphate) molecule. The only difference between this molecule and the AMP (adenosine monophosphate) molecule shown in Figure 2–20 is the addition of two phosphate groups. Although this structural difference may seem minor, it is the key to the function of ATP in living systems.

DNA. Some RNA molecules function as enzymelike catalysts (sometimes referred to as ribozymes).

The discovery of the structure and function of DNA and RNA is undoubtedly the greatest triumph thus far of the molecular approach to the study of biology. In Section 3, we shall trace the events leading to the key discoveries and consider in some detail the marvelous processes—the details of which are still being worked out—by which the nucleic acids perform their functions.

The Molecule ATP Is the Cell's Energy Currency

In addition to their role as the building blocks of nucleic acids, nucleotides have an independent and crucial function in living systems. When modified by the addition of two more phosphate groups, they are the carriers of the energy needed to power the numerous chemical reactions occurring within cells.

The energy in storage carbohydrates, such as starch and glycogen, and in lipids is like money in savings bonds or certificates of deposit—not readily accessible. The energy in glucose is like money in a checking account—accessible but not very handy for everyday transactions. The energy in modified nucleotides, however, is like money in your pocket—available immediately, in convenient amounts, and universally accepted.

The principal energy carrier for most processes in living organisms is the molecule **adenosine triphosphate,** or **ATP,** shown schematically in Figure 2–23. Note the three phosphate groups. The bonds linking these groups are relatively weak, and they can be broken quite readily by hydrolysis. The products of the most common reaction, illustrated in Figure 2–24, are **ADP (adenosine diphosphate),** a phosphate group, and energy. As this energy is released, it can be used to power other chemical reactions.

2–24

The hydrolysis of ATP. With the addition of a molecule of water to ATP, one phosphate group is removed from the molecule. The products of this reaction are ADP, a free phosphate group, and energy. About 7 kcal of energy are released for every mole of ATP hydrolyzed. With an energy input of 7 kcal per mole, the reaction can be reversed.

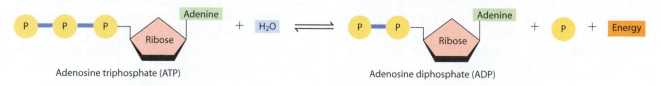

Adenosine triphosphate (ATP)   Adenosine diphosphate (ADP)

During respiration, ADP is "recharged" to ATP when glucose is oxidized to carbon dioxide and water, just as the money in your pocket is "recharged" when you cash a check or visit an automatic teller machine. In Chapter 6, we shall consider this process in more detail. For now, however, the important thing to remember is that ATP is the molecule that is directly involved in providing energy for the living cell, to which we shall turn our attention in Section 2.

Secondary Metabolites

Historically, the compounds produced by plants have been separated into primary and secondary metabolites, or products. **Primary metabolites,** by definition, are molecules that are found in all plant cells and are necessary for the life of the plant. Examples of primary metabolites are simple sugars, amino acids, proteins, and nucleic acids. **Secondary metabolites,** by contrast, are restricted in their distribution, both within the plant and among the different species of plants. Once considered waste products, secondary metabolites are now known to be important for the survival and propagation of the plants that produce them. Many serve as chemical signals that enable the plant to respond to environmental cues. Others function in the defense of the producer against herbivores, pathogens (disease-causing organisms), or competitors. Some provide protection from radiation from the sun, while still others aid in pollen and seed dispersal.

As indicated, secondary metabolites are not evenly distributed throughout the plant. Their production typically occurs in a specific organ, tissue, or cell type at specific stages of development (e.g., during flower, fruit, seed, or seedling development). Some, the **phytoalexins,** are antimicrobial compounds produced only after wounding or after attack by bacteria or fungi (see pages 62, 780, and 782). Secondary metabolites are produced at various sites within the cell and are stored primarily within vacuoles. They frequently are synthesized in one part of the plant and stored in another. Moreover, their concentration in a plant often varies greatly during a 24-hour period.

There are three major classes of secondary plant compounds: alkaloids, terpenoids, and phenolics.

Alkaloids Are Classes of Molecules That Include Morphine, Cocaine, Caffeine, Nicotine, and Atropine

Alkaloids are among the most important pharmacologically, or medicinally, active compounds. Interest in them has traditionally stemmed from their dramatic physiological or psychological effect on humans. They are bitter-tasting nitrogenous compounds that are basic (alkaline) in their chemical properties.

(a) Morphine

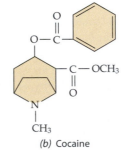
(b) Cocaine

2–25

The structures of some physiologically active alkaloids. (a) Morphine is contained in the milky juice released by slitting the seed pods of the opium poppy (Papaver somniferum); *(b) cocaine is present in the leaves of the coca*

The first alkaloid to be identified—in 1806—was **morphine,** from the opium poppy (*Papaver somniferum*). It is used today in medicine as an analgesic (pain reliever) and cough suppressant; however, excessive use of this drug can lead to strong addiction. Nearly 10,000 alkaloids have now been isolated and their structures identified, including cocaine, caffeine, nicotine, and atropine. The structures of some of these physiologically active alkaloids are shown in Figure 2–25.

Cocaine comes from coca (*Erythroxylum coca*), a shrub or small tree that is indigenous to the eastern slopes of the Andes Mountains of Bolivia and Peru. Many Incas living at the high elevations of these mountains chew coca leaves to lessen hunger pangs and fatigue while working in this harsh environment. Chewing the leaves, which contain small concentrations of cocaine, is relatively harmless compared with the smoking, snorting, or intravenous injection of cocaine. The habitual use of cocaine and its derivative "crack" can have devastating effects both physically and psychologically, and can lead to death. Cocaine has been used as an anesthetic in eye surgery and as a local anesthetic by dentists.

Caffeine, found in such plants as coffee (*Coffea arabica*), tea (*Camellia sinensis*), and cocoa (*Theobroma cacao*), is a component of popular beverages. It acts as a stimulant. The high concentrations of caffeine present in the

(c) Caffeine

(d) Nicotine

plant (Erythroxylum coca); (c) coffee (Coffea)
beans and tea (Camellia) leaves contain
caffeine; and (d) cultivated tobacco
(Nicotiana tabacum) plants contain nicotine.

developing seedlings of the coffee plant have been shown to be highly toxic and lethal to both insects and fungi. In addition, caffeine released by the seedling apparently inhibits germination of other seeds in the vicinity of the seedling, preventing the growth of competitors. This process, called **allelopathy,** will be discussed further in Chapter 32.

Nicotine, another stimulant, is obtained from the leaves of the tobacco plant *(Nicotiana tabacum).* It is a highly toxic alkaloid and has received considerable attention because of concern over the harmful effects of cigarette smoking. Nicotine is synthesized in the roots and transported to the leaves, where it is sequestered in vacuoles. It is an effective deterrent to attack by grazing herbivores and insects. Nicotine is synthesized in response to wounding, apparently functioning as a phytoalexin.

Atropine-containing extracts from the Egyptian henbane *(Hyoscyamus muticus)* were used by Cleopatra in the last century B.C. to dilate her pupils, in the hope that she would appear more alluring. During the Medieval period, European women used atropine-containing extracts from the deadly nightshade *(Atropa belladonna)* for the same purpose. Atropine is used today as a cardiac stimulant, a pupil dilator for eye examinations, and an effective antidote for some nerve gas poisoning.

Terpenoids Are Composed of Isoprene Units and Include Essential Oils, Taxol, Rubber, and Cardiac Glycosides

Terpenoids, also called terpenes, occur in all plants and are by far the largest class of secondary metabolites, with over 22,000 terpenoid compounds described. The simplest of the terpenoids is the hydrocarbon isoprene (C_5H_8). All terpenoids can be classified according to their number of isoprene units (Figure 2–26a). Familiar categories of terpenoids are the monoterpenoids, which consist of two isoprene units; sesquiterpenoids (three isoprene units); and diterpenoids (four isoprene units). A single plant may synthesize many different terpenoids, at different locations within the plant for a variety of purposes, and at different times during the course of its development.

Isoprene itself is emitted in significant quantities by the leaves of many plant species and is largely responsible for the bluish haze that hovers over wooded hills and mountains in summer (Figure 2–26b). It is also a component of smog. Isoprene is emitted only in the light. It is made in chloroplasts from carbon dioxide recently converted to organic compounds by photosynthesis. One may wonder why plants produce and discharge such large quantities of isoprene. Studies have shown that isoprene emissions are highest on hot days, indicating that isoprene production may somehow aid the plant in coping with heat.

(b)

2–26

(a) *A diverse group of compounds is formed from isoprene units. All sterols, for example, are built of six isoprene units (Figure 2–14).*
(b) *Blue haze, composed largely of isoprene, hovering over the Three Sisters Blue Mountains of Australia.*

Isoprene (C_5H_8)

(a)

Many of the monoterpenoids and sesquiterpenoids are called **essential oils** because they are highly volatile and contribute to the fragrance, or essence, of the plants that produce them. In the mint *(Mentha)* large quantities of volatile monoterpenoids (menthol and menthone) are both synthesized and stored in glandular trichomes (hairs), which are outgrowths of the epidermis. The essential oils produced by the leaves of some plants deter herbivores; some protect against attack by fungi or bacteria; others are known to be allelopathic. The terpenoids of flower fragrances attract insect pollinators to the flowers.

The diterpenoid **taxol** has attracted considerable attention because of its anti-cancer properties. It has been shown to shrink cancers of the ovary and the breast. At one time, the only source of taxol was the bark of the Pacific yew tree *(Taxus brevifolia)*. Harvesting all of the bark from a tree yields only a very small amount of taxol (a scant 300-milligram dose from a 40-foot-tall 100-year-old tree). Moreover, removal of the bark kills the tree. Fortunately, it has been found that extracts from the needles of the European yew tree *(Taxus baccata)* and *Taxus* bushes, as well as a yew fungus, can yield taxol-like compounds. Needles can be harvested without destroying the yew trees and bushes. Taxol has since been synthesized in the laboratory, but the synthesis technique remains to be refined. Even then, it may be cheaper to make commercial forms of taxol from natural sources.

The largest known terpenoid compound is **rubber,** which consists of molecules containing 400 to more than 100,000 isoprene units. Rubber is obtained commercially from the milky fluid, called *latex,* of the tropical plant *Hevea brasiliensis,* a member of the *Euphorbiaceae* family (Figure 2–27). About 1800 species of eudicotyledons have been reported to contain rubber, but only a few yield enough rubber to make them commercially valuable. In *Hevea,* rubber may constitute 40 to 50 percent of the latex. Latex is obtained from the rubber tree by making a V-shaped incision in the bark. A spout is inserted at the bottom of the incision, and the latex flows down the incision and is collected in a cup attached to the tree. The latex is processed, and the rubber is removed and pressed into sheets for shipment to factories.

Many terpenoids are poisonous, among them the *cardiac glycosides,* which are sterol derivatives that can cause heart attacks. When used medicinally, cardiac glycosides can result in a slower and strengthened heartbeat. Foxglove plants *(Digitalis)* are the principal source of the most active cardiac glycosides, digitoxin and digoxin. Cardiac glycosides synthesized by members of the dogbane order of plants *(Apocynales)* provide an effective defense against herbivores. Interestingly, some insects have learned to adapt to these toxins. An example is the caterpillar of the monarch butterfly, which feeds preferentially on milkweed *(Asclepias* spp.) and stores the cardiac glycosides safely within its body (Figure 2–28). When

2–27

Tapping of the tropical rubber tree, Hevea brasiliensis, *to obtain rubber, a component of the milky latex. These cultivated trees are being tapped by an Iban village elder on the island of Borneo, Malaysia.*

the adult butterfly leaves the host plant, the bitter-tasting cardiac glycosides protect it against predator birds. Ingestion of the cardiac glycosides causes the birds to vomit, and they soon learn to recognize and avoid other brightly colored monarchs on sight alone.

Terpenoids play a multiplicity of roles in plants. Some are photosynthetic pigments (carotenoids) or hormones (gibberellins, abscisic acid), while others serve as structural components of membranes (sterols) or as electron carriers (ubiquinone, plastoquinone), in addition to the roles already mentioned. All of these substances will be discussed in subsequent chapters.

Phenolics Include Flavonoids, Tannins, Lignins, and Salicylic Acid

The term **phenolics** encompasses a broad range of compounds, all of which have a hydroxyl group (—OH) attached to an aromatic ring (a ring of six carbons containing three double bonds). They are almost universally present in plants and are known to accumulate in all plant parts (roots, stems, leaves, flowers, and fruits). Although they represent the most studied of secondary metabolites, the function of many phenolic compounds is still unknown.

The **flavonoids,** which are water-soluble pigments present in the vacuoles of plant cells, represent the largest group of plant phenolic compounds. Flavonoids found in red wines and grape juice have received considerable attention because of their reported lowering of cholesterol levels in the blood. Over 3000 different flavonoids have been described, and they are probably

2–29

The intensely blue color of these day flowers, Commelina communis, *is the result of co-pigmentation. In the day flower, anthocyanin and flavone molecules are joined to a magnesium ion to form the blue pigment commelinin.*

2–28

(a) A monarch butterfly caterpillar feeds on milkweed, ingesting and storing the toxic compounds produced by the plant. The monarch caterpillar and also the monarch butterfly (b) thus become unpalatable and poisonous. The conspicuous coloration of caterpillar and butterfly warns would-be predators.

the most intensively studied secondary metabolites of plants. Flavonoids are divided into several classes, including the widespread anthocyanins, flavones, and flavonols. The **anthocyanins** range in color from red through purple to blue. Most of the flavones and flavonols are yellowish or ivory-colored pigments, some are colorless. The colorless flavones and flavonols can alter the color of a plant part through the formation of complexes with anthocyanins and metal ions. This phenomenon, called **co-pigmentation,** is responsible for some intensely blue flower colors (Figure 2–29).

Flower pigments act as visual signals to attract pollinating birds and bees, a role recognized by Charles Darwin and by naturalists before and after his time. Flavonoids also affect how plants interact with other organisms such as symbiotic bacteria living within plant roots, as well as microbial pathogens. For example, flavonoids released from the roots of legumes can stimulate or inhibit specific genetic responses in the different bacteria associated with them. Flavonoids may also provide protection against damage from ultraviolet radiation.

Probably the most important deterrents to herbivore feeding in angiosperms (flowering plants) are the **tannins,** phenolic compounds present in relatively high concentrations in the leaves of a wide array of woody plants. It is their astringent taste that is repellent to insects, reptiles, birds, and higher animals. Humans have made use of tannins to tan leather, denaturing the protein of the leather and protecting it from bacterial attack. Tannins are sequestered in vacuoles in order to protect the other components of the cell (Figure 2–30).

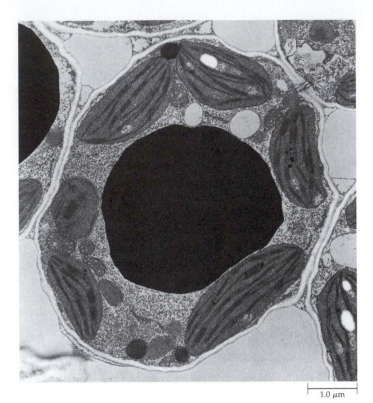

1.0 μm

2–30

Tannin-containing vacuole in a leaf cell of the sensitive plant, Mimosa pudica. *The electron-dense tannin literally fills the central vacuole of this cell.*

Lignins, unlike other phenolic compounds, are deposited in the cell wall rather than in the vacuole. Second only to cellulose as the most abundant organic compound on Earth, lignins are polymers formed from three types of monomers: *p*-coumaryl, coniferyl, and sinapyl alcohols. The relative amount of each monomer differs significantly depending on whether the lignin is from gymnosperms, woody angiosperms, or grasses. In addition, there is great variation in the monomeric composition of lignins of different species, organs, tissues, and even cell wall fractions.

The major importance of lignin is the compressive strength and stiffness it adds to the cell wall. **Lignification,** the process of lignin deposition, is believed to have played a major role in the evolution of terrestrial plants. Although unlignified cell walls can withstand substantial tensile forces, they are weak against the compressive forces of gravity. With the addition of lignin to the walls, it was possible for terrestrial plants to increase in stature and to develop branch systems capable of supporting large photosynthetic surfaces.

Lignin also waterproofs the cell wall. It therefore facilitates upward transport of water in the conducting cells of the xylem by limiting the outward movement of water from the cells. In addition, lignin assists the water-conducting cells in resisting the tension generated by the stream of water (the transpiration stream) being pulled to the top of tall plants (see Chapter 31). A further role of lignin is indicated by its deposition in response to vari-

Salicylic acid

Acetylsalicylic acid
(aspirin)

(a)

2–31

*(a) Chemical structures of salicylic acid and aspirin. (b) Willow tree (*Salix*) growing along a stream bank.*

(b)

ous types of injury and attack by fungi. The so-called "wound lignin" protects the plant from fungal attack by increasing the resistance of walls to mechanical penetration, protecting them against fungal enzyme activity, and reducing the diffusion of enzymes and toxins from the fungi into the plant. It has been suggested that lignin may have first functioned as an antifungal and antibacterial agent and only later assumed a role in water transport and mechanical support in the evolution of terrestrial plants.

Salicylic acid, the active ingredient in aspirin, was first known for its analgesic properties. It was discovered by the ancient Greeks and by Native Americans, who obtained it for pain relief from tea brewed from willow *(Salix)* bark (Figure 2–31). Only recently, however, has the action of this phenolic acid in plant tissues been discovered: it is essential for the development of systemic acquired resistance, commonly referred to as SAR. SAR develops in response to a localized attack by pathogenic bacteria, fungi, or viruses. As a result, other portions of the plant are provided with long-lasting protection against the same and unrelated pathogens. Salicylic acid also probably triggers the dramatic increases in temperature during flowering of the voodoo lily *(Sauromatum guttatum)* and other species of *Araceae*.

With this introduction to the molecular composition of plant cells, the stage has been set to turn our attention to the living cell—its structure and the activities by which it maintains itself as an entity distinct from the nonliving world surrounding it. We shall explore how the organic molecules introduced in this chapter perform their functions. It should come as no surprise that organic molecules rarely function in isolation but rather in combination with other organic molecules. The marvelous processes by which these molecules perform their functions still are not clearly understood and remain the object of intensive investigations.

Summary

Living Matter Is Composed of Only a Few of the Naturally Occurring Elements

Living organisms are composed mostly of only six elements: carbon, hydrogen, nitrogen, oxygen, phosphorus, and sulfur. The bulk of living matter is water. Most of the rest of living material is composed of organic molecules—carbohydrates, lipids, proteins, and nucleic acids. Polysaccharides, proteins, and nucleic acids are examples of macromolecules, which are made up of similar monomers joined together by dehydration synthesis

reactions to form polymers. By the reverse process, called hydrolysis, polymers can be split apart to their constituent monomers.

Carbohydrates Are Sugars and Their Polymers

Carbohydrates serve as a primary source of chemical energy for living systems and as important structural elements in cells. The simplest carbohydrates are the monosaccharides, such as glucose and fructose. Monosaccharides can be combined to form disaccharides, such as sucrose, and polysaccharides, such as starch and cellulose. These molecules can usually be broken apart by the addition of a water molecule at each linkage, a chemical reaction known as hydrolysis.

Lipids Are Hydrophobic Molecules That Play a Variety of Roles in the Cell

Lipids are another source of energy and structural material for cells. Compounds in this group—fats, oils, phospholipids, cutin, suberin, waxes, and steroids—are generally insoluble in water.

Fats and oils, also known as triglycerides, store energy. Phospholipids are modified triglycerides that are important components of cellular membranes. Cutin, suberin, and waxes are lipids that form barriers to water loss. The surface cells of stems and leaves are covered with a cuticle, composed of wax and cutin, that prevents water loss. Steroids are molecules having four interconnected hydrocarbon rings. Steroids are found in cellular membranes, and they may have other roles in the cell as well.

Proteins Are Versatile Polymers of Amino Acids

Amino acids have an amino group, a carboxyl group, a hydrogen, and a variable R group attached to the same carbon atom. Twenty different kinds of amino acids—differing in the size, charge, and polarity of the R group—are used to build proteins. By the process of dehydration synthesis, amino acids are joined by peptide bonds. A chain of amino acids is a polypeptide, and proteins consist of one or more long polypeptides.

A protein's structure can be described in terms of levels of organization. The primary structure is the linear sequence of amino acids joined by peptide bonds. The secondary structure, most commonly an alpha helix or a beta pleated sheet, is formed as a result of hydrogen bonding between amino and carboxyl groups. The tertiary structure is the folding that results from interactions between R groups. The quaternary structure results from specific interactions between two or more polypeptide chains.

Summary TABLE   Biologically Important Organic Molecules

CLASS OF MOLECULE	TYPES	SUBUNITS	MAIN FUNCTIONS	OTHER FEATURES
Carbohydrates	Monosaccharides (e.g., glucose)	Monosaccharide	Ready energy source	Carbohydrates are sugars and polymers of sugars.
	Disaccharides (e.g., sucrose)	Two monosaccharides	Transport form in plants	To identify carbohydrates, look for compounds that consist of monomers with many hydroxyl groups (—OH) and usually one carbonyl (—C≡O) group attached to the carbon skeleton. However, if the sugars are in the ring form, the carbonyl group is not evident.
	Polysaccharides	Many monosaccharides	Energy storage or structural components	
	Starch		Major energy storage in plants	
	Glycogen		Major energy storage in prokaryotes, fungi, and animals	
	Cellulose		Component of plant cell walls	
	Chitin		Component of fungal cell walls	
Lipids	Triglycerides	3 fatty acids + 1 glycerol	Energy storage	Lipids are nonpolar molecules that will not dissolve in polar solvents such as water. Thus lipids are the ideal molecule for long-term energy storage. They can be "put aside" in a cell, will not dissolve in the watery environment or "leak out" into the rest of the cell.
	Oils		Major energy storage in seeds and fruits	
	Fats		Major energy storage in animals	
	Phospholipids	2 fatty acids + 1 glycerol + 1 phosphate group	Major component of all cell membranes	Phospholipids and glycolipids are modified triglycerides with a polar group at one end. The polar "head" of the molecule is hydrophilic and thus dissolves in water; the nonpolar "tail" is hydrophobic and insoluble in water. This is the basis for their role in cell membranes, where they are arranged tail to tail.
	Cutin, suberin, and waxes	Vary; complex lipid structures	Protection	Act as waterproofing for stems, leaves, and fruits.
	Steroids	Four linked hydrocarbon rings	Component of cell membranes; hormones	A sterol is a steroid with a hydroxyl group at the carbon-position 3.
Proteins (polypeptides)	Many different types	Amino acids	Numerous; include structural, catalytic (enzymes)	Primary, secondary, tertiary, and quaternary structures.
Nucleic acids	DNA	Nucleotides	Carrier of genetic information	Each nucleotide is composed of a sugar, a nitrogenous base, and a phosphate group. ATP is a nucleotide that functions as the principal energy carrier for cells.
	RNA		Involved in protein synthesis	

Enzymes are globular proteins that catalyze chemical reactions in cells. Because of enzymes, cells are able to accelerate the rate of chemical reactions at moderate temperatures.

Nucleic Acids Are Polymers of Nucleotides

Nucleotides are complex molecules consisting of a phosphate group, a nitrogenous base, and a five-carbon sugar. They are the building blocks of the nucleic acids deoxyribonucleic acid (DNA) and ribonucleic acid (RNA), which transmit and translate the genetic information. Some RNA molecules function as catalysts.

Adenosine triphosphate (ATP) is the cell's energy currency. ATP can be hydrolyzed, releasing adenosine diphosphate (ADP), phosphate, and considerable energy. This energy can be used to drive other reactions in the cell. In the reverse reaction, ADP can be "recharged" to ATP with the addition of a phosphate group and an input of energy.

Secondary Metabolites Play a Variety of Roles Not Directly Related to the Basic Functioning of the Plant

Three main classes of secondary metabolites found in plants are alkaloids, terpenoids, and phenolics. Although the botanical functions of these substances are not clearly known, some are thought to deter predators and/or competitors. Examples of such compounds include caffeine and nicotine (alkaloids), as well as cardiac glycosides (terpenoids) and tannins (phenolics). Others, such as anthocyanins (phenolics) and essential oils (terpenoids), attract pollinators. Still others, such as phenolic lignins, are responsible for the compressive strength, stiffness, and waterproofing of the plant body. Some secondary metabolites, such as rubber (a terpenoid) and morphine and taxol (alkaloids), have important commercial or medicinal uses.

Selected Key Terms

adenosine triphosphate (ATP) p. 31

alkaloids p. 32

amino acids p. 26

carbohydrates p. 18

catalyst p. 30

cellulose p. 21

dehydration synthesis p. 20

disaccharides p. 18

enzymes p. 30

hydrolysis p. 20

hydrophilic p. 18

hydrophobic p. 22

lignins p. 36

lipids p. 18

macromolecules p. 18

monomers p. 18

monosaccharides p. 18

nucleic acids p. 18

nucleotides p. 30

organic molecules p. 18

peptide bond p. 28

phenolics p. 34

phytoalexins p. 32

polymerization p. 18

polymers p. 18

polypeptides p. 28

polysaccharides p. 18

primary metabolites p. 32

proteins p. 18

saturated fats p. 24

secondary metabolites p. 32

steroids p. 25

tannins p. 35

terpenoids p. 33

triglycerides p. 22

unsaturated fats p. 24

Questions

1. Why must starch be hydrolyzed before it can be used as an energy source or transported?

2. What advantage is there to a plant to store food energy as fructans rather than as starch? As oils rather than as starch or fructans?

3. Most plant oils are unsaturated. What is the principal difference between a saturated and an unsaturated fat or oil?

4. All amino acids have the same basic structure. What aspect of their structure do all amino acids have in common? What part of an amino acid determines its identity?

5. There are several levels of protein organization. What are those levels, and how do they differ from one another?

6. The coagulation of egg white when an egg is cooked is a common example of protein denaturation. What happens when a protein is denatured?

7. A number of insects, including the monarch butterfly, have adopted a strategy of utilizing certain secondary metabolites of plants for protection against predators. Explain.

8. Lignin, a cell wall constituent, is believed to have played a major role in the evolution of terrestrial plants. Explain in terms of all the presumed functions of lignin.

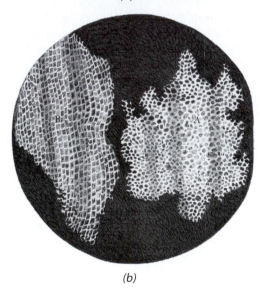

(a)

(b)

3–1

*The English microscopist Robert Hooke first used the term "cell" to refer to the small chambers he saw in magnified slices of cork. (**a**) Hooke's drawing of his microscope, reproduced from a book he published in 1665. Light from an oil lamp was directed to the specimen through a water-filled glass globe that acted as a condenser. The specimen was mounted on a pin, just below the tip of the microscope. The microscope was focused by moving it up and down using a screw held to the stand by a clamp. (**b**) Among the many other illustrations in the book was this drawing of two slices of cork.*

Chapter 3

Introduction to the Plant Cell

OVERVIEW

Try to define "life" and you will find it a difficult task. But you are in good company because biologists themselves cannot agree on a succinct, all-encompassing definition that adequately describes the richness, the diversity, and the complexity of what it means to be alive. Nevertheless, biologists have been able to agree on the fundamental unit of life—the smallest structure that has all the basic characteristics of living organisms. That fundamental unit is the cell. Mostly invisible to the naked eye, cells function as minuscule factories, obtaining and storing needed substances from the environment, processing and packaging molecules, transporting materials from place to place, obtaining energy, and doing much more. As in a factory, the various activities of the cell do not take place randomly but are associated with specific structures that interact in a highly organized manner.

This chapter begins with the discovery of cells and the development of the cell theory, followed by a description of the two major types of cells. Prokaryotic cells, such as bacteria, which are relatively simple and ancient, are compared with eukaryotic cells, such as those in plants, fungi, and animals, which are more complex and more recent. The remainder of the chapter focuses on the internal organization of plant cells, including the structure and function of membrane-bounded compartments called organelles, as well as other, nonmembranous structures, such as ribosomes and the cell wall.

CHECKPOINTS

By the time you finish reading this chapter, you should be able to answer the following questions:

1. How does the structure of a prokaryotic cell differ from that of a eukaryotic cell?

2. What are the various types of plastids, and what role(s) does each play in the cell?

3. What are the principal components of the endomembrane system, and what role does each play in that system?

4. What is meant by the cytoskeleton of the cell, and with what cellular processes is it involved?

5. How do primary cell walls differ from secondary cell walls?

In the last chapter, we progressed from atoms and small molecules to large, complex molecules, such as proteins and nucleic acids. At each level of organization, new properties appear. Water, we know, is not the sum of the properties of elemental hydrogen and oxygen, both of which are gases. Water is something more and also something different. In proteins, amino acids become organized into polypeptides, and polypeptide chains are arranged in new levels of organization—the secondary, tertiary, and, in some cases, quaternary structure of the complete protein molecule. Only at the final level of organization do the complex properties of the protein emerge, and only then can the molecule assume its function.

The characteristics of living systems, like those of atoms or molecules, do not emerge gradually as the degree of organization increases. They appear quite suddenly and specifically, in the form of the living cell—something that is more than and different from the atoms and molecules of which it is composed. Life begins where the cell begins.

Development of the Cell Theory

Cells are the fundamental units of life—both in structure and function. The smallest organisms are composed of single cells. The largest are made up of billions of cells, each of which still lives a partly independent existence. The realization that all organisms are composed of cells was one of the most important conceptual advances in the history of biology because it emphasizes the underlying sameness of all living systems. It therefore brings unity to widely varied studies involving many different kinds of organisms.

The word "cell" was first used in a biological sense some 300 years ago. In the seventeenth century, the English scientist Robert Hooke, using a microscope of his own construction, noticed that cork and other plant tissues are made up of what appeared to be small cavities separated by walls (Figure 3–1). He called these cavities "cells," meaning "little rooms." However, "cell" did not take on its present meaning—the basic unit of living matter—for more than 150 years.

In 1838, Matthias Schleiden, a German botanist, reported his observation that all plant tissues consist of organized masses of cells. In the following year, zoologist Theodor Schwann extended Schleiden's observation to animal tissues and proposed a cellular basis for all life. In 1858, the idea that all living organisms are composed of one or more cells took on an even broader significance when the pathologist Rudolf Virchow generalized that cells can arise only from preexisting cells: "Where a cell exists, there must have been a preexisting cell, just as the animal arises only from an animal and the plant only from a plant."

From the perspective provided by Darwin's theory of evolution, published in the following year, Virchow's concept takes on an even larger significance. There is an unbroken continuity between modern cells—and the organisms they compose—and the first primitive cells that appeared on Earth at least 3.5 billion years ago.

In its modern form, the **cell theory** states that (1) all living organisms are composed of one or more cells; (2) the chemical reactions of a living organism, including its energy-releasing processes and its biosynthetic reactions, take place within cells; (3) cells arise from other cells; and (4) cells contain the hereditary information of the organisms of which they are a part, and this information is passed from parent cell to daughter cell.

Prokaryotic Cells and Eukaryotic Cells

All cells share two essential features. One is an outer membrane, the **plasma membrane** (also known as the *plasmalemma,* or *cell membrane*), that isolates the cell's contents from the external environment. The other is the genetic material—the hereditary information—that directs a cell's activities and enables it to reproduce, passing on the cell's characteristics to its offspring.

The organization of the genetic material is one of the characteristics that distinguish prokaryotic cells from eukaryotic cells. In **prokaryotic cells,** the genetic material is in the form of a large, circular molecule of DNA, with which a variety of proteins are loosely associated. This molecule is known as the **chromosome. In eukaryotic cells,** by contrast, the DNA is linear and tightly bound to special proteins known as **histones,** forming a number of more complex chromosomes.

Within the eukaryotic cell, the chromosomes are surrounded by a **nuclear envelope,** made up of two membranes, that separates them from the other cell contents in a distinct **nucleus** (hence the name, *eu,* meaning "true," and *karyon,* meaning "nucleus" or "kernel"). In prokaryotes ("before a nucleus"), the chromosome is not contained within a membrane-bounded nucleus, although it is localized in a distinct region known as the **nucleoid.**

The remaining components of a cell (other than the nucleus or nucleoid and the cell wall, when present) constitute the **cytoplasm,** the outer boundary of which is the plasma membrane. The cytoplasm contains a large variety of molecules and molecular complexes. For example, both prokaryotic and eukaryotic cells contain complexes of protein and RNA, known as **ribosomes,** that play a crucial role in the assembly of protein molecules from their amino acid subunits. Eukaryotic cells also contain a variety of membrane-bounded structures called **organelles.** These specialized structures carry out particular functions within the cell. In addition, virtually all eukaryotic cells possess a complex network of protein

Viewing the Cellular World

In the three centuries since Robert Hooke first observed the structure of cork through his simple microscope (Figure 3–1), our capacity to view the cell and its contents has increased dramatically. Most cells can be seen only with the aid of a microscope. Unaided, the human eye has a resolving power of about 0.1 millimeter, or 100 micrometers (Table 3–1). This means that if one looks at two lines that are less than 100 micrometers apart, they will appear as a single line. Similarly, two dots less than 100 micrometers apart look like a single blurry dot. In order to resolve structures closer together than this, microscopes must be used.

The best light microscopes have a resolving power of 0.2 micrometer, or about 200 nanometers, and so improve on the naked eye about 500 times. A light microscope cannot do better than this because the limiting factor is the wavelength of light. The shorter the wavelength, the greater the resolution. The shortest wavelength of visible light is about 0.4 micrometer, and this sets the limit of resolution with the light microscope.

Notice that resolving power and magnification are different. Using even the best light microscope, if you take a photograph of two lines that are less than 0.2 micrometer (200 nanometers) apart, you can enlarge the image indefinitely, but the two lines will still blur together. By using more powerful lenses, you can increase the magnification, but this will not improve the resolution. You will simply see a larger blur.

The Transmission Electron Microscope Uses Electrons That Pass through the Specimen

With the transmission electron microscope (TEM), the limit of resolution imposed by light can be reduced because electrons have much shorter wavelengths than visible light. Theoretically, a TEM operating at an accelerating voltage of 100,000 volts should have a resolution of about 0.002 nanometer. Because of problems with specimen preparation, contrast, and radiation damage, however, the resolution of biological objects is more like 2 nanometers. Nonetheless, this is still 100 times better than the resolution of the light microscope.

The TEM has its disadvantages. In order to produce a beam of electrons, air must be pumped out of the microscope to

TABLE 3–1	Measurements Used in Microscopy
1 centimeter (cm)	= 1/100 meter = 0.4 inch*
1 millimeter (mm)	= 1/1,000 meter = 1/10 cm
1 micrometer (μm)**	= 1/1,000,000 meter = 1/10,000 cm
1 nanometer (nm)	= 1/1,000,000,000 meter = 1/10,000,000 cm
1 angstrom (Å)†	= 1/10,000,000,000 meter = 1/100,000,000 cm = 1/10 nm
or	
1 meter	= 10^2 cm = 10^3 mm = 10^6 μm = 10^9 nm = 10^{10} Å

*A metric-to-English conversion table is found in Appendix C.
**Micrometers were formerly known as microns (μ), and nanometers as millimicrons (mμ).
†The Ångstrom is not an accepted measurement in the International System of Units. It is included here, however, because it was widely used in microscopy in the past.

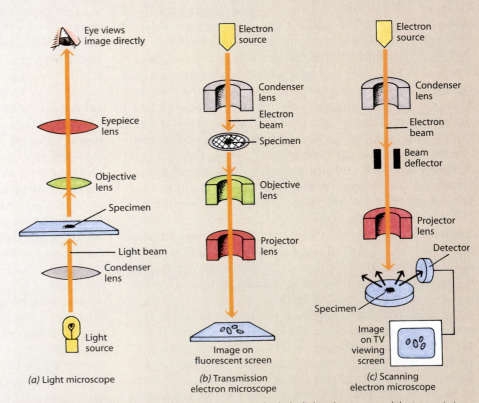

(a) Light microscope   (b) Transmission electron microscope   (c) Scanning electron microscope

A comparison of **(a)** the light microscope, **(b)** the transmission electron microscope, and **(c)** the scanning electron microscope. The focusing lenses in the light microscope are glass or quartz; those in the electron microscopes are magnetic coils. In both the light microscope and the transmission electron microscope, the illuminating beam passes through the specimen. In the scanning electron microscope, it is deflected from the surface.

create a vacuum. The specimen must be put into this vacuum, making it impossible to examine living material. Typically, the specimen is first preserved by chemical fixation and then embedded in plastic so that it can be sectioned (cut into slices) by a special cutting instrument called an ultra-microtome. Because electrons have very limited penetrating power, the block of plastic containing the specimen must be cut into exceedingly thin slices (50 to 100 nanometers). Such thin sections are effectively two-dimensional slices of tissue, which fail to convey a three-dimensional perspective. Also, it is often difficult to determine what changes may have been produced in the material in the course of its preparation. The micrographs in Figures 3–4 and 3–5 are two of the many transmission electron micrographs in this book.

The Scanning Electron Microscope Uses Electrons That Are Scattered from the Specimen's Surface

Although the resolution of the scanning electron microscope (SEM) is only about 10 nanometers, this instrument is a valuable tool for biologists. Unlike the TEM, which uses electrons that have passed through the specimen to form an image, the SEM uses electrons that are scattered or emitted from the surface of the specimen. The electron beam is focused into a fine probe, which is rapidly passed back and forth over the specimen. Complete scanning from top to bottom usually takes only a few seconds. Variations in the surface of the specimen affect the pattern in which the electrons are scattered from it. Holes and fissures appear dark, and knobs and ridges appear light. The pattern produced by the electrons is amplified and transmitted to a television monitor, providing a visual image of the specimen. Scanning electron microscopy provides vivid three-dimensional representations of the surfaces of whole cells and cellular structures that more than compensate for its limited resolution. Figure 3–18a was prepared with a scanning electron microscope.

Studies with these three types of microscopes, complemented by a variety of biochemical techniques, have provided a wealth of knowledge about the structure of cells and their dynamic processes.

filaments, called the **cytoskeleton,** that is involved in many processes. A cytoskeleton and organelles are absent in prokaryotes.

The plasma membrane of prokaryotes is surrounded by an outer **cell wall** that is manufactured by the cell itself. Some eukaryotic cells, including plant cells and fungi, have cell walls, although they differ from prokaryotic cell walls in both composition and structure. Other eukaryotic cells, including those of our own bodies and of other animals, do not have cell walls. Another feature distinguishing eukaryotic and prokaryotic cells is size: eukaryotic cells are usually larger than prokaryotic cells, the difference often being an order of magnitude.

The Fossil Record Suggests Evolutionary Relationships among Different Types of Organisms

Modern prokaryotes are represented by the archaea and bacteria (Figure 3–2), including the cyanobacteria (Figure 3–3), a group of photosynthetic bacteria that were formerly known as the blue-green algae. According to the fossil record, the earliest living organisms were comparatively simple cells, resembling present-day prokaryotes. Until eukaryotes evolved, prokaryotes were the only forms of life on this planet. Many biologists believe that the transition from the prokaryotic cell to the eukaryotic cell, a topic we shall explore in Chapter 13, was the most significant event in the history of life, second only in biological importance to the first appearance of living systems.

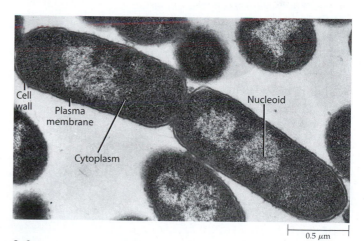

3–2

Prokaryote. Electron micrograph of cells of Escherichia coli, *a bacterium that is a common, usually harmless inhabitant of the human digestive tract. This heterotrophic (nonphotosynthetic) prokaryote is the most thoroughly studied of all living organisms. Each rod-shaped cell has a cell wall, a plasma membrane, and cytoplasm. The genetic material (DNA) is found in the less granular area in the center of each cell. This region, known as the nucleoid, is not surrounded by a membrane. The densely granular appearance of the cytoplasm is largely due to the presence of numerous ribosomes, which are involved with protein synthesis. The two cells in the center have just divided but have not yet separated completely.*

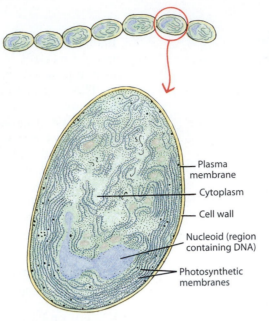
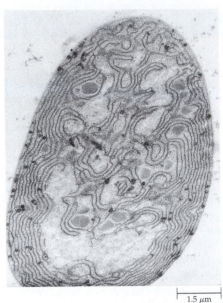

3–3

Prokaryote. Electron micrograph of a single cell from a filament of cells of a photosynthetic prokaryote, the cyanobacterium Anabaena azollae. In addition to the cytoplasmic components found in E. coli (Figure 3–2), this cell contains a series of membranes in which chlorophyll and other photosynthetic pigments are embedded. Anabaena synthesizes its own energy-rich organic compounds in chemical reactions powered by the radiant energy of the sun.

Plasma membrane

Cytoplasm

Cell wall

Nucleoid (region containing DNA)

Photosynthetic membranes

1.5 μm

3–4

Eukaryote. Electron micrograph of Chlamydomonas, a photosynthetic eukaryotic cell, which contains a membrane-bounded nucleus and numerous organelles. The most prominent organelle is the single, irregularly shaped chloroplast that fills most of the cell. It is surrounded by an envelope consisting of two membranes and is the site of photosynthesis. Other membrane-bounded organelles, the mitochondria, provide energy for cellular functions, including the flicking movements of the two flagella, only one of which is visible in this micrograph. These flicking movements propel the cell through the water. The organism's food reserves are in the form of starch grains, clustered around a structure known as the pyrenoid, which is found in the chloroplast. The cytoplasm is enclosed by the plasma membrane, outside of which is a cell wall composed of polysaccharides.

Figure 3–4 provides an example of a modern single-celled photosynthetic eukaryote, the alga *Chlamydomonas*. It is a common inhabitant of freshwater ponds. These organisms are small and bright green (because of their chlorophyll), and they move very quickly with a characteristic darting motion. Being photosynthetic, they are usually found near the water's surface, where the light intensity is greatest.

Some features that distinguish prokaryotic cells from eukaryotic cells are listed in Table 3–2.

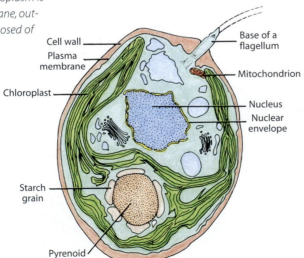

Cell wall

Plasma membrane

Chloroplast

Starch grain

Pyrenoid

Base of a flagellum

Mitochondrion

Nucleus

Nuclear envelope

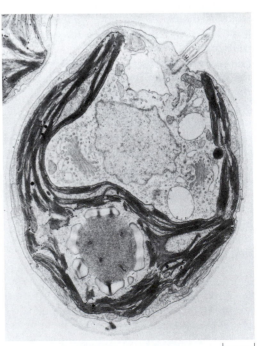

1 μm

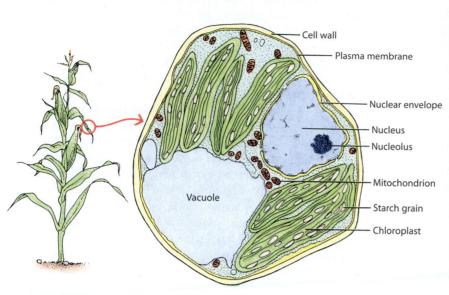

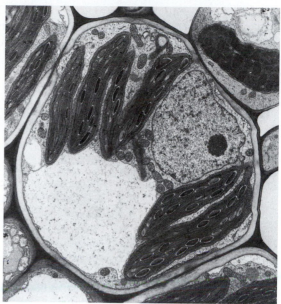

3.0 μm

3–5

Eukaryote. Electron micrograph of a cell from the leaf of a maize plant. The granular material within the nucleus is chromatin. It contains DNA associated with histone proteins. The nucleolus is the region within the nucleus *where the RNA components of ribosomes are synthesized. Note the many mitochondria and chloroplasts, all bounded by membranes. The vacuole, a fluid-filled region enclosed by a* *membrane, and the cell wall are characteristic of plant cells. As you can see, this cell closely resembles* Chlamydomonas, *shown in Figure 3–4.*

TABLE 3–2 **Comparison of Selected Features of Prokaryotic and Eukaryotic Cells**

	Prokaryotic cells	Eukaryotic cells
Cell size (length)	Generally 1 to 10 micrometers	Generally 5 to 100 micrometers; many much longer than 100 micrometers
Nuclear envelope	Absent	Present
DNA	Circular, in nucleoid	Linear, in nucleus
Organelles (e.g., mitochondria and chloroplasts)	Absent	Present
Cytoskeleton (microtubules and actin filaments)	Absent	Present
Ribosome size*	70S	80S in cytoplasm, 70S in mitochondria and plastids

*S refers to Svedberg units, which are used to express the sedimentation coefficient, a measure of how rapidly a particle sediments in an ultracentrifuge. Bacterial ribosomes and those in mitochondria and plastids are smaller than the cytoplasmic ribosomes of eukaryotic cells and therefore sediment more slowly (that is, have smaller S values).

The first multicellular organisms, as far as can be determined from the fossil record, made their appearance a mere 750 million years ago. The cells of modern multicellular organisms closely resemble those of single-celled eukaryotes. They differ from single-celled eukaryotes in that each type of cell is specialized to carry out a relatively limited function in the life of the organism. However, each remains a remarkably self-sustaining unit.

Notice how similar a cell from the leaf of a maize (corn) plant (Figure 3–5) is to *Chlamydomonas*. The plant cell is also photosynthetic, supplying its own energy needs from sunlight. However, unlike the alga, it is part of a multicellular organism and depends on other cells for water, minerals, protection from drying out, and other necessities.

The Plant Cell: An Overview

The plant cell typically consists of a more or less rigid **cell wall** and a **protoplast** (Table 3–3). The term "protoplast" is derived from the word **protoplasm,** which is used to refer to the contents of cells. A protoplast is the unit of protoplasm inside the cell wall.

A protoplast consists of **cytoplasm** and a **nucleus.** The cytoplasm, which is bounded externally by the plasma membrane, includes organelles, systems of membranes, and nonmembranous structures such as ribo-

TABLE 3–3	An Inventory of Plant Cell Components	
Cell wall	Middle lamella Primary wall Secondary wall Plasmodesmata	
Protoplast	Nucleus	Nuclear envelope Nucleoplasm Chromatin Nucleolus

	Cytoplasm	Plasma membrane (outer boundary of cytoplasm)
		Cytosol
		Organelles bounded by two membranes:
		    Plastids
		    Mitochondria
		Organelles bounded by one membrane:
		    Peroxisomes
		    Vacuoles, bounded by tonoplast
		Endomembrane system*
		(major components)
		    Endoplasmic reticulum
		    Golgi complex
		    Vesicles
		Cytoskeleton
		    Microtubules
		    Actin filaments
		Ribosomes
		Oil bodies

*The endomembrane system also includes the plasma membrane, the nuclear envelope, the tonoplast, and all other internal membranes with the exception of mitochondrial, plastid, and peroxisome membranes.

3–6

Under high magnification, cellular membranes often have a three-layered (dark-light-dark) appearance, as seen in the plasma membranes on either side of the common wall between two cells from a maize (Zea mays) leaf.

somes. The rest of the cytoplasm—the "cellular soup," or cytoplasmic matrix, in which the nucleus, various bodies, and membrane systems are suspended—is called the **cytosol.** In contrast to most animal cells, plant cells develop one or more liquid-filled cavities, or **vacuoles,** within the cytosol. The vacuole is bounded by a single membrane called the **tonoplast.**

In a living plant cell, the cytoplasm is frequently in motion. The organelles, as well as various substances suspended in the cytosol, can be observed being swept along in an orderly fashion in the moving currents. This movement is known as **cytoplasmic streaming,** or *cyclosis*, and it continues as long as the cell is alive (see essay on page 47). Cytoplasmic streaming undoubtedly facilitates the exchange of materials within the cell and between the cell and its environment. It is not known, however, whether this is a primary function of cytoplasmic streaming.

Plasma Membrane

The plasma membrane typically has a three-layered appearance, consisting of two dark layers separated by a lighter layer, in electron micrographs (Figure 3–6). The term "unit membrane" has been used to designate membranes with such an appearance.

The plasma membrane has several important functions: (1) it mediates the transport of substances into and out of the protoplast, (2) it coordinates the synthesis and assembly of the cellulose microfibrils that make up the cell wall, and (3) it receives and transmits hormonal and environmental signals involved in the control of cell growth and differentiation. The plasma membrane has the same basic structure as the internal membranes of the cell, consisting of a lipid bilayer in which globular proteins are embedded. In Chapter 4, we will discuss the plasma membrane—both its structure and its vital functions—in greater detail.

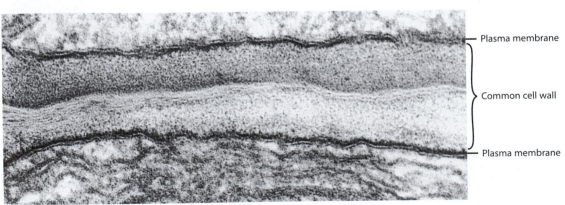

Plasma membrane

Common cell wall

Plasma membrane

Cytoplasmic Streaming in Giant Algal Cells

Much of our understanding of cytoplasmic streaming comes from work on the giant cells of green algae such as Chara *and* Nitella. *In these cells, which are 2 to 5 centimeters long, the chloroplast-containing layer of cytoplasm bordering the wall is stationary. Spirally arranged bundles of actin filaments extend for several centimeters along the cells, forming distinct "tracks" that are firmly attached to the stationary chloroplasts. The moving layer of cytoplasm, which occurs between the bundles of actin filaments and the tonoplast, contains the nucleus, mitochondria, and other cytoplasmic components.*

The generating force necessary for cytoplasmic streaming comes from an interaction between actin and myosin, a protein molecule with an ATPase-containing "head" that is activated by actin. ATPase is an enzyme that breaks down (hydrolyzes) ATP to release energy (page 31). Apparently, the organelles in the streaming cytoplasm are indirectly attached to the actin filaments by myosin molecules,

which use the energy released by ATP hydrolysis to "walk" along the actin filaments, pulling the organelles with them. Streaming always occurs from the minus to the plus ends of the actin filaments, all of which are similarly oriented within a bundle.

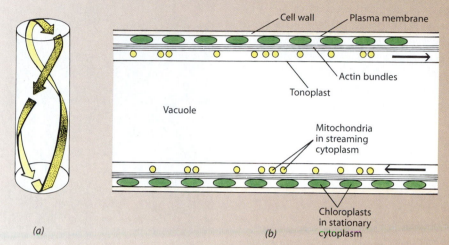

(a)

(b)

(a) *A track followed by the streaming cytoplasm in a giant algal cell.* (b) *A longitudinal section through part of the cell, showing the arrangement of stationary and streaming layers of cytoplasm. The proportions have been distorted in both diagrams for clarity.*

Nucleus

The **nucleus** is often the most prominent structure within the protoplast of eukaryotic cells. The nucleus performs two important functions: (1) it controls the ongoing activities of the cell by determining which protein molecules are produced by the cell and when they are produced, as we shall see in Chapter 11; and (2) it stores most of the cell's genetic information, passing it on to the daughter cells in the course of cell division.

The nucleus is bounded by a pair of membranes called the **nuclear envelope.** The nuclear envelope, as seen with the aid of an electron microscope, contains a large number of circular pores that are 30 to 100 nanometers in diameter (Figure 3–7). At each pore, the inner and outer membranes are joined, forming the pore's lining. The pores are not merely holes in the envelope; each has a complicated structure. Nuclear pores provide a direct passageway for the exchange of materials between the nucleus and the cytoplasm. In various places the outer membrane of the nuclear envelope may be continuous with the **endoplasmic reticulum,** a complex system of membranes that plays a central role in the synthesis of molecules needed by the cell. The nuclear envelope may be considered a specialized, locally differentiated portion of the endoplasmic reticulum.

In specially stained cells, thin threads and grains of **chromatin** can be distinguished from the **nucleoplasm,** or nuclear matrix (Figure 3–8). Chromatin is made up of DNA combined with large amounts of histone proteins. During the process of nuclear division, the chromatin becomes progressively more condensed until it is visible under the microscope as individual chromosomes. The dispersed chromatin of nondividing nuclei apparently is attached at one or more sites to the inner surface of the nuclear envelope. The hereditary information is carried in the molecules of DNA contained within the chromosomes.

Different types of organisms vary in the number of chromosomes present in their somatic (body) cells. Take, for example, the following vascular plants: *Haplopappus gracilis,* a desert annual, has 4 chromosomes per cell; *Arabidopsis thaliana,* the small flowering weed used widely for genetic research, 10; *Brassica oleracea,* cabbage, 18; *Triticum vulgare,* bread wheat, 42; and one species of the fern *Ophioglossum,* about 1250. The gametes, or sex cells, however, have only half the number of chromosomes that is characteristic of the somatic cells of each organism. The number of chromosomes in the gametes is referred to as the **haploid** ("single set") number, and that in the somatic cells as the **diploid** ("double set") number.

Polysome

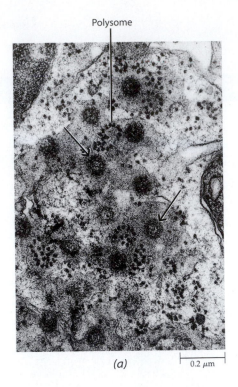

Endoplasmic reticulum

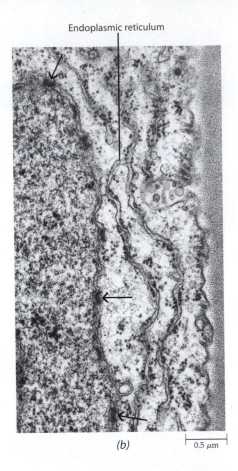

(a)   0.2 μm

(b)   0.5 μm

3–7

Nuclear pores as revealed in electron micrographs of nuclei in parenchyma cells of the seedless vascular plant Selaginella kraussiana. *In (a) the pores are shown in surface view, whereas in (b) they are shown in sectional view (see arrows). Note the polysomes (coils of ribosomes) on the surface of the nuclear envelope in (a) and the rough endoplasmic reticulum paralleling the nuclear envelope in (b).*

Often the only structures within a nucleus that are discernible with the light microscope are the spherical structures known as **nucleoli** (singular: nucleolus), one or more of which are present in each nondividing nucleus (Figure 3–5). In many diploid organisms, the nucleus contains two nucleoli, one for each haploid set of chromosomes. The nucleoli may fuse and then appear as one larger structure. Each nucleolus contains large amounts of RNA and proteins, along with large loops of DNA emanating from several chromosomes. The loops of DNA, known as nucleolar organizer regions, are the sites at which ribosomal subunits are constructed (see Figure 11–7). The ribosomal subunits are then transferred to the cytoplasm, via the nuclear pores, where they are assembled to form ribosomes. The very presence of the nucleolus is due to the accumulation of the RNA and protein molecules being packaged to form the ribosomal subunits. In fact, the size of a nucleolus is a reflection of the level of its activity.

Chloroplasts and Other Plastids

Together with vacuoles and cell walls, **plastids** are characteristic components of plant cells and are concerned with such processes as photosynthesis and storage. The principal types of plastids are chloroplasts, chromoplasts, and leucoplasts. Each plastid is bounded by an envelope consisting of two membranes. Internally, the plastid is differentiated into a system of membranes, or **thylakoids,** and a more or less homogeneous matrix, the **stroma.** The degree of development of the thylakoid system varies among plastid types.

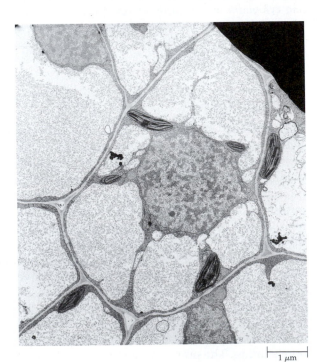

1 μm

3–8

*Parenchyma cell from a tobacco (*Nicotiana tabacum*) leaf, with its nucleus "suspended" in the middle of the cell by dense strands of cytoplasm. The less granular regions are portions of a large central vacuole that merge beyond the plane of this cell section. The dense granular substance in the nucleus is chromatin.*

3–9

A three-dimensional diagram of a chloroplast-containing plant cell. Typically, the disk-shaped chloroplasts are located in the cytoplasm along the wall, with their broad surfaces facing the surface of the wall. Most of the volume of this cell is occupied by a vacuole (bounded by the tonoplast), which is traversed by a few strands of cytoplasm. In this cell, the nucleus lies in the cytoplasm along the wall, although in some cells (note Figure 3–8) it might appear suspended by strands of cytoplasm in the center of the vacuole.

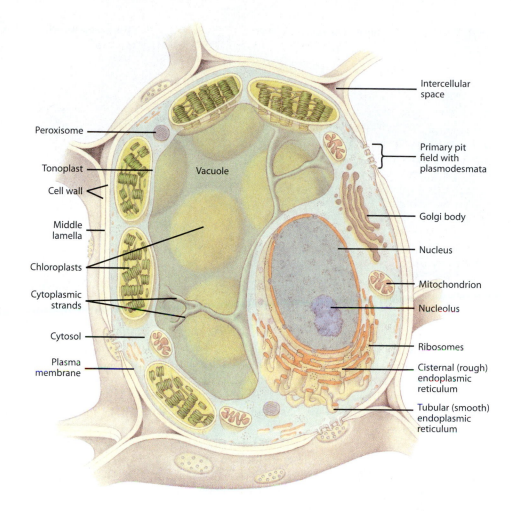

Chloroplasts Are the Sites of Photosynthesis

Mature plastids are commonly classified in part on the basis of the kinds of pigments they contain. **Chloroplasts,** the sites of photosynthesis, contain chlorophylls and carotenoid pigments. The chlorophyll pigments, which are the receptors for light energy necessary for photosynthesis, are responsible for the green color of these plastids. In plants, chloroplasts are usually disk-shaped and measure between 4 and 6 micrometers in diameter. A single mesophyll ("middle of the leaf") cell may contain 40 to 50 chloroplasts; a square millimeter of leaf contains some 500,000. The chloroplasts are usually found with their broad surfaces parallel to the cell wall (Figure 3–9). They can reorient in the cell under the influence of light—for example, gathering along the walls parallel with the leaf surface under low or medium light intensity. Under potentially damaging high light intensity, the chloroplasts can orient themselves perpendicular to the leaf surface.

The internal structure of the chloroplast is complex (Figure 3–10). The stroma is traversed by an elaborate system of thylakoids, consisting of **grana** (singular: granum)—stacks of disklike thylakoids that resemble a stack of coins—and **stroma thylakoids** (or intergranal thylakoids) that traverse the stroma between grana and interconnect them. The grana and stroma thylakoids and their internal compartments are believed to constitute a single, interconnected system. Chlorophylls and carotenoid pigments are embedded in the thylakoid membranes. These pigments are responsible for capturing light that drives photosynthesis. How this is done and how the light energy is converted to chemical energy are discussed in Chapter 7.

The chloroplasts of green algae and plants often contain starch grains and small oil bodies called plastoglobuli. The starch grains are temporary storage products and accumulate only when the alga or plant is actively photosynthesizing (Figures 3–4 and 3–5). They may be

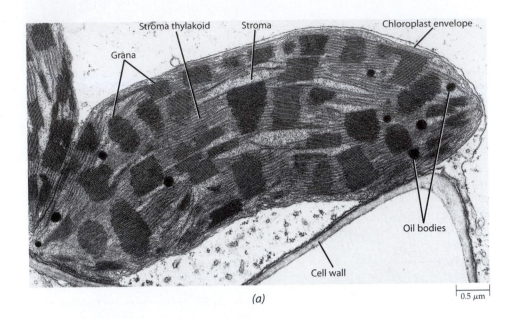

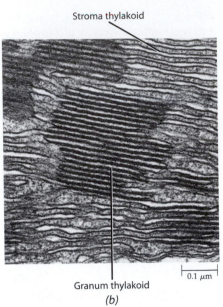

(a)

0.5 μm

(b)

0.1 μm

3–10
A chloroplast of a maize (Zea mays) *leaf.* **(a)** *Section showing grana and stroma thylakoids.* **(b)** *Detail showing a granum composed of stacks of disklike thylakoids. The thylakoids of the various grana are interconnected by other thylakoids, commonly called stroma thylakoids.*

lacking in the chloroplasts of plants that have been kept in the dark for as little as 24 hours. The starch has been broken down to supply sugar to the plant, which is unable to photosynthesize in the dark. Starch grains will often reappear after the plant has been in the light for only 3 or 4 hours.

The ability to form chloroplasts and the pigments associated with them involves the contribution of both nuclear and plastid DNA. Overall control clearly resides in the nucleus, however. Some chloroplast proteins are encoded, or specified, by plastid DNA and are synthesized within the chloroplast itself. Most of the chloroplast proteins, however, are encoded by nuclear DNA, synthesized in the cytosol, and then imported into the chloroplast.

Chloroplasts are the ultimate source of virtually all of our food supplies and of our fuel. As we shall discuss in Chapter 7, it is in the chloroplast that the light energy of the sun is converted to chemical energy and carbon dioxide is fixed to form carbohydrate. Chloroplasts are not only sites of photosynthesis; they are also involved in amino acid synthesis and fatty acid synthesis, and, as we've noted, they provide space for the temporary storage of starch.

Chromoplasts Contain Pigments Other Than Chlorophyll

Chromoplasts (Gk. *chroma*, "color") are also pigmented plastids (Figure 3–11). Of variable shape, chromoplasts lack chlorophyll but synthesize and retain carotenoid pigments, which are often responsible for the yellow, orange, or red colors of many flowers, old leaves, some fruits, and some roots, such as carrots. Chromoplasts may develop from previously existing green chloroplasts by a transformation in which the chlorophyll and internal membrane structure of the chloroplast disappear and

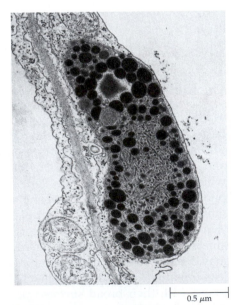

0.5 μm

3–11
Chromoplast from a petal of Forsythia, *a common garden shrub that is covered with yellow flowers in the early spring. The chromoplast contains numerous electron-dense oil bodies in which the yellow pigment is stored.*

masses of carotenoids accumulate, as occurs during the ripening of many fruits. The precise functions of chromoplasts are not well understood, although at times they act as attractants to insects and other animals. In Chapter 22, we will discuss the essential role of these attractants in the cross-pollination of flowering plants and the dispersal of fruits and seeds.

Leucoplasts Are Nonpigmented Plastids

Structurally the least differentiated of mature plastids, **leucoplasts** lack pigments and an elaborate system of inner membranes (Figure 3–12). Some leucoplasts, known as *amyloplasts*, synthesize starch (Figure 3–13), whereas others are thought to be capable of forming a variety of substances, including oils and proteins.

Proplastids Are the Precursors of Other Plastids

Proplastids are small, colorless or pale green, undifferentiated plastids that occur in meristematic (dividing) cells of roots and shoots. They are the precursors of the other, more highly differentiated plastids such as chloroplasts, chromoplasts, or amyloplasts (Figure 3–14). If the development of a proplastid into a more highly differentiated form is arrested by the absence of light, it may form one or more **prolamellar bodies,** which are semicrystalline bodies composed of tubular membranes (Figure 3–15). Plastids containing prolamellar bodies are called **etioplasts.** Etioplasts form in leaf cells of plants grown in the dark. During subsequent development of

3–12

Leucoplasts—small colorless plastids— clustered around the nucleus in an epidermal cell of a leaf from the houseplant known as a wandering Jew (Zebrina). The purple color is due to anthocyanin pigments in vacuoles of the epidermal cells seen here.

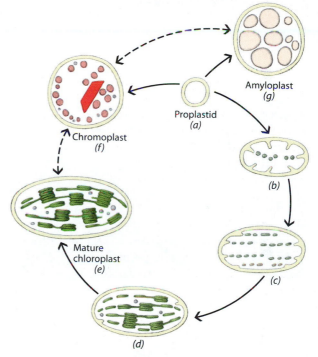

3–14

Plastid developmental cycle, beginning with the development of a chloroplast from a proplastid. (a) Initially the proplastid contains few or no internal membranes. (b)–(d) As the proplastid differentiates, flattened vesicles develop from the inner membrane of the plastid envelope and eventually align themselves into grana and stroma thylakoids. (e) The thylakoid system of the mature chloroplast appears discontinuous with the envelope. (f), (g) Proplastids may also develop into chromoplasts and leucoplasts, such as the starch-synthesizing amyloplast shown here. Note that chromoplasts may be formed from proplastids, chloroplasts, or leucoplasts. The various kinds of plastids can change from one type to another (dashed arrows).

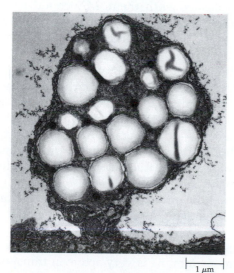

3–13

Amyloplast, a type of leucoplast, from the embryo sac of soybean (Glycine max). The round, clear bodies are starch grains. The smaller, dense bodies are oil bodies. Amyloplasts are involved with the synthesis and long-term storage of starch in seeds and storage organs, such as potato tubers.

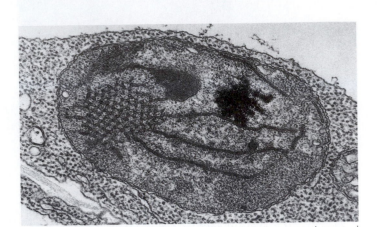

0.25 μm

3–15

Etioplast in a leaf cell of tobacco (Nicotiana tabacum) *grown in the dark. Note the semicrystalline prolamellar body (checkerboard pattern at the left). When exposed to light, the tubular membranes of the prolamellar body develop into thylakoids.*

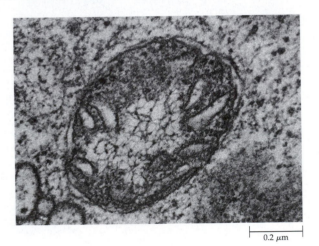

0.2 μm

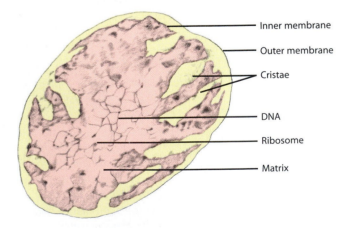

Inner membrane

Outer membrane

Cristae

DNA

Ribosome

Matrix

3–16

Mitochondrion in a leaf cell of spinach (Spinacia oleracea), *in a section revealing some strands of DNA in the nucleoid. The envelope of the mitochondrion consists of two separate membranes, each made up of a lipid bilayer. The outer membrane is smooth, but the inner membrane folds inward to form cristae, which are embedded in a dense matrix. The small particles in the matrix are ribosomes.*

etioplasts into chloroplasts in the light, the membranes of the prolamellar bodies develop into thylakoids. In nature, the proplastids in the embryos of seeds first develop into etioplasts. Upon exposure to light, the etioplasts develop into chloroplasts. The various kinds of plastids are remarkable for the relative ease with which they can change from one type to another.

Plastids are semiautonomous organelles that resemble bacteria in several ways. For example, like bacteria, plastids contain one or more *nucleoids*, clear, grana-free regions containing DNA. The DNA of the chloroplast, like that of a bacterium, exists in circular form and is not associated with histones. In addition, the ribosomes of both bacteria and plastids are about two-thirds the size of the cytoplasmic ribosomes. Moreover, protein synthesis on bacterial and plastid ribosomes is inhibited by antibiotics, such as chloramphenicol and streptomycin, that have no effect on eukaryotic ribosomes.

Plastids reproduce by fission, the process of dividing into equal halves, which is characteristic of bacteria. In meristematic cells, the division of proplastids roughly keeps pace with cell division. In mature cells, however, the greater proportion of the final plastid population may be derived from the division of mature plastids.

Mitochondria

Mitochondria, like plastids, are bounded by two membranes (Figure 3–16). The inner membrane has numerous invaginations called **cristae** (singular: crista). The cristae have the form of folds or tubules, which greatly increase the surface area available to proteins and the reactions associated with them. Mitochondria are generally smaller than plastids, measuring about half a micrometer in diameter, and exhibit great variation in length and shape.

Mitochondria are the sites of respiration, a process involving the release of energy from organic (fuel) molecules and its transfer to molecules of ATP, or adenosine triphosphate (page 31), the chief immediate source of chemical energy for all eukaryotic cells. Most plant cells contain hundreds or thousands of mitochondria, the number of mitochondria per cell being related to the cell's demand for ATP.

Mitochondria are in constant motion, turning and twisting and moving from one part of the cell to another; they also fuse and divide by fission. Mitochondria tend to congregate where energy is required. In cells in which the plasma membrane is very active in transporting materials into or out of the cell, mitochondria often can be found arrayed along the membrane surface. In motile, single-celled algae, mitochondria are typically clustered at the bases of locomotor structures known as flagella (see page 60), presumably providing energy for flagellar movement.

Mitochondria, like plastids, are semiautonomous organelles—that is, they contain the components necessary for the synthesis of some, but not all, of their own proteins. The inner membrane of the mitochondrion encloses a liquid matrix that contains proteins, RNA, DNA, small ribosomes similar to those of bacteria, and various solutes (dissolved substances). The DNA of the mitochondrion, like that of the plastid, occurs as circular molecules in one or more clear areas, the nucleoids (Figure 3–16). Thus, in plant cells genetic information is found in three different compartments: nucleus, plastid, and mitochondrion. The **nuclear genome,** the total genetic information stored in the nucleus, is far larger than the genome of either the plastid or the mitochondrion and accounts for most of the genetic information of the cell. Both plastids and mitochondria can code for some, but by no means all, of their own polypeptides.

Mitochondria and Chloroplasts Evolved from Bacteria

On the basis of the close similarity between bacteria and the mitochondria and chloroplasts of eukaryotic cells, it seems probable that mitochondria and chloroplasts originated as bacteria that found shelter within larger heterotrophic cells. These larger cells were the forerunners of the eukaryotes. The smaller cells, which contained (and still contain) all the mechanisms necessary to trap and/or convert energy from their surroundings, donated these useful capacities to the larger cells. Cells with their respiratory and/or photosynthetic "assistants" had a clear advantage over their contemporaries and undoubtedly soon multiplied at their contemporaries' expense. All but a very few modern eukaryotes contain mitochondria, and all autotrophic eukaryotes also contain chloroplasts; both seem to have been acquired by independent symbiotic events. (*Symbiosis* is a close association between two or more dissimilar organisms that may be, but is not necessarily, beneficial to each. See Chapter 13.) The smaller cells—now established as symbiotic organelles within the larger cells—obtained protection from environmental extremes. As a consequence of these symbioses, eukaryotes were able to invade the land and acidic waters, where the prokaryotic cyanobacteria are absent but where the eukaryotic green algae abound.

Peroxisomes

Peroxisomes (also called *microbodies*) are spherical organelles that are bounded by a single membrane and range in diameter from 0.5 to 1.5 micrometers. They have a granular interior, which may contain a body, sometimes crystalline, composed of protein (Figure 3–17). There are no internal membranes. In addition,

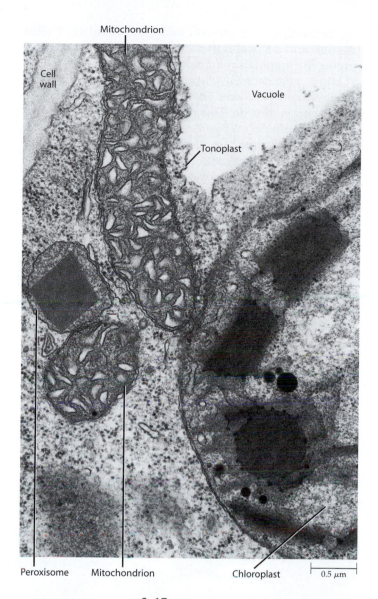

3–17

Organelles in a leaf cell of tobacco (Nicotiana tabacum). *A peroxisome with a large crystalline inclusion and a single bounding membrane may be contrasted with the two mitochondria and the chloroplast, each of which is bounded by two membranes. Due to the plane of section, the double nature of the chloroplast envelope is apparent only in the lower portion of the micrograph. A single membrane, the tonoplast, separates the vacuole from the rest of the cytoplasm.*

peroxisomes possess neither DNA nor ribosomes and, therefore, must import all of their proteins. Typically, peroxisomes are closely associated with one or two segments of endoplasmic reticulum (see page 56). Once thought to have their origin from the endoplasmic reticulum, peroxisomes are now believed to be self-replicating organelles, like plastids and mitochondria. Unlike plastids and mitochondria, however, peroxisomes must import the materials required for their replication.

Some peroxisomes play an important role in photorespiration, a process that consumes oxygen and releases carbon dioxide. This process is the reverse of what happens during photosynthesis (see Chapter 7). In green leaves, peroxisomes are closely associated with mitochondria and chloroplasts (Figure 3–17). Other peroxisomes, called *glyoxysomes*, contain the enzymes necessary for the conversion of stored fats to sucrose during germination in many seeds. Fats represent an efficient way to store carbon and energy, but sucrose is the transport form of carbon and energy in the growing seedling.

Vacuoles

Together with plastids and a cell wall, the **vacuole** is one of the three characteristic structures that distinguish plant cells from animal cells. Vacuoles are membrane-bounded regions within the cell that are filled with a liquid called **cell sap.** The single membrane surrounding the vacuole is known as the **tonoplast,** or vacuolar membrane (Figures 3–9 and 3–17). The vacuole may originate directly from the endoplasmic reticulum (see Figure

3–19), but most of the tonoplast and vacuolar proteins are derived directly from the Golgi complex, which is discussed on pages 57 and 58.

The principal component of the cell sap is water, with other components varying according to the type of plant, organ, and cell and their developmental and physiological states. In addition to inorganic ions such as Ca^{2+}, K^+, Cl^-, Na^+, and HPO_4^{2-}, vacuoles commonly contain sugars, organic acids, and amino acids. Sometimes a particular substance is present at so high a concentration that it forms crystals. Calcium oxalate crystals, which can assume several different forms, are especially common (Figure 3–18). Cell sap is usually slightly acidic. Some cell sap, such as that of the vacuoles in citrus fruits, is very acidic—hence the tart, sour taste of the fruit. In most cases, vacuoles do not synthesize the molecules they accumulate but instead receive them from other parts of the cytoplasm.

The immature plant cell typically contains numerous small vacuoles that increase in size and fuse into a single vacuole as the cell enlarges. In the mature cell, as much as 90 percent of the volume may be taken up by the vacuole, with the rest of the cytoplasm consisting of a thin peripheral layer closely pressed against the cell wall (Figures 3–8 and 3–9). By filling such a large proportion of the cell with "inexpensive" vacuolar contents, plants not only save "expensive" (in terms of energy) nitrogen-rich cytoplasmic material but also acquire a large surface between the thin layer of cytoplasm and the cell's external environment. Most of the increase in the size of the cell results from enlargement of the vacuole(s). A direct consequence of this strategy is the development of internal pressure and the maintenance of tissue rigidity, one of the principal roles of the vacuole and tonoplast (see Chapter 4).

Different kinds of vacuoles with distinct functions may be found in a single mature cell. Vacuoles are important storage compartments for primary metabolites, such as sugars and organic acids and the reserve proteins in seeds. Vacuoles also remove toxic secondary metabolites, such as nicotine and tannin, from the rest of the cytoplasm (see Figure 2–30). Such substances are sequestered permanently in the vacuoles. As mentioned in Chapter 2, the secondary metabolites contained in the vacuoles are toxic not only to the plant itself but also to pathogens, parasites, and/or herbivores, and they therefore play an important role in plant defense.

The vacuole is often a site of pigment deposition. The blue, violet, purple, dark red, and scarlet colors of plant cells are usually caused by a group of pigments known as the anthocyanins (see Chapter 2). Unlike most other plant pigments, the anthocyanins are readily soluble in water and are dissolved in the cell sap (Figure 3–12). They are responsible for the red and blue colors of many vegetables (radishes, turnips, cabbages), fruits (grapes, plums, cherries), and a host of flowers (cornflowers,

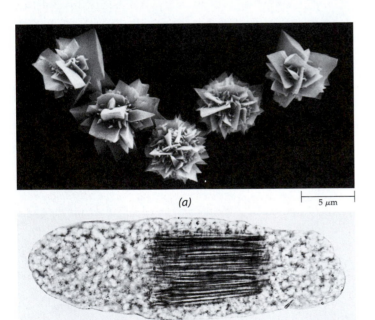

(a)   ⊢─────⊣ 5 µm

(b)   ⊢─────⊣ 50 µm

3–18

Vacuoles can contain different forms of calcium oxalate crystals.
(a) *Druses, or aggregates of crystals composed of calcium oxalate, from epidermal cells of redbud* (Cercis canadensis), *as seen with a scanning electron microscope.* **(b)** *A bundle of raphides, or needlelike crystals of calcium oxalate, in a vacuole of a leaf cell of the snake plant* (Sansevieria). *The tonoplast surrounding the vacuole is not discernible. The granular substance seen here is the cytoplasm.*

geraniums, delphiniums, roses, and peonies). Sometimes the pigments are so brilliant that they mask the chlorophyll in the leaves, as in the ornamental red maple.

Anthocyanins are also responsible for the brilliant red colors of some leaves in autumn. These pigments form in response to cold, sunny weather, when the leaves stop producing chlorophyll. As the chlorophyll that is present disintegrates, the newly formed anthocyanins are unmasked. In leaves that do not form anthocyanin pigments, the breakdown of chlorophyll in autumn may unmask the more stable yellow-to-orange carotenoid pigments already present in the chloroplasts. The most spectacular autumnal coloration develops in years when cool, clear weather prevails in the fall.

Vacuoles are also involved in the breakdown of macromolecules and the recycling of their components within the cell. Entire cell organelles, such as mitochondria and plastids, may be deposited and degraded in vacuoles. Because of this digestive activity, vacuoles are comparable in function with the organelles known as *lysosomes* that occur in animal cells.

Oil Bodies

Oil bodies, or lipid droplets, are more or less spherical structures that impart a granular appearance to the cytoplasm of a plant cell when viewed with the light micro-

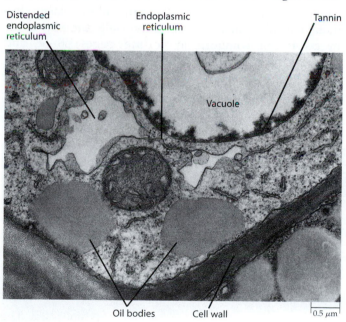

3–19

Cytoplasmic components in a parenchyma cell from the thickened stem, or corm, of the quillwort Isoetes muricata. Two oil bodies are visible just below the mitochondrion, at the center of this electron micrograph. The dense material lining the vacuole is tannin. To the upper left of the mitochondrion, a cisterna of the endoplasmic reticulum is greatly distended. Some vacuoles are thought to arise from the endoplasmic reticulum in this way.

scope. In electron micrographs, the oil bodies have an amorphous appearance (Figures 2–8 and 3–19). Oil bodies are widely distributed throughout the cells of the plant body but are most abundant in fruits and seeds. Approximately 45 percent of the weight of sunflower, peanut, flax, and sesame seeds is composed of oil. The oil provides energy and a source of carbon to the developing seedling. The oil bodies, often incorrectly described as organelles (called spherosomes), are thought to arise in the endoplasmic reticulum, one of the two main sites of lipid synthesis in plants, but they are not bounded by a membrane. The other site of lipid synthesis is the plastid.

Ribosomes

Ribosomes are small particles, only about 17 to 23 nanometers in diameter, consisting of about equal amounts of protein and RNA. Each ribosome consists of a large and a small subunit, which are produced in the nucleolus and exported to the cytoplasm where they are assembled into a ribosome. As we shall discuss in Chapter 11, ribosomes are the sites at which amino acids are linked together to form proteins. Abundant in the cytoplasm of metabolically active cells, ribosomes are found both free in the cytosol and attached to the endoplasmic reticulum. As mentioned previously, plastids and mitochondria contain smaller ribosomes, similar to those in prokaryotes.

Ribosomes actively involved in protein synthesis occur in clusters or aggregates called **polysomes,** or polyribosomes (Figure 3–20). Cells that are synthesizing

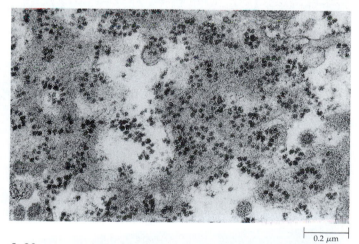

3–20

Polysomes (clusters of ribosomes) on the surface of rough endoplasmic reticulum. The endoplasmic reticulum is a network of membranes extending throughout the cytoplasm of the eukaryotic cell, dividing it into compartments and providing surfaces on which chemical reactions can take place. Polysomes are the sites at which amino acids are assembled into proteins. This electron micrograph shows a portion of a leaf cell from the fern Regnellidium diphyllum.

proteins in large quantities often contain extensive systems of polysome-bearing endoplasmic reticulum. In addition, polysomes are often attached to the outer surface of the nuclear envelope (Figure 3–7). Membrane-bound ribosomes and free ribosomes are both structurally and functionally identical, differing from one another only in the proteins they are making at any given time.

Endoplasmic Reticulum

The **endoplasmic reticulum (ER)** is a complex, three-dimensional membrane system. In sectional view, the ER appears as two parallel membranes with a narrow space, or *lumen*, between them. This profile of endoplasmic reticulum should not be confused with a single unit membrane: each of the parallel ER membranes is itself a unit membrane. The form and abundance of the ER varies greatly from cell to cell, depending upon the cell type, its metabolic activity, and its stage of development. For example, cells that store proteins have abundant **rough endoplasmic reticulum,** which consists of flattened sacs, or **cisternae** (singular: cisterna), with numerous ribosomes on their outer surface (Figure 3–20). In contrast, cells that secrete lipids have extensive systems of **smooth endoplasmic reticulum,** which lacks ribosomes and is largely tubular in form. The **tubular endoplasmic reticulum** is involved in lipid synthesis. (Note that cisternal and tubular ER are distinguished by the shape of the ER membranes, while rough and smooth ER are distinguished by the presence or absence of bound ribosomes. Rough ER is typically cisternal, however, and smooth ER is generally tubular.) Both rough and smooth forms of ER occur within the same cell and have numerous connections between them.

In many cells, an extensive network of ER, consisting of interconnected cisternae and tubules, is located just inside the plasma membrane in the peripheral, or *cortical,* cytoplasm (Figure 3–21). It has been suggested that this cortical network of ER serves as a structural element that stabilizes or anchors the cytoskeleton of the cell. The most likely function of the cortical ER appears to be in regulating the level of Ca^{2+} ions in the cytosol (see page 87). The cortical ER may therefore play a role in a host of developmental and physiological processes involving calcium.

Some electron micrographs show the rough ER to be continuous with the outer membrane of the nuclear envelope. As we have mentioned, the nuclear envelope may be considered a specialized, locally differentiated portion of the ER. When the nuclear envelope breaks into fragments during prophase of nuclear division, it becomes indistinguishable from cisternae of the rough ER. When new nuclei are formed during telophase, vesicles of ER join to form the nuclear envelopes of the two daughter nuclei. (The stages of nuclear division are the subject of Chapter 8.)

The ER functions as a communications system within the cell and as a system for channeling materials—such as proteins and lipids—to different parts of the cell. In addition, the cortical ER of adjacent plant cells is interconnected by cytoplasmic threads, called plasmodesmata, which traverse their common walls and play a role in cell-to-cell communication (see pages 66–68 and 87–89).

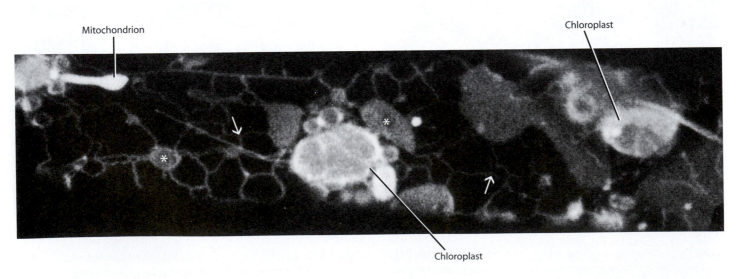

Mitochondrion

Chloroplast

Chloroplast

3–21

A portion of a filament of the moss Funaria hygrometrica showing cortical endoplasmic reticulum. The filament had been stained with a fluorescent dye and photographed using confocal laser scanning microscopy. The tubular ER network (arrows) is interspersed with cisternal ER (asterisks). The cortical ER is believed to be a general indicator of the metabolic and developmental status of a cell. Quiescent cells have less cortical ER, and developing and physiologically active cells have more.

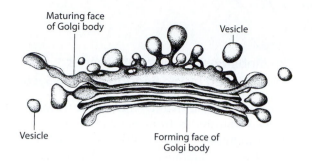

Maturing face
of Golgi body

Vesicle

Vesicle

Forming face of
Golgi body

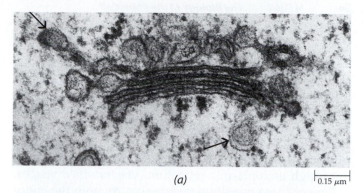

(a) 0.15 μm

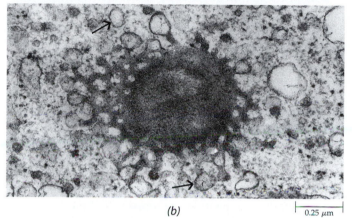

(b) 0.25 μm

3–22

*The Golgi body consists of a group of flat, membranous sacs with vesicles that bud off from the sacs. It serves as a "packaging center" for the cell and is concerned with secretory activities in eukaryotic cells. In **(a)** the cisternae of a Golgi body in a cell from the stem of the horsetail Equisetum hyemale are shown in sectional view, whereas **(b)** shows a single cisterna in a surface view. The arrows in both micrographs indicate vesicles that have pinched off from the cisternae.*

Golgi Complex

The term **Golgi complex** is used to refer collectively to all of the **Golgi bodies** (also known as dictyosomes) of a cell. Golgi bodies consist of stacks of flattened, disk-shaped sacs, or cisternae, which are often branched into complex series of tubules at their margins (Figure 3–22). The Golgi bodies of seed plants usually consist of four to eight cisternae.

The Golgi complex is a dynamic, highly polarized membrane system. Customarily, the two opposite poles of a Golgi stack are referred to as forming (or *cis*) and maturing (or *trans*) faces. The stacks between the two faces comprise the middle (or *medial*) cisternae. An additional structurally and biochemically distinct compartment, the ***trans*-Golgi network,** occurs on the maturing face of the Golgi body (Figure 3–23).

3–23

A diagrammatic representation of the endomembrane system. This drawing depicts the origin of new membranes from the rough endoplasmic reticulum. Transition vesicles pinch off a smooth-surfaced portion of the ER and carry membranes and enclosed substances to the forming face of the Golgi body. In this cell, cell wall substances are being transported stepwise across the Golgi stack to the trans-Golgi network by means of shuttle vesicles. Secretory vesicles derived from the trans-Golgi network then migrate to the plasma membrane and fuse with it, contributing new membrane to the plasma membrane and discharging their contents into the wall.

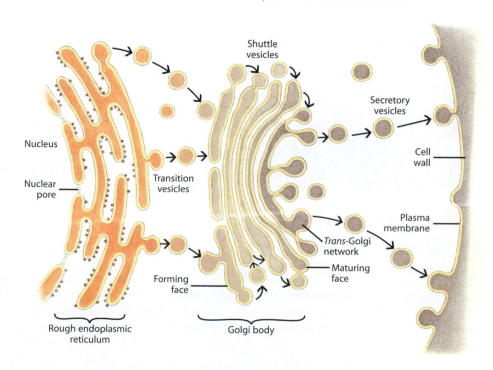

Shuttle vesicles

Secretory vesicles

Cell wall

Nucleus

Nuclear pore

Transition vesicles

Plasma membrane

Trans-Golgi network

Maturing face

Forming face

Rough endoplasmic reticulum

Golgi body

Golgi bodies are involved in secretion. In plants, most of the Golgi bodies are involved in the synthesis and secretion of noncellulosic cell wall polysaccharides. Evidence indicates that different steps in the synthesis of these polysaccharides occur in different cisternae of the Golgi body. Golgi bodies also process and secrete glycoproteins (see page 63) that are transferred to them from the rough endoplasmic reticulum, via *transition vesicles*. A flow of these transition vesicles occurs from the endoplasmic reticulum to the forming face of the Golgi body. The glycoproteins are transported stepwise across the stack to the maturing face by means of shuttle vesicles. They are then sorted in the *trans*-Golgi network for delivery to the vacuole or for secretion at the cell surface (Figure 3–23). Polysaccharides destined for secretion at the cell surface may also be sorted in the *trans*-Golgi network. A given Golgi body can process polysaccharides and glycoproteins simultaneously.

Newly formed vacuolar proteins are packaged at the *trans*-Golgi network into *coated vesicles* (Figure 3–24). The coats of these vesicles contain several proteins, including *clathrin*, a protein composed of three large and three smaller polypeptide chains that together form a three-pronged structure, called a triskelion. It is the arrangement of the triskelions on the cytoplasmic surface of the vesicles that forms their characteristic coat.

Glycoproteins and complex polysaccharides destined for secretion at the cell surface are packaged in noncoated, or smooth-surfaced, vesicles. The movement of these vesicles from the *trans*-Golgi network to the plasma membrane appears to depend upon the presence of actin filaments (see page 60). When the vesicles reach the plasma membrane, they fuse with it and discharge their contents into the wall. In enlarging cells, the vesicle membranes become incorporated into the plasma membrane, contributing to its growth. The secretion of substances from cells in vesicles is called **exocytosis.**

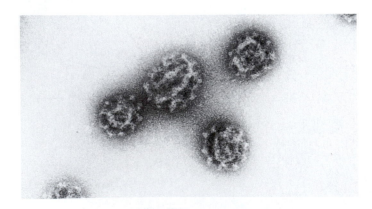

3–24

Coated vesicles isolated from bean (Phaseolus) leaves. The coated vesicles are surrounded by proteins, including clathrin. The three-pronged clathrin subunits are associated with each other to form cages around the vesicles.

The Mobility of Cellular Membranes Is Exemplified by the Endomembrane System

Membranes are dynamic, mobile structures that are frequently transferred from one compartment of the cytoplasm to another. An excellent example of the mobility of cellular membranes is provided by the **endomembrane system,** the major components of which are the *endoplasmic reticulum;* the *Golgi complex,* consisting of *Golgi bodies;* the *trans*-Golgi network; and various kinds of vesicles (Figure 3–23). Also included in the endomembrane system are the plasma membrane, the nuclear envelope, the tonoplast, and all other internal membranes, with the exception of mitochondrial, plastid, and peroxisome membranes. The membranes of the endomembrane system form a continuum, with the ER being the initial source of membranes. Transition vesicles transport new membrane material from the ER to the Golgi bodies, and secretory vesicles derived from the *trans*-Golgi network contribute to the plasma membrane. The *trans*-Golgi network also supplies vesicles that fuse with the tonoplast and thus contribute to the formation of the vacuole. The ER, Golgi complex, and *trans*-Golgi network are therefore a functional unit in which the Golgi bodies serve as the main vehicles for the transformation of endoplasmic-reticulum-like membranes into plasma-membrane-like and tonoplast-like membranes.

Cytoskeleton

As we have mentioned, virtually all eukaryotic cells possess a **cytoskeleton,** a complex network of protein filaments that extends throughout the cytosol. The cytoskeleton is intimately involved in many processes, including cell division, growth and differentiation, and the movement of organelles from one location to another within the cell. The cytoskeleton of plant cells consists of two main types of protein filament: microtubules and actin filaments. In addition, plant cells, like animal cells, may contain a third type of cytoskeletal filament, the *intermediate filament.* Little is known about the structure or role of the intermediate filament in plant cells.

Microtubules Are Cylindrical Structures Composed of Tubulin Subunits

Microtubules are long, thin cylindrical structures about 24 nanometers in diameter and of varying lengths. Each microtubule is built up of subunits of the protein called tubulin. The subunits are arranged in a helix to form 13 rows, or "protofilaments," around a hollow core (Figure 3–25a, b). Within each protofilament the subunits are oriented in the same direction, and all the protofilaments are aligned in parallel with the same polarity. The microtubule, consequently, is a polar structure, for which there

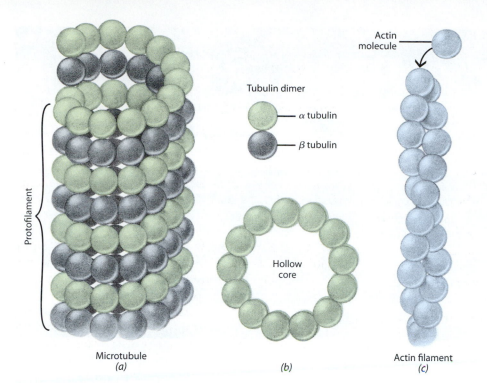

Actin molecule

Tubulin dimer

α tubulin

β tubulin

Protofilament

Hollow core

Microtubule
(a)

(b)

Actin filament
(c)

3–25
Two components of the cytoskeleton, microtubules and actin filaments, are formed from globular protein subunits. (a), (b) Microtubules are hollow tubules composed of two different types of molecules, alpha (α) tubulin and beta (β) tubulin. These tubulin molecules first come together to form soluble dimers ("two parts"), which then self-assemble into insoluble hollow tubules. The arrangement results in 13 "protofilaments" around a hollow core. (a) Longitudinal view of a short portion of a microtubule; (b) transverse section of a microtubule showing the ends of the 13 protofilaments. (c) Actin filaments consist of two linear chains of identical molecules coiled around one another to form a helix.

can be designated plus and minus ends. The plus ends grow faster than the minus ends, and the ends of the microtubules can alternate between growing and shrinking states, a behavior called *dynamic instability*. Microtubules are dynamic structures that undergo regular sequences of breakdown and re-formation into new configurations at specific points in the cell cycle (see Chapter 8). Their assembly takes place at sites within the cell known as *microtubule organizing centers*. The surface of the nucleus and portions of the cortical cytoplasm (the cytoplasm just inside the plasma membrane) have been identified as microtubule organizing centers.

Microtubules have many functions. In enlarging and differentiating cells, microtubules in the cortical cyto-

plasm, known as cortical microtubules, are involved in the orderly growth of the cell wall (Figure 3–26). These microtubules control the alignment of the cellulose microfibrils that are added to the cell wall, and the direction of cell expansion is governed, in turn, by this alignment of cellulose microfibrils in the wall. Microtubules also serve to direct secretory Golgi vesicles containing noncellulosic cell wall substances toward the developing wall. In addition, microtubules make up the spindle fibers that play a role in chromosome movement and in cell plate formation in dividing cells (see Chapter 8). Microtubules are also important components of flagella and cilia and are involved in the movement of these structures.

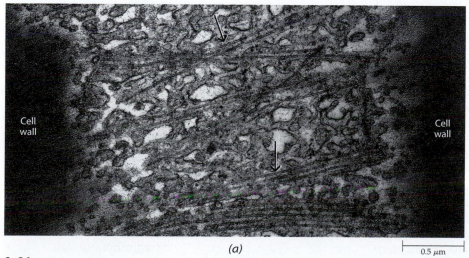

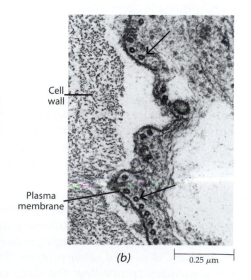

Cell wall

Cell wall

Cell wall

Plasma membrane

(a)

0.5 μm

(b)

0.25 μm

3–26
Cortical microtubules (indicated by arrows) in leaf cells of the fern Botrychium virginianum. (a) A longitudinal view of the microtubules just inside the wall and plasma membrane. (b) A transverse view of cortical microtubules, which can be seen to be separated from the wall by the plasma membrane. Cortical microtubules play a role in the alignment of cellulose microfibrils in the cell wall.

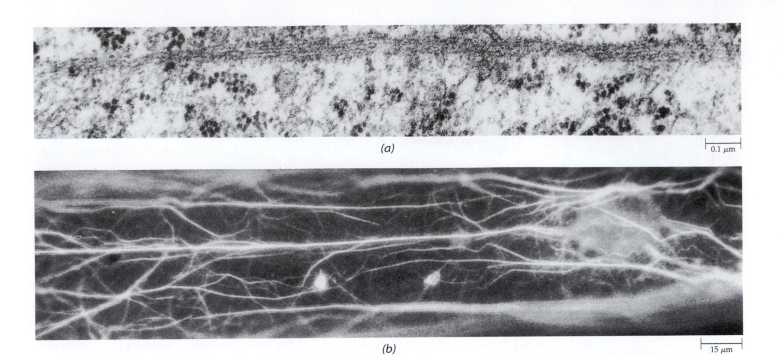

(a)

0.1 μm

(b)

15 μm

3–27

*Actin filaments. **(a)** A bundle of actin filaments as revealed in an electron micrograph of a leaf cell of maize (Zea mays). **(b)** Several bundles of actin filaments as revealed in a fluorescence micrograph of a stem hair of tomato (Solanum lycopersicum). Actin filaments are involved in a variety of activities, including cytoplasmic streaming.*

Actin Filaments Are Narrower Than Microtubules

Actin filaments, also called microfilaments, are, like microtubules, polar structures with distinct plus and minus ends. They are composed of a protein called *actin* and occur as long filaments 5 to 7 nanometers in diameter (Figure 3–25c). Actin filaments are associated spatially with microtubules and, like microtubules, form new configurations at specific points in the cell cycle. In addition to single filaments, bundles of actin filaments have been found in many plant cells (Figure 3–27).

Actin filaments are involved with a variety of activities in plant cells, including cytoplasmic streaming (see "Cytoplasmic Streaming in Giant Algal Cells" on page 47), cell wall deposition, tip growth of pollen tubes, movement of the nucleus before and following cell division, organelle movement, vesicle-mediated secretion, and organization of the ER.

Flagella and Cilia

Flagella and **cilia** (singular: flagellum and cilium) are hairlike structures that extend from the surface of many different types of eukaryotic cells. They are relatively thin and constant in diameter (about 0.2 micrometer), but they vary in length from about 2 to 150 micrometers. By convention, those that are longer or are present alone or in small numbers are usually referred to as flagella, whereas those that are shorter or occur in greater numbers are called cilia. In the following discussion, we will use the term "flagellum" to refer to both.

In some algae and other protists, flagella are locomotor structures, propelling the organisms through the water. In plants, flagella are found only in the gametes (reproductive cells) and then only in plants that have motile sperm, including the mosses, liverworts, ferns, cycads, and maidenhair tree (*Ginkgo biloba*). Some flagella, called tinsel flagella, bear one or two rows of minute lateral appendages, whereas others, called whiplash flagella, lack such structures (Figure 3–28).

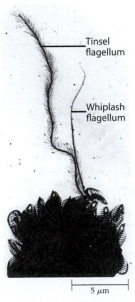

Tinsel flagellum

Whiplash flagellum

5 μm

3–28

Two types of flagella are represented in this single cell of the colonial organism Synura petersenii, *an alga. This organism has a long tinsel flagellum and a somewhat shorter whiplash flagellum.*

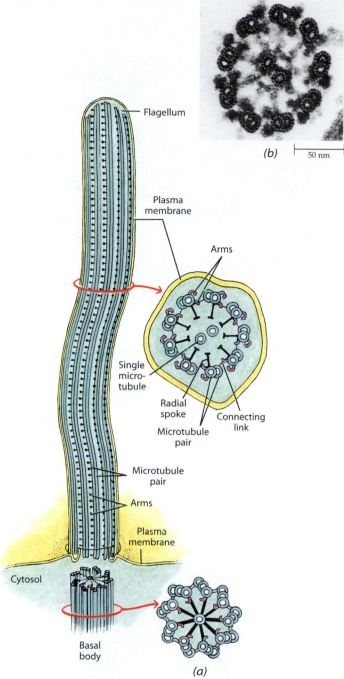

3–29

*The structure of a flagellum. **(a)** Diagram of a flagellum with its underlying basal body, and **(b)** an electron micrograph of the flagellum of* Chlamydomonas, *as seen in transverse section. Virtually all eukaryotic flagella have this same internal structure, which consists of an outer cylinder of nine pairs of microtubules surrounding two additional microtubules in the center. The "arms," the radial spokes, and the connecting links are formed from different types of protein. The basal bodies from which flagella arise have nine outer triplets, with no microtubules in the center. The "hub" of the wheel in the basal body is not a microtubule, although it has about the same diameter.*

One of the most intriguing of the discoveries made possible by the electron microscope is the internal structure of flagella. Each flagellum has a precise internal organization (Figure 3–29). An outer ring of nine pairs of microtubules surrounds two additional microtubules in the center of the flagellum proper. This basic 9-plus-2 pattern of organization is found in all flagella of eukaryotic organisms.

The movement of a flagellum originates from within the structure itself. Flagella are capable of movement even after they have been detached from cells. The movement is produced by a sliding microtubule mechanism in which the outer pairs of microtubules move past one another without contracting. As the pairs slide past one another, their movement causes localized bending of the flagellum. Sliding of the pairs of microtubules results from cycles of attachment and detachment of enzyme-containing "arms" between neighboring pairs in the outer ring (Figure 3–29).

Flagella grow out of cylinder-shaped structures in the cytoplasm known as *basal bodies,* which then form the basal portion of the flagellum. The internal structure of the basal body resembles that of the flagellum itself, except that the outer tubules in the basal body occur in triplets rather than in pairs and the two central tubules are absent.

Cell Wall

The **cell wall,** above all other characteristics, distinguishes plant cells from animal cells. Its presence is the basis of many of the characteristics of plants as organisms. The cell wall is rigid and therefore limits the size of the protoplast, preventing rupture of the plasma membrane when the protoplast enlarges following the uptake of water by the cell. The cell wall largely determines the size and shape of the cell, the texture of the tissue, and the final form of the plant organ. Cell types are often identified by the structure of their walls, reflecting the close relationship between cell wall structure and cell function.

Once regarded as merely an outer, inactive structure produced by the protoplast, the cell wall is now recognized as having specific and essential functions and to be an integral part of the plant cell. Cell walls contain a variety of enzymes and play important roles in the absorption, transport, and secretion of substances in plants. They may also, like vacuoles, serve as sites of digestive activity.

In addition, the cell wall may play an active role in defense against bacterial and fungal pathogens by receiving and processing information from the surface of the pathogen and transmitting this information to the plasma membrane of the plant cell. Through gene-activated processes (see Chapter 11), the plant cell may then become resistant to attack through the production of

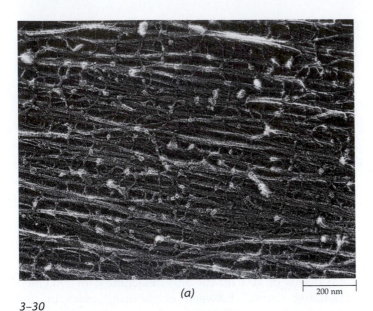

(a)

200 nm

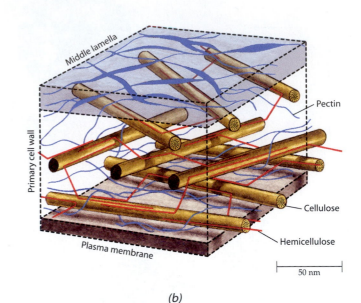

(b)

50 nm

3–30
Primary walls. (a) Surface view of the primary wall of a carrot (Daucus carota) *cell, prepared by a fast-freeze, deep-etch technique, showing cellulose microfibrils cross-linked by an intricate web of matrix molecules. (b) Schematic diagram showing how the cellulose microfi-*

brils are cross-linked into a complex network by hemicellulose molecules. The hemicellulose molecules are linked to the surface of the microfibrils by hydrogen bonds. The cellulose-hemicellulose network is permeated by a network of pectins, which are highly

hydrophilic polysaccharides. Both hemicellulose and pectin are matrix substances. The middle lamella is a pectin-rich layer that cements together the primary walls of adjacent cells.

phytoalexins (see pages 32 and 780)—antibiotics that are toxic to the pathogens—or through the synthesis and deposition of substances such as lignin (page 36), which act as barriers to invasion. Certain cell wall polysaccharides, dubbed "oligosaccharins," may even function as signal molecules, regulating plant growth and development.

Cellulose Is the Principal Component of Plant Cell Walls

The principal component of plant cell walls is **cellulose,** which largely determines their architecture. Cellulose is made up of repeating monomers of glucose attached end to end (see Figure 2–5). These long, thin cellulose molecules are united into **microfibrils** about 10 to 25 nanometers in diameter (Figure 3–30). Cellulose has crystalline properties (Figure 3–31) because of the orderly arrangement of its molecules in certain parts, the *micelles,* of the microfibrils (Figure 3–32). Cellulose microfibrils wind together to form fine threads that may coil around one another like strands in a cable. Each "cable," or macrofibril, measures about 0.5 micrometer in diameter and may reach 4 micrometers in length. Cellulose molecules wound in this fashion have a strength exceeding that of an equivalent thickness of steel.

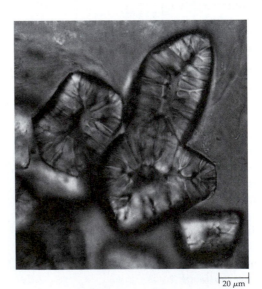

20 μm

3–31
Stone cells (sclereids) from the flesh of a pear (Pyrus communis). *Clusters of such stone cells are responsible for the gritty texture of this fruit. The stone cells have very thick secondary walls traversed by numerous simple pits, which appear as lines in the walls.*

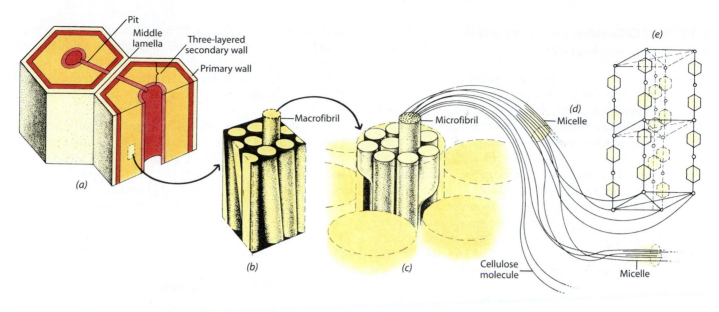

3–32

The detailed structure of a cell wall. (a) Portion of wall showing the middle lamella, primary wall, and three layers of secondary wall. Cellulose, the principal component of the cell wall, exists as a system of fibrils of different sizes. (b) The largest fibrils, macrofibrils, can be seen with the light microscope. (c) With the aid of an electron microscope, the macrofibrils can be resolved into microfibrils about 10 to 25 nanometers wide. (d) Parts of the microfibrils, the micelles, are arranged in an orderly fashion and impart crystalline properties to the wall. (e) A fragment of a micelle shows parts of the chainlike cellulose molecules in a lattice arrangement.

Cellulose Forms a Framework Interpenetrated by a Matrix of Noncellulosic Molecules

The cellulose framework of the wall is interpenetrated by a cross-linked *matrix* of noncellulosic molecules. These molecules are the polysaccharides known as hemicelluloses and pectins, as well as the structural proteins called glycoproteins (Figure 3–30).

The **hemicelluloses** vary greatly in different cell types and among the different plant groups. *Xyloglucans* (see Figure 2–7) are the principal hemicelluloses of the first-formed cell wall layers of magnoliids and eudicotyledons. *Xylans* are the principal hemicelluloses of such wall layers in monocotyledons, the other group of flowering plants. Both kinds of hemicelluloses are tightly hydrogen-bonded to the cellulose microfibrils, apparently limiting the extensibility of the cell wall by tethering adjacent microfibrils and, hence, probably playing a significant role in regulating cell enlargement.

Pectins, which are best known for their ability to form gels, are characteristic of the first-formed cell wall layers and of the intercellular substance that cements together the walls of contiguous cells in eudicotyledons and, to a lesser extent, in monocotyledons. They may be lacking entirely from subsequently formed wall layers. Pectins are highly hydrophilic polysaccharides (see Figure 2–6), and the water they introduce into the cell wall imparts plastic, or pliable, properties to the wall, a condition necessary for wall expansion. Growing primary walls are composed of about 65 percent water.

Cell walls may also contain **glycoproteins**—structural proteins—as well as enzymes. The glycoproteins, which are matrix components, comprise about 10 percent of the dry weight of many primary walls. The best characterized glycoproteins are the **extensins** (a family of hydroxyproline-rich proteins), so-named because they were originally assumed to be involved with the cell wall's extensibility. It appears, however, that the deposition of extensin might strengthen the wall, making it less extensible. A large number of enzymes have been reported in the first-formed wall layers. Such enzymes include peroxidases, phosphatases, cellulases, and pectinases.

Another important constituent of the walls of many kinds of cells is **lignin,** which adds compressive strength and bending stiffness (rigidity) to the cell wall (pages 36 and 37). It is commonly found in the walls of plant cells that have a supporting or mechanical function. Lignin, which is hydrophobic, replaces the water in the cell wall. The process by which a cell becomes lignified, that is, impregnated with lignin, begins in the intercellular substance at the corners of the cells. It then spreads to the first-formed wall layers and finally to those formed last.

Cutin, suberin, and **waxes** are fatty substances commonly found in the walls of the outer, protective tissues of the plant body. Cutin, for example, is found in the walls of the epidermis, and suberin is found in those of the secondary protective tissue, cork. Both substances occur in combination with waxes and function largely to reduce water loss from the plant (pages 24 and 25).

Many Plant Cells Have a Secondary Wall in Addition to a Primary Wall

Plant cell walls vary greatly in thickness, depending partly on the role the particular cells play in the structure of the plant and partly on the age of the individual cell. Each protoplast forms its wall from the outside inward so that the youngest layer of a given wall is the innermost portion, next to the protoplast. The cell wall layers formed first make up the **primary wall.** The region of union of the primary walls of adjacent cells is called the **middle lamella,** or intercellular substance. Many cells subsequently deposit additional wall layers. These form the **secondary wall.** If present, the secondary wall is laid down by the protoplast of the cell on the inner surface of the primary wall (Figure 3–32a).

The Middle Lamella Joins Adjacent Cells   The middle lamella is the pectin-rich layer that cements together the primary walls of adjacent cells. Frequently, it is difficult to distinguish the middle lamella from the primary wall, especially in cells that develop thick secondary walls. In such cases, the two adjacent primary walls and the middle lamella, and perhaps the first layer of the secondary wall of each cell, may be called a *compound middle lamella.*

The Primary Wall, Composed of the First-formed Wall Layers, Is Deposited while the Cell Is Increasing in Size   The primary wall is deposited before and during the growth of the plant cell. Primary walls are composed of cellulose, hemicelluloses, pectic substances, proteins (both glycoproteins and enzymes), and water. Primary walls may also contain lignin, suberin, or cutin.

Actively dividing cells commonly have only primary walls, as do most mature cells involved with such metabolic processes as photosynthesis, respiration, and secretion. These cells—that is, living cells with only primary walls—are able to lose their specialized cellular form, divide, and differentiate into new types of cells. For this reason, it is principally cells with only primary walls that are involved in wound healing and regeneration in the plant.

Usually, primary cell walls are not of uniform thickness throughout but have thin areas called **primary pit-fields** (Figure 3–33). Cytoplasmic threads, or plasmodesmata, which connect the living protoplasts of adjacent cells, are commonly aggregated in the primary pit-fields, but they are not restricted to such areas.

The Secondary Wall Is Deposited inside the Primary Wall after the Primary Wall Has Stopped Increasing in Size

Although many plant cells have only a primary wall, in others the protoplast deposits a secondary wall inside the primary wall. Secondary wall formation occurs mostly after the cell has stopped growing and the primary wall is no longer increasing in surface area. Secondary walls are particularly important in specialized cells that have a strengthening function and in those involved in the conduction of water. In these cells, the protoplast often dies after the secondary wall has been laid down.

Cellulose is more abundant in secondary walls than in primary walls, and pectins may be lacking; the secondary wall is therefore rigid and not readily stretched. The matrix of the secondary wall is composed of hemicellulose. Glycoproteins and enzymes, which are relatively abundant in primary cell walls, apparently are absent in secondary cell walls.

Frequently, three distinct layers—designated S_1, S_2, and S_3, for the outer, middle, and inner layer, respectively—can be distinguished in a secondary wall (Figure 3–34). The layers differ from one another in the orienta-

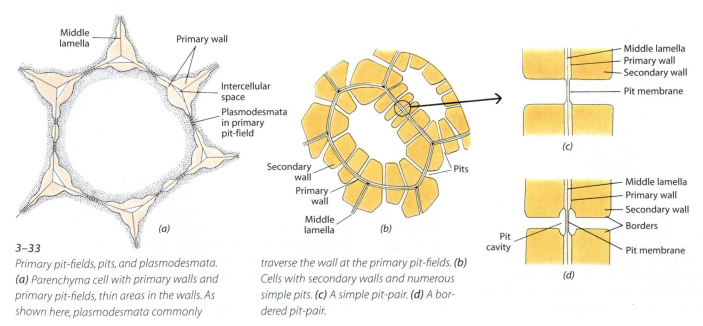

3–33

Primary pit-fields, pits, and plasmodesmata.
(a) Parenchyma cell with primary walls and primary pit-fields, thin areas in the walls. As shown here, plasmodesmata commonly traverse the wall at the primary pit-fields. (b) Cells with secondary walls and numerous simple pits. (c) A simple pit-pair. (d) A bordered pit-pair.

Cell Theory versus Organismal Theory

In its classical form, the cell theory proposed that the bodies of plants and animals are aggregations of individual, differentiated cells. The proponents of this concept believed that the activities of the whole plant or animal might be considered the summation of the activities of the individual constituent cells, with the individual cells of prime importance. This concept has been compared to the Jeffersonian theory of democracy, which considered the nation to be dependent upon and secondary in rights and privileges to its individual, constituent states.

By the latter half of the nineteenth century, an alternative to the cell theory was formulated. Known as the **organismal** *theory, it replaced some of the ideas expounded by the cell theory. The proponents of the organismal theory consider the entire organism, rather than the individual cells, to be of prime importance. The many-celled plant or animal is regarded not merely as a group of independent units, but as a more or less continuous*
mass of protoplasm, which, in the course of evolution, became subdivided into cells. The organismal theory arose in part from the results of physiological research that demonstrated the necessity for coordination of the activities of the various organs, tissues, and cells for normal growth and development of the organism. The organismal theory might be likened to the theory of government that holds that the unified nation is of prime importance, not the states of which it is formed.

The nineteenth-century German botanist Julius von Sachs concisely stated the organismal theory when he wrote, "Die Pflanze bildet Zelle, nicht die Zelle Pflanzen," which means "The plant forms cells, the cells do not form plants."

Indeed, the organismal theory is especially applicable to plants, whose protoplasts do not pinch apart during cell division, as in animal cell division, but are partitioned initially by insertion of a cell plate (see Chapter 8). Moreover, the separation of plant cells is rarely complete, the
protoplasts of contiguous cells remaining interconnected by the cytoplasmic strands known as plasmodesmata. Plasmodesmata traverse the walls and unite the entire plant body into an organic whole called the symplast, which consists of the interconnected protoplasts and their plasmodesmata. As appropriately stated by Donald Kaplan and Wolfgang Hagemann, "Instead of higher plants being federal aggregations of independent cells, they are unified organisms whose protoplasts are incompletely subdivided by cell walls."

In its modern form, the cell theory states simply that all living organisms are composed of one or more cells, that the chemical reactions of a living organism take place in cells, that cells arise from other cells, and that cells contain the hereditary information passed from parent cell to daughter cell. The cell and organismal theories are not mutually exclusive. Together, they provide a meaningful view of structure and function at cellular and organismal levels.

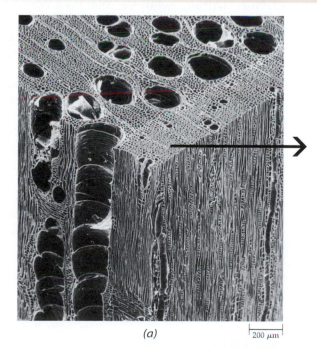

(a)

200 μm

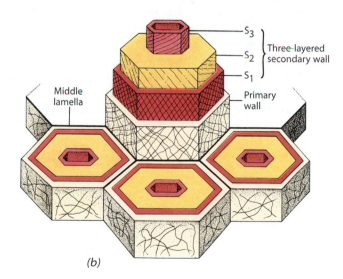

(b)

3–34

The layers of secondary cell walls. (a) Scanning electron micrograph of block of red oak (Quercus rubra) wood showing the location of the supporting cells (known as fibers) depicted in the diagram. (b) Diagram showing the organization of the cellulose microfibrils and the three layers (S_1, S_2, S_3) of the secondary wall. The different orientations of the three layers strengthen the secondary wall.

tion of their cellulose microfibrils. Such multiple wall layers are found in certain cells of the secondary xylem, or wood. The laminated structure of such secondary walls greatly increases their strength, and the cellulose microfibrils are laid down in a denser pattern than in the primary wall. The secondary walls of cells found in wood commonly contain lignin.

Whereas the Primary Wall Has Pit-fields, the Secondary Wall Has Pits When the secondary cell wall is deposited, it is not laid down over the primary pit-fields of the primary wall. Consequently, characteristic interruptions, or **pits,** are formed in the secondary wall (Figure 3–33). In some instances, pits are also formed in areas where there are no primary pit-fields. Lignified secondary walls are not permeable to water, but, with the formation of pits, at least at those sites the adjacent cells are separated only by primary walls.

A pit in a cell wall usually occurs opposite a pit in the wall of an adjoining cell. The middle lamella and two primary walls between the two pits are called the **pit membrane.** The two opposite pits plus the membrane constitute a **pit-pair.** Two principal types of pits are found in cells with secondary walls: **simple** and **bordered.** In bordered pits, the secondary wall arches over the **pit cavity.** In simple pits, there is no overarching. (See page 655 for additional information on the properties of pit membranes in tracheary elements.)

Growth of the Cell Wall Involves Interactions among Plasma Membrane, Secretory Vesicles, and Microtubules

Cell walls grow both in thickness and in surface area. Extension of the wall is a complex process under the close biochemical control of the protoplast. During growth, the primary wall must yield enough to allow an appropriate extent of expansion, while at the same time remaining strong enough to constrain the protoplast. Growth of the primary wall requires loosening of wall structure, a phenomenon influenced by some hormones (see Chapter 28). There is also an increase in protein synthesis and in respiration to provide needed energy, as well as an increase in uptake of water by the cell. Most of the new cellulose microfibrils are placed on top of those previously formed, layer upon layer, although some may be inserted into the existing wall structure.

In cells that enlarge more or less uniformly in all directions, the microfibrils are laid down in a random array, forming an irregular network. In contrast, in elongating cells, the microfibrils of the side walls are deposited in a plane at right angles (perpendicular) to the axis of elongation (Figure 3–35).

Newly deposited cellulose microfibrils run parallel to the cortical microtubules lying just beneath the plasma membrane. It is generally accepted that the cellulose microfibrils are synthesized by **cellulose synthase** complexes that occur in the plasma membrane (Figure 3–36). In seed plants, these enzyme complexes appear as rings, or rosettes, of six hexagonally arranged particles that span the membrane, with the hole in the center of each rosette occupied by a particle commonly referred to as a

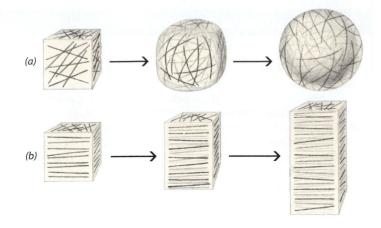

3–35

*The orientation of cellulose microfibrils within the primary wall influences the direction of cell expansion. (**a**) If the cellulose microfibrils are randomly oriented in all walls, the cell will expand equally in all directions, tending to become spherical in shape. (**b**) If the microfibrils are oriented at right angles to the ultimate long axis of the cell, the cell will expand longitudinally along that axis.*

globule. During cellulose synthesis, the complexes, which move in the plane of the membrane, extrude the microfibrils onto the outer surface of the membrane. From these, the cellulose microfibrils are integrated into the cell wall. Movement of the complexes is guided by the underlying cortical microtubules. The rosettes are inserted into the plasma membrane via secretory vesicles from the *trans*-Golgi network.

The matrix substances—hemicelluloses and pectic substances—are carried to the wall in secretory vesicles, as are the glycoproteins. The type of matrix substance synthesized and secreted by a cell at any given time depends on the stage of development. Pectins, for example, are more characteristic of enlarging cells, whereas hemicelluloses predominate in cells that are no longer enlarging.

Plasmodesmata

As mentioned previously, the protoplasts of adjacent plant cells are connected with one another by **plasmodesmata** (singular: plasmodesma). Although such structures have long been visible with the light microscope (Figure 3–37), they were difficult to interpret. Not until they could be observed with an electron microscope was their nature as cytoplasmic strands confirmed.

3–36

*Cellulose microfibrils are synthesized by enzyme complexes that move within the plane of the plasma membrane. **(a)** The enzymes are complexes of cellulose synthase that form rosettes embedded in the plasma membrane. Each enzyme rosette synthesizes cellulose from the glucose derivative UDP-glucose (uridine diphosphate glucose). The UDP-glucose molecules enter the rosette on the inner (cytoplasmic) face of the membrane, and a cellulose microfibril is extruded from the outer face of the membrane. **(b)** As the far ends of the newly formed microfibrils become integrated into the wall, the rosettes continue synthesizing cellulose, moving along a route (arrows) that parallels the cortical microtubules in the underlying cytoplasm.*

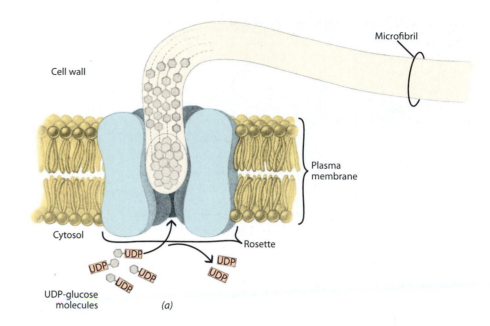

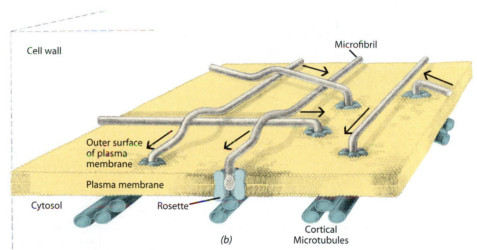

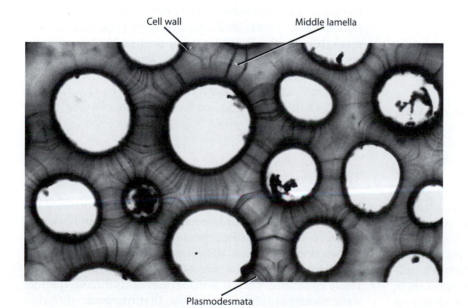

3–37

Light micrograph of plasmodesmata in the thick primary walls of persimmon (Diospyros) endosperm, the nutritive tissue within the seed. The plasmodesmata appear as fine lines extending from cell to cell across the walls. The middle lamella appears as a light line between these cells. Plasmodesmata generally are not discernible with the light microscope. The extreme thickness of the persimmon endosperm cell walls greatly increases the length of the plasmodesmata, however, making them discernible with the light microscope.

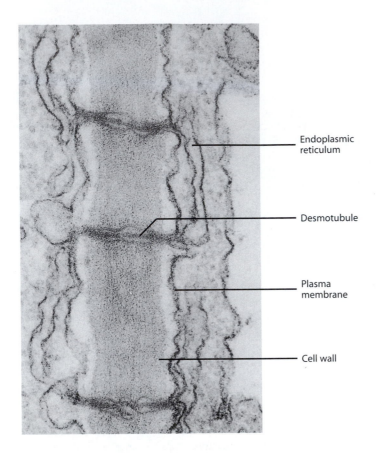

Endoplasmic
reticulum

Desmotubule

Plasma
membrane

Cell wall

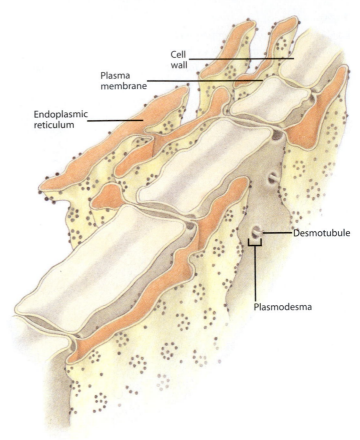

Cell
wall

Plasma
membrane

Endoplasmic
reticulum

Desmotubule

Plasmodesma

3–38

Plasmodesmata connecting two leaf cells from the cottonwood (Populus deltoides) tree. Plasmodesmata are narrow, plasma-membrane-lined channels in the walls, each traversed by a modified tubule of endoplasmic reticulum known as the desmotubule. The middle lamella between the adjacent primary walls is not discernible in the electron micrograph, as is common in such preparations. Note the continuity of the ER on either side of the wall with the desmotubules of the plasmodesmata.

Plasmodesmata may occur throughout the cell wall, or they may be aggregated in primary pit-fields or in the membranes between pit-pairs. With the electron microscope, plasmodesmata appear as narrow channels (about 30 to 60 nanometers in diameter) lined by a plasma membrane and traversed by a modified tubule of ER known as the *desmotubule* (Figure 3–38). Many plasmodesmata are formed during cell division as strands of tubular ER become trapped within the developing cell plate (see Figure 8–11). Plasmodesmata are also formed in the walls of nondividing cells. These structures provide a pathway for the transport of certain substances between cells and are discussed in greater detail in Chapter 4 (see pages 87–89).

Summary

The Cell Is the Fundamental Unit of Life

All living matter is composed of cells. Cells are extremely varied, ranging in structure and function from independent, single-celled organisms to the highly specialized, interdependent cell types found in complex multicellular plants and animals. However, cells are remarkably similar in their basic structure. All cells are bounded by an outer membrane, known as the plasma membrane. Enclosed within this membrane are the cytoplasm and the hereditary information in the form of DNA. Collectively, the contents of the cell are called protoplasm.

Cells Are of Two Fundamentally Different Types: Prokaryotic and Eukaryotic

Prokaryotic cells lack nuclei and membrane-bounded organelles. They are represented today by the archaea and bacteria, including the cyanobacteria. The prokaryotic chromosome consists of a single circular molecule of DNA localized in the nucleoid. Eukaryotic cells have DNA contained within a true nucleus and have distinct membrane-bounded organelles that perform different functions.

Plant Cells Typically Consist of a Cell Wall and a Protoplast

The protoplast consists of the cytoplasm and a nucleus. The plasma membrane is the outer boundary of the cytoplasm next to the cell wall. The cytosol, or cytoplasmic matrix, of plant cells is frequently in motion, a phenomenon known as cytoplasmic streaming.

The Nucleus Is Bounded by a Nuclear Envelope and Contains Nucleoplasm, Chromatin, and One or More Nucleoli

The nucleus is the control center of the cell and contains the cell's genetic information (the nuclear genome). It is often the most prominent structure within the protoplast. It is surrounded by a nuclear envelope composed of a pair of membranes, which may be considered a specialized, locally differentiated portion of the endoplasmic reticulum. Enclosed within the nuclear envelope is the chromatin, which consists of DNA and histone proteins. The nucleolus is the site within the nucleus where ribosomal subunits are formed.

There Are Three Main Types of Plastids: Chloroplasts, Chromoplasts, and Leucoplasts

Together with vacuoles and cell walls, plastids are characteristic components of plant cells. Each plastid is bounded by an envelope consisting of two membranes. Mature plastids are classified in part on the basis of the kinds of pigments they contain: chloroplasts contain chlorophylls and carotenoid pigments; chromoplasts contain carotenoid pigments; and leucoplasts are nonpigmented. Proplastids are the precursors of plastids.

Mitochondria Are the Sites of Respiration

Mitochondria, like plastids, are organelles bounded by two membranes. The inner membrane is folded to form an extensive inner membrane system, increasing the surface area available to enzymes and the reactions associated with them. Mitochondria are the principal sites of respiration in eukaryotic cells.

Plastids and Mitochondria Share Certain Features with Prokaryotic Cells

Both plastids and mitochondria are semiautonomous organelles containing circular molecules of DNA and ribosomes similar to those of bacteria. Plastids and mitochondria probably originated as bacteria that found shelter within larger heterotrophic cells. Plant cells contain three genomes: those of the nucleus, the plastids, and the mitochondria.

Peroxisomes Are Bounded by a Single Membrane

Unlike plastids and mitochondria, peroxisomes are organelles bounded by a single membrane. Some peroxisomes play an important role in photorespiration. Others are involved in the conversion of stored fats to sucrose during seed germination.

Vacuoles Contain Mostly Water but Perform a Variety of Functions

Plant cells are characterized by the presence of vacuoles within their cytoplasm. These organelles are membrane-bounded regions that are filled with cell sap, an aqueous solution containing a variety of salts, sugars, anthocyanin pigments, and other substances. Vacuoles play an important role in cell enlargement and the maintenance of tissue rigidity. In addition, many vacuoles are involved in the breakdown of macromolecules and the recycling of their components within the cell. The vacuole is bounded by a single membrane called the tonoplast.

Ribosomes Are the Sites of Protein Synthesis

Ribosomes, which are found both free in the cytosol and attached to the endoplasmic reticulum and the outer surface of the nuclear envelope, are the sites at which amino acids are linked together to form proteins. During protein synthesis, the ribosomes occur in clusters called polysomes.

The Endoplasmic Reticulum Is an Extensive Three-Dimensional System of Membranes with a Variety of Roles

The endoplasmic reticulum exists in two forms: rough ER, which is studded with ribosomes, and smooth ER, which lacks ribosomes. Rough ER is involved with membrane and protein synthesis, and smooth ER with lipid synthesis. Some lipids occur as oil bodies in the cytosol.

The Golgi Complex Is a Highly Polarized Membrane System Involved in Secretion

The Golgi complex consists of Golgi bodies, which are composed of stacks of flattened, disk-shaped sacs, or cisternae. Most plant Golgi bodies are involved in the synthesis and secretion of complex noncellulosic cell wall polysaccharides, which are transported to the surface of the cell in secretory vesicles derived from the *trans*-Golgi network. The vesicles also contribute to the plasma membrane and tonoplast. Golgi bodies also process and secrete glycoproteins transferred to them from rough endoplasmic reticulum via transition vesicles.

Summary TABLE   Plant Cell Components

MAJOR COMPONENTS	INDIVIDUAL CONSTITUENTS	KEY DESCRIPTIVE FEATURES	FUNCTION(S)
Cell wall		Consists of cellulose microfibrils embedded in a matrix of hemicelluloses, pectins, and glycoproteins. Lignin, cutin, suberin, and waxes may also be present.	Strengthens the cell; determines cell size and shape.
	Middle lamella	Pectin-rich layer between cells.	Cements adjacent cells together.
	Primary wall	First wall layers to form. Contains primary pit-fields.	Found in actively dividing and actively metabolizing cells.
	Secondary wall	Formed after the primary wall is laid down. Located interior to the primary wall. Contains pits.	Found in cells with strengthening and/or water-conducting function. Is rigid and thus imparts added strength.
	Plasmodesmata	Cytoplasmic strands traversing the cell wall.	Interconnect protoplasts of adjacent cells, providing a pathway for the transport of substances between cells.
Nucleus		Bounded by a pair of membranes, the nuclear envelope, and containing nucleoplasm, nucleoli, and chromatin (chromosomes) consisting of DNA and histone proteins.	Controls cellular activities. Stores genetic information.
Plasma membrane		Single membrane, forming the outer boundary of the cytoplasm.	Mediates transport of substances into and out of the cell. Site of cellulose synthesis. Receives and transmits hormonal and environmental signals.
Cytoplasm	Cytosol	The least differentiated part of the cytoplasm.	Matrix in which organelles and membrane systems are suspended and biochemical events occur.
	Plastids	Surrounded by a double-membrane envelope. Semiautonomous organelles containing their own DNA and ribosomes.	Sites of food manufacture and storage.
	Chloroplasts	Contain chlorophyll and carotenoid pigments embedded in thylakoid membranes.	Sites of photosynthesis. Involved in amino acid synthesis and fatty acid synthesis. Temporary storage of starch.
	Chromoplasts	Contain carotenoid pigments.	May function in attracting insects and other animals essential for cross-pollination and fruit and seed dispersal.
	Leucoplasts	Lack pigments entirely.	Some (amyloplasts) store starch; others form oils.
	Proplastids	Undifferentiated plastids; may form prolamellar bodies.	Precursors of other plastids.

The Mobility of Cellular Membranes Is Exemplified by the Endomembrane System

The major components of the endomembrane system are the endoplasmic reticulum, the Golgi complex, the *trans*-Golgi network, and various kinds of vesicles. Collectively, the membranes of the system form a continuum, with the ER being the initial source of membranes.

The Cytoskeleton Is Composed of Microtubules and Actin Filaments

The cytosol of eukaryotic cells is permeated by the cytoskeleton, a complex network of protein filaments, of which there are two well-characterized types in plant cells: microtubules and actin filaments. Microtubules are thin cylindrical structures of variable length and are composed of subunits of the protein tubulin. They play a role in cell division, the growth of the cell wall, and the movement of flagella. Actin filaments are composed of the protein actin. These long filaments occur singly and in bundles, and they play a role in cytoplasmic streaming.

Flagella and Cilia Extend from the Surface of Certain Cells and Cause Movement

Flagella and cilia are hairlike structures that project from the surface of many different types of eukaryotic cells, serving as locomotor structures. All flagella of eukaryotic cells have the same characteristic 9-plus-2 internal structure—that is, an outer ring of nine pairs of microtubules surrounding an inner pair of microtubules in the center of the flagellum.

MAJOR COMPONENTS	INDIVIDUAL CONSTITUENTS	KEY DESCRIPTIVE FEATURES	FUNCTION(S)
Cytoplasm	Mitochondria	Surrounded by a double-membrane envelope. The inner membrane is folded into cristae. Semiautonomous organelles, containing their own DNA and ribosomes.	Sites of cellular respiration.
	Peroxisomes	Surrounded by a single membrane. Sometimes contain crystalline protein bodies.	Contain enzymes for a variety of processes such as photorespiration and conversion of fats to sucrose.
	Vacuoles	Surrounded by a single membrane (the tonoplast); may take up most of the cell volume.	Filled with cell sap, which is mostly water. Often contain anthocyanin pigments; store primary and secondary metabolites; break down and recycle macromolecules.
	Ribosomes	Small, electron-dense particles, consisting of RNA and protein.	Sites of protein synthesis.
	Oil bodies	Have an amorphous appearance.	Sites of lipid storage, especially triglycerides.
	Endoplasmic reticulum	Network of membrane channels.	Rough ER (mainly cisternal ER) has ribosomes and thus is involved in protein synthesis. Smooth ER (mainly tubular ER) is involved in lipid synthesis. Channels materials throughout the cell.
	Golgi complex	Collective term for the cell's Golgi bodies, stacks of flattened, membranous sacs.	Processes and packages substances for secretion and for use within the cell.
	Endomembrane system	Collective term for the endoplasmic reticulum, Golgi complex, *trans*-Golgi network, plasma membrane, nuclear envelope, tonoplast, and various vesicles.	The dynamic network in which membranes and various substances are transported thoughout the cell.
	Cytoskeleton	Complex network of protein filaments.	Involved in cell division, growth, and differentiation.
	Microtubules	Dynamic, cylindrical structures composed of tubulin.	Involved in many processes, such as cell plate formation, deposition of cellulose microfibrils, and directing the movement of Golgi vesicles and chromosomes.
	Actin filaments	Dynamic, filamentous structures composed of actin.	Involved in many processes, including cytoplasmic streaming and the movement of the nucleus and organelles.

The Cell Wall Is the Major Distinguishing Feature of the Plant Cell

The cell wall determines the structure of the cell, the texture of plant tissues, and many important characteristics that distinguish plants as organisms. All plant cells have a primary wall. In addition, many also have a secondary wall, which occurs interior to the primary wall. The region between the primary walls of adjacent cells is the middle lamella, a pectin-rich layer cementing together the primary walls of adjacent cells. Cellulose is the principal component of primary and secondary walls. The cellulose microfibrils of primary walls occur in a cross-linked matrix of noncellulosic molecules, including hemicelluloses, pectins, and glycoproteins. Because of the presence of pectins, primary walls are highly hydrated, making them more plastic. Actively dividing and elongating cells commonly have only primary walls. Secondary walls contain hemicelluloses but apparently lack pectins and glycoproteins. Lignin may also be present in primary walls but is especially characteristic of cells with secondary walls. Lignin adds compressive strength and rigidity to the wall.

Plasmodesmata Are Cytoplasmic Strands Connecting the Protoplasts of Adjacent Plant Cells

The protoplasts of adjacent cells are connected to one another by cytoplasmic strands called plasmodesmata, which provide a pathway for the transport of certain substances between cells.

Selected Key Terms

actin filaments   p. 60

cell sap   p. 54

cell theory   p. 41

cellulose   p. 62

cellulose synthase   p. 66

cell wall   p. 61

chloroplasts   p. 49

chromatin   p. 47

chromoplasts   p. 50

crista/cristae   p. 52

cytoplasm   p. 41

cytoplasmic streaming   p. 46

cytoskeleton   p. 58

cytosol   p. 46

endomembrane system   p. 58

endoplasmic reticulum   p. 56

eukaryotic cells   p. 41

glycoproteins   p. 63

Golgi bodies   p. 57

Golgi complex   p. 57

granum/grana   p. 49

hemicelluloses   p. 63

leucoplasts   p. 51

microfibrils   p. 62

microtubules   p. 58

middle lamella   p. 64

mitochondria   p. 52

nucleolus/nucleoli   p. 48

nucleus   p. 47

organelles   p. 41

organismal theory   p. 65

pectins   p. 63

peroxisomes   p. 53

pits   p. 66

plasma membrane   p. 41

plasmodesma/plasmodesmata   p. 66

polysomes   p. 55

primary pit-fields   p. 64

primary wall   p. 64

prokaryotic cells   p. 41

proplastids   p. 51

protoplast   p. 45

ribosomes   p. 55

secondary wall   p. 64

stoma   p. 48

thylakoids   p. 48

tonoplast   p. 54

vacuoles   p. 54

Questions

1. What is meant by the cell theory, and what is the significance of this theory for biology?

2. What three features of plant cells distinguish them from animal cells?

3. Both plastids and mitochondria are said to be "semiautonomous" organelles. Explain.

4. In what ways do chloroplasts and mitochondria resemble bacteria?

5. Once regarded as depositories for waste products in plant cells, vacuoles now are known to play many essential roles. What are some of those roles?

6. Explain the phenomenon of autumn coloration.

7. Distinguish between rough endoplasmic reticulum and smooth endoplasmic reticulum both structurally and functionally.

8. Distinguish between microtubules and actin filaments. With what functions are each of these protein filaments associated?

9. Once regarded as merely an outer, inactive structure produced by the protoplast, the cell wall is now recognized as an integral part of the plant cell and as having a number of specific and essential functions. What are some of those functions?

10. Using the following terms, explain the process of cell wall growth and cellulose deposition in elongating cells: cellulose microfibrils, cellulose synthase complexes (rosettes), cortical microtubules, secretory vesicles, matrix substances, plasma membrane.

Chapter

4

Membrane Structure and Function

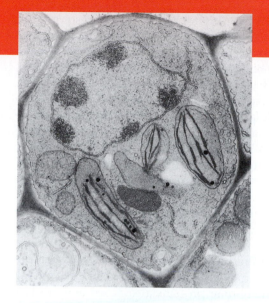

Nuclear envelope

Plasma membrane

Cell wall

Tonoplast

Peroxisome

Endoplasmic reticulum

Mitochondrion

Chloroplast

4–1

Cellular membranes. In addition to the plasma membrane, which controls the movement of substances into and out of the cell, numerous internal membranes control the passage of substances within the cell. Membranes surround the nucleus, chloroplasts and mitochondria (both of which contain internal membranes), peroxisomes, and vacuoles (each surrounded by a tonoplast). In addition, the endoplasmic reticulum (segments of which are seen here) is made up of membranes. This electron micrograph is of a leaf cell of Moricandia arvensis, *a crucifer (cabbage family).*

OVERVIEW

At first glance, the plasma membrane seems to accomplish the impossible. Among other functions, this membrane mediates the transport of substances into and out of the protoplast. Whereas some substances pass through freely, others are blocked. The plasma membrane must allow the passage of essential substances but at the same time prevent the loss of cellular constituents. How can the plasma membrane—as well as the other membranes of the cell—function so selectively?

The structure of the membrane is the key to its function. All cellular membranes contain basically the same constituents, and it is the arrangement of these components that enables the membrane to be a selectively permeable barrier. Some molecules, such as O_2, CO_2, and H_2O, readily pass through membranes, but others, such as sugars, amino acids, and inorganic ions, require the involvement of transport proteins in the membrane. Still others, such as larger molecules and solid particles, require large-scale transport processes involving vesicles.

The plasma membrane also facilitates communication between cells. "Local" communication is provided by direct cytoplasmic connections. "Long-distance" communication involves the reception and processing of chemical signals arriving from various parts of the plant.

CHECKPOINTS

By the time you finish reading this chapter, you should be able to answer the following questions:

1. What is the fluid-mosaic model of membrane structure and, in general terms, what role is played by its two major components?

2. Distinguish between diffusion and osmosis. What kinds of substances enter and leave cells by each process?

3. What are transport proteins, and of what importance are they to plant cells?

4. What are the similarities and differences between facilitated diffusion and active transport?

5. What are the roles of signal transduction and plasmodesmata in cell-to-cell communication?

All living organisms are surrounded on all sides by nonliving matter with which they constantly exchange materials. Yet living organisms differ from their nonliving surroundings in the kinds and amounts of chemical substances they contain. A tree root, for example, is chemically very different from the soil in which it grows. Without this capacity to differ, organisms would be unable to maintain the organization on which their existence depends.

In all living systems, from the smallest bacterium to the largest sequoia tree, exchanges of substances between the living organism and the nonliving world of air, soil, and water occur at the level of the individual cell. These exchanges are regulated by the plasma membrane. In multicellular organisms—ourselves, for example—the plasma membrane also regulates the exchanges of substances among the various specialized cells that make up the organism. Control of these exchanges between cells is essential to (1) protect each cell's integrity, (2) maintain the conditions at which its metabolic activities can take place, and (3) coordinate the activities of the different cells.

In addition to the plasma membrane, which controls the passage of materials into and out of the cell, there are internal membranes, such as those surrounding mitochondria, chloroplasts, and the nucleus, that control the passage of materials among compartments within the cell (Figure 4–1). In this way the cell can maintain the specialized chemical environments necessary for the processes occurring in the different cytoplasmic compartments.

Membranes also permit differences in electric potential, or voltage, to become established between the cell and its external environment and between adjacent compartments of the cell. Differences in chemical concentration (of various ions and molecules) and electric potential across membranes are forms of potential energy that are essential to many cellular processes. In fact, their existence is a criterion by which we may distinguish living systems from the surrounding nonliving environment.

In order to maintain the internal environment of the cell and its various parts, the plasma membrane and the internal membranes must perform a complex double function. Not only must they keep certain substances out while letting others in, but they must also keep certain substances in while letting others out. The capacity of a membrane to accomplish this depends on the physical and chemical properties of both the membrane and the ions and molecules that interact with the membrane.

Structure of Cellular Membranes

As mentioned in Chapter 3, cellular membranes commonly exhibit a three-layered appearance—two dark layers separated by a lighter layer—at the resolution of the electron microscope (see Figure 3–6). For a time, the two dark layers were considered to consist of protein, and the space between them of a lipid bilayer. In 1960, J. David Robertson proposed the **unit membrane model,** which emphasized a general protein-lipid-protein structure for all cellular membranes. With further study, however, it became apparent that different membranes vary considerably in their protein/lipid ratio, and that the proteins do not form continuous layers on the surface of the membrane but exist as discrete globular entities embedded in the lipid bilayer.

In 1972, S. Jonathan Singer and Garth Nicolson proposed the widely accepted **fluid-mosaic model** of membrane structure, which envisions a mosaic of proteins embedded in a fluid lipid bilayer (page 24), the components of which are in constant motion. In this model, many of the membrane proteins also move more or less freely in the fluid bilayer. As the proteins and lipid molecules move laterally within the bilayer, the proteins form different patterns, or mosaics, that vary from time to time and place to place—hence the name "fluid-mosaic" for this model of membrane structure.

There is now broad agreement that all of the membranes of the cell have the same basic structure, consisting of a lipid bilayer and globular proteins. Many membrane proteins extend across the bilayer and protrude on either side (Figure 4–2). The portion of these **transmembrane proteins** embedded in the bilayer is hydrophobic, whereas the portions exposed on either side of the membrane are hydrophilic. Transmembrane proteins and other proteins tightly bound to the membrane are called **integral proteins.** Although some of the transmembrane proteins float more or less freely in the lipid bilayer, others are anchored in place (perhaps to the cytoskeleton) and limited in mobility. Some membrane proteins do not extend into the hydrophobic interior of the lipid bilayer. Called **peripheral proteins,** they are attached to protruding portions of some of the transmembrane proteins.

The two surfaces of a membrane differ considerably in chemical composition. For example, there are two major types of lipids in the membranes of plant cells: **phospholipids** (the more abundant) and **sterols,** particularly stigmasterol (not cholesterol, which is the major sterol in animal tissues). The two layers of the bilayer have different concentrations of each. Moreover, the transmembrane proteins have definite orientations within the bilayer, and the protruding portions on each side have different amino acid compositions and tertiary structures.

On the outer surface of the plasma membrane, short-chain carbohydrates (oligosaccharides) are attached to most of the protruding proteins, forming **glycoproteins.** The carbohydrates, which form a coat on the outer surface of the plasma membrane of all eukaryotic cells, are

4–2

Fluid-mosaic model of membrane structure. The membrane is composed of a bilayer (double layer) of lipid molecules—with their hydrophobic "tails" facing inward—and large protein molecules. The proteins traversing the bilayer are a type of integral protein known as transmembrane proteins. Other proteins, called peripheral proteins, are attached to some of the transmembrane proteins. The portion of a transmembrane protein molecule embedded in the lipid bilayer is hydrophobic; the portion or portions exposed on either side of the membrane are hydrophilic. Short carbohydrate chains are attached to most of the protruding transmembrane proteins on the outer surface of the plasma membrane. The whole structure is quite fluid, and hence the proteins can be thought of as floating in a lipid "sea."

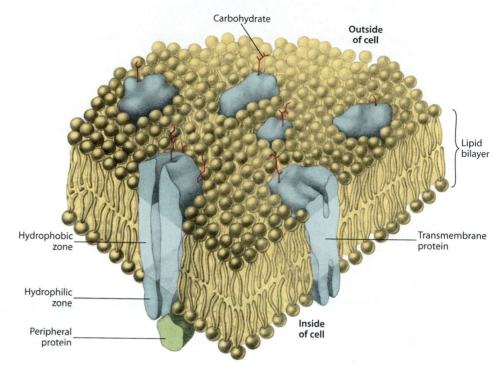

believed to play important roles in the "recognition" of molecules (such as hormones, the coat proteins of viruses, and molecules on the surfaces of bacteria) that interact with the cell.

Most membrane carbohydrates are present in the form of glycoproteins, but a small proportion are present as **glycolipids,** membrane lipids with short-chain carbohydrates attached to them. The arrangement of the carbohydrate groups on the external surface of the plasma membrane has been revealed largely by experiments using **lectins,** proteins that bind tenaciously to specific carbohydrate groups.

Two basic configurations have been identified among transmembrane proteins (Figure 4–3). One is a relatively simple rodlike structure consisting of an alpha helix embedded in the hydrophobic interior of the membrane, with less regular, hydrophilic portions extending on either side of it. The other configuration is found in large globular proteins with complex three-dimensional structures that make repeated "passes" through the membrane. In such "multipass" membrane proteins, the polypeptide chain usually spans the lipid bilayer as a series of alpha helices.

Whereas the lipid bilayer provides the basic structure and impermeable nature of cellular membranes, the proteins are responsible for most membrane functions. Most membranes are composed of 40 to 50 percent lipid (by weight) and 60 to 50 percent protein, and the amounts and types of proteins in a membrane reflect its function. Membranes involved with energy transduction (the conversion of energy from one form to another), such as the internal membranes of mitochondria and chloroplasts,

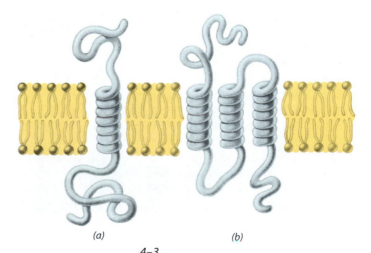

(a) *(b)*

4–3

*Two configurations of transmembrane proteins. Some extend across the lipid bilayer as a single alpha helix **(a)** and others—multipass proteins—as multiple alpha helices **(b)**. The portions of protein protruding on either side of the membrane are hydrophilic; the helical portions within the membrane are hydrophobic.*

consist of about 75 percent protein. Some membrane proteins are enzymes that catalyze membrane-associated reactions, whereas others are carriers involved in the transport of specific molecules or ions into and out of the cell or organelle. Still others act as receptors for receiving and transducing chemical signals from the cell's internal or external environment.

Movement of Water and Solutes

Of the many kinds of molecules surrounding and contained within the cell, by far the most common is water. Further, most of the other molecules and ions important in the life of the cell (Figure 4–4) are dissolved in water. Therefore, let us begin our consideration of transport across cellular membranes by looking at how water moves.

Water Moves Down a Water Potential Gradient

Except when it is locked in ice, water is constantly moving. It moves across the continents in rivers and streams, from the soil through the bodies of plants into the atmosphere, through the human body by way of the bloodstream, and into and out of living cells. In both the living and nonliving worlds, water molecules move from one place to another because of differences in potential energy. **Potential energy** is the stored energy an object—or a collection of objects, such as a collection of water molecules—possesses because of its position. The potential energy of water is usually referred to as the **water potential.**

Water moves from a region where water potential is higher to a region where water potential is lower, regardless of the reason for the difference in water potential. A simple example is water running downhill in response to gravity (Figure 4–5). Water at the top of a hill has more potential energy (that is, a higher water potential) than water at the bottom of a hill. As the water runs downhill, its potential energy is converted to kinetic energy. This, in turn, can be converted to mechanical energy doing useful work if, for example, a waterwheel is placed in the path of the moving water.

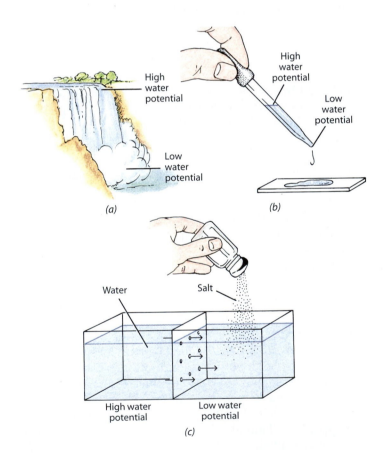

(a)   *(b)*

(c)

4–5

The three factors that most commonly determine water potential are (a) gravity, (b) pressure, and (c) the concentration of dissolved solutes. Water moves from the region of higher water potential to the region of lower water potential, regardless of the reason for the difference in water potential. Often, a combination of factors is involved. In (b), for example, the pressure applied when the bulb is squeezed is not the only factor contributing to the difference in water potential. Given the angle at which the eyedropper is held, gravity is also a factor.

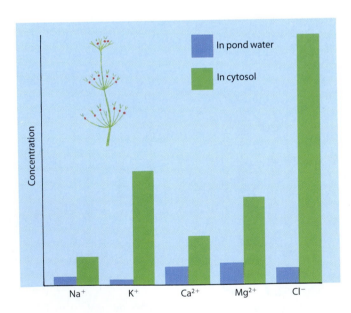

4–4

Among the most important ions in many living cells are sodium (Na^+), potassium (K^+), calcium (Ca^{2+}), magnesium (Mg^{2+}), and chloride (Cl^-). This graph shows the relative concentrations of these ions in the cytosol of the green alga Nitella and in the surrounding pond water. Such differences in ion concentration between the interior of living cells and their nonliving environment indicate that cells regulate their exchanges of materials with the surroundings. This regulation is accomplished by the plasma membrane.

Pressure is another source of water potential. If we fill an eyedropper with water and then squeeze the bulb, water will squirt out. Like water at the top of a hill, this water has been given a high water potential and will move to a lower one. Can we make the water that is running downhill move uphill by means of pressure? Obviously we can move water uphill—but only so long as the water potential produced by the pressure exceeds the water potential produced by gravity.

In solutions, water potential is affected by the concentration of dissolved particles (solutes). As the concentration of solute particles (that is, the number of solute particles per unit volume of solution) increases, the water potential decreases. Conversely, as the concentration of solute particles decreases, the water potential increases. In the absence of other factors (such as pressure) affecting the water potential, water molecules in solutions move from regions of lower solute concentration (higher water potential) to regions of higher solute concentration (lower water potential). This fact is of great importance for living systems.

The concept of water potential is useful because it enables physiologists to predict the way in which water will move under various conditions. Water potential is usually measured in terms of the pressure required to stop the movement of water—that is, the hydrostatic (water-stopping) pressure—under the particular circumstances involved. The units used to express this pressure are the bar and megapascal (MPa). A bar is approximately equal to the average pressure of the air at sea level. One megapascal is equal to 10 bars. By convention, the water potential of pure water—to be precise, at atmospheric pressure and at sea level—is set at zero. Under these same conditions, the water potential of an aqueous solution of a substance will therefore have a negative value (less than zero) because higher solute concentration results in lower water potential.

Two mechanisms are involved in the movement of water and solutes: bulk flow and diffusion. In living systems, bulk flow is the process used to move water and solutes from one part of a multicellular organism to another part. Diffusion, by contrast, plays a major role in the movement of many molecules and ions into, out of, and through cells. A particular instance of diffusion— that of water moving through a membrane that separates solutions of different concentration—is known as osmosis.

Bulk Flow Is the Overall Movement of a Fluid In **bulk flow,** the molecules of a liquid move all together and in the same direction. For example, water runs downhill by bulk flow in response to the differences in water potential at the top and the bottom of a hill. Blood moves through your blood vessels by bulk flow as a result of the water potential (blood pressure) created by the pumping of your heart. Sap—an aqueous solution of su-

crose and other solutes in the conduits of the food-conducting tissue (phloem) of plants—moves by bulk flow from the leaves to other parts of the plant body.

Diffusion Results in the Uniform Distribution of a Substance Diffusion is a familiar phenomenon. If a few drops of perfume are sprayed in one corner of a room, the scent will eventually fill the entire room even if the air is still. If a few drops of dye are put in one end of a tank of water, the dye molecules will slowly become evenly distributed throughout the tank (Figure 4–6). This process may take a day or more, depending on the size of the tank, the temperature, and the size of the dye molecules.

Why do the dye molecules move apart? If you could observe the individual dye molecules in the tank, you would see that each one moves independently and at random. Imagine a thin section through the tank, running from top to bottom. Dye molecules will move into and out of the section, some moving in one direction, some moving in the other. But you would see more dye

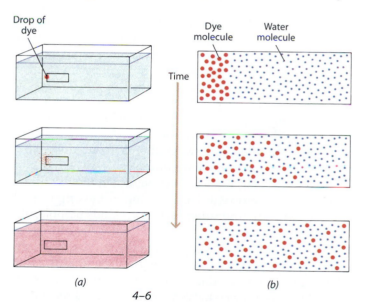

4–6

The diffusion process, as revealed by the addition of a drop of dye to a tank of water (top left). The outlined section within each panel in (a) is shown in close-up in (b). Diffusion is the result of the random movement of individual molecules (or ions), which produces a net movement of particles from a region where they are more concentrated to a region where they are less concentrated. This movement is down the concentration gradient. Notice that as the dye molecules (red) diffuse to the right, the water molecules (blue) diffuse to the left. The ultimate result is an even distribution of both types of molecules. Can you see why the net movement of molecules will slow down as equilibrium is reached?

molecules moving from the side of greater dye concentration. Why? Simply because there are more dye molecules at that end of the tank. For example, if there are more dye molecules on the left, more of them will move randomly to the right, even though there is an equal probability that any one dye molecule will move from right to left. Consequently, the *net* (overall) movement of dye molecules will be from left to right. Similarly, the net movement of water molecules is from right to left.

What happens when all the molecules are evenly distributed throughout the tank? Their even distribution does not affect the behavior of the molecules as individuals; they still move at random. But there are now as many molecules of dye and as many molecules of water on one side of the tank as on the other, and so there is no net direction of motion. There is, however, just as much individual motion (thermal agitation) as before, provided the temperature has not changed.

Substances that are moving from a region of higher concentration to a region of lower concentration are said to be moving *down a gradient,* which is somewhat analogous to flowing downhill. Diffusion occurs only down a gradient. The steeper the downhill gradient—that is, the larger the difference in concentration—the more rapid the net movement. Also, diffusion is more rapid in gases than in liquids and is more rapid at higher than at lower temperatures. Can you explain why? A substance moving toward a higher concentration of its own molecules would be moving *against a gradient,* which is somewhat analogous to being pushed uphill.

Notice that there are two gradients in our tank, one of dye molecules and one of water molecules. The dye molecules are moving in one direction, down their gradient, and the water molecules are moving in the opposite direction, down their gradient. The two kinds of molecules are moving independently of each other. In both cases, the movement is down a gradient. When the molecules have reached a state of equal distribution (that is, when there are no more gradients), they continue to move, but now there is no *net* movement in either direction. In other words, the *net transfer* of the molecules is zero; the system may be said to be in a state of **equilibrium.** Remember, that does not mean that there is no molecular or ionic motion. Molecules and ions of all kinds are in constant motion so that they are continuously being transferred from one region to another.

The concept of water potential is also useful in understanding diffusion. A high concentration of dye molecules in one region of the tank means a low concentration of water molecules there and, thus, low water potential. If the pressure is equal everywhere throughout the tank, water molecules, as they move down the gradient, are moving from a region of higher water potential to a region of lower water potential. When equilibrium is reached, water potential is equal in all parts of the tank.

The net result of diffusion is that the diffusing substance eventually becomes evenly distributed. Briefly, **diffusion** may be defined as the dispersion of substances by a movement of their ions or molecules, which tends to equalize their concentrations throughout the system. Each substance moves randomly and independently of the other(s).

Cells and Diffusion

Water, oxygen, carbon dioxide, and a few other simple molecules diffuse freely across the plasma membrane. Carbon dioxide and oxygen, which are both nonpolar, are soluble in lipids and move easily through the lipid bilayer. Despite their polarity, water molecules also move through the membrane without hindrance, apparently through momentary openings created by spontaneous movements of the membrane lipids. Other uncharged polar molecules, provided they are small enough, also diffuse through these openings. The permeability of the membrane to these solutes varies inversely with the size of the molecules, indicating that the openings are small and that the membrane acts like a sieve in this respect.

Diffusion is also the principal way in which substances move within cells. One of the major factors limiting cell size is this dependence upon diffusion, which is essentially a slow process except over very short distances. Diffusion is not an effective way to move molecules over long distances at the rates required for cellular activity. In many cells, the transport of materials is speeded up by active streaming of the cytoplasm (see Chapter 3).

Efficient diffusion requires a steep **concentration gradient,** that is, a short distance and a substantial concentration difference. (The concentration gradient is the concentration difference of a substance per unit distance.) Cells maintain such gradients by their metabolic activities. For example, in a nonphotosynthesizing cell, oxygen is used up within the cell almost as rapidly as it enters, thereby maintaining a steep gradient from outside to inside. Conversely, carbon dioxide is produced by the cell, and so a gradient from inside to outside the cell is maintained along which carbon dioxide can diffuse out of the cell. Similarly, within a cell, molecules or ions are often produced at one place and used at another. Thus a concentration gradient is established between the two regions of the cell, and the substance diffuses down the gradient from the site of production to the site of use.

Osmosis Is a Special Case of Diffusion

A membrane that permits the passage of some substances while blocking the passage of others is said to be

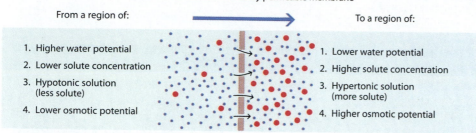

From a region of:

1. Higher water potential
2. Lower solute concentration
3. Hypotonic solution (less solute)
4. Lower osmotic potential

To a region of:

1. Lower water potential
2. Higher solute concentration
3. Hypertonic solution (more solute)
4. Higher osmotic potential

4–7

The direction of water movement in osmosis.

selectively permeable. The movement of water molecules through such a membrane is a special case of diffusion, known as **osmosis.** Osmosis results in a net transfer of water from a solution that has higher water potential (lower solute concentration) to a solution that has lower water potential (higher solute concentration).

The diffusion of water is not affected by *what* is dissolved in the water, only by *how much* is dissolved—that is, the concentration of particles of solute (molecules or ions) in the water. A small solute particle, such as a sodium ion, counts just as much as a large solute particle, such as a sugar molecule.

Two or more solutions that have equal numbers of dissolved particles per unit volume—and therefore the same water potential—are said to be **isotonic** (from the Greek *isos*, meaning "equal," and *tonos*, "tension"). There is no net movement of water across a membrane separating two solutions that are isotonic to one another, unless, of course, pressure is exerted on one side.

In comparing solutions of different concentration, the solution that has less solute (and therefore a higher water potential) is known as **hypotonic,** and the one that has more solute (a lower water potential) is known as

hypertonic. (Note that *hyper* means "more"—in this case, more particles of solute; and *hypo* means "less"—in this case, fewer particles of solute.) In osmosis, water molecules diffuse from a hypotonic solution (or from pure water), through a selectively permeable membrane, into a hypertonic solution (Figure 4–7).

Osmosis results in a buildup of pressure as water molecules continue to diffuse across the membrane into a region of lower water concentration. If water is separated from a solution by a membrane that allows the ready passage of water but not of solute in a system such as that shown in Figure 4–8, the water will move across the membrane and cause the solution to rise in the tube until an equilibrium is reached—that is, until the water potential is the same on both sides of the membrane. If enough pressure is applied on the solution in the tube by a piston, say, such as in Figure 4–8c, it is possible to prevent the net movement of water into the tube. The pressure that would have to be applied to the solution to stop water movement is called the **osmotic pressure.** The tendency of water to move across a membrane because of the effect of solutes on water potential is called the **osmotic potential** (or solute potential).

Force of gravity

Pressure applied

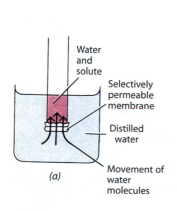

Water and solute

Selectively permeable membrane

Distilled water

Movement of water molecules

(a)

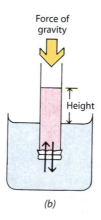

Height

(b)

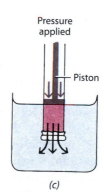

Piston

(c)

4–8

Osmosis and the measurement of the osmotic potential of a solution. (a) The tube contains a solution and the beaker contains distilled water. A selectively permeable membrane at the base of the tube permits the passage of water molecules but not of solute particles. (b) The diffusion of water into the solution causes its volume to increase, and thus the column of liquid rises in the tube.

However, the downward pressure created by the force of gravity acting on the column of solution is proportional to the height of the column and the density of the solution. Thus, as the column of solution rises in the tube, the downward pressure gradually increases until it becomes so great that it counterbalances the tendency of water to move into the solution. In other words, the water potential on the two

sides of the membrane becomes equal. At this point, there is no further net movement of water. (c) The pressure that must be applied to the piston to force the column of solution back to the level of the water in the beaker provides a quantitative measure of the osmotic potential of the solution—that is, of the tendency of water to diffuse across the membrane into the solution.

Imbibition

Water molecules exhibit a tremendous cohesiveness because of their polarity, that is, the difference in charge between one end of a water molecule and the other (see Appendix A). Similarly, because of this difference in charge, water molecules can cling (adhere) either to positively charged or to negatively charged surfaces. Many large biological molecules, such as cellulose, are polar and so attract water molecules. The adherence of water molecules is also responsible for the biologically important phenomenon called imbibition or, sometimes, hydration.

Imbibition (from the Latin imbibere, "to drink in") is the movement of water molecules into substances such as wood or gelatin that swell as a result of the accumulation of water molecules. The pressures developed by imbibition can be astonishingly large. It is said that stone for the ancient Egyptian pyramids was quarried by driving wooden pegs into holes drilled in the rock face and then soaking the pegs with water. The swelling wood created a force that split the slab of stone. In living plants, imbibition occurs particularly in seeds, which may increase to many times their original size as a result. Imbibition is essential to the germination of the seed (see Chapter 23).

The germination of seeds begins with changes in the seed coat that permit a massive uptake of water by imbibition. The embryo and surrounding structures then swell, bursting the seed coat. In the acorn on the left, photographed on a forest floor, the embryonic root emerged after the tough outer layers of the fruit split open.

Osmosis and Living Organisms

The movement of water across the plasma membrane in response to differences in water potential causes some crucial problems for living systems, particularly those in an aqueous environment. These problems vary according to whether the water potential of the cell or organism is hypotonic, isotonic, or hypertonic in relation to its environment. For example, one-celled organisms that live in salt water are usually isotonic to the medium they inhabit, which is one way of solving the problem.

Many types of cells, however, live in a hypotonic environment. In freshwater single-celled organisms, such as *Euglena*, which lacks a cell wall, the interior of the cell is hypertonic relative to the surrounding medium; consequently, water tends to move into the cell by osmosis. If too much water were to move into the cell, it could rupture the plasma membrane. In *Euglena* this is prevented by a specialized organelle known as a contractile vacuole, which collects water from various parts of the cell body and pumps it out of the cell with a rhythmic contraction.

Turgor Pressure Contributes to the Stiffness of Plant Cells

If a plant cell is placed in a hypotonic solution with a relatively high water potential, the protoplast expands and the plasma membrane stretches and exerts pressure against the cell wall. However, the plant cell does not rupture because it is restrained by the relatively rigid cell wall.

Plant cells tend to concentrate relatively strong solutions of salts within their vacuoles, and they can also accumulate sugars, organic acids, and amino acids. As a result, plant cells absorb water by osmosis and build up their internal hydrostatic pressure. This pressure against the cell wall keeps the cell **turgid**, or stiff. Consequently, the hydrostatic pressure in plant cells is commonly re-

4–9

Plasmolysis in a leaf epidermal cell. (a) Under normal conditions, the plasma membrane of the protoplast is in close contact with the cell wall. (b) When the cell is placed in a relatively concentrated sugar solution, water passes out of the cell into the hypertonic medium, the protoplast contracts slightly, and the plasma membrane moves away from the wall. (c) When immersed in a more concentrated sugar solution, the cell loses even larger amounts of water and the protoplast contracts still further. As water is lost from the vacuole, its contents become more concentrated. The widths of the arrows indicate the relative amounts of water entering or leaving the cell.

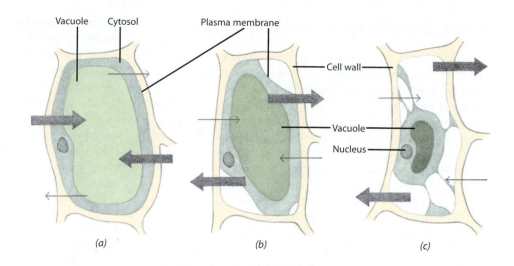

(a) (b) (c)

ferred to as turgor pressure. **Turgor pressure** is the pressure that develops in a plant cell as a result of osmosis and/or imbibition (see "Imbibition" on page 80). Equal to and opposing the turgor pressure at any instant is the inwardly directed mechanical pressure of the cell wall, called the **wall pressure.**

Turgor in the plant is especially important in the support of nonwoody plant parts. Most of the growth of a plant cell is the direct result of water uptake, with the bulk of the increase in size of the cell resulting from enlargement of the vacuoles (page 54). Before the cell can increase in size, however, there must be a loosening of the wall structure in order to decrease the resistance of the wall to turgor pressure (see Chapter 28).

Turgor is maintained in most plant cells because they generally exist in a hypotonic medium. However, if a sample of plant tissue with turgid cells is placed in a hy-

pertonic solution (for example, a sugar or salt solution), water will leave the cell by osmosis. As a result, the vacuole and rest of the protoplast will shrink, thus causing the plasma membrane to pull away from the cell wall (Figure 4–9). This phenomenon is known as **plasmolysis.** The process can be reversed if the cell is then transferred to pure water. (Because of the high water permeability of membranes, the vacuole and the rest of the protoplast are at or very near equilibrium with respect to water potential.) Figure 4–10 shows *Elodea* leaf cells before and after plasmolysis. Although the plasma membrane and the tonoplast, or vacuolar membrane, are, with few exceptions, permeable only to water, the cell walls allow both solutes and water to pass freely through them. The loss of turgor by plant cells may result in **wilting,** or drooping, of leaves and stems, or of your dinner salad.

4–10

Elodea leaf cells. (a) Turgid cells and (b) plasmolyzed cells after being placed in a relatively concentrated sucrose solution.

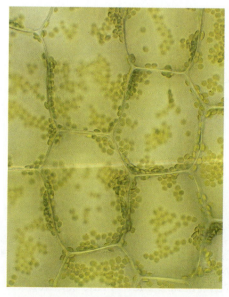

(a) (b) 25 μm

Transport of Solutes across Membranes

As mentioned previously, small nonpolar molecules (such as oxygen and carbon dioxide) and small, uncharged polar molecules (such as water) can permeate the lipid bilayer of cellular membranes freely by simple diffusion. In fact, the observation that nonpolar molecules diffuse readily across plasma membranes provided the first evidence of the lipid nature of the membrane.

By contrast, most substances required by cells are polar and require **transport proteins** to transfer them across membranes. Each transport protein is highly selective; it may accept one type of ion (such as Ca^{2+} or K^+) or molecule (such as a sugar or an amino acid) and exclude a nearly identical one. All known transport proteins have proven to be multipass transmembrane proteins (Figure 4–3b). These proteins provide the specific solutes they transport with a continuous pathway across the membrane without the solutes coming into contact with the hydrophobic interior of the lipid bilayer.

If a molecule is uncharged, its transport across the membrane is influenced only by the difference in its concentration on the two sides of the membrane (the concentration gradient). If a solute carries a net charge, however, both the concentration gradient and the total **electrical gradient**, or electric voltage across the membrane (the **membrane potential**), influence its transport. Together, the concentration gradient of a charged substance and the membrane potential are referred to as the **electrochemical gradient** of that substance. Just as differences in water potential provide the driving force for the movement of water across a membrane, the electrochemical gradient of a charged substance, say K^+, is the energy source that powers its diffusion. Plant cells typically maintain electrochemical gradients across both the plasma membrane and the tonoplast. The cytosol is electrically negative relative to both the watery environment outside the cell and the solution, known as the cell sap, inside the vacuole.

The transport of solutes across a membrane can be either passive or active, depending on the energy requirement of the process. **Passive transport** is a diffusional process in which the direction of net movement is dictated by the concentration gradient or, in the case of ions, by the electrochemical gradient. One example of passive transport is the **simple diffusion** of small nonpolar molecules across the lipid bilayer (Figure 4–11a). Most passive transport, however, requires transport pro-

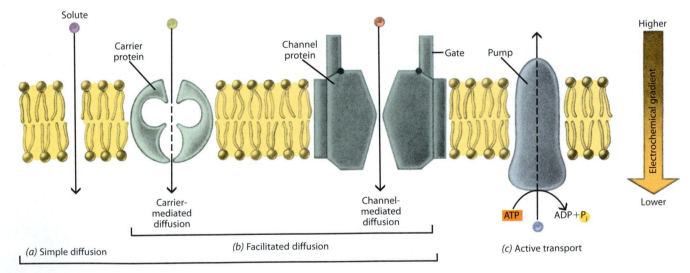

4–11

Modes of transport through the plasma membrane. **(a)** In simple diffusion, small nonpolar molecules, such as oxygen and carbon dioxide, and small uncharged, polar molecules, such as water, pass directly through the lipid bilayer, down their concentration gradient. **(b)** Facilitated diffusion occurs via either carrier proteins or channel proteins. Carrier proteins bind the specific solute and undergo conformational changes as the solute molecule is transported. Channel proteins allow selected solutes—commonly ions such as Na^+ and K^+—to pass directly through water-filled pores. Channel proteins are gated. When the gates are open, solutes pass through, but when closed, solute flow is blocked. As with simple diffusion, facilitated diffusion occurs down concentration or electrochemical gradients. Both diffusional processes are passive transport processes, which do not require energy. **(c)** Active transport, on the other hand, moves solutes against the concentration or electrochemical gradient and therefore requires an input of energy, usually supplied by the hydrolysis of ATP to ADP and P_i (see Figure 2–24). The transport proteins involved in active transport are known as pumps.

Patch-Clamp Recording in the Study of Ion Channels

The membranes of both plant and animal cells contain channel proteins that form pathways for passive ion movement. When activated, these ion channels become permeable to ions, allowing solute flow across the membrane. The opening and closing of ion channels is referred to as gating.

The first evidence for gated ion channels was obtained from electrophysiological experiments using intracellular microelectrodes. Such methods can be applied only to relatively large cells, however, and consequently the currents (that is, the ion movements) measured flow through many channels at a time. In addition, different types of channels may be opened simultaneously. Plant cells pose another problem: when a microelectrode is inserted into the protoplast, it usually penetrates both the plasma membrane and the tonoplast, so that the data reflect the collective behavior of both membranes.

The patch-clamp technique revolutionized the study of ion channels. This technique involves the electrical analysis of a very small patch of membrane of a naked (wall-less) protoplast or tonoplast. In this way, it is possible to identify a single ion-specific channel in the membrane and to study the transport of ions through that channel.

In patch-clamp experiments, a glass electrode (micropipette) with a tip diameter of about 1.0 micrometer is brought into contact with the membrane. When gentle suction is then applied, an extremely tight seal is formed between the micropipette and the membrane. With the patch still attached to the intact protoplast (a), the transport of ions can be recorded between the cytoplasm and the artificial solution that fills the micropipette. If the pipette is withdrawn from the attached protoplast, it is possible to separate only the small patch

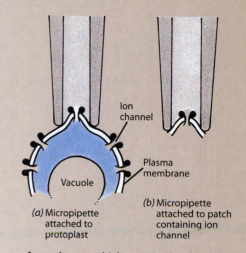

(a) Micropipette attached to protoplast

(b) Micropipette attached to patch containing ion channel

of membrane, which remains intact within the tip (b). With a detached patch it is easy to alter the composition of the solution on either or both sides of the membrane to test the effect of different solutes on channel behavior.

---

teins to facilitate the passage of ions and polar molecules across the hydrophobic interior of the membrane. This process is called **facilitated diffusion**. The solute still diffuses across the membrane in the direction dictated by the concentration gradient, but its movement is facilitated by a specific transport protein in the membrane.

Two types of transport proteins—**carrier proteins** and **channel proteins**—permit the movement of a substance across a membrane but only down the substance's electrochemical gradient, that is, they are passive transporters (Figure 4–11b). Carrier proteins bind the specific solute being transported and undergo a series of conformational changes in order to transport the solute across the membrane. Channel proteins form water-filled pores that extend across the membrane and, when open, allow specific solutes (usually inorganic ions such as Na^+, K^+, Ca^{2+}, and Cl^-) to pass through them. The conduction of ions by channel proteins involves transient, or brief, binding events within the channel. The selectivity of the channel proteins apparently is realized by these binding events. The ion channels are not open continuously. Instead, they have "gates" that open briefly and then close again, a process referred to as "gating."

Recent evidence indicates the presence of proteins that form water channels in the plasma membrane of specific animal cell types and in the tonoplast and possibly the plasma membrane of plant cells. These water channel proteins, called **aquaporins**, enhance the move-ment of water through the membranes while excluding ions and metabolites.

Carrier proteins are further categorized according to how they function. Some carriers are **uniporters**, which are simply proteins that transport only one solute from one side of the membrane to the other. All carriers involved with facilitated diffusion and all channel proteins operate as uniporters. Other carrier proteins function as **cotransport systems**, in which the transfer of one solute depends on the simultaneous or sequential transfer of a second solute. The second solute may be transported in the same direction, in which case the carrier protein is known as a **symporter**, or in the opposite direction, as in the case of an **antiporter** (Figure 4–12).

Neither simple diffusion nor facilitated diffusion is capable of moving solutes *against* a concentration gradient or an electrochemical gradient. Transport of a substance against its electrochemical gradient requires the input of additional energy and is called **active transport**. A good example of an active transport protein is a **pump** (Figure 4–11c). Pumps are driven by chemical energy (ATP), electrical energy, or light energy. In plant and fungal cells they typically are proton pumps. The proton pump is, in fact, a membrane-bound H^+-ATPase, an enzyme that uses the energy from the hydrolysis of ATP to transport protons (H^+ ions) across the plasma membrane against their gradient. The enzyme generates a large gradient of protons across the membrane that provides the

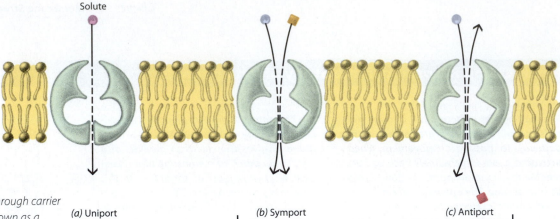

(a) Uniport

(b) Symport (c) Antiport

Cotransport systems

4–12

Three ways that solutes pass through carrier proteins. (a) In the simplest, known as a uniport, one particular solute is moved directly across the membrane in one direction. Carrier proteins involved with facilitated diffusion function as uniporters, as do all channel proteins. (b) In the type of cotransport system known as a symport, two different solutes are moved across the membrane simultaneously and in the same direction. (c) In another type of cotransport system, known as an antiport, two different solutes are moved across the membrane either simultaneously or sequentially, but in opposite directions.

4–13

Primary and secondary active transport of sucrose. (a) Primary active transport occurs when the proton pump (the enzyme H^+-ATPase) pumps protons (H^+) against their gradient. The result is a proton gradient across the membrane. (b) The proton gradient energizes secondary active transport. As the protons flow passively down their gradient, sucrose molecules are cotransported across the membrane against their gradient. The carrier protein is known as a sucrose-proton symporter.

driving force for solute uptake by all proton-coupled cotransport systems. The uptake of sucrose—the form in which sugar is commonly transported throughout the plant—depends on sucrose-proton cotransport (see page 768). The process by which the proton pump actively transports protons against their gradient is referred to as "primary active transport." The second process, which is powered by the movement of the protons down their gradient, cotransports sucrose molecules against their gradient and is known as "secondary active transport" (Figure 4–13). By this process, even neutral solutes can be accumulated to concentrations much higher than those outside the cell, simply by cotransporting with a charged molecule—a proton, for example.

The three categories of transport proteins move solutes at different speeds. The number of solute ions or molecules transported per protein per second is relatively slow with pumps (fewer than 500 per second), intermediate with carriers (500 to 10,000 per second), and most rapid with channels (10,000 to many millions per second).

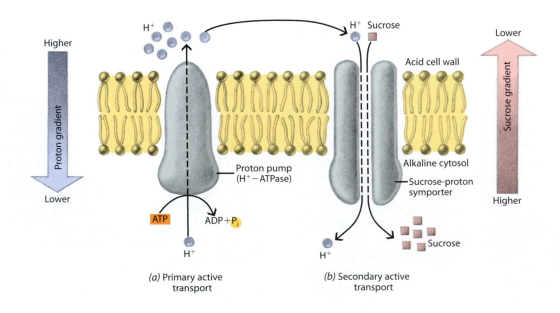

(a) Primary active transport

(b) Secondary active transport

Vesicle-Mediated Transport

The transport proteins that ferry ions and small polar molecules across the plasma membrane cannot accommodate large molecules, such as proteins and polysaccharides, or large particles, such as microorganisms or bits of cellular debris. These large molecules and particles are transported by means of vesicles or saclike structures that bud off from or fuse with the plasma membrane, a process called **vesicle-mediated transport.** As we saw in Figure 3–23, vesicles move from the *trans*-Golgi network to the surface of the cell. The hemicelluloses, pectins, and glycoproteins that form the matrix of the cell wall are carried to developing cell walls in secretory vesicles that fuse with the plasma membrane, thus releasing their contents into the wall. As mentioned previously, this process is known as **exocytosis** (Figure 4–14). The slimy substance that lubricates a growing root and helps it penetrate the soil is a polysaccharide that is transported by secretory vesicles and is released into the walls of the rootcap cells by exocytosis (see page 591). Exocytosis is not limited to the secretion of substances derived from the Golgi bodies. For example, the digestive enzymes secreted by carnivorous plants, such as the Venus flytrap and sundew (see pages 717 and 737), are carried to the plasma membrane in vesicles derived from the endoplasmic reticulum.

Transport by means of vesicles or saclike structures can also work in the opposite direction. In **endocytosis,** material to be taken into the cell induces the plasma membrane to bulge inward, producing a vesicle enclosing the substance. Three different forms of endocytosis are known: phagocytosis, pinocytosis, and receptor-mediated endocytosis.

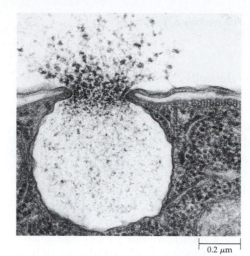

4–14

Exocytosis. A secretory vesicle, formed by a Golgi complex of the protist Tetrahymena furgasoni, *discharges mucus at the cell surface. Notice how the membrane enclosing the vesicle has fused with the plasma membrane.*

Phagocytosis ("cell eating") involves the ingestion of relatively large, solid particles, such as bacteria or cellular debris, via large vesicles derived from the plasma membrane (Figure 4–15a). Many one-celled organisms, such as amoebas, feed in this way, as do plasmodial slime molds and cellular slime molds. A unique example of phagocytosis in plants is found in the nodule-forming roots of legumes during the release of *Rhizobium* bacteria from infection threads. The bacteria are enveloped by portions of the plasma membrane of the root hairs (see page 739).

Pinocytosis ("cell drinking") involves the taking in of liquids, as distinct from solid matter (Figure 4–15b). It is the same in principle as phagocytosis. Unlike phagocytosis, however, which is carried out only by certain specialized cells, pinocytosis is believed to occur in all

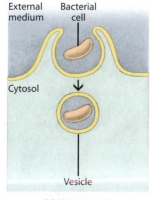

(a) Phagocytosis

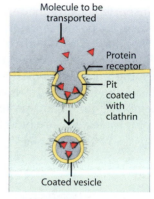

(c) Receptor-mediated endocytosis

4–15

Three types of endocytosis. (a) In phagocytosis, contact between the plasma membrane and particulate matter, such as a bacterial cell, causes the plasma membrane to extend around the particle, engulfing it in a vesicle. (b) In pinocytosis, the plasma membrane

pouches inward, forming a vesicle around liquid from the external medium that is to be taken into the cell. (c) In receptor-mediated endocytosis, the molecules to be transported into the cell must first bind to specific receptor proteins. The receptors are either localized in

indented areas of the plasma membrane known as coated pits or migrate to such areas after binding the molecules to be transported. When filled with receptors carrying their particular molecules, the pit buds off as a coated vesicle.

eukaryotic cells, as the cells continuously and indiscriminately "sip" small amounts of fluid from the surrounding medium.

In **receptor-mediated endocytosis,** currently the subject of a great deal of research, particular membrane proteins serve as receptors for specific molecules that are transported into the cell (Figure 4–15c). Until recently, receptor-mediated endocytosis was thought to be uncommon in plant cells. As indicated by studies using isolated protoplasts and intact tissues and the salts of heavy metals (such as lead, which is visible in electron micrographs), however, an endocytic cycle similar to that of animal cells exists in plant cells (Figure 4–16).

This cycle begins at specialized regions of the plasma membrane called **coated pits,** where specific receptors are localized. Coated pits are depressions of the plasma membrane coated on their inner, or cytoplasmic, surfaces with the peripheral protein *clathrin* (page 58). The substance being transported attaches to the receptors in a coated pit. Shortly thereafter (within minutes usually) the coated pit invaginates and pinches off to form a **coated vesicle.** The vesicles thus formed contain not only the substance being transported but also the receptor molecules, and they have an external coating of clathrin. Within the cell, the coated vesicles shed their coats and then fuse with some other membrane-bounded structure (for example, Golgi bodies or small vacuoles), releasing their contents in the process. As you can see in Figure 4–15c, the surface of the membrane facing the interior of a vesicle is equivalent to the surface of the plasma membrane facing the exterior of the cell. Similarly, the surface of the vesicle facing the cytoplasm is equivalent to the cytoplasmic surface of the plasma membrane.

As we noted in Chapter 3, new material needed for expansion of the plasma membrane by enlarging cells is transported, ready-made, in Golgi vesicles. During endocytosis, portions of the plasma membrane are returned to the Golgi bodies. Furthermore, current evidence indicates that portions of membranes used in forming endocytic vesicles are transported back to the plasma membrane in exocytosis. In the process, the membrane lipids and proteins, including the specific receptor molecules, are recycled.

Cell-to-Cell Communication

Thus far in our consideration of the transport of substances into and out of cells, we have assumed that individual cells exist in isolation, surrounded by a watery environment. In multicellular organisms, however, this is generally not the case. Cells are organized into **tissues,** groups of specialized cells with common functions. Tissues are further organized to form **organs,** each of which has a structure that suits it for a specific function.

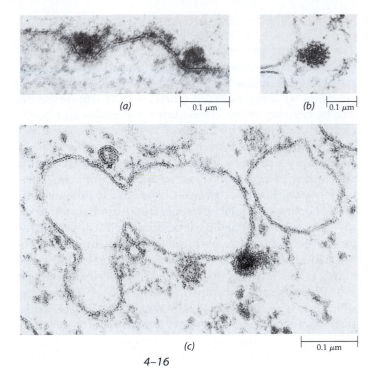

(a)   ⊢0.1 μm⊣   (b)   ⊢0.1 μm⊣

(c)   ⊢0.1 μm⊣

4–16

Endocytosis in maize (Zea mays) rootcap cells that have been exposed to a solution containing lead nitrate. (The rootcap cells form a protective covering at the tip of the root.) **(a)** *Granular deposits containing lead can be seen in two coated pits.* **(b)** *A coated vesicle with lead deposits.* **(c)** *Here, one of two coated vesicles has fused with a large Golgi vesicle where it will release its contents. This coated vesicle (dark structure) still contains lead deposits, but it appears to have lost its coat, which is located just to the right of it. The coated vesicle to its left is clearly intact.*

Signal Transduction Is the Process by which Cells Use Chemical Messengers to Communicate

As you might imagine, the successful existence of multicellular organisms depends upon the ability of the individual cells to communicate with one another so that they can collaborate to create harmonious tissues and organs, and ultimately, a properly functioning organism. This communication is accomplished in large part by means of chemical **signals**—that is, by substances that are synthesized within and then transported out of one cell and travel to another cell. In plants, the chemical signals are represented largely by hormones, chemical messengers typically produced by one cell type or tissue in order to regulate the function of cells or tissues elsewhere in the plant body (see Chapter 28). The signal molecules must be small enough to get through the cell wall easily.

The plasma membrane plays a key role in signal recognition. When the signal molecules reach the plasma membrane of the **target cell,** they may be transported into the cell by any one of the endocytic processes we have considered. Alternatively, they may remain outside the cell but bind to specific **receptors** on the outer surface of the membrane. In most cases the receptors are transmembrane proteins that become activated when they bind a signal molecule (the first messenger) and generate secondary signals, or **second messengers,** on the inside. The second messengers, which increase in concentration within the cell in response to the signal, pass the signal on by altering the behavior of selected cellular proteins, thereby triggering chemical changes within that cell. This process by which a cell converts an extracellular signal into a response is called **signal transduction.** Two of the most widely used second messengers are calcium ions and, in animals and fungi, cyclic AMP (cyclic adenosine monophosphate, a molecule formed from ATP).

The signal-transduction pathway may be divided into three steps: **reception, transduction,** and **induction** (Figure 4–17). The binding of the hormone (or of any chemical signal) to its specific receptor represents the reception step. During the transduction step, the second messenger, which is capable of amplifying the stimulus and initiating the cell's response, is formed in or released into the cytosol. The calcium ion, Ca^{2+}, has been identified as the second messenger in many plant responses. Binding of a hormone to its specific receptor triggers the release of Ca^{2+} ions stored in the vacuole into the cytosol. The Ca^{2+} ions enter the cytosol through specific Ca^{2+} channels in the tonoplast. In some plant cells, Ca^{2+} ions stored in the lumen of the endoplasmic reticulum may also be involved (page 56). The Ca^{2+} ions then combine with **calmodulin,** the principal calcium-binding protein in plant cells. The Ca^{2+}-calmodulin complex influences, or induces, numerous cellular processes, generally via activation of appropriate enzymes.

Plasmodesmata Enable Cells to Communicate

Plasmodesmata, the narrow strands of cytoplasm that interconnect the protoplasts of neighboring plant cells, are also important pathways in cell-to-cell communication. Because they are intimately interconnected by plasmodesmata, all of the protoplasts of the plant body, together with their plasmodesmata, constitute a continuum called the **symplast.** Accordingly, the movement of substances from cell to cell by means of the plasmodesmata is called **symplastic transport.** In contrast, the movement of substances in the cell wall continuum, or **apoplast,** surrounding the symplast is called **apoplastic transport.**

As mentioned in Chapter 3, a plasmodesma typically is traversed by a tubular strand of endoplasmic reticulum, called a *desmotubule*, which is continuous with the endoplasmic reticulum of the adjacent cells. The desmotubule, however, does not resemble the adjoining endoplasmic reticulum. It is much smaller in diameter and contains a central, rodlike structure. Considerable controversy has centered on the interpretation of the central rod, but today most investigators believe it represents the merger of the inner portions of the bilayers of the tightly appressed endoplasmic reticulum forming the desmotubule (Figure 4–18). If this interpretation is correct, the desmotubule lacks a lumen, or opening, and all transport via the plasmodesma would be restricted to the cytoplasmic channel surrounding the desmotubule. This channel, called the **cytoplasmic sleeve,** appears to be subdivided into several narrower channels by spoke-like structures that extend between the desmotubule and the plasma membrane lining the cytoplasmic sleeve. Interestingly, these narrower channels have about the same diameter as the openings that make up the connecting channels—known as gap junctions—between adjacent animal cells.

Plasmodesmata apparently provide a more efficient pathway between neighboring cells than the less direct,

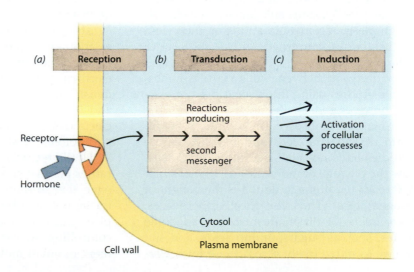

4–17

*Generalized model of a signal-transduction pathway. (**a**) Reception. A hormone (or other chemical signal) binds to a specific receptor in the plasma membrane. (**b**) Transduction. The receptor now stimulates the cell to produce a second messenger. (**c**) Induction. The second messenger enters the cytosol and activates cellular processes. In other cases, the chemical signals enter the cell and bind to specific receptors inside.*

(a) **Reception** (b) **Transduction** (c) **Induction**

Reactions producing second messenger

Activation of cellular processes

Receptor

Hormone

Cell wall

Cytosol

Plasma membrane

4–18
Plasmodesmata in cell walls of the sugarcane (Saccharum) leaf. Electron micrographs showing (a) longitudinal view of the plasmodesmata and (b) plasmodesmata in transverse view. Note that the endoplasmic reticulum is connected to the desmotubule, which contains a central rod, apparently formed by the merger of the inner portions of the ER bilayers. The cytoplasmic channel surrounding the desmotubule is called the cytoplasmic sleeve. Note the spokelike structures in the cytoplasmic sleeves of the plasmodesmata in (b).

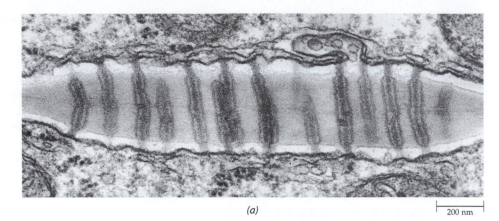

(a)   200 nm

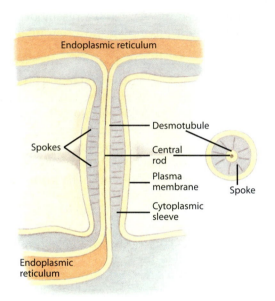

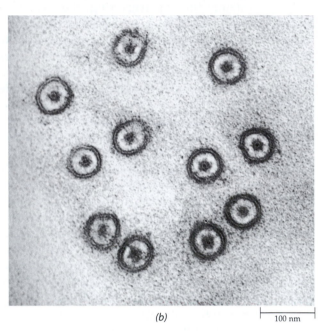

(b)   100 nm

alternative route of plasma membrane, cell wall, and plasma membrane. It is believed that cells and tissues that are far removed from direct sources of nutrients can be supplied with nutrients either by simple diffusion or by bulk flow through plasmodesmata. In addition, as we will discuss in Chapter 31, some substances are believed to move through plasmodesmata to and from the xylem and phloem, the tissues concerned with long-distance transport in the plant body.

Evidence for transport between cells via plasmodesmata comes from studies using fluorescent dyes or electric currents. The dyes, which do not easily cross the plasma membrane, can be observed moving from the injected cells into neighboring cells and beyond (Figure 4–19). Such studies have revealed that most plasmodesmata can allow the passage of molecules up to molecular weights of 800 to 1000 daltons (a dalton is the weight of a single hydrogen atom). Thus, the effective pore sizes, or *size exclusion limits*, of plasmodesmata are adequate for small solutes, such as sugars, amino acids, and signaling molecules, to move freely across these intercellu-

lar connections. Studies involving fluorescent dyes, however, have revealed the presence of barriers at various regions in the symplast. These barriers are due to differences in the size exclusion limits of the plasmodesmata at the boundaries of the various regions. Such regions are referred to as *symplastic domains*.

The passage of pulses of electric current from one cell to another can be monitored by means of receiver electrodes placed in neighboring cells. The magnitude of the electrical force is found to vary with the frequency, or density, of plasmodesmata and with the number of cells and length of cells between the injection and receiver electrodes, indicating that plasmodesmata can serve as a path for electrical signaling between plant cells.

Until recently, plasmodesmata were depicted as rather passive entities that exerted little direct influence over the substances that moved through them. That image is fast fading and being replaced by one that depicts plasmodesmata as dynamic structures capable of controlling, to varying degrees, the intercellular movement of small molecules. Moreover, recent studies indi-

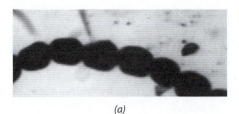

(a)

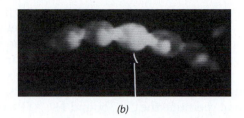

(b)

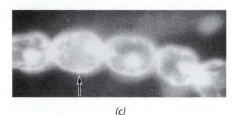

(c)

4–19

*Movement through plasmodesmata. **(a)** A hair from a stamen of Setcreasea purpurea before injection with the fluorescent dye disodium fluorescein. **(b)**, **(c)** Two and five*

minutes after the dye was injected into the cytoplasm of the cell indicated by the arrow. Notice that the dye has passed in both directions into the cytoplasm of the neighboring

cells. Since the plasma membrane is impermeable to the dye, the movement of dye from cell to cell must have occurred via plasmodesmata connecting the adjoining cells.

cate that some plasmodesmata have the capacity to transport macromolecules, such as proteins and nucleic acids, by an active energy-dependent process, thereby playing a primary role in the coordination of plant growth and development. Plant viruses, which far exceed the size exclusion limits typical of plasmodesmata, have long been known to spread infection by moving from cell to cell via plasmodesmata. Evidence shows that such viruses produce specific "movement proteins" that dramatically increase the size exclusion limit or otherwise alter the structure of the plasmodesma.

The transport of materials into and out of cells through plasmodesmata, through the transmembrane proteins, and by means of endocytosis and exocytosis appear superficially to be three quite different processes. They are fundamentally similar, however, in that they all depend on the precise, three-dimensional structure of a great variety of specific protein molecules. These protein molecules not only form channels through which transport can occur but also endow the plasma membrane with the capacity to "recognize" particular molecules. This capacity is the result of billions of years of an evolutionary process that began, as far as we are able to discern, with the formation of that first fragile film around a few organic molecules.

Summary

The plasma membrane regulates the passage of materials into and out of the cell, a function that makes it possible for the cell to maintain its structural and functional integrity. This regulation depends on interaction between the membrane and the materials that pass through it.

Membranes Consist of a Lipid Bilayer and Proteins

According to the fluid-mosaic model, the plasma membrane and other cellular membranes are composed of lipid bilayers in which proteins are embedded. In plant cells, the major types of lipids are phospholipids (the more abundant) and sterols. Different membrane proteins perform different functions; some are enzymes, others are receptors, and still others are transport proteins. The two surfaces of a membrane differ considerably in chemical composition. The exterior surface of the plasma membrane is characterized by short carbohydrate chains that are believed to play important roles in the "recognition" of molecules that interact with the cell.

Water Moves Down a Water Potential Gradient

Water is one of the principal substances passing into and out of cells. Water potential determines the direction in which the water moves; that is, the movement of water is from regions of higher water potential (lower solute concentration) to regions of lower water potential (higher solute concentration) provided that the pressure in the two regions is equal.

Water Movement Takes Place by Bulk Flow and Diffusion

Bulk flow is the overall movement of water molecules, as when water flows downhill or moves in response to pressure. Diffusion involves the independent movement of molecules and results in net movement down a gradient. Diffusion is most efficient when the distance involved is short and the concentration gradient is steep. By their metabolic activities, cells maintain steep concentration gradients of many substances across the plasma membrane and between different compartments of the cytoplasm. The rate of movement of substances within cells is increased by cytoplasmic streaming. Carbon dioxide and oxygen are two important molecules that move into and out of cells by diffusion across the membrane.

Osmosis Is a Special Case of Diffusion

Osmosis is the diffusion of water through a selectively permeable membrane—a membrane that permits the movement of water but inhibits the passage of solutes. In the absence of other forces, the movement of water by osmosis is from a region of lower solute concentration (a hypotonic medium), and therefore of higher water potential, to a region of higher solute concentration (a hypertonic medium), and so of lower water potential. The turgor (rigidity) of plant cells is a consequence of osmosis and a strong but somewhat elastic cell wall.

Small Molecules Cross Membranes via Simple Diffusion, Facilitated Diffusion, or Active Transport

Both simple diffusion and facilitated diffusion (diffusion assisted by carrier or channel proteins) are passive transport processes. If transport requires the expenditure of energy by the cell, it is known as active transport. Active transport can move substances against the concentration gradient or electrochemical gradient. This process is mediated by transport proteins known as pumps. In plant and fungal cells, an important pump is a membrane-bound H^+-ATPase enzyme.

Large Molecules and Particles Cross Membranes via Vesicle-Mediated Transport

Controlled movement of large molecules into and out of a cell also occurs by endocytosis or exocytosis, processes in which substances are transported within vesicles. Three forms of endocytosis are known: phagocytosis, in which solid particles are taken into the cell; pinocytosis, in which liquids are taken in; and receptor-mediated endocytosis, in which molecules and ions to be transported into the cell are bound to specific receptors in the plasma membrane. During exocytosis and endocytosis, portions of membranes are recycled between the Golgi bodies and the plasma membrane.

 *Summary* **TABLE**   The Movement of Substances across Membranes

MOVEMENT OF IONS AND SMALL MOLECULES

NAME OF PROCESS	MOVEMENT AGAINST OR DOWN A GRADIENT	TRANSPORT PROTEINS REQUIRED?	ENERGY SOURCE, SUCH AS ATP, REQUIRED?	SUBSTANCES MOVED	COMMENTS
Passive transport					
Simple diffusion	Down	No	No	Small nonpolar molecules (O_2, CO_2, and others)	Diffusion is the net movement of a substance down its concentration gradient.
Osmosis (a special case of diffusion)	Down	No	No	H_2O	Osmosis is the diffusion of water across a selectively permeable membrane.
Facilitated diffusion	Down	Yes	No	Polar molecules and ions	Carrier proteins undergo conformational changes to transport a specific solute. Channel proteins form water-filled pores for specific ions.
Active transport	Against	Yes	Yes	Polar molecules and ions	Often involves proton pumps. Enables cells to accumulate or expel solutes at high concentrations.

MOVEMENT OF LARGE MOLECULES AND PARTICLES (VESICLE-MEDIATED TRANSPORT)

NAME OF PROCESS	BASIC FUNCTION	EXAMPLES AND COMMENTS
Exocytosis	Releasing materials from the cell	Secretion of polysaccharides of the cell wall matrix; secretion of digestive enzymes by carnivorous plants.
Endocytosis	Taking materials into the cell	
Phagocytosis	Ingesting solids	Ingestion of bacteria, cellular debris.
Pinocytosis	Taking up liquids	Incorporation of fluid from the environment.
Receptor-mediated endocytosis	Taking up specific molecules	Molecules bind specific receptors in clathrin-coated pits, which then invaginate to form coated vesicles in the cell.

Signal Transduction Is the Process by which Cells Use Chemical Messengers to Communicate

In multicellular organisms, communication among cells is essential for coordination of the different activities of the cells in the various tissues and organs. Much of this communication is accomplished by chemical signals that either pass through the plasma membrane or interact with receptors on the membrane surface. Most receptors are transmembrane proteins that become activated when they bind a signal molecule and generate second messengers on the inside. The second messengers, in turn, amplify the stimulus and trigger the cell's response. This process is known as signal transduction.

Plasmodesmata Enable Cells to Communicate

Plasmodesmata also are important pathways in cell-to-cell communication. All of the protoplasts, together with their plasmodesmata, constitute a continuum called the symplast. The cell wall continuum surrounding the symplast is called the apoplast. Once thought of as rather passive entities through which ions and small molecules move by simple diffusion or bulk flow, plasmodesmata are now known to be dynamic structures capable of controlling the intercellular movement of various-sized molecules.

Selected Key Terms

active transport   p. 83

apoplast   p. 87

bulk flow   p. 77

carrier proteins   p. 83

channel proteins   p. 83

coated pits   p. 86

concentration gradient p. 78

cotransport systems p. 83

cytoplasmic sleeve   p. 87

desmotubule   p. 87

diffusion   p. 78

electrochemical gradient p. 82

endocytosis   p. 85

exocytosis   p. 85

facilitated diffusion p. 83

fluid-mosaic model   p. 74

hypertonic   p. 79

hypotonic   p. 79

imbibition   p. 80

integral proteins   p. 74

isotonic   p. 79

membrane potential p. 82

osmosis   p. 79

osmotic potential   p. 79

osmotic pressure   p. 79

passive transport   p. 82

peripheral protein   p. 74

phagocytosis   p. 85

pinocytosis   p. 85

plasmodesmata   p. 87

plasmolysis   p. 81

pumps   p. 83

receptor-mediated endocytosis   p. 86

second messengers p. 87

selectively permeable membrane   p. 79

signal transduction p. 87

simple diffusion   p. 82

symplast   p. 87

transmembrane proteins p. 74

transport proteins   p. 82

turgor pressure   p. 81

vesicle-mediated transport p. 85

water potential   p. 76

Questions

1. Of what value is the concept of water potential to plant physiologists?

2. Distinguish between a substance moving *down* a concentration gradient and a substance moving *against* a concentration gradient.

3. In terms of both solute concentration and water potential, distinguish among isotonic, hypotonic, and hypertonic solutions.

4. At the close of the Punic Wars, when the Romans destroyed the city of Carthage (in 146 B.C.), they are supposed to have sown the ground with salt and plowed it in. Explain, in terms of the physiological processes discussed in this chapter, why this action would make the soil barren to most plants for many years.

5. What is vesicle-mediated transport, and how does endocytosis differ from exocytosis?

6. What are the differences between phagocytosis and receptor-mediated endocytosis?

7. After the spring thaw and with April showers, homeowners often find water backing up in the basement drain because the sewer pipe is clogged with roots. Using the information you have obtained in this chapter, explain how the roots got into the sewer pipe.

8. Secondary active transport is significant to the plant because it enables a cell to accumulate even neutral solutes at concentrations much higher than those outside the cell. Using such terms as proton pump (H^+-ATPase), proton gradient, proton-coupled cotransport, sucrose-proton cotransport, primary active transport, and secondary active transport, explain how this system works.

9. Explain, in general terms, what happens in each step—reception, transduction, and induction—of the signal-transduction pathway.

10. Viruses are too large to move through plant cell walls, yet some viruses spread from intact cell to intact cell with relative ease. How is that possible?

Energetics

A chickadee feasting on the seeds of a dried sunflower head in winter. During the previous summer, energy from sunlight was captured by the then green leaves and converted into chemical energy stored in sugar and other organic molecules. Much of the sugar was transported to the developing flower heads, where it was converted into other energy-rich storage molecules by the developing seeds. Each of the many flowers of the flower head produces a single seed. Chickadees and other animals eat such seeds and other plant parts to obtain the fuel necessary to carry out their many life processes.

Chapter

5

The Flow of Energy

OVERVIEW

Energy is the capacity to do work, and in this chapter we will look at work and energy from the cell's perspective. Cells carry out various types of work, such as synthesizing molecules, moving organelles and chromosomes from place to place, and transporting substances across membranes. Each of these activities requires energy, and the cell must be able to obtain energy and then utilize it in various ways.

How do cells utilize energy in the proper amounts? Two key factors are involved. First, almost every chemical reaction in a cell requires the involvement of an enzyme. Enzymes are specific proteins that make it possible for reactions to take place with only small inputs of energy. Without enzymes, cells could not function because chemical reactions would take place too slowly, requiring the input of so much energy that the cell would be damaged.

A second key factor is ATP, the cell's "energy currency." As we will see, energy released during certain types of chemical reactions can be stored in molecules of ATP, which then can donate that energy to power other reactions. In a cell, these various types of reactions—each requiring a specific enzyme—take place sequentially in metabolic pathways that channel the flow of energy.

CHECKPOINTS

By the time you finish reading this chapter, you should be able to answer the following questions:

1. What are the first and second laws of thermodynamics, and how do they relate to living organisms?

2. Why are oxidation-reduction reactions important in biology?

3. How do enzymes catalyze chemical reactions, and what are some of the factors that influence enzyme activity?

4. How does feedback inhibition regulate cellular activities?

5. What are coupled reactions, and how does ATP function as an intermediate between exergonic and endergonic reactions?

ife here on Earth is solar-powered. Nearly every process vital to life depends on a steady flow of energy from the sun. An immense amount of solar energy—estimated to be 13×10^{23} calories per year—reaches the Earth. About 30 percent of this solar energy is immediately reflected back into space as light, just as light is reflected from the moon. About 20 percent is absorbed by Earth's atmosphere. Much of the remaining 50 percent is absorbed by the Earth itself and converted to heat. Some of this absorbed heat energy serves to evaporate the waters of the oceans, producing the clouds that, in turn, produce rain and snow. Solar energy, in combination with other factors, is also responsible for the movements of air and water that help set patterns of climate over the surface of the Earth.

Less than one percent of the solar energy reaching the Earth is captured by the cells of plants and other photosynthetic organisms and converted by them into the energy that drives nearly all the processes of life. Living systems change energy from one form to another, transforming the radiant energy of the sun into the chemical,

electrical, and mechanical energy used by living organisms (Figure 5–1).

This flow of energy is the essence of life. In fact, one way of viewing evolution is as a competition among organisms for the most efficient use of energy resources. A cell can be best understood as a complex system for transforming energy. At the other end of the biological scale, the structure of an ecosystem (that is, all the living organisms in a particular locale and the nonliving factors with which they interact) or of the biosphere itself is determined by the energy exchanges occurring among the groups of organisms within it.

In this chapter, we shall look first at the general principles governing all energy transformations. Then we shall turn our attention to the characteristic ways in which cells regulate the energy transformations that take place within living systems. In the chapters that follow, the principal and complementary processes of energy flow through the biosphere will be examined—glycolysis and respiration in Chapter 6 and photosynthesis in Chapter 7.

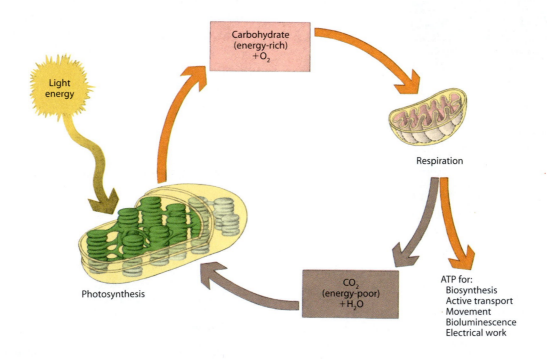

5–1

An example of the flow of biological energy. The radiant energy of sunlight is produced by nuclear fusion reactions taking place in the sun. Chloroplasts, present in all photosynthetic eukaryotic cells, capture the radiant energy of sunlight and use it to convert water and carbon dioxide into carbohydrates, such

as glucose, sucrose, or starch. Oxygen is released into the air as a product of the photosynthetic reactions. Mitochondria, present in eukaryotic cells, carry out the final steps in the breakdown of these carbohydrates and capture their stored energy in ATP molecules. This process, cellular respiration, consumes

oxygen and produces carbon dioxide and water, completing the cycling of the molecules. With each transformation, some energy is dissipated to the environment in the form of heat. Thus the flow of energy through the biosphere is one-way. It can continue only as long as there is an input of energy from the sun.

The Laws of Thermodynamics

Energy is an elusive concept. Today it is usually defined as the capacity to do work. Until about 200 years ago, heat—the form of energy most readily studied—was considered to be a separate, though weightless, substance called "caloric." An object was hot or cold depending upon how much caloric it contained; when a cold object was placed next to a hot one, the caloric flowed from the hot object into the cold one; and when metal was pounded with a hammer, it became warm because the caloric was forced to the surface. Even though the idea of a caloric substance proved incorrect, the concept turned out to be surprisingly useful.

The development of the steam engine in the latter part of the eighteenth century, more than any other single chain of events, changed scientific thinking about the nature of energy. Energy became associated with work, and heat and motion came to be seen as forms of energy. This new understanding led to the study of **thermodynamics**—the science of energy transformations—and to formulation of its laws.

The First Law States That the Total Energy in the Universe Is Constant

The **first law of thermodynamics** states, quite simply: *Energy can be changed from one form to another but cannot be created or destroyed.* In engines, for instance, chemical energy (such as in coal or gasoline) is converted to heat, or thermal energy, which is then partially converted to mechanical movements (kinetic energy). Some of the energy is converted back to heat by the friction of these movements, and some leaves the engine in the form of exhaust products. Unlike the heat in the engine or the boiler, the heat produced by friction and lost in the exhaust cannot produce "work"—that is, it cannot turn the gears—because it is dissipated into the environment. But it is nevertheless part of the total equation. In fact, engineers calculate that most of the energy consumed by an engine is dissipated randomly as heat; most engines work with less than 25 percent efficiency.

The notion of **potential energy** developed in the course of such engine-efficiency studies. A barrel of oil or a ton of coal could be assigned a certain amount of potential energy, expressed in terms of the amount of heat it would liberate when burned. The efficiency of the conversion of the potential energy to usable energy depended on the design of the energy-conversion system.

Although these concepts were formulated in terms of engines running on heat energy, they apply to other systems as well. For example, a boulder pushed to the top of a hill contains potential energy. Given a little push (the energy of activation), it rolls down the hill again, converting the potential energy to energy of motion and to heat produced by friction. As mentioned previously, water can also possess potential energy (page 76). As it moves by bulk flow from the top of a waterfall or over a dam, it can turn waterwheels that turn gears, for example, to grind maize. Thus, the potential energy of water, in this system, is converted to the kinetic energy of the wheels and gears and to heat, which is produced by the movement of the water itself as well as the turning wheels and gears.

Light is another form of energy, as is electricity. Light can be changed to electrical energy, and electrical energy can be changed to light (by letting it flow through the tungsten wire in a light bulb, for example).

The first law of thermodynamics can be stated more completely as follows: *In all energy exchanges and conversions, the total energy of the system and its surroundings after the conversion is equal to the total energy before the conversion.* A "system" can be any clearly defined entity—for example, an exploding stick of dynamite, an idling automobile engine, a mitochondrion, a living cell, a tree, a forest, or the Earth itself. The "surroundings" consist of everything outside the system.

The first law of thermodynamics can be stated as a simple bookkeeping rule: When we add up the energy income and expenditures for any physical process or chemical reaction, the books must always balance. (Note, however, that the first law does not apply to nuclear reactions, in which energy is, in fact, created by the conversion of mass to energy.)

The Second Law States That the Entropy of the Universe Is Increasing

The energy that is released as heat in an energy conversion has not been destroyed—it is still present in the random motion of atoms and molecules—but it has been "lost" for all practical purposes. It is no longer available to do useful work. This brings us to the **second law of thermodynamics**, which is the more important one, biologically speaking. It predicts the direction of all events involving energy exchanges. Thus, it has been called "time's arrow."

The second law states: *In all energy exchanges and conversions—if no energy leaves or enters the system under study—the potential energy of the final state will always be less than the potential energy of the initial state.* The second law is entirely in keeping with everyday experience (Figure 5–2). A boulder will roll downhill but never uphill. A ball that is dropped will bounce—but not back to the height from which it was dropped. Heat will flow from a hot object to a cold one and never the other way.

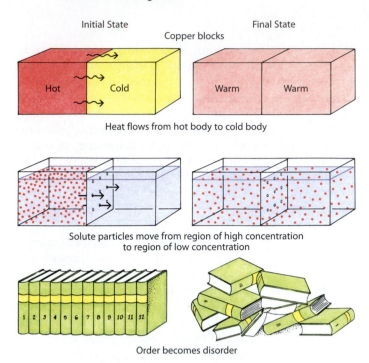

Initial State   Final State
Copper blocks

Hot   Cold   Warm   Warm

Heat flows from hot body to cold body

Solute particles move from region of high concentration
to region of low concentration

Order becomes disorder

5–2

Some illustrations of the second law of thermodynamics. In each case, a concentration of energy—in the hot copper block, in the solute particles on one side of a tank, and in the neatly organized books—is dissipated. In nature, processes tend toward randomness, or disorder. Only an input of energy can reverse this tendency and reconstruct the initial state from the final state. Ultimately, however, disorder will prevail, because the total amount of energy in the universe is finite.

A process in which the potential energy of the final state is less than that of the initial state is one that releases energy. Such a process is said to be **exergonic** ("energy-out"). Only exergonic processes can take place spontaneously. Although "spontaneously" has an explosive sound, the word says nothing about the speed of the process—just that it can take place without an input of energy from outside the system (Figure 5–3a). By contrast, a process in which the potential energy of the final state is greater than that of the initial state is one that requires energy. Such processes are said to be **endergonic** ("energy-in"). In order for an endergonic process to proceed, there must be an input of energy to the system (Figure 5–3b).

One important factor in determining whether or not a reaction is exergonic is ΔH, the change in heat content of the system, where Δ stands for change, H for heat content (the formal term for H is "enthalpy"). In general, the change in heat content is approximately equal to the

5–3

(a) In exergonic processes, whether physical or chemical, the potential energy of the final state is lower than the potential energy of the initial state, and energy is released. Exergonic processes occur spontaneously, that is, they can take place without an input of energy. (b) In endergonic processes, by contrast, the potential energy of the final state is higher than the potential energy of the initial state. Such processes cannot take place spontaneously. For them to occur, an input of energy is required.

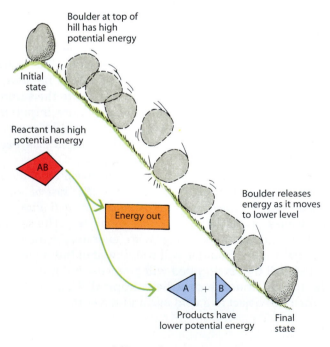

Boulder at top of hill has high potential energy

Initial state

Reactant has high potential energy

AB

Energy out

Boulder releases energy as it moves to lower level

A + B

Products have lower potential energy

Final state

(a) Exergonic processes

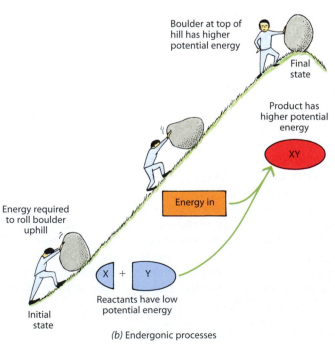

Boulder at top of hill has higher potential energy

Final state

Product has higher potential energy

XY

Energy in

Energy required to roll boulder uphill

X + Y

Reactants have low potential energy

Initial state

(b) Endergonic processes

change in potential energy. As noted in Figure A–15 in Appendix A, the energy change that occurs when glucose, for instance, is oxidized can be measured in a calorimeter and expressed in terms of ΔH. The complete oxidation of a mole* of glucose to carbon dioxide and water yields 673 kilocalories:

$$C_6H_{12}O_6 + 6O_2 \longrightarrow 6CO_2 + 6H_2O + 673 \text{ kcal}$$

Glucose   Oxygen   Carbon   Water   Heat
                      dioxide   released

$$\Delta H = -673 \text{ kcal/mole}$$

In many cases, an exergonic chemical reaction is also an exothermic reaction—that is, it gives off heat and thus has a negative ΔH.

Another factor besides the gain or loss of heat determines the direction of a process. This factor, called **entropy** (symbolized as S), is a measurement of the disorder, or randomness, of a system. Let us return to water as an example. The change from ice to liquid water and the change from liquid water to water vapor are both endothermic processes—a considerable amount of heat is absorbed from the surroundings as they occur. Yet, under the appropriate conditions, they proceed spontaneously. The key factor in these processes is the increase in entropy. In the case of the change of ice into water, a solid is being turned into a liquid, and some of the hydrogen bonds that hold the water molecules together in a crystal (ice) are being broken. As the liquid water turns to vapor, the rest of the hydrogen bonds are ruptured as the individual water molecules separate, one by one. In each case, the disorder of the system has increased.

The notion that there is more disorder associated with more numerous and smaller objects than with fewer, larger ones is in keeping with our everyday experience. If there are 20 papers on a desk, the possibilities for disorder are greater than if there are 2 or even 10. If each of the 20 papers is cut in half, the entropy of the system—the capacity for randomness—increases. The relationship between entropy and energy is also a commonplace idea. If you were to find your room tidied up and your books in alphabetical order on the shelf, you would recognize that someone had been at work—that energy had been expended. To organize the papers on a desk similarly requires the expenditure of energy.

Now let us return to the question of the energy changes that determine the course of chemical reactions. As discussed, both the change in the heat content of the system (ΔH) and the change in entropy (ΔS) contribute to the overall change in energy. This total change—which takes into account both heat and entropy—is

called the **free-energy change** and is symbolized as ΔG, after the American physicist Josiah Willard Gibbs (1839–1903), who was one of the first to integrate all of these ideas.

Keeping ΔG in mind, let us examine once again the oxidation of glucose. The ΔH of this reaction is -673 kcal/mole. The ΔG is -686 kcal/mole. (Note that these are values under standard conditions, with all reactants and products present at a concentration of one mole. The real values under actual conditions are likely to be somewhat different.) Thus, the entropy factor has contributed 13 kcal/mole to the free-energy change of the process. The changes in heat content and in entropy both contribute to the lower energy state of the products of the reaction.

The relationship among ΔG, ΔH, and entropy is given in the following equation:

$$\Delta G = \Delta H - T \Delta S$$

This equation states that the free-energy change is equal to the change in heat content (a negative value in exothermic reactions, which give off heat) minus the change in entropy multiplied by the absolute temperature T. In exergonic reactions, ΔG is always negative, but ΔH may be zero or even positive. Since T is always positive, the greater the change in entropy, the more negative ΔG will be, that is, the more exergonic the reaction will be. Therefore it is possible to state the second law in another, simpler way: *All naturally occurring processes are exergonic.*

Living Organisms Require a Steady Input of Energy

The most interesting implication of the second law, as far as biology is concerned, is the relationship between entropy and order. Living systems are continuously expending large amounts of energy to maintain order. Stated in terms of chemical reactions, living systems continuously expend energy to maintain a position far from equilibrium. If equilibrium were to occur, the chemical reactions in a cell would, for all practical purposes, stop, and no further work could be done. At equilibrium, a cell would be dead.

The universe is a closed system—that is, neither matter nor energy enters or leaves the system. The matter and energy present in the universe at the time of the "big bang" are all the matter and energy it will ever have. Moreover, after each and every energy exchange and transformation, the universe as a whole has less potential energy and more entropy than it did before. In this view, the universe is running down. The stars will flicker out, one by one. Life—any form of life on any planet—will come to an end. Finally, even the motion of individual molecules will cease. Take heart though. Even the most pessimistic among us do not believe this will occur for another 20 billion years or so.

---

*A mole is the amount of a substance, in grams, that equals its molecular weight. For example, the atomic weight of carbon is 12 and the atomic weight of oxygen is 16, so the molecular weight of CO_2 is 44. One mole of CO_2 is therefore 44 grams of CO_2.

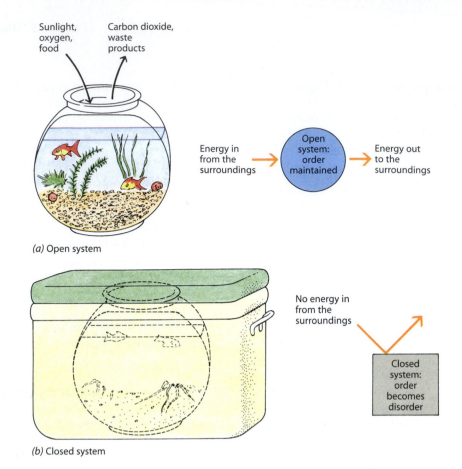

Sunlight, oxygen, food

Carbon dioxide, waste products

Energy in from the surroundings → Open system: order maintained → Energy out to the surroundings

(a) Open system

No energy in from the surroundings

Closed system: order becomes disorder

(b) Closed system

5–4

(a) A fish bowl is, like the Earth, an open system—matter and energy enter and leave the system. Sunlight passes through the glass, oxygen diffuses into the water at its surface, and food is added by a human caretaker. Heat leaves the system through the glass and through the opening at the top, carbon dioxide diffuses out of the water at its surface, and the waste products of the animals are removed when the bowl is cleaned. Although energy is lost from the system with each and every energy exchange, a steady supply of energy—mainly food—from outside the system maintains its order. (b) If, however, the fish bowl is placed in an opaque container that is sealed and insulated, it becomes a closed system, consisting of the fish bowl, its contents, and the air within the container. Neither matter nor energy can enter or leave the system. For a period of time after the container is sealed, energy continues to be converted from one form to another by the organisms in the fish bowl. With each and every conversion, however, some of the energy is given off as heat and dissipated into the water, the glass, and the air within the container. In time, the system will run down—the organisms will die and their bodies will disintegrate. The order originally present in the system will have become the disorder of individual atoms and molecules moving at random.

In the meantime, life can exist *because* the universe is running down. Although the universe as a whole is a closed system, the Earth is not. It is an open system (Figure 5–4), receiving an energy input of about 13×10^{23} calories per year from the sun. Photosynthetic organisms are specialists at capturing the light energy released by the sun as it slowly burns itself out. They use this energy to organize small, simple molecules (water and carbon dioxide) into larger, more complex molecules (sugars). In the process, the captured light energy is stored as chemical energy in the bonds of sugars and other molecules.

Living cells—including photosynthetic cells—can convert this stored energy into motion, electricity, light, and, by shifting the energy from one type of chemical bond to another, into more useful forms of chemical energy. At each transformation, energy is lost to the surroundings as heat. But before the energy captured from the sun is completely dissipated, organisms use it to create and maintain the complex organization of structures and activities that we know as life.

Oxidation-Reduction

Chemical reactions are essentially energy transformations in which energy stored in chemical bonds is transferred to other, newly formed chemical bonds. In such transfers, electrons shift from one energy level to another (see Appendix A, Electrons and Energy). In many reactions, electrons pass from one atom or molecule to another. These reactions, known as **oxidation-reduction** (or redox) **reactions**, are of great importance in living systems.

The *loss* of an electron is known as **oxidation**, and the atom or molecule that loses the electron is said to be oxidized. The reason electron loss is called oxidation is that oxygen, which attracts electrons very strongly, is often the electron acceptor.

Reduction is, conversely, the gain of an electron. Oxidation and reduction always take place simultaneously. The electron lost by the oxidized atom is accepted by another atom, which is thus reduced—hence the term "redox" (for reduction-oxidation) reactions (Figure 5–5).

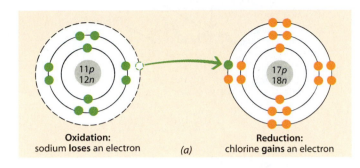

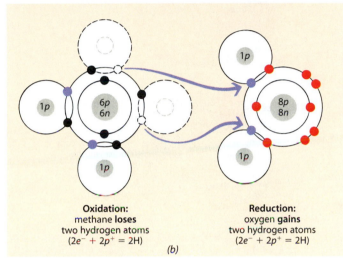

5–5

*Redox reactions. **(a)** In some oxidation-reduction reactions, such as the oxidation of sodium and the reduction of chlorine, a single electron is transferred from one atom to another. Such simple reactions typically involve elements or inorganic compounds. **(b)** In other oxidation-reduction reactions, such as the partial oxidation of methane (CH$_4$), the electrons are accompanied by protons. In these reactions, which often involve organic molecules, oxidation is the loss of hydrogen atoms, and reduction is the gain of hydrogen atoms. When an oxygen atom gains two hydrogen atoms, as shown here, the product is, of course, a water molecule.*

Redox reactions may involve only a solitary electron, as when sodium loses an electron and becomes oxidized to Na$^+$, and chlorine gains an electron and is reduced to Cl$^-$. In biological reactions, however, the electron often travels with a proton, that is, as a hydrogen atom. In such cases, oxidation involves the removal of hydrogen atoms, and reduction the gain of hydrogen atoms. For example, when glucose is oxidized, electrons and hydrogen atoms are lost by the glucose molecule and are gained by oxygen atoms, which are reduced to water:

$$C_6H_{12}O_6 + 6O_2 \longrightarrow 6CO_2 + 6H_2O + \text{Energy}$$

Glucose   Oxygen   Carbon dioxide   Water

The electrons are moved to a lower energy level, and energy is released. In other words, the oxidation of glucose is an exergonic process.

Conversely, during photosynthesis, electrons and hydrogen atoms are transferred from water to carbon dioxide, thereby oxidizing the water to oxygen and reducing the carbon dioxide to form three-carbon sugars:

$$6CO_2 + 6H_2O + \text{Energy} \longrightarrow 2C_3H_6O_3 + 6O_2$$

Carbon dioxide   Water   Three-carbon sugar   Oxygen

In this case, the electrons are moved to a higher energy level, and an input of energy is required for the reaction to take place. The reduction of carbon dioxide to sugar, in other words, is an endergonic process.

In living systems, the energy-capturing reactions (photosynthesis) and the energy-releasing reactions (glycolysis and respiration) are oxidation-reduction reactions. The complete oxidation of a mole of glucose releases 686 kilocalories of energy (that is, $\Delta G = -686$ kcal/mole). Conversely, the reduction of carbon dioxide to form the equivalent of a mole of glucose stores 686 kilocalories of energy in the chemical bonds of glucose.

If the energy released during the oxidation of glucose were to be released all at once, most of it would be dissipated as heat. Not only would it be of no use to the cell, but the resulting high temperature would destroy the cell. Mechanisms have evolved in living systems, however, that regulate these chemical reactions—and a multitude of others—in such a way that energy is stored in particular chemical bonds from which it can be released in small amounts as the cell needs it. These mechanisms, which require only a few kinds of molecules, enable cells to use energy efficiently, without disrupting the delicate balances that characterize a living system. To understand how they work, we must look more closely at the proteins known as enzymes and at the molecule known as ATP.

Enzymes

Most chemical reactions require an initial input of energy to get started. This is true even for exergonic reactions such as the oxidation of glucose or the burning of natural gas in a home furnace. The added energy increases the kinetic energy of the molecules, enabling them to collide with enough force (1) to overcome the repulsion between the electrons surrounding one molecule and those surrounding the other molecule and (2) to break existing chemical bonds within the molecules, making possible the formation of new bonds. The energy that must be possessed by the molecules in order to react is known as the **energy of activation** (Figure 5–6).

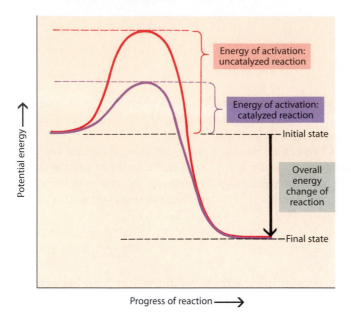

5–6

In order to react, molecules must possess enough energy—the energy of activation— to collide with sufficient force to overcome their mutual repulsion and to break existing chemical bonds. An uncatalyzed reaction requires more activation energy than a catalyzed one, such as an enzymatic reaction. The lower activation energy in the presence of the catalyst is often within the range of energy possessed by the molecules within living cells, and so the reaction can occur at a rapid rate with little or no added energy. Note, however, that the overall energy change from the initial state to the final state is the same with and without the catalyst.

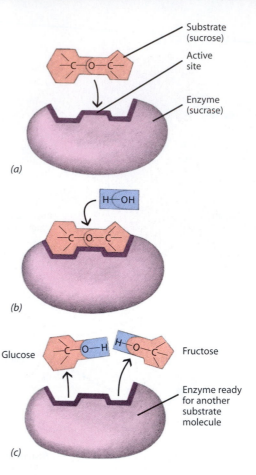

5–7

*A model of enzyme action. **(a)**, **(b)** Sucrose, a disaccharide, is hydrolyzed to yield a molecule of glucose and a molecule of fructose **(c)**. The enzyme involved in this reaction, sucrase, is specific for this process. As you can see, the active site of the enzyme fits the opposing surface of the sucrose molecule. The fit is so exact that a molecule composed, for example, of two subunits of glucose would not be affected by this enzyme.*

In the laboratory, the energy of activation is usually supplied as heat. In a cell, however, many different reactions are going on at the same time, and heat would affect all of these reactions indiscriminately. Moreover, heat would break the hydrogen bonds that maintain the structure of many of the molecules within the cell and would have other generally destructive effects. Cells avoid this problem by the use of **enzymes**, protein molecules that are specialized to serve as catalysts.

A **catalyst** is a substance that lowers the activation energy required for a reaction by forming a temporary association with the molecules that are reacting. This temporary association brings the reacting molecules close to one another and may also weaken existing chemical bonds, making it easier for new ones to form. As a result, little, if any, added energy is needed to start the reaction, and it goes more rapidly than it would in the absence of the catalyst. The catalyst itself is not permanently altered in the process, and so it can be used over and over again. (In Chinese, the word for "catalyst"

is the same as the word for "marriage broker," and the functions are indeed analogous.)

Because of enzymes, cells are able to carry out chemical reactions at great speed and at comparatively low temperatures. A single enzyme molecule may catalyze the reaction of tens of thousands of identical molecules in a second. Thus enzymes are typically effective in very small amounts.

Almost 2000 different enzymes are now known, each of them capable of catalyzing a specific chemical reaction. No cell, however, contains all the known enzymes—different types of cells synthesize different types of enzymes. The particular enzymes that a cell manufactures are a major factor in determining the biological activities and functions of that cell. A cell can carry out a given chemical reaction at a reasonable rate only if it has the enzyme that can catalyze that reaction.

The molecule on which an enzyme acts is known as its **substrate**. For example, in the reaction in Figure 5–7, sucrose is the substrate and sucrase is the enzyme.

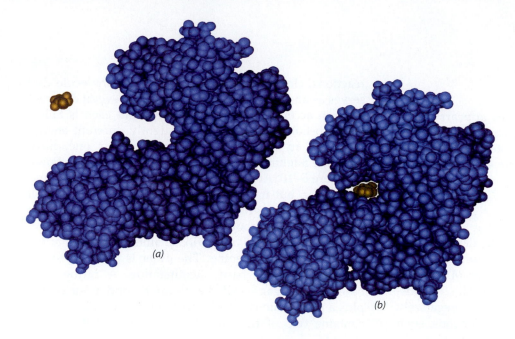

5–8
Space-filling models of the yeast enzyme hexokinase (blue) and one of its substrates, glucose (brown). Hexokinase catalyzes the first step in the breakdown of glucose in respiration (see page 111). Such models, which are produced by computer techniques, show the three-dimensional shape of the molecules. Here the glucose molecule is shown colliding with the enzyme and binding to the active site, which appears as a cleft in the side of the hexokinase molecule. (a) In the absence of glucose, hexokinase has an open cleft. (b) When glucose is bound by hexokinase, the cleft is partially closed.

An Enzyme Has an Active Site That Binds a Specific Substrate

A few enzymes are RNA molecules known as ribozymes. All other enzymes, however, are large, complex globular proteins consisting of one or more polypeptide chains (page 30). The polypeptide chains of an enzyme are folded in such a way that they form a groove or pocket on the surface (Figure 5–8). The substrate fits very precisely into this groove, which is the site of the reactions catalyzed by the enzyme. This portion of the enzyme is known as the **active site**.

The active site not only has a precise three-dimensional shape but also has exactly the correct array of charged and uncharged, or hydrophilic and hydrophobic, areas on its binding surface. If a particular portion of the substrate has a negative charge, the corresponding feature on the active site has a positive charge, and so on. Thus, the active site not only confines the substrate molecule but also orients it in the correct manner.

The amino acids involved in the active site need not be adjacent to one another on the polypeptide chains. In fact, in an enzyme with quaternary structure (page 30), these amino acids may even be on different polypeptide chains. The amino acids are brought together at the active site by the precise folding of the polypeptide chains in the molecule.

The Induced-Fit Hypothesis States That a Substrate May Cause the Active Site to Change Shape

The binding that takes place between enzyme and substrate alters the conformation of the enzyme, thus inducing an even closer fit between the active site and the substrates (Figure 5–8). This *induced fit* may place some strain on the reacting molecules and so further facilitate the reaction (Figure 5–9).

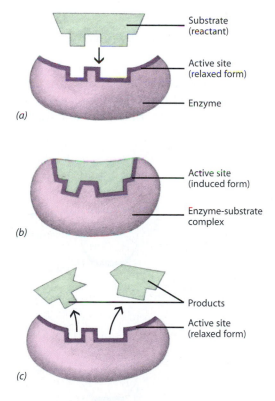

5–9
The induced-fit hypothesis. (a) The active site is believed to be flexible and (b) to adjust its shape to that of the substrate. This induces a close fit between the active site and the substrate and may also put some strain on the substrate molecule, facilitating the reaction (c). Note the induced fit exhibited by hexokinase in Figure 5–8.

Cofactors in Enzyme Action

The catalytic activity of some enzymes appears to depend only on their structure as proteins. Other enzymes, however, require one or more nonprotein components, known as **cofactors**, without which the enzymes cannot function.

Some Cofactors Are Metal Ions

Certain metal ions are cofactors for particular enzymes. For example, magnesium ion (Mg^{2+}) is required in most enzymatic reactions involving the transfer of a phosphate group from one molecule to another. The two positive charges on the magnesium ion hold the negatively charged phosphate group in position. Other ions, such as Ca^{2+} and K^+, play similar roles in other reactions. In some cases, ions serve to hold the enzyme protein in its proper three-dimensional shape.

Other Cofactors Are Organic Molecules Called Coenzymes

Nonprotein organic cofactors also play a crucial role in enzyme-catalyzed reactions. Such cofactors are called **coenzymes**. For example, in some oxidation-reduction reactions, electrons are passed along to a molecule that serves as an electron acceptor. There are several different electron acceptors in any given cell, and each is tailor-made to hold the electron at a slightly different energy level. As an example, let us look at just one, nicotinamide adenine dinucleotide (NAD^+), which is shown in Figure 5–10. At first glance, NAD^+ looks complex and unfamiliar, but if you look at it more closely, you will find that you recognize most of its component parts. The two ribose (five-carbon sugar) units are linked by a *pyrophosphate bridge*. One of the ribose units is attached to the nitrogenous base adenine. The other is attached to another nitrogenous base, nicotinamide. A nitrogenous base plus a sugar is called a *nucleoside,* and a nucleoside plus a phosphate is called a **nucleotide**. A molecule that contains two of these combinations is called a dinucleotide.

The nicotinamide ring is the active end of NAD^+—that is, the part that accepts electrons. Nicotinamide is a vitamin, which is known as niacin. Vitamins are organic compounds that are required in small quantities by many living organisms. Whereas plants can synthesize all of their required vitamins, humans and other animals cannot synthesize most vitamins and so must obtain them in their diets. Many vitamins are precursors of coenzymes or parts of coenzymes.

5–10

Nicotinamide adenine dinucleotide, an electron acceptor, in its oxidized form, NAD+, and in its reduced form, NADH. Nicotinamide is a derivative of nicotinic acid, one of the essential B vitamins. Notice how the bonding within the nicotinamide ring (shaded rectangle) shifts as the molecule changes from the oxidized to the reduced form, and vice versa. Reduction of NAD+ to NADH requires two electrons and one hydrogen ion (H+). The two electrons, however, generally travel as components of two hydrogen atoms; thus, there is one hydrogen ion "left over" when NAD+ is reduced.

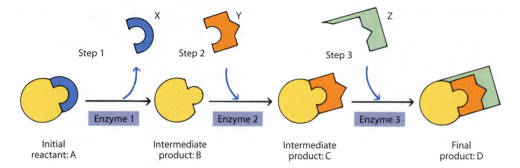

5–11

A schematic representation of a metabolic pathway. In order to produce the final product (D) from the initial reactant (A), a series of reactions is required. Each reaction is cat- alyzed by a different enzyme, and each results in a small but significant modification in the substrate molecule. If any step of the pathway is inhibited—either because of a nonfunc- tioning enzyme or because a substrate is unavailable—the pathway will shut down, and the subsequent reactions of the series will not occur.

When nicotinamide is present, our cells can use it to make NAD^+, which, like many other coenzymes, is recycled. That is, NAD^+ is regenerated when $NADH + H^+$ passes its electrons to another electron acceptor. Thus, although this coenzyme is involved in many cellular reactions, the actual number of NAD^+ molecules required is relatively small.

Some enzymes use nonprotein cofactors that remain attached to the enzyme protein. Such tightly bound cofactors (either ions or coenzymes) are referred to as *prosthetic groups*. Examples of prosthetic groups are the iron-sulfur clusters of ferredoxins (see page 138) and the pyridoxal phosphate (vitamin B_6) of some transaminases.

Metabolic Pathways

Enzymes characteristically work in series, like workers on an assembly line. Each enzyme catalyzes one small step in an ordered series of reactions that together form a **metabolic**, or **biochemical, pathway** (Figure 5–11). Different metabolic pathways serve different functions in the life of the cell. For example, one pathway may be involved in the breakdown of the polysaccharides in bacterial cell walls, another in the breakdown of glucose, and another in the synthesis of a particular amino acid.

Cells derive several advantages from this sort of arrangement. First, the groups of enzymes making up a common pathway can be segregated within the cell. Some are found in solution, as in the vacuoles, whereas others are embedded in the membranes of specialized organelles, such as mitochondria and chloroplasts. The enzymes located in membranes appear to be lined up in sequence, so the product of one reaction moves directly to the adjacent enzyme for the next reaction of the series. A second advantage is that there is little accumulation of intermediate products, since each product tends to be used up in the next reaction along the pathway. A third advantage is a thermodynamic one. If any of the reactions along the pathway are highly exergonic (that is, energy-releasing), they will rapidly use up the products of the preceding reactions, pulling those reactions forward. Similarly, the accumulation of products from the exergonic reactions will push the subsequent reactions forward by increasing the concentrations of the reactants.

Some reactions are common to two or more pathways in a cell. Frequently, however, identical reactions that occur in different pathways are catalyzed by different enzymes. Such enzymes are referred to as **isozymes**. Commonly, each isozyme is coded for by a different set of genes. Isozymes are adapted to the specific pathway and cellular location where they are used.

Regulation of Enzyme Activity

Another remarkable feature of metabolism is the extent to which each cell regulates the synthesis of the products necessary for its well-being, making them in the appropriate amounts and at the rates required. At the same time, cells avoid overproduction, which would waste both energy and raw materials. The availability of reactant molecules or of cofactors is a principal factor in limiting enzyme action, and for this reason most enzymes probably work at a rate well below their maximum.

Temperature affects enzymatic reactions. An increase in temperature increases the rate of enzyme-catalyzed

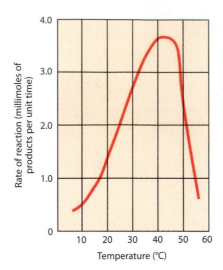

5–12

The effect of temperature on the rate of an enzyme-catalyzed reaction. The concentrations of enzyme and reacting molecules (substrate) were kept constant. The rate of the reaction, as in most metabolic reactions, approximately doubles for every 10°C rise in temperature up to about 40°C. Above this temperature, the rate decreases as temperature increases, and at about 60°C the reaction stops altogether, presumably because the enzyme is denatured.

reactions, but only to a point. As can be seen in Figure 5–12, the rate of most enzymatic reactions approximately doubles for each 10°C rise in temperature in the range 10° to 40°C but then drops off very quickly after about 40°C. The increase in reaction rate occurs because of the increased energy of the reactants. The decrease in the reaction rate above about 40°C occurs as the structure of the enzyme molecule itself begins to unfold, as heat disrupts the relatively weak forces that hold it in its specifically active shape. This unfolding of the enzyme molecule is known as *denaturation* (page 30).

The pH of the surrounding solution also affects enzyme activity. Among other factors, the conformation of an enzyme depends on the attraction and repulsion between negatively charged (acidic) and positively charged (basic) amino acids. As the pH changes, these charges change, and so the shape of the enzyme changes until it is so drastically altered that the enzyme is no longer functional. More important, probably, the charges of the active site and substrate are changed so that the binding capacity is affected. Some enzymes are frequently found at a pH that is not their optimum, suggesting that this discrepancy may not be an evolutionary oversight but a way of controlling enzyme activity.

Several enzymes of the photosynthetic pathway are regulated by changes in pH.

Living systems also have more precise ways of turning enzyme activity on and off. In each metabolic pathway there is at least one enzyme that sets the rate of the overall enzyme sequence because it catalyzes the slowest, or rate-limiting, reaction. These **regulatory enzymes** exhibit increased or decreased catalytic activity in response to certain signals. Such regulatory enzymes constantly adjust the rate of each enzyme sequence to meet changes in the cell's demands for energy and for molecules required for growth and repair of the cell. In most multienzyme systems the first enzyme of the sequence is a regulatory enzyme. By regulating a metabolic pathway at the beginning of the pathway, minimal energy is spent and metabolites are diverted for more important processes.

The most important kind of regulatory enzymes in metabolic pathways are **allosteric enzymes**. The term allosteric derives from the Greek *allos*, "other," and *stereos*, "shape" or "form." Allosteric enzymes have at least two sites: an **active site** that binds the substrate, and an **effector site** that binds the regulatory substance. When the regulatory substance is bound at the effector site, the enzyme is reversibly changed from one form to the other and the enzyme is said to be allosterically regulated.

In some multienzyme systems, the regulatory enzyme is specifically inhibited by the end product of the pathway, which accumulates when the amount of end product exceeds the cell's needs. By inhibiting the regulatory enzyme, all subsequent enzymes operate at reduced rates because their substrates become depleted. In this way the pathway's end product is brought into balance with the cell's needs. This type of regulation is called **feedback inhibition** (Figure 5–13). When more of this end product is needed by the cell, molecules of the end product will dissociate from the effector site, and activity of the regulatory enzyme will increase again.

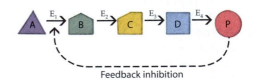

Feedback inhibition

5–13

Feedback inhibition of a metabolic pathway catalyzed by a sequence of four enzymes (E_1 through E_4). Such inhibition typically involves allosteric inhibition of the first enzyme (E_1) in the sequence by the end product (P) of the pathway. Therefore, enzyme E_1 will be more active when amounts of P are low.

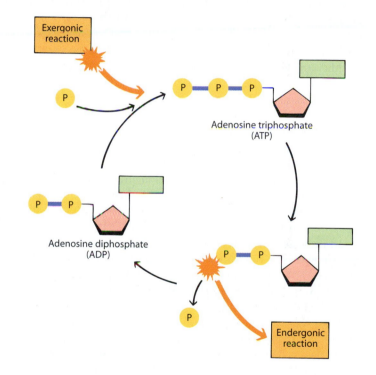

5–14

The structure of adenosine triphosphate (ATP), adenosine diphosphate (ADP), and adenosine monophosphate (AMP). A phosphoester bond links the first phosphate group to the ribose of adenosine, whereas phosphoanhydride bonds, designated by the blue bonds, link the second and third phosphate groups to the molecule. At pH 7, the phosphate groups are fully ionized.

Phosphates

Adenine

Ribose

Adenosine

AMP

ADP

ATP

The Energy Factor: ATP Revisited

All of the biosynthetic activities of the cell (and many other activities as well) require energy. A large proportion of this energy is supplied by **ATP**, the nucleotide derivative that is the cell's chief energy currency.

As discussed in Chapter 2, ATP is made up of adenine, the five-carbon sugar ribose, and three phosphate groups (Figure 5–14). These three phosphate groups, with strong negative charges, are linked to each other by phosphoanhydride bonds and, in turn, are linked to the ribose by a phosphoester bond.

To understand the role of ATP, let us briefly review the concept of the chemical bond and bond energies. Because a chemical bond is a stable configuration of electrons, reacting molecules must collide with a certain amount of energy in order to separate the bonded atoms, thus allowing the formation of new bonds. This energy is the energy of activation. Because enzymes reduce the required energy of activation to a level already possessed by a significant proportion of the reacting molecules (Figure 5–6), the reactions essential to life can proceed at an appropriate rate.

As we have seen, however, the *direction* in which a reaction proceeds is determined by the free-energy change ΔG. Only if the reaction is exergonic (negative ΔG) will it proceed at all. Yet many cellular reactions, including biosynthetic reactions—such as the formation of a disaccharide from two monosaccharide molecules—are endergonic (positive ΔG). In such reactions, the electrons forming the chemical bonds of the product are at a higher energy level than the electrons in the bonds of the starting materials. That is, the potential energy of the product is greater than the potential energy of the reactants, an apparent violation of the second law of thermodynamics. Cells circumvent this difficulty by **coupled reactions** in which endergonic reactions are linked to and driven by exergonic reactions that provide a surplus of energy. The result is that the net process is exergonic and thus able to proceed spontaneously (Figure 5–15). ATP is the molecule that most frequently serves as the intermediate between the exergonic and the otherwise endergonic reaction in such coupled reactions.

Exergonic reaction

Adenosine triphosphate (ATP)

Adenosine diphosphate (ADP)

Endergonic reaction

5–15

In living systems, endergonic reactions, such as biosynthetic reactions, are powered by the energy released in exergonic reactions to which they are coupled. In most coupled reactions, ATP is the intermediate that carries energy from one reaction to the other.

Enzymes catalyzing the hydrolysis of ATP are known as **ATPases**. A variety of different ATPases have been identified. The protein "arms" on the microtubules in flagella (see Figure 3–29), for example, are ATPase molecules, catalyzing the energy release that causes the microtubules to move past one another. Many of the proteins that move molecules and ions through cellular membranes against concentration gradients are not only transport proteins but also ATPases, releasing energy to power the transport process (page 83).

Because of its structure, the ATP molecule is well suited to this role in living systems. Energy is released from the ATP molecule when one phosphate group is removed by hydrolysis, producing a molecule of ADP (adenosine diphosphate) and a free phosphate ion:

$$ATP + H_2O \longrightarrow ADP + Phosphate + Energy$$

In the course of this reaction, about 7.3 kilocalories of energy are released per mole of ATP hydrolyzed. Removal of a second phosphate group produces AMP (adenosine monophosphate) and releases an equivalent amount of chemical energy:

$$ADP + H_2O \longrightarrow AMP + Phosphate + Energy$$

The covalent bonds—that is, the phosphoanhydride bonds—linking these two phosphates to the rest of the molecule are relatively weak bonds, and so are easily broken by hydrolysis. The new bonds formed as a result of this hydrolysis are much stronger, and so energy is released in this reaction. The fragility of the terminal anhydride bond in ATP is due to the negative charges on the three phosphates (Figure 5–14). The resulting mutual repulsions significantly weaken the bond, which is why relatively little energy is required to break the bond. Hydrolysis leaves the molecule with only two negative charges adjacent—a more stable arrangement. Hence, ADP is more stable than ATP. For the same reason, AMP is more stable still, and cellular reactions requiring an extra boost of energy will cleave both phosphates.

In most reactions within a cell, the terminal phosphate group of ATP is not simply removed but is transferred to another molecule. This addition of a phosphate group to a molecule is known as **phosphorylation**; the enzymes that catalyze such transfers are known as **kinases**. The addition of a negatively charged phosphate group destabilizes the phosphorylated compound, which, thus energized, can participate in some other metabolic reaction.

Let us look at a simple example of energy exchange involving ATP in the formation of sucrose in sugarcane. Sucrose is formed from the monosaccharides glucose and fructose. Under standard thermodynamic conditions, sucrose synthesis is strongly endergonic, requiring an input of 5.5 kilocalories for each mole of sucrose formed:

$$Glucose + Fructose + Energy \longrightarrow Sucrose + H_2O$$

However, when coupled with the breakdown of ATP, the synthesis of the sucrose is actually exergonic. During the series of reactions involved in the formation of sucrose (see Figure 2–3), two molecules of ATP are used to phosphorylate the glucose and fructose, thus energizing each of them:

$$ATP + Glucose \longrightarrow Glucose\ phosphate + ADP$$

$$ATP + Fructose \longrightarrow Fructose\ phosphate + ADP$$

They are then linked by hydrolyzing these phosphates.

The overall equation for sucrose formation from the phosphorylated monosaccharides is:

$$Glucose\ phosphate + Fructose\ phosphate \longrightarrow$$
$$Sucrose + 2\ Phosphate$$

The cell spends a total of 2×7.3 kilocalories = 14.6 kilocalories of ATP energy, and uses 5.5 kilocalories of it to form a mole of sucrose. The other 9.1 kilocalories is used to drive the reaction forward irreversibly and is ultimately released as heat. Thus, the sugarcane plant is able to form sucrose by coupling the breakdown of two molecules of ATP to the synthesis of a covalent bond between glucose and fructose.

Where does the ATP originate? As we shall see in the next chapter, energy released by the exergonic breakdown of molecules such as glucose is used to "recharge" the ADP molecule to ATP. Of course, the energy released in these reactions is originally derived from the sun as radiant energy that is converted, during photosynthesis, into chemical energy. Some of this chemical energy is stored in the ATP molecule before being converted to chemical bond energies of other organic molecules. Thus the ATP/ADP system serves as a universal energy-exchange system, shuttling between energy-releasing and energy-requiring reactions.

Summary

Life on Earth Is Dependent on the Flow of Energy from the Sun

A small fraction of this energy, captured in the process of photosynthesis, is converted to the energy that drives the many other metabolic reactions associated with living systems and from which living systems derive their order and organization.

In photosynthesis, the energy of the sun is used to forge high-energy carbon-carbon and carbon-hydrogen bonds of organic compounds. In respiration, these bonds are subsequently broken to release carbon dioxide and water from the compounds in which the bonds are present, and energy is released. Some of this energy is used to drive cellular processes, but as in machines, some energy is lost as heat in each energy-conversion step.

Living Systems Operate According to the Laws of Thermodynamics

The first law of thermodynamics states that energy can be converted from one form to another but cannot be created or destroyed. The potential energy of the initial state (or reactants) is equal to the potential energy of the final state (or products) plus the energy released in the process or reaction. The second law of thermodynamics states that, in the course of energy conversions, if no energy enters or leaves a system, the potential energy of the final state will always be less than the potential energy of the initial state. Stated differently, all natural processes tend to proceed in such a direction that the disorder, or randomness, of the universe increases. This disorder, or randomness, is known as entropy. To maintain the organization on which life depends, living systems must have a constant supply of energy to overcome the tendency toward increasing disorder.

Oxidation-Reduction Reactions Play an Important Role in the Flow of Energy

Energy transformations in cells involve the transfer of electrons from one energy level to another and, often, from one atom or molecule to another. Reactions involving the transfer of electrons from one molecule to another are known as oxidation-reduction reactions. An atom or molecule that loses electrons is oxidized; one that gains electrons is reduced. Oxidation (the loss of electrons) produces a product with less potential energy. Reduction (the addition of electrons) produces a product with a greater amount of potential energy.

Enzymes Enable Chemical Reactions to Take Place at Temperatures Compatible with Life

Enzymes are the catalysts of biological reactions, lowering the energy of activation and thus enormously increasing the rate at which reactions take place. With few exceptions, enzymes are large globular protein molecules folded in such a way that a particular group of amino acids forms an active site. The reacting molecules, known as the substrates, fit precisely into this active site. Although the shape of an enzyme may change temporarily in the course of a reaction, it is not permanently altered.

Many enzymes require cofactors, which may be metal ions or nonprotein organic molecules known as coenzymes. Coenzymes often serve as electron carriers, with different coenzymes holding electrons at slightly different energy levels.

Enzyme-catalyzed reactions take place in ordered series of steps called metabolic pathways. Each step in a pathway is catalyzed by a particular enzyme. The stepwise reactions of metabolic pathways enable cells to carry out their chemical activities with remarkable efficiency in terms of both energy and materials. Each metabolic pathway is under the tight control of one or more regulatory enzymes.

ATP Supplies the Energy for Most of the Activities of the Cell

The ATP molecule consists of the nitrogenous base adenine, the five-carbon sugar ribose, and three phosphate groups. The three phosphate groups are linked by two covalent bonds that are easily broken, each yielding some 7.3 kcal/mole. Cells are able to carry out endergonic (energy-requiring) reactions and processes by coupling them with exergonic (energy-yielding) reactions that provide a surplus of energy. Such coupled reactions usually involve ATP as the common intermediate.

Selected Key Terms

active site   p. 101

allosteric enzymes   p. 104

ATP   p. 105

ATPases   p. 105

catalyst   p. 100

coenzymes   p. 102

cofactors   p. 102

coupled reactions   p. 105

effector site   p. 104

endergonic   p. 96

energy   p. 95

energy of activation p. 99

entropy   p. 97

enzymes   p. 100

exergonic   p. 96

feedback inhibition p. 104

free-energy change (ΔG) p. 97

induced-fit hypothesis p. 101

isozymes   p. 103

metabolic pathway p. 103

oxidation-reduction reaction   p. 98

phosphorylation   p. 106

regulatory enzymes p. 104

substrate   p. 100

Questions

1. Distinguish between the following: active site/substrate; AMP/ADP/ATP; ATPases/kinases.

2. At least four types of energy conversions take place in photosynthetic cells. Name them.

3. Enzymes characteristically work in series, called metabolic pathways, like workers on an assembly line. What are some advantages gained by the cell from this sort of arrangement?

4. The laws of thermodynamics apply only to closed systems, that is, to systems in which no energy is entering and leaving. Is an aquarium a closed system? If not, could you convert it to one? A space station may or may not be a closed system, depending on certain features of its design. What would these features be? Is the Earth a closed system? How about the universe?

5. The most interesting implication of the second law of thermodynamics, as far as living systems are concerned, is the relationship between entropy and order. Explain.

6 | Respiration

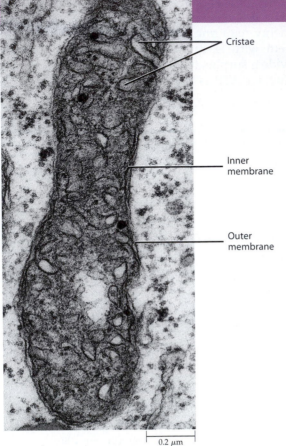

Cristae

Inner
membrane

Outer
membrane

0.2 μm

6–1

*Mitochondrion in a leaf cell of a fern
(Regnellidium diphyllum). Mitochondria are
the sites of respiration, the process by which
chemical energy is transferred from carbon-
containing compounds to ATP. Most of the
ATP is produced on the surfaces of the cristae
by enzymes embedded in these membranes.*

OVERVIEW

*Just as we discussed the flow of energy in the previous chapter,
now we focus on the flow of carbon—although in reality the two
cannot be separated. In this chapter, the emphasis is on the breakdown of molecules to obtain energy for the cell. Both plants and
animals use food molecules as building blocks to construct and
repair cells and as sources of energy. The main difference is that
plants can manufacture their own food molecules by photosynthesis, whereas animals must obtain them in their diets. Once
inside the plant or animal cell, these molecules, principally glucose, are degraded in the same step-by-step, assembly-line manner, and much of the energy originally present in the food
molecules is converted to ATP—a more versatile form of energy.*

*You will see that the assembly line is divided into four dis-
tinct stages. In glycolysis, the glucose molecule is first made
reactive by "energizing" it with ATP, and then the energized
glucose is partially degraded to a three-carbon product. The next
stage is the Krebs cycle, in which the remnants of the glucose
molecule are completely degraded to carbon dioxide and water.
Two more stages—the electron transport chain and oxidative
phosphorylation—are of critical importance. It is during these
stages that the greatest amount of ATP is formed, at least when
oxygen is present. The chapter ends with a discussion of the
events that occur when oxygen is absent and when different
types of food molecules are used.*

CHECKPOINTS

By the time you finish reading this chapter, you should be able to answer the following questions:

1. What is the overall reaction, or equation, for respiration, and what is the main function of this process?

2. What are the main events that occur during glycolysis?

3. Where in the cell does the Krebs cycle occur, and what are the products formed?

4. How does the flow of electrons in the electron transport chain result in the formation of ATP?

5. When a cell metabolizes a glucose molecule, how does the net energy yield under aerobic conditions differ from that obtained under anaerobic conditions? How do you explain the difference?

ATP is the universal energy currency in living systems. It takes part in a great variety of cellular events, from the biosynthesis of organic molecules to the flick of a flagellum, the streaming of cytoplasm, or the active transport of a molecule across the plasma membrane (Figure 6–1). In the following pages, we describe how a cell oxidizes carbohydrates and captures a portion of the released energy in the phosphoanhydride bonds of ATP. This process provides an excellent illustration both of the chemical principles described in the previous chapter and of the way in which cells carry out biochemical processes.

An Overview of Glucose Oxidation

As mentioned in Chapter 2, energy-yielding carbohydrate molecules are generally stored in plants as sucrose or starch. A necessary preliminary step to **respiration**—the complete breakdown of sugars or other organic molecules to carbon dioxide and water—is the hydrolysis of these storage molecules to monosaccharides. Respiration itself is generally considered to begin with glucose, an end product of the hydrolysis of both sucrose and starch.

The oxidation of glucose (and other carbohydrates) is complicated in detail but simple in its overall design. **Oxidation,** as we saw in the last chapter, is the loss of electrons. **Reduction** is the gain of electrons. In the oxidation of glucose, the glucose molecule is split apart, and the hydrogen atoms (that is, electrons and their accompanying protons) are removed from the carbon atoms and combined with oxygen, which is thereby reduced. The electrons go from higher energy levels to lower energy levels, and energy is released.

Glucose can be used as a source of energy under both aerobic (that is, in the presence of oxygen) and anaerobic (in the absence of oxygen) conditions. However, maximum energy yields for oxidizable organic compounds are achieved only under aerobic conditions. Consider, for example, the overall reaction for the complete oxidation of glucose:

$$C_6H_{12}O_6 + 6O_2 \longrightarrow 6CO_2 + 6H_2O + \text{Energy}$$
$$\text{Glucose} \quad \text{Oxygen} \quad \text{Carbon} \quad \text{Water}$$
$$\text{dioxide}$$

With oxygen as the ultimate electron acceptor, this reaction is highly exergonic (energy-yielding), with a ΔG of -686 kcal/mole. This reaction represents the process of respiration. When energy is extracted from organic compounds without the involvement of oxygen, the process is called **fermentation,** which is discussed later in this chapter.

Respiration involves four distinct stages: glycolysis, the Krebs cycle, the electron transport chain, and oxidative phosphorylation (Figure 6–2). In **glycolysis,** the six-carbon glucose molecule is broken down to a pair of three-carbon molecules of pyruvic acid or **pyruvate.** (Pyruvic acid dissociates, producing pyruvate and a hydrogen ion. Pyruvic acid and pyruvate exist in equilibrium, and the two terms are frequently used interchangeably.) In the **Krebs cycle,** the pyruvate molecules are further broken down to carbon dioxide, and the resulting electrons are passed to the **electron transport chain.** In **oxidative phosphorylation,** the energy that is released as electrons move through the electron transport chain is used to form ATP from ADP and phosphate.

As the glucose molecule is oxidized, some of its energy is extracted in a series of small, discrete steps and is stored in the phosphoanhydride bonds of ATP. In accord with the second law of thermodynamics, however, most of its energy is dissipated as heat energy.

Glycolysis

As mentioned previously, in glycolysis (from *glyco,* meaning "sugar," and *lysis,* meaning "splitting"), the six-carbon glucose molecule is split into two molecules of pyruvate (Figure 6–3). Glycolysis occurs in a series of ten steps, each catalyzed by a specific enzyme. This series of reactions is carried out by virtually all living cells, from bacteria to the eukaryotic cells of plants and animals. Glycolysis is an anaerobic process that occurs in the cytosol. Biologically, glycolysis may be considered a primitive process, in that it most likely arose before the appearance of atmospheric oxygen and before the origin of cellular organelles.

The glycolytic pathway is shown in detail in Figure 6–4. It illustrates the principle that biochemical processes of a living cell proceed in small sequential steps, each catalyzed by a specific enzyme. As the series of reactions is discussed, notice how the carbon skeleton of the molecule is disassembled as its atoms are rearranged step by step. It is not intended that you memorize these steps; simply follow them closely. Note especially the formation of ATP from ADP and phosphate and the formation of NADH from NAD$^+$ (see Figure 5–10). **ATP** and **NADH** represent the cell's net energy harvest from glycolysis. In Chapter 7, we will see that reactions 4 through 7 also occur in the Calvin cycle, a portion of the photosynthetic process. This repetition illustrates a principle of biochemical evolution: pathways do not arise entirely anew; rather, a few new reactions are added to an existing set to make a "new" pathway.

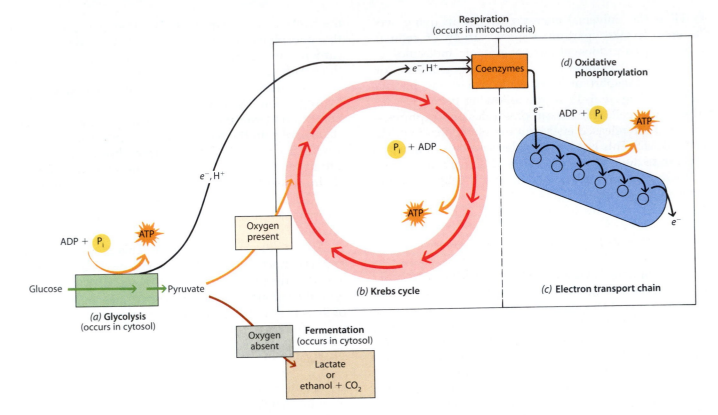

6–2

An overview of the oxidation of glucose. The complete oxidative breakdown of glucose is called respiration. Glycolysis is considered the first of four stages. The Krebs cycle, the electron transport chain, and oxidative phosphorylation are the other three stages. (*a*) In glycolysis, glucose is split into pyruvate. A small amount of ATP is synthesized from ADP and phosphate, and a few electrons (e⁻) and their accompanying protons (H⁺) are transferred to coenzymes that function as electron carriers. (*b*) In the presence of oxygen (aerobic pathway), the pyruvate is fed into the Krebs cycle. In the course of this cycle, additional ATP is synthesized and more electrons and protons are transferred to coenzymes. (*c*) The coenzymes then transfer the electrons to an electron transport chain in which the electrons drop, step by step, to lower energy levels, generating a proton gradient. (*d*) The proton gradient is then used to drive the formation of considerably more ATP. This process is called oxidative phosphorylation. At the end of the electron transport chain, the electrons reunite with protons and combine with oxygen to form water.

In the absence of oxygen (anaerobic pathway), the pyruvate is converted to either lactate or ethanol. This process, known as fermentation, produces no additional ATP, but it does regenerate the coenzymes that are necessary for glycolysis to continue.

6–3

In glycolysis, the six-carbon glucose molecule is split, via a series of ten reactions, into two molecules of a three-carbon compound known as pyruvate. In the course of glycolysis, four hydrogen atoms are removed from the original glucose molecule.

6–4

The steps of glycolysis. Each step is catalyzed by a specific enzyme, as indicated. **P** *= phosphate group.*

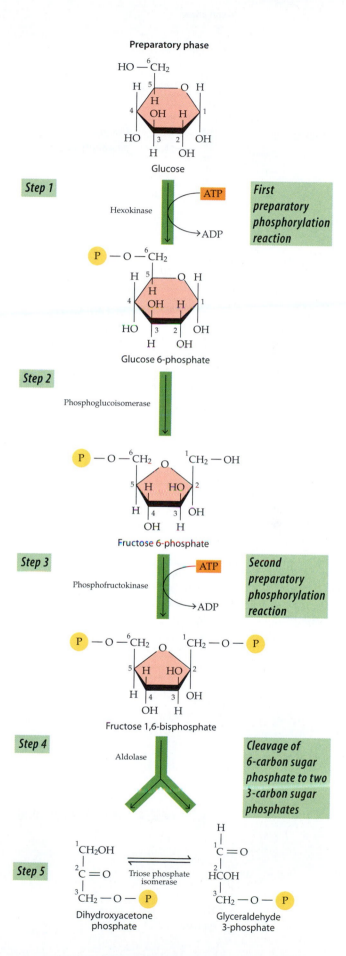

Step 1 The preparatory phase in glycolysis requires an input of energy. The energy is supplied by coupling these steps to the ATP/ADP system. In the first step—the first preparatory reaction—the terminal phosphate group is transferred from an ATP molecule to the carbon in the sixth position of the glucose molecule, to make glucose 6-phosphate. The reaction of ATP with glucose to produce glucose 6-phosphate and ADP is an exergonic reaction. Some of the energy it releases is conserved in the chemical bond linking the phosphate to the glucose molecule, thereby energizing the glucose molecule. This reaction is catalyzed by a specific enzyme (hexokinase; see Figure 5–8).

Step 2 In this step, the molecule is rearranged, again with the help of a specific enzyme (phosphoglucoisomerase). The six-sided ring characteristic of glucose becomes a five-sided fructose ring. (As shown in Figure 2–3, glucose and fructose have the same types and numbers of atoms—$C_6H_{12}O_6$—but differ in the arrangement of these atoms.) This reaction is driven by the accumulation of glucose 6-phosphate from step 1 and by the depletion of fructose 6-phosphate as it enters step 3.

Step 3 In this step—the second preparatory reaction—fructose 6-phosphate gains a second phosphate by investment of another ATP. The added phosphate is bonded to the first carbon, producing fructose 1,6-bisphosphate—that is, fructose with phosphates in the carbon-1 and carbon-6 positions. Note that in the course of the reactions thus far, two molecules of ATP have been converted to two molecules of ADP and no energy has been recovered. The energy yield thus far, in other words, is −2 ATP.

Step 4 This is the cleavage step from which glycolysis derives its name. The six-carbon sugar molecule is split in half, producing two three-carbon molecules, glyceraldehyde 3-phosphate and dihydroxyacetone phosphate.

Step 5 Glyceraldehyde 3-phosphate and dihydroxyacetone phosphate can be interconverted by the enzyme triose phosphate isomerase. However, because the glyceraldehyde 3-phosphate is used up in subsequent reactions, the net result is the conversion of dihydroxyacetone phosphate into glyceraldehyde 3-phosphate. Thus, *the products of subsequent steps must be counted twice to account for the fate of each glucose molecule.* With the completion of step 5, the preparatory phase is complete.

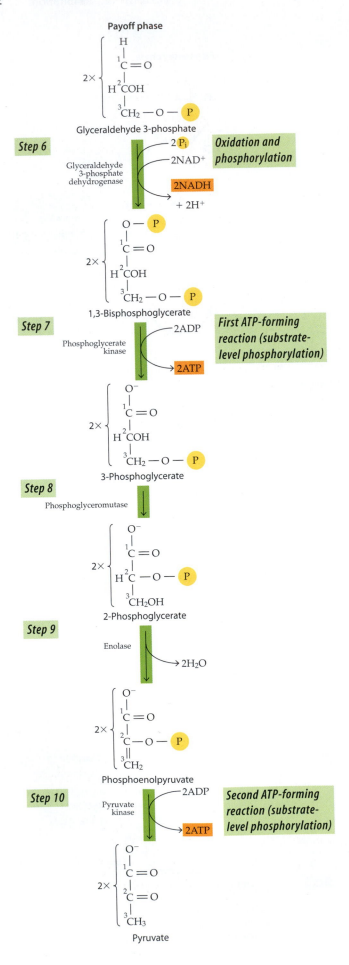

Payoff phase

Glyceraldehyde 3-phosphate

Step 6 — Glyceraldehyde 3-phosphate dehydrogenase — $2 P_i$ / $2NAD^+$ → **2NADH** $+ 2H^+$

Oxidation and phosphorylation

1,3-Bisphosphoglycerate

Step 7 — Phosphoglycerate kinase — $2ADP$ → **2ATP**

First ATP-forming reaction (substrate-level phosphorylation)

3-Phosphoglycerate

Step 8 — Phosphoglyceromutase

2-Phosphoglycerate

Step 9 — Enolase → $2H_2O$

Phosphoenolpyruvate

Step 10 — Pyruvate kinase — $2ADP$ → **2ATP**

Second ATP-forming reaction (substrate-level phosphorylation)

Pyruvate

Step 6 In the first reaction of the payoff phase, glyceraldehyde 3-phosphate molecules are oxidized—that is, hydrogen atoms with their electrons are removed—and NAD^+ is reduced to NADH and H^+ (a total of two molecules of NADH and two H^+ ions per molecule of glucose—one for each of the two molecules of glyceraldehyde 3-phosphate generated per molecule of glucose). This is the first reaction in which the cell harvests energy, stored in the form of the high-energy compound NADH. Some of the energy from this oxidation reaction is also conserved in the attachment of a phosphate group to the carbon-1 position of the molecule, forming 1,3-bisphosphoglycerate. The properties of this new bond are similar to those of the phosphoanhydride bonds of ATP. (The designation P_i indicates inorganic phosphate, which is available as a phosphate ion in solution in the cytosol.)

Step 7 The bond energy of phosphate released from the 1,3-bisphosphoglycerate molecule is used to phosphorylate a molecule of ADP (a total of two molecules of ATP are formed per molecule of glucose). This is a highly exergonic reaction (that is, its ΔG has a large negative value), in effect, and "pulls" all the previous reactions forward. The formation of ATP by the enzymatic transfer of a phosphate group from a metabolic intermediate to ADP is referred to as **substrate-level phosphorylation.**

Step 8 The remaining phosphate group is transferred from the carbon-3 position to the carbon-2 position of the glycerate molecule.

Step 9 In this step, a molecule of water is removed from the three-carbon compound. As a consequence of the rearrangement of electrons and energy in the molecule, a high-energy phosphorylated compound (phosphoenolpyruvate) is formed.

Step 10 The phosphate group of phosphoenolpyruvate is transferred to a molecule of ADP, forming another molecule of ATP. (Again, a total of two molecules of ATP are formed per molecule of glucose by substrate-level phosphorylation.) This is also a highly exergonic reaction, and it pulls forward the preceding two reactions (steps 8 and 9).

Summary of Glycolysis

The complete sequence of glycolysis begins with one molecule of glucose (Figure 6–5). Energy enters the sequence at steps 1 and 3 by the transfer of a phosphate group from an ATP molecule—one at each step—to the sugar molecule. The six-carbon molecule splits at step 4, and after step 5 the pathway yields energy. At step 6, two molecules of NAD^+ are reduced to two molecules of NADH

6–5

A summary of the two phases of glycolysis. The preparatory phase requires an energy investment of 2 ATP. This stage ends with the splitting of the six-carbon sugar molecule into two three-carbon molecules. The payoff phase produces an energy yield of 4 ATP and 2 NADH—a substantial return on the original investment. The net ATP yield is therefore 2 molecules of ATP per molecule of glucose. Carbohydrates other than glucose, including glycogen, starch, various disaccharides, and a number of monosaccharides, can undergo glycolysis once they have been converted to glucose 6-phosphate or fructose 6-phosphate.

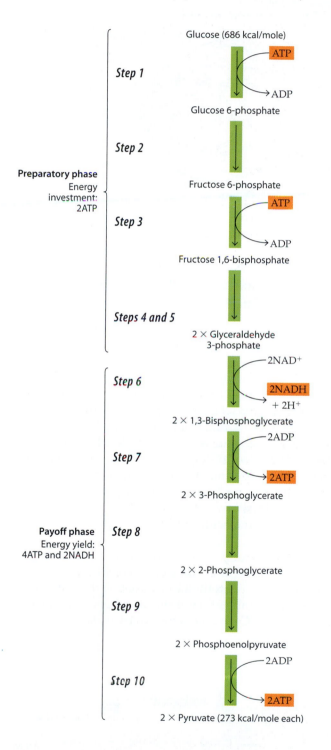

NADH, thereby storing some of the energy from the oxidation of glyceraldehyde 3-phosphate as the high-energy electrons of the reduced coenzyme. At steps 7 and 10, two molecules of ADP take energy from the system, forming additional phosphoanhydride bonds to become two molecules of ATP per molecule of glyceraldehyde 3-phosphate—or four molecules of ATP per molecule of glucose. Two of the four ATP are, in effect, replacements for the two ATP used in steps 1 and 3. The net ATP yield is only two molecules of ATP per molecule of glucose.

Glycolysis (from glucose to pyruvate) can be summarized by the overall equation:

$$\text{Glucose} + 2\text{NAD}^+ + 2\text{ADP} + 2\text{P}_i \longrightarrow$$
$$2\text{ Pyruvate} + 2\text{NADH} + 2\text{H}^+ + 2\text{ATP} + 2\text{H}_2\text{O}$$

Thus, one glucose molecule is converted to two molecules of pyruvate. The *net* harvest—the energy yield—is two molecules of ATP and two molecules of NADH per molecule of glucose. Two moles of pyruvate have a total energy content of about 546 kilocalories, compared with 686 kilocalories stored in a mole of glucose. A large portion (about 80 percent) of the energy stored in the original glucose molecule is therefore still present in the two pyruvate molecules.

Also note that, under aerobic conditions, the two molecules of NADH can yield additional ATP molecules in the mitochondria when used as electron donors to the electron transport chain of the aerobic pathway (see page 116).

The Aerobic Pathway

Pyruvate is a key intermediate in cellular energy metabolism because it can be utilized in one of several pathways. Which pathway it follows depends in part upon the conditions under which metabolism takes place and in part upon the specific organism involved. In some cases, the pathway is determined by the particular tissue in the organism. The principal environmental factor is the availability of oxygen. In the presence of oxygen, pyruvate is oxidized completely to carbon dioxide, and glycolysis is but the initial stage of respiration. This aerobic pathway results in the complete oxidation of glucose and a much greater ATP yield than can be achieved by glycolysis alone. In eukaryotic cells, the reactions of the aerobic pathway take place within mitochondria.

Structure of a mitochondrion. A mito-chondrion is surrounded by two mem-branes, as shown in this three-dimensional diagram. The inner membrane folds *inward, forming the cristae. Many of the enzymes and electron carriers involved in respiration are present within the inner membrane.*

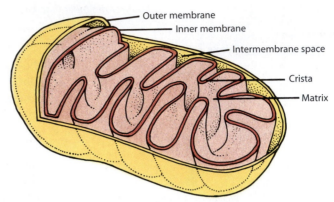

The Structure of the Mitochondrion Provides the Key to Its Function

A mitochondrion, as we saw in Chapter 3, is surrounded by two membranes, the inner one convoluted inwardly into folds called **cristae** (Figures 6–1 and 6–6). Within the inner compartment of the mitochondrion, bathing the cristae, is a liquid **matrix.** The matrix contains water, enzymes, coenzymes, phosphates, and other molecules and ions involved in respiration. All but one of the enzymes of the Krebs cycle are found in solution within the matrix. One Krebs-cycle enzyme and the components of the electron transport chain are present within the inner membrane and are therefore prominent features of the cristae. Thus, a mitochondrion resembles a self-contained chemical factory.

The outer membrane of the mitochondrion is permeable to most small molecules, and the solution in the space between the inner and outer membranes is therefore similar in composition to the cytosol. The inner membrane, however, permits the passage of only certain molecules, such as pyruvate, ADP, and ATP. It restrains the passage of other molecules and ions, including H^+ ions (protons). As we shall see, this selective permeability of the inner membrane is critical to the ability of the mitochondrion to harness the power of respiration for the production of ATP.

A Preliminary Step: Pyruvate Enters the Mitochondrion and Is Both Oxidized and Decarboxylated

Pyruvate passes from the cytosol, where it is produced by glycolysis, to the matrix of the mitochondrion, crossing the outer and inner membranes in the process. Pyruvate, however, is not used directly in the Krebs cycle. Within the mitochondrion, pyruvate is both oxidized and decarboxylated—that is, electrons are removed and CO_2 is split out of the molecule. In the course of this exergonic reaction, a molecule of NADH is produced from NAD^+ (Figure 6–7). The two pyruvate molecules derived from the original glucose molecule have now been oxidized to two acetyl ($—CH_3CO$) groups. In addition, two molecules of CO_2 have been liberated, and two molecules of NADH have been formed from NAD^+.

Each acetyl group is temporarily attached to **coenzyme A (CoA)**—a large molecule, consisting of a nucleotide linked to pantothenic acid, one of the B-complex vitamins. The combination of the acetyl group and CoA, known as **acetyl CoA,** enters the Krebs cycle.

The three-carbon pyruvate molecule is oxidized and decarboxylated to form the two-carbon acetyl group, which is attached to coenzyme A as acetyl CoA. The oxidation *of the pyruvate molecule is coupled to the reduction of NAD^+ to NADH. Acetyl CoA is the form in which carbon atoms derived from glucose enter the Krebs cycle.*

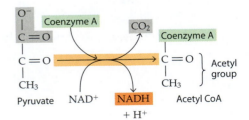

The Krebs Cycle Oxidizes the Acetyl Groups of the Acetyl CoA Molecules

The Krebs cycle is named in honor of Sir Hans Krebs, whose research group was largely responsible for its elucidation. Krebs postulated this metabolic pathway in 1937 and later received a Nobel Prize in recognition of his brilliant work. The Krebs cycle can also be called the citric acid cycle or the tricarboxylic acid (TCA) cycle because it begins with the formation of citric acid or citrate, which has three carboxylic acid groups.

The Krebs cycle always begins with acetyl CoA, its only real substrate. Upon entering the Krebs cycle (Figure 6–8), the two-carbon acetyl group is combined with a four-carbon compound (oxaloacetate) to produce a six-carbon compound (citrate). The coenzyme A is released to combine with a new acetyl group when another molecule of pyruvate is oxidized. In the course of the cycle, two of the six carbons are removed and oxidized to CO_2, and oxaloacetate is regenerated—thus literally making this series of reactions a cycle. Each turn around the cycle uses up one acetyl group and regenerates one molecule of oxaloacetate, which is then ready to begin the Krebs cycle again.

6–8

In the course of the Krebs cycle, two carbons enter as the acetyl group of acetyl CoA and two carbons are oxidized to carbon dioxide; the hydrogen atoms are passed to the coenzymes NAD^+ and FAD. As in glycolysis, a specific enzyme is involved at each step.

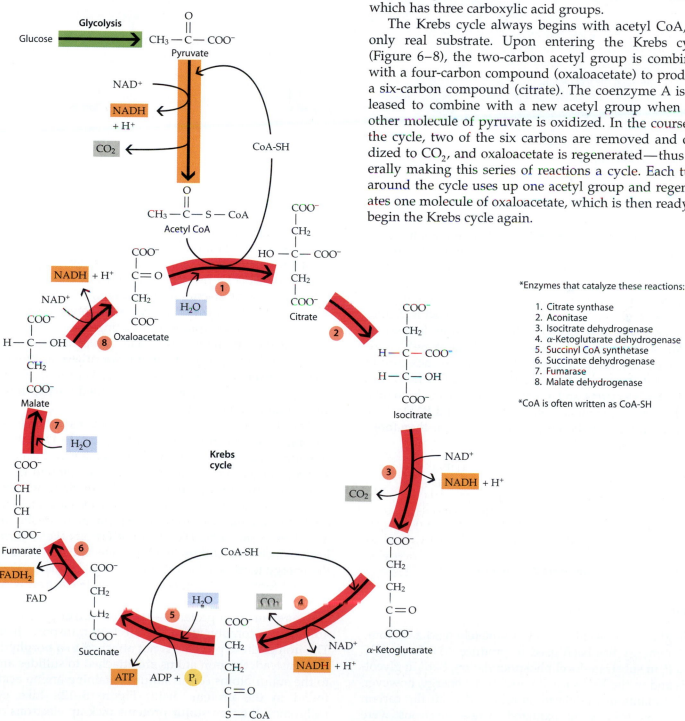

*Enzymes that catalyze these reactions:

1. Citrate synthase
2. Aconitase
3. Isocitrate dehydrogenase
4. α-Ketoglutarate dehydrogenase
5. Succinyl CoA synthetase
6. Succinate dehydrogenase
7. Fumarase
8. Malate dehydrogenase

*CoA is often written as CoA-SH

6–9

The coenzyme flavin adenine dinucleotide in its oxidized form (FAD) and its reduced form (FADH₂). Riboflavin is a vitamin (vitamin B₂) made by all plants and many microorganisms but required in the diet of most animals. It is a pigment and, in its oxidized form, is bright yellow. A related electron acceptor, flavin mononucleotide (FMN), consists of riboflavin with one phosphate group. It accepts electrons from NADH in the electron transport chain.

In the course of these steps, some of the energy released by the oxidation of the carbon atoms is used to convert ADP to ATP (one molecule per cycle; another case of substrate-level phosphorylation), but most is used to reduce NAD^+ to NADH (three molecules per cycle). In addition, some of the energy is used to reduce a second electron carrier—the coenzyme known as flavin adenine dinucleotide (FAD) (Figure 6–9). One molecule of **FADH₂** is formed from FAD in each turn of the cycle. Oxygen is not directly involved in the Krebs cycle; the electrons and protons removed in the oxidation of carbon are all accepted by NAD^+ and FAD.

The overall equation for the Krebs cycle is therefore:

Oxaloacetate + Acetyl CoA + $3H_2O$ + ADP + P_i + $3NAD^+$ + FAD $\longrightarrow$

Oxaloacetate + $2CO_2$ + CoA + ATP + 3NADH + $3H^+$ + FADH₂

In the Electron Transport Chain, Electrons Removed from the Glucose Molecule Are Transferred to Oxygen

The glucose molecule is now completely oxidized. Some of its energy has been used to produce ATP from ADP and P_i in substrate-level phosphorylation, both in glycolysis and in the Krebs cycle. Most of the energy, however, still remains in the electrons removed from the carbon atoms as they were oxidized. These electrons were passed to the electron carriers NAD^+ and FAD and are still at a high energy level in NADH and FADH₂.

In the next stage of respiration, these high-energy electrons of NADH and FADH₂ are passed step-by-step to the low energy level of oxygen. This stepwise passage is made possible by the **electron transport chain** (Figure 6–10), a series of electron carriers, each of which holds the electrons at a slightly lower energy level. Each carrier is capable of accepting or donating one or two electrons. Each component of the chain can accept electrons from the preceding carrier and transfer them to the following carrier in a specific sequence. With one exception, all of these carriers are embedded in the inner mitochondrial membrane.

The electron carriers of the electron transport chain of the mitochondria differ from NAD^+ and FAD in their chemical structures. Some of them belong to a class of electron carriers known as **cytochromes**—protein molecules with an iron-containing porphyrin ring, or heme group, attached (Figure 6–11). Cytochromes pick up electrons on their iron atoms, which can be reversibly reduced from the ferric (Fe^{3+}) to the ferrous (Fe^{2+}) form. Each cytochrome differs in its protein structure and in the energy level at which it holds the electrons. In their reduced form, cytochromes carry a single electron without a proton.

Non-heme iron proteins—the **iron-sulfur proteins**—are another component of the electron transport chain. The iron of these proteins is not attached to a porphyrin ring; instead, the iron atoms are attached to sulfides and to the sulfur atoms of the sulfur-containing amino acids found in the protein chain (Figure 6–12). Like cytochromes, the iron-sulfur proteins pick up electrons on their iron atoms and therefore carry electrons but not protons.

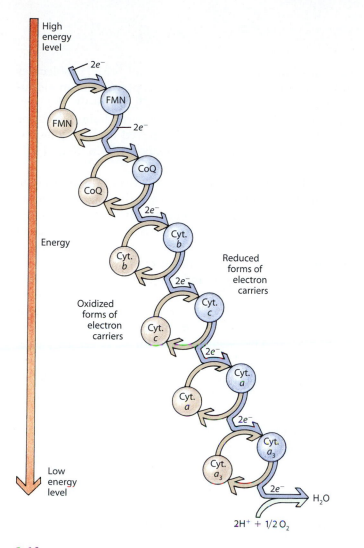

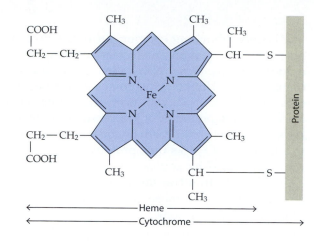

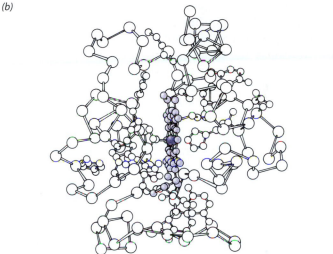

6–10

A schematic representation of the electron transport chain. The mole-cules shown here—flavin mononucleotide (FMN), coenzyme Q (CoQ), and cytochromes b, c, a, and a₃—are the principal electron carriers of the chain. At least nine other molecules also function as intermediates between these electron carriers.

Electrons carried by NADH enter the chain when they are transferred to FMN, which is thus reduced (blue). Almost instantaneously, FMN passes the electrons on to CoQ. In the process, FMN returns to its oxi-dized form (gray), ready to receive another pair of electrons, and CoQ is reduced. CoQ then passes the electrons on to the next carrier and returns to its oxidized form, and so on down the line. As the electrons move down the chain, they drop to successively lower energy levels. The elec-trons are ultimately accepted by oxygen, which combines with protons (hydrogen ions) to form water.

Electrons carried by FADH₂ are at a slightly lower energy level than those carried by NADH. They enter the electron transport chain farther down the line at CoQ and generate only 2 ATP per coenzyme molecule versus the yield of 3 ATP per molecule for NADH.

6–11

Cytochromes are molecules that participate in electron transfer in mitochondria. (a) Each cytochrome contains an atom of iron held in a nitrogen-containing porphyrin ring. The porphyrin ring with its atom of iron is known as heme. Each iron atom accepts one electron and is reduced from Fe^{3+} to Fe^{2+}. The cyto-chrome shown here is cytochrome c. (b) The overall structure of the cytochrome c mole-cule, showing the position of the heme group (color) within the globular protein.

6–12

Postulated arrangement of iron (Fe) and sulfur (S) atoms in the reactive center of an iron-sulfur protein. The iron-sulfur center shown here con-sists of four iron atoms and eight sulfur atoms. The iron-sulfur centers are attached to cysteine (cys) groups of the protein. Iron-sulfur proteins are involved in electron transfer.

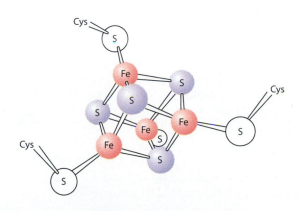

The most abundant components of the electron transport chain are quinone molecules. In mitochondria the quinone is *ubiquinone*, also called **coenzyme Q (CoQ)** (Figure 6–13). Unlike cytochromes and iron-sulfur proteins, a quinone can accept or donate either one or two electrons. This permits CoQ to serve as an intermediate between two-electron carriers and one-electron carriers. In addition, CoQ picks up a proton along with each electron it carries—the equivalent of a hydrogen atom. By alternating electron transfer between carriers that carry electrons only and those that carry hydrogen atoms, CoQ can shuttle protons across the inner mitochondrial membrane. For example, every time a quinone molecule accepts an electron from a cytochrome, it also picks up a proton (H^+) from the mitochondrial matrix. When the quinone gives up its electron to the next carrier, such as a cytochrome, the proton is released into the intermembrane space. Because the electron carriers are oriented in the mitochondrial membrane so that protons are always picked up on the matrix side of the membrane and released into the intermembrane space, a proton gradient is generated across the inner mitochondrial membrane. The importance of this gradient for aerobic ATP production is discussed later. Because CoQ is small and hydrophobic, it can move freely within the lipid bilayer of the membrane and thus can shuttle electrons between other, less mobile carriers.

At the "top" (most energetic end) of the electron transport chain are the electrons held by NADH and $FADH_2$. For every molecule of glucose oxidized, the yield of the Krebs cycle was two molecules of $FADH_2$ and six molecules of NADH, and the oxidation of pyruvate to acetyl CoA yielded two molecules of NADH. Recall that an additional two molecules of NADH were produced in glycolysis; in the presence of oxygen, electrons from these NADH molecules are transported into the mitochondrion. Electrons from all NADH molecules are transferred to the electron acceptor flavin mononucleotide (abbreviated FMN; see Figure 6–9), the first component of the electron transport chain. Electrons from $FADH_2$ molecules are transferred to CoQ, which is farther down the electron transport chain than FMN (Figure 6–10).

As electrons flow along the electron transport chain from higher to lower energy levels, the released energy is harnessed and used to generate the proton gradient. The proton gradient, in turn, drives the formation of ATP from ADP and P_i in a process known as **oxidative phosphorylation.** At the end of the chain, the electrons are accepted by oxygen and combine with protons (hydrogen ions) to produce water. Each time one pair of electrons passes from NADH to oxygen, enough protons are "pumped" across the membrane to generate three molecules of ATP. Each time a pair of electrons passes from $FADH_2$, which holds them at a slightly lower energy level than NADH, enough protons are "pumped" to form two molecules of ATP.

Oxidative Phosphorylation Is Achieved by the Chemiosmotic Coupling Mechanism

Until the early 1960s, the mechanism of oxidative phosphorylation was one of the most baffling puzzles in all of cell biology. As a result of the insight and experimental creativity of the British biochemist Peter Mitchell (1920–1992)—and the subsequent work of many other investigators—much of the puzzle has now been solved. Oxidative phosphorylation depends on a gradient of protons (H^+ ions) across the mitochondrial membrane and the subsequent use of the potential energy stored in that gradient to form ATP from ADP and phosphate.

As shown in Figure 6–14, the components of the electron transport chain are arranged sequentially in the inner membrane of the mitochondrion. Most of the electron carriers are tightly associated with proteins embedded in the membrane, forming three distinct complexes. Within each complex, the electron carriers are held in the proper positions in relation to one another.

As noted above, the protein complexes are also proton pumps. As the electrons drop to lower energy levels

6–13

Oxidized and reduced forms of coenzyme Q. The oxidized form of CoQ is reduced by accepting electrons from a donor in the electron transport chain. The protons picked up by CoQ come from the matrix side of the inner mitochondrial membrane. Coenzyme Q is also known as "ubiquinone," a reflection of the ubiquity of this compound—it occurs in virtually all eukaryotic cells.

Coenzyme Q,
oxidized (CoQ)

Coenzyme Q,
reduced (CoQH₂)

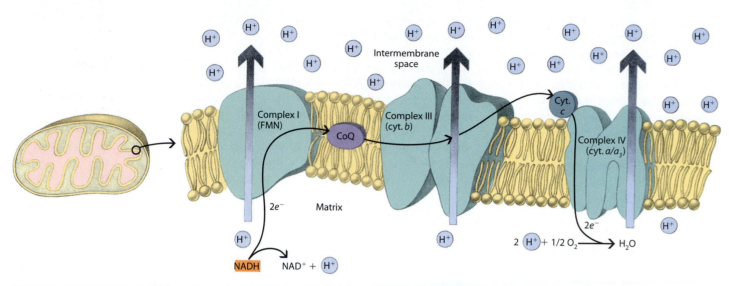

6–14

The arrangement of the components of the electron transport chain in the inner membrane of the mitochondrion. Three complex protein structures (here labeled I, III, and IV) are embedded in the membrane. They contain electron carriers and the enzymes required to catalyze the transfer of electrons from one carrier to the next. Complex I contains the electron carrier FMN, which receives two electrons from NADH and passes them to CoQ. CoQ, located in the lipid interior of the membrane, ferries electrons from Complex I to Complex III, which contains cytochrome b. From Complex III, the electrons move to cytochrome c, a peripheral membrane protein that shuttles back and forth between Complexes III and IV. The electrons then move through cytochromes a and a$_3$, located in Complex IV, back into the matrix where they combine with protons (H$^+$ ions) and oxygen, forming water.

Complex II, another complex protein structure (not shown here) embedded in the inner mitochondrial membrane, contains FAD. Electrons are passed from succinate (in the Krebs cycle) to FAD, forming FADH$_2$, and then to CoQ. Thus, electrons from FADH$_2$ enter the electron transport chain at CoQ. Complex II is not part of the transfer of electrons from NADH to O$_2$.

As the electrons make their way down the electron transport chain, protons are pumped through the three protein complexes from the matrix to the intermembrane space. This transfer of protons from the matrix side of the inner mitochondrial membrane to the other side establishes the proton gradient that powers the synthesis of ATP.

during their transit through the electron transport chain, the released energy is used by the protein complexes to pump protons from the mitochondrial matrix into the intermembrane space. It is thought that for each pair of electrons moving down the electron transport chain from NADH to oxygen, about 10 protons are pumped out of the matrix.

As we noted earlier, the inner membrane of the mitochondrion is impermeable to protons. Thus, the protons that are pumped into the intermembrane space cannot easily move back across the membrane into the matrix. The result is a concentration gradient of protons across the inner membrane of the mitochondrion, with a much higher concentration of protons in the intermembrane space than in the matrix.

Like a boulder at the top of a hill or water at the top of a falls, the difference in the concentration of protons between the intermembrane space and the matrix represents potential energy. This potential energy results not only from the actual concentration difference (more hydrogen ions outside the matrix than inside) but also from the difference in electric charge (more positive charges outside than inside). The potential energy is thus in the form of an **electrochemical gradient.** It is available to power any process that provides a channel allowing the protons to flow down the gradient back into the matrix.

Such a channel is provided by a large enzyme complex known as **ATP synthase** (Figure 6–15). This enzyme complex, which is embedded in the inner membrane of the mitochondrion, has binding sites for ATP and ADP. It also has an inner channel, or pore, through which protons can pass. When protons flow through this channel, moving down the electrochemical gradient from the intermembrane space back into the matrix, the energy that is released powers the synthesis of ATP from ADP and phosphate.

6–15

*The ATP synthase complex. **(a)** This enzyme complex consists of two major portions, F_o, which is contained within the inner membrane of the mitochondrion, and F_1, which extends into the matrix. Binding sites for both ATP and ADP are located on the F_1 portion, which consists of nine separate protein subunits. A channel, or pore, connecting the intermembrane space to the mitochondrial matrix, passes through the entire complex. When protons flow through this channel, moving down the electrochemical gradient, ATP is synthesized from ADP and phosphate. **(b)** The knobs protruding from the vesicles in this electron micrograph are the F_1 portions of ATP synthase complexes. The F_o portions to which they are attached are embedded in the membrane and are not visible. The vesicles were prepared by disrupting the inner mitochondrial membrane with ultrasonic waves. When the membrane is disrupted in this way, the fragments immediately reseal, forming closed vesicles. The vesicles are, however, inside out. The outer surface here is the surface that faces the matrix in the intact mitochondrion.*

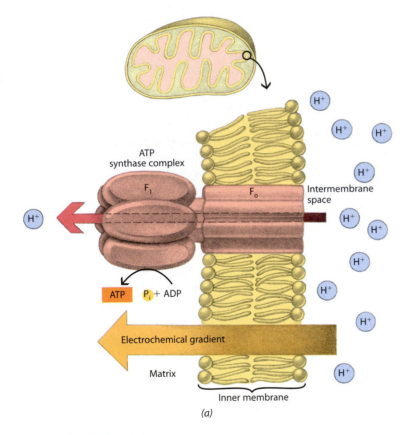

(a)

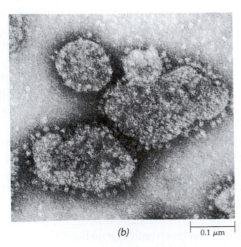

(b)   0.1 μm

Note that ATP synthase functions in a manner that is opposite to that of the proton pump H^+-ATPase, described in Chapter 4. There, ATP was *used* as an energy source to pump protons against their electrochemical gradient. ATP synthase, on the other hand, uses the energy of protons moving down their gradient to *produce* ATP.

This mechanism of ATP synthesis, summarized in Figure 6–16, is known as **chemiosmotic coupling.** The term "chemiosmotic," coined by Peter Mitchell, reflects the fact that the production of ATP in oxidative phosphorylation includes both chemical processes (the "chemi" portion of the term) and transport processes across a selectively permeable membrane (the "osmotic" portion of the term). As we have seen, two distinct events take place in chemiosmotic coupling: (1) a proton gradient is established across the inner membrane of the mitochondrion, and (2) potential energy stored in the gradient is used to generate ATP from ADP and phosphate.

Chemiosmotic power also has other uses in living systems. For example, it provides the power that drives the rotation of bacterial flagella. In photosynthetic cells, as we shall see in the next chapter, it is involved in the formation of ATP using energy supplied to electrons by the sun. It can also be used to power other transport processes. In the mitochondrion, for example, the energy stored in the proton gradient is also used to carry other substances through the inner membrane. Both phosphate and pyruvate are carried into the matrix by membrane proteins that simultaneously transport protons down the gradient.

The Overall Energy Harvest Involves NADH and FADH$_2$ As Well As ATP

We are now in a position to see how much of the energy originally present in the glucose molecule has been recovered in the form of ATP. The "balance sheet" for ATP yield given in Figure 6–17 may help you keep track of the discussion that follows.

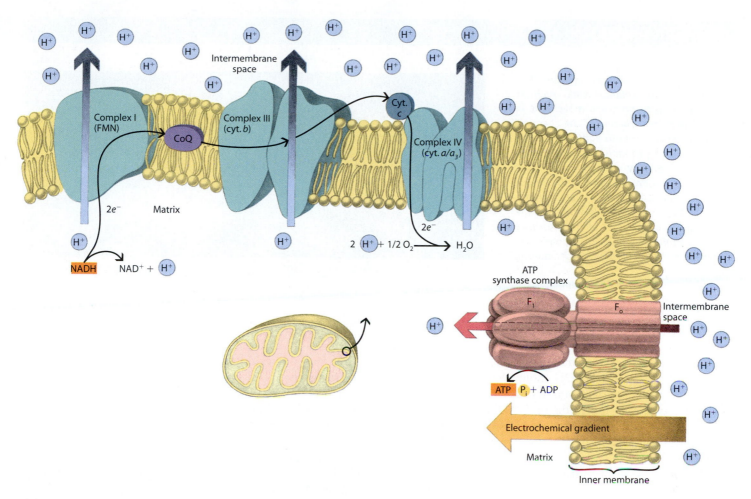

6–16

A summary of the chemiosmotic synthesis of ATP in the mitochondrion. As electrons pass down the electron transport chain, which forms a part of the inner mitochondrial membrane, protons are pumped out of the mitochondrial matrix into the intermembrane space. This creates an electrochemical gradient. The subsequent movement of protons down the gradient as they pass through the ATP synthase complex provides the energy by which ATP is regenerated from ADP and phosphate. Current evidence suggests that three protons flow through the ATP synthase complex for each molecule of ATP formed.

6–17

A summary of the net energy yield from the complete oxidation of one molecule of glucose.

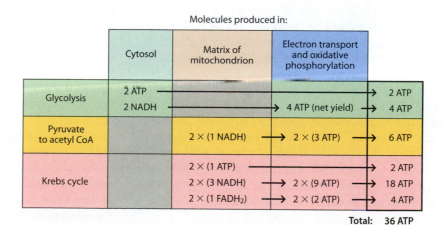

		Molecules produced in:		
	Cytosol	Matrix of mitochondrion	Electron transport and oxidative phosphorylation	
Glycolysis	2 ATP			2 ATP
	2 NADH		4 ATP (net yield)	4 ATP
Pyruvate to acetyl CoA		2 × (1 NADH)	2 × (3 ATP)	6 ATP
Krebs cycle		2 × (1 ATP)		2 ATP
		2 × (3 NADH)	2 × (9 ATP)	18 ATP
		2 × (1 $FADH_2$)	2 × (2 ATP)	4 ATP
			Total:	36 ATP

Bioluminescence

In most organisms, the energy stored in ATP is used for cellular work, such as biosynthesis, transport, and movement. In some, however, a fraction of the chemical energy is reconverted to light energy. Bioluminescence is probably an accidental by-product of energy exchanges in most luminescent organisms, such as the fungus Mycena lux-coeli shown here, photographed by its own light. However, in some luminescent organisms, bioluminescence serves a useful function. For example, the firefly uses flashes of light as its mating signal.

Glycolysis takes place in the cytosol and in the presence of oxygen yields 2 molecules of ATP directly plus 2 molecules of NADH per molecule of glucose. The electrons held by these 2 NADH molecules are transported across the mitochondrial membrane at a "cost" of one ATP molecule per molecule of NADH. Thus, the net yield from reoxidation of the 2 NADH molecules is only 4 molecules of ATP, rather than the 6 that would otherwise be expected.

The conversion of pyruvate to acetyl CoA occurs in the matrix of the mitochondrion, yielding 2 molecules of NADH for each molecule of glucose. When the electrons held by the 2 NADH molecules pass down the electron transport chain, enough protons are pumped across the mitochondrial membrane to synthesize 6 ATP.

The Krebs cycle also occurs in the matrix of the mitochondrion, yielding 2 molecules of ATP, 6 of NADH, and 2 of $FADH_2$. The passage down the electron transport chain of the electrons held by these NADH and $FADH_2$ molecules drives the pumping of enough protons across the membrane to yield 22 ATP. Thus, for each molecule of glucose, the total yield of the Krebs cycle is 24 ATP.

As the balance sheet shows (Figure 6–17), the net yield from a single molecule of glucose is 36 molecules of ATP. All but 2 of the 36 molecules of ATP have come from reactions in the mitochondrion, and all but 4 involve the oxidation of NADH or $FADH_2$ along the electron transport chain coupled with oxidative phosphorylation.

The total difference in free energy (ΔG) between the reactants (glucose and oxygen) and the products (carbon dioxide and water) is −686 kilocalories per mole. The terminal phosphoanhydride bonds of the 36 ATP molecules account for about 263 kilocalories (7.3 × 36) per mole of glucose. In other words, about 38 percent of the energy is conserved in the ATP. The remainder is lost as heat.

Other Substrates for Respiration

Thus far we have regarded glucose as the main substrate for respiration. It is important to note, however, that fats and proteins can also be converted to acetyl CoA and enter the Krebs cycle (see Figure 6–19). For a fat, a triglyceride molecule is first hydrolyzed to glycerol and three fatty acids. Then, beginning at the carboxyl end of the fatty acids, two-carbon acetyl groups are successively removed as acetyl CoA by a process called *beta oxidation*. A molecule such as oleic acid (see Figure 2–9), which contains 18 carbon atoms, yields nine molecules of acetyl CoA that can be oxidized by the Krebs cycle. Proteins are similarly broken down into their constituent amino acids, and the amino groups are removed. Some of the residual carbon skeletons are converted to Krebs cycle intermediates such as α-ketoglutarate, oxaloacetate, and fumarate, and thereby enter the cycle.

Anaerobic Pathways

In most eukaryotic cells (as well as in most bacteria), pyruvate usually follows the aerobic pathway and is completely oxidized to carbon dioxide and water. However, in the absence or shortage of oxygen, pyruvate is not the end product of glycolysis. Under these conditions, the NADH produced during the oxidation of glyceraldehyde 3-phosphate cannot pass its electrons to O_2 via the electron transport chain, but must still be reoxidized to NAD^+. Without this reoxidation, glycolysis would soon stop because the cell would run out of NAD^+ as an electron acceptor.

In many bacteria, fungi, protists, and animal cells, this oxygenless, or anaerobic, process results in the formation of lactate. This process is therefore called **lactate fermentation.** In yeast and most plant cells, however, pyruvate is broken down to ethanol (ethyl alcohol) and carbon dioxide. This anaerobic process is therefore called **alcohol fermentation.** In both cases, the two electrons (and one proton) from NADH are transferred to what was the middle carbon of pyruvate. In the case of alcohol fermentation, however, the reoxidation of NADH is preceded by the release of carbon dioxide (decarboxylation) (Figure 6–18).

Thermodynamically, lactate fermentation and alcohol fermentation are similar. In both, the NADH is reoxidized, and the energy yield for glucose breakdown is limited to the net gain of 2 ATP molecules produced during glycolysis. The complete, balanced equation for the fermentation of glucose can be written as follows:

$$\text{Glucose} + 2\text{ADP} + 2\text{P}_i \longrightarrow$$
$$2 \text{ Ethanol} + 2\text{CO}_2 + 2\text{ATP} + 2\text{H}_2\text{O}$$

or

$$\text{Glucose} + 2\text{ADP} + 2\text{P}_i \longrightarrow 2 \text{ Lactate} + 2\text{ATP} + 2\text{H}_2\text{O}$$

During alcohol fermentation, approximately 7 percent of the total available energy of the glucose molecule—about 52 kilocalories per mole—is released, with about 93 percent remaining in the two alcohol molecules. However, if one considers the efficiency with which the anaerobic cell conserves much of those 52 kilocalories as ATP (7.3 kilocalories per mole of ATP, or 14.6 kilocalories per mole of glucose), the efficiency of energy conserva-

tion is about 26 percent. As noted in Chapter 5, the efficiency of human-made machines seldom exceeds 25 percent, and that of respiration, as we have seen in this chapter, is about 38 percent.

The fact that glycolysis does not require oxygen suggests that the glycolytic sequence evolved early, before free oxygen was present in the atmosphere. Presumably, primitive one-celled organisms used glycolysis (or something very much like it) to extract energy from the organic molecules they absorbed from their watery surroundings. Although the anaerobic pathways generate only two molecules of ATP for each glucose processed, this low yield was and is adequate for the needs of many organisms or parts of organisms. The root systems of rice plants in flooded rice paddies often carry out extensive fermentation to provide energy for the growth and metabolism of the roots.

The Strategy of Energy Metabolism

The various pathways by which different organic molecules are broken down to yield energy are known collectively as **catabolism.** They are also central to the biosynthetic processes of life. These processes, known collectively as **anabolism,** are the pathways by which cells synthesize the diversity of molecules that constitute a living organism. Because many of these molecules,

6–18

(a) The two-step process by which pyruvate is converted anaerobically to ethanol. In the first step, carbon dioxide is released. In the second, NADH is oxidized and acetaldehyde is reduced. Most of the energy of the glucose remains in the alcohol, which is the principal end product of the sequence. However, by regenerating NAD+, these steps allow glycolysis to continue, with its small but sometimes vital yield of ATP. (b) An example of anaerobic glycolysis. Ancient Egyptian wall paintings, such as the one shown here, are the earliest historical record of wine-making. They have been dated to about 5000 years ago. However, recently discovered pottery fragments, stained with wine, suggest that the Sumerians had mastered the art of wine-making at least 500 years before the Egyptians. The grapes were picked and then crushed by foot, and the juice was collected in jugs and allowed to ferment, producing wine. In modern wine-making, pure yeast cultures are added to relatively sterile grape juice for fermentation, rather than relying on the yeasts carried on the grapes.

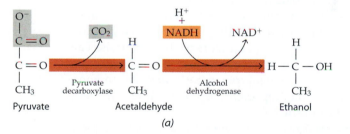

(a)

(b)

6–19

An outline of some major pathways of catabolism and anabolism in the living cell. Catabolic pathways (arrows pointing down) are exergonic. A significant portion of the energy released in these pathways is captured in the synthesis of ATP. Anabolic pathways (arrows pointing up) are endergonic. The energy that powers the reactions in these pathways is supplied primarily by ATP and NADH.

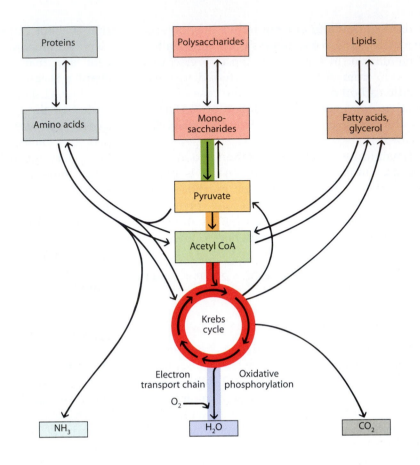

such as proteins and lipids, can be broken down and fed into the central pathway of glucose metabolism, you might guess that the reverse process can occur—namely, that the various intermediates of glycolysis and the Krebs cycle can serve as precursors for biosynthesis. This is indeed the case, as outlined in Figure 6–19. Hence, the Krebs cycle plays a critical role in both catabolic and anabolic processes and represents a major "hub" of metabolic activities in the cell.

For the reactions of the catabolic and anabolic pathways to occur, there must be a steady supply of organic molecules that can be broken down to yield not only energy but also building-block molecules. Without a supply of such molecules, the metabolic pathways cease to function and the organism dies. Heterotrophic cells (including the heterotrophic cells of plants, such as root cells) are dependent on external sources—specifically, autotrophic cells—for the organic molecules that are essential for life. Autotrophic cells, however, are able to synthesize their own energy-rich organic molecules from simple inorganic molecules and an external energy source. These molecules supply energy as well as building-block molecules.

By far the most important autotrophic cells are the photosynthetic cells of algae and plants. In the next chapter, we shall examine how these cells capture the energy of sunlight and use it to synthesize the monosaccharide molecules on which life on this planet depends.

Summary

Respiration, or the Complete Oxidation of Glucose, Is the Chief Source of Energy in Most Cells

As glucose is broken down in a series of sequential enzyme-catalyzed reactions, some of the energy released is packaged in the form of terminal phosphoanhydride bonds in ATP and the rest is lost as heat.

In Glycolysis, Glucose Is Split into Pyruvate

The first phase in the breakdown of glucose is glycolysis, in which the six-carbon glucose molecule is split into two three-carbon molecules of pyruvate; two molecules of ATP and two of NADH are formed. This reaction occurs in the cytosol.

The Krebs Cycle Completes the Metabolic Breakdown of Glucose to Carbon Dioxide

In the course of respiration, the three-carbon pyruvate molecules are broken down within the mitochondrion to two-carbon acetyl groups, which then enter the Krebs cycle as acetyl CoA. In the Krebs cycle, each acetyl group is oxidized in a series of reactions to yield two additional molecules of carbon dioxide, one molecule of ATP, and four molecules of reduced electron carriers (three NADH and one $FADH_2$). With two turns of the cycle, the carbon atoms derived from the glucose molecule are completely oxidized.

In the Electron Transport Chain, the Flow of Electrons Is Coupled to the Pumping of Protons across the Inner Mitochondrial Membrane and the Synthesis of ATP

The next stage of respiration is the electron transport chain, which involves a series of electron carriers and enzymes embedded in the inner membrane of the mitochondrion. Along this series of electron carriers, the high-energy electrons carried by NADH and $FADH_2$ pass "downhill" to oxygen. The large quantity of free energy released during the passage of electrons down the electron transport chain powers the pumping of protons (H^+ ions) out of the mitochondrial matrix. This creates a gradient of potential energy across the inner membrane of the mitochondrion. When the protons pass through the ATP synthase complex as they flow down the gradient back into the matrix, the energy released is used to form ATP from ADP and phosphate. This process, known as chemiosmotic coupling, is the mechanism by which oxidative phosphorylation is accomplished.

In the course of the aerobic breakdown of the glucose molecule to CO_2 and H_2O, 36 molecules of ATP are generated, most of them in the mitochondrion in the final stage of respiration, oxidative phosphorylation.

Fermentation Reactions Occur under Anaerobic Conditions

In the absence or shortage of oxygen, pyruvate produced by glycolysis may be converted either to lactate (in many bacteria, fungi, and animal cells) or to ethanol and carbon dioxide (in yeasts and most plant cells). These anaerobic processes—called fermentation—yield 2 ATP for each glucose molecule.

The Krebs Cycle Is the "Metabolic Hub" for the Breakdown and Synthesis of Many Different Types of Molecules

Although glucose is regarded as the main substrate for respiration in most cells, fats and proteins can also be converted to molecules that can enter the respiratory sequence at various steps. The various pathways by which organic molecules are broken down to yield energy are known collectively as catabolism. The biosynthetic processes of life are known collectively as anabolism.

Selected Key Terms

acetyl CoA p. 114

alcohol fermentation p. 123

anabolism p. 123

anaerobic pathways p. 122

ATP synthase p. 119

catabolism p. 123

chemiosmotic coupling p. 120

coenzyme A (CoA) p. 114

coenzyme Q (CoQ) p. 118

cristae p. 114

cytochromes p. 116

electrochemical gradient p. 119

electron transport chain p. 116

$FADH_2$ p. 116

fermentation p. 109

glycolysis p. 109

iron-sulfur proteins p. 116

Krebs cycle p. 109

lactate fermentation p. 122

mitochondrial matrix p. 114

NADH p. 109

oxidative phosphorylation p. 118

pyruvate p. 109

respiration p. 109

substrate-level phosphorylation p. 112

Questions

1. Distinguish between substrate-level phosphorylation and oxidative phosphorylation. Where do these processes occur in the cell, in relation to respiration?

2. Sketch the structure of a mitochondrion. Describe where the various stages in the complete breakdown of glucose take place in relation to mitochondrial structure. What molecules and ions cross the mitochondrial membranes during these processes?

3. One might say that two distinct events take place in chemiosmotic coupling. What are those events? What are some uses of chemiosmotic power in living systems?

4. Certain chemicals function as "uncoupling" agents when they are added to respiring mitochondria. The passage of electrons down the electron transport chain to oxygen continues, but no ATP is formed. One of these agents, the antibiotic valinomycin, is known to transport K^+ ions through the inner membrane into the matrix. Another, 2,4-dinitrophenol, transports H^+ ions through the membrane. How do these substances prevent the formation of ATP?

5. With some strains of yeast, fermentation stops before the sugar is exhausted, usually at an alcohol concentration in excess of 12 percent. What is a plausible explanation?

Chapter

7 | Photosynthesis, Light, and Life

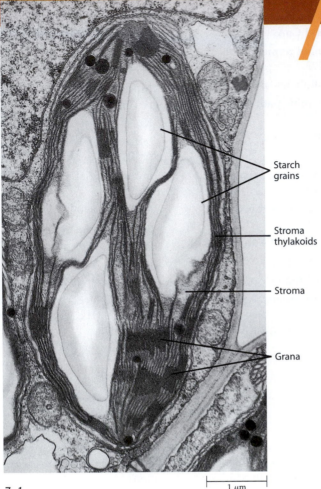

7–1

Photosynthesis in eukaryotic organisms occurs in chloroplasts. Shown here is a chloroplast from a mesophyll cell of the pigweed (Amaranthus retroflexus) *leaf. The light-capturing reactions occur in the internal membranes, or thylakoids, where the chlorophylls and other pigments are embedded. Many of the thylakoids occur in disklike stacks called grana. The thylakoids of different grana are interconnected via stroma thylakoids. The series of reactions by which the captured light energy is used to synthesize carbon-containing compounds occurs in the stroma, the material surrounding the thylakoids. During periods of intense photosynthesis, some of the carbohydrate is stored temporarily in the chloroplast as grains of starch. At night, sucrose is produced from the starch and exported from the leaf to other parts of the plant, where it is eventually used for the manufacture of other molecules needed by the plant.*

Starch grains

Stroma thylakoids

Stroma

Grana

1 μm

OVERVIEW

From a human perspective, photosynthesis is the most important process on Earth. During photosynthesis, plants, algae, and photosynthetic bacteria are able to "harness the sun," using the radiant energy to convert simple molecules—carbon dioxide and water—into complex organic molecules that can be used by plants and animals alike as sources of energy and molecular building blocks. Moreover, photosynthesis releases oxygen (O_2) into the air we breathe, and it is this oxygen that plays such an important role in cellular respiration and the accompanying synthesis of ATP. Thus, without photosynthesis, plants and animals, including humans, would suffocate and starve.

After a discussion of the nature of light and the role of pigments, this chapter introduces the two series of reactions involved in photosynthesis. In the first series, called the light reactions (or energy-transduction reactions), molecules of chlorophyll absorb light energy and use it to synthesize ATP as well as NADPH. These products then function in the second series of reactions, termed carbon fixation, to convert CO_2 molecules present in the air into carbohydrates. The oxygen so important to aerobic organisms is released as a waste product in photosynthesis when water molecules are split apart in the light reactions.

CHECKPOINTS

By the time you finish reading this chapter, you should be able to answer the following questions:

1. What is the role of light in photosynthesis, and what properties of light suggest that it is a wave? A particle?

2. What are the principal pigments involved in photosynthesis, and why are leaves green?

3. What are the main products of the energy-transduction, or light, reactions of photosynthesis?

4. What are the main products of the carbon-fixation reactions of photosynthesis, and why is the name "dark reactions" misleading for this series of reactions?

5. What are the main events associated with each of the two photosystems in the light reactions, and what is the difference between antenna pigments and reaction center pigments?

6. What are the principal differences between the C_3, C_4, and CAM pathways for carbon fixation? What features do they have in common?

In the previous chapter, we described the breakdown of carbohydrates to yield the energy required for the many kinds of activities carried out by living systems. In the pages that follow, we will complete the circle by describing the way in which light energy from the sun is captured and converted to chemical energy.

This process—photosynthesis—is the route by which virtually all energy enters our biosphere. Each year more than 250 billion metric tons of sugar are produced worldwide by photosynthetic organisms. The importance of photosynthesis, however, extends far beyond the sheer weight of this product. Without this flow of energy from the sun, channeled largely through the chloroplasts of eukaryotic cells, the pace of life on this planet would swiftly diminish and then would virtually cease altogether, as dictated by the inexorable second law of thermodynamics.

Photosynthesis: A Historical Perspective

The importance of photosynthesis in the economy of nature was not recognized until comparatively recent times. Aristotle and other Greeks, observing that the life processes of animals were dependent upon the food they ate, thought that plants derived all of their food from the soil.

More than 350 years ago, in one of the first carefully designed biological experiments ever reported, the Belgian physician Jan Baptista van Helmont (ca. 1577–1644) offered the first experimental evidence that soil alone does not nourish the plant. Van Helmont grew a small willow tree in an earthenware pot, adding only water to the pot. At the end of 5 years, the willow had increased in weight by 74.4 kilograms, whereas the soil had decreased in weight by only 57 grams. On the basis of these results, van Helmont concluded that all the substance of the plant was produced from the water and none from the soil! Van Helmont's conclusions, however, were too broad.

Toward the end of the eighteenth century, the English clergyman/scientist Joseph Priestley (1733–1804) reported that he had "accidently hit upon a method of restoring air that had been injured by the burning of candles." On 17 August 1771, Priestley "put a [living] sprig of mint into air in which a wax candle had burned out and found that, on the 27th of the same month, another candle could be burned in this same air." The "restorative which nature employs for this purpose," he stated, was "vegetation." Priestley extended his observations and soon showed that air "restored" by vegetation was not "at all inconvenient to a mouse." Priestley's experiments offered the first logical explanation of how air remained "pure" and able to support life despite the burning of countless fires and the breathing of many animals. When he was presented with a medal for his discovery, the citation read in part: "For these discoveries we are assured that no vegetable grows in vain . . . but cleanses and purifies our atmosphere." Today we would explain Priestley's experiments simply by saying that plants take up the CO_2 produced by combustion or exhaled by animals, and that animals inhale the O_2 released by plants.

Shortly thereafter, the Dutch physician Jan Ingenhousz (1730–1799) confirmed Priestley's work and showed that the air was "restored" only in the presence of sunlight and only by the green parts of plants. In 1796, Ingenhousz suggested that carbon dioxide is split in photosynthesis to yield carbon and oxygen, with the oxygen then released as gas. Subsequently, the proportion of carbon, hydrogen, and oxygen atoms in sugars and starches was found to be about one atom of carbon per molecule of water (CH_2O), as the word "carbohydrate" indicates. Thus, in the overall reaction for photosynthesis,

$$CO_2 + H_2O + \text{Light energy} \longrightarrow (CH_2O) + O_2$$

it was generally assumed that the carbohydrate came from a combination of water molecules and the carbon atoms of carbon dioxide and that the oxygen was released from the carbon dioxide. This entirely reasonable hypothesis was widely accepted; but, as it turned out, it was quite wrong.

The investigator who upset this long-held theory was C. B. van Niel of Stanford University. Van Niel, then a graduate student, was investigating the activities of different types of photosynthetic bacteria (Figure 7–2). One

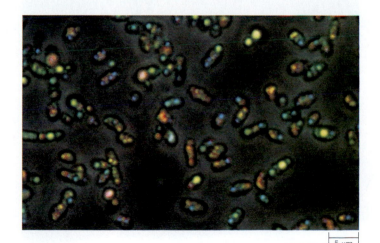

5 μm

7–2

Purple sulfur bacteria. These present-day bacteria reduce carbon to carbohydrates during photosynthesis but do not release oxygen. In these cells, hydrogen sulfide (H_2S) plays the same role that water plays in the photosynthetic process of plants. The hydrogen sulfide is split, and the sulfur accumulates as globules, visible within these cells.

particular group of such bacteria—the purple sulfur bacteria—reduces carbon to carbohydrates during photosynthesis but does not release oxygen. The purple sulfur bacteria require hydrogen sulfide for their photosynthetic activity. In the course of photosynthesis, globules of sulfur accumulate inside the bacterial cells (Figure 7–2). Van Niel found that the following reaction takes place during photosynthesis in these bacteria:

$$CO_2 + 2H_2S \xrightarrow{\text{Light}} (CH_2O) + H_2O + 2S$$

This finding was simple and did not attract much attention until van Niel made a bold extrapolation. He proposed the following generalized equation for photosynthesis:

$$CO_2 + 2H_2A \xrightarrow{\text{Light}} (CH_2O) + H_2O + 2A$$

In this equation, H_2A represents an oxidizable substance, such as hydrogen sulfide or free hydrogen. In algae and green plants, however, H_2A is water (Figure 7–3). In short, van Niel proposed that water, *not* carbon dioxide, was the source of the oxygen in photosynthesis.

In 1937, Robin Hill showed that, when exposed to light, isolated chloroplasts were able to produce O_2 in the absence of CO_2. This light-driven release of O_2 in the absence of CO_2—called the **Hill reaction**—occurred only when the chloroplasts were illuminated and provided with an artificial electron acceptor. This finding supported van Niel's proposal made six years earlier.

More convincing evidence that the O_2 released in photosynthesis is derived from H_2O came in 1941, when investigators used a heavy isotope of oxygen ($^{18}O_2$) to trace the oxygen from water to oxygen gas:

$$CO_2 + 2H_2^{18}O \xrightarrow{\text{Light}} (CH_2O) + H_2O + {}^{18}O_2$$

Thus, in the case of algae and green plants, in which water serves as the electron donor, a complete, balanced equation for photosynthesis can be written as follows:

$$3CO_2 + 6H_2O \xrightarrow{\text{Light}} C_3H_6O_3 + 3O_2 + 3H_2O$$

Although glucose commonly is represented as the carbohydrate product of photosynthesis in summary equations, in reality, very little glucose is generated in photosynthesizing cells. The more immediate carbohydrate products are trioses (three-carbon sugars).

As noted above, light was discovered to be required for the process we now call photosynthesis about 200 years ago. In fact, photosynthesis occurs in two stages, only one of which actually requires light. Evidence for this two-stage process was first presented in 1905 by the English plant physiologist F. F. Blackman, as the result of experiments in which he measured the individual and combined effects of changes in light intensity and temperature on the rate of photosynthesis. These experiments showed that photosynthesis has both a light-dependent stage and a light-independent stage.

In Blackman's experiments, the light-independent reactions increased in rate as the temperature was increased, but only up to about 30°C, after which the rate began to decrease. From this evidence it was concluded that these reactions were controlled by enzymes, since this is the way enzymes are expected to respond to temperature (see Figure 5–12). This conclusion has since been proven to be correct.

7–3

The bubbles on the leaves of this submerged pondweed, Elodea, *are bubbles of oxygen, one of the products of photosynthesis. Van Niel was the first to propose that the oxygen produced in photosynthesis came from the splitting of water rather than the breakdown of carbon dioxide.*

The Nature of Light

Over 300 years ago, the English physicist Sir Isaac Newton (1642–1727) separated light into a spectrum of visible colors by passing it through a prism. In this way, Newton showed that white light actually consists of a number of different colors, ranging from violet at one end of the spectrum to red at the other. The separation of colors is possible because light of different colors is bent (refracted) at different angles in passing through a prism.

In the nineteenth century, the British physicist James Clerk Maxwell (1831–1879) demonstrated that light is

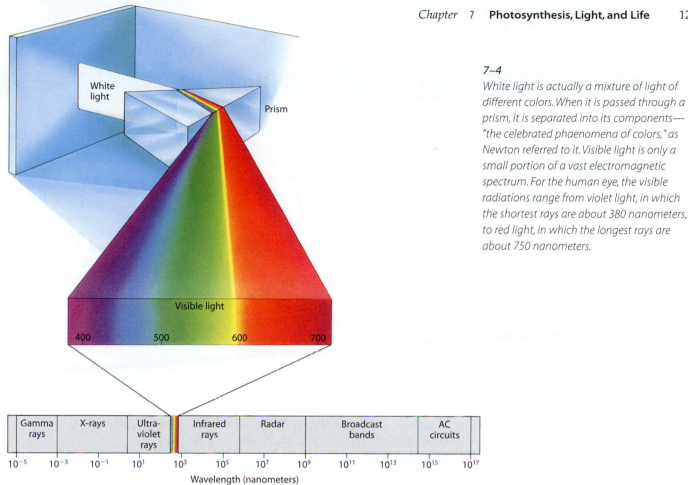

White light

Prism

Visible light

400 500 600 700

| Gamma rays | X-rays | Ultra-violet rays | Infrared rays | Radar | Broadcast bands | AC circuits |

10^{-5} 10^{-3} 10^{-1} 10^{1} 10^{3} 10^{5} 10^{7} 10^{9} 10^{11} 10^{13} 10^{15} 10^{17}

Wavelength (nanometers)

7–4

White light is actually a mixture of light of different colors. When it is passed through a prism, it is separated into its components— "the celebrated phaenomena of colors," as Newton referred to it. Visible light is only a small portion of a vast electromagnetic spectrum. For the human eye, the visible radiations range from violet light, in which the shortest rays are about 380 nanometers, to red light, in which the longest rays are about 750 nanometers.

but a small part of a vast continuous spectrum of radiation, the **electromagnetic spectrum** (Figure 7–4). All the radiations in this spectrum travel in waves. The **wavelengths**—that is, the distances from the crest of one wave to the crest of the next—range from those of gamma rays, which are measured in fractions of a nanometer (1 nanometer $= 10^{-9}$ meter), to those of low-frequency radio waves, which are measured in kilometers (1 kilometer $= 10^{3}$ meters $= 0.6$ mile). Radiation of each particular wavelength has a characteristic amount of energy associated with it. The shorter the wavelength, the greater the energy; conversely, the longer the wavelength, the lower the energy. Within the spectrum of visible light, violet light has the shortest wavelength and red light has the longest. The shortest rays of violet light have almost twice the energy of the longest rays of red light.

Light Has the Properties of Waves and Particles

By 1900 it had become clear that the **wave model** of light was not adequate. The key observation, a very simple one, was made in 1888: when a zinc plate is exposed to ultraviolet light, it acquires a positive charge. The metal becomes positively charged because the light energy dislodges electrons from the metal atoms. Subsequently it was discovered that this **photoelectric effect,** as it is known, can be produced in all metals. Every metal has a

critical maximum wavelength for the effect; the light, or other radiation, must be of that particular wavelength or shorter (that is, more energetic) for the effect to occur.

With some metals, such as sodium, potassium, and selenium, the critical wavelength is within the spectrum of visible light, and as a consequence, visible light striking the metal can set up a moving stream of electrons (an electric current). Exposure meters, television cameras, and the electric eyes that open doors at supermarkets and airline terminals all operate on this principle of turning light energy into electrical energy.

What, then, is the problem with the wave model of light? Simply this: The wave model predicts that the brighter the light—that is, the stronger, or more intense, the beam—the greater the force with which the electrons would be dislodged from a metal. Whether or not light can eject the electrons of a particular metal, however, depends only on the wavelength of the light, not on its intensity. A very weak beam of the critical wavelength is effective, whereas a stronger (brighter) beam of a longer wavelength is not. Furthermore, increasing the intensity of light increases the number of electrons dislodged but not the velocity at which they are ejected from the metal. To increase the velocity, one must use a shorter wavelength of light. Nor is it necessary for energy to accumulate in the metal. With even a dim beam of the critical wavelength, an electron may be emitted the instant the light hits the metal.

The Fitness of Light

Light, as Maxwell showed, is only a tiny band in a continuous spectrum. From the physicist's point of view, the difference between light and darkness—so dramatic to the human eye—is only a few nanometers of wavelength, or, expressed differently, a small amount of energy. Why is it that this tiny portion of the electromagnetic spectrum is responsible for vision, for phototropism (the curving of an organism toward light), for photoperiodism (the seasonal changes that take place in an organism with the changing length of day and night), and for photosynthesis, on which all life depends? Is it an amazing coincidence that all these biological activities depend on these same wavelengths?

George Wald of Harvard University, one of the foremost experts on the subject of light and life, says no. He thinks that if life exists elsewhere in the universe, it is probably dependent upon this same fragment of the electromagnetic spectrum. Wald bases this conjecture on two points.

*First, living things are composed of large, complicated molecules held in special configurations and relationships to one another by hydrogen bonds and other weak bonds. Radiation of even slightly higher energies (shorter wavelengths) than the energy of violet light breaks these bonds and so disrupts the structure and function of the molecules. DNA molecules, for example, are particularly vulnerable to such disruption. Radiations with wavelengths less than 200 nanometers—that is, with still higher energies—drive electrons out of atoms to create ions; hence, it is called **ionizing radiation**. On the other hand, radiations with wavelengths longer than those of the visible band—that is, with less energy than red light—are absorbed by water, which makes up the bulk of all living things on Earth. When such radiation is absorbed by organic molecules, it causes them to increase their motion (increasing heat), but it does not trigger changes in their electron configurations. Only those radiations within the range of visible light have the property of exciting molecules—that is, of moving electrons into higher energy levels—and so of producing chemical and, ultimately, biological change.*

The second reason the visible band of the electromagnetic spectrum was "chosen" by living things is simply that it is what is available. Most of the radiation reaching the surface of the Earth from the sun is within this range. Most of the higher-energy wavelengths are absorbed by the oxygen and ozone high in the atmosphere. Much infrared radiation is screened out by water vapor and carbon dioxide before it reaches the Earth's surface.

This is an example of what has been termed "the fitness of the environment." The suitability of the environment for life and that of life for the physical world are exquisitely interrelated. If they were not, life could not exist.

To explain such phenomena, the **particle model** of light was proposed by Albert Einstein in 1905. According to this model, light is composed of particles of energy called **photons**, or quanta of light. The energy of a photon (a quantum of light) is inversely proportional to its wavelength—the longer the wavelength, the lower the energy. Photons of violet light, for example, have almost twice the energy of photons of red light, the longest visible wavelength.

The wave model of light permits physicists to describe certain aspects of its behavior mathematically, whereas the photon model permits another set of mathematical calculations and predictions. These two models are no longer regarded as opposing one another; rather, they are complementary, in the sense that both are required for a complete description of the phenomenon we know as light.

The Role of Pigments

For light energy to be used by living systems, it must first be absorbed. A substance that absorbs light is known as a **pigment.** Some pigments absorb all wavelengths of light and so appear black. Most pigments, however, absorb only certain wavelengths and transmit or reflect the wavelengths they do not absorb (Figure 7–5). The light absorption pattern of a pigment is known

Sunlight

Sunlight

7–5

When light strikes a colored object, some wavelengths are absorbed, whereas others are transmitted or reflected. The colors we perceive are the wavelengths of light that are transmitted or reflected. For example, a ripe tomato appears red because it reflects light in the red portion of the spectrum; light in all the other portions of the visible spectrum is absorbed. Similarly, the leaves of the tomato plant appear green because they reflect light in the green portion of the spectrum.

as the **absorption spectrum** of that substance. **Chlorophyll,** the pigment that makes leaves green, absorbs light principally in the violet and blue wavelengths and also in the red; because it reflects green light, it appears green.

An **action spectrum** demonstrates the relative effectiveness of different wavelengths of light for a specific light-requiring process, such as photosynthesis or flowering. Similarities between the absorption spectrum of a pigment and the action spectrum of a light-requiring process provide evidence that the pigment is responsible for that particular process (Figure 7–6). One line of evidence that chlorophyll is the principal pigment involved in photosynthesis is the similarity between its absorption spectrum and the action spectrum for photosynthesis (Figure 7–7).

When chlorophyll molecules (or other pigment molecules) absorb light, electrons are temporarily boosted to a higher energy level, called the **excited state.** As the electrons return to their lower energy level, or ground level, the energy released has three possible fates. The first possibility is that the energy is converted to heat or to some combination of heat and light of longer wavelength, a phenomenon known as **fluorescence.** Fluorescence occurs, however, only when light energy is absorbed by isolated chlorophyll molecules in solution. A second possibility is that the energy—but not the electron—may be transferred from the excited chlorophyll molecule to a neighboring chlorophyll molecule, exciting the second molecule and allowing the first one to return to its ground state. This process is known as **resonance energy transfer,** and it may be repeated to a third, a

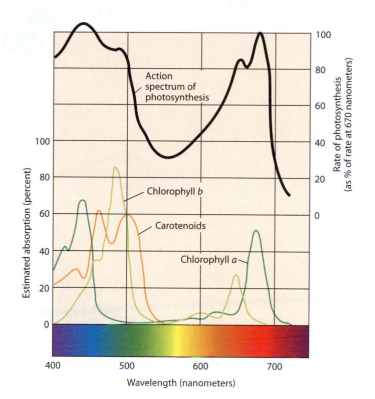

7–6

The action spectrum for photosynthesis (upper curve) and the absorption spectra for chlorophyll a, chlorophyll b, and carotenoids (lower curves) in a plant chloroplast. Note the relationship between the action spectrum of photosynthesis and the absorption spectra for chlorophyll a, chlorophyll b, and carotenoids, all of which absorb light used in photosynthesis.

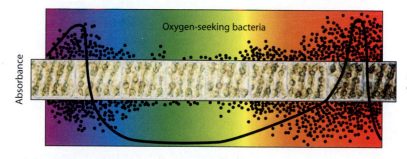

7–7

Correlating action and absorption spectra. Results of an experiment performed in 1882 by T. W. Engelmann revealed the action spectrum of photosynthesis in the filamentous alga Spirogyra. Like investigators working more recently, Engelmann used the rate of oxygen production to measure the rate of photosynthesis. Unlike his successors, however, he lacked sensitive electronic devices for detecting oxygen. As his oxygen indicator, he chose motile bacteria that are attracted by

oxygen. He replaced the mirror and diaphragm usually used to illuminate objects under view in his microscope with a "microspectral apparatus," which, as its name implies, projected a tiny spectrum of colors onto the slide under the microscope. Then he arranged a filament of algal cells parallel to the spread of the spectrum. The oxygen-seeking bacteria congregated mostly in the areas where the violet and red wavelengths

fell upon the algal filament. As you can see, the action spectrum for photosynthesis paralleled the absorption spectrum of chlorophyll (as indicated by the heavy black line). Engelmann concluded that photosynthesis depends on the light absorbed by chlorophyll. This is an example of the sort of experiment that scientists refer to as "elegant," not only brilliant but also simple in design and conclusive in its results.

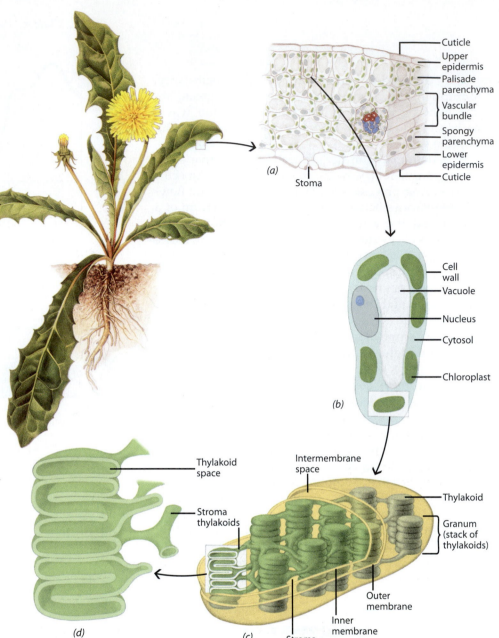

7–8

*Journey into the leaf of a dandelion (Taraxacum officinale). **(a)** The inner tissue of the leaf, called mesophyll, is specialized for photosynthesis. The mesophyll of the dandelion leaf consists of elongated, columnar-shaped cells (palisade parenchyma beneath the upper epidermis) and irregularly shaped cells (spongy parenchyma), all of which contain numerous chloroplasts and are exposed for much of their surface to intercellular (air) spaces. Oxygen, carbon dioxide, and other gases, including water vapor, enter the leaf largely through special openings, the stomata (singular: stoma). These gases fill the intercellular spaces, entering and leaving the leaf and cells by diffusion. Water and minerals taken up by the root enter the leaf in the water-conducting tissue (xylem) of the vascular bundles, or veins, of the leaf. Sugars, the products of photosynthesis, move out of the leaf by way of the food-conducting tissue (phloem) of the vascular bundles, traveling to nonphotosynthetic parts of the plant. **(b)** The chloroplasts occur in the narrow layer of protein-rich cytoplasm that surrounds a large central vacuole. **(c)** The three-dimensional structure of a chloroplast and **(d)** the arrangement of the pigment-containing thylakoid membranes. The stacks of disklike thylakoids, known as grana, are interconnected by the thylakoids, known as stroma thylakoids, that traverse the stroma.*

fourth, or more chlorophyll molecules. The third possibility is that the high-energy electron itself may be transferred to a neighboring molecule (an electron acceptor) that is part of an electron transport chain, leaving an "electron hole" in the excited chlorophyll molecule.

During the process of photosynthesis in intact chloroplasts, both the second and third possibilities—namely, transfer of energy of the excited chlorophyll to a neighboring chlorophyll and transfer of the high-energy electron itself to a neighboring electron acceptor—are useful energy-releasing events.

As you know from Chapter 3, in eukaryotic cells photosynthesis takes place in the chloroplast, and chloroplast structure plays a key role in these energy transfers (Figures 7–1 and 7–8). The chlorophyll molecules themselves are embedded in the thylakoids of the chloroplast.

The Main Photosynthetic Pigments Are the Chlorophylls, the Carotenoids, and the Phycobilins

There are several kinds of chlorophyll, which differ from one another both in the details of their molecular structure and in their specific absorption properties. **Chlorophyll** *a* occurs in all photosynthetic eukaryotes and in the cyanobacteria. Not surprisingly, chlorophyll *a* is essential for the oxygen-generating photosynthesis carried out by organisms of these groups (Figure 7–9).

Plants, green algae, and euglenoid algae also contain the pigment **chlorophyll** *b,* which has a slightly different absorption spectrum than chlorophyll *a*. Chlorophyll *b* is an **accessory pigment**—a pigment that is not directly involved in photosynthetic energy transduction but serves to broaden the range of light that can be used in photosynthesis (Figure 7–6). When a molecule of chlorophyll *b* absorbs light, the energy is ultimately transferred to a molecule of chlorophyll *a*, which then transforms it into chemical energy during the course of photosynthesis. In the leaves of most green plants, chlorophyll *a* generally constitutes about three-fourths of the total chlorophyll content, with chlorophyll *b* accounting for the remainder.

Chlorophyll *c* takes the place of chlorophyll *b* in some groups of algae, most notably the brown algae and the diatoms (Chapter 17). The photosynthetic bacteria (other than cyanobacteria) contain either **bacteriochlorophyll,** which is found in purple bacteria, or **chlorobium chlorophyll,** which occurs in green sulfur bacteria. These bacteria cannot extract electrons from water and thus do not evolve oxygen. Chlorophylls *b* and *c* and the photosynthetic pigments of purple bacteria and green sulfur bacteria are simply chemical variations of the basic structure shown in Figure 7–9.

Two other classes of pigments that are involved in the capture of light energy are the **carotenoids** and the **phycobilins.** The energy absorbed by these accessory pigments must be transferred to chlorophyll *a*; like chlorophylls *b* and *c*, these accessory pigments cannot substitute for chlorophyll *a* in photosynthesis. Although the carotenoid pigments can help collect light of different wavelengths, their principal function is that of an anti-oxidant, preventing photooxidative damage to the chlorophyll molecules. Without the carotenoids, there would be no photosynthesis in the presence of oxygen.

Carotenoids are red, orange, or yellow lipid-soluble pigments found in all chloroplasts and in cyanobacteria. Like the chlorophylls, the carotenoid pigments of chloroplasts are embedded in the thylakoid membranes. Two groups of carotenoids—**carotenes** and **xanthophylls**—are normally present in chloroplasts (Figure 7–10). The beta-carotene found in plants is the principal source of the vitamin A required by humans and other animals. In green leaves, the color of the carotenoids is usually masked by that of the much more abundant chlorophylls, but in temperate regions of the world carotenoids

7–9

Chlorophyll a contains a magnesium ion held in a nitrogen-containing porphyrin ring, highlighted in blue. Attached to the ring is a long hydrocarbon chain, forming a hydrophobic tail that serves to anchor the molecule to specific hydrophobic proteins of the thylakoid membranes. Chlorophyll b differs from chlorophyll a in having a —CHO *group in place of the* —CH₃ *group indicated in gray. Alternating single and double bonds (known as conjugated bonds), such as those in the porphyrin ring of chlorophylls, are common among pigments (see also Figure 7–10).*

become visible when the chlorophylls break down in autumn.

The third class of accessory pigments, the phycobilins, are found in the cyanobacteria and in the chloroplasts of the red algae. Unlike the carotenoids, the phycobilins are water-soluble.

The Reactions of Photosynthesis

Today, the many reactions that occur during photosynthesis are divided into two major processes: **energy-transduction reactions** and **carbon-fixation reactions.** Commonly, the energy-transduction reactions are referred to as the **light reactions** (or *light-dependent reactions*) because of the importance of light in these reactions. Traditionally, the carbon-fixation reactions

The chemical structures shown at the top of the page:

(a) beta-Carotene

(b) Vitamin A (retinol)

(c) Retinal

(d) Zeaxanthin

7–10

Related carotenoids. **(a)** *The pigment beta-carotene gives carrots, sweet potatoes, and other yellow vegetables their characteristic color.* **(b)** *Most animals, including humans, can enzymatically convert a molecule of beta-carotene into two molecules of vitamin A (retinol) by cleaving the beta-carotene molecule at the point indicated by the arrow in* **(a)** *and adding an —H and —OH to each end.* **(c)** *Oxidation of vitamin A yields retinal, the pigment essential for vision.* **(d)** *Zeaxanthin (a xanthophyll) is the pigment responsible for the yellow color of maize (Zea mays) kernels. Unlike the carotenes, xanthophylls contain oxygen. Note the conjugated bonds (alternating single and double bonds) in the carbon chains, which are similar to the conjugated bonds of the porphyrin ring of the chlorophylls (Figure 7–9).*

have been referred to as the *dark reactions* or *light-independent reactions*. These terms are misleading, however, because carbon fixation depends crucially upon the chemical energy harvested during the light reactions to reduce carbon. Moreover, as long as that energy is available, the carbon-fixation reactions can occur in either light or dark.

In the light reactions, light energy is used to form ATP from ADP and to reduce electron-carrier molecules, notably the coenzyme $NADP^+$. $NADP^+$ is similar in structure to NAD^+ (see Figure 5–10)—it has an extra phosphate on one of the riboses—but its biological role is distinctly different. Its reduced form, NADPH, is used by cells to provide energy for biosynthetic pathways. NADH, as we saw in Chapter 6, transfers its electrons to the electron transport chain. In addition, water molecules are split and O_2 is liberated (Figure 7–11).

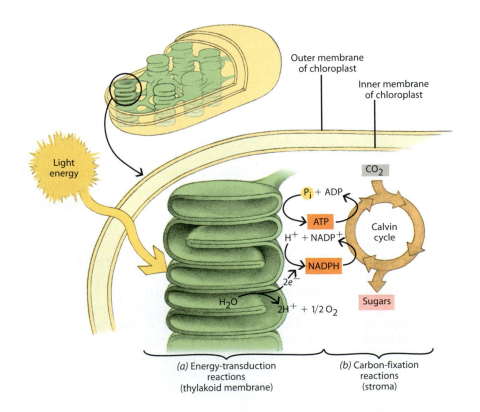

(a) Energy-transduction reactions (thylakoid membrane)

(b) Carbon-fixation reactions (stroma)

7–11

An overview of photosynthesis. Photosynthesis takes place in two stages: the energy-transduction reactions and the carbon-fixation reactions. **(a)** *In the energy-transduction reactions, light energy absorbed by chlorophyll a molecules in the thylakoid membrane is used indirectly to power the synthesis of ATP. Simultaneously, in the interior of the thylakoid, water is split into oxygen gas and hydrogen atoms (electrons and protons). The electrons are ultimately accepted by $NADP^+$ and H^+, producing NADPH.* **(b)** *In the carbon-fixation reactions, which occur in the stroma of the chloroplast, sugars are synthesized from carbon dioxide and the hydrogen carried by NADPH. This process is powered by the ATP and NADPH produced in the energy-transduction reactions. As we shall see, it involves a series of reactions, known as the Calvin cycle, that are repeated over and over.*

During the carbon-fixation reactions, the energy of ATP is used to link carbon dioxide covalently to an organic molecule, and the reducing power of NADPH is then used to reduce the newly fixed carbon atoms to a simple sugar. In the process, the chemical energy of ATP and NADPH is used to synthesize molecules suitable for transport (sugar) and storage (starch). At the same time, a carbon skeleton is generated from which all other organic molecules can be built. This conversion of carbon dioxide into organic compounds is known as **carbon fixation,** or **CO_2 fixation.**

Two Photosystems Are Involved in the Light Reactions

In the chloroplast (Figures 7–1, 7–8, and 7–11), the chlorophyll and other pigment molecules are embedded in the thylakoids in discrete units of organization called **photosystems** (Figure 7–12). Each photosystem includes an assembly of about 250 to 400 pigment molecules and consists of two closely linked components: an **antenna complex** and a **reaction center.** The antenna complex consists of pigment molecules that gather light energy and "funnel" it to the reaction center. The reaction center is made up of a complex of proteins and chlorophyll molecules that enable light energy to be converted into chemical energy. Within the photosystems, the chlorophyll molecules are bound to specific chlorophyll-binding membrane proteins and held in place to allow efficient capture of light energy.

All of the pigments within a photosystem are capable of absorbing photons, but only one pair of special chlorophyll *a* molecules per photosystem can actually use the energy in the photochemical reaction. This special pair of chlorophyll *a* molecules is situated at the core of the reaction center of the photosystem. The other pigment molecules, called **antenna pigments** because they are part of the light-gathering network, are located in the antenna complex. In addition to chlorophyll, varying amounts of carotenoid pigments are also located in each antenna complex.

Light energy absorbed by a pigment molecule anywhere in the antenna complex is transferred from one pigment molecule to the next by resonance energy transfer until it reaches the reaction center, with its special pair of chlorophyll *a* molecules (Figure 7–13). When either of the two chlorophyll *a* molecules of the reaction center absorbs energy, one of its electrons is boosted to a higher energy level and is transferred to an electron-acceptor molecule to initiate electron flow. The chlorophyll molecule is thus oxidized and positively charged.

Two different kinds of photosystems, Photosystem I and Photosystem II, are linked together by an electron transport chain. The photosystems were numbered in order of their discovery. In **Photosystem I,** the special chlorophyll *a* molecules of the reaction center are known

7–12

Internal surface of a thylakoid, as viewed by the freeze-fracture technique. The particles embedded within the membrane are thought to be photosystems, the structural units involved in the light reactions.

as P_{700}. The "P" stands for pigment and the subscript "700" designates the optimal absorption peak in nanometers. The reaction center of **Photosystem II** also contains a special form of chlorophyll *a*. Its optimal absorption peak is at 680 nanometers, and accordingly it is called P_{680}.

In general, Photosystem I and Photosystem II work together simultaneously and continuously. As we shall see, however, Photosystem I can operate independently.

In the Light Reactions, Electrons Flow from Water to Photosystem II to Photosystem I to NADP⁺

Figure 7–14 shows the present model of how the two photosystems work together. According to this model, light energy incident upon Photosystem II is absorbed by molecules of P_{680} in the reaction center, either directly or indirectly via one or more of the antenna molecules. When a P_{680} molecule is excited, its energized electron is transferred to a primary acceptor molecule, which transfers its extra electron to a secondary acceptor molecule and so on down the electron transport chain. By an incompletely understood reaction, the electron-deficient P_{680} molecule is able to replace its electrons one at a time by extracting them from water molecules. As soon as four electrons have been removed from two water molecules, requiring four photons, the two water molecules are split, yielding four electrons, four protons, and oxygen gas:

$$2H_2O \longrightarrow 4e^- + 4H^+ + O_2$$

135

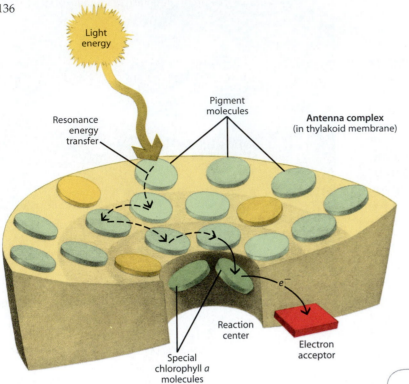

Light energy

Resonance energy transfer

Pigment molecules

Antenna complex (in thylakoid membrane)

Reaction center

Special chlorophyll *a* molecules

e^-

Electron acceptor

7–13

Energy transfer during photosynthesis. Light energy absorbed by a pigment molecule anywhere in the antenna complex passes by resonance energy transfer from one pigment molecule to another until it reaches one of two special chlorophyll a molecules at the reaction center. When a chlorophyll a molecule at the reaction center absorbs the energy, one of its electrons is boosted to a higher energy level and is transferred to an electron-acceptor molecule.

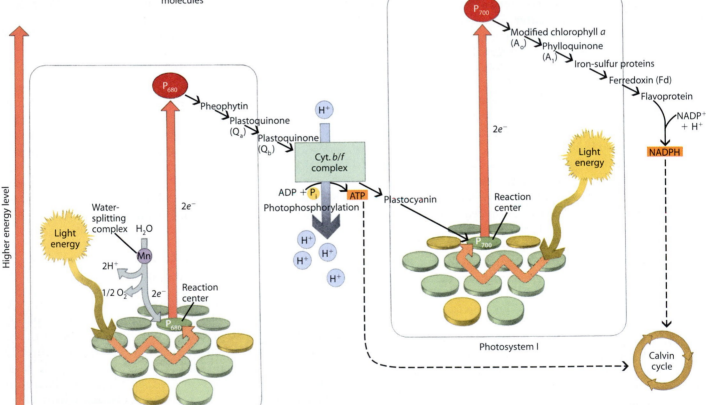

Higher energy level

P_{680} → Pheophytin → Plastoquinone (Q_a) → Plastoquinone (Q_b)

Water-splitting complex

H_2O

Mn

$2H^+$

$1/2\ O_2$

$2e^-$

Light energy

Reaction center

P_{680}

$2e^-$

H^+

Cyt. *b/f* complex

ADP + P_i → ATP

Photophosphorylation

H^+ H^+ H^+ H^+

Plastocyanin

Photosystem II

P_{700} → Modified chlorophyll *a* (A_o) → Phylloquinone (A_1) → Iron-sulfur proteins → Ferredoxin (Fd) → Flavoprotein

$NADP^+ + H^+$

NADPH

$2e^-$

Light energy

Reaction center

P_{700}

Photosystem I

Calvin cycle

7–14

How the two photosystems work together: noncyclic electron flow and photophosphorylation. This zigzag scheme (Z scheme) shows the pathway of electron transfer from H_2O to $NADP^+$ that occurs during noncyclic electron flow, as well as the energy relationships. To raise the energy of electrons derived from H_2O to the energy level required to reduce $NADP^+$

to NADPH, each electron must be energized twice (broad red arrows) by photons absorbed in Photosystems I and II. After each excitation step, the high-energy electrons flow "downhill" via the electron-carrier chains shown (black arrows). Protons are pumped across the thylakoid membrane into the thylakoid lumen during the water-splitting

reaction and during electron transfer through the cytochrome b/f complex, producing the proton gradient that is central to ATP formation (see Figure 7–15 for the details of this process). The formation of ATP by noncyclic electron flow is called noncyclic photophosphorylation.

Chemiosmotic Coupling in Chloroplasts and Mitochondria

Although there are similarities between the chemiosmotic processes in chloroplasts and mitochondria—the occurrence of similar ATP synthase complexes and of electron transport chains that pump protons across a membrane as the electrons are passed along a series of carriers—there are some noteworthy differences as well. For example, in the mitochondrion, the high-energy electrons passed down the electron transport chain are obtained through the oxidation of food molecules. The chemical energy from the food molecules is used to make ATP. In the chloroplast, the photosystems capture light energy and use it to drive electrons to the top of the electron transport chain. Hence, the chloroplast does not need food to manufacture ATP. It uses light energy, which is transformed to chemical energy, to make ATP.

The spatial orientation of the ATP synthase complexes and the direction of proton flow also differ in chloroplasts and mitochondria. In the mitochondrion, protons are pumped out of the matrix into the intermembrane space. In the chloroplast, protons are pumped from the stroma into the thylakoid space. Thus, the intermembrane space serves as the reservoir of hydrogen ions that powers the ATP synthase complexes in the inner mitochondrial membrane, while the thylakoid space performs the same role for the ATP synthase complexes in the thylakoid membrane.

The proton gradient across the thylakoid membrane is substantial. As protons are pumped out of the stroma into the thylakoid space, the pH in the thylakoid space drops to about 5, while increasing to about 8 in the stroma. The resultant gradient of about three pH units represents a proton motive force of about 200 millivolts (mV) across the thylakoid membrane. Nearly all of this motive force is contributed by the pH gradient rather than by a membrane potential. It is this proton motive force that drives ATP synthesis by the ATP synthase complex in the thylakoid membrane.

The mitochondrial matrix, like the stroma of the chloroplast, also has a pH of about 8. The pH of the intermembrane space, however, is only about 7, or the same pH as that of the cytosol. Most of the proton motive force in the mitochondrion, therefore, is created by the resulting membrane potential rather than by a pH gradient. Note that in both chloroplast and mitochondrion, the catalytic end of the ATP synthase complex is located in a solution full of enzymes and with a pH of about 8. It is there (stroma and matrix, respectively) that the organelles' ATP is made.

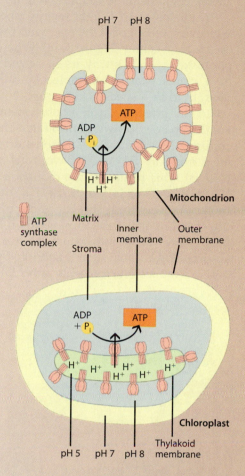

Comparison of the flow of protons and the position of the ATP synthase complexes in the membranes of mitochondria and chloroplasts.

Compartments of the organelles with similar pH have been colored the same.

This light-dependent oxidative splitting of water molecules is called **photolysis**. The water-splitting enzyme complex that catalyzes the photolysis of water is located on the inside of the thylakoid membrane, and the protons are released into the lumen of the thylakoid, or thylakoid space, not directly into the stroma of the chloroplast. Thus, the photolysis of water molecules contributes to the generation of a proton gradient across the thylakoid membrane—the sole means by which ATP is generated during photosynthesis. Manganese is an essential cofactor for the oxygen-evolving mechanism.

The components of the electron transport chain between the two photosystems resemble those of the electron transport chain of respiration: quinones, cytochromes, and iron-sulfur proteins are involved. In addition, photosynthetic electron transport involves the copper-containing protein plastocyanin and a chlorophyll derivative, known as pheophytin, in which the central magnesium atom has been replaced by two hydrogen atoms. The electron transport chain is so arranged that the transfer of electrons between successive carriers in the chain is coupled obligatorily to the pumping of protons across the thylakoid membrane, thereby generating the electrochemical proton gradient that is capable of driving ATP synthesis. ATP synthase complexes, embedded in the thylakoid membrane, provide a channel through which the protons can flow down the gradient, back into the stroma of the chloroplast. As they do so, the potential energy of the gradient drives the synthesis of ATP from ADP and P_i. This process is entirely analogous to proton-driven ATP synthesis in the mitochondrion but is called **photophosphor-**

ylation here to emphasize that light provides the energy used to maintain the proton gradient. Thus, chloroplasts and mitochondria generate ATP by the same basic mechanism: chemiosmotic coupling (Figure 7–15).

In Photosystem I, light energy photoexcites antenna molecules, which pass the energy to the P_{700} molecules at the reaction center (Figure 7–14). When a P_{700} molecule is photoexcited in this way, its energized electron is passed to a primary acceptor molecule called A_o, which is thought to be a special chlorophyll with a function similar to the pheophytin of Photosystem II. The electrons are then passed downhill through a chain of carriers, including phylloquinone (A_1), iron-sulfur proteins, including ferredoxin (Fd), and a flavoprotein, to the coenzyme $NADP^+$. This results in the reduction of the $NADP^+$ to NADPH and oxidation of the P_{700} molecule. The electrons removed from the P_{700} molecule are replaced by the electrons that have moved down the electron transport chain from Photosystem II. Two photons must be absorbed by Photosystem II and two by Photosystem I in order to reduce one molecule of $NADP^+$ to NADPH and

to produce one atom of oxygen—one-half of an O_2 molecule—from H_2O.

Thus, in the light, electrons flow continuously from water through Photosystems II and I to $NADP^+$, resulting in the oxidation of water to oxygen (O_2) and the reduction of $NADP^+$ to NADPH. This unidirectional flow of electrons from water to $NADP^+$ is called **noncyclic electron flow,** and the ATP production that occurs is called **noncyclic photophosphorylation.**

The free energy change (ΔG) for the reaction

$$H_2O + NADP^+ \longrightarrow NADPH + H^+ + \tfrac{1}{2}O_2$$

is 51 kilocalories per mole. The energy of 700-nanometer light is about 40 kilocalories per mole of photons. Because four photons are required to boost two electrons to the level of NADPH, about 160 kilocalories are available. Approximately one-third (51/160) of the available energy is captured as NADPH. The total energy harvest from noncyclic electron flow (based on the passage of 6 pairs of electrons from H_2O to $NADP^+$) is 6 ATP and 6 NADPH.

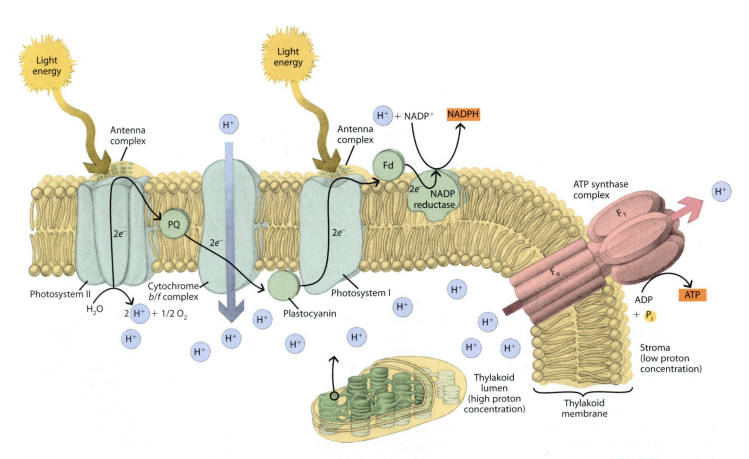

7–15

The chemiosmotic coupling mechanism of photophosphorylation. Electrons move from H_2O through Photosystem II, the intermediate chain of electron carriers, Photosystem I, and finally to $NADP^+$. (PQ stands for a pool of
plastoquinones, mobile electron carriers, which are lipid soluble.) Protons are pumped from the stroma into the thylakoid lumen by the flow of electrons through the chain of carriers between Photosystem II and Photosystem I.
This creates an electrochemical proton gradient, which drives ATP synthesis as the protons flow down this gradient through an ATP synthase complex in the membrane and reenter the stroma.

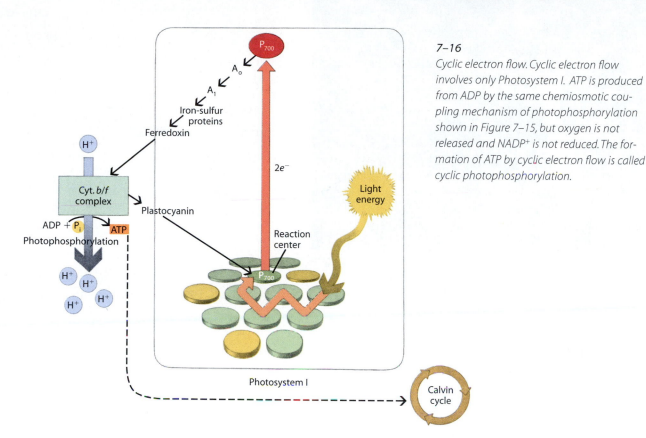

7–16
Cyclic electron flow. Cyclic electron flow involves only Photosystem I. ATP is produced from ADP by the same chemiosmotic coupling mechanism of photophosphorylation shown in Figure 7–15, but oxygen is not released and $NADP^+$ is not reduced. The formation of ATP by cyclic electron flow is called cyclic photophosphorylation.

Cyclic Photophosphorylation Generates Only ATP

As mentioned previously, Photosystem I can work independently of Photosystem II. In this process, called **cyclic electron flow,** energized electrons are transferred from P_{700} to A_o, as before (Figure 7–16). Instead of being passed downhill to $NADP^+$, however, the electrons are shunted to an acceptor in the electron transport chain between Photosystems I and II. The electrons then pass downhill through that chain back into the reaction center of Photosystem I, driving the transport of protons across the thylakoid membrane and hence powering the generation of ATP. Because this process involves a cyclic flow of electrons, it is called **cyclic photophosphorylation.** It is believed that the most primitive photosynthetic mechanisms worked in this way, and this is apparently the way in which some bacteria carry out photosynthesis. Photosynthetic eukaryotes (plants and algae) are also able to synthesize ATP by cyclic electron flow. However, no water is split, no O_2 is evolved, and no NADPH is formed. The only product is ATP.

As we have seen, the total energy harvest from noncyclic electron flow (based on the passage of 6 pairs of electrons from H_2O to $NADP^+$) is 6 ATP and 6 NADPH. Yet, the carbon-fixation reactions that we are about to encounter inherently require more ATP than NADPH. Cyclic photophosphorylation is therefore an ongoing necessity to meet the needs of the Calvin cycle, as well as to drive a host of other energy-requiring processes within the chloroplast.

The Carbon-Fixation Reactions

In the second series of photosynthetic reactions, the ATP and NADPH generated by the light reactions are used to fix and reduce carbon and to synthesize simple sugars. Carbon is available to photosynthetic cells in the form of carbon dioxide. For algae and cyanobacteria, this carbon dioxide is found dissolved in the surrounding water. In most plants, carbon dioxide reaches the photosynthetic cells through special openings, called stomata, in the leaves and green stems (Figure 7–17).

In the Calvin Cycle, CO_2 Is Fixed via a Three-Carbon Pathway

In many plant species, the reduction of carbon occurs exclusively in the stroma of the chloroplast by means of a series of reactions often called the **Calvin cycle** (named after its discoverer, Melvin Calvin, who received a Nobel Prize in 1961 for his work on the elucidation of this pathway). The Calvin cycle is analogous to other metabolic cycles (page 115) in that, by the end of each turn of the cycle, the starting compound is regenerated. The starting (and ending) compound in the Calvin cycle is a five-carbon sugar with two phosphate groups known as **ribulose 1,5-bisphosphate (RuBP).**

The Calvin cycle occurs in three stages. The first stage begins when carbon dioxide enters the cycle and is enzymatically combined with, or "fixed" (bonded cova-

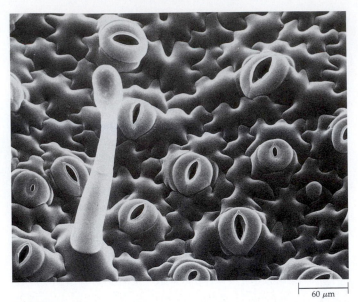

7–17

Scanning electron micrograph of stomata on the lower surface of a tobacco (Nicotiana tabacum) leaf. A plant can open or close its stomata as needed; in this micrograph, most of them are open. It is through the stomata that the carbon dioxide required for photosynthesis diffuses into the interior of the leaf and the oxygen produced as a by-product of photosynthesis diffuses out. The structure extending from the surface at the lower left is a leaf hair.

7–18

The first stage of the Calvin cycle. Melvin Calvin and his coworkers, Andrew A. Benson and James A. Bassham, briefly exposed photosynthesizing algae to radioactive carbon dioxide ($^{14}CO_2$), then killed the cells in boiling alcohol and separated the various ^{14}C-containing compounds by two-dimensional paper chromatography. They found that the various intermediates became radioactively labeled. They deduced that radioactive carbon is covalently linked to a molecule of ribulose 1,5-bisphosphate (RuBP). The resulting six-carbon compound then immediately splits to form two molecules of 3-phosphoglycerate (PGA). The radioactive carbon atom, indicated here in orange, appears in one of the two molecules of PGA.

lently) to, RuBP. The resultant six-carbon compound, an unstable enzyme-bound intermediate, is immediately hydrolyzed to generate two molecules of 3-phosphoglycerate, or 3-phosphoglyceric acid (PGA) (Figure 7–18). Each PGA molecule—the first detectable product of the Calvin cycle—contains three carbon atoms. Hence, the Calvin cycle is also known as the **C₃ pathway.**

RuBP carboxylase/oxygenase, often called **Rubisco** for short, is the enzyme that catalyzes this crucial initial reaction. (The enzyme's oxygenase activity is discussed later in this chapter.) Rubisco is undoubtedly the world's most abundant enzyme; by some estimates it may account for as much as 40 percent of the total soluble protein of most leaves.

In the second stage of the cycle, 3-phosphoglycerate is reduced to glyceraldehyde 3-phosphate, or 3-phosphoglyceraldehyde (PGAL) (Figure 7–19). This occurs in

7–19

The second stage of the Calvin cycle involves the conversion of 3-phosphoglycerate (PGA) to glyceraldehyde 3-phosphate (PGAL) in two steps. (a) In the first step of the sequence, the enzyme 3-phosphoglycerate kinase in the stroma catalyzes the transfer of phosphate from ATP to PGA, yielding 1,3-bisphosphoglycerate. (b) In the second step, NADPH donates electrons in a reduction catalyzed by glyceraldehyde 3-phosphate dehydrogenase, producing PGAL. In addition to its role as an intermediate in CO_2 fixation, PGAL has several possible fates in the plant cell. It may be oxidized via glycolysis for energy production or used for hexose synthesis.

two steps that are essentially the reversal of the corresponding steps in glycolysis, with one exception: the nucleotide cofactor for the reduction of 1,3-bisphosphoglycerate is NADPH, not NADH. Note that it takes the fixation of three molecules of CO_2 to three molecules of ribulose 1,5-bisphosphate to form six molecules of glyceraldehyde 3-phosphate.

In the third stage of the cycle, five of the six molecules of glyceraldehyde 3-phosphate are used to regenerate three molecules of ribulose 1,5-bisphosphate, the starting material.

The complete cycle is summarized in Figure 7–20. As in every metabolic pathway, each step in the Calvin cycle is catalyzed by a specific enzyme. At each full turn of the cycle, a molecule of carbon dioxide enters the cycle and is reduced, and a molecule of RuBP is regenerated. Three turns of the cycle, with the introduction of

three atoms of carbon, are necessary to produce one molecule of glyceraldehyde 3-phosphate, the phosphorylated form of the $C_3H_6O_3$ in the last equation on page 128. The overall equation for the production of one molecule of glyceraldehyde 3-phosphate is:

$$3CO_2 + 9ATP + 6NADPH + 6H^+ \longrightarrow$$
$$\text{Glyceraldehyde 3-phosphate} + 9ADP + 8P_i +$$
$$6NADP^+ + 3H_2O$$

(Note again that the Calvin cycle requires more ATP than NADPH, hence the need for the ATP generated by cyclic photophosphorylation.)

The immediate product of the cycle is **glyceraldehyde 3-phosphate,** the primary molecule transported from the chloroplast to the cytosol of the cell. This same triose phosphate ("triose" means a three-carbon sugar) is formed when the fructose 1,6-bisphosphate molecule is

7–20

Summary of the Calvin cycle. At each full "turn" of the cycle, one molecule of carbon dioxide (CO_2) enters the cycle. Three turns are summarized here—the number required to make one molecule of glyceraldehyde 3-phosphate (PGAL), the equivalent of one molecule of a three-carbon sugar. The energy that drives the Calvin cycle is provided in the form of ATP and NADPH, produced by the light reactions of photosynthesis.

(a) Stage 1: Fixation. The cycle begins at the upper left when three molecules of ribulose 1,5-bisphosphate (RuBP), a five-carbon compound, are combined with three molecules of carbon dioxide. This produces three molecules of an unstable intermediate that splits apart immediately, yielding six molecules of 3-phosphoglycerate (PGA), a three-carbon compound. (b) Stage 2: Reduction. The six molecules of PGA are reduced to six molecules of glyceraldehyde 3-phosphate (PGAL). (c) Stage 3: Regeneration of acceptor. Five of the six PGAL molecules are combined and rearranged to form three five-carbon molecules of RuBP. The one "extra" molecule of PGAL represents the net gain from the Calvin cycle. PGAL serves as the starting point for the synthesis of sugars, starch, and other cellular components.

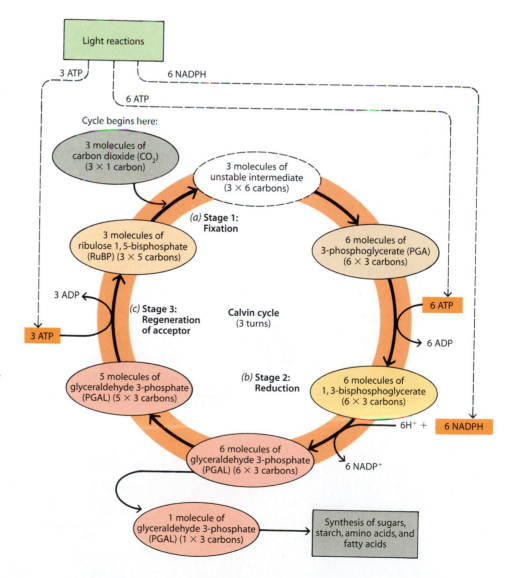

split at the fourth step in glycolysis, and it is interconvertible with another triose phosphate, dihydroxyacetone phosphate (see Figure 6–4). Using the energy provided by hydrolysis of the phosphate bonds, the first four steps of glycolysis can be reversed to form glucose from glyceraldehyde 3-phosphate.

Most Fixed Carbon Is Converted to Sucrose or Starch

As we have mentioned, although glucose commonly is represented as the carbohydrate product of photosynthesis in summary equations, in reality, very little free glucose is generated in photosynthesizing cells. Rather, most of the fixed carbon is converted either to **sucrose,** the major transport sugar in plants, or to **starch,** the major storage carbohydrate in plants (see Chapter 2).

Much of the glyceraldehyde 3-phosphate produced by the Calvin cycle is exported to the cytosol, where, through a series of reactions, it is converted to sucrose. Most of the glyceraldehyde 3-phosphate that remains in the chloroplast is converted to starch, which is stored temporarily during the light period as starch grains in the stroma (Figure 7–1). At night, sucrose is produced from starch and exported from the leaf via the vascular bundles to other parts of the plant.

Photorespiration Occurs When Rubisco Binds O_2 Instead of CO_2

In the presence of ample carbon dioxide, the enzyme Rubisco catalyzes the carboxylation of ribulose 1,5-bisphosphate with great efficiency. Under such conditions, the thermodynamic efficiency for the Calvin cycle is close to 90 percent, and the maximum overall thermodynamic efficiency for photosynthesis about 33 percent. (Most of the light energy is lost in the generation of ATP and NADPH by the light reactions.)

As mentioned previously, however, Rubisco is not absolutely specific for CO_2 as a substrate. Oxygen competes with CO_2 at the active site, and Rubisco catalyzes the condensation of O_2 with RuBP to form one molecule of 3-phosphoglycerate and one of **phosphoglycolate** (Figure 7–21). This is the enzyme's oxygenase activity, as reflected in its name: RuBP carboxylase/oxygenase. No carbon is fixed during this reaction, and energy must be expended to salvage the carbons from phosphoglycolate, which is not a useful metabolite.

7–21
Reactions catalyzed by RuBP carboxylase/oxygenase (Rubisco). (a) Rubisco's carboxylase activity—fixation of CO_2 in the Calvin cycle—is favored by high CO_2 and low oxygen concentrations (see also Figure 7–18). (b) The oxygenase action of Rubisco also occurs to a significant extent, especially in the presence of low CO_2 and high oxygen concentrations (normal atmospheric concentrations). The oxygenase action of Rubisco lessens the efficiency of photosynthesis since only one molecule of 3-phosphoglycerate (PGA) is formed from ribulose 1,5-bisphosphate (RuBP) rather than two, as in Rubisco's carboxylase activity. The oxygenase activity of Rubisco combined with the salvage pathway (see Figure 7–22) consumes O_2 and releases CO_2, a process called photorespiration.

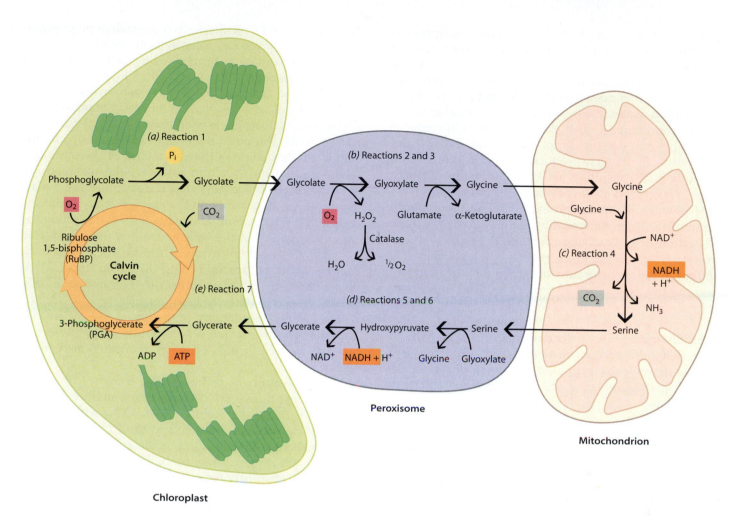

Chloroplast

Peroxisome

Mitochondrion

7–22

The pathway by which phosphoglycolate formed during photorespiration is salvaged by conversion into serine and thus to 3-phosphoglycerate (PGA), which is fed into the Calvin cycle. **(a)** *Reaction 1: Phosphoglycolate is dephosphorylated in chloroplasts to form glycolate.* **(b)** *Reactions 2 and 3: In peroxisomes, the glycolate is oxidized to glyoxylate, which is then transaminated to glycine.*

(c) *Reaction 4: In mitochondria, two glycine molecules condense to form serine and the CO_2 that is released during photorespiration.* **(d)** *Reactions 5 and 6: In peroxisomes, the serine is transaminated to hydroxypyruvate, which is then reduced to glycerate. The glycerate then enters the chloroplasts.* **(e)** *Reaction 7: The glycerate is phosphorylated to PGA,*

which rejoins the Calvin cycle. Oxygen is consumed at two points in the photorespiratory pathway, once in the chloroplast (the oxygenase activity of Rubisco) and once in the peroxisome (oxidation of glycolate to glyoxylate). CO_2 is released at one point in the mitochondrion (condensation of two glycine molecules to form one molecule of serine).

The salvage pathway is long and employs three cellular organelles: chloroplast, peroxisome, and mitochondrion (Figure 7–22). The pathway involves, in part, the conversion of two molecules of phosphoglycolate into a molecule of the amino acid serine (which has three carbons) and a molecule of CO_2. The oxygenase activity of Rubisco combined with the salvage pathway *consumes* O_2 and *releases* CO_2, a process called **photorespiration.** Unlike mitochondrial respiration (often referred to as "dark respiration" to distinguish it from photorespiration, which occurs only in the light), photorespiration is a wasteful process, yielding neither ATP nor NADH. In

some plants, as much as 50 percent of the carbon fixed in photosynthesis may be reoxidized to carbon dioxide during photorespiration. Apparently the evolution of Rubisco produced an active site that is not able to discriminate between CO_2 and O_2, perhaps because much of its evolution occurred before O_2 was an important component of the atmosphere.

The condensation of O_2 with RuBP occurs concurrently with CO_2 fixation under today's atmospheric conditions, with an atmosphere consisting of 21 percent O_2 and only 0.036 percent (360 parts per million, or ppm) CO_2. Moreover, conditions that can alter the CO_2 to O_2

ratio in favor of O_2 and, hence, enhance photorespiration are quite common. Carbon dioxide is not continuously available to the photosynthesizing cells of a plant. As we have seen, it enters the leaf by way of the stomata, specialized pores that open and close depending on, among other factors, water stress. When a plant is subjected to hot, dry conditions, it must close its stomata to conserve water. This cuts off the supply of carbon dioxide and also allows the oxygen produced by photosynthesis to accumulate. The resulting low carbon dioxide and high oxygen concentrations favor photorespiration.

Also, when plants are growing in close proximity to one another, the air surrounding the leaves may be quite still, with little gas exchange between the immediate environment and the atmosphere as a whole. Under such conditions, the concentration of carbon dioxide in the air closest to the leaves may be rapidly reduced to low levels by the photosynthetic activities of the plant. Even if the stomata are open, the concentration gradient between the outside of the leaf and the inside may be so slight that little carbon dioxide diffuses into the leaf. Meanwhile, oxygen accumulates, favoring photorespiration and greatly reducing the photosynthetic efficiency of the plants.

The Four-Carbon Pathway Is a Solution to Photorespiration

The Calvin cycle is not the only pathway used in the carbon-fixation reactions. In some plants, the first detectable product of CO_2 fixation is not the three-carbon molecule 3-phosphoglycerate, but rather the four-carbon molecule oxaloacetate, which is also an intermediate in the Krebs cycle. Plants that employ this C_4 **pathway,** along with the Calvin cycle, are commonly called C_4 **plants** (for four-carbon), as distinct from the C_3 **plants,** which use *only* the Calvin cycle. The C_4 pathway is also referred to as the Hatch–Slack pathway after

M. D. Hatch and C. R. Slack, two Australian plant physiologists who played key roles in its elucidation.

The oxaloacetate is formed when carbon dioxide is fixed to phosphoenolpyruvate (PEP) in a reaction catalyzed by the enzyme **PEP carboxylase,** which is found in the cytosol of mesophyll cells of C_4 plants (Figure 7–23). The oxaloacetate is then reduced to malate or converted, with the addition of an amino group, to the amino acid aspartate in the chloroplast of the same cell. These steps, using CO_2 from adjacent air spaces, occur in **mesophyll cells.** The next step is a surprise: the malate (or aspartate, depending on the species) moves from the mesophyll cells to **bundle-sheath cells** surrounding the vascular bundles of the leaf, where it is decarboxylated to yield CO_2 and pyruvate. The CO_2 then enters the Calvin cycle by reacting with RuBP to form PGA. Meanwhile, the pyruvate returns to the mesophyll cells, where it reacts with ATP to regenerate PEP (Figure 7–24). Hence, the anatomy of the leaves of C_4 plants establishes a *spatial separation* between the C_4 pathway and the Calvin cycle, which occur in two different types of cells.

The two primary carboxylating enzymes of photosynthesis use different forms of the carbon dioxide molecule as a substrate. Rubisco uses CO_2, whereas PEP carboxylase uses the hydrated form of carbon dioxide, the bicarbonate ion (HCO_3^-), as its substrate. PEP carboxylase has a high affinity for bicarbonate. Hence, it operates very efficiently even when the concentration of its substrate is quite low.

Typically, the leaves of C_4 plants are characterized by an orderly arrangement of the mesophyll cells around a layer of large bundle-sheath cells, so that together the two form concentric layers around the vascular bundle (Figure 7–25). This wreathlike arrangement has been termed **Kranz anatomy** (*Kranz* is the German word for "wreath"). In some C_4 plants, the chloroplasts of the mesophyll cells have well-developed grana while those

7–23

Carbon fixation by the C_4 pathway. Carbon dioxide is "fixed" to phosphoenolpyruvate (PEP) by the enzyme PEP carboxylase. PEP carboxylase uses the hydrated form of CO_2, which is HCO_3^- (bicarbonate ion). Depending on the species, the resulting oxaloacetate is either reduced to malate or transaminated to aspartate through the addition of an amino group (—NH$_2$). The malate or aspartate moves into the bundle-sheath cells where CO_2 is released for use in the Calvin cycle. Notice that PEP contains a phosphoanhydride bond. Like ATP, PEP is a high-energy compound.

7–24

The pathway for carbon fixation in the maize plant (Zea mays), a C_4 plant. Carbon dioxide is first fixed in mesophyll cells as oxaloacetate, which is rapidly converted to malate. The malate is then transported to bundle-sheath cells, where the CO_2 is released to enter the Calvin cycle, ultimately yielding sugars and starch. Pyruvate returns to the mesophyll cell for regeneration of phosphoenolpyruvate (PEP). Hence, there is a spatial separation between the C_4 pathway, which occurs in the mesophyll cells, and the Calvin cycle, which takes place in the bundle-sheath cells.

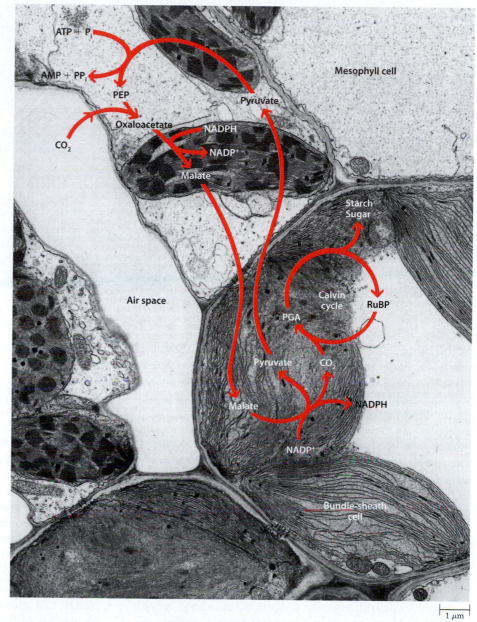

7–25

Transverse section of a portion of a maize (Zea mays) leaf. As is typical of C_4 plants, the vascular bundles (composed of xylem and phloem) are surrounded by large, chloroplast-containing bundle-sheath cells that are, in turn, surrounded by a layer of mesophyll cells. The C_4 pathway takes place in the mesophyll cells; the Calvin cycle occurs in the bundle-sheath cells. Seen here are four vascular bundles—one large and three small. The uptake of sugar from the mesophyll occurs largely in the small bundles. The large bundles are involved primarily with the export of the sugar from the leaf to other parts of the plant.

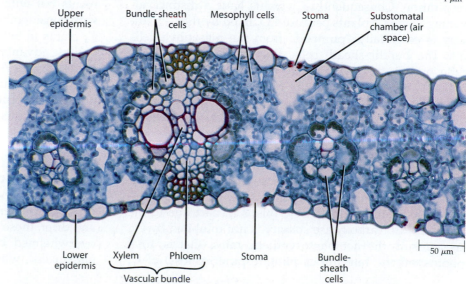

7–26

Electron micrograph showing portions of chloroplasts in a mesophyll cell (above) and a bundle-sheath cell (below) of a maize (Zea mays) leaf. Compare the well-developed grana of the mesophyll cell chloroplast with the poorly developed grana of the bundle-sheath cell chloroplast. Note the plasmodesmata in the wall between these two cells. The intermediates of photosynthesis move from one cell to the other via the plasmodesmata in this C_4 plant.

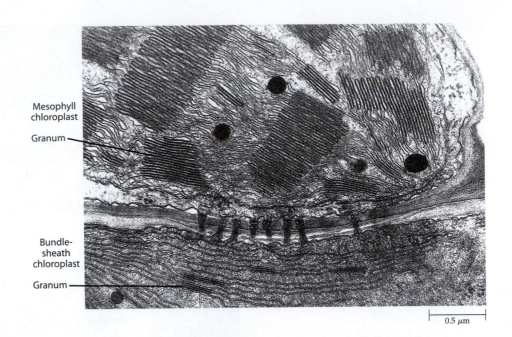

Mesophyll chloroplast

Granum

Bundle-sheath chloroplast

Granum

0.5 μm

of the bundle-sheath cells have either poorly developed grana or none at all (Figure 7–26). In addition, when photosynthesis is occurring, the bundle-sheath chloroplasts commonly form larger and more numerous starch grains than the mesophyll chloroplasts.

Photosynthesis Is Generally More Efficient in C_4 Plants Than in C_3 Plants Fixation of CO_2 in C_4 plants has a larger energy cost than in C_3 plants. For each molecule of CO_2 fixed in the C_4 pathway, a molecule of PEP must be regenerated at the cost of two phosphate groups of ATP (Figure 7–24). Therefore, C_4 plants need five ATP molecules to fix one molecule of CO_2, whereas C_3 plants need only three. One might well ask why C_4 plants have evolved such an energetically expensive method of providing CO_2 to the Calvin cycle.

High CO_2 and low O_2 concentrations limit photorespiration. Consequently, C_4 plants have a distinct advantage over C_3 plants because CO_2 fixed by the C_4 pathway is essentially "pumped" from the mesophyll cells into the bundle-sheath cells, thus maintaining a high ratio of CO_2 to O_2 at the site of Rubisco activity. This high CO_2:O_2 ratio favors the carboxylation of RuBP. In addition, since both the Calvin cycle and photorespiration are localized in the inner, bundle-sheath layer of cells, any CO_2 liberated by photorespiration into the outer, mesophyll layer can be refixed by the C_4 pathway that operates there. The CO_2 liberated by photorespiration can thus be prevented from escaping from the leaf. Moreover, compared with C_3 plants, C_4 plants are superior utilizers of available CO_2; this is in part due to the fact that PEP carboxylase activity is not inhibited by O_2. As a result, the net photosynthetic rates (that is, total photosynthetic rate minus photorespiratory loss) of C_4

grasses, for example, can be two to three times the rates of C_3 grasses under the same environmental conditions. In short, the gain in efficiency from the elimination of photorespiration in C_4 plants more than compensates for the energetic cost of the C_4 pathway. Maize (*Zea mays*), sugarcane (*Saccharum officinale*), and sorghum (*Sorghum vulgare*) are examples of C_4 grasses. Wheat (*Triticum aestivum*), rye (*Secale cereale*), oats (*Avena sativa*), and rice (*Oryza sativa*) are examples of C_3 grasses.

C_4 plants evolved primarily in the tropics and are especially well adapted to high light intensities, high temperatures, and dryness. The optimal temperature range for C_4 photosynthesis is much higher than that for C_3 photosynthesis, and C_4 plants flourish even at temperatures that would eventually be lethal to many C_3 species. Because of their more efficient use of carbon dioxide, C_4 plants can attain the same photosynthetic rate as C_3 plants but with smaller stomatal openings and, hence, with considerably less water loss. The preponderance of C_4 plants in hotter, drier climates may be an expression of these advantages of C_4 photosynthesis at high temperatures (Figure 7–27). In addition, C_4 plants have three to six times less Rubisco than C_3 plants, and the overall leaf nitrogen content in C_4 plants is less than in C_3 plants; hence, C_4 plants are able to use nitrogen more efficiently than C_3 plants.

A familiar example of the competitive capacity of C_4 plants is seen in lawns in the summertime. In most parts of the United States, lawns consist mainly of C_3 grasses, such as Kentucky bluegrass (*Poa pratensis*) and creeping bent (*Agrostis tenuis*). As the summer days become hotter and drier, these dark green, fine-leaved grasses are often overwhelmed by rapidly growing crabgrass (*Digitaria sanguinalis*), which disfigures the lawn as its yellowish-green, broader-leafed plants slowly take over. Crabgrass, you will not be surprised to learn, is a C_4 plant.

All of the plants now known to utilize C_4 photosynthesis are flowering plants (angiosperms), including at least 19 families, 3 of which are monocots and 16 eudicots. No family has been found, however, that contains only C_4 species. This pathway has undoubtedly arisen independently many times in the course of evolution.

Species have been discovered in several genera with photosynthetic characteristics intermediate between those of C_3 and C_4 species. These so-called **C_3–C_4 intermediates,** which are characterized by Kranz-like leaf anatomy, partially suppressed photorespiration, and a reduced sensitivity to O_2, are considered by some plant biologists as evidence of stepwise evolution of the C_4 pathway from C_3 ancestors.

Plants Having Crassulacean Acid Metabolism Can Fix CO_2 in the Dark

Another strategy for CO_2 fixation has evolved independently in many succulents, including cacti and stonecrops. Because it was first recognized in stonecrops (family *Crassulaceae*), it is called **crassulacean acid metabolism (CAM).** Plants that take advantage of CAM photosynthesis are called **CAM plants.** CAM plants, like C_4 plants, use both C_4 and C_3 pathways. In CAM plants, however, there is a *temporal separation*—a separation in time—rather than a spatial separation of the two pathways (Figure 7–28).

Plants are considered to have CAM if their photosynthetic cells have the ability to fix CO_2 in the dark through the activity of PEP carboxylase in the cytosol. The initial carboxylation product is oxaloacetate, which is immediately reduced to malate. The malate so formed is stored as malic acid in the vacuole and is detectable as a sour taste. During the following light period, the malic acid is recovered from the vacuole, decarboxylated, and the CO_2 transferred to RuBP of the Calvin cycle *within the same cell* (Figure 7–29). Thus, a structural precondition of all CAM plants is the presence of cells that have both (1) large vacuoles, in which the malic acid can be temporarily stored in aqueous solution, and (2) chloroplasts, where the CO_2 obtained from the malic acid can be transformed into carbohydrates.

CAM plants are largely dependent upon nighttime accumulation of CO_2 for their photosynthesis because their stomata are closed during the day, retarding water loss. This is obviously advantageous in the conditions of high light intensity and water stress under which most CAM plants live. If all atmospheric CO_2 uptake in a CAM plant occurs at night, the water-use efficiency of that plant can be many times greater than that of a C_3 or C_4 plant. Typically, a CAM plant loses 50 to 100 grams of

7–27

Distribution of C_4 grasses. Analyses of the geographic distribution of C_4 species in North America have shown that they are generally most abundant in climates with high temperatures. Differences appear to exist, however, between monocots and eudicots in the particular type of high temperature favored. For example, C_4 grasses are most abundant in regions having the highest temperatures during the growing season. In contrast, C_4 eudicots are most abundant in regions having the greatest aridity during the growing season. The numbers shown here indicate the percentages of the total number of grass species having the C_4 pathway in each of 32 local floras in North America. The highest percentages are found in those regions with the highest temperatures during the growing season.

Sugarcane
C$_4$ plant

Pineapple
CAM plant

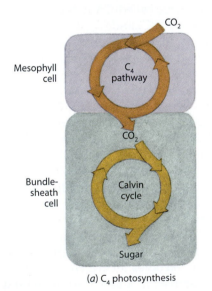

Mesophyll
cell

Bundle-
sheath
cell

Stage 1:
Initial fixation
of CO$_2$ to form
4-carbon acids

Stage 2:
Release of CO$_2$
to Calvin
cycle

(a) C$_4$ photosynthesis

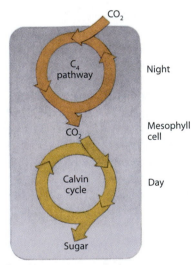

Night

Mesophyll
cell

Day

(b) CAM photosynthesis

7–28

Comparison of C$_4$ and CAM photosynthesis. C$_4$ and CAM plants utilize both C$_4$ and C$_3$ (Calvin cycle) pathways, with CO$_2$ initially being incorporated into four-carbon acids in the C$_4$ pathway. Subsequently, the CO$_2$ is transferred to the C$_3$ pathway, or Calvin cycle. (a) In C$_4$ plants, the two pathways occur in different cells; hence, they are said to be spatially separated (see Figure 7–24). **(b)** In CAM plants, by contrast, they are temporally separated, functioning at different times (see Figure 7–29). The C$_4$ pathway, or initial fixation of CO$_2$, occurs at *night, and the C$_3$ pathway functions during the day. The stomata of C$_4$ plants are open during the day and closed at night, while those of CAM plants are closed during the day and open at night.*

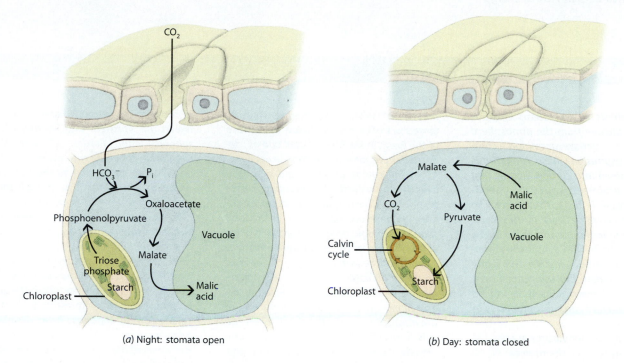

(a) Night: stomata open   (b) Day: stomata closed

7–29

Crassulacean acid metabolism (CAM) involves the formation of malic acid at night and its disappearance during daytime. CAM plants are therefore known as plants that taste sour at night and sweet during the day. (a) CO_2 is first fixed at night when the stomata are open. At night, starch from the chloroplast is broken down as far as phosphoenolpyruvate (PEP).

CO_2, hydrated to form HCO_3^- (bicarbonate ion), reacts with PEP to form oxaloacetate, which then is reduced to malate. Most of the malate is pumped into the vacuole and stored there as malic acid. (b) During the daytime the malic acid is recovered from the vacuole and decarboxylated, producing CO_2 and pyruvate. The CO_2 enters the Calvin cycle, where it is

refixed by Rubisco. Much of the pyruvate may be converted to sugars and starch by reverse glycolysis (note that the pathway shown in Figure 6–4 can proceed in both directions). Stomatal closure during the daytime prevents loss of water and of CO_2 released by decarboxylation of malate.

water for every gram of CO_2 gained, compared with 250 to 300 grams for C_4 plants and 400 to 500 grams for C_3 plants. During periods of prolonged drought, some CAM plants can keep their stomata closed both night and day, maintaining low metabolic rates by refixing the CO_2 produced by respiration.

Among the vascular plants, CAM is more widespread than C_4 photosynthesis. It has been reported in at least 23 families of flowering plants, mostly eudicots, including such familiar houseplants as the maternity plant, wax plant, and snake plant, a monocot. Not all CAM plants are highly succulent (fleshy); two examples of less succulent ones are the pineapple and Spanish "moss," both members of the family *Bromeliaceae* (monocots). Some nonflowering plants have also been reported to show CAM activity, including the bizarre gymnosperm *Welwitschia mirabilis* (see Figure 20–43 on page 492), the aquatic species of quillwort (*Isoetes*) (see Figure 19–19 on page 442), and some ferns. *Welwitschia*, however, fixes CO_2 predominantly in the way a C_3 plant does.

Each of the Carbon-Fixation Mechanisms Has Its Advantages and Disadvantages in Nature

The type of photosynthetic mechanism used by a plant is important, but it is not the only factor that determines where the plant lives. All three mechanisms—C_3, C_4, and CAM photosynthesis—have advantages and disadvantages, and a plant can compete successfully only when the benefits of its type of photosynthesis outweigh other factors. For example, although C_4 plants generally tolerate higher temperatures and drier conditions than C_3 species, C_4 plants may not compete successfully at temperatures below 25°C. This is, in part, because they are more sensitive to cold than C_3 species. In addition, as discussed, CAM plants conserve water by closing their stomata during the day, a practice that severely reduces their ability to take in and fix CO_2. Hence, CAM plants grow slowly and compete poorly with C_3 and C_4 species under conditions other than extreme aridity. Thus, to some extent, each photosynthetic type of plant has limitations imposed by its own photosynthetic mechanism.

The Carbon Cycle

In photosynthesis, living systems incorporate carbon dioxide from the atmosphere into organic, carbon-containing compounds. In respiration, these compounds are broken down into carbon dioxide and water. These processes, viewed on a worldwide scale, result in the **carbon cycle**. The principal photosynthesizers in this cycle are plants, phytoplankton, marine algae, and cyanobacteria. They synthesize carbohydrates from carbon dioxide and water and release oxygen into the atmosphere. About 100 billion metric tons of carbon per year are bound into carbon compounds by photosynthesis.

Some of the carbohydrates are used by the photosynthesizers themselves. Plants release carbon dioxide from their roots and leaves, and phytoplankton, marine algae, and cyanobacteria release carbon dioxide into the water where it maintains an equilibrium with the carbon dioxide of the air. Some 42,000 billion metric tons of carbon are "stored" as dissolved carbon dioxide in the oceans, and some 740 billion metric tons in the atmosphere. Some of the carbohydrates are used by animals that feed on live plants, on algae, and on one another, releasing carbon dioxide. An enormous amount of organic carbon is contained in the dead bodies of plants and other organisms as well as in discarded leaves and shells, feces, and other waste materials. All these materials settle into the soil or sink to the ocean floors, where in many cases they are consumed by decomposers—small invertebrates, bacteria, and fungi. Carbon dioxide is also released by these processes into the reservoir of the air and oceans. A vast amount of carbon—an estimated 60,000 billion metric tons—is fixed in the form of limestone (calcium carbonate). Limestone deposits represent the shells of former marine creatures. The carbon reenters the cycle if and when the limestone deposits are exposed to the atmosphere (by geologic uplift), and the process of weathering begins. Another large store of carbon lies below the surface of the Earth in the form of coal and oil, deposited there some 300 million years ago.

In the course of years and averaged over the entire Earth, the natural processes of photosynthesis and respiration essentially balance one another. Over the long span of geologic time, the carbon dioxide concentration of the atmosphere has varied, but for the last 10,000 years—at least up to the Industrial Revolution—it has remained relatively constant. By volume, carbon dioxide is a very small proportion of the atmosphere, only about 0.036 percent. It is important, however, because carbon dioxide, along with water vapor, methane, and other so-called greenhouse gases, absorbs infrared radiation. While allowing sunlight to pass through to Earth's surface, it blocks heat (infrared radiation) from being radiated back into space. Thus, the atmosphere warms up. Some of the heat from the atmosphere is transferred to the oceans and raises their temperature as well. As the atmosphere and oceans warm, the overall temperature of the Earth rises. Because the carbon dioxide and other gases trap the sun's radiation in much the same way as the glass of a greenhouse, global warming produced in this manner is known as the **greenhouse effect**. Since 1850, the concentration of carbon dioxide in the atmosphere has climbed from about 270 parts per million (ppm) to about 365 ppm today, owing mainly to our use of fossil fuels, such as coal, oil, and natural gas, to our plowing of the soil, and to our destruction and burning of forests, particularly in the tropics. During at least the last two decades of the twentieth century, the carbon dioxide content of the atmosphere has been rising at a rate of about 0.4 percent per year.

Although 95 percent of carbon dioxide from fossil fuels is released in the Northern Hemisphere, only a 3 ppm difference exists in atmospheric carbon dioxide levels between the Northern and Southern Hemispheres. Using models of atmospheric circulation together with estimates of carbon dioxide absorption by the oceans, researchers were unable to account for the very small difference in atmospheric carbon dioxide concentration between the two hemispheres. Strong evidence exists that the missing carbon is found in the boreal, or northern coniferous, forests of North America and Eurasia, pointing to these forests as major factors in the global carbon balance. (Interestingly, in black spruce forests, the moss layer assimilates and stores as much carbon dioxide as do the stems of the trees.) Whereas increasing concern over tropical deforestation and depletion of trees in temperate forests has slowed these practices, little has been done to deter the threat to the boreal forests.

At the beginning of the 1980s, a major study by the Environmental Protection Agency of the United States predicted that the increase in the carbon dioxide "blanket" would significantly increase the average temperatures on Earth, beginning by the turn of the century. A number of subsequent studies, coupled with worldwide temperatures in the 1980s and early 1990s that were higher than in any decade since records have been kept, have now convinced not only most members of the scientific community but also political leaders throughout the world that the increase is real and has already begun. If the warming trend continues, by the middle of the twenty-first century Earth's average temperature may increase by 1.5° to 4.5°C (2.5° to 7.5°F).

The consequences of this increase cannot be known with certainty. In some parts of the world, there may be lengthened growing seasons, increased precipitation, and, in conjunction with increased levels of carbon dioxide available to plants, greater agricultural productivity. In other parts of

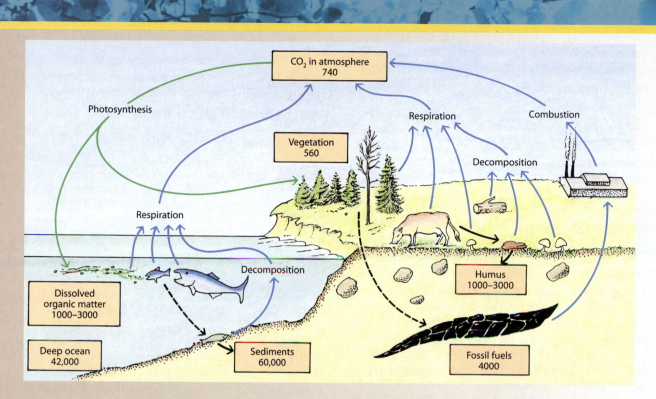

The carbon cycle. The arrows indicate the movement of carbon atoms. The numbers are all estimates of the amount of carbon, expressed in billions of metric tons, stored in various reservoirs as of 1988. The amount of carbon released by respiration and combustion has begun to exceed the amount fixed by photosynthesis. The amount of carbon present in the atmosphere is currently increasing by 3 billion metric tons each year.

the world, however, precipitation may be reduced, lowering crop yields and, in already arid areas, accelerating the spread of the great deserts of the world. Not all plant species respond in the same way to high levels of carbon dioxide. C_3 plants might be expected to respond dramatically to higher carbon dioxide levels with increased photosynthesis and growth as photorespiration is effectively minimized. The response of C_4 plants might not be so dramatic as they lose their competitive advantage over the C_3 plants. Rises in sea level, resulting from the melting of polar ice, pose a potential threat not only to human inhabitants of coastal regions but also to the multitude of marine organisms that live or reproduce in shallow waters at the continents' edges.

Although it now appears that global warming is the inevitable consequence of past and present human activities, national and international efforts are underway to develop strategies for agriculture, energy use, and manufacturing that will slow—and perhaps eventually reverse—the process. At the 1992 United Nations Earth Summit in Rio de Janeiro, for example, leaders of the world's industrialized countries signed the Framework Convention on Climate Change, with a goal of stabilizing carbon dioxide emissions by the year 2000. Subsequently, at the Kyoto Climate Conference, which was held in Kyoto, Japan, in December 1997, a Protocol was adopted to reduce greenhouse emissions. If fully implemented, the Kyoto Protocol has the potential "to redirect the Earth from the path of an overheating climate and to a safer world." It remains to be seen whether we have awakened in time.

Summary

In photosynthesis, light energy is converted to chemical energy and carbon is "fixed" into organic compounds. A complete, balanced equation for this reaction can be written as follows:

$$3CO_2 + 6H_2O \xrightarrow{\text{Light}} C_3H_6O_3 + 3O_2 + 3H_2O$$

Light Is Absorbed by Pigments Organized into Photosystems

The first step in photosynthesis is the absorption of light energy by pigment molecules. The pigments involved in eukaryotic photosynthesis include the chlorophylls and the carotenoids, which are packed in the thylakoids of chloroplasts as photosynthetic units called photosystems. Light absorbed by pigment molecules boosts their electrons to a higher energy level. Because of the way the pigment molecules are arranged in the photosystems, they are able to transfer this energy to special pigment molecules at the reaction centers. Most photosynthetic organisms contain two photosystems, Photosystem I and Photosystem II.

The many reactions that occur during photosynthesis are divided into two major processes: the energy-transduction reactions, or light reactions, and the carbon-fixation reactions.

In the Light Reactions, Electrons Flow from Water to Photosystem II, down an Electron Transport Chain to Photosystem I, and Finally to NADP+

In the currently accepted model of the light reactions, light energy enters Photosystem II, where it is trapped by pigment molecules and passed to the P_{680} chlorophyll molecules of the reaction center. Energized electrons are transferred from P_{680} to an electron acceptor. As the electrons are removed from P_{680}, they are replaced by low-energy electrons from water molecules, and oxygen is produced (photolysis). Pairs of electrons then pass downhill to Photosystem I along an electron transport chain. This passage generates a proton gradient that drives the synthesis of ATP from ADP and phosphate (photophosphorylation). Meanwhile, light energy absorbed in Photosystem I is passed to the P_{700} chlorophyll molecules of the Photosystem I reaction center. The energized electrons are ultimately accepted by the coenzyme molecule NADP+, and the electrons removed from P_{700} are replaced by the electrons from Photosystem II. The energy yield from the light-dependent reactions is stored in the molecules of NADPH and in the ATP formed by photophosphorylation. Photophosphorylation also occurs in cyclic electron flow, a process that does not require Photosystem II. The only product of cyclic electron flow is ATP. This extra ATP is required by the Calvin cycle, which uses ATP and NADPH in a 3:2 ratio.

In the Electron Transport Chain, Electron Flow Is Coupled to Proton Pumping and ATP Synthesis via a Chemiosmotic Mechanism

Like oxidative phosphorylation in mitochondria, photophosphorylation in chloroplasts is a chemiosmotic process. As electrons flow down the electron transport chain from Photosystem II to Photosystem I, protons are pumped from the stroma into the thylakoid lumen, creating a gradient of potential energy. As protons flow down this gradient from the thylakoid lumen back into the stroma, passing through an ATP synthase, ATP is formed.

In the Calvin Cycle, CO2 Is Fixed via a Three-Carbon Pathway

In the carbon-fixation reactions, which take place in the stroma of the chloroplast, the NADPH and ATP produced in the light reactions are used to reduce carbon dioxide to organic carbon. The Calvin cycle is responsible both for the initial fixation of carbon dioxide and for the subsequent reduction of the newly fixed carbon. In the Calvin cycle, a molecule of carbon dioxide combines with the starting compound, a five-carbon sugar called ribulose 1,5-bisphosphate (RuBP), to form two molecules of the three-carbon compound 3-phosphoglycerate (PGA). The PGA is then reduced to the three-carbon molecule glyceraldehyde 3-phosphate (PGAL). At each turn of the cycle, one carbon atom enters the cycle. Three turns of the cycle produce one molecule of glyceraldehyde 3-phosphate. At each turn of the cycle, RuBP is regenerated. Most of the fixed carbon is converted to either sucrose or starch.

The Carbon-Fixation Pathway in C4 Plants Is a Solution to the Problem of Photorespiration

Plants in which the Calvin cycle is the only carbon-fixation pathway, and in which the first detectable product of CO_2 fixation is PGA, are called C_3 plants. In the so-called C_4 plants, carbon dioxide is initially fixed to phosphoenolpyruvate (PEP) to yield oxaloacetate, a four-carbon compound, in the mesophyll cells of the leaf. The oxaloacetate is rapidly converted to malate (or to aspartate, depending upon the species), which moves from the mesophyll cells to the bundle-sheath cells. There the malate is decarboxylated and the CO_2 enters the Calvin cycle by reacting with RuBP to form PGA. Thus, the C_4 pathway takes place in the mesophyll cells, but the Calvin cycle occurs in bundle-sheath cells.

C_4 plants are more efficient utilizers of CO_2 than C_3 plants, in part because PEP carboxylase is not inhibited by O_2. Thus, C_4 plants can attain the same photosynthetic rate as C_3 plants, but with smaller stomatal openings and, hence, with less water loss. In addition, C_4 plants are more competitive than C_3 plants at high temperatures.

CAM Plants Can Fix CO_2 in the Dark

Crassulacean acid metabolism (CAM) occurs in many succulent plants. In CAM plants, the fixation of CO_2 to phosphoenolpyruvate to form oxaloacetate occurs at night, when the stomata are open. The oxaloacetate is rapidly converted to malate, which is stored overnight in the vacuole as malic acid. During the daytime, when the stomata are closed, the malic acid is recovered from the vacuole and the fixed CO_2 is transferred to RuBP of the Calvin cycle. The C_4 pathway and the Calvin cycle occur within the same cells in CAM plants; hence, these two pathways, which are spatially separated in C_4 plants, are temporally separated in CAM plants.

Selected Key Terms

absorption spectrum p. 131

accessory pigment p. 133

action spectrum p. 131

antenna complex p. 135

bundle-sheath cells p. 144

C_3 plants, C_4 plants p. 144

C_3-C_4 intermediates p. 147

Calvin cycle, or three-carbon (C_3) pathway p. 139

CAM plants p. 147

carbon cycle p. 150

carbon fixation, or CO_2 fixation p. 139

carotenes p. 133

carotenoids p. 133

chlorophyll p. 133

crassulacean acid metabolism (CAM) p. 147

cyclic electron flow p. 139

cyclic photophosphorylation p. 139

electromagnetic spectrum p. 129

four-carbon (C_4) pathway p. 144

glyceraldehyde 3-phosphate p. 141

greenhouse effect p. 150

Kranz anatomy p. 144

light reactions p. 133

mesophyll cells p. 144

noncyclic electron flow p. 138

noncyclic photophosphorylation p. 138

PEP carboxylase p. 144

phosphoglycolate p. 142

photolysis p. 137

photons p. 130

photophosphorylation p. 137

photorespiration p. 143

Photosystem I, Photosystem II p. 135

phycobilins p. 133

pigment p. 130

reaction center p. 135

resonance energy transfer p. 131

ribulose 1,5-bisphosphate (RuBP) p. 139

RuBP carboxylase/oxygenase (Rubisco) p. 140

xanthophylls p. 133

Questions

1. Explain how the Hill reaction and the use of $^{18}O_2$ provided evidence for van Niel's proposal that water, not carbon dioxide, is the source of the oxygen evolved in photosynthesis.

2. What is the relationship between the absorption spectrum of a pigment and the action spectrum of that same pigment?

3. As excited electrons return to ground level, the energy released has three possible fates. What are those fates, and which two are useful energy-releasing events in photosynthesis?

4. Distinguish between noncyclic and cyclic electron flow and photophosphorylation. What products are produced by each? Why is cyclic photophosphorylation essential to the Calvin cycle?

5. Carbon dioxide accounts for only 0.036 percent of the air under today's atmospheric conditions, yet overall that is sufficient for all of the photosynthesis that takes place. Conditions can arise, however, under which the concentration of CO_2 is inadequate for photosynthesis to take place. What are some of those conditions?

6. By means of a labeled diagram, explain the term Kranz anatomy.

7. In what ways do C_4 plants have an advantage over C_3 plants?

8. Whereas the C_4 pathway and the Calvin cycle (C_3 pathway) are *spatially separated* in C_4 plants, in CAM plants these two pathways are *temporally separated*. Explain.

9. CAM plants are said to taste sweet during the day and sour at night. Explain why.

10. What is photophosphorylation, and what is the relationship between this process and the thylakoid membrane?

11. How is Rubisco involved in both photosynthesis and photorespiration, and why is photorespiration detrimental to a plant?

Genetics and Evolution

SECTION 3

Maize, or corn (Zea mays), is one of the three most important crop plants in the world. Maize has long been an important organism for genetic studies. The speckled pattern on some of the kernels in this painting is the result of genetic transposition, the study of which brought American geneticist Barbara McClintock a Nobel Prize.

Chapter

The Reproduction of Cells

OVERVIEW

When referring to the process by which cells reproduce, we sometimes say that "cells multiply by dividing." Mathematically this sounds peculiar, but biologically it's correct. Cells reproduce by cell division, the process by which the "parent" cell is partitioned in such a way that its component parts are apportioned between the two resulting "daughter" cells. This chapter and the next focus on cell division in eukaryotic cells. We will examine the ways the genetic material—the DNA of the chromosomes— becomes apportioned to the daughter cells. In this chapter we will focus on somatic (body) cell division, which results in daughter cells that are genetically identical to the parent cell. In the next chapter, we will consider the type of cell division that gives rise to reproductive cells, each of which is genetically different from its parent cell.

Cell division is part of the total life history of the cell—what is known as the cell cycle. During the somatic cell cycle, the DNA is replicated, the duplicate copies are apportioned into two daughter nuclei, and the cytoplasm is divided in such a way that each daughter nucleus ends up in its own cell. This process is so fundamental to living organisms that we often fail to realize its importance until things go wrong. When problems occur in the cell cycle, cells may divide in an uncontrolled manner, resulting in a cancer-like growth. Thus, it is important to understand the cell cycles of both plants and animals.

CHECKPOINTS

By the time you finish reading this chapter, you should be able to answer the following questions:

1. How does cell division in prokaryotes differ from that in eukaryotes? How are the two processes similar?

2. What is the cell cycle, and what key events occur in the G_1, S, G_2, G_0, and M phases?

3. What is the significance of mitosis? What events occur during each of the four mitotic phases?

4. What role does the mitotic spindle play in positioning and then separating sister chromatids during mitosis?

5. What roles do the phragmosome, the phragmoplast, and the cell plate play during cytokinesis?

Cells reproduce by a process known as **cell division,** in which the contents of a cell are divided between two new **daughter cells.** In one-celled organisms, such as bacteria and many protists, cell division increases the number of individuals in a population. In many-celled organisms, such as plants and animals, cell division, together with cell enlargement, is the means by which the organism grows. It is also the means by which injured or worn-out tissues are repaired or replaced.

An individual cell grows by absorbing substances from the environment and using these materials to synthesize new structural and functional molecules. When such a cell reaches a certain size and metabolic state, it divides. The two daughter cells, each about half the size of the original parent cell, or mother cell, then begin growing again. In one-celled organisms, cell division may occur every day, every few hours, or even every few minutes, producing a succession of identical organisms.

The new cells are similar both in structure and function to the parent cell and to one another. They are similar, in part, because each new cell typically receives about half of the parent cell's cytoplasm. More important, in terms of structure and function, each new cell inherits an exact replica of the genetic, or hereditary, information of the parent cell. Therefore, before cell division can occur, all of the genetic information present in the parent cell nucleus must be faithfully replicated.

Cell Division in Prokaryotes

The distribution of exact replicas of the genetic information is relatively simple in prokaryotic cells. In such cells, most of the genetic information is in the form of a single, long, circular molecule of DNA, which is associated with a variety of proteins. This molecule, the **bacterial chromosome,** is duplicated before cell division, producing two daughter chromosomes. Each of the two daughter chromosomes is attached to a different site on the interior of the plasma membrane. As the cell grows and the membrane elongates, the chromosomes gradually are separated (Figure 8–1). When the cell has approximately doubled in size and the chromosomes are completely separated, the plasma membrane and cell wall grow inward, dividing the cell in two. This type of cell division is called **binary fission,** meaning "dividing in two."

Cell Division in Eukaryotes

In eukaryotic cells, the problem of allocating the genetic material is much more complex. The nucleus of a typical eukaryotic cell contains about a thousand times more DNA than a prokaryotic cell. This DNA is present as linear molecules, forming a number of distinct chromo-

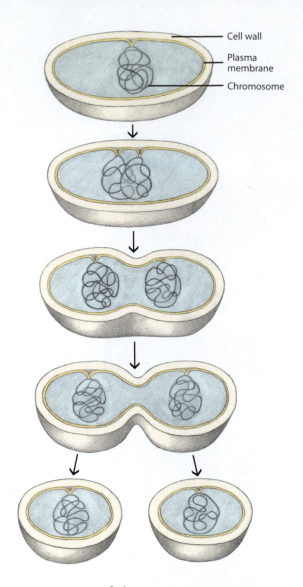

Cell wall

Plasma membrane

Chromosome

8–1

Cell division in a bacterium. Attachment of the chromosome to an inward fold of the plasma membrane ensures that one duplicated chromosome is distributed to each daughter cell as the plasma membrane elongates.

somes. For instance, bread wheat (*Triticum vulgare*) cells have 42 chromosomes. When these cells divide, each daughter cell has to receive one copy—and only one copy—of each of the 42 chromosomes.

The solution to this problem is, as we shall see, ingenious and elaborate. In a series of steps known collectively as **mitosis,** or nuclear division, a complete set of previously duplicated chromosomes is allocated to each of the two daughter nuclei. Mitosis is usually followed by **cytokinesis,** a process that divides the entire cell into two new cells. Each new cell contains not only a nucleus with a full chromosome complement but also approximately half of the cytoplasm of the parent cell.

Although mitosis and cytokinesis are the two events most commonly associated with the reproduction of eukaryotic cells, they represent only a relatively small portion of a larger process, the cell cycle.

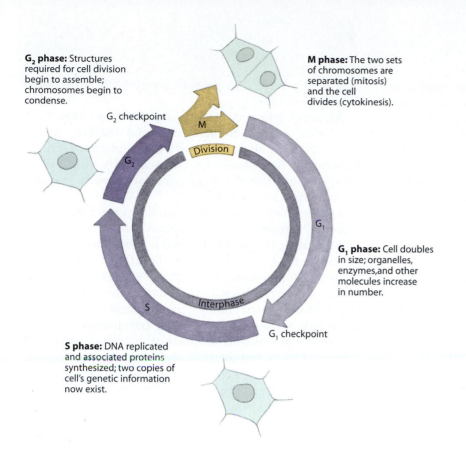

G₂ phase: Structures required for cell division begin to assemble; chromosomes begin to condense.

M phase: The two sets of chromosomes are separated (mitosis) and the cell divides (cytokinesis).

G₁ phase: Cell doubles in size; organelles, enzymes, and other molecules increase in number.

S phase: DNA replicated and associated proteins synthesized; two copies of cell's genetic information now exist.

8–2

The cell cycle. Cell division, which consists of mitosis (the division of the nucleus) and cytokinesis (the division of the cytoplasm), takes place after the completion of the three preparatory phases (G₁, S, and G₂) of interphase. Progression of the cell cycle is mainly controlled at two checkpoints, one at the end of G₁ and the other at the end of G₂. After the G₂ phase comes mitosis, which is usually followed by cytokinesis. Together, mitosis and cytokinesis constitute the M phase of the cell cycle. In cells of different species or of different tissues within the same organism, the various phases occupy different proportions of the total cycle.

The Cell Cycle

Dividing eukaryotic cells pass through a regular, repeated sequence of events known as the **cell cycle** (Figure 8–2). Completion of the cycle requires varying periods of time from a few hours to several days, depending on both the type of cell and external factors, such as temperature and available nutrients. Commonly, the cell cycle is divided into interphase and mitosis. **Interphase** precedes mitosis. It is a period of intense cellular activity, during which elaborate preparations are made for cell division, including chromosome duplication. Interphase can be divided into three phases, which are designated G₁, S, and G₂. Mitosis and cytokinesis together are referred to as the **M phase** of the cell cycle.

Some cell types pass through successive cell cycles during the life of the organism. This group includes the one-celled organisms and certain cell types in plants and animals. For example, the plant cells called **initials,** along with their immediate derivatives, or sister cells, constitute the **apical meristems** located at the tips of roots and shoots (Figure 8–3). Meristems are perpetually young tissues, and most of the cell divisions in plants occur in and near the meristems. Initials may pause in their progress around the cell cycle in response to environmental factors, as during winter dormancy, and resume proliferation at a later time. This specialized resting, or dormant, state is often called the **G₀ phase** (G-zero phase).

In a multicellular organism, it is of critical importance that cells divide at a sufficient rate to produce as many cells as are needed for normal growth and development.

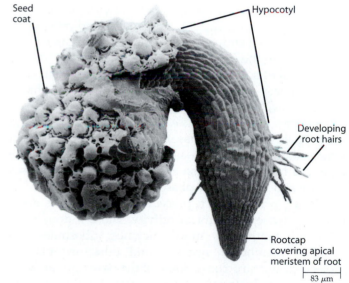

Seed coat

Hypocotyl

Developing root hairs

Rootcap covering apical meristem of root

83 μm

8–3

A germinating Arabidopsis seed, with part of the stem (hypocotyl) and the root of the embryo protruding from the seed coat. The rest of the embryo, including the cotyledons (seed leaves) and apical meristem of the shoot, are enclosed within the seed coat. The resumption of growth by the embryo depends in large part on the presence of apical meristems at the tips of the roots and stems. Meristems are perpetually embryonic tissues. Those of the tips of roots and shoots are involved with extension of the plant body.

In addition, the cell must have mechanisms that can sense whether certain conditions have been met before proceeding to the next phase. For example, replication of the DNA and synthesis of its associated proteins must be completed before the cell can proceed from the G_2 phase to mitosis and cytokinesis.

The nature of the control or controls that regulate the cell cycle is currently the subject of intense research and apparently is basically similar in all eukaryotic cells. In the typical cell cycle, progression is mainly controlled at two crucial transition points, called **checkpoints**—one at the end of G_1 and another at the end of G_2 (Figure 8–2). It is at the G_1 checkpoint that the control system either arrests the cycle or triggers a process that will initiate the S phase. At the G_2 checkpoint, the control system again either arrests the cycle or initiates mitosis.

Let us now examine some of the events that take place during each of the three phases of interphase.

Interphase

Before a cell can begin mitosis and actually divide, it must replicate its DNA and synthesize more of the proteins associated with the DNA in the chromosomes. It must also produce a supply of organelles and other cytoplasmic components adequate for two daughter cells, and assemble the structures needed to carry out mitosis and cytokinesis. These preparatory processes occur during interphase—that is, during the G_1, S, and G_2 phases of the cell cycle (Figure 8–2).

The key process of DNA replication occurs during the **S phase** (synthesis phase) of the cell cycle, a time in which many of the DNA-associated proteins, most notably the histones, are also synthesized. G (gap) phases precede and follow the S phase.

The **G_1 phase,** which precedes the S phase, is a period of intense biochemical activity. The cell doubles in size and synthesizes more enzymes, ribosomes, organelles, membrane systems, and other cytoplasmic molecules and structures. Some of the structures are synthesized entirely *de novo* ("from scratch") within the cell. These include microtubules, actin filaments, and ribosomes, all of which are composed, at least in part, of protein subunits. Membranous structures, such as Golgi bodies, various vesicles, and vacuoles, are apparently derived either directly or indirectly from the endoplasmic reticulum, which is renewed and enlarged by the synthesis of lipid and protein molecules.

In those cells that contain **centrioles**—that is, in most eukaryotic cells except those of fungi and plants—the centrioles begin to separate from one another and to duplicate. Centrioles, which are structures identical to the basal bodies of cilia and flagella (page 61), are surrounded by a cloud of amorphous material called the **centrosome.** The centrosome is also duplicated, so that each of the duplicated pairs of centrioles is surrounded by a centrosome.

Mitochondria and plastids are produced only from previously existing mitochondria and plastids. In the case of plastids, their precursors also replicate. Each mitochondrion and plastid has its own DNA, which is organized much like the single circular chromosome of a bacterial cell (see page 41).

During the **G_2 phase,** which follows the S phase and precedes mitosis, the final preparations for cell division occur. During this phase, the cell begins to assemble the structures required not only for allocating a complete set of chromosomes to each daughter nucleus, but also for dividing the cytoplasm and separating the daughter nuclei. In centriole-containing cells, duplication of the centriole pair is completed, with the two mature centriole pairs lying just outside the nuclear envelope, somewhat separated from one another. By the end of interphase, the newly duplicated chromosomes, which are dispersed in the nucleus, begin to condense but are difficult to distinguish from the nucleoplasm.

Cell Division in Plants

Two events that occur in interphase are unique to plants. First, before mitosis can begin, the nucleus must migrate to the center of the cell, if it is not already there. This migration appears to start in the G_1 phase just before DNA replication and is especially discernible in plant cells with large vacuoles. In such cells, the nucleus initially becomes anchored in the center of the cell by cytoplasmic strands (Figure 8–4). Gradually, the strands merge to form a transverse sheet of cytoplasm that bisects the cell in the plane where it will ultimately divide. This sheet, called the **phragmosome,** contains both microtubules and actin filaments involved with its formation.

Second, in addition to the migration of the nucleus to the center of the cell, one of the earliest signs that a plant cell is about to divide is the appearance of a narrow, ringlike band of microtubules that lies just beneath the plasma membrane (see the essay on facing page). This relatively dense band of microtubules encircles the nucleus in a plane corresponding to the equatorial plane of the future mitotic spindle (see Figure 8–12). Because the band of microtubules appears during G_2, just before the first phase of mitosis (prophase), it is called the **preprophase band.** Actin filaments are aligned in parallel with the preprophase band microtubules. The preprophase band disappears after initiation of the mitotic spindle, long before initiation of the cell plate in telophase, the last phase of mitosis. Yet, as the cell plate forms, it grows outward to fuse with the cell wall of the parent cell precisely at the zone previously occupied by the preprophase band.

Immunofluorescence Microscopy

Immunofluorescence microscopy has been used to reveal the location of microtubules at different stages of cell division. This technique has the capacity to provide three-dimensional views. Here root-tip cells of onion (Allium cepa) were stained with specially prepared fluorescent antibodies to tubulin, the protein that makes up microtubules (page 58). The fluorescent antibodies attach to the tubulin, and when photographed by fluorescence microscopy, the antibodies reveal the location of the microtubules. **(a)** *Prior to the formation of the preprophase band of microtubules, most of the microtubules lie just beneath the plasma membrane. This is the cortical array of microtubules that is involved in the orientation of cellulose microfibrils.* **(b)** *A preprophase band of microtubules (arrowheads) encircles each cell just before prophase. Other microtubules (arrows), forming the preprophase spindle, outline* the nuclear envelope, which itself is not visible. The lower right-hand cell is at a later stage of preprophase than that above, and the nucleus-associated microtubules merge above and below at the two regions that will become the poles of the mitotic spindle. **(c)** *The mitotic spindle at metaphase.* **(d)** *During telophase, new microtubules form a phragmoplast, in which cell plate formation takes place.*

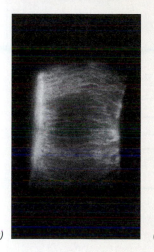

(a)

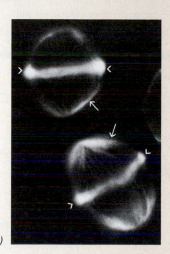

(b)

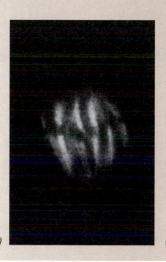

(c)

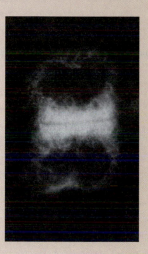

(d)

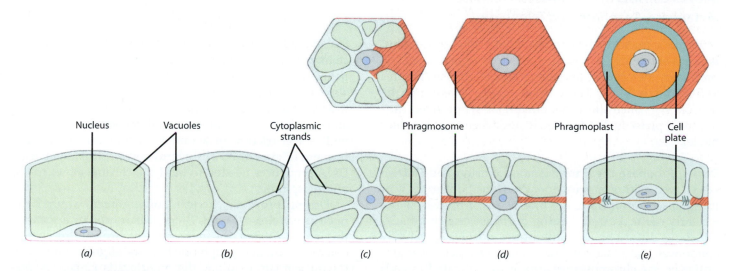

Nucleus   Vacuoles   Cytoplasmic strands   Phragmosome   Phragmoplast   Cell plate

(a)   (b)   (c)   (d)   (e)

8–4

Diagrammatic representation of some stages of cell division in a highly vacuolate cell. **(a)** *Initially, the nucleus lies along one wall of the cell, which contains a large central vacuole.* **(b)** *Strands of cytoplasm penetrate the vacuole, providing a pathway for the* nucleus to migrate to the center of the cell. **(c)** *The nucleus has reached the center of the cell and is suspended there by numerous cytoplasmic strands. Some of the strands have begun to merge to form the phragmo-* some through which cell division will take place. **(d)** *The phragmosome, which forms a layer that bisects the cell, is fully formed.* **(e)** *When mitosis is completed, the cell will divide in the plane occupied by the phragmosome.*

8–5

Mitosis, a diagrammatic representation with four chromosomes. (a) During early prophase, the chromosomes become visible as long threads scattered throughout the nucleus. (b) As prophase continues, the chromosomes shorten and thicken until each can be seen to consist of two threads (chromatids) attached to each other at their centromeres. (c) By late prophase kinetochores develop on both sides of each chromosome at the centromere. Finally, the nucleolus and nuclear envelope disappear. (d) Metaphase begins with the appearance of the spindle in the area formerly occupied by the nucleus. During metaphase, the chromosomes migrate to the equatorial plane of the spindle. At full metaphase (shown here) the centromeres of the chromosomes lie on that plane. (e) Anaphase begins as the centromeres of the sister chromatids separate. The sister chromatids, now called daughter chromosomes, then move to opposite poles of the spindle. (f) Telophase begins when the daughter chromosomes have completed their migration.

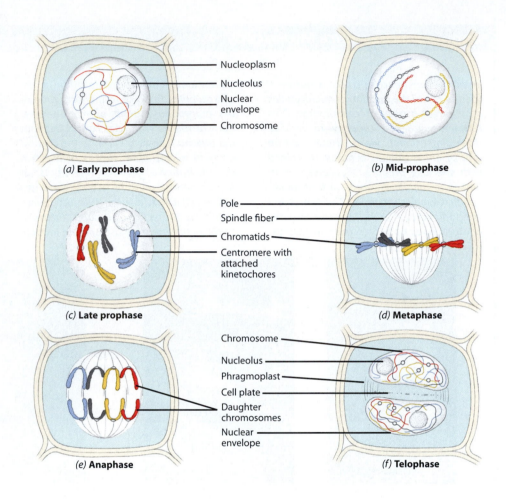

Nucleoplasm
Nucleolus
Nuclear envelope
Chromosome

(a) **Early prophase**

(b) **Mid-prophase**

Pole
Spindle fiber
Chromatids
Centromere with attached kinetochores

(c) **Late prophase**

(d) **Metaphase**

Chromosome
Nucleolus
Phragmoplast
Cell plate
Daughter chromosomes
Nuclear envelope

(e) **Anaphase**

(f) **Telophase**

Mitosis Consists of Four Phases: Prophase, Metaphase, Anaphase, and Telophase

Mitosis, or nuclear division, is a continuous process but is conventionally divided into four major phases: prophase, metaphase, anaphase, and telophase (Figures 8–5 through 8–7). These four phases make up the process by which the genetic material synthesized during the S phase is divided equally between two daughter nuclei.

During Prophase the Chromosomes Shorten and Thicken
As viewed with a microscope, the transition from G_2 of interphase to **prophase,** the first phase of mitosis, is not a clearly defined event. During prophase, the chromatin, which is diffuse in the interphase nucleus, gradually condenses into well-defined chromosomes. Initially, however, the chromosomes appear as elongated threads scattered throughout the nucleus. (The threadlike appearance of the chromosomes when they first become visible is the source of the name "mitosis"; *mitos* is the Greek word for "thread.")

As prophase advances, the threads shorten and thicken, and as the chromosomes become more distinct, it becomes evident that each is composed of not one but two threads coiled about one another. During the preceding S phase, each chromosome duplicated itself; hence, each chromosome now consists of two **sister chromatids.** By late prophase, after further shortening, the two chromatids of each chromosome lie side by side and nearly parallel, joined together along their length, with a constriction in a unique region called the **centromere** (Figure 8–8). The centromeres consist of specific DNA sequences needed to bind the chromosomes to the mitotic spindle.

During prophase, a clear zone appears around the nuclear envelope (Figure 8–6a). Microtubules appear in this zone. At first the microtubules are randomly oriented, but by late prophase they are aligned parallel to the nuclear surface along the spindle axis. Called the *preprophase spindle,* this is the earliest manifestation of mitotic spindle assembly, and it forms while the preprophase band is still present.

Toward the end of prophase, the nucleolus gradually becomes indistinct and disappears. Either simultaneously or shortly afterward, the nuclear envelope breaks down, marking the end of prophase.

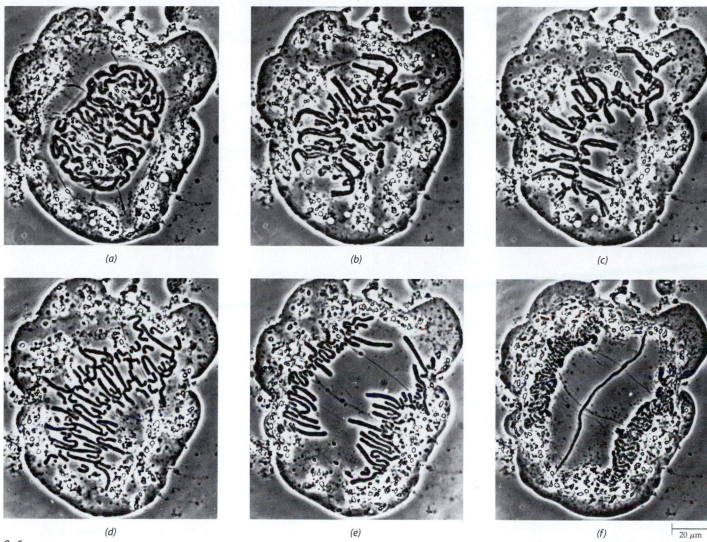

(a) *(b)* *(c)*

(d) *(e)* *(f)* `20 μm`

8–6

Mitosis as seen with phase-contrast optics in a living cell of the African blood lily (Haemanthus katherinae). The spindle is barely discernible in these cells, which have been flattened to show all of the chromosomes more clearly. (a) Late prophase: The chromosomes have condensed. A clear zone has developed around the nucleus. (b) Late

prophase–early metaphase: The nuclear envelope has disappeared, and the ends of some of the chromosomes are protruding into the cytoplasm. (c) Metaphase: The chromosomes are arranged with their centromeres on the equatorial plane. (d) Mid-anaphase: The sister chromatids (now called daughter chromosomes) have separated and

are moving to opposite poles of the spindle. (e) Late anaphase. (f) Telophase–cytokinesis: The daughter chromosomes have reached the opposite poles, and the two chromosome masses have begun the formation of two daughter nuclei. Cell plate formation is nearly complete.

During Metaphase, the Chromosomes Become Aligned on the Equatorial Plane of the Mitotic Spindle   The second phase of mitosis is metaphase. **Metaphase** begins as the **mitotic spindle,** a three-dimensional structure that is widest in the middle and tapers toward its poles, appears in the area formerly occupied by the nucleus (Figure 8–9). The spindle consists of *spindle fibers,* which are bundles of microtubules that gradually replace those of the preprophase band (see Figure 8–12c). The early

events of metaphase are among the most dynamic features of mitosis. With the sudden breakdown of the nuclear envelope, some of the spindle microtubules become attached to, or "captured" by, specialized protein complexes called **kinetochores** (Figure 8–8). These structures develop on both sides of each chromosome at the centromere, so that each chromatid has its own kinetochore. These microtubules are called **kinetochore microtubules.**

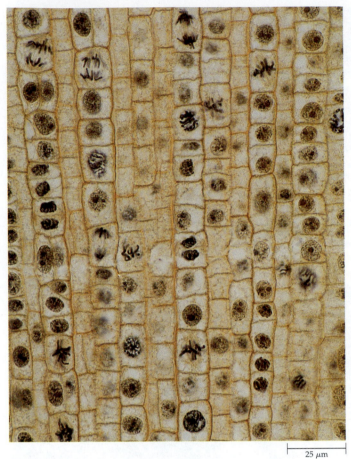

8–7

Photomicrograph of dividing cells in an onion (Allium) root tip. By comparing these cells with the phases of mitosis illustrated in Figures 8–5 and 8–6, you should be able to identify the various mitotic phases shown in this micrograph.

25 μm

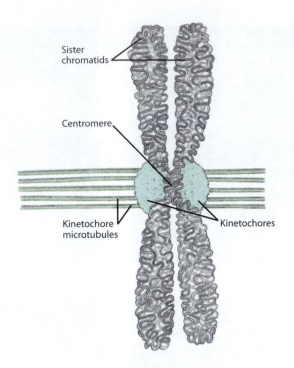

8–8

A diagram of a fully condensed chromosome. The chromosomal DNA was replicated during the S phase of the cell cycle. Each chromosome now consists of two identical parts, called sister chromatids. The centromere, the constricted area in the center, is the site of attachment of the two chromatids. The kinetochores are protein-containing structures, one on each chromatid, associated with the centromere. Attached to the kinetochores are microtubules that form part of the spindle.

As soon as the first microtubule becomes attached to one of the kinetochores, the chromosome begins to move toward the pole of the captured microtubule. Soon, one or more microtubules are captured by the opposing kinetochore, and the chromosome is pulled toward the opposite pole. This tug-of-war continues until the kinetochore microtubules have aligned the chromosome midway between the spindle poles so that the centromere lies on the equatorial plane of the spindle. When the chromosomes have all moved to the equatorial plane, or metaphase plate, the cell has reached *full metaphase*. The chromatids are now in position to separate.

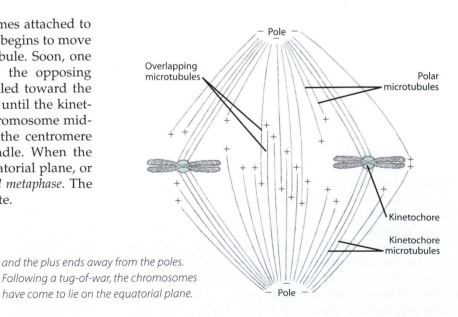

8–9

Mitotic spindle at metaphase, consisting of kinetochore microtubules and overlapping polar microtubules. Note that the minus ends of the microtubules are at or near the poles and the plus ends away from the poles. Following a tug-of-war, the chromosomes have come to lie on the equatorial plane.

The Mitotic Spindle Consists of a Highly Organized Array of Kinetochore Microtubules and Polar Microtubules The mitotic spindle is a complex, highly organized apparatus that consists of two major classes of microtubules, the kinetochore microtubules and the **polar microtubules**, which are not attached to kinetochores (Figure 8–9). All spindle microtubules are oriented with one end (the minus end) at or near one of the poles and the other end (the plus end) pointing away from the pole. Some of the polar microtubules are relatively short, but most are long enough to overlap with polar microtubules from the opposite pole. Hence, the mitotic spindle contains an array of microtubules that consists of two halves. Actin filaments are intermingled among the microtubules of the spindle and form an elastic "cage" around the spindle during mitosis.

As mentioned previously, the assembly of microtubules takes place at specific sites known as *microtubule organizing centers* (page 59). In animals and many protists, the microtubule organizing center responsible for the formation of the mitotic spindle is the centrosome. During mitosis, a centrosome is found at each pole of the spindle, and the spindle microtubules can be seen emanating from the centrosomes. In plant cells, the microtubule organizing centers provide for the assembly of the spindle, even though the spindle poles are poorly defined.

During Anaphase, the Sister Chromatids Separate and, As Daughter Chromosomes, Move to Opposite Poles of the Spindle The most rapid phase of mitosis, **anaphase** (Figure 8–6e), begins abruptly with the simultaneous separation of all of the sister chromatids at the centromeres. The sister chromatids are now called **daughter chromosomes.** As the kinetochores of the daughter chromosomes move toward opposite poles, the arms of the chromosomes seem to drag behind. As anaphase continues, the two identical sets of chromosomes move rapidly toward the opposite poles of the spindle.

The separation of the chromatids and the moving apart of the daughter chromosomes are the consequences of two independent processes mediated by the spindle. In the first, the poleward movement of the chromatids is accompanied by shortening of the kinetochore microtubules. In the second, the poles themselves move apart as the polar microtubules increase in length. By the end of anaphase, the two identical sets of chromosomes have moved to opposite poles (Figure 8–6e).

To ensure that sister chromatids will separate from one another at anaphase, it is crucial that the sister chromatids are attached to microtubules extending from opposite poles, and that they separate from each other by their movement toward opposite poles. As the daughter chromosomes move apart, the kinetochore microtubules shorten by the loss of tubulin subunits, primarily at the kinetochore ends. It would seem, therefore, that move-

ment of the chromosomes toward the poles results solely from shortening of the microtubules. Evidence indicates, however, that **motor proteins** (such as dynein), which are part of the kinetochore, use the energy of ATP to pull the chromosomes along their attached microtubules. Hence, the motor proteins provide the driving force in separation of the daughter chromosomes, whereas shortening of the microtubules at their kinetochore ends is a secondary event.

During Telophase, the Chromosomes Lengthen and Become Indistinct During **telophase** (Figure 8–6f), the separation of the two identical sets of chromosomes is made final as nuclear envelopes are organized around each set. The membranes of these nuclear envelopes are derived from vesicles. The spindle apparatus disappears. In the course of telophase, the chromosomes become increasingly indistinct, elongating to become slender threads again. Nucleoli also re-form at this time. When these processes are done and the chromosomes have once more disappeared from view, mitosis is complete, and the two daughter nuclei have entered interphase.

The two daughter nuclei produced during mitosis are genetically equivalent to one another and to the nucleus that divided to produce them. This is important, for the nucleus is the control center of the cell, as is described in more detail in Chapter 11. The nucleus contains coded instructions that specify the production of proteins, many of which mediate cellular processes by acting as enzymes whereas others serve directly as structural elements in the cell. This hereditary blueprint is faithfully transmitted to the daughter cells, and its precise distribution is ensured in eukaryotic organisms by the organization of chromosomes and their division during the process of mitosis.

The duration of mitosis varies with the tissue and the organism involved. However, prophase is always the longest phase and anaphase is always the shortest. In a root tip (Figure 8–7), the relative lengths of time for each of the four phases are as follows: prophase, 1 to 2 hours; metaphase, 5 to 15 minutes; anaphase, 2 to 10 minutes; and telophase, 10 to 30 minutes. By contrast, the duration of interphase in a root tip ranges from 12 to 30 hours.

Cytokinesis in Plants Occurs by the Formation of a Phragmoplast and a Cell Plate

As we have noted, *cytokinesis*—the division of the cytoplasm—typically follows mitosis. In most organisms, cells divide by ingrowth of the cell wall, if present, and constriction of the plasma membrane, a process that cuts through the spindle fibers. In all plants (bryophytes and vascular plants) and in a few algae, cell division occurs by formation of a **cell plate** (Figures 8–10 through 8–12).

8–10

In plant cells, separation of the daughter chromosomes is followed by formation of a cell plate, which completes the separation of the dividing cells. Here numerous Golgi vesicles can be seen fusing in an early stage of cell plate formation. The two groups of chromosomes on either side of the developing cell plate are at telophase. Arrows point to portions of the nuclear envelope reorganizing around the chromosomes.

Chromosomes   Golgi vesicles fusing to form cell plate   Chromosomes

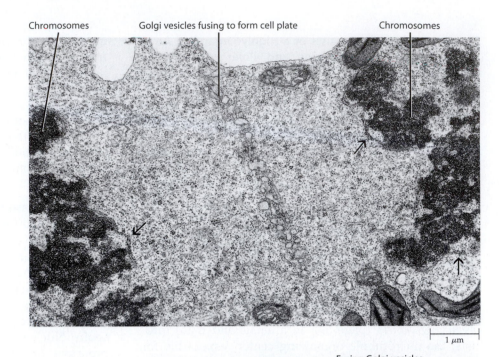

1 μm

8–11

Progressive stages of cell plate formation in root cells of lettuce (Lactuca sativa), *showing association of the endoplasmic reticulum with the developing cell plate and the origin of plasmodesmata. (a) A relatively early stage of cell plate formation, with numerous small, fusing Golgi vesicles and loosely arranged elements of tubular (smooth) endoplasmic reticulum. (b) An advanced stage of cell plate formation, revealing a persistent close relationship between the endoplasmic reticulum and fusing vesicles. Strands of tubular endoplasmic reticulum become trapped during cell plate consolidation. (c) Mature plasmodesmata, which consist of a plasma-membrane-lined channel and a tubule, the desmotubule, of endoplasmic reticulum.*

Fusing Golgi vesicles

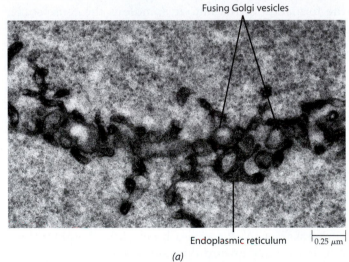

Endoplasmic reticulum   0.25 μm

(a)

Fusing Golgi vesicles

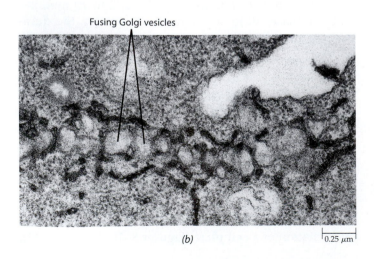

0.25 μm

(b)

Endoplasmic reticulum   Desmotubule

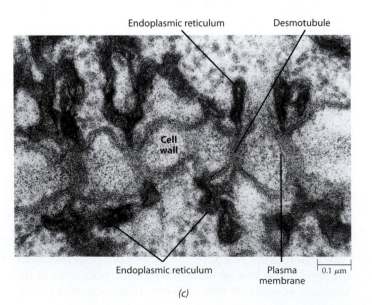

Cell wall

Endoplasmic reticulum   Plasma membrane   0.1 μm

(c)

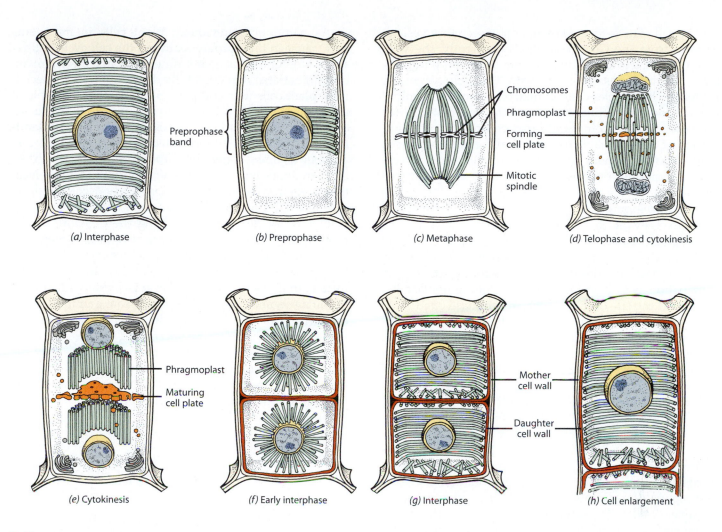

Chromosomes

Phragmoplast

Forming cell plate

Mitotic spindle

Preprophase band

(a) Interphase   *(b)* Preprophase   *(c)* Metaphase   *(d)* Telophase and cytokinesis

Phragmoplast

Maturing cell plate

Mother cell wall

Daughter cell wall

(e) Cytokinesis   *(f)* Early interphase   *(g)* Interphase   *(h)* Cell enlargement

8–12

Changes in the distribution of microtubules during the cell cycle and cell wall formation during cytokinesis. (a) During interphase, and in enlarging and differentiating cells, the microtubules lie just inside the plasma membrane. (b) Just before prophase, a ringlike band of microtubules, the preprophase band, encircles the nucleus in a plane corresponding to the equatorial plane of the future mitotic spindle. (c) During metaphase, the micro-

tubules form the mitotic spindle. (d) During telophase, microtubules are organized into a phragmoplast between the two daughter nuclei. The cell plate, made up of fusing Golgi vesicles guided into position by the phragmoplast microtubules, forms at the equator of the phragmoplast. (e) As the cell plate matures in the center of the phragmoplast, the phragmoplast and developing cell plate grow outward until they reach the wall of the

dividing cell. (f) During early interphase, microtubules radiate outward from the nuclear envelope into the cytoplasm. (g) Each sister cell forms its own primary wall. (h) With enlargement of the daughter cells (only the upper one is shown here), the mother cell wall is torn. In (g) and (h) the microtubules once more lie just inside the plasma membrane, where they play a role in the orientation of newly forming cellulose microfibrils.

In early telophase, an initially barrel-shaped system of microtubules called the **phragmoplast** forms between the two daughter nuclei. The phragmoplast, like the mitotic spindle that preceded it, is composed of microtubules that form two opposing arrays on either side of the division plane. Also, like the microtubules of the spindle halves, those of the phragmoplast halves overlap. The cell plate is initiated as a disk suspended in the phragmoplast (Figure 8–12d). At this stage, the phrag-

moplast does not extend to the walls of the dividing cell. Phragmoplast microtubules disappear where the cell plate has formed but are successively regenerated at the margins of the cell plate. The cell plate—preceded by the phragmoplast—grows outward until it reaches the walls of the dividing cell, completing the separation of the two daughter cells. In cells with large vacuoles, the phragmoplast and cell plate are formed within the phragmosome (Figure 8–4).

The cell plate is formed by fusion of secretory vesicles derived from the Golgi complex. Apparently, the vesicles are directed to the division plane by the phragmoplast microtubules, possibly with the help of motor proteins. The vesicles contain matrix molecules, hemicelluloses and/or pectins, that form the cell plate. When the vesicles fuse, their membranes contribute to the formation of the plasma membrane on either side of the cell plate. Plasmodesmata are formed at this time as segments of smooth endoplasmic reticulum are entrapped between the fusing vesicles (Figure 8–11).

The developing cell plate fuses with the parent cell wall precisely at the zone demarcated earlier by the preprophase band. Actin filaments have been found bridging the gap between the leading edge of the phragmoplast and the cell wall, possibly providing an explanation as to how the expanding phragmoplast "remembers" the site of the former preprophase band.

Following formation of the middle lamella, each protoplast deposits a primary wall next to the middle lamella. In addition, each daughter cell deposits a new layer of primary wall around the entire protoplast. This new wall is continuous with the wall at the cell plate. The original wall of the parent cell stretches and ruptures as the daughter cells enlarge (Figure 8–12h).

Because of the presence of cell walls, several aspects of cell division are unique to plants. In animals, cell migration, or movement, plays an important role in **morphogenesis,** the formation and differentiation of tissues and organs. Constraints imposed by the plant cell wall, however, essentially eliminate the ability of the cell to move and also restrict cell expansion. Consequently, the organization of cells within a plant part largely reflects the planes of cell division and the directions of cell expansion that have occurred during the plant's development. The microtubule components of the cytoskeleton play important roles in the alignment of microfibrils and, hence, in the direction of cell expansion (page 66).

Cell Division and the Reproduction of the Organism

In many one-celled organisms, mitosis is the key event in reproduction. It is the means by which exact replicas of the chromosomes are transmitted from parent to offspring. Mitosis plays the same essential role in the reproduction of some large organisms. Plants may produce new individuals from roots or runners. Reproduction in which exact replicas of the chromosomes are transmitted from parent to offspring by mitosis is known as **asexual reproduction.** It is also sometimes called vegetative reproduction because it is particularly common among plants.

As mentioned previously, plant cells in meristems may pause in their progress around the cell cycle and resume proliferation at a later time. Cells in a meristem are undifferentiated, but even fully differentiated, or specialized, plant cells with complete protoplasts are capable of reentering the cell cycle. Cells specialized for storage, secretion, or photosynthesis, for example, must first dedifferentiate. Once dedifferentiated, they can divide any number of times, depending upon conditions, and then redifferentiate into cell types appropriate to their location in the root or stem. These processes occur regularly during wound healing, as when roots are injured during cultivation of the soil, and during the development of roots on the stems of plant cuttings placed in water.

All multicellular organisms produce new cells—and thus grow—by mitosis and cytokinesis, and some produce entirely new individuals in this fashion. Most multicellular organisms, however, arise from a unique single cell—the fertilized egg, or zygote. This cell is formed by the fusion of two cells, the sperm and the egg. How are these cells formed? And what process ensures that the new organism receives a full complement of chromosomes? In the next chapter, we shall turn our attention to these questions, as we consider the process of meiosis.

Summary

Prokaryotic Cells and Eukaryotic Cells Can Divide to Produce Daughter Cells Similar to the Parent Cell

Cells reproduce by a process known as cell division, in which the cellular contents are apportioned between two new daughter cells. The new cells are structurally and functionally similar both to the parent cell and to one another because each new cell inherits an exact replica of the genetic information of the parent cell.

Prokaryotic Cells Divide by the Process of Binary Fission

Cell division in prokaryotes is a relatively simple process, in which the two daughter chromosomes are attached to different sites on the interior of the plasma membrane. As the membrane elongates, the chromosomes are separated. The plasma membrane and cell wall then grow inward, dividing the parent cell in two.

In Eukaryotic Cells, Cell Division Occurs by Mitosis and Cytokinesis

Cell division in eukaryotes is a more complex process. Eukaryotic cells contain a large amount of genetic material (DNA) organized into a number of different chromosomes. Dividing eukaryotic cells pass through a regular sequence of events known as the cell cycle, consisting of four major phases: M, which consists of mitosis and cytokinesis, G_1, S, and G_2. The G_1, S, and G_2 phases—known collectively as interphase—are the preparatory phases of the cycle. During the G_1 phase, the cell doubles in size. This size increase is accompanied by an increase in cytoplasmic molecules and structures. DNA replication occurs only during the S phase, resulting in duplication of the chromosomes. During the G_2 phase, the structures required for mitosis and cytokinesis are assembled. Progression through the cycle is mainly controlled at two crucial checkpoints, one at the transition from G_1 to S and the other at the transition from G_2 to M.

During Prophase, the Duplicated Chromosomes Shorten and Thicken

When the cell is in interphase, the chromosomes are in an uncoiled state and are difficult to distinguish from the nucleoplasm. Mitosis in plant cells is preceded by migration of the nucleus to the center of the cell and the appearance of the preprophase band, a dense band of microtubules that marks the equatorial plane of the future mitotic spindle. As prophase of mitosis begins, the chromatin gradually condenses into well-defined chromosomes, each chromosome consisting of identical strands called sister chromatids, held together at the centromere. Simultaneously, the spindle begins to form.

Metaphase, Anaphase, and Telophase Result in the Duplicated Chromosomes Being Distributed Equally into New Nuclei

Prophase ends with the breakdown of the nuclear envelope and the disappearance of the nucleolus. During metaphase, the chromatid pairs, maneuvered by kinetochore microtubules of the mitotic spindle, come to lie in the center of the cell, with their centromeres on the equatorial plane. During anaphase, the sister chromatids separate, and each chromatid, now an independent chromosome, moves toward an opposite pole. During telophase, the separation of the two identical sets of chromosomes is made final as nuclear envelopes are formed around each set. Nucleoli also re-form at this time.

Mitosis Is Generally Followed by Cytokinesis, the Division of the Cytoplasm

In plants and certain algae, the cytoplasm is divided by a cell plate that begins to form during mitotic telophase. The developing cell plate arises from fusing Golgi vesicles guided to the division plane by microtubules of the phragmoplast, an initially barrel-shaped system of microtubules that forms between the two daughter nuclei in early telophase.

Summary TABLE The Cell Cycle in Plants

PHASES OF THE CELL CYCLE			KEY EVENTS OCCURRING IN THE PHASE
Interphase	G_1 phase		Cell doubles in size; organelles and other structures increase in number; enzymes and other proteins are synthesized; nucleus begins to migrate to center of cell.
	S phase		DNA is replicated; DNA-associated proteins are synthesized.
	G_2 phase		Structures required for mitosis and cytokinesis are assembled; preprophase band appears; chromosomes begin to condense.
M phase	Mitosis	Prophase	Chromosomes continue to condense; preprophase band disappears; preprophase spindle appears; kinetochores develop; nucleolus disappears; nuclear envelope breaks down.
		Metaphase	Mitotic spindle appears; kinetochores "capture" some spindle microtubules; chromosomes move to the equatorial plane.
		Anaphase	Sister chromatids separate, and the resulting daughter chromosomes move to opposite poles of the cell.
		Telophase	Nuclear envelopes form around each set of identical chromosomes; nucleoli re-form; spindle apparatus disappears; chromosomes elongate and become indistinct. Phragmosome forms, and the cell plate begins to develop.
	Cytokinesis		Cell plate, preceded by phragmoplast, forms between the two nuclei and grows outward. Middle lamella appears, and primary wall is deposited entirely around each cell. Original cell wall is stretched and ruptured.

Selected Key Terms

anaphase   p. 163

apical meristems   p. 157

bacterial chromosome
p. 156

binary fission   p. 156

cell cycle   p. 157

cell plate   p. 163

centromere   p. 160

cytokinesis   p. 156

daughter chromosomes
p. 163

G₀ phase   p. 157

G₁ phase   p. 158

G₂ phase   p. 158

initials   p. 157

interphase   p. 157

kinetochore microtubules
p. 161

kinetochores   p. 161

M phase   p. 157

metaphase   p. 161

mitosis   p. 156

mitotic spindle   p. 161

morphogenesis   p. 166

motor proteins   p. 163

phragmoplast   p. 165

phragmosome   p. 158

polar microtubules
p. 163

preprophase band   p. 158

prophase   p. 160

sister chromatids   p. 160

S phase   p. 158

telophase   p. 163

Questions

1. Most of the cell divisions in plants occur in and near meristems. What is a meristem? What kinds of cells make up meristems? What is the so-called G_0 phase?

2. In the typical cell cycle there are checkpoints. What are these checkpoints? What purpose do they serve?

3. Distinguish between the centromere and the kinetochores.

4. What is the preprophase band? What role does it play in plant cell division?

5. What are the relative durations of prophase, metaphase, anaphase, and telophase?

6. In animals, cell migration, or movement, plays an important role in morphogenesis. What is meant by morphogenesis, and why does cell migration not play a similar role in plants?

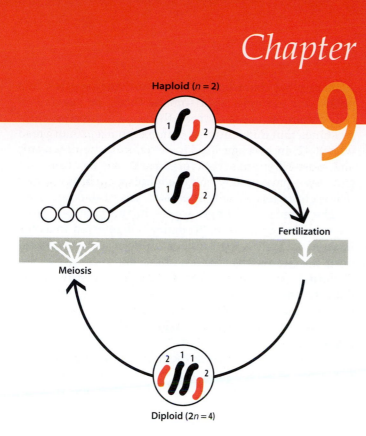

Haploid (n = 2)

Fertilization

Meiosis

Diploid (2n = 4)

9–1

Sexual reproduction is characterized by two events: the halving of the number of chromosomes (meiosis) and the coming together of the gametes (fertilization). Following meiosis, there is a single set of chromosomes—that is, the haploid number (n); here n = 2. Following fertilization, there is a double set of chromosomes—that is, the diploid number (2n).

Chapter 9

Meiosis and Sexual Reproduction

OVERVIEW

When you think of sex, think of meiosis. Sexual reproduction involves meiosis—a type of nuclear division that is a variation on mitosis. Recall that mitosis results in daughter cells that have exactly the same number and type of chromosomes as the parent. During meiosis, however, the chromosome number is reduced by half. That is, the resulting daughter cells have half the number of chromosomes present in the parent cell.

Why is sexual reproduction dependent on this reduction in chromosome number? In plants, as in humans, sexual reproduction involves the union of gametes—sperm and egg. When gametes fuse, the chromosome number of the resulting cell is twice that of either gamete. Without meiosis, the fusing of sperm and egg would result in a doubling of the chromosome number generation after generation, something that does not happen. Thus, one function of meiosis is to produce gametes having half the chromosome number of the parent cell. Then, when the gametes fuse, the chromosome number of the parent cell is reestablished. In contrast with mitosis, meiosis also results in genetic diversity. Gametes are not only genetically different from the parent cell, they are also genetically different from each other. Meiosis is part of the reason that the offspring produced by sexual reproduction differ from their parents, whether those parents are conifers, flowering plants, humans, or algae.

CHECKPOINTS

By the time you finish reading this chapter, you should be able to answer the following questions:

1. What is the relationship between haploid and diploid chromosome numbers and meiosis and fertilization?

2. What is the principal difference between zygotic meiosis, gametic meiosis, and sporic meiosis?

3. How is a sporophyte different from a gametophyte, and what is meant by the term "alternation of generations"?

4. What are the events that occur during crossing-over, and why is this process so important?

5. What are the main events that occur during meiosis I? How is meiosis I different from meiosis II?

6. What are the advantages and disadvantages of sexual and asexual reproduction?

Many organisms can get by without sex. As noted in the previous chapter, mitosis is the key event in the reproduction of many unicellular organisms, transmitting exact replicas of the parent's chromosomes to its offspring. Asexual reproduction, therefore, results in offspring that are genetically identical to the parent organism. Many unicellular eukaryotes are also capable of sexual reproduction. Sexual reproduction is the principal mode of reproduction of multicellular eukaryotes. Sexual reproduction involves the mixing of genomes of two individuals and, in contrast to asexual reproduction, results in offspring that differ genetically from one another and from both of their parents.

Sexual reproduction involves a regular alternation between two critical events: **meiosis** and **fertilization**. Meiosis is a special kind of nuclear division that is believed to have evolved from mitosis and that uses much of the same cellular apparatus. Fertilization is the means by which the different genetic combinations of the two parents are brought together to form the genetic identity of the offspring.

Although some eukaryotic organisms do not reproduce sexually, they probably once did but lost this capacity in the course of their evolutionary history. Later in this chapter, we shall consider why sexual reproduction is virtually universal among eukaryotes.

Haploid and Diploid

To understand meiosis, we must look once more at the chromosomes, focusing this time on their numbers. As mentioned in Chapter 3, every organism has a chromosome number characteristic of its particular species. In a potato, for example, each somatic (body or vegetative) cell contains 46 chromosomes, the same number of chromosomes found in a human being; in maize, 20; in bread wheat, 42; in cabbage, 18. However, in these organisms and most other eukaryotic organisms, the sex cells—or **gametes**—have exactly half the number of chromosomes that is characteristic of the somatic cells of the organism.

The number of chromosomes in the gametes (from the Greek word *gamein*, "to marry") is referred to as the **haploid** ("single set") number. And, as you might expect, the number in the somatic cells is known as the **diploid** ("double set") number. Cells that have more than two sets of chromosomes are said to be **polyploid** ("many sets"). Polyploid cells rarely occur in animals, but the somatic cells of a wide variety of flowering plants are polyploid.

In biological shorthand, the haploid number is designated as *n* and the diploid number as 2*n*. In potatoes and humans, for example, *n* = 23 and 2*n* = 46. When a sperm fertilizes an egg, the two haploid nuclei fuse, *n* + *n* = 2*n*, and the diploid number is restored (Figure 9–1). A diploid cell produced by the fusion of two gametes is known as a **zygote** (from the Greek *zygōtos*, "joined together").

In every diploid cell, each chromosome has a partner. The members of a pair of chromosomes are known as **homologous chromosomes,** or **homologs.** The two homologs resemble each other in size and shape and also, as we shall see, in the kinds of hereditary information they contain. One homolog comes from the gamete of one parent, and its partner is from the gamete of the other parent. After fertilization, both homologs are present in the zygote (Figure 9–2).

9–2

*Homologous chromosomes. **(a)** During or prior to gamete formation, individual homologs (members of a homologous pair) are parceled out by meiosis so that a haploid (n) gamete carries only one member of each homologous pair. **(b)** At fertilization, the chromosomes in the sperm nucleus and the egg nucleus come together in the diploid (2n) zygote, producing once again pairs of chromosomes. Each pair consists of one homolog from one parent (paternal chromosome) and one from the other parent (maternal chromosome).*

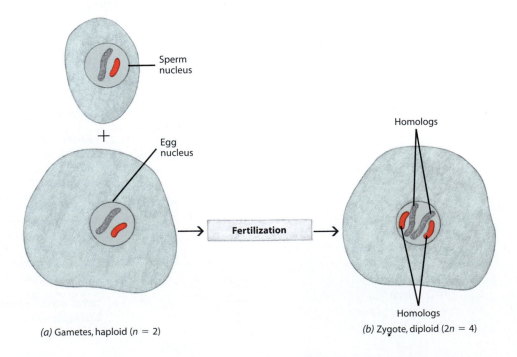

Sperm nucleus

+

Egg nucleus

Fertilization

Homologs

Homologs

(a) Gametes, haploid (*n* = 2)

(b) Zygote, diploid (2*n* = 4)

In meiosis, the diploid set of chromosomes, which contains the two homologs of each pair, is reduced to a haploid set, which contains only one homolog of each pair. Meiosis thus counterbalances the effects of fertilization, ensuring that the number of chromosomes remains constant from generation to generation. As we shall see, meiosis is also a source of new combinations within the chromosomes themselves.

Meiosis, the Life Cycle, and Diploidy

The first eukaryotic organisms were probably haploid and asexual, but once sexual reproduction was established among them, the stage was set for the evolution of diploidy. It seems likely that this condition first arose when two haploid cells combined to form a diploid zygote; such an event probably took place repeatedly. Presumably the zygote then divided immediately by meiosis (a process called **zygotic meiosis**), thus restoring the haploid condition (Figure 9–3a). In organisms with this simple kind of life cycle (some algae and all fungi), the zygote is the only diploid cell.

By "accident"—an accident that occurred in a number of separate evolutionary lines—some of these diploid zygotes divided mitotically instead of meiotically and, as a consequence, produced an organism that was composed of diploid cells, with meiosis occurring later. In animals, the water molds, and some green and brown algae, this delayed meiosis results in the production of gametes—eggs and sperm—and is therefore called **gametic meiosis**. If they come together, these gametes fuse, an event that immediately restores the diploid state (Figure 9–3b). Therefore, in animals and

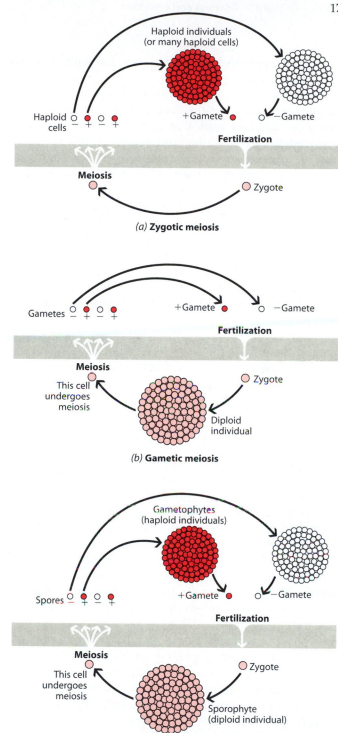

(a) **Zygotic meiosis**

(b) **Gametic meiosis**

(c) **Sporic meiosis** (alternation of generations)

9–3

Diagrams of the principal types of life cycles. In these diagrams, the diploid phase of the cycle takes place below the broad bar, and the haploid phase occurs above it. The four white arrows signify the products of meiosis; the single white arrow represents the fertilized egg, or zygote. **(a)** In zygotic meiosis, the zygote divides by meiosis to form four haploid cells. Each of these cells divides by mitosis to produce either more haploid cells or a multicellular haploid individual that eventually gives rise to gametes by differentiation. This type of life cycle is found in Chlamydomonas and a number of other algae and in the fungi. **(b)** In gametic meiosis, the haploid gametes are formed by meiosis in a diploid individual and fuse to form a diploid zygote that divides to produce another diploid individual. This type of life cycle is characteristic of most animals and some protists (Oomycota, the water molds), as well as some green and brown algae (for example, Fucus, a brown alga). **(c)** In sporic meiosis, the sporophyte, or diploid individual, produces haploid spores as a result of meiosis. These spores do not function as gametes but undergo mitotic division. This gives rise to multicellular haploid individuals (gametophytes), which eventually produce gametes that fuse to form diploid zygotes. These zygotes, in turn, differentiate into diploid individuals. This kind of life cycle, known as alternation of generations, is characteristic of plants and many algae.

some other organisms, gametes are the only haploid cells.

In plants, meiosis **(sporic meiosis)** results in the production of spores **(meiospores),** not gametes. Meiospores are haploid cells that can divide mitotically to produce a multicellular haploid organism; this is in contrast to gametes, which can develop only following fusion with another gamete. Multicellular haploid organisms that appear in alternation with diploid forms are found in plants, in many algae (among them certain brown, red, and green algae), and in a few other groups of organisms. Such organisms exhibit a type of life cycle known as an **alternation of generations** (Figure 9–3c). Among the plants and algae, the haploid, gamete-producing generation is called the **gametophyte,** and the diploid, spore-producing generation is called the **sporophyte.**

In some algae—most of the red algae, many of the green algae, a few of the brown algae—the haploid and diploid forms, or generations, are similar in external appearance. Such types of life cycles are said to exhibit an alternation of **isomorphic** (from the Greek for "equal form") **generations.**

In contrast, there are some life cycles in which the haploid and diploid forms are not identical. During the history of these groups, mutations (inheritable changes in the genetic message) occurred that were expressed in only one generation. In this way, the gametophyte and sporophyte became noticeably different from one another, and an alternation of **heteromorphic** ("different form") **generations** originated. Such life cycles are characteristic of plants and of some brown and red algae.

In the bryophytes (mosses, liverworts, and hornworts), the gametophyte is the dominant form. It is the nutritionally independent form and is usually larger than the sporophyte, which may be more complex structurally. In the vascular plants (familiar examples are the conifers and flowering plants), on the other hand, the sporophyte—the tree or flowering plant itself—is the dominant form; it is much larger and more complex than the gametophyte, which is nutritionally dependent on the sporophyte in nearly all groups.

Diploidy permits the storage of more genetic information and so perhaps allows a more subtle expression of the organism's genetic background in the course of development. This may be why the sporophyte is the large, complex, and nutritionally independent generation in vascular plants. One of the clearest evolutionary trends in this group, which predominates in most terrestrial habitats, is the increasing dominance of the sporophyte and suppression of the gametophyte. Among the flowering plants, the female gametophyte is a microscopic body that consists of only seven cells; the male gametophyte consists of only three cells. Both of these gametophytes are nutritionally dependent on the sporophyte.

The Process of Meiosis

As we saw in Chapter 8, mitosis consists of a single nuclear division and results in the formation of two daughter nuclei. Each of these nuclei receives an exact copy of the parent cell's chromosomes. Meiosis, by contrast, consists of two successive nuclear divisions (meiosis I and meiosis II), producing a total of four nuclei. Each of these four daughter nuclei contains half the number of chromosomes present in the original nucleus. Moreover, each daughter nucleus receives just one member of each pair of homologous chromosomes.

The key events in meiosis—on which all else depends—occur in the interphase preceding meiosis and in prophase at the beginning of the first meiotic division (prophase I; "I" is used to indicate all phases of meiosis I). During this interphase, as with the interphase preceding mitosis, the chromosomes are duplicated. By the beginning of meiosis, therefore, each chromosome consists of two identical sister chromatids held together at the centromere.

Early in prophase I, after the chromosomes are duplicated, the homologous chromosomes come together in pairs (Figure 9–4a). Once contact is made at any point between two homologs, pairing extends along the length of the chromatids, such that the same portions of the homologous chromosomes lie next to each other. Since each chromosome consists of two identical chromatids, the pairing of the homologous chromosomes actually involves four chromatids. The paired homologous chromosomes are called **bivalents,** or **tetrads** (from the Greek *tetra,* meaning "four"). Use of the term "tetrad" emphasizes that there are four chromatids per bivalent.

At this point, a crucial process occurs that alters the genetic makeup of the chromosomes. This process, known as **crossing-over,** involves the exchange of segments of a chromosome with corresponding segments from its homologous chromosome (Figure 9–4b). At the sites of crossing-over, portions of the chromatids of one homolog are broken and exchanged with corresponding portions of the chromatids of the other homolog. The breaks are resealed, and the result is that the sister chromatids of a single homolog no longer contain identical genetic material (Figure 9–4c). The X-like configurations that occur during crossing-over are called **chiasmata** (singular: chiasma) (Figure 9–5). The maternal homolog now contains portions of the paternal homolog, and vice versa. Thus, crossing-over is an important mechanism for **genetic recombination,** that is, for recombining the genetic material from the two parents. Moreover, the chiasmata prevent the homologs from separating during the rest of prophase I and are used to help orient homologs toward opposite poles at metaphase I, ensuring the correct segregation of the homologs at anaphase I. Hence, the chiasmata play a role analogous to that of the centromere during an ordinary mitotic division.

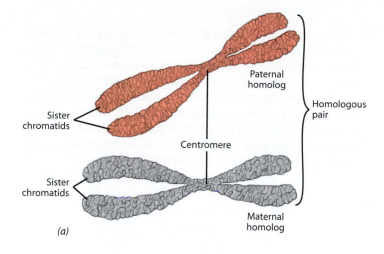

(a)

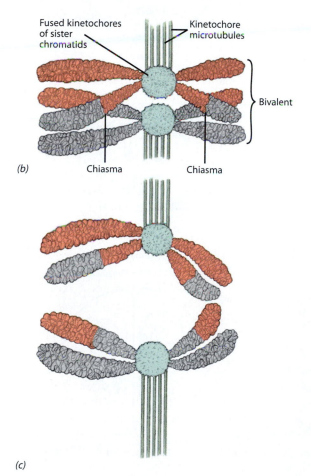

(b)

(c)

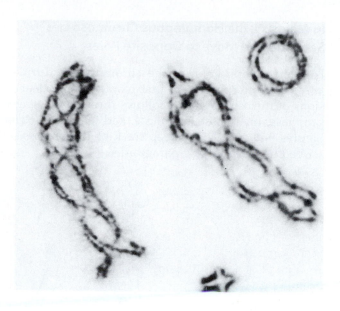

9–5

Variations in the number of chiasmata can be seen in the paired chromosomes of a grasshopper, Chorthippus parallelus, *n = 4.*

Under normal circumstances, each pair of homologs is held together by at least one chiasma. Failure of chiasmata to form between homologous chromosomes results in failure of the homologs to separate at anaphase I and can result in haploid cells that contain too few or too many chromosomes, a phenomenon known as **nondisjunction.**

The Phases of Meiosis

As we have already noted, meiosis consists of two successive nuclear divisions, which are designated as meiosis I and meiosis II. In meiosis I, homologous chromosomes pair and then separate from one another. In meiosis II, the chromatids of each homolog separate. The cells in which meiosis occurs are called **meiocytes.**

9–4

(a) *A homologous pair of chromosomes, prior to meiosis. One member of the pair is of paternal origin, and the other of maternal origin. Each of these chromosomes has duplicated and consists of two sister chromatids connected at the centromere.* **(b)** *In prophase of the first meiotic division, the two homologs come together and become closely associated with one another. The paired homologous chromosomes are called a bivalent. A homologous pair consists of four chromatids and is therefore also known as a tetrad. Within the tetrad, chromatids of the two homologs intersect at a number of points, making possible the exchange of chromatid segments. This phenomenon is known as crossing-over, and the locations at which it occurs are called chiasmata.* **(c)** *The result of crossing-over is a recombination of the genetic material of the two homologs. The sister chromatids of each homolog are no longer identical. Kinetochore microtubules attached to the fused kinetochores of sister chromatids separate the homologs at anaphase I.*

In Meiosis I, the Homologous Chromosomes Separate and Move to Opposite Poles

Following duplication of the chromosomes during the preceding interphase, the first of the two nuclear divisions of meiosis begins. It follows through the stages of prophase, metaphase, anaphase, and telophase. Refer to Figures 9–6 and 9–7 to keep track of the processes described in the following paragraphs.

In **prophase I**, the longest and most complex phase of meiosis, the chromosomes—present in the diploid number—first become visible as long, slender threads. As in mitosis, the chromosomes have already duplicated during the preceding interphase. Consequently each chromosome, at the beginning of prophase I, consists of two identical chromatids attached at the centromere. At this early stage of meiosis, however, each chromosome appears to be single rather than double.

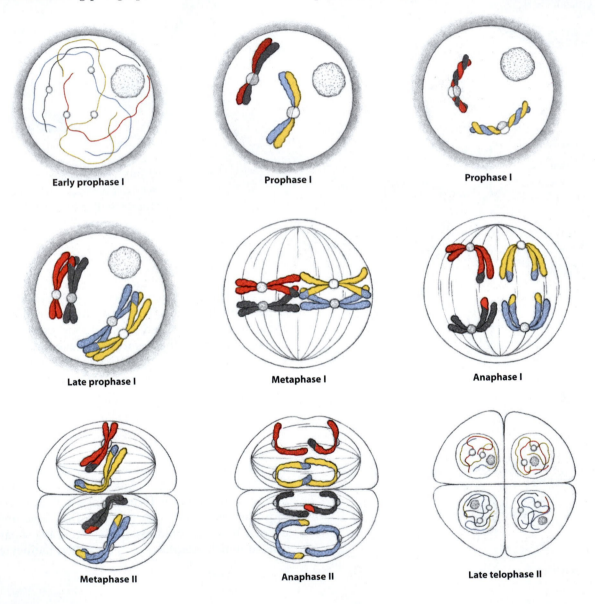

Early prophase I   **Prophase I**   **Prophase I**

Late prophase I   **Metaphase I**   **Anaphase I**

Metaphase II   **Anaphase II**   **Late telophase II**

9–6

Meiosis, a diagrammatic representation with two pairs of chromosomes. Not all stages are shown.
Prophase I: *The chromosomes become visible as elongated threads, homologous chromosomes come together in pairs, the pairs coil round one another, and the paired chromosomes become very short.*
Metaphase I: *The paired chromosomes*

move into position on the metaphase plate with their centromeres evenly distributed on either side of the equatorial plane of the spindle.
Anaphase I: *The paired chromosomes separate and move to opposite poles.*
   The second meiotic division is essentially the same as ordinary mitosis.
Metaphase II: *The chromosomes are lined*

up at the equatorial plane with their centromeres lying on the plane.
Anaphase II: *The centromeres separate, and the chromatids separate and move toward opposite poles of the spindle.*
Telophase II: *The chromosomes have completed their migration. Four new nuclei, each with the haploid number of chromosomes, are formed.*

Early Prophase I. Chromosomes appear as threads. Each thread is actually double-stranded, composed of two identical chromatids.

Late Prophase I. Homologous chromosomes pair. This is a crucial point of difference between meiosis and mitosis. Chiasmata are visible.

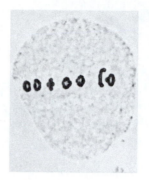

Metaphase I. Bivalents are now lined up randomly at the equatorial plane of the cell, with their centromeres evenly distributed on either side of the plane.

Anaphase I. The homologous chromosomes have separated from each other and are moving toward opposite poles of the cell.

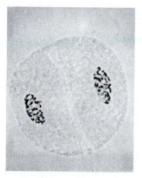

Telophase I. The chromosomes are regrouped at each pole, and the cell is dividing to form two cells.

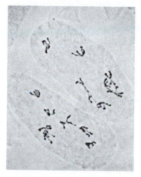

Prophase II. The chromosomes are reappearing. Each still consists of two chromatids. Because of crossing-over, the chromatids are no longer identical to each other.

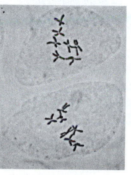

Metaphase II. The chromosomes are lined up at the equatorial plane of the cell, with their centromeres on the plane.

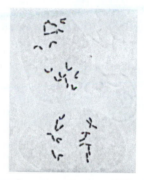

Anaphase II. The centromere of each chromosome has divided, and the chromatids—now chromosomes—are moving toward opposite poles.

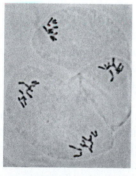

Telophase II. The chromosomes have now completely separated, and new cell walls are forming.

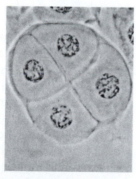

Tetrad. New plasma membranes and cell walls form as the process of cytokinesis is completed. These four haploid cells will become pollen grains. These four haploid cells (known as a tetrad) will become pollen grains.

9–7

Meiosis in lily (Lilium regale), n = 12.

During the course of prophase I, the paired threads become more and more tightly contracted, causing the chromosomes to shorten and thicken. With the aid of an electron microscope, it is possible to identify a densely staining *axial core,* consisting mainly of proteins, in each chromosome (Figure 9–8a). Both sister chromatids are attached to the axial core. During mid-prophase, the axial cores of a pair of homologous chromosomes approach each other to within 0.1 micrometer, forming a **synaptonemal complex** (Figure 9–8b). The pairing is very precise, beginning at one or more sites along the length of the chromosomes and proceeding in such a way that the corresponding portions of the homologous chromosomes lie next to one another. The synaptonemal complex consists of a long zipperlike protein structure that connects the two axial cores of the two homologous chromosomes of each bivalent. The pairing of the homologous chromosomes of each bivalent, called **synapsis,** is a necessary part of meiosis. The process cannot occur in haploid cells because such homologs are not present.

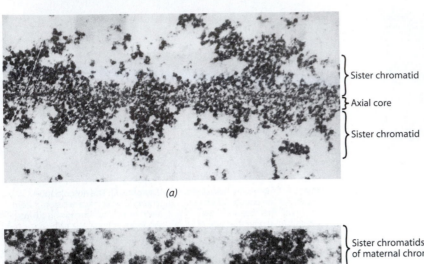

(a)

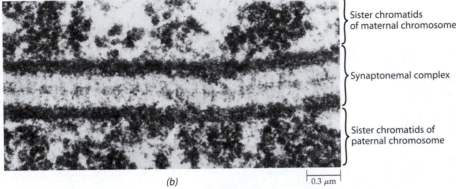

(b)

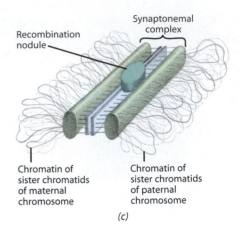

(c)

9–8

(a) Portion of a chromosome of Lilium *early in prophase I, prior to pairing with its homologous chromosome. Note the dense axial core, with sister chromatid material above and below it. This core, which consists mainly of proteins, may arrange the genetic material of the chromosome in preparation for pairing and genetic exchange. (b) Synaptonemal complex, consisting of a zipperlike protein structure connecting the two axial cores of the two homologous chromosomes, in a bivalent of* Lilium. *(c) Portion of a typical synaptonemal complex, which appears to be essential for crossing-over. Recombination nodules are protein complexes that appear to mediate the process.*

During the time that the synaptonemal complex exists, crossing-over occurs. Portions of the chromatids break apart and rejoin with corresponding segments from their homologous chromatids. As noted, the result is chromatids with a different representation of genes than originally.

The synaptonemal complex appears to be essential for crossing-over to occur, although it probably is not the agent that brings about the recombination event. Instead, additional protein complexes, called *recombination nodules,* are thought to mediate the crossing-over process (Figure 9–8c).

As prophase I proceeds, the synaptonemal complex ceases to exist. Eventually, the nuclear envelope breaks down. The nucleolus usually disappears as RNA synthesis is temporarily suspended. Finally, the homologous chromosomes appear to repulse one another. Their chromatids are held together at the chiasmata, however. These chromatids separate very slowly. As they separate, some of the chiasmata slip toward the end of the chromosome arm. One or more chiasmata may occur in each arm of the chromosome, or only one in the entire bivalent. The appearance of a particular bivalent can vary widely, depending on the number of chiasmata present (Figure 9–5).

In **metaphase I,** the spindle—an axis of microtubules similar to that which functions in mitosis—becomes conspicuous (Figure 9–9). As meiosis proceeds, micro-

tubules become attached to the kinetochores of the chromosomes of each bivalent. These paired chromosomes then move to the equatorial plane of the cell, where they line up randomly in a configuration characteristic of metaphase I. The centromeres of the paired chromosomes line up on opposite sides of the equatorial plane, the kinetochores on sister chromatids apparently having fused so that their attached microtubules all point in the same direction. In contrast, in mitotic metaphase, as we have seen, the centromeres of the individual chromosomes line up directly on the equatorial plane, with the kinetochore microtubules of sister chromatids pointing in opposite directions.

Anaphase I begins when the homologous chromosomes separate and begin to move toward opposite poles. Notice again the contrast with mitosis. In mitotic anaphase, the centromeres separate and the sister chromatids separate. In meiotic anaphase I, on the other hand, the centromeres do not separate and the sister chromatids remain together; it is the homologs that separate. Because of the exchanges of chromatid segments

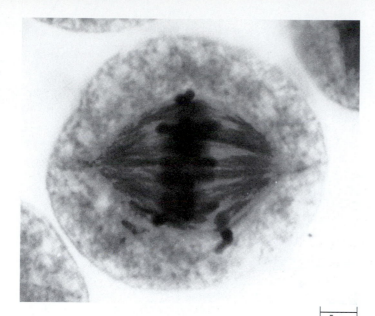

9–9

The spindle in a pollen mother cell of wheat
(Triticum aestivum) *during metaphase I of*
meiosis.

that result from crossing-over, however, the chromatids
are not identical, as they were at the onset of meiosis.

In **telophase I,** the coiling of the chromosomes re-
laxes and the chromosomes become elongated and once
again indistinct. New nuclear envelopes begin to form,
as telophase gradually changes into interphase. Finally,
the spindle disappears, the nucleoli are re-formed, and
protein synthesis commences again. In many organisms,
however, no interphase intervenes between meiotic divisions I and II. In these organisms, the chromosomes pass
more or less directly from telophase I to prophase II of
the second meiotic division without cytokinesis taking
place. In either case, no DNA replication occurs between
meiosis I and meiosis II.

In Meiosis II, the Chromatids of Each Homolog Separate and Move to Opposite Poles

At the beginning of the second meiotic division, the
chromatids are still attached by their centromeres. This
division resembles a mitotic division: the nuclear envelope (if one re-formed during telophase I) becomes dis-
organized, and the nucleolus disappears by the end of
prophase II. At **metaphase II,** a spindle again becomes
obvious, and the chromosomes—each consisting of two
chromatids—line up with their centromeres on the
equatorial plane. At **anaphase II,** the centromeres separate and are pulled apart, and the newly separated chromatids, now called daughter chromosomes, move to
opposite poles. At **telophase II,** new nuclear envelopes
and nucleoli are organized, and the contracted chromosomes relax as they fade into an interphase nucleus.
Walls develop about each new nucleus and cytoplasm.
Thus, cells are formed with the haploid number of
chromosomes.

Meiosis Produces Genetic Variability

The end result of the two nuclear divisions of meiosis is
that each cell has only half as many chromosomes as the
original diploid nucleus. But more important are the genetic consequences of the process. At metaphase I, the
orientation of the bivalents is random; that is, the chromosomes are randomly distributed between the two
new nuclei. If the original diploid cell had two pairs of
homologous chromosomes, $n = 2$, there are four possible ways in which they could be distributed among the
haploid cells (Figure 9–10). If $n = 3$, there are 8 possibilities; if $n = 4$, there are 16. The general formula is 2^n. In
potatoes and human beings, $n = 23$, and so the number

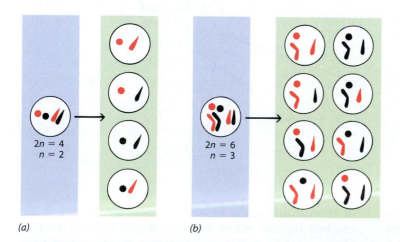

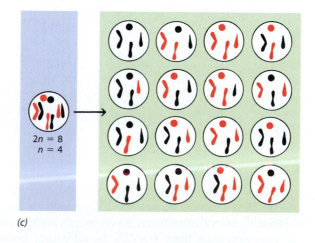

(a)   (b)   (c)

9–10

The different chromosome combinations
possible in the gametes of three organisms
with relatively few chromosomes. Red repre-

sents chromosomes of paternal origin, and
black represents chromosomes of maternal
origin. (**a**) *An organism in which* n = 2.

(**b**) *An organism in which* n = 3. (**c**) *An*
organism in which n = 4.

Mitosis

Meiosis

9–11
A comparison of the main features of mitosis and meiosis.

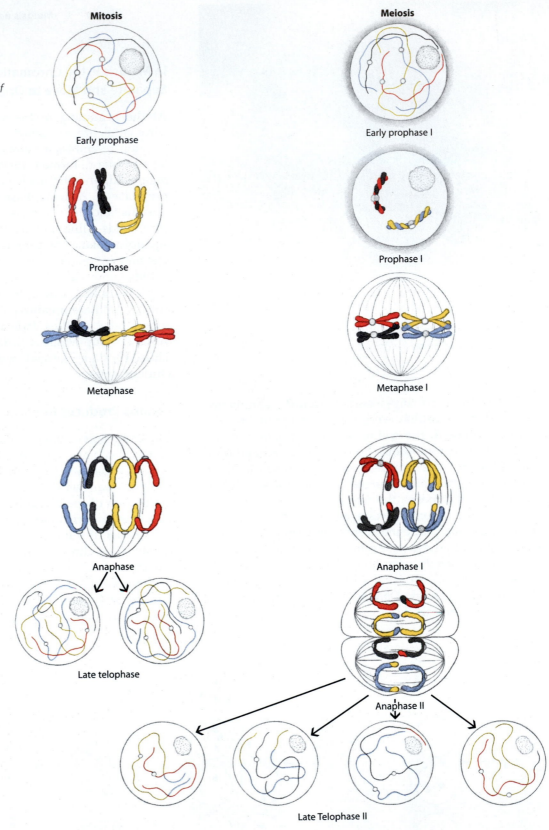

Early prophase

Prophase

Metaphase

Anaphase

Late telophase

Early prophase I

Prophase I

Metaphase I

Anaphase I

Anaphase II

Late Telophase II

of possible combinations is 2^{23}, which is equal to 8,388,608. Many organisms have much higher numbers of chromosomes than $n = 23$. In addition, because of crossing-over, each chromosome usually contains segments that have been derived from both parents. Clearly, the potential for genetic variability is enormous.

As the number of chromosomes increases, the chance of reconstituting the same set that was present in the original diploid nucleus becomes smaller and smaller. Quite apart from this, the existence of at least one chiasma in each bivalent makes it almost impossible that any cell produced by meiosis could be genetically the same as any one of those that fused to produce the diploid line of cells undergoing meiosis.

To sum up, meiosis differs from mitosis in three fundamental ways (Figure 9–11):

1. Two nuclear divisions are involved in meiosis and only one in mitosis, yet in both meiosis and mitosis the DNA is replicated only once.

2. Each of the four nuclei produced in meiosis is haploid, containing only one-half the number of chromosomes—that is, only one member of each pair of homologous chromosomes—present in the original diploid nucleus from which it was produced. By contrast, each of the two nuclei produced during mitosis has the same number of chromosomes as the original nucleus.

3. Each of the nuclei produced by meiosis contains different gene combinations from the others, whereas the nuclei produced by mitosis have identical gene combinations.

In meiosis, nuclei *different* from the original nucleus are produced, in contrast with mitosis, which produces nuclei with chromosome complements *identical* to those of the original nucleus. The genetic and evolutionary consequences of the behavior of chromosomes in meiosis are profound. Because of meiosis and fertilization, the populations of diploid organisms that occur in nature are far from uniform; instead, they consist of individuals that differ from one another in many characteristics.

Asexual Reproduction: An Alternative Strategy

Asexual reproduction (also known as vegetative reproduction) results in progeny that are identical to their single parent. In eukaryotes, there is a wide variety of means of asexual reproduction, ranging from the development of an unfertilized egg cell to division into almost equal parts of the parent organism. In all such cases, however, the new organisms are the product of mitosis and are therefore genetically identical to the parent.

Vegetative reproduction is common in plants and is accomplished in many different ways (see "Vegetative Reproduction: Some Ways and Means" on page 180). Often plants reproduce both sexually and asexually, thus hedging their evolutionary bets (Figure 9–12), but many species reproduce only asexually. Even among these, however, it is clear that their ancestors were capable of sexual reproduction and that, therefore, vegetative reproduction represents an alternative—a "choice" made in response to evolutionary pressure for extreme uniformity. This choice, if rigidly adhered to, severely restricts the ability of the population to adapt to differing conditions. Lacking recombination and genetic variability, the population cannot adjust to changing conditions as readily as populations that can reproduce sexually.

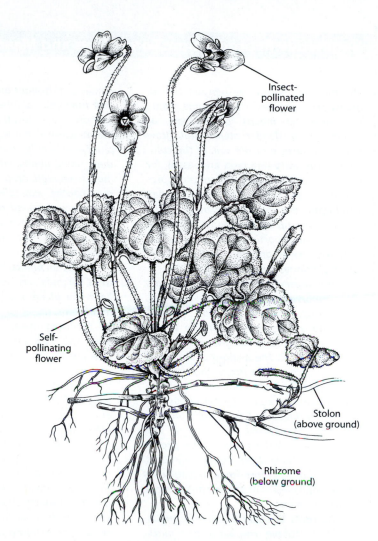

9–12

Violets reproduce both sexually and asexually. The large flowers are cross-pollinated by insects, and the seeds may be carried some distance from the parent plant by ants. The smaller flowers, closer to the ground, are self-pollinating and never open. Seeds from these flowers drop close to the parent plant and produce plants that are genetically similar to the parent. Presumably such plants are better able (on the average) to grow successfully near the parent. Both of these forms of reproduction are sexual and involve genetic recombination. Asexual reproduction in violets occurs when horizontally creeping stems (known as stolons when they are above ground and rhizomes when they are below ground) produce new, genetically identical plants close to the parent.

Insect-pollinated flower

Self-pollinating flower

Stolon (above ground)

Rhizome (below ground)

Vegetative Reproduction: Some Ways and Means

The forms of vegetative reproduction in plants are many and varied. Some plants reproduce by means of runners, or stolons—long slender stems that grow along the surface of the soil. In the cultivated strawberry (Fragaria ananassa), for example, leaves, flowers, and roots are produced at every other node on the runner. Just beyond the second node, the tip of the runner turns up and becomes thickened. This thickened portion first produces adventitious roots and then a new shoot, which continues the runner.

Underground stems, or rhizomes, are also important reproductive structures, particularly in grasses and sedges. Rhizomes invade areas near the parent plant, and each node can give rise to a new flowering shoot. The noxious character of many weeds results from this type of growth pattern, and many garden plants, such as irises, are propagated almost entirely from rhizomes. Corms, bulbs, and tubers are specialized for storage and reproduction. White potatoes are propagated artificially from tuber segments, each with one or more "eyes." It is the eyes of the "seed pieces" of potato that give rise to the new plant.

The roots of some plants—for example, cherry, apple, raspberry, and blackberry—produce "suckers," or sprouts, which give rise to new plants. Commercial varieties of banana do not produce seeds and are propagated by suckers that develop from buds on underground stems. When the root of a dandelion is broken, as it may be if one attempts to pull it from the ground, each root fragment may give rise to a new plant.

In a few species, even the leaves are reproductive. One example is the house-plant Kalanchoë daigremontiana, *familiar to many people as the "maternity plant," or "mother of thousands." The common*

names of this plant are based on the fact that numerous plantlets arise from meristematic tissue located in notches along the margins of the leaves. The maternity plant is ordinarily propagated by means of these small plants, which, when they are mature enough, drop to the soil and take root. Another example of vegetative propagation is provided by the walking fern (Asplenium rhizophyllum), in which young plants form where the leaf tips touch the ground.

In certain plants, including some citrus trees, orchids, certain grasses (such as Kentucky bluegrass, Poa pratensis), and dandelions, the embryos in the seeds may be produced asexually from the parent plant. This is a kind of vegetative reproduction known as apomixis. The seeds produced in this way give rise to individuals that are genetically identical to the parent because fertilization is not required to produce an apomictic embryo, providing another instance of asexual reproduction.

In general, asexual reproduction allows the exact replication of individuals that are particularly well suited to a certain environment or habitat. This suitability may include characteristics that are viewed as desirable in a cultivated plant or those that facilitate survival under a particular set of environmental conditions.

(a)

(b)

(c)

(a) The strawberry (Fragaria ananassa) is propagated asexually by stolons. Strawberry plants also produce flowers and reproduce sexually. (b) Kalanchoë daigremontiana, showing small plants that have arisen in the notches along the margins of the leaf. (c) The walking fern, Asplenium rhizophyllum, showing the way the leaves root at their tips and produce new plants. In this way, the fern is capable of forming large colonies of genetically identical plants.

Advantages of Sexual Reproduction

Sexual reproduction has a high selective advantage. As we have seen, it occurs only in eukaryotic organisms and involves a regular alternation between meiosis and fertilization. One of its most significant features is that the mechanism produces an infinite array of genetic diversity in natural populations and, to a certain extent, helps to maintain this diversity. As such, it provides the basic mechanism of evolution. In theory, sexual reproduction is unnecessary if an organism is particularly well in tune with its environment. What is needed in such a situation is the accurate reproduction of a particular "winning combination," generally speaking. In fact, however, natural populations have to keep adjusting to a constantly changing environment, and those that are able to invade new environments in competition with others will have an advantage.

Another measure of the evolutionary advantage of sexual reproduction is provided by the amount of energy and other resources that it requires. In angiosperms, sexual reproduction requires not only the production of gametes, often in great excess, but also the development of flowers and various other devices that enhance the possibilities—which still are often tenuous—of fertilization of these gametes. The preponderance of sexual reproduction among eukaryotes today may be viewed as evidence for the success of sexual reproduction over asexual reproduction.

The advantages of sexual reproduction were summarized aptly in 1932 by Nobel laureate Hermann J. Müller:

> There is no basic biological reason why reproduction, variation and evolution cannot go on indefinitely without sexuality or sex; therefore sex is not, in an absolute sense, a necessity, it is a "luxury." It is, however, highly desirable and useful, and so it becomes necessary in a relativistic sense, when our competitor-species are also endowed with sex, for sexless beings, although often at a temporary advantage, can not keep up the pace set by sexual beings in the evolutionary race, and when readjustments are called for, they must eventually lose out.

Summary

Sexual Reproduction Involves Meiosis and Fertilization

Sexual reproduction involves a special kind of nuclear division called meiosis. Meiosis is the process by which the chromosomes are reassorted and cells are produced that have the haploid number (*n*) of chromosomes. The other principal component of sexual reproduction is fertilization, the coming together of haploid cells to form a zygote. Fertilization restores the diploid number (2*n*) of chromosomes. There are characteristic differences among major groups of organisms as to when these events take place in the life cycle.

Sexual Life Cycles May Involve Zygotic, Gametic, or Sporic Meiosis

In the evolution of organisms, diploidy evolved subsequent to the process of sexual reproduction. In some algae and all fungi, the zygote formed by fertilization divides immediately by meiosis (zygotic meiosis). From early cycles of this sort, more complex life cycles involv-

ing diploid phases were derived on a number of occasions, when the zygote divided by mitosis. If the haploid cells produced by meiosis function immediately as gametes (gametic meiosis), the result is the type of life cycle found in animals and in some groups of protists. If the haploid cells divide by mitosis (sporic meiosis), as in many algae, all plants, and a few other organisms, they are considered spores (meiospores), and the diploid generation that gives rise to such spores is called the sporophyte. The haploid generation, which develops from the spores by mitosis, is called the gametophyte; it eventually produces gametes by mitosis. In this alternation of generations, if the gametophyte and sporophyte in a particular life cycle are approximately equal in size and complexity, the generations are said to be isomorphic. If, however, the gametophyte and sporophyte differ widely in size and complexity, the generations are said to be heteromorphic.

Meiosis Involves Two Sequential Nuclear Divisions and Results in a Total of Four Nuclei (or Cells), Each of Which Has the Haploid Number of Chromosomes

In the first meiotic division (meiosis I), paired homologous chromosomes undergo crossing-over and eventually separate. First, homologous chromosomes pair lengthwise to form bivalents (also known as tetrads). The chromosomes are double, each consisting of two chromatids. Chiasmata form between the chromatids of the homologs. These chiasmata are the visible evidence of crossing-over—the exchange of chromatid segments between homologous chromosomes. The bivalents line up at the equatorial plane in a random manner but with the centromeres of the paired chromosomes on either side of the plane. In this way, the chromosomes from the female parent and those from the male parent are reassorted during anaphase I. This reassortment, together with crossing-over, ensures that all the products of meiosis differ from the parental set of chromosomes and from each other. In this way, meiosis permits the expression of the variability that is stored in the diploid genotype.

In the second meiotic division (meiosis II), the chromosomes divide as in mitosis.

Sexual Reproduction Results in Diversity, Whereas Asexual Reproduction Does Not

Asexual reproduction results in progeny that are identical to their single parent, whereas sexual reproduction produces an infinite array of genetic diversity in natural populations. Lacking recombination and genetic variability, asexually produced plants are not able to adjust as readily to changing conditions as are plants of the same species produced sexually.

Summary TABLE   Comparison of the Main Features of Mitosis and Meiosis*

MITOSIS (IN SOMATIC CELLS)	MEIOSIS (IN CELLS IN THE SEXUAL CYCLE)
One cell division, resulting in two daughter cells	Two cell divisions, resulting in four products of meiosis
Chromosome number per nucleus maintained (e.g., for a diploid cell)	Chromosome number halved in the products of meiosis
Normally, no pairing of homologs	Full synapsis of homologs at prophase I
Normally, no chiasmata	At least one chiasma per homologous pair
Centromeres divide at anaphase	Centromeres do not divide at anaphase I but do at anaphase II
Conservative process: daughter cells' genotypes identical with parental genotype	Promotes variation among the products of meiosis
Cell undergoing mitosis can be diploid or haploid	Cell undergoing meiosis is diploid

* After: Anthony J. F. Griffiths, Jeffrey H. Miller, David T. Suzuki, Richard C. Lewontin, William M. Gelbart. 1996. *An Introduction to Genetic Analysis,* 6/e. W.H. Freeman and Company, New York. (Figure 3–57)

Selected Key Terms

alternation of generations p. 172

bivalent, or tetrad   p. 172

chiasmata   p. 172

crossing-over   p. 172

diploid   p. 170

fertilization   p. 170

gametes   p. 170

gametic meiosis   p. 171

gametophyte   p. 172

genetic recombination p. 172

haploid   p. 170

heteromorphic generations p. 172

homologous chromosome, or homolog   p. 170

isomorphic generations p. 172

meiocytes   p. 173

meiosis   p. 170

meiospores   p. 172

polyploid   p. 170

sporic meiosis   p. 172

sporophyte   p. 172

synapsis   p. 175

synaptonemal complex p. 175

zygote   p. 170

zygotic meiosis   p. 171

Questions

1. What are the two critical events in sexual reproduction in eukaryotes?

2. In life cycles with an alternation of generations, there are sporophyte and gametophyte generations. What is the first cell of each of the generations?

3. What is the difference between an alternation of isomorphic and an alternation of heteromorphic generations?

4. Distinguish between chiasma and crossing-over; between synaptonemal complex and synapsis.

5. What is the principal difference between anaphase I and anaphase II?

6. Although Müller referred to sex as a "luxury," he considered it to be "highly desirable and useful" and ultimately "necessary." Why necessary?

7. In what ways is meiosis different from mitosis?

Chapter

10

Genetics and Heredity

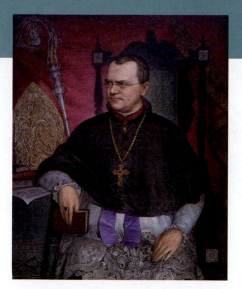

10–1

Gregor Mendel (1822–1884). As a teenager, Mendel received a thorough training in agricultural science. He then spent two years at the University of Vienna, where he studied physics, chemistry, mathematics, and botany. In 1843, at 21, he entered the Augustinian Monastery in Brünn in the Austro-Hungarian Empire (now Brno in the Czech Republic). Although Mendel published only two scientific papers during his lifetime, he performed breeding experiments with a variety of plants until his election in 1871 as abbot of the monastery. Unfortunately, almost all of Mendel's papers relating to his scientific work were destroyed shortly before or after his death in 1884.

OVERVIEW

This chapter deals with two of the most fundamental questions in biology: What is the nature of inherited traits, and how are these traits transmitted from one generation to the next? To answer these questions we first travel back to the 1800s and follow the reasoning of Gregor Mendel as he conducted the experiments that laid the foundation for modern genetics. We next examine some of the more recent variations on these Mendelian themes and some of the ways in which the hereditary material—the genes— can be altered by mutations. As you will see, even though Mendel did not know about meiosis, this process—as described in the previous chapter—is crucial for our understanding of the principles he developed.

The second half of the chapter deals with the nature of the gene itself, and thus we turn our attention to DNA. Once again we travel back in time, in this case to follow the reasoning of Watson and Crick as they pieced together the evidence to determine the structure of DNA. From structure we move to function and ask how DNA is able to replicate itself. As you know from Chapter 8, one of the key features of cellular reproduction is the replication of the DNA and the resulting formation of identical sister chromatids during interphase. At the end of this chapter, we examine the mechanism of this replication as well as the repair of errors that may occur during the process.

CHECKPOINTS

By the time you finish reading this chapter, you should be able to answer the following questions:

1. What were the major findings of Gregor Mendel, and what were the unique aspects of his experimental method that contributed to his success?

2. How is it possible for a trait to be visible in the parents but not in the offspring? What type of test could you conduct to verify your answer?

3. What are linked genes? In what way is the concept of linkage at odds with the principle of independent assortment?

4. What are the different kinds of mutations, and how do mutations affect the evolution of a population of organisms?

5. What is meant by cytoplasmic inheritance?

6. How does DNA replication occur?

Ever since people first started to look at the world around them, they have puzzled and wondered about heredity. Why is it that the offspring of all living things—whether dandelions, dogs, aardvarks, or oak trees—always resemble their parents and never some other species? Although biological inheritance—heredity—has been the object of wonder since early in human history, only rather recently have we begun to understand how it works. In fact, the scientific study of heredity, known as **genetics,** did not really begin until the second half of the nineteenth century.

The Concept of the Gene

At about the same time that Darwin was writing *The Origin of Species,* Gregor Mendel was carrying out a series of experiments that provided the first useful answers to the basic questions about heredity (Figure 10–1). His work, pursued from 1856 to 1863 in a quiet monastery garden in what was then the city of Brünn in the Austro-Hungarian Empire, marks the beginning of modern genetics. Mendel's work was largely ignored, however, until after his death.

By Mendel's time, breeding experiments with domestic plants and animals had shown that both parents contribute to the characteristics of their offspring. Further, it was known that these contributions are carried in the gametes—that is, in the sperm and eggs.

Mendel's great achievement was to demonstrate that inherited characteristics are determined by discrete factors that are passed from one generation to the next and are parceled out separately (reassorted) in each generation. These discrete factors, which Mendel called *Elemente,* eventually came to be known as **genes.**

Mendel's Experimental Method Contributed to His Success

The principal subject for Mendel's experiments in heredity was the common garden pea *(Pisum sativum).* It was a good choice, in part because the reproductive structures of the pea flower are entirely enclosed by petals, even when they are mature (Figure 10–2). Consequently, the flower normally self-pollinates; that is, sperm cells from the flower's own pollen fertilize its egg cells. Although the plants could be crossbred experimentally, accidental crossbreeding was unlikely to occur to confuse the experimental results. As Mendel stated in his original paper, "The value and utility of any experiment are determined by the fitness of the material to the purpose for which it is used."

Mendel's choice of the pea plant for his experiments was not original. However, he was the first to formulate the fundamental principles of heredity, where others had failed, because of his approach to the problem. His

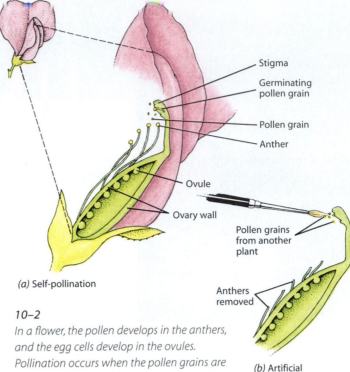

(a) Self-pollination

Anthers removed

(b) Artificial cross-pollination

10–2

In a flower, the pollen develops in the anthers, and the egg cells develop in the ovules. Pollination occurs when the pollen grains are transferred from anther to stigma. The pollen grains then germinate, forming pollen tubes, which carry the sperm to the eggs. When the nuclei of sperm and egg unite, the fertilized egg, or zygote, develops into an embryo within the ovule. The ovule matures to form the seed and the ovary wall to form the fruit, which in the pea is the pod.

Pollination in most species of flowering plants involves the transfer (often by an insect) of pollen from one plant to the stigma of another plant. This is called cross-pollination. (a) In the pea flower, however, the stigma and anthers are completely enclosed by petals, and the flower, unlike most, does not open until after fertilization has taken place. Thus self-pollination is the normal course of events in the pea flower—that is, pollen is deposited on the stigma of the same flower in which it is produced. Note that each pea in a pod represents an independent fertilization event. (b) In his cross-breeding experiments, Mendel pried open the flower bud before the pollen matured and removed the anthers with tweezers, preventing self-pollination. Then he artificially cross-pollinated the flower by dusting the stigma with pollen collected from another plant.

approach had five important components. *First,* he tested a very specific hypothesis in a series of logical experiments. He planned his experiments carefully and imaginatively, choosing for study only characteristics with clear-cut and consistent hereditary differences. *Second,* before he made a cross between two different kinds of peas, he obtained true-breeding lines for each of the traits in which he was interested, traits that appeared unchanged from one generation to the next. *Third*, he studied not only the offspring of the first generation but also those of subsequent generations and their crosses. *Fourth*, and most important, he counted the different types of offspring resulting from each cross and then analyzed the results mathematically (see essay, page 188). Even though his mathematical approach was simple, the idea that biological problems could be studied quantitatively was startlingly new. *Fifth*, and not least, he kept accurate records, organizing his data in such a way that his results could be evaluated simply and objectively. The experiments themselves were described so clearly that they could be repeated and checked by other scientists, as eventually they were.

The Principle of Segregation

The interpretation of Mendel's results was surprisingly clear. Mendel selected for study seven traits, each of which appeared in two conspicuously different forms in different varieties of plants. (The complete list of traits is given in Table 10–1.) Crosses between individuals that differ in a single trait, such as those Mendel carried out, are called **monohybrid** crosses; those that involve two traits are called **dihybrid** crosses. When Mendel crossed true-breeding plants with contrasting characteristics, he observed that, in every case, all of the progeny showed only one of the two traits, while the contrasting characteristic could not be seen at all in the first generation (now known in biological shorthand as the **F₁ genera-**

tion for "first filial generation"). For example, the seeds of all the progeny of the cross between yellow-seeded plants and green-seeded plants were as yellow as those of the yellow-seeded parent. Mendel called the characteristic for yellow seed, as well as the other characteristics that appeared in the F₁ generation, **dominant.** Characteristics that did not appear in the F₁ generation he called **recessive.** When plants of the F₁ generation were allowed to self-pollinate (Figure 10–3), the recessive characteristic reappeared in the second, or **F₂,** generation, along with the dominant characteristic, in ratios of approximately 3 dominant to 1 recessive (Table 10–1). Thus, the hereditary factors determining characteristics were still present in the F₁, but were masked.

TABLE 10–1 Results of Mendel's Experiments with Pea Plants

Trait	Original Crosses Dominant	× Recessive	Second Filial Generation (F₂) Dominant	Recessive
Seed form	Round	× Wrinkled	5474	1850
Seed color	Yellow	× Green	6022	2001
Flower position	Axial	× Terminal	651	207
Flower color	Purple	× White	705	224
Pod form	Inflated	× Constricted	882	299
Pod color	Green	× Yellow	428	152
Stem length	Tall	× Dwarf	787	277

10–3

An outline of one of Mendel's experiments. (a) A true-breeding yellow-seeded pea plant was crossed with a true-breeding green-seeded plant by removing pollen from the anthers of flowers on one plant and transferring it to the stigmas of flowers on the other plant. These plants are the parental (P) generation. (b) Pea pods containing only yellow seeds developed from the fertilized flowers. These peas (seeds) and the plants that grew from them when they were planted constitute the F₁ generation. When the F₁ plants flowered, they were not disturbed and were allowed to self-pollinate. (c) Pea pods developing from the self-pollinated flowers contained both yellow and green peas (the F₂ generation) in an approximate ratio of 3:1. That is, about 3/4 were yellow and 1/4 were green.

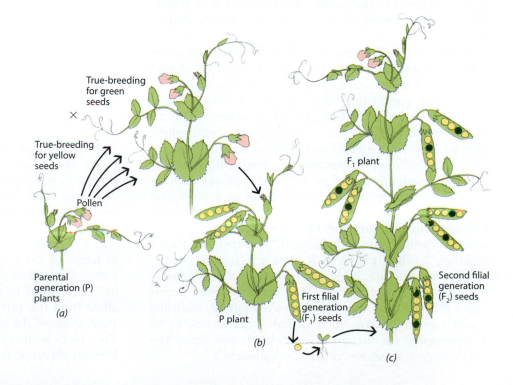

True-breeding for green seeds

True-breeding for yellow seeds

Pollen

Parental generation (P) plants

(a)

P plant

First filial generation (F₁) seeds

(b)

F₁ plant

Second filial generation (F₂) seeds

(c)

How do the recessive characteristics disappear so completely and then reappear again, and always in such constant 3:1 proportions? It was in answering this question that Mendel made his greatest contribution. He saw that the appearance and disappearance of alternative characteristics, as well as their constant proportions in the F$_2$ generation, could be explained if hereditary characteristics were determined by discrete (separable) factors. These factors, Mendel realized, must have occurred in the F$_1$ plants in pairs, with one member of each pair inherited from the maternal parent and the other from the paternal parent. The paired factors separated again when the adult F$_1$ plants produced gametes, resulting in two kinds of gametes, with one member of the pair in each.

The hypothesis that every individual carries pairs of factors for each trait and that the members of a pair segregate (separate from each other) during the formation of gametes has come to be known as Mendel's first law, or the **principle of segregation.** This phenomenon can be easily understood in terms of meiosis, a process that was unknown at the time Mendel was carrying out his experiments. Mendel never saw a chromosome.

Segregation Involves the Separation of Alleles

We now recognize that any given gene—such as those for seed color and flower color—can exist in different forms. The different forms of a gene are called **alleles.** Alleles occupy the same site, or **locus,** on homologous chromosomes. Hence, each diploid cell has two alleles for each gene, one on each of the homologous chromosomes. The alleles are represented in biological shorthand by letters. By convention, capital letters are used for the alleles for dominant characteristics and lowercase, or small, letters for the alleles for recessive characteristics.

Consider a cross between a white-flowered plant and a purple-flowered plant. The allele for purple flower color, which is dominant, is indicated by the capital letter *W.* The contrasting allele for white flower color, which is recessive, is indicated by the lowercase letter *w.* In the *true-breeding strains* of garden peas with which Mendel worked, white-flowered individuals had the genetic constitution, or **genotype,** *ww.* Purple-flowered individuals had the genotype *WW.* Individuals such as these, which have two identical alleles at a particular locus on their homologous chromosomes, are said to be **homozygous** for that gene. (The term is from the Greek *homos,* meaning "same" or "similar," and *zygōtos,* "joined together.") When plants with these contrasting characteristics are crossed, every individual in the F$_1$ generation receives a *W* allele from the purple-flowered parent and a *w* allele from the white-flowered parent, and thus has the genotype *Ww.* Such an individual is said to be **heterozygous** for the gene for flower color (from the Greek *heteros,* meaning "other" or "different").

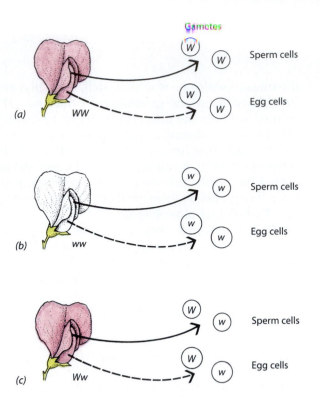

10–4

*An example of the segregation of alleles in gamete formation. **(a)** In a pea plant homozygous for purple flowers (WW), all of the gametes, both sperm and egg cells, will contain a purple-flower (W) allele. **(b)** Similarly, in a pea plant homozygous for white flowers (ww), all of the gametes will contain a white-flower (w) allele. **(c)** However, in a pea plant that is heterozygous for flower color, half of the sperm cells will contain a W allele and half will contain a w allele. Similarly, half of the egg cells will contain a W allele and half will contain a w allele.*

The letters W and w are used here because of a convention by which geneticists commonly derive the allele symbol from the first letter of the name of the form that is less common in nature (in this case, white pea flowers).

In the course of meiosis, a heterozygous individual forms two kinds of gametes, *W* and *w,* which will be present in equal proportions (Figure 10–4). As indicated in Figure 10–5, the *W* and *w* gametes derived, respectively, from the two parents will recombine to form, on average, one *WW* individual, one *ww* individual, and two *Ww* individuals for every four offspring produced. In terms of their appearance, or **phenotypes,** both the heterozygous *Ww* individuals and the homozygous *WW* individuals will be purple-flowered. The products of the allele from the purple-flowered parent are sufficient to mask those of the allele from the white-flowered parent. This, then, is the basis for the 3:1 phenotypic ratios that Mendel observed (Table 10–1).

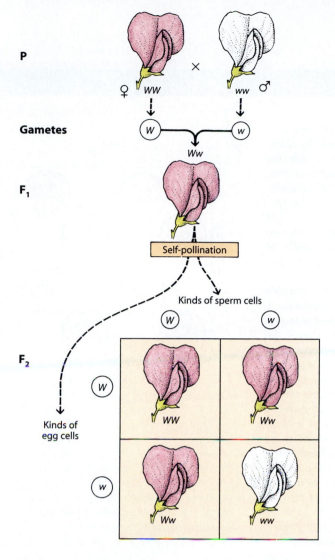

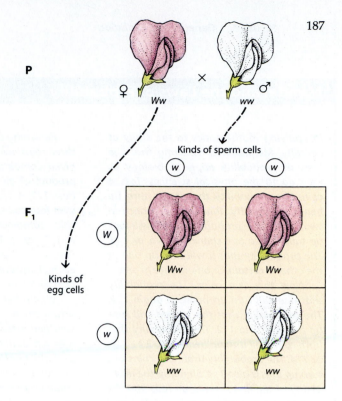

10–5

Mendel's principle of segregation, as exempli-fied in the F₁ and F₂ generations following a cross between two homozygous parent (P) pea plants, one with two dominant alleles for purple flowers (WW) and the other with two recessive alleles for white flowers (ww). The female symbol ♀ identifies the plant that contributes the egg cells (female gametes), and the male symbol ♂ indicates the plant that contributes the sperm cells (male gametes).

The phenotype of the offspring in the F₁ generation is purple, but note that the geno-type is Ww. The F₁ heterozygote produces four kinds of gametes, ♀ W, ♀ w, ♂ W, ♂ w, in equal proportions. When this plant self-pollinates, the W and w egg and sperm cells combine randomly to form, on the average, 1/4 WW (purple), 2/4 (or 1/2) Ww (purple), and 1/4 ww (white) offspring. It is this under-lying 1:2:1 genotypic ratio that accounts for the phenotypic ratio of 3 dominants (purple) to 1 recessive (white).

10–6

A testcross. For a pea flower to be white, the plant must be homozygous for the recessive allele (ww). But a purple pea flower can be produced by a plant with either a Ww or a WW genotype. How can you determine the genotype of a purple-flowered plant? Geneticists solve this problem by crossing such plants with homozygous recessives. This sort of experiment is known as a testcross. As shown here, a 1:1 phenotypic ratio in the F₁ generation (that is, one purple to one white flower) indicates that the purple-flowering parent used in the testcross must have been heterozygous. What would have been the result if the plant being tested had been homozygous for the purple-flower allele?

How can one tell whether the genotype of a plant with purple flowers is *WW* or *Ww*? As shown in Figure 10–6, one can tell by crossing such a plant with a white-flowered (homozygous recessive, *ww*) plant and count-ing the progeny of the cross. If the plant being tested is heterozygous, the progeny will have a 1:1 phenotypic ratio. Mendel performed just this kind of experiment, which is now known as a **testcross**—the crossing of an individual showing a dominant characteristic with a sec-ond individual that is homozygous recessive for that trait.

One of the simplest ways to predict the types of off-spring that will be produced in a cross is to diagram the cross as shown in Figures 10–5 and 10–6. This sort of checkerboard diagram is known as a *Punnett square*, after the English geneticist who first used it for the analysis of genetically determined traits.

Mendel and the Laws of Probability

In applying mathematics to the study of heredity, Mendel was asserting that the laws of probability apply to biology as they do to the physical sciences. Toss a coin. The probability that it will turn up heads is fifty-fifty, that is, one chance in two, or 1/2. The probability that it will turn up tails is also one chance in two, or 1/2. The probability that it will turn up one or the other is certain, or one chance in one.

Now toss two coins. The probability that one will turn up heads is again 1/2. The probability that the second will turn up heads is also 1/2. The probability that both coins will turn up heads is 1/2 × 1/2, or 1/4. The probability that both coins will turn up tails is also 1/4. Similarly, the probability of the first turning up tails and the second turning up heads is 1/2 × 1/2, and the probability of the second turning up tails and the first heads is 1/2 × 1/2. We can diagram these four possible outcomes of a two-coin toss in a Punnett square (see figure), which indicates that the combination in each square has an equal probability of occurring.

*The two-coin toss provides an example of what is known as the **product rule of probability**. This rule states that the probability of two independent events occurring together is simply the probability of one occurring alone multiplied by the probability of the other occurring alone. For instance, in Mendel's experiment diagrammed in Figure 10–5, the probability that a gamete produced by an F_1 plant of Ww genotype will carry the W allele is 1/2, and the probability that it will carry the w allele is 1/2. Therefore the probability of any specific combination of the two alleles in the offspring—for example, WW or ww—is 1/2 × 1/2, or 1/4. The appearance of the recessive phenotype in one-fourth of the offspring in the F_2 generation undoubtedly indicated to Mendel that he was dealing with a simple case of the laws of probability.*

Returning to our coin toss, if there were three coins involved, the probability of any given combination would be simply the product of all three individual probabilities: 1/2 × 1/2 × 1/2, or 1/8. Similarly, with four coins, the probability of any specific combination is 1/2 × 1/2 × 1/2 × 1/2, or 1/16. The Punnett square in Figure 10–7 expresses the probability of each of any one of four possible phenotype combinations.

*When there is more than one possible arrangement of the events producing a specified outcome, the individual probabilities are added. For instance, what is the probability of throwing a head and a tail, in either order? There are two ways you could do this: by throwing a head first and then a tail (HT), or a tail first and then a head (TH). The probability of throwing a head first and then a tail (HT) is 1/2 × 1/2, or 1/4. The probability of throwing a tail first and then a head (TH) is also 1/2 × 1/2, or 1/4. Thus, the probability that a head and a tail will be thrown, in whichever order, is the sum of their individual probabilities: (1/2 × 1/2) + (1/2 × 1/2) = 1/4 + 1/4 = 1/2. This is known as the **sum rule of probability**.*

In the cross diagrammed in Figure 10–5, a heterozygote is produced by either Ww or wW. The probability of a heterozygote in the F_2 generation is the sum of the probability of each of the two possible combinations: 1/4 + 1/4 = 1/2.

The sum rule of probability, like the product rule, applies in more complex cases as well. For example, if you were asked the probability of throwing two heads and a tail, the answer would be 3/8. Three combinations are possible: HHT, HTH, and THH. For each of these combinations, the probability is 1/2 × 1/2 × 1/2 = 1/8, that is, the product of three independent throws. Thus, the probability of throwing two heads and a

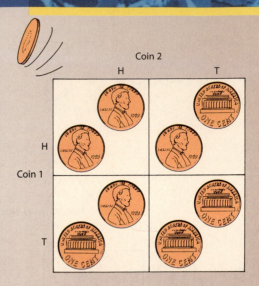

Coin 2

If you toss two coins 4 times, it is unlikely that you will get the precise results diagrammed here. However, if you toss two coins 100 times, you will come close to the proportions predicted in the Punnett square, and if you toss two coins 1000 times, you will be very close indeed. As Mendel knew, the ratio of dominants to recessives in the F_2 generation might well not have been so clearly visible if he had been dealing with a small sample. The larger the sample, however, the more closely it will conform to the results predicted by the laws of probability.

tail is the sum of the probabilities of each of the three possible combinations: 1/8 + 1/8 + 1/8 = 3/8.

Notice that in planning his experiments, Mendel made several assumptions: (1) of the male gametes produced, one-half contain one paternal allele and one-half contain the other paternal allele for each gene; (2) of the female gametes produced, one-half contain one maternal allele and one-half contain the other maternal allele for each gene; (3) the male and female gametes combine at random. Thus, the laws of probability could be employed— an elegant marriage of biology and mathematics.

The Principle of Independent Assortment

In a second series of experiments, Mendel studied hybrids that involved two pairs of contrasting characteristics; that is, he carried out dihybrid crosses. For example, he crossed a strain of garden pea with round and yellow seeds with a strain that had wrinkled and green seeds. As Table 10–1 shows, the alleles for round seeds and yellow seeds are both dominant, and those for wrinkled and green seeds recessive. All the seeds of the F_1 generation were round and yellow. When the F_1 seeds were planted and the flowers allowed to self-pollinate, 556 F_2 seeds were produced. Of these, 315 seeds showed the two dominant characteristics—round and yellow—and 32 combined the recessive characteristics, wrinkled and green. All the rest of the seeds produced were unlike those of either parent: 101 were wrinkled and yellow, and 108 were round and green. Totally new combinations of characteristics had appeared.

This experiment did not, however, contradict Mendel's previous results. If the two traits, seed color and seed shape, are considered independently, round and wrinkled still appeared in a ratio of approximately 3:1 (423 round to 133 wrinkled), and so did yellow and green (416 yellow to 140 green). But the seed shape and seed color characteristics, which had originally been combined in a certain way (round only with yellow and wrinkled only with green), behaved as if they were entirely independent of one another (yellow could now be found with wrinkled and green with round).

Figure 10–7 shows the basis for the results of the dihybrid experiments. In a cross involving two pairs of dominant and recessive alleles, with each pair on a different chromosome, the ratio of distribution of phenotypes is 9:3:3:1. The fraction 9/16 represents the proportion of the F_2 progeny expected to show the two dominant characteristics, 1/16 the proportion expected to show the two recessive characteristics, and 3/16 and 3/16 the proportions expected to show the alternative combinations of dominant and recessive characteristics. In the example shown, one parent carries both dominant alleles and the other carries both recessive alleles. Suppose that each parent carried one recessive and one dominant allele. Would the results be the same? If you are not sure of the answer, try making a diagram of the possibilities using a Punnett square, as was done in Figure 10–7.

From these experiments, Mendel formulated his second law, the **principle of independent assortment.** This law states that the two alleles of a gene assort, or segregate, independently of the alleles of other genes. Please note that the principle of segregation deals strictly with alleles of one gene, while the principle of independent assortment considers relationships between genes.

Discovery of the Chromosomal Basis of Mendel's Laws

Mendel's experiments were first reported in 1865 before a small group of people at a meeting of the Brünn Society of Natural History and published the following year in the *Proceedings* of the Society. Although the journal was circulated to libraries all over Europe, Mendel's work was ignored for 35 years. Then, in 1900, his paper was independently rediscovered by three scientists, Hugo de Vries, Carl Correns, and Erich von Tschermak, each from a different European country. Each of the scientists had performed similar breeding experiments and

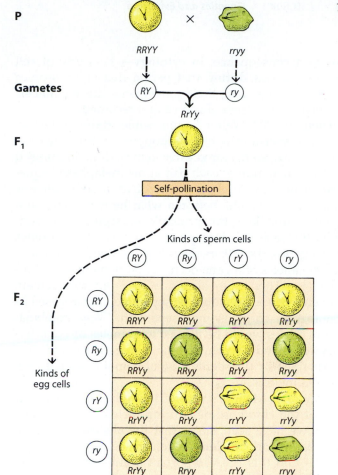

Phenotypes:

 9 Round yellow

 3 Round green

 3 Wrinkled yellow

 1 Wrinkled green

10–7

One of the experiments from which Mendel derived his principle of independent assortment. A plant homozygous for round (RR) and yellow (YY) peas is crossed with a plant having wrinkled (rr) and green (yy) peas. The F_1 peas are all round and yellow, but notice how the characteristics will, on the average, appear in the F_2 generation. Of the 16 possible combinations in the offspring, nine show the two dominant characteristics (round and yellow), three show one combination of dominant and recessive (round and green), three show the other combination (wrinkled and yellow), and one shows the two recessives (wrinkled and green). This 9:3:3:1 distribution of the phenotypes is always the expected result from a cross involving two independently assorting genes, each with one dominant and one recessive allele in each of the parents.

was searching the literature for related work. Each found that much of his own work had been anticipated in Mendel's brilliant analysis.

During the 35 years that Mendel's work remained in obscurity, great improvements were made in microscopy and, as a consequence, in **cytology**—the study of cell structure. It was during that period that chromosomes were discovered and their movements during mitosis and meiosis were first observed and recorded.

Then, in 1902, Walter Sutton, while studying the formation of sperm cells in grasshoppers, noticed that the chromosomes were paired early in meiosis I and guessed that the orientation of each pair at the metaphase I equatorial plane was purely a matter of chance. Sutton was struck by the parallel between what he was seeing and Mendel's first law, the principle of segregation. Suddenly, the facts fell into place. Suppose chromosomes carried genes, the factors described by Mendel, and suppose the alleles of a gene occurred on homologous chromosomes. Then the alleles would always remain independent and so would be separated at anaphase I as homologous chromosomes separated. New combinations of alleles would be formed as gametes fused at fertilization. Mendel's principle of segregation could thus be explained by the segregation of homologous chromosomes at meiosis.

How does Mendel's second law relate to the movement of the chromosomes during meiosis? This principle, as we have seen, states that the alleles of different genes assort independently. We can see that this can be true if—and this is an important point—the genes are on different pairs of homologous chromosomes, as shown in Figure 10–8.

On the basis of these parallels, Sutton proposed that the factors described by Mendel are carried on the chromosomes. Although Sutton's proposal was widely accepted as a working hypothesis, proof of the physical location of the gene depended on further studies. That proof was provided by A. H. Sturtevant, then a student working with fruit flies *(Drosophila melanogaster)* in the laboratory of T. H. Morgan at Columbia University. Sturtevant's studies confirmed not only that genes are located on chromosomes, as Sutton had hypothesized, but also that the genes have fixed positions in a linear sequence.

Linkage

Knowing that genes are located on chromosomes, one can readily guess that if two different genes are located relatively close together on the same pair of chromosomes, they generally will not segregate independently. Such genes, which are usually inherited together, are called **linked genes.**

The linkage of genes was first discovered by the English geneticist William Bateson and his coworkers in 1905, while they were studying the genetics of the sweet pea *(Lathyrus odoratus).* These scientists crossed a doubly homozygous recessive strain of sweet peas that had red petals and round pollen grains with a second strain (re-

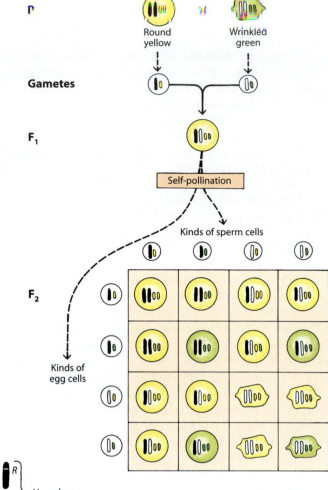

10–8

The chromosome distributions in Mendel's cross of round yellow and wrinkled green peas, according to Sutton's hypothesis. Although the pea has 14 chromosomes (n = 7), only 4 (two pairs) are shown here, the homologous pair carrying the alleles for round or wrinkled and the homologous pair carrying the alleles for yellow or green. (This selection of specific pairs of homologous chromosomes is analogous to Mendel's selection of specific traits to study.) One parent plant is homozygous for the dominants, and the other parent plant is homozygous for the recessives. Therefore, the gametes of one parent can contain only R and Y, and the gametes of the other parent can contain only r and y. The F_1 generation, therefore, must be Rr and Yy. When a cell of this generation undergoes meiosis, R is separated from r and Y from y when the members of each homologous pair separate at anaphase I. Because the alleles of the two genes are on different pairs of homologous chromosomes, they are assorted independently in meiosis. Four different types of haploid egg nuclei are possible, as the diagram reminds us, and also four different types of haploid sperm nuclei. These can combine in 4 × 4, or 16 different ways, as illustrated in the Punnett square.

sembling the wild form of the species) that had purple petals and long pollen grains. All the F_1 progeny had purple petals and long pollen grains, showing these traits to be dominant. When the F_1 was self-pollinated, they obtained the following characteristics in the F_2 generation:

4831	purple	long
390	purple	round
393	red	long
1338	red	round

If the genes for flower color and pollen shape were on the same chromosome, there should have been only two types of progeny. If the genes were on two different chromosomes, there should have been four progeny types in a ratio of 3910:1304:1304:434, or 9:3:3:1. Clearly, the experimental results were not in a 9:3:3:1 ratio.

The explanation for the results obtained is that the two genes are "linked" on one chromosome but are sometimes exchanged between homologous chromosomes during crossing-over (Figure 10–9). It is now known that crossing-over—the breakage and rejoining of chromosomes that results in the appearance of chiasmata—occurs in prophase I of meiosis (see Figure 9–4). The greater the distance between two genes on a chromosome, the greater the chance that a crossing-over will occur between them. The closer together two genes are, the greater will be their tendency to assort together in meiosis and, thus, the greater is their "linkage." Chromosome maps can be constructed based on the frequency of crossing-over between the genes and therefore on the extent to which alleles are exchanged. Such **genetic,** or **linkage, maps** provide an approximation of the positions of the genes on the chromosomes (Figure 10–10).

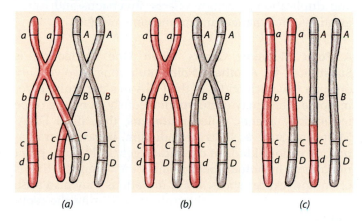

(a)   (b)   (c)

10–9

Exchange of alleles during crossing-over. (a), (b) Crossing-over takes place when breaks occur in the chromatids of homologous chromosomes early in meiosis I. Alleles are exchanged when the broken end of each chromatid joins with the chromatid of a homologous chromosome. (c) As a result of crossing-over, two of the chromosomes have allele combinations different from those of the original chromosomes.

10–10

A portion of a linkage map of chromosome 1 of a tomato (Solanum lycopersicum), showing relative positions of genes on the chromosome. Each locus is flanked by drawings of the variant phenotype that first identified that genetic locus (to the right) and the appropriate normal phenotype (to the left). Distances between loci are shown in "map units," units based on the frequency of crossing-over, or recombination, between genes. If two genes are more than 50 map units apart, the frequency of recombination is high enough that they appear to assort independently.

(a)

(b)

10–11
(a) *Hugo de Vries, shown standing next to* Amorphophallus titanum, *a member of the same family as the calla lily. The plant, a native of the Sumatran jungles, has one of the most massive inflorescences, or flower clusters, of any of the angiosperms. This picture was taken in the arboretum of the Agricultural College at Wageningen, Holland, in 1932.* **(b)** Oenothera glazioviana, *the evening primrose. De Vries reported on mutations he observed in this organism.*

Mutations

The studies on independent assortment described above depend on the existence of differences between the alleles of a gene. How do such differences arise? The first answer to this question was provided by Hugo de Vries, a Dutch geneticist (Figure 10–11a).

Mutations Are Changes in the Genetic Makeup of an Individual

In 1901, de Vries studied the inheritance of characteristics in a kind of evening primrose (*Oenothera glazioviana*) that was abundantly established on the coastal dunes of Holland. He found that, although the patterns of heredity in this plant were generally well ordered and predictable, a characteristic occasionally appeared that had not been observed previously in either parental line. De Vries hypothesized that this new characteristic was the phenotypic expression of a change in a gene. Moreover, according to his hypothesis, the changed gene would then be passed along to subsequent generations just as the other genes were. De Vries spoke of this hereditary change in one of the alleles of a gene as a mutation and of the organism carrying it as a mutant (Figure 10–11b).

Ironically, we now know that only two of roughly 2000 changes in the evening primrose that were observed by de Vries were due to a mutation, as defined by him. All of the rest were due to new genetic combinations or to the presence of extra chromosomes rather than to actual abrupt changes in any particular gene.

Today, any change in the hereditary state of an organism is called a **mutation.** Such changes may occur at the level of the gene (de Vries's concept of a mutation) or at the level of the chromosome. In **gene mutations**—also called **point mutations**—an allele of a gene changes, becoming a different allele. In **chromosome mutations,** segments of chromosomes, whole chromosomes, or even entire sets of chromosomes change. In the next few paragraphs, we shall briefly discuss point mutations and the four principal kinds of chromosome mutations: deletions and duplications, position effects, inversions and translocations, and changes in chromosome number.

A Point Mutation Occurs When One Nucleotide Is Substituted for Another   Point mutations involve only one or a few nucleotides of the DNA in a particular chromosome. They may occur spontaneously in nature or be induced by agents, called **mutagens,** that affect DNA. Mutagens, such as ionizing radiation, ultraviolet radiation, or various kinds of chemicals, generally cause point mutations. In human beings and other animals, this is often how cancer starts. Point mutations may also arise by the rare mispairings that can occur during the replication of DNA.

Deletions and Duplications Involve the Removal or Insertion of Nucleotides   Chromosomal mutations called **deletions** occur when segments of a chromosome are deleted, as by X-rays. This usually results in a change in the characteristics of the organism. Many deletions apparently occur in nature as the result of unequal crossing-over, when a chromosome segment breaks off and is not replaced by the corresponding segment from its homolog. Unequal crossing-over may also result in **duplications,** in which the same chromosome segment occurs twice. Most of the time, however, crossing-over is very accurate.

Genes May Move from One Location to Another Whereas genes usually occur in a fixed position on the chromosomes, they may on rare occasions move around. In bacteria, **plasmids**—small circular molecules of DNA separate from the main chromosome—may enter the chromosome at places where they share a common sequence of nucleotides. (The consequences of such a process for bacterial and viral evolution are discussed in Chapter 14, and the consequences for genetic engineering in Chapter 28.) In both bacteria and eukaryotes, genes may move as small segments of DNA from one location to another in the chromosomes. These movable genetic elements, known as **transposons** or "jumping genes," were first detected in maize by Barbara McClintock in the late 1940s (Figure 10–12; see also page 154). On the basis of these studies, Dr. McClintock was awarded a Nobel Prize in 1984. Whether by plasmids or transposons, the changed position of the genes involved may disrupt the action of their new neighbors, or vice versa, and lead to effects—known as position effects—that we recognize as mutations.

Pieces of Chromosomes May Be Inverted or Moved to Another Chromosome If two breaks occur in the same chromosome, the chromosomal segment between the breaks may rotate 180 degrees and reenter its chromosome with its sequence of genes oriented in the direction opposite to the original one. Such a change in chromosomal sequence is called an **inversion.** Another sort of change that is common in certain groups of plants involves the exchange of parts between two nonhomologous chromosomes to produce a chromosome mutation. Such a change in position is called a **translocation.** Translocations are often reciprocal, meaning a segment from one chromosome is exchanged with a segment from another, nonhomologous one, so that the two translocation chromosomes are formed simultaneously. In the case of either an inversion or a translocation, the genes on the chromosomal segment involved may be expressed differently in their new environment.

Entire Chromosomes May Be Lost or Duplicated Mutationlike effects may also be associated with changes in chromosome number, which occur spontaneously and rather commonly but are usually eliminated promptly. Whole chromosomes may, under certain circumstances, be added to or subtracted from the basic set, a condition known as **aneuploidy. Polyploidy,** the duplication of whole sets of chromosomes, may also occur. (The evolu-

10–12
Barbara McClintock holding an ear of corn, or maize, of the type she used in her landmark studies on transposons. Her work was largely unheralded until decades after her investigations were published.

tionary importance of polyploidy in plants is discussed in Chapter 12.) In any of these cases, associated alterations of the phenotype usually occur.

Changes in chromosome number can be induced by chemicals such as colchicine, an alkaloid derived from the autumn crocus plant *(Colchicum autumnale).* This substance is an inhibitor of microtubule assembly and, hence, of mitotic spindle formation. If cells of meristematic tissue are treated with colchicine, chromosome duplication can take place but metaphase and anaphase cannot. When the colchicine is removed to allow cell multiplication to resume, the resultant cells will contain greater numbers of chromosomes than their precursors.

Mutations Provide the Raw Materials for Evolutionary Change

When a mutation occurs in a predominantly haploid organism, such as the fungus *Neurospora* or a bacterium, the phenotype associated with this mutation is immediately exposed to the environment. If favorable, the number of organisms with the mutation tends to increase in the population as a result of natural selection (see page

239). If unfavorable, the mutant is quickly eliminated from the population. Some mutations may be more nearly neutral in their effects and then may persist by chance alone, but most have either negative or positive effects on the organism in which they occur. In a diploid organism, the situation is very different. Each chromosome and every gene is present in duplicate, and a mutation on one of the homologs, even if it would be unfavorable in a double dose, may have much less effect or even be advantageous when present in a single dose. For this reason, such a mutation may persist in the population. The mutant gene may eventually alter its function, or the selective forces on the population may change in such a way that the effects of the mutant gene become advantageous.

Whether mutations are harmful or neutral, the capacity to mutate is extremely important, for it allows the individuals in a species to vary and to adapt to changing conditions. Mutations thus provide the raw material for evolutionary change. Mutations in eukaryotes occur spontaneously at a rate of about 1 mutant gene at a given locus per 200,000 cell divisions. This, together with recombination, provides variation of the sort that is necessary for evolutionary change through natural selection.

Broadening the Concept of the Gene

Alleles Undergo Interactions That Affect the Phenotype

As the early studies in genetics proceeded, it soon became apparent that dominant and recessive characteristics are not always clear-cut, as in the seven traits studied by Mendel. Interactions can and do occur between the alleles of a gene that affect the phenotype.

Incomplete Dominance Produces Intermediate Phenotypes In cases of **incomplete dominance,** the phenotype of the heterozygote is intermediate between those of the parent homozygotes. For example, in snapdragons, a cross between a red-flowered plant and a white-flowered plant produces a plant that has pink flowers. As we shall see in Chapter 11, genes dictate the structure of proteins—in this case, proteins involved in pigment synthesis in the cells of the flower petals. For this heterozygote, the pigment produced in a flower petal cell by one allele is not completely masked by the action of the other allele. When the F_1 generation is allowed to self-pollinate, the characteristics segregate again, the result in the F_2 generation being one red-flowered (homozygous) plant to two pink-flowered (heterozygous) plants to one white-flowered (homozygous) plant (Figure 10–13). Thus, the alleles themselves remain discrete and unaltered, conforming to Mendel's principle of segregation.

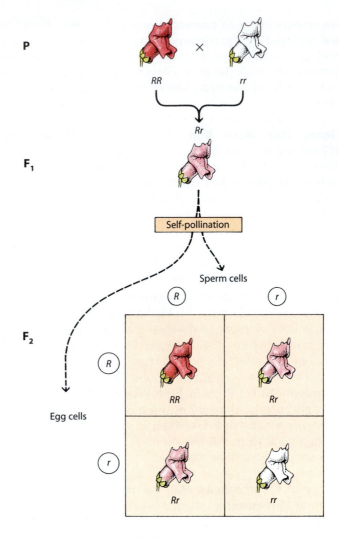

10–13

A cross between a red (RR) snapdragon and a white (rr) snapdragon. This looks very much like the cross between a purple- and a white-flowering pea plant shown in Figure 10–5, but there is a significant difference because, in this case, neither allele is dominant. The flower of the heterozygote is a blend of the two colors.

Some Genes Have Multiple Alleles Although any individual diploid organism can have only two alleles of any given gene, it is possible that more than two forms of a gene may be present in a population of organisms. When three or more alleles exist for a given gene, they are referred to as **multiple alleles.** The genes that control self-sterility (that is, genes that prevent self-pollination) in certain flowering plants can have large numbers of alleles. It is estimated, for example, that some populations of red clover have hundreds of alleles of the self-sterility gene. Fortunately for Mendel, peas are self-pollinators.

Gene Interactions Also Occur among Alleles of Different Genes

In addition to interactions between alleles of the same gene, interactions also occur among the alleles of different genes. Indeed, most of the characteristics (both structural and chemical) that constitute the phenotype of an organism are the result of the interaction of two or more distinct genes.

Epistasis Occurs When One Gene Interacts with Another

In some cases, one gene may interfere with or mask the effect of another. This type of interaction is called **epistasis** ("standing upon"). A classic example of epistasis occurs in a variety of sweet pea that produces either purple or white flowers. In order to produce purple flowers, a plant must have at least one dominant allele of each of two different genes. The dominant alleles *A* and *B* must both be present because they code for the production of enzymes that catalyze two separate reactions in the production of the purple pigment. If a plant is homozygous for either *a* or *b*, the corresponding reaction will not take place. As a result, no purple pigment will be produced, and the flowers will be white.

Certain Traits Are Controlled by Several Genes

Some traits, such as size, shape, weight, yield, and metabolic rate, are not the result of interactions between one, two, or even several genes. Instead, they are the cumulative result of the combined effects of many genes. This phenomenon is known as **polygenic inheritance.**

A trait affected by a number of genes does not show a clear difference between groups of individuals—such as the differences tabulated by Mendel. Instead, it shows a gradation known as **continuous variation.** If you make a graph of differences among individuals for any trait affected by a number of genes, none of which is dominant over the other, you will get a bell-shaped curve, with the mean, or average, usually falling in the center of the curve (Figure 10–14).

The first experiment that illustrated the way in which many genes can interact to produce a continuous pattern of variation in plants was carried out with wheat by the Swedish scientist H. Nilson-Ehle. Table 10–2 shows the phenotypic effects of various combinations of two genes, each with two alleles, that act together to control the intensity of color in wheat kernels.

A Single Gene Can Have Multiple Effects on the Phenotype

Although many genes have a single phenotypic effect, a single gene can often have multiple effects on the phenotype of an organism, affecting a number of apparently unrelated traits. This phenomenon is known as **pleiotropy** ("many turnings"). Mendel encountered pleiotropic genes during his breeding experiments with peas. When he crossed peas with purple flowers, brown seeds, and a dark spot on the axils of the leaves with a variety having white flowers, light seeds, and no spot on the axils, the same traits for flower, seeds, and leaves always stayed together as a unit. The inheritance of these traits can be accounted for by a single gene, which visibly influences several traits.

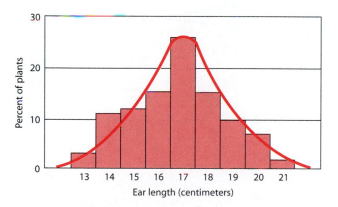

10–14

Distribution of ear length of the Black Mexican variety of maize (Zea mays). This is an example of a phenotypic characteristic that is determined by the interaction of a number of genes. Such a characteristic shows continuous variation. If the variation is plotted as a curve, the curve is bell-shaped, with the mean, or average, falling in the center of the curve.

TABLE 10–2	The Genetic Control of Color in Wheat Kernels		
Parents	$R_1R_1R_2R_2$ × $r_1r_1r_2r_2$		
	(Dark red) (White)		
F₁	$R_1r_1R_2r_2$ (Medium red)		
F₂	**Genotype**	**Phenotype**	
1	$R_1R_1R_2R_2$	Dark red	
2 } 4	$R_1R_1R_2r_2$	Medium-dark red	
2	$R_1r_1R_2R_2$	Medium-dark red	
4 } 6	$R_1r_1R_2r_2$	Medium red	15 red
1	$R_1R_1r_2r_2$	Medium red	to
1	$r_1r_1R_2R_2$	Medium red	1 white
2 } 4	$R_1r_1r_2r_2$	Light red	
2	$r_1r_1R_2r_2$	Light red	
1	$r_1r_1r_2r_2$	White	

A case of particular interest to plant breeders is the relationship between yield in wheat and the presence or absence of awns (long, slender bristles) on the lemmas (bracts) associated with wheat flowers. The same gene affects both. Wheat with awns have greater yields than wheat without awns, providing a method for judging potential yield without waiting for a mature crop.

The Inheritance of Some Characteristics Is under the Control of Genes Located in Plastids and Mitochondria

In Chapter 3, we noted that plastids and mitochondria contain their own DNA and encode for some of their own proteins. Hence, not all of the characteristics of the cell are controlled exclusively by the DNA of the chromosomes located in the nucleus. The inheritance of characteristics under the control of genes located in the cytoplasm—strictly speaking, in plastids and mitochondria—is known as **cytoplasmic inheritance.**

In most organisms, including angiosperms, or flowering plants, most cytoplasmically inherited characteristics are maternally inherited, that is, they are determined solely by the female parent. Any one of several phenomena may be responsible for **maternal inheritance.** In some angiosperms, for instance, the immediate precursor (the generative cell) of the sperm cells may receive no plastids when the microspore (immature pollen grain) divides to produce the larger tube cell and smaller generative cell (Figure 10–15). In some orchids, the generative cell receives neither plastids nor mitochondria. In several species of angiosperms the generative cells receive both plastids and mitochondria, but one or both types of organelles degenerate before the generative cell divides to produce the sperm cells. In other cases, plastids and mitochondria are excluded or removed from generative or sperm cells. Even though plastids and mitochondria may be present in sperm cells at the time of fertilization, it is still possible that they will not be transmitted into the egg. In such cases only the sperm nucleus enters the egg cell, the entire sperm cytoplasm being excluded from it.

Some conspicuous characteristics are due to cytoplasmic inheritance involving chloroplasts. Among these are the variegated or mottled leaves of certain plants that are prized as landscape plants and houseplants for their attractive foliage. In variegated coleus and hostas, for example, the lighter patches each contain cells that developed from a cell containing only mutant (nongreen) plastids. Cytoplasmic inheritance involving mitochondria includes cytoplasmic male sterility, a maternally inherited trait that prevents the production of pollen but does not affect female fertility. The cytoplasmic male sterility phenotype has been widely used in the commercial production of F_1 hybrid seed (for example, in maize, onions, carrots, beets, and petunias) because it is unnecessary to remove the stamens to prevent self-pollination before making crosses.

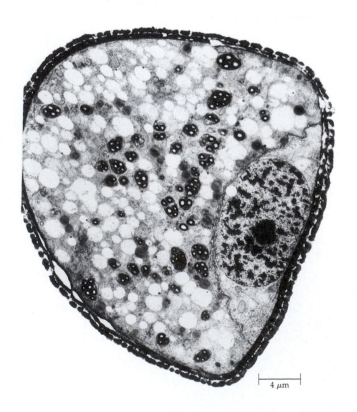

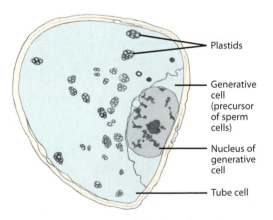

10–15

Transmission electron micrograph of a Tulbaghia violacea *(family* Liliaceae) *pollen grain showing the newly formed generative cell, which will develop into sperm cells. Note that the generative cell lacks plastids, which are numerous within the larger tube cell. The tube cell will produce a pollen tube that will carry the sperm cells to the egg apparatus. Only the sperm cells or their nuclei will enter the egg cell.*

The Phenotype Is the Result of the Genotype Interacting with the Environment

The expression of a gene is always the result of its interaction with the environment. To take a common example, a seedling may have the genetic capacity to be green, to flower, and to fruit, but it will never turn green if it is kept in the dark, and it may not flower and fruit unless certain precise environmental requirements are met.

The water buttercup is an especially striking example. It grows with part of the plant body submerged in water and part floating on the surface of the water. Although the leaves are genetically identical, the broad, floating leaves differ markedly in both form and physiology from the finely divided leaves that develop under water (Figure 10–16).

Temperature often affects gene expression. Primrose plants that are red-flowered at room temperature are white-flowered above 30°C (86°F).

The expression of a gene may be altered not only by factors in the external environment but also by factors in the internal environment of the organism throughout plant development. These factors include temperature, pH, ion concentrations, hormones, and a multitude of other influences, including the action of other genes.

The Chemical Basis of Heredity

By the early 1940s, the existence of genes and the fact that they are carried in the chromosomes were no longer in doubt. A turning point in the history of genetics came when scientists began to focus on the question of how it is possible for the chromosomes to carry what the scientists realized must be an enormous amount of very complex information.

The chromosomes, like all other parts of a living cell, are composed of atoms arranged into molecules. Some scientists, a number of them eminent in the field of genetics, thought it would be impossible to understand the complexities of heredity in terms of the structure of "lifeless" chemicals. Others thought that if the chemical structure of the chromosomes were understood, we could then come to understand how chromosomes function as the bearers of the genetic information. This insight marked the beginning of the vast range of investigations that we know as **molecular genetics.**

The Chemistry of the Gene: DNA versus Protein

Early chemical analyses revealed that the eukaryotic chromosome consists of both **deoxyribonucleic acid (DNA)** and protein, in about equal amounts. Once it be-

10–16

The water buttercup, Ranunculus peltatus, *grows with part of the plant body submerged in water. Leaves growing above the water are broad, flat, and lobed. The genetically identical underwater leaves are thin and finely divided, appearing almost rootlike. These differences are thought to be related to differences in the turgor (page 80) of the immature leaf cells in the two environments. The degree of turgor affects the expansion of the cell walls and thus the ultimate size of the cells.*

came clear to researchers that chromosomes carried the genetic information, the problem became one of deciding whether the protein or the DNA plays this essential role. By the early 1950s, a great deal of evidence for the role of DNA as the genetic material had accumulated. It was not until the discovery of the structure of DNA, however, that its genetic role came to be understood.

The Structure of DNA

In the early 1950s, a young American scientist, James Watson, went to Cambridge, England, on a research fellowship to study problems of molecular structure. There, at the Cavendish Laboratory, he met physicist Francis Crick. Both were interested in DNA, and they soon began to work together to solve the problem of its molecular structure. They did not do experiments in the usual sense but rather undertook to examine all the data about DNA provided by others and to unify them into a meaningful whole.

DNA Consists of Nucleotides, Each Containing One of Four Nitrogenous Bases

By the time Watson and Crick began their studies of DNA, a lot of information on the subject had already accumulated. It was known that the DNA molecule is very large, very long and thin, and composed of four different kinds of molecules called nucleotides (page 30). Each nucleotide contains a phosphate group, the sugar deoxyribose, and one of four bases: **adenine, guanine, cytosine,** and **thymine.** Two of the bases, adenine and guanine, are similar to one another in structure and are called **purines.** The other two bases, cytosine and thymine, are also similar to one another in structure and are called **pyrimidines.**

In 1950, Linus Pauling had shown that proteins sometimes take the form of a helix (page 28) and that the helical structure is maintained by hydrogen bonding between successive turns of the helix. Pauling had even suggested that the structure of DNA might be similar. Subsequent X-ray studies by Rosalind Franklin and Maurice Wilkins at Kings College, London, provided strong evidence that the DNA molecule was indeed a giant helix. Finally, data obtained by Erwin Chargaff indicated that the ratio of DNA nucleotides containing thymine to those containing adenine is approximately 1:1, and that the ratio of nucleotides containing guanine to those containing cytosine is also approximately 1:1 (Table 10–3).

DNA Exists in the Form of a Double Helix

From these data, Watson and Crick attempted to construct a model of DNA that would fit the known facts and explain the biological role of DNA. In order to carry the vast amount of genetic information, the molecules should be heterogeneous and varied. Also, there must be some way to replicate readily and with great precision so that faithful copies could be passed from cell to cell and from parent to offspring, generation after generation.

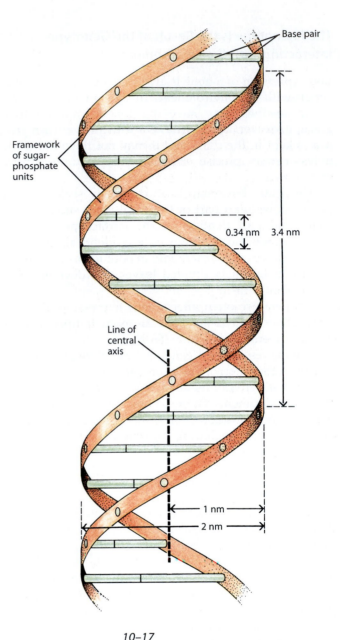

10–17

The double-stranded helical structure of DNA, as first presented in 1953 by Watson and Crick. The framework of the helix is composed of the sugar-phosphate units of the nucleotides. The rungs are formed by the four nitrogenous bases adenine and guanine (the purines) and thymine and cytosine (the pyrimidines). Each rung consists of a pair of bases. Knowledge of the distances in nanometers (nm) shown here was crucial in establishing the detailed structure of the DNA molecule. These distances were determined from X-ray diffraction photographs of DNA taken by Rosalind Franklin.

TABLE 10–3   Composition of DNA in Several Species

Source	Purines		Pyrimidines	
	Adenine	Guanine	Cytosine	Thymine
Human being	30.4%	19.6%	19.9%	30.1%
Ox	29.0	21.2	21.2	28.7
Salmon sperm	29.7	20.8	20.4	29.1
Wheat germ	28.1	21.8	22.7	27.4
Escherichia coli	24.7	26.0	25.7	23.6
Sheep liver	29.3	20.7	20.8	29.2

After Erwin Chargaff, *Essays on Nucleic Acids,* 1963.

By piecing together the various data, Watson and Crick were able to deduce that DNA is not a single-stranded helix, as are many proteins, but is instead a huge, entwined double helix. If you were to take a ladder and twist it into the shape of a helix, keeping the rungs perpendicular to the sides, a crude model of a double helix would be formed (Figure 10–17). The two sides of the ladder are made up of alternating sugar molecules and phosphate groups. The rungs of the ladder are formed by the nitrogenous bases—adenine (A), thymine (T), guanine (G), and cytosine (C)—one base for each sugar-phosphate, with two bases forming each rung. The paired bases meet in the interior of the helix and are joined by hydrogen bonds, the relatively weak bonds that Pauling had demonstrated in his studies of protein structure.

As Watson and Crick worked their way through the data, they assembled actual tin-and-wire models of the molecules, testing where each piece would fit into the three-dimensional puzzle (Figure 10–18). As they worked with the models, they noticed that the nucleotides along any one strand of the double helix could be assembled in any order, such as ATGCGTACATT and so on. Because a DNA molecule may be several thousand nucleotides long, a wide variety of arrangements is possible.

The most exciting discovery came when Watson and Crick set out to construct the matching DNA strand. In doing this, they encountered an interesting and important restriction: purines could not pair with purines and pyrimidines could not pair with pyrimidines, and adenine could pair only with thymine, and guanine could pair only with cytosine. Only these two combinations of nitrogenous bases form the correct hydrogen bonds. Adenine forms two hydrogen bonds with thymine, and guanine forms three hydrogen bonds with cytosine.

Now look again at Table 10–3. The Watson-Crick model explains in a simple and logical way the base composition of DNA—that is, that the amount of A equals that of T and the amount of C equals that of G. Perhaps the most important property of the model is that the two strands are **complementary;** that is, each strand contains a sequence of bases that is complementary to the sequence of bases in the other strand.

The double-stranded structure of a small portion of a DNA molecule is shown in Figure 10–19. In each strand, the phosphate group that joins two deoxyribose molecules is attached to one sugar at the 5' position (carbon 5 of deoxyribose) and to the other sugar at the 3' position (the third carbon in the ring of deoxyribose). This configuration gives each strand a 5' end and a 3' end. Moreover, the two strands run in opposite directions—that is, the direction from the 5' end to the 3' end of one strand is opposite that of the other strand. The strands are said to be **antiparallel.**

In what could be considered one of the great understatements of all time, Watson and Crick wrote in their original brief publication: "It has not escaped our notice that the specific pairing we have postulated immediately suggests a possible copying mechanism for the genetic material." In 1962, nine years after their hypothesis was published, Watson, Crick, and Wilkins shared a Nobel Prize in recognition of their landmark studies.

(a)

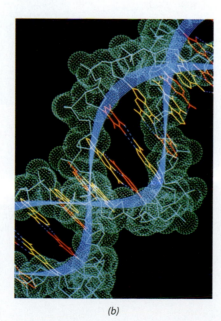

(b)

10–18

(a) James Watson (left) and Francis Crick in 1953, with one of their models of DNA. At the time they announced their discovery of the DNA structure, Watson was 23 and Crick was 34. (b) A computer-generated model of a portion of a DNA molecule. The sugar-phosphate backbone is indicated by the blue ribbons and green dots. The purines are shown in yellow, and the pyrimidines in red. The hydrogen bonds linking the base pairs are represented by blue dashed lines.

In 1993, 40 years after their discovery, Watson remarked, "The molecule is so beautiful. Its glory was reflected on Francis and me. I guess the rest of my life has been spent trying to prove that I was almost equal to being associated with DNA, which has been a hard task." Crick replied, "We were upstaged by a molecule."

10–19

The double-stranded structure of a small portion of a DNA molecule. Each nucleotide consists of a phosphate group, a deoxyribose sugar, and a purine or pyrimidine base.

Note the repetitive sugar-phosphate-sugar-phosphate sequence that forms the backbone of each strand of the molecule. Each phosphate group is attached to the 5' carbon of one sugar subunit and to the 3' carbon of the sugar subunit in the adjacent nucleotide. Each strand of the DNA molecule thus has a 5' end and a 3' end, determined by these 5' and 3' carbons. The strands are antiparallel—that is, the direction from the 5' to the 3' end of one strand is opposite to that of the other strand.

The strands are held together by hydrogen bonds (represented here by dashed lines) between the bases. Notice that adenine and thymine form two hydrogen bonds, whereas guanine and cytosine form three. Because of these bonding requirements, adenine can pair only with thymine, and guanine can pair only with cytosine. Thus the order of bases along one strand determines the order of bases along the other strand.

The sequence of bases varies from one DNA molecule to another. It is customarily written as the sequence in the 5' to 3' direction of one of the strands. Here, using the strand on the left, the sequence is TCAG.

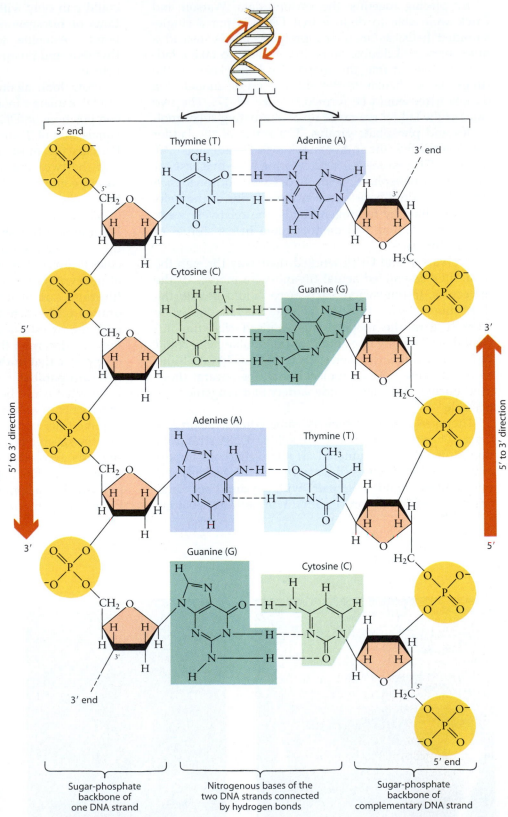

Sugar-phosphate backbone of one DNA strand

Nitrogenous bases of the two DNA strands connected by hydrogen bonds

Sugar-phosphate backbone of complementary DNA strand

DNA Replication

An essential property of the genetic material is the ability to provide for exact copies of itself. As Watson and Crick initially observed, a mechanism by which DNA can replicate itself is implicit in the double and complementary structure of the DNA helix.

At the time of DNA replication, the molecule "unzips," with the paired bases separating as the hydrogen bonds are broken. As the two strands separate, they act as **templates,** or guides, for the synthesis of two new strands. Each strand directs the synthesis of the new complementary strand along its length (Figure 10–20), using the raw materials of the cell. If a T is present on the original strand (the template), only an A can fit in the adjacent location of the new strand; a G will pair only with a C, and so on. In this way, each strand forms a copy of its original partner strand, and two exact replicas of the molecule are produced. The age-old question of how hereditary information is duplicated and passed on, generation after generation, had, in principle, been answered.

DNA Replication Is Bidirectional

Replication of DNA is a process that occurs only once in each cell generation, during the S phase of the cell cycle (page 158). It is the essential event in the duplication of chromosomes. In most eukaryotic cells, DNA replication leads ultimately to mitosis, but in cells that give rise to meiospores or to gametes it leads to meiosis. It is a remarkably rapid process. For example, in humans and other mammals, the rate of synthesis is about 50 nucleotides per second. In prokaryotes, it is even faster—about 500 nucleotides per second.

The principle of DNA replication, in which each strand of the double helix serves as a template for the formation of a new strand, is relatively simple and easy to understand. However, the actual process by which the cell accomplishes replication is considerably more complex. Like other biochemical reactions of the cell, DNA replication requires a number of different enzymes, each catalyzing a particular step of the process. The identification of the principal enzymes, their precise functions, and the sequence of events in replication has required a number of years and the efforts of many scientists working in different laboratories. Although our understanding is still incomplete, the general outlines of the process are now clear.

Initiation of DNA replication always begins at a specific nucleotide sequence known as the **origin of replication.** Initiation requires special initiator proteins and enzymes known as **helicases,** which break the hydrogen bonds linking the complementary bases at the origin of replication, opening up the helix so replication can occur. Single-strand binding proteins keep the two

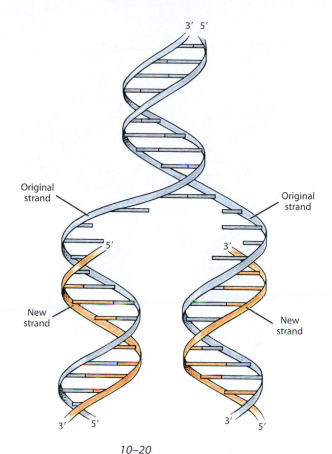

10–20

Replication of the DNA molecule as predicted by the Watson–Crick model. The strands separate as the paired bases separate at the hydrogen bonds. Each of the original strands then serves as a template along which a new, complementary strand forms from nucleotides available in the cell. Subsequent research has resulted in modification of some of the details of this process, as we shall see shortly, but the underlying principle is unchanged.

strands separated and allow the enzymes necessary for synthesis to bind. The synthesis of new strands is catalyzed by enzymes known as **DNA polymerases.**

If DNA caught in the act of replication is viewed under the electron microscope, the localized regions of synthesis appear as "eyes," or **replication bubbles.** At either end of a bubble, where the existing strands are being separated and the new, complementary strands are being synthesized, the molecule appears to form a Y-shaped structure. This is known as a **replication fork.** The two replication forks move in opposite directions away from the origin (Figure 10–21), and thus replication is said to be **bidirectional.**

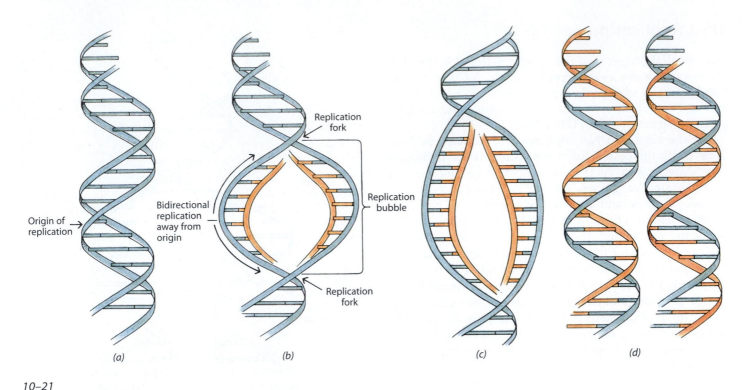

Origin of
replication

Bidirectional
replication
away from
origin

Replication
fork

Replication
bubble

Replication
fork

(a)   (b)   (c)   (d)

10–21

*An overview of DNA replication. **(a)** The two strands of the DNA molecule separate at the origin of replication as a result of the action of special initiator proteins and enzymes. **(b)**, **(c)** The two replication forks move away from the origin of replication in opposite directions, forming a replication bubble that expands in both directions (bidirectionally). **(d)** When synthesis of the new DNA strands is complete, the two double-stranded chains separate into two new double helixes. Each helix consists of one old strand and one new strand.*

In prokaryotes, there is a single origin of replication in the chromosome. In eukaryotes, by contrast, there are many replication origins in each chromosome. Replication proceeds along the linear chromosome as each bubble expands bidirectionally (Figure 10–22) until it meets an adjacent bubble.

DNA polymerase synthesizes new DNA strands only in the 5′ to 3′ direction. Along one of the original strands, therefore, DNA can be synthesized continuously in the 5′ to 3′ direction as a single unit. This newly synthesized DNA is known as the **leading strand.** Because of DNA's antiparallel nature and the directionality of DNA polymerase, the other strand is synthesized, again in the 5′ to 3′ direction, as a series of fragments. These fragments of what is known as the **lagging strand** are each individually synthesized in a direction *opposite* to the overall direction of replication. As the fragments, known as **Okazaki fragments,** lengthen, they are eventually joined together by an enzyme called **DNA ligase.** The complex process of DNA replication is summarized in Figure 10–23.

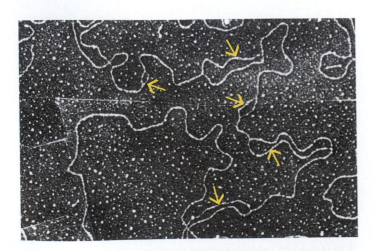

10–22

Replication of eukaryotic chromosomes is initiated at multiple origins. The individual replication bubbles spread until ultimately they meet and join. In this electron micrograph of a replicating chromosome in an embryonic cell of the fruit fly (Drosophila), the replication bubbles are indicated by the arrows.

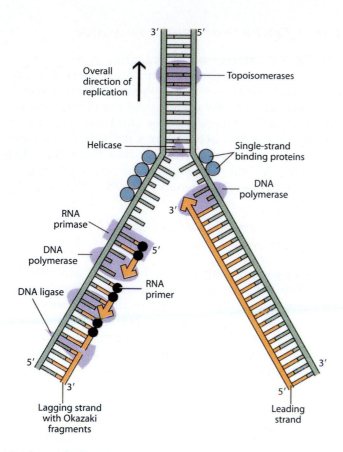

10–23

A summary of DNA replication. The two strands of the DNA double helix separate, and new complementary strands are synthesized by DNA polymerase in the 5' to 3' direction, using the original strands as templates.

Each new strand begins with a short RNA primer formed by the enzyme RNA primase. Synthesis of the leading strand, which requires a single RNA primer (not shown in diagram), is continuous. However, synthesis of the lagging strand, which occurs in a direction opposite to the overall direction of replication, is discontinuous. The short DNA molecules of the lagging strand, each requiring individual RNA primers, are called Okazaki fragments. Synthesis of each Okazaki fragment ends when the DNA polymerase runs into the RNA primer attached to the 5' end of the previous fragment. Following replacement of the RNA primer of the previous Okazaki fragment by DNA nucleotides, the fragment is joined to the growing strand by the enzyme DNA ligase.

Enzymes called topoisomerases prevent DNA tangling during replication, and the two strands, separated at the replication fork by enzymes known as helicases, are stabilized by single-strand binding proteins.

The Problem of the Ends of Linear DNA

DNA replication in the circular DNA molecules of prokaryotes is finished when the leading and lagging strands complete the circle. The leading strand simply continues to grow in the 5' to 3' direction until its 3' end reaches and is joined to the 5' end of the lagging strand of the same complementary strand coming around the other way. With the linear DNA of eukaryotes, however, there is no way for the DNA polymerases to polymerize the last few nucleotides of the lagging strand because there is no 3' end ahead of it. As a result, with each round of replication, the replicated DNA molecules become progressively shorter. If this trend were to continue, the chromosome with its essential genes would gradually be lost.

Eukaryotic chromosomes counter this loss through the presence of special end-sequences called **telomeres.** Although the ends of chromosomes usually do shorten with replication of the DNA molecule, the telomeres protect the nontelomeric DNA from degradation. On occasion, the telomeres are lengthened as DNA of the telomeres is synthesized by a special DNA polymerase called **telomerase.**

A function for telomeres was proposed in 1939 by Barbara McClintock, who reported that broken chromosomes in maize frequently fuse with other broken chromosomes. Because intact chromosomes with telomeres do not fuse with one another, she hypothesized that the telomeres protect chromosomes from end-to-end fusion, a role that has come to be accepted.

Errors in Replication Are Usually Corrected by DNA Polymerases

Occasional mistakes can be made in replication and damage can be done to DNA, for example, by ultraviolet radiation and chemicals. DNA polymerases are remarkably accurate, however, so that few mistakes occur during replication. Moreover, if a mistake is made, the enzyme reverses itself, and the incorrect nucleotide is quickly replaced with the correct one. In addition, many different mechanisms have evolved to repair damaged DNA. The bottom line is that most DNA is faithfully preserved, but mistakes—mutations—occasionally happen. Some mutations may be harmful and lead to death of the organism, but a few of them may be beneficial and provide the raw material for natural selection.

The Energetics of DNA Replication

The nucleotides required for DNA replication are synthesized by the same biochemical pathways as the nucleotides required for other functions of the cell. However, they are supplied not in the form of the monophosphates but rather as triphosphates. For example, adenine is supplied as deoxyadenosine triphosphate (dATP). The only difference between ATP and dATP is in the sugar subunit—in ATP (page 31) it is ribose, and in dATP it is deoxyribose. Similarly, guanine is supplied as dGTP, cytosine as dCTP, and thymine as dTTP.

Just as ATP provides energy to power a variety of biosynthetic reactions, dATP, dGTP, dCTP, and dTTP provide energy to power the synthesis of DNA. As each nucleotide is attached to the growing DNA strand, the "extra" pyrophosphate group is removed and immediately split into two phosphate ions. The energy released in this process powers the reactions mediated by the DNA polymerases.

DNA As a Carrier of Information

You will recall that a necessary property of the genetic material is the capacity to carry a vast amount of information. In the DNA molecule, the information is carried in the sequence of the bases, and *any* sequence of bases is possible. The number of paired bases ranges from about 5000 for the simplest virus up to an estimated 200 billion in lily. In a single human cell, with "only" 6 billion paired bases, the DNA contains an amount of information equal to some 600,000 printed pages averaging 500 words each, the equivalent of a library of about a thousand books. Quite clearly, the DNA molecule has the capacity to store the necessary genetic information.

We have come a long way in our understanding of the mechanism of heredity and of the ways in which the characteristics of organisms are produced. Much more remains to be learned, however, and the field will remain one of central importance to the science of botany for many years to come.

Summary

Mendel's Experiments Provide the Foundation of Modern Genetics

Gregor Mendel's breeding experiments in pea plants revealed that hereditary characteristics are determined by discrete factors (now called genes) that occur in pairs. According to Mendel's principle of segregation, the members of each pair segregate (that is, separate from one another) during gamete formation. When two gametes come together in fertilization, the offspring receives one member of each pair from each parent. The members of a given pair may be the same (homozygous) or they may be different (heterozygous). Different forms for the same gene are known as alleles.

Alleles Can Be Dominant or Recessive

The genetic makeup of an organism is its genotype, while its observable characteristics are its phenotype. An allele that is expressed in the phenotype of a heterozygous individual to the exclusion of the other allele is a dominant allele. An allele whose effects are concealed in the phenotype of a heterozygous individual is a recessive allele. In crosses involving two individuals heterozygous for the same gene, the expected ratio of dominant to recessive phenotypes in the offspring is 3:1.

Unlinked Genes Assort Independently during Meiosis

Mendel's second principle—independent assortment—applies to the behavior of two or more different genes. This principle states that, during gamete formation, the alleles of one gene segregate independently of the alleles of another gene. When organisms heterozygous for each of two independently assorting genes are crossed, the expected phenotypic ratio in the offspring is 9:3:3:1.

Genes Are Carried on Chromosomes

Sutton was among the first to notice the analogy between the behavior of the chromosomes at meiosis and the segregation and assortment of the factors described by Mendel. On the basis of this observation, Sutton proposed that genes are carried on chromosomes.

Genes That Occur on the Same Chromosome Are Said to Be Linked

Further studies by Bateson showed that although some genes assort independently, adhering to Mendel's second principle, others tend to remain together. Such genes, located relatively close together on a chromosome, are said to be linked. Linkage maps, based on the amount of crossing-over that occurs between genes, can give an approximation of the location of genes on chromosomes.

Mutations Are Changes in the Genetic Makeup of an Individual

Mutations are random changes in the genotype. Different mutations of a single gene increase the diversity of alleles of the gene in a population. As a consequence, mutation provides the variability among organisms that is the raw material for evolution. There are a number of different kinds of mutations: point mutations, deletions, duplications, position effects, inversions, translocations, and changes in chromosome number.

Alleles Can Be Incompletely Dominant and/or Exist in More than Two Forms

Many traits are inherited according to the patterns revealed by Mendel. However, for others—perhaps the majority—the patterns are more complex. Although many alleles interact in a dominant-recessive manner, some show varying degrees of incomplete dominance, with intermediate phenotypes. In a population of organisms, multiple alleles of a single gene may exist, but only two alleles can be present in any given diploid individual.

Different Genes Can Also Interact with One Another

Novel phenotypes may result from gene interactions, or genes may affect one another in an epistatic manner, such that one hides the effect of the other. The phenotypic expression of many traits is influenced by a number of genes. This phenomenon is known as polygenic inheritance. Such traits typically show continuous variation in a population, as represented by a bell-shaped curve. Conversely, a single gene can affect two or more superficially unrelated traits. This property of a gene is known as pleiotropy.

Some Genes Are Located in the Cytoplasm

Although most of the traits inherited by the cell are contributed by nuclear genes, some inherited traits are controlled by genes located in the cytoplasm—specifically, in plastids and mitochondria. This phenomenon is known as cytoplasmic inheritance.

The Phenotype Is the Result of the Genotype Interacting with the Environment

Gene expression is also affected by factors in the external and internal environments. Variations in the expression of particular alleles may result from environmental influences, interactions with other genes, or both.

Watson and Crick Deduced That DNA Is a Double Helix

The Watson and Crick model of DNA is a double-stranded helix, shaped like a twisted ladder. The two sides of the ladder are composed of repeating subunits consisting of a phosphate group and the five-carbon sugar deoxyribose. The "rungs" are made up of paired nitrogenous bases. There are four bases in DNA—the purines adenine (A) and guanine (G) and the pyrimidines thymine (T) and cytosine (C). A can pair only with T, and G only with C, so a "rung" always consists of one purine and one pyrimidine. The four bases are the four "letters" used to spell out the genetic message. The paired bases are joined by hydrogen bonds.

When DNA Replicates, Each Strand Is Used As a Template to Synthesize a Complementary Strand

When the DNA molecule replicates, the two strands separate locally, breaking at the hydrogen bonds that otherwise hold them together. Each strand acts as a template for the formation of a new, complementary strand from nucleotides available in the cell. The addition of nucleotides to the new strands is catalyzed by DNA polymerases. A variety of other enzymes also play key roles in the replication process.

DNA Replication Is Bidirectional

Replication begins at a particular nucleotide sequence on the chromosome, the origin of replication. It proceeds bidirectionally, by way of two replication forks that move in opposite directions. The circular DNA molecules of prokaryotes have only one origin of replication; replication is completed when the leading and lagging strands complete the circle. By contrast, the linear DNA molecules of eukaryotes have many replication origins, and the DNA polymerases are unable to add the last few nucleotides of the lagging strand. Consequently, the DNA molecules tend to become progressively shorter with each replication. Eukaryotic chromosomes counter this loss through the presence of special end-sequences called telomeres.

Errors in DNA Replication Are Usually Corrected by DNA Polymerases

DNA is continually suffering damage, ranging from mispairing of bases to chromosome breakage. Many different enzymes and mechanisms have evolved for repairing the damaged DNA, and the process of maintaining the cell's genetic integrity is therefore remarkably accurate.

DNA Is the Carrier and Transmitter of Genetic Information

The role of DNA as the carrier and transmitter of the genetic information is now universally accepted. With the discovery of the mechanism by which the living cell replicates its DNA, the question of how the hereditary information is faithfully transmitted from parent cell to daughter cell, generation after generation, was answered.

Selected Key Terms

alleles   p. 186

aneuploidy   p. 193

chromosome mutations
p. 192

deletions   p. 192

dihybrid   p. 185

DNA ligase   p. 202

DNA polymerases
p. 201

dominant   p. 185

duplications   p. 193

epistasis   p. 195

F₁, F₂ generations   p. 185

gene, or point, mutations
p. 192

genes   p. 184

genetic, or linkage, maps
p. 191

genotype   p. 186

heterozygous   p. 186

homozygous   p. 186

incomplete dominance
p. 194

inversion   p. 193

linked genes   p. 190

locus   p. 186

monohybrid   p. 185

multiple alleles   p. 194

mutagens   p. 192

mutation   p. 192

Okazaki fragments
p. 202

phenotype   p. 186

pleiotropy   p. 195

polygenic inheritance
p. 195

polyploidy   p. 193

recessive   p. 185

telomeres   p. 203

testcross   p. 187

translocation   p. 193

transposons   p. 193

Questions

1. Distinguish between the following terms: gene/allele; genotype/phenotype; epistasis/pleiotropy; leading strand/lagging strand.

2. Why is a homozygous recessive always used in a testcross?

3. Explain how the movement of chromosomes during meiosis relates to Mendel's two principles, or laws.

4. Explain what is meant by cytoplasmic inheritance; maternal inheritance.

5. A pea plant that breeds true for round, green seeds ($RRyy$) is crossed with a plant that breeds true for wrinkled, yellow seeds ($rrYY$). Each parent is homozygous for one dominant characteristic and for one recessive characteristic. (a) What is the genotype of the F_1 generation? (b) What is the phenotype? (c) The F_1 seeds are planted and their flowers are allowed to self-pollinate. Draw a Punnett square to determine the ratios of the phenotypes in the F_2 generation. How do the results compare with those of the experiment shown in Figure 10–7?

6. In Jimson weed, the allele for violet petals (W) is dominant over the allele for white petals (w), and the allele for prickly capsules (S) is dominant over the allele for smooth capsules (s). A plant with white petals and prickly capsules was crossed with one that had violet petals and smooth capsules. The F_1 generation was composed of 47 plants with white petals and prickly capsules, 45 plants with white petals and smooth capsules, 50 plants with violet petals and prickly capsules, and 46 plants with violet petals and smooth capsules. What were the genotypes of the parents?

7. How are "complementary bases" and "antiparallel strands" involved in the structure of DNA?

8. Shown here is the sequence of bases in the 5' to 3' direction in one strand of a hypothetical DNA molecule. Identify the sequence of bases in the complementary strand.

   5' – A – A – G – T – T – T – G – G – T – T – A – C – T – T – G – 3'
   3' – – – – – – – – – – – – – – – 5'

9. Distinguish among the following: origin of replication, replication bubble, and replication fork.

10. Eukaryotic cells are grown for a number of generations in a medium containing thymine labeled with tritium, a radioactive isotope of hydrogen (^3H). The cells are then removed from the radioactive medium, placed in an ordinary, nonradioactive medium, and allowed to divide. Studies of the distribution of the radioactive isotope are made after each generation to determine the presence or absence of radioactive material in the chromatids.

    Before the cells are placed in the nonradioactive medium, all the chromatids contain ^3H. After one generation in the nonradioactive medium, all of the chromatids still contain radioactive ^3H. (a) Assuming each chromatid contains a single DNA molecule, explain the results. (b) Is this consistent with the Watson-Crick hypothesis of DNA replication? (c) What would be the distribution of the ^3H after two generations in the nonradioactive medium? Why? (*Hint:* You may find it helpful to look at Figure 10–21 as you think about these questions.)

Chapter 11

Gene Expression

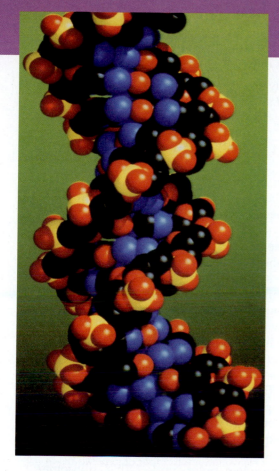

11–1
Computer-generated model of DNA.

OVERVIEW

How does DNA store hereditary information? And how is this information used to control the activities of the cell and thus the traits of an individual, such as the flower colors that Mendel studied? In this chapter, you will find that the genetic information encoded in the sequence of nitrogenous bases in the DNA is used to make proteins. Recall that proteins, particularly enzymes, play vital roles in the cell and that each protein consists of a specific linear sequence of amino acids that determines how that protein will function in the cell. As we will see, the sequence of bases in the DNA controls cellular activities by determining the sequence of amino acids in proteins.

The second part of the chapter deals with the regulation of gene activity. A plant grows from a single cell—the zygote—as the result of mitotic divisions, and therefore all its vegetative parts are genetically identical. Root cells, for example, contain the genes to make leaves, and leaf cells contain the genes to make roots. Leaves, however, do not normally grow out of roots, so the leaf genes must somehow be inactive in root cells but active in leaf cells. Although there is still much uncertainty about the mechanisms by which genes are switched on and off, we will examine some models proposed for prokaryotic cells and some of the factors regulating gene activity in eukaryotic cells.

The last part of the chapter presents information on recombinant DNA technology and the various techniques by which genes are manipulated for human purposes.

CHECKPOINTS

By the time you finish reading this chapter, you should be able to answer the following questions:

1. What is the nature of the genetic code, and in what way is it "universal"?

2. What are the main steps in the transcription of RNA from DNA?

3. Where does translation occur in a eukaryotic cell, and what are the major steps that occur in the process?

4. How do eukaryotic and prokaryotic chromosomes differ?

5. What are some of the factors regulating gene expression in eukaryotes?

6. How can recombinant DNA technology be used to clone a gene of interest?

A s we saw in the last chapter, Watson and Crick—by constructing a model of the DNA molecule—established the chemical nature of the gene and suggested the mechanism for its replication. The DNA molecule (Figure 11–1) carries the instructions for manufacturing cell structures and for enabling the cell to grow, develop, and function. The DNA molecule also transmits those instructions to new cells and organisms. Yet, two fundamental questions were left unanswered: How are instructions encoded in the DNA molecule, and how are they carried out?

The search for the answer to these questions led to ribonucleic acid (RNA), a molecule similar to DNA (page 31). RNA's involvement in gene expression was long suspected because cells that are synthesizing large amounts of protein invariably contain large amounts of RNA. Moreover, unlike DNA, which is found mostly in the nucleus, RNA is found mostly in the cytoplasm, where protein synthesis takes place. Also, both prokaryotic and eukaryotic cells that make large amounts of protein have numerous ribosomes, and ribosomes contain substantial RNA.

Additional evidence for the involvement of RNA in gene expression came from studies with viruses. When a bacterial cell is infected by a DNA-containing bacteriophage (a bacterial virus), RNA is synthesized from the viral DNA before viral protein synthesis begins. Also, some viruses consist of only RNA and protein. These clues all indicated that RNA as well as DNA contained information about protein structure.

Like DNA, RNA is a long-chain macromolecule of nucleic acid, but it differs from DNA in three important properties (Figure 11–2).

1. In the nucleotides of RNA, the sugar component is ribose rather than deoxyribose.

2. The nitrogenous base thymine is found in DNA but does not occur in RNA. Instead, RNA contains a closely related pyrimidine, **uracil (U).** Uracil, like thymine, pairs only with adenine.

3. RNA is usually single-stranded and does not form a regular helical structure.

From DNA to Protein: The Role of RNA

It turned out that not one but three kinds of RNA play roles as intermediaries in the steps that lead from DNA to proteins: messenger RNA, transfer RNA, and ribosomal RNA.

As we shall see in this chapter, DNA serves as a template for the synthesis of **messenger RNA (mRNA)** molecules, a process called transcription. Transcription follows rules similar to those of replication and is catalyzed by the enzyme **RNA polymerase.** The role of mRNA is to carry the genetic message from the DNA to the ribosomes, which are composed of **ribosomal RNA (rRNA)** and proteins.

The ribosomes are the actual sites of protein synthesis. The role of the **transfer RNA (tRNA)** molecules, each of which is specific for a particular amino acid, is to translate the coded nucleotide sequence of mRNA into the amino acid sequence of protein. The tRNA molecules carry the amino acids to the ribosomes, where they are attached to a growing polypeptide. This ribosome-mediated synthesis of a polypeptide is called translation. It is the nucleotide sequence in the mRNA that determines the amino acid sequence in the protein.

Thus, RNA is synthesized from DNA, and RNA is used to synthesize protein. Three processes of information transfer can now be recognized: **replication** (the syn-

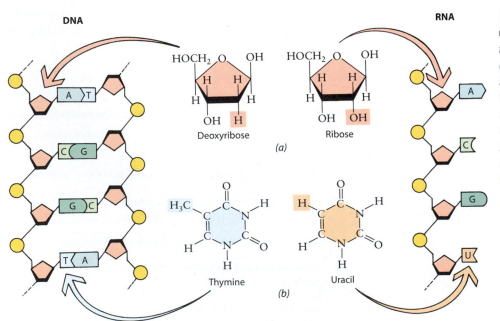

DNA   **RNA**   *11–2*

*Chemically, RNA is very similar to DNA, but there are two differences in its nucleotides. **(a)** One difference is in the sugar component. Instead of deoxyribose, RNA contains ribose, which has an additional oxygen atom. **(b)** The other difference is that instead of thymine, RNA contains the closely related pyrimidine uracil (U). Uracil, like thymine, pairs only with adenine. A third difference concerns the structure of RNA. Unlike DNA, RNA is usually single-stranded and does not form a regular helix.*

Deoxyribose   Ribose

(a)

Thymine   Uracil

(b)

11–3

The three processes of information transfer: replication, transcription, and translation. Replication of the DNA occurs only once in each cell cycle (page 158), during the S phase prior to mitosis or meiosis. Transcription and translation, however, occur repeatedly throughout the interphase portions of the cell cycle.

thesis of DNA), **transcription** (the synthesis of mRNA, which is a copy of a portion of one strand of the double-stranded DNA helix), and **translation** (the synthesis of a polypeptide directed by the nucleotide sequence of the mRNA) (Figure 11–3). The principle of unidirectional flow from gene to mRNA molecule to protein is known as the **central dogma of molecular biology,** a term coined by Francis Crick.

The Genetic Code

The identification of mRNA as the working copy of the genetic information that dictates the sequence of amino acids in proteins still left unresolved the big question: How is this accomplished? Proteins contain 20 different kinds of amino acids, but DNA and RNA each contain only four different kinds of nucleotides. Somehow, these nucleotides constitute a **genetic code** for the amino acids.

As it turned out, the idea of a code was useful not only as a dramatic metaphor but also as a working analogy. Scientists, seeking to understand how the sequence of nucleotides stored in the double helix of DNA could specify the quite dissimilar structures of protein molecules, approached the problem with methods used by cryptographers in deciphering codes.

As we saw in Chapter 2, the primary structure of a particular kind of protein molecule consists of a specific linear arrangement of the 20 different kinds of amino acids. Similarly, there are four different kinds of nucleotides, arranged in a specific linear sequence in a DNA molecule. If each nucleotide "coded" for one amino acid, only four amino acids could be specified by the four bases. If two nucleotides specified one amino acid, there could be a maximum number, using all possible arrangements of the nucleotides, of 4×4, or 16—still not quite enough to code for all 20 amino acids. Therefore, following the code analogy, at least three nucleotides in sequence must specify each amino acid. This would provide for $4 \times 4 \times 4$, or 64, possible combinations, or **codons**—clearly more than enough.

The three-nucleotide, or triplet, codon was widely and immediately adopted as a working hypothesis. Its existence, however, was not actually demonstrated until the code was finally broken, a decade after Watson and Crick first presented their DNA structure. The scientists who performed the initial, crucial experiments were Marshall Nirenberg and his colleague Heinrich Matthaei, both of the National Institutes of Health. Utilizing the extracted contents of *Escherichia coli* cells, radioactively labeled amino acids, and mRNAs from a variety of sources, they defined the first code word—UUU for phenylalanine—and provided a method for defining the others. As a result of these and similar experiments, subsequently performed in a number of laboratories, the mRNA codons for all of the amino acids were soon worked out. Of the 64 possible triplet combinations, 61 specify particular amino acids and 3 are stop signals. With 61 combinations coding only 20 amino acids, you can see that there must be more than one codon for many of the amino acids; hence, the genetic code is said to be **degenerate,** or redundant. As can be seen in Figure 11–4, codons specifying the same amino acid often differ only in the third nucleotide.

Second letter

First letter (5′ end)	U	C	A	G	Third letter (3′ end)
U	UUU UUC phe / UUA UUG leu	UCU UCC UCA UCG ser	UAU UAC tyr / UAA stop UAG stop	UGU UGC cys / UGA stop UGG trp	U C A G
C	CUU CUC CUA CUG leu	CCU CCC CCA CCG pro	CAU CAC his / CAA CAG gln	CGU CGC CGA CGG arg	U C A G
A	AUU AUC ile AUA / AUG met	ACU ACC ACA ACG thr	AAU AAC asn / AAA AAG lys	AGU AGC ser / AGA AGG arg	U C A G
G	GUU GUC GUA GUG val	GCU GCC GCA GCG ala	GAU GAC asp / GAA GAG glu	GGU GGC GGA GGG gly	U C A G

11–4

The genetic code, consisting of 64 codons (triplet combinations of mRNA bases) and their corresponding amino acids. (For the names and structures of the 20 amino acids, see Figure 2–15.) Of the 64 codons, 61 specify particular amino acids. The other three codons are stop signals, which cause the polypeptide chain to terminate. Since 61 triplets code for 20 amino acids, there obviously must be "synonyms"; for example, leucine (leu) has six codons. Each codon, however, specifies only one amino acid.

The Genetic Code Is Universal

One of the most remarkable discoveries of molecular biology is that the genetic code is nearly identical in all organisms, with only minor exceptions. Under proper circumstances, genes from bacteria can function perfectly in plant cells. Also, plant genes can be introduced into bacteria, where plant proteins can then be produced by the protein synthesis "machinery" of the bacterial cell. Not only does this observation provide a striking demonstration of the fact that all life on Earth has had a common origin, it also provides the basis for the techniques of genetic engineering, which hold such promise (see Chapter 28).

Protein Synthesis

Now that we know about the genetic code, we can address the question of how the information encoded in the DNA and transcribed into mRNA is subsequently translated into a specific amino acid sequence of a polypeptide chain. The principles of protein synthesis are basically similar in prokaryotic and eukaryotic cells, although there are some differences in detail. First, we shall focus on the process as it takes place in prokaryotic cells, using *E. coli* as our model.

Messenger RNA Is Synthesized from a DNA Template

As mentioned previously, instructions for protein synthesis are encoded in nucleotide sequences in the DNA and are copied, or transcribed, into molecules of mRNA following the same base-pairing rules that govern DNA replication. Each new mRNA molecule is transcribed from one of the two strands of the DNA helix (Figure 11–5). Transcription is catalyzed by RNA polymerase. The single-stranded mRNA molecules produced in this process range from less than 500 to over 10,000 nucleotides in length. Specific nucleotide sequences of the DNA, called **promoters,** are the binding sites for the RNA polymerase and thus determine where RNA synthesis starts and which DNA strand is used as a template. Once an RNA polymerase molecule attaches to a promoter and the DNA double helix is opened, transcription, or RNA synthesis, can begin. Transcription ends after the RNA polymerase transcribes a special sequence in the DNA called the **terminator.**

As we have mentioned, protein synthesis requires, in addition to mRNA molecules, the two other types of RNA: transfer RNA and ribosomal RNA. These molecules, which are transcribed from their own specific genes in the DNA of the cell, differ both structurally and functionally from mRNA.

Each Transfer RNA Carries an Amino Acid

Transfer RNA (tRNA) molecules are sometimes called "the dictionary of the language of life" because of the role they play in translating the nucleotide sequence of mRNA into the amino acid sequence of protein. Each tRNA molecule is relatively small, consisting of about 80 nucleotides that form a long, single strand that folds back on itself (Figure 11–6). There are over 60 different tRNA molecules in every cell, at least one for each of the 20 amino acids found in proteins.

Each tRNA molecule has two important attachment sites. One of these sites, the **anticodon,** consists of a sequence of three nucleotides that binds to the codon on an mRNA molecule. The other site, at the 3′ end of the tRNA molecule, attaches to a particular amino acid. The tRNA with its attached amino acid is called an **aminoacyl-tRNA.** Attachment of the tRNA molecules to their amino acids is brought about by enzymes known as **aminoacyl-tRNA synthetases.** There are at least 20 dif-

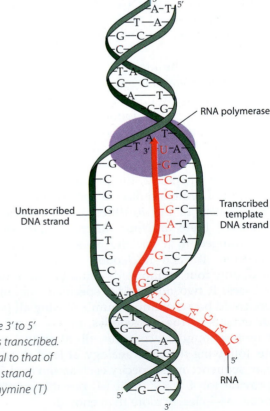

11–5

RNA transcription. At the point of attachment of the enzyme RNA polymerase, the DNA opens up, and as the RNA polymerase moves along the DNA molecule, the two strands of the DNA molecule separate. Nucleotide building blocks are assembled in RNA in a 5′ to 3′ direction. Note that the RNA strand is comple- *mentary—not identical—to the 3′ to 5′ template strand from which it is transcribed. Its sequence is, however, identical to that of the untranscribed (5′ to 3′) DNA strand, except for the replacement of thymine (T) by uracil (U).*

11–6

The structure of tRNA. Each tRNA molecule consists of about 80 nucleotides linked together in a single chain. The chain always terminates in a CCA sequence at its 3' end. An amino acid links to its specific tRNA molecule at this end. Some nucleotides are the same in all tRNAs; these are shown in gray. The other nucleotides vary according to the particular tRNA. The unlabeled boxes represent unusual modified nucleotides characteristic of tRNA molecules.

Some of the nucleotides are hydrogen-bonded to one another, as indicated by the dashed lines. In some regions, the unpaired nucleotides form loops. The loop on the right in this diagram is thought to play a role in binding the tRNA molecule to the surface of the ribosome. Three of the unpaired nucleotides in the loop at the bottom of the diagram (brown) form the anticodon. They serve to "plug in" the tRNA molecule to an mRNA codon.

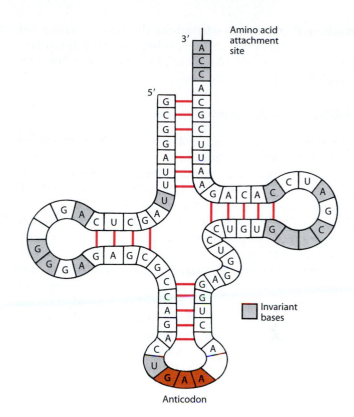

ferent aminoacyl-tRNA synthetases, one or more for each amino acid. Powered by energy released in ATP hydrolysis, these enzymes are key elements in the translation of the genetic message. They determine which amino acid will be associated with which tRNA (and thus with which anticodon).

Ribosomal RNA Is Associated with Protein to Form a Ribosome

As the name implies, ribosomal RNA (rRNA) is associated with the ribosomes, which are large complexes of RNA and protein molecules. Functionally, ribosomes are protein-synthesizing machines to which tRNA molecules bind in precise relationship to the mRNA molecules so as to read accurately the genetic message encoded in the mRNA. Ribosomes consist of two subunits, one small and the other large, each composed of specific rRNA molecules and proteins (Figure 11–7). The smaller subunit has an **mRNA-binding site**. The larger subunit has three sites where tRNA can bind: an **A (aminoacyl) site,** where the incoming amino acid-bearing tRNA usually

(a) Model based on electron micrographs

(b) Diagrammatic representation

mRNA-binding site

mRNA

11–7

In both prokaryotes and eukaryotes, ribosomes consist of two subunits, one large and one small. Each subunit is composed of specific rRNA and protein molecules. The size and density of both the subunits and the whole ribosome are greater in eukaryotes than prokaryotes. (a) One view of the three-dimensional structure of the E. coli ribosome, as revealed by electron micrographs. (b) A schematic diagram of another view of the E. coli ribosome. In protein synthesis, the ribosome moves along an mRNA molecule that is threaded between the two subunits.

binds; a **P (peptidyl) site,** where the tRNA bearing the growing polypeptide chain resides; and an **E (exit) site,** from which the tRNAs leave the ribosome after they have released their amino acids.

mRNA Is Translated into Protein

The synthesis of protein is known as "translation" because it involves the transfer of information from one language (a sequence of nucleotides) to another (a sequence of amino acids). In most cells, protein synthesis consumes more energy than any other biosynthetic process. The major stages in translation are initiation, elongation of the polypeptide chain, and chain termination (Figure 11–8).

Initiation begins when the smaller ribosomal subunit attaches to a strand of mRNA near its 5′ end, exposing its first codon, the *initiation codon*. The anticodon on the first tRNA then pairs with the initiation codon of the mRNA in an antiparallel fashion: the initiation codon usually is (5′)–AUG–(3′) and the tRNA anticodon is (3′)–UAC–(5′). The initiator tRNA, which binds to the AUG codon, carries a modified form of the amino acid methionine (fMet), the amino acid that starts the polypeptide chain. The larger ribosomal subunit then attaches to the smaller subunit, resulting in the fMet-tRNA being bound to the P site and the A site being made available for an incoming aminoacyl-tRNA. The energy for this step is provided by the hydrolysis of guanosine triphosphate (GTP).

At the beginning of the **elongation** stage, the second codon of the mRNA is positioned opposite the vacant ribosomal A site. A tRNA with an anticodon complementary to the second mRNA codon binds to the mRNA and, with its amino acid, occupies the A site of the ribosome. With both P and A sites occupied, an enzyme, **peptidyl transferase,** which is part of the larger ribosomal subunit, forges a peptide bond between the two amino acids, attaching the first amino acid (fMet) to the second. The first tRNA is released from the ribosome via the E site and reenters the cytoplasmic tRNA pool. The ribosome then moves one more codon down the mRNA molecule. Consequently, the second tRNA, to which the fMet and second amino acid are now attached, is transferred from the A to the P position. A third aminoacyl-tRNA moves into the A position opposite the third codon on the mRNA, and the step is repeated. Over and over, the P position accepts the tRNA bearing the growing polypeptide chain, and the A position accepts the tRNA bearing the new amino acid that will be added to the chain. As the ribosome moves along the mRNA strand, the initiator portion of the mRNA molecule is freed, and another ribosome can form an initiation complex with it. A group of ribosomes translating the same mRNA molecule is known as a **polyribosome,** or **polysome** (Figure 11–9).

11–8

*Three stages in protein synthesis. (**a**) Initiation. The smaller ribosomal subunit attaches to the 5′ end of the mRNA molecule. The first, or initiator, tRNA molecule, bearing the modified amino acid fMet, binds to the AUG initiation codon on the mRNA molecule. The larger ribosomal subunit locks into place, with the tRNA occupying the P (peptidyl) site. The A (aminoacyl) and E (exit) sites are vacant.*

*(**b**) Elongation. A second tRNA with its attached amino acid moves into the A site, and its anticodon binds to the mRNA. A peptide bond is formed between the two amino acids brought together at the ribosome. At the same time, the bond between the first amino acid and its tRNA is broken. The ribosome moves along the mRNA chain in a 5′ to 3′ direction, and the second tRNA, with the dipeptide attached, is moved to the P site from the A site as the first tRNA is released from the ribosome via the E site. A third tRNA moves into the A site, and another peptide bond is formed. The growing peptide chain is always attached to the tRNA that is moving from the A site to the P site, and the incoming tRNA bearing the next amino acid always occupies the A site. This step is repeated over and over until the polypeptide is complete.*

*(**c**) Termination. When the ribosome reaches a termination codon (in this example, UGA), a release factor occupies the A site. The polypeptide is cleaved from the last tRNA, the tRNA is released from the P site, and the two subunits of the ribosome separate.*

11–9

Clusters of ribosomes. All the ribosomes in a specific cluster are reading the same molecule of mRNA. Such groups are called polyribosomes, or polysomes.

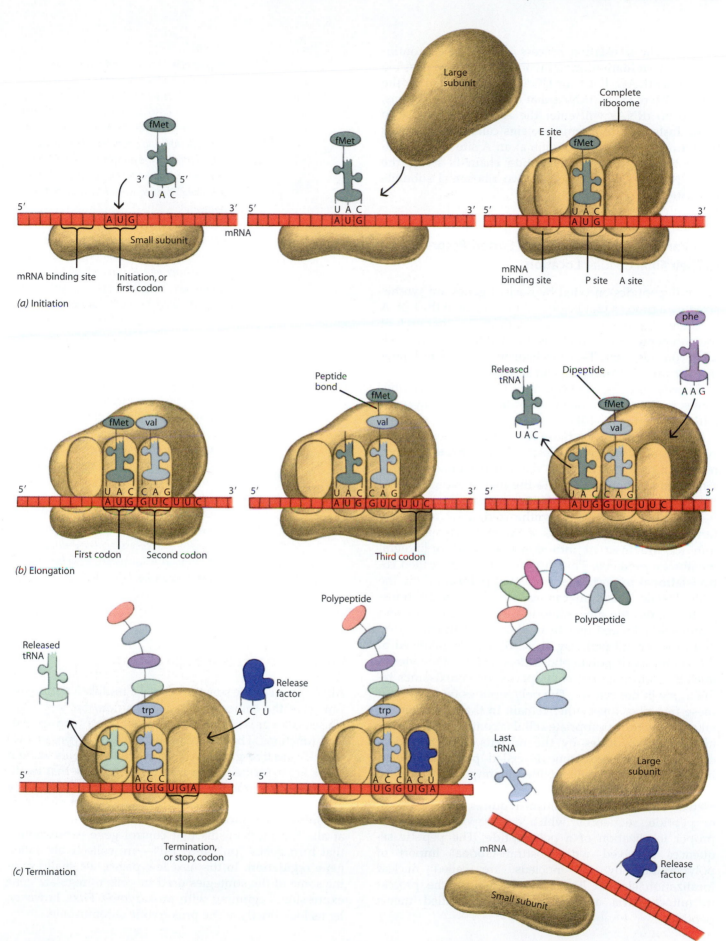

(a) Initiation

(b) Elongation

(c) Termination

The cyclic translation process just described continues until **termination** occurs when one of three possible *stop codons* (UAG, UAA, or UGA) is encountered on the mRNA. There are no tRNAs that recognize these codons, and so no tRNAs will enter the A site in response to them. Instead, cytoplasmic proteins called *release factors* bind directly to any stop codon at an A site on the ribosome, the completed polypeptide chain is freed, the last tRNA is released, and the two ribosomal subunits separate.

In Eukaryotes, Polypeptides Are Sorted According to Their Final Cellular Location

The polypeptides encoded by nuclear genes are synthesized by a process that begins in the cytosol of the cell. A mechanism is required, therefore, to ensure that each of the polypeptides eventually is directed to the correct cellular compartment. This mechanism is called **polypeptide (or protein) targeting** and **sorting.**

Shortly after polypeptide synthesis begins on ribosomes in the cytosol, one of two divergent pathways is followed (Figure 11–10).

In the first pathway, those ribosomes involved in the synthesis of polypeptides destined for the endoplasmic reticulum or for membranes derived from endoplasmic reticulum become attached to the membrane of the endoplasmic reticulum early in the translational process. As the polypeptides are synthesized, they are transferred across the membrane of the endoplasmic reticulum (or are inserted into it, in the case of integral membrane proteins). This process is therefore called **cotranslational import.** The completed polypeptide in the endoplasmic reticulum can remain there or be transported across the membrane to the Golgi complex and various vesicles and thus to another destination.

In the second pathway, those ribosomes involved in the synthesis of polypeptides destined for the cytosol, nucleus, mitochondria, chloroplasts, or peroxisomes remain free in the cytosol. The polypeptides released from these free ribosomes either remain in the cytosol or are taken up by the appropriate cellular component. The uptake of polypeptides by the nucleus, mitochondrion, chloroplast, or peroxisome is called **posttranslational import** because the polypeptides are imported into these cellular components *after* translation has been completed. The import mechanism requires specific signal, or peptide, sequences, which are responsible for the proper localization of a polypeptide. The peptide sequences involved with posttranslational import of polypeptides by the nucleus are called nuclear localization signals. Those involved with the process in mitochondria and chloroplasts are called transit peptides.

11–10

Polypeptide targeting and sorting. (a) Synthesis of all polypeptides encoded by nuclear genes begins in the cytosol. The large and small ribosomal subunits associate with each other and with the 5' end of an mRNA molecule, forming a functional ribosome that begins making the polypeptide. When the developing polypeptide is about 30 amino acids long, it enters one of two alternative pathways.

(b) If destined for any of the compartments of the endomembrane system (page 58), the polypeptide becomes attached to an endoplasmic reticulum membrane and is transferred into its lumen (the interior space) as synthesis continues. This process is called cotranslational import. The completed polypeptide then either remains in the endoplasmic reticulum or is transported via the Golgi complex and various vesicles to its final destination. Integral membrane proteins are inserted into the membrane of the endoplasmic reticulum as they are made.

(c) If the polypeptide is destined for the cytosol or for import into the nucleus, mitochondria, chloroplasts, or peroxisomes, its synthesis continues in the cytosol. When synthesis is complete, the polypeptide is released from the ribosome and either remains in the cytosol or is imported into an organelle. This process is called posttranslational import. Polypeptide uptake by the nucleus occurs via the nuclear pores.

Regulating Gene Expression

Although all of the somatic cells of a multicellular organism have the same genetic composition, the organism consists of a great many different cell types with specialized functions. The difference between cell types is a result of selective gene expression—that is, only certain genes are expressed. As a result the enzymes that mediate cell differentiation are expressed selectively. In a given cell, some genes are expressed continuously, others only as their products are needed, and still others not at all. The mechanisms that control gene expression—that turn genes "on" and "off"—are collectively called **gene regulation.** In the next few pages, we shall examine some of the strategies used by cells to regulate gene expression, beginning with prokaryotes. First, however, let us look briefly at the prokaryotic chromosome.

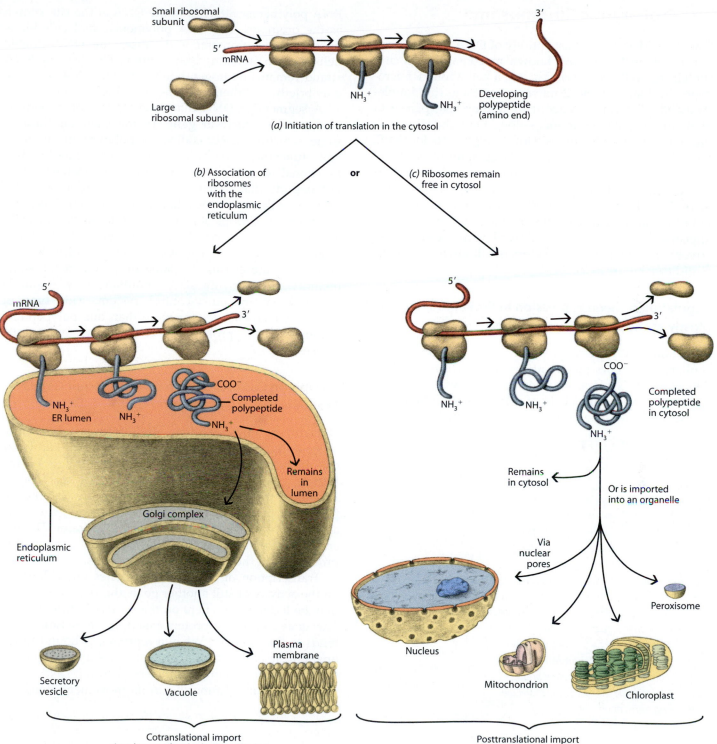

Small ribosomal subunit

5'
mRNA

Large ribosomal subunit

3'

NH_3^+

NH_3^+

Developing polypeptide (amino end)

(a) Initiation of translation in the cytosol

or

(b) Association of ribosomes with the endoplasmic reticulum

(c) Ribosomes remain free in cytosol

5'
mRNA

3'

NH_3^+
ER lumen

NH_3^+

COO^-
Completed polypeptide

NH_3^+

Remains in lumen

Golgi complex

Endoplasmic reticulum

Secretory vesicle

Vacuole

Plasma membrane

Cotranslational import into lumen of endoplasmic reticulum, followed by transport to final destination

5'

3'

NH_3^+

NH_3^+

COO^-

NH_3^+

Completed polypeptide in cytosol

Remains in cytosol

Or is imported into an organelle

Via nuclear pores

Nucleus

Mitochondrion

Chloroplast

Peroxisome

Posttranslational import into various organelles

The Prokaryotic Chromosome

Once considered to consist entirely of DNA, the prokaryotic chromosome is now known to contain RNA and protein as well. In *Escherichia coli*, a cell about 1 micrometer in diameter and 2 micrometers long, the double-stranded DNA contains about 4.7 million base pairs and is about 1.6 millimeters long—long enough to encircle the cell more than 400 times! How is it possible for the *E. coli* cell to accommodate such a large amount of DNA? The circular DNA of the bacterial chromosome is supercoiled and folded into a number of loops, which appear to be held in place by the RNA and protein. As seen with the electron microscope, the bacterial chromosome appears as a tangled mass of threadlike material in a relatively clear region of the cytoplasm known as the nucleoid (see Figure 3–2).

Regulation of Gene Expression in Prokaryotes Occurs Mostly at the Level of Transcription

Regulation of gene expression in *E. coli* and other prokaryotes occurs mostly at the level of transcription. As we have seen, transcription in prokaryotes involves the synthesis of a molecule of mRNA along a template strand of DNA. The process begins when the enzyme RNA polymerase attaches to the DNA at the site known as the promoter. The RNA polymerase molecule binds tightly to the promoter and causes the DNA double helix to open, initiating transcription. The growing RNA strand remains hydrogen-bonded to the DNA template only briefly, and then separates as a single strand.

A segment of DNA that codes for a polypeptide is known as a **structural gene.** In the bacterial chromosome, structural genes coding for polypeptides with related functions often occur together in sequence. Such functional groups might include, for instance, two polypeptide chains that together constitute a particular enzyme or three enzymes that work in a single biochemical pathway. Groups of genes coding for such molecules are typically transcribed into a single mRNA strand (Figure 11–11). Thus, a group of polypeptides that are needed by the cell at the same time and in the same quantity can be synthesized simultaneously, a simple and efficient inventory-control system. This type of mRNA, which encodes for more than one polypeptide, is found only in prokaryotes and in plant and animal viruses.

An Operon Consists of a Group of Adjacent Genes That Function as a Regulatory Unit

Our current understanding of the regulation of transcription in prokaryotes rests upon a model, known as the **operon model,** proposed some years ago by the French scientists François Jacob and Jacques Monod. According to this model, an **operon** (Figure 11–12) comprises the promoter, one or more structural genes, and another DNA sequence known as the **operator.** The operator is a sequence of nucleotides located between the promoter and the structural gene or genes.

Transcription of the structural genes often depends on the activity of still another gene, the **regulator,** which may be located anywhere on the bacterial chromosome. This gene codes for a protein called the **repressor,** which binds to the operator. When a repressor is bound to the operator, it obstructs the promoter. As a consequence, RNA polymerase either cannot bind to the DNA molecule or, if bound, cannot begin its movement along the

11–11

In prokaryotes, transcription often results in an mRNA molecule that contains coding sequences for several different polypeptide chains, with the sequences separated by stop and start codons. In this diagram, the stop and start codons are adjacent, but sometimes they are separated by as many as 100 to 200 nucleotides. The 5' end of the mRNA molecule has a short leader sequence, and the 3' end has a trailer sequence; neither of these sequences codes for protein. Translation starts at the leading (5') end of the mRNA and usually begins while the rest of the molecule is still being transcribed.

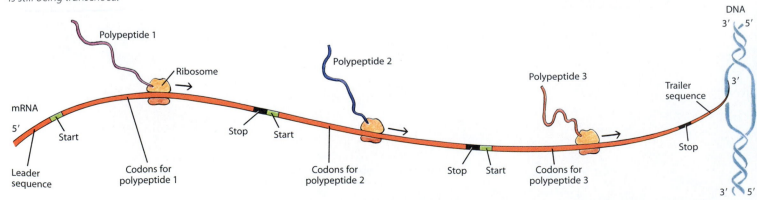

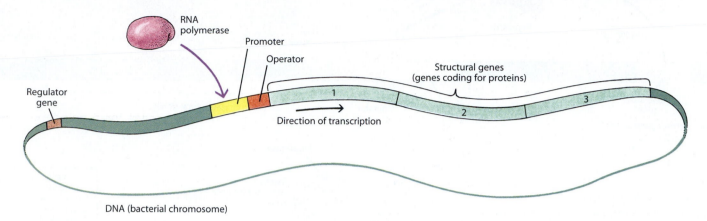

11–12

A schematic representation of an operon. An operon consists of a promoter, an operator, and structural genes (that is, genes that code for proteins, often enzymes, that work sequentially in a particular biochemical pathway).

The promoter, which precedes the operator, is the binding site for RNA polymerase. The operator is the site at which a repressor can bind; the repressor may overlap the promoter, the first structural gene, or both.

Another gene involved in operon function is the regulator, which codes for the repressor. Although the regulator may be adjacent to the operon, in most cases it is located elsewhere on the bacterial chromosome.

molecule. The result in either case is the same: no mRNA transcription occurs. However, when the repressor is removed, transcription may begin.

The capacity of the repressor to bind to the operator and thus to block protein synthesis depends, in turn, on yet another molecule that either activates or inactivates the repressor for that particular operon (Figure 11–13). A molecule that activates a repressor is known as a **corepressor,** and one that inactivates a repressor is known as an **inducer.** For example, when lactose (milk sugar) is present in the growth medium, the first step in its metabolism by bacterial cells produces a closely related sugar, allolactose. Allolactose is an inducer—it binds to and inactivates the repressor, removing it from the oper-

11–13

In an operon system, the transcription of mRNA—and thus the synthesis of proteins—is regulated by interaction of a repressor with either an inducer or a corepressor. Active repressors block RNA polymerase from transcribing mRNA. (a) In some systems, such as the lac operon, the repressor molecule is active in binding to the operator site until it combines with the inducer. (b) In other systems, such as the trp operon, the repressor is not active until it combines with the corepressor.

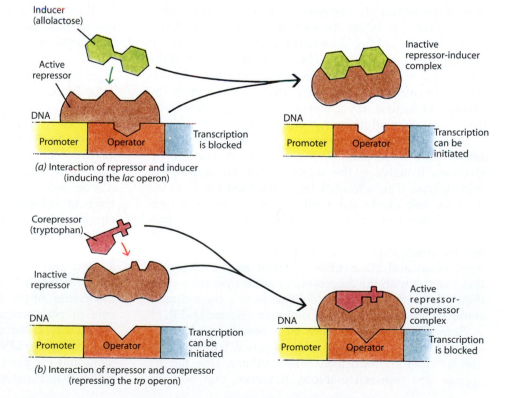

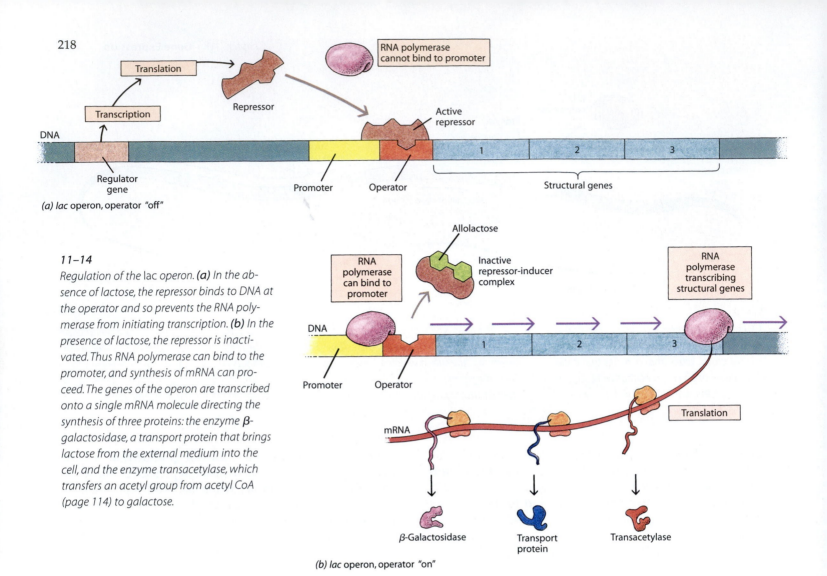

11–14

Regulation of the lac operon. (a) In the absence of lactose, the repressor binds to DNA at the operator and so prevents the RNA polymerase from initiating transcription. (b) In the presence of lactose, the repressor is inactivated. Thus RNA polymerase can bind to the promoter, and synthesis of mRNA can proceed. The genes of the operon are transcribed onto a single mRNA molecule directing the synthesis of three proteins: the enzyme β-galactosidase, a transport protein that brings lactose from the external medium into the cell, and the enzyme transacetylase, which transfers an acetyl group from acetyl CoA (page 114) to galactose.

ator of the lactose (*lac*) operon. As a consequence, RNA polymerase can begin its movement along the DNA molecule, transcribing the structural genes of the operon into mRNA (Figure 11–14). The enzymes involved in the metabolism of lactose are produced, and the bacterium is therefore able to use lactose as an energy source. In the absence of lactose, the enzymes are not produced and the cell can expend its energy elsewhere.

An example of a corepressor is provided by the amino acid tryptophan. When present in the growth medium, it activates the repressor for the tryptophan (*trp*) operon. The activated repressor then binds to the operator and blocks the synthesis of the unneeded enzymes. Since tryptophan is readily available to the bacterium, it does not need to expend the energy to produce the enzymes required for tryptophan synthesis. Both tryptophan and allolactose—as well as the molecules that interact with the repressors of other operons—exert their effects by causing a change in the three-dimensional shape of the repressor molecule (Figure 11–13).

Some 75 different operons have now been identified in *E. coli*, comprising 260 structural genes. Some are, like the *lac* operon, inducible, while others, like the *trp* operon, are repressible. Note, however, that both in-

ducible and repressible systems are examples of negative control, since both involve repressors that turn off transcription.

The Eukaryotic Chromosome

In the early days of molecular genetics (the late 1950s and into the 1960s), it was tempting to think that the eukaryotic chromosome might turn out to be simply a large-scale version of the *E. coli* chromosome. However, it soon became clear that there are many important differences between the bacterial chromosome and the chromosomes of eukaryotes—some expected and some very surprising. These differences include: (1) a far greater quantity of DNA in eukaryotic chromosomes; (2) a great deal of repetition in the sequence of this DNA, with much of it lacking any apparent function; (3) a close association of the DNA with proteins that play a major role in chromosome structure; (4) considerably more complexity in the organization of the protein-coding sequences of the DNA and the regulation of their expression; and (5) gene regulation that is primarily positive rather than negative, as in prokaryotes.

(a)

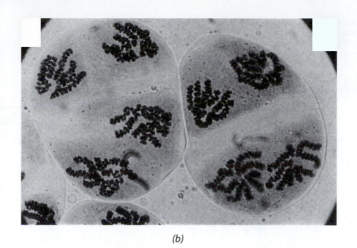

(b)

11–15

(a) Wake-robin (Trillium erectum) *flowers in the early spring. (b) The second anaphase of meiosis during microspore formation in T. erectum. Separation of the fully condensed,* *clearly visible chromosomes is nearly complete. Each of the newly forming nuclei will possess a haploid set of chromosomes, containing a total of over 30 meters of DNA.*

DNA is an "exquisitely thin filament," in the words of E. J. DuPraw, who calculated that a length sufficient to reach from the Earth to the sun would weigh only half a gram. The DNA of each eukaryotic chromosome is thought to be in the form of a single, linear molecule. If you were impressed with the size of the *E. coli* genome and the length of its DNA, consider the genome size and length of DNA in *Trillium*, a common spring wildflower (Figure 11–15). (Genome size is usually expressed in numbers of base pairs found in a haploid set of chromosomes.) A haploid set of *Trillium* chromosomes contains 1×10^{11} (100 billion) base pairs. Each diploid cell thus contains about 68 meters of DNA! We are only beginning to understand what *Trillium* does with all that DNA.

Chromosomes Contain Histone Proteins

Packaging of the DNA in eukaryotic cells is more complicated than in prokaryotic cells. As we noted in Chapter 3, the combination of DNA and its associated proteins in eukaryotic chromosomes is known as **chromatin** ("colored threads") because of its staining properties. Chromatin is more than half protein, and the most abundant proteins belong to a class of small polypeptides known as **histones.** Histones are positively charged (basic) and so are attracted to the negatively charged (acidic) DNA. They are always present in chromatin and are synthesized in large amounts during the S phase of the cell cycle. The histones are primarily responsible for the folding and packaging of DNA.

There are five distinct types of histones. They are present in enormous quantities—about 60 million molecules per cell of each of four of the types (H2A, H2B, H3, and H4) and about 30 million molecules per cell of the fifth type (H1). With the exception of H1, the amino acid sequences of the histones are remarkably similar in widely diverse groups of organisms. For instance, the se-

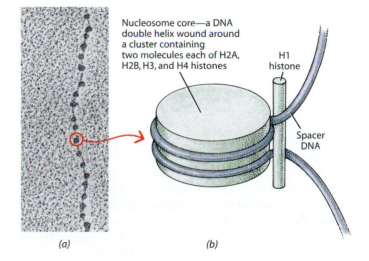

(a)   (b)

11–16

Nucleosomes. (a) Electron micrograph of nucleosomes and connecting threads of DNA from a chicken erythrocyte (red blood cell). The nucleosomes—the structures that look like beads on a string—are each approximately 10 nanometers in diameter. (b) Each nucleosome consists of a protein core made up of four different kinds of histones (two molecules each of H2A, H2B, H3, and H4), around which the DNA helix is wound twice. Nucleosomes are separated from each other by spacer, or linker, DNA. Histone H1 binds to the outer surface of the nucleosomes.

quences of the H4 histones of cows and peas differ by only two of 102 amino acids.

The fundamental packing units of chromatin are the **nucleosomes** (Figure 11–16), which resemble beads on a string. Each nucleosome consists of a core of eight histone molecules (two each of H2A, H2B, H3, and H4), around which the DNA filament is wrapped twice, like thread around a spool. Histone H1 lies on the DNA, outside the nucleosome core. When a fragment of DNA is

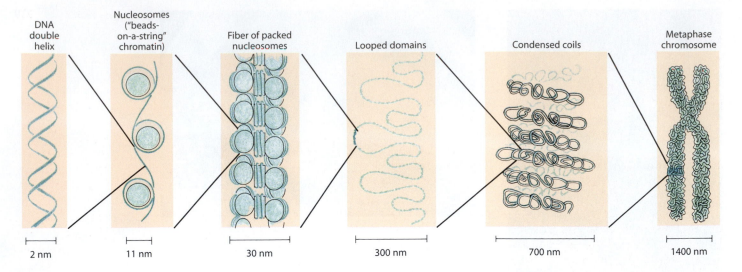

DNA double helix	Nucleosomes ("beads-on-a-string" chromatin)	Fiber of packed nucleosomes	Looped domains	Condensed coils	Metaphase chromosome
2 nm	11 nm	30 nm	300 nm	700 nm	1400 nm

11–17

Stages in the folding of chromatin, culminating in a fully condensed metaphase chromosome. The model is derived from electron micrographs of chromatin at different degrees of condensation. According to present evidence, each chromatid of a replicated chromosome contains a single molecule of double-stranded DNA and is, by weight, about 60 percent protein.

wrapped up in a nucleosome, it is about one-sixth the length it would be if fully extended.

As shown in Figure 11–17, additional packing of the nucleosomes produces a fiber that is about 30 nanometers in diameter. Further condensation of this fiber produces a series of loops, known as looped domains. The looped domains also coil until, ultimately, clusters of neighboring looped domains condense into the compact metaphase chromosomes that become visible with the light microscope during mitosis and meiosis.

Other proteins associated with the chromosome are the enzymes involved in DNA and RNA synthesis, regulatory proteins, and a large number and variety of molecules that have not yet been isolated and identified. Unlike the histones, these molecules vary from one cell type to another.

Regulation of Gene Expression in Eukaryotes

As we saw earlier, regulation of gene expression in prokaryotes typically involves turning genes on and off in response to changes in the nutrients available in the environment. In eukaryotes, especially multicellular eukaryotes, the problems of regulation are very different. A multicellular organism usually starts life as a fertilized egg, the zygote. The zygote divides repeatedly by mitosis and cytokinesis, producing many cells. At some stage these cells begin to differentiate, becoming, in a flowering plant, for example, epidermal cells, photosynthetic cells, storage cells, water-conducting cells (tracheary elements), food-conducting cells (sieve elements), and so forth. Each cell type, as it differentiates, begins to produce characteristic proteins that distinguish it structurally and functionally from other types of cells.

There is evidence, however, that all of the genetic information originally present in the zygote is also present in every diploid cell of the organism. This is especially clear for plants, in which differentiated cells with complete protoplasts have been shown to have the capacity to dedifferentiate and divide, and their progeny to differentiate into virtually any cell type. Individual cells may even regenerate an entire plant genetically similar to the original plant (a phenomenon known as totipotency, which is discussed in Chapter 28). Thus, although the original differentiated cell produced only its characteristic proteins—and not proteins characteristic of other cell types—it may later produce other proteins that change its characteristics. (Of course, differentiated cells of different types contain many proteins in common, consistent with the many functions that all cells carry out, such as the Krebs cycle and the processes of replication, transcription, and translation.) It becomes apparent that differentiation of the cells of a multicellular organism depends on the activation of certain groups of genes and the inactivation of others.

Chromosome Condensation Is an Important Factor in Gene Expression

Many lines of evidence indicate that the degree of condensation of the DNA of the chromosome, as shown by chromatin staining, plays a major role in the regulation of gene expression in eukaryotic cells. Staining reveals two types of chromatin: **euchromatin,** the more open chromatin, which stains weakly, and **heterochromatin,** the more condensed chromatin, which stains strongly. During interphase, heterochromatin remains tightly condensed, but euchromatin becomes less condensed. Transcription takes place during interphase, when the euchromatin is less condensed and is therefore accessible to molecules of RNA polymerase.

Some regions of heterochromatin are constant from cell to cell and are never expressed. An example is the highly condensed chromatin located in the centromere region of each chromosome. This region, which does not code for protein, is believed to play a structural role in the movement of the chromosomes during mitosis and meiosis. Similarly, no transcription takes place at the telomeres, the protective ends of the chromosomes (page 203).

The degree of condensation of other regions of the chromatin, however, varies from one type of cell to another within the same organism. This most likely reflects the fact that different types of cells synthesize different proteins and thus require the transcription of different segments of the DNA. Moreover, as a cell differentiates during embryonic development, the proportion of heterochromatin to euchromatin increases as the cell becomes more specialized. Segments of DNA that will not be needed by the differentiated cell are effectively "silenced."

Specific Binding Proteins Regulate Gene Expression

In eukaryotes, as in prokaryotes, transcription is regulated by proteins that bind to specific sites on the DNA molecule. Working in conjunction with other proteins called transcription factors that directly or indirectly affect the initiation of transcription, these proteins rearrange the nucleosomes, allowing the transcriptional machinery access to the DNA. Many of these proteins and their binding sites have now been identified, and it is increasingly clear that this level of transcriptional control is far more complex in multicellular eukaryotes than in prokaryotes. Recent evidence indicates that histones may also play a role in the regulation of gene expression by selectively exposing the genes to transcription.

A gene in a multicellular organism appears to respond to the sum of many different regulatory proteins, some tending to turn the gene on and others to turn it off. The sites at which these proteins bind may be hundreds or even thousands of base pairs away from the promoter sequence at which RNA polymerase binds and transcription begins. This, as you might expect, adds to the difficulty of identifying the regulatory molecules and also of understanding exactly how they exert their effects.

The DNA of the Eukaryotic Chromosome

Early studies of the DNA of eukaryotic cells revealed two surprising facts. First, with a few exceptions, the amount of DNA per cell is the same for every diploid cell of any given species (which is not surprising), but the variations among different species are enormous (Table 11–1). Second, in every eukaryotic cell, there is what appears to be a great excess of DNA, or at least of DNA whose functions are unknown. It is estimated that in eukaryotic cells less than 10 percent of all the DNA codes for proteins. In humans, it may even be as little as 1 percent. By contrast, prokaryotes use their DNA very thriftily. Except for regulatory or signal sequences, virtually all of their DNA is expressed at some time in the organism's life.

Continuing research is revealing the organization of eukaryotic DNA, as well as some of the functions of the long chains of nucleotides that do not encode protein.

In Eukaryotic DNA, Many Nucleotide Sequences Are Repeated

Eukaryotic genomes contain large numbers of copies of apparently nonessential nucleotide sequences that do not code for protein. The presence of a seemingly excess amount of DNA accounts, at least in part, for the large amounts of DNA in organisms such as the salamander and *Trillium* (Table 11–1).

Two major categories of repeated DNA can be recognized: tandemly repeated DNA and interspersed repeated DNA. The essential feature of **tandemly repeated DNA** is that the repeated copies are arranged one after the other in a row, that is, in tandem series. The repeated

TABLE 11–1	DNA Content of Selected Bacterial and Eukaryotic Genomes	
Genome	Approximate Number of Base Pairs (kb)*	Number of Chromosomes (haploid)
Bacteria (circular chromosomes)		
Mycoplasma hominis	760	
Bacillus	3000	
E. coli	4700	
Eukaryotes (linear chromosomes)		
Fungi		
Saccharomyces cerevisiae (yeast)	13,500	16
Animals		
Drosophila melanogaster (fruit fly)	165,000	4
Homo sapiens (human being)	3,000,000	23
Amphiuma sp. (salamander)	76,500,000	14
Plants		
Arabidopsis thaliana	100,000	10
Zea mays (maize)	4,500,000	10
Pisum sativum (pea)	5,000,000	7
Trillium	100,000,000	5

*kb = kilobase; 1 kb is a thousand base pairs.

unit in tandemly repeated DNA may range from as few as one base pair to as many as 2000 base pairs.

A subcategory of tandemly repeated DNA is **simple-sequence repeated DNA** (originally called satellite DNA), which consists of sequences with fewer than 10 base pairs that are typically present in enormous quantities. Simple-sequence DNA segments are thought to be vital to chromosome structure. Long blocks of short repetitive sequences occur around the centromeres, which play an important role in chromosome movement. Indeed, these segments of simple-sequence repeated DNA may actually be the centromere. The ends of the chromosomes, or telomeres, which protect the chromosomes from degradation with each round of replication, also consist of simple-sequence repeats.

Interspersed repeated DNA units are dispersed throughout the DNA, rather than being arranged in tandem, and they tend to be hundreds or even thousands of base pairs long. About 20 to 40 percent of the DNA of most multicellular organisms consists of interspersed repeated sequences. The dispersed units, which may number in the hundreds of thousands, are similar to one another, rather than identical.

Although the roles of interspersed repeated sequences are unknown, most such sequences are believed to have originated from transposable DNA sequences, or transposons (page 193), and to represent a kind of molecular parasite. Transposons can multiply and move around the genome to numerous other sites, leaving copies of themselves there.

Some Nucleotide Sequences Are Not Repeated

Single-copy DNA, unlike repeated DNA, is present in only one or a few copies per genome. Depending on the species, it constitutes anywhere from 50 to 70 percent of the DNA of the organism. Essentially, all of the protein-coding genes belong to this class of DNA. However, only a very small proportion of single-copy DNA appears to be transcribed into RNA. Some of the noncoding single-copy DNA is accounted for by long stretches of spacer DNA (Figure 11–16), which is never transcribed into RNA. The remainder has been found in a most unexpected location.

Most Structural Genes Consist of Introns and Exons

One of the great surprises in the study of eukaryotic DNA was the discovery that the protein-coding sequences of genes are usually not continuous but are instead interrupted by noncoding sequences—that is, by nucleotide sequences that are not translated into protein. These noncoding interruptions within a gene are known as intervening sequences, or **introns.** The coding sequences—the sequences that *are* translated into protein—are called **exons.**

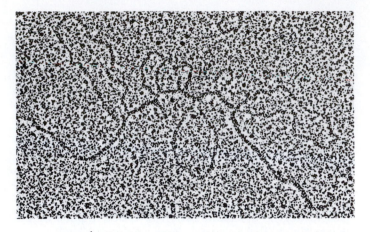

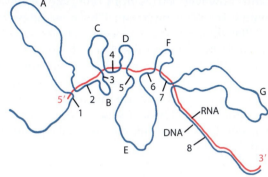

11–18

This electron micrograph reveals the results of an experiment in which a single strand of DNA containing the gene coding for ovalbumin was hybridized with the messenger RNA for ovalbumin. The complementary sequences of the DNA and mRNA are held together by hydrogen bonds. There are eight such sequences (the exons labeled 1 through 8 in the accompanying diagram). Some segments of the DNA do not have corresponding mRNA segments and so loop out from the hybrid. These are the seven introns, labeled A through G. Only the exons are translated into protein.

The presence of introns in the genes of eukaryotes was reported almost simultaneously in 1977 by several groups of investigators. Electron micrographs of hybrids between mRNA molecules and segments of DNA known to contain the genes coding for these mRNA molecules showed that there was not a perfect match between the mRNA molecules and the genes from which they were transcribed (Figure 11–18). The nucleotide sequences of the genes were much longer than the complementary mRNA molecules found in the cytosol.

It is now known that most, but not all, structural genes of multicellular eukaryotes contain introns. The introns are part of the newly transcribed RNA molecules, but they are snipped out before translation occurs. The number of introns per gene varies widely. Introns have also been found in genes coding for transfer RNAs and ribosomal RNAs. Prokaryotic genes, on the other hand, contain no introns.

Transcription and Processing of mRNA in Eukaryotes

Transcription in eukaryotes is the same, in principle, as in prokaryotes. It begins with the attachment of an RNA polymerase to a particular nucleotide sequence on the DNA molecule. The enzyme then moves along the molecule, using the 3′ to 5′ strand as a template for the synthesis of RNA molecules as shown in Figure 11–5. The transcribed RNA molecules (rRNAs, tRNAs, and mRNAs) then play their various roles in the translation of the encoded genetic information into protein.

Despite these basic similarities, there are significant differences between prokaryotes and eukaryotes in transcription, translation, and the events that occur between these two processes. One important difference is that eukaryotic genes are not grouped so that two or more structural genes are transcribed onto a single RNA molecule. In eukaryotes, each structural gene is transcribed separately, and its transcription is under separate controls.

Another important difference is that in eukaryotes, unlike prokaryotes, the DNA is separated from the sites of protein synthesis in the cytoplasm by the nuclear envelope. In eukaryotic cells, therefore, transcription and translation are separated both in time and space. After transcription is completed in the nucleus, the mRNA transcripts of eukaryotes are intensively modified before they are transported to the cytoplasm, the site of translation.

Even before transcription is completed, while the newly forming mRNA strand is only about 20 nucleotides long, a "cap" of an unusual nucleotide is added to its leading (5′) end. This cap is necessary for the binding of the mRNA to the eukaryotic ribosome. After transcription has been completed and the molecule released from the DNA template, special enzymes add a string of adenine nucleotides, known as the poly-A tail, to the trailing (3′) end of the molecule.

Before the modified mRNA molecules leave the nucleus, the introns are removed, and the exons are spliced together to form a single, continuous molecule (Figure 11–19). The splicing mechanism is exceedingly precise. As you can see, the addition or deletion of even a single nucleotide would shift the reading of the triplet codons so that completely different amino acids would be coded for. Consequently, an entirely new, probably nonfunctional, protein would be produced.

A number of instances have now been found in which identical mRNA transcripts are processed in more than one way. Such alternative splicing can result in the formation of different functional polypeptides from RNA molecules that were originally identical. In such cases, an intron may become an exon, or vice versa. Thus, the more that is learned about eukaryotic DNA

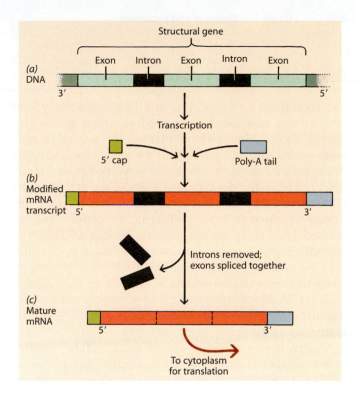

11–19

A summary of the stages in the processing of an mRNA transcribed from a structural gene of a eukaryote. (a) The genetic information encoded in the DNA is transcribed into an RNA copy. (b) This copy is then modified with the addition of a cap at the 5′ end and a poly-A tail at the 3′ end. (c) The introns are snipped out, and the exons are spliced together. The mature mRNA then passes to the cytoplasm, where it is translated into protein.

and its expression, the more difficult it becomes to define "gene" or "intron" or "exon."

As we noted earlier, molecular biologists once thought that the eukaryotic chromosome might be simply a large-scale version of the prokaryotic chromosome. That turned out not to be the case. As we have seen, the structure and organization of the chromosome, the regulation of gene expression, and the processing of mRNA molecules are all much more complex in eukaryotes than in prokaryotes. For many years it also seemed reasonable to believe that the chromosomes of eukaryotes were stable. Recombinations were produced by crossing-over, of course, but it was thought that—except for occasional point mutations and mistakes in meiosis—the structure of each chromosome was essentially fixed and unchanging. Perhaps the greatest surprise of all has been the discovery that this too is not the case. In both prokaryotes and eukaryotes, segments of DNA can move from one place to another within a chromosome, they can move into and out of chromosomes, they can move from one chromosome to another—and sometimes even from one organism to another.

Recombinant DNA Technology

For thousands of years, humans have selectively bred both plants and animals to produce more and higher quality foods. The seeds of plants with desirable traits were saved and used for the following year's crop, in the hope that the desirable traits would reappear. It was not until after the discovery of Mendel's work, in which he demonstrated that inherited characteristics are determined by discrete factors—that is, by genes—and described how these factors behave during a cross, that scientifically conducted programs of selective breeding became possible. Such breeding programs often require years to obtain desirable results. Moreover, they depend exclusively on naturally occurring genetic recombination within an interbreeding population, or species, for genetic variability.

Today, it is possible to obtain a nucleotide sequence of an isolated fragment of DNA from virtually *any* organism, recombine it with the DNA of a carrier (a plasmid or a virus), and insert it into the cells of any other organism. Using this **recombinant DNA** technology, geneticists are able to create novel genotypes, a feat that would be impossible to achieve using traditional genetic techniques. This technology not only allows individual genes to be inserted into organisms in a way that is both precise and simple, it also allows gene transfer between species that otherwise are incapable of hybridizing with one another—for example, the transfer of a disease-resistance gene in petunia into the DNA of a tomato plant.

Restriction Enzymes Are Used to Make Recombinant DNA

Recombinant DNA technology is based in large part on the ability to cut DNA molecules from different sources precisely into specific pieces and to combine those pieces to produce new combinations. This procedure depends on the existence of **restriction enzymes** *(restriction endonucleases)*, which recognize specific sequences of double-stranded DNA known as recognition sequences. These sequences are typically four to six nucleotides long and are always symmetrical (that is, one strand is identical to the other strand when read in the opposite direction).

Restriction enzymes cut DNA within or near their particular recognition sequences. Most restriction enzymes make straight cuts, but some cut through the two strands a few nucleotides apart, leaving what are called **sticky ends** (Figure 11–20). The DNA strands at the two sticky ends are complementary, and they can therefore pair with one another and rejoin through the action of DNA ligase (see Figure 10–23). More important, these sticky ends can join with any other segment of DNA— from any number of sources—that has been cut by the same restriction enzyme and therefore has complementary sticky ends. This property makes it possible to create virtually unlimited recombinations of genetic material, since the ability of DNA fragments to recombine is independent of the original sources of the fragments.

How does one make and manipulate recombinant DNA? The basic procedure is to use restriction enzymes to cut up DNA from a **donor organism.** Consisting of one to several genes each, the DNA fragments (also called *restriction fragments*) are combined with small DNA molecules, such as bacterial plasmids, which can replicate autonomously when introduced into bacterial cells. The plasmids act as **vectors** for the foreign DNA fragments. Together with their inserts, the plasmids are examples of recombinant DNA because they consist of DNA originating from two different sources—a donor

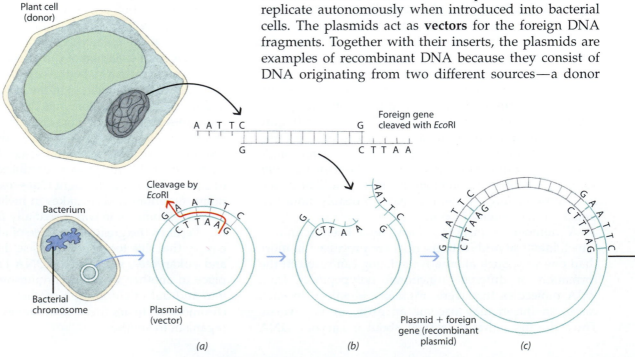

Plant cell (donor)

Foreign gene cleaved with *Eco*RI

Bacterium

Cleavage by *Eco*RI

Bacterial chromosome

Plasmid (vector)

Plasmid + foreign gene (recombinant plasmid)

(a)   (b)   (c)

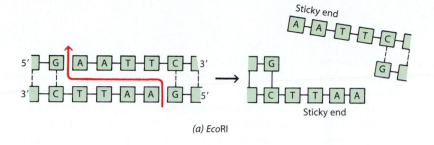

(a) EcoRI

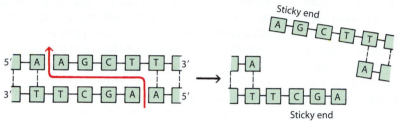

(b) HindIII

11–20

The DNA nucleotide sequences recognized by two widely used restriction enzymes, (a) EcoRI and (b) HindIII. These two enzymes cut the DNA so that sticky ends result. A sticky end can reattach to its complementary sequence at the end of any DNA molecule that has been cut by the same restriction enzyme. Restriction enzymes are named for the bacteria from which they are obtained by combining the first letter of the genus with the first two letters of the specific epithet. EcoRI is from E. coli, and HindIII is from Hemophilus influenzae.

and a bacterium. The recombinant DNA, now existing as recombinant plasmids, is then taken up by bacterial cells, which are said to be **transformed.** If a bacterial cell containing a recombinant plasmid is allowed to divide and grow into a colony with millions of cells, each cell will contain the same recombinant plasmid with the same DNA insert. This process of amplifying, or generating, many identical DNA fragments is called **DNA cloning,** or gene cloning. Figure 11–21 gives an overview of this procedure, which uses plasmids as cloning vectors. A virus, usually a bacteriophage, can also be used as a cloning vector.

Screening Processes Are Used to Identify the Colonies Carrying the Recombinant Plasmid

Even though all of the bacteria in a population may be exposed to recombinant plasmids, not all of them will be transformed. It is essential, therefore, to be able to identify the bacteria that contain the recombinant plasmids. This process is referred to as "screening." Selective screening is accomplished by using genetically modified plasmid vectors that contain antibiotic-resistance genes. One such plasmid carries the gene *amp*^{R}, which confers resistance to the antibiotic ampicillin. Bacteria carrying recombinant plasmids with this gene, for example, will be able to grow in the presence of ampicillin. Hence, the transformed bacteria will survive, and the rest will die.

11–21

The use of plasmids in DNA cloning. (a) The plasmid is cleaved by a restriction enzyme. In this example, the enzyme is EcoRI, which cleaves the plasmid at the sequence (5')–GAATTC–(3'), leaving sticky ends exposed. (b) These ends, consisting of TTAA and AATT sequences, can join with any other segment of DNA that has been cleaved by the same enzyme. Thus it is possible to splice a foreign gene—say, from a plant cell—into the plasmid (c). (In this illustration, the length of the GAATTC sequences is exaggerated and the lengths of the other portions of both the foreign gene and the plasmid are compressed.) (d) When plasmids incorporating a foreign gene are released into a medium in which bacteria are growing, they are taken up by some of the bacterial cells. (e) As these cells multiply, the recombinant plasmids replicate. The result is an increasing number of cells, all making copies of the same plasmid. (f) The recombinant plasmids can then be separated from the other contents of the cells and treated with EcoRI to release the copies of the cloned gene.

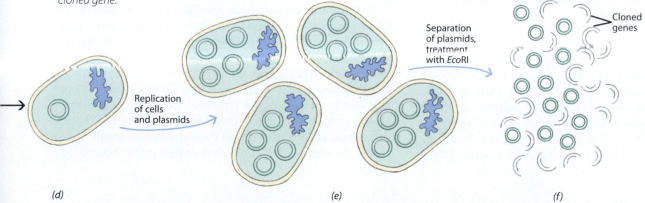

(d)

Replication of cells and plasmids

(e)

Separation of plasmids, treatment with EcoRI

Cloned genes

(f)

An alternative screening process involves the use of vectors that carry the gene *lacZ*. The *lacZ* gene encodes for β-galactosidase, the enzyme that hydrolyzes the sugar lactose. Such plasmids have a single recognition sequence for the restriction enzyme used, and the recognition sequence lies within the *lacZ* gene. Bacteria containing a plasmid with the *lacZ* gene intact will form blue colonies when grown on solid culture media that are supplemented with a modified sugar called X-gal. Colonies that have had a foreign DNA fragment inserted into the recognition sequence of the plasmid will appear white because the *lacZ* gene has been disrupted, preventing it from producing β-galactosidase. In some cases, both antibiotic resistance (selective screening) and color identification (nonselective screening) methods are used in tandem (Figure 11–22).

DNA Libraries Can Be Either Genomic or Complementary

The procedure described above for the cloning of DNA is sometimes called "shotgun cloning." This approach involves cutting the entire genome of an organism into a large number of restriction fragments and then introducing them randomly into a large number of bacterial cells or viruses. Such a collection of cloned DNA is called a **genomic library** because it contains cloned fragments of most, if not all, of the genome.

Another kind of DNA library is the **cDNA, or complementary DNA, library.** Complementary DNA is synthetic DNA made from messenger RNA, using a special enzyme called **reverse transcriptase.** This enzyme is obtained from animal viruses called retroviruses. Using mRNA as a template, reverse transcriptase catalyzes the synthesis of a single strand of cDNA by reverse transcription. The single-stranded DNA is then converted into double-stranded DNA using DNA polymerase. Because it is made from mRNA, cDNA is devoid of introns. This means that, unlike genomic DNA, cDNA from eukaryotes will contain only protein-coding sequences and can be translated into functional protein in bacteria, which are unable to remove introns.

The Polymerase Chain Reaction Can Be Used to Amplify Segments of DNA

The **polymerase chain reaction (PCR)** is a technique by which any piece of DNA can be greatly amplified in a relatively short period of time. Literally millions of copies of a DNA segment can be made by PCR in just a few hours. The procedure involves adding a short primer at each end of a selected DNA sequence, separating the two strands of the double helix by heating, and exposing them to the temperature-resistant DNA

11–22

Overview of DNA cloning using a bacterial plasmid as a vector. (a) Cells of E. coli containing plasmids with the ampR and lacZ genes are treated to release the plasmids. The ampR gene confers resistance to ampicillin, an antibiotic, and the lacZ gene produces β-galactosidase, an enzyme that hydrolyzes lactose, a sugar. In addition, plant cells containing the gene of interest are treated to release their DNA. (b) The plasmids and plant DNA are both treated with the same restriction enzyme. The recognition sequence for the enzyme is within the lacZ gene on the plasmid, so the gene is disrupted. For each plant cell, DNA is cut into many fragments, one of which contains the gene of interest. Treatment by the enzyme results in sticky ends on both the plasmids and the plant DNA fragments. (c) Combining the treated plasmids and fragments results in recombinant plasmids containing the gene of interest, as well as recombinant plasmids containing other plant genes. Insertion occurs by base-pairing the plasmid sticky ends with the complementary sticky ends of the plant DNA fragments. (d) The plasmid solution, consisting of recombinant plasmids as well as intact plasmids, is mixed with bacteria lacking the ampR and lacZ genes. Uptake of plasmids will occur in some of the bacteria by transformation. (e) The bacteria are grown on solid culture media with ampicillin and X-gal, a modified sugar that turns blue when digested by β-galactosidase. Bacteria containing plasmids will grow and form colonies on such media because they are resistant to the ampicillin. Bacteria containing recombinant plasmids have a nonfunctional lacZ gene; therefore those bacteria will form white colonies. Bacterial colonies containing plasmids without a disrupted lacZ gene will appear blue because they can produce β-galactosidase. (f) Each white colony consists of a clone of bacteria carrying identical recombinant plasmids. Additional selecting is needed to determine whether or not a white colony is made up of transformed bacteria carrying the plant gene of interest.

polymerase known as *Taq* polymerase, which catalyzes growth from the DNA primers. (If DNA is heated, the hydrogen bonds holding the two strands together are broken and the strands separate, the three-dimensional structure of the molecule is lost, and the molecule is said to be denatured.) *Taq* polymerase is obtained from the bacterium *Thermus aquaticus*, which thrives in hot springs. Hence, it is not denatured by the temperatures used to denature the DNA.

In the polymerase chain reaction, both strands of the DNA are copied simultaneously. After completion of the replication of the segment between the two primers, the two newly formed double-stranded DNA molecules are then heated. This causes the strands to separate into four single strands, and a second cycle of replication is carried out by lowering the temperature in the presence of all the components necessary for polymerization. If this procedure is repeated for 20 cycles, amplifications of up to a millionfold can be achieved within a few hours.

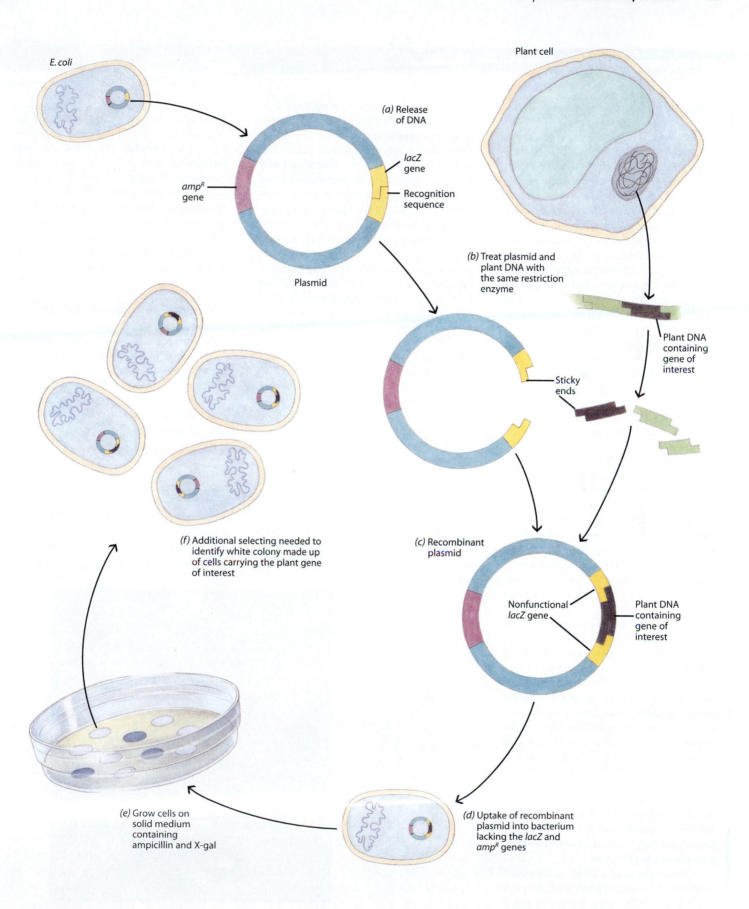

E. coli

Plant cell

(a) Release of DNA

amp^R gene

lacZ gene

Recognition sequence

Plasmid

(b) Treat plasmid and plant DNA with the same restriction enzyme

Plant DNA containing gene of interest

Sticky ends

(c) Recombinant plasmid

Nonfunctional *lacZ* gene

Plant DNA containing gene of interest

(f) Additional selecting needed to identify white colony made up of cells carrying the plant gene of interest

(e) Grow cells on solid medium containing ampicillin and X-gal

(d) Uptake of recombinant plasmid into bacterium lacking the *lacZ* and *amp^R* genes

Arabidopsis thaliana: The Model Plant

Arabidopsis thaliana, a small weed of the mustard family (Brassicaceae), has become the model experimental organism in the study of plant molecular genetics. Among the attributes that make it particularly suited for both classical and molecular genetic research are: (1) Its short generation time. Only four to six weeks are required to obtain mature plants, each of which has the potential to produce more than 10,000 seeds. (2) Its small size. Arabidopsis is such a small plant that literally dozens can be grown in a small pot, requiring only moist soil and fluorescent light for rapid growth. (3) Its adaptability. Arabidopsis plants grow well on sterile, biochemically defined media. In addition, Arabidopsis cells have been grown in culture and plants have been regenerated from such cells. (4) Typically, it self-fertilizes. This allows new mutations to be made homozygous with minimal effort. Many mutations have been identified in Arabidopsis, including visible ones useful as markers in genetic mapping. Such maps provide an approximation of the actual positions of genes on the chromosomes. (5) Its relatively small genome and small amount of repetitive DNA, which simplify the task of identifying and isolating genes. (6) Its susceptibility to infection by the bacterium Agrobacterium tumefaciens, which carries plasmids capable of being recombined with foreign genes (see page 692).

None of the crop plants share all of these traits with Arabidopsis. Typical crop plants have generation times of several months and require a great deal of space for growth in large numbers. In addition, those that have been used for recombinant DNA studies have large genomes and large amounts of repetitive DNA. Arabidopsis has joined such organisms as E. coli, yeast, the fruit fly, and mice as vehicles in the molecular biologist's quest to gain insight into basic life processes. A goal of scientists affiliated with the Multinational Coordinated Arabidopsis thaliana Genome Research Project is to characterize and completely sequence the Arabidopsis genome by the year 2000.

The genetic approach to solving basic problems in plant growth and development is currently being applied in major research laboratories around the world.

The strategy involves screening large populations of plants that have been treated with mutagenic agents that randomly cause disruptions in individual genes. The researcher looks for plants that show some alteration relative to the nonmutagenized plants. The geneticist refers to these alterations as mutant phenotypes. Through genetic analysis it can be determined that a mutant phenotype is due to a defect in a single gene within the plant's genome. As an example, plant biologists long have recognized a series of single gene mutations that cause a dwarf phenotype resulting from reduction in internodal elongation. Chemical analyses of tissues from these mutants have revealed that, in many cases, the mutant plants are deficient in one or more of the group of plant hormones known as gibberellins, which are discussed in Chapter 28. Thus, it appears that some dwarf phenotypes are due to specific disruptions of genes that code for enzymes involved in the biosynthesis of gibberellins. The study of plants that are defective in a single gene can provide important information about the role that the gene and its protein product play in plant growth and development. For example, many gibberellin-deficient mutants have reduced height but show normal development of leaves and flowers, indicating that the missing gibberellins are involved in internodal elongation but not in these other developmental processes.

With a model organism such as Arabidopsis, one can take advantage of the increasingly sophisticated body of information available on the structure of the plant genome. Plant biologists have already identified some of the genes that control development of the embryo, seedling, root, and flower. In addition, specific models of developmental regulation have been proposed for each of these developmental processes.

Arabidopsis thaliana, (at left, below) which has become the object of investigation of hundreds of researchers worldwide.

Various stages in the development of the Landsberg erecta strain of Arabidopsis thaliana. (a) 14 days after planting. At this stage, the shoot contains a rosette of 7 to 8 leaves. (b) 21 days *after planting. The plant has already undergone transition from a vegetative to reproductive phase. (c) 37 days after planting. The plant has produced a great many flowers, and fruit and seed development is well under way. (d) 53 days after planting. The plant is undergoing senescence.*

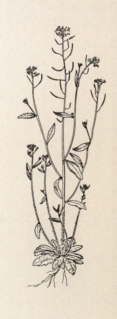

(a)   (b)   (c)   (d)

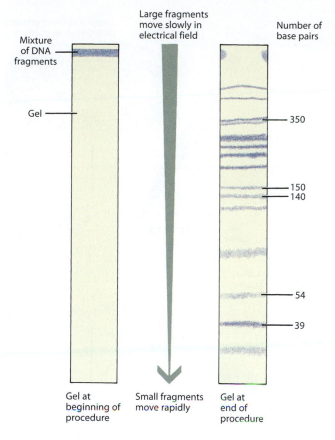

Large fragments move slowly in electrical field

Mixture of DNA fragments

Gel

Number of base pairs

— 350

— 150
— 140

— 54

— 39

Gel at beginning of procedure

Small fragments move rapidly

Gel at end of procedure

11–23

Electrophoresis can be used to separate fragments of DNA. In electrophoresis, the electrical field separates molecules both by charge and by length, or size. Smaller molecules move faster than larger ones. A mixture of DNA fragments containing different numbers of base pairs can be cleanly separated according to size. The gel is then sliced into sections, and the separate, purified fragments are washed out of the gel unharmed. This separation procedure is important in many aspects of recombinant DNA work.

DNA Sequencing Has Revealed the Genomes of Organisms

With the development of techniques for cutting DNA molecules into smaller pieces and making multiple copies of those pieces, it is now possible to determine the nucleotide sequence of any isolated gene. One of the most important features of restriction enzymes is that different enzymes cut DNA molecules at different sites. Cutting a DNA molecule with one restriction enzyme produces one particular set of short DNA fragments. Cutting an identical DNA molecule with a different restriction enzyme produces a different set of short DNA fragments. The fragments of each set can be separated from one another by electrophoresis on the basis of their lengths, or sizes (Figure 11–23), and cloned into multiple copies.

These copies can then be analyzed to determine the exact nucleotide sequence of each fragment. The methods used in these analyses involve complex series of chemical or enzymatic reactions. However, most DNA sequencing is now carried out automatically by DNA-sequencing machines, which use dideoxyribonucleotides (nucleotides in which the sugar subunit has two fewer oxygens than ribose) tagged with a different colored fluorescent dye for each of the four nucleotides. This approach greatly reduces the time required to determine the nucleotide sequences of short fragments.

Because the sets of fragments produced by different restriction enzymes overlap, the information obtained from sequencing the different sets can be pieced together like a puzzle to reveal the entire sequence of a DNA molecule (Figure 11–24). Complete sequences are now known for organellar DNA (from both mitochondria and chloroplasts), for some viruses, and also for the whole genomes of *E. coli*; a representative of the *Archaea*,

(a)

T – T | A – A
A – A | T – T

Restriction enzyme 1

C | A – T – G
G – T – A | C

Restriction enzyme 2

(b) DNA fragments

A – T – G – T – T – A – A – C – T – T – A – A – C

C – A – A – T – T – G – A – A – T – T – G – T – A

A – A – C – A – T – G – C – A – T – G – T – T
T – T – G – T – A – C – G – T – A – C – A – A

T – T – A – A – C
A – A – T – T – G – T – A

A – A – C – T – T
T – T – G – A – A

A – T – G
C

T – T
A – A

A – T – G – C
C – G – T – A

A – A – C – A – T – G
T – T – G – T – A – C

(c) Sequenced DNA molecule

T – T – A – A – C – A – T – G – C – A – T – G – T – T – A – A – C – T – T – A – A – C – A – T – G
A – A – T – T – G – T – A – C – G – T – A – C – A – A – T – T – G – A – A – T – T – G – T – A – C

11–24

A simplified example of DNA sequencing. Identical samples of the DNA molecule to be sequenced are treated with different restriction enzymes that cut the DNA at different sites. **(a)** *One sample is treated with one enzyme (restriction enzyme 1), producing one set of fragments, and another sample is treated with another enzyme (restriction enzyme 2), producing a different set of fragments. The fragments of each set are then separated from one another, cloned, and analyzed, revealing the nucleotide sequence of each individual fragment* **(b)**. *As you can see, the fragments produced by the two restriction enzymes overlap, making it possible to determine the nucleotide sequence of the molecule as a whole* **(c)**.

Methanococcus jannaschii; and the yeast *Saccharomyces cerevisiae.* Work is well under way on the human genome and that of *Arabidopsis* (see the essay on page 228) as well.

Nucleic Acid Hybridization Is Used to Locate Specific DNA Segments

Before a DNA segment of interest—such as a particular gene or portion of a gene—can be cloned, sequenced, or manipulated, it must first be located and isolated. The chromosomes of even the simplest eukaryotic cells contain an enormous quantity of DNA. Locating a specific segment of that DNA is like trying to find the proverbial needle in a haystack. The "magnet" that scientists use is **nucleic acid hybridization,** one of the earliest methods for studying DNA and RNA molecules.

The foundation of hybridization is the base-pairing properties of the nucleic acids. As you know, when dissolved DNA molecules are heated gently, the hydrogen bonds between the strands are broken and the strands separate. When the solution is cooled slowly, the hydrogen bonds re-form, reconstituting the double helix.

When DNA molecules from different sources are mixed together and heated, the strands separate and undergo random collisions. If two strands with nearly complementary sequences find each other as the solution is cooled, they will form a hybrid double helix. The extent to which segments from two samples associate, or hybridize, and the speed with which they do so provide an estimate of the similarity between the nucleotide sequences.

This procedure has been adapted to locate specific nucleotide sequences in either DNA or RNA. A **nucleic acid probe** can be prepared by incorporating a radioactive isotope into a short segment of single-stranded DNA or RNA that is complementary to the nucleotide sequence of interest. For example, suppose that one wants to identify the gene that codes for a specific enzyme. If the nucleotide sequence is known for that gene, the researcher can prepare a short, radioactively labeled strand of DNA complementary to the part of the DNA sequence needed for the production of the enzyme. The labeled probe then is used to "tag," or label, the colonies containing the complementary DNA. Since the probe is radioactively labeled, it can be detected by X-ray film (Figure 11–25). Alternatively, the probe can be labeled with a fluorescent dye (Figure 11–26).

Probes may be mRNA molecules, DNA fragments produced by restriction enzymes, complementary DNAs synthesized from RNA by reverse transcriptase, or nucleotide sequences synthesized in the laboratory. After a probe has been prepared, it can be used to seek out segments of DNA or RNA that contain the complementary nucleotide sequence. Labeled RNA molecules are routinely used to find corresponding DNA segments, and vice versa.

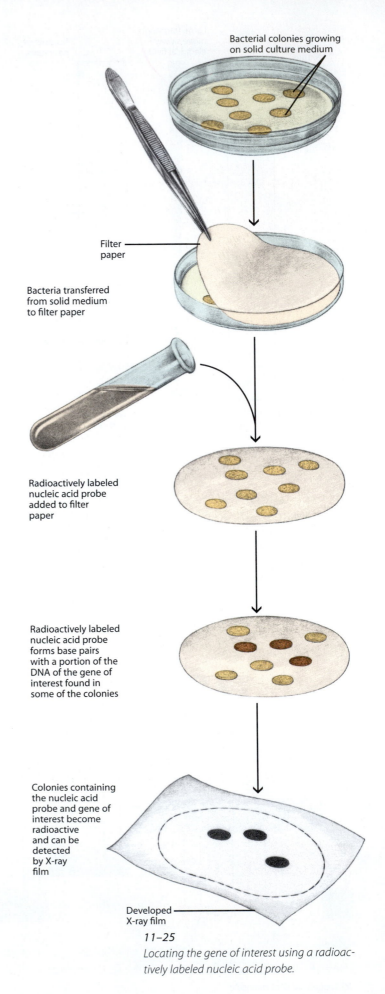

Bacterial colonies growing on solid culture medium

Filter paper

Bacteria transferred from solid medium to filter paper

Radioactively labeled nucleic acid probe added to filter paper

Radioactively labeled nucleic acid probe forms base pairs with a portion of the DNA of the gene of interest found in some of the colonies

Colonies containing the nucleic acid probe and gene of interest become radioactive and can be detected by X-ray film

Developed X-ray film

11–25

Locating the gene of interest using a radioactively labeled nucleic acid probe.

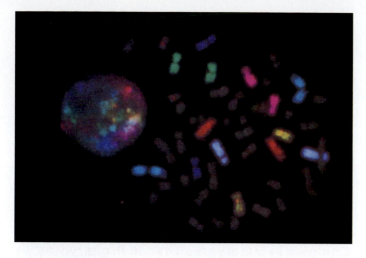

11–26

Chromosome preparation in which several homologous pairs have been labeled with chromosome-specific fluorescent probes. The specific labeling reveals the position of the chromosomes in the intact nucleus (left).

Several Techniques Are Available to Locate Genes of Interest

As mentioned previously, the entire nucleotide sequences for the genomes of several organisms are already known, and those for humans and at least one plant—*Arabidopsis*—will be obtained within the next decade. A greater challenge will be the identification of all of the individual genes and the functions of the proteins coded for by those genes. Several techniques are currently being used to match up DNA sequences with the appropriate protein product.

If a scientist is working with a protein of known function, that protein can sometimes be purified and a portion of the amino acid sequence determined by chemical methods. The amino acid sequence is then used to predict the corresponding nucleotide sequence, which in turn can be synthesized, labeled with radioactive phosphate, and used as a probe on genomic or cDNA libraries to identify the appropriate clone.

A second method of identifying genes uses a genetic approach. A mutation that affects a specific biochemical process can be localized on the genetic linkage map and matched with DNA that maps at the same location (see Chapter 10). For example, if the recessive mutation that causes purple pea flowers to become white were due to a mutation in an enzyme responsible for the production of purple pigment, the gene that codes for that enzyme could be identified by co-localizing the mutation and a previously cloned fragment of DNA on the genetic linkage map for the pea.

Using the methods described above, molecular biologists have identified gene sequences representing hundreds of different proteins of known function. These sequences have been compiled in computer databases that are accessible to the entire scientific community.

Because organisms share common ancestry, similar genes from different organisms often show a high degree of sequence similarity at both the protein and DNA levels. A sequence of DNA of unknown function can be aligned against the existing sequence database using sophisticated computer software. A match often provides information on the function of the unknown coding sequence. Known gene sequences from one organism can also be used as probes to identify similar genes from the organisms of interest.

As more information on the structure and function of individual genes is accumulated from the combined work of biochemists, geneticists, and molecular biologists, the expanding database makes it increasingly easy to match up DNA sequences with the processes. However, the goal of identifying all of the genes contained within an organism's genome is a daunting one that will occupy researchers for years to come.

Some of the applications of recombinant DNA technology, broadly referred to as **biotechnology**, are described in Chapter 28. There we discuss genetic engineering in plants—that is, how foreign genes are transferred via a bacterial plasmid to a plant's genome, creating a plant whose genome has been modified by externally applied, new DNA.

Summary

DNA Contains Encoded Hereditary Information

Genetic information is encoded in the sequence of nucleotides in molecules of DNA, and these, in turn, determine the sequence of amino acids in molecules of protein.

In the Process of Transcription, mRNA Is Synthesized Using DNA as a Template

The genetic information in the DNA is not expressed directly but is transferred via messenger RNA (mRNA). The long molecules of mRNA are assembled by complementary base-pairing along one strand of the DNA helix. This process, called transcription, is catalyzed by the enzyme RNA polymerase. Each sequence of three nucleotides in the coding region of the mRNA molecule is the codon for a specific amino acid.

The Genetic Code Is a Triplet Code

The genetic code has been deciphered; that is, it is now known which amino acid is called for by a given mRNA codon. Of the 64 possible triplet combinations of the four-lettered nucleotide code, 61 specify particular amino acids and 3 are termination codons. With 61 combinations coding for 20 amino acids, there is more than one codon for many amino acids.

In the Process of Translation, the Information Encoded in a Strand of mRNA Is Used to Synthesize a Specific Protein

Protein synthesis—translation—takes place at the ribosomes. A ribosome is formed from two subunits, one large and one small, each consisting of characteristic ribosomal RNAs (rRNAs) complexed with specific proteins. Also required for protein synthesis is another group of RNA molecules, known as transfer RNA (tRNA). These small molecules can carry an amino acid on one end, and they have a triplet of bases, the anticodon, on a central loop at the opposite end of the three-dimensional structure. The tRNA molecule is the adapter that pairs the correct amino acid with each mRNA codon during protein synthesis. There is at least one kind of tRNA molecule for each kind of amino acid found in proteins.

In bacteria, even as the mRNA strand is still being transcribed, ribosomes are attaching, one after another, near its free end. At the point where the strand of mRNA is in contact with a ribosome, tRNAs are bound temporarily to the mRNA strand. This binding takes place by complementary base-pairing between the mRNA codon and the tRNA anticodon. Each tRNA molecule carries the specific amino acid called for by the mRNA codon to which the tRNA attaches. Thus, following the sequence originally dictated by the DNA, the amino acid units are brought into line one by one and, as peptide bonds form between them, are linked into a polypeptide chain. A summary of protein synthesis in a bacterial cell is depicted in Figure 11–27.

The Operon Consists of a Group of Adjacent Genes That Function as a Regulatory Unit

The essential genetic information of prokaryotes, of which *E. coli* is the best-studied example, is encoded in a circular double-stranded molecule of DNA. A principal means of genetic regulation in prokaryotes is the operon system. An operon contains not only a linear sequence of genes coding for a group of functionally related proteins but also adjacent DNA sequences known as the promoter and operator. Transcription from the operon is controlled by the promoter, the binding site for RNA polymerase, and the operator, the binding site for the repressor. When the repressor is attached to the DNA molecule at the operator, RNA polymerase cannot initiate the transcription of RNA. When the repressor is inactivated, RNA polymerase can attach to the DNA, permitting transcription and protein synthesis to take place.

Eukaryotic Chromosomes Contain Histone Proteins

The eukaryotic chromosome differs in many ways from the chromosome of prokaryotes. Its DNA is always associated with proteins—primarily histones—that play a major role in chromosome structure. The DNA molecule wraps around histone cores to form nucleosomes, which are the basic packaging units of eukaryotic DNA.

Regulation of Gene Expression Is More Complex in Eukaryotes Than in Prokaryotes

During development of multicellular eukaryotes, different groups of genes are activated or inactivated in different types of cells. Gene expression is correlated with the degree of condensation of the chromosome. A variety of specific regulatory proteins is also thought to play key roles in the regulation of gene expression.

In Eukaryotes, Most Structural Genes Contain Introns and Exons

Not all of the DNA in a gene specifies a protein sequence. In the process of transcription, some of the sequences of DNA that are transcribed into RNA are introns that must be removed before the transcript can be used as an mRNA molecule. These mRNA segments are excised from the mRNA before it reaches the cytoplasm. The remaining mRNA segments, transcribed from portions of the DNA known as exons, are spliced together in the nucleus before the mRNA moves into the cytoplasm.

In Eukaryotic DNA, Many Nucleotide Sequences Are Repeated

In addition to introns, eukaryotic genomes contain large numbers of copies of other apparently excess and "meaningless" DNA. There are two major categories of repeated DNA: tandemly repeated DNA and interspersed repeated DNA. A subcategory of tandemly repeated DNA, called simple-sequence repeated DNA, occurs at the centromeres and telomeres (the tips) of chromosomes. Single-copy DNA makes up 50 to 70 percent of eukaryotic chromosomal DNA, but only a very small proportion of it appears to be transcribed into RNA.

Transcription in Eukaryotes Differs from That in Prokaryotes in a Number of Respects

In eukaryotes, transcription involves many regulatory proteins. Also, structural genes are not grouped in operons as they often are in prokaryotes. The transcription of each gene is regulated separately, and each gene produces an RNA transcript containing the encoded infor-

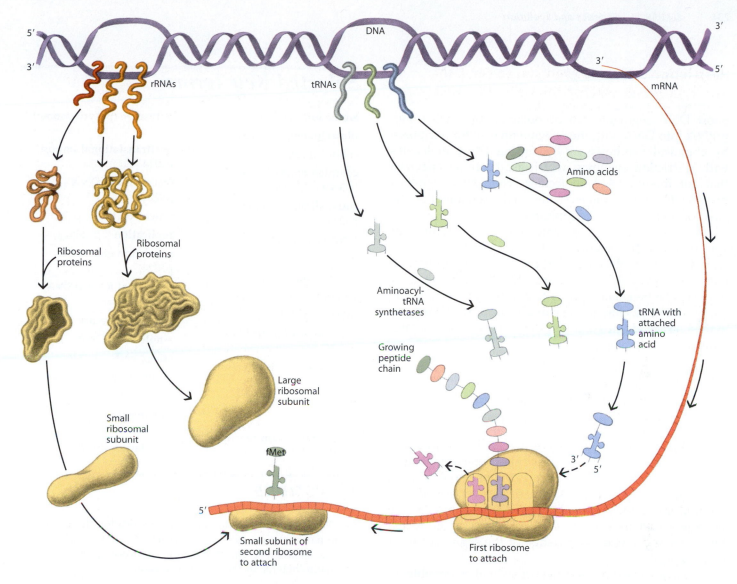

11–27

A summary of protein synthesis in a bacterial cell. Three different kinds of rRNA molecules are transcribed from the DNA of the chromosome. These combine with specific proteins to form the ribosomal subunits. At least 32 different kinds of tRNA molecules are also transcribed from the DNA. These molecules are so structured that each can be attached at one end (by an aminoacyl-tRNA synthetase) to a specific amino acid. Each contains an anticodon that is complementary to an mRNA codon for that particular amino acid.

Protein synthesis begins as an mRNA strand is being transcribed from its DNA template. When the leading (5') end of the mRNA strand attaches to a small ribosomal subunit, an initiator tRNA carrying fMet plugs into the initiation codon of the mRNA. Addition of a large ribosomal subunit completes the initiation complex. The second amino acid is brought in by a tRNA molecule with an anticodon complementary to the next codon in the mRNA strand. The tRNA plugs in momentarily to that codon, a peptide bond is formed between the first and second amino acids, and the first tRNA molecule is released. As the ribosome moves along the mRNA strand, this process is repeated over and over, as the amino acids are brought into line one by one, following the exact order originally dictated by the DNA from which the mRNA was transcribed.

E. coli can synthesize as many as 3000 proteins.

mation for a single product. RNA transcripts are processed in the nucleus to produce the mature mRNA molecules that move from the nucleus to the cytoplasm. This processing includes the removal of introns and the splicing together of exons. Alternative splicing of identical RNA transcripts in different types of cells can produce different mRNA molecules and thus different polypeptides.

Recombinant DNA Technology Is Used to Create Novel Genotypes

Recombinant DNA technology includes methods for (1) obtaining DNA segments short enough to be analyzed and manipulated, (2) obtaining large quantities of identical DNA segments, (3) determining the exact sequence of nucleotides in a DNA segment, and (4) locating and identifying specific DNA segments of interest.

Restriction Enzymes Are Used to Cut DNA into Fragments Having Sticky Ends

Short DNA segments can be obtained by transcribing mRNA into DNA with the enzyme reverse transcriptase, by chemical synthesis, or by cutting DNA molecules with restriction enzymes, which are bacterial enzymes that cut foreign DNA molecules. The DNA segments produced by restriction enzymes can be separated on the basis of their size by electrophoresis.

Different restriction enzymes cut DNA at different specific nucleotide sequences. Instead of cutting the two strands of the molecule straight across, some restriction enzymes leave sticky ends. Any DNA cut by such an enzyme can be joined readily to another DNA molecule cut by the same enzyme. The discovery of restriction enzymes has made possible the development of recombinant DNA technology.

DNA Cloning and the Polymerase Chain Reaction Are Used to Produce Large Quantities of Identical DNA Segments

In cloning, the segments to be copied are introduced into bacterial cells by means of plasmids or bacteriophages, which function as vectors. Once in the bacterial cell, the vector and the foreign DNA it carries are replicated, and the multiple copies can be harvested from the cells. The polymerase chain reaction is a much more rapid process but requires a greater knowledge of the segment that is to be copied.

The availability of multiple copies makes possible, in turn, the determination of the exact order of nucleotides in a DNA segment. By combining sequencing information for sets of short segments produced by different restriction enzymes, molecular biologists can determine the complete sequence of a long DNA segment (such as an entire gene).

A Specific Nucleic Acid Probe Is Used to Locate Specific DNA Segments

Segments of DNA of interest for study and manipulation can be identified by nucleic acid hybridization, using single-stranded probes labeled with radioactive isotopes or fluorescent dyes. This technique is based on the capacity of a single strand of RNA or DNA to combine, or hybridize, with another strand with a complementary nucleotide sequence.

Selected Key Terms

anticodon   p. 210

biotechnology   p. 231

codon   p. 209

complementary DNA (cDNA)   p. 226

cotranslational import p. 214

DNA sequencing   p. 229

exon   p. 222

genetic code   p. 209

genomic library   p. 226

intron   p. 222

messenger RNA (mRNA) p. 208

nucleic acid hybridization p. 230

nucleosome   p. 219

operon   p. 216

polymerase chain reaction (PCR) p. 226

polysome (polyribosome) p. 212

posttranslational import p. 214

recombinant DNA p. 224

repeated DNA   p. 221

replication   p. 208

restriction enzyme p. 224

ribosomal RNA (rRNA) p. 208

RNA polymerase   p. 208

single-copy DNA   p. 222

sticky ends   p. 224

structural gene   p. 216

transcription   p. 209

transfer RNA (tRNA) p. 208

translation   p. 209

Questions

1. Distinguish among the following: replication/transcription/translation; mRNA/tRNA/rRNA; A site/P site/E site. Distinguish between codons/anticodons; euchromatin/heterochromatin; introns/exons.

2. Why is the genetic code said to be degenerate?

3. Most of the DNA of a bacterial cell codes for messenger RNA, and most of the RNA produced by the cell is mRNA. Yet analysis of the RNA content of a cell reveals that, typically, rRNA accounts for about 80 percent of the cellular RNA and tRNA makes up most of the rest. Only about 2 percent is normally mRNA. How do you explain these findings?

4. What amino acid is carried by the tRNA molecule shown in Figure 11–6? How do you know?

5. "There are significant differences between prokaryotes and eukaryotes in transcription, translation, and the events that occur between these two processes." Explain.

6. What are restriction enzymes? What are their uses in recombinant DNA technology?

7. Describe the role of sticky ends in recombinant DNA technology. How are sticky ends produced? What enzyme is required to complete the recombination?

8. Suppose you treated a DNA molecule with a particular restriction enzyme and obtained five fragments, which you separated and cloned into multiple copies. Using the multiple copies, you then sequenced the five fragments. What would you do next to establish their sequence in the original molecule?

Chapter

12

The Process of Evolution

12–1

Charles Darwin in 1840, four years after he returned from his voyage on HMS Beagle. In his later book, The Voyage of the Beagle, Darwin made the following comments about his selection as ship's naturalist for the voyage: "Afterwards, on becoming very intimate with Fitz Roy [the captain of the Beagle], I heard that I had run a very narrow risk of being rejected on account of the shape of my nose! He . . . was convinced that he could judge of a man's character by the outline of his features; and he doubted whether anyone with my nose could possess sufficient energy and determination for the voyage. But I think he was afterwards well satisfied that my nose had spoken falsely."

OVERVIEW

The theory of evolution has been called the greatest unifying principle in biology—in part because it helps us answer so many questions about the world around us. For example, why are there so many different types of organisms on Earth? Why do organisms live where they do? How do new kinds of organisms originate? First formulated by the naturalist Charles Darwin, the theory of evolution now draws on Mendelian genetics, molecular biology, and mathematics.

Although we can define evolution in different ways, it is helpful to focus first on the evolution of populations. At this level of organization, evolution is simply the change in the genetic makeup of a population over time. We examine the processes that result in an increase in favorable traits and a decrease in unfavorable traits, such that, over time, the population becomes better adapted to its environment.

We then take a close look at the level of the species and examine the mechanisms by which new kinds of organisms arise. This process, called speciation, helps us understand how organisms as diverse as pine trees, dandelions, and asparagus could have originated from a single ancestral plant. The chapter ends with a discussion of some current models of evolutionary change.

CHECKPOINTS

By the time you finish reading this chapter, you should be able to answer the following questions:

1. What is Darwin's theory of evolution? What is the difference between natural selection and artificial selection?

2. How is the Hardy–Weinberg equilibrium important for studying evolution? What are the four agents, other than natural selection, that can change the frequencies of alleles in a population, and how do these changes occur?

3. Evolutionarily speaking, what is the advantage of sexual reproduction?

4. What are some ways in which organisms adapt to their physical environments?

5. How does the biological species concept differ from the morphological species concept?

6. How are new species formed? What mechanisms keep closely related species from interbreeding?

In 1831, as a young man of 22, Charles Darwin (Figure 12–1) set out on a five-year voyage as ship's naturalist on a British navy ship, HMS *Beagle*. The book he wrote about the journey, *The Voyage of the Beagle*, not only is a classic work of natural history but also provides us with insight into the experiences that led directly to Darwin's proposal of his theory of evolution by natural selection.

At the time of Darwin's historic voyage, most scientists—and nonscientists as well—still believed in the theory of "special creation." According to this idea, the many different kinds of living organisms were each created (or otherwise came into existence) in their present form. Some scientists, such as Jean Baptiste de Lamarck (1744–1829), had proposed theories of **evolution,** the process by which the earliest forms on Earth have been transformed into the vast diversity of life forms seen today. Lamarck and others, however, could not convincingly explain the mechanism by which the process occurred.

Darwin's theory brought about a great intellectual revolution because he provided painstakingly gathered evidence that all living things are descended from a single common ancestor and that the diversity of living things is accounted for by modifications that accumulated over time. The mechanism he proposed—**natural selection**—is the process by which certain "favorable" modifications become more and more common from one generation to the next.

Particularly important in the genesis of Darwin's ideas were the plants and animals he observed during a stay of some five weeks in the Galápagos Islands, an archipelago that lies in equatorial waters some 950 kilometers off the coast of Ecuador (Figures 12–2 and 12–3). There he made two important observations. First, he noted that the plants and animals found on the islands, although distinctive, were similar to those on the nearby South American mainland. If each kind of plant and animal had been created separately and was unchangeable, as was then generally believed, why did the plants and animals of the Galápagos not resemble those of Africa, for example, rather than those of South America? Or indeed, why were they not utterly unique, unlike organisms anywhere else on Earth? Second, people familiar with the islands pointed out variations that occurred from island to island in such organisms as the giant tortoises, after which the islands were named (*galápagos* is a Spanish word for "tortoise"). Sailors who took these tortoises on board and kept them as sources of fresh meat on their voyages could generally tell which island any particular tortoise had come from (Figure 12–4). If the Galápagos tortoises had been specially created, why did they not all look alike?

12–2

The Beagle's *voyage around South America. The ship left England in December 1831 and arrived at Bahia, Brazil, in late February of 1832. About 3¹/₂ years were spent along the coast of South America, surveying and making inland explorations. The stop at the Galápagos Islands was for slightly more than a month, and during that brief time Darwin made the wealth of observations that were to change the course of the science of biology. The remainder of the voyage, across the Pacific to New Zealand and Australia, across the Indian Ocean to the Cape of Good Hope, back to Bahia once more, and at last home to England, occupied another year.*

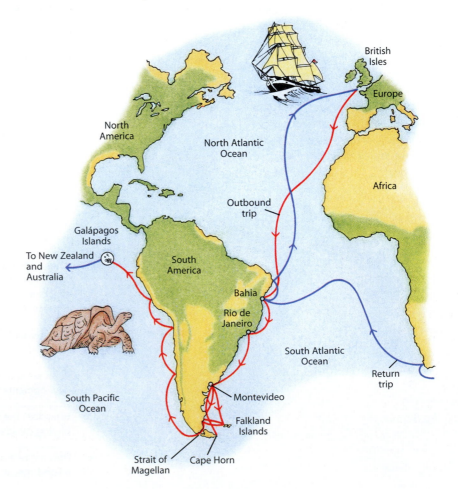

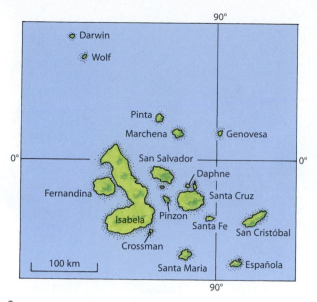

12–3

The Galápagos Archipelago, some 950 kilometers west of the coast of Ecuador, consists of 13 principal volcanic islands and many smaller islets and rocks. These islands have been called "a living laboratory of evolution." "One is astonished," wrote Charles Darwin in 1837, "at the amount of creative force . . . displayed on these small, barren, and rocky islands. . . ."

Darwin's Theory

Not long after his return to England, Darwin read an essay by the Reverend Thomas Malthus that had appeared in 1798. In this essay, Malthus warned that the human population was increasing so rapidly that it would soon be impossible to feed all of Earth's inhabitants.

Darwin saw that Malthus's conclusion—that food supply and other factors hold populations in check—is true for all species, not just the human one. For example, Darwin calculated that a single breeding pair of elephants, which are among the slowest reproducers of all animals, would, if all their offspring lived long and, in turn, reproduced normally, produce a standing population of 19 million elephants in 750 years. Yet the average number of elephants remained the same over the years. So, although a single breeding pair could have in theory produced 19 million descendants, it did in fact produce an average of only two. Although the numbers of every species should tend to become enormously large, the population remains approximately constant. Darwin recognized that there must be a struggle for existence within all populations. He concluded that the individuals that will survive and reproduce and those that will not must be determined to a significant degree by the interaction between the variations among individual organisms and the environment. Some variations enable individuals to adapt, to survive longer, and to produce more offspring than other individuals. Darwin called these variations "favorable" and argued that, over time, inherited favorable variations tended, through natural selection, to become more common in a population.

Natural selection, according to Darwin, was a process analogous to the type of selection exercised by breeders

Darwin began to wonder if all the tortoises and other strange plants and animals of the Galápagos might be related to organisms that existed on the mainland of South America. Once they reached this remote archipelago, these organisms might have spread slowly from island to island, changing bit by bit in response to local conditions and eventually becoming distinct variants that could be distinguished by a human observer.

(a)

(b)

12–4

A distinguishing feature of the Galápagos tortoise is the shape of its shell, or carapace, which varies according to its island of origin. (a) Tortoises found on islands with comparatively lush vegetation are characterized by a

domed shell. The domed shell protects the tortoise's soft parts as it makes its way through the thick undergrowth. This tortoise is a native of Isabela Island. (b) Tortoises found on arid islands, on which the typical vegeta-

tion is thornbush and cactus, are characterized by a saddleback shell. The high arch at the front of the shell enables the tortoise to reach upward in search of food. This tortoise was photographed on the island of Española.

| Kale | Brussels sprouts | Broccoli | Kohlrabi | Cabbage | Cauliflower |

12–5

Six vegetables produced from a single species of plant (Brassica oleracea), a member of the mustard family. They are the result of selection for leaves (kale), lateral buds (brussels sprouts), *flowers and stem (broccoli), stem (kohlrabi), enlarged terminal buds (cabbage), and flower clusters (cauliflower). Kale most resembles the ancestral wild plant. Artificial selection, as* *practiced by plant and animal breeders, gave Darwin the clue to the concept of natural selection.*

of domesticated plants and animals (Figure 12–5). He called this process **artificial selection.** He asked "If man can by patience select variations useful to him, why, under changing and complex conditions of life, should not variations useful to nature's living products often arise, and be preserved and selected?" In natural selection—the major force in evolution—the action of the environment takes the place of human choice. As individuals with certain hereditary characteristics survive and reproduce, and as individuals with other hereditary characteristics are eliminated, the population will slowly change.

In artificial selection, breeders can concentrate their efforts on one or a few characteristics of interest, such as fruit size or animal weight. In natural selection, however, the entire organism must be "fit" in terms of the total environment in which it lives. In other words, it is the entire phenotype that is subject to natural selection.

Evolutionary changes that result in the differentiation of major groups of organisms, such as plants or fungi, obviously require long periods of time. Thus, it is no coincidence that the conclusions of the geologist Charles Lyell, who demonstrated that the Earth was much older than previously thought, had a profound influence on Darwin. In developing his theory of evolution, Darwin needed such an ancient Earth as a stage on which to view the unfolding of the diversity of living things. The many distinctive kinds of fossils (relics of long-dead organisms) that were being discovered—fossils that, with increasing age, became more and more different from living organisms—also provided evidence important to the development of the theory of evolution (Figure 12–6).

The essential difference between Darwin's formulation and that of any of his predecessors is the central role he gave to variation. Others had thought of variations as mere disturbances in the overall design, whereas Darwin saw that variations among individuals are the raw material of the evolutionary process. Which individuals will survive and reproduce and which will not is determined to a significant degree by the interaction between these variations and the environment. Some variations enable individuals to survive longer and produce more offspring than other individuals. Given sufficient time, natural selection can lead to the accumulation of changes that differentiate groups of organisms from one another. Species arise, Darwin proposed, when differences among individuals within a group are gradually converted, over many generations, into differences between groups. The continuation of such divergences over millions of years resulted in the great diversity of organisms currently found on Earth.

As originally formulated by Darwin, the theory of evolution had a major weakness: the absence of any valid mechanism to explain heredity. Although Mendel was at work on his experiments with pea plants at the time Darwin was writing *On the Origin of Species*, his paper was not delivered until 1865 and did not enter the mainstream of biological thinking until early in the twentieth century. The subsequent development of genetics made it possible to answer three questions that Darwin was never able to resolve: (1) how inherited characteristics are transmitted from one generation to the next; (2) why inherited characteristics are not "blended out" but can disappear and reappear in later generations; and (3) how the variations arise on which natural selection acts.

(a)

(b)

(c)

12–6

Various kinds of fossils. Fossils occur in a variety of forms, and commonly are found in sedimentary rocks, which are formed by the accumulation of rock particles. (a) Some plant parts are flattened as the sediments accumulate, and all that is left is a thin carbonaceous (carbon-containing) film, such as this cycad leaf dating from the Jurassic period (145 to 208 million years ago). A great diver- *sity of cycads were contemporary with the dinosaurs; about 140 species of cycads survive today. (b) These 225-million-year-old logs in Petrified National Park, Arizona, are examples of replacement fossils. Such fossils are formed when ground water containing dissolved minerals permeates logs buried in sediment. As the water passes through the logs, the* *wood is replaced by the dissolved minerals and is turned to stone—that is, it is petrified. (c) An Acacia flower entombed approximately 40 million years ago in amber (hardened tree resin) from the Dominican Republic. The numerous stamens seen here consist of long filaments topped by small, rounded anthers.*

The Concept of the Gene Pool

A new branch of biology, **population genetics,** emerged from the synthesis of Mendelian principles with Darwinian evolution. A **population** can be defined as a localized group of individuals belonging to the same species. For now, we can think of a **species** as a group of populations that have the potential to interbreed in nature.

A population is unified and defined by its **gene pool,** which is simply the sum total of all the alleles of all the genes of all the individuals in the population. From the viewpoint of the population geneticist, each individual organism is only a temporary vessel, holding a small sampling of the gene pool for a moment in time. Population geneticists are interested in gene pools, the changes in their composition over time, and the forces causing these changes.

In natural populations, some alleles increase in frequency from generation to generation, and others decrease. (The frequency of an allele is simply the proportion of that allele in a population in relation to all the alleles of the same gene.) If an individual has a favorable combination of alleles in its genotype, it is more likely to survive and reproduce. As a consequence, its alleles are likely to be present in an increased proportion in the next generation. Conversely, if the combination of alleles is not favorable, the individual is less likely to survive and reproduce. Representation of its alleles in

the next generation will be reduced or perhaps eliminated. Evolution is the result of such accumulated changes in the gene pool over time.

In the context of population genetics, the **fitness** of an individual does not mean physical well-being or optimal adaptation to the environment. The only criterion—the sole measure—of an individual's fitness is the relative number of surviving offspring, that is, the extent to which the alleles in an individual's genotype are present in succeeding generations.

The Behavior of Genes in Populations: The Hardy–Weinberg Law

In the early 1900s, biologists raised an important question, alluded to earlier, about the maintenance of variability in populations. How, they asked, can both dominant and recessive alleles remain in populations? Why don't dominants simply drive out recessives? For example, if in a population of plants with contrasting flower color, the allele for purple color is dominant over the allele for white color, why are not all the flowers purple? The question is clearly important to understanding evolution, because it is the genetic variability of populations that provides the "raw material" on which the process of natural selection operates. This question was answered in 1908 by G. H. Hardy, an English mathematician, and G. Weinberg, a German physician.

Working independently, Hardy and Weinberg showed that in a large population in which random mating occurs and in the absence of forces that change the proportions of alleles (discussed below), the original ratio of dominant alleles to recessive alleles will be retained from generation to generation. In other words, the **Hardy–Weinberg law,** as it is now known, states that the proportions, or **frequencies,** of the alleles and genotypes in a population's gene pool remain constant, or at equilibrium, from generation to generation unless acted upon by agents other than sexual recombination. To demonstrate this, they examined the behavior of alleles in an idealized *nonevolving* population in which five conditions hold:

1. No mutations. Mutations alter the gene pool by changing one allele into another.

2. Isolation from other populations. The movement of individuals—with the transfer of their alleles—into or out of the population can change gene pools.

3. Large population size. If the population is large enough, the laws of probability apply—that is, it is highly unlikely that chance alone can alter the frequencies, or relative proportions, of alleles.

4. Random mating. The Hardy–Weinberg equilibrium will hold only if an individual of any genotype chooses its mates at random from the population.

5. No natural selection. Natural selection alters the gene pool as certain alleles become more frequent in the population and others become less frequent.

Consider a single gene that has only two alleles, *A* and *a*. Hardy and Weinberg demonstrated mathematically* that if the five conditions listed above are met, the frequencies of alleles *A* and *a* in the population will not change from generation to generation. Moreover, the frequencies of the three possible combinations of these alleles—the genotypes *AA*, *Aa*, and *aa*—will not change from generation to generation. In other words, the gene pool will be in a steady state—an equilibrium—with respect to these alleles.

This equilibrium is expressed by the Hardy–Weinberg equation:

$$p^2 + 2pq + q^2 = 1$$

In this equation, the letter *p* designates the frequency of one allele at a specific locus (for example, *A*), and the letter *q* designates the frequency of the other allele (*a*). The sum of *p* plus *q* must always equal 1 (that is, 100 percent of the alleles of that particular gene in the gene pool). The expression p^2 designates the frequency of individuals homozygous for one allele (*AA*), q^2 the frequency of individuals homozygous for the other allele (*aa*), and $2pq$ the frequency of heterozygotes (*Aa*).

*This mathematical demonstration and its application in specific examples are given in Appendix B.

The Hardy–Weinberg Equilibrium Provides a Standard for Detecting Evolutionary Change

The Hardy–Weinberg equilibrium and its mathematical formulation have proved as valuable a foundation for population genetics as Mendel's principles have been for classical genetics. At first glance, this seems hard to understand, since the five conditions specified for a gene pool in equilibrium—that is, a nonevolving population—are seldom likely to be met in a natural population. How, then, can the Hardy–Weinberg equation be useful? An analogy from physics may be helpful. Newton's first law says that a body remains at rest or maintains a constant velocity when not acted upon by an external force. In the real world, bodies are always acted upon by external forces, but this first law is an essential premise for examining the nature of such forces. It provides a standard against which to measure.

Similarly, the Hardy–Weinberg equation provides a standard against which we can measure the changes in allele frequencies that are always occurring in natural populations. Without the Hardy–Weinberg equation, we would not be able to detect change, determine its magnitude and direction, or uncover the forces responsible for it.

The Agents of Change

In order for evolution to occur, the frequencies of alleles or genotypes in a population must deviate from the Hardy–Weinberg equilibrium. At the population level, evolution may be defined as a generation-to-generation change in a population's genetic structure. Such small-scale generation-to-generation change in the frequency of a population's alleles is referred to as **microevolution.**

According to modern evolutionary theory, natural selection is the major force in changing allele frequencies. Let us first consider some other agents that can change the frequencies of alleles in a population. There are four: mutations, gene flow, genetic drift, and nonrandom mating.

Mutations Provide the Raw Material for Evolutionary Change

From the point of view of population genetics, **mutations** are inheritable changes in the genotype (Figure 12–7). As we learned earlier, a mutation may involve the substitution of one or a few nucleotides in a DNA molecule or changes in whole chromosomes, segments of chromosomes, or even entire sets of chromosomes (pages 192–193). Most mutations occur "spontaneously"—meaning simply that we do not know the factors that triggered them. Mutations are generally said to occur at random, or by chance. This does not mean that

12–7

A mutant red delicious apple. The mutant allele that is responsible for the golden color arose in a cell of the flower's ovary wall, which subsequently developed into the fleshy part of the apple. The seeds of this apple would produce red apples because they were unaffected by the mutation. Golden delicious apples, however, were discovered on a mutated branch of a tree that produced red delicious apples. The seeds of these golden apples, which carried the mutation, produced trees bearing golden delicious apples.

mutations occur without cause but rather that the events triggering them are independent of their subsequent effects. Although the rate of mutation can be influenced by environmental factors, the specific mutations produced are independent of the environment—and independent of their potential for subsequent benefit or harm to the organism and its offspring.

Although the rate of spontaneous mutation is generally low, mutations provide the raw material for evolutionary change because they provide the variation on which other evolutionary forces act. However, because mutation rates are so low, mutations probably never determine the direction of evolutionary change.

Gene Flow Is the Movement of Alleles into or out of a Population

Gene flow can occur as a result of immigration or emigration of individuals of reproductive age. In the case of plants, it can also occur through the movement of gametes via pollen between populations.

Gene flow can introduce new alleles into a population or it can change existing allele frequencies. Its overall effect is to decrease the difference between populations. Natural selection, by contrast, is more likely to increase differences, producing populations more suited for local conditions. Thus, gene flow often counteracts natural selection.

The possibilities of gene flow between natural populations of most plant species diminish rapidly with distance. Although pollen can be dispersed great distances at times, the chances of its falling on a receptive stigma at any great distance are slight. For several kinds of insect-pollinated plants that grow in temperate regions, a gap of only 300 meters may effectively isolate two populations. Rarely will more than 1 percent of the pollen that reaches a given individual come from this far away. In plants whose pollen is spread by wind, very little falls more than 50 meters from the parent plant under normal circumstances.

Genetic Drift Refers to Changes That Occur Due to Chance

As we stated previously, the Hardy–Weinberg equilibrium holds true only if the population is large. This qualification is necessary because the equilibrium depends on the laws of probability. Consider, for example, an allele, say *a*, that has a frequency of 1 percent. In a population of 1 million individuals, 20,000 *a* alleles would be present in the gene pool. (Remember that each diploid individual carries two alleles for any given gene. In the gene pool of this population there are 2 million alleles for this particular gene, of which 1 percent, or 20,000, are allele *a*.) If a few individuals in this population were destroyed by chance before leaving offspring, the effect on the frequency of allele *a* would be negligible.

In a population of 50 individuals, however, the situation would be quite different. In this population, it is likely that only one copy of allele *a* would be present. If the lone individual carrying this allele failed to reproduce or were destroyed by chance before leaving offspring, allele *a* would be completely lost. Similarly, if 10 of the 49 individuals homozygous for allele *A* were lost, the frequency of *a* would jump from 1 in 100 to 1 in 80.

This phenomenon, a change in the gene pool that takes place as a result of chance, is **genetic drift.** Population geneticists and other evolutionary biologists generally agree that genetic drift plays a role in determining the evolutionary course of small populations. Its relative importance, however, as compared with that of natural selection, is a matter of debate. There are at least two situations—the founder effect and the bottleneck effect—in which genetic drift has been shown to be important.

The Founder Effect Occurs When a Small Population Colonizes a New Area A small population that becomes separated from a larger one may or may not be genetically representative of the larger population from which it was derived (Figure 12–8). Some rare alleles may be overrepresented or, conversely, may be completely absent in the small population. An extreme case would be

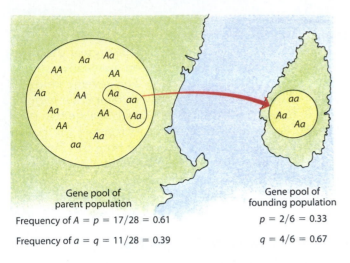

Gene pool of
parent population

Gene pool of
founding population

Frequency of $A = p = 17/28 = 0.61$ $p = 2/6 = 0.33$

Frequency of $a = q = 11/28 = 0.39$ $q = 4/6 = 0.67$

12–8

The founder effect. When a small subset of a population founds a new colony (for example, on a previously uninhabited island), the allele frequencies within the founding group may be different from those within the parent population. Thus the gene pool of the new population will have a different composition than the gene pool of the parent population.

the initiation of a new population by a single plant seed. As a consequence, when and if the small population increases in size, it will continue to have a different genetic composition—a different gene pool—from that of the parent group. This phenomenon, a type of genetic drift, is known as the **founder effect.**

The Bottleneck Effect Occurs When Environmental Factors Suddenly Decrease Population Size   The **bottleneck effect** is another type of situation that can lead to genetic drift. It occurs when a population is drastically reduced in numbers by an event, such as an earthquake, flood, or fire, that may have little or nothing to do with the usual forces of natural selection. A population bottleneck is likely not only to eliminate some alleles entirely but also to cause others to become overrepresented in the gene pool.

Nonrandom Mating Decreases the Frequency of Heterozygotes

Disruption of the Hardy–Weinberg equilibrium can also be produced by nonrandom mating. Typically, the members of a population mate more often with close neighbors than with more distant ones. Thus, within the large population, neighboring individuals tend to be closely related. Such nonrandom mating promotes **inbreeding,** the mating of closely related individuals. An extreme

form of nonrandom mating that is particularly important in plants is self-pollination (as in the pea plants Mendel studied).

Inbreeding and self-pollination tend to increase the frequencies of homozygotes in a population at the expense of the heterozygotes. Take, for example, Mendel's pea plants in which only two alleles are involved in flower color, *W* (purple) and *w* (white). When *WW* plants and *ww* plants self-pollinate, all of their progeny will be homozygous. When *Ww* plants self-pollinate, however, only half of their progeny will be heterozygous. With succeeding generations, there will be a decrease in the frequency of heterozygotes, with a corresponding increase in the frequencies of the two homozygotes. Note that although nonrandom mating, as exemplified by the pea plants, can change the ratio of genotypes and phenotypes in a population, the frequencies of the alleles in question remain the same.

Preservation and Promotion of Variability

Sexual Reproduction Produces New Genetic Combinations

By far the most important method by which eukaryotic organisms promote variation in their offspring is sexual reproduction. Sexual reproduction produces new genetic combinations in three ways: (1) by independent assortment at the time of meiosis (see Figure 9–10), (2) by crossing-over with genetic recombination (see Figure 9–4), and (3) by the combination of two different parental genomes at fertilization. At every generation, alleles are assorted into new combinations.

In contrast, consider organisms that reproduce only asexually—that is, by processes that involve mitosis and cytokinesis but not meiosis. Except when a mutation has occurred in the duplication process, the new organism will exactly resemble its only parent. In the course of time, various clones may form, each carrying one or more mutations, but potentially favorable combinations are unlikely to accumulate in one genotype. The only advantage to the organism of sexual reproduction, from a strictly scientific viewpoint, appears to be the promotion of variation by the production of new combinations of alleles among the offspring. Why such variation is advantageous to the individual organism is a matter of controversy.

Various Mechanisms Promote Outbreeding

Many means have evolved by which new genetic combinations are promoted in sexually reproducing populations. Among plants, various mechanisms ensure that the sperm-bearing pollen from flowers on one plant is

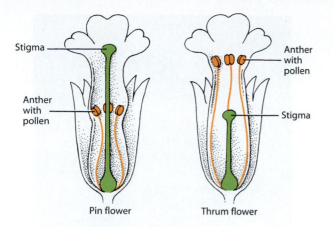

12–9

Diagrams of two types of flowers ("pin" and "thrum") of the same species of primrose. Notice that the pollen-bearing anthers of the pin flower and the pollen-receiving stigma of the thrum flower are both situated about halfway up the length of the flower and that the pin stigma is level with the thrum anthers. An insect foraging for nectar deep inside these flowers would collect pollen on different areas of its body, so that thrum pollen would be deposited on pin stigmas, and vice versa. Self-pollination is therefore inhibited.

TABLE 12–1	Protection of Recessive Alleles by Diploidy			
Frequency of Allele *w* in Gene Pool	Genotype Frequencies			Percentage of Allele *w* in Heterozygotes
	WW	*Ww*	*ww*	
0.9	0.01	0.18	0.81	10
0.1	0.81	0.18	0.01	90
0.01	0.9801	0.0198	0.0001	99

delivered to the stigmas of flowers on a different plant. In some plants, such as the holly and the date palm, male flowers are on one tree and female flowers on another. In others, such as the avocado, the pollen of a particular plant matures at a time when its own stigma is not receptive. In some species, anatomical arrangements inhibit self-pollination (Figure 12–9).

Some plants have genes for self-sterility. Typically, such a gene has multiple alleles—s^1, s^2, s^3, and so on. A plant carrying the s^1 allele cannot fertilize a plant with an s^1 allele. One with an s^1s^2 genotype cannot fertilize any plant with either of those alleles, and so forth. In one population of about 500 evening primrose plants, 37 different self-sterility alleles were found, and it has been estimated that there may be hundreds of alleles for self-sterility in red clover. A plant with a rare self-sterility allele is more likely to be able to fertilize another plant than is a plant with a common self-sterility allele. As a consequence, the self-sterility system strongly encourages variability in a population. Selection for the rare allele makes it more common, whereas more common alleles become rarer.

Diploidy Enables Recessive Alleles to Be Stored

Another factor in the preservation of variability in eukaryotes is diploidy. In a haploid organism, genetic variations are immediately expressed in the phenotype and are therefore exposed to the selection process. In a diploid organism, however, such variations may be

stored as recessives, as with the allele for white flowers in Mendel's pea plants (Table 12–1). As the table reveals, the lower the frequency of allele *w*, the smaller the proportion of it exposed in the *ww* homozygote becomes. The removal of the allele by natural selection slows down accordingly.

Heterozygotes May Have a Selective Advantage over Homozygotes

Recessive alleles, even ones that are harmful in the homozygous state, may not only be sheltered in the heterozygous state but sometimes may actually be selected for. This phenomenon, in which the heterozygotes produce more offspring than either type of homozygote, is known as **heterozygote advantage.** It is another way that genetic variability is preserved.

Probably the best example of heterozygote advantage in plants is found in the crossing of inbred lines of crop plants, with the resultant increase in vigor and productivity of the **hybrids**—by definition, the offspring of genetically dissimilar parents. Hybrids exhibiting such **hybrid vigor** have played an important role in increasing the yield of crop plants throughout the world, most notably of maize. Hybridization was the first systematic application of genetic principles to crop breeding. The discovery of hybrid vigor in maize dates to 1908, when the American plant breeder G. H. Shull discovered that crossing inbred lines produced hybrids that quadrupled the yield (from about 20 to 80 bushels per acre!) of the inbred lines.

Responses to Selection

In the course of the controversies that led to the synthesis of evolutionary theory with Mendelian genetics, some biologists argued that natural selection could serve only to eliminate the "less fit." As a consequence, it would tend to reduce the genetic variation in a population and thus reduce the potential for further evolution. Modern population genetics has demonstrated that this is not the case. Natural selection can, in fact, be a critical factor in preserving and promoting genetic variability in a population.

In general, only the phenotype is being selected, in the sense that it is the relationship between the phenotype and the environment that determines how many offspring an individual will contribute to the next generation. Because nearly all characteristics of individual organisms are determined by the interactions of many genes, phenotypically similar individuals can have many different genotypes.

When some feature, such as tallness, is strongly selected for, there is an accumulation of alleles that contribute to this feature and an elimination of those alleles that work in the opposite direction. But selection for a polygenic characteristic is not simply the accumulation of one set of alleles and the elimination of others. Gene interactions, such as epistasis and pleiotropy (pages 195–196), are of fundamental importance in determining the course of selection in a population.

Bear in mind, also, that the phenotype is not determined solely by the interactions of the multitude of alleles making up the genotype. The phenotype is also a product of the interaction of the genotype with the environment in the course of the individual's life (Figure 12–10).

Evolutionary Changes in Natural Populations May Occur Rapidly

Under certain circumstances, the characteristics of populations may change rapidly, often in response to a rapidly changing environment. Particularly during the past few centuries, the influence of human beings in many areas has been so great that some populations of other organisms have had to adjust rapidly in order to survive. Evolutionary biologists have been particularly interested in examples of such rapid changes—"evolution in action"—because the principles involved are presumed to be the same as those that govern changes (even though much slower changes) in populations generally.

In plants, strong selective forces have been observed to produce rapid changes in natural populations. For example, plants in the grazed part of an experimental pasture in Maryland were much shorter than those in the ungrazed part, and it was thought that this might be due to the direct action of grazing. This hypothesis was tested by digging up some of the plants from both parts of the pasture and growing them together in a garden. It was assumed that if the short plants were short merely because they had been grazed, they would soon become tall in the absence of grazing. Some of them did, but plants of white clover (*Trifolium repens*), Kentucky bluegrass (*Poa pratensis*), and orchard grass (*Dactylis glomerata*) remained short, indicating that these populations had been modified genetically by the selective force of grazing, over a period of only two or three centuries. A similar example is illustrated in Figure 12–11.

(a)

(b)

12–10
Jeffrey pines (Pinus jeffreyi) *usually grow tall and straight, as in* **(a)**. *Environmental forces, however, can alter the normal growth patterns, as shown in* **(b)**. *This tree is growing on a mountaintop in Yosemite National Park, California, where it is exposed to strong, constant winds.*

In Wales, the tailings around a number of abandoned lead mines are rich in lead (up to 1 percent) and zinc (up to 0.03 percent)—substances that are toxic to most plants at these concentrations. Because of the presence of these metals, the tailings are often nearly devoid of plant life. Observing that one species of grass, *Agrostis tenuis*, was colonizing these areas, scientists took some of the mine plants and other *Agrostis* from nearby pastures and grew them together either in normal soil or in mine soil. In the normal soil, the mine plants were slower growing and smaller than the pasture plants. On the mine soil, how-

12–11

Prunella vulgaris is a common herb of the mint family; it is widespread in woods, meadows, and lawns in temperate regions of the world. Most populations consist of erect plants, such as those shown in (a), which grow in open, often some-what moist, grassy places throughout the cooler regions of the world. Populations found in lawns, however, always consist of prostrate plants, such as those shown in (b), growing in Berkeley, California. Erect plants of P. vulgaris cannot survive in lawns because they are damaged by mowing and do not have the capability for resprouting low branches from the base, which would be necessary for survival. When lawn plants are grown in an experimental garden, some remain prostrate whereas others grow erect. The prostrate habit is determined geneti-cally in the first group and environmentally in the second group.

(a)

(b)

12–12

Bent grass (Agrostis tenuis) growing in the foreground on the tailings of abandoned lead mines rich in lead, photographed in Wales. Such tailings are often devoid of plant life because the lead is present at levels toxic to plants. The lead-tolerant bent grass plants growing here are descendants of a popula-tion of plants that gradually adapted to the lead in the tailings.

ever, the mine plants grew normally but the pasture plants did not grow at all. Half the pasture plants in mine soil were dead in three months and had misshapen roots that were rarely more than 2 millimeters long. But a few of the pasture plants (3 of 60) showed some resis-tance to the effects of the metal-rich soil. They were doubtless genetically similar to the plants originally se-lected in the development of the lead-resistant strain of *Agrostis.* The mines were no more than 100 years old and so the lead-resistant strain had developed in a relatively short period of time. The resistant plants had been se-lected from the genetically variable pasture plants, some of whose seeds had fallen on the mine soil, germinated, and produced seeds that could survive there. Then, due to natural selection, a distinct strain of *Agrostis* was con-stituted (Figure 12–12).

The Result of Natural Selection: Adaptation

Natural selection results in **adaptation,** a term with sev-eral meanings in biology. First, it can mean a state of being adjusted to the environment. Every living organ-ism is adapted in this sense, just as Abraham Lincoln's legs were, as he remarked, "just long enough to reach the ground." Second, adaptation can refer to a particular characteristic that aids in the adjustment of an organism to its environment. Third, adaptation can mean the evo-lutionary process, occurring over the course of many generations, that produces organisms better suited to their environment.

Natural selection involves interactions between indi-vidual organisms, their physical environment, and their biological environment—that is, other organisms. In many cases, the adaptations that result from natural selection can be clearly correlated with environmental factors or with the selective forces exerted by other organisms.

Clines and Ecotypes Are Reflections of Adaptation to the Physical Environment

Developmental plasticity is the tendency of individuals to vary over time in response to different environmental conditions, or for genetically identical organisms to differ in response to different environmental stimuli. Such plasticity is much greater in plants than in animals because the indeterminate (unrestricted) growth pattern that is characteristic of plants can be modified more easily to produce strikingly different expressions of a particular genotype. Environmental factors, as every gardener knows, can cause profound differences in the phenotypes of many species of plants. Leaves that develop in the shade, for example, may be thinner and larger than those that develop in the sun.

Sometimes phenotypic variation within the same species follows a geographic distribution and can be correlated with gradual changes in temperature, humidity, or some other environmental condition. A gradual change of this kind in the characteristics of populations of an organism is called a **cline**. Many species exhibit north-south clines of various traits. Plants, for example, growing in the south often have slightly different requirements for flowering or for ending dormancy than the same kind of plants growing in the north, although they may all belong to the same species.

Clines are frequently encountered in organisms that live in the sea, where the temperature often rises or falls very gradually with changes in latitude. Clines are also characteristic of organisms that occur in such areas as the eastern United States, where rainfall gradients may extend over thousands of kilometers. When populations of plants are sampled along a cline, the differences are often proportional to the distance between the populations.

A species that occupies many different habitats may appear to be slightly different in each one. Each group of distinct phenotypes is known as an **ecotype**. Are the differences among ecotypes determined entirely by the environment, or do these differences represent adaptation resulting from the action of natural selection on genetic variation?

Of particular interest in the study of ecotypes is the work of Jens Clausen, David Keck, and William Hiesey with the perennial herb *Potentilla glandulosa*, which ranges through a wide variety of climatic zones in California. These scientists established experimental gardens at three localities in California at sites where native populations of *P. glandulosa* occur: (1) Stanford, located between the inner and outer Coast Ranges, at 30 meters elevation, with warm temperate weather and predominant winter rainfall; (2) Mather, on the western slope of the Sierra Nevada, at 1400 meters elevation, with long, cold, snowy winters and hot, mostly dry summers; and (3) Timberline, east of the crest of the Sierra Nevada at

roughly the same latitude as the two other stations but at 3050 meters elevation, with very long and cold, snowy winters and short, cool, rather dry summers (Figure 12–13).

When *P. glandulosa* plants from numerous locations were grown side-by-side in the gardens set up at the three stations, four distinct ecotypes became apparent. The morphological, or structural, characteristics of each ecotype were correlated with their physiological responses, which in turn were critical to the survival of each ecotype in its native environment.

For example, the Coast Ranges ecotype consists of plants that grew actively in both winter and summer when cultivated at Stanford, which lies within their native range. Plants of this ecotype decreased in size but survived at Mather, outside their native range, even though they were subjected to about five months of cold winter weather. At Mather they became winter-dormant, but they stored enough food during their growing season to carry them through the long, unfavorable winter. At Timberline, plants of the Coast Ranges ecotype failed to survive, almost invariably dying during the first winter. The short growing season at this high elevation did not permit them to store enough food to survive the long winter. Other species that occur in California's Coast Ranges produce ecotypes that have physiological responses comparable to those of *P. glandulosa*. Indeed, strains of unrelated plant species that occur together naturally in a given location are often more similar to one another physiologically than they are to other populations of their own species.

The physiological and morphological characteristics of ecotypes, as in *P. glandulosa*, usually have a complex genetic basis, involving dozens (or, in some cases, perhaps even hundreds) of genes. Sharply defined ecotypes are characteristic of regions, like western North America, where the breaks between adjacent habitats are sharply defined. On the other hand, when the environment changes more gradually from one habitat to another, the characteristics of the plants that occur in that region may do likewise.

Ecotypes Differ Physiologically   In order to understand why ecotypes flourish where they do, we must understand the physiological basis for their ecotypic differentiation. For example, when Scandinavian strains of goldenrod (*Solidago virgaurea*) from shaded habitats and from exposed habitats were grown under different light intensities, they showed differences in their photosynthetic response. The plants from shaded environments grew rapidly under low light intensities, whereas their growth rate was markedly retarded under high light intensities. In contrast, plants from exposed habitats grew rapidly under conditions of high light intensity, but much less well at low light levels.

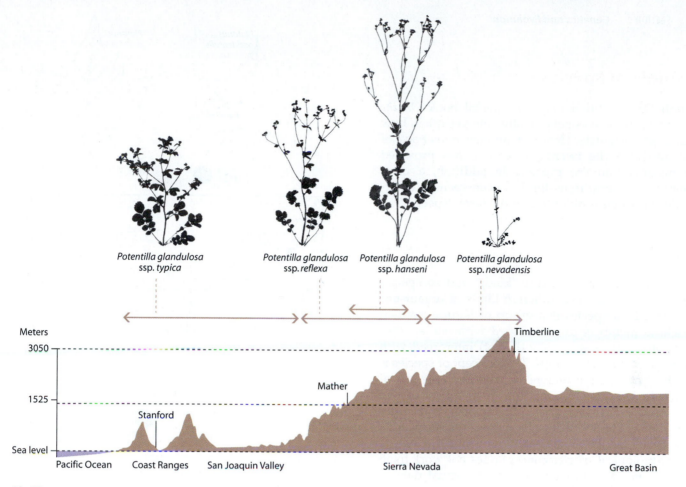

Potentilla glandulosa
ssp. typica

Potentilla glandulosa
ssp. reflexa

Potentilla glandulosa
ssp. hanseni

Potentilla glandulosa
ssp. nevadensis

Meters

Timberline

3050

1525

Mather

Stanford

Sea level

Pacific Ocean   Coast Ranges   San Joaquin Valley   Sierra Nevada   Great Basin

12–13

Ecotypes of Potentilla glandulosa, *a relative of the strawberry. Plants from a number of populations of* P. glandulosa *were collected at 38° north latitude from the Pacific Ocean to Timberline and transplanted to experimental gardens at Stanford, Mather, and Timberline, all also at about 38° north latitude. The plants were propagated asexually so that genetically alike individuals could be grown at the three climatically unlike gardens. When grown side-by-side, four distinct ecotypes became apparent. The four ecologically distinct ecotypes were correlated with differences in their morphology, especially flower and leaf characteris-tics. Many of these characteristics were passed on to the next genera-tion, indicating that these differences are genotypic as well as phenotypic.*

Each ecotype is distributed over a range of altitudes. Where the ranges of two ecotypes overlap, the two ecotypes grow in different environments. Representatives of the four ecotypes at the flowering stage and their approximate ranges are shown. Each of the four eco-types has been given a subspecies name.

In another experiment, arctic and alpine populations of the widespread herb *Oxyria digyna* were studied using strains from an enormous latitudinal range that extended southward from Greenland and Alaska to the mountains of California and Colorado. Plants of northern populations had more chlorophyll in their leaves, as well as higher respiration rates at all temperatures, compared with plants from farther south. High-elevation plants from near the southern limits of the species' range carried out photosynthesis more efficiently at high light intensities than did low-elevation plants from farther north. Thus the respective ecotypes could function better in their own habitats, characterized, for example, by high light intensities in high mountain habitats and lower intensities in the far north. The existence of *O. digyna* over such a wide area and such a wide range of ecological conditions is made possible, in part, by differences in metabolic potential among its various populations.

Coevolution Results from Adaptation to the Biological Environment

When populations of two or more species interact so closely that each exerts a strong selective force on the other, simultaneous adjustments occur that result in **co-evolution**. One of the most important, in terms of sheer number of species and individuals involved, is the co-evolution of flowers and their pollinators, which is described in Chapter 22. Another example of coevolution involves the monarch butterfly and the milkweed plants (page 34).

The Origin of Species

Although Darwin titled his monumental book *On the Origin of Species*, he was never really able to explain how species might originate. However, an enormous body of work, mostly in the twentieth century, has provided many insights into the process. In addition, a great amount of time and discussion has been spent attempting to develop a clear definition of the term "species."

What Is a Species?

In Latin, **species** simply means "kind," and so species are, in the simplest sense, different kinds of organisms. More precisely, a species is a group of natural populations whose members can interbreed with one another but cannot (or at least usually do not) interbreed with members of other such groups. This concept of species is called the *biological species concept*. The key criterion of this definition is **genetic isolation**: if members of one species freely exchanged genes with members of another species, they could no longer retain those unique characteristics that identify them as different kinds of organisms. The biological species concept does not work in all situations, and it is difficult to apply to actual data in nature. Several alternative species concepts have therefore been proposed. In practice, "biological species" are generally identified purely on an assessment of their morphological, or structural, distinctness. In fact, most species recognized by taxonomists have been designated as such based on anatomical and morphological criteria. This practical approach has been referred to as the *morphological species concept*.

The inability to form fertile hybrids has often been used as a basis for defining species. However, this criterion is not generally applicable. In some groups of plants—particularly such long-lived woody plants as trees and shrubs—very morphologically distinct species often can form fertile hybrids with one another. Take the case of the sycamores *Platanus orientalis* and *Platanus occidentalis*, which have been isolated from one another in nature for at least 50 million years. *P. orientalis* is native from the eastern Mediterranean region to the Himalayas, while *P. occidentalis* is native to eastern North America (Figure 12–14). *P. orientalis* has been widely cultivated in southern Europe since Roman times, but it cannot be grown in northern Europe away from the moderating influence of the sea. After the European discovery of the New World, *P. occidentalis* was brought into cultivation in the colder portions of northern Europe, where it flourished. About 1670, these two very distinct trees produced intermediate and fully fertile hybrids when they were cultivated together in England. Called the London plane, the hybrid (*Platanus* × *hybrida*) is capable of growing in regions with cold winters and is now grown

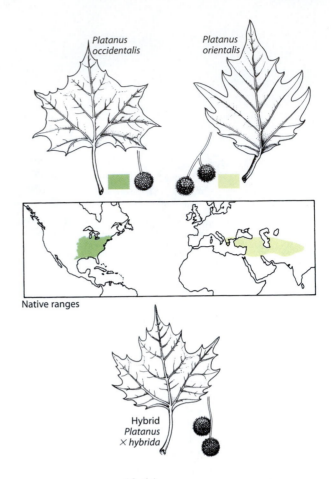

12–14

The distribution of two species of sycamore, Platanus orientalis, *which is native to areas from the eastern Mediterranean region to the Himalayas, and* P. occidentalis, *which is native to North America. The fully fertile hybrid,* Platanus × hybrida, *is the London plane, a sturdy tree suitable for growing along city streets.*

as a street tree in New York City and throughout the temperate regions of the world.

One might argue that *P. orientalis* and *P. occidentalis* had no potential to interbreed in nature. But, then, there are many groups, such as the birches, oaks, and willows, in which many species freely interbreed in nature. Does that mean that each of these groups of interbreeding species should be considered a single species, and that what appear as morphologically distinct species should be considered subspecies (subdivisions of a species)? It is unlikely that most plant taxonomists would be willing to go that far.

From an evolutionary perspective, a species is a population or group of organisms reproductively united but very probably changing as it moves through space and time. Splinter groups reproductively isolated from the population as a whole can undergo sufficient change that they become new species. This process is known as **speciation.** Occurring repeatedly in the course of more than 3.5 billion years, it has given rise to the diversity of organisms that have lived in the past and that live today.

How Does Speciation Occur?

By definition, the members of a species share a common gene pool that is effectively separated from the gene pools of other species. A central question, then, is, how does one pool of genes split off from another to begin a separate evolutionary journey? A subsidiary question is, how do two species, often very similar to one another, inhabit the same place at the same time and yet remain reproductively isolated?

According to current thinking, speciation is most commonly the result of the geographic separation of a population of organisms; this process is known as **allopatric** ("other country") **speciation.** Under certain circumstances, speciation may also occur without geographic isolation, in which case it is known as **sympatric** ("same country") **speciation.**

Allopatric Speciation Involves the Geographic Separation of Populations

Every widespread species that has been carefully studied has been found to contain geographically representative populations that differ from each other to a greater or lesser extent. Examples are the ecotypes of *Potentilla glandulosa* and the strains of *Oxyria digyna*. A species composed of such geographic variants is particularly susceptible to speciation if geographic barriers arise, preventing gene flow.

Geographic barriers are of many different types (Figure 12–15). Islands are frequently sites for the development of new species, and set the stage for the sudden (in geologic time) diversification of a group of organisms that share a common ancestor, often itself newly evolved. The sudden diversification of such a group of organisms is called **adaptive radiation.** It is associated with the opening up of a new biological frontier that may be as vast as the land or the air, or, as in the case of the Galápagos tortoises, as small as an archipelago. Adaptive radiation results in the almost simultaneous formation of many new species in a wide range of habitats. (See "Adaptive Radiation in Hawaiian Tarweeds" on pages 250–251.)

Differentiation on islands is particularly striking because, in the absence of competition, organisms seem more likely to produce highly unusual forms than do related species on the continents. In island localities, the characteristics of plants and animals may change more rapidly than on the mainland, and features that are never encountered elsewhere may arise. Similar clusters of species may also arise in mainland areas, of course, and may involve spectacular differentiation.

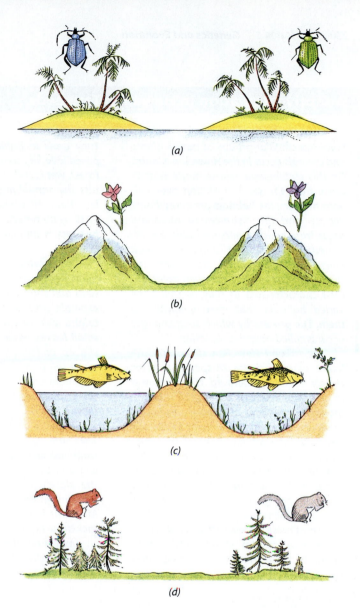

12–15
Four different geographic barriers that can lead to speciation may be (a) islands; (b) mountaintops; (c) ponds, lakes, or even oceans; and (d) isolated clumps of vegetation.

Sympatric Speciation Occurs without Geographic Separation

A well-documented mechanism by which new species are produced through sympatric speciation—that is, when there is no geographic isolation—is **polyploidy.** By definition, polyploidy is an increase in the number of chromosomes beyond the typical diploid ($2n$) complement. Polyploidy may arise as a result of nondisjunction (page 173) during mitosis or meiosis, or it may be generated when the chromosomes divide properly during mitosis or meiosis but cytokinesis does not subsequently occur. Polyploid individuals can be produced deliberately in the laboratory by use of the drug colchicine, which disrupts microtubule formation and hence prevents separation of chromosomes during mitosis (page 161).

Adaptive Radiation in Hawaiian Tarweeds

Some spectacular groups of native plants and animals occur in the Hawaiian Islands. On this archipelago, whose major islands arose from the sea in isolation over millions of years, the habitats are exceedingly diverse, and the distances to mainland areas from which plants and animals migrated to colonize the islands are great. Those organisms that did reach the Hawaiian Islands often changed greatly in their characteristics as they occupied the varied habitats that were available to them. The process by which such changes occur is called adaptive radiation.

The striking Hawaiian silversword alliance, which includes 28 species in three closely related genera of the sunflower family, Asteraceae, is one of the most remarkable examples of adaptive radiation in plants. These species belong to the tarweed subtribe Madiinae, most of whose members are found in California and adjacent regions. The 28 species belong to the genus Argyroxiphium (silverswords) and to two other genera also found only in Hawaii, Dubautia and Wilkesia. The species of these genera range in habit from small, mat-forming shrubs and rosette plants to large trees and climbing vines.

They grow in habitats as diverse as exposed lava, dry scrub, dry woodland, moist forest, wet forest, and bogs. In these habitats, the annual precipitation ranges from less than 40 centimeters to more than 1230 centimeters. The rainiest of these habitats is among the wettest places on Earth.

This enormous variation in habitats is paralleled by significant variation in leaf sizes and shapes among these plants. For example, species of Dubautia that grow in bright, dry habitats usually have very small leaves, while those that inhabit the shaded understories of wet forests have much larger ones. The silversword, Argyroxiphium sandwicense, which grows on the dry alpine slopes of Haleakala Crater on the island of Maui, has leaves that are covered with a dense, silvery mat of hairs. These hairs apparently provide protection from intense solar radiation and aid in the conservation of moisture. The leaves of the closely related greensword, Argyroxiphium grayanum, which also occurs on Maui but in wet forest and bog habitats, lack these hairs.

Important physiological differences likewise characterize the species of

The silversword, Argyroxiphium sandwicense, *a remarkable plant that grows on the exposed, upper cinder slopes of Haleakala Crater on the island of Maui, where the plants are exposed to high levels of solar radiation and low humidities.*

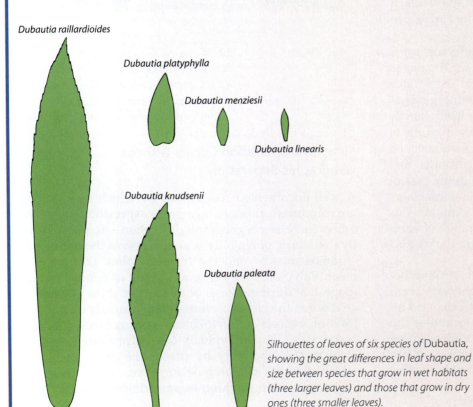

Dubautia raillardioides

Dubautia platyphylla

Dubautia menziesii

Dubautia linearis

Dubautia knudsenii

Dubautia paleata

Silhouettes of leaves of six species of Dubautia, showing the great differences in leaf shape and size between species that grow in wet habitats (three larger leaves) and those that grow in dry ones (three smaller leaves).

Dubautia reticulata. *Species of the genus* Dubautia *(which includes 21 of the estimated 28 species of Hawaiian tarweeds) range from trees and shrubs to lianas (woody vines) and small, matted plants that are scarcely woody.* Dubautia reticulata, *shown here, grows in the wet forests of Maui, where it may reach a height of 8 meters or more and develop a trunk with a diameter of nearly 0.5 meter.*

Wilkesia gymnoxiphium. *This bizarre, yuccalike plant occurs only on the island of Kauai, where it is restricted to dry scrub vegetation along the fringes of Waimea Canyon. Kauai is the oldest of the main islands of the Hawaiian Archipelago. Robert Robichaux, shown here, is studying the physiological ecology of this fascinating group of plants.*

species was closely related (and similar) to modern Californian species of the tarweed genera Madia and Raillardiopsis.

Evolutionary patterns, reconstructed from DNA sequence data, indicate that the most recent common ancestor of the modern silversword alliance species occurred on the oldest high island of Kauai. The many species of this group that occur outside Kauai owe their existence to interisland dispersal of founders from older to progressively younger islands, followed by

Dubautia scabra. *This low, matted, herbaceous member of the genus is found in moist to wet habitats on several of the islands in the Hawaiian Archipelago. It is believed to be the progenitor of several additional species on the younger islands of the group.*

speciation that involved major ecological shifts—the hallmarks of adaptive radiation. Similar patterns have been confirmed recently for several other endemic lineages of Hawaiian organisms.

Hawaiian tarweeds, which must meet very different kinds of environmental challenges. For example, species of Dubautia that occur in dry habitats have a much greater tolerance to water stress than those that occur in moist to wet habitats. The species that grow in dry habitats have leaves with more elastic cell walls and under dry conditions are able to maintain higher turgor pressures than their relatives.

Despite their great diversity in appearance and the very different habitats in which they grow, all 28 species of these three genera are very closely related to one another. Any two of them can be hybridized, as far as we know, and all of the hybrids are at least partly fertile. Additionally, experimental hybrids between Hawaiian and Californian tarweeds have been produced, underscoring their very close relationship. The entire group of three genera appears to have evolved in isolation on the Hawaiian Islands following the arrival of a single, original colonist from western North America. Molecular studies have revealed that this ancestral

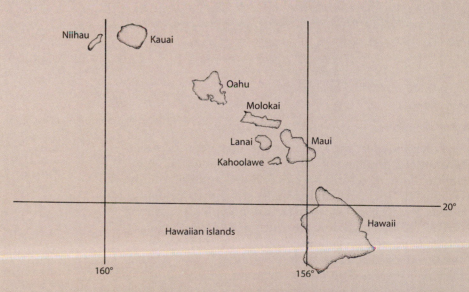

Niihau

Kauai

Oahu

Molokai

Lanai   Maui

Kahoolawe

20°

Hawaiian islands   Hawaii

160°   156°

The Hawaiian Islands. *The oldest of the main islands, Kauai, includes some rocks that are as much as 6 million years old; the youngest, Hawaii, is still being formed. The Hawaiian Archipelago is gradually moving northwest with the Pacific Plate, the older islands gradually being eroded below the surface of the sea and the younger ones continuously being formed, apparently as they pass over a "hot spot," which is a thin spot in the Earth's crust through which lava erupts. Thus we know that there were islands in approximately the present position of Hawaii much more than 6 million years ago.*

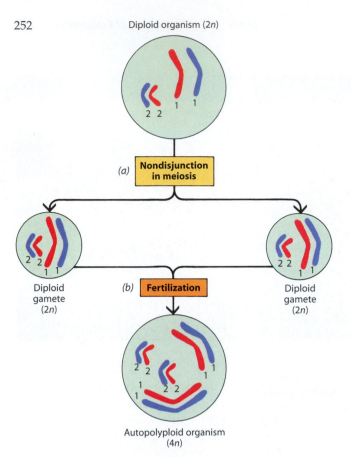

12–16

*Autopolyploidy. Polyploidy within individual organisms can lead to the formation of new species. **(a)** If the chromosomes of a diploid organism do not separate during meiosis (nondisjunction), diploid (2n) gametes may result. **(b)** Union of two such gametes, produced either by the same individual or by different individuals of the same species, will produce an autopolyploid, or tetraploid (4n), individual. Although this individual may be capable of sexual reproduction, it will be reproductively isolated from the diploid parent species.*

Polyploidy leading to the formation of new species as a result of a doubling of chromosome number within individual organisms of a species is called **autopolyploidy** (Figure 12–16). Such individuals are called **autopolyploids.** Sympatric speciation by autopolyploidy was first discovered by Hugo de Vries, during his investigation of the genetics of the evening primrose (*Oenothera lamarckiana*), a diploid species with 14 chromosomes. Among his plants appeared an unusual variant, which, upon microscopic examination, was found to be tetraploid (4*n*), with 28 chromosomes. De Vries was unable to breed the tetraploid primrose with the diploid primrose, because of problems with pairing of the chromosomes during meiosis. The tetraploid (an autopolyploid) was a new species, which de Vries named *Oenothera gigas* (*gigas*, meaning "giant").

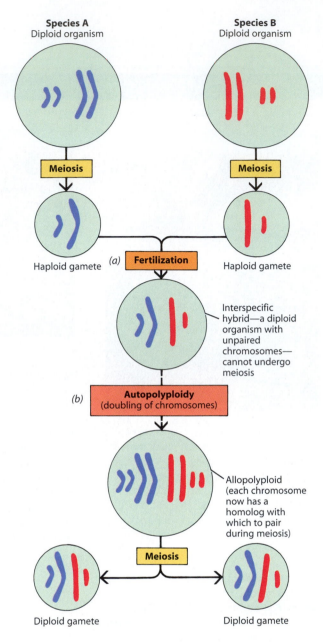

12–17

*Allopolyploidy. **(a)** An organism that is a hybrid between two different species—an interspecific hybrid—and is produced from two haploid (n) gametes can grow normally because mitosis is normal. It cannot reproduce sexually, however, because the chromosomes cannot pair at meiosis. **(b)** If autopolyploidy subsequently occurs and the chromosome number doubles, the chromosomes can pair at meiosis. As a result, the hybrid—an allopolyploid—can produce viable diploid (2n) gametes and is now a new species capable of reproducing sexually.*

A much more common mode of polyploidy is **allopolyploidy,** which results from a cross between two different species, producing an **interspecific hybrid** (Figure 12–17). Such hybrids are usually sterile because the chromosomes cannot pair at meiosis (having no ho-

12–18

Flowering heads of Tragopogon *(goat's beard). Three highly fertile diploid (2n = 12) species introduced from Europe,* **(a)** Trago-pogon dubius, **(b)** Tragopogon porrifolius, *and* **(c)** Tragopogon pratensis, *were all well established in southeastern Washington and adjacent Idaho by 1930. These species hybridized easily, resulting in highly sterile diploid (F₁) interspecific hybrids:* **(d)** Tragopogon dubius × porrifolius, **(e)** Tragopogon porrifolius × pratensis, *and* **(f)** Tragopogon dubius × pratensis. *In 1949, four small populations of* Tragopogon *were discovered that were clearly different from the diploid hybrids and were immediately sus-pected of being two newly originated poly-ploid species. The suspected polyploids* **(g), (h)** *differed in none of their characters from the diploid hybrids* **(d), (f)** *except that they were much larger in every way and were obviously fertile, with flower heads containing many developing fruits. It was soon confirmed that these populations were tetraploid (2n = 24) species and were named* **(g)** Tragopogon mirus *and* **(h)** Tragopogon miscellus. *These tetraploid species are among the few allo-polyploids whose time of origin is known with a high degree of certainty.* **(i)** *Note that this inflorescence of* Tragopogon porrifolius × pratensis *is a highly sterile diploid hybrid of a generation later than the F₁ seen here as* **(e)**.

mologs), a necessary step for producing viable gametes. If, however, autopolyploidy then occurs in such a sterile hybrid and the resulting cells divide by future mitosis and cytokinesis, they eventually produce a new individ-ual asexu-ally. That individual—an allopolyploid—will have twice as many chromosomes as its parent. As a consequence, it is reproductively isolated from its parental line. However, its chromosomes—now dupli-cated—can pair, meiosis can occur normally, and fertility is restored. It is a new species capable of sexual reproduc-tion.

Hybridization and sympatric speciation through polyploidy are important, well-established phenomena in plants. It is clear that they have played significant roles in the evolution of flowering plants. Recent esti-mates are that from 47 percent to over 70 percent of flowering plants are polyploid. Furthermore, 80 percent of the species in the grass family are estimated to be polyploid, and a large number of the major food crops are polyploid, including wheat, sugarcane, potatoes, sweet potatoes, and bananas.

Some polyploids that originated as weeds in habitats associated with the activities of human beings have been spectacularly successful. Probably the best-documented examples are two species of goat's beard, *Tragopogon mirus* and *Tragopogon miscellus*, which are products of al-lopolyploid speciation (Figure 12–18). Both arose during the last hundred years in the Palouse region of south-eastern Washington and adjacent Idaho, following the introduction and naturalization of their Old World pro-genitors, *Tragopogon dubius*, *Tragopogon porrifolius*, and *Tragopogon pratensis*. All three of the Old World species are diploid (2n = 12), and each crosses readily with both of the others, forming F₁ hybrids that are highly sterile. In 1949, two tetraploid (2n = 24) hybrids were discov-ered that are fairly fertile and have increased substan-tially in the Palouse region since their discovery. Both *T. mirus* and *T. miscellus* have been reported in Arizona, and *T. miscellus* also occurs in Montana and Wyoming. The progenitors of *T. mirus* are *T. dubius* and *T. porri-folius*, and those of *T. miscellus* are *T. dubius* and *T. praten-sis*. The origin of the allopolyploids of *Tragopogon* has recently been confirmed by molecular analysis of ribosom-al RNA genes.

12–19

Polyploidy has been investigated extensively among grasses of the genus Spartina, *which grow in salt-marsh habitats along the coasts of North America and Europe. (a) This salt marsh is on the coast of Great Britain. (b) A* Spartina *hybrid. (c)* Spartina maritima, *the native European species of salt-marsh grass,*

has 2n = 60 chromosomes, shown here from a cell in meiotic anaphase I. (d) Spartina alterniflora *is a North American species with 2n = 62 chromosomes (there are 30 bivalents and 2 unpaired chromosomes), shown here from a cell in meiotic metaphase I. (e) A vigorous polyploid,* S. anglica, *arose sponta-*

neously from a hybrid between these species and was first collected in the early 1890s. This polyploid, which has 2n = 122 chromosomes, shown here from a cell in meiotic anaphase I, is now extending its range through the salt marshes of Great Britain and other temperate countries.

One of the best known polyploids, the origin of which is associated with human activity, is a salt-marsh grass of the genus *Spartina* (Figure 12–19). One native species, *S. maritima,* occurs in marshes along the coasts of Europe and Africa. A second species, *S. alterniflora,* was introduced into Great Britain from eastern North America in about 1800, spreading from where it was first planted and forming large but local colonies.

In Britain, the native *S. maritima* is short in stature whereas *S. alterniflora* is much taller, frequently growing to 0.5 meter and occasionally to 1 meter or even more in height. Near the harbor at Southampton, in southern England, both the native species and the introduced species existed side by side throughout the nineteenth century. In 1870, botanists discovered a sterile hybrid between these two species that reproduced vigorously by rhizomes. Of the two parental species, *S. maritima* has a somatic chromosome number of $2n = 60$ and *S. alterniflora* has $2n = 62$; the hybrid, owing perhaps to some minor meiotic misdivision, also has $2n = 62$ chromosomes. This sterile hybrid, which was named *Spartina × townsendii,* still persists. About 1890, a vigorous seed-producing polyploid, named *S. anglica,* was derived naturally from it. This fertile polyploid, which had a diploid chromosome number of $2n = 122$ (one chromosome pair was evidently lost), spread rapidly along the coasts of Great Britain and northwestern France. It is often

planted to bind mud flats, and such use has contributed to its further spread.

One of the most important polyploid groups of plants is the genus *Triticum,* the wheats. The most commonly cultivated crop in the world, bread wheat, *T. aestivum,* has $2n = 42$ chromosomes. Bread wheat originated at least 8000 years ago, probably in central Europe, following the natural hybridization of a cultivated wheat with $2n = 28$ chromosomes and a wild grass of the same genus with $2n = 14$ chromosomes. The wild grass probably occurred spontaneously as a weed in the fields where wheat was being cultivated. The hybridization that gave rise to bread wheat probably occurred between polyploids that arose from time to time within the populations of the two ancestral species.

It is likely that the desirable characteristics of the new, fertile, 42-chromosome wheat were easily recognized, and it was selected for cultivation by the early farmers of Europe when it appeared in their fields. One of its parents, the 28-chromosome cultivated wheat, had itself originated following hybridization between two wild 14-chromosome species in the Near East. Species of wheat with $2n = 28$ chromosomes are still cultivated, along with their 42-chromosome derivative. Such 28-chromosome wheats are the chief grains used in macaroni products because of the desirable agglutinating properties of their proteins.

12–20

Sympatric speciation. Helianthus anomalus *is the result of a cross between two other distinct species of sunflower,* Helianthus annuus *and* Helianthus petiolaris. *Individually, the three species are easily distinguished from one another.* H. anomalus, *for example, has few petals, which are wider than those in* H. petiolaris *and* H. annuus, *and has small leaves with short petioles (leaf stalks).* H. petiolaris *is named for its long, thin petioles, and* H. annuus *has large leaves with thick petioles.*

Helianthus
annuus

Helianthus
petiolaris

Helianthus
anomalus

An example of sympatric speciation not involving polyploidy is provided by the anomalous sunflower (*Helianthus anomalus*), the product of interbreeding between two other distinct species of sunflower, the common sunflower (*Helianthus annuus*) and the petioled sunflower (*Helianthus petiolaris*) (Figure 12–20). All three species occur widely in the western United States. Molecular evidence indicates that *H. anomalus* arose by **recombination speciation,** a process in which two distinct species hybridize, the mixed genome of the hybrid becoming a third species that is genetically (reproductively) isolated from its ancestors.

First-generation hybrids of *H. annuus* and *H. petiolaris* are semisterile, a condition apparently due to unfavorable interactions between the genomes of the parental species that result in difficulties during meiosis in the hybrid. Over several generations, however, full fertility is achieved in the hybrids as the gene combinations become rearranged. Individuals with the newly arranged genomes are compatible with one another—that is, *H. anomalus* with *H. anomalus*—but are incompatible with *H. annuus* and *H. petiolaris*.

In a recent study undertaken by Loren H. Rieseberg and coworkers, the genomic composition of three experimentally produced hybrid lineages involving crosses between *H. annuus* and *H. petiolaris* was compared with that of the naturally occurring *H. anomalus*. Surprisingly, by the fifth generation, the genomic composition of all three lineages was remarkably similar to that of the nat-

urally occurring *H. anomalus*. In addition, in all three experimentally produced hybrid lineages, the fertility was uniformly high (greater than 90 percent). It has been hypothesized that certain combinations of genes from *H. annuus* and *H. petiolaris* consistently work better together and, hence, are always found together in surviving hybrids. This study has been characterized as "a first-ever re-creation of a new species."

Sterile Hybrids May Become Widespread If They Are Able to Reproduce Asexually

Even if hybrids are sterile, as in the horsetail hybrid *Equisetum* × *ferrissii*, they may become widespread providing they are able to reproduce asexually (Figure 12–21). In some groups of plants, sexual reproduction is combined with frequent asexual reproduction, so that recombination occurs but successful genotypes can be multiplied exactly (see Figure 9–12).

An outstanding example of such a system is the extremely variable Kentucky bluegrass (*Poa pratensis*), which in one form or another occurs throughout the cooler portions of the Northern Hemisphere. Occasional hybridization with a whole series of related species has produced hundreds of distinct races of this grass, each characterized by a form of asexual reproduction called **apomixis,** in which seeds are formed but they contain embryos that are produced independent of fertilization. Consequently, the embryos are genetically identical to

12–21

*One of the most abundant and vigorous of the horsetails (see Figure 19–25) found in North America is **(a)** Equisetum × ferrissii, a completely sterile hybrid of Equisetum hyemale and Equisetum laevigatum. Horsetails propagate readily from small fragments of underground stems, and the hybrid maintains itself over its wide range through such vegetative propagation. **(b)** Range of E. × ferrissii and those of its parental species.*

(a)

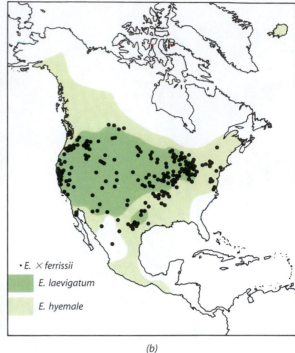

- *E. × ferrissii*
  E. laevigatum
  E. hyemale

(b)

the parent. Apomixis occurs in the ovule or immature seed, with the apomictic embryo being formed by one of two different pathways, depending upon the species. Of the angiosperms, over 300 species from more than 35 families have been described as apomictic. Among them are the *Poaceae* (grasses), *Asteraceae* (composites), and *Rosaceae* (rose family).

In apomictic species, or in species with well-developed vegetative reproduction, the individual strains may be particularly successful in specific habitats. In addition, such asexually propagated strains do not require **outcrossing** (cross-pollination between individuals of the same species) and therefore often do well in environments such as high mountains, where pollination by insects may be uncertain.

Maintaining Reproductive Isolation

Once speciation has occurred, the now-separate species can live together without interbreeding, despite the fact that some are so similar phenotypically that only an expert can tell them apart. What factors operate to maintain this **reproductive,** or **genetic, isolation** of closely related species?

Isolating mechanisms may be conveniently divided into two categories: **prezygotic mechanisms** (premating mechanisms), which prevent the formation of hybrid zygotes, and **postzygotic mechanisms** (postmating mechanisms), which prevent or limit gene exchange even after hybrid zygotes have been formed. One of the most significant things about postzygotic isolating mechanisms is that, in nature, they are rarely tested. The prezygotic mechanisms alone usually prevent any hybridization.

One type of prezygotic mechanism in plants is microhabitat isolation. For example, both scarlet oak (*Quercus coccinea*) and black oak (*Quercus velutina*) occur through-

out the eastern United States (Figure 12–22), and the two species, which are wind-pollinated, form fertile hybrids readily in cultivation. Nevertheless, hybrids between these two species are rare in nature. In general, scarlet oaks are found in relatively moist, low areas with acidic soil, whereas black oaks are found in drier, well-drained habitats. Although these two species occur in the same general area, their separation in these two different habitats virtually precludes cross-fertilization. Only where the environment has been disturbed—as by burning or cutting of the trees—do hybrids become more common.

Among other prezygotic mechanisms that can prevent the formation of hybrids between plant species that occur together are seasonal differences in time of flowering. If two species do not flower together, they will not hybridize in nature even when they grow side by side. Alternatively, they may be pollinated by different kinds of insects or other animals, and pollen may rarely be transferred between them. They will hybridize only when a pollinator mistakenly visits the "wrong" flower.

The Origin of Major Groups of Organisms

As knowledge about the ways in which species may originate has become better developed, evolutionary biologists have turned their attention to the origin of genera and higher taxonomic groups of organisms. As suggested by the essay on adaptive radiation (pages 250–251), genera may originate through the same kinds of evolutionary processes that are responsible for the origin of species. If a particular species has a distinctive adaptation to a habitat that is very different from the habitat in which its progenitor grows, the adapted species may come to be very different from that progeni-

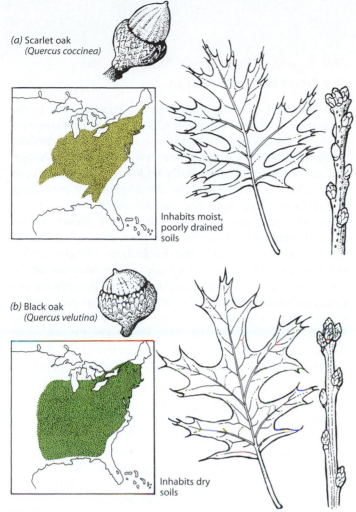

(a) Scarlet oak
(Quercus coccinea)

Inhabits moist, poorly drained soils

(b) Black oak
(Quercus velutina)

Inhabits dry soils

12–22

*Microhabitat isolation, a prezygotic mechanism that prevents inbreeding. Although these two species of oaks occur throughout the eastern United States, they are isolated reproductively, largely by their specific habitat requirements. **(a)** Scarlet oak* (Quercus coccinea) *inhabits relatively moist, poorly drained acidic soils, and **(b)** black oak* (Quercus velutina) *inhabits drier, well-drained soils.*

tor. It may gradually give rise to many new species and come to constitute a novel evolutionary line, a process called phyletic change. Gradually, it may become so distinct that this line may be classified as a new genus, family, or even class of organisms. (The levels of classification, or taxonomic groups, are discussed in Chapter 13.) Accordingly, no special mechanisms would be necessary to account for the origin of taxonomic groups above the level of species, that is, for **macroevolution**—only discontinuity in habitat, range, or way of life, and the accumulation of many small changes in the frequencies of alleles in gene pools. This is the **gradualism model** of evolution.

Although the fossil record documents many impor-

tant stages in evolutionary history, there are numerous gaps, and gradual transitions of fossil forms are rarely found. Instead, new forms representing new species appear rather suddenly (in geologic terms) in strata, apparently persist unchanged for their tenure on Earth, and then disappear from the rocks as suddenly as they appeared. For many years this discrepancy between the model of slow phyletic change and the poor documentation of such change in much of the fossil record was attributed to the imperfection of the record itself. Darwin, in *On the Origin of Species*, noted that the geologic record is ". . . a history of the world imperfectly kept, and written in a changing dialect; of this history we possess the last volume alone, relating only to two or three countries. Of this volume, only here and there a short chapter has been preserved; and of each page, only here and there a few lines."

In 1972, two young scientists, Niles Eldredge and Stephen Jay Gould, ventured the proposal that perhaps the fossil record is not so imperfect after all. Both Eldredge and Gould had backgrounds in geology and invertebrate paleontology, and both were impressed with the fact that there was very little evidence for a gradual phyletic change in the fossil species they studied.

Typically a species would appear "abruptly" in fossil-bearing strata, last 5 million to 10 million years, and disappear, apparently not much different from when it first appeared. Another species, related but distinctly different, would take its place, persist with little change, and disappear "abruptly." Suppose, Eldredge and Gould argued, that these long periods of little or no change, followed by what appear to be gaps in the fossil record, are not flaws in the record but *are* the record, the evidence of what really happens.

Eldredge and Gould proposed that species undergo most of their morphological modification as they first diverge from their progenitors, and then change little even as they give rise to additional species. In other words, long periods of gradual change or of no change at all (periods of equilibrium) are punctuated by periods of rapid change, that is, of rapid speciation. This theory is called the **punctuated equilibrium model** of evolution.

How could new species make such "sudden" appearances? Eldredge and Gould turned to allopatric speciation for the answer. If new species formed principally in small populations isolated from parent populations and occurred rapidly (in thousands rather than millions of years), and if the new species then outcompeted the old species, taking over their geographic range, the resulting fossil pattern would be the one observed.

The punctuated equilibrium model has stimulated a vigorous and continuing debate among biologists, a reexamination of evolutionary mechanisms as currently understood, and a reappraisal of the evidence. Perhaps populations change more rapidly at some times than at others, particularly in periods of environmental stress.

Regularly, new studies are published in support of gradualism or of punctuated equilibrium, but a consensus on either model has not been achieved. These activities have at times been misinterpreted as a sign that Darwin's theory is "in trouble." In fact, they indicate that evolutionary biology is alive and well and that scientists are doing what they are supposed to be doing—asking questions. Darwin, we think, would have been delighted.

Summary

Darwin Proposed a Theory of Evolution by Natural Selection

Charles Darwin was not the first to propose a theory of evolution, but his theory differed from others in that it envisioned evolution as a two-part process, depending upon (1) the existence in nature of inheritable variations among organisms and (2) the process of natural selection by which some organisms, by virtue of their inheritable variations, leave more surviving progeny than others. Darwin's theory is regarded as the greatest unifying principle in biology.

Population Genetics Is the Study of Gene Pools

Population genetics is a synthesis of the Darwinian theory of evolution with the principles of Mendelian genetics. For the population geneticist, a population is an interbreeding group of organisms, defined and united by its gene pool (the sum of all the alleles of all the genes of all the individuals in the population). Evolution is the result of accumulated changes in the composition of the gene pool.

The Hardy–Weinberg Law States That in an Ideal Population, the Frequency of Alleles Will Not Change over Time

The Hardy–Weinberg law describes the steady state in allele and genotype frequencies that would exist in an ideal, nonevolving population, in which five conditions are met: (1) no mutation, (2) isolation from other populations, (3) large population size, (4) random mating, and (5) no natural selection. The Hardy–Weinberg equilibrium demonstrates that the genetic recombination that results from meiosis and fertilization cannot, in itself, change the frequencies of alleles in the gene pool. The mathematical expression of the Hardy–Weinberg equilibrium provides a quantitative method for determining the extent and direction of change in allele and genotype frequencies.

Five Agents Cause Gene Frequencies in a Gene Pool to Change

The principal agent of change in the composition of the gene pool is natural selection. Other agents of change include mutation, gene flow, genetic drift, and nonrandom mating. Mutations provide the raw material for change, but mutation rates are usually so low that mutations, in themselves, do not determine the direction of evolutionary change. Gene flow, the movement of alleles into or out of the gene pool, may introduce new alleles or alter the proportions of alleles already present. It often has the effect of counteracting natural selection. Genetic drift is the phenomenon in which certain alleles increase or decrease in frequency, and sometimes even disappear, as a result of chance events. Circumstances that can lead to genetic drift, which is most likely to occur in small populations, include the founder effect and the bottleneck effect. Nonrandom mating causes changes in the proportions of genotypes but may or may not affect allele frequencies.

Various Processes Preserve and Promote Variability

Sexual reproduction is the most important factor promoting genetic variability in populations. Mechanisms that favor outbreeding further promote variability, including self-sterility alleles in plants. Variability is preserved by diploidy, which shelters rare, recessive alleles from selection. In cases of heterozygote advantage, the heterozygote is selected over either homozygote, thus maintaining both the recessive and the dominant alleles in the population. One of the best examples of heterozygote advantage in plants is hybrid vigor.

Natural Selection Acts on the Phenotype, Not the Genotype

Only the phenotype is accessible to selection. Similar phenotypes can result from very different combinations of alleles. Because of epistasis and pleiotropy, single alleles cannot be selected in isolation. Selection affects the whole genotype.

The Result of Natural Selection Is the Adaptation of Populations to Their Environment

Evidence of adaptation to the physical environment can be seen in gradual variations that follow a geographic distribution (cline) and in distinct groups of phenotypes (ecotypes) of the same species occupying different habitats. Adaptation to the biological environment results from the selective forces exerted by interacting species of organisms on each other (coevolution).

Species Are Usually Defined on the Basis of Genetic Isolation

A species is commonly defined as a group of natural populations whose members can interbreed with one another but cannot (or at least usually do not) interbreed with members of other such groups. In practice, most species are generally identified purely on an assessment of their morphological, or structural, distinctness.

In order for speciation—the formation of new species—to occur, populations that formerly shared a common gene pool must be reproductively isolated from one another and subsequently subjected to different selection pressures.

Allopatric Speciation Involves the Geographic Separation of Populations, while Sympatric Speciation Occurs among Organisms Living Together

Two principal modes of speciation are recognized, allopatric ("other country") and sympatric ("same country"). Allopatric speciation occurs in geographically isolated populations. Islands are frequently sites for sudden diversification and the development of new species from a common ancestor, a pattern of speciation that is called adaptive radiation. Sympatric speciation, which does not require geographic isolation, occurs principally in plants through polyploidy, often coupled with hybridization. Hybrid populations derived from two species are common in plants, especially trees and shrubs. Even if hybrids are sterile, they may become widespread by asexual means of reproduction, including apomixis, in which seeds are formed but with embryos that are produced independent of fertilization.

The Gradualism and Punctuated Equilibrium Models Are Used to Explain Evolution of Major Groups of Organisms

Carried out over time, the same processes responsible for the evolution of species may give rise to genera and other major groups. This is the gradualism model of evolution. Paleontologists have presented evidence for an additional pattern of evolution known as punctuated equilibrium. They propose that new species are formed during bursts of rapid speciation among small isolated populations, that the new species outcompete many of the existing species (which become extinct), and that, in turn, the new species abruptly become extinct.

Selected Key Terms

adaptation   p. 245

adaptive radiation   p. 249

allopatric speciation   p. 249

allopolyploidy   p. 252

apomixis   p. 255

artificial selection   p. 238

autopolyploidy   p. 252

bottleneck effect   p. 242

cline   p. 246

coevolution   p. 247

Darwin's theory   p. 237

developmental plasticity   p. 246

ecotype   p. 246

founder effect   p. 242

gene flow   p. 241

gene pool   p. 239

genetic drift   p. 241

gradualism model of evolution   p. 257

Hardy–Weinberg law   p. 240

hybrid vigor   p. 243

macroevolution   p. 257

microevolution   p. 240

mutations   p. 240

natural selection   p. 236

polyploidy   p. 249

population genetics   p. 239

punctuated equilibrium model of evolution   p. 257

recombination speciation   p. 255

reproductive (genetic) isolation   p. 256

speciation   p. 248

species   p. 248

sympatric speciation   p. 249

Questions

1. Explain the influence of Thomas Malthus and Charles Lyell on the development of Darwin's theory of evolution.

2. How did Darwin's concept of evolution differ primarily from that of his predecessors? What was the major weakness in Darwin's theory?

3. What is meant by developmental plasticity? Why is developmental plasticity much greater in plants than in animals?

4. Distinguish between each of the following: cline/ecotype; microevolution/macroevolution; allopatric speciation/sympatric speciation; autopolyploidy/allopolyploidy.

5. Define genetic isolation. Why is it such an important factor in speciation?

6. When two distinct species of plants hybridize, their hybrid offspring usually are sterile. Why? Explain how such a sterile hybrid can give rise to a new species capable of sexual reproduction.

Diversity

SECTION 4

Yellow lady's slipper (Cypripedium calceo-lus), *an orchid, blooming in early summer in a Wisconsin wood. Orchidaceae is the largest family of flowering plants, with some 20,000* species, the majority of which grow in the tropics. As is true of most orchids, especially those of temperate climates, the lady's slipper is threatened with extinction.

Systematics: The Science of Biological Diversity

OVERVIEW

When you pause to examine a flower, shrub, or tree, you may very well wonder: "What's the name of that plant?" Such a question—arising out of a simple curiosity to identify organisms in the world around us—has intrigued people as far back as Aristotle, and no doubt before. In this chapter, you will find that the seemingly trivial process of naming an organism is, in fact, part of a highly organized system for establishing genetic relationships and identifying evolutionary trends.

Because people commonly name plants and other organisms in the language of their country, there will be almost as many names for the same organism as there are languages. For botanists and other biologists, this multitude of names represents a significant barrier to the sharing of information. Therefore, in addition to the "common names" that vary from country to country, each organism also has a "scientific name"—a two-word Latin name that identifies it precisely to the world at large.

Not only does a scientific name provide a universal "identity card" for an organism, but it also gives clues about the relationships of one organism to another. Thus, after describing the rules and rationale behind the scientific naming of organisms, we broaden the scope to discuss the different characteristics used for classifying organisms into groups. This is followed by an overview of the major groups of organisms and the hypothesized mechanism by which eukaryotic organisms evolved from prokaryotes.

CHECKPOINTS

By the time you finish reading this chapter, you should be able to answer the following questions:

1. What is the binomial system of nomenclature?

2. Why is the term "hierarchical" used to describe taxonomic categories, and what are the principal categories between the levels of species and kingdom?

3. What is cladistic analysis, and how is a cladogram constructed?

4. What evidence is there for the existence of the three major domains, or groups, of living organisms?

5. What are the four kingdoms of eukaryotes, and what are the major identifying characteristics of each?

In the previous section, we considered the mechanisms by which evolutionary change occurs. Now we turn our attention to the products of evolution, that is, to the multitude of different kinds, or species, of living organisms—estimated at over 30 million—that share our biosphere today. The scientific study of biological diversity and its evolutionary history is called **systematics.**

Taxonomy and Hierarchical Classification

An important aspect of systematics is **taxonomy**—the identifying, naming, and classifying of species. Modern biological classification began with the eighteenth-century Swedish naturalist Carl Linnaeus (Figure 13–1),

13–1

The eighteenth-century Swedish professor, physician, and naturalist Carl Linnaeus (1707–1778), who devised the binomial system for naming species of organisms and established the major categories that are used in the hierarchical system of biological classification. When he was 25, Linnaeus spent five months exploring Lapland for the Swedish Academy of Sciences; he is shown here wearing his Lapland collector's outfit.

whose ambition was to name and describe all of the known kinds of plants, animals, and minerals. In 1753, Linnaeus published a two-volume work entitled *Species Plantarum* ("The Kinds of Plants") in which he described each species in Latin by a sentence limited to twelve words. He regarded these descriptive Latin phrase names, or **polynomials,** as the proper names for the species, but in adding an important innovation devised earlier by Caspar Bauhin (1560–1624), Linnaeus made permanent the **binomial** ("two-term") **system** of nomenclature. In the margin of the *Species Plantarum,* next to the "proper" polynomial name of each species, he wrote a single word. This word, when combined with the first word of the polynomial—the **genus** (plural: genera)—formed a convenient "shorthand" designation for the species. For example, for catnip, which was formally named *Nepeta floribus interrupte spicatus pedunculatis* (meaning "*Nepeta* with flowers in an interrupted pedunculate spike"), Linnaeus wrote the word "cataria" (meaning "cat-associated") in the margin of the text, thus calling attention to a familiar attribute of the plant. He and his contemporaries soon began calling this species *Nepeta cataria,* and this Latin name is still used for this species today.

The convenience of this new system was obvious, and the cumbersome polynomial names were soon replaced by binomial names. The earliest binomial name applied to a particular species has priority over other names applied to the same species later. The rules governing the application of scientific names to plants are embodied in the *International Code of Botanical Nomenclature.* Codes also exist for animals (*International Code of Zoological Nomenclature*) and microbes (*International Code of Nomenclature of Bacteria*).

The Species Name Consists of the Genus Name Plus the Specific Epithet

A species name consists of two parts. The first is the name of the genus—also called the generic name—and the second is the **specific epithet.** For catnip, the generic name is *Nepeta,* the specific epithet is *cataria,* and the species name is *Nepeta cataria.*

A generic name may be written alone when one is referring to the entire group of species making up that genus. For instance, Figure 13–2 shows three species of the violet genus, *Viola.* A specific epithet is meaningless, however, when written alone. The specific epithet *biennis,* for example, is used in conjunction with dozens of different generic names. *Artemisia biennis,* a kind of wormwood, and *Lactuca biennis,* a species of wild lettuce, are two very different members of the sunflower family, and *Oenothera biennis,* an evening primrose, belongs to a different family altogether. Because of the danger of confusing names, a specific epithet is always preceded by the name or the initial letter of the genus that includes it:

(a)

(b)

(c)

13–2

Three members of the violet genus. (a) The common blue violet, Viola papilionaceae, *which grows in temperate regions of eastern North America as far west as the Great Lakes. (b)* Viola tricolor, *a yellow-flowered violet. (c) Pansy,* Viola tricolor var. hortensis, *an*

annual, cultivated strain of a mostly perennial species that is native to western Europe. These photographs indicate the kinds of differences in flower color and size, leaf shape and margin, and other features that distinguish the

species of this genus, even though there is an overall similarity between all of them. There are about 500 species of the genus Viola; *most of them are found in temperate regions of the Northern Hemisphere.*

for example, *Oenothera biennis* or *O. biennis*. Names of genera and species are printed in italics or are underlined when written or typed.

If a species is discovered to have been placed in the wrong genus initially and must then be transferred to another genus, the specific epithet moves with it to the new genus. If there is already a species in that genus that has that particular specific epithet, however, an alternative name must be found.

Each species has a **type specimen,** usually a dried plant specimen housed in a museum or herbarium, which is designated either by the person who originally named that species or by a subsequent author if the original author failed to do so (Figure 13–3). The type specimen serves as a basis for comparison with other specimens in determining whether they are members of the same species or not.

13–3

Type specimen of the angiosperm Podandrogyne formosa *(family Capparidaceae), found in Costa Rica and western Panama. This specimen was collected by Theodore S. Cochrane and described by him in a paper published in the journal* Britonnia *(volume 30, pages 405–410, 1978).*

The Members of a Species May Be Grouped into Subspecies or Varieties

Certain species consist of two or more subspecies or varieties (some botanists consider varieties to be sub-categories of subspecies, and others regard them as equivalent). All of the members of a subspecies or variety of a given species resemble one another and share one or more features not present in other subspecies or varieties of that species. As a result of these subdivisions, although the binomial name is still the basis of classification, the names of some plants and animals may consist of three parts. The subspecies or variety that includes the type specimen of the species repeats the name of the species, and all the names are written in italics or underlined. Thus the peach tree is *Prunus persica* var. *persica*, whereas the nectarine is *Prunus persica* var. *nectarina*. The repeated *persica* in the name of the peach tree tells us that the type specimen of the species *P. persica* belongs to this variety, abbreviated "var."

Organisms Are Grouped into Broader Taxonomic Categories Arranged in a Hierarchy

Linnaeus (and earlier scientists) recognized three king-doms—plant, animal, and mineral—and until very recently the **kingdom** was the most inclusive unit used in biological classification. In addition, several hierarchical taxonomic categories were added between the levels of genus and kingdom: genera were grouped into **families,** families into **orders,** and orders into **classes.** The Swiss-French botanist Augustin-Pyramus de Candolle (1778–1841), who invented the word "taxonomy," added another category—**division**—to designate groups of classes in the plant kingdom. Hence, the divisions became the largest inclusive groups of the plant kingdom. In this hierarchical system—that is, of groups within groups, with each group ranked at a particular level—the taxonomic group at any level is called a **taxon** (plural: taxa). The level at which it is ranked is called a **category.** For example, genus and species are categories, and *Prunus* and *Prunus persica* are taxa within those categories.

At the XV International Botanical Congress in 1993, the *International Code of Botanical Nomenclature* made the term **phylum** (plural: phyla) nomenclaturally equivalent to division. "Phylum" has long been used by zoologists for groups of classes. In addition, the Code recommended the practice of italicizing all taxonomic names, not just the names of genera and species. We have adopted that practice and the use of the term "phylum" instead of "division" in this edition. All taxa at the genus level or higher are capitalized.

Regularities in the form of the names for the different taxa make it possible to recognize them as names at that level. For example, names of plant families end in *-aceae*,

with a very few exceptions. Older names are allowed as alternatives for a few families, such as *Fabaceae*, the pea family, which may also be called by the older name, *Leguminosae*; *Apiaceae*, the parsley family, also known as *Umbelliferae*; and *Asteraceae*, the sunflower family, also known as *Compositae*. Names of plant orders end in *-ales*.

Sample classifications of maize (*Zea mays*) and the commonly cultivated edible mushroom (*Agaricus bisporus*) are given in Table 13–1.

Classification and Phylogeny

As mentioned previously, taxonomy is only one aspect of systematics. For Linnaeus and his immediate successors, the goal of taxonomy was the revelation of the grand, unchanging design of creation. After publication of Darwin's *On the Origin of Species* in 1859, however, differences and similarities among organisms came to be

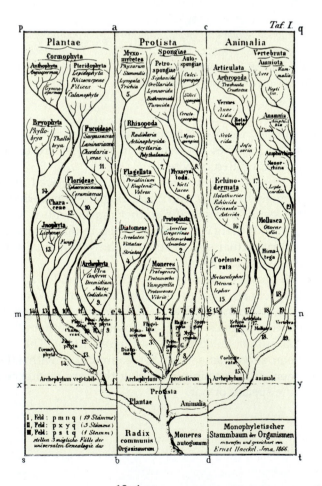

13–4

One of the earliest phylogenetic trees appeared in The History of Creation, *by the German naturalist Ernst Haeckel, published in 1866. The terminology reflects the system of classification in use at the time.*

TABLE 13-1

Biological Classification. Notice how much you can tell about an organism when you know its place in the system. The descriptions here do not define the various categories but tell you something about their characteristics. The kingdoms *Plantae* and *Fungi* belong to the domain *Eukarya*.

Category	Taxon	Description
Maize		
Kingdom	*Plantae*	Organisms that are primarily terrestrial, with chlorophylls *a* and *b* contained in chloroplasts, spores enclosed in sporopollenin (a tough wall substance), and nutritionally dependent multicellular embryos.
Phylum	*Anthophyta*	Vascular plants with seeds and flowers; ovules enclosed in an ovary, pollination indirect; the angiosperms.
Class	*Monocotyledones*	Embryo with one cotyledon; flower parts usually in threes; many scattered vascular bundles in the stem; the monocots.
Order	*Commelinales*	Monocots with fibrous leaves; reduction and fusion in flower parts.
Family	*Poaceae*	Hollow-stemmed monocots with reduced greenish flowers; fruit a specialized achene (caryopsis); the grasses.
Genus	*Zea*	Robust grasses with separate staminate and carpellate flower clusters; caryopsis fleshy.
Species	*Zea mays*	Maize, or corn.
Edible Mushroom		
Kingdom	*Fungi*	Nonmotile, multinucleate, heterotrophic, absorptive organisms in which chitin predominates in the cell walls.
Phylum	*Basidiomycota*	Dikaryotic fungi that form a basidium bearing four spores (basidiospores); the *Basidiomycetes, Teliomycetes,* and *Ustomycetes*.
Class	*Basidiomycetes*	Fungi that produce basidiomata, or "fruiting bodies," and club-shaped, aseptate basidia that line gills or pores; the hymenomycetes.
Order	*Agaricales*	Fleshy fungi with radiating gills or pores.
Family	*Agaricaceae*	*Agaricales* with gills.
Genus	*Agaricus*	Dark-spored soft fungi with a central stalk and gills free from the stalk.
Species	*Agaricus bisporus*	The common edible mushroom.

seen as products of their evolutionary history, or **phylogeny.** Biologists now wanted classifications to be not only informative and useful but also an accurate reflection of the evolutionary relationships among organisms. The evolutionary relationships among organisms have often been diagrammed as **phylogenetic trees,** which depict the genealogic relationships between taxa as hypothesized by a particular investigator (Figure 13–4). Like other hypotheses, certain aspects of phylogenetic trees can be tested and revised as necessary. Examples of

such testing would be detailed study of the fossil record and examination of the structural and molecular characteristics of living organisms.

In a classification scheme that accurately reflects phylogeny, every taxon is, ideally, **monophyletic.** This means that the members of a taxon, at whatever categorical level, be it genus, family, or order, should all be descendants of a single common ancestral species. Thus, a genus should consist of all species descended from the most recent common ancestor—and only of species de-

Convergent Evolution

Comparable selective forces, acting on plants growing in similar habitats but different parts of the world, often cause totally unrelated species to assume a similar appearance. The process by which this happens is known as **convergent evolution**.

Let us consider some of the adaptive characteristics of plants growing in desert environments—fleshy, columnar stems (which provide the capacity for water storage), protective spines, and reduced leaves. Three fundamentally different families of flowering plants—the spurge fam-

ily (Euphorbiaceae), the cactus family (Cactaceae), and the milkweed family (Asclepiadaceae)—have members that have evolved these features. The cactuslike representatives of the spurge and milkweed families shown here evolved from leafy plants that look quite different from one another.

Native cacti occur (with the exception of one species) exclusively in the New World. The comparably fleshy members of the spurge and milkweed families occur mainly in desert regions in Asia and espe-

cially Africa, where they play an ecological role similar to that of the New World cacti.

Although the plants shown here—(a) *Euphorbia*, a member of the spurge family; (b) *Echinocereus*, a cactus; (c) *Hoodia*, a fleshy milkweed—have CAM photosynthesis, all three are related to and derived from plants that have only C_3 photosynthesis (page 147). This indicates that the physiological adaptations involved in CAM photosynthesis also arose as a result of convergent evolution.

(a)

(b)

(c)

scended from that ancestor. Similarly, a family should consist of all genera descended from a more distant common ancestor—and only of genera descended from that ancestor.

Although this ideal, which results in **natural taxa,** sounds relatively straightforward, it is often difficult to attain. In many cases, biologists do not know enough about the evolutionary history of the organisms to establish taxa that are, with a reasonable degree of certainty, monophyletic. However, where relationships are unknown or uncertain, it may be more practical to create an **artificial taxon.** Thus, as we shall see, some widely accepted taxa contain members descended from more than one ancestral line. Such taxa are said to be **polyphyletic.** Other taxa exclude one or more descendants of a common ancestor. These taxa are said to be **paraphyletic.**

Homologous Features Have a Common Origin, and Analogous Features Have a Common Function but Different Evolutionary Origins

Systematics is, to a great extent, a comparative science. It groups organisms into taxa from the categorical levels of genus through phylum based on similarities in structure

and other characters. From Aristotle on, however, biologists have recognized that superficial similarities are not useful criteria for taxonomic decisions. To take a simple example, birds and insects should not be grouped together simply because both have wings. A wingless insect (such as an ant) is still an insect, and a flightless bird (such as the kiwi) is still a bird.

A key question in systematics is the origin of a similarity or difference. Does the similarity of a particular feature reflect inheritance from a common ancestor, or does it reflect adaptation to similar environments by organisms that do not share a common ancestor? A related question arises concerning differences between organisms: Does a difference reflect separate evolutionary histories, or does it reflect instead the adaptations of closely related organisms to very different environments? As we shall see in later chapters, foliage leaves, cotyledons, bud scales, and floral parts have quite different functions and appearances, but all are modifications of the same type of organ, namely, the leaf. Such structures, which have a common origin but not necessarily a common function, are said to be **homologous** (from the Greek *homologia*, meaning "agreement"). These are the features upon which evolutionary classification systems are ideally constructed.

By contrast, other structures, which may have a similar function and superficial appearance, have an entirely different evolutionary background. Such structures are said to be **analogous** and are the result of convergent evolution (see the essay on the facing page). Thus the wings of a bird and those of an insect are analogous, not homologous. Similarly, the spine of a cactus (a modified leaf) and the thorn of a hawthorn (a modified stem) are analogous, not homologous. Distinguishing between homology and analogy is seldom so simple, and generally requires detailed comparison as well as evidence from other features of the organisms under study.

Methods of Classification

The Traditional Method Is Based on a Comparison of Outward Similarities

Traditionally, the classification of a recently discovered organism and its phylogenetic relationship to other organisms has been assessed on the basis of its overall outward similarities to other members of that taxon. Phylogenetic trees constructed by traditional methods rarely include detailed considerations of comparative information. Instead, they reflect the relatively intuitive consideration and weighing of a large number of factors. The resulting trees often contain information about both the sequence in which branchings occurred and the extent of the subsequent biological changes (Figure 13–5). Although this approach has produced many useful results, it is based to a large extent on the investigator's opinion of the relative importance of the various factors being taken into account in determining the classification. Therefore, it is not surprising that very different classifications have sometimes been proposed for the same groups of organisms.

The Cladistic Method Is Based on Phylogeny

The most widely used method of classifying organisms today is known as **cladistics,** or **phylogenetic analysis,** because it explicitly seeks to understand phylogenetic relationships. The approach focuses on the branching of one lineage from another in the course of evolution. It attempts to identify monophyletic groups, or **clades,** which can be defined by the possession of unique features (sometimes called *shared derived character states*), as opposed to the possession of more widespread features. These widespread features can be interpreted as *preexisting*, or *ancestral, character states*. The two types of character states usually are distinguished from one another by comparison with one or more **outgroups,** that is, with closely related taxa outside the group that is being analyzed.

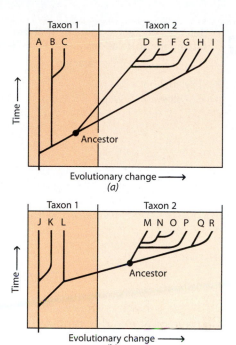

13–5

The evolutionary history of a group of related organisms can be represented by a phylogenetic tree constructed by using traditional methods. The vertical locations of the branching points indicate when particular taxa diverged from one another. The horizontal distances indicate how much the taxa have diverged, taking into account a number of different characteristics.

The two diagrams shown here represent the evolutionary histories of two different groups of taxa, labeled A through I in (a) and J through R in (b). In (a), the ancestor of taxa D through I is included in taxon 1 because of its close resemblance to taxa B and C. In (b), the ancestor of taxa M through R is placed in taxon 2, because of its close resemblance to taxon M. In each case, taxa 1 and 2 would themselves be members of a taxon at a higher categorical level, which would probably include other taxa as well.

The result of cladistic analysis is a **cladogram,** which provides a graphical representation of a working model, or hypothesis, of branching sequences. These hypotheses can then be tested by attempting to incorporate additional plants or characters that may or may not conform to the predictions of the model.

To see how a cladogram is constructed, let us consider four different groups of plants: mosses, ferns, pines, and oaks. For each of the plant groups, we have selected four homologous characters to be analyzed

TABLE 13–2	Selected Characters Used in Analyzing the Phylogenetic Relationships of Four Plant Taxa			
	Characters*			
Taxon	Xylem and Phloem	Wood	Seeds	Flowers
Mosses	−	−	−	−
Ferns	+	−	−	−
Pines	+	+	+	−
Oaks	+	+	+	+

*The character state "present" (+) is the derived condition; the character state "absent" (−) is the ancestral condition.

pothesized to form a monophyletic group. Figure 13–6b shows how further resolution is obtained as information about other features is added.

How does one interpret the cladogram in Figure 13–6b? To begin, note that cladograms do not indicate that one group gave rise to another, as in many phylogenetic trees constructed by the traditional method. Rather, they imply that groups terminating adjacent branches (the branch points are called **nodes**) shared a common ancestor. The cladogram of Figure 13–6b tells us that oaks shared a more recent common ancestor with pines than with ferns, and are more closely related to pines than to ferns. The relative positions of plants on the cladogram indicate their relative times of divergence.

A fundamental principle of cladistics is that a cladogram should be constructed in the simplest, least complicated, and most efficient way. This principle is called the **principle of parsimony.** When conflicting cladograms are constructed from the data at hand, the one with the greatest number of statements of homology and the fewest of analogy is preferred.

(Table 13–2). To keep matters simple, the characters are considered to have only two different states: present (+) and absent (−).

Through their possession of embryos, mosses are known to be related to the other three plant groups, which also have embryos. However, mosses lack many features that the other three plants share—for example, xylem and phloem, and many characters not shown here. The mosses can be used as the outgroup and can be considered to have diverged earlier than the other taxa from a common ancestor. Accordingly, the mosses can be used to determine whether features shared among ferns, pines, and oaks can potentially be used to define a clade. For example, seeds are not present in mosses and can therefore be hypothesized as a potential shared derived feature that would support uniting pines and oaks as a monophyletic group. Applying this argument to our few characters results in the "absent" character state being consistently recognized as the ancestral condition and a "present" character state as the derived condition.

Figure 13–6a shows how one might sketch a cladogram based on the presence or absence of the vascular tissues xylem and phloem. Inasmuch as the ferns, pines, and oaks all have xylem and phloem, they can be hy-

Molecular Systematics

Until relatively recently, classification by any methodology was based largely on comparative morphology and anatomy. During the past decade, however, plant systematics has been revolutionized by the application of molecular techniques. The techniques most widely used are those for determining both the sequence of amino acids in proteins and of nucleotides in nucleic acids—sequences that are genetically determined. Molecular data are different from data obtained from traditional sources in several important ways: in particular, they are easier to quantify, potentially provide many more characters for phylogenetic analysis, and allow comparison of organisms that are morphologically very different. With the development of molecular techniques, it has become possible to compare organisms at the most basic level—the gene. The drawbacks with molecular data are that they can rarely be obtained from fossils and that homologies are sometimes very difficult to assess.

13–6

Cladograms showing phylogenetic relationships between ferns, pines, and oaks, indicating the shared characters that support the patterns of relationships. (a) A cladogram based on the presence or absence of xylem and phloem. (b) Further resolution of the relationships, based on additional information regarding the presence or absence of wood, seeds, and flowers.

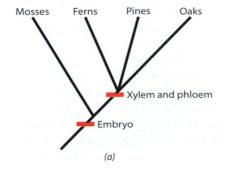

(a)

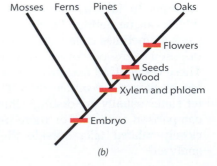

(b)

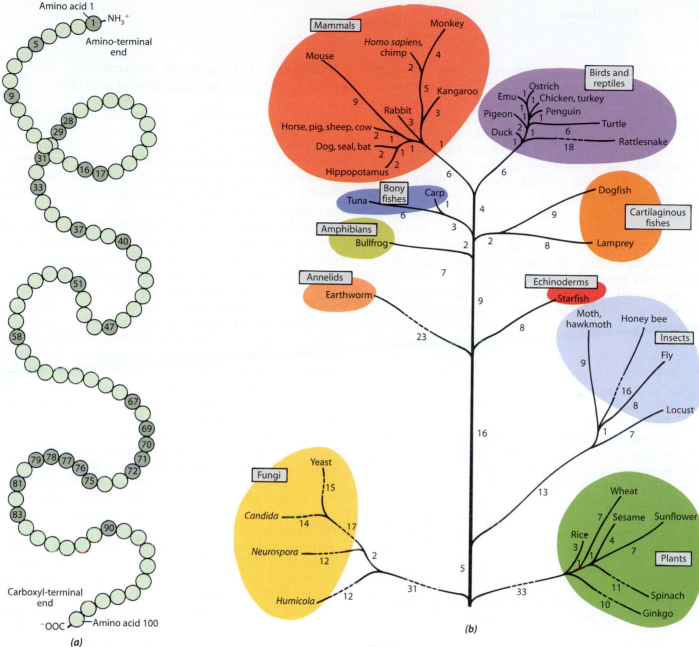

(a)

(b)

13-7

The use of amino acid sequences of homologous proteins to determine evolutionary relationships. This method assumes that the greater the number of amino acid differences between homologous proteins of any two organisms, the more distant their evolutionary relationship. Conversely, the fewer the differences, the closer their relationship.

One of the most frequently sequenced proteins is cytochrome c, a carrier in the electron transport chain. **(a)** In cytochrome c molecules from the more than 60 species that have been studied thus far, 27 of the amino acids are identical (dark blue-green). **(b)** The main branches of an evolutionary tree based on comparisons of amino acid sequences of cytochrome c molecules. The numbers indicate the number of amino acids by which each cytochrome c differs from the cytochrome c at the nearest branch point. Dashes indicate that a line has been shortened and is therefore not to scale. Although based on comparisons of a single type of protein molecule, this tree is in fairly good agreement with evolutionary trees constructed by more conventional means, which take into account a variety of data.

A Comparison of Amino Acid Sequences Provides a Molecular Clock

Among the first proteins to be analyzed in taxonomic studies was cytochrome c, one of the carriers of the electron transport chain (page 116). Cytochrome c molecules from a great variety of organisms were sequenced, making it possible to determine the number of amino acids by which the molecules of various organisms differ. The number of similarities and differences between the amino acid constituents of different organisms was then used to evaluate their evolutionary relationship: the smaller the number of differences, the closer the relationship between any two organisms. Figure 13-7 illustrates a phylogeny of eukaryotic organisms based on cytochrome c data. The results conform fairly well, but not perfectly, with phylogenies constructed by more traditional methods.

As data on protein variations accumulated, it became clear that although protein structure is a useful parameter of evolutionary relationships, there are difficulties in interpreting the results. Some biologists maintained that differences in protein structure resulted in functional differences among the molecules. Other biologists contended that amino acid changes occur regularly and randomly as the result of mutation—and that they do not represent the result of a selection process. These changes are seen as merely marking off the passage of time, like grains of sand trickling through an hourglass, or the decay of radioisotopes, or the ticking of a clock. From this viewpoint, the amino acid differences in the homologous proteins of different groups of organisms do not represent functional differences. Instead, they represent the differences in the number of amino acid substitutions that have occurred in the homologous proteins since the lineages branched apart. This viewpoint led to the concept of a **molecular clock,** which, simply stated, uses the rate at which proteins (or nucleic acids) shared by different groups of organisms changed over time as an indication of when those groups diverged from a common ancestor.

Although the use of homologous proteins to estimate evolutionary relationships has been largely abandoned, amino acid sequences from 57 different enzymes have recently been used to determine the divergence times of prokaryotes and eukaryotes. The results of this study indicate that prokaryotic organisms and eukaryotic organisms last shared a common ancestor about 2 billion years ago. These results conflict with an earlier assessment, based on ribosomal RNA sequences, that prokaryotes and eukaryotes evolved from a common ancestor over a relatively short time in the planet's history and were already in existence 3.5 billion years ago.

A Comparison of Nucleotide Sequences Provides Evidence for Three Domains of Life

Nucleic acid sequencing is technically far easier than protein sequencing, dealing as it does with only four different nucleotides, as compared with 20 different amino acids. Also, it assesses similarities and differences more sensitively, since changes in nucleotides may not be reflected in amino acids because of the many synonyms in the genetic code (page 209).

As the sequences of nucleic acids from a variety of species have been determined, the information has been entered into computer banks. It is therefore possible to make detailed comparisons among large numbers of taxa. Such comparisons have demonstrated the value of nucleic acid sequences in systematic studies. For example, analyzing the sequences of the small-subunit ribosomal RNA provided the first evidence that the living world is divided into three major groups, or domains— *Bacteria, Archaea,* and *Eukarya* (Figure 13–8). As discussed further in Chapter 14, the **Bacteria** are the prokaryotes considered to be true bacteria, and the **Archaea** are prokaryotes able to live in extreme environments. The *Eukarya* include all eukaryotes.

The first full sequencing of a genome—that is, the DNA—for a representative of the *Archaea, Methanococcus jannaschii,* has recently been completed (Figure 13–9). The results of this DNA sequencing further support the presence of three domains of life and indicate that the *Archaea* and *Eukarya* shared a common evolutionary pathway independent of the lineage of *Bacteria.* Direct sequencing of ribosomal RNA has also provided data for studying plant phylogeny and for testing hypotheses on the origin of the angiosperms (see Chapter 22).

The most comprehensive studies of seed plant phy-

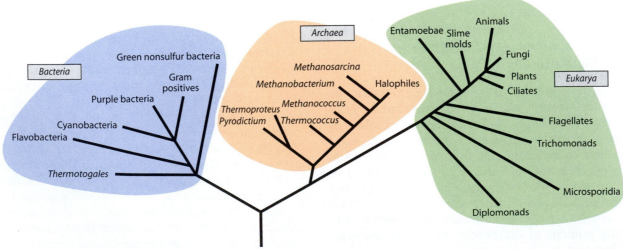

13–8

A universal evolutionary tree as determined by comparing sequences of ribosomal RNA. The data support the division of the living world into three domains, two of which *consist of prokaryotic organisms (Bacteria and Archaea) and one of eukaryotic organisms (Eukarya). The universal ancestor of all cells is found at the base of the tree. Animals,* *fungi, and plants each constitute a separate kingdom in the domain Eukarya. The remaining groups in the Eukarya belong to the kingdom Protista.*

logeny have been based on the variation in nucleotide sequences of the *rbc*L gene found in the DNA of the chloroplast. The *rbc*L gene, which codes for the large subunit of the Rubisco enzyme of the Calvin cycle (page 140), is especially well suited for analysis of such a broad group of plants. Not only is it a slow-evolving, single-copy gene, but it lacks introns and is large enough (about 1500 base pairs) to preserve a significant number of phylogenetically informative characters.

Molecular data alone may not provide the most accurate account of phylogenetic relationships. Some systematists think that all available evidence—molecular, morphological, anatomical, ultrastructural, developmental, and fossil—should be taken into consideration in assessing phylogenetic relationships.

The Major Groups of Organisms:
Bacteria, Archaea, and *Eukarya*

In Linnaeus's time, as we mentioned earlier, three kingdoms were recognized—animals, plants, and minerals—and until fairly recently it was common to classify every living thing as either an animal or a plant. Kingdom *Animalia* included those organisms that moved and ate things and whose bodies grew to a certain size and then stopped growing. Kingdom *Plantae* comprised all living things that did not move or eat and that grew indefinitely. Thus the fungi, algae, and bacteria, or prokaryotes, were grouped with the plants, and the protozoa—the one-celled organisms that ate and moved—were classified as animals. Lamarck, Cuvier, and most other eighteenth- and nineteenth-century biologists continued to place all organisms in one or the other of these kingdoms. This old division into plants and animals is still widely reflected in the organization of college textbooks, including this one. That is why, in addition to plants, we include algae, fungi, and prokaryotes in this text.

In the twentieth century, new data began to emerge. This was partly a result of improvements in the light microscope and, subsequently, the development of the electron microscope. It was also due, in part, to the application of biochemical techniques to studies of differences and similarities among organisms. As a result, the number of groups recognized as constituting different kingdoms increased. The new techniques revealed, for example, the fundamental differences between prokaryotic and eukaryotic cells. These differences were sufficiently great to warrant placing the prokaryotic organisms in a separate kingdom, *Monera*. Since then, as we have just noted, studies of ribosomal RNA sequences have revealed the two distinct lineages of prokaryotic organisms—*Bacteria* and *Archaea*—in addition to the distinct lineage of eukaryotic organisms—*Eukarya* (Figure 13–10).

Utilizing this molecular evidence, the system of classification adopted in this book recognizes the three **domains**, *Bacteria, Archaea,* and *Eukarya*. The protists, fungi, plants, and animals are now regarded as kingdoms within the domain *Eukarya*. Thus, the domain, not the kingdom, is the highest level of taxonomic categories. Table 13–3 summarizes some of the major differences that distinguish the three domains.

0.5 µm

13–9

Micrograph of Methanococcus jannaschii, *the first* Archaea *whose entire genome has been fully sequenced. Its genome consists of a circular chromosome made up of DNA. Completed in 1996, the sequencing showed that two-thirds of the organism's genes consist of sequences never seen before by researchers. Arrows indicate where the cell has ruptured, spilling its contents.*

TABLE 13–3	Some Major Distinguishing Features of the Three Domains of Life*		
Characteristic	Bacteria	Archaea	Eukarya
Cell type	Prokaryotic	Prokaryotic	Eukaryotic
Nuclear envelope	Absent	Absent	Present
Number of chromosomes	1	1	More than 1
Chromosome configuration	Circular	Circular	Linear
Organelles (mitochondria and plastids)	Absent	Absent	Present (in all but a few)
Cytoskeleton	Absent	Absent	Present
Chlorophyll-based photosynthesis	Yes	No	Yes

*Note that some features listed apply to only certain representatives of a certain domain.

13–10

*Representatives of the three domains. Electron micrographs of **(a)** a prokaryote, the cyanobacterium* Anabaena *(domain Bacteria); **(b)** a prokaryote, the archaea* Methanothermus fervidus *(domain Archaea); and **(c)** a eukaryotic cell, from the leaf of a sugarbeet* (Beta vulgaris) *(domain Eukarya). The cyanobacterium is a common inhabitant of ponds, while Methanothermus, which is adapted to high temperatures, grows optimally at 83° to 88°C. Note the greater complexity of the eukaryotic cell, with its conspicuous nucleus and chloroplasts, and its much larger size (note the scale markers).*

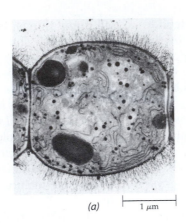

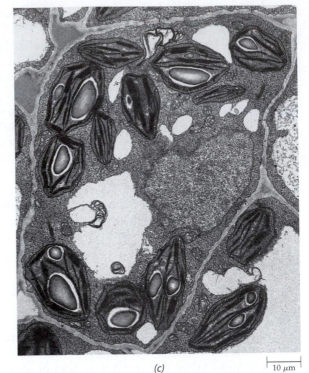

(a)   1 μm

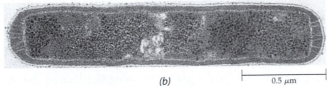

(b)   0.5 μm

(c)   10 μm

Origin of the Eukaryotes

One of the most momentous series of events to take place in the evolution of life on Earth was the transformation of relatively simple prokaryotic cells into intricately organized eukaryotic cells. You will recall from Chapter 3 that eukaryotic cells are typically much larger than prokaryotes, and their DNA, which is more highly structured, is enclosed within a nuclear envelope. In addition to an internal cytoskeleton, eukaryotic cells differ further from prokaryotic cells in their possession of mitochondria and, in plants and algae, chloroplasts that are about the size of a prokaryotic cell.

The Serial Endosymbiotic Theory Provides a Hypothesis for the Origin of Mitochondria and Chloroplasts

Both mitochondria and chloroplasts are believed to be the descendants of bacteria that were taken up and adopted by some ancient **host cell.** This concept for the origin of mitochondria and chloroplasts is known as the **serial endosymbiotic theory,** with the prokaryotic ancestors of mitochondria and chloroplasts as the **endosymbionts.** An endosymbiont is an organism that lives within another, dissimilar organism. The process by which eukaryotic cells originated is termed *serial* endosymbiosis because the events did not occur simultaneously—mitochondria are believed to pre-date chloroplasts.

The Endomembrane System Is Thought to Have Evolved from Portions of the Plasma Membrane   Endosymbiosis has had a profound influence upon diversification of the eukaryotes. Most experts believe that the process of establishing an endosymbiotic relationship was preceded by the evolution of some prokaryotic host cell into a primitive **phagocyte** (meaning "eating cell")—a cell capable of engulfing large particles such as bacteria (Figure 13–11). It is likely that the ancestral host cell was a wall-less heterotroph living in an environment that provided it with food. Such cells would need a flexible plasma membrane capable of enveloping bulky extracellular objects by folding inward. This endocytosis would be followed by the breakdown of the food particles within vacuoles derived from the plasma membrane. The plasma membrane would have been rendered flexible by the incorporation of sterols, and the development of a

13–11

*Diagrammatic representation of the origin of the photosynthetic eukaryotic cell of a heterotrophic prokaryote. **(a)** Most prokaryotes contain a rigid cell wall, so it is likely that an initial step in the transformation of a prokaryote to a eukaryotic cell was loss of the prokaryote's ability to form a cell wall. **(b), (c)** This free-living, naked form now had the*

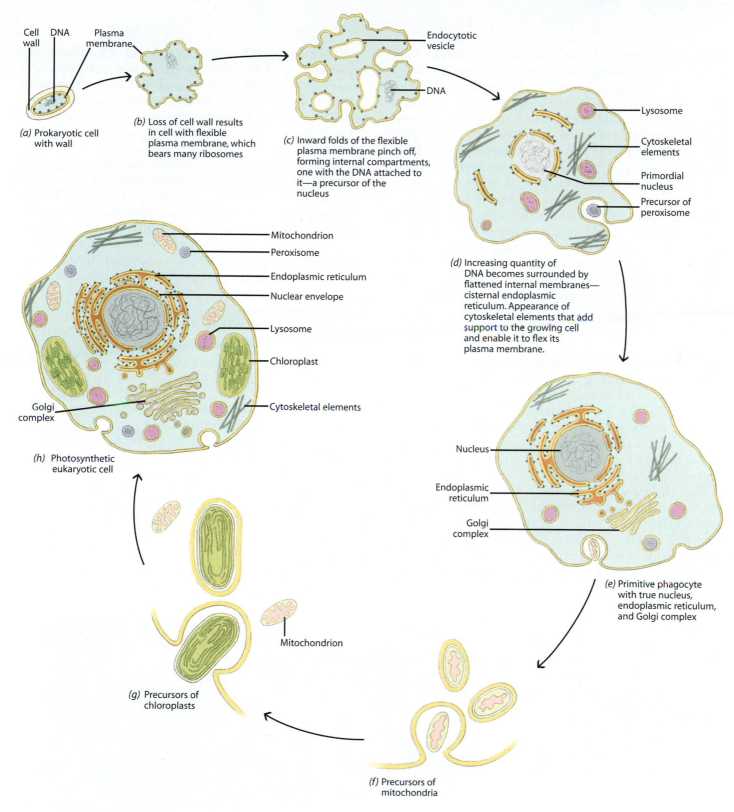

Cell wall **DNA** **Plasma membrane**

(a) Prokaryotic cell with wall

(b) Loss of cell wall results in cell with flexible plasma membrane, which bears many ribosomes

Endocytotic vesicle

DNA

(c) Inward folds of the flexible plasma membrane pinch off, forming internal compartments, one with the DNA attached to it—a precursor of the nucleus

Lysosome

Cytoskeletal elements

Primordial nucleus

Precursor of peroxisome

(d) Increasing quantity of DNA becomes surrounded by flattened internal membranes— cisternal endoplasmic reticulum. Appearance of cytoskeletal elements that add support to the growing cell and enable it to flex its plasma membrane.

Mitochondrion
Peroxisome
Endoplasmic reticulum
Nuclear envelope
Lysosome
Chloroplast
Cytoskeletal elements

Golgi complex

(h) Photosynthetic eukaryotic cell

Nucleus

Endoplasmic reticulum

Golgi complex

(e) Primitive phagocyte with true nucleus, endoplasmic reticulum, and Golgi complex

Mitochondrion

(g) Precursors of chloroplasts

(f) Precursors of mitochondria

ability to increase in size, change shape, and engulf extracellular objects by infolding of the plasma membrane (endocytosis). **(d), (e)** Internalization of a patch of the plasma membrane to which DNA was attached was the probable precursor of the nucleus. The primitive phagocyte eventually acquired a true nucleus containing an increased quan-

tity of DNA. A cytoskeleton must also have developed to provide inner support for the wall-less cell and to play a role in movement, of both the cell itself and its internal components. **(f)** The mitochondria of the eukaryotic cell had their origin as bacterial endosymbionts, which ultimately transferred most of their DNA to the host's nucleus. **(g)** Chloro-

plasts also are believed to be the descendants of bacteria. They ultimately transferred most of their DNA to the host's nucleus. **(h)** The photosynthetic eukaryotic cell contains a complex endomembrane system and a variety of other internal structures such as the peroxisomes, mitochondria, and chloroplasts depicted here.

cytoskeleton (especially of microtubules) would have provided the mechanism necessary for capturing food or prey and carrying it inward by endocytosis. The host cell's lysosomes (membrane-bounded vesicles that contain degradative enzymes) would fuse with the food vacuoles, breaking down their contents into usable organic compounds. The intracellular membranes derived from the plasma membrane gradually compartmentalized the host cells, forming what is known as the endomembrane system of the eukaryotic cell (page 58).

The genesis of the nucleus—the principal feature of eukaryotic cells—could also have begun by infolding of the plasma membrane. You will recall that in prokaryotes the circular DNA molecule, or chromosome, is attached to the plasma membrane. Infolding of this portion of the plasma membrane could have resulted in enclosure of the DNA within an intracellular sac, the primordial nucleus (Figure 13–11).

Mitochondria and Chloroplasts Are Thought to Have Evolved from Bacteria That Were Phagocytized   The stage is set. A phagocyte now exists that can prey on bacteria. The phagocyte, however, still lacks mitochondria. The next step is for the phagocyte not to digest the bacterial precursors of the mitochondria (or chloroplasts) but to adopt them, establishing a symbiotic ("living together") relationship.

The green *Vorticella* shown in Figure 13–12 is an example of a modern protist that establishes endosymbioses with certain species of the green alga *Chlorella*. The algal cells remain intact within the host cells as endosymbionts, providing photosynthetic products useful to the heterotrophic host. In return, the algae receive es-

sential mineral nutrients from the host. There are many examples of prokaryotic (bacterial) and eukaryotic endosymbionts in other protists, as well as in the cells of some 150 genera of freshwater and marine invertebrate animals. Algal endosymbionts, including those occurring in the polyps of reef-building corals, increase host productivity and survival (see Chapter 16).

Transformation of an endosymbiont into an organelle usually involved loss of the endosymbiont's cell wall (if any existed) and other unneeded structures. In the course of evolution, the DNA of the endosymbiont and many of its functions were gradually transferred to the host's nucleus. Hence, the genomes of modern mitochondria and chloroplasts are quite small compared with the nuclear genome. Although the mitochondrion or chloroplast cannot live outside a eukaryotic cell, both are self-replicating organelles that have retained many of the characteristics of their prokaryotic ancestors.

Relationships within the Eukaryotes Are Not Always Well Defined

In contrast to the sharp distinction between the *Bacteria*, *Archaea*, and *Eukarya* domains, the relationships within the *Eukarya* are more complex and the divisions between eukaryotic kingdoms much less clear-cut. For convenience, the predominantly unicellular phyla and some of the multicellular lines associated with them are grouped in the kingdom *Protista*, from which the three kingdoms consisting essentially of multicellular organisms— *Plantae*, *Animalia*, and *Fungi*—have evolved. While plants, animals, and fungi are probably monophyletic groups, *Protista* is paraphyletic.

13–12

Endosymbiosis in Vorticella. *(a) Each bell-shaped cell of the protozoan* Vorticella *contains numerous cells of the autotrophic, endosymbiotic alga* Chlorella. *(b) Electron micrograph of a* Vorticella *containing cells of* Chlorella. *Each algal cell is found in a separate vacuole (perialgal vacuole) bounded by a single membrane. The protozoan provides the algae with protection and mineral nutrients, while the algae produce carbohydrates that serve as nourishment for their heterotrophic host cell.*

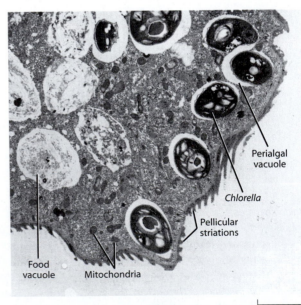

Perialgal vacuole

Chlorella

Pellicular striations

Food vacuole

Mitochondria

5 μm

(a)   (b)

TABLE 13–4 Classification of Living Organisms Included in This Book. (See Appendix D for summary descriptions of these groups.)

Prokaryotic Domains

Bacteria	Bacteria
Archaea	Archaea

Eukaryotic Domain

Eukarya

Kingdom *Fungi*	Fungi	Phylum *Chytridiomycota* (chytrids)
		Phylum *Zygomycota* (zygomycetes)
		Phylum *Ascomycota* (ascomycetes)
		Phylum *Basidiomycota* (*Basidiomycetes, Teliomycetes,* and *Ustomycetes*)
Kingdom *Protista*	Heterotrophic protists	Phylum *Myxomycota* (plasmodial slime molds)
		Phylum *Dictyosteliomycota* (cellular slime molds)
		Phylum *Oomycota* (water molds)
	Photosynthetic protists ("algae")	Phylum *Euglenophyta* (euglenoids)
		Phylum *Cryptophyta* (cryptomonads)
		Phylum *Rhodophyta* (red algae)
		Phylum *Dinophyta* (dinoflagellates)
		Phylum *Haptophyta* (haptophytes)
		Phylum *Bacillariophyta* (diatoms)
		Phylum *Chrysophyta* (chrysophytes)
		Phylum *Phaeophyta* (brown algae)
		Phylum *Chlorophyta* (green algae)
Kingdom *Plantae*	Bryophytes	Phylum *Hepatophyta* (liverworts)
		Phylum *Anthocerophyta* (hornworts)
		Phylum *Bryophyta* (mosses)
	Vascular plants	
	Seedless vascular plants	Phylum *Psilotophyta* (psilotophytes)
		Phylum *Lycophyta* (lycophytes)
		Phylum *Sphenophyta* (horsetails)
		Phylum *Pterophyta* (ferns)
	Seed plants	Phylum *Cycadophyta* (cycads)
		Phylum *Ginkgophyta* (ginkgo)
		Phylum *Coniferophyta* (conifers)
		Phylum *Gnetophyta* (gnetophytes)
		Phylum *Anthophyta* (angiosperms)

The Eukaryotic Kingdoms

The following is a synopsis of the four kingdoms included in the domain *Eukarya* (see Table 13–4, which does not include the kingdom *Animalia*).

The Kingdom *Protista* Includes Unicellular, Colonial, and Simple Multicellular Eukaryotes

Protista (Figure 13–13) is here considered to comprise all organisms traditionally regarded as protozoa (one-celled "animals"), which are heterotrophic, as well as all algae, which are autotrophic. Also included in the kingdom *Protista* are some heterotrophic groups of organisms that have traditionally been placed with the fungi—including the water molds and their relatives (phylum *Oomycota*), the cellular slime molds (phylum *Dictyosteliomycota*), and the plasmodial slime molds (phylum *Myxomycota*).

The reproductive cycles of protists are varied but typically involve both cell division and sexual reproduction. Protists may be motile by 9-plus-2 flagella or cilia or by amoeboid movement, or they may be nonmotile. Red algae and some predominantly unicellular groups of protists included in this book are discussed in Chapter 16, and two major groups of algae—green algae and brown algae (and related organisms)—are discussed in Chapter 17. Green algae are clearly very closely related to plants.

(a)

(b)

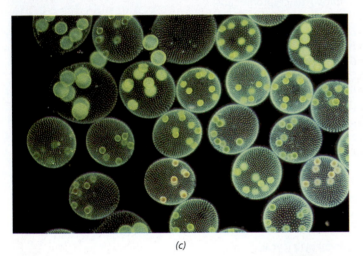

(c)

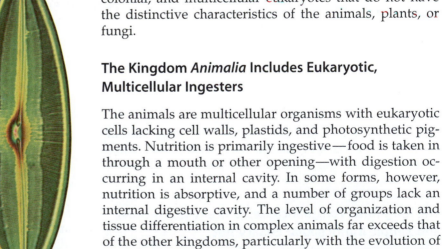

(d)

13–13

Protists. (a) Plasmodium of a plasmodial slime mold, Physarum *(phylum Myxomycota), growing on a leaf. (b)* Postelsia palmiformis, *the "sea palm" (phylum Phaeophyta), growing on exposed intertidal rocks off Vancouver Island, British Columbia. (c)* Volvox, *a motile colonial green alga (phylum Chlorophyta). (d) A red alga (phylum Rhodophyta). (e) A pennate diatom (phylum Bacillariophyta) showing the intricately marked shell characteristic of this group.*

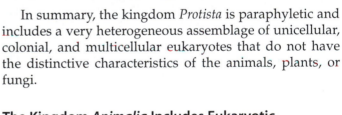

(e)

In summary, the kingdom *Protista* is paraphyletic and includes a very heterogeneous assemblage of unicellular, colonial, and multicellular eukaryotes that do not have the distinctive characteristics of the animals, plants, or fungi.

The Kingdom *Animalia* Includes Eukaryotic, Multicellular Ingesters

The animals are multicellular organisms with eukaryotic cells lacking cell walls, plastids, and photosynthetic pigments. Nutrition is primarily ingestive—food is taken in through a mouth or other opening—with digestion occurring in an internal cavity. In some forms, however, nutrition is absorptive, and a number of groups lack an internal digestive cavity. The level of organization and tissue differentiation in complex animals far exceeds that of the other kingdoms, particularly with the evolution of complex sensory and neuromotor systems. The motility of the organism (or, in sessile forms, of its component parts) is based on contractile fibrils. Reproduction is predominantly sexual. Animals are not discussed in this book, except in relation to some of their interactions with plants and other organisms.

The Kingdom *Fungi* Includes Eukaryotic, Multicellular Absorbers

Members of the kingdom *Fungi*, which are nonmotile, filamentous eukaryotes that lack plastids and photosynthetic pigments, absorb their nutrients from either dead or living organisms (Figure 13–14). The fungi have traditionally been grouped with plants, but there is no longer any doubt that the fungi are an independent evolutionary line. Moreover, comparisons of ribosomal RNA sequences indicate that the fungi are more closely related to animals than to plants. Apparently animals and fungi share a unique evolutionary history, their most recent common ancestor being a flagellated protist similar to modern choanoflagellates. Choanoflagellates are characterized by flattened mitochondrial cristae and a posterior flagellum. Aside from their filamentous growth habit,

fungi have little in common with any of the protist groups that have been classified as algae. Typically, for example, the cell walls of fungi include a matrix of chitin. The structures in which fungi form their spores are often complex. Fungal reproductive cycles, which can also be quite complex, typically involve both sexual and asexual processes. Fungi are discussed in Chapter 15.

The Kingdom *Plantae* Includes Eukaryotic, Multicellular Photosynthesizers

Plants—with three phyla of bryophytes (mosses, liverworts, and hornworts) and the nine living, or extant, phyla of vascular plants—constitute a kingdom of photosynthetic organisms adapted for a life on the land (Figure 13–15). Their ancestors were specialized green algae. All plants are multicellular and are composed of

13–14

Fungi. **(a)** *Red blanket lichen* (Herpothallon sanguineum) *growing on a tree trunk.* **(b)** *A white coral fungus (family* Clavariaceae). **(c)** *Mushrooms (genus probably* Mycena)*, with dew droplets, growing in a rainforest in Peru.* **(d)** *An earthball,* Scleroderma aurantium.

(a)

(b)

(c)

(d)

(a)

(b)

(c)

(d)

(e)

13–15

Plants. **(a)** Sphagnum, *the peat moss (phylum Bryophyta), forms extensive bogs in cold and temperate regions of the world.* **(b)** Marchantia *is
by far the most familiar of the thallose liverworts (phylum Hepatophyta). It is a widespread, terrestrial genus that grows on moist soil and
rocks.* **(c)** *Diphasiastrum digitatum, a "club moss" (phylum Lycophyta).*
(d) *Wood horsetail,* Equisetum sylvaticum *(phylum Sphenophyta).* **(e)**
Bulblet fern, Cystopteris bulbifera *(phylum Pterophyta).* **(f)** *The dandelion,* Taraxacum officinale, *and* **(g)** *fishhook cactus,* Mammillaria microcarpa, *are eudicots (phylum Anthophyta).* **(h)** *Foxtail barley,* Hordeum
jubatum, *and* **(i)** Cymbidium *orchids are monocots (phylum
Anthophyta).* **(j)** *Sugar pine,* Pinus lambertiana, *and incense cedar,*
Calocedrus decurrens, *are both conifers (phylum Coniferophyta).*

(f)

(g)

(h)

(i)

(j)

eukaryotic cells that contain vacuoles and are surrounded by cell walls that contain cellulose. Their principal mode of nutrition is photosynthesis, although a few plants have become heterotrophic. Structural differentiation occurred during the evolution of plants on land, with trends toward the evolution of organs specialized for photosynthesis, anchorage, and support. In more complex plants, such organization has produced specialized photosynthetic, vascular, and covering tissues. Reproduction in plants is primarily sexual, with cycles of alternating haploid and diploid generations. In the more advanced members of the kingdom, the haploid generation (the gametophyte) has been reduced during the course of evolution. The bryophytes are discussed in Chapter 18 and the vascular plants in Chapters 19 through 22.

Summary

Systematics, the scientific study of biological diversity, encompasses taxonomy—the identifying, naming, and classifying of species—and phylogenetics—discerning the evolutionary interrelationships among organisms.

Organisms Are Named with a Binomial and Grouped into Taxonomic Categories Arranged in a Hierarchy

Organisms are designated scientifically by a name that consists of two words—a binomial. The first word in the binomial is the name of the genus, and the second word, the specific epithet, combined with the name of the genus, completes the name of the species. Species are sometimes subdivided into subspecies or varieties. Genera are grouped into families, families into orders, orders into classes, classes into phyla, and phyla into kingdoms. Based on recent taxonomic studies, kingdoms are further grouped into domains.

Organisms Are Classified Phylogenetically Using Homologous, Rather than Analogous, Features

In classifying organisms into the categories of genus through domain, systematists seek to group the organisms in ways that reflect their phylogeny (evolutionary history). In a phylogenetic system, every taxon should be, ideally, monophyletic—that is, every taxon should consist only of organisms descended from a common ancestor. A major principle of such classification is that the similarities should be homologous—that is, the result of common ancestry—rather than the result of convergent evolution.

The frequently intuitive traditional methods of classifying organisms have been largely replaced by more explicit cladistic methods, which attempt to understand branching sequences (genealogy) based on the possession of shared derived character traits.

A Comparison of the Molecular Composition of Organisms Can Be Used to Predict Their Evolutionary Relationships

New techniques in molecular systematics are providing a relatively objective and explicit method of comparing organisms at the most basic level of all, the gene. A large body of evidence indicates that under certain circumstances protein and nucleic acid molecules are molecular clocks, with changes in composition reflecting the time that has elapsed since different groups of organisms diverged from one another. Both amino acid and nucleotide sequencing are making valuable contributions to more accurate classification schemes that reflect an improved understanding of biological diversity and its evolutionary history.

Organisms Are Classified into Two Prokaryotic Domains and One Eukaryotic Domain, Which Consists of Four Kingdoms

In this text—based largely on data obtained from the sequencing of the small-subunit ribosomal RNA—living organisms are grouped into three domains, *Bacteria*, *Archaea*, and *Eukarya*. The *Bacteria* and *Archaea* are two distinct lineages of prokaryotic organisms. The *Archaea* are more closely related to the *Eukarya* than they are to

the *Bacteria*. *Eukarya* consist entirely of eukaryotes. The *Protista*, *Fungi*, *Plantae*, and *Animalia* are kingdoms within the *Eukarya*.

Selected Key Terms

analogy   p. 267

Archaea   p. 270

artificial taxon   p. 266

Bacteria   p. 270

binomial system   p. 262

category   p. 264

clades   p. 267

cladistics   p. 267

cladogram   p. 267

convergent evolution   p. 266

domains   p. 271

endosymbiosis   p. 272

Eukarya   p. 270

genus   p. 262

homology   p. 266

molecular clock   p. 270

molecular systematics   p. 268

monophyletic   p. 265

outgroup   p. 267

paraphyletic   p. 266

phylogenetic tree   p. 265

phylogeny   p. 265

polyphyletic   p. 266

principle of parsimony   p. 268

serial endosymbiotic theory   p. 272

specific epithet   p. 262

systematics   p. 262

taxon   p. 264

taxonomy   p. 262

type specimen   p. 263

Questions

1. Distinguish between or among the following: category/taxon; monophyletic/polyphyletic/paraphyletic; host/endosymbiont.

2. Identify which of the following are categories and which are taxa: undergraduates; the faculty of The Pennsylvania State University; the Green Bay Packers; major league baseball teams; the U.S. Marine Corps; the family Robinson.

3. A key question in systematics is the origin of a similarity or difference. Explain.

4. How does a cladogram differ from a phylogenetic tree constructed by the traditional method?

5. It is generally thought that similarities and differences in homologous nucleotide sequences provide a more sensitive indicator of the evolutionary distance between organisms than similarities and differences in the amino acid sequences of homologous proteins. Explain why this should be so.

6. Explain the role of endosymbiosis in the origin of eukaryotic cells.

14

Prokaryotes and Viruses

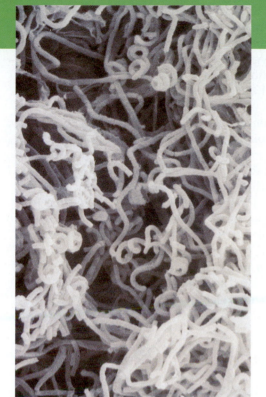

14–1

A common prokaryote, the filamentous actinomycete known as Streptomyces sca- *bies. Actinomycetes are abundant in soil, where they are largely responsible for the "moldy" odor of damp soil and decaying material.* Streptomyces scabies *is the bacterium that causes potato scab disease.*

OVERVIEW

The most abundant organisms on Earth—that is, the prokaryotes, which make up the domains Bacteria *and* Archaea*—are usually invisible to the naked eye. But small does not mean insignificant. Prokaryotes, for example, brought oxygen to the atmosphere of the early Earth and are responsible for decomposing dead organisms and recycling their chemical elements. In addition, prokaryotes supply commercially important products such as antibiotics, and they provide scientists with model systems for studying the ways in which organisms survive and even thrive in hostile environments. On the other hand, certain prokaryotes attain notoriety because of the problems they cause, most notably diseases of plants and animals.*

Viruses—other microscopic agents of disease discussed in this chapter—do not belong to any domain or kingdom and thus are something of a paradox. By themselves, viruses are metabolically inert entities. Yet, once inside a living cell, a virus can reproduce by taking over the cell's genetic machinery.

CHECKPOINTS

By the time you finish reading this chapter, you should be able to answer the following questions:

1. What is the basic structure of a prokaryotic cell?

2. How do prokaryotes reproduce, and in what ways does genetic recombination take place in prokaryotes?

3. In what ways are the cyanobacteria ecologically important?

4. Metabolically, what are the principal differences between the cyanobacteria and the purple and green bacteria?

5. How do the mycoplasmas differ from all other *Bacteria*?

6. Physiologically, what are the three large groups of *Archaea*?

7. What is the basic structure of a virus? How do viruses reproduce?

Of all organisms, the prokaryotes are the simplest structurally, smallest physically, and the most abundant worldwide. Even though individuals are microscopically small, the total weight of prokaryotes in the world is estimated to exceed that of all other living organisms combined. In the sea, for example, prokaryotes make up an estimated 90 percent or more of the total weight of living organisms. In a single gram (about $\frac{1}{28}$ of an ounce) of fertile agricultural soil, there may be 2.5 billion individuals. About 2700 species of prokaryotes are currently recognized, but many more await discovery.

The prokaryotes are, in evolutionary terms, the oldest organisms on Earth. The oldest known fossils are chainlike prokaryotes found in rocks from Western Australia, dated at about 3.5 billion years old (see Figure 1–2). Although some present-day prokaryotes phenotypically resemble these ancient organisms, none of the prokaryotes living today are primitive. Rather, they are modern organisms that have succeeded in adapting to their particular environments.

The prokaryotes are, in fact, the most dominant and successful forms of life on Earth. Their success is undoubtedly due to their great metabolic diversity and their rapid rate of cell division. Growing under optimal conditions, a population of *Escherichia coli*, probably the best-known prokaryote, can double in size and divide every 20 minutes. Prokaryotes can survive in many environments that support no other form of life. They occur in the icy wastes of Antarctica, in the dark depths of the ocean, in the near-boiling waters of natural hot springs (Figure 14–2), and in the superheated water found near undersea vents. Some prokaryotes are among the very few modern organisms that can survive without free oxygen, obtaining their energy by anaerobic processes (page 122). Oxygen is lethal to some types, whereas others can exist with or without oxygen.

In Chapter 13, we emphasized that there are two distinct lineages of prokaryotes, the *Bacteria* and the *Archaea*. At the molecular level, these two domains, while both prokaryotic, are as evolutionarily distinct from one another as either is from the *Eukarya*. We begin by considering some features largely shared by prokaryotes, while noting any differences that may exist between the two domains (see Table 13–3). We then turn our attention specifically to *Bacteria* and then to *Archaea*. Finally, we look briefly at the viruses. Viruses are not cells and, hence, they have no metabolism of their own. A virus consists primarily of a genome (either DNA or RNA) that replicates itself within a living host cell by directing the genetic machinery of that cell to synthesize viral nucleic acids and proteins.

Characteristics of the Prokaryotic Cell

Prokaryotes lack an organized nucleus bounded by a nuclear envelope (see page 41). Instead, a single circular, or continuous, molecule of DNA, associated with non-histone proteins, is localized in a region of the cell called the **nucleoid.** In addition to its so-called "chromosome," a prokaryotic cell may also contain one or more smaller extrachromosomal pieces of circular DNA, called **plasmids,** that replicate independently of the cell's chromosome.

The cytoplasm of most prokaryotes is relatively unstructured, although it often has a fine granular appearance due to its many ribosomes—as many as 10,000 in a single cell. These are smaller (70S in size) than the cytoplasmic ribosomes (80S in size) of eukaryotes (see Table 3–2). Prokaryotes occasionally contain **inclusions,** distinct granules consisting of storage material. They lack a cytoskeleton. The cytoplasm does not contain any membrane-bounded organelles, nor is it generally divided or compartmentalized by membranes. The principal exception occurs in the cyanobacteria, which contain an extensive system of membranes (thylakoids) bearing chlorophyll and other photosynthetic pigments (see Figure 14–10). Purple bacteria also have extensive membranes for their photosynthetic apparatus.

14–2
Aerial view of a very large boiling hot spring, Grand Prismatic Spring, in Yellowstone National Park in Wyoming. Thermophilic ("heat-loving") bacteria thrive in such hot springs. Carotenoid pigments of thickly growing thermophilic bacteria, including cyanobacteria, color the runoff channels a brownish orange.

The Plasma Membrane Serves as a Site for the Attachment of Various Molecular Components

The plasma membrane of a prokaryotic cell is formed from a lipid bilayer and is similar in chemical composition to that of a eukaryotic cell. However, except in the mycoplasmas (the smallest free-living cells known), the plasma membranes of prokaryotes lack cholesterol or other sterols. In aerobic prokaryotes, the plasma membrane incorporates the electron transport chain found in the mitochondrial membrane of eukaryotic cells, lending further support to the serial endosymbiotic theory (page 272). In the photosynthetic purple and green bacteria (but not the cyanobacteria), the sites of photosynthesis are found in the plasma membrane. In the photosynthetic purple bacteria and aerobes with large energy requirements, the membrane is often extensively convoluted, greatly increasing its working surface. Also, the membrane apparently contains specific attachment sites for the DNA molecules, ensuring separation of the replicated chromosomes at cell division (page 156).

The Cell Wall of Most Prokaryotes Contains Peptidoglycans

The protoplasts of almost all prokaryotes are surrounded by a cell wall, which gives the different types their characteristic shapes. Many prokaryotes have rigid walls, some have flexible walls, and only the mycoplasmas have no cell walls at all.

The cell walls of prokaryotes are complex and contain many kinds of molecules not present in eukaryotes. Except for the *Archaea*, the walls of prokaryotes contain complex polymers known as **peptidoglycans**, which are primarily responsible for the mechanical strength of the wall. Thus, peptidoglycan has been dubbed a "signature molecule" for differentiating species of *Bacteria* from species of *Archaea*.

Bacteria can be divided into two major groups on the basis of the capacity of their cell walls to retain the dye known as crystal violet. Those whose cell walls retain the dye are called **gram-positive,** whereas those that do not are called **gram-negative,** after Hans Christian Gram, the Danish microbiologist who discovered the distinction. Gram-positive and gram-negative bacteria differ markedly in the structure of their cell walls. In gram-positive bacteria, the wall, which ranges from 10 to 80 nanometers in thickness, has a homogeneous appearance and consists of as much as 90 percent peptidoglycan (Figure 14–3a). In gram-negative bacteria, the wall consists of two layers: an inner peptidoglycan layer, only 2 to 3 nanometers thick, and an outer layer of lipopolysaccharides and proteins (Figure 14–3b). The molecules of the outer layer are arranged in the form of a bilayer, about 7 to 8 nanometers in thickness, similar in structure to the plasma membrane. Gram staining is widely used to identify and classify bacteria, because it reflects a fundamental difference in the architecture of the cell wall.

14–3

Bacterial cell walls. (a) The wall of a gram-positive bacterium consists of a homogeneous layer of mostly peptidoglycan, seen here as the inner dark band. The outer dark layer in this electron micrograph represents a layer of surface proteins. (b) In a gram-negative bacterium, a layer of peptidoglycan is sandwiched between the plasma membrane and an outer layer of lipopolysaccharides and proteins, similar in composition to the plasma membrane.

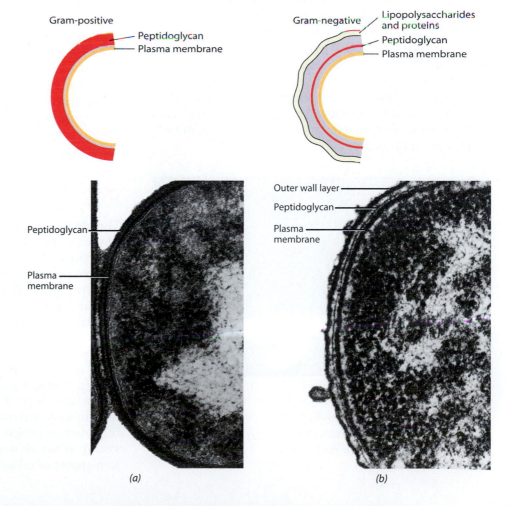

Gram-positive
— Peptidoglycan
— Plasma membrane

Gram-negative
— Lipopolysaccharides and proteins
— Peptidoglycan
— Plasma membrane

Peptidoglycan

Plasma membrane

Outer wall layer
Peptidoglycan
Plasma membrane

(a)

(b)

Many prokaryotes secrete slimy or gummy substances on the surface of their walls. Most of these substances consist of polysaccharides; a few consist of proteins. Commonly known as a "capsule," the general term for these layers is **glycocalyx.** The glycocalyx plays an important role in infection by allowing certain pathogenic bacteria to attach to specific host tissues. It is believed that the glycocalyx may also protect the bacteria from desiccation.

Prokaryotes Store Various Compounds in Granules

A wide variety of prokaryotes—both *Bacteria* and *Archaea*—contain inclusion bodies, or storage granules, consisting of **poly-β-hydroxybutyric acid,** a lipidlike compound, which serves as a depository for carbon and energy. Another storage compound of prokaryotes is **glycogen,** a starchlike substance. Glycogen granules are usually smaller than poly-β-hydroxybutyric acid granules.

Prokaryotes Have Distinctive Flagella

Many prokaryotes are motile, and their ability to move independently is usually due to long, slender appendages known as **flagella** (singular: flagellum) (Figure 14–4). Lacking microtubules and a plasma membrane, these flagella differ greatly from those of eukaryotes (page 60). Each prokaryotic flagellum is composed of subunits of a protein called flagellin, which are arranged into chains that are wound in a triple helix (three chains) with a hollow central core. Bacterial flagella grow at the tip. Flagellin molecules formed in the cell pass up through the hollow core and are added at the far end of the chains. In some species, the flagella are distributed over the entire cell surface; in others, they occur in tufts at one end of the cell.

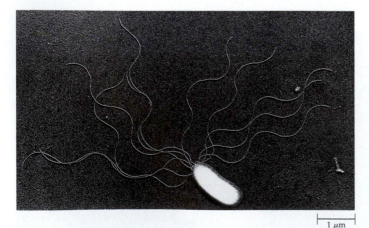

14–4

Flagella on Pseudomonas marginalis, *a common bacterium that is widespread in soils. It causes a soft-rot disease found mainly in fleshy and leafy vegetables.*

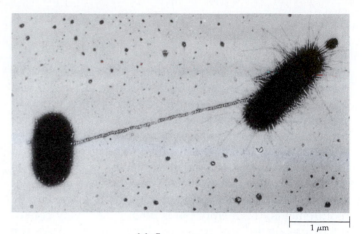

1 μm

14–5

Electron micrograph of conjugating E. coli *cells. The elongated donor cell at the right of the micrograph is connected to the more rotund recipient cell by a long pilus, which is necessary for conjugation. Numerous short fimbriae are visible on the donor cell.*

Fimbriae and Pili Are Involved in Attachment

Fimbriae and pili—the two terms are often used interchangeably—are filamentous structures assembled from protein subunits in much the same way as the filaments of flagella. **Fimbriae** (singular: fimbria) are much shorter, more rigid, and typically more numerous than flagella (Figure 14–5). The functions of fimbriae are not known for certain, but they may serve to attach the organism to a food source or other surfaces.

Pili (singular: pilus) are generally longer than fimbriae, and only one or a few are present on the surface of an individual (Figure 14–5). Some pili are involved in the process of conjugation between prokaryotes, serving first to connect the two cells and then, by retracting, to draw them together for the actual transfer of DNA.

Diversity of Form

The oldest method of identifying prokaryotes is by their physical appearance. Prokaryotes exhibit considerable diversity of form, but many of the most familiar species fall into one of three categories (Figure 14–6). A prokaryote with a cylindrical shape is called a **rod,** or **bacillus** (plural: bacilli); spherical ones are called **cocci** (singular: coccus); and long curved, or spiral, rods are called **spirilla** (singular: spirillum). Cell shape is a relatively constant feature in most species of prokaryotes.

In many prokaryotes, the cells remain together, producing filaments, clusters, or colonies that also have a distinctive shape. For example, cocci and rods may adhere to form chains, and such behavior is characteristic of particular genera. Bacilli usually separate after cell division. When they do remain together, they form long, thin chains of cells, as are found in the filamentous acti-

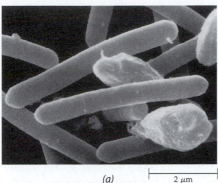

(a)   2 μm

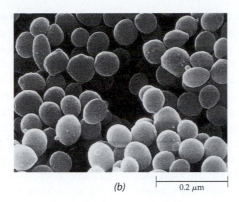

(b)   0.2 μm

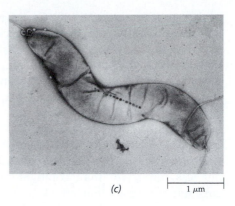

(c)   1 μm

14–6

*The three major forms of prokaryotes are: bacilli, cocci, and spirilla. **(a)** Clostridium botulinum, the source of the toxin that causes deadly food poisoning, or botulism, is a bacillus, or rod-shaped bacterium. The saclike structures are endospores, which are resistant to heat and cannot easily be destroyed. Bacilli are responsible for many plant*

*diseases, including fire blight of apples and pears (caused by Erwinia amylovora) and bacterial wilt of tomatoes, potatoes, and bananas (caused by Pseudomonas solanacearum). **(b)** Many prokaryotes, such as Micrococcus luteus, shown here, take the shape of spheres. Among the cocci are Streptococcus lactis, a common milk-*

*souring agent and Nitrosococcus nitrosus, a soil bacterium that oxidizes ammonia to nitrites. **(c)** Spirilla, such as Magnetospirillum magnetotacticum, are less common than bacilli and cocci. Flagella can be seen at either end of this cell, which was isolated from a swamp. The string of dark magnetic particles orient the cell in the Earth's magnetic field.*

0.25 μm

14–7

One of the diverse types of prokaryotes. This is a fruiting body of Chondromyces crocatus, a myxobacterium, or gliding bacterium. Each fruiting body, which may contain as many as 1 million cells, consists of a central stalk that branches to form clusters of myxospores. Normally, the myxobacteria are rods that glide together along slime tracks and eventually form the kinds of bodies shown here.

nomycetes (Figure 14–1). The gram-negative rods of the myxobacteria aggregate and construct complex fruiting bodies, within which some cells become converted into resting cells called myxospores (Figure 14–7). The myxospores are more resistant to drying, UV radiation, and heat than the vegetative cells and are of survival value to the bacterium.

Reproduction and Gene Exchange

Most prokaryotes reproduce by a simple type of cell division called **binary fission** (page 156). In some forms, reproduction is by budding or by fragmentation of filaments of cells. As they multiply, prokaryotes, barring mutation, produce clones of genetically identical cells. Mutations do occur, however. It has been estimated that in a culture of *E. coli* that has divided 30 times, about 1.5 percent of the cells are mutants. Mutations, combined with a rapid generation time, are responsible for the extraordinary adaptability of prokaryotes.

Further adaptability is provided by the genetic recombinations that take place as a result of conjugation, transformation, and transduction. Evidence is rapidly accumulating that such genetic recombinations are quite common in nature. All three mechanisms of gene transfer occur together in certain bacteria as well as archaea.

Conjugation has been characterized as the prokaryotic version of sex. This form of mating takes place when a pilus produced by the donor cell comes in contact with the recipient cell (Figure 14–5). This "sex pilus" then retracts, pulling the two cells together so that a tube, called a conjugation bridge, can form between them. A portion of the donor chromosome then passes through this bridge to the recipient cell. Conjugation is the mechanism used by plasmids to transfer copies of themselves to a new host. Conjugation can transfer genetic information between distantly related organisms; for example, plasmids can be transferred between bacteria and fungi as well as between bacteria and plants. **Transformation** occurs when a prokaryote takes up free, or naked, DNA from the environment. The free DNA may have been left behind by an organism that died. Because DNA is not chemically stable outside cells, transformation is probably less important than conjugation. **Transduction** occurs when viruses that attack bacteria—viruses known as bacteriophages (see page 297)—bring with them DNA they have acquired from their previous host.

Endospores

Certain species of bacteria have the capacity to form **endospores,** which are dormant resting cells (Figure 14–8). This process, called sporulation, has been studied extensively in the genera *Bacillus* and *Clostridium*. It characteristically occurs when a population of cells begins to use up its food supply.

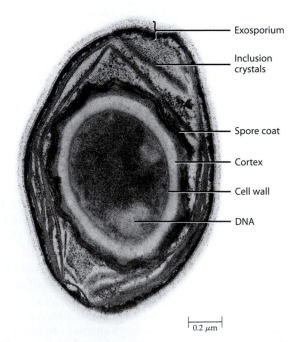

Exosporium

Inclusion crystals

Spore coat

Cortex

Cell wall

DNA

0.2 μm

14–8

Mature free spore of Bacillus megaterium. *The outermost layer is the enveloping exosporium, made up of a faint peripheral layer and a dark basal layer. Underlying the exosporium are large inclusion crystals. The spore proper is covered with a protein spore coat. Beneath the spore coat is a thick peptidoglycan cortex, which is essential for the unique resistance properties of bacterial spores. Interior to the cortex is a thin cell wall, also of peptidoglycan, that covers a dehydrated spore protoplast.*

Endospore formation greatly increases the capacity of the prokaryotic cell to survive. Endospores are exceedingly resistant to heat, radiation, and chemical disinfectants, primarily due to their dehydrated protoplasts. The endospores of *Clostridium botulinum,* the organism that causes often-fatal blood poisoning, are not destroyed by boiling for several hours. In addition, endospores can remain viable (that is, they can germinate and develop into vegetative cells) for a very long time. For example, endospores recovered in 7000-year-old fractions of sediment cores from a Minnesota lake proved to be viable. More remarkably, ancient endospores preserved in the gut of an extinct bee trapped in amber were also reported to be revivable. The amber—and hence the endospores—was estimated to be 25 to 40 million years old.

Metabolic Diversity

Prokaryotes Are Autotrophs or Heterotrophs

Prokaryotes exhibit tremendous metabolic diversity. Although some are autotrophs (meaning literally "self-feeding")—organisms that use carbon dioxide as their sole source of carbon—most are **heterotrophs**—organisms requiring organic compounds as a carbon source. The vast majority of heterotrophs are **saprophytes** (Gk. *sapros*, "rotten," or "putrid"), which obtain their carbon from dead organic matter. Saprophytic bacteria and fungi are responsible for the decay and recycling of organic material in the soil; indeed, they are the recyclers of the biosphere.

50 μm

14–9

Thiothrix, *a colorless sulfur bacterium.* Thiothrix *obtains energy by the oxidation of hydrogen sulfide,* H_2S. *The chains of cells, each filled with particles of sulfur, are attached to the substrate at the base (center of micrograph) and so form a characteristic rosette.*

Among the autotrophs are those that obtain their energy from light. These organisms are referred to as **photosynthetic autotrophs.** Some autotrophs, known as **chemosynthetic autotrophs,** are able to use inorganic compounds, rather than light, as an energy source (Figure 14–9). The energy is obtained from the oxidation of reduced inorganic compounds containing nitrogen, sulfur, or iron or from the oxidation of gaseous hydrogen.

Prokaryotes Vary in Their Tolerance of Oxygen and Temperature

Prokaryotes vary in their need for, or tolerance of, oxygen. Some species, called **aerobes,** require oxygen for respiration. Others, called **anaerobes,** lack an aerobic pathway and therefore cannot use oxygen as a terminal electron acceptor. There are two kinds of anaerobes: **strict anaerobes,** which are killed by oxygen and, hence, can live only in the absence of oxygen, and **facultative anaerobes,** which can grow in the presence or absence of oxygen.

Prokaryotes also vary with regard to the range of temperatures under which they can grow. Some have a low optimum temperature (that is, a temperature at which growth is most rapid). These organisms, called **psychrophiles,** can grow at 0°C or lower and can survive indefinitely when it is much colder. At the other extreme are the **thermophiles** and **extreme thermophiles,** which have high and very high temperature optima, respectively. Thermophilic prokaryotes, with growth optima below 80°C, are common inhabitants of hot springs. Some extreme thermophiles have optimal growth temperatures greater than 100°C and have even been found growing in the 140°C water near deep-sea vents. Because their heat-stable enzymes are capable of catalyzing biochemical reactions at high temperatures, thermophiles and extreme thermophiles are being intensively investigated for use in industrial and biotechnological processes.

Prokaryotes Play a Vital Role in the Functioning of the World Ecosystem

The autotrophic bacteria make major contributions to the global carbon balance. The role of certain bacteria in fixing atmospheric nitrogen—that is, in incorporating nitrogen gas into nitrogen compounds—is likewise of major biological significance (see page 290 and Chapter 30). Through the action of decomposers, materials incorporated into the bodies of once-living organisms are degraded and released and made available for successive generations. More than 90 percent of the CO_2 production in the biosphere, other than that associated with human activities, results from the metabolic activity of bacteria and fungi. The abilities of certain bacteria to decompose toxic natural and synthetic substances such as petroleum, pesticides, and dyes may lead to their widespread use in cleaning up dangerous spills and toxic dumps, when the techniques of using these bacteria are better developed.

Some Prokaryotes Cause Disease

In addition to their ecological roles, bacteria are important as agents of disease in both animals and plants. Human diseases caused by bacteria include tuberculosis, cholera, anthrax, gonorrhea, whooping cough, bacterial pneumonia, Legionnaires' disease, typhoid fever, botulism, syphilis, diphtheria, and tetanus. A clear link has now been established between stomach ulcers and infection with *Helicobacter pylori* and between coronary heart disease and *Chlamydia pneumoniae*, which have been found growing in arteries. About 80 species of bacteria, including many strains that appear to be identical but differ in the species of plant they infect, have been found to cause diseases of plants. Many of these diseases are highly destructive; some of them are described later in this chapter.

Some Prokaryotes Are Used Commercially

Industrially, bacteria are sources of a number of important antibiotics: streptomycin, aureomycin, neomycin, and tetracycline, for example, are produced by actinomycetes. Bacteria are also widely used commercially for the production of drugs and other substances, such as vinegar, various amino acids, and enzymes. The production of almost all cheeses involves bacterial fermentation of the sugar lactose into lactic acid, which coagulates milk proteins. The same kinds of bacteria that are responsible for the production of cheese are also responsible for the production of yogurt and for the lactic acid that preserves sauerkraut and pickles.

Bacteria

Phylogenetic analysis, employing sequencing of ribosomal RNA, reveals that there are twelve major lineages, or kingdoms, of *Bacteria*. The *Bacteria* are a diverse group of organisms. They range from the most ancient lineage of extreme thermophilic chemosynthetic autotrophs that oxidize gaseous hydrogen or reduce sulfur compounds to the lineages of photosynthetic autotrophs, represented by the cyanobacteria and the purple and green bacteria. The bacteria selected for individual treatment here are those we consider to be of special evolutionary and ecological importance.

Cyanobacteria Are Important from Ecological and Evolutionary Perspectives

The **cyanobacteria** deserve special emphasis because of their great ecological importance, especially in the global carbon and nitrogen cycles, as well as their evolutionary significance. They represent one of the major evolutionary lines of *Bacteria*. Photosynthetic cyanobacteria have chlorophyll *a*, together with carotenoids and other, unusual accessory pigments known as **phycobilins.** There are two kinds of phycobilins: **phycocyanin,** a blue pigment, and **phycoerythrin,** a red one. Within the cells of cyanobacteria are numerous layers of membranes, often parallel to one another (Figure 14–10). These membranes are photosynthetic thylakoids that resemble those found in chloroplasts—which, in fact, correspond in size to the entire cyanobacterial cell. The main storage product of the cyanobacteria is glycogen. Cyanobacteria are believed to have given rise through symbiosis to at least some eukaryotic chloroplasts. In biochemical and structural detail, cyanobacteria are especially similar to the chloroplasts of red algae (see page 358).

Many cyanobacteria produce a mucilaginous envelope, or sheath, that binds groups of cells or filaments together. The sheath is often deeply pigmented, particularly in species that sometimes occur in terrestrial habitats. The colors of the sheaths in different species include light gold, yellow, brown, red, emerald green, blue, violet, and blue-black. Despite their former name ("blue-green algae"), therefore, only about half of the species of cyanobacteria are actually blue-green in color.

Cyanobacteria often form filaments and may grow in large masses 1 meter or more in length. Some cyanobacteria are unicellular, a few form branched filaments, and a very few form plates or irregular colonies (Figure 14–11). Following division of the cyanobacterial cell, the resulting subunits may then separate to form new colonies. In addition, the filaments may break into fragments called **hormogonia** (singular: hormogonium). As in other filamentous or colonial bacteria, the cells of cyanobacteria are usually joined only by their walls or by mucilaginous sheaths, so that each cell leads an independent life.

Some filamentous cyanobacteria are motile, gliding and rotating around the longitudinal axis. Short segments that break off from a cyanobacterial colony can glide away from their parent colony at rates as rapid as 10 micrometers per second. This movement may be connected with the extrusion of mucilage through small pores in the cell wall, together with the production of contractile waves in one of the surface layers of the wall. Some cyanobacteria exhibit intermittent jerky movements.

Cyanobacteria Can Live in a Wide Variety of Environments

Although more than 7500 species of cyanobacteria have been described and given names, there may actually be as few as 200 distinct, free-living nonsymbiotic species. Like other bacteria, cyanobacteria sometimes grow under extremely inhospitable conditions, from the water of hot springs to the frigid lakes of Antarctica, where they sometimes form luxuriant mats 2 to 4 centimeters thick in water beneath more than 5 meters of permanent ice. However, cyanobacteria are absent in acidic waters, where eukaryotic algae are often abundant. The greenish color of some polar bears in zoos is due to the presence of colonies of cyanobacteria that develop within the hollow hairs of their fur.

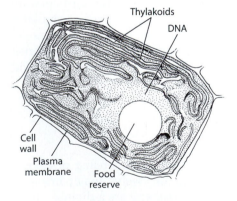

14–10

A cell of the cyanobacterium Anabaena cylindrica. *Photosynthesis takes place in the chlorophyll-containing membranes—the thylakoids—within the cell. The three-dimensional quality of this electron micrograph is due to the freeze-fracturing technique used in the preparation of the tissue.*

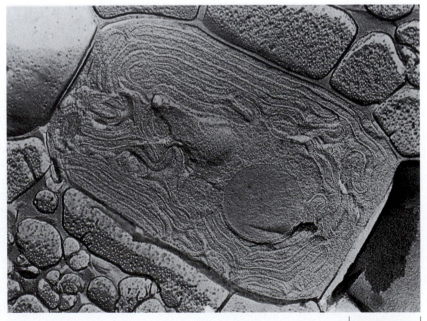

1 μm

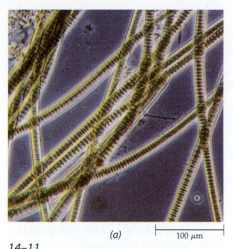

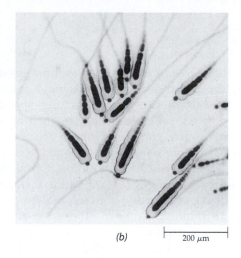

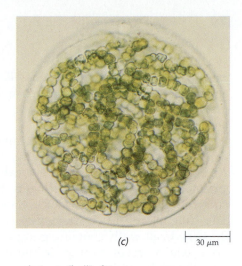

(a)   100 µm   (b)   200 µm   (c)   30 µm

14–11

Three common genera of cyanobacteria. **(a)** Oscillatoria, in which the only form of reproduction is by means of fragmentation of the filament. **(b)** Calothrix, a filamentous form

with a basal heterocyst (see page 290). Calothrix is capable of forming akinetes—enlarged cells that develop a resistant outer envelope—just above the heterocysts. **(c)** A

gelatinous "ball" of Nostoc commune, containing numerous filaments. These cyanobacteria occur frequently in freshwater habitats.

Layered chalk deposits called **stromatolites** (Figure 14–12), which have a continuous geologic record covering 2.7 billion years, are produced when colonies of cyanobacteria bind calcium-rich sediments. Today, stromatolites are formed only in a few places, such as shallow pools in hot, dry climates. Their abundance in the fossil record is evidence that such environmental conditions were prevalent in the past, when cyanobacteria played the decisive role in elevating the level of free oxygen in the atmosphere of the early Earth. More ancient stromatolites (3 billion years or older), produced in an oxygen-free environment, probably were made by purple and green bacteria. Recent evidence suggests that some stromatolites may have been produced solely by geologic processes.

Many marine cyanobacteria occur in limestone (calcium carbonate) or lime-rich substrates, such as coralline algae (see page 359) and the shells of mollusks. Some freshwater species, particularly those that grow in hot springs, often deposit thick layers of lime in their colonies.

Cyanobacteria Form Gas Vesicles, Heterocysts, and Akinetes The cells of cyanobacteria living in freshwater or marine habitats—especially those that inhabit the surface layers of the water, in the community of microscopic organisms known as the **plankton**—commonly contain bright, irregularly shaped structures called **gas vesicles.** These vesicles provide and regulate the buoyancy of the organisms, thus allowing them to float at certain levels in the water. When numerous cyanobacteria become unable to regulate their gas vesicles properly—for example, because of extreme fluctuations of temperature or oxygen supply—they may float to the surface of the body of water and form visible masses called "blooms." Some cyanobacteria that form blooms secrete chemical substances that are toxic to other organisms,

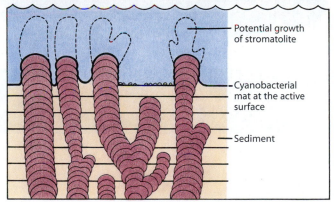

Potential growth of stromatolite

Cyanobacterial mat at the active surface

Sediment

14–12

Stromatolites are produced when flourishing colonies of cyanobacteria bind calcium carbonate into domed structures, like the ones shown in the diagram and photograph, or others with a more intricate form. Such structures are abundant in the fossil record but today are being formed in only a few, highly suitable environments, such as the tidal flats of Hamlin Pool in Western Australia shown here.

causing large numbers of deaths. The Red Sea apparently was given its name because of the blooms of planktonic species of *Trichodesmium*, a red cyanobacterium.

Many genera of cyanobacteria can fix nitrogen, converting nitrogen gas to ammonium, a form in which the nitrogen is available for biological reactions. In filamentous cyanobacteria, **nitrogen fixation** often occurs within **heterocysts,** which are specialized, enlarged cells (Figure 14–13). Heterocysts are surrounded by thickened cell walls containing large amounts of glycolipid, which serves to impede the diffusion of oxygen into the cell. Within a heterocyst, the cell's internal membranes are reorganized into a concentric or reticulate pattern. Heterocysts are low in phycobilins, and they lack Photosystem II, so the cyclic photophosphorylation that occurs in these cells does not result in the evolution of oxygen (page 139). The oxygen that is present is either rapidly reduced by hydrogen, a by-product of nitrogen fixation, or expelled through the wall of the heterocyst. Nitrogenase, the enzyme that catalyzes the nitrogen-fixing reactions, is sensitive to the presence of oxygen, and nitrogen fixation therefore is an anaerobic process. Heterocysts have small plasmodesmatal connections—microplasmodesmata—with adjacent vegetative cells. It is via these microplasmodesmata that the products of nitrogen fixation are transported from the heterocyst to the vegetative cells and the products of photosynthesis move from the vegetative cells to the heterocyst.

Among cyanobacteria that fix nitrogen are free-living species such as *Trichodesmium*, which occurs in certain tropical oceans. *Trichodesmium* accounts for about a quarter of the total nitrogen fixed there, an enormous amount. Symbiotic cyanobacteria are likewise very important in nitrogen fixation. In the warmer parts of Asia, rice can often be grown continuously on the same land without the addition of fertilizers because of the presence of nitrogen-fixing cyanobacteria in the rice paddies (Figure 14–14). Here the cyanobacteria, especially members of the genus *Anabaena* (Figure 14–13), often occur in association with the small, floating water fern *Azolla*, which forms masses on the paddies.

Cyanobacteria occur as symbionts within the bodies of some sponges, amoebas, flagellated protozoa, diatoms, green algae that lack chlorophyll, other cyanobacteria, mosses, liverworts, vascular plants, and oomycetes, in addition to their familiar role as the photosynthetic partner in many lichens (see Chapter 15). Some symbiotic cyanobacteria lack a cell wall, in which case they function as chloroplasts. The symbiotic cyanobacterium divides at the same time as the host cell by a process similar to chloroplast division. Bacteria similar to cyanobacteria appear to have given rise to at least some chloroplasts.

In addition to the heterocysts, some cyanobacteria form resistant spores called **akinetes,** which are enlarged cells surrounded by thickened envelopes (Figure 14–13b). Like the endospores formed by other bacteria,

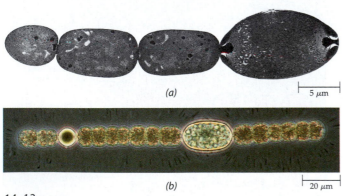

(a)   5 μm

(b)   20 μm

14–13

Filament of Anabaena. *(a) This electron micrograph shows a chain of cells held together along incompletely separated walls. The first cell on the right end of the chain is a heterocyst, in which nitrogen fixation takes place. The gelatinous matrix of this filament was destroyed during preparation of the specimen for electron microscopy. (b) In this preparation, the gelatinous matrix is barely discernible as striations extending outward from the cell surfaces. The third cell from the left is a heterocyst.* Anabaena, *like the* Calothrix *shown in Figure 14–11b, forms akinetes (large oval body toward the right).*

14–14

Women planting rice in a paddy in Perak, Malaysia. In Southeast Asia, rice can often be grown on the same land continuously without the addition of fertilizers because of nitrogen fixation by Anabaena azollae, *which grows in the tissues of the water fern* Azolla *growing in the rice paddies.*

akinetes are resistant to heat and drought and thus allow the cyanobacterium to survive during unfavorable periods.

Purple and Green Bacteria Have Unique Types of Photosynthesis

The purple and green bacteria together represent the second major group of photosynthetic bacteria. The overall photosynthetic process and the photosynthetic pigments utilized by these bacteria differ from those used by the cyanobacteria. Whereas cyanobacteria produce oxygen during photosynthesis, the purple and green bacteria do not. In fact, purple and green bacteria can grow in light only under anaerobic conditions because pigment synthesis in these organisms is repressed by oxygen. Cyanobacteria employ chlorophyll a and two photosystems in their photosynthetic process. By contrast, purple and green bacteria employ several different types of bacteriochlorophyll, which differ in certain respects from chlorophyll, and only one photosystem (Figure 14–15). The photosystems present in purple and green bacteria appear to be ancestors of individual photosystems—Photosystem II and Photosystem I, respectively—present in photosynthetic autotrophs, including plants, algae, and cyanobacteria.

The colors characteristic of photosynthetic bacteria are associated with the presence of several accessory pigments that function in photosynthesis. In two groups of purple bacteria, these pigments are yellow and red carotenoids. In the cyanobacteria, as we have seen, the pigments are the red and blue phycobilins, which are not found in purple and green bacteria.

The purple and green bacteria are subdivided into those that use mostly sulfur compounds as electron donors and those that do not. In the purple sulfur and green sulfur bacteria, sulfur compounds play the same role in photosynthesis that water plays in organisms containing chlorophyll a (page 128).

Purple or green sulfur bacterium:

$$CO_2 + 2H_2S \xrightarrow{\text{Light}} (CH_2O) + H_2O + 2S$$

Carbon dioxide Hydrogen sulfide   Carbohydrate   Water   Sulfur

Cyanobacterium, alga, or plant:

$$CO_2 + 2H_2O \xrightarrow{\text{Light}} (CH_2O) + H_2O + O_2$$

Carbon dioxide   Water   Carbohydrate   Water   Oxygen

The purple nonsulfur and green nonsulfur bacteria, which are able to use sulfide at only low levels, also use organic compounds as electron donors. These compounds include alcohols, fatty acids, and a variety of other organic substances.

The prokaryote that became an endosymbiont of eukaryotes and evolved into mitochondria is thought to have been a relative of the purple nonsulfur bacteria. This conclusion is based both on the similar metabolic features of mitochondria and purple nonsulfur bacteria and on comparisons of the base sequences in their small-subunit ribosomal RNAs.

Because of their requirement for H_2S or a similar substrate, the purple and green sulfur bacteria are able to grow only in habitats that contain large amounts of decaying organic material, recognizable by the sulfurous odor. In these bacteria, as in the closely related colorless sulfur bacterium *Thiothrix*, elemental sulfur may accumulate as deposits within the cell (Figure 14–9).

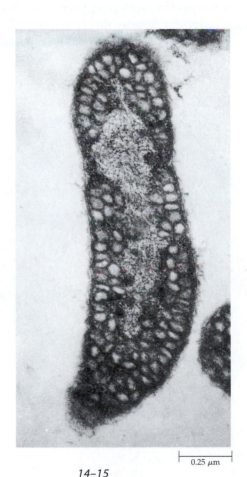

0.25 μm

14–15

A purple nonsulfur bacterium, Rhodospirillum rubrum. *The structures that resemble vesicles are intrusions of the plasma membrane, which contains the photosynthetic pigments. This cell, with its many membrane intrusions, has a very high content of bacteriochlorophyll. It is from a culture that was grown in dim light. In cells grown in bright light, the membrane intrusions are less extensive because there is a decreased need for photosynthetic pigments.*

Prochlorophytes Contain Chlorophylls *a* and *b* and Carotenoids

The **prochlorophytes** are a group of photosynthetic prokaryotes that contain chlorophylls *a* and *b*, as well as carotenoids, but do not contain phycobilins. Thus far, only three genera of prochlorophytes have been discovered. The first of these is *Prochloron*, which lives only along tropical seashores, as a symbiont within colonial sea squirts (ascidians). The cells of *Prochloron* are roughly spherical in shape and contain an extensive system of thylakoids (Figure 14–16).

The two other known prochlorophyte genera are *Prochlorothrix* and *Prochlorococcus*. *Prochlorothrix*, which is filamentous, has been found growing in several shallow lakes in the Netherlands. *Prochlorococcus* has been found deep in the euphotic zone—the zone into which sufficient light penetrates for photosynthesis to occur—of the open oceans. Prochlorococci are extremely small cocci.

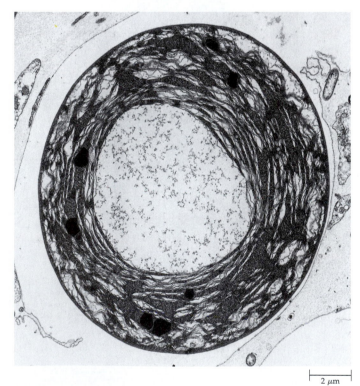

2 μm

14–16

A single cell of the bacterium Prochloron, *showing the extensive system of thylakoids.* Prochloron *is a photosynthetic bacterium with chlorophylls a and b and carotenoids, the same pigments found in the green algae and plants. The prochlorophytes resemble both cyanobacteria (because they are prokaryotic and contain chlorophyll a) and the chloroplasts of green algae and plants (because they contain chlorophyll b instead of phycobilins).*

Mycoplasmas Are Wall-less Organisms That Live in a Variety of Environments

Mycoplasmas and mycoplasmalike organisms are bacteria that lack cell walls. Usually about 0.2 to 0.3 micrometer in diameter, they are probably the smallest organisms capable of independent growth. The genome of the mycoplasmas is likewise small, only one-fifth that of *E. coli* and other common prokaryotes. Because they lack a cell wall and, consequently, rigidity, mycoplasmas can assume various forms. In a single culture, a mycoplasma may range from small rods to highly branched filamentous forms.

Mycoplasmas may be free-living in soil and sewage, parasites of the mouth or urinary tract of humans, or pathogens in animals and plants. Among the plant-pathogenic mycoplasmas are the **spiroplasmas,** long spiral or corkscrew-shaped cells less than 0.2 micrometer in diameter, which are motile even though they lack flagella (Figure 14–17). They are motile by means of a rotary motion or a slow undulation. Some spiroplasmas have been cultured on artificial media, including *Spiroplasma citri*, which causes citrus stubborn disease. Symptoms of stubborn disease, such as bunchy upright growth of twigs and branches, are slow to develop and difficult to detect. The disease, which is widespread, is difficult to control. In California and some Mediterranean countries, stubborn is probably the greatest threat to the production of grapefruit and sweet oranges. *Spiroplasma citri* has also been isolated from corn (maize) plants suffering from corn stunt disease.

Mycoplasmalike Organisms Cause Diseases in Plants

Mycoplasmalike organisms (MLOs), or phytoplasmas, have been identified in more than 200 distinct plant dis-

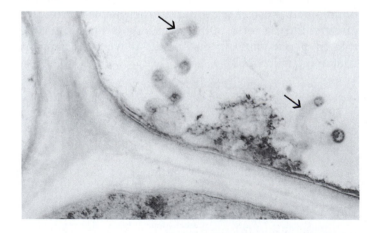

14–17

*Spiroplasmas (arrows) in a sieve tube of a maize (*Zea mays*) plant that was suffering from corn stunt disease.* Spiroplasma citri *causes both corn stunt disease and citrus stubborn disease.*

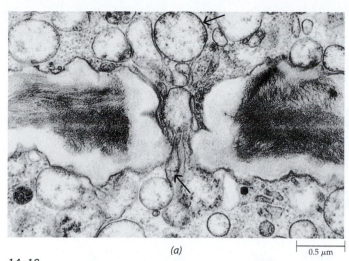

14–18

(a) *Mycoplasmalike organisms (arrows) apparently traversing a sieve-plate pore in a young inflorescence of a coconut palm* (Cocos nucifera) *affected by lethal yellowing disease.* **(b)** *A devastated grove of coconut palms—now looking like telephone poles— in Jamaica. Lethal yellowing has been responsible for the death of palms of many genera in southern Florida and elsewhere.*

eases affecting several hundred genera of plants. Some of these diseases are very destructive, for example, X-disease of peach, which can render a tree commercially worthless in 2 to 4 years, and pear decline, so-called because it commonly causes a slow, progressive weakening and ultimate death of pear trees. Aster yellows, another MLO-caused disease, results in a general yellowing (chlorosis) of the foliage and infects a wide variety of host crops, ornamentals, and weeds. Carrots are among the host crops suffering the greatest losses, commonly 10 to 25 percent and as high as 90 percent. Lethal yellowing of coconut is also caused by an MLO (Figure 14–18).

In flowering plants, MLOs are generally confined to the conducting elements of the phloem known as sieve tubes. Most MLOs are believed to move passively from one sieve-tube member to another through the sieve-plate pores, as the sugar solution is transported in the phloem. Motile spiroplasmas, which also are found in sieve tubes, may be able to move actively in phloem tissue. Most MLOs and spiroplasmas are transmitted from plant to plant by insect vectors that acquire the pathogen while feeding on an infected plant.

Plant-Pathogenic Bacteria Cause a Wide Variety of Diseases

In addition to those already mentioned, many more economically important diseases of plants are caused by bacteria, contributing substantially to the one-eighth of crops worldwide that are lost to disease annually. Almost all plants can be affected by bacterial diseases, and many of these diseases can be extremely destructive.

Virtually all plant-pathogenic bacteria are gram-negative, and all but *Streptomyces*, which is filamentous, are rod-shaped. They are **parasites**—symbionts that are harmful to their hosts. The symptoms caused by plant-pathogenic bacteria are quite varied, with the most common appearing as spots of various sizes on stems, leaves, flowers, and fruits (Figure 14–19). Almost all such bacterial spots are caused by members of two closely related genera, *Pseudomonas* and *Xanthomonas*.

Some of the most destructive diseases of plants—such as blights, soft rots, and wilts—are also caused by bacteria. Blights are characterized by rapidly developing necroses (dead, discolored areas) on stems, leaves, and flowers. Fire blight in apples and pears, caused by *Erwinia amylovora*, is a widespread, economically important disease that can kill young trees within a single season. Fire blight was so destructive to pear trees that by the 1930s their cultivation on a commercial scale had been eliminated from almost all parts of the United States except the Northwest. Bacterial soft rots occur most commonly in the fleshy storage tissues of vegetables, such as potatoes or carrots, as well as in fleshy fruits, such as tomatoes and eggplants, and succulent stems or leaves, such as cabbage or lettuce. The most destructive soft rots are caused by bacteria of the genus *Erwinia*, with heavy losses occurring in the postharvest period.

Bacterial vascular wilts affect mainly herbaceous plants. The bacteria invade the vessels of the xylem, where they multiply. They interfere with the movement of water and inorganic nutrients by producing high-molecular-weight polysaccharides, which results in the wilting and death of the plants. The bacteria commonly degrade portions of the vessel walls and can even cause the vessels to rupture. Once the walls have ruptured, the bacteria then spread to the adjacent parenchyma tissues,

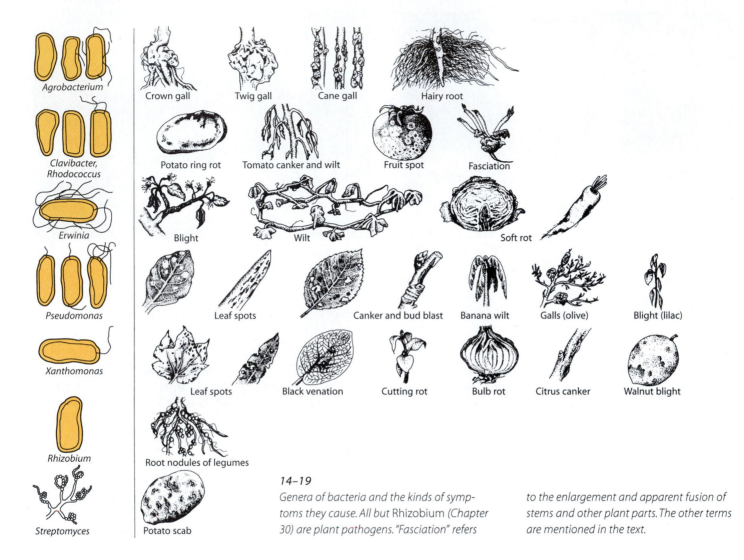

14–19

Genera of bacteria and the kinds of symptoms they cause. All but Rhizobium *(Chapter 30) are plant pathogens. "Fasciation" refers to the enlargement and apparent fusion of stems and other plant parts. The other terms are mentioned in the text.*

where they continue to multiply. In some bacterial wilts, the bacteria ooze to the surface of the stems or leaves through cracks formed over cavities filled with cellular debris, gums, and bacterial cells. More commonly, however, the bacteria do not reach the surface of the plant until the plant has been killed by the disease.

Among the most important examples of wilts are bacterial wilt of alfalfa, tomato, and bean plants (each caused by different species of *Clavibacter*); bacterial wilt of cucurbits, such as squashes and watermelons (caused by *Erwinia tracheiphila*); and black vein of crucifers, such as cabbage (caused by *Xanthomonas campestris*). The most economically important wilt disease of plants, however, is caused by *Pseudomonas solanacearum*. It affects more than 40 different genera of plants, including such major crops as bananas, peanuts, tomatoes, potatoes, eggplants, and tobacco, to name a few. This disease occurs worldwide in tropical, subtropical, and warm temperate areas.

The members of another genus of *Bacteria*, known as *Agrobacterium*, cause a cancer-like disease known as crown gall in plants, "gall" being a general term for swelling. Because of their role in genetic engineering, these *Bacteria* will be discussed in Chapter 28.

Archaea

There are four phylogenetically distinct lineages of *Archaea*, exhibiting enormous physiological diversity. On the basis of that diversity, the *Archaea* can be divided into three large groups—extreme halophiles, methanogens, and extreme thermophiles—and one small group represented by a thermophile that lacks a cell wall. Until very recently, the archaea were generally thought of as noncompetitive relics inhabiting hostile environments that were of little importance to global ecology. It is now known, however, that the archaea are present in less hostile environments such as soil. Archaea also constitute a major component of the oceanic picoplankton (organisms smaller than 1 micrometer), possibly outnumbering all other oceanic organisms.

14–20

Extreme halophilic archaea growing in evaporating ponds of seawater, as seen in an aerial view taken near San Francisco Bay, California. The ponds yield table salt, as well as other salts of commercial value. As the water evaporates and the salinity increases, the halophiles multiply harmlessly, forming massive growths, or "blooms," that turn the seawater pink or reddish purple.

The Extreme Halophiles Are the "Salt-Loving" Archaea

The **extreme halophilic archaea** are a diverse group of prokaryotes that occur everywhere in nature where the salt concentration is very high—in places such as Great Salt Lake and the Dead Sea, as well as in ponds where seawater is left to evaporate, yielding table salt (Figure 14–20). Extreme halophiles have a very high requirement for salt, with most of them requiring 12 to 23 percent salt (sodium chloride, NaCl) for optimal growth. Their cell walls, ribosomes, and enzymes are stabilized by the sodium ion, Na^+.

All extreme halophiles are chemoorganotrophs (heterotrophs that obtain their energy from the oxidation of organic compounds), and most species require oxygen. In addition, certain species of extreme halophiles exhibit a light-mediated synthesis of ATP that does not involve any chlorophyll pigments. Among them is *Halobacterium halobium*, the prevalent species of *Archaea* in the Great Salt Lake. Although the high concentration of salt in their environment limits the availability of O_2 for respiration, such extreme halophiles are able to supplement their ATP-producing capacity by converting light energy into ATP using a protein called bacteriorhodopsin, which is found in the plasma membrane.

The Methanogens Are the Methane-Producing Archaea

The **methanogens** are a unique group of prokaryotes—the only ones that produce methane (Figure 14–21). All are strictly anaerobic and will not tolerate even the slightest exposure to oxygen. Methanogens can produce methane (CH_4) from hydrogen (H_2) and carbon dioxide (CO_2), with the needed electrons being derived from the H_2 and the CO_2 serving as both carbon source and electron acceptor.

All methanogens use ammonium (NH_4^+) as a nitrogen source, and a few can fix nitrogen. Methanogens are common in sewage-treatment plants, in bogs, and in the ocean depths. In fact, most of the natural gas reserves now used as fuel were produced by the activities of methane-producing prokaryotes in the past. Methanogens are also found in the digestive tracts of cattle and other ruminant (cud-chewing) animals, where they make possible the digestion of cellulose. It is estimated that a cow belches about 50 liters of methane a day while chewing the cud.

1 μm

14–21

Scanning electron micrograph of a methane-producing archaea from the digestive tract of a cud-chewing animal. Cells such as those shown here produce methane and carbon dioxide. Methanogens are strict anaerobes and can therefore live only in the absence of oxygen—a condition prevailing on the young Earth but occurring today only in isolated environments.

14–22

A steaming hot spring in Yellowstone National Park, with hydrogen sulfide–rich steam rising to the surface of the Earth. This is a typical habitat for extreme thermophilic archaea, where, because of the heat and acidity, higher forms of life cannot develop. The archaea form a mat around the spring.

The Extreme Thermophiles Are the "Heat-Loving" Archaea

The **extreme thermophilic archaea** contain representatives of the most heat-loving of all known prokaryotes. The membranes and enzymes of these archaea are unusually stable at high temperatures: all have temperature optima above 80°C, and some grow at temperatures over 110°C. Most species of extreme thermophiles metabolize sulfur in some way, and, with only a few exceptions, they are strict anaerobes. These archaea are inhabitants of hot, sulfur-rich environments, such as the hot springs and geysers found in Iceland, Italy, New Zealand, and Yellowstone National Park (Figure 14–22). As noted, extreme thermophilic archaea also grow near deep-sea hydrothermal vents and cracks in the ocean floor from which geothermally superheated water is emitted (page 287).

Thermoplasma Is an Archaea Lacking a Cell Wall

A fourth group of *Archaea* consists of a single known genus, *Thermoplasma*, containing a single species, *T. acidophilum*. *Thermoplasma* resembles the mycoplasmas (page 292) in lacking a cell wall and being very small; individuals vary from spherical (0.3 to 2 micrometers in diameter) to filamentous. *Thermoplasma* is only known to occur in acidic, self-heating coal refuse piles in southern Indiana and western Pennsylvania, in sites within the piles where the temperatures range from 32° to 80°C— the sort of very unusual habitat where archaea seem to thrive.

Because archaea often grow in places that resemble habitats characteristic of the early Earth, it has been suggested that the methanogens might have originated more than 3 billion years ago, when there was an atmosphere rich in CO_2 and H_2, and that they might have per-

sisted in certain habitats similar to those in which they first evolved.

Viruses

The existence of viruses was first recognized just over 100 years ago when Dutch, German, and Russian scientists showed that tobacco mosaic disease could be produced on plants with a liquid that had passed through a filter used to trap bacteria. Until the 1930s, viruses were considered to be extremely small bacteria or even enzymes or proteins. Evidence against this point of view began to accumulate in 1933, when Wendell Stanley, a biochemist working at the Rockefeller Institute for Medical Research, prepared an extract from diseased tobacco plants that contained the tobacco mosaic virus (TMV) and purified it. In the presence of high salt concentrations, the purified virus precipitated in the form of crystals, thus behaving more like a chemical than an organism. When these needlelike crystals were redissolved and the solution applied to a tobacco leaf, the characteristic symptoms of tobacco mosaic disease appeared. Thus the virus had retained its ability to infect tobacco plants even after it had been crystallized, unlike any known organism.

Practically every kind of organism is now known to be infected by distinct viruses, and it is clear that a tremendous diversity of viruses exists. A virus typically is associated with a specific type of host and is usually discovered and studied because it causes a disease in that host.

In humans, viruses are responsible for many infectious diseases, including chicken pox, measles, mumps, influenza, colds (often complicated by secondary bacterial infections), infectious hepatitis, polio, rabies, herpes, AIDS, and deadly hemorrhagic fevers (such as those caused by Ebola virus and Hantaan virus). Viruses also

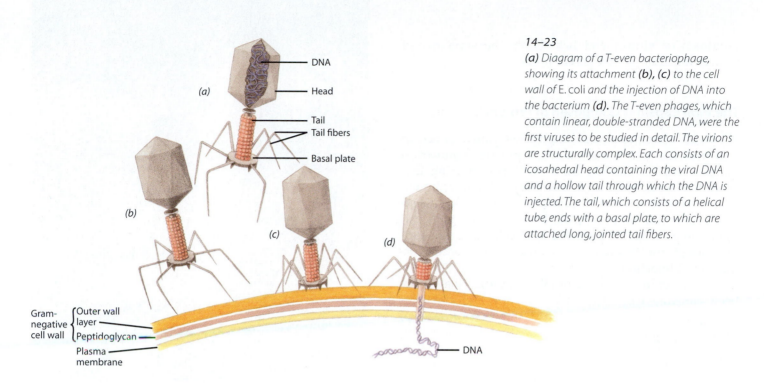

(a) Diagram of a T-even bacteriophage, showing its attachment (b), (c) to the cell wall of E. coli and the injection of DNA into the bacterium (d). The T-even phages, which contain linear, double-stranded DNA, were the first viruses to be studied in detail. The virions are structurally complex. Each consists of an icosahedral head containing the viral DNA and a hollow tail through which the DNA is injected. The tail, which consists of a helical tube, ends with a basal plate, to which are attached long, jointed tail fibers.

attack other animals, as well as prokaryotes (including mycoplasmas), protists, fungi, and plants. Viruses cause many plant diseases. In fact, about one-fourth of all known viruses attack and cause diseases of plants. Some of the viruses infecting plants are quite similar to those infecting animals. The main differences are in the proteins on the cell surface, which restrict a virus to a narrow ecological niche. Once inside a cell, viral gene expression and replication require specific interactions with other host components. Subsequent spread of viruses between plant cells requires yet other virus-host interactions. Any of these steps can be blocked by incompatibilities between virus and host components.

Viruses that infect bacteria are called **bacteriophages** ("bacteria eaters") or simply **phages** (Figure 14–23). Studies of bacteriophages were very important in the early development of molecular biology. Increasingly powerful approaches have also been developed for studying animal and plant viruses.

Experiments with bacteriophages have revealed many important and previously unappreciated features of eukaryotic cell biology, including RNA splicing and many aspects of DNA replication, transcription, and translation. Thus viruses have been and continue to be very important sources of insight into the general biology of cells. Viruses also serve as gene-transfer systems and have become indispensable tools of modern biotechnology, as discussed in Chapter 28.

Viruses Are Noncellular Structures Consisting of Protein and DNA or RNA

Viruses consist of a nucleic acid genome surrounded by protein, which protects the genome in an external environment and helps the virus attach to the next cell or host. Viruses may survive on their own outside a living cell, but increase in number only by replication in appropriate susceptible host cells. Outside the cell a virus is a submicroscopic particle that is metabolically inert. The virus particle, commonly called the **virion,** is the structure by which the virus genome is carried from one cell to another.

Viral genomes are composed of either RNA or DNA—double-stranded or single-stranded—depending on the virus. Viruses lack plasma membranes, cytoplasm with ribosomes, and enzymes necessary for protein synthesis and energy production. Viral genes, however, encode the information for the production of enzymes involved in nucleic acid replication and protein processing. Viral genomes are highly organized sequences of nucleotides with distinct genes and regulatory sequences. The regulatory sequences in DNA viruses often function in ways similar to those that regulate genes in cellular DNA. In RNA viruses, however, the mechanisms of regulation of gene expression often differ from the DNA-based pathways common in cells. The entire nucleotide sequence has been determined for hundreds of viruses and has shown the similarity of viruses infecting very different hosts.

The Viral Capsid Is Composed of Protein Subunits

All viruses have one or more proteins, called capsid or coat proteins, which assemble in a precise symmetrical manner to form the **capsid,** a shell-like covering that protects the nucleic acid. Some viruses also have an envelope of lipid molecules interspersed with proteins on the outer surface of the capsid. The surface proteins and lipids help recognize potential host cells and also provide targets for the immune response of animals combating a viral infection.

The structures of a large number of viruses have now been elucidated by electron microscopy and crystallography. The tobacco mosaic virus—the first to be seen, in 1939—is a rodlike particle about 300 nanometers long and 15 nanometers in diameter. The RNA of over 6000 nucleotide bases forms a single strand that fits into a groove in each of over 2000 identical protein molecules arranged with helical symmetry much like a coiled spring (Figure 14–24). Most plant viruses, like tobacco mosaic virus, are single-stranded RNA viruses. The few plant viruses with a double-stranded RNA genome all belong to the family *Reoviridae*. Rice gall dwarf disease is caused by a reovirus, specifically a phytoreovirus that is spread from plant to plant by leafhoppers.

The most common shape of a virus is an icosahedron, a 20-sided structure, in which the capsids are assembled from 180 or more protein molecules arranged in a symmetry similar to that of a geodesic dome (Figure 14–25). The size of most of the icosahedral plant viruses is about 30 nanometers in diameter. Viruses having this shape include tobacco ringspot, cucumber mosaic, cowpea chlorotic mottle (Figure 14–25a), and Tulare apple mosaic viruses (Figure 14–26a). Other plant viruses are bullet-shaped (Figure 14–26b) or appear as long, flexible rods or threads (Figure 14–26c).

Viruses Multiply by Taking Over the Genetic Machinery of the Host Cell

The transmission, or spread, of viruses from diseased to healthy plants most commonly involves insect vectors, such as aphids, leafhoppers, or whiteflies, with sucking mouthparts called stylets. Such insects that have fed on diseased plants carry viruses on their stylets and transmit the virus to a healthy plant when they feed upon it. Except for a few of the plant RNA viruses (see below), plant viruses do not actually infect their vectors, that is, the organisms that transmit them. Instead, plant viruses are simply carried in a very specific manner, and each type of plant virus is transmitted only by a particular

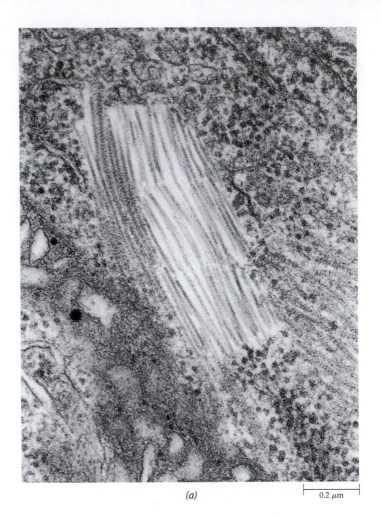

(a)   0.2 μm

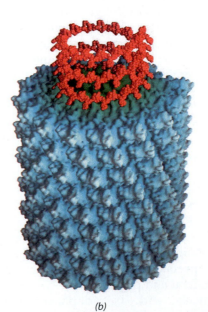

(b)

14–24

Tobacco mosaic virus (TMV). (a) Electron micrograph showing virus particles in a mesophyll cell of a tobacco leaf. (b) A portion of the TMV particle, as determined by X-ray crystallography. The single-stranded RNA, shown here in red, fits into the grooves of the protein subunits that assemble into the helical arrangement of the capsid.

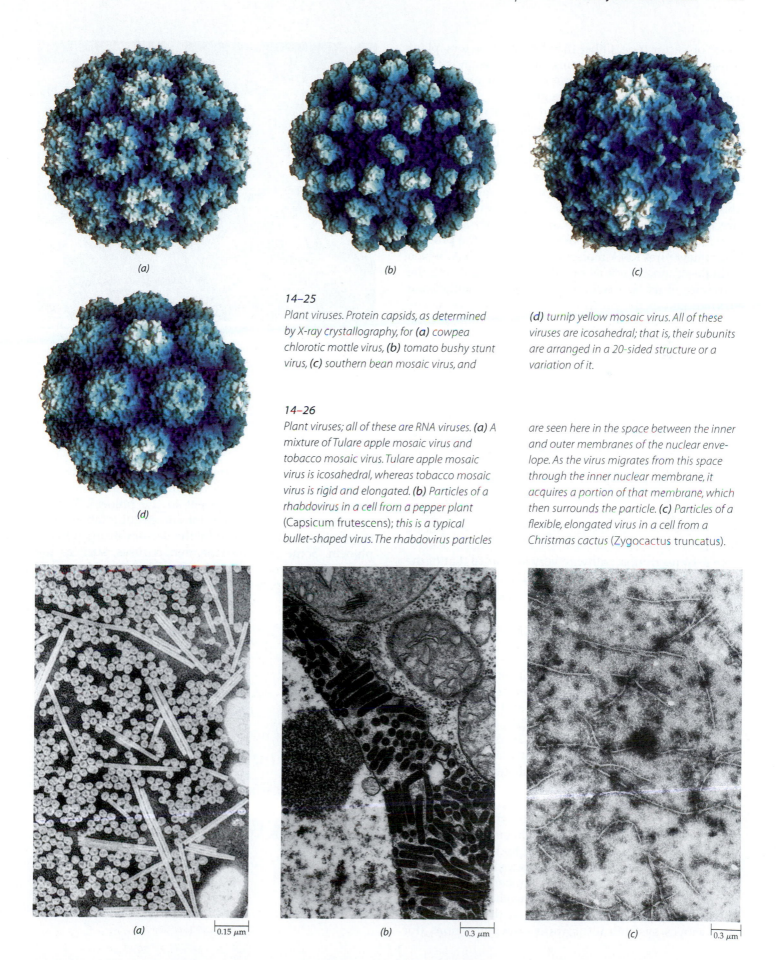

(a)

(b)

(c)

14–25

*Plant viruses. Protein capsids, as determined by X-ray crystallography, for **(a)** cowpea chlorotic mottle virus, **(b)** tomato bushy stunt virus, **(c)** southern bean mosaic virus, and* **(d)** *turnip yellow mosaic virus. All of these viruses are icosahedral; that is, their subunits are arranged in a 20-sided structure or a variation of it.*

(d)

14–26

*Plant viruses; all of these are RNA viruses. **(a)** A mixture of Tulare apple mosaic virus and tobacco mosaic virus. Tulare apple mosaic virus is icosahedral, whereas tobacco mosaic virus is rigid and elongated. **(b)** Particles of a rhabdovirus in a cell from a pepper plant (Capsicum frutescens); this is a typical bullet-shaped virus. The rhabdovirus particles are seen here in the space between the inner and outer membranes of the nuclear envelope. As the virus migrates from this space through the inner nuclear membrane, it acquires a portion of that membrane, which then surrounds the particle. **(c)** Particles of a flexible, elongated virus in a cell from a Christmas cactus (Zygocactus truncatus).*

(a) |0.15 μm|

(b) |0.3 μm|

(c) |0.3 μm|

vector or vector type. Other than by insect vectors, plant viruses can enter the plant only through wounds made mechanically or by transmission into an ovule via a pollen tube of an infected pollen grain.

Once inside a host cell, a virion sheds its capsid (and, if one is present, its envelope), freeing its nucleic acid. Within the cell, the RNA or DNA of the virus then multiplies by taking over the cell's genetic machinery, thus producing the nucleic acids and proteins necessary to assemble additional virus particles. In a single-stranded RNA virus, such as tobacco mosaic virus, the viral RNA directs the formation of a complementary strand of RNA, which then serves as the template for the production of new viral RNA molecules. The viral RNA, which is single-stranded, acts as messenger RNA, utilizing the ribosomes of its host cell and directing the synthesis of enzymes and of the protein subunits of the capsid. The new RNA strands and capsid protein subunits are assembled into complete virus particles within the host cell.

Three types of plant viruses, the geminiviruses, the badnaviruses, and the caulimoviruses, have DNA as their genetic material. Geminiviruses are small, spherical particles that almost always exist as connected pairs (Figure 14–27). Bean golden mosaic is a disease of bean plants caused by a geminivirus. It is spread from plant to plant by whiteflies and occurs mainly in tropical climates. Another such geminivirus causes maize (corn) streak, which is spread by leafhoppers and has the smallest known genome of any virus. Badnaviruses and caulimoviruses are larger viruses that replicate their DNA by means of an RNA intermediate, using a reverse transcription pathway very similar to that of the human immunodeficiency virus (HIV) and other retroviruses.

Within the Plant, Viruses Move via Plasmodesmata, and Some Travel in the Phloem

Some viruses are confined to a relatively small area of the initial site of infection, whereas others move throughout the plant body and are said to be **systemic.** Short-distance movement from cell to cell occurs through plasmodesmata (Figure 14–28). Such movement is facilitated by viral-encoded proteins called **movement proteins,** which apparently bring about an increase in the size of the plasmodesmata and allow passage of the virus. Cell-to-cell movement via plasmodesmata is a slow process. In a leaf, for instance, the virus moves about 1 millimeter, or 8 to 10 parenchyma cells, per day.

Although some viruses appear to be restricted in their movement to parenchyma cells and their interconnecting plasmodesmata, a large number of viruses are rapidly transported in the conducting channels, or sieve tubes, of the phloem tissue. Once in the phloem, the virus moves systemically toward growing regions (for

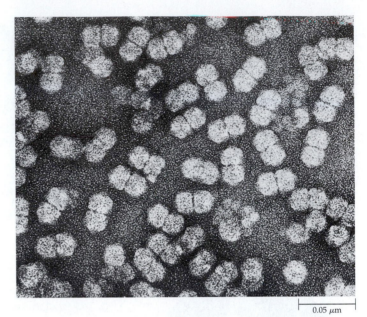

0.05 μm

14–27

Purified geminivirus from the grass Digitaria, *negatively stained in 2 percent aqueous uranyl acetate. The geminiviruses, which have DNA as their genetic material, typically occur in pairs.*

example, shoot tips and root tips) and regions of storage, such as rhizomes and tubers, where the virus reenters the parenchyma cells adjacent to the phloem. Viruses that depend on the phloem for successful establishment of infection are introduced by the vector directly into the phloem. Some phloem-dependent viruses, such as the beet yellows virus, seem to be limited to the phloem and a few adjacent parenchyma cells. Tobacco mosaic virus,

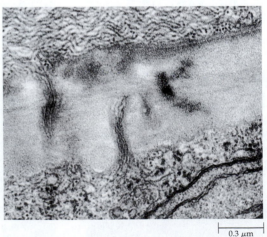

0.3 μm

14–28

Beet yellows virus particles in plasmodesmata, moving from a sieve-tube member of the phloem (above) to its sister cell, a companion cell (below).

by contrast, invades all kinds of cells. Upon entering the sieve tubes, the viruses are transported to other parts of the plant along with the sugars and other substances present in the sieve-tube sap.

Viruses Cause a Variety of Plant Diseases

Well over 2000 kinds of plant diseases are known to be caused by the more than 600 different kinds of plant viruses identified. Viral diseases greatly reduce the productivity of many kinds of agricultural and horticultural crops. Worldwide losses due to viral diseases are estimated at about $15 billion annually.

Often the only symptom produced by virus infection is reduced growth rate, resulting in various degrees of dwarfing or stunting of the plant. The most obvious symptoms are usually those that appear on leaves, where the viruses interfere with chlorophyll production, thus affecting photosynthesis. Mosaics and ring spots are the most common symptoms produced by systemic viruses. In mosaic diseases, light green, yellow, or white areas—ranging in size from small flecks to large stripes—appear intermingled with the normal green of leaves and fruit (Figure 14–29). In ring spot diseases, chlorotic (yellow) or necrotic (dead tissue) rings appear on the leaves and sometimes also on the stems and fruits. Less common viral diseases include leaf roll (potato leaf roll), yellows (beet yellows), dwarf (barley yellow dwarf), canker (cherry black canker), and tumor (wound tumor) (Figure 14–30). The yellow blotches or borders on the leaves of some prized horticultural varieties may be caused by viruses, and the variegated appearance of some flowers is the result of viral infections that are passed on from generation to generation in vegetatively propagated plants (Figure 14–31).

14–29

Tobacco leaf infected with tobacco mosaic virus.

14–30

(a) Tumors produced by the wound tumor virus in sweet clover (Melilotus alba). *(b)* Wound tumor virus particles (arrows) visible in an electron micrograph of a cell of the host plant. *(c)* The virus is transmitted by the clover leafhopper (Agallia constricta). This electron micrograph shows a clover epidermal cell. Viruses are being produced in the honeycomblike area at the upper left. Individual viruses can be seen in the dark area below.

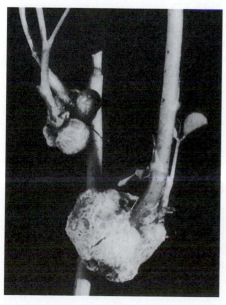

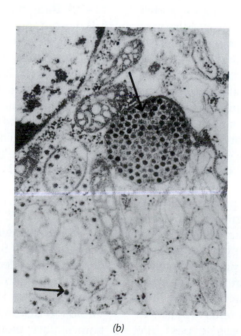

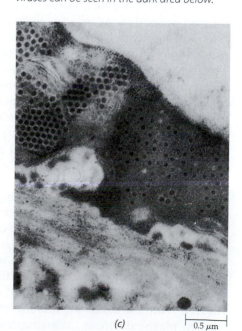

(a)   (b)   (c)   0.5 μm

14–31

Streaked flowers of Rembrandt tulips. The streaking is caused by a viral infection transmitted directly from plant to plant. The infection weakens the plant, and it subsequently dies. Current practice is to achieve streaking by selecting for transposon effects (page 193).

Viral Diseases Are Controlled in Various Ways   Because viral diseases greatly reduce the productivity of many different kinds of crops worldwide, or decrease the desirability of the product, much effort is being expended to find effective ways to control these diseases. Chemicals sprayed onto infected plants are not effective, and vector control is often not practical and efficient. For the vegetatively propagated plants, including potatoes and many flower and fruit crops, meristem culture has been effective. The virus often does not invade the growing tip, so a new plant grown in tissue culture from the tip dissected away from the parent plant will often be virus-free. Such plants are then used as "mother" plants from which cuttings are produced to supply growers with virus-free plants. Eliminating seedborne virus is also effective in many crops, including peas, beans, barley, and lettuce. Samples of seeds are grown, and if the virus is present, the entire seed lot is discarded.

Another approach to the control of plant viruses has focused on the ability of a host plant to prevent the virus from replicating or from moving within the plant body, thus making the host plant resistant. As mentioned previously, viral-encoded movement proteins appear to modify plasmodesmata, allowing viruses to pass into adjacent cells. Some movement proteins display substantial host specificity and can only support significant cell-to-cell spread in certain hosts. Interestingly, in at least some cases, such misadapted movement proteins are competent for initial cell-to-cell spread in nonhost plants but lead to rapid induction of host resistance responses that block infection of further cells. Other types of plant resistance involve preventing the initial replication of the virus, blocking long-distance movement, or responding with a hypersensitive reaction that limits the virus by killing cells near the point of infection.

Resistance is known to be conditioned by specific host genes, and one such gene, the *N* gene from tobacco, has recently been cloned and characterized. Several other host resistance genes have been genetically mapped to chromosomal locations, and still others are widely used in developing varieties of crops that resist viruses. Viruses, like many other plant pathogens, often mutate and diversify to overcome these resistance genes. Plants can be genetically engineered, however, to resist virus infection by inserting any of several pieces of the viral nucleic acid into the host's genome. Some sequences work by completely blocking virus replication, whereas others greatly decrease the ability of the virus to move from cell to cell. Squash was the first crop to be marketed with genetically engineered resistance to viruses, but several others are expected to be used in cases where there is no other means of viral disease control.

Viroids: Other Infectious Particles

Viroids are the smallest known agents of infectious disease. They consist of small, circular, single-stranded molecules of RNA and lack capsids of any kind (Figure 14–32). Ranging in size from 246 base pairs (the coconut cadang-cadang viroid) to 375 base pairs (the citrus exocortis viroid), viroids are much smaller than the smallest viral genomes. Although the viroid RNA is a single-stranded circle, it can form a secondary structure that resembles a short double-stranded molecule with closed ends. The viroid RNA contains no protein-encoding genes, and hence it is totally dependent on the host for its replication. The viroid RNA molecule appears to be replicated in the nucleus of the host cell, where it apparently mimics DNA and allows the host cell RNA polymerase to replicate it. Viroids may cause their symptoms by interfering with gene regulation in the infected host cell.

The term "viroid" was first used by Theodor O. Diener of the U.S. Department of Agriculture in 1971, to describe the infectious agent causing the spindle tuber disease of potato. Potatoes infected by the potato spindle tuber viroid, or PSTVd (the d is added to the initials of viroids to distinguish them from abbreviations for virus names), are elongated (spindle-shaped) and gnarled.

14–32

Electron micrograph of potato spindle tuber viroid (arrows) mixed with portions of the double-stranded DNA molecule of a bacteriophage. This micrograph illustrates the tremendous difference in size between a viroid and a virus.

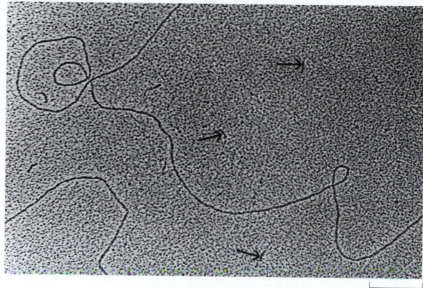

0.25 μm

They sometimes have deep crevices in their surfaces. Viroids have been identified as the cause of several economically important plant diseases. The coconut cadang-cadang viroid (CCCVd), for example, has resulted in the death of millions of coconut trees in the Philippines over the last half century, and the chrysanthemum stunt viroid (CSVd) nearly destroyed the chrysanthemum industry in the United States in the early 1950s.

The Origin of Viruses

Because of the simplicity of viral genomes and viral structure, some earlier investigators thought that viruses might represent the direct descendants of the self-replicating units from which the first cells evolved. Other investigators think that this line of descent is unlikely, for viruses exist only by virtue of their ability to utilize the genetic machinery of their host cells. They compete with the nucleic acids of these cells and take over the host genetic and metabolic activities in directing the formation of new viral particles. Therefore, viruses may have come into being after the evolution of cells in which the genetic code was already established.

Viruses may have originated as segments of host genomic material that escaped and acquired the ability to replicate independently in a cell and be transmitted to another cell. Acquisition of a protein that could protect the nucleic acid in the transmission process, and facilitate recognition of new host cells, was key to the evolution of viruses. Viruses can evolve with astonishing rapidity in response to strong selection pressures. It is likely that novel viruses are still evolving today from both bacteria and eukaryotes.

Summary

The *Bacteria* and *Archaea* Are the Two Prokaryotic Domains

The prokaryotes are the smallest and structurally simplest organisms. In evolutionary terms, they are also the oldest organisms on Earth, and they consist of two distinct lineages, the domains *Bacteria* and *Archaea*. Prokaryotes lack an organized nucleus and membrane-bounded cellular organelles. Most of their genetic material is incorporated into a single circular molecule of double-stranded DNA, which replicates before cell division. Often, additional small pieces of circular DNA, known as plasmids, are also present. Except for mycoplasmas, all prokaryotes have rigid cell walls. In the *Bacteria*, this wall is composed mainly of peptidoglycan. Gram-negative *Bacteria*, which have walls that do not retain the dye known as crystal violet, have an outer layer of lipopolysaccharides and proteins over the peptidoglycan layer. Many prokaryotes secrete slimy or gummy substances on the surface of their walls, forming a layer called the glycocalyx, or capsule. A wide variety of prokaryotes—both *Bacteria* and *Archaea*—contain granules of poly-β-hydroxybutyric acid and glycogen, which are food storage compounds.

Prokaryotic cells may be rod-shaped (bacilli), spherical (cocci), or spiral (spirilla). All prokaryotes are unicellular, but, if the cell wall does not divide completely following cell division, the daughter cells may adhere in groups, in filaments, or in solid masses. Many prokaryotes have flagella and, hence, are motile; the rotation of the flagella moves the cell through the medium. Lacking microtubules, the flagella of prokaryotes differ greatly from those of eukaryotes. Prokaryotes may also have fimbriae or pili.

Most prokaryotes reproduce by simple cell division, also called binary fission. Mutations, combined with the rapid generation time of prokaryotes, are responsible for their extraordinary adaptability. Further adaptability is provided by the genetic recombinations that take place as a result of conjugation, transformation, and transduction. Certain species of bacteria have the capacity to form endospores, dormant resting cells that allow the cell to survive unfavorable conditions.

Prokaryotes Exhibit Tremendous Metabolic Diversity

Although some are autotrophs, most prokaryotes are heterotrophs. The vast majority of heterotrophs are saprophytes and, together with the fungi, they are the recyclers of the biosphere. Some autotrophs—the photosynthetic autotrophs—obtain their energy from light. Other autotrophs obtain their energy from the reduction of inorganic compounds and are called chemosynthetic autotrophs. A number of genera play important roles in the cycling of nitrogen, sulfur, and carbon. Of all living organisms, only certain bacteria are capable of nitrogen fixation. Without bacteria, life on Earth as we know it would not be possible.

Some prokaryotes are aerobic, others are strict anaerobes, and still others are facultative anaerobes. Prokaryotes also vary with regard to the range of temperatures at which they grow, ranging from those that can grow at 0°C or lower (psychrophiles) to those that can grow at temperatures greater than 100°C (extreme thermophiles).

The *Bacteria* Include Pathogenic and Photosynthetic Organisms

Many bacteria are important pathogens in both plants and animals. One distinctive group of bacteria, the mycoplasmas (and mycoplasmalike organisms), which lack a cell wall and are very small, include a number of disease-causing organisms.

Photosynthetic bacteria can be divided into three major groups: the cyanobacteria, the prochlorophytes, and the purple and green bacteria. The cyanobacteria and prochlorophytes contain chlorophyll *a*, the same molecule that occurs in all photosynthetic eukaryotes. In addition, the prochlorophytes contain chlorophyll *b*, but lack phycobilins, which are present in cyanobacteria. Many genera of cyanobacteria can fix nitrogen. Clearly

Bacteria seem to have been involved in the symbiotic origin of chloroplasts. Early in the history of eukaryotes, a symbiotic event similar to that occurring in the origin of chloroplasts seems to have given rise to mitochondria; purple nonsulfur bacteria apparently were the group involved in this event.

The *Archaea* Are Physiologically Diverse Organisms That Occupy a Wide Variety of Habitats

The *Archaea* can be divided into three large groups: extreme halophiles, methanogens, and extreme thermophiles. A fourth group is represented by a single genus, *Thermoplasma*, which lacks a cell wall. Once believed to occupy primarily hostile environments, the *Archaea* are now known to constitute a major component of the oceanic picoplankton.

Viruses Are Noncellular Structures Consisting of DNA or RNA Surrounded by a Coat of Protein

Viruses possess genomes that replicate within a living host by directing the genetic machinery of that cell to synthesize viral nucleic acids and proteins. Viruses contain either RNA or DNA—single-stranded or double-stranded—surrounded by an outer protein coat, or capsid, and sometimes also by a lipid-containing envelope. Viruses are comparable in size to large macromolecules and come in a variety of shapes. Most are spherical, with icosahedral symmetry, and a number of others are rod-shaped, with helical symmetry.

Viruses and Viroids Cause Diseases in Plants and Animals

Viruses are responsible for many diseases of humans and other animals, and also for more than 2000 known kinds of plant diseases. The transmission of viruses from diseased to healthy plants most commonly involves insect vectors. Once inside a host cell, the virus particle, or virion, sheds its capsid, freeing its nucleic acid. Most plant viruses are RNA viruses. In single-stranded RNA viruses, such as tobacco mosaic virus, the viral RNA directs the formation of a complementary strand of RNA, which then serves as a template for the production of new viral RNA molecules. Utilizing the ribosomes of the host cell, the viral RNA directs the synthesis of capsid proteins. The new RNA strands and capsid proteins then are assembled into complete virions within the host cell.

Short-distance movement of viruses from cell to cell within host plants occurs through plasmodesmata. Such movement is facilitated by viral-encoded proteins called movement proteins. A large number of plant viruses move systemically throughout the plant in the phloem.

Viroids, the smallest known agents of infection, consist of small, circular, single-stranded molecules of RNA. Unlike viruses, viroids lack protein coats. They are thought to interfere with gene regulation in infected host cells, where they occur mainly within the nucleus.

Selected Key Terms

aerobes   p. 287

akinetes   p. 290

anaerobes   p. 287

bacilli   p. 284

bacteriophage, or phage   p. 297

binary fission   p. 285

capsid   p. 298

chemosynthetic autotrophs   p. 287

cocci   p. 284

conjugation   p. 285

cyanobacteria   p. 288

endospores   p. 286

facultative anaerobes   p. 287

fimbriae   p. 284

glycocalyx   p. 284

gram-negative   p. 283

gram-positive   p. 283

halophiles   p. 295

heterocysts   p. 290

heterotrophs   p. 286

hormogonia   p. 288

methanogens   p. 295

movement proteins   p. 300

mycoplasmalike organisms (MLOs)   p. 292

mycoplasmas   p. 292

nitrogen fixation   p. 290

nucleoid   p. 282

peptidoglycans   p. 283

photosynthetic autotrophs   p. 287

pili   p. 284

plankton   p. 289

plasmid   p. 282

prochlorophytes   p. 292

psychrophiles   p. 287

saprophytes   p. 286

spirilla   p. 284

spiroplasmas   p. 292

strict anaerobes   p. 287

stromatolites   p. 289

thermophiles   p. 287

transduction   p. 286

transformation   p. 286

virion   p. 297

viroids   p. 302

Questions

1. Distinguish between the following: gram-positive bacterium/gram-negative bacterium; fimbria/pilus; endospore/akinete; virus/viroid.

2. Pear decline is so called because it causes a slow, progressive weakening and ultimate death of the pear tree. It is a systemic disease caused by an MLO, or phytoplasma. What is meant by a "systemic disease," and by what pathway do MLOs move through the tree?

3. What genetic factors contribute to the extraordinary adaptability of prokaryotes to a broad range of environmental conditions?

4. What are some of the ways by which plant viral diseases may be controlled?

5. One might argue that viruses should be considered living organisms. By what criteria might viruses be considered alive?

Chapter

15 | Fungi

15–1
Mycelia of several different fungi growing under the bark of a fallen tree. Decomposition by fungi and bacteria makes it possible for the organic material incorporated into the bodies of organisms to be recycled in the ecosystem.

OVERVIEW

Fungi are literally everywhere on Earth. They absorb their food by secreting digestive enzymes into the immediate environment. These enzymes catalyze the breakdown of large food molecules into molecules small enough to be absorbed into the fungal cell. For this reason, fungi usually grow within or on top of their food supply.

This absorptive mode of nutrition is responsible for both the praises and curses earned by fungi. On the one hand, fungi are extremely beneficial because the digestive enzymes they secrete are responsible for decomposing dead plants and animals, making it possible to recycle chemical elements. However, fungi also secrete enzymes that digest materials valuable to us, causing mold on food, mildew on fabrics, dry rot on wood, and athlete's foot. Commercial use of fungi and their enzymes produces wine, beer, certain cheeses, and various antibiotics.

We first examine the groups that make up the kingdom **Fungi** *and then explore the important symbiotic relationships of fungi with other organisms.*

CHECKPOINTS

By the time you finish reading this chapter, you should be able to answer the following questions:

1. In what ways do the fungi differ from all other life forms? In other words, what are the distinctive characteristics of the *Fungi?*

2. From what type of organism is it thought that fungi evolved?

3. What are the distinguishing characteristics of the *Chytridiomycota, Zygomycota, Ascomycota,* and *Basidiomycota?*

4. What are the deuteromycetes, and what is their relationship to other groups of fungi?

5. What is a yeast, and what is the relationship of yeasts to the other groups of fungi?

6. What kinds of symbiotic relationships exist between fungi and other organisms?

The fungi are heterotrophic organisms that were once considered to be primitive or degenerate plants lacking chlorophyll. It is now clear, however, that the only characteristic fungi share with plants—other than those common to all eukaryotes—is a multicellular growth form. (A few fungi, including yeasts, are unicellular.) Recent molecular evidence strongly suggests that fungi are more closely related to animals than to plants. As we shall see, the fungi are a form of life so distinctive from all others that they have been assigned their own kingdom—the kingdom *Fungi.*

Over 70,000 species of fungi have been identified thus far, with some 1700 new species discovered each year. Conservative estimates for the total number of species exceed 1.5 million, placing the fungi second only to insects in this regard. The largest living organism on Earth today may be an individual of the tree root-rot fungus *Armillaria ostoyae,* which encompasses more than 600 hectares (1500 acres) of forest near Mount Adams in the state of Washington. This fungus is estimated to be 400 to 1000 years old. A close relative of *A. ostoyae, Armillaria gallica,* has been found occupying 15 hectares (35 acres) in northern Michigan. This "humongous fungus," as it has been dubbed, is estimated to be at least 1500 years old.

The Importance of Fungi

Fungi Are Important Ecologically As Decomposers

The ecological impact of the fungi cannot be overestimated. Together with the heterotrophic bacteria, fungi are the principal decomposers of the biosphere (Figure 15–1). Decomposers are as necessary to the continued existence of the world as are food producers. Decomposition releases carbon dioxide into the atmosphere and returns nitrogenous compounds and other materials to the soil, where they can be used again—recycled—by plants and eventually by animals. Estimates are that, on average, the top 20 centimeters of fertile soil contains nearly 5 metric tons of fungi and bacteria per hectare (2.47 acres). Some 500 known species of fungi, representing a number of distinct groups, are marine, breaking down organic material in the sea just as their relatives do on land.

As decomposers, fungi often come into direct conflict with human interests. A fungus makes no distinction between a rotting tree that has fallen in the forest and a fence post; the fungus is just as likely to attack one as the other. Equipped with a powerful arsenal of enzymes that break down organic substances, including lignin and cellulose of wood, fungi are often nuisances and are sometimes highly destructive. Fungi attack cloth, paint, leather, waxes, jet fuel, petroleum, wood, paper, insulation on cables and wires, photographic film, and even the coating of the lenses of optical equipment—in fact,

15–2
The common mold Rhizopus *growing on strawberries.*

almost any conceivable substance. Although individual species of fungi are highly specific to particular substrates, as a group they attack virtually anything. Everywhere, they are the scourge of food producers, distributors, and sellers alike, for they grow on bread, fresh fruits (Figure 15–2), vegetables, meats, and other products. Fungi reduce the nutritional value, as well as the palatability, of such foodstuffs. In addition, some produce very toxic substances known as **mycotoxins** on certain plant materials.

Fungi Are Important Medically and Economically As Pests, Pathogens, and Producers of Certain Chemicals

The importance of fungi as commercial pests is enhanced by their ability to grow under a wide range of conditions. Some strains of *Cladosporium herbarum,* which attacks meat in cold storage, can grow at a temperature as low as −6°C. In contrast, one species of *Chaetomium* grows optimally at 50°C and survives even at 60°C.

Many fungi attack living organisms, rather than dead ones, and sometimes do so in surprising ways (see "Predaceous Fungi" on page 333). They are the most important causal agents of plant diseases. Well over 5000 species of fungi attack economically valuable crop and garden plants, as well as trees and many wild plants. Other fungi—over 150 species have been identified—cause serious diseases in domestic animals and humans.

Although fungal infections of humans are most common in tropical regions of the world, an alarming increase has occurred in the numbers of individuals infected with fungi in all regions of the world. This increase is due in part to the growing population of individuals with compromised immune systems, such as AIDS patients in hospitals. Nearly 40 percent of all deaths from hospital-acquired infections in the mid-1980s were found to be due not to bacteria or viruses but to fungi. Around 80 percent of AIDS deaths are due to pneumonia caused by *Pneumocystis carinii*, long thought to be a protozoan. Evidence now points toward classifying *P. carinii* as a fungus, most probably a chytrid. Another serious fungal pathogen of AIDS patients is *Candida*, which causes thrush and other infections of the mucous membranes.

The qualities that make fungi such important pests can also make them commercially valuable. Certain yeasts, such as *Saccharomyces cerevisiae*, are useful because they produce ethanol and carbon dioxide, which play a central role in baking, brewing, and winemaking. Other fungi provide the distinctive flavors and aromas of specific kinds of cheese. The commercial use of fungi in industry is growing, and many antibiotics—including penicillin, the first antibiotic to be used widely—are produced by fungi. Dozens of different kinds of fungi (mushrooms) are eaten regularly by humans, and some of them are cultivated commercially. The ability of fungi to break down substances is leading to investigations into the use of fungi in toxic waste cleanup programs. The white rot fungus *Phanerochaete chrysosporium*, which survives by digesting woody material, has been very effective in the degradation of toxic organic compounds.

A striking example of the potential value of compounds derived from fungi is cyclosporin, a "wonder drug" isolated from the soil-inhabiting fungus *Tolypocladium inflatum*. Cyclosporin suppresses the immune reactions that cause rejection of organ transplants, but without the undesirable side effects of other drugs used for this purpose. This remarkable drug became available in 1979, making it possible to resume organ transplants, which had essentially been abandoned. As a result of cyclosporin, successful organ transplants have today become almost commonplace.

Fungi Form Important Symbiotic Relationships

The kinds of relationships between fungi and other organisms are extremely diverse. For example, at least 80 percent of all vascular plants form mutually beneficial associations, called **mycorrhizae,** between their roots and fungi. These associations, which are discussed further beginning on page 340, play a critical role in plant nutrition. Lichens, many of which occupy extremely hostile habitats, are symbiotic associations between fungi and either algal or cyanobacterial cells (see page 334). Symbiotic relationships also exist between fungi and insects. In one such relationship, the fungi, which produce cellulase and other enzymes needed for digestion of plant material, are cultivated by ants in fungus "gardens." The ants supply the fungus with leaf cuttings and anal droppings, and the ants eat nothing but the fungus. Neither the fungus nor the ants can exist without the other. Other symbiotic relationships involve a great variety of fungi, known as **endophytes,** that live inside the leaves and stems of apparently healthy plants. Many of these fungi produce toxic secondary metabolites that appear to protect their hosts against pathogenic fungi and attack by insects and grazing mammals (see "From Pathogen to Symbiont: Fungal Endophytes" on page 335).

TABLE 15–1   Major Characteristics of Fungal Phyla

Phylum	Representatives	Nature of Hyphae	Method of Asexual Reproduction	Type of Sexual Spore	Common Plant Diseases
Chytridiomycota (790 species)	*Allomyces, Coelomomyces*	Aseptate, coenocytic	Zoospores	None	Brown spot of corn, crown wart of alfalfa, black wart of potato
Zygomycota (1060 species)	*Rhizopus* (common bread mold), *Glomus* (endomycorrhizal fungus)	Aseptate, coenocytic	Nonmotile spores	Zygospore (in zygosporangium)	Soft rot of various plant parts
Ascomycota (32,300 species)	*Neurospora*, powdery mildews, *Morchella* (edible morels), *Tuber* (truffles)	Septate	Budding, conidia (nonmotile spores), fragmentation	Ascospore	Powdery mildew, brown rot of stone fruits, chestnut blight, Dutch elm disease
Basidiomycota (22,244 species)	Mushrooms (*Amanita*, poisonous; *Agaricus*, edible), stinkhorns, puffballs, shelf fungi, rusts, smuts	Septate with dolipore	Budding, conidia (nonmotile spores, including urediniospores), fragmentation	Basidiospore	Black stem rust of wheat and other cereals, white pine blister rust, common corn smut, loose smut of oats, *Armillaria* root rot

15–3

Fungi. (a) The chytrid Polyphagus euglenae *parasitizing a* Euglena *cell. The cytoplasm of the rounded-up* Euglena *cell is degraded. (b) A flower fly* (Syrphus) *that has been killed by the fungus* Entomophthora muscae, *a zygomycete. (c) A common morel,* Morchella esculenta, *an ascomycete. The morels are among the most prized of the edible fungi. (d) A mushroom,* Hygrocybe aurantiosplendens, *a species of* Basidiomycetes. *A mushroom is made up of densely packed hyphae, collectively known as the mycelium.*

Today most mycologists recognize four phyla of fungi: *Chytridiomycota,* the chytrids; *Zygomycota,* zygomycetes; *Ascomycota,* ascomycetes; and *Basidiomycota, Basidiomycetes, Teliomycetes,* and *Ustomycetes* (Figure 15–3; Table 15–1). Reinclusion of the chytrids in the kingdom *Fungi* is very recent. About 15 to 20 years ago, the chytrids were included in the kingdom *Fungi,* but fungal systematists began to emphasize the absence of flagellated cells as a requirement for membership in the kingdom. Although chytrids were known to share many characteristics with the fungi, their formation of flagellated cells (zoospores) disqualified them as fungi. Instead, chytrids were placed in the kingdom *Protista.* Nevertheless, recent evidence obtained from comparisons of protein and nucleic acid sequences has tipped the scales in favor of placing the chytrids in the kingdom *Fungi.* We have therefore included the *Chytridiomycota* in the *Fungi,* but it should be noted that the systematics of the fungi is in a state of flux.

(a)

(b)

(c)

(d)

Biology and Characteristics of Fungi

Most Fungi Are Composed of Hyphae

Fungi are primarily terrestrial. Although some fungi are unicellular, most are filamentous, and structures such as mushrooms consist of a great many such filaments, packed tightly together (Figure 15–3c, d). Fungal filaments are known as **hyphae,** and a mass of hyphae is called a **mycelium** (Figure 15–1). Growth of hyphae occurs at their tips, but proteins are synthesized throughout the mycelium. The hyphae grow rapidly. An individual fungus may produce more than a kilometer of new hyphae within 24 hours. (The words "mycelium" and **mycology**—the study of fungi—are derived from the Greek word *mykēs,* meaning "fungus.")

The hyphae of most species of fungi are divided by partitions, or crosswalls, called **septa** (singular: septum). Such hyphae are said to be **septate.** In other species, septa typically occur only at the bases of reproductive structures (sporangia and gametangia) and in older, highly vacuolated portions of hyphae. Hyphae lacking septa are said to be **aseptate,** or **coenocytic,** which means "contained in a common cytoplasm" or multinucleate. In most fungi, the septa are perforated by a central pore so that the protoplasts of adjacent cells are essentially continuous from cell to cell. In members of the *Ascomycota,* the pores are usually unobstructed (Figure 15–4) and usually large enough to allow nuclei, which are quite small, to squeeze through. Such mycelia are therefore functionally coenocytic. The nuclei of fungal hyphae are haploid.

All fungi have cell walls. The cell walls of plants and many protists are built on a framework of cellulose microfibrils, interpenetrated by a matrix of noncellulosic molecules, such as hemicelluloses and pectic substances. In fungi, the cell wall is composed primarily of another polysaccharide—**chitin**—which is the same material found in the hard shells, or exoskeletons, of arthropods, such as insects, arachnids, and crustaceans. Chitin is more resistant to microbial degradation than is cellulose.

With their rapid growth and filamentous form, fungi have a relationship to their environment that is very different from that of any other group of organisms. The surface-to-volume ratio of fungi is very high, so they are in as intimate a contact with the environment as are the bacteria. Usually no somatic part of a fungus is more than a few micrometers from its external environment, being separated from it only by a thin cell wall and the plasma membrane. (The terms **somatic** and **soma,** from the Greek *sōma,* meaning "body," are equivalent to the term "vegetative" for plants.) With its extensive mycelium, a fungus can have a profound effect on its surroundings—for example, in binding soil particles together. Hyphae of individuals of the same species often fuse, thus increasing the intricacy of the network.

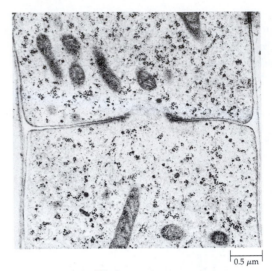

15–4

Electron micrograph of a septum between two cells in the ascomycete Gibberella acuminata. *The large globular structures are mitochondria, and the tiny dark granules are ribosomes. This specimen was thin-sectioned through the central pore region of a septum.*

Fungi Are Heterotrophic Absorbers

Because their cell walls are rigid, fungi are unable to engulf small microorganisms or other particles. Typically a fungus will secrete enzymes onto a food source and then absorb the smaller molecules that are released. Fungi absorb food mostly at or near the growing tips of their hyphae.

All fungi are heterotrophic. In obtaining their food, they function either as saprophytes (living on organic materials from dead organisms), as parasites, or as mutualistic symbionts (see page 334). Some fungi, mainly yeasts, obtain their energy by fermentation, producing ethyl alcohol from glucose. Glycogen is the primary storage polysaccharide in some fungi, as it is in animals and bacteria. Lipids serve an important storage function in other fungi.

Specialized hyphae, known as **rhizoids,** anchor some kinds of fungi to the substrate. Parasitic fungi often have similar specialized hyphae, called **haustoria** (singular: haustorium), which absorb nourishment directly from the cells of other organisms (Figure 15–5).

Fungi Have Unique Variations of Mitosis and Meiosis

One of the most characteristic features of the fungi involves nuclear division. In fungi, the processes of meio-

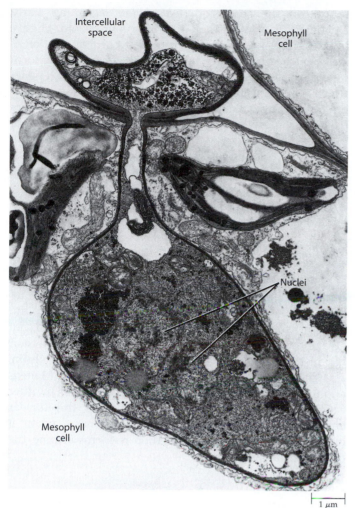

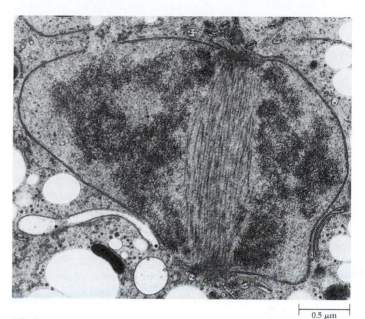

15–6

Electron micrograph of metaphase nucleus of Arthuriomyces peckianus, a rust fungus, showing the spindle inside the nucleus and the two spindle pole bodies at either end of the spindle. Spindle pole bodies, which are microtubule organizing centers, are characteristic of the Zygomycota, Ascomycota, and Basidiomycota.

15–5

Electron micrograph of a haustorium of Melampsora lini, a rust fungus, growing in a leaf cell of flax (Linum usitatissimum). In the intercellular space at the top of the micrograph is the haustorial mother cell. The narrow penetration hypha leads to the large, bulbous haustorium within the lower mesophyll cell.

sis and mitosis are different from those that occur in plants, animals, and many protists. In most fungi, the nuclear envelope does not disintegrate and re-form but is constricted near the midpoint between the two daughter nuclei. In others, it breaks down near the mid-region. In most fungi, the spindle forms within the nuclear envelope, but in some *Basidiomycota* it appears to form within the cytoplasm and move into the nucleus. Except for the chytrids, all fungi lack centrioles, but they form unique structures called **spindle pole bodies,** which appear at the spindle poles (Figure 15–6). Both spindle pole bodies and centrioles function as microtubule organizing centers during mitosis and meiosis.

Fungi Reproduce Both Asexually and Sexually

Fungi reproduce through the formation of spores that are produced either sexually or asexually. Except for the chytrids, nonmotile spores are the characteristic means of reproduction in fungi. Some spores are dry and very small. They can remain suspended in the air for long periods, thus being carried to great heights and for great distances. This property helps to explain the very wide distributions of many species of fungi. Other spores are slimy and stick to the bodies of insects and other arthropods, which may then spread them from place to place. The spores of some fungi are propelled ballistically into the air (see "Phototropism in a Fungus" on page 315). The bright colors and powdery textures of many types of molds are due to the spores. Some fungi never produce spores.

The most common method of asexual reproduction in fungi is by means of spores, which are produced either in **sporangia** (singular: sporangium) or from hyphal cells called **conidiogenous cells.** The spores produced by conidiogenous cells occur singly or in chains and are called **conidia** (singular: conidium). The sporangium is a saclike structure, the entire contents of which are converted into one or more—usually many—spores. Some fungi also reproduce asexually by fragmentation of their hyphae.

Sexual reproduction in fungi consists of three distinct phases: plasmogamy, karyogamy, and meiosis. The first two phases are phases of syngamy, or fertilization. **Plasmogamy** (the fusion of protoplasts) precedes **karyogamy** (the fusion of nuclei). In some species karyogamy follows plasmogamy almost immediately, whereas in others the two haploid nuclei do not fuse for some time, forming a **dikaryon** ("two nuclei"). Karyogamy may not take place for several months or even years. During that time the pairs of nuclei may divide in tandem, producing a dikaryotic mycelium. Eventually, the nuclei fuse to form a diploid nucleus, which sooner or later undergoes meiosis, reestablishing the haploid condition. Sexual reproduction in most fungi results in the formation of specialized spores such as zygospores, ascospores, and basidiospores.

It is important to emphasize that the diploid phase in the life cycle of a fungus is represented only by the zygote. Meiosis typically follows formation of the zygote; in other words, meiosis in fungi is zygotic (see Figure 9–3a). In addition, the gametes, when produced by fungi, are similar in appearance and size. Such gametes are referred to as **isogametes.** In most fungi, the nuclei act as the gametes. The general name of the gamete-producing structures is **gametangium** (plural: gametangia).

Evolution of the Fungi

As noted earlier, we include four phyla—*Chytridiomycota, Zygomycota, Ascomycota,* and *Basidiomycota*—in the kingdom *Fungi.* It now appears that the chytrids are the most primitive fungi and that the flagellated condition—represented by the flagellated zoospores—is a primitive character retained by the chytrids after they evolved from flagellated protists. Moreover, there is considerable molecular evidence that both animals and fungi diverged from a common ancestor, most likely a colonial protist resembling a choanoflagellate (Figure 15–7). If, in fact, the earliest fungi were aquatic flagellated organisms, the progenitors of the *Zygomycota, Ascomycota,* and *Basidiomycota* probably lost their flagellated stages fairly early in their evolutionary history. Although the *Fungi* appear to be a monophyletic lineage with a closer relationship to animals than to any other kingdom, relationships within the *Fungi* are far from certain.

Fungi have been around for a long time, but their history is poorly understood. The oldest fossils resembling fungi are represented by aseptate filaments from the Lower Cambrian, about 544 million years ago. They are believed to have been reef saprophytes. Fossil fungi morphologically similar to the chytrid *Allomyces* were found on the stems of *Aglaophyton major,* an Early Devonian plant more than 400 million years old. Highly branched hyphae were also found in cortical cells of *A. major.* Such fungi, which belong to the *Zygomycota,* form mycorrhizae, specifically **endomycorrhizae,** which penetrate cells (see page 341). They are one of the few symbiotic plant-fungus associations in the fossil record and are believed to have played a major role in the evolution of plants. The oldest known fossil *Ascomycota* are found in rocks of Silurian age (438 million years), and the oldest known *Basidiomycota* come from the Upper Devonian (380 million years ago).

Chytrids: Phylum *Chytridiomycota*

The *Chytridiomycota,* or chytrids, are a predominantly aquatic group, consisting of about 790 species. Soils from ditches and the banks of ponds and streams are also inhabited by chytrids, and some chytrids are even found in desert soils. Chytrids are varied not only in form, but also in the nature of their sexual interactions and in their life histories. The cell walls of chytrids contain chitin, and like other fungi, the chytrids store glycogen. Meiosis and mitosis in chytrids resemble these processes in other fungi in that they are intranuclear; that is, the nuclear envelope remains intact until late telophase, when it breaks in a median plane and then re-forms around the daughter nuclei.

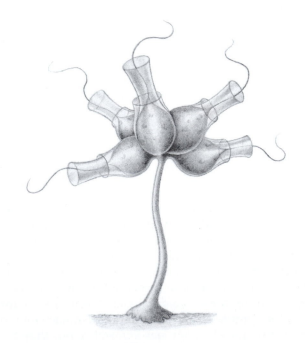

15–7

A choanoflagellate, a colonial protist that many zoologists believe is related to the ancestor of both animals and fungi.

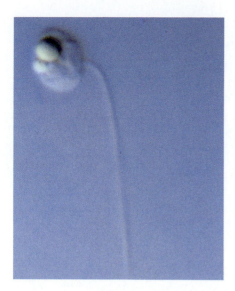

15–8
A uniflagellated zoospore of the chytrid Polyphagus euglenae. *The chytrids are distinguished from other fungi primarily by their characteristic motile zoospores and gametes.*

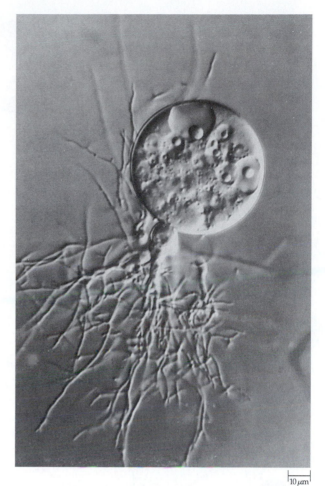

10 μm

15–9
Chytridium confervae, *a common chytrid, as seen with the aid of differential interference contrast system optics. Note the slender rhizoids extending downward.*

Almost all chytrids are coenocytic, with few septa at maturity. They are distinguished from other fungi primarily by their characteristic motile cells (zoospores and gametes), most of which have a single, posterior, whiplash flagellum (Figure 15–8). Some chytrids are simple, unicellular organisms that do not develop a mycelium. In them, the whole organism is transformed into a reproductive structure at the appropriate time. Other chytrids have slender rhizoids that extend into the substrate and serve as an anchor (Figure 15–9). Some species are parasites of algae, protozoa, and aquatic oomycetes and of the spores, pollen grains, or other parts of plants. Some other chytrid species are saprophytic on such substrates as dead insects.

Several species of chytrids are plant pathogens, including *Physoderma maydis* and *Physoderma alfalfae,* which cause minor diseases known, respectively, as brown spot of corn and crown wart of alfalfa. *Synchytrium endobioticum* causes a disease of potatoes known as black wart disease, which is a serious problem in regions of Europe and Canada.

Chytrids exhibit a variety of modes of reproduction. Some species of *Allomyces,* for example, have an alternation of isomorphic generations like that shown in Figure 15–10, whereas in other species the alternating generations are heteromorphic—the haploid and diploid individuals do not resemble one another closely. Alternation of generations is characteristic of plants and of many algae but is otherwise found only in *Allomyces,* in one other closely related genus of chytrids, and in a very few

heterotrophic protists not considered in this book. In terms of its life cycle, morphology, and physiology, *Allomyces* is the best-known chytrid.

Phylum *Zygomycota*

Most zygomycetes live on decaying plant and animal matter in the soil, while some are parasites of plants, insects, or small soil animals. Still others form symbiotic associations—endomycorrhizae—with plants, and a few occasionally cause severe infections in humans and domestic animals. There are approximately 1060 described species of zygomycetes. Most of them have coenocytic hyphae, within which the cytoplasm can often be seen streaming rapidly. Zygomycetes can usually be recognized by their profuse, rapidly growing hyphae, but some of them also exhibit a unicellular, yeastlike form of growth under certain conditions. Asexual reproduction by means of haploid spores pro-

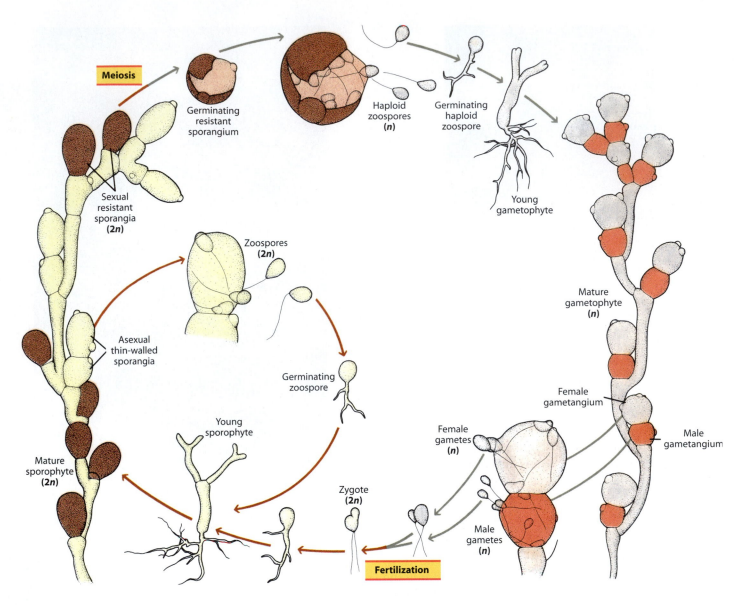

Meiosis

Germinating
resistant
sporangium

Haploid
zoospores
(*n*)

Germinating
haploid
zoospore

Sexual
resistant
sporangia
(**2n**)

Young
gametophyte

Zoospores
(**2n**)

Mature
gametophyte
(*n*)

Asexual
thin-walled
sporangia

Germinating
zoospore

Young
sporophyte

Female
gametangium

Male
gametangium

Mature
sporophyte
(**2n**)

Female
gametes
(*n*)

Zygote
(**2n**)

Male
gametes
(*n*)

Fertilization

15–10

In the life cycle of the chytrid Allomyces arbusculus, *there is an alternation of isomorphic generations. The haploid and diploid individuals are indistinguishable until they begin to form reproductive organs. The haploid individuals (gametophytes) produce approximately equal numbers of colorless female gametangia and orange male gametangia (right). The male gametes, which* are about half the size of the female gametes, are attracted by sirenin, a hormone produced by the female gametes. The zygote loses its flagella and germinates to produce a diploid individual. This sporophyte forms two kinds of sporangia. The first are asexual sporangia—colorless, thin-walled structures that release diploid zoospores—which in turn germinate and repeat the diploid generation. The second kind are sexual sporangia—thick-walled, reddish-brown structures that are able to withstand severe environmental conditions. After a period of dormancy, meiosis occurs in these sexual, resistant sporangia, resulting in the formation of haploid zoospores. These zoospores develop into gametophytes, which produce gametangia at maturity.

Phototropism in a Fungus

Over the millennia, the fungi have evolved a variety of methods that ensure wide dispersal of their spores. One of the most ingenious is found in Pilobolus, *a zygomycete that grows on dung. The sporangiophores of this fungus, which attain a height of 5 to 10 millimeters, are positively phototropic—that is, they grow toward the light. An expanded region of the sporangiophore located just below the sporangium (known appropriately as the subsporangial swelling) functions as a lens, focusing the sun's rays on a photoreceptive area at its base. Light focused elsewhere promotes maximum growth of the sporangiophore on the side away from the light, causing the sporangiophore to curve toward the light.*

The vacuole in the subsporangial swelling contains a high concentration of solutes, which results in water moving into it by osmosis. Eventually, the turgor pressure becomes so great that the swelling splits, shooting the sporangium in the direction of the light. The initial velocity may approach 50 kilometers per hour, and the sporangium may travel a distance greater than 2 meters. Considering that the sporangia are only about 80 micrometers in diameter, this is an enormous distance. This mechanism is adapted to shoot spores away from the dung—where animals do not feed—and into the grass where they can be eaten by herbivores and excreted in the fresh dung to continue the cycle anew.

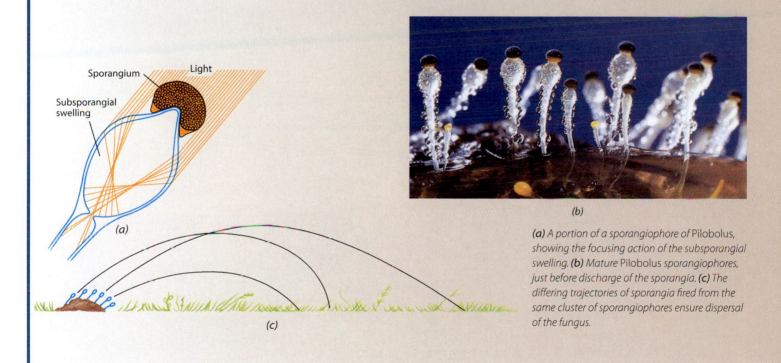

(a) A portion of a sporangiophore of Pilobolus, *showing the focusing action of the subsporangial swelling. (b) Mature* Pilobolus *sporangiophores, just before discharge of the sporangia. (c) The differing trajectories of sporangia fired from the same cluster of sporangiophores ensure dispersal of the fungus.*

duced in specialized sporangia borne on the hyphae is almost universal in zygomycetes.

One of the best-known and most familiar members of this phylum is *Rhizopus stolonifer,* a black mold that forms cottony masses on the surface of moist, carbohydrate-rich foods such as bread and similar substances that are exposed to air (Figure 15–2). This organism is also a serious pest of stored fruits and vegetables. The life cycle of *R. stolonifer* is illustrated in Figure 15–11. The mycelium of *Rhizopus* is composed of several distinct kinds of haploid hyphae. Most of the mycelium consists of rapidly growing, coenocytic hyphae that grow through the substrate, absorbing nutrients. From them, arching hyphae called **stolons** are formed. The stolons form rhizoids wherever their tips come into contact with the substrate. From each of these points, a sturdy, erect branch arises, which is called a **sporangiophore** ("sporangium bearer") because it produces a spherical sporangium at its apex. Each sporangium begins as a swelling, into which a number of nuclei flow. The sporangium is eventually isolated by the formation of a septum. The protoplasm within is cleaved, and a cell wall forms around each of the asexually produced nuclei to form spores. As the sporangium wall matures, it becomes black, giving the mold its characteristic color. With the breaking of the sporangium wall, the spores are liberated. Each spore can germinate to produce a new mycelium, completing the asexual cycle.

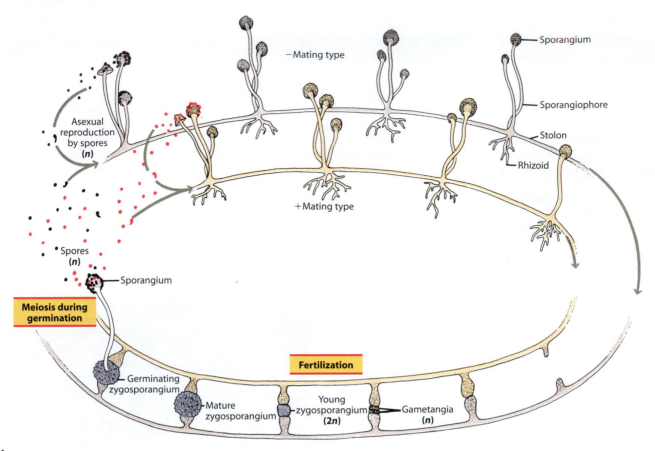

Sporangium

Sporangiophore

Stolon

Rhizoid

−Mating type

+Mating type

Asexual reproduction by spores (*n*)

Spores (*n*)

Sporangium

Meiosis during germination

Fertilization

Germinating zygosporangium

Mature zygosporangium

Young zygosporangium (2*n*)

Gametangia (*n*)

15–11

In Rhizopus stolonifer, *as in most other zygomycetes, asexual reproduction by means of haploid spores is the chief mode of reproduction. Less frequently, sexual reproduction occurs. The spores are formed in sporangia, whose black walls give the mold its characteristic color. In this common species, sexual reproduction involves genetically differentiated mating strains, which have traditionally been labeled + and − types. (Although the mating strains are morphologically indistinguishable from one another, they are shown here in two colors.) Sexual reproduction results in the formation of a resting spore called a zygospore, which develops within a zygosporangium. The zygosporangium in* Rhizopus *develops a thick, rough, black coat, and the zygospore remains dormant, often for several months.*

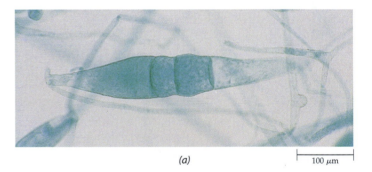

(a)   100 μm

(b)   100 μm

15–12

The zygomycete Rhizopus stolonifer, *a black mold. (a) Gametangia, the gamete-producing structures, are in the process of fusing to produce a zygospore. (b) The zygospore develops within a thick-walled zygosporangium.*

The phylum *Zygomycota* is named for its chief characteristic—the formation of sexually produced resting spores called **zygospores,** which develop within thick-walled structures called **zygosporangia** (Figure 15–12). The zygospores often remain dormant for long periods. Sexual reproduction in *R. stolonifer* requires the presence of two physiologically distinct mycelia, designated + and − strains. When two compatible individuals are in close proximity, hormones are produced that cause out-growths of the hyphae to come together and develop into gametangia. Species such as *R. stolonifer* that require + and − strains for sexual reproduction are said to be **heterothallic,** whereas self-fertile species are called **homothallic.**

In any event, the gametangia become separated from the rest of the fungal body by the formation of septa (Figure 15–11). The walls between the two touching gametangia dissolve, and the two multinucleate protoplasts come together. Following plasmogamy (the fusion of the two multinucleate gametangia), the + and − nuclei pair, and a thick-walled zygosporangium is produced. Inside the zygosporangium, the paired + and − nuclei fuse (karyogamy) to form diploid nuclei, which develop into a single multinucleate zygospore. At the time of germination, the zygosporangium cracks open, and a sporangiophore emerges from the zygospore. Meiosis occurs at the time of germination so that the spores produced asexually within the new sporangium are haploid. When these spores germinate, the cycle begins again.

Only two genera of zygomycetes commonly cause disease in living plants and living plant tissue. One of them is *Rhizopus*, which causes soft rot of many flowers, fleshy fruits, seeds, bulbs, and corms. The other is *Choanephora*, which causes a soft rot of squash, pumpkin, okra, and pepper.

One of the most important groups of zygomycetes includes *Glomus* and related genera, which always grow in intimate association with the roots of plants, forming endomycorrhizae. Another group of zygomycetes that has great ecological significance, the order *Entomophthorales*, is parasitic on insects and other small animals (see Figure 15-3b). Species of this order, most of which reproduce by means of a terminal, asexual spore that is discharged at maturity, are being increasingly used in the biological control of insect pests of crops.

The trichomycetes, a third group of zygomycetes, have an intriguing relationship with arthropods. These fungi are found in aquatic insect larvae, millipedes, crayfish, and even crustaceans from undersea thermal vents. Trichomycetes are seldom seen because they occur in the gut of the host organism, where some are thought to provide vitamins to the animal.

Phylum *Ascomycota*

The ascomycetes, which comprise about 32,300 described species, include a number of familiar and economically important fungi. Most of the blue-green, red, and brown molds that cause food spoilage are ascomycetes. Ascomycetes are also the cause of a number of serious plant diseases, including the powdery mildews, which primarily attack leaves; brown rot of stone fruits (caused by *Monilinia fructicola*); chestnut blight (caused by the fungus *Cryphonectria parasitica*, which was accidentally introduced into North America from northern China); and Dutch elm disease (caused by *Ophiostoma ulmi* and *O. novo-ulmi*, which were introduced from northern Europe and the region of Romania and Ukraine, respectively). Many yeasts are also ascomycetes, as are the edible morels and truffles (Figure 15–13). Many new families and thousands of additional species of ascomycetes—some undoubtedly of great economic importance—await discovery and scientific description.

15–13

Ascomycetes. (a) Eyelash cup, Scutellinia scutellata. *(b) The highly prized, edible ascoma of a black truffle,* Tuber melanosporum. *In the truffles, this spore-bearing structure is produced below ground and remains closed, liberating its ascospores only when the ascoma decays or is broken open by digging animals. Truffles are mycorrhizal (see page 342), mainly on oaks and hazelnuts, and are searched for by specially trained dogs and pigs. The pigs used are sows because truffles emit a chemical that mimics the pheromone from the saliva of a hog. Recently, truffles have been cultivated commercially on a small scale by inoculating the roots of seedling host plants with their spores.*

(a)

(b)

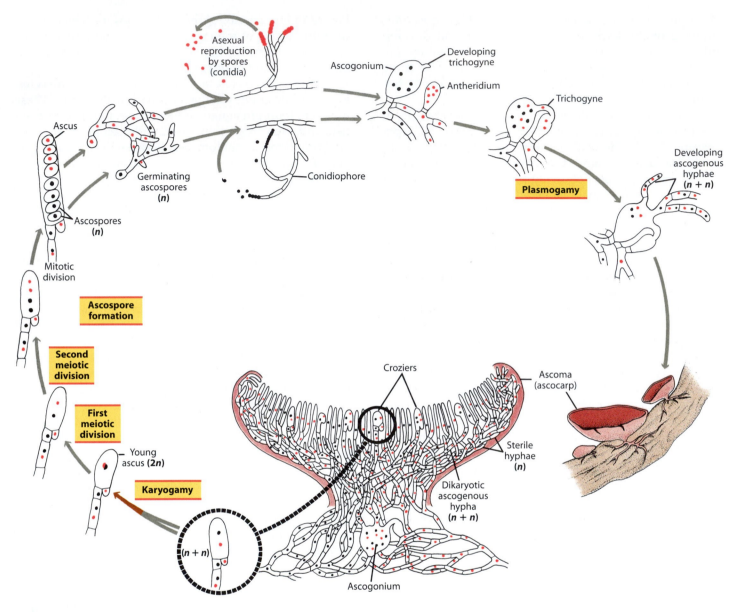

15–14

The typical life cycle of an ascomycete. Asexual reproduction occurs by way of specialized spores, known as conidia, which are usually multinucleate. Sexual reproduction

involves the formation of asci and ascospores. Plasmogamy produces fused protoplasts with as yet unfused nuclei, designated n + n.

Karyogamy is followed immediately by meiosis in the ascus, resulting in the production of ascospores.

Ascomycetes, with the exception of the unicellular yeasts, have filamentous growth forms. In general, their hyphae have perforated septa (Figure 15–4), which allow the cytoplasm and nuclei to move from one cell to the next. The hyphal cells of the vegetative mycelium may be either uninucleate or multinucleate. Some ascomycetes are homothallic, others heterothallic.

The life cycle of an ascomycete is diagrammed in Figure 15–14. In most species of this phylum, asexual reproduction is by the formation of usually multinucleate

conidia. The conidia are formed from conidiogenous cells (Figure 15–15), which are borne at the tips of modified hyphae called **conidiophores** ("conidia bearers"). Unlike zygomycetes, which produce spores internally within a sporangium, ascomycetes produce their asexual spores externally as conidia.

Sexual reproduction in ascomycetes always involves the formation of an **ascus** (plural: asci), a saclike structure within which haploid **ascospores** are formed following meiosis. Because the ascus resembles a sac, the

15–15

Conidia are the characteristic asexual spores of ascomycetes; they are usually multinucleate. These electron micrographs show stages in the formation of conidia in Nomuraea rileyi, which infects the velvetbean caterpillar. **(a)** *Scanning electron micrograph of conidia at various stages of development.* **(b)** *Transmission electron micrograph of conidia.*

(a)

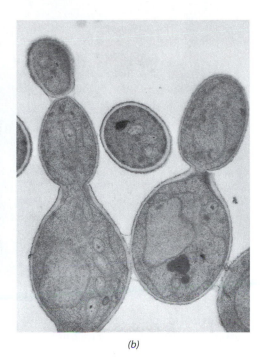

(b)

15–16

Asci and ascospores. **(a)** *An electron micrograph showing two asci of Ascodesmis nigricans in which ascospores are maturing.* **(b)** *Ascoma of Erysiphe aggregata, showing the enclosed asci and ascospores. This completely enclosed type of ascoma is called a cleistothecium.* **(c)** *An ascoma of Coniochaeta, showing the enclosed asci and ascospores. Note the small pore at the top. This sort of ascoma, with a small opening, is known as a perithecium.*

ascomycetes commonly are referred to as the "sac fungi." Both asci and ascospores are unique structures that distinguish the ascomycetes from all other fungi (Figure 15–16). Ascus formation usually occurs within a complex structure composed of tightly interwoven hyphae—the **ascoma** (plural: ascomata), or ascocarp. Many ascomata are macroscopic. An ascoma may be open and more or less cup-shaped (an *apothecium*; Figure 15–13a), closed and spherical (a *cleistothecium*; Figure 15–16b), or spherical to flask-shaped with a small pore through which the ascospores escape (a *perithecium*; Figure 15–16c). The asci usually develop on the inner surface of the ascoma. The layer of asci is usually called the **hymenium**, or **hymenial layer** (Figure 15–17).

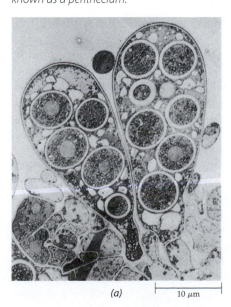

(a) 10 μm

(b) 25 μm

(c) 100 μm

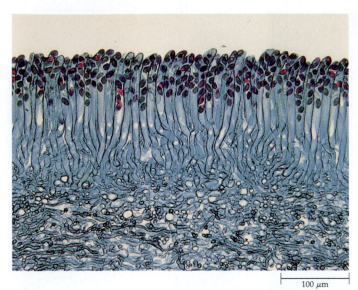

100 μm

15–17
A stained thin section through the hymenial layer of a morel (Morchella), showing asci with ascospores.

15–18
Life cycle of a mushroom (phylum Basidiomycota, a hymenomycete). Monokaryotic, primary mycelia are produced from basidiospores and give rise to dikaryotic, secondary mycelia, often following the fusion of different mating types, in which case the mycelia are heterokaryotic. Dikaryotic, tertiary mycelia form the basidioma, within which basidia form on the hymenia that line the gills, ultimately releasing up to billions of basidiospores.

In the life cycle of an ascomycete (Figure 15–14, top left corner), the mycelium is initiated with the germination of an ascospore on a suitable substrate. Soon after, the mycelium begins to reproduce asexually by forming conidia. Many crops of conidia are produced during the growing season, and it is the conidia that are primarily responsible for propagating and disseminating the fungus.

Sexual reproduction, involving ascus formation, occurs on the same mycelium that produces conidia. The formation of multinucleate gametangia called **antheridia** (the male gametangia) and **ascogonia** (the female gametangia) precedes sexual reproduction. The male nuclei of the antheridium pass into the ascogonium via the **trichogyne,** which is an outgrowth of the ascogonium. Plasmogamy—the fusion of protoplasts—has now taken place. The male nuclei may then pair with the genetically different female nuclei within the common cytoplasm, but they *do not yet fuse with them.* **Ascogenous hyphae** now begin to grow out of the ascogonium. As they continue to develop, compatible pairs of nuclei migrate into them, and cell division occurs in such a way that the resultant cells are invariably **dikaryotic,** which means they contain two compatible haploid nuclei. (Monokaryotic cells contain only one nucleus.)

The asci form near the tips of the dikaryotic, ascogenous hyphae. Commonly, it is the apical cell of the dikaryotic hypha that grows to form a hook, or **crozier.** In this hooked cell, the two nuclei divide in such a way that their spindle fibers are parallel to each other. Two of the daughter nuclei are close to one another at the top of the hook; one of the remaining two daughter nuclei is

near the tip, and the other is near the basal septum of the hook. Two septa are then formed; these divide the hook into three cells, of which the middle one becomes the ascus. It is in this middle cell that karyogamy occurs: the two nuclei fuse to form a diploid nucleus (zygote), the only diploid nucleus in the life cycle of the ascomycetes. Soon after karyogamy, the young ascus begins to elongate. The diploid nucleus then undergoes meiosis, which is generally followed by one mitotic division, producing an ascus with eight nuclei. These haploid nuclei are then cut off in segments of the cytoplasm to form ascospores. In most ascomycetes, the ascus becomes turgid at maturity and finally bursts, releasing its ascospores explosively into the air in the cup fungi and some of the perithecium-forming species. The ascospores are generally propelled about 2 centimeters from the ascus, but some species propel them as far as 30 centimeters. This initiates their airborne dispersal.

Phylum *Basidiomycota*

Phylum *Basidiomycota,* the last of the four phyla of fungi to be discussed, includes some of the most familiar fungi. Among the 22,300 distinct species of this phylum are the mushrooms, toadstools, stinkhorns, puffballs, and shelf fungi, as well as two important groups of plant pathogens, the rusts and smuts. Members of the *Basidiomycota* play a central role in the decomposition of plant litter, often constituting two-thirds of the living biomass (not including animals) in the soil.

A diagram of the life cycle of a mushroom will provide a convenient reference point as we proceed with our discussion (Figure 15–18). The *Basidiomycota* are distinguished from other fungi by their production of **basidiospores,** which are borne outside a club-shaped spore-producing structure called the **basidium** (plural: basidia) (Figure 15–19). In nature, most *Basidiomycota* reproduce primarily through the formation of basidiospores.

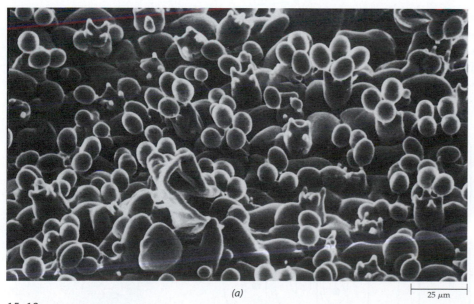

Plasmogamy

Germinating
basidiospores and
monokaryotic, primary
mycelium
(*n*)

Basidiospores (*n*)

Basidium

Basidiospore
formation

Second
meiotic
division

First
meiotic
division

Young
basidium
(2*n*)

Karyogamy

(*n* + *n*)

(*n* + *n*)

Dikaryotic,
secondary
mycelium
(*n* + *n*)

Young
basidioma

Basidioma
(basidiocarp)

Dikaryotic,
tertiary
mycelium
(*n* + *n*)

Secondary mycelium
(*n* + *n*)

Tertiary mycelium
(*n* + *n*)

Secondary mycelium
(*n* + *n*)

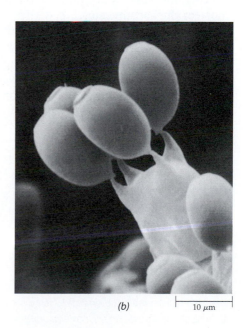

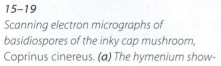

(a)

25 μm

(b)

10 μm

15–19

*Scanning electron micrographs of
basidiospores of the inky cap mushroom,
Coprinus cinereus. (a) The hymenium show-* *ing numerous basidia frozen at the time of
basidiospore release. (b) The top of a basid-* *ium, with four basidiospores, each attached
to a stalklike sterigma.*

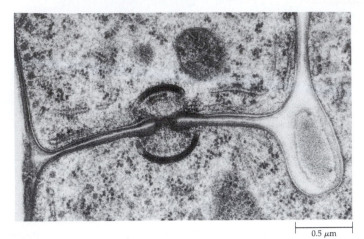

0.5 μm

15–20

A dolipore septum, shown in Auricularia auricula, a common wood-decay species of Basidiomycetes. Such septa are common in Basidiomycetes. Each dolipore septum is perforated by a pore. Parenthesomes are visible on either side of the dolipore.

The mycelium of the *Basidiomycota* is always septate, but the septa are perforated. In many species, the pore of the septum has an inflated doughnut-like or barrel-shaped margin called a **dolipore**. Any fungus with dolipore septa belong to the *Basidiomycota*. On either side of the dolipore may be membranous caps called **parenthesomes** because in profile they resemble a pair of parentheses (Figure 15–20). Many *Basidiomycota*, including the rusts and smuts, have septa that resemble those of the ascomycetes.

In most species of *Basidiomycota*, the mycelium passes through two distinct phases—monokaryotic and dikaryotic—during the life cycle of the fungus. When it germinates, a basidiospore produces a mycelium that may be multinucleate initially. Septa are soon formed, however,

and the mycelium is divided into **monokaryotic** (uninucleate) cells. This mycelium also is referred to as the **primary mycelium.** Commonly, the dikaryotic mycelium is produced by fusion of monokaryotic hyphae from different mating types (in which case it is heterokaryotic), resulting in formation of a **dikaryotic** (binucleate), or **secondary, mycelium,** since karyogamy does not immediately follow plasmogamy.

The apical cells of the dikaryotic mycelium usually divide by the formation of **clamp connections** (Figure 15–21). These clamp connections, which ensure the allocation of one nucleus of each type to the daughter cells, are found only in the *Basidiomycota*, although as many as 50 percent of the species may not form them.

The mycelium that forms the **basidiomata** (singular: basidioma)—fleshy, basidiospore-producing bodies, such as mushrooms and puffballs—also is dikaryotic. It is called the **tertiary mycelium.** The formation of the basidiomata may require light and low CO_2 levels, both of which signal to the mycelium that it is "outside" its substrate. As it forms the basidiomata, the tertiary mycelium becomes differentiated into specialized hyphae that play different functions within the basidiomata.

The *Basidiomycota* can be divided into three classes: *Basidiomycetes, Teliomycetes,* and *Ustomycetes.* The *Basidiomycetes* include all fungi that produce basidiomata, such as mushrooms, shelf fungi, and puffballs. Neither the *Teliomycetes* (the rusts) nor the *Ustomycetes* (the smuts) form basidiomata. Instead, these fungi produce their spores in masses called **sori** (singular: sorus). The basidiomata, which are characteristic of the *Basidiomycetes,* are analogous to the ascomata of the ascomycetes.

15–21

Clamp connections. (a) In the Basidiomycetes, dikaryotic hyphae characteristically are distinguished by the formation of clamp connections during cell division in tips of hyphae. They presumably ensure the proper distribution of the two genetically distinct types of nuclei in the basidioma. Two septa form to divide the parent cell into two daughter cells. (b) Electron micrograph of a clamp connection and characteristic septa in a hypha of Auricularia auricula.

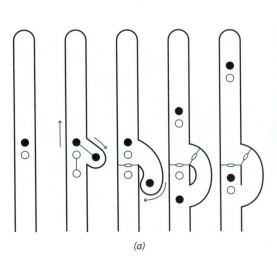

(a)

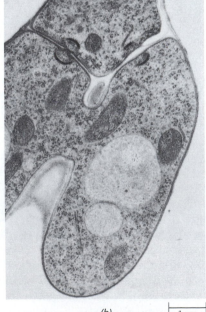

(b)   1 μm

(a)

(b)

(c)

15–22

Hymenomycetes. **(a)** *The fly agaric,* Amanita muscaria. *The mushrooms are at various stages of growth. Among the characteristics of this genus of mushrooms, many members of which are poisonous, are the scales on the cap, the ring on the stalk, and the cup, or volva, around its base.* **(b)** Polyporus arcularius, *a polypore. The polypores lack the gills found in most kinds of mushrooms. In P. arcularius, spores are shed through diamond-shaped pores.* **(c)** *A shelf fungus,* Ganoderma applanatum. *Shelf fungi are wood-rotting fungi.* **(d)** *An edible tooth fungus,* Hericium coralloides. *The hymenium, an outer spore-bearing layer of basidia, is borne on the surface of the downwardly directed teeth.*

(d)

Class *Basidiomycetes* Includes the Hymenomycetes and the Gasteromycetes

The class *Basidiomycetes* includes the edible and poisonous mushrooms, coral fungi, tooth fungi, and shelf or bracket fungi (Figure 15–22). These *Basidiomycetes* commonly are referred to as hymenomycetes, because they produce their basidiospores on a distinct fertile layer, the **hymenium,** which is exposed before the spores are mature (Figure 15–23). Another group of *Basidiomycetes,* the so-called gasteromycetes (literally the "stomach fungi"), include forms in which no distinct hymenium is visible at the time the basidiospores are released. Among the

more familiar gasteromycetes are the stinkhorns, earth-stars, false puffballs, bird's-nest fungi, and puffballs (see Figure 15–27). Most *Basidiomycetes* have club-shaped, aseptate (internally undivided) basidia, usually bearing four basidiospores, each on a minute projection called a **sterigma** (plural: sterigmata) (Figures 15–19b and 15–23). Others—the jelly fungi (Figure 15–24)—have septate (internally divided) basidia like those of the rusts and smuts.

The structure that one recognizes as a mushroom or toadstool is a basidioma (Figure 15–18). ("Mushroom" is sometimes popularly used to designate the edible forms of basidiomata, and "toadstool" is used to designate the

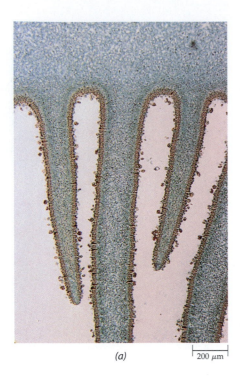

(a)   ⊢ 200 μm ⊣

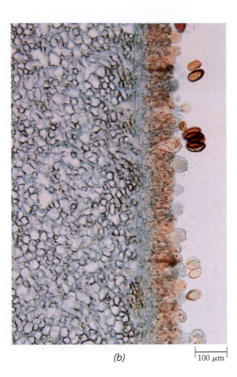

(b)   ⊢ 100 μm ⊣

(c)   ⊢ 50 μm ⊣

15–23

Stained sections through the gills of Coprinus, a common mushroom, at progressively higher magnifications. The hymenial layer is stained darker in each of these preparations. (a) Outlines of some of the gills. (b) Developing basidia and basidiospores are shown in a section through the hymenial layer. (c) Nearly mature basidiospores attached to basidia by sterigmata.

inedible ones, but mycologists do not recognize such a distinction and use only the term "mushroom." In this book, all such forms are referred to as mushrooms; this does not mean that they are all edible.) The mushroom generally consists of a **pileus,** or **cap,** that sits atop a **stipe,** or **stalk.** The masses of hyphae in the basidiomata usually form distinct layers. Early in its development—the "button" stage—the mushroom may be covered by a membranous tissue that ruptures as the mushroom enlarges. In some genera, remnants of this tissue are visible as patches on the upper surface of the cap and as a cup, or **volva,** at the base of the stipe (Figure 15–22a). In many hymenomycetes, the lower surface of the cap consists of radiating strips of tissue called **gills** (Figure 15–23), which are lined with the hymenium. In other members of the class, the hymenium is located elsewhere; for example, in the tooth fungi (Figure 15–22d) the hymenium covers downwardly directed pegs. In the shelf fungi and polypores (Figure 15–22b, c), the hymenium lines vertical tubes that open as pores.

As mentioned previously, in hymenomycetes, basidia form in well-defined hymenia that are exposed before the basidiospores are mature. Each basidium develops from a terminal cell of a dikaryotic hypha. Soon after the young basidium enlarges, karyogamy occurs. This is followed almost immediately by meiosis of each diploid nucleus, resulting in the formation of four haploid nuclei (Figure 15–18). Each of the four nuclei then migrates into a sterigma, which enlarges at its tip to form a uninucleate, haploid basidiospore. At maturity, the basidiospores are discharged forcibly from the basidioma but depend on the wind for dispersal. The reproductive capacity of a single mushroom is tremendous, with billions of spores being produced by a single basidioma. This reproductive capacity is essential. Each species occupies a narrow niche in the environment, and the chance that a given spore will land on a substrate suitable for germination and growth is slim.

In relatively uniform habitats, such as lawns and fields, the mycelium from which mushrooms are produced spreads underground, growing downward and

15–24

A jelly fungus growing on a dead tree limb in the Amazon forest of Brazil. Jelly fungi form septate basidia. For this and other reasons, many mycologists no longer group the various jelly fungi with the hymenomycetes.

15–25

A "fairy ring" formed by the mushroom Marasmius oreades. *Some fairy rings are estimated to be up to 500 years old. Because of the exhaustion of key nutrients, the grass immediately inside such a ring is often stunted and lighter green than the grass outside the ring.*

outward and forming a ring of mushrooms on the edge of the colony. This ring may grow as large as 30 meters in diameter. In an open area, the mycelium expands evenly in all directions, dying at the center and sending up basidiomata at the outer edges, where it grows most actively because this is the area in which the nutritive material in the soil is most abundant. As a consequence, the mushrooms appear in rings, and as the mycelium grows, the rings become larger. Such circles of mushrooms are known in European folk legends as "fairy rings" (Figure 15–25).

The best known hymenomycetes are the gill fungi, including *Agaricus campestris,* the common field mushroom. The closely related *Agaricus bisporus* is one of the few mushrooms cultivated commercially. It is now grown in more than 70 countries, and the value of the world crop exceeds $14 billion. Together with the Oriental shiitake mushroom, *Lentinula edodes, A. bisporus* makes up about 86 percent of the world's mushroom crop. Other mushrooms are also being cultivated, and some are gathered in large quantities in nature. An alarm has been sounded that mushrooms are declining both in total numbers of species and in the quantity of individual species in forests of Europe and the Pacific Northwest of the United States. If this trend continues, it could result in a dramatic decline in the health of the trees that are dependent on mycorrhizal fungi for nutrient uptake as well as disruption of the nutrient cycle in the ecosystem. The cause of the decline has not been identified, but pollutants such as nitrates are suspected.

The gill fungi also include many poisonous mushrooms. The genus *Amanita* includes the most poisonous of all mushrooms, as well as some that are edible. Just a few bites of the "destroying angel," *A. virosa,* can be fatal. Some *Basidiomycetes* contain chemicals that cause hallucinations in humans who eat them (Figure 15–26).

The gasteromycetes (Figure 15–27) are characterized by the fact that their basidiospores mature inside the basidiomata and are not discharged forcibly from them. Once considered a separate class, *Gasteromycetes,* this group is now known to be polyphyletic in nature. The basidiomata of the gasteromycetes possess a distinct outer covering, called the **peridium,** that varies from almost papery thin in some species to thick and rubbery or leathery in others. In some species the peridium opens naturally when the spores are mature; in others, it re-

(a)

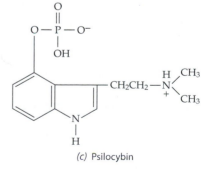

(b)

(c) Psilocybin

15–26

Mushrooms figure prominently in the religious ceremonies of several groups of Indians in southern Mexico and Central America. The Indians eat certain hymenomycetes for their hallucinogenic qualities. (a) One of the most important of these mushrooms is Psilocybe mexicana, *shown here growing in a pasture near Huautla de Jiménez, Oaxaca, Mexico.*

(b) The shaman María Sabina is shown eating Psilocybe *in the course of a midnight religious ceremony. (c) Psilocybin, the chemical responsible for the colorful visions experienced by those who eat these "sacred" mushrooms, is a structural analogue of LSD and mescaline (see Figure 22–50a).*

15–27

Gasteromycetes. **(a)** *Puffballs,* Calostoma cinnabarina. *Raindrops cause the thin outer layer, or peridium, of the puffball to dimple, forcing out a puff of air, mixed with spores, through the opening.* **(b)** *The netted stinkhorn,* Dictyophora duplicata. *The basidiospores are released in a foul-smelling, sticky mass at the top of the fungus. Flies visit it for food and spread the spores, which adhere to their legs and bodies in great numbers.* **(c)** *White-egg bird's-nest fungus,* Crucibulum laeve. *The round structures ("eggs") of the basidiomata (the "nests") of these fungi contain the basidiospores, which are splashed out and dispersed by raindrops.* **(d)** *Earthstar,* Geastrum saccatum, *showing one fully opened individual and two others in earlier stages of development. The outer layers of the peridium fold back in this genus, raising the spore mass above the dead leaves.*

(a)

(b)

(c)

(d)

mains permanently closed, with the spores being liberated only after it has been ruptured through the action of an external agent.

Stinkhorns (Figure 15–27b) have a remarkable morphology. They develop underground as leathery, egg-shaped structures. At maturity, they differentiate into an elongating stalk and a pileus, or cap, bearing a **gleba,** which is the fertile portion of the basidioma. The gleba forms an unpleasant-smelling, sticky mass of spores that attracts flies and beetles, which disperse the spores.

The puffballs are familiar gasteromycetes. At maturity, the interior of a puffball dries up, and it releases a cloud of spores when struck (Figure 15–27a). Some giant puffballs may reach 1 meter in diameter and may produce several trillion basidiospores. The bird's-nest fungi (Figure 15–27c) begin their development like puffballs, but the disintegration of much of their internal structure leaves them looking like miniature bird's nests.

Class *Teliomycetes* Includes the Rusts

The class *Teliomycetes* consists of the fungi commonly referred to as rusts, approximately 7000 species of which have been described. Unlike the *Basidiomycetes*, the rusts do not form basidiomata. As mentioned previously, their spores occur in masses called sori (Figure 15–28). They do, however, form dikaryotic hyphae and basidia, which are septate like those of the jelly fungi (members of the class *Basidiomycetes*). As plant pathogens, the rusts are of tremendous economic importance, causing billions of dollars of damage to crops throughout the world each year. Among the more serious rust diseases are black stem rust of cereals, white pine blister rust, coffee rust, cedar-apple rust, and peanut rust.

The life cycles of many rusts are complex, and these pathogens are a constant challenge to the plant pathologist whose task it is to keep them under control. Until recently, the rusts were thought to be obligate parasites on vascular plants, but now several species are grown in artificial culture. Some smuts are also able to complete their development under laboratory conditions.

15–28

Orange sori of the blister rust (Kuehneola uredinis) on a blackberry leaf photographed in San Mateo County, California.

An example of a rust life cycle is provided by *Puccinia graminis*, the cause of black stem rust of wheat. Numerous strains of *P. graminis* exist, and, in addition to wheat, they parasitize other cereals such as barley, oats, and rye, and various species of wild grasses. *Puccinia graminis* is a continuous source of economic loss for wheat growers. In a single year, the losses in Minnesota, North Dakota, South Dakota, and the prairie provinces of Canada amounted to nearly 8 million metric tons. As early as A.D. 100, Pliny described wheat rust as "the greatest pest of the crops." Today plant pathologists combat black stem rust largely by breeding resistant wheat varieties, but mutation and recombination in the rust make any advantage short-lived.

Puccinia graminis is **heteroecious;** that is, it requires two different hosts to complete its life cycle (see Figure 15–29 on the following two pages). **Autoecious** parasites, in contrast, require only one host. *Puccinia graminis* can grow indefinitely on its grass host, but there it reproduces only asexually. In order for sexual reproduction to take place, the rust must spend part of its life cycle on barberry (*Berberis*) and part on a grass. One method of attempting to eliminate this rust has been to eradicate barberry bushes. For example, the crown colony of Massachusetts passed a law ordering "whoever . . . hath any barberry bushes growing in his or their land . . . shall cause the same to be extirpated or destroyed on or before the thirteenth day of June, A.D. 1760."

Infection of the barberry occurs in the spring (Figure 15–29, upper left), when uninucleate basidiospores infect the plant by forming haploid mycelia, which first develop **spermogonia,** primarily on the upper surfaces of the leaves. The form of *P. graminis* that grows on barberry consists of separate + and − strains, so that the basidiospores and the spermogonia derived from them are either + or −. Each spermogonium is a flask-shaped

pustule lined by cells that form sticky, uninucleate cells called **spermatia.** The mouth of the spermogonium is surrounded by a brush of orange, stiff, unbranched, pointed hairs, the **periphyses,** which hold droplets of sugary, sweet-smelling nectar. The nectar, which is attractive to flies, contains the spermatia. Among the periphyses, branched **receptive hyphae** are also to be found. Flies visit the spermogonia and feed on the nectar. In moving from one spermogonium to another, they transfer spermatia. If a + spermatium of one spermogonium comes in contact with the − receptive hypha of another spermogonium, or vice versa, plasmogamy occurs and dikaryotic hyphae are produced. Aecial initials are produced from the dikaryotic hyphae that extend downward from the spermogonium. **Aecia** are then formed primarily on the lower surface of the leaf, where they produce chains of **aeciospores.** The dikaryotic aeciospores must then infect the wheat; they will not grow on barberry.

The first external manifestation of infection on the wheat is the appearance of rust-colored, linear streaks on the leaves and stems (the red stage). These streaks are **uredinia,** containing unicellular, dikaryotic **urediniospores.** Urediniospores are produced throughout the summer and reinfect the wheat; they are also the primary means by which the wheat rust has spread throughout the wheat-growing regions of the world. In late summer and early fall, the red-colored sori gradually darken and become **telia** with two-celled dikaryotic **teliospores** (the black stage). The teliospores are overwintering spores, which infect neither wheat nor barberries. Shortly after they are formed, karyogamy takes place, and the teliospores overwinter in the diploid state. Meiosis actually begins immediately but is arrested in prophase I. In early spring, prior to germination, meiosis is completed in the short, curved basidia that emerge from the two cells of the teliospore. Septa are formed between the resultant nuclei, which then migrate into the sterigmata and develop into basidiospores. Thus, the year-long cycle is completed.

In certain regions, the life cycle of wheat rust can be short-cut through the persistence of the uredinial state when actively growing plant tissues are available throughout the year. In the North American plains, urediniospores from winter wheat in the southwestern states and Mexico drift north to southern Manitoba. Later generations scatter westward to Alberta, and finally there is a southern drift at the end of summer, apparently moving along the eastern flank of the Rocky Mountains, and so back to the wintering grounds. Under such circumstances, wheat rust does not depend on barberry to persist, so that barberry eradication was not an effective tool for controlling wheat rust in this area. In contrast, the spread of urediniospores from south to north is prevented in Eurasia, where there are extensive east-west mountain ranges, and barberry *is* necessary for survival of the pathogen.

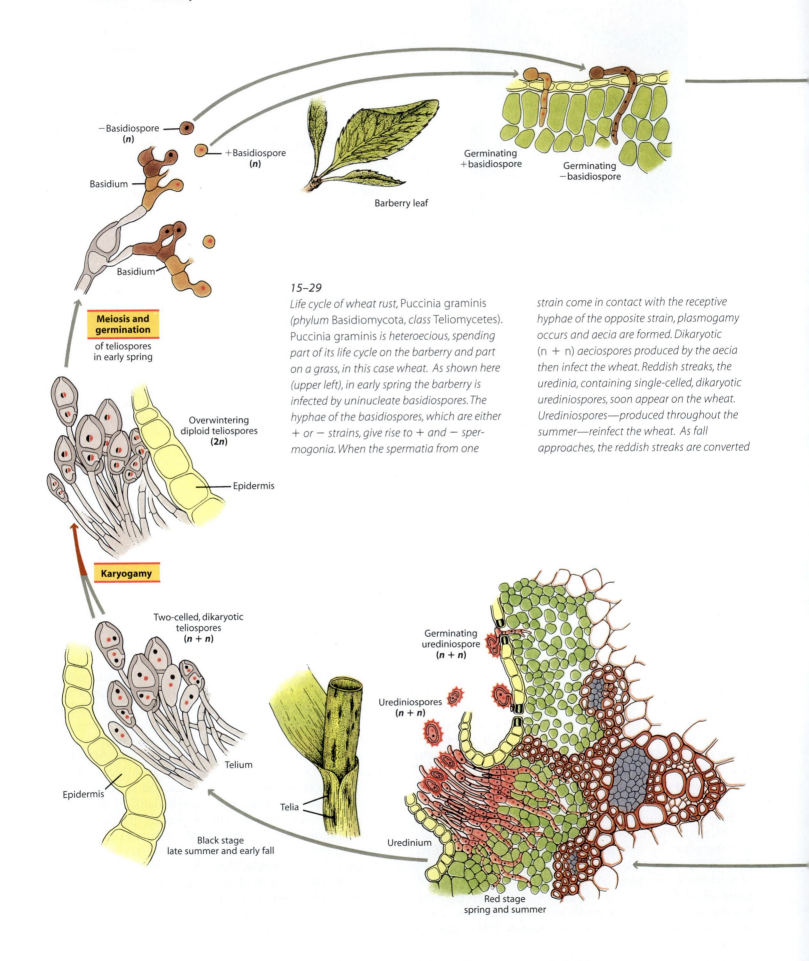

−Basidiospore (*n*)

+Basidiospore (*n*)

Basidium

Basidium

Barberry leaf

Germinating +basidiospore

Germinating −basidiospore

Meiosis and germination

of teliospores in early spring

Overwintering diploid teliospores (2*n*)

Epidermis

Karyogamy

Two-celled, dikaryotic teliospores (*n + n*)

Telium

Epidermis

Telia

Black stage late summer and early fall

Uredinium

Germinating urediniospore (*n + n*)

Urediniospores (*n + n*)

Red stage spring and summer

15–29

Life cycle of wheat rust, Puccinia graminis *(phylum* Basidiomycota, *class* Teliomycetes*). Puccinia graminis is heteroecious, spending part of its life cycle on the barberry and part on a grass, in this case wheat. As shown here (upper left), in early spring the barberry is infected by uninucleate basidiospores. The hyphae of the basidiospores, which are either + or − strains, give rise to + and − sper-mogonia. When the spermatia from one strain come in contact with the receptive hyphae of the opposite strain, plasmogamy occurs and aecia are formed. Dikaryotic (n + n) aeciospores produced by the aecia then infect the wheat. Reddish streaks, the uredinia, containing single-celled, dikaryotic urediniospores, soon appear on the wheat. Urediniospores—produced throughout the summer—reinfect the wheat. As fall approaches, the reddish streaks are converted*

Spermogonia on upper surface of leaf

Aecia on lower surface of leaf

—Spermatia (*n*)

Plasmogamy

—Receptive hyphae (*n*)

+Spermogonium

—Spermogonium

Aecium

Aeciospores (*n* + *n*)

to dark-colored telia containing teliospores, which initially are dikaryotic. Shortly after the teliospores are formed, the two nuclei in each half of the teliospore fuse (karyogamy), and the teliospores, which infect neither host, overwinter in the diploid state. In early spring, as the two cells of the teliospore germinate, the diploid nuclei complete meiosis. Each cell gives rise to a basidium and four haploid basidiospores.

Dispersal of aeciospores to wheat in spring

Wheat

Uredinia

Germinating aeciospore on wheat

Class *Ustomycetes* Includes the Smuts

All members of the *Ustomycetes* are parasites of flowering plants and are commonly referred to as smuts. The name "smut" refers to the sooty or smutty appearance of the black, dusty masses of teliospores, which are the characteristic resting spores of the smut fungi. Approximately 1070 species of *Ustomycetes* have been described. Most smuts form septate basidia. Economically, the smuts are very important. They attack approximately 4000 species of flowering plants, including both food crops and ornamentals. Three of the better known smut fungi are *Ustilago maydis,* the cause of common corn smut (Figure 15–30); *Ustilago avenae,* which causes loose smut of oats; and *Tilletia tritici,* the cause of bunt or stinking smut of wheat.

The life cycle of a smut, which is autoecious (requiring only one host), is considerably simpler than that of *Puccinia graminis.* Take for example the *Ustilago maydis* life cycle. Infections by spores of *U. maydis* remain localized, producing sori or large tumors. The most conspicuous tumors, or galls, occur on the ear of corn (maize) where the kernels become much enlarged and unsightly due to the development within them of a massive mycelium. A dikaryotic mycelium eventually gives rise to the thick-walled teliospores, in which karyogamy and meiosis take place.

Upon germination, the teliospore gives rise to a four-celled basidium. Two + and two − haploid, uninucleate basidiospores are formed, one from each of the four cells of the basidium (*U. maydis,* like *P. graminis,* is heterothallic). The basidiospores may infect maize plants directly or give rise by budding to populations of uninucleate cells called **sporidia,** which can also infect maize plants. Upon germination the basidiospores or sporidia produce either a + or − mycelium. When mycelia of opposite strains come into contact, plasmogamy occurs, producing a dikaryotic (*n* + *n*) mycelium, most cells of which change into teliospores.

Yeasts

A **yeast,** by definition, is simply a unicellular fungus that reproduces primarily by budding. Some fungi exhibit both unicellular and filamentous growth forms, shifting from one form to the other under changing environmental conditions. In many of these fungi the filamentous form is the phase in which the fungus spends most of its life cycle. Others exist primarily as yeasts, including the most familiar of yeasts, *Saccharomyces cerevisiae.* At one time, *S. cerevisiae* was thought to exist only in the unicellular phase (Figure 15–31a). It came as a great surprise, therefore, when quite by accident it was found that *S. cerevisiae* also exhibits a filamentous phase, which is triggered by starvation (Figure 15–31b). Generally,

15–30

Corn smut is a familiar plant disease in which the fungus Ustilago maydis *produces black, dusty-looking masses of spores in ears of corn. When young and white, these are cooked and eaten in Mexico and Central America, where they are regarded as a delicacy.* Ustilago *is a smut, a member of the class* Teliomycetes *of the phylum* Basidiomycota.

most laboratory cultures provide all of the nutrients essential for *S. cerevisiae* to continue its existence as a yeast. Apparently the filamentous phase is *S. cerevisiae*'s way of foraging for food.

The yeasts are not a formal taxonomic group. The yeast growth form is exhibited by a broad range of unrelated fungi encompassing the *Zygomycota, Ascomycota,* and *Basidiomycota.* There are at least 80 genera of yeasts, with approximately 600 known species. Most yeasts are ascomycetes, but at least a quarter of the genera are *Basidiomycota.*

A common form of reproduction exhibited by yeasts is **budding,** that is, the production of a small outgrowth, the bud, from the parent cell. Each yeast cell, therefore, can be regarded as a conidiogenous cell. Budding is, of course, an asexual method of reproduction. Some yeasts multiply asexually only in the haploid condition. Each haploid cell is capable of serving as a gamete, and at times two haploid cells may fuse to form a diploid cell, or zygote, which functions as an ascus (Figure 15–31c). (There is no dikaryotic phase in yeasts.)

Meiosis occurs within the ascus in ascomycomycetous yeasts. Usually four ascospores are produced per ascus, although in some species meiosis is followed by one or more mitotic divisions resulting in greater numbers of ascospores. In other yeasts, such as *S. cerevisiae,* meiosis is sometimes delayed and the zygote divides mitotically to form a population of diploid cells that reproduce

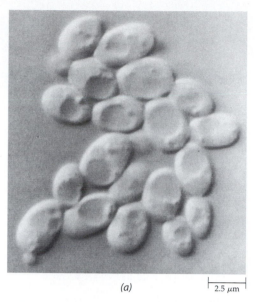

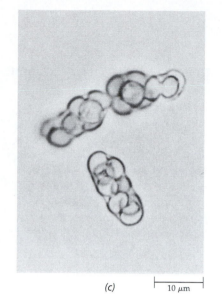

(a) | 2.5 μm | (c) | 10 μm |

15–31
Yeasts. (a) Single celled and (b) filamentous forms of bread yeast, Saccharomyces cerevisiae. *Yeasts commonly reproduce by budding (lower left in each micrograph), an asexual process. (c) Asci, with eight ascospores each, of* Schizosaccharomyces octosporus.

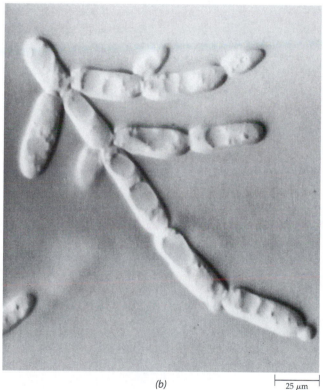

(b) | 25 μm |

come directly from the grape, but most arise from the direct action of the yeast (Figure 15–32). Most of the yeasts important in the production of wine, cider, sake, and beer are strains of *S. cerevisiae*, although other species also play a role. Most lager beer, for example, is made using *Saccharomyces carlsbergensis*. *Saccharomyces cerevisiae* is now virtually the only species used in baking bread (Figure 15–31a). Some species of yeast are important as human pathogens, causing such diseases as thrush (caused by *Candida albicans*) and cryptococcosis, the latter caused by a species of *Basidiomycetes* (*Cryptococcus neoformans*) that has the growth habit of a yeast when it grows as a human pathogen.

asexually by budding. Thus, such yeasts have both haploid and diploid budding stages. Diploid cells may ultimately undergo meiosis and revert to the haploid condition. In still other yeasts, the ascospores fuse in pairs immediately after they are formed; the ascospore is the only haploid cell in the life cycle, which is predominantly diploid.

As mentioned earlier, yeasts are utilized by vintners as a source of ethanol, by bakers as a source of carbon dioxide, and by brewers as a source of both substances. Many domestically useful strains of yeast have been developed by selection and breeding, and the techniques of genetic engineering are now being used to improve these strains further through the addition of useful genes from other organisms. Some of the flavors of wine

15–32
In some areas of the world, yeasts that are present on grapes are still used to produce wine. When the grapes are crushed, the yeast mix with the juice and then break down the glucose in the juice to alcohol.

15–33
Penicillium *and* Aspergillus—*two of the common genera of deuteromycetes.* **(a)** *A culture of* Penicillium notatum, *the original penicillin-producing fungus, showing the distinctive colors produced during growth and spore development.* **(b)** *A culture of* Aspergillus fumigatus, *a fungus that causes respiratory disease in humans. Notice the concentric growth pattern produced by successive "pulses" of spore production.*

(a)

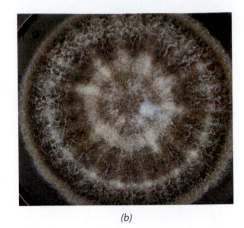
(b)

A number of yeasts, most notably *S. cerevisiae*, have become important laboratory organisms for genetic research. This yeast has become the organism of choice for studies of the metabolism, molecular genetics, and development of eukaryotic cells, and for chromosome studies. Haploid cells of *S. cerevisiae* have 16 chromosomes, and the DNA sequences of all 16 have been determined. *Saccharomyces cerevisiae* is thus the first eukaryote whose genome has been completely sequenced. In the future, such detailed knowledge, coupled with the ease of manipulating the genetics of yeasts growing on defined media, will doubtless greatly increase the importance of yeasts for scientific investigations and industrial applications.

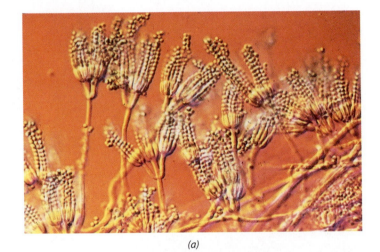
(a)

Deuteromycetes

The deuteromycetes, or conidial fungi, are an artificial assemblage of about 15,000 distinct species of fungi for which only the asexually reproducing state is known or in which the sexual reproductive features are not used as the basis of classification (Figures 15–33 and 15–34). They have often been called "Fungi Imperfecti" because they were thought of as being second- (deutero-) class or "imperfect" members among sexually reproducing— "perfect"—organisms. Considering that many of the deuteromycetes are abundant, flourishing organisms, the term "imperfect" is somewhat misleading.

In some of the fungi classified as deuteromycetes, the sexual state of the life cycle has apparently been lost in the course of evolution. In others, the sexual state may simply not have been discovered yet, and in still others,

(b)

15–34
The conidiogenous cells and conidiophores— the specialized hyphae that bear the conidia—of deuteromycetes are used in their classification. **(a)** Penicillium *(brushlike) and* **(b)** Aspergillus *(tightly clumped and arising* *from the swollen top of the conidiophore). Note the long chains of small, dry conidia.*

Predaceous Fungi

Among the most highly specialized of the fungi are the predaceous fungi, which have developed a number of mechanisms for capturing small animals they use as food. Although microscopic fungi with such habits have been known for many years, recently it has been learned that a number of species of gilled fungi also attack and consume the small roundworms known as nematodes. The oyster mushroom, Pleurotus ostreatus, for example, grows on decaying wood (a, b). Its hyphae secrete a substance that anesthetizes nematodes, after which the hyphae envelop and penetrate these tiny worms. The fungus apparently uses them primarily as a source of nitrogen, thus supplementing the low levels of nitrogen that are present in wood.

Some of the microscopic deuteromycetes secrete on the surface of their hyphae a sticky substance in which passing protozoa, rotifers, small insects, or other animals become stuck (c). More than 50 species of this group trap or snare nematodes. In the presence of these roundworms, the fungal hyphae produce loops that swell rapidly, closing the opening like a noose when a nematode rubs against its inner surface. Presumably the stimulation of the cell wall increases the amount of osmotically active material in the cell, causing water to enter the cells and increase their turgor pressure. The outer wall then splits, and a previously folded inner wall expands as the trap closes.

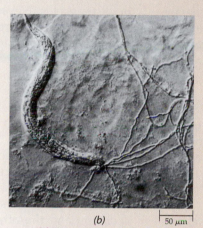
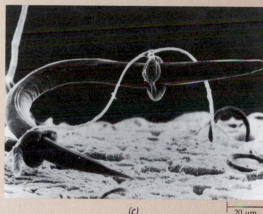

(a)

(b)   50 μm

(c)   20 μm

(a) The oyster mushroom, Pleurotus ostreatus.
(b) Hyphae of the oyster mushroom, which produce a substance that anesthetizes, converging on the mouth of an immobilized nematode.

(c) The predaceous deuteromycete Arthrobotrys anchonia *has trapped a nematode. The trap consists of rings, each comprising three cells, which when triggered swell rapidly to about three times their original size in 0.1 second and strangle the nematode. Once the worm has been trapped, fungal hyphae grow into its body and digest it.*

it may not be used as the principal basis of classification because of the close resemblance between the conidium-producing features and those of other deuteromycetes. Thus for some members of the well-known deuteromycete genera *Aspergillus* and *Penicillium,* the sexual state is known, but the species are classified as members of these genera because of their overall resemblance to the other species. On the basis of their overall characteristics, most deuteromycetes are clearly ascomycetes, reproducing only by means of conidia. A few deuteromycetes are *Basidiomycetes* or zygomycetes, the former marked by the septa and clamp connections that are characteristic of the *Basidiomycetes.* New molecular techniques are now supplementing traditional morphological features in the evaluation of the relationship between particular deuteromycetes and their relatives that form sexual states. When the sexual and asexual states of a fungus are identified, the entire organism in all of its forms is given the name of the sexual state.

Many fungi exhibit the phenomenon of **heterokaryosis,** in which genetically different nuclei occur together in a common cytoplasm. The nuclei may differ from one another because of mutation or because of the fusion of genetically distinct hyphae. Since genetically different nuclei may occur in different proportions in different parts of a mycelium, these sectors may have different properties.

Among the deuteromycetes, as well as some other fungi, genetically distinct haploid nuclei occasionally fuse. Within the resulting diploid nuclei, the chromosomes may associate, recombination may follow, and genetically novel haploid nuclei may form. Restoration of the haploid condition does not involve meiosis. Instead, it results from a gradual loss of chromosomes, a process called *haploidization.* This genetic phenomenon, in which plasmogamy, karyogamy, and haploidization occur in sequence, is known as **parasexuality;** it was discovered in *Aspergillus,* a deuteromycete. Within the hyphae of this common fungus, there is one diploid nucleus, on average, for every 1000 haploid nuclei. Parasexual cycles may add considerably to genetic and evolutionary flexibility in fungi that lack a true sexual cycle.

Among the deuteromycetes are many organisms of great economic importance. For example, a number of the most important plant pathogens are deuteromycetes. Anthracnose diseases of plants, which cause lesions and blackening, are generally caused by deuteromycetes. In addition, an often fatal disease of dogwoods (*Cornus florida*) that was detected over a wide area of the eastern United States in the late 1980s is caused by the deuteromycete *Discula destructiva*.

Other members of the group produce effects that are valued by humans. For example, certain *Penicillium* species give some types of cheese the appearance, flavor, odor, and texture so highly prized by gourmets. Roquefort, Danish blue, Stilton, and Gorgonzola are all ripened by *Penicillium roqueforti*. Another species, *Penicillium camemberti*, gives Camembert and Brie cheeses their special qualities.

Soy paste (miso) is produced by fermenting soybeans with *Aspergillus oryzae*, and soy sauce by fermenting soybeans with a mixture of *A. oryzae* and *Aspergillus soyae*, as well as lactic acid bacteria. *Aspergillus oryzae* is also important for the initial steps in brewing sake, the traditional alcoholic beverage of Japan; *Saccharomyces cerevisiae* is important later in the process.

The ubiquitous soil deuteromycete *Trichoderma* has many commercial applications. For example, cellulose-degrading enzymes produced by *Trichoderma* are used by clothing manufacturers to give jeans a "stone-washed" look. The same enzymes are added to some household laundry detergents to help remove fabric nubs. *Trichoderma* is also used by farmers in the biological control of other fungi that attack crops and forest trees.

Many important antibiotics are produced by deuteromycetes. The first antibiotic was discovered by Sir Alexander Fleming, who noted in 1928 that a strain of *Penicillium* that had contaminated a culture of *Staphylococcus* growing on a nutrient agar plate had completely halted the growth of the bacteria. Ten years later, Howard Florey and his associates at Oxford University purified penicillin and later came to the United States to promote the large-scale production of the drug. Production of penicillin increased enormously with increasing demand during World War II. Penicillin is effective in curing a wide variety of bacterial diseases, including pneumonia, scarlet fever, syphilis, gonorrhea, diphtheria, and rheumatic fever.

Not all substances produced by deuteromycetes are useful to us. For example, the aflatoxins are potent causative agents for liver cancer in humans. These highly carcinogenic mycotoxins, which show their effects at concentrations as low as a few parts per billion, are secondary metabolites produced by certain strains of *Aspergillus flavus* and *Aspergillus parasiticus*. Both of these fungi frequently grow in stored food products, especially peanuts, maize, and wheat. In tropical countries, aflatoxins have been estimated to contaminate at least 25 percent of the food. Aflatoxins have been detected occasionally in maize harvested in the United States, although strong efforts have been made to detect and destroy contaminated maize.

One group of deuteromycetes, the dermatophytes (Gk. *derma*, "skin," + *phyton*, "plant"), are the cause of ringworm, athlete's foot, and other fungal skin diseases. Such diseases are especially prevalent in the tropics. The pathogenic stages of these fungi are asexual, but most of these organisms have now been correlated with species of ascomycetes. Nevertheless, they continue to be classified as deuteromycetes on the basis of their disease-causing forms. During World War II, more soldiers had to be sent back from the South Pacific because of skin infections than because of wounds received in battle.

Symbiotic Relationships of Fungi

Symbiosis—"living together"—is a close and long-term association between organisms of different species. Some symbiotic relationships, typically disease-causing, are **parasitic.** One species (the parasite) benefits from the association and the other (the host) is harmed. Although many fungi are parasites, other fungi are involved in symbiotic relationships that are **mutualistic**—that is, the association is beneficial to both organisms. Two of these mutualistic symbioses—lichens and mycorrhizae—have been and are of extraordinary importance in enabling photosynthetic organisms to become established in previously barren terrestrial environments.

A Lichen Consists of a Mycobiont and a Photobiont

A lichen is a mutualistic symbiotic association between a fungal partner and a population of unicellular or filamentous algal or cyanobacterial cells. The fungal component of the lichen is called the **mycobiont** (Gk. *mykēs*, "fungus," + *bios*, "life"), and the photosynthetic component is called the **photobiont** (Gk. *phōto-*, "light," + *bios*, "life"). The scientific name given to the lichen is the name of the fungus. About 98 percent of the lichen-forming fungi belong to the *Ascomycota*, the remainder to the *Basidiomycota*. Lichens are polyphyletic. Recent DNA evidence indicates that they have evolved independently on at least five occasions, and it is likely that they evolved independently many more times.

About 13,250 species of lichen-forming fungi have been described, representing almost half of all known ascomycetes. Some 40 genera of photobionts are found in combination with these ascomycetes. The most frequent

From Pathogen to Symbiont: Fungal Endophytes

The leaves and stems of plants are often riddled with fungal hyphae. Such fungi are called endophytes, a general term used for a plant or fungus growing within a plant. Although some fungal endophytes cause disease symptoms in the plants they inhabit, others produce no such effects. Instead, they protect host plants from their natural enemies: herbivores and, in some cases, pathogenic microbes.

In many species of grasses, endophytic fungi infect the flowers of the host and proliferate in the seeds. Eventually, a substantial mass develops throughout the stems and leaves of the mature grass plant, with the fungal hyphae growing between the host cells. A good example of such a relationship is provided by tall fescue, Festuca arundinacea, a grass that covers more than 15,000 square kilometers (35 million acres) of lawns, fields, and pastures in the United States, especially in the East and Midwest. Tall fescue plants that are free of fungus provide good forage for livestock, but cattle feeding on in-

fected plants become lethargic and stop grazing, often panting and drooling excessively. If the animals are not moved to other forage, they become feverish, gain weight slowly, produce little milk, have difficulty in conceiving, develop gangrene, and eventually die, if they have not been slaughtered by that time for their salvage value. In the 1970s, scientists at the University of Kentucky discovered that these symptoms were associated with the endophytic ascomycete Sphacelia typhina. In relationships with fungal endophytes, which are common in grasses, the host plant gains protection from its enemies (including, unfortunately for us, cattle), and the fungus obtains all of its nutritional needs from the grass. The deterrent effects on herbivores occur because the fungi produce alkaloids—bitter, nitrogen-rich compounds that are abundant in some plants. Alkaloids have physiological effects on humans and other animals (page 32).

Some of the alkaloids produced by an-

other fungus, Claviceps purpurea, which infects rye (Secale cereale) and other grasses, are identical to those produced by Sphacelia. The rye-infecting fungus, which replaces infected grain with a hard mycelial mat, or sclerotium, causes a plant disease called ergot. The sclerotium—also called an ergot—contains lysergic acid amide (LDA), a precursor of lysergic acid diethylamide (LSD). LSD was first discovered in studies of the alkaloids of C. purpurea. Domestic animals and people who eat the infected grain develop a disease called ergotism, which is often accompanied by gangrene, nervous spasms, psychotic delusions, and convulsions. It occurred frequently during the Middle Ages, when it was known as St. Anthony's fire. It has been suggested that the widespread accusations of witchcraft in Salem Village (now Danvers) and other communities in Massachusetts and Connecticut in 1692, which led to a number of executions, may have resulted from an outbreak of convulsive ergotism. In one epidemic in Europe in the year 994, more than 40,000 people died. As recently as 1951, there was an outbreak of ergotism in a small French village, in which 30 people became temporarily insane, believing that they were pursued by demons or snakes. Five of the villagers died. Some of the ergot alkaloids have the property of enhancing muscle constriction and have been used for more than 400 years to hasten uterine contraction during childbirth. Other ergot alkaloids constrict the veins and are used in the treatment of migraine. Because of these useful medical properties, efforts are now under way to produce improved strains of C. purpurea and to grow them commercially.

Sclerotia of Claviceps purpurea, *which causes the plant disease known as ergot, are seen here growing on stalks of rye (Secale cereale).*

are the green algae *Trebouxia, Pseudotrebouxia,* and *Trentepohlia,* and the cyanobacterium *Nostoc.* About 90 percent of all lichens have one of these four genera as their photobiont. A number of lichens incorporate two photobionts—a green alga and a cyanobacterium. Different species of the algal genus can serve as photo-

bionts in a single lichen species. In addition, a single fungal species can form lichens with different algae or cyanobacteria.

Lichens are able to live in some of the harshest environments on Earth, and consequently they are extremely widespread (Figure 15–35). They occur from arid desert

(a)

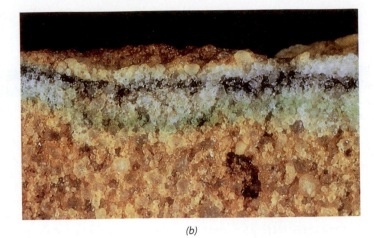

(b)

15–35
In this seemingly lifeless, dry region of Antarctica **(a)**, lichens live just beneath the exposed surface of the sandstone. **(b)** In the fractured rock, the colored bands are distinct zones biologically. The white and black zones are formed by a lichen, while the lower green zone is produced by a nonlichenized unicellular green alga. The air temperatures in this part of Antarctica rise almost to freezing in the summer and probably fall to −60°C in winter.

(a)

regions to the Arctic and grow on bare soil, tree trunks, sunbaked rocks, fence posts, and windswept alpine peaks all over the world (Figures 15–36 and 15–37). Some lichens are so tiny that they are almost invisible to the unaided eye; others, like the reindeer "mosses," may cover kilometers of land with ankle-deep growth. One species, *Verrucaria serpuloides*, is a permanently submerged marine lichen. Lichens are often the first colonists of newly exposed rocky areas. In Antarctica, there are more than 350 species of lichens (Figure 15–35) but only two species of vascular plants; seven species of lichens actually occur within 4 degrees of the South Pole! Although widely distributed, individual lichen species usually occupy fairly specific substrates, such as the surfaces or interiors of rocks, soil, leaves, and bark. Some lichens provide the substrate for other lichens and parasitic fungi that may be closely related to the lichen being parasitized.

(b)

(c)

15–36
Crustose and foliose lichens. **(a)** A crustose ("encrusting") lichen, Caloplaca saxicola, growing on a bare rock surface in central California. **(b)** Crustose and foliose ("leafy") lichens growing on the gravestone of Roland Thaxter (1858–1932), renowned mycologist from Harvard University. **(c)** Parmelia perforata, a foliose ("leafy") lichen, forms part of a hummingbird's nest on a dead tree branch in Mississippi.

(a)

(b)

(c)

(d)

15–37

Several fruticose ("shrubby") lichens. (a) Goldeye lichen, Teloschistes chrysophthalmus. *(b) British soldiers,* Cladonia cristatella, *are 1 to 2 centimeters tall. (c) Witch's hair* (Alectoria sarmentosa), *a hanging lichen that often occurs in masses on the limbs of trees. Although superficially similar to* Alectoria *and occupying a similar ecological niche, "Spanish moss," which is common throughout the southern United States, is not a lichen but a flowering plant—a member of the pineapple family (Bromeliaceae). (d)* Cladonia subtenuis, *commonly called "reindeer moss," is actually a lichen. Lichens of this group, which are abundant in the Arctic, concentrated radioactive substances following atmospheric nuclear tests and nuclear reactor accidents. Reindeer feeding on the lichens concentrated these radioactive substances still further and passed them on to the humans and other animals that consumed them or their products, especially milk and cheese.*

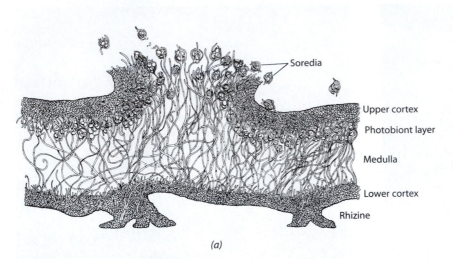

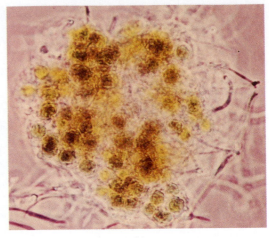

(a)

(b)

15–38

*A stratified lichen. (**a**) A cross section of the lichen* Lobaria verrucosa, *shown here releasing soredia consisting of hyphae wrapped around cyanobacteria. The simplest lichens consist of a crust of fungal hyphae entwining colonies of algae or cyanobacteria. More complex lichens, however, exhibit a definite growth form with a characteristic internal structure. The lichen shown here has four distinct layers: (1) the upper cortex, a protective surface of heavily gelatinized fungal hyphae; (2) the photobiont layer, which, in* Lobaria, *consists of cyanobacterial cells and loosely interwoven, thin-walled hyphae; (3) the medulla, consisting of loosely packed, weakly gelatinized hyphae, which makes up about two-thirds of the thickness of the thallus and appears to serve as a storage area; and (4) the lower cortex, which is thinner than the upper one and is covered with fine projections (rhizines) that attach the lichen to its substrate. (**b**) A soredium, which is composed of fungal hyphae and cells of the photobiont.*

In almost all cases, the fungus makes up most of the thallus and apparently plays the major role in determining the form of the lichen. There are two general types of lichen thalli. In one, the cells of the photobiont are more or less evenly distributed throughout the thallus; in the other, the cells of the photobiont form a distinct layer within the thallus (Figure 15–38). Three major growth forms are recognized among the latter, stratified lichens: **crustose,** which is flattened and adheres firmly to the substrate, having a "crusty" (encrusting) appearance; **foliose,** which is leaflike; and **fruticose,** which is erect and often branched and "bushy" (Figures 15–36 and 15–37).

The colors of lichens range from white to black, through shades of red, orange, brown, yellow, and green, and these organisms contain many unusual chemical compounds. Many lichens are used as sources of dyes; for example, the characteristic color of Harris tweed originally resulted from the treatment of the wool with a lichen dye. Many lichens have also been used as medicines, components of perfumes, or minor sources of food. Some species are being investigated for their ability to secrete anti-tumor compounds.

Lichens commonly reproduce by simple fragmentation, by the production of special powdery propagules known as **soredia** (Figure 15–38), or by small outgrowths known as **isidia.** Fragments, soredia, and isidia, which all contain both fungal hyphae and algae or cyanobacteria, act as small dispersal units to establish the lichen in new localities. The fungal component of the lichen produces ascospores, conidia, or basidiospores that are typical of its taxonomic group. If the fungus is an ascomycete, it may form ascomata, which are similar to those of other ascomycetes, except that in lichens the ascomata may endure and produce spores slowly but continuously over a number of years. Regardless, any of these spores may form new lichens when they germinate and come into contact with the appropriate green algae or cyanobacteria.

The Survival of Lichens Is Due to Their Ability to Dry Out Very Rapidly   How can the lichens survive under environmental conditions so severe that they exclude any other form of life? At one time, it was thought that the secret of a lichen's success was that the fungal tissue protected the alga or cyanobacterium from drying out. Actually, one of the chief factors in their survival seems to be the fact that they dry out very rapidly. Lichens are frequently very desiccated in nature, with a water content ranging from only 2 to 10 percent of their dry weight. When a lichen dries out, photosynthesis ceases. In this state of "suspended animation," even blazing sunlight or great extremes of heat or cold can be endured by some species of lichens. Cessation of photosynthesis depends, in large part, on the fact that the upper cortex of the lichen becomes thicker and more opaque when dry, cutting off the passage of light energy. A wet lichen may be damaged or destroyed by light intensities or temperatures that do not harm a dry lichen.

The lichen reaches its maximum vitality, as judged by the rate of photosynthesis, after it has been soaked with water and has begun to dry. Its rate of photosynthesis reaches a peak when the water content is 65 to 90 percent of the maximum the organism can hold; below this level, if the lichen continues to lose water, the rate of photosynthesis decreases. In many environments, the water content of the lichen varies markedly in the course of a day, with most photosynthesis taking place only during a few hours, usually in the early morning after wetting by fog or dew. As a consequence, lichens have an extremely slow rate of growth, their radius increasing at rates ranging from 0.1 to 10 millimeters a year. Calculated on this basis, some mature lichens have been estimated to be as much as 4500 years old. They achieve their most luxuriant growth along seacoasts and on fog-shrouded mountains. The oldest known fossil lichens are found in the Early Devonian (400 million years old).

The Nature of the Relationship between Mycobiont and Photobiont Considerable debate has surrounded the nature of the relationship between mycobiont and photobiont, that is, whether the relationship is parasitic or mutualistic symbiosis. As noted by David L. Hawksworth, the issue is in reality one of scale. At the cellular level, individual photobiont cells can be considered parasitized by the mycobionts whose hyphae adhere closely to the phycobiont cell walls (Figure 15–39). These hyphae typically form haustoria or **appressoria,** which are specialized hyphae that penetrate photobiont cells by means of pegs. These structures are involved with the transfer of carbohydrates and nitrogen compounds (in the case of nitrogen-fixing cyanobacteria such as *Nostoc*) from the photobiont to the fungus. In addition, the mycobiont controls the division rate of its photobiont. The mycobiont in turn provides the photobiont with a suitable physical environment in which to grow and absorbs needed minerals from the air in the form of dust or from rain. At the whole lichen level, the association clearly is mutualistic as neither partner can flourish in niches in which they occur in nature without the other. Today mutualism generally is judged on the basis of the dual functional unit, and therefore lichen symbiosis is considered to be mutualistic.

Although at one time it was thought that separate dispersal of mycobiont and photobiont was uncommon, there now is evidence that independent dispersal and resynthesis of the lichen is common in nature in some species. Several lichenologists have developed methods for growing many photobionts and mycobionts separately in defined media and resynthesizing the lichen. The mechanism by which the germinating fungal spore recognizes the appropriate alga or cyanobacterium is not well understood. When grown together in culture, the fungus seems first to bring the photobiont partner under control and then to establish the characteristic form of the mature lichen (Figure 15–39).

Lichens Are Important Ecologically Lichens clearly play an important role in ecosystems. Mycobionts produce large numbers of secondary metabolites called **lichen acids,** which sometimes amount to 40 percent or more of the dry weight of a lichen. These metabolites are known to play a role in the biogeochemical weathering of rock

15–39

Scanning electron micrographs of early stages in the interaction between fungal and algal components of British soldier lichens (Cladonia cristatella) *in laboratory culture. The photosynthetic component in this lichen (Figure 15–37b) is* Trebouxia, *a green alga.* **(a)** *An algal cell surrounded by fungal hyphae.* **(b)** *Penetration of an algal cell by a fungal haustorium (arrow).* **(c)** *Mixed groups of fungal and algal components developing into the mature lichen.*

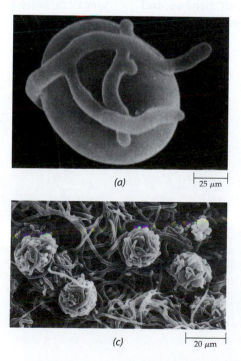

(a) 25 μm

(c) 20 μm

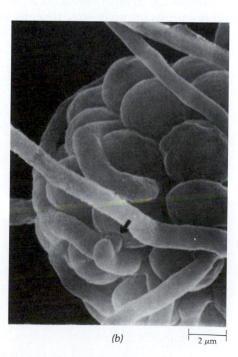

(b) 2 μm

and soil formation. The lichens trap the newly formed soil, making it possible for a succession of plants to grow.

Lichens containing a cyanobacterium are of special importance because they contribute fixed nitrogen to the soil. Such lichens are an important factor in supplying nitrogen in many ecosystems, including the old-growth forests of the Pacific Northwest of the United States, some tropical forests, and certain desert and tundra sites.

Because they have no means of excreting the elements that they absorb, some lichens are particularly sensitive to toxic compounds. The toxins cause the limited amount of chlorophyll in their algal or cyanobacterial cells to deteriorate. Lichens are very sensitive indicators of the toxic components—particularly sulfur dioxide—of polluted air, and they are being used increasingly to monitor atmospheric pollutants, especially around cities. Since lichens containing cyanobacteria are especially sensitive to sulfur dioxide, air pollution can substantially limit nitrogen fixation in natural communities, causing long-range changes in soil fertility. Both the state of health of the lichens and their chemical composition are used to monitor the environment. For example, analysis of lichens can detect the distribution of heavy metals and other pollutants around industrial sites. Fortunately, many lichens have the ability to bind heavy metals outside their cells and thus escape damage themselves.

When nuclear tests were being conducted in the atmosphere, lichens were used to monitor the fallout. Now, lichens provide a useful way to monitor the contamination by radioactive substances following events such as the explosion at the Chernobyl nuclear power plant in 1986.

There are many interactions between lichens and other groups of organisms. Lichens are eaten by a number of vertebrate and invertebrate animals. They are an important winter food source for reindeer and caribou in the far-north regions of North America and Europe and are eaten by mites, insects, and slugs. Lichens are dispersed by birds that use lichens in their nests (Figure 15–36c). The nests of flying squirrels can contain up to 98 percent lichens. Some lichens have been found to have important functions as antibiotics.

Mycorrhizae Are Mutualistic Associations between Fungi and Roots

The most prevalent and possibly the most important mutualistic symbiosis in the plant kingdom is the **mycorrhiza,** which literally means "fungus root." Mycorrhizae are intimate and mutually beneficial symbiotic associations between fungi and roots, and they occur in the vast majority of vascular plants, both wild and cultivated. The few families of flowering plants that usually lack mycorrhizae include the mustard family *(Brassicaceae)* and the sedge family *(Cyperaceae)*.

Mycorrhizal fungi benefit their host plants by increasing the plants' ability to capture water and essential elements (see Chapter 30), especially phosphorus. Increased absorption of zinc, manganese, and copper—three other essential nutrients—has likewise been demonstrated. For many forest trees, if seedlings are grown in a sterile nutrient solution and then transplanted to a grassland soil, they grow poorly, and many eventually die from malnutrition (Figure 15–40). Mycorrhizal fungi also provide protection against attack by pathogenic fungi and nematodes (small roundworms). In return for these benefits, the fungal partner receives from the host plant carbohydrates and vitamins essential for its growth.

15–40

Mycorrhizae and tree nutrition. Nine-month-old seedlings of white pine (Pinus strobus) were raised for two months in a sterile nutrient solution and then transplanted to prairie soil. The seedlings on the left were transplanted directly. The seedlings on the right were grown for two weeks in forest soil containing fungi before being transplanted to the grassland soil.

15–41

Endomycorrhizae. Glomus versiforme, *a zygomycete, is shown here growing in association with the roots of leeks,* Allium porrum. ***(a)*** *Arbuscules (highly branched structures) and vesicles (dark, oval structures). Arbuscules predominate in young infections, with vesicles becoming common later.* ***(b)*** *Arbuscules growing inside a leek root cell.*

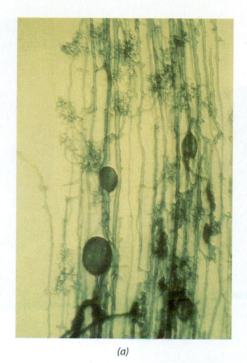 

(a)   (b)

Endomycorrhizae Penetrate Root Cells There are two major types of mycorrhizae: **endomycorrhizae,** which penetrate the root cells, and **ectomycorrhizae,** which surround the root cells. Of these two, endomycorrhizae are by far the more common, occurring in about 80 percent of all vascular plants. The fungal component is a zygomycete (order *Glomales*), with fewer than 200 species involved in such associations worldwide. Thus, endomycorrhizal relationships are not highly specific. The fungal hyphae penetrate the cortical cells of the plant root, where they form highly branched structures called **arbuscules** (Figures 15–41 and 15–42) and in some cases terminal swellings called **vesicles.** The arbuscules do not enter the protoplast but greatly invaginate the plasma membrane, increasing its surface area and presumably facilitating the bidirectional transfer of metabolites and nutrients by the two mycorrhizal partners. Most, or perhaps all, exchange between plant and fungus takes place at the arbuscules. Vesicles may also occur between host cells and are thought to function as storage compartments for the fungus. Endomycorrhizae are often called vesicular-arbuscular (or V/A) mycorrhizae. The hyphae extend out into the surrounding soil for several centimeters and thus greatly increase the potential for the absorption of water and the uptake of phosphates and other essential nutrients.

15–42

Endomycorrhizae. Scanning electron micrograph showing arbuscules of Glomus etunicatum, *a zygomycete, in cortical cells of a sugar maple* (Acer saccharum) *root. The protoplasts, which typically envelop the hyphae, are not apparent.*

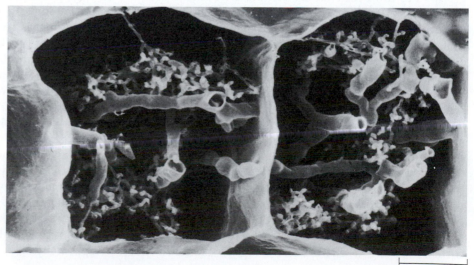

10 μm

15–43
Ectomycorrhizae. A section through the extensive ectomycorrhizae of a lodgepole pine (Pinus contorta) *seedling. The seedling extends about 4 centimeters above the soil surface.*

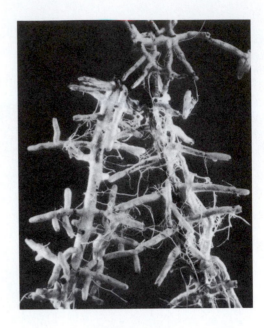

15–44
Ectomycorrhizae from a western hemlock (Tsuga heterophylla)*. In such ectomycorrhizae, the fungus commonly forms a sheath of hyphae, called a fungal mantle, around the root. Hormones secreted by the fungus cause the root to branch. This growth pattern and the hyphal sheath impart a characteristic branched and swollen appearance to the ectomycorrhizae. The narrow mycelial strands extending from the mycorrhizae function as extensions of the root system.*

Ectomycorrhizae Surround but Do Not Penetrate Root Cells   Ectomycorrhizae (Figures 15–43 and 15–44) are characteristic of certain groups of trees and shrubs that are found primarily in temperate regions. Among these groups are the beech family *(Fagaceae)*, which includes the oaks; the willow, poplar, and cottonwood family *(Salicaceae);* the birches *(Betulaceae);* the pine family *(Pinaceae);* and certain groups of tropical trees that form dense stands of only one or a few species. The trees that grow at timberline (the upper altitudes and latitudes of tree growth) in different parts of the world—such as pines in the northern mountains, *Eucalyptus* in Australia, and southern beech *(Nothofagus;* see Figure 22–10)—almost always are ectomycorrhizal. The ectomycorrhizal association apparently makes the trees more resistant to the harsh, cold, dry conditions that occur at the limits of tree growth.

In ectomycorrhizae, the fungus surrounds but does not penetrate living cells in the roots. In the conifers, the hyphae grow between the cells of the root epidermis and cortex, forming a characteristic highly branched network, the **Hartig net** (Figure 15–45), which eventually surrounds many of the cortical and epidermal cells. In the roots of most angiosperms colonized by ectomycorrhizal fungi, the epidermal cells are triggered to enlarge primarily at right angles to the surface of the root, thickening the root rather than extending it, and the Hartig net is confined to this layer (Figure 15–45b). The Hartig net functions as the interface between the fungus and the plant. In addition to the Hartig net, ectomycorrhizae are characterized by a **mantle,** or sheath, of hyphae that covers the root surface. Mycelial strands extend from the mantle into the surrounding soil (Figure 15–44). Typically, root hairs do not develop on ectomycorrhizae, and the roots are short and often branched. Ectomycorrhizae are mostly formed with *Basidiomycetes,* including many genera of mushrooms, but some ectomycorrhizae involve associations with ascomycetes, including truffles *(Tuber)* and possibly the edible morels *(Morchella).* At least 5000 species of fungi are involved in ectomycorrhizal associations, often with a high degree of specificity.

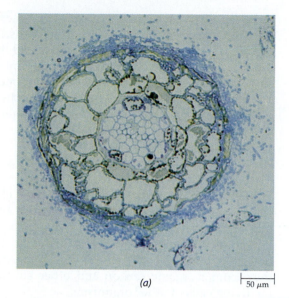

(a) | 50 μm

15–45

Ectomycorrhizae. (a) Transverse section of an ectomycorrhiza of Pinus. *The hyphae of the fungus form a mantle around the root and also penetrate between the epidermal and cortical cells, where they form the characteristic Hartig net. (b) Longitudinal section of ectomycorrhiza of birch (*Betula alleghaniensis). *The fungus ensheathes the root, forming a mantle around it. The Hartig net is confined to the layer of radially enlarged epidermal cells.*

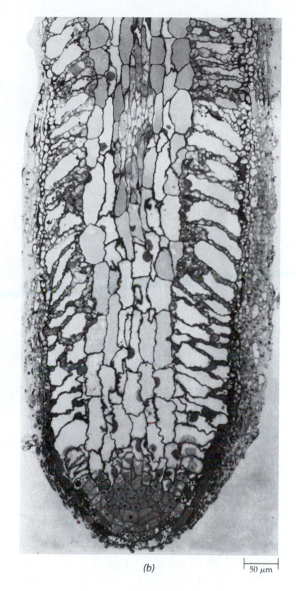

(b) | 50 μm

Other Kinds of Mycorrhizae Are Found in the Heather and Orchid Families Two other kinds of mycorrhizae are those characteristic of the heather family (*Ericaceae*) and a few closely related groups, and those associated with the orchids (*Orchidaceae*). In *Ericaceae*, the fungal hyphae form an extensive, loosely organized web over the root surface. Rather than extending the absorptive surface significantly, the principal role of the fungus is to release enzymes into the soil to break down certain compounds and make them available to the plant. The mycorrhizae of *Ericaceae* seem to function primarily in enhancing nitrogen, rather than phosphorus, uptake by the plants. This allows *Ericaceae* to colonize the kinds of infertile, acidic soils where they are especially frequent. Both ascomycetes and *Basidiomycetes* are involved in mycorrhizal associations with *Ericaceae*.

In nature, orchid seeds germinate only in the presence of suitable fungi. This mycorrhizal association is unique in that the fungus, which is internal, also supplies its host with carbon, at least when the host is a seedling. The fungi in such associations are mostly *Basidiomycetes*, with more than 100 species involved.

Mycorrhizae Were Probably Important in the History of Vascular Plants A study of the fossils of early plants has revealed that endomycorrhizal associations were as frequent then as they are in their descendants (Figure 15–46). This finding led K. A. Pirozynski and D. W. Malloch to suggest that the evolution of mycorrhizal associations may have been a critical step in allowing colonization of the land by plants. Given the poorly developed soils that would have been available at the time of the first colonization, the role of mycorrhizal fungi (probably zygomycetes) may have been of crucial significance, particularly in facilitating the uptake of phosphorus and other nutrients. A similar relationship has been demonstrated among contemporary plants colonizing extremely nutrient-poor soils, such as slag

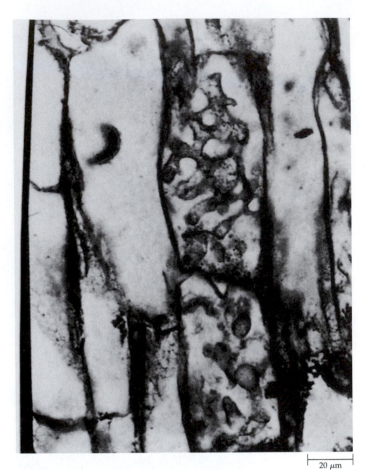

15–46

Silicified roots of a gymnosperm from the Triassic period of Antarctica, showing well-developed arbuscules. Endomycorrhizae were characteristic of the earliest plants for which fossils are known.

heaps: those individuals with endomycorrhizae have a much better chance of surviving. Thus it may have been not a single organism but rather a symbiotic association of organisms, comparable to a lichen, that initially invaded the land.

Summary

Fungi Are Ecologically and Economically Important

Fungi, together with the heterotrophic bacteria, are the principal decomposers of the biosphere, breaking down organic products and recycling carbon, nitrogen, and other components to the soil and air. As decomposers, they often come into direct conflict with human interests, attacking almost any conceivable substance. Most fungi are saprophytes; that is, they live on organic mate-

rial from dead organisms. Many fungi attack living organisms, however, and are important causal agents of diseases in plants, domestic animals, and humans. A number of fungi are economically important to humans as destroyers of foodstuffs and other organic materials. The kingdom *Fungi* also includes yeasts, cheese molds, edible mushrooms, and *Penicillium* and other producers of antibiotics.

Most Fungi Are Composed of Hyphae

Fungi are rapidly growing organisms that characteristically form filaments called hyphae, which may be septate or aseptate. In most fungi, the hyphae are highly branched, forming a mycelium. Parasitic fungi often have specialized hyphae (haustoria) by means of which they extract organic carbon and other substances from the living cells of other organisms.

Fungi Are Absorbers That Reproduce by Means of Spores

Fungi, which are almost all terrestrial, reproduce by means of spores, which are usually wind-dispersed. No motile cells are formed at any stage of the fungal life cycle, except by chytrids. The primary component of the fungal cell wall is chitin. Typically, a fungus secretes enzymes onto a food source and then absorbs the small molecules that are released. Glycogen is the primary storage polysaccharide.

The Diploid Phase in the Life Cycle of a Fungus Is Represented Only by the Zygote

The fungi are isogamous (gametes similar in size and shape), and meiosis in fungi is zygotic—that is, the zygote divides by meiosis to form four haploid cells.

There Are Four Phyla of Fungi

The kingdom *Fungi* includes four phyla: *Chytridiomycota*, *Zygomycota*, *Ascomycota*, and *Basidiomycota*, as well as an artificial assemblage known as the deuteromycetes. There is considerable evidence that animals and fungi diverged from a common ancestor, most likely a colonial protist resembling a choanoflagellate.

The *Chytridiomycota* Form Flagellated, Motile Cells

The *Chytridiomycota* are a predominantly aquatic group, and the only group of fungi with motile reproductive cells (zoospores and gametes). Most chytrids are coenocytic, with few septa at maturity. Some species are parasitic and others saprobic. Several are plant pathogens causing minor diseases such as brown spot of corn and crown wart of alfalfa.

The *Zygomycota* Form Zygospores in Zygosporangia

The zygomycetes (phylum *Zygomycota*) mostly have coenocytic mycelia. Their asexual spores are generally formed in sporangia—saclike structures in which the entire contents are converted to spores. The zygomycetes are so named because they form resting spores called zygospores during sexual reproduction. The zygospores develop within thick-walled structures called zygosporangia.

The *Ascomycota* Form Ascospores in Asci

Phylum *Ascomycota*, the members of which are commonly called ascomycetes, has some 32,300 named, distinct species, more than any other group of fungi. The distinguishing characteristic of ascomycetes is the ascus—a saclike structure in which the meiotic (sexual) spores, known as ascospores, are formed. In the ascomycete life cycle, the protoplasts of male and female gametangia fuse, and the female gametangia produce specialized hyphae that are dikaryotic (each compartment containing a pair of haploid nuclei). The ascus forms near the tip of a dikaryotic hypha. Ascospores are generally forcibly expelled. The asci are incorporated into complex spore-producing bodies called ascomata. Typically, asexual reproduction is by formation of usually multinucleate spores known as conidia.

The *Basidiomycota* Form Basidiospores Borne on Basidia

The phylum *Basidiomycota* includes many of the largest and most familiar fungi. They include mushrooms, puffballs, shelf fungi, stinkhorns, and others, as well as the rusts and smuts, which are important plant pathogens. The distinguishing characteristic of this phylum is the production of basidia. The basidium is produced at the tip of a dikaryotic hypha and is the structure in which meiosis occurs. Each basidium typically produces four basidiospores, and these are the principal means of reproduction in the *Basidiomycota*.

The Mushrooms, Rusts, and Smuts Are Representatives of the Three Classes of *Basidiomycota*

The *Basidiomycota* can be divided into three classes: *Basidiomycetes*, *Teliomycetes*, and *Ustomycetes*. In the *Basidiomycetes*, the basidia are incorporated into complex spore-producing bodies called basidiomata. In the hy-menomycetes, which include the mushrooms and shelf fungi, the basidiospores are produced on a distinct fertile layer, the hymenium. This layer often lines the gills or tubes of the hymenomycetes and is exposed before the forcibly discharged spores are mature. In the gasteromycetes, which include the stinkhorns and puffballs, the basidiospores mature inside the basidiomata and are not discharged forcibly from them. The members of the classes *Teliomycetes* and *Ustomycetes*, the rusts and smuts, respectively, do not form basidiomata. Instead, rusts and smuts have septate basidia, as do the jelly fungi, which are members of the class *Basidiomycetes*. All *Basidiomycetes* other than the jelly fungi form aseptate basidia.

Yeasts Are Unicellular Fungi

Yeasts are unicellular fungi that typically reproduce by budding, an asexual method of reproduction. Some fungi exhibit both unicellular and filamentous growth forms, shifting from one form to the other under changing environmental conditions. Most yeasts are ascomycetes, which reproduce sexually through the production of ascospores.

Fungi with No Known Sexual State Are Placed in the Deuteromycetes

An artificial group of fungi—the deuteromycetes, also known as Fungi Imperfecti—includes many thousands of species with no known sexual cycle. Most of these deuteromycetes are allied with ascomycetes, but some are *Basidiomycetes* or zygomycetes.

Lichens Consist of a Mycobiont and a Photobiont

Lichens are mutualistic symbiotic partnerships between a fungal partner (the mycobiont) and a population of either green algae or cyanobacteria (the photobiont). About 98 percent of the fungal partners belong to the *Ascomycota*, the rest to the *Basidiomycota*. The fungi receive carbohydrates and nitrogen compounds from their photosynthetic partners and provide the photobiont with a suitable physical environment in which to grow. The ability of the lichen to survive under adverse environmental conditions is related to its ability to withstand desiccation and remain dormant when dry.

Mycorrhizae Are Mutualistic Associations between Fungi and Roots

Mycorrhizae—symbiotic associations between plant roots and fungi—characterize all but a few families of vascular plants. Endomycorrhizae, also called vesicular-arbuscular mycorrhizae, in which the fungal partners are zygomycetes, occur in about 80 percent of all kinds of vascular plants. In such associations, the fungus actually penetrates the cortical cells of the host, but does not enter the protoplasts. In the second major type of mycorrhizal association, ectomycorrhizae, the fungus does not penetrate the host cells but forms a sheath, or mantle, that surrounds the roots and also a network (the Hartig net) that grows around the cortical cells. Mostly *Basidiomycetes,* but also some ascomycetes, are involved in ectomycorrhizal associations. Mycorrhizal associations are important in obtaining phosphorus and other nutrients for the plant and in providing organic carbon for the fungus. Such associations were characteristic of the first plants to invade the land.

Selected Key Terms

aeciospores   p. 327

aecium   p. 327

antheridium   p. 320

arbuscule   p. 341

ascogonium   p. 320

ascoma   p. 319

ascospores   p. 318

ascus   p. 318

autoecious   p. 327

basidioma   p. 322

basidiospores   p. 320

basidium   p. 320

coenocytic   p. 310

conidiophore   p. 318

conidia   p. 311

dolipore   p. 322

endophytes   p. 308

heteroecious   p. 327

hymenium   p. 319

mycobiont   p. 334

mycology   p. 310

mycorrhizae   p. 340

mycotoxins   p. 307

parasexuality   p. 333

photobiont   p. 334

septate   p. 310

septum   p. 310

soredia   p. 338

sori   p. 322

spermogonia   p. 327

sporangiophore   p. 315

stolons   p. 315

telia   p. 327

teliospores   p. 327

uredinia   p. 327

urediniospores   p. 327

zygosporangium   p. 316

zygospores   p. 316

Questions

1. Distinguish between the following: hyphae/mycelium; somatic/vegetative; rhizoids/haustoria; plasmogamy/karyogamy; sporangium/gametangium; heterothallic/homothallic; dikaryotic/monokaryotic; parasitic/mutualistic; arbuscules/vesicles; endomycorrhizae/ectomycorrhizae.

2. "The fungi are of enormous importance both ecologically and economically." Elaborate upon this statement both in general terms and with specific reference to each of the major groups of fungi, including the yeasts, lichens, and mycorrhizal fungi.

3. How might one determine, on the basis of hyphal structure alone, whether a given fungus is a member of the *Zygomycota, Ascomycota,* or *Basidiomycota*?

4. What do zygospores, ascospores, and basidiospores have in common? Zoospores, conidia, aeciospores, and urediniospores?

5. Many fungi produce antibiotics. What do you think the function of the antibiotics might be for the fungi that produce them?

6. In the life cycle of a mushroom, three types of hyphae, or mycelia, can be recognized: primary, secondary, and tertiary. How do these three types of mycelia relate to one another, and how do they fit into the life cycle?

7. "Both the state of health of the lichens and their chemical composition are used to monitor the environment." Explain.

8. "The most prevalent and probably the most important mutualistic symbiosis in the plant kingdom is the mycorrhiza." Explain.

Chapter 16

Protista I: Euglenoids, Slime Molds, Cryptomonads, Red Algae, Dinoflagellates, and Haptophytes

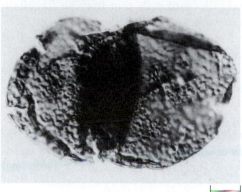

0.2 mm   0.2 mm

0.1 mm

16–1

Three very different kinds of acritarchs that occurred together about 700 million years ago in the seas of what is now the Grand Canyon region of Arizona. Acritarchs, which first appeared in the fossil record about 1.5 billion years ago, were evidently eukaryotes, as indicated by their size and by the complexity of their cell walls. Acritarchs, which were very diverse, were probably planktonic autotrophs. Some closely resemble the resting zygotes of dinoflagellates (see page 365), whereas others resemble certain green algae.

OVERVIEW

In this chapter and the next we visit the kingdom Protista, which encompasses a wide variety of organisms, including unicellular, colonial, and simple multicellular forms. This kingdom may be regarded as a "catchall" assemblage of organisms that simply do not fit well into any of the other kingdoms.

The organisms considered in this chapter include two phyla of entirely heterotrophic organisms once treated as fungi and five phyla of algae. In general, algae are simple, photosynthetic organisms, although many are heterotrophic. Some, like the euglenoids, the cryptomonads, and the dinoflagellates, are largely unicellular. Others, like the haptophytes, are unicellular or colonial, and the red algae are mostly multicellular. The two phyla of heterotrophic organisms, on the other hand, exhibit strikingly different growth habits. The plasmodial slime molds grow mostly as large, multinucleate masses of protoplasm, whereas the cellular slime molds usually exist as unicellular amoeba-like cells that can aggregate to form a mass that resembles a multicellular organism.

As you will see, many of the organisms discussed in this chapter have tremendous ecological importance and provide clues concerning the evolution of eukaryotic cells and the origin of the other kingdoms of eukaryotic organisms.

CHECKPOINTS

By the time you finish reading this chapter, you should be able to answer the following questions:

1. What criteria are used to place an organism in the kingdom *Protista?* How is this kingdom different from the other four kingdoms of organisms?

2. How are algae of ecological importance?

3. In what ways are the cryptomonads and euglenoids similar? Why is it difficult to classify these organisms on the basis of the mechanisms by which they obtain their food?

4. What features distinguish the plasmodial from the cellular slime molds? Why are these organisms not considered to be algae?

5. What are the distinctive cellular characteristics of the red algae? In what ways is the life history of red algae unusual?

6. What are the distinctive features of the phylum *Haptophyta,* and how are haptophytes important in food webs?

With approximately 70 percent of its surface covered by water, Earth is known as the "water planet." This abundance of water provided the watery habitat in which life began, some 3.5 billion years ago, with the first appearance of prokaryotes. DNA evidence suggests that the first eukaryotes appeared between 2.5 and 1 billion years ago, at least a billion years after the first occurrence of prokaryotes.

Until recently, the oldest eukaryotic fossils known were the **acritarchs,** which first appeared in the fossil record some 1.5 billion years ago (Figure 16–1). Recently, however, eukaryotic fossils have been retrieved from sediments in China that are 1.7 billion years old. These fossils are said to resemble the modern aquatic eukaryotes known as brown algae. An even more ancient eukaryotic fossil, a 0.5-meter-long, thin fossil named *Grypania,* has been found in rocks that are 2.1 billion years old. Described as a photosynthetic eukaryote, *Grypania* may be the earliest known eukaryote. Although these interpretations are somewhat controversial, they suggest that the first eukaryotes are even more ancient than previously thought.

Today a diverse array of the descendants of these early eukaryotes—protists—populates the oceans and marine shorelines, as well as freshwater lakes, ponds, and streams. A few protists are able to live in terrestrial habitats, but their principal realm remains the water.

Biologists have grouped the protists together into the kingdom *Protista,* which contains eukaryotes that are not clearly assignable to the kingdoms *Fungi, Plantae,* or *Animalia* (see Figure 13–13). Most biologists would agree that fungi, plants, and animals are derived from ancient protists and that the study of modern protists sheds light on the origin of these important groups. In addition to the evolutionary importance of protists, some cause diseases of plants or animals, whereas others are of great ecological significance.

Protist groups covered in this book (in this and the following chapter) include photosynthetic organisms that function ecologically like plants—that is, as primary producers using light energy to manufacture their own food (see Chapter 32). Photosynthetic protists occur in the phyla made up of algae. Among the algae, the green algae are particularly significant because plants are derived from an ancestor that, if still alive, would be classified as a green alga. In addition to these autotrophs, we will describe some colorless, heterotrophic protists, including oomycetes. These heterotrophic protists are very closely related to the algae. We will also describe some protists, the mixotrophs, that display both photosynthesis and heterotrophy. The algae include autotrophic, heterotrophic, and mixotrophic eukaryotes. Also considered are two groups of protists—the cellular slime molds and the plasmodial slime molds—that are not algae. These two groups, together with the oomycetes, have traditionally been the concern of the *mycologist,* or student of fungi. Although these heterotrophic protists are not closely related to fungi, they continue to be described by terminology used for fungi. Similarities and differences among the protist groups discussed in this chapter are summarized in Table 16–1.

Protists exhibit an amazing array of body types that include amoeba-like cells; single cells—with or without cell walls—that may or may not have flagella; colonies consisting of aggregations of cells that may be flagellate or not; branched or unbranched filaments; one- or two-cell-thick sheets; tissue that resembles some tissues of plants and animals; and multinucleate masses of protoplasm with or without cell walls. Protists vary in size from the smallest known eukaryote—a tiny green alga—to 30-meter-long brown algae known as kelps. Both extremely small and very large sizes confer protection against being eaten by aquatic herbivores. Many protists reproduce sexually and have complex life histories, but some reproduce only by asexual means. All three types of sexual life cycles—zygotic, sporic, and gametic (see Figure 9–3)—occur among protists. It is common for different phases of the life history of a single protist species to be very different in size and appearance.

Studies of nuclear-encoded ribosomal RNA gene sequences have clarified the relationships of the various protist groups to one another. In this chapter, we will consider the groups believed to have appeared earliest: euglenoids, plasmodial and cellular slime molds, cryptomonads, red algae, dinoflagellates, and haptophyte algae. In the next chapter, we will discuss the protists that appeared somewhat more recently: water molds, diatoms, chrysophyte algae, brown algae, and the green algae.

Ecology of the Algae

In all bodies of water, minute photosynthetic cells and tiny animals occur as suspended **plankton** (Gk. *planktos,* "wandering"). The planktonic algae and the cyanobacteria, which together constitute the **phytoplankton,** are the beginning of the food chain for the heterotrophic organisms that live in the ocean and in bodies of fresh water. The heterotrophic plankton—**zooplankton**—consists mainly of tiny crustaceans, the larvae of many different phyla of animals, and many heterotrophic protists and bacteria.

In the sea, most small and some large fish, as well as most of the great whales, feed on the plankton, and still larger fish feed on the smaller ones. In this way, the "great meadow of the sea," as the phytoplankton is sometimes called, can be likened to the meadows of the land, serving as the source of nourishment for het-

TABLE 16–1   Comparative Summary of Characteristics of *Protista* I

Phylum	Number of Species	Photosynthetic Pigments	Carbohydrate Food Reserve	Flagella	Cell Wall Component	Habitat
Euglenophyta (euglenoids)	900	Most have none, or chlorophylls *a* and *b*; carotenoids	Paramylon	Usually 2; often unequal, with 1 forward and 1 behind or reduced to stub; extend apically; various hairs	No cell wall; have a flexible or rigid pellicle of proteinaceous strips beneath the plasma membrane	Mostly freshwater, some marine
Myxomycota (plasmodial slime molds)	700	None	Glycogen	Usually 2; apical; unequal; in reproductive cells only, whiplash	None on plasmodium	Terrestrial
Dictyosteliomycota (dictyostelids, or cellular slime molds)	50	None	Glycogen	None (amoeboid)	Cellulose	Terrestrial
Cryptophyta (cryptomonads)	200	None, or chlorophylls *a* and *c*; phycobilins; carotenoids	Starch	2; unequal; subapical; tinsel (1 with one row, 1 with two rows of hairs)	No cell wall; have a stiff layer of proteinaceous plates beneath plasma membrane	Marine and freshwater; cold waters
Rhodophyta (red algae)	4000–6000	Chlorophyll *a*; phycobilins; carotenoids	Floridean starch	None	Cellulose microfibrils embedded in matrix (usually galactans); deposits of calcium carbonate in many	Predominantly marine, about 100 freshwater species; many tropical species
Dinophyta (dinoflagellates)	2000–4000	None in many, or chlorophylls *a* and *c*; carotenoids, mainly peridinin	Starch	None (except in gametes) or 2, dissimilar; lateral (1 transverse, 1 longitudinal); both have hairs	Have a layer of vesicles beneath the plasma membrane with or without cellulose plates	Mostly marine, many freshwater; some in symbiotic relationships
Haptophyta (haptophyte algae)	300	Chlorophylls *a* and *c*; carotenoids, especially fucoxanthin	Chrysolaminarin	None or 2; equal or unequal; most have no hairs; most have haptonema	Scales of cellulose; scales of calcified organic material in some	Great majority marine, a few freshwater

erotrophic organisms. Floating or swimming single-celled or colonial chrysophytes, diatoms, green algae, and dinoflagellates are the most important organisms at the base of freshwater food chains. In marine waters, unicellular or colonial haptophytes, dinoflagellates, and diatoms are the most important eukaryotic members of the marine phytoplankton and therefore essential to the support of marine animal life (Figure 16–2).

0.2 mm

16–2

Marine phytoplankton. The organisms shown here are dinoflagellates and filamentous and unicellular diatoms.

Marine phytoplankton is being increasingly used as "fodder" to support the growth of shrimp, shellfish, and other marketable seafood products. Seaweeds can be grown in aquatic farming operations to produce edible and industrially useful products (see "Algae and Human Affairs" on page 380). Both types of applications of algae are examples of *mariculture,* by which marine organisms are cultivated in their natural environment, and are analogous to terrestrial agricultural systems.

In both marine and fresh waters relatively free of serious human disturbance, phytoplankton populations are usually held in check by seasonal climate changes, nutrient limitation, and predation. When humans pollute aquatic systems, however, some algae may be released from these constraints, and their populations grow to undesirable, "bloom" proportions. In the ocean, some of these blooms are known as "red tides" or "brown tides" because the water becomes colored by large numbers of algal cells containing red or brown accessory pigments. Algal blooms correlate with the release of large quantities of toxic compounds into the water. These toxins, which may have evolved as a defense against protist or animal predators, can cause human illness and massive die-offs of fishes, birds, or aquatic mammals (see "Red Tides/Toxic Blooms" on page 364). In recent years the frequency of toxic marine blooms has increased globally, although only a few dozen species of phytoplankton are toxic. Some ecologists link this increase with the worldwide decline in water quality caused by increased human populations.

Algae, which are abundant on this watery planet, are important in carbon cycling (see "The Carbon Cycle" on page 150). They are able to transform carbon dioxide

(CO_2)—a so-called "greenhouse gas" that contributes to global warming—into carbohydrates by photosynthesis and into calcium carbonate by calcification. Large amounts of organic carbon and calcium carbonate are incorporated into algae and have been transported to the ocean bottom. Today, the marine phytoplankton absorbs about one-half of all the CO_2 that results from human activities such as the burning of fossil fuels. The extent to which such carbon is transported to the ocean depths, and thus does not contribute to global temperature increases, is a matter of controversy.

The mechanism by which some phytoplanktonic organisms reduce the amount of CO_2 in the atmosphere is by favoring the formation of calcium carbonate as they fix CO_2 during photosynthesis. The calcium carbonate is deposited as tiny scales covering the phytoplankton. The CO_2 removed from the water by this calcification process and by photosynthesis is replaced by atmospheric CO_2, creating a suction effect, also known as "CO_2 drawdown." Phytoplankton covered with calcium carbonate, upon settling to the ocean floor, gave rise to much of the famous White Cliffs of Dover and to the economically important North Sea oil deposits. Several kinds of red, green, and brown seaweeds can also become encrusted with calcium carbonate. The effect of calcification of these seaweeds on the global carbon cycle is not as well understood as the effect of phytoplankton calcification.

Some marine phytoplankton, particularly haptophytes and dinoflagellates, produce significant amounts of a sulfur-containing organic compound that aids in regulating osmotic pressure within their cells. A volatile compound derived from it is excreted by the cells and subsequently is converted to sulfur oxides in the atmosphere, which increase cloud cover and, hence, reflectiveness. Sulfur oxides, which are also generated by burning fossil fuels, contribute to acid rain and have a cooling effect on the climate. Scientists who try to predict future climates must take such cooling effects into account, along with the climate-warming effects of the greenhouse gases.

In this chapter we will consider the following, largely unrelated protist phyla: *Euglenophyta,* the euglenoids; *Myxomycota* and *Dictyosteliomycota,* the plasmodial and cellular slime molds; *Cryptophyta,* the cryptomonads; *Rhodophyta,* the red algae; *Dinophyta,* the dinoflagellates; and *Haptophyta,* the haptophytes. All of these groups of organisms and those considered in the chapter that follows are of great ecological importance.

Euglenoids: Phylum *Euglenophyta*

Flagellates—protists bearing flagella—known as euglenoids make up the phylum *Euglenophyta.* There are 900 known species of euglenoids. Molecular evidence suggests that the earliest euglenoids were phagocytes (parti-

16–3

Euglena. *(a) Electron micrograph showing two large storage bodies of paramylon and the helically arranged proteinaceous strips of the pellicle. A red eyespot, or stigma, is visible at the upper end of the cell. (b) The structure of* Euglena, *as interpreted from electron micrographs.*

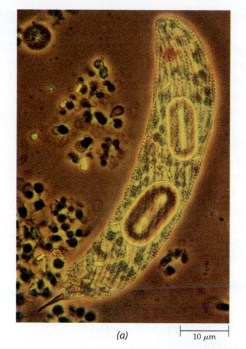

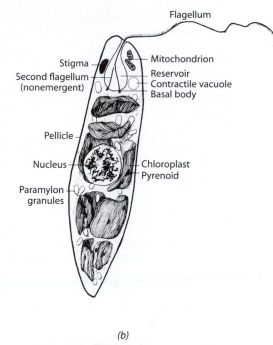

(a)   10 μm   (b)

cle eaters). About one-third of the genera, including the common *Euglena,* contain chloroplasts. The similarities between the chloroplasts of the euglenoids and those of the green algae—both possess chlorophylls *a* and *b* together with several carotenoids—suggest that the euglenoid chloroplasts were derived from endosymbiotic green algae. About two-thirds of the genera are colorless heterotrophs that rely upon particle feeding or absorption of dissolved organic compounds. These feeding habits, along with a more general requirement for vitamins, explain why many of the euglenoids occur in fresh waters that are rich in organic particles and compounds.

The structure of *Euglena* (Figure 16–3) illustrates many of the features typical of euglenoids in general. With one exception (the colonial genus *Colacium*), euglenoids are unicellular. *Euglena,* like most euglenoids, does not have a cell wall or any other rigid structure covering the plasma membrane. The genus *Trachelomonas* (Figure 16–4), however, has a wall-like covering made of iron and manganese minerals. The plasma membrane of euglenoids is supported by an array of helically arranged proteinaceous strips, which are in the cytoplasm, immediately beneath the plasma membrane. These strips form a structure called the **pellicle,** which may be flexible or rigid. *Euglena's* flexible pellicle allows the cell to change its shape, facilitating movement in muddy habitats where flagellar swimming is difficult. Swimming *Euglena* cells have a single, long flagellum that emerges from the base of an anterior depression called the **reservoir,** and a second, nonemergent flagellum (Figure 16–3). A swelling at the base of the emergent flagellum, together with a nearby, distinctive red **eyespot,** or **stigma,** in the cytoplasm, constitute a light-sensing system in euglenoids.

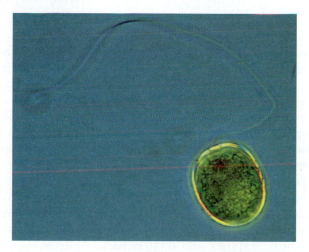

16–4

Trachelomonas, a widespread euglenoid. Its green-pigmented cell is encased in a rigid, often ornamented wall-like covering called a lorica, which may vary from clear to black-brown in color. A single flagellum protrudes through a pore at the apex of the lorica. Reproduction is accomplished by mitosis.

A **contractile vacuole** collects excess water from all parts of the euglenoid cell. The water is discharged from the cell via the reservoir. After each discharge, a new contractile vacuole is formed by coalescence of small vesicles. Contractile vacuoles are commonly observed in freshwater protists, which need to eliminate the excess water that accumulates as a result of osmosis. If the water is not removed, the cells will burst.

In contrast to the chloroplasts of green algae, euglenoid plastids do not store starch. Instead, granules of a unique polysaccharide known as **paramylon** form in the cytoplasm. Euglenoid plastids often resemble plastids of many green and other algae in their possession of a protein-rich region called a **pyrenoid,** which is the site of Rubisco and some other enzymes involved in photosynthesis (page 140).

Euglenoids reproduce by mitosis and lengthwise cytokinesis, continuing to swim while they divide. The nuclear envelope remains intact during mitosis as it does in many other protists and in fungi. This suggests that an intact mitotic nuclear envelope is a primitive feature and that nuclear envelopes that break down during mitosis, as they do in plants, animals, and various protists, are a derived characteristic. Sexual reproduction and meiosis do not seem to occur in euglenoids, suggesting that these processes had not yet evolved when this group diverged from the main lineage of protists.

16–5
Plasmodium of a plasmodial slime mold, Physarum, *growing on a tree trunk.*

Plasmodial Slime Molds: Phylum *Myxomycota*

The plasmodial slime molds, or myxomycetes, are a group of about 700 species that seems to have no direct relationship to the cellular slime molds, the fungi, or any other group. Although referred to as molds, as are certain fungi, molecular evidence demonstrates that neither the plasmodial slime molds nor cellular slime molds are closely related to the fungi. When conditions are appropriate, the plasmodial slime molds exist as thin, streaming multinucleate masses of protoplasm that creep along in amoeboid fashion. Lacking a cell wall, this "naked" mass of protoplasm is called a **plasmodium.** As these plasmodia travel, they engulf and digest bacteria, yeast cells, fungal spores, and small particles of decaying plant and animal matter. Plasmodia can also be cultured on media that do not contain particulate matter, suggesting that plasmodia can also obtain food by absorption of dissolved organic compounds.

The plasmodium may grow to weigh as much as 20 to 30 grams and, because slime molds are spread out thinly, this amount can cover an area of up to several square meters (Figure 16–5). The plasmodium, with its many nuclei, is not divided by cell walls. As it grows, the nuclei divide repeatedly and synchronously; that is, all the nuclei in a plasmodium divide at the same time. Centrioles are present, and mitosis is similar to that of plants and higher animals, although the chromosomes are very small.

Typically, the moving plasmodium is fan-shaped, with flowing protoplasmic tubules that are thicker at the base of the fan and spread out, branch, and become thinner toward their outer ends. The tubules are composed of slightly solidified protoplasm through which more liquefied protoplasm flows rapidly. The foremost edge of the plasmodium consists of a thin film of gel separated from the substrate by only a plasma membrane and a slime sheath.

Plasmodial slime molds have a life history that involves sexual reproduction and therefore may be among the earliest divergent protists to have acquired sex. Plasmodial growth continues as long as an adequate food supply and moisture are available. Generally, when either of these is in short supply, the plasmodium migrates away from the feeding area. At such times, plasmodia may be found crossing roads or lawns, climbing trees, or in other unlikely places. In many species, when the plasmodium stops moving, it divides into a large number of small mounds. The mounds are similar in size and volume, and so their formation is probably controlled by chemical effects within the plasmodium. The life cycle of a typical plasmodial slime mold is summarized in Figure 16–6. Each mound produces a sporangium, usually borne at the tip of a stalk. The mature sporangium is often extremely ornate (Figure 16–7). The protoplasm of the young sporangium contains many nuclei, which increase in number by mitosis. Progressively, the protoplasm is cleaved into a large number of spores, each containing a single diploid nucleus. Meiosis then occurs, giving rise to four haploid nuclei per spore. Three of the four nuclei disintegrate, however, leaving each spore with a single haploid nucleus. In some members of this group, discrete sporangia are not produced, and the entire plasmodium may develop either into a **plasmodiocarp** (Figure 16–7c), which retains the former shape of the plasmodium, or into an **aethalium** (Figure 16–7d), in which the plasmodium forms a large mound that is essentially a single large sporangium.

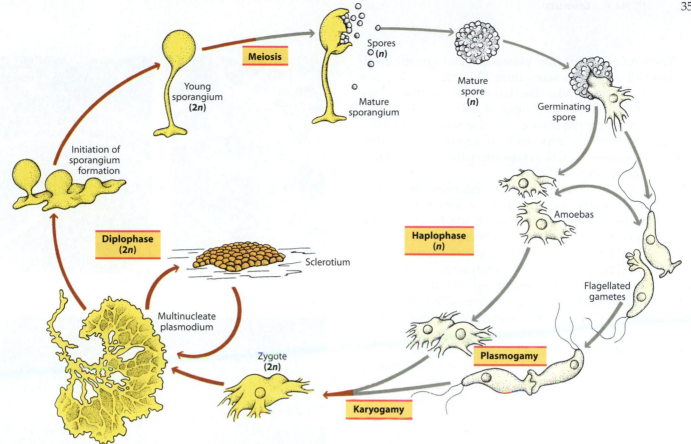

Meiosis

Young sporangium **(2n)**

Spores **(n)**

Mature sporangium

Mature spore **(n)**

Germinating spore

Initiation of sporangium formation

Amoebas

Diplophase (2n)

Haplophase (n)

Sclerotium

Flagellated gametes

Multinucleate plasmodium

Zygote **(2n)**

Plasmogamy

Karyogamy

16–6

Life cycle of a typical myxomycete. Sexual reproduction in the plasmodial slime molds consists of three distinct phases: plasmogamy, karyogamy, and meiosis. Plasmogamy is the union of two protoplasts, which may be amoebas or flagellated gametes derived from germinated spores; it brings two haploid nuclei together in the

same cell. Karyogamy is the fusion of these two nuclei, resulting in the formation of a diploid zygote and initiation of the so-called diplophase of the life cycle. The plasmodium is a multinucleate, free-flowing mass of protoplasm that can pass through a silk cloth or a piece of filter paper and remain unchanged.

Plasmodia often form a hardened stage, the sclerotium, in nature and are able to survive dry periods well in this condition. In any event, the active plasmodium may ultimately form sporangia. Within the developing sporangia, meiosis restores the haploid condition and thus initiates the haplophase of the life cycle.

16–7

Spore-producing structures in plasmodial slime molds. (a) Sporangium of Arcyria nutans. (b) Sporangia of Stemonitis splendens. (c) Plasmodiocarp of Hemitrichia serpula. (d) Aethalia of Lycogala growing on bark.

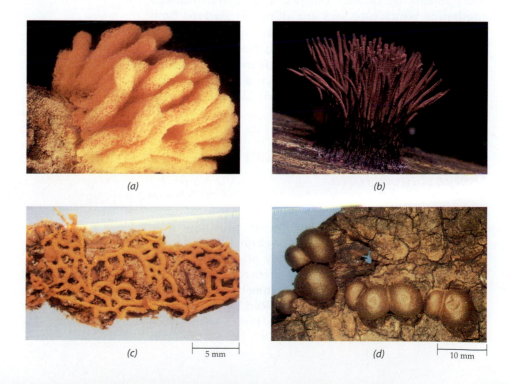

(a)

(b)

(c)   5 mm

(d)   10 mm

When habitats dry out, plasmodia can quickly form an encysted stage, the **sclerotium.** Sclerotia can be seen easily on firewood piles because these organisms are often brightly colored in shades of yellow and orange. They are of great importance for the survival of plasmodial slime molds, especially in fast-drying habitats such as dead cacti or soil in deserts, where these organisms are abundant.

The spores of myxomycetes are also resistant to environmental extremes and can be very long-lived. Some have germinated after being kept in the laboratory for more than 60 years. Thus, spore formation in this group seems to make possible not only genetic recombination but also survival under adverse conditions.

Under favorable conditions, the spores split open and the protoplast slips out (Figure 16–6). The protoplast may remain amoeboid, or it may develop one to four whiplash flagella. The amoeboid and flagellated stages are readily interconvertible. The amoebas feed by the ingestion of bacteria and organic material and multiply by mitosis and cell cleavage. If the food supply is used up or conditions are otherwise unfavorable, an amoeba may cease moving about, become round, and secrete a thin wall to form a **microcyst.** These microcysts can remain viable for a year or more, resuming activity when favorable conditions return.

After a period of growth, plasmodia appear in the amoeba population. Their appearance is governed by a number of factors, including cell age, environment, density of amoebas, and cyclic adenosine monophosphate (cyclic AMP, or cAMP). These factors play roles similar to those in the cellular slime mold *Dictyostelium discoideum*, discussed in the following section. One method of plasmodium formation is by the fusion of gametes. The gametes are usually genetically different from one another, in which case they are ultimately derived from different haploid spores. These gametes are simply some of the amoebas or flagellated cells that now have a new role. In some species and strains, however, the plasmodium is known to form directly from a single amoeba. Such plasmodia are usually haploid, like the amoebas from which they arose.

Cellular Slime Molds: Phylum *Dictyosteliomycota*

The cellular slime molds, or dictyostelids—a group of about 50 species in four genera—are probably more closely related to the amoebas (phylum *Rhizopoda*, not considered in this book) than to any other group. They are common inhabitants of most litter-rich soils, where they usually exist as free-living amoeba-like cells, or **myxamoebas.** These myxamoebas feed on bacteria by phagocytosis (Figure 16–8). Unlike fungi, with which

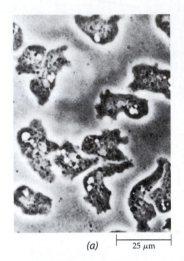
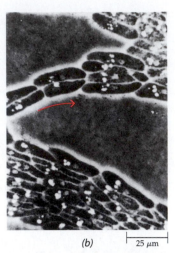

(a)   ⊢ 25 μm   (b)   ⊢ 25 μm

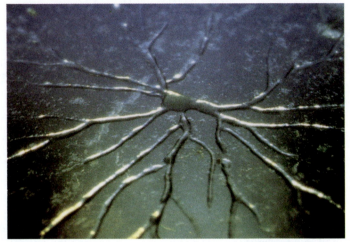

(c)

16–8

Life cycle of the cellular slime mold Dictyostelium discoideum. **(a)** *The feeding stage of the myxamoebas. The light gray area in the center of each cell is the nucleus, and the white areas are contractile vacuoles.* **(b, c)** *Myxamoebas aggregating. The direction in which the stream is moving is indicated by an arrow.* **(d)** *Migrating pseudoplasmodium, formed of many myxamoebas. Each sluglike mass deposits a thick slime sheath, which collapses behind it.* **(e–g)** *At the end of the migration, the pseudoplasmodium gathers itself into a mound and begins to rise vertically, differentiating into a stalk and a mass of spores, as seen in these scanning electron micrographs.*

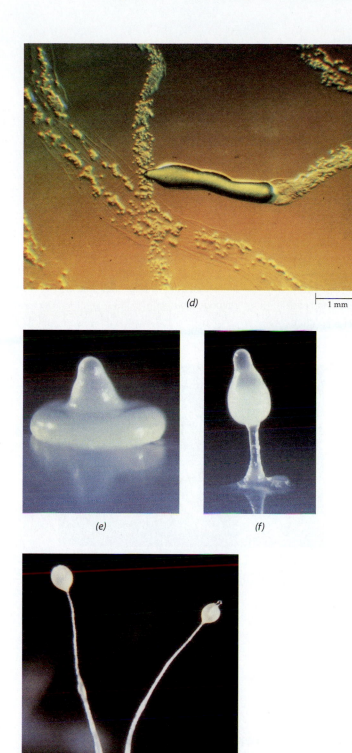

(d)

1 mm

(e)

(f)

(g)

0.5 mm

they were once grouped, the cellular slime molds have cellulose-rich cell walls during part of their life cycle and undergo the type of mitosis found in plants and animals, in which the nuclear envelope breaks down. In addition, again unlike the fungi, cellular slime molds have centrioles.

Dictyostelium discoideum has become an important model system for laboratory study of eukaryotic gene expression and developmental processes. One such process is programmed cell death, or **apoptosis,** normal cell death in which the cell activates a death program and kills itself. *Dictyostelium* reproduces by cell division and shows little morphological differentiation until it exhausts the available supply of bacteria. In response to starvation, the cellular slime molds produce asexual spores, but not in a simple way. Individual cells first aggregate to form a motile sluglike mass. The uninucleate, haploid myxamoebas retain their individuality in the mass, which usually contains 10,000 to 125,000 individuals. This mass, called a **pseudoplasmodium** (Figure 16–8d), or **slug,** migrates to a new place before differentiating and releasing spores. Thus, it avoids releasing new spores into habitats depleted of bacteria.

The myxamoebas aggregate by **chemotaxis,** migrating toward a source of cAMP, which is secreted by the starved myxamoebas. The cAMP diffuses away from the cells, establishing a concentration gradient along which the surrounding cells move toward the cells that are secreting the cAMP. The secreting cells, in turn, are stimulated to emit a new "pulse" of cAMP after a period of about five minutes. At least three waves of cells are recruited in this fashion. Binding of the cAMP signal to receptors in the plasma membrane triggers massive rearrangement of actin filaments, enabling the myxamoeba to crawl toward the source of cAMP. As the cells accumulate at the aggregation center, their plasma membranes become sticky; this causes them to adhere to each other and results in formation of a pseudoplasmodium.

The eventual developmental fate of an individual cell is determined at an early stage by its position in the aggregation, which appears to be controlled by the stage of the cell cycle the individual myxamoeba was in when aggregation began. Cells that divide between about 1.5 hours before and 40 minutes after the onset of starvation enter the aggregation last. When migration ceases, cells in the forward, or anterior, end of the slug become the stalk cells of the developing spore-bearing body. The stalk cells become coated with cellulose, providing rigidity to the stalk, and then die by apoptosis. Meanwhile, the posterior cells of the pseudoplasmodium move to the top of the stalk and become dormant spores. Eventually, the spores are dispersed. If spores fall on a warm, damp surface, they germinate. Each spore releases a single myxamoeba, and the cycle is repeated.

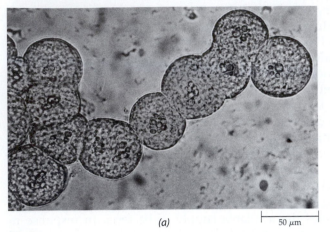 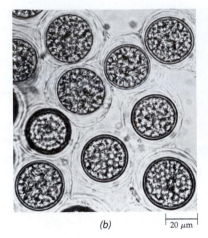 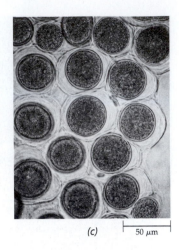

(a)   (b)   `20 µm`   (c)   `50 µm`
        `50 µm`

16–9

*The process of macrocyst formation in
Dictyostelium mucoroides. (a) Each zygote,
or giant cell, is beginning to engulf the myx-*
*amoebas around it. (b) Giant cells have
engulfed all of the myxamoebas, and each
has laid down a cellulose wall. (c) Mature*
*macrocysts; at this stage they appear
homogeneous.*

Reproduction involving asexual spores is common in the cellular slime molds. Sexual reproduction also occurs frequently and results in the formation of walled zygotes called **macrocysts.** The macrocysts are formed by aggregations of myxamoebas that are smaller than those involved in the formation of slugs. In addition, these aggregations are rounded in outline, rather than being elongate (Figure 16–9). During the formation of a macrocyst, two haploid myxamoebas first fuse, forming a single large myxamoeba, the zygote, which becomes actively phagocytic. The zygote continues to feed voraciously until all of the surrounding myxamoebas have been engulfed, becoming a giant cell in the process. At this stage, a thick cell wall, rich in cellulose, is laid down around the giant cell, and a mature macrocyst is formed. Within the macrocyst, the zygote—the only diploid cell of the life cycle—undergoes meiosis and several mitotic divisions before germination and the release of numerous, haploid myxamoebas.

16–10

*Cryptomonas, a cryptomonad. Each of these
unicellular algae has two, slightly unequal
flagella (not apparent here) of about the
same length as the cell. Both flagella emerge
from an anterior depression of the cell.*

Cryptomonads: Phylum *Cryptophyta*

The cryptomonads are chocolate-brown, olive, blue-green, or red, fast-growing, single-celled flagellates that occur in marine and fresh waters (Figure 16–10). There are 200 known species of cryptomonads (Gk. *kryptos,* "hidden," + *monos,* "alone"). They are aptly named because their small size (3 to 50 micrometers) often makes them inconspicuous. They occur primarily in cold or subsurface waters, are not easily preserved, and are readily eaten by aquatic herbivores. Cryptomonads are particularly rich in polyunsaturated fatty acids, which are essential to the growth and development of zooplankton. Cryptomonads are ecologically important phytoplanktonic algae because of their edibility, and be-

cause they are frequently the dominant algae in lakes and coastal waters when diatom and dinoflagellate populations subside on a seasonal basis.

Cryptomonads are like euglenoids not only in requiring certain vitamins but also in having both pigmented photosynthetic members and colorless phagocytic members that consume particles such as bacteria. Cryptomonads provide some of the best evidence that colorless eukaryotic hosts can obtain chloroplasts from endosymbiotic eukaryotes. The chloroplasts of cryptomonads and certain other algae have four bounding membranes. Evidence indicates that the cryptomonads arose through the fusion of two different eukaryotic cells, one het-

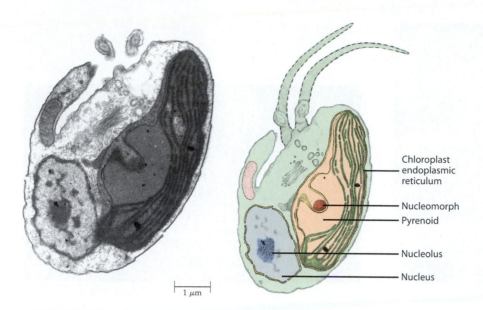

Chloroplast
endoplasmic
reticulum

Nucleomorph
Pyrenoid

Nucleolus

Nucleus

1 μm

16–11
Electron micrograph of a cryptomonad, showing the nucleus and the nucleomorph, which is wedged in a cleft in the chloroplast. The nucleomorph is considered to be a vestigial nucleus belonging to the endosymbiont—a red algal cell—engulfed by a heterotrophic host. The outer of the four membranes surrounding the chloroplast is the chloroplast endoplasmic reticulum.

erotrophic and the other photosynthetic, establishing a **secondary endosymbiosis.** In addition to chlorophylls *a* and *c* and carotenoids, some cryptomonad chloroplasts contain a phycobilin, either phycocyanin or phycoerythrin. These water-soluble accessory pigments are otherwise known only in cyanobacteria and red algae, providing evidence for the origin of the cryptomonad chloroplast.

The outer of the four membranes surrounding the cryptomonad chloroplast is continuous with the nuclear envelope and is called the **chloroplast endoplasmic reticulum** (Figure 16–11). The space between the second and third chloroplast membranes contains starch grains and the remains of a reduced nucleus, complete with three linear chromosomes and a nucleolus with typical eukaryotic RNA. The reduced nucleus, called a **nucleomorph** (meaning "looks like a nucleus"), is interpreted as the remains of the nucleus of a red algal cell that was ingested and enslaved for its photosynthetic capabilities by a heterotrophic host. The endosymbiont resembles the chloroplasts of other algae in that most of its genes have been transferred to the host's nucleus so that it is no longer capable of independent existence.

Red Algae: Phylum *Rhodophyta*

Red algae are particularly abundant in tropical and warm waters, although many are found in the cooler regions of the world. There are 4000 to 6000 known species in 680 or so genera, only a very few of which are unicellular—such as *Cyanidium*, one of the only organisms capable of growth in acidic hot springs—or microscopic filaments. The vast majority of red algae are more structurally complex, macroscopic seaweeds. Fewer than 100 different species of red algae occur in fresh water (Figure 16–12), but in the sea the number of species is greater

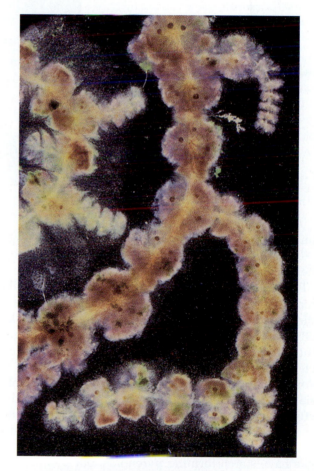

16–12
The simple, filamentous, freshwater red alga Batrachospermum moniliforme. *The soft, gelatinous, branched axes of this freshwater red alga are most frequently found in cold streams, ponds, and lakes, where they occur throughout the world.*

16–13

Marine red algae. (a) In Bonnemaisonia hamifera, the basically filamentous structure of the red algae is clearly evident. The branched filaments of this red alga are hooked, enabling it to cling to other sea-weeds. (b) Jointed coralline algae in a tidal pool in California. (c) The reef-stabilizing, crustose coralline red alga Porolithon cras-pedium. (d) Irish moss (Chondrus crispus), an important source of carrageenan.

(a)

(b)

(c)

(d)

than that of other types of seaweeds combined (Figure 16–13). Red algae usually grow attached to rocks or to other algae, but there are a few floating forms.

The chloroplasts of red algae contain phycobilins, which mask the color of chlorophyll *a* and give red algae their distinctive color. These pigments are particularly well suited to the absorption of the green and blue-green light that penetrates into deep water, where red algae are well represented. Biochemically and structurally, the chloroplasts of red algae closely resemble the cyanobacteria from which they were almost certainly derived, directly following endosymbiosis. Some red algae have lost most or all of their pigments and grow as parasites upon other red algae. Chloroplasts of a few primitive red algae have starch-forming pyrenoids, but pyrenoids appear to have been lost prior to the origin of the more complex forms.

Cells of Red Algae Have Some Unique Features

Red algae are unusual among the algae, and unique among algal phyla, in having neither centrioles nor flagellated cells. Whether they lost centrioles and flagella

at some ancient time or their ancestors never had these structures is not known. In place of centrioles, which occur in many other eukaryotes, the red algae have microtubule organizing centers called **polar rings.** The main food reserves of the red algae are granules of **floridean starch,** which are stored in the cytoplasm. Floridean starch, a unique molecule that resembles the amylopectin portion of starch, is actually more like glycogen than starch (page 20).

The cell walls of most red algae include a rigid, inner component consisting of microfibrils of cellulose or of another polysaccharide and an outer mucilaginous layer, usually a sulfated polymer of galactose, such as agar or carrageenan (see "Algae and Human Affairs" on page 380). It is this mucilaginous layer that gives red algae their characteristic flexible, slippery texture. Continuous production and sloughing of the mucilage helps red algae rid themselves of other organisms that might colonize their surfaces and reduce their exposure to sunlight.

In addition, certain red algae deposit calcium carbonate in their cell walls. The function of algal calcification is uncertain. One hypothesis is that calcification helps algae to obtain carbon dioxide from the water for photo-

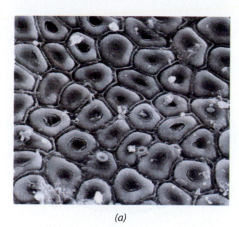

(a)   (b)

16–14

*Coralline algae. **(a)** Scanning electron micrograph of an unidentified species of crustose red alga from a depth of 268 meters on a seamount in the Bahamas, nearly 100 meters below the lowest limits established for any other photosynthesizing organism. At this depth, the light intensity was estimated as 0.0005 percent of its value at the ocean surface. The alga formed patches about a meter across, covering about 10 percent of the rock surface. When tested in the laboratory, this alga was found to be about 100 times more efficient than its shallow-water relatives in capturing and using light energy. **(b)** Purple crustose coralline algae from the same seamount.*

synthesis. Many of the calcified red algae are especially tough and stony and constitute the family *Corallinaceae*, the **coralline algae.** Calcification explains the occurrence of possible fossil coralline algae that are more than 700 million years old. Coralline algae are common throughout the oceans of the world, growing on stable surfaces that receive enough light, including 268-meter-deep seabed rocks (Figure 16–14). Other habitats include tidal pool rocks, on which jointed coralline red algae grow (Figure 16–13b), and the surf-pounded shoreward surfaces of coral reefs, where crust-forming (crustose) coralline algae (Figure 16–13c) help to stabilize reef structure. Large areas of diverse coral reefs around the world owe their survival, in part, to the architectural strength conferred by coralline algae. In recent years, a bright orange bacterium that causes a lethal disease of coralline algae has been spreading throughout the South Pacific, endangering thousands of kilometers of reef.

Many red algae produce unusual toxic terpenoids (page 33) that may assist in deterring herbivores. Some red algal terpenoids have anti-tumor activity and are currently being tested for possible use as anti-cancer drugs.

A few genera, such as *Porphyra,* have cells densely packed together into one- or two-layered sheets (Figure 16–15). However, most red algae are composed of filaments that are often densely interwoven and held to-

16–15

The red alga, Porphyra nereocystis, *the life history of which includes both bladelike and filamentous phases. **(a)** The bladelike phase, which is a gametophyte, produces both female gametangia, called carpogonia (reddish cell aggregates on left), and male gametangia, called spermatangia (cell aggregates on right). **(b)** After fertilization, the diploid carpospores give rise to a system of branched filaments. Meiosis occurs during germination of the spores (conchospores) produced by the filamentous phase. These haploid cells then grow into the bladelike gametophyte.*

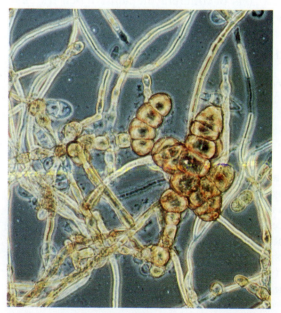

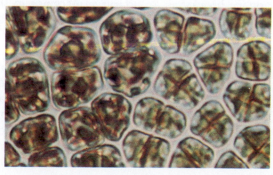

(a)   (b)

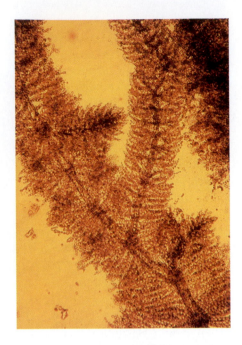

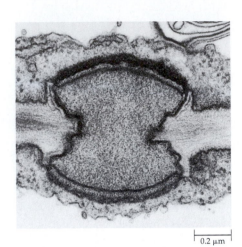

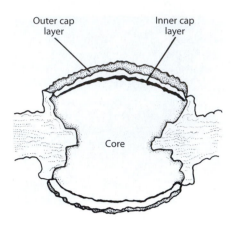

16–17

A pit connection in the red alga Palmaria. *Pit connections are distinct, lens-shaped plugs that form between the cells of red algae as the cells divide. These connections are also formed frequently between the cells of adja-*

cent filaments that come into contact with each other, linking together the bodies of red algae. Pit-connection cores are protein, and their outer cap layers are, at least in part, polysaccharide.

16–16

Batrachospermum sirodotia, *showing the whorls of lateral branches.*

gether by the mucilaginous layer, which has a rather firm consistency. Growth in filamentous red algae is initiated by a single, dome-shaped apical cell that cuts off segments sequentially to form an axis. This axis, in turn, forms whorls of lateral branches (Figure 16–16). In most red algae, the cells are interconnected by primary **pit connections** (Figure 16–17) that develop at cytokinesis. Many red algae are multiaxial—that is, made up of many coherent filaments forming a three-dimensional body. In such forms, the filaments are interconnected by the formation of secondary pit connections. These pits form between cells of different filaments when the filaments come into contact with one another.

Red Algae Have Complicated Life Histories

Many red algae reproduce asexually by discharging spores, called **monospores,** into the water. If conditions are suitable, monospores may attach to a substrate. By repeated mitoses, a new seaweed similar to the monospore-producing parent is produced. Sexual reproduction also occurs widely among the multicellular red algae and can involve very complex life histories.

The simplest type of sexual life history in red algae involves an alternation of generations between two separate, multicellular forms of the same species—a haploid, gamete-producing **gametophyte** and a diploid, spore-producing **sporophyte.** The gametophyte produces **spermatangia** (singular: spermatangium), structures that generate and release nonmotile **spermatia** (singular:

spermatium), or male gametes, which are carried to female gametes by water currents. The female gamete, or egg, is the lower, nucleus-containing portion of a structure known as the **carpogonium** (plural: carpogonia), which is borne on the same gametophyte as the spermatangia and remains attached to it. The carpogonium develops a protuberance, the **trichogyne,** for reception of spermatia. When a spermatium comes in contact with a trichogyne, the two cells fuse. The male nucleus then migrates down the trichogyne to the female nucleus and fuses with it. The resultant diploid zygote then produces a few diploid **carpospores,** which are released from the parent gametophyte into the water. If they survive, the carpospores attach to a surface and grow into the sporophyte, which produces haploid spores by sporic meiosis. If these haploid spores survive, they in turn attach to a surface and grow into gametophytes, thus completing the life cycle.

Experts believe that the red algae acquired an alternation of two multicellular generations early in their evolutionary history as an adaptive response to their lack of flagellated male gametes. Such nonflagellated gametes cannot swim toward female gametes, as do the flagellated male gametes of some other protists, animals, and some plants. Hence, fertilization may be more a matter of chance, with the consequence that zygote formation may be relatively rare. Alternation of generations is regarded as an adaptation that increases the number and genetic diversity of the progeny resulting from each individual fertilization event, or zygote. This is because a multicellular sporophyte can produce many more spores—and more diverse haploid spores—than could a single, meiotic zygote nucleus. Alternation of two multicellular generations also occurs in several other protist groups, such as green and brown seaweeds (discussed in Chapter 17), and in bryophytes and vascular plants (discussed in Chapters 18 to 21), where it may have similar ecological and genetic benefits.

A further evolutionary advance has occurred in most red algae. Rather than immediately producing spores, the zygote nucleus divides repeatedly by mitosis, generating a third multicellular life-cycle phase, the diploid **carposporophyte** generation. The carposporophyte generation remains attached to its parental gametophyte and probably receives organic nutrients from it. These nutrients help to support rapid proliferation of cells by mitosis. When the carposporophyte reaches its mature size, mitosis occurs in apical cells, giving rise to carpospores. The carpospores are released into the water, settle onto a substrate, and grow into separate diploid sporophytes.

In many red algae, a mitotically produced copy of the diploid zygote nucleus is transferred to another cell of the gametophyte. This cell, known as an **auxiliary cell**, serves as a host and nutritional source for repeated mitoses by the adopted nucleus. Proliferation of diploid filaments from the auxiliary cell generates a carposporophyte and carpospores. In many forms, multiple copies of the diploid zygote nucleus are carried by the growth of long tubular cells throughout the algal body and are deposited into many additional auxiliary cells. Each diploid nucleus then produces many carposporophytes, which release very large numbers of carpospores into the water. In one case, each zygote nucleus is known to result in the release of some 4500 carpospores. Each carpospore is capable of growing into a usually freeliving, multicellular diploid generation, called the **tetrasporophyte.** Meiosis occurs in specialized cells of the tetrasporophyte, called **tetrasporangia.** Each of the tetraspores produced can germinate into a new gametophyte, if conditions are favorable. *Polysiphonia* provides an example of this kind of life cycle (Figure 16–18).

The life history of most red algae thus consists of three phases: (1) a haploid gametophyte; (2) a diploid phase, called a carposporophyte; and (3) another diploid phase, called a tetrasporophyte. The red algal carposporophyte generation is regarded as an additional way to increase the genetic products of sexual reproduction when fertilization rates are low. Alternation of generations involving three multicellular generations is unique to the red algae. The ability to produce many carposporophytes with the resultant larger numbers of carpospores and potentially huge numbers of tetraspores, all from a single zygote, has helped the red algae to conquer the sexual disability imposed by their lack of flagella.

In most red algae, the gametophyte and tetrasporophyte generations resemble one another closely and are therefore said to be isomorphic, as in *Polysiphonia*. Coralline algae also have isomorphic life cycles. However, an increasing number of heteromorphic life cycles are also being discovered. In these species, the tetrasporophytes either are microscopic and filamentous or consist of a thin crust that is tightly attached to a rock substrate. *Phycologists*—scientists who study algae—speculate that differences in appearance have selective advantages in responding to seasonal changes or other environmental variations. The development of techniques for cultivating algae in the laboratory has led to the discovery that what appear to be distinct species are in some cases actually alternating generations of the same species.

The most important discovery of this kind resulted from British phycologist Kathleen Drew Baker's laboratory cultivation research. She discovered that the inconspicuous red filaments of *Conchocelis*, which grows in seashells, are the diploid phase of the edible, bladeforming seaweed *Porphyra* (the haploid phase). Her work, published in the journal *Nature* in 1949, has formed the basis for the modern, billion-dollar-per-year nori production industry in Japan, Korea, and China (see the essay on page 380). In gratitude, a memorial park was established in Drew Baker's honor in Kumamoto Prefecture, Japan, where she is annually revered with a ceremony dedicated to "the mother of the sea."

Dinoflagellates: Phylum *Dinophyta*

Molecular systematic data indicate that the dinoflagellates are closely related to ciliate protozoa such as *Paramecium* and *Vorticella* (see Figure 13–12). They are also closely related to the sporozoa, which include malarial and other parasites of animals and humans.

Most dinoflagellates are unicellular biflagellates (Figure 16–19). Some 2000 to 4000 species are known, many of them abundant and highly productive members of the marine phytoplankton; others occur in fresh water. Dinoflagellates are unique in that their flagella beat within two grooves. One groove encircles the body like a belt, and the second groove is perpendicular to the first. The beating of the flagella in their respective grooves causes the dinoflagellate to spin like a top as it moves. The encircling flagellum is ribbon-like. There are also numerous nonmotile dinoflagellates, but they typically produce reproductive cells having flagella in grooves, from which their relationship to other dinoflagellates is deduced.

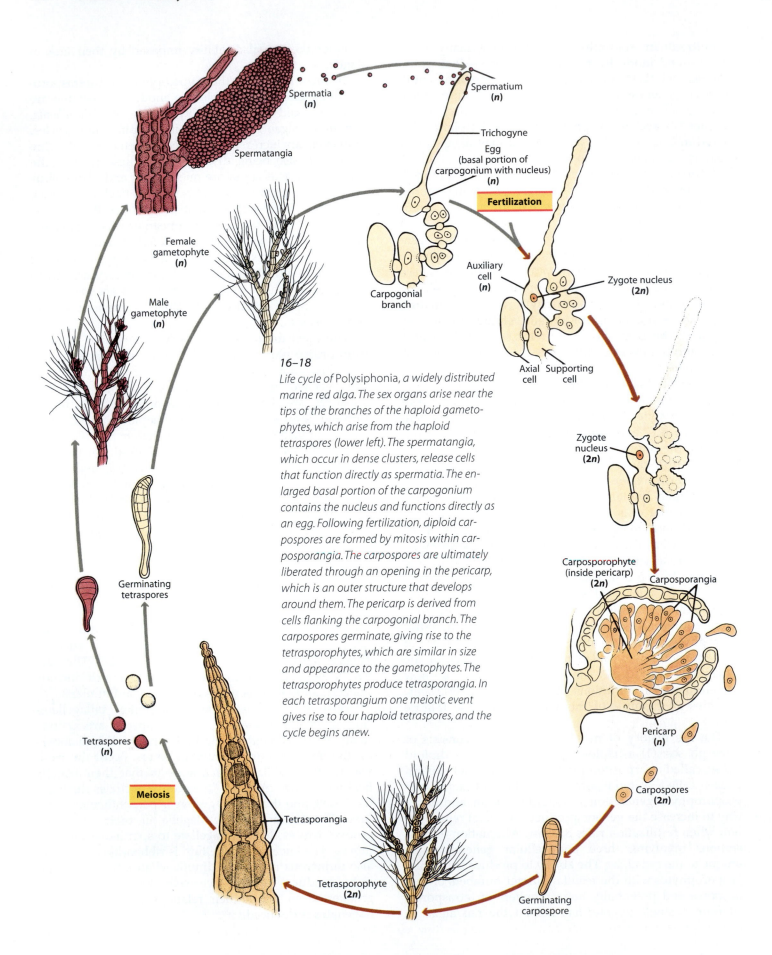

16–18

Life cycle of Polysiphonia, *a widely distributed marine red alga. The sex organs arise near the tips of the branches of the haploid gametophytes, which arise from the haploid tetraspores (lower left). The spermatangia, which occur in dense clusters, release cells that function directly as spermatia. The enlarged basal portion of the carpogonium contains the nucleus and functions directly as an egg. Following fertilization, diploid carpospores are formed by mitosis within carposporangia. The carpospores are ultimately liberated through an opening in the pericarp, which is an outer structure that develops around them. The pericarp is derived from cells flanking the carpogonial branch. The carpospores germinate, giving rise to the tetrasporophytes, which are similar in size and appearance to the gametophytes. The tetrasporophytes produce tetrasporangia. In each tetrasporangium one meiotic event gives rise to four haploid tetraspores, and the cycle begins anew.*

Spermatia (*n*)

Spermatium (*n*)

Trichogyne

Egg (basal portion of carpogonium with nucleus) (*n*)

Spermatangia

Fertilization

Female gametophyte (*n*)

Auxiliary cell (*n*)

Zygote nucleus (*2n*)

Carpogonial branch

Axial cell   Supporting cell

Male gametophyte (*n*)

Zygote nucleus (*2n*)

Carposporophyte (inside pericarp) (*2n*)

Carposporangia

Germinating tetraspores

Pericarp (*n*)

Tetraspores (*n*)

Carpospores (*2n*)

Meiosis

Tetrasporangia

Tetrasporophyte (*2n*)

Germinating carpospore

16–19

The "armor" of some dinoflagellates consists of cellulose plates in vesicles inside the plasma membrane. The vesicles of genera that appear to be naked may or may not contain cellulose plates.

Dinoflagellates are unusual, though not unique, in having permanently condensed chromosomes. This feature, along with some unusual aspects of mitosis, was once thought to indicate that dinoflagellates were quite primitive. Current opinion is that dinoflagellates are a highly derived protist group. The chief method of reproduction of dinoflagellates is by longitudinal cell division, with each daughter cell receiving one of the flagella and a portion of the wall, or theca. Each daughter cell then reconstructs the missing parts in a very intricate sequence.

Many of the dinoflagellates are bizarre in appearance, with stiff cellulose plates forming the **theca,** which often looks like a strange helmet or part of an ancient coat of armor (Figures 16–19 and 16–20). The cellulose plates of the wall are within vesicles just inside the outermost membrane of the cell. Dinoflagellates in the open ocean often have large, elaborate sail-like thecal plates that aid in flotation. Other dinoflagellates have very thin or no cellulose plates and therefore do not appear to have a theca.

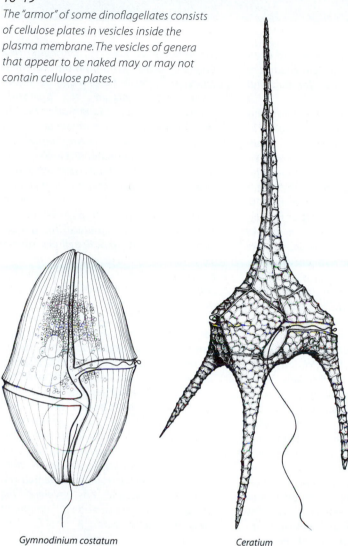

Gymnodinium costatum

Ceratium

Exuviaella

Gymnodinium neglectum

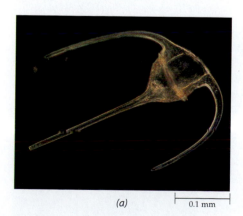

(a)   0.1 mm

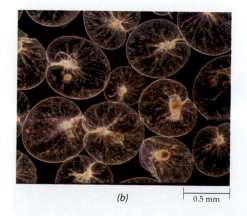

(b)   0.5 mm

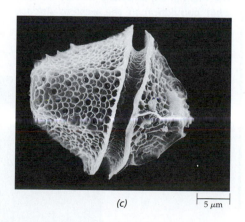

(c)   5 μm

16–20

*Dinoflagellates. **(a)** Ceratium tripos, an armored dinoflagellate. **(b)** Noctiluca scintil-* *lans, a bioluminescent marine dinoflagellate. **(c)** Gonyaulax polyedra, the dinoflagellate* *responsible for the spectacular red tides along the coast of southern California.*

Red Tides/Toxic Blooms

In late August 1987, the west coast of Florida was ravaged by a major red tide incident—one of dozens that are known to have occurred there over the past 150 years. Hundreds of thousands of dead fish littered the beaches, and millions of tourist dollars were lost. The organism responsible for most red tides and their associated effects in the Gulf of Mexico is the dinoflagellate Gymnodinium breve. Following a period of rapid dinoflagellate reproduction, a toxic bloom formed in which G. breve was so abundant that the sea was colored reddish brown. Environmental factors that favor such blooms include warm surface temperatures, high nutrient content in the water, low salinity (which often occurs during rainy periods), and calm seas. Thus rainy weather followed by sunny weather in the summer months is often associated with red tide outbreaks.

Two months later, G. breve invaded estuaries along the North Carolina coast, areas where it had never before been observed. The dinoflagellates became so abundant that the waters turned yellowish, devastating the tourist industry in this region. At least 41 cases of respiratory, gastrointestinal, or neurological illness were reported in people who swam in these waters—all associated with the toxins produced by the dinoflagellate bloom. Gymnodinium is thought to have traveled north from Florida in the Gulf Stream; with it came schools of a fish called menhaden, which consumed large amounts of the dinoflagellates. Bottle-nosed dolphins ate the menhaden, and fully half of the dolphin population in the western Atlantic died. The fish had not been injured by the toxic dinoflagellates in their guts, but the dolphins, eating the fish whole, had been poisoned. In their weakened condition, dolphins proved easy prey for bacterial and viral diseases.

Red tides have been around for a long time. Records of their presence are found in the Old Testament and Homer's Iliad. There is considerable concern, however, that their incidence is increasing and spreading around the globe. In 1990, there were twice as many outbreaks as in 1970. Ecologists are not sure whether this increase is merely an upturn in a natural cycle or the beginning of a serious global epidemic. The increased frequency is correlated with increased nutrients in coastal waters originating from runoff from human development, the heavy use of fertilizers, and livestock farms.

The most recently identified toxin-producing dinoflagellate is Pfiesteria piscicida (see page 366), which has been responsible for massive fish kills in the Neuse and Pamlico Rivers in North Carolina. The toxins produced by Pfiesteria are powerful enough to tear bleeding sores in fish. Special care must be taken by people working with the toxins, which have been reported to cause nausea, vomiting, headaches, burning eyes, memory lapses, breathing difficulty, mood swings, and impaired speech. Pfiesteria is the first known toxin-producing dinoflagellate with multiple stages (at least 24), including flagellated, amoeboid, and cyst stages with transition forms between each of these (see Figure 16–22).

Other dinoflagellates are responsible for the formation of red tides in different areas. Thus Gonyaulax tamarensis is the organism involved along the northeastern Atlantic coast, from the Canadian Maritime provinces to southern New England, while Gymnodinium catenella causes red tides at times along the Pacific Coast from Alaska to California, Ptichodiscus brevis in the Gulf of Mexico, and Protogonyaulax tamarensis in the North Sea off the coast of Northumberland in the United Kingdom. Over 40 marine species of dinoflagellates have been identified as producing toxins that kill birds and mammals, render shellfish toxic, or produce a widespread tropical fish-poisoning disease called ciguatera. The poisons produced by some dinoflagellates, such as G. catenella, are extraordinarily powerful nerve toxins. The chemical nature and biological activity of most of these toxins are relatively well known.

When shellfish, such as mussels and clams, ingest toxic dinoflagellates or other organisms, they accumulate and concentrate the toxins. Depending on the species of toxic organisms the shellfish have consumed, they themselves may become dangerously toxic to the people who eat them. Along the Atlantic Coast, fisheries are commonly closed in summer, and people regularly suffer from poisoning after consuming mussels, oysters, scallops, or clams taken from certain regions. Research with Pfiesteria piscicida has revealed that the dinoflagellates can attack humans not only indirectly, through shellfish, but directly.

(a) Fish killed by Pfiesteria. *(b)* Gymnodinium breve, the unarmored dinoflagellate responsible for the outbreaks of red tide along the west coast of Florida. The curved, transverse flagellum lies in a groove encircling the organism. Its longitudinal flagellum, only a portion of which is visible, extends from the middle of the organism toward the lower left. The apical groove at the top is an identifying characteristic of Gymnodinium.

(a)

(b)

Many Dinoflagellates Ingest Solid Food Particles or Absorb Dissolved Organic Compounds

About half of all dinoflagellates lack a photosynthetic apparatus and hence obtain their nutrition either by ingesting solid food particles or by absorbing dissolved organic compounds. Even many pigmented—photosynthetic—and heavily armored dinoflagellates can feed in these ways. Some feeding dinoflagellates extrude a tubular process known as a peduncle, which can suction organic materials into the cell. The peduncle is retracted into the cell when feeding is finished.

Most pigmented dinoflagellates typically contain chlorophylls *a* and *c*, which are generally masked by carotenoid pigments, including **peridinin,** which is similar to fucoxanthin, an accessory pigment typical of chrysophytes (see Chapter 17). The presence of peridinin supports the hypothesis that the chloroplasts of many dinoflagellates were derived from ingested chrysophytes by endosymbiosis, as described in Chapter 13. Other dinoflagellates have green or blue-green plastids obtained from ingested green algae or cryptomonads. The carbohydrate food reserve in dinoflagellates is starch, which is stored in the cytoplasm.

Pigmented dinoflagellates occur as symbionts in many other kinds of organisms, including sponges, jellyfish, sea anemones, tunicates, corals, octopuses and squids, gastropods, turbellarians, and certain protists. In the giant clams of the family *Tridachnidae,* the dorsal surface of the inner lobes of the mantle may appear chocolate-brown as a result of the presence of symbiotic dinoflagellates. When they are symbionts, the dinoflagellates lack armored plates and appear as golden spherical cells called **zooxanthellae** (Figure 16–21).

Zooxanthellae are primarily responsible for the photosynthetic productivity that makes possible the growth of coral reefs in tropical waters, which are notoriously nutrient-poor. Coral tissues may contain as many as 30,000 symbiotic dinoflagellates per cubic millimeter, primarily in cells that line the gut of the coral polyps. Amino acids produced by the polyps stimulate the dinoflagellates to produce glycerol instead of starch. The glycerol is used directly for coral respiration. Because the dinoflagellates require light for photosynthesis, the corals containing them grow mainly in ocean waters less than 60 meters deep. Many of the variations in the shapes of coral are related to the light-gathering properties of different geometric arrangements. This relationship is somewhat similar to the ways in which various branching patterns of trees serve to maximally expose their leaves to sunlight.

During Periods of Low Nutrient Levels, Dinoflagellates Form Resting Cysts

Under conditions that do not allow continued population growth, such as low nutrient levels, dinoflagellates may produce nonmotile resting cysts that drift to the lake or ocean bottom, where they remain viable for years. Ocean currents may transport these benthic (meaning "bottom of the water") cysts to other locations. When conditions become favorable, the cysts may germinate, reviving the population of swimming cells. Cyst production, movement, and germination explain many aspects of the ecology and geography of toxic, dinoflagellate blooms. They explain why blooms do not necessarily occur at the same site every year, and why blooms are associated with nutrient pollution of the ocean with sewage and agricultural runoff. In addition, they explain why blooms appear to move from one location to another from year to year.

Sexual reproduction has been found in a number of species of dinoflagellates. Dinoflagellate zygotes form distinctive thick, chemically inert, ornamented cell walls that resemble some ancient fossil acritarchs (Figure 16–1). The life histories of dinoflagellates may be very complex, involving multiple forms, some of which resemble amoebas. As is the case for certain red algae, by growing the organisms in laboratory culture, it is possible to show that morphologically different protists may be components of the life history of a single species. Understanding life histories is important in understanding the roles of phytoplankton, including dinoflagellates, in food webs and in toxic bloom formation.

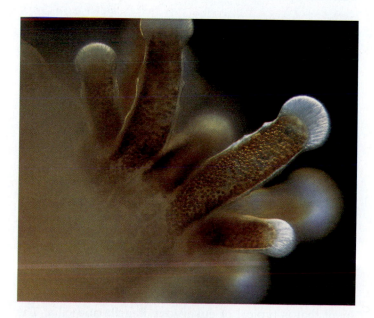

16–21

Zooxanthellae, the symbiotic form of dinoflagellates, shown here in a tentacle of a coral animal. These symbionts are responsible for much of the productivity of coral reefs.

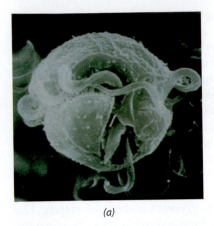

(a)

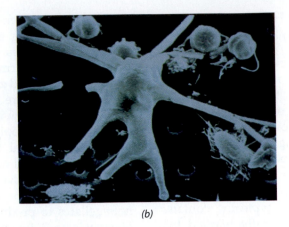

(b)

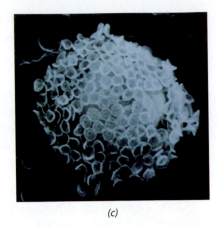

(c)

16–22

Three of the known stages in the complex life cycle of Pfiesteria piscicida, *a dinoflagellate. (a) A biflagellated cell showing flagella and the peduncle, which is used in feeding.*

(b) One of the amoeboid stages, which dominate the life cycle. (c) An amoeboid cyst stage. The biflagellated cyst can transform into an amoeboid stage within minutes. Pfiesteria

piscicida *has been termed an "ambush predator" because it releases a toxin that kills fish and then consumes the sloughed tissue of these fish.*

Many Dinoflagellates Produce Toxic or Bioluminescent Compounds

About 20 percent of all known dinoflagellate species produce one or more highly toxic compounds that are economically and ecologically significant (see essay on page 364). Dinoflagellate toxins may provide protection from predation, but at least one recently discovered dinoflagellate, *Pfiesteria piscicida,* uses its deadly toxin in a "hit and run" feeding strategy (Figure 16–22). The presence of fish, such as the menhaden, stimulates benthic cysts to germinate into swimming *Pfiesteria* cells that produce a toxin that paralyzes the fishes' respiratory systems, causing death by suffocation. As the fish decay, the dinoflagellates extend their peduncles and feed frenetically upon bits of fish flesh. When the food has been consumed, the phantom dinoflagellates rapidly return to the more obscure benthic cyst stage, sometimes within a period of just two hours. As a result, the cause of massive fish kills can be difficult to ascertain unless water samples are taken at the very beginning of the killing event. Ocean nutrient pollution is related to fish kills in that excess phosphorus in the water stimulates at least one phase in the life cycle of *Pfiesteria.*

Marine dinoflagellates are also famous for their bioluminescent capabilities (Figure 16–20b). These are responsible for the attractive sparkling of ocean waters that is commonly observed at night as boats or swimmers agitate the water. When dinoflagellate cells are disturbed, a series of well-understood biochemical events results in a reaction involving luciferin and the enzyme luciferase that, as in fireflies and other organisms, creates a brief flash of light (see Chapter 28). Bioluminescence is

thought to serve as protection against predators, such as copepods, small crustaceans that are the most numerous members of the zooplankton. One hypothesis is that the dinoflagellates' light flashes directly disrupt feeding by startling the predators. Another hypothesis suggests a more indirect process—copepods that have fed upon luminescent dinoflagellates become more visible to the fish that feed upon them.

Haptophytes: Phylum *Haptophyta*

The phylum *Haptophyta* consists of a diverse array of primarily marine phytoplankton, though a few freshwater and terrestrial forms are known. The phylum consists of unicellular flagellates, colonial flagellates, and nonmotile single cells and colonies. There are about 300 known species in 80 genera, but new species are being discovered continuously. Haptophyte species diversity is highest in the tropics.

The most distinctive feature of the haptophyte algae is the **haptonema** (Gk. *haptein,* "to fasten"; relates to sense of touch). The haptonema is a threadlike structure that extends from the cell along with two flagella of equal length (Figure 16–23). It is structurally distinct from a flagellum. Although microtubules are present in the haptonema, they do not have the 9-plus-2 arrangement that is typical of eukaryotic flagella and cilia. The haptonema can bend and coil but cannot beat like a flagellum. In some cases, it allows the haptophyte cell to catch prey food particles, functioning somewhat like a fishing rod. In other cases, it seems to help the cells sense and avoid obstacles.

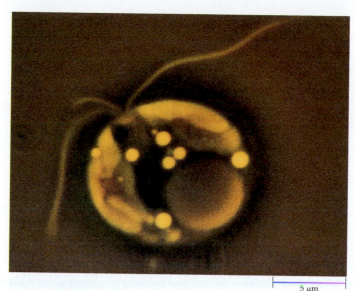

16–23

The haptophyte Prymnesium parvum. *Most haptophytes have two smooth and nearly equal flagella. Many also possess a haptonema (the smaller appendage shown here), which can bend or coil but cannot beat as do the flagella.*

Another characteristic of haptophyte algae is the presence of small, flat scales on the outer surface of the cell (Figure 16–24). These scales are composed of organic material or calcified organic material. The calcified scales are known as **coccoliths,** and the 12 or more families of organisms that are decorated with coccoliths are known as the coccolithophorids. Coccoliths are of two types. Those produced inside the cell—in Golgi vesicles—are transported to the cell's exterior. The other type of coccolith is generated outside the cell. Coccoliths form the basis for a continuous fossil record back to their first appearance in the Late Triassic, some 230 million years ago.

Most haptophytes are photosynthetic, with chlorophyll *a* and some variation of chlorophyll *c*. At least one nonphotosynthetic representative is known. Some have the accessory pigment fucoxanthin, in common with chrysophytes and diatoms (see Chapter 17). Other haptophytes lack fucoxanthin.

In common with the cryptomonads, the plastids of haptophytes are surrounded by a chloroplast endoplasmic reticulum, which is continuous with the nuclear envelope. As in cryptomonads, the chloroplast endoplasmic reticulum is evidence that the plastids were acquired by secondary endosymbiosis. Sexual reproduction and alternation of heteromorphic generations occur in haptophytes, but the chromosome levels and life histories of many forms are as yet unknown.

Marine haptophytes are significant components of food webs, serving both as producers and, even though most are autotrophic, as consumers. As consumers, they graze on small particles such as cyanobacteria or absorb dissolved organic carbon. They are an important means by which organic carbon and two-thirds of the oceans' calcium carbonate are transported to the deep ocean. In addition, they are important producers of sulfur oxides connected with acid rain (page 350). The gelatinous, colonial stage of *Phaeocystis* dominates the phytoplankton of the marginal ice zone in polar regions and contributes some 10 percent of the atmospheric sulfur compounds that are generated by phytoplankton. In addition, its gelatin contributes significant organic carbon to the water. In all of the oceans, especially at midlatitudes, *Emiliania huxleyi* can form blooms covering thousands of square kilometers of ocean. The two haptophyte genera *Chrysochromulina* and *Prymnesium* are notorious for forming marine toxic blooms that kill fishes and other marine life.

In the next chapter, we will turn our attention to the remaining protist phyla traditionally studied by botanists. Special emphasis is placed on the green algae, which include representatives considered to be more closely related than any other living organisms to the ancestors of bryophytes and vascular plants.

16–24

Haptophyte algae. **(a)** Emiliania huxleyi, *a coccolithophorid. This is the most widespread and abundant of the estimated 300 species of this group of extremely minute algae. The platelike scales covering cells of the coccolithophorids consist of calcium carbonate.* **(b)** *A fluorescence micrograph of a young colony of* Phaeocystis, *stained with acridine orange. The cells are embedded in a polysaccharide mucilage. In temperate oceans, such as the North Sea, massive* Phaeocystis *blooms clog fishing nets and wash up onto beaches, producing meter-thick mounds of foam.*

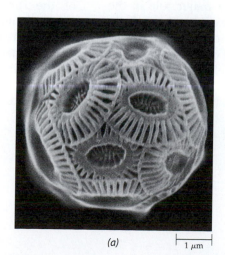

(a)

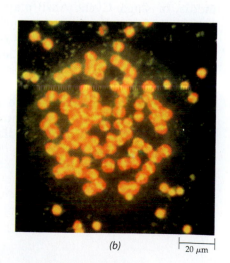

(b)

Summary

The Kingdom *Protista* Includes a Variety of Autotrophic and Heterotrophic Organisms

Protists are eukaryotic organisms that are not included in the fungal, plant, or animal kingdoms. Those considered in this book include both photosynthetic (autotrophic) organisms, or algae, and heterotrophic organisms once treated as fungi. The latter include the *Myxomycota* and *Dictyosteliomycota*, as well as the *Oomycota*, which are considered in the next chapter.

The algae include photosynthetic protists as well as colorless relatives. Algae are important components of aquatic food webs and, together with the cyanobacteria, constitute the phytoplankton. The algae also play significant roles in global carbon and sulfur cycling.

The Plasmodium of the *Myxomycota* Is a "Naked" Mass of Protoplasm

The plasmodial slime molds, phylum *Myxomycota*, may exist as streaming, multinucleate masses of protoplasm called plasmodia, which are usually diploid. These plasmodia typically form sporangia in which diploid spores are formed. Meiosis occurs within each of the spores, and three of the resulting nuclei disintegrate, leaving one haploid nucleus in each spore. Under favorable conditions, the spores split open, producing amoebas, which may become flagellated. These amoebas or flagellated cells may function as gametes. Plasmodium formation often, but not always, follows fusion of gametes.

The Pseudoplasmodium of the *Dictyosteliomycota* Is an Aggregation of Myxamoebas

The cellular slime molds, phylum *Dictyosteliomycota*, are a group of amoeba-like organisms that aggregate together at one stage of their life cycle to form pseudoplasmodia, or slugs. Cyclic AMP plays a major role in the aggregation of individual myxamoebas into these slugs, which undergo complex patterns of differentiation and programmed cell death that have led to their being used as models in this field. Flagellated cells are not known. Reproduction involving asexual spores is common. Sexual reproduction involves the formation of walled zygotes called macrocysts. In contrast to fungi, but like many other protists, the plasmodial and cellular slime molds ingest particulate food.

Algae Obtain Nutrients in a Variety of Ways

Although many algae are capable of photosynthesis, uptake and utilization of dissolved organic compounds is also common. Particle feeding occurs in at least some members of the euglenoids, cryptomonads, dinoflagellates, and haptophytes. Phagocytosis is regarded as a method by which the ancestors of the pigmented members of these algal groups probably acquired their chloroplasts.

The Euglenoids, Phylum *Euglenophyta*, Occur Primarily in Fresh Water and Are Unicellular

About one-third of the euglenoids are photosynthetic and have chloroplasts containing chlorophylls *a* and *b* and several carotenoids. The plastids do not store starch; rather, granules of a unique polysaccharide called paramylon are stored in the cytoplasm. Euglenoids lack a cell wall but have a series of helically arranged proteinaceous strips, called a pellicle, which lies just beneath the plasma membrane. The cells contain a contractile vacuole and bear flagella. Sexual reproduction is unknown in euglenoids.

Cryptomonads, Phylum *Cryptophyta*, and Dinoflagellates, Phylum *Dinophyta*, Typically Occur in Both Freshwater and Marine Habitats

The cryptomonads are single-celled flagellates, which appear to have arisen through the fusion of two different eukaryotic cells, one heterotrophic and the other photosynthetic. In addition to chlorophylls *a* and *c* and carotenoids, some cryptomonad chloroplasts contain either phycocyanin or phycoerythrin, water-soluble pigments otherwise known only in cyanobacteria and red algae.

The dinoflagellates are unicellular biflagellates. Many are bizarre in appearance, with stiff cellulose plates forming a wall, or theca. Other dinoflagellates have very thin or no cellulose plates at all. About half of the dinoflagellates are photosynthetic, containing chlorophylls *a* and *c*, which are generally masked by carotenoid pigments, including peridinin. Some marine dinoflagellates produce toxic compounds and/or harmful red tides. Other marine dinoflagellates serve as food-producing endosymbionts within the cells of reef-forming corals.

Red Algae, Phylum *Rhodophyta,* Are a Large Group Particularly Common in Warmer Marine Waters

Red algae usually grow attached to a substrate, and some grow at great depths (down to 268 meters). The chloroplasts contain phycobilins, which mask the color of chlorophyll *a* and give the red algae their distinctive color. The chloroplasts of the red algae are biochemically and structurally very similar to the cyanobacteria from which they very likely were derived. Red algae are notable for their complex life cycles, which commonly involve alternation of three distinct generations—gametophyte, carposporophyte, and tetrasporophyte. Red algae are also sources of useful, valuable carbohydrates such as agar and carrageenan.

The Haptophytes, Phylum *Haptophyta,* Are Primarily Marine Phytoplankton

The most distinctive feature of the haptophytes is the haptonema, a threadlike structure that seems to help the cell to sense and avoid obstacles or to catch prey food particles. Most haptophytes are photosynthetic, with chlorophylls *a* and *c.* They are ecologically important in both the carbon and sulfur cycles on a worldwide basis.

Questions

1. Expound upon the phytoplankton as the "great meadow of the sea."

2. By means of a simple, labeled diagram, explain the structure-function relationships of *Euglena.*

3. "When the going gets tough, the tough get going." Describe how each of the following organisms adapt to tough times, such as periods of inadequate nutrient or moisture levels: plasmodial slime molds, cellular slime molds, dinoflagellates.

4. Of what advantage is the diploid carposporophyte generation to the red algae?

5. What do such organisms as *Gymnodinium breve, Pfiesteria piscicida,* and *Gonyaulax tamarensis* have in common?

Selected Key Terms

acritarchs   p. 348

aethalium   p. 352

apoptosis   p. 355

carposporophyte   p. 361

chloroplast endoplasmic reticulum   p. 357

contractile vacuole p. 351

coralline algae   p. 359

eyespot, or stigma p. 351

floridean starch   p. 358

gametophyte   p. 360

haptonema   p. 366

monospores   p. 360

myxamoebas   p. 354

nucleomorph   p. 357

pellicle   p. 351

phytoplankton   p. 348

pit connections   p. 360

plasmodiocarp   p. 352

plasmodium   p. 352

pseudoplasmodium p. 355

pyrenoid   p. 352

sclerotium   p. 354

secondary endosymbiosis p. 357

sporophyte   p. 360

tetrasporophyte   p. 361

theca   p. 363

zooplankton   p. 348

zooxanthellae   p. 365

Chapter

Protista II: Heterokonts and Green Algae

17–1

Multicellular photosynthetic organisms anchored themselves to rocky shores early in the course of their evolution. These kelp, seen at low tide on the rocks at Botanical Beach on Vancouver Island, British Columbia, are brown algae (Phaeophyta), a group in which multicellularity evolved independently of other groups of organisms.

OVERVIEW

Algae are crucially important components of aquatic ecosystems, producing oxygen and serving as food for aquatic animals. In this chapter, we continue our study of the algae. We first examine the heterokonts, which consist of four phyla whose members have two flagella of different length and structure. The heterokonts include three mostly photosynthetic phyla—algae of various types—and one entirely heterotrophic phylum, the oomycetes.

We then turn our attention to the green algae—phylum Chlorophyta—whose motile members have flagella of similar length and structure. In this phylum you will find organisms with almost every type of body plan, including unicellular, colonial, filamentous, and flat, tissuelike forms. Chlorophytes are important evolutionarily because some ancestral members are thought to have given rise to the bryophytes and vascular plants, setting the stage for the story continued in the following chapters.

CHECKPOINTS

By the time you finish reading this chapter, you should be able to answer the following questions:

1. What do all phyla of heterokonts have in common?

2. How do the oomycetes differ from the other heterokonts, and what are some important plant diseases caused by oomycetes?

3. What is unique about the cell walls of diatoms? What effect does this feature have on the diatom life cycle, generation after generation?

4. What are the basic characteristics of the brown algae?

5. What characteristics of the green algae have led botanists to conclude that the green algae are the protist group from which plants evolved?

6. How does the mode of cell division in the *Chlorophyceae* differ from that of the other classes of green algae?

The open sea, the shore, and the land are the three life zones that make up our biosphere. Of these zones, the sea and the shore are more ancient. Here algae play a role comparable to the role played by plants in the far younger terrestrial world. Often, algae are also dominant in freshwater habitats—ponds, streams, and lakes—where they may be the most important contributors to the productivity of these ecosystems. Everywhere they grow, algae play an ecological role comparable to that of the plants in land habitats.

Along rocky shores can be found larger, more complex seaweeds, such as the red algae described in the previous chapter and members of the brown and green algae. At low tide, it is easy to observe fairly distinct banding patterns or layers that reflect the positions of seaweed species in relation to their ability to survive exposure (Figure 17–1). Seaweeds of this intertidal zone are subjected twice a day to large fluctuations of humidity, temperature, salinity, and light, in addition to the pounding action of the surf and forceful, abrasive water motions. Polar seaweeds must endure months of darkness under the sea ice. Seaweeds are also prey to a host of herbivores as well as microbial pathogens. Their complex biochemistries, structures, and life histories reflect adaptation to these physical and biological challenges.

Anchored offshore beyond the zone of waves, massive brown kelps form forests that provide shelter for a rich diversity of fishes and invertebrate animals, some of which are valued human foods. Many large carnivores, including sea otters and tuna, find food and refuge in kelp beds, such as those occurring off the coast of California. The kelps themselves, along with some red algal species, are harvested by humans for extraction of industrial products (see "Algae and Human Affairs" on page 380).

In this chapter we continue our study of the protists by examining the green and brown algae, as well as three groups of protists that are closely related to the brown algae. Green algae include the smallest known eukaryotes, which are unicellular or colonial members of the phytoplankton. They also include attached filaments, macroscopic sheets, and tissuelike and coenocytic (multinucleate) forms. The brown algae have more complex structures than the green algae. Generally, they are attached to a substrate, and they range from microscopic branched filaments to 50-meter-long giant kelps. Chrysophytes and diatoms include many unicellular and colonial members of marine and freshwater phyto plankton. Oomycetes, formerly thought to be related to the fungi, are now known to be more closely related to the chrysophytes, diatoms, and brown algae. Oomycetes

are microscopic, heterotrophic residents of both aquatic and terrestrial habitats. Similarities and differences among these protist groups are summarized in Table 17–1.

The Heterokonts

Electron microscopists have long suspected that oomycetes, chrysophytes, diatoms, brown algae, and certain other groups (not discussed in this book) are closely related, based on the common occurrence of similar flagella. These organisms are known as **heterokonts** (meaning "different flagella") because their flagella differ in length and ornamentation. The flagella occur in pairs, with one flagellum long and ornamented with distinctive hairs (tinsel) and the other shorter and smooth (whiplash), as shown in Figure 3–28. Molecular sequence analyses have confirmed the suspicion, once based only on these unique flagella, that oomycetes, chrysophytes, diatoms, and brown algae are indeed closely related.

Molecular studies have further revealed that: (1) the fungus-like heterokonts (including oomycetes) diverged relatively early; (2) the pigmented heterokonts are derived from a single, common ancestor; and (3) there was an early separation of the pigmented forms into two clades. One of these pigmented lineages consists of the diatoms, the other includes the rest of the pigmented heterokonts.

Oomycetes: Phylum *Oomycota*

The phylum *Oomycota*, with about 700 species, is a distinctive group, commonly called oomycetes. In common with dinoflagellates and many green algae, the cell walls of these organisms are composed largely of cellulose or cellulose-like polymers. The oomycetes range from unicellular to highly branched, coenocytic, and filamentous forms. The latter somewhat resemble the hyphae that are characteristic of the fungi, and for this reason, oomycetes have been grouped with fungi in the past.

Most species of oomycetes can reproduce both sexually and asexually. Asexual reproduction among the oomycetes is by means of motile zoospores, which have two flagella that characterize the heterokonts—one tinsel, which is distinctive, and one whiplash. Sexual reproduction is oogamous, meaning that the female gamete is a relatively large, nonflagellated egg, and the male gamete is noticeably smaller and flagellate (Figure 17–2).

TABLE 17–1   Comparative Summary of Characteristics of *Protista* II

Phylum	Number of Species	Photosynthetic Pigments	Carbohydrate Food Reserve	Flagella	Cell Wall Component	Habitat
Oomycota (including water molds)	700	None	Glycogen	2; in zoospores and male gametes only; apical or lateral; 1 tinsel (two rows of hairs) forward, 1 whiplash behind	Cellulose or cellulose-like	Marine, freshwater, and terrestrial (need water)
Bacillariophyta (diatoms)	100,000	None, or chlorophylls *a* and *c*; carotenoids, mainly fucoxanthin	Chrysolaminarin	None or 1; only in male gametes of centric type; apical; tinsel (two rows of hairs)	Silica	Marine and freshwater
Chrysophyta (chrysophytes)	1000	None, or chlorophylls *a* and *c*; carotenoids, mainly fucoxanthin	Chrysolaminarin	None or 2; apical; tinsel (two rows of hairs) forward, whiplash behind	None or silica scales; cellulose also in scales of some	Predominantly freshwater, a few marine
Phaeophyta (brown algae)	1500	Chlorophylls *a* and *c*; carotenoids, mainly fucoxanthin	Laminarin, mannitol (transported)	2; only in reproductive cells; lateral; tinsel (two rows of hairs) forward, whiplash behind	Cellulose embedded in matrix of mucilaginous algin; plasmodesmata in some	Almost all marine; mostly temperate and polar, flourish in cold ocean waters
Chlorophyta (green algae)	17,000	Chlorophylls *a* and *b*; carotenoids	Starch	None or 2 (or more); apical or subapical; equal or unequal; whiplash (some with hairs)	Glycoproteins, noncellulose polysaccharides or cellulose; plasmodesmata in some	Mostly aquatic, freshwater or marine; many in symbiotic relationships

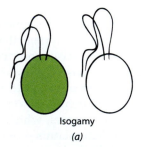

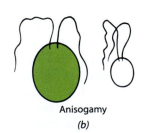

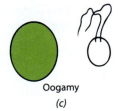

17–2

Types of sexual reproduction, based on gamete form. **(a)** *Isogamy—the gametes are equal in size and shape.* **(b)** *Anisogamy—one gamete, conventionally termed male, is smaller than the other.* **(c)** *Oogamy—one gamete, usually the larger, is nonmotile and female.*

Oogamy is also characteristic of some of the brown and green algae covered in this chapter and of bryophytes and vascular plants, as well as the red algae described in Chapter 16.

In the oomycetes, one to many eggs are produced in a structure called an **oogonium** (plural: oogonia), and an **antheridium** (plural: antheridia) contains numerous male nuclei (Figure 17–3). Fertilization results in the formation of a thick-walled zygote, the **oospore,** for which this phylum is named. The oospore serves as a resting stage that can tolerate stressful conditions. When conditions improve, the oospore germinates. Other heterokonts, as well as many green algae that live in similar habitats, also produce resistant resting stages that result directly from sexual reproduction.

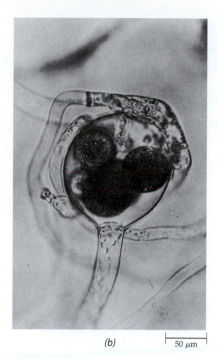

(a) ⊢50 μm⊣   (b) ⊢50 μm⊣

17–3

*Achlya ambisexualis, an oomycete that reproduces both asexually and sexually. **(a)** Empty sporangium with zoospores encysted about its opening, a distinctive feature of Achlya. **(b)** Sex organs, showing fertilization tubes extending from the antheridium through the wall of the oogonium to the eggs. Fertilization results in the formation of thick-walled zygotes known as oospores.*

The Water Molds Are Aquatic Oomycetes

One large group of the phylum *Oomycota* is aquatic. The members of this group—the so-called water molds—are abundant in fresh water and are easy to isolate from it. Most of them are saprophytic, living on the remains of dead plants and animals, but a few are parasitic, including species that cause diseases of fishes and fish eggs.

In some water molds, such as *Saprolegnia* (Figure 17–4), sexual reproduction can occur with male and female sex organs borne on the same individual; in other words, these water molds are **homothallic.** Other water molds, such as some species of *Achlya* (Figure 17–3), are **heterothallic**—male and female sex organs are borne by different individuals, or if they are borne on one individual, that individual is genetically incapable of fertilizing itself. Both *Saprolegnia* and *Achlya* can reproduce sexually and asexually.

Some Terrestrial Oomycetes Are Important Plant Pathogens

Another group of oomycetes is primarily terrestrial, although the organisms still form motile zoospores when liquid water is available. Among this group—the order

Peronosporales—are several forms that are economically important. One of these is *Plasmopara viticola*, which causes downy mildew in grapes. Downy mildew was accidentally introduced into France in the late 1870s on grape stock from the United States that had been imported because of its resistance to other diseases. The mildew soon threatened the entire French wine industry. It was eventually brought under control by a combination of good fortune and skillful observation. French vineyard owners in the vicinity of Medoc customarily put a distasteful mixture of copper sulfate and lime on vines growing along the roadside to discourage passersby from picking the grapes. A professor from the University of Bordeaux who was studying the problem of downy mildew noticed that these plants were free from symptoms of the disease. After conferring with the vineyard owners, the professor prepared his own mixture of chemicals—the Bordeaux mixture—which was made generally available in 1882. The Bordeaux mixture was the first chemical used in the control of a plant disease.

Another economically important member of this group is the genus *Phytophthora* (meaning "plant destroyer"). *Phytophthora*, with about 35 species, is a particularly important plant pathogen that causes widespread destruction of many crops, including cacao, pineapples, tomatoes, rubber, papayas, onions, strawberries, apples, soybeans, tobacco, and citrus. A widespread member of this genus, *Phytophthora cinnamomi*, which occurs in soil, has killed or rendered unproductive millions of avocado trees in southern California and elsewhere. It has also destroyed tens of thousands of hectares of valuable eucalyptus timberland in Australia. The zoospores of *P. cinnamomi* are attracted to the plants they infect by chemicals exuded by the roots. This oomycete also produces resistant spores that can survive for up to six years in moist soil. Extensive breeding efforts are now under way with avocados and other susceptible crops to produce strains resistant to this oomycete.

The best-known species of *Phytophthora*, however, is *Phytophthora infestans* (Figure 17–5), the cause of the late blight of potatoes, which produced the great potato famine of 1846–1847 in Ireland. The population of Ireland, which had risen from 4.5 million people in 1800 to about 8.5 million in 1845, fell to 6.5 million people in 1851 as a result of this famine. About 800,000 people starved to death, and great numbers of people emigrated, many of them to the United States. Virtually the entire Irish potato crop was wiped out in a single week in the summer of 1846. This was a disaster for the Irish peasants, who depended almost entirely on potatoes for their food. Each adult consumed between 4 and 6 kilograms (nearly 9 pounds and more) of potatoes every day—the amount required to supply sufficient quantities of protein to keep alive. Even today, *Phytophthora infestans* is still a serious pest of the potato crop.

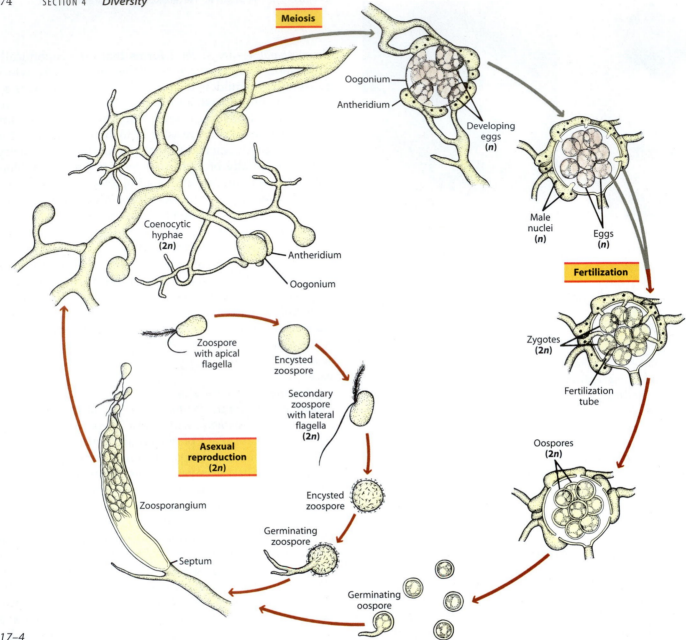

Meiosis

Oogonium

Antheridium

Developing eggs (*n*)

Male nuclei (*n*)

Eggs (*n*)

Coenocytic hyphae (**2n**)

Antheridium

Oogonium

Zoospore with apical flagella

Encysted zoospore

Secondary zoospore with lateral flagella (**2n**)

Asexual reproduction (2n)

Zoosporangium

Septum

Encysted zoospore

Germinating zoospore

Germinating oospore

Fertilization

Zygotes (**2n**)

Fertilization tube

Oospores (**2n**)

17–4

Life cycle of Saprolegnia, *an oomycete. The mycelium of this water mold is diploid. Reproduction is mainly asexual (lower left). Biflagellated zoospores released from a sporangium, known as a zoosporangium because it produces zoospores, swim for a while and then encyst. Each eventually gives rise to a secondary zoospore, which also encysts and then germinates to produce a new mycelium.*

During sexual reproduction, oogonia and antheridia are formed, in this species, on the same hypha (upper left). Meiosis occurs within these structures. The oogonia are enlarged cells in which a number of spherical eggs are produced. The antheridia develop from the tips of other filaments of the same individual and produce numerous male nuclei. In mating, the antheridia grow toward the oogonia and develop fertilization tubes, which penetrate the oogonia, as seen in Achlya in Figure 17–3b.

Male nuclei travel down the fertilization tubes to the female nuclei and fuse with them. Following each nuclear fusion, a thick-walled zygote—the oospore—is produced. On germination, the oospore develops into a hypha, which eventually produces a zoosporangium, beginning the cycle anew.

Also deserving of mention is the genus *Pythium*, the members of which are soil-inhabiting organisms found all over the world. *Pythium* species are the most important causes of **damping-off** diseases of seedlings. They attack a wide variety of economically important crops and can even be a very serious problem in turf grasses used for golf courses and football fields. Some *Pythium* species attack and rot seeds in the field and may destroy seedlings either before they emerge from the soil (pre-emergence damping-off) or afterward (postemergence damping-off). Postemergence damping-off, in which the young seedling rots near the soil line and then falls over, is a special problem in greenhouses, where large numbers of seedlings are grown in dense stands. Many home gardeners encounter this problem in the spring when renewing their flower beds with recently purchased annuals.

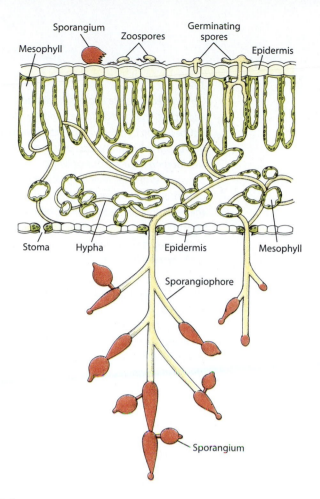

17–5

Phytophthora infestans, *the cause of the late blight of potatoes. The cells of the potato leaf are shown in green. In the presence of water and low temperatures, either of two events can occur. Zoospores can be released from the sporangia and swim to the germination site (as shown here), or the sporangia can germinate directly through a germ tube.*

Diatoms: Phylum *Bacillariophyta*

The diatoms are unicellular or colonial organisms that are exceedingly important components of the phytoplankton (Figure 17–6). It has been estimated that marine planktonic diatoms account for as much as 25 percent of the total primary production on Earth. Diatoms, especially very small forms, account for the greatest biomass and species diversity of phytoplankton in polar waters. Diatoms are a primary source of food for aquatic animals in both marine and freshwater habitats. Species such as *Thalassiosira pseudonana* are commonly used as food in marine cultures, or mariculture, of economically valuable bivalves, such as oysters. Diatoms provide essential carbohydrates, fatty acids, sterols, and vitamins to the animals.

It is estimated that there are 250 genera and 100,000 living species of diatoms, and many phycologists believe that the number may be a great deal higher. There are also thousands of extinct species, known from the remains of silica-containing cell walls. Diatoms first appeared about 250 million years ago. They first became abundant in the fossil record some 100 million years ago, during the Cretaceous period. Many of the fossil species are identical to those still living today, which indicates an unusual persistence through geologic time.

17–6

Diatoms. (a) A selected array of marine diatoms, as seen with a light microscope. (b) Scanning electron micrograph of one half of an Entogonia *frustule. (c)* Licmophora flagellata, *a stalked pennate diatom, as seen with a light microscope. (d) Scanning electron micrograph of* Cyclotella meneghiniana, *a centric diatom that occurs in brackish water.*

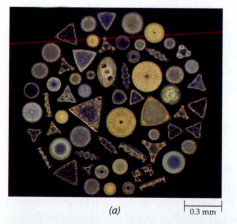

(a) 0.3 mm

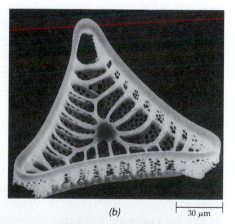

(b) 30 μm

(c)

(d) 5 μm

There are often tremendous numbers of individuals in small areas. For example, over 30 to 50 million individuals of the freshwater genus *Achnanthes* may occur on 1 square centimeter of a submerged rock in the streams of North America. Large numbers of species likewise may occur together. In two small samples of mud from the ocean near Beaufort, North Carolina, for example, 369 species of diatoms were identified. Most species of diatoms occur in plankton, but some are bottom dwellers or grow on other algae or on submerged plants.

The Walls of Diatoms Consist of Two Halves

Diatoms differ from *Chrysophyta* in lacking flagella, except for some male gametes, and in the unique structure of their two-part cell walls. Known as **frustules,** the walls, which are made of polymerized, opaline silica ($SiO_2 \cdot nH_2O$), consist of two overlapping halves. The two halves fit together like a laboratory Petri dish. Electron microscopy has shown that fine tracings on the diatom frustules are actually composed of a large number of minute, intricately shaped depressions, pores, or passageways, some of which connect the living protoplasm within the frustule to the outside environment (Figure 17–6b, d). Species can be distinguished by differences in frustule ornamentation. In most cases, both halves of the frustule have exactly the same ornamentation, but in some cases, the ornamentation may differ.

On the basis of symmetry, two types of diatoms are recognized: the **pennate** diatoms, which are bilaterally symmetrical (Figure 17–6c), and the **centric** diatoms, which are radially symmetrical (Figure 17–6d). Centric diatoms, which have a larger surface-to-volume ratio than pennate ones and consequently float more easily, are more abundant than pennate in large lakes and marine habitats.

Reproduction in Diatoms Is Mainly Asexual, Occurring by Cell Division

When cell division takes place, each daughter cell receives one half of the frustule of its parental cell and constructs a new half (Figure 17–7). As a consequence, one of the two new cells is typically somewhat smaller than the parental cell, and after a long series of cell divisions, the size of the diatoms in the resulting population will often have declined. In some diatom populations, when the size decreases to a critical level, sexual reproduction occurs. The cells that develop from division of the zygote typically regain maximum size for the species. In some other cases, sexual reproduction is triggered by changes in the physical environment.

The sexual life history of diatoms is gametic, like that of animals and certain brown and green seaweeds described later in this chapter. When sexual reproduction occurs in the centric diatoms, it is oogamous. The male

17–7

Generalized life cycle in a centric diatom. Reproduction in the diatoms is mainly asexual, occurring by cell division. The cell walls, or frustules, in all diatoms are composed of two overlapping portions, one of which encloses the other. When cell division takes place, each daughter cell (bottom left) receives one half of the parental frustule (bottom right) and constructs a new frustule half. The existing half is always the larger part of the silica wall, with the new half fitting inside it. Thus one daughter cell of each new pair tends to be smaller than the parental cell from which it is derived.

In some species, the frustules are expandable and are enlarged by the growing protoplasm within them. In other species, however, the frustules are more rigid. Thus, in a population, the average cell size decreases through successive cell divisions. When the individuals of these species have decreased in size to about 30 percent of the maximum diameter, sexual reproduction may occur (top). Certain cells function as male gametangia, and each produces sperm through meiosis. Other cells function as female gametangia. In these female gametangia, two or three of the four products of meiosis are nonfunctional, so that one or two eggs are produced per cell. This is an example of gametic meiosis (see Figure 9–3b). After fertilization, the resulting auxospore, or zygote, expands to the full size characteristic of the species. The walls formed by the auxospore are often different from those of the asexually reproducing cells of the same species. Once the auxospore is mature, it divides and produces new frustule halves with intricate markings typical of the asexually reproducing cells.

gametes, which may have a single, tinsel flagellum, are the only flagellated cells found in diatoms at any stage of their life cycle. In the pennate diatoms, sexual reproduction is isogamous, and both male and female gametes are nonflagellate. Both types of sexual reproduction result in empty frustules that readily sediment (meaning "to sink down"). Researchers have observed that mass sexual reproduction of marine diatoms can result in the formation of layers of silica in southern ocean sediments.

Unfavorable conditions, such as low levels of mineral nutrients, can cause marine coastal or benthic diatoms to form resting stages. The resting cells have heavy frustules, which allow them to sink readily to the bottom if

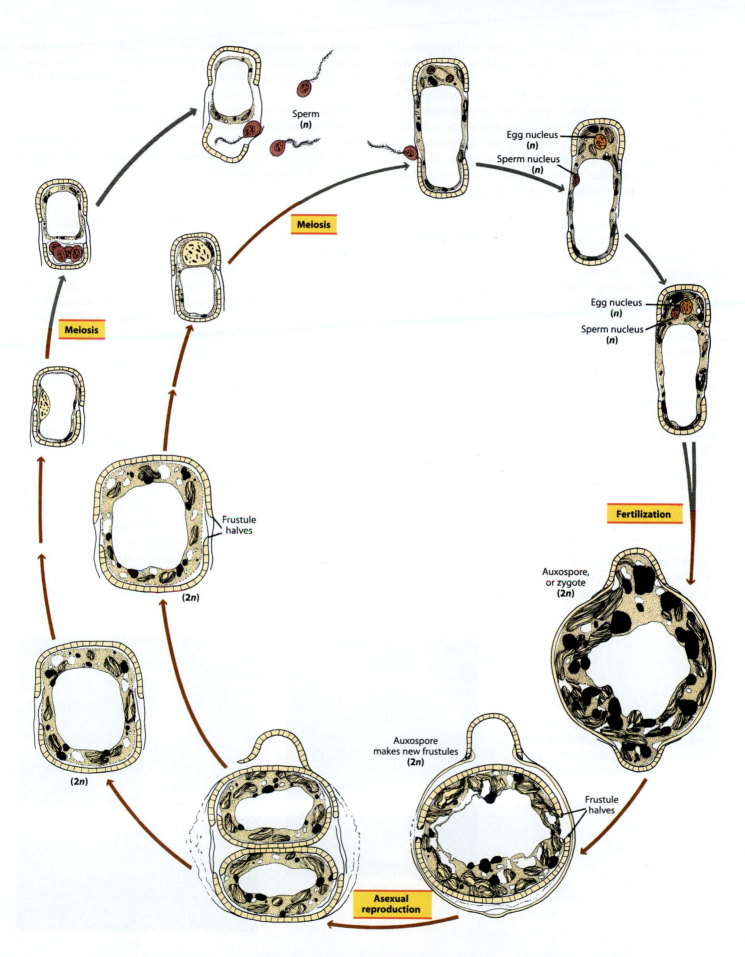

Sperm
(*n*)

Egg nucleus
(*n*)

Sperm nucleus
(*n*)

Meiosis

Egg nucleus
(*n*)

Sperm nucleus
(*n*)

Meiosis

Frustule
halves

Fertilization

(2*n*)

Auxospore,
or zygote
(2*n*)

(2*n*)

Auxospore
makes new frustules
(2*n*)

Frustule
halves

(2*n*)

**Asexual
reproduction**

they are not already there. These cells will germinate when nutrient conditions improve. Diatoms are frequently most abundant in the spring and fall, when upwelling in the oceans or wind-driven turnover of stratified lakes occurs. These processes resuspend sufficient silica for diatom growth. When the silica becomes depleted, diatom blooms give way to dominance by other phytoplankton that do not require silica. Diatoms may bloom beneath the winter ice cover of lakes because herbivorous animals do not actively feed during the cold season.

The silica frustules of diatoms have accumulated in ocean sediments over millions of years, forming the fine, crumbly substance known as diatomaceous earth. This substance is used as an abrasive in silver polish and as a filtering and insulating material. In the Santa Maria, California, oil fields there is a subterranean deposit of diatomaceous earth that is 900 meters thick, and near Lompoc, California, more than 270,000 metric tons of diatomaceous earth are quarried annually for industrial use.

The most conspicuous features within the protoplast of diatoms are the brownish plastids that contain chlorophylls *a* and *c* as well as **fucoxanthin,** a golden-brown carotenoid. There are usually two large plastids in the cells of pennate diatoms, whereas centric diatoms have numerous discoid plastids. Diatom reserve storage materials include lipids and the water-soluble polysaccharide **chrysolaminarin,** which is stored in vacuoles. Chrysolaminarin is similar to the laminarin found in brown algae.

Although most species of diatoms are autotrophic, some are heterotrophic, absorbing dissolved organic carbon. These heterotrophic species are primarily pennate diatoms that live on the bottom of the sea in relatively shallow habitats. A few diatoms are obligate heterotrophs. They lack chlorophyll and thus are not capable of producing their own food through photosynthesis. On the other hand, some diatoms, lacking their characteristic frustules, live symbiotically in large marine protozoa (order *Foraminifera*) and provide organic carbon to their hosts. Certain diatoms are associated with production of the neurotoxin domoic acid, which causes amnesiac shellfish poisoning in humans.

Chrysophytes: Phylum *Chrysophyta*

Chrysophytes are primarily unicellular or colonial organisms that are abundant in fresh and salt water throughout the world (Figure 17–8). There are a few plasmodial, filamentous, and tissuelike forms and about 1000 known species. Some chrysophytes are colorless, whereas others have chlorophylls *a* and *c,* the color of which is largely masked by an abundance of fucoxanthin. The golden color of this pigment gives rise to the name "chrysophyte" (Gk. *chrysos,* "gold," + *phyton,* "plant"). An individual pigmented cell usually contains one or two large chloroplasts. As in the diatoms, the carbohydrate food reserve is chrysolaminarin. It is stored in a vacuole usually found in the posterior of the cell.

Several chrysophytes are known to ingest bacteria and other organic particles. Individual cells of the freshwater, pigmented, motile colony *Dinobryon* can each consume about 36 bacteria per hour. *Dinobryon* and its chrysophyte relatives are the major consumers of bacteria in some of the cooler lakes of North America. *Poterioochromonas* can ingest motile algal cells that are two to three times larger in diameter than itself. Its cell volume can expand as much as 30 times to accommo-

17–8

Representative chrysophytes. (a) Scanning electron micrograph of a mature resting cyst of Dinobryon cylindricum. *The hooked, cylindrical collar contains a pore through which the amoeboid cell emerges when it is ready to germinate. Surface spines represent ornamentation. (b) Scanning electron micrograph of silica scales and flagella of* Synura *cells in a colony. The scales are arranged in very regular overlap patterns in spiral rows.*

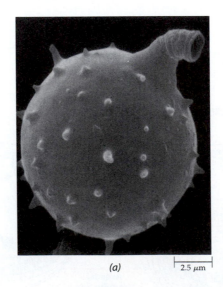

(a)   2.5 μm

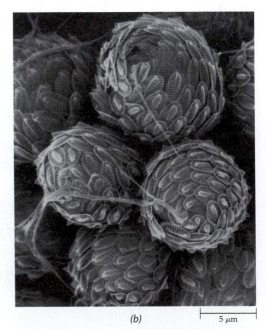

(b)   5 μm

date the food it has eaten. *Uroglena americana,* another relative, seems to require prey food as a source of an essential phospholipid. Such forms can be dominant components of the phytoplankton in temperate lakes of low nutrient content (oligotrophic). The ability to consume particulate food provides an advantage under such low-nutrient conditions.

Some chrysophytes have walls made of interwoven cellulosic fibrils that may be impregnated with minerals. Others are wall-less, and some look very much like amoebas with plastids. The members of one group, known as the synurophytes after the motile, colonial genus *Synura* (Figure 17–8b), are covered with overlapping, ornamented silica scales. These are produced inside the cells in vesicles and are then transported to the outside of the cell. The presence of a scaly enclosure prevents these algae from feeding on particles. In cold, acidic lakes, their scales may persist in the sediments and provide useful information about past habitats.

Reproduction in most chrysophytes is asexual and, for some forms, involves zoospore formation. Sexual reproduction is also known for some species. Characteristic resting cysts are commonly formed at the end of the growing season. Sometimes, but not always, cysts are the result of sexual reproduction. In some groups the resting cysts contain silica, which, like silica-containing scales, can sediment and form valuable records of past ecological conditions.

The marine chrysophytes *Heterosigma* and *Aureococcus* have generated toxic "brown tides" that have caused millions of dollars worth of damage to the shellfish and salmon fisheries. Some freshwater chrysophytes can also form blooms and are blamed for unpleasant tastes and odors that arise from their excretion of organic compounds into drinking water.

Brown Algae: Phylum *Phaeophyta*

The brown algae, an almost entirely marine group, include the most conspicuous seaweeds of temperate, boreal (or northern), and polar waters. Although there are only about 1500 species, the brown algae dominate rocky shores throughout the cooler regions of the world (Figure 17–9). Most people have observed seashore rocks covered with rockweeds, the common name for members of the brown algal order *Fucales.* The larger brown algae of the order *Laminariales,* a number of which form extensive beds offshore, are called kelps. In clear water, brown algae flourish from low-tide level to a depth of 20 to 30 meters. On gently sloping shores, they may extend 5 to 10 kilometers from the coastline.

(a)

17–9

Brown algae. (a) Bull kelp (Durvillea antarctica) *exposed at low tide off a rocky coast in New Zealand. (b) Detail of the kelp* Laminaria, *showing holdfasts, stipes, and the bases of several fronds. (c) Rockweed* (Fucus vesiculosus) *densely covers many rocky shores that are exposed at low tide. When submerged, the air-filled bladders on the blades carry them up toward the light. Photosynthetic rates of frequently exposed marine algae are one to seven times as great in air as in water, whereas the rates are higher in water for those rarely exposed. This difference accounts in part for the vertical distribution of seaweeds in intertidal areas.*

(b)

(c)

Algae and Human Affairs

People of various parts of the world, especially in the Far East, eat both red and brown algae. Kelps ("kombu") are eaten regularly as vegetables in China and Japan. They are sometimes cultivated but are mainly harvested from natural populations. Porphyra ("nori"), a red alga, is eaten by many inhabitants of the north Pacific Basin and has been cultivated in Japan, Korea, and China for centuries (page 359). Various other red algae are eaten on the islands of the Pacific and also on the shores of the North Atlantic. Seaweeds are generally not of high nutritive value as a source of carbohydrates because humans, like most other animals, lack the enzymes necessary to break down most of the materials in cell walls, such as cellulose and the protein-rich intercellular matrix. Seaweeds do, however, provide necessary salts, as well as a number of important vitamins and trace elements, and so are valuable supplementary foods. Some green algae, such as Ulva, or sea lettuce, are also eaten as greens.

In many northern temperate regions, kelp is harvested for its ash, which is rich in sodium and potassium salts and is therefore valued for industrial processes. Kelp is also harvested regionally and used directly for fertilizer.

Alginates, which are a group of substances derived from kelps, such as Macrocystis, are widely used as thickening agents and colloid stabilizers in the food, textile, cosmetic, pharmaceutical, paper, and welding industries. Off the west coast of the United States, Macrocystis kelp beds can be harvested several times a year by cropping them just below the water surface.

One of the most useful, direct commercial applications of any alga is the preparation of agar, which is made from a mucilaginous material extracted from the cell walls of a number of genera of red algae. Agar is used to make the capsules that contain vitamins and drugs, as well as for dental-impression material, as a base for cosmetics, and as a culture medium for bacteria and other microorganisms. Purified agarose is the gel often used in electrophoresis in biochemical experimentation. Agar is also employed as an anti-drying agent in bakery goods, in the preparation of rapid-setting jellies and desserts, and as a temporary preservative for meat and fish in tropical regions. Agar is produced in many parts of the world, but Japan is the principal manufacturer. A similar algal colloid called carrageenan is used in preference to agar for the stabilization of emulsions, such as paints, cosmetics, and dairy products. In the Philippines, the red alga Eucheuma is cultivated commercially as a source of carrageenan.

(a)

(b)

(c)

(a) A forest of giant kelp (Macrocystis pyrifera) growing off the coast of California. (b) Harvesting the seaweed Nudaria by hand from submerged ropes in Japan. (c) A kelp harvester operating in the nearshore waters of California. Cutting racks at the rear of the ship are lowered 3 meters below the water's surface, and the ship moves backward, cutting through the kelp canopy. The harvested kelp is moved via conveyor belts to a collecting bin on board the ship.

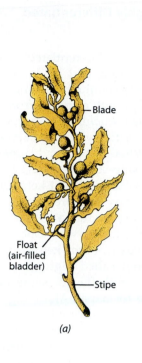

—Blade

Float
(air-filled
bladder)

—Stipe

(a)

(b)

(a) The brown alga Sargassum *has a complex pattern of organization.* Sargassum, *like* Fucus, *is a member of the order* Fucales *and has a life cycle like that shown in Figure 17–14.* *(b)* *Two species of this genus, which lack sexual reproduction, form the great free-floating masses of the Sargasso Sea.*

Even in the tropics, where the brown algae are less common, there are immense floating masses of *Sargassum* (Figure 17–10) in such areas as the Sargasso Sea (so named because of the abundance of *Sargassum*) in the Atlantic Ocean northeast of the Caribbean islands. *Sargassum muticum* and some other brown algae can form nuisance growths when introduced into non-native areas and can seriously interfere with mariculture operations. *Sargassum* can also outcompete and replace members of the *Laminariales* and *Fucales*, which are regarded as critical or keystone species in their communities.

The Basic Form of a Brown Alga Is a Thallus

Though they are a monophyletic group, brown algae range in size from microscopic forms to the largest of all seaweeds, the kelps, which are as much as 60 meters long and weigh more than 300 kilograms. The basic form of a brown alga is a **thallus** (plural: thalli)—a simple, relatively undifferentiated vegetative body. The thalli range in complexity from simple branched filaments (Figure 17–11), to aggregations of branched filaments that are called pseudoparenchyma because they look tissuelike, to authentic tissues (Figure 17–12). As in some green algae and plants, adjacent cells are typically linked by plasmodesmata. Unlike plant plasmodesmata, those of brown algae do not seem to have desmotubules connecting the endoplasmic reticulum of adjacent cells (see page 87).

In the past, brown algae have been classified into orders on the basis of thallus structure. However, recent molecular and cellular studies have revealed that thallus organization is not a good indicator of brown algal relationships. Closely related species of brown algae may have quite different thallus organization, and unrelated genera may appear structurally similar. The expression of similar morphologies by unrelated species is a response to similar selective pressures. This phenomenon, which also occurs in the green algae and the angiosperms, is known as convergent evolution (see page 266).

The Pigment Fucoxanthin Gives the Brown Algae Their Characteristic Color

Brown algal cells typically contain numerous disk-shaped, golden-brown plastids that are similar both biochemically and structurally to the plastids of chrysophytes and diatoms, with which they probably had a common origin. In addition to chlorophylls *a* and *c*, the chloroplasts of brown algae also contain various carotenoids, including an abundance of the xanthophyll fucoxanthin, which gives the members of this phylum their characteristic dark-brown or olive-green color. The reserve storage material in brown algae is the carbohydrate laminarin, which is stored in vacuoles. Molecular analyses suggest that there are two major lineages of brown algae: those with starch-producing pyrenoids in

17–11

Ectocarpus, *a brown alga that has simple branched filaments. This micrograph of E. siliculosus shows unilocular sporangia (the short, rounded, light-colored structures) and plurilocular sporangia (the longer, dark-colored structures), which are borne on sporophytes. Meiosis, resulting in haploid zoospores, takes place within the unilocular sporangia. Diploid zoospores are formed in the plurilocular sporangia. Ectocarpus occurs in shallow water and estuaries throughout the world, from cold Arctic and Antarctic waters to the tropics.*

their plastids, including *Ectocarpus* (Figure 17–11), and those without pyrenoids, such as *Laminaria* and its relatives. These two groups also consistently differ in the structure of their sperm. Oogamy and a heteromorphic type of life cycle occur only in the group that lacks pyrenoids.

The Kelps Have the Most Highly Differentiated Bodies among the Algae

Large kelps, such as *Laminaria,* are differentiated into regions known as the holdfast, stipe, and blade, with a meristematic region located between the blade and stipe (Figure 17–9b). The pattern of growth resulting from this type of meristematic activity is particularly important in the commercial use of *Macrocystis,* which is harvested along the California coast. When the older blades are harvested at the surface by kelp-cutting boats, *Macrocystis* is able to regenerate new blades. Giant kelps, such as *Macrocystis* and *Nereocystis,* may be more than 60 meters long. They grow very rapidly so that a considerable amount of material is available for harvest. One of the most important products derived from the kelps is a mucilaginous intercellular material called **algin,** which is important as a stabilizer and emulsifier for some foods and for paint and as a coating for paper. Algin, together with cellulose in inner cell wall layers, provides the flexibility and toughness that allow seaweeds to withstand mechanical stress imposed by waves and currents. Algin, which also helps to reduce drying when the seaweeds are exposed by low tides, increases buoyancy and helps to slough off organisms that attempt to colonize the algal blades.

The internal structure of kelps is complex. Some of them have, in the center of the stipe, elongated cells that are modified for food conduction. These cells resemble the food-conducting cells in the phloem of vascular plants, including the presence of sieve plates (Figure 17–12b). The cells are able to conduct food material rapidly—at rates as high as 60 centimeters per hour—from the blades at the water surface to the poorly illuminated stipe and holdfast regions far below. Lateral translocation from the outer photosynthetic layers to the inner cells takes place in many relatively thick kelps. **Mannitol** is the primary carbohydrate that is translocated, along with amino acids.

17–12

Some brown algae, such as the giant kelp Macrocystis integrifolia, *have evolved sieve tubes comparable to those found in food-conducting tissue of vascular plants. (a) Longitudinal section of part of a stipe, with sieve tubes, the relatively wide elements in the middle of the micrograph. The sieve-tube members are joined end-on-end by the sieve plates, which appear as narrow cross walls here. (b) Cross section showing a sieve plate.*

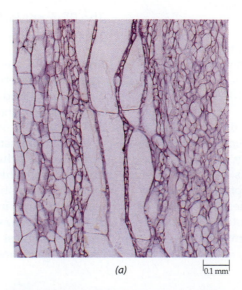

(a)   0.1 mm

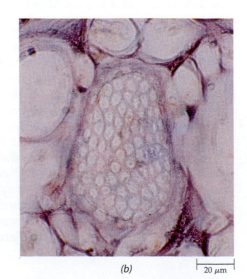

(b)   20 μm

The rockweed *Fucus* (Figure 17–9c) is a dichotomously branching brown alga that has air bladders near the ends of its blades. The pattern of differentiation of *Fucus* otherwise resembles that of the kelps. *Sargassum* (Figure 17–10) is related to *Fucus*. Some species of *Sargassum* remain attached, whereas in others the individuals form floating masses in which the holdfasts have been lost. Both forms occur within certain species. *Fucus* and *Sargassum* and some other brown algae grow by means of repeated divisions from a single apical cell and not from a meristem located within the body, as is characteristic of the kelps.

The Life Cycles of Most Brown Algae Involve Sporic Meiosis

The life cycles of most brown algae involve an alternation of generations and, therefore, sporic meiosis (see Figure 9–3c). The gametophytes of the more primitive brown algae, such as *Ectocarpus*, produce multicellular reproductive structures called **plurilocular gametangia**. They may function as male or female gametangia or produce flagellated haploid spores that give rise to new gametophytes. The diploid sporophytes produce both **plurilocular** and **unilocular sporangia** (Figure 17–11). The plurilocular sporangia form diploid zoospores that produce new sporophytes. Meiosis takes place within the unilocular sporangia, producing haploid zoospores that germinate to produce gametophytes. Unilocular sporangia, along with algin and plasmodesmata, are defining features of the brown algae.

In *Ectocarpus*, the gametophyte and sporophyte are similar in size and appearance (isomorphic). Many of the larger brown algae, including the kelps, undergo an alternation of heteromorphic generations—a large sporophyte and a microscopic gametophyte, as in the common kelp *Laminaria* (Figure 17–13). In *Laminaria*, the unilocular sporangia are produced on the surface of the mature blades. Half of the zoospores that the sporangia produce have the potential to grow into male gametophytes and half into female gametophytes. According to one hypothesis, the plurilocular gametangia borne on these gametophytes have become modified during the course of their evolution into one-celled antheridia and one-celled oogonia. Each antheridium releases a single sperm, and each oogonium contains a single egg. The fertilized egg in *Laminaria* remains attached to the female gametophyte and develops into a new sporophyte. In several genera of brown algae, the female gametes attract the male gametes by organic compounds.

Fucus and its close relatives have a gametic life cycle (see Figure 9–3b and Figure 17–14), as do the diatoms and certain green seaweeds. Understanding the evolutionary pressures that stimulated the origin of gametic life cycles in these protists may illuminate the early ap-

pearance of the gametic life cycle in our own metazoan (Gk. *meta*, "between," + *zōion*, "animal") lineage. *Fucus* and other brown seaweeds may contain large amounts of phenolic compounds that discourage herbivores. Other brown seaweeds tend to produce terpenes for the same purpose (page 33). In some cases, these compounds also have anti-microbial or anti-tumor activities, prompting investigations of their potential use in human medicine.

Green Algae: Phylum *Chlorophyta*

The green algae, including at least 17,000 species, are diverse in structure and life history. Although most green algae are aquatic, they are found in a wide variety of habitats, including on the surface of snow (Figure 17–15), on tree trunks, in the soil, and in symbiotic associations with lichens, freshwater protozoa, sponges, and coelenterates. Some green algae—such as species of the unicellular genera *Chlamydomonas* and *Chloromonas*, found growing on the surface of snow, and *Trentepohlia*, a filamentous alga that grows on rocks and tree trunks or branches—produce large amounts of carotenoids that function as a shield against intense light. These accessory pigments give the algae an orange, red, or rust color. Most aquatic green algae are found in fresh water, but a number of groups are marine. Many green algae are microscopic, although some of the marine species are large. *Codium magnum* of Mexico, for example, sometimes attains a breadth of 25 centimeters and a length of more than 8 meters.

The *Chlorophyta* resemble plants in several important characteristics. They contain chlorophylls *a* and *b*, and they store starch, their food reserve, inside plastids. Green algae and plants are the only two groups to do so. Some, but not all, green algae are like plants in having firm cell walls composed of cellulose, hemicelluloses, and pectic substances. In addition, the microscopic structure of the flagellated reproductive cells in some green algae resembles that of plant sperm. Many other biochemical details, such as the production of phytochrome (see Chapter 29) in at least two genera of green algae, also indicate that there is a very close relationship between the two groups. For these reasons, green algae are believed to be the protist group from which plants evolved.

Molecular and cellular studies of green algae have revealed that at least two multicellular lineages have evolved separately from unicellular flagellates. Each lineage includes a diverse mixture of morphological types. Unicellular forms, colonies, unbranched filaments, branched filaments, and other forms may be found.

Traditional classification systems group green algae according to their outward structure. Unicellular flagel-

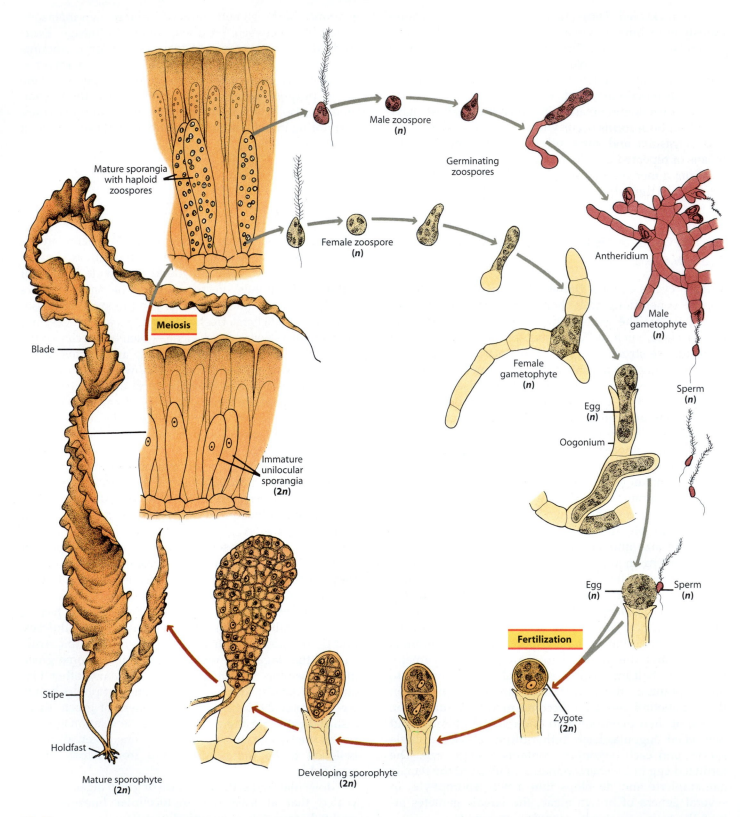

Mature sporangia with haploid zoospores

Male zoospore (*n*)

Germinating zoospores

Female zoospore (*n*)

Antheridium

Male gametophyte (*n*)

Female gametophyte (*n*)

Sperm (*n*)

Egg (*n*)

Oogonium

Egg (*n*)   Sperm (*n*)

Meiosis

Blade

Immature unilocular sporangia (**2n**)

Fertilization

Zygote (**2n**)

Developing sporophyte (**2n**)

Stipe

Holdfast

Mature sporophyte (**2n**)

17–13

Life cycle of the kelp Laminaria, *an example of sporic meiosis. Like many of the brown algae,* Laminaria *has an alternation of heteromorphic generations in which the sporophyte is conspicuous. Motile haploid zoospores are produced in the sporangia following meiosis (upper left). From these zoospores grow the microscopic, filamentous gametophytes, which in turn produce the motile sperm and nonmotile eggs. In some other brown algae, the sporophyte and gametophyte are often similar; they have an alternation of isomorphic generations.*

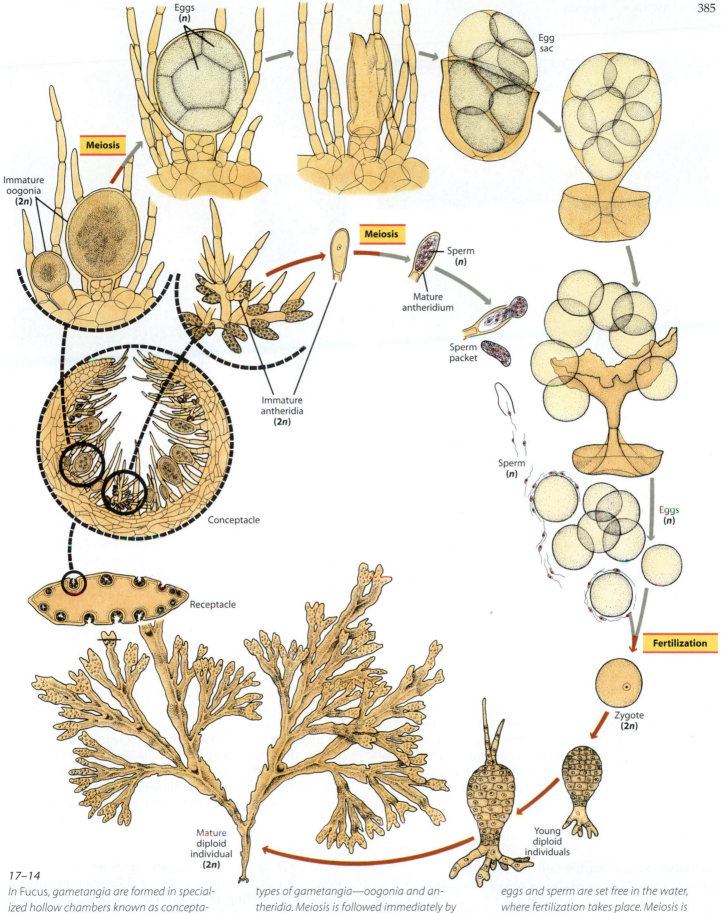

Eggs (n)

Meiosis

Immature oogonia (2n)

Meiosis

Egg sac

Sperm (n)

Mature antheridium

Sperm packet

Immature antheridia (2n)

Sperm (n)

Eggs (n)

Conceptacle

Fertilization

Receptacle

Zygote (2n)

Young diploid individuals

Mature diploid individual (2n)

17–14

In *Fucus*, gametangia are formed in specialized hollow chambers known as conceptacles, which are found in fertile areas called receptacles at the tips of the branches of diploid individuals (lower left). There are two types of gametangia—oogonia and antheridia. Meiosis is followed immediately by mitosis to give rise to 8 eggs per oogonium and 64 sperm per antheridium. Eventually the eggs and sperm are set free in the water, where fertilization takes place. Meiosis is gametic, and the zygote grows directly into the new diploid individual.

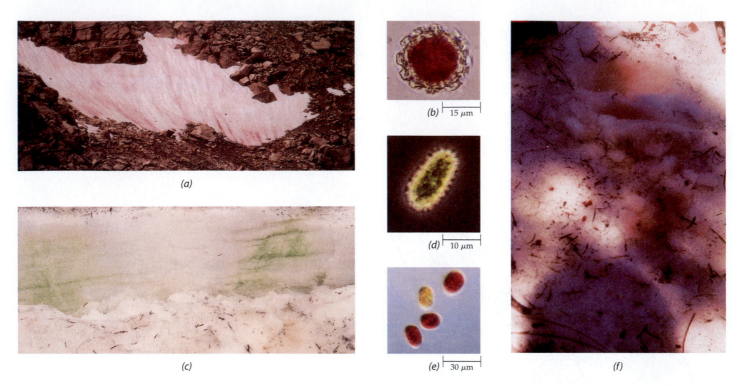

(a)

(b) $\overline{\text{ 15 }\mu m}$

(c)

(d) $\overline{\text{ 10 }\mu m}$

(e) $\overline{\text{ 30 }\mu m}$

(f)

17–15

Snow algae. Snow algae are unique because of their tolerance to temperature extremes, acidity, high levels of irradiation, and minimal nutrients for growth. (a) In many parts of the world, as in alpine areas from northern Mexico to Alaska, the presence of large numbers of snow algae produces "red snow" during the summer. This photograph was taken near Beartooth Pass in Montana. (b) Dormant zygote of the snow alga Chlamydomonas nivalis. *The red color results from carotenoids*

that serve to protect the chlorophyll in the zygote.

(c) Green snow occurs just below the surface, usually near tree canopies in alpine forests. It is widespread, occurring as far south as Arizona and as far north as Alaska and Quebec. This photograph was taken at Cayuse Pass, Mt. Rainier National Park, Washington. (d) Resting zygote of the alga Chloromonas brevispina, *found in green snow.*

(e) Three brilliant orange resting zygotes

of Chloromonas granulosa, *the alga responsible for orange snow. The single yellow zygote will change to orange as it matures. (f) Orange snow, associated with trees in forested alpine regions, receives more irradiation than green snow and less than red snow. Orange snow, which is not as common as the other two kinds of colored snow, occurs from Arizona and New Mexico to Alaska. This photograph was taken at Bill Williams Mountain in Arizona.*

lates are grouped together, filamentous types are grouped together, and so forth. As we observed earlier in the brown algae, related green algae cannot always be recognized by their outward structure. However, evidence of a relationship is revealed by ultrastructural studies of mitosis, cytokinesis, and reproductive cells, as well as molecular similarities. This new information has resulted in a new systematic alignment of the green algae into several classes, three of which—the *Chlorophyceae*, the *Ulvophyceae*, and the *Charophyceae*—are discussed in this book.

Cell Division and Motile Cell Differences Exist among Classes of Green Algae

The members of the largest class of green algae, the freshwater *Chlorophyceae*, have a unique mode of cytokinesis involving a **phycoplast** (Figure 17–16). In these

algae, the daughter nuclei move toward one another as the nonpersistent mitotic spindle collapses, and a new system of microtubules, the phycoplast, develops parallel to the plane of cell division. Presumably, the role of the phycoplast is to ensure that the **cleavage furrow,** resulting from infolding of the plasma membrane, will pass between the two daughter nuclei. The nuclear envelope persists throughout mitosis. In motile cells of the *Chlorophyceae*, there is a cross-shaped pattern of four narrow bands of microtubules known as **flagellar roots,** which are associated with the basal bodies (centrioles) of the flagella (Figure 17–17).

In other green algal classes, spindles may remain present throughout cytokinesis until they are disrupted, either by furrowing or by growth of a cell plate. A cell plate originates in the central region of the cell and grows outward to its margins. Some of the members of the class *Charophyceae* produce a new cytokinetic micro-

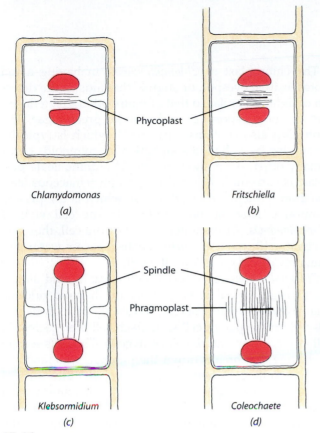

Phycoplast

Chlamydomonas
(a)

Fritschiella
(b)

Spindle

Phragmoplast

Klebsormidium
(c)

Coleochaete
(d)

17–16

Cytokinesis in two classes of the phylum Chlorophyta. *(a), (b) In the class* Chlorophyceae, *the mitotic spindle is nonpersistent and the daughter nuclei, which are relatively near one another, are separated by a phycoplast. Cytokinesis in* Chlamydomonas *is by furrowing, while cytokinesis in* Fritschiella *occurs by cell plate formation. (c) In simpler members of the class* Charophyceae, *such as* Klebsormidium, *the mitotic spindle is persistent and the daughter nuclei are relatively far apart. Cytokinesis occurs by furrowing. (d) Advanced charophytes like* Coleochaete *and* Chara *have a plantlike phragmoplast, and cytokinesis occurs by cell plate formation as in plants.* Ulvophyceae, *like* Charophyceae, *also have a persistent spindle, but do not have a phragmoplast or a cell plate.*

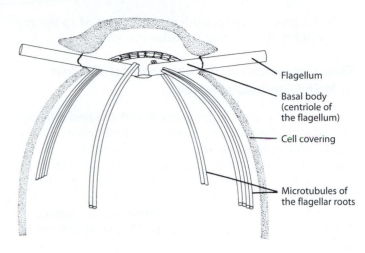

Flagellum

Basal body (centriole of the flagellum)

Cell covering

Microtubules of the flagellar roots

17–17

Diagram of the cross-shaped arrangement of four narrow bands of microtubules known as flagellar roots. These flagellar roots are associated with the flagellar basal bodies (centrioles) and are characteristic of green algae of the class Chlorophyceae.

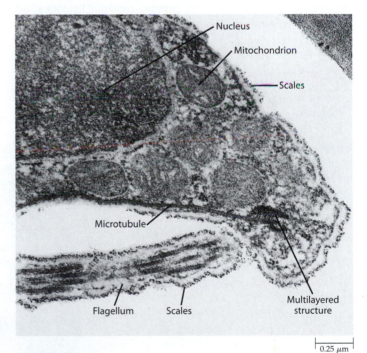

Nucleus

Mitochondrion

Scales

Microtubule

Flagellum   Scales

Multilayered structure

0.25 μm

17–18

Electron micrograph of the anterior portion of a motile sperm of the green alga Coleochaete *(class* Charophyceae*). Shown here is the multilayered structure, which is associated with the flagellar root system at the base of the flagellum. The multilayered structure is also characteristic of the sperm of bryophytes and some vascular plants, and it is one of the features linking these plants with* Charophyceae, *their ancestors. As seen here, a layer of microtubules extends from the multilayered structure down into the posterior end of the cell and serves as a cytoskeleton for these wall-less cells. The flagellar and plasma membranes are covered by a layer of small scales.*

tubular system, the **phragmoplast,** which is nearly identical to that present in plants. The microtubules in a phragmoplast are oriented perpendicular to the plane of cell division (Figure 17–16).

The flagellated cells of the *Charophyceae* are distinct from those of the other classes in that they have an asymmetrical flagellar root system of microtubules (Figure 17–18). One role of the flagellar root system is to provide the means by which a flagellum is anchored in place. Quite often, an entity known as a multilayered structure is associated with one of the flagellar roots. The type of flagellar root is often an important taxonomic character. The flagellar root system of the *Charophyceae*, with its multilayered structure, is very similar to that found in the sperm of bryophytes and some vascular plants. For these and other reasons, including biochemical and molecular similarities, *Charophyceae* is the group of living green algae considered to be closest to the ancestry of bryophytes and vascular plants.

Class *Chlorophyceae,* the Chlorophytes, Consists of Mainly Freshwater Species

The class *Chlorophyceae* includes flagellated and nonflagellated unicellular algae, motile and nonmotile colonial algae, filamentous algae, and algae consisting of flat sheets of cells. The members of this class live mainly in fresh water, although a few unicellular, planktonic species occur in coastal marine waters. A few *Chlorophyceae* are essentially terrestrial, occurring in habitats such as on snow (Figure 17–15), in soil, or on wood.

Chlamydomonas* Is an Example of a Motile Unicellular *Chlorophyceae The common freshwater green alga *Chlamydomonas* (Figure 17–19), a unicellular form with two equal flagella, has been widely used as a model system for molecular studies of the genes regulating photosynthesis and other cell processes. Molecular analyses have revealed that *Chlamydomonas* is a polyphyletic group; that is, it consists of several distinct lineages, all of which coincidentally are unicellular with two equal flagella.

The chloroplast of *Chlamydomonas,* in having a red photosensitive eyespot, or stigma, that aids in the detection of light, is similar to that of many other green flagellates and of zoospores of multicellular green algae. The chloroplast also contains a pyrenoid, which is typically surrounded by a shell of starch. Similar pyrenoids occur in many other green algal species. The uninucleate protoplast is surrounded by a thin glycoproteinaceous (carbohydrate-protein) cell wall, inside which is the plasma membrane. There is no cellulose in the cell wall of *Chlamydomonas.* At the anterior end of the cell, there are two contractile vacuoles, which collect excess water and ultimately discharge it from the cell.

Chlamydomonas reproduces both sexually and asexually. During asexual reproduction, the haploid nucleus usually divides by mitosis to produce up to 16 daughter cells within the parent cell wall. Each cell then secretes a wall around itself and develops flagella. The cells secrete an enzyme that breaks down the parental wall, and the daughter cells can then escape, although fully formed daughter cells are often retained for some time within the parent cell wall.

Sexual reproduction in *Chlamydomonas* involves the fusion of individuals belonging to different mating types (Figure 17–20). The vegetative cells are induced to form gametes by nitrogen starvation. The gametes, which re-

17–19

Chlamydomonas, *a unicellular green alga. The stigma was not preserved in this cell, and only the bases of the flagella can be seen in this electron micrograph. The nucleus of this uninucleate flagellate contains a prominent nucleolus. A shell of starch surrounds the pyrenoid, which is located within the chloroplast.*

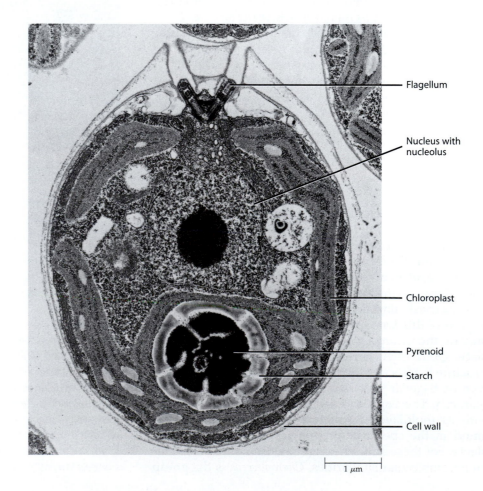

Flagellum

Nucleus with nucleolus

Chloroplast

Pyrenoid

Starch

Cell wall

1 μm

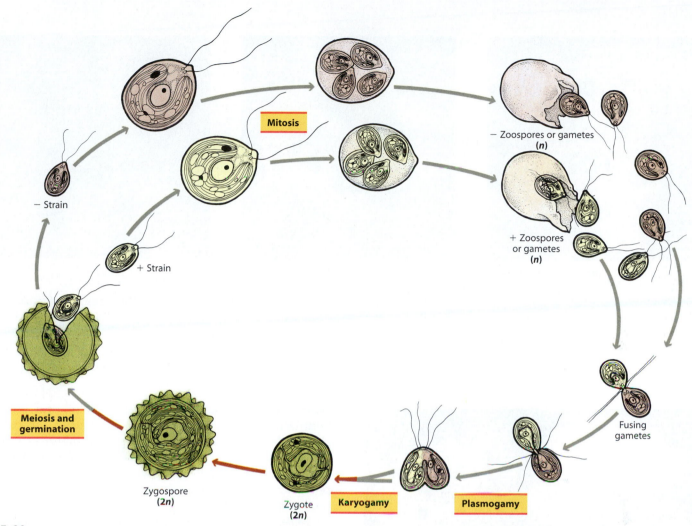

Mitosis

– Zoospores or gametes
(*n*)

– Strain

+ Strain

+ Zoospores
or gametes
(*n*)

Fusing
gametes

**Meiosis and
germination**

Zygospore
(*2n*)

Zygote
(*2n*)

Karyogamy

Plasmogamy

17–20

Life cycle of Chlamydomonas. *Sexual reproduction occurs when gametes of different mating types come together, cohering at first by their flagellar membranes and then by a slender protoplasmic thread—the conjugation tube (lower right). The protoplasts of the two cells fuse completely (plasmogamy), followed by the union of their nuclei (karyogamy). A thick wall is then formed around the diploid zygote, known at this point as a zygospore. After a period of dormancy, meiosis occurs, followed by germination, and four haploid cells emerge. Asexual reproduction of the haploid individuals by cell division is the most frequent mode of reproduction.*

semble the vegetative cells, first become aggregated in clumps. Within these clumps, pairs are formed that stick together, first by their flagellar membranes and later by a slender protoplasmic thread—the conjugation tube—that connects them at the base of their flagella. As soon as this protoplasmic connection is formed, the flagella become free, and one or both pairs of flagella propel the partially fused gametes through the water. The protoplasts of the two gametes fuse completely (plasmogamy), followed by fusion of their nuclei (karyogamy), forming the zygote. Soon the four flagella shorten and eventually disappear, and a thick cell wall forms around the diploid zygote. This thick-walled, resistant zygote, or zygospore, then undergoes a period of dormancy. Meiosis occurs at the end of the dormant period, resulting in the production of four haploid cells,

each of which develops two flagella and a cell wall. These cells can either divide asexually or mate with a cell of another mating strain to produce a new zygote. Thus, *Chlamydomonas* exhibits zygotic meiosis (see Figure 9–3a), and the haploid phase is the dominant phase in its life cycle.

Volvox* Is the Most Spectacular of the Motile Colonial *Chlorophyceae The class *Chlorophyceae* also includes colonies of various shapes and sizes that are aggregates of flagellated cells resembling *Chlamydomonas*. Examples of such colonies are *Gonium, Pandorina, Eudorina,* and *Volvox* (Figures 17–21 and 17–22). Like *Chlamydomonas, Volvox* has been the subject of much laboratory investigation. With the exception of their zygotes, *Chlamydomonas* and *Volvox* are both haploid. Therefore, mutant

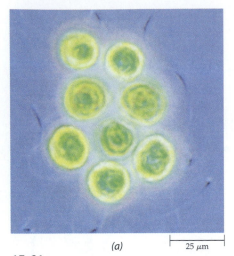

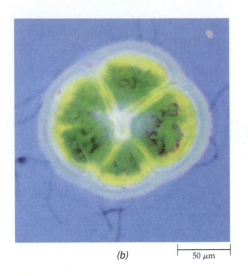

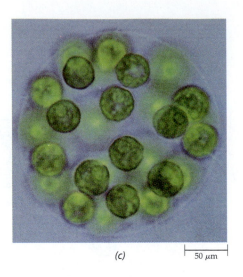

(a)   25 μm   *(b)*   50 μm   *(c)*   50 μm

17–21

Several colonial Chlorophyceae. *(a)* Gonium, *(b)* Pandorina, *and (c)* Eudorina. *In these algae, cells similar to those of*

Chlamydomonas *adhere in a gelatinous matrix to form multicellular colonies propelled by the beating of the flagella of the*

individual cells. Varying degrees of cellular specialization are found in different genera.

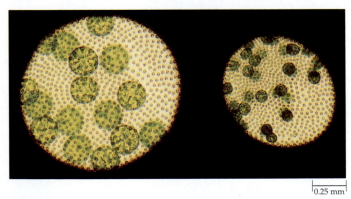

0.25 mm

17–22

Two spheroids (individuals) of Volvox carteri. *The approximately 2000 small cells that dot the periphery of the spheroids are the* Chlamydomonas-*like somatic cells. In V. carteri, somatic cells are not interconnected in mature colonies as they are in some members of the genus. A few cells are capable of becoming reproductive, either sexually or asexually. The spheroid at the left is asexual. Mitotic division has already occurred, producing 16 juvenile spheroids, each of which will eventually digest a passageway out of the parent colony and swim away.*

The spheroid at the right is sexual. When heat-shocked, a female spheroid produces egg-bearing sexual female individuals and a male spheroid produces sperm-laden sexual male individuals. The zygotes produced by fertilization are heat- and desiccation-resistant.

genes are not masked by dominant alleles, and mutations affecting development can be readily detected. Hundreds of mutant strains with specific, inheritable developmental defects have been isolated and are being studied in order to learn how specific genes regulate photosynthesis, sexual reproduction, cellular differentiation, and other cell processes.

Volvox consists of a hollow sphere, the spheroid, made up of a single layer of 500 to 60,000 vegetative, biflagellated cells that serve primarily a photosynthetic function, and a small number of larger, nonflagellated reproductive cells. The specialized reproductive cells undergo repeated mitoses to form many-celled, juvenile spheroids, which "hatch" from the parental spheroid by releasing an enzyme that dissolves the transparent parental matrix. Interestingly, as the spheroids first develop, all of the flagella face the hollow center, so the colony must turn inside out before it can become motile.

Sexual reproduction in *Volvox* is always oogamous (Figure 17–2). In all species that have been studied, sexual reproduction is synchronized within the population of colonies by a sexual inducer molecule, a glycoprotein with a molecular weight of about 30,000. This inducer molecule is produced by a spheroid that has itself become sexual by some other, as yet poorly understood, mechanism. One male colony of *V. carteri* may produce enough inducer to induce sexual reproduction in over half a billion other colonies.

The Class *Chlorophyceae* Also Includes Nonmotile Unicellular Members   One such member is *Chlorococcum*, which is very commonly found in the microbial flora of soils (Figure 17–23). There is a very large number of uni-

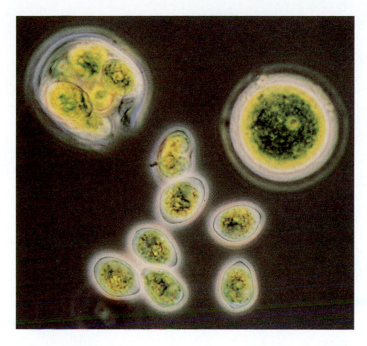

17–23

Chlorococcum echinozygotum. *At the upper left is a cell filled with asexual zoospores, which have formed mitotically within the cell. The smaller cells in the middle are biflagellate zoospores (the flagella are not visible). A nonmotile vegetative cell is at the upper right.*

17–24

(a) A stained portion of the "water net," Hydrodictyon, *a colonial member of the* Chlorophyceae. **(b)** *A higher magnification of a portion of* Hydrodictyon reticulatum.

cellular soil algal genera that superficially resemble *Chlorococcum* but that can be distinguished on the basis of cellular, reproductive, and molecular features. *Chlorococcum* and its relatives reproduce asexually by producing biflagellated zoospores, which are released from the parental cell. Sexual reproduction is accomplished by the release of flagellated gametes, which fuse in pairs to form zygotes. Meiosis is zygotic, as it is in all members of the *Chlorophyceae*.

Some *Chlorophyceae* Are Nonmotile Colonies   Nonmotile colonial members of the *Chlorophyceae* include *Hydrodictyon*, the "water net" (Figure 17–24). Under favorable conditions, it forms massive surface blooms in ponds, lakes, and gentle streams. Each colony consists of many large, cylindrical cells arranged in the form of a large, lacy, hollow cylinder. Initially uninucleate, each cell eventually becomes multinucleate. At maturity, each cell contains a large central vacuole and peripheral cytoplasm containing the nuclei and a large reticulate (resembling a net) chloroplast with numerous pyrenoids. *Hydrodictyon* reproduces asexually through the formation of large numbers of uninucleate, biflagellated zoospores in each cell of the net. The zoospores are not released from parental cells but, rather amazingly, group themselves into geometric arrays of four to nine (most typically six) within the cylindrical parent cell. Zoospores then lose their flagella and form the component cells of daughter mini-nets. These are eventually released from the parent cell and grow into large mature nets by dramatic cell enlargement. In view of this mode of reproduction, it is easy to see how *Hydrodictyon* can form such conspicuous blooms in nature. Sexual reproduction in *Hydrodictyon* is isogamous, and meiosis is zygotic, as in all sexually reproducing *Chlorophyceae*.

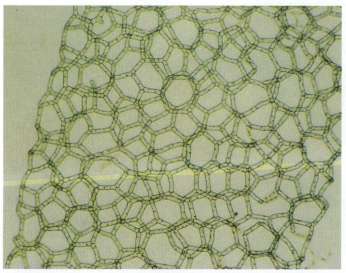

(a)

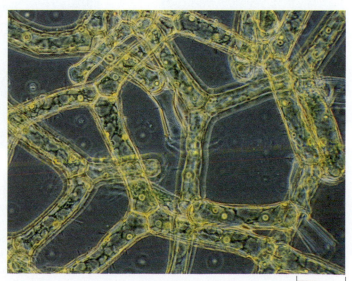

(b)   50 μm

There Are Also Filamentous and Parenchymatous *Chlorophyceae* *Oedogonium* is an example of an unbranched filamentous member of the *Chlorophyceae.* Filaments begin their development attached to underwater substrates by a holdfast, but massive growths may later break away to form noticeable floating blooms in lakes. The mode of cell division in *Oedogonium* results in the formation of characteristic "caps" or annular scars with each cell division (Figure 17–25). Thus, these scars reflect the number of divisions that have occurred in a given cell.

Asexual reproduction in *Oedogonium* takes place by means of zoospore formation, with a single zoospore produced per cell. Each zoospore has a crown of about 120 flagella. Sexual reproduction is oogamous (Figure 17–26). Each antheridium produces two multiflagellated sperm, and each oogonium produces a single egg. Like the brown algae, *Oedogonium* uses water-soluble chemical attractants to bring sperm to eggs. Meiosis in *Oedogonium* is zygotic, as in all *Chlorophyceae.*

The branched filamentous and parenchymatous, or tissuelike, *Chlorophyceae* include algae that have the most complex structures found in the class. Their cells can be specialized with respect to particular functions or positions in the algal body and, like plant cells, are sometimes connected by plasmodesmata. *Stigeoclonium* consists of branched filaments (Figure 17–27), and *Fritschiella* is composed of subterranean rhizoids—a parenchymatous prostrate system near the soil surface—and two kinds of erect branches (Figure 17–28). *Fritschiella* grows on damp surfaces, such as tree trunks, moist walls, and leaf surfaces.

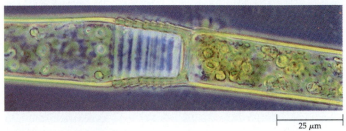

17–25

Oedogonium, *an unbranched, filamentous member of the* Chlorophyceae. *A section of vegetative filament showing annular scars.*

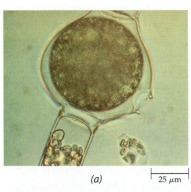

(a)   25 μm

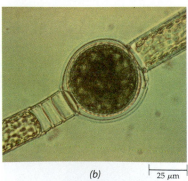

(b)   25 μm

17–26

Sexual reproduction in Oedogonium *is oogamous. Each oogonium produces a single egg, whereas each antheridium produces two multiflagellated sperm. (a) An oogonium of* Oedogonium cardiacum, *with a large, round egg and, at the lower right, a sperm cell. (b) An oogonium of* Oedogonium foveolatum, *with a young zygote and, to the left, a chain of empty antheridia.*

17–27

Stigeoclonium, *a branched, filamentous member of the* Chlorophyceae *that was formerly considered to belong to the genus* Ulothrix (Ulvophyceae). *Branching has just begun in this young individual, with a dark holdfast at its base and a rhizoid extending downward from just above the holdfast.*

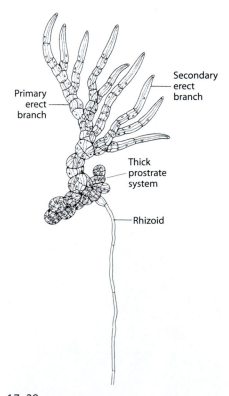

17–28

Fritschiella, *a terrestrial member of the* Chlorophyceae. *In its adaptation to a terrestrial habitat,* Fritschiella *has independently evolved some of the features that are characteristic of plants.*

Class *Ulvophyceae,* the Ulvophytes, Consists of Mainly Marine Species

The *Ulvophyceae* are primarily marine, but a few important representatives occur in fresh water, probably having moved there from the marine habitat at some time in the past. Ulvophytes may be filamentous or composed of flat sheets of cells, or they may be macroscopic and multinucleate. Ulvophytes have a closed mitosis in which the nuclear envelope persists; the spindle is persistent through cytokinesis.

The flagellated cells of ulvophytes are scaly or naked like those of *Charophyceae,* but in contrast to motile charophytes, ulvophytes are nearly radially symmetrical and have apical, forward-directed flagella like those of the chlorophytes. The flagellated cells of *Ulvophyceae* may have two, four, or many flagella, like *Chlorophyceae.* Those of *Charophyceae* are always biflagellate. Ulvophytes are the only green algae that have an alternation of generations with sporic meiosis or a diploid, dominant life history involving gametic meiosis.

One evolutionary line of *Ulvophyceae* consists of filamentous algae with large, multinucleate septate cells. One of the genera of this group, *Cladophora* (Figure 17–29), is widespread in both salt water and fresh waters, sometimes forming nuisance blooms in the latter habitat. Its filaments commonly grow in dense mats, which are either free-floating or attached to rocks and vegetation. The filaments elongate and branch near the ends. Each cell contains many nuclei and a single, peripheral, netlike chloroplast with many starch-forming pyrenoids. Marine species of *Cladophora* have an alternation of isomorphic generations. Most of the freshwater species, however, do not have an alternation of generations, apparently having lost this characteristic during the course of their transition from marine to fresh waters.

(a) (b) 25 μm

(c) 25 μm

(d)

17–29

Cladophora, *a member of the class* Ulvophyceae, *is widespread in marine and freshwater habitats. The marine species, like most* Ulvophyceae, *have an alternation of generations, while the freshwater species do not.* **(a)** *Branched filaments of* Cladophora. **(b)** *Part of an individual cell, showing the netlike chloroplast.* **(c)** *Commencement of branching at the apical end of a cell.* **(d)** *An individual of* Cladophora *growing in a sluggish stream in California.*

17–30

Sea lettuce, Ulva, *a common member of the class* Ulvophyceae *that grows on rocks, pilings, and similar places in shallow seas worldwide.*

A second kind of growth habit among the *Ulvophyceae* is that of *Ulva*, commonly known as sea lettuce (Figure 17–30). This familiar alga is common along temperate seashores throughout the world. Individuals of *Ulva* consist of a glistening, flat thallus that is two cells thick and a meter or more long in exceptionally large individuals. The thallus is anchored to the substrate by a holdfast produced by extensions of the cells at its base. Each cell of the thallus contains a single nucleus and chloroplast. *Ulva* is anisogamous (Figure 17–2) and has an alternation of isomorphic generations like that of many other *Ulvophyceae* (Figure 17–31).

The siphonous marine algae, characterized by very large, branched, coenocytic cells that are rarely septate, constitute additional evolutionary lines in the *Ulvo-*

phyceae (Figure 17–32). These algae, which are very diverse, develop as a result of repeated nuclear division without the formation of cell walls. Cell walls are produced only in the reproductive phases of the siphonous green algae.

Codium, mentioned on page 383, is a member of this group. It is a spongy mass of densely intertwined coenocytic filaments (Figure 17–32a). Strains of *Codium fragile* may be quite weedy, and nuisance growths of this seaweed are spreading in quiet waters of the temperate zone. *Ventricaria* (also known as *Valonia*), which is common in tropical waters, has been widely used in studies of cell walls and in physiological experiments requiring large amounts of cell sap. *Ventricaria* appears to be unicellular but is actually a large, multinucleate vesicle at-

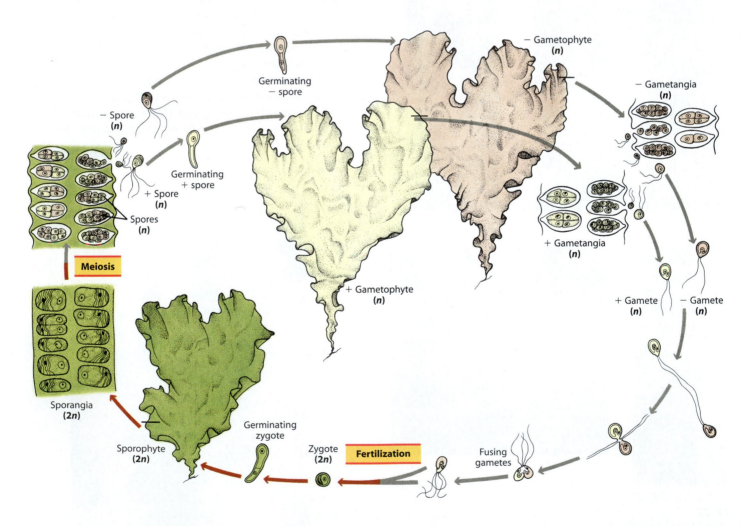

17–31

In the sea lettuce, Ulva, we can see the reproductive pattern known as an alternation of generations, in which one generation produces spores (left), the other gametes (right).

The haploid (n) gametophyte produces haploid isogametes, and the gametes fuse to form a diploid (2n) zygote. A sporophyte, a multicellular body in which all the cells are

diploid, develops from the zygote. The sporophyte produces haploid spores by meiosis. The haploid spores develop into haploid gametophytes, and the cycle begins again.

17–32

Four genera of siphonous green algae of the class Ulvophyceae. *(a) A species of* Codium, *abundant along the Atlantic coast.* **(b)** Ventricaria, *common in tropical waters; individuals are often about the size of a hen's* egg. *(c)* Acetabularia, *the "mermaid's wine glass," a mushroom-shaped siphonous green alga. The siphonous green alga in the background is* Dasycladus. *This photograph was taken in the Bahamas.* **(d)** Halimeda, *a* siphonous green alga that is often dominant in reefs in warmer waters throughout the world. This alga produces distasteful compounds that retard grazing by fishes and other marine herbivores.

tached to the substrate by several rhizoids (Figure 17–32b). It grows to the size of a hen's egg. Another well-known siphonous green alga is *Acetabularia* (Figure 17–32c), which has been widely used in experiments on the genetic basis of differentiation. The siphonous green algae are primarily diploid. The gametes are the only haploid cells in the life cycle.

The chloroplasts of some siphonous green algae, including *Codium*, are harvested by sea slugs (nudibranchs), which are marine mollusks that lack shells (Figure 17–33). The sea slugs eat the algae, and the algal chloroplasts persist in the cells that line the animal's respiratory chamber. In the presence of light, these chloro-

plasts carry on photosynthesis so efficiently that individuals of the nudibranch *Placobranchus ocellatus* are reported to evolve more oxygen than they consume.

Halimeda and related genera of siphonous green algae are notable for their calcified cell walls. When these algae die and disintegrate, they play a major role in the generation of the white carbonate sand that is so characteristic of tropical waters. A number of genera of algae, including *Halimeda* (Figure 17–32d), contain secondary metabolites that significantly reduce feeding by herbivorous fishes. The fronds of *Halimeda* expand and grow rapidly at night, building up toxic compounds that deter herbivores, which are active mainly during the day.

17–33

Placobranchus, *a sea slug. The tissues of this animal contain chloroplasts, which it obtains by eating certain green algae.* (a) *Ordinarily, flaplike structures called parapodia are folded over the animal's back, hiding the chloroplast-containing tissues.* (b) *When the parapodia are spread apart, however, the deep green tissues become visible. The chloroplasts carry on photosynthesis so efficiently that, within a 24-hour cycle of light and darkness, some of the animals produce more oxygen than they consume.*

(a)   1 cm

(b)

Within an hour after sunrise, the fronds turn from white to green as chloroplasts from the lower portions of the thallus migrate upward and begin to photosynthesize rapidly.

Class *Charophyceae*, the Charophytes, Includes Members That Closely Resemble Plants

Charophyceae consist of unicellular, colonial, filamentous, and parenchymatous genera. Their relationship with one another, and to plants, is revealed by many fundamental structural, biochemical, and genetic similarities. These include the presence of asymmetrical flagellated cells, some of which have distinctive multilayered structures (Figure 17–18). Other similarities are breakdown of the nuclear envelope at mitosis, persistent spindles or phragmoplasts at cytokinesis, the presence of phytochrome, and other molecular features.

Spirogyra (Figure 17–34) is a well-known genus of unbranched, filamentous *Charophyceae* that often forms frothy or slimy floating masses in bodies of fresh water. Each filament is surrounded by a watery sheath. The name *Spirogyra* refers to the helical arrangement of the one or more ribbon-like chloroplasts with numerous pyrenoids found within each uninucleate cell. Asexual reproduction occurs in *Spirogyra* by cell division and fragmentation. There are no flagellated cells at any stage of the life cycle, but, as noted earlier, flagellated reproductive cells do occur in other genera of *Charophyceae*. During sexual reproduction in *Spirogyra*, a conjugation tube forms between two filaments. The contents of the two cells that are joined by the tube serve as isogametes. Fertilization may occur in the tube, or one of the gametes may migrate into the other filament, where fertilization then takes place. The zygotes become surrounded by thick walls that contain **sporopollenin**. This protective substance is the most resistant biopolymer known. Sporopollenin enables the zygotes to survive harsh con-

(a)   25 μm

(b)   25 μm

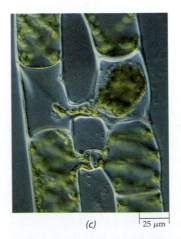

(c)   25 μm

(d)   25 μm

17–34

Sexual reproduction in Spirogyra. (a), (b) *The formation of conjugation tubes between the cells of adjacent filaments.* (c) *The contents of the cells of the − strain (on the left) pass* through these tubes into the cells of the + strain. (d) *Fertilization occurs within these cells. The resulting zygote develops a thick, resistant cell wall and is termed a zygospore.* *The vegetative filaments of* Spirogyra *are haploid, and meiosis occurs during germination of the zygospores, as it does in all* Charophyceae.

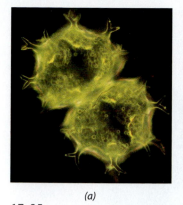  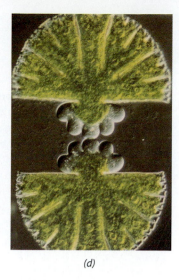

(a)　(b)　(c)　(d)

17–35

Desmids. The desmids are a group of thousands of species of unicellular, freshwater Charophyceae. Many desmids are constricted, which gives them a very charac- *teristic appearance. (a) Xanthidium armatum. (b) Euastrum affine. (c) Micrasterias radiosa. (d) Cell division in Micrasterias thomasiana.* *The smaller half of each daughter individual will grow to the size of the larger half, forming a full-sized desmid.*

ditions for long periods of time before germinating when conditions improve. Meiosis is zygotic, as in all members of the *Charophyceae*.

The desmids are a large group of freshwater green algae related to *Spirogyra*. Like *Spirogyra*, they lack flagellated cells. Some desmids are filamentous, but most are unicellular. Most desmid cells consist of two sections, or semi-cells, joined by a narrow constriction—the isthmus—between them (Figure 17–35). Desmid cell division and sexual reproduction are very similar to those of *Spirogyra*. There are thousands of desmid species. They are most abundant and diverse in peat bogs and ponds that are poor in mineral nutrients. Some are associated with distinctive and possibly symbiotic bacteria, which live in the mucilaginous sheaths.

The *Coleochaetales* and *Charales* Have Plantlike Characteristics　Two orders of charophycean green algae, the *Coleochaetales* and the *Charales*, resemble plants more closely than do other charophytes in the details of their cell division and sexual reproduction. These orders have

a plantlike microtubular phragmoplast operating during cytokinesis. Like plants, they are oogamous, and their sperm are ultrastructurally similar to those of bryophytes. Plants were probably derived from an extinct member of the *Charophyceae* that in many respects resembled members of the *Coleochaetales* and *Charales*.

The order *Coleochaetales* includes both branched filamentous genera and discoid (disk-shaped) genera that grow by division of apical or peripheral cells (Figure 17–36a). *Coleochaete*, which grows on the surface of submerged rocks or freshwater plants, has uninucleate vegetative cells that each contain one large chloroplast with an embedded pyrenoid. Very similar chloroplasts and pyrenoids occur in some of the hornworts, a group of bryophytes that is discussed in Chapter 18. *Coleochaete*, like a number of charophytes, reproduces asexually by zoospores that are formed singly within cells. Sexual reproduction is oogamous. The zygotes, which remain attached to the parental thallus, stimulate the growth of a layer of cells that covers the zygotes (Figure 17–36b). In at least one species, these parental cells have wall in-

 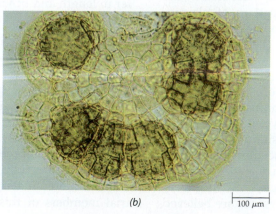

(a)　0.25 µm　(b)　100 µm

17–36

(a) Coleochaete grows on rocks and on stems of aquatic flowering plants in shallow lake waters. (b) Individuals of this species of Coleochaete consist of a parenchymatous disk, which is generally one cell thick. The large cells are zygotes, which are protected by a cellular covering. The hair cells extending from the disk are ensheathed at the base; Coleochaete means "sheathed hair." These hairs are thought to discourage aquatic animals from feeding on the alga.

17–37

(a) Chara, a stonewort (class Charophyceae) that grows in shallow waters of temperate lakes. (b) Chara showing gametangia. The top structure is an oogonium, and that below is an antheridium.

(a)

(b)   200 μm

growths similar to those occurring at the gametophyte-sporophyte junction of bryophytes (see Chapter 18) and other plants. These specialized cells, called transfer cells, are believed to function in nutrient transport between the gametophyte and sporophyte. The wall ingrowths suggest how the distinctive life cycle of plants and the sporophyte generation might have evolved in the long-extinct ancestors of modern plants.

The order *Charales* includes about 250 living species of distinctive green algae found primarily in fresh water or sometimes in brackish water. Modern forms such as *Chara* (Figure 17–37) are commonly known as the stoneworts, because some of them have heavily calcified cell walls. Calcification of the distinctive reproductive structures of ancient relatives has resulted in a good fossil record extending back to Late Silurian times (about 410 million years ago).

The *Charales*, like *Coleochaete* and plants, exhibit apical growth. In addition, the thallus of the *Charales* is differentiated into nodal and internodal regions. The tissue organization in the nodal regions resembles the parenchyma of plants, as does the pattern of plasmodesmatal connections. From the nodal regions arise whorls of branches. In some species, files of cells grow over the central, filamentous axis, creating a thicker, stronger thallus. The sperm of *Charales* are produced in multicellular antheridia that are more complex than those found in any other group of protists. Their eggs are borne in oogonia enclosed by several long, tubular, twisted cells. These cells are in a position analogous to the distinctive female gametangia of seedless plants, and they may serve similar functions. Sperm are the only flagellated cells in the *Charales* life cycle, and they very closely resemble bryophyte sperm. Zygotes are believed

to germinate by meiosis, though this has been difficult to study because zygotes are enclosed in very tough walls that include sporopollenin. Sporopollenin is also a component of the walls of plant spores and pollen and is responsible for the widespread occurrence of these cells in the fossil record.

Summary

Heterokonts Have One Long Tinsel Flagellum and One Short Whiplash Flagellum

Brown algae (phylum *Phaeophyta*) are closely related to a diverse array of microscopic protists, including oomycetes (phylum *Oomycota*), diatoms (phylum *Bacillariophyta*), and chrysophytes (phylum *Chrysophyta*). Collectively, these organisms are known as heterokonts because characteristically they have two flagella that differ in length and ornamentation.

Phylum *Oomycota* Includes Both Aquatic and Terrestrial Heterotrophs That Form Thick-Walled Oospores

The oomycetes range from unicellular forms to highly branched coenocytic filamentous forms. The cell walls are composed largely of cellulose or cellulose-like polymers. Asexual reproduction is by means of zoospores. Sexual reproduction involves a large, immobile egg and a small, motile male gamete. Three of the partly terrestrial members of this phylum are *Phytophthora*, a very important causative agent of plant diseases, including late blight of Irish potatoes; *Plasmopara viticola*, which causes downy mildew of grapes; and *Pythium*, which causes damping-off diseases of seedlings.

Diatoms Belong to the Phylum *Bacillariophyta*

The diatoms are unicellular or colonial organisms that occur in fresh as well as in marine waters. They are important components of the phytoplankton and are unique for their two-part siliceous cell walls.

Chrysophytes Belong to Phylum *Chrysophyta*

Chrysophytes are colorless or gold-pigmented, unicellular or colonial organisms that are abundant in fresh waters and in the oceans of the world. Some chrysophytes feed on bacteria and other organic particles, whereas others cannot feed because they are coated by overlapping silica scales.

Brown Algae Belong to Phylum *Phaeophyta*

The brown algae include the largest and most complex of marine algae and are the most conspicuous seaweeds of temperate, boreal, and polar waters. In many types, the vegetative body is well differentiated into holdfast, stipe, and blade. Some have food-conducting tissues that approach, in their complexity, those of vascular plants. Giant brown kelps form ecologically important marine forests, and these algae are the source of industrially important polysaccharides. The life cycles of most brown algae involve an alternation of generations.

Chlorophyceae, Ulvophyceae, and *Charophyceae* Are Classes of the Phylum *Chlorophyta*

The green algae include a wide diversity of freshwater and terrestrial forms, in addition to seaweeds. Several classes of green algae have been defined on the basis of cell division, reproductive cell structure, and other features. The *Ulvophyceae* are primarily marine, and alternation of generations occurs in some forms. The *Chlorophyceae* are primarily freshwater species, and two members of this group, *Chlamydomonas* and *Volvox*, are important laboratory model systems. The *Chlorophyceae* have a unique mode of cytokinesis—the mitotic spindle collapses at telophase, and a phycoplast develops parallel to the plane of cell division. The freshwater *Charophyceae* are the green algae that are most closely related to bryophytes and vascular plants. The modern genus *Coleochaete* and the order *Charales* have cell division, reproductive, and other features that link them particularly closely to the ancestry of plants.

Selected Key Terms

algin p. 382
antheridium p. 372
centric p. 376
chrysolaminarin p. 378
cleavage furrow p. 386
damping-off p. 374
flagellar roots p. 386
frustules p. 376
fucoxanthin p. 378
heterokonts p. 371
heterothallic p. 373
homothallic p. 373
mannitol p. 382

oogonium p. 372
oospore p. 372
pennate p. 376
phragmoplast p. 387
phycoplast p. 386
plurilocular gametangia p. 383
plurilocular sporangia p. 383
sporopollenin p. 396
thallus p. 381
unilocular sporangia p. 383

Questions

1. Distinguish between each of the following: oogonium/ antheridium; homothallic/heterothallic; pennate/centric; phycoplast/phragmoplast.

2. Identify the plant diseases caused by each of the following oomycetes: *Plasmopara viticola, Phytophthora infestans,* and *Pythium* sp.

3. The diatoms may be characterized as "the algae that live in glass houses." Explain.

4. What pigments do the diatoms, chrysophytes, and brown algae have in common? Which of these pigments is responsible for the color of these algae?

5. The kelps have the most highly differentiated bodies among the algae. How so?

6. *Fucus* has a life cycle that is, in some ways, similar to our own. Explain.

7. Distinguish among the three classes of *Chlorophyta*: *Chlorophyceae, Ulvophyceae,* and *Charophyceae.*

8. What features of the modern genus *Coleochaete* and the order *Charales* link them closely to the ancestry of plants?

Chapter

18

Bryophytes

18–1
Dense growth of the moss Fissidens *on limestone rocks in a waterfall. This photograph was taken in a nature reserve just west of Yalta on the Crimean Peninsula in Ukraine.*

OVERVIEW

With the bryophytes—liverworts, hornworts, and mosses—we see the important evolutionary move from the water to the land. Such a move involved solving a variety of problems—the most crucial of which was how to avoid drying out. The gametes of bryophytes are enclosed in multicellular protective structures—an antheridium encloses the sperm and an archegonium encloses the egg. But a vestige of their aquatic algal ancestors persists in that sperm must still swim through water to reach the egg. As you will see, fertilization takes place within the archegonium, and the resulting embryo is nourished and protected by the maternal plant. The growing embryo (the young sporophyte) never becomes independent of the parent (the gametophyte), and in fact one of the characteristics of the bryophytes is that the free-living gametophyte is almost always larger than the sporophyte. Meiosis quickly occurs in the sporophyte, and the resulting spores are dispersed into the environment, where they germinate and give rise to new gametophytes.

CHECKPOINTS

By the time you finish reading this chapter, you should be able to answer the following questions:

1. What are the general characteristics of the bryophytes? In other words, what does it take to be a bryophyte?

2. What are the three phyla of bryophytes? How are they similar to and different from one another?

3. How does sexual reproduction occur in bryophytes? What are the principal parts of the resulting sporophyte of most bryophytes?

4. How can the three types of liverworts be distinguished from one another?

5. What are the distinguishing features of the hornworts?

6. What are the distinguishing features of each of the three classes of mosses?

Bryophytes—liverworts, hornworts, and mosses—are small "leafy" or flat plants that most often grow in moist locations in temperate and tropical forests or along the edges of streams and wetlands (Figure 18–1). Bryophytes are not confined, however, to such habitats. Many species of mosses are found in relatively dry deserts, and several can form extensive masses on dry, exposed rocks that can become very hot (Figure 18–2).

Mosses sometimes dominate the terrain to the exclusion of other plants over large areas north of the Arctic Circle. Mosses are also the dominant plants on rocky slopes above timberline in the mountains, and a significant number of mosses are able to withstand the long periods of severe cold on the Antarctic continent (Figure 18–3). A few bryophytes are aquatic, and some are even found on rocks splashed by ocean waves, although none are truly marine.

Bryophytes contribute significantly to plant biodiversity and are also important in some parts of the world for the large amounts of carbon they store, thereby playing a significant role in the global carbon cycle (page 150). Increasing evidence indicates that the first plants were much like modern, or extant, bryophytes, and even today, together with lichens, bryophytes are important initial colonizers of bare rock and soil surfaces. Like the lichens, some bryophytes are remarkably sensitive to air pollution, and they are often absent or represented by only a few species in highly polluted areas.

The Relationships of Bryophytes to Other Groups

In many respects, bryophytes are transitional between the charophycean green algae, or charophytes (see Chapter 17), and the vascular plants, which are discussed in Chapters 19 through 22. In the last chapter, we considered some of the features shared by charophytes and plants. Both contain chloroplasts with well-devel-

18–2

A dry-land moss, Tortula obtusissima, *that lives on and around limestone in the central plateau of Mexico. Having no roots, the plants obtain their moisture directly from the external environment in the form of dew or rain. They can recover physiologically from complete dryness in less than five minutes.*

(a)

(b)

18–3

(a) At about 3000 meters elevation on Mount Melbourne, Antarctica, the daily temperatures in summer mainly range from −10° to −30°C. In this incredibly harsh environment, botanists from New Zealand discovered patches of a moss of the genus Campylopus *(b), growing in the bare areas visible in the photograph, where volcanic activity produces temperatures that may reach 30°C. The* growth of Campylopus *in this locality demonstrates the remarkable dispersal powers of mosses as well as their ability to survive in harsh habitats.*

oped grana, and both have motile cells that are asymmetrical, with flagella that extend from the side rather than the end of the cell. During the cell cycle, both charophytes and plants exhibit breakdown of the nuclear envelope at mitosis and persistent spindles or phragmoplasts during division of the cytoplasm (cytokinesis). In addition, you may recall that, among the charophytes, the *Charales* and *Coleochaetales* appear to be more closely related to the plants than any others. In *Coleochaete*, the zygotes are retained within the parental thallus and, in at least one species of *Coleochaete*, the cells covering the zygotes develop wall ingrowths. These covering cells apparently function as transfer cells involved with the transport of sugars to the zygotes. Like *Coleochaete*, all plants are oogamous, that is, they have an egg cell that is fertilized by sperm.

Bryophytes and vascular plants share a number of characters that distinguish them from the charophytes. These include: (1) the presence of male and female gametangia, called **antheridia** and **archegonia**, respectively, with a protective layer called a sterile jacket layer; (2) retention of both the zygote and the developing multicellular embryo, or young sporophyte, within the archegonium or the female gametophyte; (3) the presence of a multicellular diploid sporophyte, which results in an increased number of meioses and an amplification of the number of spores that can be produced following each fertilization event; (4) multicellular sporangia consisting of a sterile jacket layer and internal spore-producing (**sporogenous**) tissue; and (5) spores with walls containing **sporopollenin,** which resists decay and drying. Charophytes lack all of these shared bryophyte and vascular plant characters, which are correlated with the existence of plants on land.

Living bryophytes lack the water- and food-conducting (vascular) tissues called xylem and phloem, respectively, which are present in vascular plants. Although some bryophytes have specialized conducting tissues, the cell walls of the bryophyte water-conducting cells are not lignified, as are those of the vascular plants. Also, there are differences in the life cycles of bryophytes and vascular plants, both of which exhibit alternating heteromorphic gametophytic and sporophytic generations. In the bryophytes, the gametophyte is dominant and free-living and the sporophyte is small, permanently attached to and nutritionally dependent upon its parental gametophyte. By contrast, the sporophyte of vascular plants is the dominant generation and is also ultimately free-living. In addition, the bryophyte sporophyte is unbranched and bears only a single sporangium, whereas the sporophytes of extant vascular plants are branched and bear many more sporangia. Vascular plant sporophytes therefore produce a great many more spores than do the sporophytes of bryophytes.

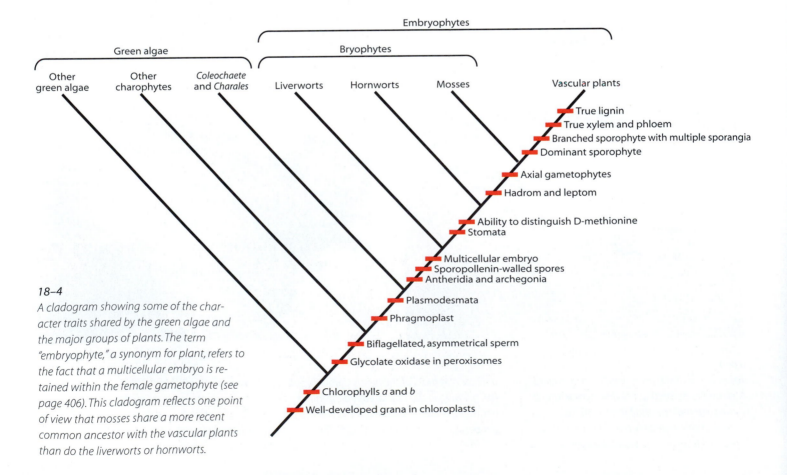

18–4

A cladogram showing some of the character traits shared by the green algae and the major groups of plants. The term "embryophyte," a synonym for plant, refers to the fact that a multicellular embryo is retained within the female gametophyte (see page 406). This cladogram reflects one point of view that mosses share a more recent common ancestor with the vascular plants than do the liverworts or hornworts.

In the past, the relationship of bryophytes to vascular plants was controversial. Today, however, nucleotide sequence data and recent fossil discoveries, together with classical morphological characters and newly described ultrastructural features, reveal that bryophytes include the earliest of the extant plant groups to have diverged within a monophyletic plant lineage (Figure 18–4). Modern bryophytes can therefore provide important insights into the nature of the earliest plants and the process by which vascular plants evolved. A comparison of the structure and reproduction of extant bryophytes with those of ancient fossils and modern vascular plants can show how various features of vascular plants may have evolved.

Comparative Structure and Reproduction of Bryophytes

Some bryophytes, namely the hornworts and certain liverworts, are described as "thalloid" because their gametophytes, which are generally flat and dichotomously branched (forking repeatedly into two equal branches), are **thalli** (singular: thallus). Thalli are undifferentiated bodies, or bodies not differentiated into roots, leaves, and stems. Such thalli are often relatively thin, which may facilitate the uptake of both water and CO_2. Some bryophyte gametophytes have specialized adaptations on their upper surface for increasing CO_2 permeability while at the same time reducing water loss. The surface pores of the liverwort *Marchantia* are one such example

(Figure 18–5). The gametophytes of some liverworts (leafy liverworts) and the mosses are said to be differentiated into "leaves" and "stems," but it could be argued that these are not true leaves and stems because they occur in the gametophytic generation and do not contain xylem and phloem. However, the thalli of certain liverworts and mosses do contain centrally located strands of cells that appear to have conducting functions. Such cells may be similar to ancient evolutionary precursors of phloem and lignified vascular tissues. Inasmuch as the terms "leaf" and "stem" are commonly used when referring to the leaflike and stemlike structures of the gametophytes of leafy liverworts and mosses, this practice will be followed in this book.

Surface layers reminiscent of the cuticles commonly found on surfaces of the true leaves and stems of vascular plants also occur on the surfaces of some bryophytes. The cuticle of sporophytes is closely correlated with the presence of stomata, which function primarily in the regulation of gas exchange. The pores seen in some bryophyte gametophytes, such as those of *Marchantia*, are considered to be analogous to stomata (Figure 18–5). The biochemistry and evolution of the bryophyte cuticle are poorly understood, however, primarily because bryophyte cuticles are more difficult to remove for chemical analysis than are cuticles of vascular plants.

The gametophytes of both thalloid and leafy bryophytes are generally attached to the substrate, such as soil, by **rhizoids** (Figure 18–5). The rhizoids of mosses are multicellular, each consisting of a linear row of cells, whereas those of liverworts and hornworts are unicellular. The rhizoids of bryophytes generally serve only to anchor the plants, since absorption of water and inorganic ions commonly occurs directly and rapidly throughout the gametophyte. Mosses, in particular, often have special hairs and other structural adaptations that aid in external water transport and absorption by leaves and stems. In addition, bryophytes often harbor fungal or cyanobacterial symbionts that may aid in acquisition of mineral nutrients. Rootlike organs are lacking in the bryophytes.

18–5

Surface pores of Marchantia, *a thalloid liverwort. (a) Transverse section of the gametophyte of* Marchantia. *Numerous chloroplast-bearing cells are evident in the upper layers, and there are several layers of colorless cells below them, as well as rhizoids that anchor the plant body to the substrate. Pores permit the exchange of gases in the air-filled chambers that honeycomb the upper photosynthetic layer. The specialized cells that surround each pore are usually arranged in four or five superimposed circular tiers of four cells each, and the whole structure is barrel-shaped. Under dry conditions, the cells of the lowermost tier, which usually protrude into the chamber, become juxtaposed and retard water loss, whereas under moist conditions they separate. Thus the pores serve a function similar to that of the stomata of vascular plants. (b) A scanning electron micrograph of a pore on the dorsal surface of a gametophyte of* Marchantia.

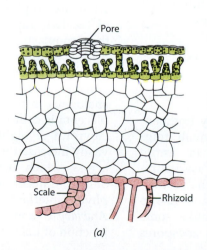

(a)

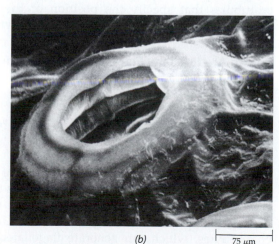

(b)

75 μm

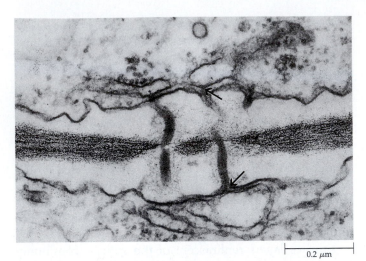

0.2 μm

18–6

Longitudinal view of plasmodesmata in the liverwort Monoclea
gottschei. *Note (arrows) that the desmotubule in the plasmodesma on
the right is continuous with the endoplasmic reticulum in the cytosol.*

The cells of bryophyte tissues are interconnected by
plasmodesmata. Bryophyte plasmodesmata are similar
to those of vascular plants in possessing an internal component known as the desmotubule (Figure 18–6). The
desmotubule is derived from a segment of tubular endoplasmic reticulum that becomes entrapped in developing cell plates during cytokinesis (see Figure 8–11).
Certain charophytes also possess plasmodesmata, but
the degree to which they are similar to those of
bryophytes and represent the ancient forerunners of
plant plasmodesmata is still under study (page 398). The
cells of most bryophytes resemble those of vascular
plants in having many small, disk-shaped plastids. All of
the cells of some hornwort species, and the apical
and/or reproductive cells of many bryophytes, by contrast, have only a single large plastid per cell. This characteristic is believed to be an evolutionary holdover from
ancestral green algae, which, like modern *Coleochaete,*
probably contained only a single large plastid per cell.
During cell division, the cells of bryophytes and vascular plants produce preprophase bands consisting of microtubules that specify the position of the future cell
wall. Such bands are lacking in the charophytes.

Sperm Are the Only Flagellated Cells Produced by Bryophytes, and They Require Water to Swim to the Egg

Many bryophytes can reproduce asexually by fragmentation (vegetative propagation), whereby small fragments, or pieces, of tissue produce an entire
gametophyte. Another widespread means of asexual reproduction in both liverworts and mosses is the production of **gemmae**—multicellular bodies that give rise to
new gametophytes (see Figure 18–15). Unlike some
charophytes, which can generate flagellated zoospores

for asexual reproduction, sperm are the only flagellated
cells produced by bryophytes. Loss of ability to produce
zoospores, which are likely to be less useful on land than
in the water, is probably correlated with the absence of
centrioles from the spindles of bryophytes and other
plants (page 158). Mitosis in certain liverworts and hornworts shows features that are intermediate between
those of charophytes and vascular plants, suggesting
evolutionary stages leading to the absence of centrioles
in plant mitosis.

Sexual reproduction in bryophytes involves production of antheridia and archegonia, often on separate
male or female gametophytes. In some species, sex is
known to be controlled by the distribution at meiosis of
distinctive sex chromosomes. In fact, sex chromosomes
in plants were first discovered in bryophytes. The spherical or elongated antheridium is commonly stalked and
consists of a sterile jacket layer, one cell thick, that surrounds numerous **spermatogenous cells** (Figure 18–7a).
The "jacket" layer of cells is said to be "sterile" because
it cannot produce sperm. Each spermatogenous cell
forms a single biflagellated sperm that must swim
through water to reach the egg located inside an
archegonium. Liquid water is therefore required for fertilization in bryophytes.

The archegonia of bryophytes are flask-shaped, with
a long neck and a swollen basal portion, the **venter,**
which encloses a single egg (Figure 18–7b). The outer
layer of cells of the neck and venter forms the sterile protective layer of the archegonium. The central cells of the
neck, the **neck canal cells,** disintegrate when the egg is
mature, resulting in a fluid-filled tube through which the
sperm swim to the egg. During this period, chemicals
are released that attract sperm. After fertilization, the zygote remains within the archegonium where it is nourished by sugars, amino acids, and probably other
substances provided by the maternal gametophyte. This
form of nutrition is known as **matrotrophy** ("food derived from the mother"). Thus supplied, the zygote undergoes repeated mitotic divisions, generating the
multicellular embryo (Figure 18–8), which eventually
develops into the mature sporophyte (Figure 18–9).
There are no plasmodesmatal connections between cells
of the two adjacent generations. Nutrient transport is
thus apoplastic—that is, nutrients move along the cell
walls. This transport is facilitated by a **placenta** that occurs at the interface between the sporophyte and
parental gametophyte (Figure 18–10) and is therefore
analogous to the placenta of mammals. The bryophyte
placenta is composed of transfer cells with an extensive
labyrinth of highly branched cell wall ingrowths that
vastly increase the surface area of the plasma membrane
across which active nutrient transport takes place.
Similar transfer cells occur at the gametophyte-sporophyte interface of vascular plants (for example,
Arabidopsis and soybean) and at the haploid-diploid
junction of *Coleochaete* (page 397). The occurrence of pla-

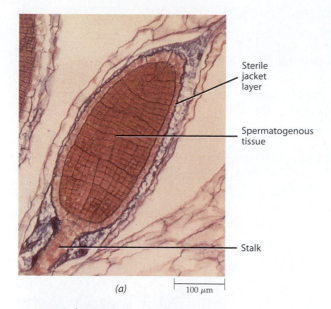

Sterile
jacket
layer

Spermatogenous
tissue

Stalk

(a)   100 µm

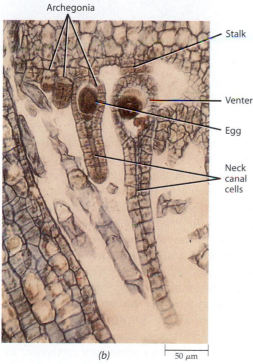

Archegonia

Stalk

Venter

Egg

Neck
canal
cells

(b)   50 µm

18–7

*Gametangia of Marchantia, a liverwort. **(a)** A developing antheridium, consisting of a stalk and sterile—that is, non-sperm-forming— jacket layer enclosing spermatogenous tissue. The spermatogenous tissue develops into spermatogenous cells, each of which forms a single sperm propelled by two flagella.*

* **(b)** Several archegonia at different stages of development. An egg is contained in the venter, a swollen portion at the base of each flask-shaped archegonium. When the egg is mature, the neck canal cells disintegrate, creating a fluid-filled tube through which the biflagellated sperm swim to the egg in response to chemical attractants. In Marchantia, the archegonia and antheridia are borne on different gametophytes.*

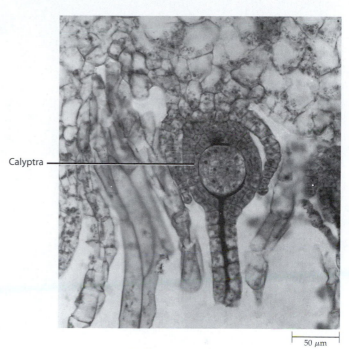

Calyptra

50 µm

18–8

An early stage in development of the embryo, or young sporophyte, of Marchantia. *Here the young sporophyte is nothing more than an undifferentiated spherical mass of cells within the enlarged venter, or calyptra.*

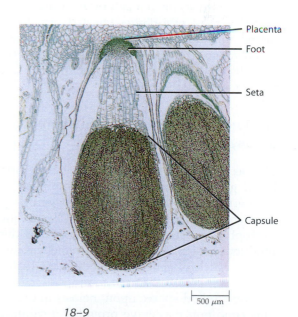

Placenta

Foot

Seta

Capsule

500 µm

18–9

A nearly mature sporophyte of Marchantia. *At this stage of development the foot, seta, and capsule, or sporangium, of the sporophyte are distinct. The placenta occurs at the interface between the foot and gametophyte and consists of transfer cells of both sporophyte and gametophyte.*

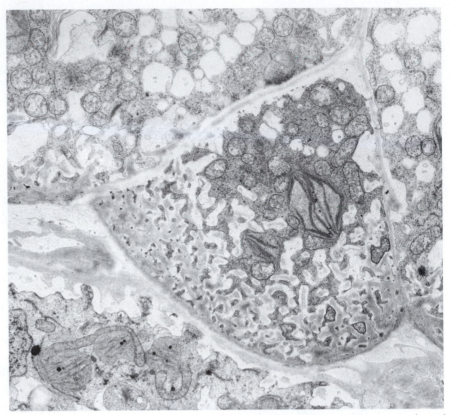

18–10

The gametophyte-sporophyte junction (placenta) in the liverwort Carrpos monocarpos. *Extensive wall ingrowths develop in the single cell layer of transfer cells in the sporophyte. There are several layers of transfer cells in the gametophyte, but their wall ingrowths are not as highly branched as those of the sporophyte layer. Numerous chloroplasts and mitochondria are present in the placental cells of both generations.*

1 μm

cental cells in *Coleochaete* suggests that matrotrophy had already evolved in the charophyte ancestors of plants.

As the bryophyte embryo develops, the venter undergoes cell division, keeping pace with the growth of the young sporophyte. The enlarged venter of the archegonium is called a **calyptra.** At maturity the sporophyte of most bryophytes consists of a **foot,** which remains embedded in the archegonium, a **seta,** or stalk, and a **capsule,** or **sporangium** (Figure 18–9). The transfer cells at the junction between the foot and archegonium constitute the placenta.

The Term "Embryophytes" Is an Appropriate Synonym for Plants

The occurrence of a multicellular, matrotrophic embryo in all groups of plants, from bryophytes through angiosperms, is the basis for the term **embryophytes** as a synonym for plants (Figure 18–4). The advantage of matrotrophy and the plant placenta is that they fuel the production of a many-celled diploid sporophyte, each cell of which is genetically equivalent to the fertilized egg. These cells can be used to produce many genetically diverse haploid spores upon meiosis in the sporangium. This condition may have provided a significant advantage to early plants as they began to occupy the land. Production of greater numbers of spores per fertilization event may also have helped compensate for low fertilization rates when water became scarce. The sporophytic generation of plants is thought to have evolved from a zygote, such as those produced by charophytes, in which meiosis had been delayed until after at least a few mi-

totic divisions had occurred. The more mitotic divisions that occur between fertilization and meiosis, the larger the sporophyte that can be formed and the greater the number of spores that can be produced. Throughout the evolutionary history of plants, there has been a tendency for sporophytes to become increasingly larger in relation to the size of the gametophyte generation.

The sporophyte epidermis of hornworts and mosses typically contains stomata—each bordered by two guard cells—that resemble the stomata of vascular plants. The stomata apparently have similar functions. One such function is to aid in the uptake of CO_2 by the sporophyte for photosynthesis. Another function is to generate a flow of water and nutrients between the sporophyte and gametophyte induced by the loss of water vapor through the stomata, and, of course, when closed, the stomata retard water loss. Liverwort sporophytes, which are typically smaller and more ephemeral than those of mosses and hornworts, lack stomata. The epidermal cell walls of the moss and liverwort sporophytes are impregnated with decay-resistant phenolic materials that may protect developing spores. Those of the hornwort sporophyte are covered with a protective cuticle.

The Sporopollenin Walls of Bryophyte Spores Have Survival Value

Bryophyte spores, like those of all other plants, are encased in a substantial wall impregnated with the most decay- and chemical-resistant biopolymer known, sporopollenin. The sporopollenin walls enable bryophyte spores to survive dispersal through the air from one moist site to another. The spores of charophytes, which are typically dispersed in water, are not enclosed by a sporopollenin wall. Charophyte zygotes, however, are lined with sporopollenin and can therefore tolerate exposure and microbial attack, remaining viable for long periods. The sporopollenin-walled spores of plants are thought to have originated from charophyte zygotes by change in the timing of sporopollenin deposition.

Bryophyte spores germinate to form juvenile developmental stages, which in mosses are called **protonemata** (singular: protonema; from the Greek, *prōtos*, "first," and *nēma*, "thread"). From the protonemata develop gametophytes and gametangia. Protonemata, which are characteristic of all mosses, are also found in some liverworts, but not in hornworts.

The bryophytes are grouped into three phyla: *Hepatophyta* (the liverworts), *Anthocerophyta* (the hornworts), and *Bryophyta* (the mosses). In addition, there is an enigmatic bryophyte known as *Takakia* (Figure 18–11). In the past, *Takakia* was classified as a liverwort, but now it is considered to be a very divergent moss. The relative order of evolutionary divergence of these four groups is controversial. There is some disagreement as to which group represents the earliest divergent bryophytes and which is most closely related to the vascular plants. With the availability of more molecular sequence information and integration of information from the fossil record, however, it seems most likely that the liverworts diverged first and mosses are more closely related to vascular plants.

18–11

Takakia ceratophylla, a bryophyte that for many years was known only from its unusual gametophytes. On the left is a female gametophyte with an attached immature sporophyte; on the right is a male gametophyte with orange antheridia.

Liverworts: Phylum *Hepatophyta*

Liverworts, or hepatics, are a group of about 6000 species of plants that are generally small and inconspicuous, although they may form relatively large masses in favorable habitats, such as moist, shaded soil or rocks, tree trunks, or branches. A few kinds of liverworts grow in water. The name "liverwort" dates from the ninth century, when it was thought, because of the liver-shaped outline of the gametophyte in some genera, that these plants might be useful in treating diseases of the liver. According to the medieval "Doctrine of Signatures," the outward appearance of a body signaled the possession of special properties. The Anglo-Saxon ending *-wort* (originally *wyrt*) means "herb"; it appears as a part of many plant names in the English language.

Most liverwort gametophytes develop directly from spores, but some genera first form a protonema-like filament of cells, from which the mature gametophyte develops. Gametophytes continue to grow from an apical meristem. There are three major types of liverworts that can be differentiated on the basis of structure and grouped into two clades (page 267). One clade consists of the complex thalloid liverworts, which have internal tissue differentiation. The other clade contains the leafy liverworts and the simple thalloid types, which consist of ribbons of relatively undifferentiated tissue.

Complex Thalloid Liverworts Include *Riccia*, *Ricciocarpus*, and *Marchantia*

Thalloid liverworts can be found on moist, shaded banks and in other suitable habitats, such as flowerpots in a cool greenhouse. The thallus, which is about 30 cells thick at the midrib and approximately 10 cells thick in the thinner portions, is sharply differentiated into a thin, chlorophyll-rich upper (dorsal) portion and a thicker, colorless lower (ventral) portion (Figure 18–5a). The lower surface bears rhizoids, as well as rows of scales. The upper surface is often divided into raised regions, each with a large pore that leads to an underlying air chamber (Figure 18–5b).

The sporophyte structure of *Riccia* and *Ricciocarpus* is among the simplest seen in liverworts (Figure 18–12). *Ricciocarpus*, which grows in water or on damp soil, is bisexual—that is, both sex organs arise on the same plant. Some species of *Riccia* are aquatic, although most are terrestrial. *Riccia* gametophytes may be either unisexual or bisexual. In both *Riccia* and *Ricciocarpus*, the sporophytes are deeply embedded within the dichotomously branched gametophytes and consist of little more than a sporangium. No special mechanism for spore dispersal occurs in these sporophytes. When the portion of the gametophyte containing mature sporophytes dies and decays, the spores are liberated.

18–12
Riccia, *one of the simplest of liverworts.* **(a)** *The system of branching of* Riccia *gameto-phytes is dichotomous, that is, the main and subsequent axes fork into two branches.* **(b)** *The sporophyte, which is embedded within the gametophyte, consists solely of a spherical capsule.*

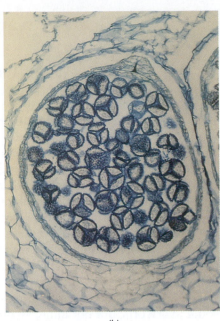

(a)   (b)

One of the most familiar of liverworts is *Marchantia*, a widespread genus that grows on moist soil and rocks (see Figure 13–15b). Its dichotomously branched gameto-phytes are larger than those of *Riccia* and *Ricciocarpus.* Unlike the latter two genera, in which the sex organs are distributed along the dorsal surface of the thallus, *Marchantia* has its gametangia borne on specialized structures called **gametophores.**

The gametophytes of *Marchantia* are unisexual, and the male and female gametophytes can be readily distinguished by their distinctive gametophores. The antheridia are borne on disk-headed gametophores called **antheridiophores,** whereas the archegonia are borne on umbrella-headed gametophores called **archegoniophores** (Figure 18–13). In *Marchantia,* the sporophyte generation consists of a foot, a short seta, and a capsule (Figure 18–9). In addition to spores, the mature spo-

rangium contains elongate cells called **elaters,** which have helically arranged hygroscopic (moisture-absorbing) wall thickenings (Figure 18–14). The walls of elaters are sensitive to slight changes in humidity, and, after the capsule dehisces (dries out and opens) into a number of petal-like segments, the elaters undergo a twisting action that helps disperse the spores.

Fragmentation is the principal means of asexual reproduction in liverworts, but another widespread mechanism is the production of gemmae. In *Marchantia,* the gemmae are produced in special cuplike structures—called **gemma cups**—located on the dorsal (upper) surface of the gametophyte (Figure 18–15). The gemmae are dispersed primarily by splashes of rain.

The life cycle of *Marchantia* is illustrated in Figure 18–17 (pages 410 and 411).

(a)

(b)

18–13
Gametophytes of Marchantia. *The antheridia* **(a)** *and archegonia* **(b)** *are elevated on stalks—the antheridiophores and archego-niophores, respectively—above the thallus.*

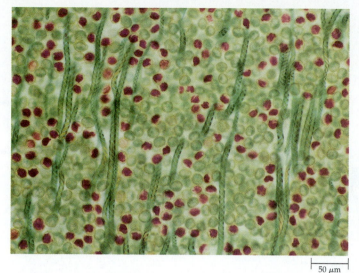

18–14

Mature spores (red spheres) and elaters (green strands) from a capsule of Marchantia.

(a)

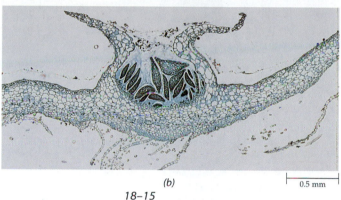

(b) 0.5 mm

18–15

Gemma cups. **(a)** *Gametophytes of* Marchantia, *with gemma cups containing gemmae. The gemmae appear as more or less spherical pieces of tissue. The gemmae are splashed out by the rain and may then grow into new gametophytes, each genetically identical to the parent plant from which it was derived by mitosis.* **(b)** *Longitudinal section of a gemma cup. The gemmae are the dark, more or less lens-shaped structures.*

Leafy Liverworts Have a Distinctive Leaf Structure and/or Arrangement

The leafy liverworts are a very diverse group that includes more than 4000 of the 6000 species of the phylum *Hepatophyta* (Figure 18–16). The leafy liverworts are especially abundant in the tropics and subtropics, in regions of heavy rainfall or high humidity, where they grow on the leaves and the bark of trees, as well as on other plant surfaces (Figure 18–18, page 412). There are probably many tropical species that have not yet been described. Leafy liverworts are also well represented in temperate regions. The plants are usually well branched and form small mats.

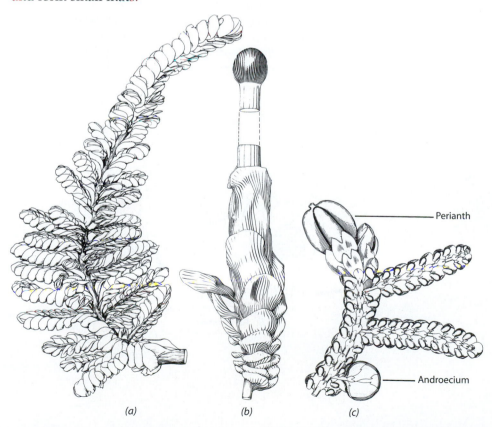

Perianth

Androecium

(a)   *(b)*   *(c)*

18–16

Leafy liverworts. **(a)** Clasmatocolea puccionana, *showing the characteristic arrangement of the leaves.* **(b)** *The end of a branch of* Clasmatocolea humilis. *The capsule and the long stalk of the sporophyte are visible.* **(c)** *A portion of a branch of* Frullania, *showing the characteristic arrangement of its leaves. The antheridia are contained within the androecium. The archegonium and developing sporophyte are contained within the perianth.*

409

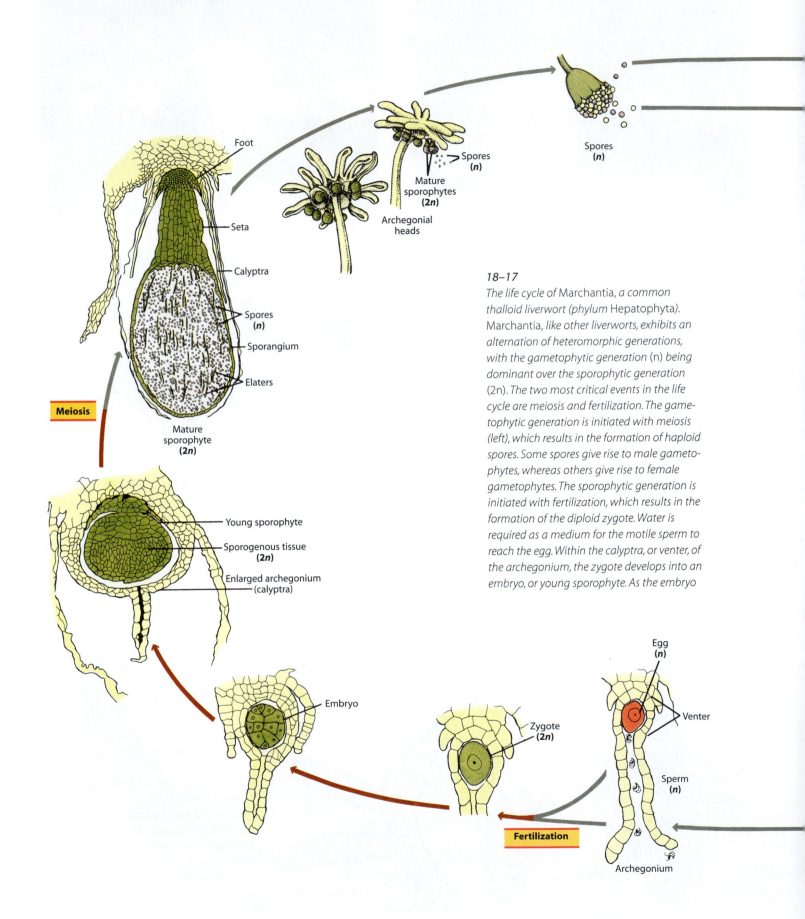

Foot

Seta

Calyptra

Spores
(*n*)

Sporangium

Elaters

Meiosis

Mature
sporophyte
(*2n*)

Young sporophyte

Sporogenous tissue
(*2n*)

Enlarged archegonium
(calyptra)

Embryo

Spores
(*n*)

Mature
sporophytes
(*2n*)

Archegonial
heads

Spores
(*n*)

Zygote
(*2n*)

Egg
(*n*)

Venter

Sperm
(*n*)

Fertilization

Archegonium

18–17

The life cycle of Marchantia, *a common
thalloid liverwort (phylum* Hepatophyta).
Marchantia, *like other liverworts, exhibits an
alternation of heteromorphic generations,
with the gametophytic generation* (n) *being
dominant over the sporophytic generation*
(2n). *The two most critical events in the life
cycle are meiosis and fertilization. The gametophytic generation is initiated with meiosis
(left), which results in the formation of haploid
spores. Some spores give rise to male gametophytes, whereas others give rise to female
gametophytes. The sporophytic generation is
initiated with fertilization, which results in the
formation of the diploid zygote. Water is
required as a medium for the motile sperm to
reach the egg. Within the calyptra, or venter, of
the archegonium, the zygote develops into an
embryo, or young sporophyte. As the embryo*

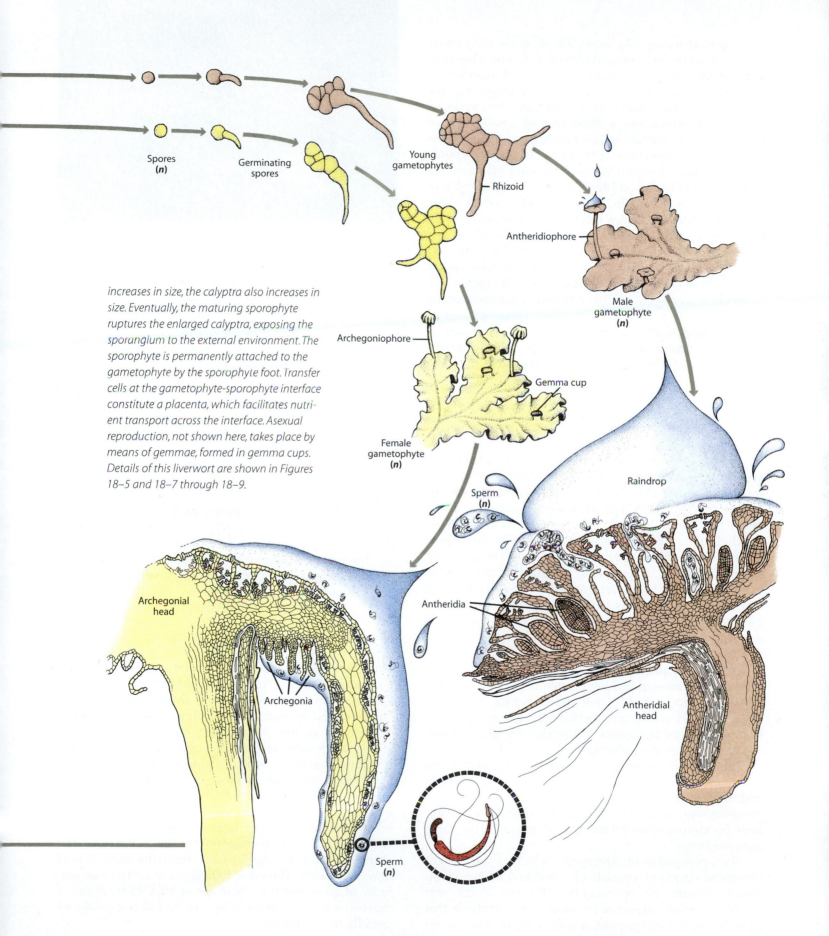

Spores
(**n**)

Germinating
spores

Young
gametophytes

Rhizoid

Antheridiophore

Male
gametophyte
(**n**)

increases in size, the calyptra also increases in size. Eventually, the maturing sporophyte ruptures the enlarged calyptra, exposing the sporangium to the external environment. The sporophyte is permanently attached to the gametophyte by the sporophyte foot. Transfer cells at the gametophyte-sporophyte interface constitute a placenta, which facilitates nutrient transport across the interface. Asexual reproduction, not shown here, takes place by means of gemmae, formed in gemma cups. Details of this liverwort are shown in Figures 18–5 and 18–7 through 18–9.

Archegoniophore

Gemma cup

Female
gametophyte
(**n**)

Sperm
(**n**)

Raindrop

Archegonial
head

Archegonia

Antheridia

Antheridial
head

Sperm
(**n**)

Liverwort leaves, like moss leaves, generally consist of only a single layer of undifferentiated cells. One way to distinguish liverworts from mosses is that moss leaves are usually equal in size and spirally arranged around the stem, whereas many liverworts have two rows of equal-sized leaves and a third row of smaller leaves along the lower surface of the gametophyte. Most moss leaves splay outward from the stem in three dimensions, but a few have leaves flattened into one plane, as do many liverworts. In addition, moss leaves sometimes have a thickened "midrib," but liverwort leaves lack this structure. Moss leaves are most often entire, in contrast to liverwort leaves, which can be highly lobed or dissected. In *Frullania*, a common liverwort that grows on bark, the leaves consist of a large dorsal lobe and a small, helmet-shaped ventral lobe (Figure 18–16c).

In the leafy liverworts, the antheridia generally occur on a short side branch with modified leaves known as the **androecium.** The developing sporophyte, as well as the archegonium from which it develops, is characteristically surrounded by a tubular sheath known as the **perianth** (Figure 18–16c).

Hornworts: Phylum *Anthocerophyta*

Hornworts constitute a small phylum of about 100 species. The members of the genus *Anthoceros* are the most familiar of the six genera. The gametophytes of hornworts (Figure 18–19a) superficially resemble those of the thallose liverworts, but there are many features that indicate a relatively distant relationship. For example, the cells of most species usually have a single large chloroplast with a pyrenoid, as in the green alga *Coleochaete*. Some hornwort species have cells containing many small chloroplasts lacking pyrenoids, as do most plant cells, but even in these hornworts the apical cell contains a single plastid, reflecting the ancestral condition (page 404).

Hornwort gametophytes are often rosettelike, and their dichotomous branching is often not apparent. They are usually about 1 to 2 centimeters across. *Anthoceros* has extensive internal cavities that are often inhabited by cyanobacteria of the genus *Nostoc* that fix nitrogen and supply it to their host plants.

The gametophytes of some species of *Anthoceros* are unisexual, whereas others are bisexual. The antheridia and archegonia are sunken on the dorsal surface of the gametophyte, with the antheridia clustered in chambers. Numerous sporophytes may develop on the same gametophyte.

The sporophyte of *Anthoceros*, which is an upright elongated structure, consists of a foot and a long, cylindrical capsule, or sporangium (Figures 18–19 and 18–20). A unique aspect of hornwort sporophytes is that early in their development, a meristem, or zone of ac-

18–18

A leafy liverwort growing on the leaf of an evergreen tree in the rainforest of the Amazon Basin, near Manaus, Brazil.

tively dividing cells, develops between the foot and the sporangium. This basal meristem remains active as long as conditions are favorable for growth. As a result, the sporophyte continues to elongate for a prolonged period of time. It is green, having several layers of photosynthetic cells. It is also covered with a cuticle and has stomata (Figure 18–19c). The presence of stomata on the sporophytes of hornworts and mosses is regarded as evidence of an important evolutionary link to the vascular plants. Maturation of the spores, and, ultimately, dehiscence of the sporangium, begins near its tip and extends toward the base as the spores mature (Figure 18–19b). Among the spores are sterile, elongate, often multicellular structures that resemble the elaters of liverworts. The dehiscing sporangium splits longitudinally into ribbonlike halves.

Mosses: Phylum *Bryophyta*

Many groups of organisms contain members that are commonly called "mosses"—reindeer "mosses" are lichens, scale "mosses" are leafy liverworts, while club "mosses" and Spanish "moss" belong to different groups of vascular plants. Sea "moss" and Irish "moss" are algae. The genuine mosses, however, are members of the phylum *Bryophyta*, which consists of three classes: *Sphagnidae* (the peat mosses), *Andreaeidae* (the granite mosses), and *Bryidae* (often referred to as the "true mosses"). These groups are very distinct from one another, differing in many important features. Molecular and other information suggests that the peat mosses and granite mosses diverged earlier from the main line of moss evolution. The class *Bryidae* contains the vast majority of moss species. There are at least 9500 species of mosses, with new forms being discovered constantly, especially in the tropics.

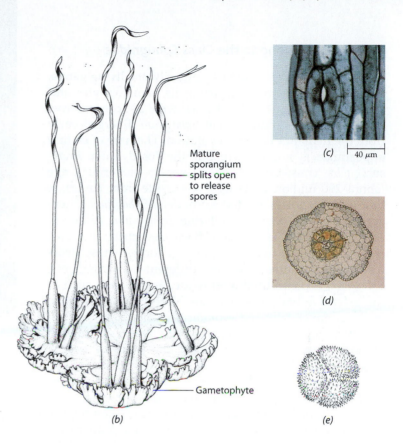

(a)

(b)

(c) | 40 μm

(d)

Mature sporangium splits open to release spores

Gametophyte

(e)

18–19

Anthoceros, *a hornwort.* **(a)** *A gametophyte with attached (elongate) sporophytes.* **(b)** *When mature, the sporangium splits, and the spores are released.* **(c)** *A stoma; stomata are* abundant on the sporophytes of the hornworts, which are green and photosynthetic. **(d)** *Developing spores, visible in the center of this cross section of a sporangium, and* **(e)** mature spores still held in a tetrad, a group of four spores—three of which are visible here—formed from a spore mother cell by meiosis.

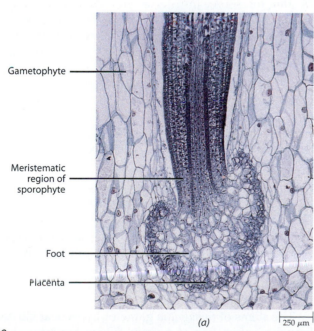

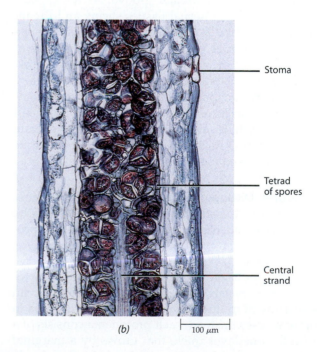

Gametophyte

Meristematic region of sporophyte

Foot

Placenta

(a) | 250 μm

Stoma

Tetrad of spores

Central strand

(b) | 100 μm

18–20

Anthoceros. **(a)** *Longitudinal section of the lower portion of a sporophyte, showing its foot embedded in the tissue of the gameto-* phyte. **(b)** *Longitudinal section of a portion of a sporangium, showing tetrads of spores with elater-like structures among them. The central* strand of tissue in the lower part of the sporangium consists of tissue that may function in conduction.

Peat Mosses Belong to the Class *Sphagnidae*

The class *Sphagnidae* contains primarily one living genus, *Sphagnum.* Distinctive features of its gametophytes and sporophytes (Figure 18–21), as well as comparative DNA sequences, indicate that *Sphagnum* diverged early from the main line of moss evolution. The time of its first appearance is not known, but the fossil order *Proto-sphagnales,* consisting of several genera of Permian age (about 290 million years ago; see inside front cover), is clearly very closely related to modern *Sphagnum.* There are at least 150 species of living *Sphagnum.* Although more than 300 species have been described, many of these are thought to be polymorphic forms whose structure differs somewhat in different environments. *Sphagnum* is distributed worldwide, in wet areas such as the extensive bog regions of the Northern Hemisphere, and is commercially and ecologically valuable.

Sexual reproduction in *Sphagnum* involves formation of antheridia and archegonia at the ends of special branches located at the tips of the moss gametophyte. Fertilization takes place in late winter, and four months later mature spores are discharged from sporangia.

Among mosses, the sporophytes of *Sphagnum* (Figure 18–21a) are quite distinctive. The red to blackish-brown capsules are nearly spherical and are raised on a stalk, the **pseudopodium,** which is part of the gametophyte and may be up to 3 millimeters long. The sporophyte has a very short seta, or stalk. Spore discharge in *Sphagnum* is spectacular (Figure 18–21c). At the top of the capsule is a lidlike **operculum,** separated from the rest of the capsule by a circular groove. As the capsule matures and dries, its internal tissues shrink and the internal air pressure builds to several bars, similar to pressures found in the tires of trailer trucks. The operculum is eventually blown off with an audible click, and the escaping gas carries a cloud of spores out of the capsule.

Asexual reproduction by fragmentation is very common. Young branches and stem pieces that break off from the gametophyte and injured leaves can regenerate new gametophytes. As a result, *Sphagnum* then forms large, densely packed clumps.

Three Features Distinguish the *Sphagnidae* from Other Mosses   The most distinctive differences between the class *Sphagnidae* and other mosses are its unusual protonema, the peculiar morphology of its gametophyte, and its explosive operculum mechanism. The protonema—the first stage of development of the gametophyte—of the *Sphagnidae* does not consist of an extensive array of multicellular, branched filaments as in most other mosses. Instead, each protonema consists of a plate of cells, one layer thick, that grows by a marginal meristem, most of whose cells can divide in one of only two possible directions. In these respects, the protonema of *Sphagnum* is remarkably similar to the disk-shaped

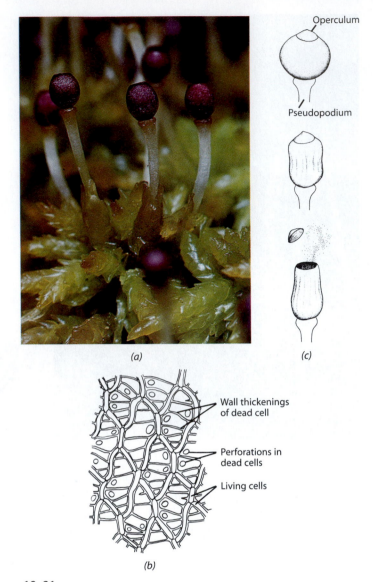

(a)   (b)   (c)

18–21

A peat moss, Sphagnum. **(a)** *A gametophyte, with many attached sporophytes. Some of the capsules, such as the two at the front, have already discharged their spores.* **(b)** *Structure of a leaf. Large, dead cells are surrounded by smaller living cells, rich in chloroplasts.* **(c)** *Dehiscence of a capsule. As the capsule dries, it contracts, changing from a spherical to a cylindrical shape. This change in shape causes compression of trapped gas within the capsule. When the compressed gas reaches a pressure of about 5 bars, the pressure inside the capsule explosively blows off the operculum, releasing a cloud of spores.*

thalli of *Coleochaete* (page 397). The erect gametophyte arises from a budlike structure that grows from one of the marginal cells (Figure 18–22). This structure contains an apical meristem that divides in three directions, forming leaf and stem tissues.

The stems of *Sphagnum* gametophytes bear clusters of branches, often five at a node, which are more densely tufted near the tips of the stem, resulting in a moplike head. Both branches and stems bear leaves, but stem leaves often have little or no chlorophyll, whereas most

18–22
Young leafy gametophyte of Sphagnum.

The Ecology of *Sphagnum* Is of Worldwide Importance

Sphagnum-dominated peatlands occupy more than 1 percent of the Earth's surface, an enormous area equal to about one-half that of the United States. *Sphagnum* is thus one of the most abundant plants in the world. Peatlands are of particular importance in the global carbon cycle because peat stores very large amounts (about 400 gigatons, or 400 billion metric tons, on a global basis) of organic carbon that is not readily decayed to CO_2 by microorganisms. Peat is formed from the accumulation and compression of the mosses themselves, as well as the sedges, reeds, grasses, and other plants that grow among them. In Ireland and some other northern regions, dried peat is burned and used widely as industrial fuel, as well as for domestic heating. Ecologists are concerned that global warming brought about by increasing amounts of CO_2 and other gases in the atmosphere—due in large part to human activities—might result in oxidation of peatland carbon. This could further increase CO_2 levels and global temperatures.

branch leaves are green. The leaves are one cell layer thick and have a distinctive cell pattern unlike that of any other moss. Each leaf consists of large dead cells with circular wall thickenings, surrounded by narrow, green or occasionally red-pigmented, living cells containing several discoid plastids (Figure 18–21b). The walls of dead leaf cells, and the outermost stem cells, are perforated so that they readily become filled with water. As a result, the water-holding capacity of the peat mosses is up to 20 times their dry weight. By comparison, cotton absorbs only four to six times its dry weight. Both living and dead cell walls of *Sphagnum* are impregnated with decay-resistant phenolic compounds and have antiseptic properties. In addition, the peat mosses contribute to the acidity of their own environment by releasing H^+ ions; in the center of the bogs the pH is often less than 4—very unusual for a natural environment.

Because of their superior absorptive and antiseptic qualities, *Sphagnum* mosses have been used as diaper material by native peoples and, in Europe from the 1880s to World War I, as dressings for wounds and boils. *Sphagnum* is still very widely used in horticulture as a packing material for plant roots, as a planting medium, and as a soil additive. Gardeners mix peat moss with soil to increase the water-holding capacity of the soil and make it more acidic. The harvesting and processing of *Sphagnum* from peat bogs for these purposes is a multimillion dollar industry and is of ecological concern because it can result in severe degradation of some wetlands. Efforts are under way to develop techniques for regenerating peatlands because of their ecological importance.

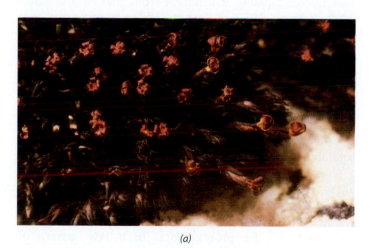

(a)

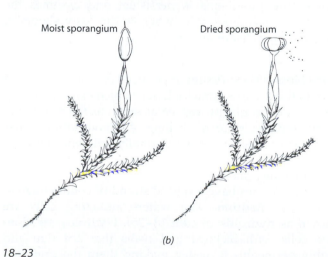

Moist sporangium   Dried sporangium

(b)

18–23
(a) Andreaea rothii *growing on granite rock in Devon, England.* **(b)** *The* Andreaea *sporangium (or capsule) contracts and splits as it dries out, allowing the spores to fall out under dry conditions.*

Granite Mosses Belong to the Class *Andreaeidae*

The genus *Andreaea* consists of about 100 species of small, blackish-green or dark reddish-brown tufted rock mosses (Figure 18–23), which in their own way are as peculiar as *Sphagnum. Andreaea* occurs in mountainous or Arctic regions, often on granite rocks—hence its common name. A second genus, *Andreaeobryum,* is restricted to northwestern Canada and adjacent Alaska and grows primarily on calcareous (calcium-containing) rocks. In this group, the protonema is unusual in having two or more rows of cells, rather than one row as in most mosses. The rhizoids are unusual in consisting of two rows of cells. The minute capsules are marked by four vertical lines of weaker cells along which the capsule splits. The capsule remains intact above and below these dehiscence lines. The resulting four valves are very sensitive to the humidity of the surrounding air, opening when it is dry—the spores can be carried far by the wind under such circumstances—and closing when it is moist. This mechanism of spore discharge, by means of slits in the capsule, is different from that of any other moss (Figure 18–23). Recent evidence indicates that *Takakia* (page 407) is closely allied to *Andreaea* and *Andreaeobryum.*

"True Mosses" Belong to the Class *Bryidae*

The class *Bryidae* contains most of the species of mosses. In this group of mosses—the "true mosses"—the branching filaments of the protonemata are composed of a single row of cells and resemble filamentous green algae (Figure 18–24). They can usually be distinguished from the green algae, however, by their slanted cross walls. Leafy gametophytes develop from minute budlike structures on the protonemata. In a few genera of mosses the protonema is persistent and assumes the major photosynthetic role, whereas the leafy shoots of the gametophyte are minute.

Many Mosses Have Tissues Specialized for Water and Food Conduction   Moss gametophytes, which exhibit varying degrees of complexity, can be as small as 0.5 millimeter or 50 centimeters or more long. All have multicellular rhizoids, and the leaves are normally only one cell layer thick except at the midrib (which is lacking in some genera). In many mosses, the stems of the gametophytes and sporophytes have a central strand of conducting tissue called **hadrom.** The water-conducting cells are known as **hydroids** (Figure 18–25). Hydroids are elongate cells with inclined end walls that are thin and highly permeable to water, making them the preferred pathways for water and solutes. Hydroids resemble the water-conducting tracheary elements of vascular plants because both lack a living protoplast at maturity (see

18–24

A "true moss." Protonema of a moss with a budlike structure, from which the leafy gametophyte will develop. Protonemata are the first stage of the gametophyte generation of mosses and some liverworts. They often resemble filamentous green algae.

Chapter 24). Unlike tracheary elements, however, hydroids lack specialized, lignin-containing wall thickenings. In some moss genera, food-conducting cells, also known as **leptoids,** surround the strand of hydroids (Figure 18–25). The food-conducting tissue is called **leptom.** The leptoids are elongate cells that have some structural and developmental similarities to the food-conducting sieve elements of seedless vascular plants (see page 427). At maturity, both cell types have inclined end walls with small pores and living protoplasts with degenerate nuclei. The conducting cells of mosses—the hydroids and leptoids—apparently are similar to those of certain fossil plants known as **protracheophytes,** which may represent an intermediate stage in the evolution of vascular plants, or **tracheophytes** (see page 434).

Sexual Reproduction in Mosses Is Similar to That of Other Bryophytes   The sexual cycle of mosses is similar to that of liverworts and hornworts in that it involves production of male and female gametangia, (Figure 18–26), an unbranched matrotrophic (maternally nourished) sporophyte, and specialized spore dispersal processes (see the moss life cycle in Figure 18–27, pages 418 and 419.)

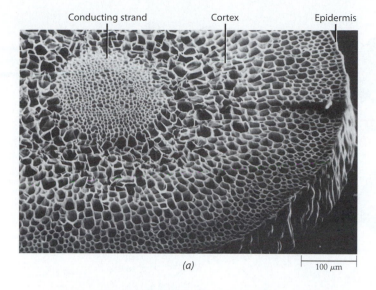

Conducting strand   Cortex   Epidermis

(a)   100 µm

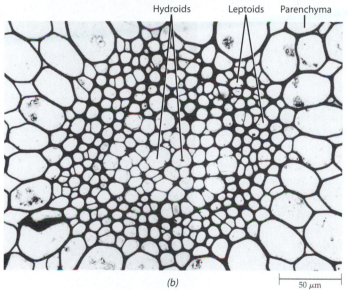

Hydroids   Leptoids   Parenchyma

(b)   50 µm

Hydroid   Leptoid   Parenchyma

(c)   20 µm

18–25

Conducting strands in the seta, or stalk, of a sporophyte of the moss
Dawsonia superba. *(a) General organization of the seta as seen in
transverse section with the scanning electron microscope. (b) Transverse
section showing the central column of water-conducting hydroids
surrounded by a sheath of food-conducting leptoids and the
parenchyma of the cortex. (c) Longitudinal section of a portion of the
central strand, showing (from left to right) hydroids, leptoids, and
parenchyma.*

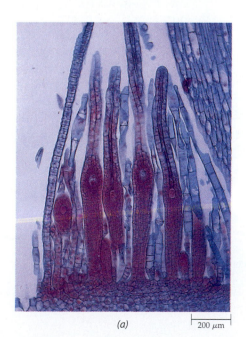

(a)   200 µm

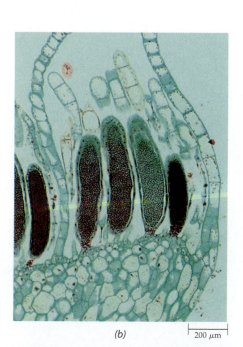

(b)   200 µm

18–26

Gametangia of Mnium, *a unisexual moss.
(a) Longitudinal section through an archegonial head showing the pink-stained archegonia surrounded by sterile structures called
paraphyses. (b) Longitudinal section through
an antheridial head showing antheridia
surrounded by paraphyses.*

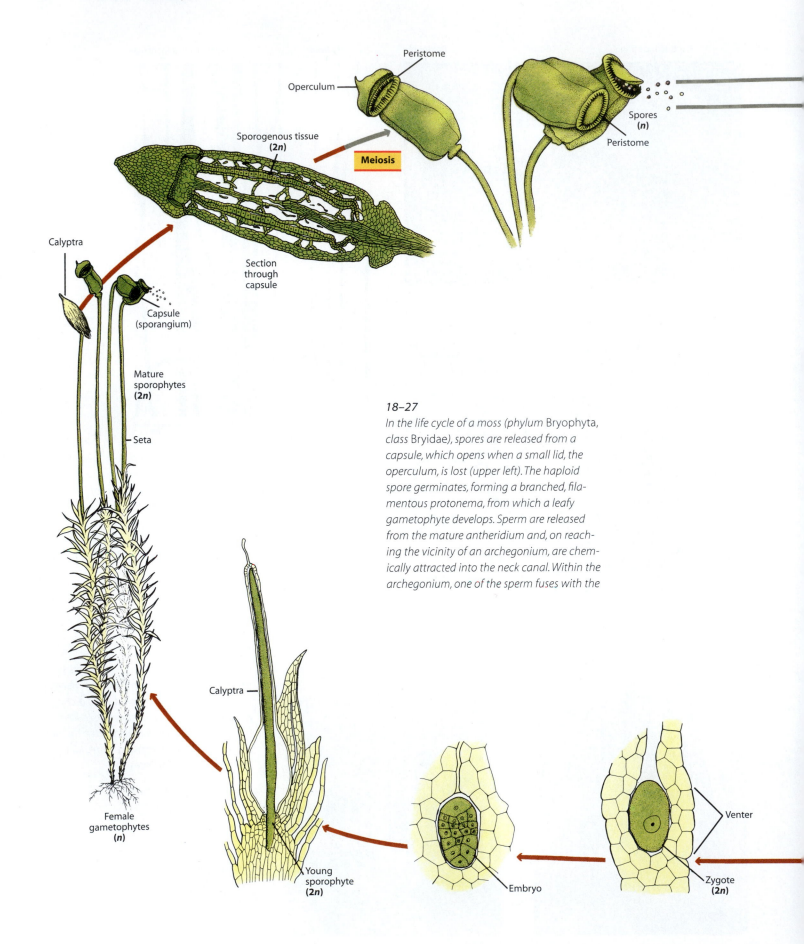

Peristome

Operculum

Spores (*n*)

Peristome

Sporogenous tissue (**2n**)

Meiosis

Calyptra

Section through capsule

Capsule (sporangium)

Mature sporophytes (**2n**)

Seta

18–27

In the life cycle of a moss (phylum Bryophyta, class Bryidae*), spores are released from a capsule, which opens when a small lid, the operculum, is lost (upper left). The haploid spore germinates, forming a branched, fila-mentous protonema, from which a leafy gametophyte develops. Sperm are released from the mature antheridium and, on reach-ing the vicinity of an archegonium, are chem-ically attracted into the neck canal. Within the archegonium, one of the sperm fuses with the*

Calyptra

Female gametophytes (*n*)

Young sporophyte (**2n**)

Embryo

Venter

Zygote (**2n**)

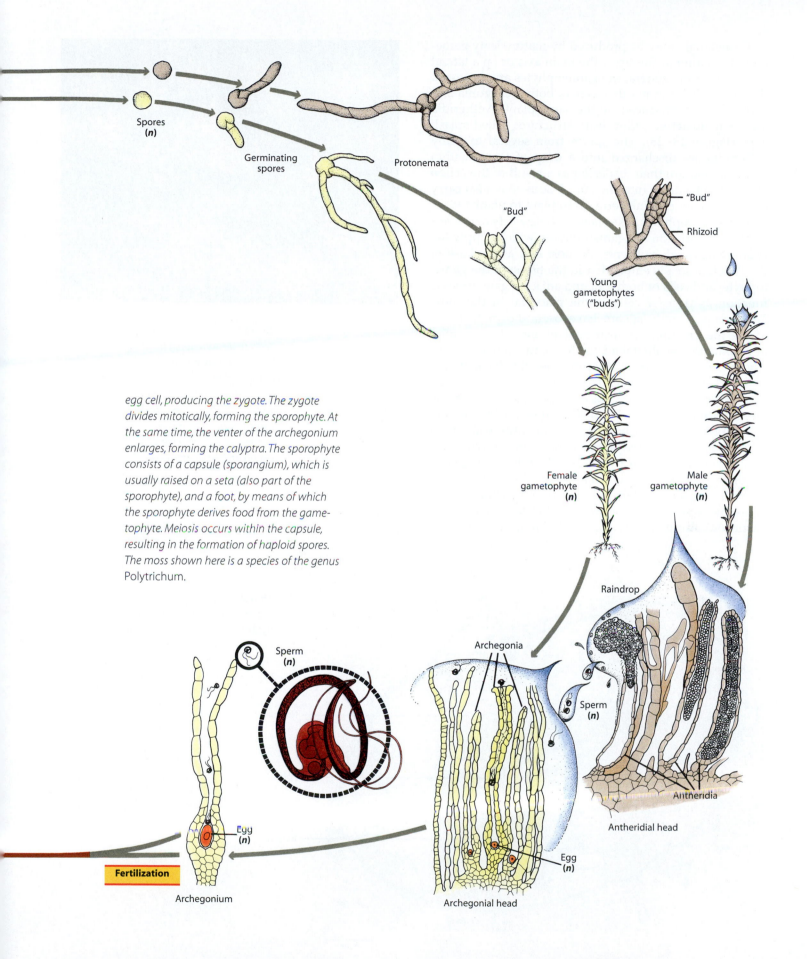

Spores
(*n*)

Germinating
spores

Protonemata

"Bud"

"Bud"

Rhizoid

Young
gametophytes
("buds")

egg cell, producing the zygote. The zygote
divides mitotically, forming the sporophyte. At
the same time, the venter of the archegonium
enlarges, forming the calyptra. The sporophyte
consists of a capsule (sporangium), which is
usually raised on a seta (also part of the
sporophyte), and a foot, by means of which
the sporophyte derives food from the gametophyte. Meiosis occurs within the capsule,
resulting in the formation of haploid spores.
The moss shown here is a species of the genus
Polytrichum.

Female
gametophyte
(*n*)

Male
gametophyte
(*n*)

Raindrop

Sperm
(*n*)

Archegonia

Sperm
(*n*)

Antheridia

Antheridial head

Egg
(*n*)

Egg
(*n*)

Fertilization

Archegonium

Archegonial head

Gametangia may be produced by mature leafy gametophytes, either at the tip of the main axis or on a lateral branch. In some genera, the gametophytes are unisexual (Figure 18–26), but in other genera both archegonia and antheridia are produced by the same plant. Antheridia are often clustered within leafy structures called splash cups (Figure 18–28). The sperm from several to many antheridia are discharged into a drop of water within each cup and are then dispersed as a result of the action of raindrops falling into the cup. Insects may also carry drops of water, rich in sperm, from plant to plant.

Moss sporophytes, like those of hornworts and liverworts, are borne on the gametophytes, which supply the sporophytes with nutrients. A short foot at the base of the stalklike seta is embedded in the tissue of the gametophyte, and cells of both foot and adjacent gametophyte function as transfer cells in the placenta. In the moss *Polytrichum*, simple sugars have been shown to move across the junction between generations. The capsules, or sporangia, usually take 6 to 18 months to reach maturity in temperate species and are generally elevated on a seta into the air, thus facilitating spore dispersal. Some mosses produce brightly colored sporangia that attract insects. Setae may reach 15 to 20 centimeters in length in a few species, but may be very short or entirely absent in other species. The setae of many moss sporophytes contain a central strand of hydroids, which in some genera is surrounded by leptoids (Figure 18–25). Stomata are normally present on the epidermis of moss sporophytes. Some moss stomata, however, are bordered by only a single, doughnut-shaped guard cell (Figure 18–29).

18–28

Leafy male gametophytes of the moss Polytrichum piliferum, *showing the mature antheridia clustered together in cup-shaped heads. The sperm are discharged into drops of water held within these leafy heads and are then splashed out of them by raindrops, sometimes reaching the vicinity of archegonia on other gametophytes (Figure 18–27).*

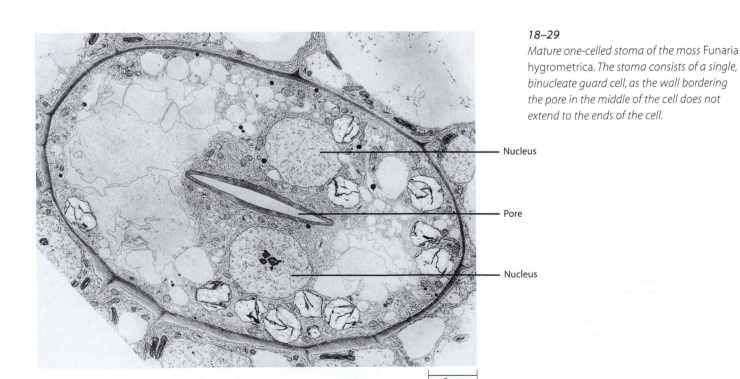

18–29

Mature one-celled stoma of the moss Funaria hygrometrica. *The stoma consists of a single, binucleate guard cell, as the wall bordering the pore in the middle of the cell does not extend to the ends of the cell.*

Nucleus

Pore

Nucleus

5 μm

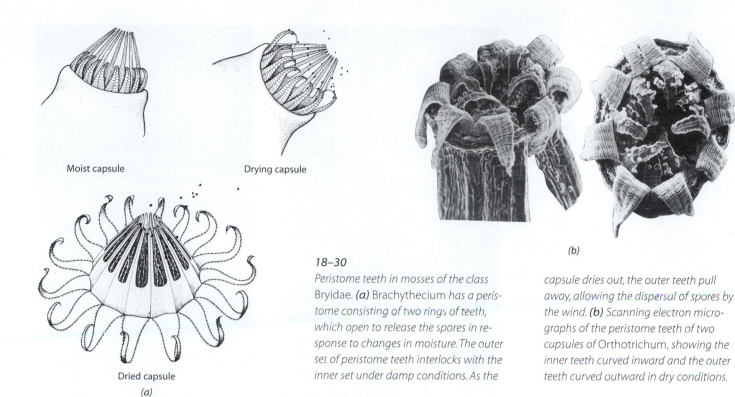

18–30

Peristome teeth in mosses of the class Bryidae. (a) Brachythecium has a peristome consisting of two rings of teeth, which open to release the spores in response to changes in moisture. The outer set of peristome teeth interlocks with the inner set under damp conditions. As the *capsule dries out, the outer teeth pull away, allowing the dispersal of spores by the wind. (b) Scanning electron micrographs of the peristome teeth of two capsules of Orthotrichum, showing the inner teeth curved inward and the outer teeth curved outward in dry conditions.*

Moist capsule

Drying capsule

Dried capsule

(a)

(b)

Generally, the cells of the young and maturing sporophyte contain chloroplasts and carry out photosynthesis. When a moss sporophyte is mature, however, it gradually loses its ability to photosynthesize and turns yellow, then orange, and finally brown. The calyptra, derived from the archegonium, is commonly lifted upward with the capsule as the seta elongates. Prior to spore dispersal, the protective calyptra falls off, and the operculum of the capsule bursts off, revealing a ring of teeth—the **peristome**—surrounding the opening (Figure 18–30). The teeth of the peristome are formed by the splitting, along a zone of weakness, of a cellular layer near the end of the capsule. In most mosses, the teeth uncurl slowly when the air is relatively dry and curl up again when it is moist. The movements of the teeth expose the spores, which are gradually released. A capsule sheds up to 50 million haploid spores, each of which is capable of giving rise to a new gametophyte. The peristome is a characteristic of the class *Bryidae* and is lacking in the other two classes of mosses. Distinctive features of the peristomes of different moss groups are used in classifying and identifying mosses.

Asexual reproduction usually occurs by fragmentation as virtually any portion of the moss gametophyte is capable of regeneration. However, some mosses produce specialized asexual reproductive structures.

Mosses Exhibit "Cushiony" or "Feathery" Growth Patterns

Two patterns of growth are common among *Bryidae* (Figure 18–31). In the "cushiony" mosses, the gametophytes are erect and little branched, usually bearing terminal sporophytes. In the second pattern, the gametophytes are highly branched and "feathery," the plants are creeping, and the sporophytes are borne laterally. This second type of growth pattern is commonly found in those mosses that hang in masses from the branches of trees in rainforests and tropical cloud forests. Organisms such as these, which grow upon other organisms but are not parasitic on them, are called **epiphytes.** Trees provide a diverse array of microhabitats that are occupied by moss and other bryophyte species. Among these microhabitats are tree bases and buttress roots, fissures and ridges of bark, irregular surfaces of twigs, depressions at the bases of branches, and the surfaces of leaves.

A number of moss genera and species are highly endemic, that is, restricted to very limited geographic areas. Many of the endemic mosses occur as epiphytes in high-altitude temperate and tropical cloud forests, where the bryophyte biodiversity is poorly known. Bryophytes also have important but poorly catalogued interactions with a variety of invertebrates, some of which live, breed, and feed preferentially in mosses. Some experts are concerned that growth of human populations may drastically alter natural environments, leading to extensive loss of bryophyte species and associated animals before many of the organisms are even described.

18–31

The two common growth forms found in the gametophytes of different genera of mosses in the class Bryidae. *(a)* "Cushiony" *form, in which the gametophytes are erect and have few branches, shown here in* Polytrichum juniperinum. *Sporophytes can be seen rising above the gametophytes. Each consists of a spore capsule atop a long, slender seta.* *(b)* "Feathery" *form, with matted, creeping gametophytes, shown here in* Thuidium abietinum.

(a)

(b)

Summary **TABLE**   Comparative Summary of Characteristics of Bryophyte Phyla

PHYLUM	NUMBER OF SPECIES	GENERAL CHARACTERISTICS OF GAMETOPHYTE	GENERAL CHARACTERISTICS OF SPOROPHYTE	HABITATS
Hepatophyta (liverworts)	6000	Dominant and free-living generation; both thalloid and leafy genera; pores in some thalloid types; unicellular rhizoids; most cells have numerous chloroplasts; many produce gemmae; protonema stage in some; growth from apical meristem	Small and nutritionally dependent on gametophyte; unbranched; consists of little more than sporangium in some genera, and of foot, short seta, and sporangium in others; phenolic materials in epidermal cell walls; lacks stomata	Mostly moist temperate and tropical; a few aquatic; often as epiphytes
Anthocerophyta (hornworts)	100	Dominant and free-living generation; thalloid; unicellular rhizoids; most have single chloroplast per cell	Small and nutritionally dependent on gametophyte; unbranched; consists of foot and long, cylindrical sporangium, with a meristem between foot and sporangium; cuticle; stomata; no specialized conducting tissues	Moist temperate and tropical
Bryophyta (mosses)	9500	Dominant and free-living generation; leafy; multicellular rhizoids; most cells have numerous chloroplasts; many produce gemmae; protonema stage that grows by marginal meristem followed by further growth from an apical meristem in *Sphagnum*; growth by apical meristem only in *Bryidae*; some species have leptoids and nonlignified hydroids	Small and nutritionally dependent on gametophyte; unbranched; consists of foot, long seta, and sporangium in *Bryidae*; phenolic materials in epidermal cell walls; stomata; some species have leptoids and nonlignified hydroids	Mostly moist temperate and tropical; some Arctic and Antarctic; many in dry habitats; a few aquatic

Summary

Plants Most Likely Evolved from a Charophyte Ancestor

Plants, collectively known as embryophytes, seem to have been derived from a charophycean green alga. The two groups share many unique features, including a phragmoplast and cell plate at cytokinesis. Molecular and other evidence strongly suggests that plants are descended from a single common ancestor and that bryophytes include the earliest living plants to have diverged from the main line of plant evolution. The earliest plants were probably similar in several respects to modern bryophytes. The characteristics they share are thalli constructed of tissues produced by an apical meristem, a life history involving alternation of heteromorphic generations, protective walled gametangia, matrotrophic embryos, and sporopollenin-walled spores.

The Bryophytes Are the Liverworts, Hornworts, and Mosses

The bryophytes consist of three phyla of structurally rather simple, small plants. Their gametophytes are always nutritionally independent of the sporophytes, whereas the sporophytes are permanently attached to the gametophytes and nutritionally dependent upon them for at least some time early in embryo development. Thus, the gametophyte is the dominant generation. The male sex organs, antheridia, and female sex organs, archegonia, both have protective jacket layers. Each archegonium contains a single egg, whereas each antheridium produces numerous sperm. The biflagellated sperm are free-swimming and require water to reach the egg. The sporophyte is typically differentiated into a foot, a seta, and a capsule, or sporangium. Maturing hornwort and moss sporophytes are green and become less nutritionally dependent on their gametophytes than those of liverworts, which usually remain completely dependent on their gametophyte.

The Bryophytes Differ from One Another in the Presence or Absence of Stomata and Conducting Tissues and in Types of Meristems

Liverworts (phylum *Hepatophyta*) differ from mosses and hornworts in lacking stomata. Hornworts (phylum *Anthocerophyta*) have a unique basal meristem and lack specialized conducting tissue. Mosses (phylum *Bryophyta*) have, at least in some groups, both specialized conducting tissue and stomata that resemble those of vascular plants. The conducting tissue of mosses, when present, consists of hydroids, water-conducting cells, and leptoids, food-conducting cells.

Bryophytes Are Important Ecologically

Bryophytes are particularly abundant and diverse in temperate rainforests and tropical cloud forests. The moss *Sphagnum* occupies more than 1 percent of the Earth's surface, is economically valuable, and plays an essential role in the global carbon cycle.

Selected Key Terms

antheridia p. 402	**matrotrophy** p. 404
archegonia p. 402	**neck canal cells** p. 404
calyptra p. 406	**operculum** p. 414
capsule, or **sporangium** p. 406	**peristome** p. 421
elaters p. 408	**placenta** p. 404
embryophytes p. 406	**protonema/protonemata** p. 407
foot p. 406	**rhizoids** p. 403
gametophores p. 408	**seta** p. 406
gemma cups p. 408	**spermatogenous cells** p. 404
gemmae p. 404	**sporopollenin** p. 402
hadrom p. 416	**tracheophytes** p. 416
hydroids p. 416	**venter** p. 404
leptoids p. 416	
leptom p. 416	

Questions

1. By means of a simple, labeled diagram, outline a generalized life cycle of a bryophyte. Explain why it is referred to as an alternation of heteromorphic generations.

2. What evidence is there in support of a charophyte ancestry for plants?

3. Bryophytes and vascular plants share a number of characters that distinguish them from charophytes and that adapt them for existence on land. What are those characters?

4. In your opinion, which of the bryophytes has the most highly developed sporophyte? Which has the most highly developed gametophyte? In each case, give the reasons for your answer.

5. Judging from the cladogram in Figure 18–4, what group of bryophytes is most closely related to vascular plants? What characters shared by vascular plants are lacking in bryophytes?

6. Describe the structural modifications related to water absorption in *Sphagnum*. Why is *Sphagnum* of such great ecological importance?

Chapter

19

Seedless Vascular Plants

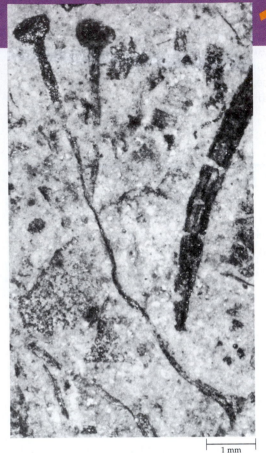

19–1
Cooksonia, the oldest known vascular plant, consisted of dichotomously branching axes. This fossil, found in New York State, is from the Late Silurian (414–408 million years ago). Its leafless aerial stems ranged up to about 6.5 centimeters long and terminated in sporangia, or spore-producing structures. These small plants probably lived in moist environments such as mud flats.

OVERVIEW

In this chapter you will learn not only about the ferns, but also about their closest relatives, often referred to as the "fern allies." All have two characteristics in common: they lack seeds but have vascular tissues, the xylem and phloem, that transport water, minerals, and sugars throughout the plant. Because of their vascular tissues, vascular plants can grow taller than the bryophytes discussed in the previous chapter. And as we will see in the following three chapters, the vascular plants became the dominant plants on Earth.

From the fossil record, we can piece together some of the major steps in the evolution of this group. The earliest vascular plants consisted only of stems, but from this humble beginning we can trace both the evolution of roots and the evolution of leaves from simple scalelike structures to large, complex fronds. Associated with these adaptations came modifications in the types of spores produced, the arrangement of vascular tissues, and the size of the gametophyte—all of which are key events in the story that will be continued in subsequent chapters.

CHECKPOINTS

By the time you finish reading this chapter, you should be able to answer the following questions:

1. What "pivotal steps" in the early history of plant evolution contributed to the success of vascular plants in their occupation of the land?

2. What explanations are there for the evolutionary origin of microphylls and megaphylls? Which groups of seedless vascular plants have microphylls? Which have megaphylls?

3. What is meant by homospory and heterospory? What are the relative sizes of the gametophytes produced by homosporous and heterosporous plants?

4. What are the characteristics of each of the following phyla of seedless vascular plants: *Rhyniophyta, Zosterophyllophyta, Lycophyta, Trimerophytophyta, Psilotophyta, Sphenophyta,* and *Pterophyta*? Which of these are exclusively fossil phyla?

5. In terms of their structure and method of development, how do eusporangia differ from leptosporangia? Which ferns are eusporangiate? Which are leptosporangiate?

Plants, like all living organisms, had aquatic ancestors. The story of plant evolution is therefore inseparably linked with their progressive occupation of the land and their increasing independence from water for reproduction. In this chapter, we shall first discuss the general features of vascular plant evolution—features linked with life on land—and then describe the features of the seedless vascular plants, those with the most generalized features. This chapter will tell the story of the club mosses, horsetails, ferns, and other groups of vascular plants that retain some of the simpler features characteristic of the earliest members of this great group of plants.

Evolution of Vascular Plants

In the previous chapter, we noted that the bryophytes and vascular plants share a number of important characters, and that together these two groups of plants—both of which have multicellular embryos—form a monophyletic lineage, the embryophytes. You will recall that it has been hypothesized that the ultimate origin of this lineage can be traced to a *Coleochaete*-like organism. Both bryophytes and vascular plants have a basically similar life cycle—an alternation of heteromorphic generations—in which the gametophyte differs from the sporophyte. An important characteristic of bryophytes, however, is the dominant gametophyte, whereas in vascular plants the sporophyte dominates (Figure 19–1). Thus, the occupation of the land by the bryophytes was undertaken with emphasis on the gamete-producing generation, which requires water to enable its motile sperm to swim to the eggs. This requirement for water undoubtedly accounts for the small size and ground-hugging form of most bryophyte gametophytes.

Relatively early in the history of plants, the evolution of efficient fluid-conducting systems, consisting of xylem and phloem, solved the problem of water and food transport throughout the plant body—a serious problem for any large organism growing on land. The ability to synthesize lignin, which is incorporated into the cell walls of supporting and water-conducting cells, was also a pivotal step in the evolution of plants. It has been suggested that the earliest plants could stand erect only by means of turgor pressure, which limited not only the environments in which they could live but also the stature of such plants. Lignin adds rigidity to the walls, making it possible for the vascularized sporophyte—the dominant generation of vascular plants—to reach great heights. Vascular plants are also characterized by the capacity to branch profusely through the activity of apical meristems located at the tips of stems and branches. In the bryophytes, on the other hand, increase in the length of the sporophyte is subapical; that is, it occurs below

the tip of the stem. In addition, each bryophyte sporophyte is unbranched and produces a single sporangium. By contrast, the branched sporophytes of vascular plants produce multiple sporangia. Picture a pine tree—a single individual—with its numerous branches and many cones, each of which contains multiple sporangia, and below it a carpet of moss gametophytes—many individuals—each bearing a single unbranched sporophyte tipped by a single sporangium.

The belowground and aboveground parts of the sporophytes of early vascular plants differed little from one another structurally, but ultimately the ancient vascular plants gave rise to more specialized plants with a more highly differentiated plant body. These plants consisted of roots, which function in anchorage and absorption of water and minerals, and of stems and leaves, which provide a system well suited to the demands of life on land—namely, the acquisition of energy from sunlight, of carbon dioxide from the atmosphere, and of water. Meanwhile, the gametophytic generation underwent a progressive reduction in size and gradually became more protected and nutritionally dependent upon the sporophyte. Finally, seeds evolved in one evolutionary line. **Seeds** are structures that provide the embryonic sporophyte with nutrients and that also help protect it from the rigors of life on land—thus providing a means of withstanding unfavorable environmental conditions. Obviously, the seedless vascular plants lack seeds. Moreover, the gametophytes of most seedless vascular plants, like those of the bryophytes, are free-living, and water is required in the environment for their motile sperm to swim to the eggs.

Because of their adaptations for existence on land, the vascular plants have been ecologically successful and are the dominant plants in terrestrial habitats. They were already numerous and diverse by the Devonian period (408–362 million years ago; see inside of front cover.) (Figure 19–2). There are nine phyla with living representatives. In addition, there are several phyla that consist entirely of extinct vascular plants. In this chapter we shall describe some of the characteristic features of vascular plants and discuss seven phyla of seedless vascular plants, three of which are extinct. In Chapters 20 through 22 we shall discuss the seed plants, which include five phyla with living representatives.

Organization of the Vascular Plant Body

The sporophytes of early vascular plants were dichotomously branched (evenly forked) axes that lacked roots and leaves. With evolutionary specialization, morphological and physiological differences arose between various parts of the plant body, bringing about the differentiation of roots, stems, and leaves—the organs of

19–2

By the Early Devonian, some 408 to 387 million years ago, small leafless plants with simple vascular systems were growing up- right on land. It is thought that their pioneer- ing ancestors were bryophyte-like plants, seen here near the water at the center, that in- vaded land sometime in the Ordovician (510
to 439 million years ago). The vascular colo- nizers shown are, from left to center, very tiny Cooksonia *with rounded sporangia,* Zosterophyllum *with clustered sporangia, and* Aglaophyton *with solitary, elongated sporangia. During the Middle Devonian (387 to 374 million years ago), larger plants with*
more complex features became established. Seen here on the right are, from back to front, Psilophyton, *a robust trimerophyte with plentiful sterile and fertile branchlets, and two lycophytes with simple microphyllous leaves,* Drepanophycus *and* Protolepidodendron.

the plant (Figure 19–3). Collectively, the roots make up the **root system,** which anchors the plant and absorbs water and minerals from the soil. The stems and leaves together make up the **shoot system,** with the stems rais- ing the specialized photosynthetic organs—the leaves— toward the sun. The vascular system conducts water and minerals to the leaves and the products of photosynthe- sis away from the leaves to other parts of the plant.

The different kinds of cells of the plant body are organ- ized into tissues, and the tissues are organized into still larger units called tissue systems. Three tissue systems, dermal, vascular, and ground, which occur in all organs of the plant, are continuous from organ to organ and re- veal the basic unity of the plant body. The **dermal tissue system** makes up the outer, protective covering of the

plant. The **vascular tissue system** comprises the conduc- tive tissues—**xylem** and **phloem**—and is embedded in the **ground tissue system** (Figure 19–3). The principal differences in the structures of root, stem, and leaf lie in the relative distribution of the vascular and ground tis- sue systems, as will be discussed in Section 5.

Primary Growth Involves the Extension of Roots and Stems, and Secondary Growth Increases Their Thickness

Primary growth may be defined as the growth that oc- curs relatively close to the tips of roots and stems. It is initiated by the apical meristems and is primarily in-

19-3

A diagram of a young sporophyte of the club moss Lycopodium lagopus, *which is still attached to its subterranean gametophyte. The dermal, vascular, and ground tissues are shown in transverse sections of (a) leaf, (b) stem, and (c) root. In all three organs, the dermal tissue system is represented by the epidermis, and the vascular tissue system, consisting of xylem and phloem, is embedded in the ground tissue system. The ground tissue in the leaf—in* Lycopodium, *a microphyll—is represented by the mesophyll, and in the stem and root by the cortex, which surrounds a solid strand of vascular tissue, or protostele. The leaf is specialized for photosynthesis, the stem for support of the leaves and for conduction, and the root for absorption and anchorage.*

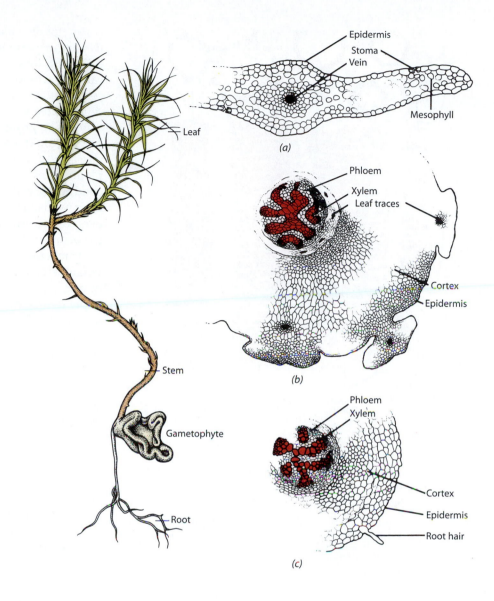

volved with extension of the plant body—often the vertical growth of a plant. The tissues arising during primary growth are known as **primary tissues,** and the part of the plant body composed of these tissues is called the **primary plant body.** Ancient vascular plants, and many contemporary ones as well, consist entirely of primary tissues.

In addition to primary growth, many plants undergo additional growth that thickens the stem and root; such growth is termed **secondary growth.** It results from the activity of lateral meristems, one of which, the **vascular cambium,** produces **secondary vascular tissues** known as secondary xylem and secondary phloem (see Figure 27–6). The production of secondary vascular tissues is commonly supplemented by the activity of a second lateral meristem, the **cork cambium,** which forms a **periderm,** composed mostly of cork tissue. The periderm replaces the epidermis as the dermal tissue system of the

plant. The secondary vascular tissues and periderm make up the **secondary plant body.** Secondary growth appeared in the Middle Devonian period, about 380 million years ago, in several unrelated groups of vascular plants.

Tracheary Elements—Tracheids and Vessel Elements—Are the Conducting Cells of the Xylem

Sieve elements, the conducting cells of the phloem, have soft walls and often collapse after they die, so they are rarely well preserved as fossils. In contrast, **tracheary elements,** the conducting cells of the xylem, have distinctive, lignified wall thickenings (Figure 19–4) and are frequently well preserved in the fossil record. Because of their various wall patterns, the tracheary elements provide valuable clues to the interrelationships of the different groups of vascular plants.

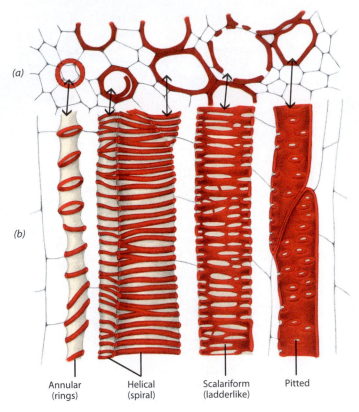

Annular (rings)   Helical (spiral)   Scalariform (ladderlike)   Pitted

19–4

Tracheary elements, the conducting cells of the xylem. A portion of the stem of Dutchman's pipe (Aristolochia) in (a) transverse and (b) longitudinal views, showing some of the distinctive types of wall thickenings exhibited by tracheary elements. Here, the wall thickenings vary, left to right, from those elements formed earliest in the development of the plant to those formed more recently.

In fossil vascular plants from the Silurian and Devonian periods, the tracheary elements are elongate cells with long, tapering ends. Such tracheary elements, called **tracheids,** were the first type of water-conducting cell to evolve; they are the only type of water-conducting cell in most vascular plants other than angiosperms and a peculiar group of gymnosperms, the *Gnetophyta* (see page 490). Tracheids not only provide channels for the passage of water and minerals, but in many modern plants they also provide support for stems. Water-conducting cells are rigid, mostly because of the lignin in their walls. This rigidity made it possible for plants to evolve an upright habit and eventually for some of them to become trees.

Tracheids are more primitive (less specialized) than **vessel elements**—the principal water-conducting cells in angiosperms. Vessel elements apparently evolved independently from tracheids in several groups of vascular plants. This is an excellent example of convergent evolution—the independent development of similar structures by unrelated or only distantly related organisms (see the essay on page 266).

Vascular Tissues Are Located in the Vascular Cylinders, or Steles, of Roots and Stems

The primary vascular tissues—primary xylem and primary phloem—and, in some vascular plants, a central column of ground tissue known as the **pith** make up the central cylinder, or **stele,** of the stem and root in the primary plant body. Several types of steles are recognized, among them the protostele, the siphonostele, and the eustele (Figure 19–5).

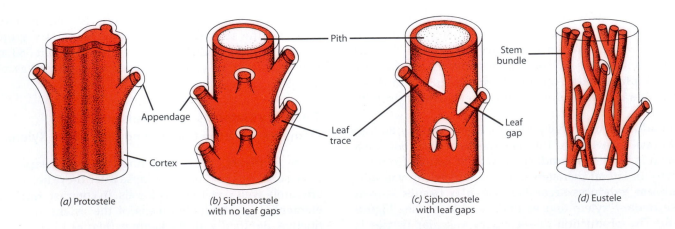

(a) Protostele   (b) Siphonostele with no leaf gaps   (c) Siphonostele with leaf gaps   (d) Eustele

19–5

Steles. (a) A protostele, with diverging traces of appendages (leaves or leaf precursors), the evolutionary precursors of leaves. (b) A siphonostele with no leaf gaps; the vascular *traces leading to the leaves simply diverge from the solid cylinder. This sort of siphonostele is found in* Selaginella, *among other plants. (c) A siphonostele with leaf gaps,* *commonly found in ferns. (d) A eustele, found in almost all seed plants. Siphonosteles and eusteles appear to have evolved independently from protosteles.*

The **protostele**—the simplest and most ancient type of stele—consists of a solid cylinder of vascular tissue in which the phloem either surrounds the xylem or is interspersed within it (Figures 19–3 and 19–5a). It is found in the extinct groups of seedless vascular plants discussed below, as well as in the *Psilotophyta* and *Lycophyta* (composed primarily of club mosses) and in the juvenile stems of some other living groups. In addition, it is the type of stele found in most roots.

The **siphonostele**—the type of stele found in the stems of most species of seedless vascular plants—is characterized by a central pith surrounded by the vascular tissue (Figure 19–5b). The phloem may form only outside the cylinder of xylem or on both sides of it. In the siphonosteles of ferns, the departure from the stem of the vascular strands leading to the leaves—the **leaf traces**—generally is marked by gaps—**leaf gaps**—in the siphonostele (as in Figure 19–5c). These leaf gaps are filled with parenchyma cells just like those that occur within and outside the vascular tissue of the siphonostele. Although the leaf traces in seed plants are associated with parenchymatous areas reminiscent of leaf gaps, these areas are generally not considered to be homologous to leaf gaps. Therefore, we will refer to these areas in seed plants as **leaf trace gaps.**

If the primary vascular cylinder consists of a system of discrete strands around a pith, as it does in almost all seed plants, the stele is called a **eustele** (Figure 19–5d). Comparative studies of living and fossil vascular plants have suggested that the eustele of seed plants evolved directly from a protostele. Eusteles appeared first among the progymnosperms, a group of spore-bearing plants that are discussed in Chapter 20 (pages 470–472). Siphonosteles evidently evolved independently from protosteles. This relationship indicates that none of the groups of seedless vascular plants with living representatives gave rise to any living seed plants.

Roots and Leaves Evolved in Different Ways

Although the fossil record reveals little information on the origins of roots as we know them today, they must have evolved from the lower, often subterranean, portions of the axis of ancient vascular plants. For the most part, roots are relatively simple structures that seem to have retained many of the ancient structural characteristics no longer present in the stems of modern plants.

Leaves are the principal lateral appendages of the stem. Regardless of their ultimate size or structure, they arise as protuberances (leaf primordia) from the apical meristem of the shoot. From an evolutionary perspective, there are two fundamentally distinct kinds of leaves—microphylls and megaphylls.

Microphylls are usually relatively small leaves that contain only a single strand of vascular tissue (Figure 19–6a). Microphylls are typically associated with stems

possessing protosteles and are characteristic of the lycophytes. The leaf traces leading to microphylls are not associated with gaps, and there is usually only a single vein in each leaf. Even though the name *microphyll* means "small leaf," some species of *Isoetes* have fairly long leaves (see Figure 19–19). In fact, certain Carboniferous and Permian lycophytes had microphylls a meter or more in length.

According to different theories, microphylls may have evolved as superficial lateral outgrowths of the stem (Figure 19–7a) or from the sterilization of sporangia in lycophyte ancestors. According to one theory, microphylls began as small, scalelike or spinelike outgrowths, called enations, devoid of vascular tissue. Gradually, rudimentary leaf traces developed, which initially extended only to the base of the enation. Finally, the leaf traces extended into the enation, resulting in formation of the primitive microphyll.

Most **megaphylls,** as the name implies, are larger than most microphylls. With few exceptions, they are associated with stems that have either siphonosteles or eusteles. The leaf traces leading to the megaphylls from siphonosteles and eusteles are associated with leaf gaps and leaf trace gaps, respectively (Figure 19–6b). Unlike the microphylls, the blade, or **lamina,** of most megaphylls has a complex system of branching veins.

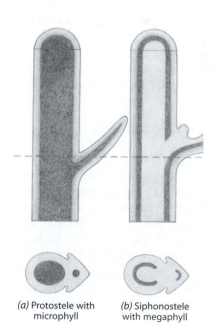

(a) Protostele with microphyll (b) Siphonostele with megaphyll

19–6

Microphylls and megaphylls. Longitudinal and transverse sections through (a) a stem with a protostele and a microphyll and (b) a stem with a siphonostele and a megaphyll, emphasizing the nodes, or regions where the leaves are attached. Note the presence of pith and a leaf gap in the stem with a siphonostele and their absence in the stem with a protostele. Microphylls are characteristic of lycophytes, while megaphylls are found in all other vascular plants.

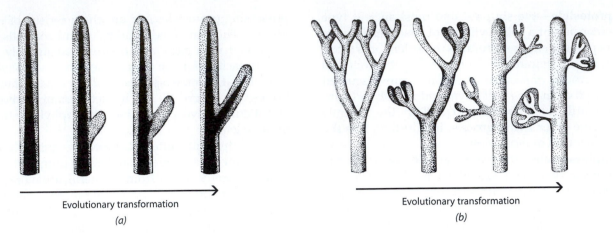

Evolutionary transformation
(a)

Evolutionary transformation
(b)

19–7

*Evolution of microphylls and megaphylls. (a)
According to one widely accepted theory,*

*microphylls evolved as outgrowths, called
enations, of the main axis of the plant.*

*(b) Megaphylls evolved by fusion of branch
systems.*

It seems likely that megaphylls evolved from entire branch systems by a series of steps similar to that shown in Figure 19–7b. The earliest plants had a leafless, dichotomously branching axis, without distinction between axis and megaphylls. Unequal branching resulted in more aggressive branches "overtopping" the weaker ones. The subordinated, overtopped lateral branches represented the beginning of leaves, and the more aggressive portions became stemlike axes. This was followed by flattening out, or "planation," of the lateral branches. The final step was fusion, or "webbing," of the separate lateral branches to form a primitive lamina.

produce bisexual gametophytes—that is, gametophytes that bear both antheridia and archegonia. Studies have revealed, however, that the gametophytes of diploid species of homosporous ferns are functionally unisexual. For example, if a sperm from a bisexual gametophyte were to fertilize an egg from that same gametophyte, the resulting sporophyte would be homozygous for all loci. Genetic studies indicate, however, that the sporophytes of most ferns are heterozygous, so they cannot have resulted from self-fertilization. Even when bisexual gametophytes of diploid species were isolated and "forced" to interbreed, they failed to produce new sporophytes.

Reproductive Systems

As mentioned previously, all vascular plants are oogamous—that is, they have large nonmotile eggs and small sperm that swim or are conveyed to the egg. In addition, all vascular plants have an alternation of heteromorphic generations, in which the sporophyte, the dominant phase of the life cycle, is larger and structurally much more complex than the gametophyte (Figure 19–8). Oogamy is clearly favored in plants, since only one of the kinds of gametes must navigate across a hostile environment outside the plant.

Homosporous Plants Produce Only One Kind of Spore, whereas Heterosporous Plants Produce Two Types

Early vascular plants produced only one kind of spore as a result of meiosis; such vascular plants are said to be **homosporous.** Among living vascular plants, homospory is found in the psilotophytes, sphenophytes (horsetails), some of the lycophytes, and almost all ferns. Upon germination, such spores have the potential to

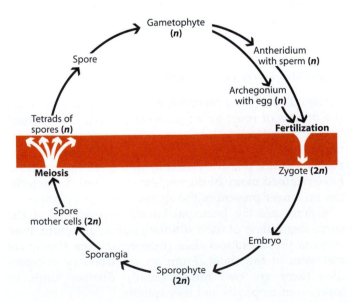

19–8

*Generalized life cycle of a vascular plant, in
which the sporophyte is the dominant phase
of the life cycle.*

Thus, rather than fertilizing their own eggs, the sperm produced by bisexual gametophytes usually fertilize the eggs of neighboring, genetically different gametophytes. Moreover, in most natural populations, although any one gametophyte may produce both antheridia and archegonia, both sex organs are not mature at the same time.

Heterospory—the production of two types of spores in two different kinds of sporangia—is found in some of the lycophytes, in a few ferns, and in all seed plants. Heterospory arose many times in unrelated groups during the evolution of vascular plants. It was common as early as the Devonian period, with the earliest record from about 370 million years ago. The two types of spores are called **microspores** and **megaspores,** and they are produced in **microsporangia** and **megasporangia,** respectively. The two types of spores are differentiated on the basis of function and not necessarily relative size. Microspores give rise to male gametophytes (microgametophytes), and megaspores give rise to female gametophytes (megagametophytes). Both of these types of unisexual gametophytes are much reduced in size compared with the gametophytes of homosporous vascular plants. Another difference is that in heterosporous plants, the gametophytes develop within the spore wall (endosporic development), while in homosporous plants the gametophytes develop outside the spore wall (exosporic development).

Over Evolutionary Time, the Gametophytes of Vascular Plants Have Become Smaller and Simpler

The relatively large gametophytes of homosporous plants are independent of the sporophyte for their nutrition, although the subterranean gametophytes of some species—such as those of *Psilotum* (see Figure 19–22) and several genera of club mosses (*Lycopodiaceae*)—are heterotrophic, depending on endomycorrhizal fungi for their nutrients. Other genera of club mosses, like most ferns and the horsetails, have free-living, photosynthetic gametophytes. In contrast, the gametophytes of many heterosporous vascular plants, and especially those of the seed plants, are dependent on the sporophyte for their nutrition.

While the initial stages of plant evolution from *Coleochaete*-like ancestors involved elaboration and modification of both the gametophyte and sporophyte, within vascular plants the evolution of the gametophyte is characterized by an overall trend toward reduction in size and complexity, and angiosperm gametophytes are the most reduced of all (see page 503). The mature megagametophyte of angiosperms commonly consists of only seven cells, one of them an egg cell. When mature, the microgametophyte contains only three cells, and two of them are sperm. Archegonia and antheridia, which are found in all seedless vascular plants, have apparently been lost in the lineage leading to angiosperms. All

but a few gymnosperms, which will be discussed in the next chapter, produce archegonia but lack antheridia. In the seedless vascular plants, the motile sperm swim through water to the archegonium. These plants must therefore grow in habitats where water is at least occasionally plentiful. In angiosperms and in most gymnosperms, entire microgametophytes (**pollen grains**) are carried to the vicinity of the megagametophyte. This transfer of the pollen grains is called **pollination.** Germination of the pollen grains produces special structures called **pollen tubes,** through which motile sperm (in cycads and *Ginkgo*) swim to or nonmotile sperm (in conifers and *Gnetales*) are transferred to the egg to achieve fertilization.

The Phyla of Seedless Vascular Plants

Several groups of seedless vascular plants flourished during the Devonian, of which the three most important have been recognized as the *Rhyniophyta, Zosterophyllophyta,* and *Trimerophytophyta.* All three groups were extinct by about the end of the Devonian, 360 million years ago. All three phyla consisted of seedless plants that were relatively simple in structure. A fourth phylum of seedless vascular plants, *Progymnospermophyta,* or progymnosperms, will be discussed in Chapter 20 because the members of this group may have been ancestral to the seed plants, both gymnosperms and angiosperms (Figure 19–9). In addition to these extinct phyla, we shall discuss in this chapter the *Psilotophyta, Lycophyta, Sphenophyta,* and *Pterophyta,* the four phyla of seedless vascular plants that have living representatives.

The overall pattern of diversification of plants may be interpreted in terms of the successive rise to dominance of four major plant groups that largely replaced the groups that were dominant earlier. In each instance, numerous species evolved in the groups that were rising to dominance. The major groups involved are as follows:

(1) Early vascular plants—characterized by relatively small stature and a simple and presumably primitive morphology. These plants included the rhyniophytes, zosterophyllophytes, and trimerophytes (Figure 19–10), which were dominant from the mid-Silurian to the mid-Devonian, from about 425 to 370 million years ago (Figure 19–2).

(2) Ferns, lycophytes, sphenophytes, progymnosperms. These more complex groups were dominant from the Late Devonian period through the Carboniferous period (see Figures 19–11 and 20–1), from about 375 to about 290 million years ago (see "Coal Age Plants" on pages 456 and 457).

(3) Seed plants arose starting in the Late Devonian period, at least 380 million years ago, and evolved many new lines by the Permian period. Gymnosperms dominated land floras throughout most of the Mesozoic era until about 100 million years ago.

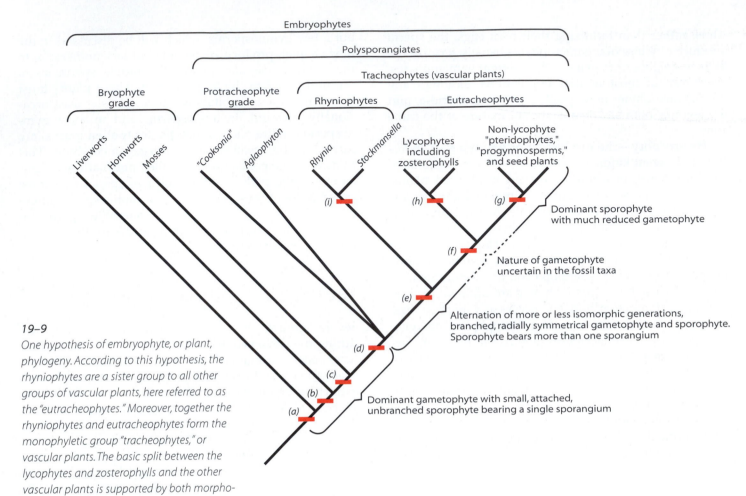

19–9

One hypothesis of embryophyte, or plant, phylogeny. According to this hypothesis, the rhyniophytes are a sister group to all other groups of vascular plants, here referred to as the "eutracheophytes." Moreover, together the rhyniophytes and eutracheophytes form the monophyletic group "tracheophytes," or vascular plants. The basic split between the lycophytes and zosterophylls and the other vascular plants is supported by both morphological and molecular data.

Characters are as follows: (a) Embryophytes: multicellular embryos; antheridia and archegonia; cuticle; preprophase band microtubules. (b) Hornwort-moss plus tracheophyte clade: ability to distinguish D-methionine; stomata. (c) Moss plus polysporangiate clade: axial gametophyte; terminal gametangia; persistent and internally differentiated sporophyte. (d) Polysporangiates: branched sporophyte with multiple sporangia; dominant sporophyte; alternation of more or less isomorphic generations. (e) Tracheophytes: tracheids with internal annular or helical wall thickenings. (f) Eutracheophytes: distinctive tracheids found in zosterophylls and lycophytes and comparable to the earliest-formed elements in some modern seedless vascular plants. (g) Non-lycophyte clade: pitted water-conducting cells; longitudinal dehiscence, or opening, of sporangia; sporangia arranged in terminal clusters. (h) Lycophyte clade: kidney-shaped sporangia; sporangia lateral on short stalks. (i) Rhyniophytes: separation or isolation layer at base of sporangium; sporangium attached directly to main axis or terminal on short branch.

(4) Flowering plants, which appeared in the fossil record about 130 million years ago. This phylum became abundant in most parts of the world within 30 to 40 million years and has remained dominant ever since.

Phylum *Rhyniophyta*

The earliest known vascular plants that we understand in detail belong to the phylum *Rhyniophyta*, a group that dates back to the mid-Silurian, at least 425 million years ago. The group became extinct in the mid-Devonian (about 380 million years ago). Earlier vascular plants were probably similar; their remains go back at least another 15 million years. Rhyniophytes were seedless plants, consisting of simple, dichotomously branching axes, or stems, with terminal sporangia. Their plant bodies were not differentiated into roots, stems, and leaves, and they were homosporous. The name of the phylum comes from the good representation of these primitive plants as fossils preserved in chert near the village of Rhynie, in Scotland.

Among the first rhyniophytes to be described is *Rhynia gwynne-vaughanii*. Probably a marsh plant, it consisted of an upright, dichotomously branched aerial system attached to a dichotomously branched rhizome (underground stem) system with rhizoids. Among the

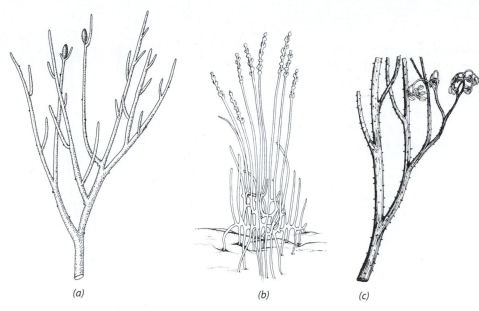

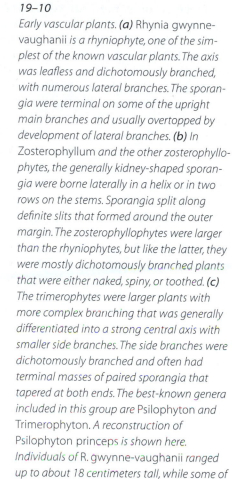

(a)   (b)   (c)

distinctive features of *R. gwynne-vaughanii* are the numerous lateral branches that arose from the dichotomized axes (Figure 19–10a) and the short branches on which the sporangia were often borne. The aerial branch system, which was about 18 centimeters in height, was covered with a cuticle and bore stomata. Lacking leaves, the aerial axes served as the photosynthetic organs.

The internal structure of *R. gwynne-vaughanii* was similar to that of many of today's vascular plants. A

19–10

Early vascular plants. (a) Rhynia gwynne-vaughanii *is a rhyniophyte, one of the simplest of the known vascular plants. The axis was leafless and dichotomously branched, with numerous lateral branches. The sporangia were terminal on some of the upright main branches and usually overtopped by development of lateral branches. (b) In* Zosterophyllum *and the other zosterophyllophytes, the generally kidney-shaped sporangia were borne laterally in a helix or in two rows on the stems. Sporangia split along definite slits that formed around the outer margin. The zosterophyllophytes were larger than the rhyniophytes, but like the latter, they were mostly dichotomously branched plants that were either naked, spiny, or toothed. (c) The trimerophytes were larger plants with more complex branching that was generally differentiated into a strong central axis with smaller side branches. The side branches were dichotomously branched and often had terminal masses of paired sporangia that tapered at both ends. The best-known genera included in this group are* Psilophyton *and* Trimerophyton. *A reconstruction of* Psilophyton princeps *is shown here. Individuals of* R. gwynne-vaughanii *ranged up to about 18 centimeters tall, while some of the trimerophytes were up to 1 meter or more in height. See also Figure 19–2.*

19–11

Reconstruction of a Carboniferous period swamp forest. See also Figure 20–1.

single layer of superficial cells—the epidermis—surrounded the photosynthetic tissue of the cortex, and the center of the axis consisted of a solid strand of xylem surrounded by one or two layers of phloemlike cells. The tracheids were different from those of most vascular plants, and although they had internal thickenings, they share some features with the water-conducting cells of mosses.

Probably the best known plant found in the Rhynie chert is the plant originally called *Rhynia major* (Figure 19–12). A more robust plant than *R. gwynne-vaughanii*, reaching heights of about 50 centimeters, it consisted of an extensive dichotomously branched rhizome system with a limited number of upright stems that branched dichotomously. All axes terminated in sporangia. For over 60 years, *R. major* was considered to be a vascular plant. Then it was shown that the cells forming the central strand of conducting tissue lack the wall thickenings typical of tracheids. Rather than being tracheids, these cells are more like the hydroids of modern mosses. For this reason, this fossil plant has been removed from the genus *Rhynia* and assigned to a new genus, *Aglaophyton*. *Aglaophyton major*, with its branched axes and multiple sporangia, may represent an intermediate stage—known as a **protracheophyte**—in the evolution of vascular plants and should probably not be retained in the phylum *Rhyniophyta*.

Cooksonia, a rhyniophyte that is believed to have inhabited mud flats, has the distinction of being the oldest known vascular plant (see Figures 1–7 and 19–1). Specimens of *Cooksonia* have been found in Wales, Scotland, Bohemia, Canada, and the United States. *Cooksonia* is the smallest and simplest vascular plant known from the fossil record. Its slender, leafless aerial stems ranged up to about 6.5 centimeters long and terminated in globose sporangia. Tracheids have been identified in the central region of the axes of *Cooksonia pertoni* from the Lower Devonian. Plants similar in form to *C. pertoni* occur in the Silurian, but it is questionable whether they contained vascular tissues. The genus *Cooksonia* may contain early simple fossil plants of diverse relationships. Some of these plants may be associated with the protracheophyte *Aglaophyton*, but others are almost certainly true vascular plants. *Cooksonia* had become extinct by the Lower Devonian period, about 390 million years ago.

Evidence from the Rhynie chert and the Lower Devonian of Germany indicates that the gametophytes of plants such as *Aglaophyton* and, by implication, *Rhynia* and *Cooksonia* (among others) were relatively large, branched structures. Some of these gametophytes apparently had water-conducting cells, cuticles, and stomata. Hence, some of these plants had an alternation of isomorphic generations, in which sporophyte and gametophyte were basically similar except for their sporangia and gametangia, respectively.

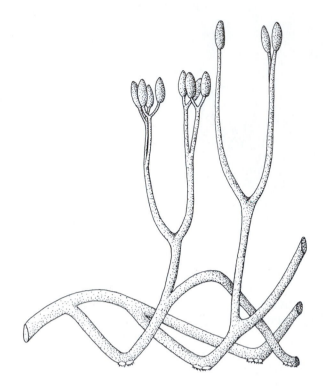

19–12

Aglaophyton major, *previously known as* Rhynia major *when it was considered to be a vascular plant. Its central strand of conducting tissue lacks tracheids but contains cells similar to the hydroids of mosses.* Aglaophyton major *may be an intermediate stage, known as a protracheophyte, in the evolution of vascular plants.*

Phylum *Zosterophyllophyta*

The fossils of a second phylum of extinct seedless vascular plants—the *Zosterophyllophyta*—have been found in strata from the Early to Late Devonian period, from about 408 to 370 million years ago. Like the rhyniophytes, the zosterophyllophytes, or zosterophylls, were leafless and dichotomously branched. The aerial stems were covered with a cuticle, but only the upper ones contained stomata, indicating that the lower branches may have been embedded in mud. In *Zosterophyllum*, it has been suggested that the lower branches frequently produced lateral branches that forked into two axes, one that grew upward, the other downward (Figure 19–10b). The downward-growing branches may have functioned like a root, permitting the plant to spread outwardly from the center by providing support. The zosterophylls are so named because of their general resemblance to the modern seagrass *Zostera*, marine angiosperms that superficially resemble grasses.

Unlike those of the rhyniophytes, the globose or kidney-shaped sporangia of the zosterophylls were borne laterally on short stalks. These plants were homosporous. The internal structure of the zosterophylls was essentially similar to that of the rhyniophytes, except that in the zosterophylls the first xylem cells to mature were located around the periphery of the xylem strand and the last to mature were located in the center. This process, known as centripetal differentiation, is the opposite of the centrifugal differentiation found in the rhyniophytes.

The zosterophylls were almost certainly the ancestors of the lycophytes. The sporangia of zosterophylls and early lycophytes are very similar and in both groups were also borne laterally. The xylem in both phyla also differentiated centripetally.

Phylum *Lycophyta*

The 10 to 15 living genera and approximately 1000 living species of *Lycophyta* are the representatives of an evolutionary line that extends back to the Devonian period. The progenitors of the lycophytes were almost certainly early zosterophylls (Figure 19–9). There are a number of orders of lycophytes, and at least three of the extinct ones included small to large trees. The three orders of living lycophytes, however, consist entirely of herbs; each order includes a single family. All lycophytes, living and fossil, possess microphylls, and this type of leaf, which shows relatively little diversity in form, is highly

characteristic of the phylum. Tree lycophytes were among the dominant plants of the coal-forming forests of the Carboniferous period (see pages 456 and 457). Most lines of woody lycophytes—those lycophytes that exhibited secondary growth—became extinct before the end of the Paleozoic era, 248 million years ago.

The Club Mosses Belong to the Family *Lycopodiaceae*

Perhaps the most familiar living lycophytes are the club mosses, family *Lycopodiaceae* (see Figure 13-15c). All but two genera of living lycophytes belong to this family, most members of which were formerly grouped in the collective genus "*Lycopodium*." Seven of these genera are represented in the United States and Canada, but most of the estimated 400 species in the family are tropical. The taxonomic boundaries of the predominantly tropical genera of this family are poorly understood, and as many as 15 genera may ultimately be recognized. *Lycopodiaceae* extend from Arctic regions to the tropics, but they rarely form conspicuous elements in any plant community. Most tropical species, many of which belong to the genus *Phlegmariurus*, are epiphytes and thus rarely seen, but several of the temperate species form mats that may be evident on forest floors. Because they are evergreen, they are most noticeable in winter.

The sporophyte of most genera of *Lycopodiaceae* consists of a branching rhizome from which aerial branches and roots arise (Figure 19–14, pages 436 and 437). Both stem and root are protostelic (Figure 19–13). The microphylls of *Lycopodiaceae* are usually spirally arranged, but

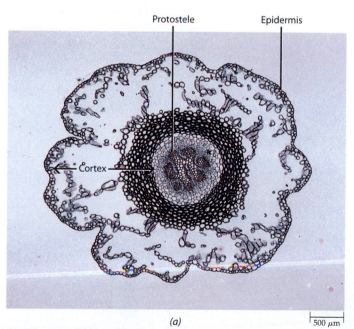

(a)

500 μm

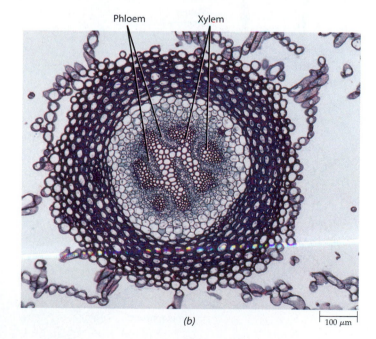

(b)

100 μm

19–13
Both the stem and root of Lycopodiaceae are protostelic. **(a)** Transverse section of Diphasiastrum complanatum *stem,* showing mature tissues. Note the large air spaces in the cortex, which surrounds the central protostele. **(b)** Detail of protostele of D. complanatum, *showing xylem and phloem. See also Figure 19–3.*

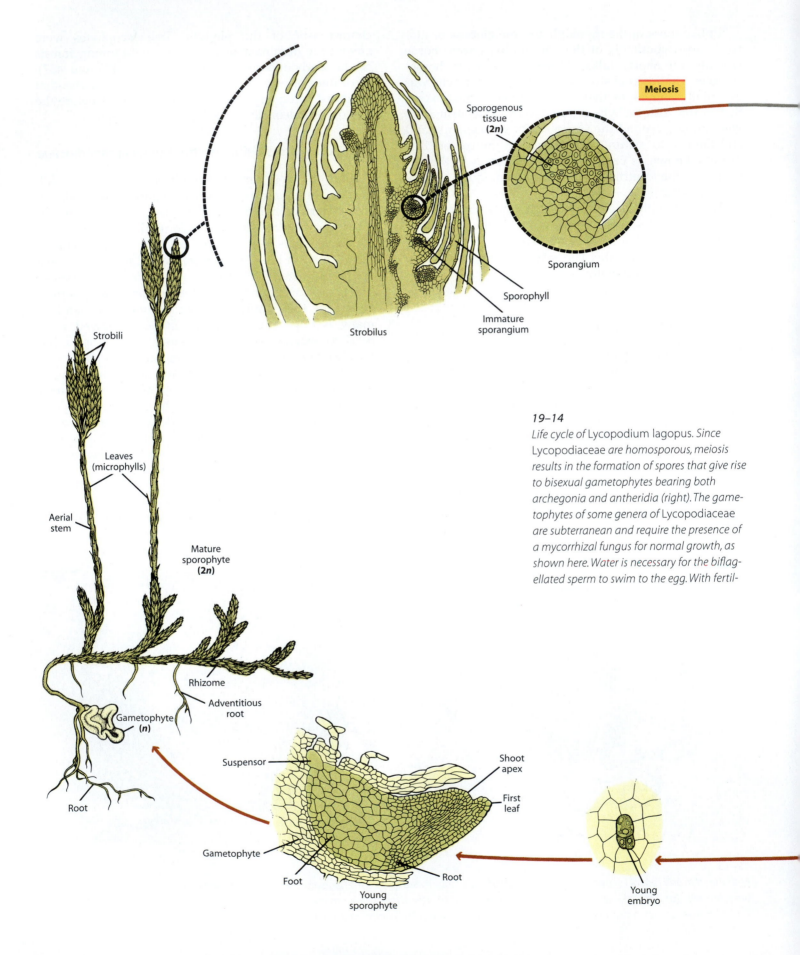

Meiosis

Sporogenous tissue (**2n**)

Sporangium

Sporophyll

Immature sporangium

Strobilus

Strobili

Leaves (microphylls)

Aerial stem

Mature sporophyte (**2n**)

Rhizome

Adventitious root

Gametophyte (**n**)

Root

Suspensor

Shoot apex

First leaf

Gametophyte

Foot

Root

Young sporophyte

Young embryo

19–14

Life cycle of Lycopodium lagopus. *Since Lycopodiaceae are homosporous, meiosis results in the formation of spores that give rise to bisexual gametophytes bearing both archegonia and antheridia (right). The gametophytes of some genera of Lycopodiaceae are subterranean and require the presence of a mycorrhizal fungus for normal growth, as shown here. Water is necessary for the biflagellated sperm to swim to the egg. With fertil-*

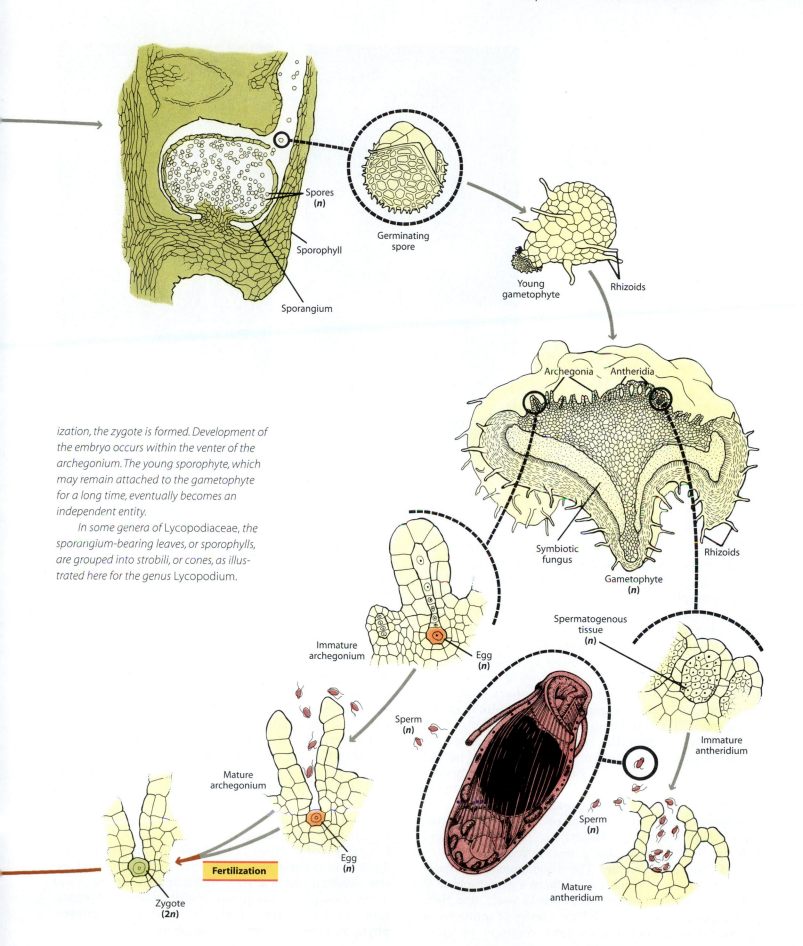

Spores
(n)

Sporophyll

Germinating
spore

Sporangium

Young
gametophyte

Rhizoids

Archegonia   Antheridia

Symbiotic
fungus

Rhizoids

Gametophyte
(n)

*ization, the zygote is formed. Development of
the embryo occurs within the venter of the
archegonium. The young sporophyte, which
may remain attached to the gametophyte
for a long time, eventually becomes an
independent entity.*

*In some genera of Lycopodiaceae, the
sporangium-bearing leaves, or sporophylls,
are grouped into strobili, or cones, as illustrated here for the genus Lycopodium.*

Immature
archegonium

Egg
(n)

Spermatogenous
tissue
(n)

Sperm
(n)

Immature
antheridium

Mature
archegonium

Egg
(n)

Sperm
(n)

Fertilization

Mature
antheridium

Zygote
(2n)

19–15

Sporophylls and strobili. (a) Huperzia lucidula *is a representative of those genera of*
Lycopodiaceae *that lack differentiated
strobili. The sporangia (small yellow structures
along stem) are borne in the axils of fertile
microphylls known as sporophylls. Areas of
fertile sporophylls alternate with regions of
sterile microphylls. (b) The terminal branches
in* Lycopodium lagopus *are terminated by
sporophylls grouped into strobili.*

(a)

(b)

they appear opposite or whorled in some members of the group. *Lycopodiaceae* are homosporous; the sporangia occur singly on the upper surface of fertile microphylls called **sporophylls,** which are modified leaves or leaflike organs that bear sporangia. In *Huperzia* (Figure 19–15a) and *Phlegmariurus*, the sporophylls are similar to ordinary microphylls and are interspersed among the sterile microphylls. In the other genera of *Lycopodiaceae* found in the United States and Canada, including *Diphasiastrum* (see Figure 13–15c) and *Lycopodium*, nonphotosynthetic sporophylls are grouped into **strobili,** or cones, at the ends of the aerial branches (Figure 19–15b).

Upon germination, the spores of *Lycopodiaceae* give rise to bisexual gametophytes that, depending on the genus, are either green, irregularly lobed masses (*Lycopodiella, Palhinhaea,* and *Pseudolycopodiella,* among the genera that occur in the United States and Canada) or subterranean, nonphotosynthetic, mycorrhizal structures (*Diphasiastrum, Huperzia, Lycopodium,* and *Phlegmariurus,* among the genera represented in the United States and Canada). The development and maturation of archegonia and antheridia in a gametophyte of *Lycopodiaceae* may require from 6 to 15 years, and their gametophytes may even produce a series of sporophytes in successive archegonia as they continue to grow.

Despite the fact that the gametophytes are bisexual, it is known that, in at least some species of lycophytes, self-fertilization rates are very low. The gametophytes of these species predominantly cross-fertilize.

Water is required for fertilization in *Lycopodiaceae*. The biflagellated sperm swim through water to the archegonium. Following fertilization, the zygote develops into an embryo, which grows within the venter of the archegonium. The young sporophyte may remain attached to the gametophyte for a long time, but eventually it becomes independent. The life cycle of *Lycopodium lagopus*, a representative of those *Lycopodiaceae* that have a subterranean, mycorrhizal gametophyte and form a strobilus, is illustrated in Figure 19–14.

Among the genera of *Lycopodiaceae* found in the United States and Canada, *Huperzia* (Figure 19–15a), the fir mosses, consists of 7 species; *Lycopodium* (Figure 19–14), the tree club mosses, of 5; *Diphasiastrum* (Figure 13–15c), the club mosses and running pines, of 11; and *Lycopodiella* of 6. These genera, and the others now recognized in *Lycopodiaceae*, differ from one another in various technical characteristics, including the arrangement of the sporophylls, the presence of rhizomes and organization of the vegetative body, the nature of the gametophyte, and basic chromosome numbers.

(a)

(b)

(c)

(d)

19–16

Representative Selaginella. *(a)* Selaginella lepidophylla, *the resurrection plant, a plant that becomes completely dried out when water is not available but quickly revives following a rain. This plant was growing in Big Bend National Park, in Texas. (b)* Selaginella rupestris, *the rock spikemoss, with strobili. (c)* Selaginella kraussiana, *a prostrate, creeping plant. Adventitious roots can be seen arising from the stems. (d)* Selaginella willdenovii, *from the Old World tropics. Shade-loving, it climbs to 7 meters and has peacock-blue leaves with a metallic sheen. Note the clearly evident rhizomes.*

The Resurrection Plant Belongs to the Family *Selaginellaceae*

Among the living genera of lycophytes, *Selaginella,* the only genus of the family *Selaginellaceae,* has the most species, about 700. Most of them are tropical in distribution. Many grow in moist places, although a few inhabit deserts, becoming dormant during the driest part of the year. Among the latter is the so-called resurrection plant, *Selaginella lepidophylla,* which occurs in Texas, New Mexico, and Mexico (Figure 19–16a).

Basically, the herbaceous sporophyte of *Selaginella* is similar to that of some *Lycopodiaceae* in that it bears microphylls and its sporophylls are arranged in strobili (Figure 19–16b). Unlike *Lycopodiaceae,* however, *Selaginella* has a small, scalelike outgrowth, called a **ligule,** near the base of the upper surface of each microphyll and sporophyll (Figure 19–17). The stem and root are protostelic (Figure 19–18).

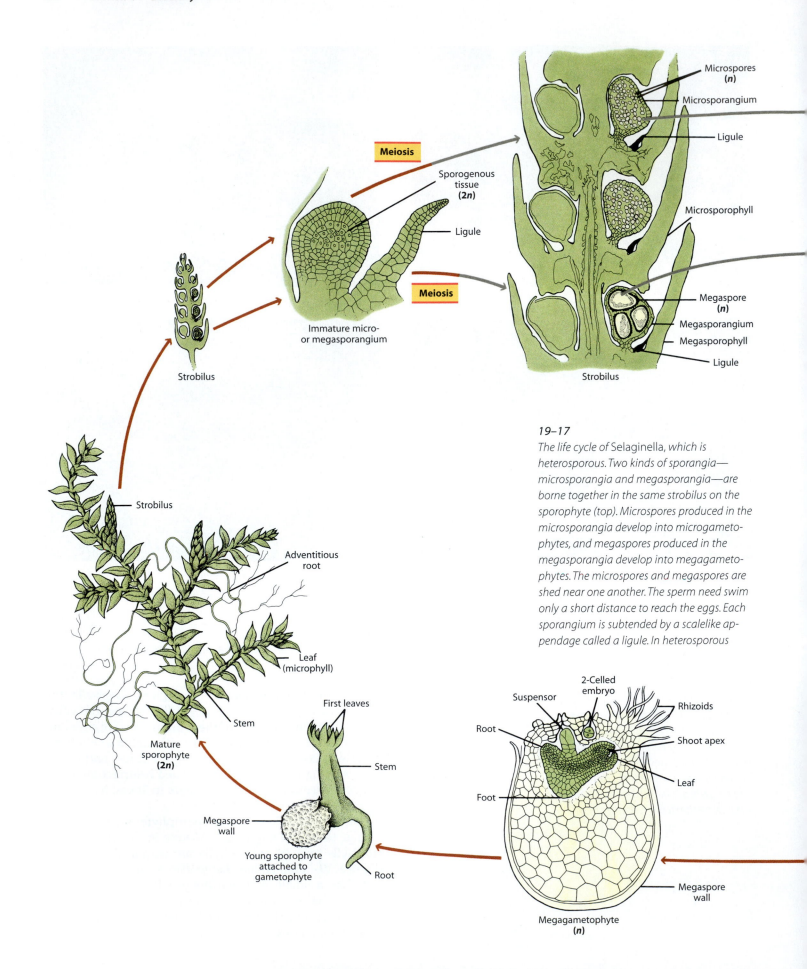

Meiosis

Microspores (*n*)

Microsporangium

Ligule

Sporogenous tissue (**2n**)

Ligule

Microsporophyll

Meiosis

Megaspore (*n*)

Megasporangium

Megasporophyll

Ligule

Immature micro- or megasporangium

Strobilus

Strobilus

Strobilus

19–17

The life cycle of Selaginella, *which is heterosporous. Two kinds of sporangia— microsporangia and megasporangia—are borne together in the same strobilus on the sporophyte (top). Microspores produced in the microsporangia develop into microgameto- phytes, and megaspores produced in the megasporangia develop into megagameto- phytes. The microspores and megaspores are shed near one another. The sperm need swim only a short distance to reach the eggs. Each sporangium is subtended by a scalelike ap- pendage called a ligule. In heterosporous*

Adventitious root

Leaf (microphyll)

Stem

Mature sporophyte (**2n**)

First leaves

Stem

Megaspore wall

Young sporophyte attached to gametophyte

Root

Suspensor

2-Celled embryo

Rhizoids

Root

Shoot apex

Leaf

Foot

Megaspore wall

Megagametophyte (*n*)

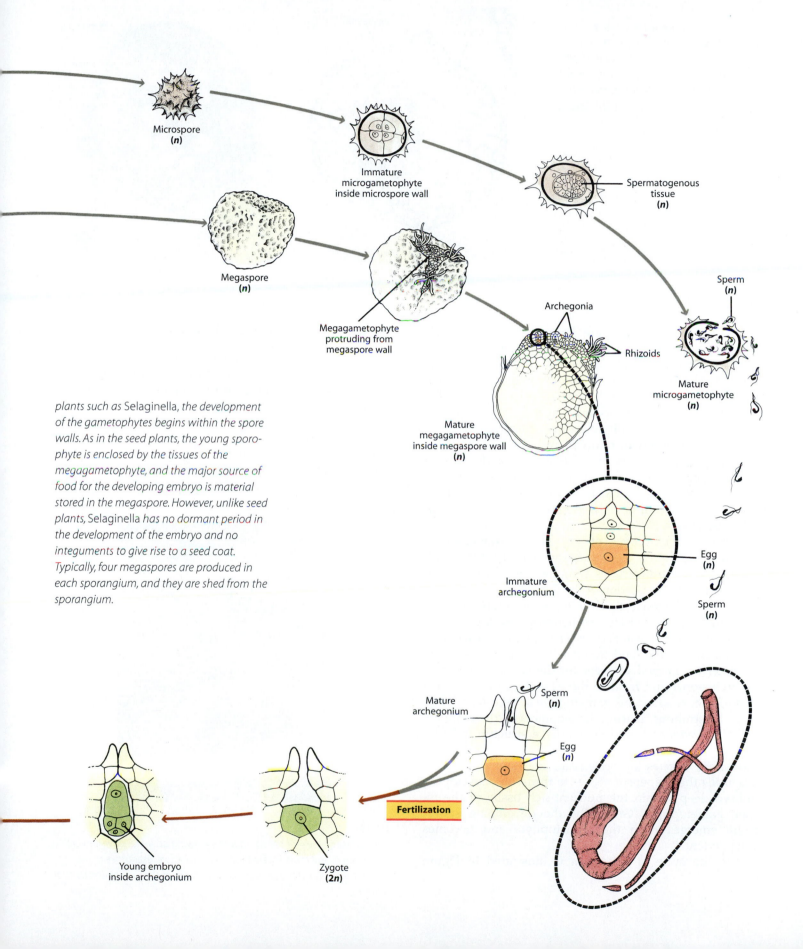

Microspore
(*n*)

Immature
microgametophyte
inside microspore wall

Spermatogenous
tissue
(*n*)

Megaspore
(*n*)

Megagametophyte
protruding from
megaspore wall

Sperm
(*n*)

Archegonia

Rhizoids

Mature
microgametophyte
(*n*)

Mature
megagametophyte
inside megaspore wall
(*n*)

Egg
(*n*)

Immature
archegonium

Sperm
(*n*)

plants such as Selaginella, *the development
of the gametophytes begins within the spore
walls. As in the seed plants, the young sporophyte is enclosed by the tissues of the
megagametophyte, and the major source of
food for the developing embryo is material
stored in the megaspore. However, unlike seed
plants,* Selaginella *has no dormant period in
the development of the embryo and no
integuments to give rise to a seed coat.
Typically, four megaspores are produced in
each sporangium, and they are shed from the
sporangium.*

Mature
archegonium

Sperm
(*n*)

Egg
(*n*)

Fertilization

Young embryo
inside archegonium

Zygote
(**2n**)

19–18.

Selaginella *protostele.* (**a**) *Transverse section of stem, showing mature tissues. The protostele is suspended in the middle of the hollow stem by elongate cortical cells (endodermal cells), called trabeculae. Only a portion of each trabecula can be seen here.* (**b**) *Detail of protostele.*

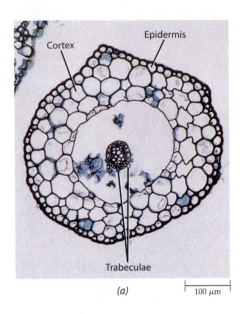

Cortex

Epidermis

Trabeculae

(a)   100 μm

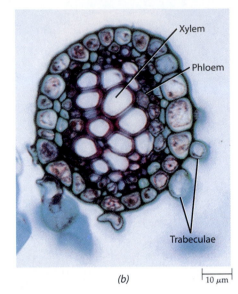

Xylem

Phloem

Trabeculae

(b)   10 μm

Whereas *Lycopodiaceae* are homosporous, *Selaginella* is heterosporous, with unisexual gametophytes. Each sporophyll bears a single sporangium on its upper surface. Megasporangia are borne by **megasporophylls,** and microsporangia are borne by **microsporophylls.** Both kinds of sporangia occur in the same strobilus.

The male gametophytes (microgametophytes) in *Selaginella* develop within the microspore, and they lack chlorophyll. At maturity the male gametophyte consists of a single prothallial, or vegetative, cell and an antheridium, which gives rise to many biflagellated sperm. The microspore wall must rupture in order for the sperm to be liberated.

During development of the female gametophyte (megagametophyte), the megaspore wall ruptures, and the gametophyte protrudes through the rupture to the outside. This is the portion of the female gametophyte in which the archegonia develop. It has been reported that the female gametophytes sometimes develop chloroplasts, although most *Selaginella* gametophytes derive their nutrition from food stored within the megaspores.

Water is required in order for the sperm to swim to the archegonia and fertilize the eggs. Commonly, fertilization occurs after the gametophytes have been shed from the strobilus. During development of the embryos in both *Lycopodiaceae* and *Selaginella*, a structure called a **suspensor** is formed. Although inactive in *Lycopodiaceae* and some species of *Selaginella*, in other species of *Selaginella* the suspensor serves to thrust the developing embryo deep within the nutrient-rich tissue of the female gametophyte. Gradually, the developing sporophyte emerges from the gametophyte and becomes independent.

The life cycle of *Selaginella* is illustrated in Figure 19–17.

The Quillworts Belong to the Family *Isoetaceae*

The only genus of the family *Isoetaceae* is *Isoetes*, the quillworts. *Isoetes* is the nearest living relative of the ancient tree lycophytes of the Carboniferous (see page 456). Plants of *Isoetes* may be aquatic, or they may grow in pools that become dry at certain seasons. The sporophyte of *Isoetes* consists of a short, fleshy underground stem (corm) bearing quill-like microphylls on its upper surface and roots on its lower surface (Figure 19–19). In *Isoetes*, each leaf is a potential sporophyll.

19–19

Isoetes storkii. *View of the sporophyte showing quill-like leaves (microphylls), stem, and roots.* Isoetes *is the last living representative of the group that included the extinct tree lycophytes of the Carboniferous coal swamps.*

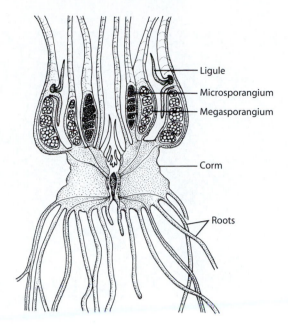

19–20

Diagram of a vertical section of an Isoetes *plant. Leaves are borne on the upper surface, and roots on the lower surface, of a short, fleshy underground stem, or corm. Some leaves (megasporophylls) bear megasporangia, and others (microsporophylls) bear microsporangia. The microsporophylls are located nearer the center of the plant.*

Like *Selaginella*, *Isoetes* is heterosporous. The megasporangia are borne at the base of megasporophylls, and the microsporangia are borne at the base of microsporophylls, similar to the megasporophylls but located nearer the center of the plant (Figure 19–20). A ligule is present just above the sporangium of each sporophyll.

One of the distinctive features of *Isoetes* is the presence of a specialized cambium that adds secondary tissues to the corm. Externally the cambium produces only parenchyma tissue, whereas internally it produces a peculiar vascular tissue consisting of sieve elements, parenchyma cells, and tracheids in varying proportions.

Some species of *Isoetes* (assigned to another genus, *Stylites*, by some) from high elevations in the tropics have the unique characteristic of obtaining their carbon for photosynthesis from the sediment in which they grow rather than from the atmosphere. The leaves of these plants lack stomata, have a thick cuticle, and carry on essentially no gas exchange with the atmosphere. Like at least some of the other species of *Isoetes* in which the plants dry out for part of the year, these species have CAM photosynthesis (page 147).

Phylum *Trimerophytophyta*

The phylum *Trimerophytophyta*, which probably evolved directly from the rhyniophytes, most likely contains plants of diverse evolutionary relationships and seems to represent the ancestral stock of the ferns, the progymnosperms, and perhaps the horsetails as well. The trimerophytes, which were larger and more complex plants than the rhyniophytes or zosterophyllophytes (Figure 19–10c), first appeared in the Early Devonian period about 395 million years ago and had become extinct by the end of the mid-Devonian, about 20 million years later—a relatively short period of existence.

Although generally larger and evolutionarily more specialized than the rhyniophytes, trimerophytes still lacked leaves. Branching, however, was more complex, with the main axis forming lateral branch systems that dichotomized several times. The trimerophytes, like the rhyniophytes and zosterophyllophytes, were homosporous. Some of their smaller branches terminated in elongate sporangia, while others were entirely vegetative. Besides their more complex branching pattern, the trimerophytes had a more massive vascular strand than the rhyniophytes. Together with a broad band of thick-walled cells in the cortex, the large vascular strand probably was capable of supporting fairly large plants, over a meter in height. As in the rhyniophytes, the xylem of the trimerophytes differentiated centrifugally. The name of the phylum comes from the Greek words *tri-*, *meros*, and *phyton*, meaning "three-parted plant," because of the organization of the secondary branches into three rows in the genus *Trimerophyton*.

Phylum *Psilotophyta*

The phylum *Psilotophyta* includes two living genera, *Psilotum* and *Tmesipteris*. *Psilotum*, the whisk fern, is tropical and subtropical in distribution. In the United States, it occurs in Alabama, Arizona, Florida, Hawaii, Louisiana, North Carolina, and Texas, as well as Puerto Rico, and it is a common greenhouse weed. *Tmesipteris* is restricted in distribution to Australia, New Caledonia, New Zealand, and other regions of the South Pacific. Although both genera are very simple plants that resemble the rhyniophytes in some aspects of their basic structure, it seems likely that they may reflect an early-diverging lineage related to living ferns. Their simple structure appears to have resulted from reduction from more complex ancestors.

Psilotum is unique among living vascular plants in that it lacks both roots and leaves. The sporophyte consists of a dichotomously branching aerial portion with small scalelike outgrowths and a branching underground portion, or a system of rhizomes with many rhi-

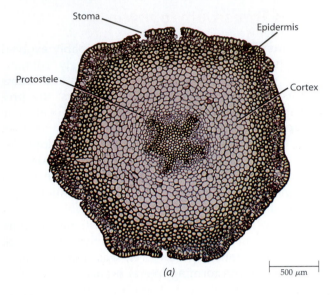

19–21
Psilotum nudum. *(a) Transverse section of stem, showing mature tissues. (b) Detail of protostele, showing xylem and phloem.*

zoids (see the mature sporophyte at the left in Figure 19–24). A symbiotic fungus—an endomycorrhizal zygomycete (page 341)—is present in the outer cortical cells of the rhizomes. *Psilotum* has a protostele (Figure 19–21).

Psilotum is homosporous, the spores being produced in sporangia that are generally aggregated in groups of three on the ends of short, lateral branches. Upon germination, the spores give rise to bisexual gametophytes, which resemble portions of the rhizome (Figure 19–22).

19–22
(a) The subterranean gametophyte of Psilotum nudum. *The gametophytes of psilotophytes are bisexual; that is, they bear both antheridia and archegonia. (b)* Psilotum nudum, *known locally as the moa plant, growing on a 1955 lava flow on the island of Hawaii. The yellow sporangia are clearly evident.*

(a)

(b)

19–23

(a) Tmesipteris parva *growing on the trunk of the tree fern* Cyathea australis *in New South Wales, Australia.* *(b)* Tmesipteris lanceolata, *in New Caledonia, an island of the southwest Pacific.*

(a)   (b)

Like the rhizome, the subterranean gametophyte contains a symbiotic fungus. In addition, some gametophytes contain vascular tissue. The sperm of *Psilotum* are multiflagellated and require water to swim to the egg. Initially the sporophyte is attached to the gametophyte by a foot, a structure that absorbs nutrients from the gametophyte. Eventually the sporophyte becomes detached from the foot, which remains embedded in the gametophyte. The life cycle of *Psilotum* is illustrated in Figure 19–24 (pages 446 and 447).

Tmesipteris grows as an epiphyte on tree ferns and other plants (Figure 19–23) and in rock crevices. The leaflike appendages of *Tmesipteris* are larger than the scalelike outgrowths of *Psilotum,* but in other respects *Tmesipteris* is essentially similar to *Psilotum.*

Phylum *Sphenophyta*

Like the *Lycophyta,* the *Sphenophyta* extend back to the Devonian period. The sphenophytes reached their maximum abundance and diversity later in the Paleozoic era, about 300 million years ago. During the Late Devonian and Carboniferous periods, they were represented by the calamites (see page 456), a group of trees that reached 18 meters or more in height, with a trunk that could be more than 45 centimeters thick. Today the *Sphenophyta* are represented by a single herbaceous

genus, *Equisetum* (Figure 19–25), which consists of 15 species. Since *Equisetum* is essentially identical to *Equisetites,* a plant that appeared about 300 million years ago, in the Carboniferous period, *Equisetum* may be the oldest surviving genus of plants on Earth.

The species of *Equisetum* are known as the "horsetails"; they are widespread in moist or damp places, by streams, and along the edge of woods. Horsetails are easily recognized because of their conspicuously jointed stems and rough texture. The small, scalelike leaves, or microphylls, are whorled at the nodes. When present, the branches arise laterally at the nodes and alternate with the leaves. The internodes (the portions of the stems between successive nodes) are ribbed, and the ribs are tough and strengthened with siliceous deposits in the epidermal cells. Horsetails have been used to scour pots and pans, particularly in colonial and frontier times, and have thus earned the name "scouring rushes." The roots are adventitious, arising at the nodes of the rhizomes, which are important in vegetative propagation.

The aerial stems of *Equisetum* arise from branching underground rhizomes, and, although the plants may die back during unfavorable seasons, the rhizomes are perennial. The aerial stem is complex anatomically (Figure 19–26). At maturity, its internodes contain a hollow pith surrounded by a ring of smaller canals called **carinal canals.** Each of these smaller canals is associated with a strand of xylem and phloem.

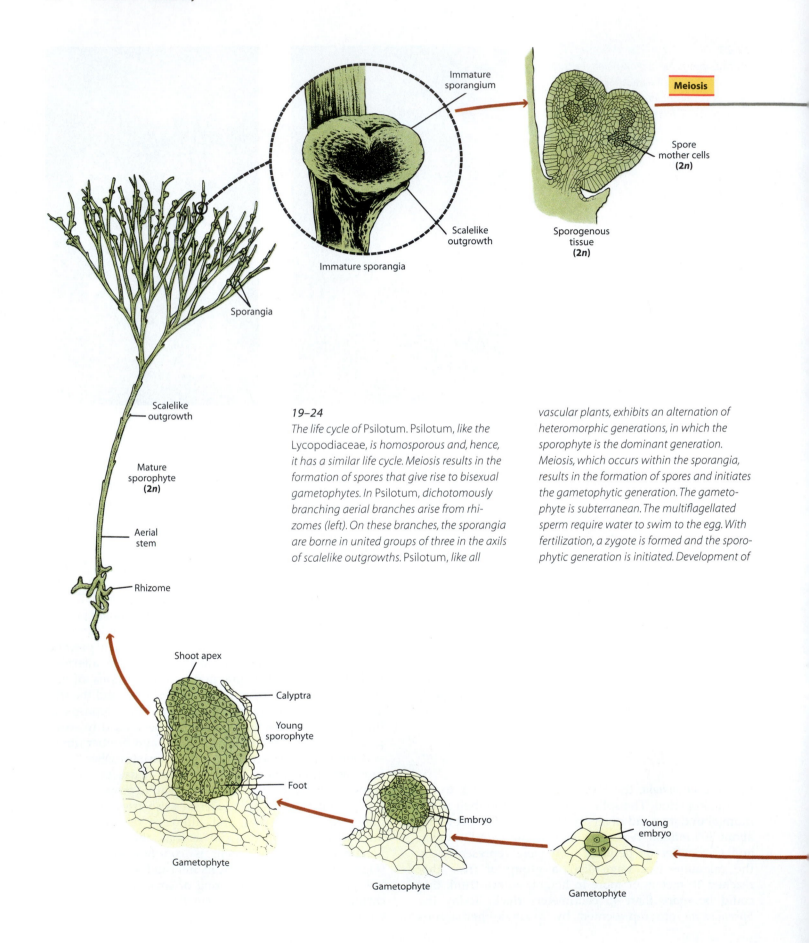

19–24
The life cycle of Psilotum. Psilotum, *like the* Lycopodiaceae, *is homosporous and, hence, it has a similar life cycle. Meiosis results in the formation of spores that give rise to bisexual gametophytes. In* Psilotum, *dichotomously branching aerial branches arise from rhizomes (left). On these branches, the sporangia are borne in united groups of three in the axils of scalelike outgrowths.* Psilotum, *like all vascular plants, exhibits an alternation of heteromorphic generations, in which the sporophyte is the dominant generation. Meiosis, which occurs within the sporangia, results in the formation of spores and initiates the gametophytic generation. The gametophyte is subterranean. The multiflagellated sperm require water to swim to the egg. With fertilization, a zygote is formed and the sporophytic generation is initiated. Development of*

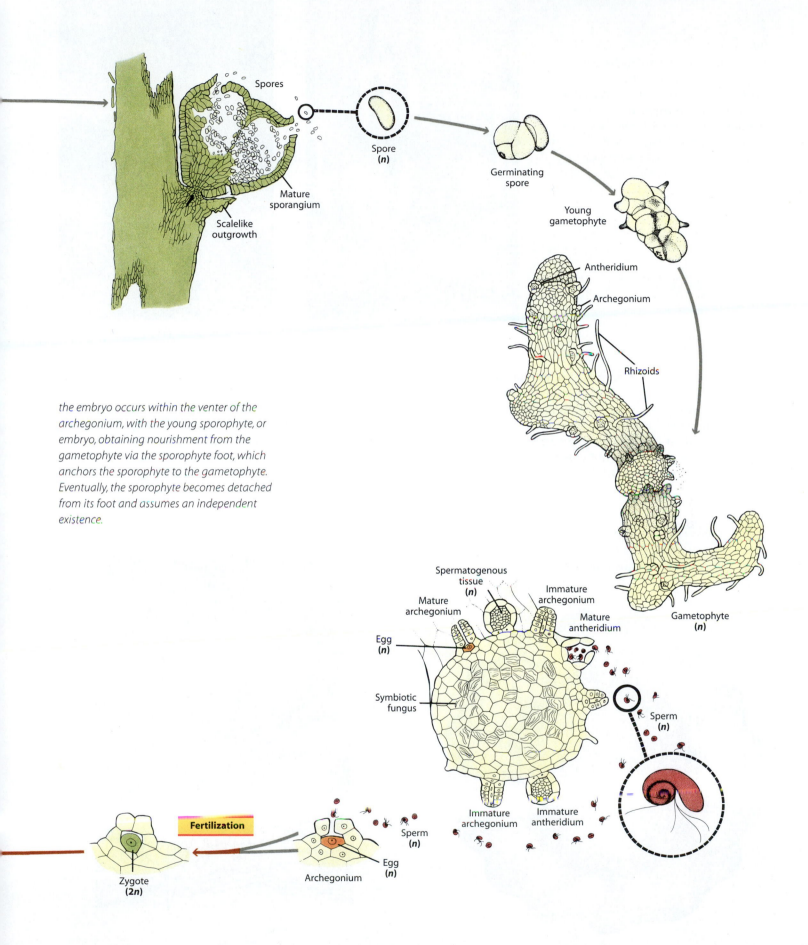

Spores

Mature
sporangium

Scalelike
outgrowth

Spore
(*n*)

Germinating
spore

Young
gametophyte

Antheridium

Archegonium

Rhizoids

the embryo occurs within the venter of the
archegonium, with the young sporophyte, or
embryo, obtaining nourishment from the
gametophyte via the sporophyte foot, which
anchors the sporophyte to the gametophyte.
Eventually, the sporophyte becomes detached
from its foot and assumes an independent
existence.

Spermatogenous
tissue
(*n*)

Mature
archegonium

Immature
archegonium

Egg
(*n*)

Mature
antheridium

Gametophyte
(*n*)

Symbiotic
fungus

Sperm
(*n*)

Immature
archegonium

Immature
antheridium

Fertilization

Sperm
(*n*)

Zygote
(**2n**)

Archegonium

Egg
(*n*)

19–25
Equisetum. **(a)** *A species of* Equisetum *in which there are separate fertile and vegetative shoots. The fertile shoots essentially lack chlorophyll and are very different in appearance from the vegetative shoots. Each fertile shoot has a terminal strobilus. Notice the whorls of scale-like leaves at each node.* **(b)** *Branching vegetative shoots of* Equisetum arvense.

(a)

(b)

Equisetum is homosporous. Sporangia are borne in groups of five to 10 along the margins of small umbrella-like structures known as **sporangiophores** (sporangia-bearing branches), which are clustered into strobili at the apex of the stem (Figures 19–25a and 19–29). The fertile stems of some species do not contain much chlorophyll. In these species, the fertile stems are sharply distinct from the vegetative stems, often appearing before the latter early in the spring (Figure 19–25). In other species of *Equisetum*, the strobili are borne at the tips of otherwise vegetative stems (see Figure 13–15d). When the spores are mature, the sporangia contract and split along their inner surface, releasing numerous spores. Elaters—thickened bands that arise from the outer layer of the

19–26
Stem anatomy of Equisetum. **(a)** *Transverse section of stem, showing mature tissues.* **(b)** *Detail of vascular strand, showing xylem and phloem.*

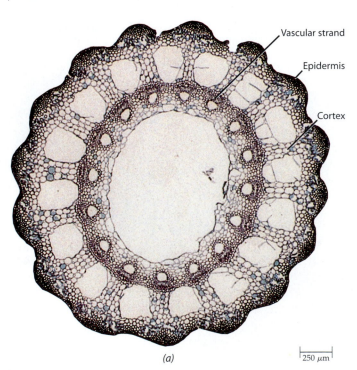

Vascular strand

Epidermis

Cortex

(a)

250 μm

Phloem   Xylem

Carinal
canal   Endodermis

(b)   50 μm

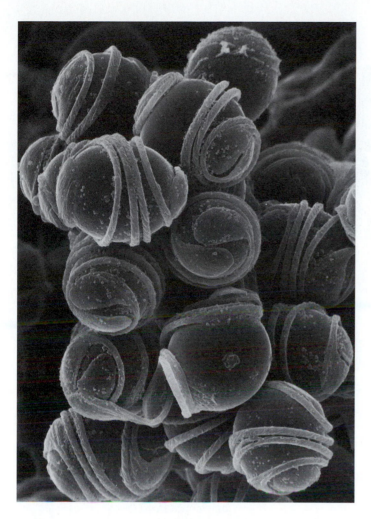

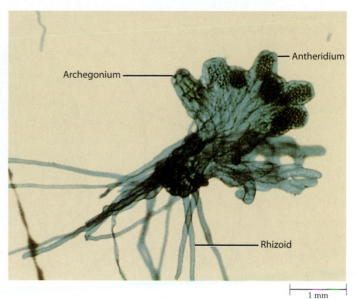

19–28

Bisexual gametophyte of Equisetum, showing male and female gametangia, antheridia and archegonia, respectively. Rhizoids can be seen extending from the lower surface of the gametophyte.

19–27

Spores of the horsetail Equisetum arvense, as seen in a scanning electron micrograph. Shown here are moist spores wrapped tightly by thickened bands, known as elaters, which are attached to the spore walls. As the spores dry, the elaters uncoil, helping to disperse the spores from the sporangium.

spore wall—coil when moist and uncoil when dry, thus playing a role in spore dispersal (Figures 19–27 and 19–29). These are quite distinct from the elaters that aid in spore dispersal in *Marchantia*. There, the elaters are elongated cells with helically arranged wall thickenings (see Figure 18–14).

The gametophytes of *Equisetum* are green and free-living and range in diameter from a few millimeters to 1 centimeter or even 3 to 3.5 centimeters in some species. Gametophytes become established mainly on mud that has recently been flooded and is rich in nutrients. The gametophytes, which reach sexual maturity in three to five weeks, are either bisexual or male (Figure 19–28). In bisexual gametophytes, the archegonia develop before the antheridia; this developmental pattern increases the probability of cross-fertilization. The sperm are multi-flagellated and require water to swim to the eggs. The eggs of several archegonia on a single gametophyte may be fertilized and develop into embryos, or young sporophytes.

The life cycle of *Equisetum* is illustrated in Figure 19–29 (pages 450 and 451).

Phylum *Pterophyta*

Ferns have been relatively abundant in the fossil record from the Carboniferous period to the present (see pages 456 and 457, and Figure 20–1). Today, ferns number about 11,000 species; they are the largest group of plants other than the flowering plants and are the most diverse in both form and habit (Figure 19–30).

The diversity of ferns is greatest in the tropics, where about three-fourths of the species are found. Here, not only are there many species of ferns, but ferns are abundant in many plant communities. Only about 380 species of ferns occur in the United States and Canada, whereas about 1000 occur in the small tropical country of Costa Rica in Central America. Approximately a third of all species of tropical ferns grow upon the trunks or branches of trees as epiphytes (Figure 19–30).

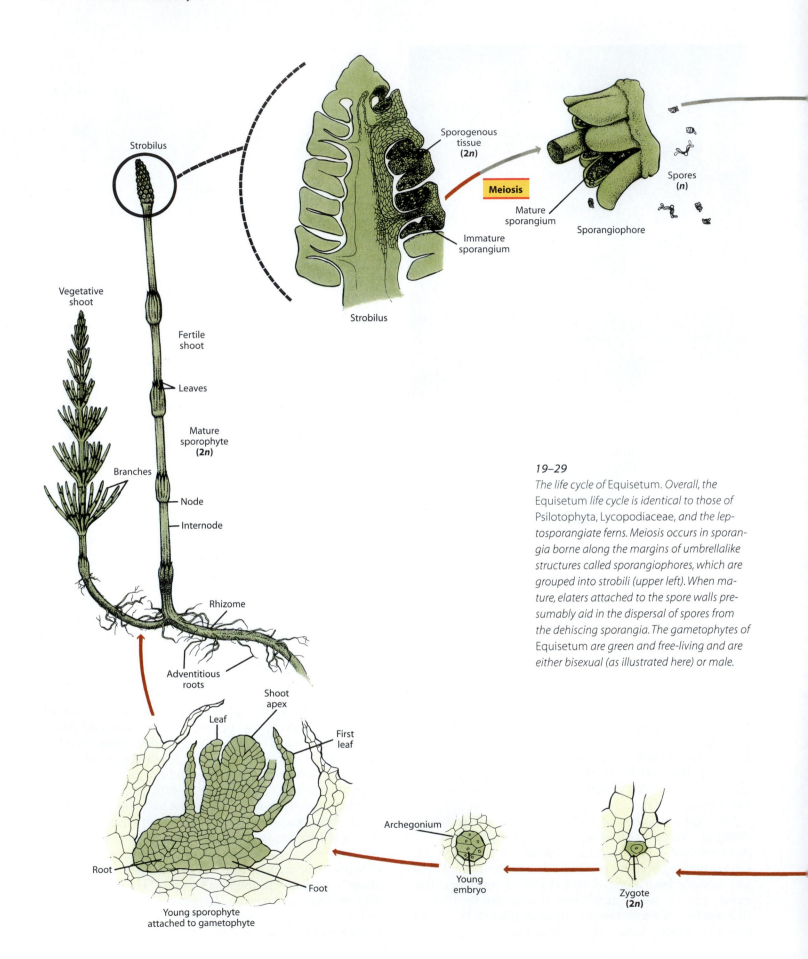

Strobilus

Sporogenous tissue (**2n**)

Mature sporangium

Immature sporangium

Strobilus

Meiosis

Spores (**n**)

Sporangiophore

Vegetative shoot

Fertile shoot

Leaves

Mature sporophyte (**2n**)

Branches

Node

Internode

Rhizome

Adventitious roots

Shoot apex

Leaf

First leaf

Root

Foot

Young sporophyte attached to gametophyte

Archegonium

Young embryo

Zygote (**2n**)

19–29

The life cycle of Equisetum. *Overall, the* Equisetum *life cycle is identical to those of* Psilotophyta, Lycopodiaceae, *and the leptosporangiate ferns. Meiosis occurs in sporangia borne along the margins of umbrellalike structures called sporangiophores, which are grouped into strobili (upper left). When mature, elaters attached to the spore walls presumably aid in the dispersal of spores from the dehiscing sporangia. The gametophytes of* Equisetum *are green and free-living and are either bisexual (as illustrated here) or male.*

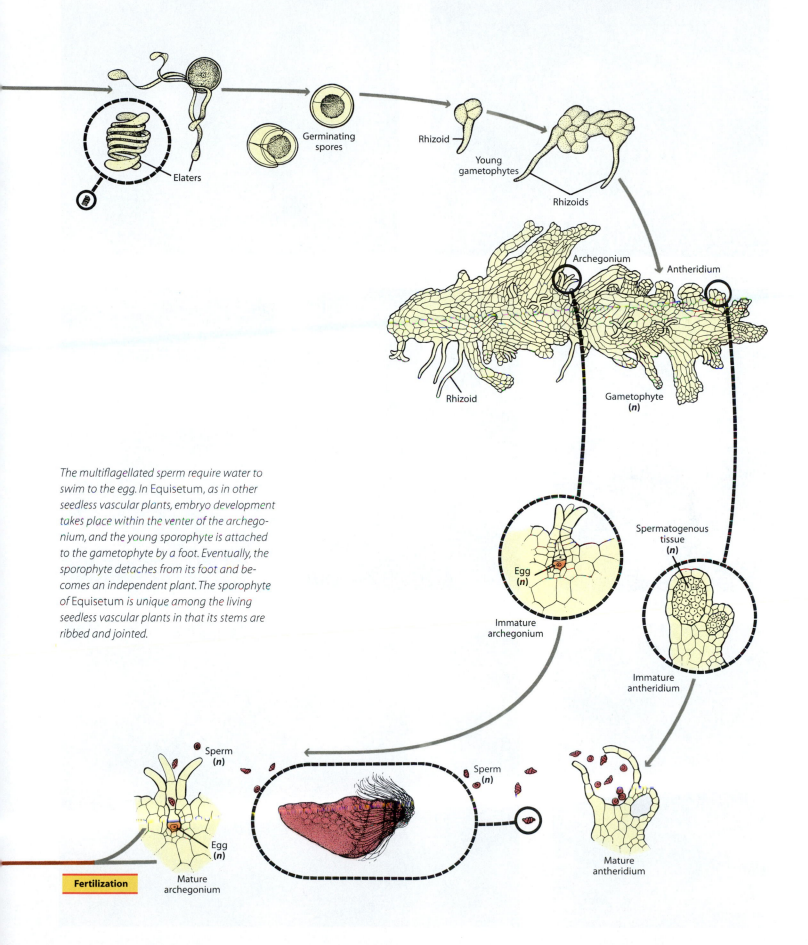

Germinating spores

Elaters

Rhizoid

Young gametophytes

Rhizoids

Archegonium

Antheridium

Rhizoid

Gametophyte (*n*)

The multiflagellated sperm require water to swim to the egg. In Equisetum, *as in other seedless vascular plants, embryo development takes place within the venter of the archegonium, and the young sporophyte is attached to the gametophyte by a foot. Eventually, the sporophyte detaches from its foot and becomes an independent plant. The sporophyte of* Equisetum *is unique among the living seedless vascular plants in that its stems are ribbed and jointed.*

Egg (*n*)

Immature archegonium

Spermatogenous tissue (*n*)

Immature antheridium

Sperm (*n*)

Sperm (*n*)

Fertilization

Egg (*n*)

Mature archegonium

Mature antheridium

(a)

(b)

(c)

(d)

(e)

(f)

(g)

19–30

The diversity of ferns, as illustrated by a few genera of the largest order of ferns, Filicales. (a) Lindsaea, Volcán Barba, Costa Rica. (b) A tree fern, Cyathea, at Monteverde, Costa Rica. (c) Plagiogyria, with distinct fertile and vegetative leaves, Volcán Poás, Costa Rica. (d) Elaphoglossum, with thick, undivided leaves, near Cuzco, Peru. (e) Asplenium septentrionale, a small fern that occurs all around the Northern Hemisphere, growing on metal-rich soil near a lead-silver mine in Wales. (f) Pleopeltis polypodioides, growing as an epiphyte on a juniper trunk in Arkansas. (g) Hymenophyllum species, one of the filmy ferns, so-called because of their delicate leaves. Filmy ferns occur as epiphytes primarily in tropical rainforests or wet temperate regions.

Some ferns are very small and have undivided leaves. *Lygodium*, a climbing fern, has leaves with a long, twining rachis (an extension of the leaf stalk, or petiole) that may be up to 30 meters or more in length. Some tree ferns (Figure 19–30b), such as those of the genus *Cyathea*, have been recorded to reach heights of more than 24 meters and to have leaves 5 meters or more in length. Although the trunks of such tree ferns may be 30 centimeters or more thick, their tissues are entirely primary in origin. Most of this thickness is the fibrous root mantle; the true stem is only four to six centimeters in diameter. The herbaceous genus *Botrychium* (see Figure 19–32a) is the only living fern known to form a vascular cambium.

In terms of the structure and method of development of their sporangia, ferns may be classified as either **eusporangiate** or **leptosporangiate** (Figure 19–31). The distinction between these two types of sporangia is important for understanding relationships among vascular plants. In a **eusporangium,** the parent cells, or initials, are located at the surface of the tissue from which the sporangium is produced (Figure 19–31a). These initials divide by the formation of walls parallel to the surface, resulting in the formation of an inner and an outer series of cells. The outer cell layer, by further divisions in both planes, builds up the several-layered wall of the spo-

19–31

Development and structure of the two principal types of fern sporangia. (a) The eusporangium originates from a series of superficial parent cells, or initials. Each eusporangium develops a wall two or more layers thick (although at maturity the inner wall layers may be crushed) and a high number of spores. (b) The leptosporangium originates from a single initial cell, which first produces a stalk and then a capsule. Each leptosporangium gives rise to a relatively small number of spores.

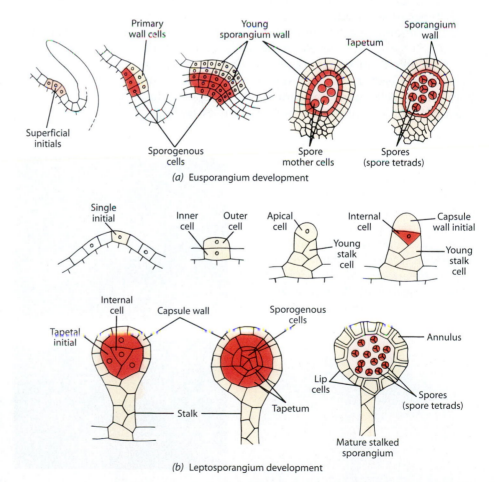

(a) Eusporangium development

(b) Leptosporangium development

rangium. The inner layer gives rise to a mass of irregularly oriented cells from which the spore mother cells ultimately arise. In many eusporangia, the inner wall layers are stretched and compressed during the course of development, so that the walls may apparently consist of a single layer of cells at maturity. Eusporangia, which are larger than leptosporangia and contain many more spores, are characteristic of all vascular plants—including those we have considered thus far—except for the leptosporangiate ferns.

In contrast to the multicellular origin of eusporangia, **leptosporangia** arise from a single superficial initial cell, which divides transversely or obliquely (Figure 19–31b). The inner of the two cells produced by this division may either contribute cells that produce a large part of the sporangial stalk or remain inactive and play no role in the further development of the sporangium, which is the more common condition. By a precise pattern of divisions, the outer cell ultimately gives rise to an elaborate, stalked sporangium, with a globose capsule having a wall that is one cell thick. Within this wall is a nutritive structure two cell layers thick called the **tapetum.** The inner mass of the leptosporangium eventually differentiates into spore mother cells, which undergo meiosis to produce four spores each. After it nourishes the young dividing cells within the sporangium, the tapetum is deposited around the spores, creating ridges, spines, and other types of surface features that are often characteristic for individual families and genera. The spores are exposed following the development of a crack in the so-called *lip cells* of the sporangium. The sporangia are stalked, and each contains a special layer of unevenly thick-walled cells called an **annulus.** As the sporangium dries out, contraction of the annulus causes tearing in the middle of the capsule. The sudden explosion and snapping back of the annulus to its original position then result in a catapultlike discharge of the spores. In eusporangia, the stalks are more massive and, while there may be preformed lines of dehiscence, there is no annulus and no catapultlike discharge of spores.

Most living ferns are homosporous; heterospory is restricted to two orders of living water ferns (see Figure 19–39), which will be discussed further below. A few extinct ferns also were heterosporous.

We shall now consider as examples three very different kinds of ferns: (1) the orders *Ophioglossales* and *Marattiales*, as examples of eusporangiate ferns; (2) the *Filicales*, or homosporous leptosporangiate ferns; and (3) the water ferns, orders *Marsileales* and *Salviniales*, the heterosporous leptosporangiate ferns.

The Orders *Ophioglossales* and *Marattiales* Are Eusporangiate Ferns

Of the three genera of the order *Ophioglossales*, *Botrychium*, the grape ferns (Figure 19–32a), and *Ophioglossum*, the adder's tongues (Figure 19–32b), are widespread in the north temperate region. In both of these genera, a single leaf typically is produced each year from the stem. Each leaf consists of two parts: (1) a vegetative portion, or blade, which is deeply dissected in *Botrychium* and undivided in most species of *Ophioglossum*, and (2) a fertile segment. In *Botrychium*, the fertile segment is dissected in the same way as the vegetative portion and bears two rows of eusporangia on the outermost segments. In *Ophioglossum*, the fertile portion is undivided and bears two rows of sunken eusporangia.

19–32

Representatives of the two genera of Ophioglossales *that occur in North America.*
(a) Botrychium parallelum. *In the genus* Botrychium, *the lower, vegetative portion of the leaf is divided. This is the only fern genus to form a vascular cambium.* **(b)** *In* Ophioglossum, *the lower portion of the leaf is undivided. In both genera, the erect, fertile, upper part of the leaf is sharply distinct from the vegetative portion.*

(a)

(b)

The gametophytes of *Botrychium* and *Ophioglossum* are subterranean, tuberous, elongate structures with numerous rhizoids; they have endophytic fungi and resemble the gametophytes of *Psilotophyta*. In *Botrychium*, the gametophytes usually possess a dorsal ridge in which the antheridia are embedded, with the archegonia generally located along the sides of the ridge. In the nature of their gametophytes, the structure of their leaves, and several other anatomical details, the *Ophioglossales* are sharply distinct from other living ferns and are clearly an early diverging and distinct group. Unfortunately, the group has no well-established fossil record before about 50 million years ago. One member of the *Ophioglossales*, *Ophioglossum reticulatum*, has the highest chromosome number known in any living organism, with a diploid complement of about 1260 chromosomes.

The only other order of ferns that has eusporangia, the tropical *Marattiales*, is an ancient group with a fossil record that extends back to the Carboniferous period. The members of this order resemble more familiar groups of ferns more closely than they do *Ophioglossales*. *Psaronius*, the extinct tree fern illustrated in Figure 20–1, was a member of this order. The six living genera of *Marattiales* include about 200 species.

Filicales Is an Order of Homosporous Leptosporangiate Ferns

Nearly all familiar ferns are members of the larger order *Filicales*, with at least 10,500 species. About 35 families and 320 genera are recognized in the order. *Filicales* differ from *Ophioglossales* and *Marattiales* in being leptosporangiate, and from the water ferns, which we shall discuss next, in being homosporous. All ferns other than *Ophioglossales* and *Marattiales*, in fact, are leptosporangiate, and very few have the subterranean gametophytes with endophytic fungi that are characteristic of the *Ophioglossales* and *Marattiales*. Clearly, leptosporangia and the other distinctive features of most ferns are specialized characters since they occur nowhere else among the vascular plants, including *Ophioglossales* and *Marattiales*, which share more features in common with other groups of ancient plants.

Most garden and woodland ferns of temperate regions have siphonostelic rhizomes (Figure 19–33) that produce new sets of leaves each year. The fern embryo produces a true root, but this soon withers, and the rest of the roots are adventitious—that is, they arise from the rhizomes near the bases of the leaves. The leaves, or **fronds,** are megaphylls and represent the most conspicuous part of the sporophyte. Their high surface-to-volume ratio allows them to capture sunlight much more effectively than the microphylls of the lycophytes. The ferns are the only seedless vascular plants to possess well-developed megaphylls. Commonly, the fronds are compound; that is, the lamina is divided into leaflets, or **pinnae,** which are attached to the **rachis,** an extension of the leaf stalk, or petiole. In nearly all ferns, the young leaves are coiled (circinate); they are commonly referred to as "fiddleheads" (Figure 19–34). This type of leaf de-

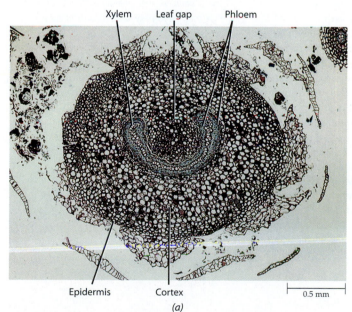

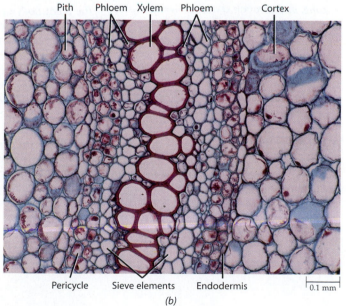

19–33

The anatomy of fern rhizomes. **(a)** *Adiantum, or maidenhair fern. Transverse section of a rhizome, showing the siphonostele. Note the* wide leaf gap. **(b)** *Transverse section of part of the vascular region of a rhizome of the tree fern Dicksonia. The phloem is composed* mainly of sieve elements; the xylem is composed entirely of tracheids.

Coal Age Plants

The amount of carbon dioxide used in photosynthesis is about 100 billion metric tons annually, about a tenth of the total carbon dioxide present in the atmosphere. The amount of carbon dioxide returned as a result of oxidation of these living materials is about the same, differing only by 1 part in 10,000. This very slight imbalance is caused by the burying of organisms in sediment or mud under conditions in which oxygen is excluded and decay is only partial. This accumulation of partially decayed plant material is known as peat (page 415). The peat may eventually become covered with sedimentary rock and thus placed under pressure. Depending on time, temperature, and other factors, peat may become compressed into soft or hard coal, one of the so-called fossil fuels.

During certain periods in the Earth's history, the rate of fossil-fuel formation was greater than at other times. One such time was the Carboniferous period, which extended from about 362 to 290 million years ago (see Figures 19–11 and 20–1). The lands were low, covered by shallow seas or swamps, and, in what are now temperate regions of Europe and North America, conditions were favorable for year-round growth. These regions were tropical to subtropical, with the equator then arcing across the Appalachians, over northern Europe, and through Ukraine. Five groups of plants dominated the swamplands, and three of them were seedless vascular plants—lycophytes, sphenophytes (calamites), and ferns. The other two were gymnospermous types of seed plants—seed ferns (Pteridospermales) and cordaites (Cordaitales).

Lycophyte Trees

For most of the "Age of Coal" in the late Carboniferous period (Pennsylvanian), lycophyte trees dominated the coal-forming swamps. Most of these plants grew to heights of 10 to 35 meters and were sparsely branched (a). After the plant attained most of its total height, the trunk branched dichotomously. Successive branching produced progressively smaller branches until, finally, the tissues of the branch tips lost their ability to grow further. The branches bore long microphylls. The tree lycophytes were largely supported by a massive periderm surrounding a relatively small amount of xylem.

Like Selaginella and Isoetes, lycophyte trees were heterosporous, and their sporophylls were aggregated into cones. Some of these trees produced structures analogous to seeds.

As the swamplands began to dry up and the climate in Euramerica began to change toward the end of the Carboniferous period, the lycophyte trees vanished almost overnight, geologically speaking. The only remaining living relative of the group is the genus Isoetes. Herbaceous lycophytes essentially similar to Lycopodium and Selaginella existed in the Carboniferous period, and representatives of some of them have survived to the present; there are 10 to 15 living genera.

Calamites

Calamites, or giant horsetails, were plants of treelike proportions, reaching heights of 18 meters or more (see Figure 20–1). Like the plant body of Equisetum, that of the calamites consisted of a branched aerial portion and an underground rhizome system. In addition, the leaves and branches were whorled at the nodes. Even the stems were remarkably similar to those of Equisetum, except for the presence of secondary xylem in the calamites, which accounted for much of the great diameter of the stems (trunks up to one-third of a meter in diameter). The similarities between calamites and living Equisetum are so strong that they are now regarded as belonging to the same order.

The fertile appendages, or sporangio-

(a)

(a) Scientists believe that once a lycophyte tree was stabilized by its shallow, forking, rootlike axes, it pushed rapidly skyward. These underground stigmarian axes produced spirally arranged rootlets, seen here as slender projections emerging from the forest floor. From left to right, a young leafy form, a pole-like juvenile, and a giant adult of 35 meters. (b) One of the most interesting gymnosperm groups is the seed ferns, a large, artificial group of primitive seed-bearing plants that appeared in the Late Devonian and flourished for about 125 million years. Fossils of these bizarre plants are common in rocks of Carboniferous age and have been well known to paleobotanists for a century or more. Their vegetative parts are so fernlike that for many years

(b)

(c)

they were grouped with the ferns. This drawing is a reconstruction of the Carboniferous seed fern Medullosa noei. The plant was about 5 meters tall. **(c)** Tip of young branch of the primitive conifer-like Cordaites, with long, straplike leaves.

phores, of the calamites were aggregated into cones. Although most were homosporous, a few giant horsetails were heterosporous. Like most of the lycophyte trees, the giant horsetails declined in importance toward the end of the Paleozoic but persisted in much reduced form through the Mesozoic and Tertiary. Today they are the horsetails and are represented by only one genus, Equisetum.

Ferns

Many of the ferns represented in the fossil record are recognizable as members of today's primitive fern families. The "Age of Ferns" in the late Carboniferous period was dominated by tree ferns such as Psaronius, one of the Marattiales—a eusporangiate group. Up to 8 meters tall, Psaronius had a stele that expanded toward the apex; the stele was covered below with adventitious roots, which played the key role in supporting the plant. The stem of Psaronius ended in an aggregate of large, pinnately compound fronds (see Figure 20–1).

Seed Plants

The two remaining plant groups that dominated the tropical lowlands of Euramerica were the seed ferns and the cordaites. The fossil plants that are usually grouped as seed ferns are probably of diverse evolutionary relationships. The remains of seed ferns are common fossils in rocks of Carboniferous age (b). Their large, pinnately compound fronds were so fernlike that these plants were long regarded as ferns. Then in 1905, F. W. Oliver and D. H. Scott demonstrated that they bore seeds and so were gymnosperms. Many species were small, shrubby, or scrambling plants.

Other probable seed ferns were tall, woody trees. The fronds of seed ferns were borne at the top of their stem, or trunk, with microsporangia and seeds borne on

them. The seed ferns survived into the Mesozoic era.

The cordaites were widely distributed during the Carboniferous period both in swamps and in drier environments. Although some members of the order were shrubs, many were tall (15 to 30 meters), highly branched trees that perhaps formed extensive forests. Their long (up to 1 meter), straplike leaves were spirally arranged at the tips of the youngest branches (c). The center of the stem was occupied by a large pith, and a vascular cambium gave rise to a complete cylinder of secondary xylem. The root system, located at the base of the plant, also contained secondary xylem. The plants bore pollen-bearing cones and seed-bearing, conelike structures on separate branches. The cordaites persisted into the Permian period (286 to 248 million years ago), the drier and cooler period that followed the Carboniferous, but were apparently extinct by the beginning of the Mesozoic.

In Conclusion

The dominant tropical coal-swamp plants of the Carboniferous period in Euramerica—the lycophyte trees—became extinct during the Late Paleozoic, a time of increasing tropical drought. Only the herbaceous relatives of the tree lycophytes and horsetails of the Carboniferous period continued to flourish and exist today, as do several groups of ferns that appeared in the Carboniferous period. Both the seed ferns and the cordaites eventually disappeared. Only one group of Carboniferous gymnosperms, the conifers (not a dominant group at the time), survived and went on to produce new types during the Permian period. The living conifers are discussed in detail in Chapter 20.

(a)

(b)

(c)

(d)

19–34

"Fiddleheads" of the ostrich fern (Matteuccia struthiopteris). Ostrich fern fiddleheads are gathered commercially in upper New England and New Brunswick; they are marketed fresh, canned, and frozen. These fiddleheads, which taste somewhat like crisp asparagus, should be picked when they are less than 15 centimeters long. The fiddleheads of many ferns are considered toxic. The ostrich fern is the most widely planted fern around house foundations in the eastern United States and Canada.

19–35

Sori are clusters of sporangia found on the undersides of leaves of ferns. (a) In Dennstaedtia punctilobula and other ferns of this genus, the sori are bare. (b) In the bracken fern (Pteridium aquilinum) shown here, as well as in the maidenhair ferns (Adiantum), the sori are located along the margins of the leaf blades, which are rolled back over them. (c) In the evergreen wood

fern (Dryopteris marginalis), the sori, which are also located near the margins of the leaf blades, are completely covered by kidney-shaped indusia. (d) In Onoclea sensibilis, the sori are enfolded by globular lobes of the pinna (leaflet) and therefore not visible. After overwintering, the lobes separate slightly, and the spores are released early in the spring, often over the snow.

velopment is known as **circinate vernation.** Uncoiling of the fiddlehead results from more rapid growth on the inner than on the outer surface of the leaf early in development and is mediated by the hormone auxin (page 671), produced by the young pinnae on the inner side of the fiddlehead. This type of vernation protects the delicate embryonic leaf tip during development. Both fiddleheads and rhizomes are usually clothed with either hairs or scales, both of which are epidermal outgrowths; the characteristics of these structures are important in fern classification.

The sporangia of *Filicales,* all of which are homosporous, occur on the margins or lower surfaces of the leaves, on specially modified leaves, or on separate stalks. The sporangia commonly occur in clusters called **sori** (singular: sorus) (Figures 19–35 and 19–36), which

may appear as yellow, orange, brownish, or blackish lines, dots, or broad patches on the lower surface of a frond. In many genera, the young sori are covered by specialized outgrowths of the leaf, the **indusia** (singular: indusium), which may shrivel when the sporangia are ripe and ready to shed their spores. The shape of the sorus, its position, and the presence or absence of an indusium are important characteristics in the taxonomy of the *Filicales.*

The spores of ferns in the *Filicales* give rise to free-living, bisexual gametophytes, which are often found in moist places, such as the sides of pots in greenhouses. The gametophyte typically develops rapidly into a flat, heart-shaped, usually membranous structure, the **prothallus,** that has numerous rhizoids on its central lower surface. Both antheridia and archegonia develop on the

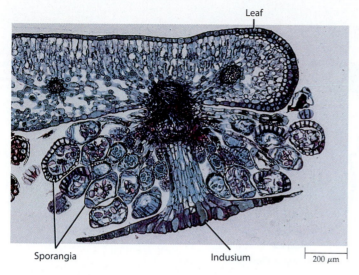

Leaf

Sporangia | Indusium | 200 μm

19–36

Cyrtomium falcatum, a homosporous fern. Transverse section of a leaf, showing a sorus on the lower surface. The sporangia are in different stages of development and are protected by an umbrellalike indusium.

(a)

(b)

(c)

ventral surface of the prothallus. The antheridia occur more typically among the rhizoids, while the archegonia are usually formed near the notch, an indentation at the anterior end of the gametophyte. The order of appearance of these gametangia is controlled genetically and can be mediated by special chemicals produced by the gametophytes. The timing of the appearance of the gametangia can influence whether the breeding system is primarily inbreeding or outcrossing. Water is required for the multiflagellated sperm to swim to the eggs.

Early in its development, the embryo, or young sporophyte, receives nutrients from the gametophyte through a foot. Development is rapid, and the sporophyte soon becomes an independent plant, at which time the gametophyte disintegrates.

The life cycle of one of the *Filicales* is shown in Figure 19–38 (pages 460 and 461).

Typically, the sporophyte is the perennial stage in ferns, and the small, thalloid gametophyte is short-lived. Remarkably, the strap-shaped or filamentous gametophytes of some species of ferns, including three genera with six tropical species found in the southern Appalachians, persist indefinitely without ever producing sporophytes. In addition, these species have never yet been induced to produce sporophytes in the laboratory (Figure 19–37). They reproduce by vegetative out-

19–37

In some ferns from widely scattered parts of the world, the gametophytes reproduce asexually and persist; sporophytes are not formed, either in the field or in the laboratory. These photographs show two of the three fern genera known to exhibit this habit in the eastern United States. (a) Typical habitat of persistent gametophytes of Vittaria and Trichomanes, Ash Cave, Hocking County, Ohio. (b) Trichomanes gametophytes, Lancaster County, Pennsylvania. (c) Vittaria gametophytes, Franklin County, Alabama.

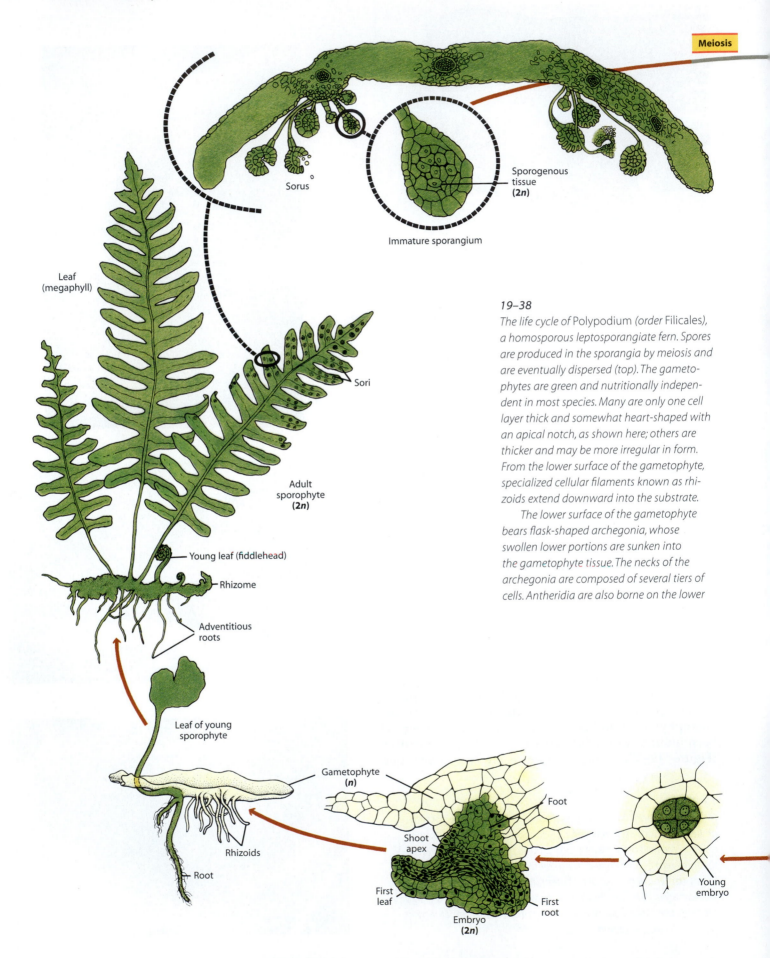

Meiosis

Sorus

Sporogenous
tissue
(**2n**)

Immature sporangium

Leaf
(megaphyll)

Sori

Adult
sporophyte
(**2n**)

Young leaf (fiddlehead)

Rhizome

Adventitious
roots

Leaf of young
sporophyte

Rhizoids

Root

Gametophyte
(**n**)

Foot

Shoot
apex

First
leaf

First
root

Embryo
(**2n**)

Young
embryo

19–38

*The life cycle of Polypodium (order Filicales),
a homosporous leptosporangiate fern. Spores
are produced in the sporangia by meiosis and
are eventually dispersed (top). The gametophytes are green and nutritionally independent in most species. Many are only one cell
layer thick and somewhat heart-shaped with
an apical notch, as shown here; others are
thicker and may be more irregular in form.
From the lower surface of the gametophyte,
specialized cellular filaments known as rhizoids extend downward into the substrate.*

*The lower surface of the gametophyte
bears flask-shaped archegonia, whose
swollen lower portions are sunken into
the gametophyte tissue. The necks of the
archegonia are composed of several tiers of
cells. Antheridia are also borne on the lower*

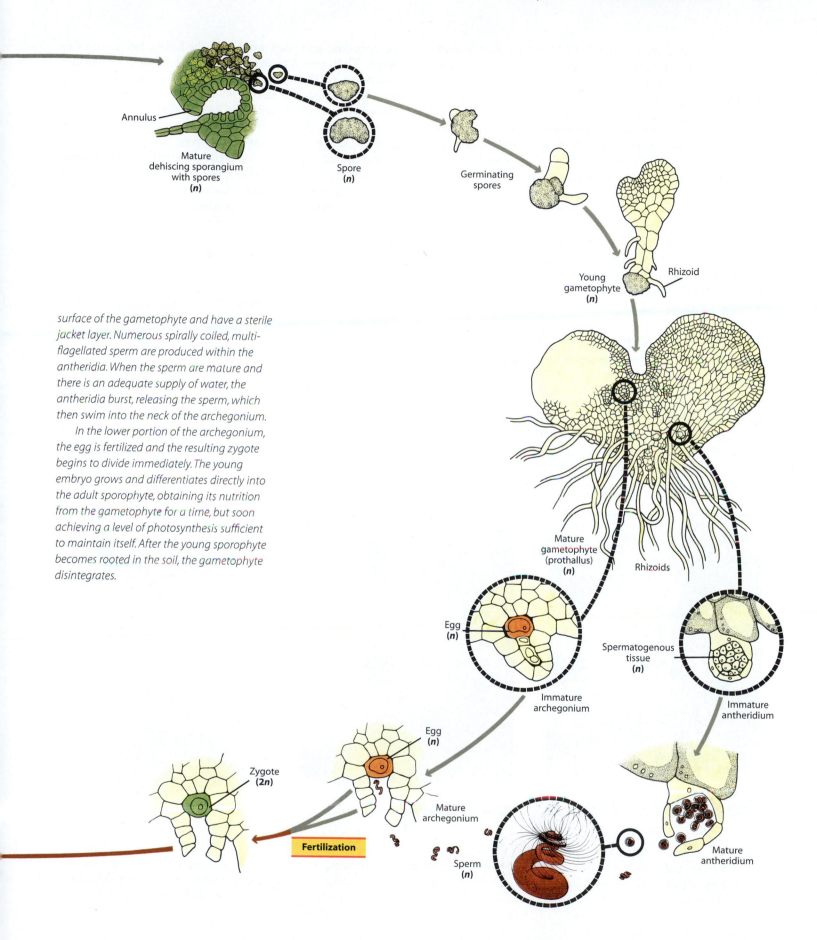

Annulus

Mature
dehiscing sporangium
with spores
(*n*)

Spore
(*n*)

Germinating
spores

Young
gametophyte
(*n*)

Rhizoid

surface of the gametophyte and have a sterile
jacket layer. Numerous spirally coiled, multiflagellated sperm are produced within the
antheridia. When the sperm are mature and
there is an adequate supply of water, the
antheridia burst, releasing the sperm, which
then swim into the neck of the archegonium.

In the lower portion of the archegonium,
the egg is fertilized and the resulting zygote
begins to divide immediately. The young
embryo grows and differentiates directly into
the adult sporophyte, obtaining its nutrition
from the gametophyte for a time, but soon
achieving a level of photosynthesis sufficient
to maintain itself. After the young sporophyte
becomes rooted in the soil, the gametophyte
disintegrates.

Mature
gametophyte
(prothallus)
(*n*)

Rhizoids

Egg
(*n*)

Spermatogenous
tissue
(*n*)

Immature
archegonium

Immature
antheridium

Egg
(*n*)

Zygote
(*2n*)

Mature
archegonium

Fertilization

Sperm
(*n*)

Mature
antheridium

growths called gemmae that fall off and are blown away to found new colonies. These ferns appear to be distinct from the other, sporophyte-producing species of their respective genera, as judged by differences in their enzymes, and probably should be treated as distinct species. Such situations are common in mosses and are being discovered in ferns much more widely than was previously expected. Populations of perennial, free-living gametophytes of *Trichomanes speciosum,* recently discovered in the Elbsandsteingebirge (a mountain range shared by Germany and the Czech Republic), are estimated to be over 1000 years old. The possibility exists that they are relics of former populations that included both sporophytes and gametophytes. The extinction of the sporophytes possibly occurred as a result of climatic changes during the glacial intervals of the last 2 million years.

The Water Ferns—Orders *Marsileales* and *Salviniales*—Are Heterosporous Leptosporangiate Ferns

The water ferns constitute two orders, *Marsileales* and *Salviniales.* Although they are structurally very different from each other, recent evidence from molecular analyses indicates that the two orders were derived from a common terrestrial ancestor. All water ferns are het-

erosporous, and they are the only living heterosporous ferns. There are five genera of water ferns. The slender rhizomes of the three genera of *Marsileales,* including *Marsilea* (which has about 50 species), grow in mud, on damp soil, or often with the leaves floating on the surface of water (Figure 19–39a). The leaves of *Marsilea* resemble those of a four-leaf clover. Drought-resistant, bean-shaped reproductive structures called **sporocarps,** which may remain viable even after 100 years of dry storage, germinate when placed in water to produce chains of sori, each bearing series of megasporangia and microsporangia (Figure 19–39b). The extremely specialized gametophytes and the heterospory of *Marsileales* are the primary reasons that we recognize these ferns as a distinct order.

The two genera of *Salviniales, Azolla* (see Figure 30–11) and *Salvinia* (Figure 19–39c), are small plants that float on the surface of water. Both genera produce their sporangia in sporocarps that are quite different in structure from those of *Marsileales.* In *Azolla,* the tiny, crowded, bilobed leaves are borne on slender stems. A pouch that forms on the upper, photosynthetic lobe of each leaf is inhabited by colonies of the cyanobacterium *Anabaena azollae.* The lower, smaller lobe of each leaf is often nearly colorless. Because of the nitrogen-fixing abilities of the *Anabaena, Azolla* is important in maintaining the fertility of rice paddies and of certain natural

(a)

(b)

(c)

19–39
Water ferns. The two very distinct orders of water ferns are the only living heterosporous ferns. (a) Marsilea polycarpa, with its leaves floating on the surface of the water, photographed in Venezuela. (b) Marsilea, *showing the germination of a sporocarp, with chains of sori. Each sorus contains a series of megasporangia and microsporangia. (c) Salvinia, with two floating leaves and one* *feathery dissected, submerged leaf at each node. These two genera are representatives of the orders Marsileales and Salviniales, respectively.*

ecosystems. The undivided leaves of *Salvinia,* which are up to 2 centimeters long, are borne in whorls of three on the floating rhizome. One of the three leaves hangs down below the surface of the water and is highly dissected, resembling a mass of whitish roots. These "roots," however, bear sporangia, which reveals that they are actually leaves. The two upper leaves, which float on the water, are covered by hairs that protect their surface from getting wet, and the leaves float back to the surface if they are temporarily submerged.

Summary

Vascular plants are characterized by the possession of the vascular tissues xylem and phloem, and they exhibit an alternation of heteromorphic generations in which the sporophyte is large and complex and is the nutritionally independent phase.

The Primary Vascular Tissues Are Arranged in Steles of Three Basic Types

The plant bodies of many vascular plants consist entirely of primary tissues. Today, secondary growth is confined largely to the seed plants, although it occurred in several unrelated fossil groups of seedless vascular plants. The primary vascular tissues and associated ground tissues exhibit three basic arrangements: (1) the protostele, which consists of a solid core of vascular tissue; (2) the siphonostele, which contains a pith surrounded by vascular tissue; and (3) the eustele, which consists of a system of strands surrounding a pith, with the strands separated from one another by ground tissue.

Roots and Leaves Evolved in Different Ways

Roots evolved from the underground portions of the ancient plant body. Leaves originated in more than one way. Microphylls, single-veined leaves whose leaf traces

Summary TABLE — A Comparison of Some of the Main Features of the Seedless Vascular Plants

PHYLUM	DICHOTO-MOUSLY BRANCHED?	DIFFERENTIATED INTO ROOTS, STEMS, AND LEAVES?	HOMOSPOROUS OR HETEROSPOROUS	TYPE OF LEAVES	TYPE OF STELE	SPORANGIA	MISCELLANEOUS CHARACTERISTICS
Rhyniophyta (rhyniophytes)	Often	Stem only	Homosporous	None	Protostele	Terminal	Exclusively fossils; likely ancestors of trimerophytes
Zosterophyllophyta (zosterophyllophytes)	Often	Stem only	Many homosporous; some heterosporous	None	Protostele	Lateral	Exclusively fossils; closely related to lycophytes
Lycophyta (lycophytes)	Some are more or less dichotomous	Yes	*Lycopodiaceae* homosporous; *Selaginellaceae* and *Isoetaceae* heterosporous	Microphyll	Most with protostele or modified protostele	On or in the axils of sporophylls	Members of the *Selaginellaceae* and *Isoetaceae* have ligules; many extinct representatives
Trimerophytophyta (trimerophytes)	Most are not	Stem only	Homosporous	None	Protostele	Terminal on ultimate dichotomies	Exclusively fossils; likely ancestors of ferns, progymnosperms, and perhaps horsetails
Psilotophyta (psilotophytes)	Yes	Stem only	Homosporous	None	Protostele	Lateral	Resemble rhyniophytes in aspects of structure; *Psilotum* and *Tmesipteris* only modern genera
Sphenophyta (sphenophytes)	No	Yes	Homosporous; some fossils heterosporous	Microphyll through reduction	Eustele-like siphonostele	On sporangiophores in strobili	Represented today by single genus, *Equisetum,* the horsetails
Pterophyta (ferns)	No	Yes	All homosporous except for *Marsileales* and *Salviniales,* which are heterosporous	Megaphyll	Protostele in some; siphonostele or more complex types in others	On sporophylls; some clustered in sori	*Ophioglossales* and *Marattiales* eusporangiate; *Filicales, Marsileales,* and *Salviniales* leptosporangiate

are not associated with leaf gaps, evolved either as superficial lateral outgrowths of the stem or from sterile sporangia. They are associated with protosteles and are characteristic of the lycophytes. Megaphylls, leaves with complex venation, evolved from branch systems. They are associated with siphonosteles and eusteles. In the siphonosteles of ferns, the leaf traces are associated with leaf gaps.

Vascular Plants May Be Either Homosporous or Heterosporous

Homosporous vascular plants produce only one type of spore, which has the potential to give rise to a bisexual gametophyte. Heterosporous plants produce microspores and megaspores, which germinate and give rise to male gametophytes and female gametophytes, respectively. The gametophytes of heterosporous plants are much reduced in size, compared with those of homosporous plants. In the history of the vascular plants, heterospory has evolved several times. There has been a long, continuous evolutionary trend toward a reduction in the size and complexity of the gametophyte, which culminated in the angiosperms. Seedless vascular plants have archegonia and antheridia. All but a few gymnosperms have archegonia, but both archegonia and antheridia have been lost in all angiosperms.

Seedless Vascular Plants Exhibit an Alternation of Heteromorphic Generations

The life cycles of the seedless vascular plants all represent modifications of an essentially similar alternation of heteromorphic generations in which the sporophyte is dominant and free-living. The gametophytes of the homosporous species are independent of the sporophyte for their nutrition. Although potentially bisexual, producing both antheridia and archegonia, these gametophytes are functionally unisexual. The gametophytes of heterosporous species are unisexual, much reduced in size, and, except for the few genera of heterosporous ferns, dependent on stored food derived from the sporophyte for their nutrition. All of the seedless vascular plants have motile sperm, and the presence of water is necessary for the sperm to swim to the eggs.

The Oldest Fossils of Vascular Plants Belong to the Phylum *Rhyniophyta*

Vascular plants go back at least 430 million years; the earliest ones about which we have many structural details belong to the phylum *Rhyniophyta*, the oldest fossils of which are from the mid-Silurian period, about 425 million years ago. Some fossils, once considered rhynio-

phytes, have conducting cells similar to bryophyte hydroids rather than to tracheids. These plants, which had branched axes and multiple sporangia, may represent an intermediate stage in the evolution of vascular plants. They are called protracheophytes. The plant bodies of the rhyniophytes and other contemporary plants were simple, dichotomously branching axes lacking roots and leaves. With evolutionary specialization, morphological and physiological differences arose between various parts of the plant body, bringing about the differentiation of root, stem, and leaf.

The Living Seedless Vascular Plants Are Classified in Four Phyla

The phyla of seedless vascular plants are the *Lycophyta* (including *Lycopodium*, *Selaginella*, and *Isoetes*), the *Psilotophyta* (*Psilotum* and *Tmesipteris*), the *Sphenophyta* (*Equisetum*), and the *Pterophyta* (ferns). Most are homosporous, but heterospory is exhibited by *Selaginella*, *Isoetes*, and the water ferns (*Salviniales* and *Marsileales*).

The *Lycophyta* are believed to have evolved from the *Zosterophyllophyta*, a phylum of entirely extinct vascular plants. The *Trimerophytophyta*, another phylum of entirely extinct vascular plants, apparently represent the ancestral stock of the ferns, the progymnosperms, and perhaps the sphenophytes as well.

Psilotophytes differ from other living vascular plants in their lack of leaves (with the possible exception of *Tmesipteris*) and roots. Lycophytes are characterized by microphylls. The members of the other phyla have megaphylls.

Two orders of ferns (*Ophioglossales* and *Marattiales*) possess eusporangia, like those of other seedless vascular plants. In eusporangia, the walls are several cell layers thick, and a number of cells participate in the initial stages of sporangial development. Other ferns—*Filicales* and the two orders of water ferns—form leptosporangia, specialized structures in which the wall consists of a single layer of cells that develop from a single initial cell.

Two of the four phyla of seedless vascular plants that contain living representatives, the lycophytes and sphenophytes, extend back to the Devonian period. Among the seedless vascular plants, only the ferns, which first appear in the fossil record in the Carboniferous period, are represented by a large number of living species, about 11,000.

Five groups of vascular plants dominated the swamplands of the Carboniferous period ("Age of Coal"), and three of them were seedless vascular plants—lycophytes, sphenophytes, and ferns. The other two were gymnosperms—the seed ferns and the cordaites.

Selected Key Terms

annulus   p. 454

eusporangium   p. 453

eustele   p. 429

frond   p. 455

heterosporous   p. 431

homosporous   p. 430

indusium   p. 458

leaf gap   p. 429

leaf trace gap   p. 429

leptosporangium   p. 454

megaphyll   p. 429

megasporangium   p. 431

megaspores   p. 431

microphyll   p. 429

microsporangium
p. 431

microspores   p. 431

phloem   p. 426

prothallus   p. 458

protostele   p. 429

protracheophyte   p. 434

seeds   p. 425

sieve elements   p. 427

siphonostele   p. 429

sorus   p. 458

sporangiophore   p. 448

sporocarp   p. 462

sporophyll   p. 438

tapetum   p. 454

tracheary elements
p. 427

tracheids   p. 428

vessel elements   p. 428

xylem   p. 426

Questions

1. What basic structural features do the *Rhyniophyta*, *Zosterophyllophyta*, and *Trimerophytophyta* have in common?

2. The vessel elements and heterospory present in several unrelated groups of vascular plants represent excellent examples of convergent evolution. Explain.

3. With the use of simple, labeled diagrams, describe the structure of the three basic types of steles.

4. Compare the life cycle of a moss with that of a homosporous leptosporangiate fern.

5. What is coal? How was it formed? What plants were involved in its formation?

6. The bryophytes often are referred to as the "amphibians of the plant kingdom," but that characterization might also be applied to the seedless vascular plants. Can you explain why?

7. What is the probable evolutionary origin of the *Lycophyta*? What evidence is there for this view?

20–1

Tropical swamps of the Late Carboniferous were dominated by several genera of giant lycophyte trees that, when mature, formed a forest canopy of airy, diffusely branched crowns, seen here in the background. These trees had massive trunks (left foreground and elsewhere) that were stabilized in the swampy mire by long stigmarian axes, from which numerous rootlets, possibly photosynthetic, extended. Note also some low-growing lycophytes with cones, in the center fore-

ground. Flooded and boggy areas of the forest floor favored other lycophyte types, such as branchless Chaloneria (center and right foreground), and horsetail types, such as the shrub Sphenophyllum (right bottom corner) and the Christmas-tree-shaped Diplocalamites (far right midground).

Less wet, or slightly elevated ground, as seen here at the left, fostered mixed vegetation, including early conifers, ground-cover ferns, tall tree ferns, and seed ferns. Among

the seed plants were, from front to back, Cordaixylon, a shrubby conifer relative with strap-shaped leaves; Callistophyton, a scrambling seed fern growing at the base of the largest lycophyte tree; and Psaronius, an early tree fern (left background). Elsewhere, robust plants like the upright umbrella-shaped seed fern Medullosa are seen occupying sunnier, more disturbed sites opened up by previous channel flooding (right midground and background).

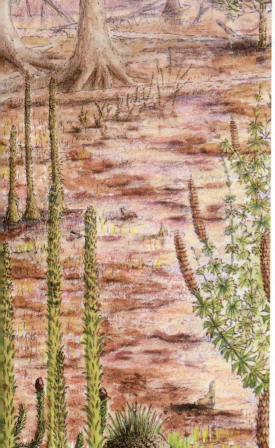

Gymnosperms

OVERVIEW

In this chapter we will examine the gymnosperms, a series of evolutionary lines of seed-bearing plants that include such common trees as pine, spruce, fir, hemlock, and cedar. Conifers, such as these trees, are a major source of lumber and paper pulp. They also include the tallest and oldest of trees.

We begin our journey through the gymnosperms with a discussion of the evolutionary origin of the ovule, the structure that develops into a seed. A marvelous innovation, the seed—consisting of a seed coat, an embryo, and stored food—replaced the spore as the unit of dispersal, and provided seed plants with an enormous advantage over the seedless vascular plants. As the name implies (the Greek word gymnos *means "naked," and* sperma *means "seed"), the seeds of gymnosperms lack the protection of an enveloping structure such as the fruit wall that encloses the seeds of angiosperms, or flowering plants.*

Another major advantage that both gymnosperms and angiosperms have over all other plants is their independence from water as a medium of transport of the sperm to the egg. In gymnosperms, the partly developed male gametophyte—the pollen grain—is transferred bodily to the vicinity of the female gametophyte within the ovule, after which it produces a pollen tube. Although not originally a sperm conveyor, the pollen tube eventually evolved into a conveyor of nonmotile sperm to the eggs of the female gametophyte within the ovule.

CHECKPOINTS

By the time you finish reading this chapter, you should be able to answer the following questions:

1. What is a seed, and why was the evolution of the seed such an important innovation for plants?

2. From which group of plants is it hypothesized that seed plants evolved? Why?

3. How do the mechanisms by which sperm reach the eggs differ in gymnosperms and seedless vascular plants?

4. What are the distinguishing features of the four phyla of living gymnosperms?

5. In what ways do gnetophytes resemble angiosperms?

One of the most dramatic innovations to arise during the evolution of the vascular plants was the seed. Seeds are one of the principal factors responsible for the dominance of seed plants in today's flora—a dominance that has become progressively greater over a period of several hundred million years. The reason is simple: the seed has great survival value. The protection that a seed affords the enclosed embryo, and the stored food that is available to that embryo at the critical stages of germination and establishment, give seed plants a great selective advantage over their free-sporing relatives and ancestors, that is, over plants that shed their spores.

Evolution of the Seed

All seed plants are heterosporous, producing megaspores and microspores that give rise respectively to megagametophytes and microgametophytes—but that trait is not unique to seed plants. As we have discussed in Chapter 19, some seedless vascular plants are also

20–2
Longitudinal section of an ovule, which consists of a megasporangium (nucellus) enveloped by an integument with an opening, the micropyle, at its apical end. A single functional megaspore is retained within the megasporangium (not shed) and will give rise to a megagametophyte, which is retained within the megasporangium. Following fertilization, the ovule matures into a seed, which becomes the unit of dispersal.

heterosporous. The production of seeds is, however, a particularly extreme form of heterospory that has been modified to form an ovule, the structure that develops into the seed. Indeed, a **seed** is simply a mature ovule containing an embryo. The immature ovule consists of a megasporangium surrounded by one or two additional layers of tissue, the **integuments** (Figure 20–2).

Several events led to the evolution of an ovule, including: (1) retention of the megaspores within the megasporangium, which is fleshy and called the **nucellus** in seed plants—in other words, the megasporangium no longer releases the spores; (2) reduction in the number of megaspore mother cells in each megasporangium to one; (3) survival of only one of the four megaspores produced by the spore mother cell, leaving a single functional megaspore in the megasporangium; (4) formation of a highly reduced megagametophyte inside the single functional megaspore—that is, formation of an endosporic (within the wall) megagametophyte that is no longer free living—which is retained within the megasporangium; (5) development of the embryo, or young sporophyte, within the megagametophyte retained within the megasporangium; (6) formation of an integument that completely envelops the megasporangium except for an opening at its apex called the **micropyle;** and (7) modification of the apex of the megasporangium to receive microspores or pollen grains. Related to these events is a basic shift in the unit of dispersal from the megaspore to the seed, the integumented megasporangium containing the mature embryo.

The Fossil Record Provides Clues to Ovule Evolution

The exact order in which these events occurred is unknown because of the incompleteness of the fossil record. They occurred fairly early in the history of vascular plants, however, because the oldest ovules or seeds are from the Late Devonian (about 365 million years ago). One of these early seed plants is *Elkinsia polymorpha* (Figure 20–3). The ovule of *Elkinsia* consisted of a nucellus and four or five integumentary lobes with little or no fusion between lobes, which led some paleobotanists to name these structures "preovules." The integumentary lobes curved inward at their tips, forming a ring around the apex of the nucellus. The apex of the nucellus was modified into a barrel-like structure for the reception of pollen. The ovules were surrounded by dichotomously branched, sterile structures called cupules.

A slightly younger Late Devonian seed plant is *Archaeosperma arnoldii* (Figure 20–4). Only the apical portion of the integument of *Archaeosperma* was divided into lobes, which formed a rudimentary micropyle. The integuments of ovules apparently evolved through the gradual fusion of integumentary lobes until the only opening left was the micropyle (Figure 20–5).

20–3

Reconstruction of a fertile branch of the Late Devonian plant Elkinsia polymorpha, *showing its ovules. Each ovule was overtopped by a dichotomously branched, sterile structure called a cupule. Note the more or less free lobes of the integument.*

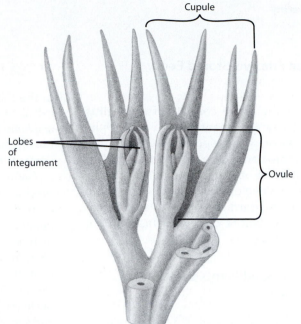

20–4

(a) Reconstruction of a fertile branch of the Late Devonian plant Archaeosperma arnoldii, *showing four ovules. The cupules, which partly enclose the ovules, are arranged in pairs, and each cupule contains two flask-shaped ovules about 4 millimeters long. The apex of each integument was lobed.* **(b)** *Diagram of the ovule, showing the position of a megaspore tetrad. The three aborted megaspores are found at the top of the functional megaspore. The lobes of the integument form a rudimentary micropyle. The question marks indicate the presumed position of the nucellus.* **(c)** *A megaspore. This fossil, from Pennsylvania, is the oldest known seedlike structure—about 360 million years in age.*

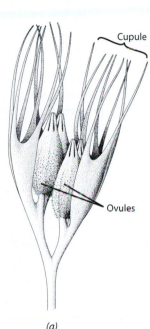

(a)

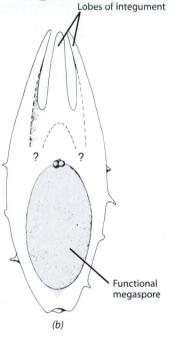

(b)

(c)

20–5

Seedlike structures in a number of Paleozoic plants, showing some of the potential stages in the evolution of the integument. **(a)** *In* Genomosperma kidstonii *(the Greek word* genomein *means "to become," and* sperma *means "seed"), eight fingerlike projections arise at the base of the megasporangium and are separated for their entire length.* **(b)** *In* Genomosperma latens *the integumentary lobes are fused from the base of the megasporangium for about a third of their length.* **(c)** *In* Eurystoma angulare *fusion is almost complete, while* **(d)** *in* Stamnostoma huttonense *it is complete, with only the micropyle remaining open at the top.*

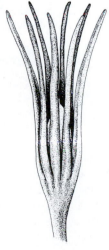

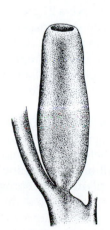

Genomosperma kidstonii
(a)

Genomosperma latens
(b)

Eurystoma angulare
(c)

Stamnostoma huttonense
(d)

A Seed Consists of an Embryo, Stored Food, and a Seed Coat

In modern seed plants the ovule consists of a nucellus enveloped by one or two integuments with a micropyle. When the ovules of most gymnosperms are ready for fertilization, the nucellus contains a megagametophyte composed of nutritive tissue and archegonia. After fertilization, the integuments develop into a **seed coat,** and a seed is formed. In most modern seed plants, an embryo develops within the seed before dispersal. In addition, all seeds contain stored food.

There Are Five Phyla of Seed Plants with Living Representatives

Seed plants arose starting in the Late Devonian period, at least 365 million years ago. During the next 50 million years, a wide array of seed-bearing plants evolved, many of which are grouped together as the so-called seed ferns, while others are recognized as cordaites and conifers (see "Coal Age Plants" on pages 456 and 457).

The seed plants, all of which typically possess megaphylls (generally large leaves with several to many veins, but modified to needles or scales in some groups), include five phyla with living representatives: the *Cycadophyta,* the *Ginkgophyta,* the *Coniferophyta,* the *Gnetophyta,* and the *Anthophyta.* The phylum *Anthophyta* comprises the angiosperms, or flowering plants, and the remaining four phyla are often grouped together as the gymnosperms.

The oldest fossil angiosperms are only about 135 million years old, from the early part of the Cretaceous period. Although the angiosperms—overwhelmingly the most successful vascular plants at the present time—actually may be somewhat older than our current understanding of the fossil record indicates, they are still relative newcomers in the broad picture of vascular plant evolution. Because they are such a large and important group, the angiosperms are considered in detail in Chapters 21 and 22.

The gymnosperms do not constitute an evolutionary line that is equivalent to the angiosperms. Instead, the gymnosperms represent a series of evolutionary lines of seed-bearing plants that lack the distinctive characteristics of the angiosperms. Although there are only about 720 species of living gymnosperms—compared with some 235,000 species of angiosperms—individual gymnosperm species are often dominant over wide areas.

Before beginning our discussion of seed plants, we shall briefly examine one more group of seedless vascular plants—the progymnosperms. They are discussed here, rather than in Chapter 19, because they are the likely progenitors of seed plants.

Progymnosperms

In the Late Paleozoic era, there existed a group of plants called the progymnosperms (phylum *Progymnospermophyta*), which had characteristics intermediate between those of the seedless vascular trimerophytes and those of the seed plants. Although the progymnosperms reproduced by means of freely dispersed spores, they produced secondary xylem (wood) remarkably similar to that of living conifers (Figure 20–6). The progymnosperms were unique among woody Devonian plants in that they also produced secondary phloem. Both the progymnosperms and Paleozoic ferns probably evolved from the more ancient trimerophytes (see Figure 19–10c), from which they differed primarily by having more elaborate and more highly differentiated branch systems and correspondingly more complex vascular systems.

In the progymnosperms, the most important evolutionary advance over both the trimerophytes and the ferns is the presence of a *bifacial vascular cambium*—that is, one that produces both secondary xylem and secondary phloem. Vascular cambia of this type are characteristic of seed plants and apparently evolved first in the progymnosperms.

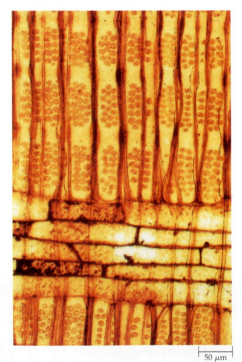

50 μm

20–6
Radial view of the secondary xylem, or wood, of the progymnosperm Callixylon newberryi. *This fossil wood, with its regular series of pitted tracheids, is remarkably similar to the wood of certain conifers.*

20–7
Reconstruction of a portion of the branch system of Triloboxylon ashlandicum, an Aneurophyton-*type progymnosperm. The main axis bears vegetative branches at the top and bottom and fertile organs with sporangia in between.*

20–8
Reconstruction of the progymnosperm Archaeopteris, *which is common in the fossil record of eastern North America. Specimens* of Archaeopteris *attained heights of 20 meters or more, and some of them seem to have formed forests.*

20–9
Reconstruction of a frondlike lateral branch system of the progymnosperm Archaeopteris macilenta. *Fertile leaves can be seen bearing maturing sporangia (seen here as brown) on centrally located primary branches.*

One kind of progymnosperm, the *Aneurophyton* type, which occurred in the Devonian period approximately 362 to 380 million years ago, was characterized by complex three-dimensional branching (Figure 20–7) and had a solid cylinder of vascular tissue, or protostele. In many respects these progymnosperms were similar to the more complex trimerophytes, but they also resemble some of the early seed ferns, which has led some paleobotanists to suggest that the branch systems of the *Aneurophyton*-type progymnosperms may have been the precursors of the fernlike leaves of early seed ferns.

A second major kind of progymnosperm, the *Archaeopteris* type, also appeared in the Devonian period, about 370 million years ago, and extended into the Mississippian period, about 340 million years ago (Figure 20–8). In this group, the lateral branch systems were flattened in one plane and bore laminar structures considered to be leaves (Figure 20–9). A eustele—that is, an

arrangement of vascular tissues in discrete strands around a pith—apparently evolved in this group of progymnosperms and is a strong similarity linking this group with the living seed plants (page 429). The larger branches of *Archaeopteris*-type progymnosperms had a pith. Although most progymnosperms were homosporous, some species of *Archaeopteris* were heterosporous.

Fossil logs of *Archaeopteris*, called *Callixylon*, may be up to a meter or more in diameter and 10 meters long, indicating that at least some species of this group were large trees. They appear to have formed extensive forests in some regions. As the reconstruction in Figure 20–8 suggests, individuals of *Archaeopteris* may have resembled conifers in their branching patterns.

Evidence accumulated during the past several decades strongly indicates that the seed plants evolved from the progymnosperms following the appearance of the seed in what now seems to have been the common

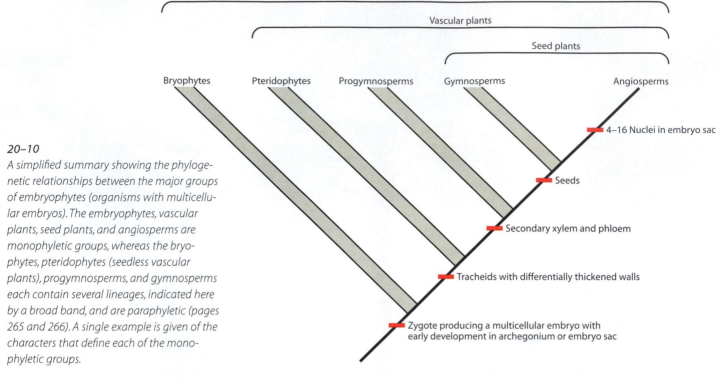

Embryophytes

Vascular plants

Seed plants

Bryophytes | Pteridophytes | Progymnosperms | Gymnosperms | Angiosperms

4–16 Nuclei in embryo sac

Seeds

Secondary xylem and phloem

Tracheids with differentially thickened walls

Zygote producing a multicellular embryo with early development in archegonium or embryo sac

20–10
A simplified summary showing the phylogenetic relationships between the major groups of embryophytes (organisms with multicellular embryos). The embryophytes, vascular plants, seed plants, and angiosperms are monophyletic groups, whereas the bryophytes, pteridophytes (seedless vascular plants), progymnosperms, and gymnosperms each contain several lineages, indicated here by a broad band, and are paraphyletic (pages 265 and 266). A single example is given of the characters that define each of the monophyletic groups.

ancestor of all seed plants (Figure 20–10). However, many problems still remain to be solved in developing a more detailed understanding of the early evolution of seed plants.

Extinct Gymnosperms

Two groups of extinct gymnosperms—the seed ferns (phylum *Pteridospermophyta*) and the *Cordaitales*, primitive coniferlike plants—were discussed and illustrated in Chapter 19 (pages 456 and 457). The seed ferns, or pteridosperms, are a very diverse and highly unnatural group that range in age from the Devonian to the Jurassic. They range in form from Late Devonian slender, branched plants with ovules, such as *Elkinsia* and *Archaeosperma*, to Carboniferous plants, such as *Medullosa*, that had the appearance of tree ferns (see Figure 20–1). Several groups of extinct Mesozoic plants are also sometimes included in the seed ferns. Exactly how these different groups of seed ferns relate to the living gymnosperms remains uncertain.

Another group of extinct gymnosperms—the cycadeoids, or *Bennettitales*—consisted of plants with palmlike leaves, somewhat resembling the living cycads (Figure 20–11a; see page 486). The *Bennettitales* are an enigmatic group of Mesozoic gymnosperms that disappeared from the fossil record during the Cretaceous. Some paleobotanists believe that the *Bennettitales* may have been members of the same evolutionary line as angiosperms. Currently, it is still uncertain exactly where the *Bennettitales* fit in phylogenetically. The *Bennettitales* lived at the same time as the extinct cycads, and both groups produced very similar leaves through most of

the Mesozoic. Reproductively, however, *Bennettitales* were distinct from the cycads in several respects, including the presence of flowerlike reproductive structures that were bisexual in some species (Figure 20–11b).

Living Gymnosperms

There are four phyla of gymnosperms with living representatives: *Cycadophyta* (cycads), *Ginkgophyta* (maidenhair tree, or ginkgo), *Coniferophyta* (conifers), and *Gnetophyta* (gnetophytes). The name *gymnosperm*, which literally means "naked seed," points to one of the principal characteristics of the plants belonging to these four phyla—namely, that their ovules and seeds are exposed on the surface of sporophylls and analogous structures.

With few exceptions, the female gametophyte of gymnosperms produces several archegonia. As a result, more than one egg may be fertilized, and several embryos may begin to develop within a single ovule—a phenomenon known as **polyembryony.** In most cases, however, only one embryo survives and therefore relatively few fully developed seeds contain more than one embryo.

In the seedless vascular plants, water is required for the motile, flagellated sperm to reach and fertilize the eggs. In the gymnosperms, however, water is not required as a medium of transport of the sperm to the eggs. Instead, the partly developed male gametophyte, the **pollen grain,** is transferred bodily (usually passively, by the wind) to the vicinity of a female gametophyte within an ovule. This process is called **pollination.** After pollination, the endosporic male gametophyte produces a tubular outgrowth, the **pollen tube.** The male gameto-

(a)

20–11
Bennettitales. *(a) Reconstruction of Wielandiella, an extinct gymnosperm from the Triassic. Wielandiella exhibits a forked branching pattern. A single strobilus, or cone, is borne at each fork. (b) A diagrammatic reconstruction of the bisporangiate, or bisexual, strobilus of Williamsoniella coronata from the Jurassic. The strobilus consists of a central ovulate receptacle surrounded by a whorl of microsporophylls bearing microsporangia containing microspores, which develop into male gametophytes (pollen grains). Hairy bracts enclose the reproductive parts.*

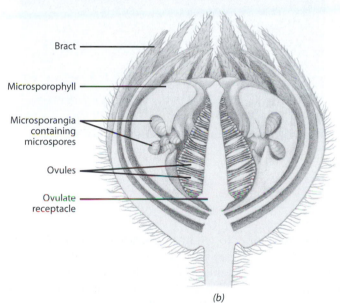

Bract

Microsporophyll

Microsporangia containing microspores

Ovules

Ovulate receptacle

(b)

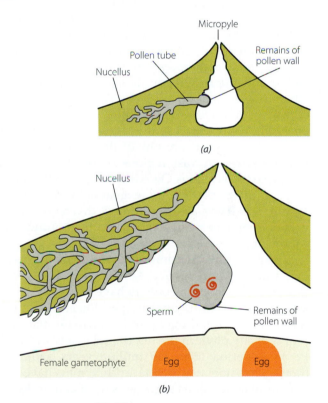

(a)

(b)

20–12
Development of the male gametophyte of Ginkgo biloba. (a) Early in its development the pollen tube grows by tip growth and begins to form what will become a highly branched haustorial structure. The pollen tube in Ginkgo grows intercellularly in the nucellus. (b) Late in development, the basal end of the pollen tube enlarges into a saclike structure that contains the two multiflagellated sperm. Subsequently, the basal end of the pollen tube ruptures, releasing the two sperm, which then swim to the eggs contained in the archegonia of the female gametophyte.

phytes of gymnosperms and other seed plants do not form antheridia.

In the conifers and gnetophytes, the sperm are nonmotile, and the pollen tubes convey them directly to the archegonia. In the cycads and *Ginkgo*, fertilization is transitional between the condition found in ferns and other seedless plants, in which free-swimming sperm occur, and the condition found in other seed plants, which have nonmotile sperm. The male gametophytes of cycads and *Ginkgo* produce a pollen tube, but this does not penetrate the archegonium (Figure 20–12). Instead, it is haustorial and may grow for several months in the tissue of the nucellus where it appears to play a role in absorbing nutrients. Eventually, the pollen grain bursts in the vicinity of the archegonium, releasing multiflagellated, swimming sperm cells (see Figure 20–39). The sperm then swim to an archegonium, and one of them fertilizes the egg.

In conifers, gnetophytes, and angiosperms the pollen tube conveys the sperm to the egg cell. With this innovation, seed plants were no longer dependent on the presence of free water to ensure fertilization—a necessity for all seedless plants. The presence of haustorial pollen tubes in *Ginkgo* and cycads suggests that, originally, the pollen tube evolved to absorb nutrients for the production of sperm by the male gametophyte during its growth within the ovule. From this perspective, the conveyance of nonmotile sperm by a pollen tube that grows directly to an egg can be seen as a later evolutionary modification of a structure initially developed for another purpose.

Phylum *Coniferophyta*

By far the most numerous, most widespread, and most ecologically important of the gymnosperm phyla living today are the *Coniferophyta*, which comprise some 50 genera with about 550 species. The tallest vascular plant, the redwood *(Sequoia sempervirens)* of coastal California and southwestern Oregon, is a conifer. Redwood trees attain heights of up to 117 meters and trunk diameters in excess of 11 meters. The conifers, which also include pines, firs, and spruces, are of great commercial value. Their stately forests are one of the most important natural resources in vast regions of the north temperate zone (see "Jobs versus Owls" in Chapter 33). During the Early Tertiary period, some genera were more widespread than they are now, and a diverse conifer flora was present across huge expanses on all of the northern continents.

The history of the conifers extends back at least to the Late Carboniferous period, some 300 million years ago. The leaves of modern conifers have many drought-resistant features, which may bring ecological advantages in certain habitats and may also be related to the diversification of the phylum during the relatively dry and cold Permian period (290–245 million years ago). At that time, increasing worldwide aridity may have favored structural adaptations such as those of conifer leaves.

Pines Are Conifers with a Unique Leaf Arrangement

The pines (genus *Pinus*), which include perhaps the most familiar of all gymnosperms (Figure 20–13), dominate broad stretches of North America and Eurasia and are widely cultivated even in the Southern Hemisphere. There are about 90 species of pines, all of which are characterized by an arrangement of the leaves that is unique among living conifers. In pine seedlings, the needlelike leaves are spirally arranged and borne singly on the stems (Figure 20–14). After a year or two of growth, a pine begins to produce its leaves in bundles, or fascicles, each of which contains a specific number of leaves (nee-

20–13
Longleaf pines, Pinus palustris, *growing in North Carolina.*

dles)—from one to eight, depending on the species (Figure 20–15). These fascicles, wrapped at the base by a series of short, scalelike leaves, are actually short shoots in which the activity of the apical meristem is suspended. Thus, a fascicle of needles in a pine is morphologically a **determinate** (restricted in growth) branch. Under unusual circumstances, the apical meristem within a fascicle of needles in a pine may be reactivated and grow into a new shoot with **indeterminate** growth, or sometimes may even produce roots and grow into an entire pine tree (Figure 20–16).

The leaves of pines, like those of many other conifers, are impressively suited for growth under conditions where water may be scarce or difficult to obtain (Figure 20–17). A thick cuticle covers the epidermis, beneath which are one or more layers of compactly arranged, thick-walled cells—the hypodermis. The stomata are sunken below the surface of the leaf. The mesophyll, or ground tissue of the leaf, consists of parenchyma cells with conspicuous wall ridges that project into the cells, increasing their surface area. Commonly, the mesophyll is penetrated by two or more resin ducts. One vascular bundle, or two bundles side by side, are found in the center of the leaf. The vascular bundles, made up of xylem and phloem, are surrounded by transfusion tis-

(a)

20–15

Bristlecone pine, Pinus longaeva, *in the White Mountains of California. Branch showing fascicles of five needles each, a mature ovulate cone on the right, and a young ovulate cone on the left. The individual needles of this species may remain functional for up to 45 years. It is also the longest-lived tree (see Figure 27–28).*

(b)

20–14

(a) Seedlings of longleaf pine, Pinus palustris, *in Georgia, showing the long leaves in fascicles, or bundles, of three. (b) A seedling of pinyon pine,* Pinus edulis, *showing spirally arranged juvenile leaves and a young taproot system. The mature leaves of this species are borne in fascicles of two needles each.*

20–16

*One-year-old Monterey pines (*Pinus radiata*) grown from rooted fascicles of needles. This experiment demonstrates that a fascicle of pine needles is actually a short shoot in which the activity of the apical meristem has been suspended but can be regenerated.*

sue, composed of living parenchyma cells and short, nonliving tracheids. The transfusion tissue is believed to conduct materials between the mesophyll and the vascular bundles. A single layer of cells known as the endodermis surrounds the transfusion tissue, separating the transfusion tissue from the mesophyll.

Most pine species retain their needles for two to four years, and the overall photosynthetic balance of a given plant depends on the health of several years' crops of needles. In bristlecone pine (*Pinus longaeva*), the longest-lived tree (Figure 20–15), the needles are retained for up to 45 years and remain photosynthetically active the en-

200 μm

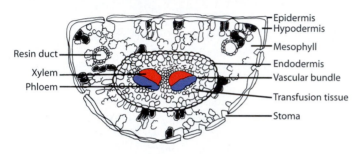

Resin duct

Xylem
Phloem

Epidermis
Hypodermis

Mesophyll

Endodermis

Vascular bundle

Transfusion tissue

Stoma

20–17

Pinus. *Transverse section of needle, showing
mature tissues.*

tire time. Because the leaves of pines and other ever-
greens function for more than one season, they are exposed to possible damage by drought, freezing, or air
pollution for much longer than the leaves of deciduous
plants, which are replaced each year.

In the stems of pines and other conifers, secondary
growth begins early and leads to the internal formation
of substantial amounts of secondary xylem, or wood
(Figure 20–18). Secondary xylem is produced toward the
inside of the vascular cambium, and secondary phloem
is produced toward the outside. The xylem of conifers
consists primarily of tracheids, while the phloem consists of sieve cells, which are the typical food-conducting
cells of gymnosperms (see page 574). Both kinds of tissues are traversed radially by narrow rays. With the initiation of secondary growth, the epidermis is eventually
replaced with a periderm, which has its origin in the
outer layer of cortical cells. As secondary growth continues, subsequent periderms are produced by active cell
division deeper in the bark.

The Pine Life Cycle Extends over a Period of Two Years
As you read this account of reproduction in pines, you
may find it useful to refer, from time to time, to the pine
life cycle (Figure 20–22) on pages 478 and 479.

The microsporangia and megasporangia in pines and
most other conifers are borne in separate cones, or strobili, on the same tree. Ordinarily, the microsporangiate
(pollen-producing) cones are borne on the lower branches
of the tree and the megasporangiate, or *ovulate*, cones are

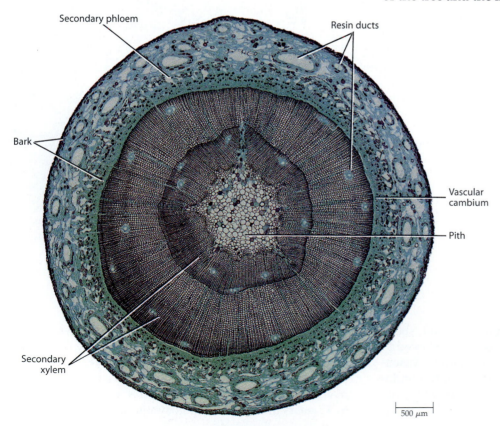

Secondary phloem

Resin ducts

Bark

Vascular
cambium

Pith

Secondary
xylem

500 μm

20–18

Pinus. *Cross section of stem, showing secondary xylem and secondary phloem separated from one another by vascular
cambium. All of the tissues outside the vascular cambium, including the phloem, make up
the bark.*

20–19

Monterey pine, Pinus radiata. *Microsporangiate cones shedding pollen, which is blown about by the wind. Some of the pollen reaches the vicinity of the ovules in the ovulate cones and then germinates, producing pollen tubes and eventually bringing about fertilization.*

borne on the upper branches. In some pines, they are borne on the same branch, with the ovulate cones closer to the tip. Since the pollen is not normally blown straight upward, the ovulate cones are usually pollinated by pollen from another tree, thus enhancing outcrossing.

Microsporangiate cones in the pines are relatively small, usually 1 to 2 centimeters long (Figure 20–19). The microsporophylls (Figure 20–20) are spirally arranged and more or less membranous. Each bears two microsporangia on its lower surface. A young microsporangium contains many **microsporocytes,** or **microspore mother cells.** In early spring the microsporocytes undergo meiosis and each produces four haploid microspores. Each microspore develops into a winged pollen grain, consisting of two **prothallial cells,** a **generative cell,** and a **tube cell** (Figure 20–21). This four-celled pollen grain is the immature male gametophyte. It is at this stage that the pollen grains are shed in enormous quantities; some are carried by the wind to the ovulate cones.

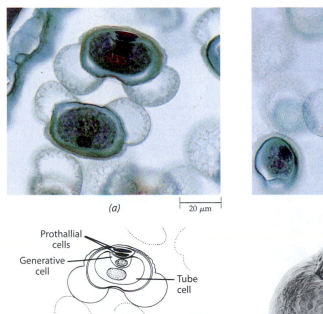

(a) 20 μm

(b) 20 μm

Prothallial cells
Generative cell
Tube cell

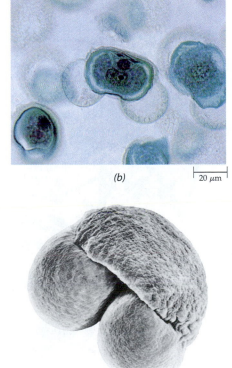

(c) 10 μm

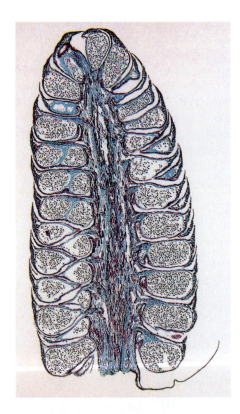

20–20

Pinus. *Longitudinal view of a microsporangiate (pollen-producing) cone, showing microsporophylls and microsporangia containing mature pollen grains.*

20–21

Pinus. **(a)** *Pollen grains with enclosed immature male gametophytes. Each gametophyte consists of two prothallial cells, a relatively small generative cell, and a relatively large tube cell.* **(b)** *A somewhat older pollen grain. Here the prothallial cells, which have no*
apparent function, have degenerated. **(c)** *Scanning electron micrograph of a pine pollen grain, with its two bladder-shaped wings. When the pollen grain germinates, the pollen tube emerges from the lower end of the grain between the wings.*

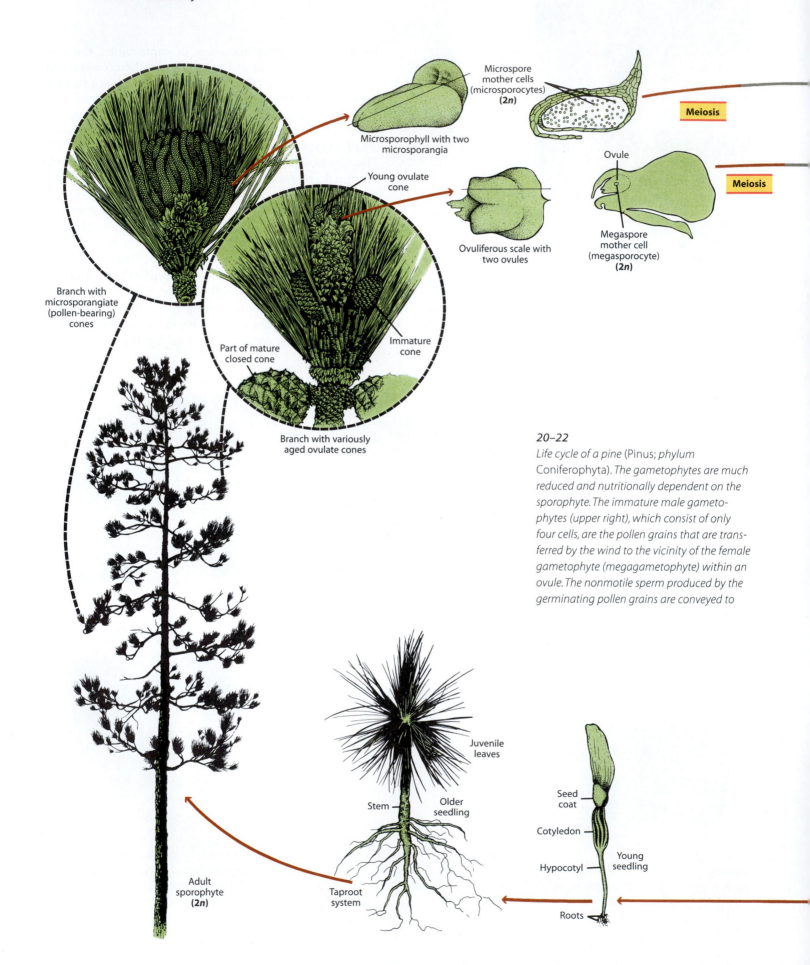

Microspore
mother cells
(microsporocytes)
(**2n**)

Meiosis

Microsporophyll with two
microsporangia

Ovule

Meiosis

Young ovulate
cone

Ovuliferous scale with
two ovules

Megaspore
mother cell
(megasporocyte)
(**2n**)

Branch with
microsporangiate
(pollen-bearing)
cones

Part of mature
closed cone

Immature
cone

Branch with variously
aged ovulate cones

20–22

*Life cycle of a pine (Pinus; phylum
Coniferophyta). The gametophytes are much
reduced and nutritionally dependent on the
sporophyte. The immature male gametophytes (upper right), which consist of only
four cells, are the pollen grains that are transferred by the wind to the vicinity of the female
gametophyte (megagametophyte) within an
ovule. The nonmotile sperm produced by the
germinating pollen grains are conveyed to*

Juvenile
leaves

Stem

Older
seedling

Seed
coat

Cotyledon

Young
seedling

Hypocotyl

Adult
sporophyte
(**2n**)

Taproot
system

Roots

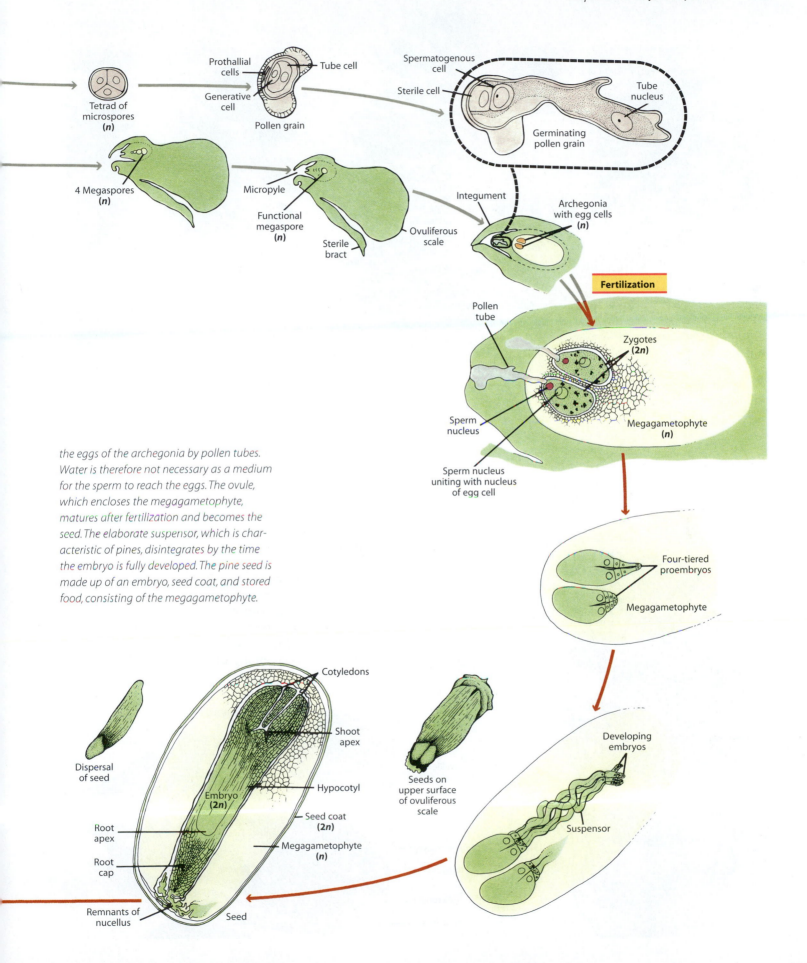

Prothallial cells

Tube cell

Tetrad of microspores
(*n*)

Generative cell

Pollen grain

Spermatogenous cell

Sterile cell

Tube nucleus

Germinating pollen grain

4 Megaspores
(*n*)

Micropyle

Functional megaspore
(*n*)

Sterile bract

Integument

Ovuliferous scale

Archegonia with egg cells
(*n*)

Fertilization

Pollen tube

Zygotes
(**2***n*)

Sperm nucleus

Sperm nucleus uniting with nucleus of egg cell

Megagametophyte
(*n*)

the eggs of the archegonia by pollen tubes. Water is therefore not necessary as a medium for the sperm to reach the eggs. The ovule, which encloses the megagametophyte, matures after fertilization and becomes the seed. The elaborate suspensor, which is characteristic of pines, disintegrates by the time the embryo is fully developed. The pine seed is made up of an embryo, seed coat, and stored food, consisting of the megagametophyte.

Four-tiered proembryos

Megagametophyte

Cotyledons

Shoot apex

Dispersal of seed

Embryo
(**2***n*)

Hypocotyl

Seed coat
(**2***n*)

Megagametophyte
(*n*)

Root apex

Root cap

Seeds on upper surface of ovuliferous scale

Developing embryos

Suspensor

Remnants of nucellus

Seed

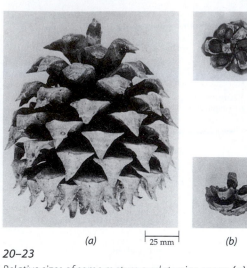

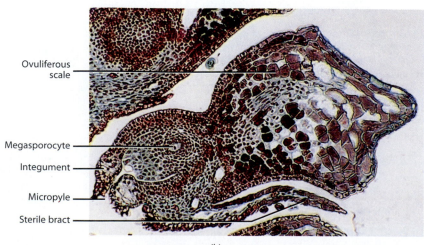

20–23

Relative sizes of some mature ovulate pine cones. **(a)** *Digger pine,* Pinus sabiniana. **(b)** *A pinyon pine,* Pinus edulis, *in top and side views. The wingless, edible seeds of this and certain other pines are called "pine nuts."* **(c)** *Sugar pine,* Pinus lambertiana. **(d)** *Yellow pine,* Pinus ponderosa. **(e)** *Eastern white pine,* Pinus strobus. **(f)** *Red pine,* Pinus resinosa.

25 mm

(a)   *(b)*   *(c)*   *(d)*   *(e)*   *(f)*

20–24

Pinus. **(a)** *Longitudinal view of young ovulate cone, showing its complex structure.* **(b)** *Detail of a portion of the cone. Note the megasporocyte (megaspore mother cell) surrounded by the nucellus. (See also the pine life cycle, Figure 20–22.)*

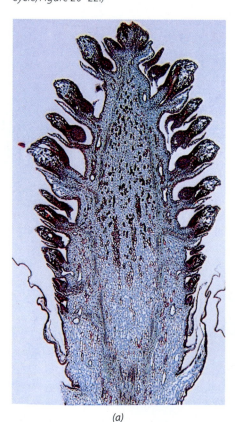

(a)

The ovulate cones of pines are much larger and more complex in their structure than are the pollen-bearing cones (Figure 20–23). The **ovuliferous scales** (cone scales), which bear the ovules, are not simply megasporophylls. Instead, they are entire modified determinate branch systems properly known as **seed-scale complexes.** Each seed-scale complex consists of the ovuliferous scale—which bears two ovules on its upper surface—and a subtending sterile bract (Figure 20–24). The scales are arranged spirally around the axis of the cone. (The ovulate cone is, therefore, a compound structure, whereas the microsporangiate cone is a simple one, in which the microsporangia are directly attached to the microsporophylls.) Each ovule consists of a multicellular nucellus (the megasporangium) surrounded by a mas-

Ovuliferous scale

Megasporocyte

Integument

Micropyle

Sterile bract

(b)

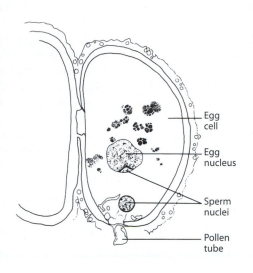

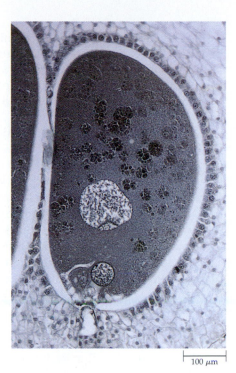

100 μm

20–25

Red ovulate cones of lodgepole pine (Pinus contorta) at the tip of a tall central branch in late spring of their first year, the time at which pollination occurs. One-year-old ovulate cones are visible at the base of this branch. Orange, pollen-bearing microsporangiate cones cluster around the shorter branches.

20–26

Pinus. *Fertilization: union of a sperm nucleus with the egg nucleus. The second sperm* *nucleus (below) is nonfunctional; it will eventually disintegrate.*

sive integument with an opening, the micropyle, facing the cone axis (Figure 20–24). Each megasporangium contains a single **megasporocyte,** or **megaspore mother cell,** which ultimately undergoes meiosis, giving rise to a linear series of four megaspores. However, only one of these megaspores is functional; the three nearest the micropyle soon degenerate.

Pollination in pines occurs in the spring (Figure 20–25). At this stage, the scales of the ovulate cone are widely separated. As the pollen grains settle on the scales, many adhere to pollination drops, which exude from the micropylar canals at the open ends of the ovules. As the pollination drops evaporate, they contract, carrying the pollen grains through the micropylar canal and into contact with the nucellus. At its micropylar end, the nucellus is slightly suppressed. The pollen grains come to lie in this shallow cavity. After pollination, the scales grow together and help protect the developing ovules. Shortly after the pollen grain comes in contact with the nucellus, it germinates to form a pollen tube. At this time, meiosis has not yet occurred in the megasporangium.

About a month after pollination, the four megaspores are produced, only one of which develops into a megagametophyte. The development of the megagame-tophyte is sluggish. It often does not begin until some six months after pollination, and even then may require another six months for completion. In the early stages of megagametophyte development, mitosis proceeds without immediate cell wall formation. About 13 months after pollination, when the female gametophyte contains some 2000 free nuclei, cell wall formation begins. Then, approximately 15 months after pollination, archegonia, usually two or three in number, differentiate at the micropylar end of the megagametophyte, and the stage is set for fertilization.

About 12 months earlier, the pollen grain germinated, producing a pollen tube that slowly digested its way through the tissues of the nucellus toward the developing megagametophyte. About a year after pollination, the generative cell of the four-celled male gametophyte undergoes division, giving rise to two kinds of cells—a **sterile cell** (stalk cell) and a **spermatogenous cell** (body cell). Subsequently, before the pollen tube reaches the female gametophyte, the spermatogenous cell divides, producing two sperm. The male gametophyte, or germinating pollen grain, is now mature. Recall that seed plants do not form antheridia.

Some 15 months after pollination, the pollen tube reaches the egg cell of an archegonium, where it discharges much of its cytoplasm and both of its sperm into the egg cytoplasm (Figure 20–26). One sperm nucleus unites with the egg nucleus, and the other degenerates.

Commonly, the eggs of all archegonia are fertilized and begin to develop into embryos (the phenomenon of polyembryony). Only one embryo usually develops fully, but about 3 to 4 percent of pine seeds contain more than one embryo and produce two or three seedlings upon germination.

During early embryogeny, four tiers of cells are produced near the lower end of the archegonium. Each of the four cells of the uppermost tier (that is, the tier farthest from the micropylar end of the ovule) begins to form an embryo. Simultaneously the four cells of the tier below the embryos, the suspensor cells, elongate greatly and force the four developing embryos through the wall of the archegonium and into the female gametophyte. Thus, a second type of polyembryony is found in the pine life cycle. Once again, however, usually only one of the embryos develops fully. During embryogeny, the integument develops into a seed coat.

The conifer seed is a remarkable structure, for it consists of a combination of two different diploid sporophytic generations—the seed coat (and remnants of the nucellus) and the embryo—and one haploid gametophytic generation (Figure 20–27). The gametophyte serves as a food reserve or nutritive tissue. The embryo consists of a hypocotyl-root axis, with a rootcap and apical meristem at one end and an apical meristem and several (generally eight) cotyledons, or seed leaves, at the other. The integument consists of three layers, of which the middle layer becomes hard and serves as the seed coat.

The seeds of pines are often shed from the cones during the autumn of the second year following the initial appearance of the cones and pollination. At maturity, the cone scales separate; the winged seeds of most species flutter through the air and are sometimes carried considerable distances by the wind. In some species of pines, such as lodgepole pines (*Pinus contorta*), the scales do not separate until the cones are subjected to extreme heat (see "The Great Yellowstone Fire" in Chapter 32). When a forest fire sweeps rapidly through the pine grove and burns the parent trees, most of the fire-resistant cones are only scorched. These cones open, releasing the seed crop accumulated over many years, and reestablish the species. In other species of pines, including the limber pine (*Pinus flexilis*), whitebark pine (*Pinus albicaulis*), and pinyon pines of western North America, as well as a few similar species of Eurasia, the wingless, large seeds are harvested, transported, and stored for later eating by large, crowlike birds called nutcrackers. The birds miss many of the seeds they store, aiding in the dispersal of the pines.

The pine life cycle is summarized in Figure 20–22.

Other Important Conifers Occur throughout the World

Although other conifers lack the needle clusters of pines, and may also differ in a number of relatively minor details of their reproductive systems, the living conifers form a fairly homogeneous group. In most conifers other than the pines, the reproductive cycle takes only a year; that is, the seeds are produced in the same season as the ovules are pollinated. In such conifers, the time between pollination and fertilization usually ranges from three days to three or four weeks, instead of 15 months or so.

Among the important genera of conifers other than

20–27

Pinus. *Longitudinal section of a seed. The hard protective seed coat (here removed) and embryo represent successive sporophyte (2n) generations, with a gametophyte generation intervening. A remnant of the nucellus (megasporangium) forms a papery shell around the gametophyte.*

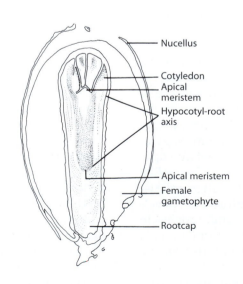

500 μm

(a)

(b)

20–28
Two genera of the pine family (Pinaceae). *(**a**)
Ovulate cone of balsam fir* (Abies balsamea).
*The upright cones, which are 5 to 10 centime-
ters long, do not fall to the ground whole, as
they do in the pines. Instead, these cones
shatter and fall apart while still attached,
scattering the winged seeds. (**b**) European
larch* (Larix decidua). *The needlelike leaves of
larch, which are borne singly on both long
shoots and short branch shoots, are spirally
arranged. Unlike most conifers, the larches are
deciduous; that is, they shed their leaves at
the end of each growing season.*

pines are the firs (*Abies;* Figure 20–28a), larches (*Larix;*
Figure 20–28b), spruces (*Picea*), hemlocks (*Tsuga*),
Douglas firs (*Pseudotsuga*), cypresses (*Cupressus;* Figure
20–29), and junipers (*Juniperus;* Figure 20–30), often misleadingly called "cedars" in North America. In the yews
(family *Taxaceae*), the ovules are not borne in cones but
are solitary and surrounded by a fleshy, cuplike structure—the **aril** (Figure 20–31a).

One of the most interesting groups of conifers is the
family *Araucariaceae*, the living members of which occur
naturally only in the Southern Hemisphere. Until recently, only two surviving genera—*Agathis* and
Araucaria—were known to exist (see the essay on page
488). A species of *Araucaria*, called Panama pine, is one
of the most valuable timber trees in South America.
Some species of *Araucaria*, such as the monkey-puzzle
tree (*Araucaria araucana*) of Chile and the Norfolk Island
pine (*Araucaria heterophylla*), are frequently cultivated in
areas where the climate is mild (Figure 20–32). Seedlings
of Norfolk Island pines are also grown as houseplants.

Another interesting group of conifers is the family
Taxodiaceae. Taxodiaceous woods are found in the
Triassic, and a number of fossils (leaves and cones) date
from the Middle Jurassic (165–185 million years ago).
The *Taxodiaceae* are represented today by widely scattered species that are the remnants of populations that
were much more widespread during the Tertiary period
(Figure 20–33). One of the most remarkable of these is
the coast redwood, *Sequoia sempervirens*, the tallest living
plant. The famous "big tree," *Sequoiadendron giganteum*
(Figure 20–34), which forms spectacular, widely scattered groves along the western slope of the Sierra
Nevada of California, and the bald cypresses (*Taxodium*)
of the southeastern United States and Mexico (Figure
20–35) also belong to this family.

20–29
*In the subglobose cones of the cypresses, the
scales are crowded together, as in this Gowen
cypress* (Cupressus goveniana). *The small
trees of this species—only about 6 meters tall
at maturity—are extremely local, being found
only near Monterey, California.*

20–30

The common juniper (Juniperus communis) has spherical ovulate cones like those of the cypresses, but in juniper the scales are fleshy and fused together. Juniper "berries" give gin its distinctive taste and aroma.

20–32

Norfolk Island pine (Araucaria heterophylla).

(a)

(b)

20–31

The conifers of the yew family (Taxaceae) have seeds that are surrounded by a fleshy cup—the aril. The arils attract birds and other animals, which eat them and thus spread the seeds. (a) Members of the genus Taxus, the yews—which occur around the Northern Hemisphere—produce red, fleshy ovulate structures. (b) Sporophylls and microsporangia of the pollen-bearing cones of a yew. Ovulate cones and pollen-bearing cones are found on separate individuals. The seeds and leaves of yews contain a toxic substance and are an important cause of poisoning by plants in children in the United States, although fatalities are extremely rare.

20–33
Fossil branchlet of Metasequoia, about 50 million years old. The accompanying map shows the geographic distribution of some living and fossil members of the redwood family (Taxodiaceae).

▲ Fossil *Sequoia* (coast redwood)

▲ Fossil *Metasequoia* (dawn redwood)

20–34
The "big trees" (Sequoiadendron giganteum) of the western slope of California's Sierra Nevada are the largest of gymnosperms. The largest specimen, the General Sherman sequoia, is more than 80 meters high and estimated to weigh at least 2500 metric tons. The largest of living animals, the blue whale, pales by comparison. Blue whales rarely exceed 35 meters in length and 180 metric tons in weight.

20–35
The bald cypress (Taxodium distichum) is a deciduous member of the redwood family that grows in the swamps of the southeastern United States. Like the larch, it is one of the few conifers that sheds its leaves (actually leafy shoots) at the end of each growing season. In this autumn photograph, the leaves have begun to change color. Spanish "moss" (Tillandsia usneoides), actually a flowering plant related to the pineapple, is seen hanging in masses from the branches of these trees.

20–36
*The dawn redwood (Metasequoia glyp-
tostroboides). This tree, growing in Hubei
Province in central China, is more than 400
years old.*

Like most of the living genera of *Taxodiaceae*, *Metasequoia*, the dawn redwood, was much more widespread in the Tertiary period than it is now (Figure 20–36). *Metasequoia* occurred widely across Eurasia and was the most abundant conifer in western and Arctic North America from the Late Cretaceous period to the Miocene epoch (from about 90 to about 15 million years ago). It survived both in Japan and in eastern Siberia until a few million years ago. The genus *Metasequoia* was first described from fossil material by the Japanese paleobotanist Shigeru Miki in 1941 (Figure 20–33a). Three years later the Chinese forester Tsang Wang, from the Central Bureau of Forest Research of China, visited the village of Mo-tao-chi in remote Sichuan Province in

south-central China. There he discovered a huge tree of a kind he had never seen before. The natives of the area had built a temple around the base of the tree. Tsang collected specimens of the tree's needles and cones, and studies of these samples revealed that the fossil *Metasequoia* had "come to life." In 1948, paleobotanist Ralph Chaney of the University of California at Berkeley led an expedition down the Yangtze River and across three mountain ranges to valleys where dawn redwoods were growing, the last remnant of the once great *Metasequoia* forest. In 1980, about 8000 to 10,000 trees still existed in the *Metasequoia* valley, about 5000 of which had diameters over 20 centimeters. Unfortunately, the trees were not reproducing there, because the seeds were being harvested for cultivation and because of the lack of a suitable habitat in which seedlings could become established. Thousands of seeds have been distributed widely, however, and this "living fossil" can now be seen growing in parks and gardens all over the world.

The Other Living Gymnosperm Phyla: *Cycadophyta, Ginkgophyta,* and *Gnetophyta*

Cycads Belong to the Phylum *Cycadophyta*

The other groups of living gymnosperms are remarkably diverse and scarcely resemble one another at all. Among them are the cycads, phylum *Cycadophyta*, which are palmlike plants found mainly in tropical and subtropical regions. These unique plants, which appeared at least 250 million years ago during the Permian period, were so numerous in the Mesozoic era, along with the superficially similar *Bennettitales*, that this period is often called the "Age of Cycads and Dinosaurs." Living cycads comprise 11 genera, with about 140 species. *Zamia pumila*, found commonly in the sandy woods of Florida, is the only cycad native to the United States (Figure 20–37).

Most cycads are fairly large plants; some reach 18 meters or more in height. Many have a distinct trunk that is densely covered with the bases of shed leaves. The functional leaves characteristically occur in a cluster at the top of the stem; thus the cycads resemble palms. (Indeed, a common name for some cycads is "sago palms.") Unlike palms, however, cycads exhibit true, if sluggish, secondary growth from a vascular cambium; the central portion of their trunks consists of a great mass of pith. Cycads are often highly toxic, with both neurotoxins and carcinogenic compounds abundant. They harbor cyanobacteria and make important contributions of fixed nitrogen to the areas where they occur.

The reproductive units of cycads are more or less reduced leaves with attached sporangia that are loosely or

20-37

Male and female plants of Zamia pumila, the only species of cycad native to the United States. The stems are mostly or entirely underground and, along with the storage roots, were used by Native Americans as food and as a source of starch. The two large gray cones in the foreground are ovulate cones; the smaller brown cones are microsporangiate cones.

tightly clustered into conelike structures near the apex of the plant. The pollen and ovulate cones of cycads are borne on different plants (Figure 20–38). The pollen tubes formed by the male gametophytes of cycads are typically unbranched or only slightly branched. In most cycads, growth of the pollen tube results in significant destruction of the nucellar tissue. Before fertilization, the basal end of the male gametophyte swells and elongates, bringing the sperm close to the eggs. The basal end then ruptures, and the released multiflagellated sperm swim to the eggs (Figure 20–39). Each male gametophyte produces two sperm.

The role of insects in the pollination of cycads is of special interest. Beetles of several groups have frequently been found to be associated with the male cones, and less frequently with the female cones, of the members of several genera of cycads. For example, weevils (family *Curculionidae*) of the genus *Rhopalotria* carry out their entire life cycles upon and within male cones of *Zamia* and also visit the female cones. Pollen-eating beetles, although not weevils, have certainly been present throughout the history of the cycads. It seems reasonable to assume that there has been a long relationship between the members of the two groups. It is less certain that wind pollination plays an important role in the pollination of cycads, despite the general assumption that it does so.

(a)

(b)

20-38

(a) Encephalartos ferox, a cycad native to Africa. Shown here is a female plant with ovulate cones. (b) A female plant of Cycas siamensis. Several megasporophylls have been removed to reveal seeds on the upper surfaces of other megasporophylls.

20–39
Sexual reproduction in cycads and Ginkgo *is unusual in combining motile sperm with pollen tubes. (a) The sperm of the cycad* Zamia pumila, *shown here, swim by virtue of an estimated 40,000 flagella. (b) Sperm are transported to the vicinity of the egg cells in the ovule by means of a pollen tube (see Figure 20–12).*

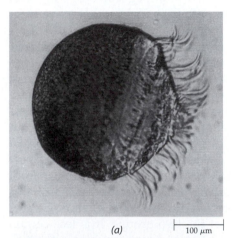

(a)   100 μm

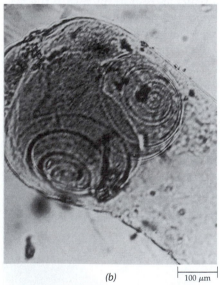

(b)   100 μm

Ginkgo biloba Is the Only Living Member of the Phylum *Ginkgophyta*

The maidenhair tree, *Ginkgo biloba*, is easily recognized by its fan-shaped leaves with their openly branched, dichotomous (forking) pattern of veins (Figure 20–40). It is an attractive and stately, but slow-growing, tree, which may reach a height of 30 meters or more. The leaves on the numerous slowly growing spur, or short, shoots of *Ginkgo* are more or less entire, whereas those on the long shoots and seedlings are often deeply lobed. Unlike most other gymnosperms, *Ginkgo* is deciduous; its leaves turn a beautiful golden color before falling in autumn.

Wollemia nobilis:

While loss of biological diversity is one of the world's most serious environmental problems, it is remarkable to discover surviving populations of an evolutionary line thought to be long extinct. This is more dramatic when the discovery is a tree up to 40 meters tall surviving in deep sandstone canyons within 150 kilometers of Sydney, Australia's largest city.

The "Wollemi Pine" was discovered in late 1994 and has been named Wollemia nobilis, *commemorating the Wollemi National Park in which it occurs, as well as both David Noble, the Wildlife Officer who discovered it, and the stature of the trees. It is one of the world's rarest plant species, because fewer than 40 plants are known, growing in two small groves. It is a member of the Araucariaceae, an ancient family of conifers.*

Araucariaceae had a worldwide distribution and their greatest diversity in the Jurassic and Cretaceous periods, between 200 and 65 million years ago, but became extinct in the Northern Hemisphere in the Late Cretaceous at about the time of the extinction of the dinosaurs. The two other surviving genera of Araucariaceae, Araucaria *and* Agathis, *have Southern Hemisphere distributions in lands that were formerly parts of the ancient supercontinent Gondwana.*

Wollemia *is characterized by having the adult foliage leaves arranged in four rows, juvenile leaves that are very different from those of the adult, a distinctive bark type, and cones terminal on the branches of the same tree—the females on the upper branches and the males on the lower branches. The seeds are not winged. Undeveloped axillary buds buried under the bark may remain dormant for long periods before eventually developing as new lateral shoots or shoots from the base. These buds increase the chance of surviving environmental disasters such as storm, fire, or rock falls.*

Botanists at the Royal Botanic Gardens in Sydney are investigating features of Wollemia *and are comparing gene sequences of the DNA of this and related groups, while horticultural researchers are studying its propagation from seeds, cuttings, and tissue cultures. Because of its rarity and since it is a handsome tree,*

A Newly Discovered Living Fossil

researchers aim to propagate *Wollemia* and make it as widely available as possible in cultivation, first in other botanic gardens for conservation and eventually in general horticulture. It will, however, take years of work before it can be made widely available.

Comparisons with fossils have involved detailed features of the leaf epidermis and the pollen, because these structures are often the best preserved parts of plant fossils. Such studies have confirmed that a pollen type found in southern Australia, New Zealand, and Antarctica, and originally thought to belong to an entirely different family, indeed represents the *Wollemia* evolutionary line. That particular fossil pollen type, which was widespread from 94 to 30 million years ago, declined greatly at about the same time as rainforests became more restricted in Australia. Some plants apparently persisted, however, and the most recent pollen fossils have been recovered from marine oil drilling on the southern Australian continental shelf, in deposits only 2 million years old.

As the third living genus of Araucariaceae, *Wollemia* has provided DNA for studies of evolutionary relationships, clarifying aspects of the evolution of Araucaria and Agathis and their relationships to other conifers. The gene sequences confirm the conclusion that its closest relative is Agathis, but that it is appropriately recognized as a separate genus.

Wollemia reflects the dramatic worldwide history of floristic change on Earth. It is a member of an ancient plant family, an example of a lineage that survived the Late Cretaceous extinction of many plant and animal groups. Also, the distribution of Araucariaceae in Australia and South America reflects past contacts between the Southern Hemisphere land masses. Most recently, the habitats of *Wollemia*—together with its extreme rarity—appear to illustrate a late stage in the widespread replacement by flowering plants of large areas that were once dominated by conifers.

As a rare species, *Wollemia* requires conservation, and its survival shows the importance of national parks in conserving biodiversity. Its habitat is a rainforest

of broad-leaved evergreen trees of *Ceratopetalum* and *Doryphora*, over which it rises as an emergent tall tree, an example of rainforest providing environmentally stable habitats that preserve species that retain many features of ancient plant groups.

(c)

(a)

(d)

(b)

(a) Tall trees of Wollemia nobilis, *growing in a deep sandstone canyon. (b) Globular shaped ovulate cones and small, brown-colored male cones. (c) Ovulate cone and a branch with leaves arranged in four rows. (d) The mature leaves of* Wollemia *bear a close resemblance to the fossil leaves of the* Araucariaceae *from Jurassic deposits in eastern Australia. (e) Cone scales and seeds of* Wollemia *placed next to a fossil cone scale from the same Jurassic deposit.*

(e)

Ginkgo biloba is the sole living survivor of a genus that has changed little for more than 150 million years and is the only living member of the phylum *Ginkgophyta*. The living species shares features with other genera of gymnosperms that range back to the Early Permian period some 270 million years ago. There are probably no wild stands of *Ginkgo* anywhere in the world, but the tree was preserved in temple grounds in China and Japan. Introduced into other parts of the world, it has been an important feature of the parks and gardens of the temperate regions of the world for about 200 years. *Ginkgo* is especially resistant to air pollution and so is commonly cultivated in urban parks and along streets.

Like the cycads, *Ginkgo* bears its ovules and microsporangia on different individuals. The ovules of *Ginkgo* are borne in pairs on the end of short stalks and ripen to produce fleshy-coated seeds in autumn (Figure 20–40b). The rotting flesh of the *Ginkgo* seed coat is infamous for its vile odor, which is derived mainly from the presence of butanoic and hexanoic acids. These are the same fatty acids found in rancid butter and Romano cheese. The kernel of the seed (that is, the female gametophytic tissue and the embryo), however, has a fishy taste and is a prized delicacy in China and Japan.

In *Ginkgo*, fertilization within the ovules may not occur until after they have been shed from their parent tree. As in cycads, the male gametophyte forms an extensively branched, haustorium-like system that develops from an initially unbranched pollen tube (see Figure 20–12). Growth of the pollen tube within the nucellus is strictly intercellular, without any apparent damage to the adjacent nucellar cells. Eventually, the basal end of this system develops into a saclike structure that, at maturity, contains two large multiflagellated sperm (Figure 20–12b). Rupture of the saclike portion of the pollen tube releases these sperm to swim to the eggs within the female gametophyte of the ovule.

The Phylum *Gnetophyta* Contains Members with Angiosperm-like Features

The gnetophytes comprise three living genera and about 70 species of very unusual gymnosperms: *Gnetum*, *Ephedra*, and *Welwitschia*. *Gnetum*, a genus of about 30 species, consists of trees and climbing vines with large, leathery leaves that closely resemble those of eudicotyledons (Figure 20–41). It is found throughout the moist tropics.

Most of the approximately 35 species of *Ephedra* are profusely branched shrubs with inconspicuous, small, scalelike leaves (Figure 20–42). With its small leaves and apparently jointed stems, *Ephedra* superficially resembles *Equisetum*. Most species of *Ephedra* inhabit arid or desert regions of the world.

(a)

(b)

20–40

(a) Ginkgo biloba, *the maidenhair tree. This tree was given its English name because of the resemblance between its leaves and the leaflets of maidenhair fern.* *(b)* Ginkgo *leaves and fleshy seeds attached to spur shoots.*

(a)

(b)

(c)

20–41

The large, leathery leaves of the tropical gnetophyte Gnetum resemble those of certain eudicots. The species of Gnetum grow as shrubs or woody vines in tropical or subtropical forests. **(a)** Megasporangiate inflores- cences and leaves of Gnetum gnemon. **(b)** Microsporangiate inflorescence and leaves and **(c)** fleshy seeds with leaves of Gnetum, photographed in the Amazon Basin of southern Venezuela.

(a)

(b)

(d)

(c)

20–42

Ephedra is the only one of the three living genera of Gnetophyta found in the United States. **(a)** A male shrub of Ephedra viridis, in California. It is a densely branched shrub that has scalelike leaves, like other members of the genus. **(b)** Microsporangiate strobili of E. viridis. Note the scalelike leaves on the stem. **(c)** Ephedra trifurca, in Arizona, with microsporangiate strobili. **(d)** Female plant of E. viridis with seeds.

491

(a)

(b)

(c)

20–43

The gnetophyte Welwitschia mirabilis, *found only in the Namib Desert of Namibia and adjacent regions of southwestern Africa.* Welwitschia *produces only two leaves, which continue to grow for the life of the plant. As growth continues, the leaves break off at the* tips and split lengthwise; thus older plants appear to have numerous leaves. *(a)* A large, seed-producing plant. *(b)* Microsporangiate strobili. *(c)* Ovulate strobili; the insect is a fire bug, sucking sap from the strobili. Welwitschia *is dioecious (see page 501).*

Welwitschia is probably the most bizarre vascular plant (Figure 20–43). Most of the plant is buried in sandy soil. The exposed part consists of a massive, woody, concave disk that produces only two strap-shaped leaves that split lengthwise with age. The cone-bearing branches arise from meristematic tissue on the margin of the disk. *Welwitschia* grows in the coastal desert of southwestern Africa, in Angola, Namibia, and South Africa.

Although the genera of *Gnetophyta* are clearly related to one another and are appropriately placed together, they differ greatly in their characteristics. These genera do, however, have many angiosperm-like features, such as the similarity of their strobili to some angiosperm inflorescences (flower clusters), the presence of very similar vessels in their xylem, and the lack of archegonia in *Gnetum* and *Welwitschia*. Current analysis favors the idea that the last two features, while similar to those in angiosperms, were independently derived in *Gnetophyta* and angiosperms.

Recently, there has been increasing support, on the basis of careful morphologically based comparisons and macromolecular analyses, for the idea that gnetophytes are the group of gymnosperms most closely related to the angiosperms. In 1990, it was reported that double fertilization—a process involving the fusion of a second sperm nucleus with a nucleus of the female gametophyte—also occurs frequently in *Ephedra*. Thus double fertilization, hitherto considered unique to angiosperms, may actually have been present in the common ancestor of angiosperms and gnetophytes. Unlike flowering plants, however, where double fertilization produces a distinctive embryo-nourishing tissue called endosperm (in addition to an embryo), the second fertilization event in *Ephedra* (and also in *Gnetum*) produces extra embryos. As is the case with all gymnosperms, in *Ephedra* and *Gnetum* a large female gametophyte serves to nourish the developing embryo within the seed. None of the living gnetophytes, however, could possibly be an ancestor of any angiosperm—each of the three living genera of gnetophytes has its own unique specializations. Interestingly, the reproductive structures of at least some species of all three genera of gnetophytes produce nectar and are visited by insects. Wind pollination is clearly important—at least in *Ephedra*—but insects also play a role in the pollination of these plants.

Summary

Seed plants consist of five phyla with living representatives. One of these, the overwhelmingly successful angiosperms (phylum *Anthophyta*), is characterized by a unique set of features. The remaining four, which lack these specializations, are commonly grouped together as gymnosperms.

A Seed Develops from an Ovule

In addition to producing seeds, all seed plants bear megaphylls. The prerequisites of the seed habit include heterospory; retention of a single megaspore; development of the embryo, or young sporophyte, within the megagametophyte; and integuments. All seeds consist of a seed coat derived from the integument(s), an embryo, and stored food. In gymnosperms, the stored food is provided by the haploid female gametophyte.

Seed Plants Most Likely Evolved from the Progymnosperms

The oldest known seedlike structures occur in strata of the Late Devonian period, about 365 million years old. The likely progenitors of the gymnosperms and angiosperms are the progymnosperms, an extinct group of seedless Paleozoic vascular plants. Among the major extinct groups of gymnosperms are the seed ferns (phylum *Pteridospermophyta*), a diverse and unnatural group, and the cycadeoids, or *Bennettitales*, which had leaves resembling those of cycads but had very different reproductive structures.

All Gymnosperms Have the Same Basic Life Cycle

Living gymnosperms comprise four phyla: *Cycadophyta, Ginkgophyta, Coniferophyta,* and *Gnetophyta.* Their life cycles are essentially similar: an alternation of heteromorphic generations with large, independent sporophytes and greatly reduced gametophytes. The ovules (megasporangia plus integuments) are exposed on the surfaces of the megasporophylls or analogous structures. At maturity the female gametophyte of most gymnosperms is a multicellular structure with several archegonia. The male gametophytes develop inside pollen grains. Antheridia are lacking in all seed plants. In gymnosperms, the male gametes, or sperm, arise directly from the spermatogenous cell. Except for the cycads and *Ginkgo,* which have flagellated sperm, the sperm of seed plants are nonmotile.

Pollination and Pollen Tube Formation Eliminate the Need for Water for the Sperm to Reach the Egg

In seed plants, water is not necessary in order for the sperm to reach the eggs. Instead, the sperm are conveyed to the eggs by a combination of pollination and pollen tube formation. Pollination in gymnosperms is the transfer of pollen from microsporangium to mega-

Summary TABLE · A Comparison of Some of the Main Features of the Gymnosperm Phyla with Living Representatives

PHYLUM	REPRESENTATIVE GENUS OR GENERA	TYPE OF TRACHEARY ELEMENT(S)	PRODUCE MOTILE SPERM?	POLLEN TUBE A TRUE SPERM CONVEYOR?	TYPE OF LEAVES PRODUCED	MISCELLANEOUS FEATURES
Cycadophyta (cycads)	*Cycas* and *Zamia*	Tracheids	Yes	No	Palmlike	Ovulate and microsporangiate cones simple and on separate plants
Ginkgophyta (maidenhair tree)	*Ginkgo*	Tracheids	Yes	No	Fan-shaped	Ovules and microsporangia on separate plants; fleshy-coated seeds
Coniferophyta (conifers)	*Abies, Picea, Pinus,* and *Tsuga*	Tracheids	No	Yes	Most needlelike or scalelike	Ovulate and microsporangiate cones on same plant; ovulate cones compound; pine needles in fascicles
Gnetophyta (gnetophytes)	*Ephedra, Gnetum,* and *Welwitschia*	Tracheids and vessel elements	No	Yes	*Ephedra:* small scalelike leaves; *Gnetum:* relatively broad, leathery leaves arranged in pairs; *Welwitschia:* two enormous, strap-shaped leaves	Ovulate and microsporangiate cones compound; borne on separate plants, except for some species of *Ephedra;* have several angiosperm-like features; leaves borne in opposite pairs

sporangium (nucellus). Fertilization occurs when one sperm of the male gametophyte (germinated pollen grain) unites with the egg, which in most gymnosperms is located in an archegonium. The second sperm has no apparent function (except perhaps in *Gnetum* and *Ephedra*), and it disintegrates. After fertilization in seed plants, each ovule develops into a seed.

There Are Four Phyla of Gymnosperms with Living Representatives

The conifers (phylum *Coniferophyta*) are the largest and most widespread phylum of living gymnosperms, with about 50 genera and some 550 species. They dominate many plant communities throughout the world, with pines, firs, spruces, and other familiar trees over wide stretches of the north. Living cycads (phylum *Cycadophyta*) consist of 11 genera and about 140 species, mainly tropical but extending away from the equator in warmer regions. Cycads are palmlike plants with trunks and sluggish secondary growth. There is only one living species of the phylum *Ginkgophyta*, the maidenhair tree *(Ginkgo biloba)*, which is known only from cultivation. The three genera of the phylum *Gnetophyta* are the closest living relatives of the angiosperms (phylum *Anthophyta*).

Questions

1. One of the most important evolutionary advances in the progymnosperms is the presence of a bifacial vascular cambium. What is a bifacial vascular cambium, and where is it found besides in the progymnosperms?

2. In what way do the *Bennettitales*, or cycadeoids, resemble cycads? How do they differ from the cycads?

3. The potential for polyembryony occurs twice in the pine life cycle. Explain.

4. Diagram and label the components of each of the following: a pine ovule with a mature megagametophyte; a mature pine microgametophyte (germinated pollen grain with sperm); and a mature pine seed.

5. Evidence exists in the cycads and *Ginkgo* that the first pollen tubes were haustorial structures and not true sperm conveyors. Explain.

Selected Key Terms

aril p. 483

determinate p. 474

generative cell p. 477

indeterminate p. 474

integument p. 468

megasporocytes, or megaspore mother cells p. 481

micropyle p. 468

microsporocytes, or microspore mother cells p. 477

nucellus p. 468

ovuliferous scales p. 480

pollen grain p. 472

pollen tube p. 472

pollination p. 472

polyembryony p. 472

prothallial cells p. 477

seed p. 468

seed coat p. 470

seed-scale complex p. 480

spermatogenous cell p. 481

sterile cell p. 481

tube cell p. 477

Introduction to the Angiosperms

21–1

A giant eucalyptus, or red tingle (Eucalyptus jacksonii), growing in the Valley of the Giants in southwestern Australia. Note the man standing in the burnt-out base of this enormous angiosperm.

OVERVIEW

Of all the plants, angiosperms—the flowering plants—are the ones that most directly affect our lives. The grains, fleshy fruits, and vegetables we eat are angiosperms, and the cotton and linen in the clothes we wear are from angiosperms as well.

The most obvious characteristic of a flowering plant is, of course, the flower. The flower contains the sexual reproductive parts of the plant and thus is of crucial importance not only for the formation of offspring but for the evolution of the species as a whole. This chapter focuses first on the basic structure of the flower. Emphasis then shifts to the formation of both the male and female gametophytes and an examination of pollination and fertilization, two processes that have features unique to angiosperms. Last to be discussed in the reproductive process are the conversions of the ovule into a seed and the surrounding ovary into a fruit. The chapter concludes with a comparison of outcrossing and self-pollination, including the different strategies angiosperms have evolved that ensure reproductive success.

CHECKPOINTS

By the time you finish reading this chapter, you should be able to answer the following questions:

1. What is a flower, and what are its principal parts?

2. What are some of the variations that exist in flower structure?

3. By what processes do angiosperms form microgametophytes (male gametophytes)? How are these processes both similar to and different from those that give rise to megagametophytes (female gametophytes)?

4. What is the structure, or composition, of the mature male gametophyte in angiosperms? Of the mature female gametophyte?

5. What is meant by "double fertilization" in angiosperms, and what are the products of this phenomenon?

6. What are some of the conditions that promote outcrossing in angiosperms, and under what circumstances might self-pollination be more advantageous than outcrossing?

Angiosperms—the flowering plants—make up much of the visible world of modern plants. Trees, shrubs, lawns, gardens, fields of wheat and corn, wildflowers, fruits and vegetables on the grocery shelves, the bright splashes of color in a florist's window, the geranium on a fire escape, duckweed and water lilies, eel grass and turtle grass, saguaro cacti and prickly pears—wherever you are, flowering plants are there also.

Diversity in the Phylum *Anthophyta*

Angiosperms make up the phylum *Anthophyta*, which includes about 235,000 species and is thus, by far, the largest phylum of photosynthetic organisms. In their vegetative features, angiosperms are enormously diverse. In size, they range from species of *Eucalyptus* trees well over 100 meters tall with trunks nearly 20 meters in girth (Figure 21–1), to some duckweeds, which are simple, floating plants often scarcely 1 millimeter long (Figure 21–2). Some angiosperms are vines that climb high into the canopy of the tropical rainforest, while others are epiphytes that grow in that canopy. Many angiosperms, such as cacti, are adapted for growth in extremely arid regions. For over 100 million years, the flowering plants have dominated the land.

In terms of their evolutionary history, the angiosperms are a group of seed plants with special character-istics: flowers, fruits, and distinctive life-cycle features that differ from those of all other plants. In this chapter, we outline these characteristics and place them in perspective, and in the following chapter, we discuss the evolution of the angiosperms. In Section 5, we consider in some detail the structure and development of the angiosperm plant body (sporophyte).

Angiosperms share so many features that do not exist in other plants that it is clear they were derived from a single common ancestor (a monophyletic group). Current investigations, based on an improved understanding of the fossil record, as well as better methods of analysis and comparisons of nucleic acid sequences, have led to a clearer recognition of the major evolutionary lines of angiosperms. Two of the major classes into which the group is divided are the *Monocotyledones* (monocots), with about 65,000 species (Figure 21–3), and the *Eudicotyledones* (eudicots), with about 165,000 species (Figure 21–4). Monocots are distinctive, including such familiar plants as the grasses, lilies, irises, orchids, cattails, and palms. The more diverse eudicots include almost all of the familiar trees and shrubs (other than the conifers) and many of the herbs (nonwoody plants). Other groups of archaic flowering plants that are neither monocots nor eudicots will be discussed in Chapter 22. Formerly these plants were grouped with the eudicots as "dicots," but we now know that this is an artificial system of classification that simply overemphasizes the dis-

(a)

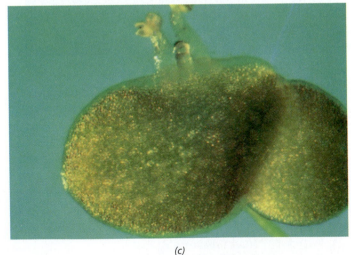

(c)

(b)

21–2

The duckweeds (family Lemnaceae*) are the smallest flowering plants. (**a**) A honeybee resting on a dense floating mat of three species of duckweed. The larger plants are* Lemna gibba, *about 2 to 3 millimeters long; the smaller ones are two species of* Wolffia, *up to 1 millimeter long. (**b**) A flowering plant of* Wolffia borealis *with a circular concave* stigma *(looking like a tiny doughnut) and a minute anther just above it, both protruding from a central cavity. The whole plant is less than 1 millimeter long. (**c**) Flowering plant of* Lemna gibba; *two stamens and a style protrude from a pocket on the upper surface of the plant.*

(a)

(b)

(c)

21–3

*Monocots. **(a)** A member of the palm family, the coconut palm* (Cocos nucifera), *growing in Tehuantepec, Oaxaca, Mexico. A coconut is a drupe, not a nut (see Chapter 22).*

(b) *Flowers and fruits of the banana plant* (Musa × paradisiaca). *The banana flower has an inferior ovary, and the tip of the fruit*

*bears a large scar left by the fallen flower parts. **(c)** Rice* (Oryza sativa) *is a member of the grass family.*

21–4

*Eudicots. **(a)** Saguaro cactus* (Carnegiea gigantea). *The cacti, of which there are about 2000 species, are almost exclusively a New World family. The thick, fleshy stems, which*

*store water, contain chloroplasts and have taken over the photosynthetic function of the leaves. **(b)** Round-lobed hepatica* (Hepatica americana), *which flowers in deciduous woodlands in the early spring. The flowers*

*have no petals but have six to ten sepals and numerous spirally arranged stamens and carpels. **(c)** California poppy* (Eschscholzia californica) *is the state flower of California and is protected by law.*

(a)

(b)

(c)

(a)

	Main Differences between Monocots and Eudicots	
Characteristic	**Eudicots**	**Monocots**
Flower parts	In fours or fives (usually)	In threes (usually)
Pollen	Triaperturate (having three pores or furrows)	Monoaperturate (having one pore or furrow)
Cotyledons	Two	One
Leaf venation	Usually netlike	Usually parallel
Primary vascular bundles in stem	In a ring	Complex arrangement
True secondary growth, with vascular cambium	Commonly present	Rare

TABLE 21–1

(b)

(c)

tinctiveness of monocots from other angiosperms. The major features of monocots and eudicots are indicated in Table 21–1.

In terms of their mode of nutrition, all but a few angiosperms are free-living, but both parasitic and saprophytic forms do exist (Figure 21–5). There are about 200 species of parasitic monocots and about 2800 species of parasitic eudicots, including the mistletoes (see Figure 22–44), *Cuscuta* (Figure 21–5a), and *Rafflesia* (Figure 21–5b). Parasitic flowering plants form specialized absorptive organs called haustoria that penetrate the tissues of their hosts. Saprophytes, in contrast, obtain their nutrition from decaying organic matter; Indian pipe (Figure 21–5c) is an example. Mycorrhizal fungi may also play an important role in the nutrition of saprophytes, in some cases transferring carbohydrates from adjacent, photosynthetic organisms.

The Flower

The flower is a determinate (growth of limited duration) shoot that bears sporophylls—sporangium-bearing leaves (Figure 21–6). The name "angiosperm" is derived

21–5

Parasitic and saprophytic angiosperms. These plants have little or no chlorophyll; they obtain their food as a result of the photosynthesis of other plants. (a) Dodder (Cuscuta salina), a parasite that is bright orange or yellow. Dodder is a member of the morning glory family (Convolvulaceae). (b) The world's largest flower, Rafflesia arnoldii, on Mount Sago, Sumatra. Plants of this genus are parasitic on the roots of a member of the grape family (Vitaceae). *(c) Indian pipe (Monotropa uniflora), a saprophyte that obtains its food from the roots of other plants via the fungal hyphae associated with its roots. There are more than 3000 species of parasitic angiosperms, representing 17 families.*

21–6

Parts of a lily (Lilium henryi) *flower. (a) An intact flower. In some flowers, such as lilies, the sepals and petals are similar to one another, and the perianth parts—the sepals and petals together—may then be referred to as tepals. Note that the sepals are attached to the receptacle below the petals. (b) Two tepals and two stamens have been removed to reveal the ovary. The gynoecium consists of the ovary, style, and stigma. The stamen consists of the filament and anther. Note that the sepals, petals, and stamens are attached to the receptacle below the ovary, which is composed, in the lily flower, of three fused carpels. Such a flower is said to be hypogynous.*

from the Greek words *angeion,* meaning "vessel," and *sperma,* meaning "seed." The definitive structure of the flower is the carpel—the "vessel." The **carpel** contains the ovules, which develop into seeds after fertilization.

Flowers may be clustered in various ways into aggregations called **inflorescences** (Figures 21–7 and 21–8). The stalk of an inflorescence or of a solitary flower is known as a **peduncle,** while the stalk of an individual flower in an inflorescence is called a **pedicel.** The part of the flower stalk to which the flower parts are attached is termed the **receptacle.**

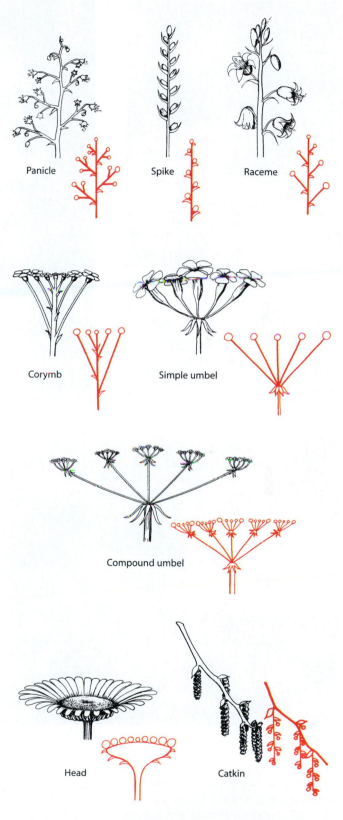

21–7

Illustrations of some of the common types of inflorescences found in the angiosperms, accompanied by simplified diagrams (in color).

(a)

(b)

(c)

(d)

(e)

21–8

Inflorescences of (a) shooting star (Dodecatheon meadia), (b) butter-and-eggs (Linaria vulgaris), (c) lupine (Lupinus diffusus), (d) bluebells (Mertensia virginica), and (e) water hemlock (Cicuta maculata). Using Figure 21–7 as a guide, can you identify the types of inflorescences shown here?

The Flower Consists of Sterile and Fertile, or Reproductive, Parts Borne on the Receptacle

Many flowers include two sets of sterile appendages, the **sepals** and **petals,** which are attached to the receptacle below the fertile parts of the flower, the **stamens** and **carpels.** The sepals arise below the petals, and the stamens arise below the carpels. Collectively, the sepals form the **calyx,** and the petals form the **corolla.** The sepals and petals are essentially leaflike in structure. Commonly the sepals are green and relatively thick, and the petals are brightly colored and thinner, although in many flowers the members of both whorls (a whorl is a circle of flower parts of one kind) are similar in color and texture. Together, the calyx (the sepals) and corolla (the petals) form the **perianth.**

The stamens—collectively the **androecium** ("house of man")—are microsporophylls. In all but a few living angiosperms, the stamen consists of a slender stalk, or **filament,** upon which is borne a two-lobed **anther** containing four microsporangia, or **pollen sacs,** in two pairs—a characteristic defining feature of angiosperms.

The carpels—collectively the **gynoecium** ("house of woman")—are megasporophylls that are folded lengthwise, enclosing one or more ovules. A given flower may contain one or more carpels, which may be separate or fused together, in part or entirely. Sometimes the individual carpel or the group of fused carpels is called a **pistil.** The word "pistil" comes from the same root as "pestle," the instrument with a similar shape that pharmacists use for grinding substances into a powder in a mortar.

In most flowers, the individual carpels or groups of fused carpels are differentiated into a lower part, the **ovary,** which encloses the ovules; a middle part, the **style,** through which the pollen tubes grow; and an upper part, the **stigma,** which receives the pollen. In some flowers, a distinct style is absent. If the carpels are fused, there may be a common style, or each carpel may retain a separate one. The common ovary of such fused carpels is generally (but not always) partitioned into two or more **locules**—chambers of the ovary that contain the ovules. The number of locules is usually related to the number of carpels in the gynoecium.

The Ovules Are Attached to the Ovary Wall at the Placenta

The portion of the ovary where the ovules originate and to which they remain attached until maturity is called the **placenta.** The arrangement of the placentae—known as the **placentation**—and consequently of the ovules varies among different groups of flowering plants (Figure 21–9). In some, the placentation is *parietal;* that is, the ovules are borne on the ovary wall or on extensions of it. In other flowers, the ovules are borne on a central column of tissue in a partitioned ovary with as

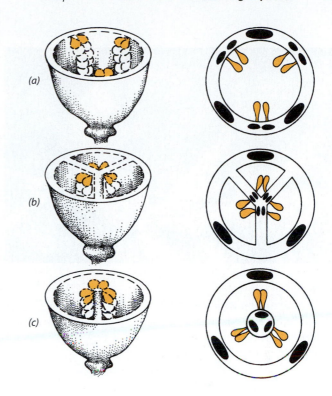

21–9
Types of placentation, with the ovules indicated in color. (a) Parietal. (b) Axile. (c) Free central. Basal and apical placentation, a single ovule at the base or apex of a unilocular ovary, are not shown here. The vascular bundles are shown as solid structures in the ovary walls.

many locules as there are carpels. This is *axile* placentation. In still others, the placentation is *free central,* the ovules being borne on a central column of tissue not connected by partitions with the ovary wall. And finally, in some flowers a single ovule occurs at the base or apex of a unilocular ovary. This is known as *basal* or *apical* placentation. These differences are important in the classification of the flowering plants.

There Are Many Variations in Flower Structure

The majority of flowers include both stamens and carpels, and such flowers are said to be **perfect** (bisexual). If either stamens or carpels are missing, the flower is **imperfect** (unisexual) and, depending on the part that is present, the flower is said to be either **staminate** or **carpellate** (or pistillate) (Figure 21–10). If both staminate and carpellate flowers occur on the same plant, as in maize (see Figure 22–32a, b) and the oaks, the species is said to be **monoecious** (from the Greek words *monos,* "single," and *oikos,* "house"). If staminate and carpellate flowers are found on separate plants, the species is said to be **dioecious** ("two houses"), as in the willows and hemp (*Cannabis sativa*).

21–10

Staminate and carpellate flowers of a tan-bark oak (Lithocarpus densiflora), of the family Fagaceae. Most members of this family, including the true oaks (Quercus), are monoecious, meaning that the staminate and carpellate flowers are separate but are borne on the same tree.

Any one of the floral whorls—sepals, petals, stamens, or carpels—may be lacking in the flowers of certain groups. Flowers with all four floral whorls are called **complete** flowers. If any whorl is lacking, the flower is said to be **incomplete.** Thus an imperfect flower is also incomplete, but not all incomplete flowers are imperfect.

The particular arrangement of the floral parts may be spiral on a more or less elongated receptacle, or similar parts—such as petals—may be attached in a whorl. The parts may be united with other members of the same whorl *(connation)* or with members of other whorls *(adnation).* An example of adnation is the union of stamens with the corolla (stamens adnate to the corolla), which is fairly common. When the floral parts of the same whorl are not joined, the prefixes *apo-* (meaning "separate") or *poly-* may be used to describe the condition. When the parts are connate, either *syn-* or *sym-* ("together") is used. For example, in an aposepalous or polysepalous calyx, the sepals are not joined; in a synsepalous one, they are.

In addition to this variation in arrangement of flower parts (spiral or whorled), the level of insertion of the sepals, petals, and stamens on the floral axis varies in relation to the ovary or ovaries (Figure 21–11). If the sepals, petals, and stamens are attached to the receptacle below the ovary, as they are in lily, the ovary is said to be **superior** (Figure 21–6). In other flowers the sepals, petals, and stamens apparently are attached near the top of the ovary, which is **inferior.** Intermediate conditions, in which part of the ovary is inferior, also occur in a number of kinds of plants.

In terms of their points of insertion, the perianth and stamens are said to be **hypogynous**—situated on the receptacle beneath the ovary and free from it and from the calyx (Figure 21–6); **epigynous**—arising from the top of the ovary (Figure 21–12); or **perigynous**—with the stamens and petals adnate to the calyx and thus forming a short tube *(hypanthium)* arising from the base of the ovary (Figure 21–13).

Finally, mention should be made of symmetry in flower structure. In some flowers, the members of the different whorls of the flower are made up of members of similar shape that radiate from the center of the flower and are equidistant from each other; they are radially symmetrical. Such flowers—tulips are an example—are said to be **regular,** or actinomorphic (from the Greek root *aktin-,* "ray"). In other flowers, one or more members of at least one whorl are different from other members of the same whorl, and such flowers are generally bilaterally symmetrical. Bilaterally symmetrical flowers—for example, snapdragons—are said to be **irregular,** or zygomorphic (Gk. *zygon,* "yoke" or "pair"). Some regular flowers have irregular color patterns, which give their visitors an image similar to that of a structurally irregular flower.

21–11

Types of flowers in common families of eudicots, showing differences in the position of the ovary. (a) In Ranunculaceae, the buttercup family, for example, the sepals, petals, and stamens are attached below the ovary and there is no fusion; such flowers are said to be hypogynous. (b) In contrast, many Rosaceae, such as cherries, have superior ovaries, with the sepals, petals, and stamens fused together to form a cup-shaped extension of the receptacle called the hypanthium. Such flowers are said to be perigynous. (c) The flowers of other plants, for example Apiaceae, the parsley family, have inferior ovaries; that is, the sepals, petals, and stamens appear to be attached above the ovaries. Such flowers are said to be epigynous.

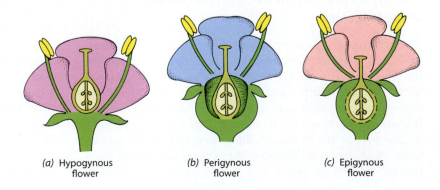

(a) Hypogynous flower   (b) Perigynous flower   (c) Epigynous flower

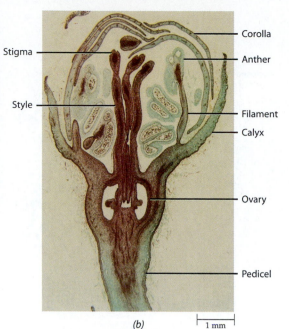

(a)

(b)

⊢ 1 mm ⊣

21–12
Apple (Malus sylvestris) flowers, (a) and (b), exhibit epigyny—their sepals, petals, and stamens apparently arise from the top of the ovary. In (b) the flower is nearly open, but the stamens are not yet erect.

The Angiosperm Life Cycle

Angiosperm gametophytes are much reduced in size— more so than those of any other heterosporous plants, including the other seed plants (gymnosperms). The mature microgametophyte consists of only three cells. The mature megagametophyte (embryo sac), which is retained for its entire existence within the tissues of the sporophyte, or specifically of the ovule, consists of only seven cells in most kinds of angiosperms. Both antheridia and archegonia are lacking. Pollination is indirect; that is, pollen is deposited on the stigma, after which the pollen tube grows through or on the surface of tissues of the carpel to convey two nonmotile sperm to the female gametophyte. After fertilization, the ovule develops into a seed, which is enclosed in the ovary. At the same time, the ovary (and sometimes additional structures associated with it) develops into a fruit.

21–13
Cherry (Prunus) flowers, (a) and (b), exhibit perigyny—their sepals (calyx), petals (corolla), and stamens are attached to a hypanthium. In (b) the filaments of the stamens are bent and crowded in the hypanthium because the flower has not yet opened.

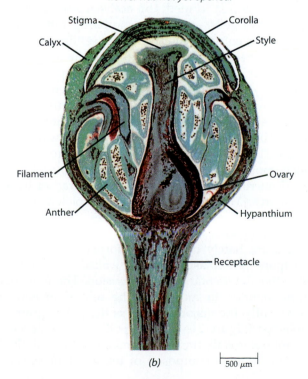

(a)

(b)

⊢ 500 μm ⊣

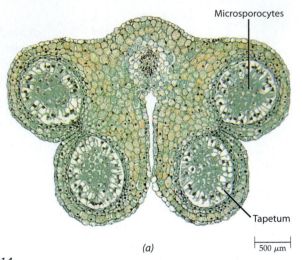

Microsporocytes

Tapetum

(a)

500 μm

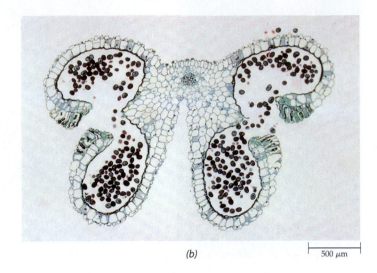

(b)

500 μm

21–14

*Two transverse sections of lily (Lilium) an-
thers. (a) Immature anther, showing the four
pollen sacs containing microsporocytes*

*surrounded by the nutritive tapetum.
(b) Mature anther containing pollen grains.*

*The partitions between adjacent pollen sacs
break down during dehiscence, as shown here.*

Microsporogenesis and Microgametogenesis Culminate in the Formation of Sperm

Two distinct processes—microsporogenesis and microgametogenesis—lead to the formation of the microgametophyte. Microsporogenesis is the formation of microspores (single-celled pollen grains) within the microsporangia, or pollen sacs, of the anther. Microgametogenesis is the development of the microgametophyte within the pollen grain up to the three-celled stage of development.

When it is first formed, the anther consists of a uniform mass of cells, except for the partly differentiated epidermis. Eventually, four groups of fertile, or **sporogenous,** cells become discernible within the anther. Each group is surrounded by several layers of sterile cells. The sterile cells develop into the wall of the pollen sac, which includes nutritive cells that supply food to the developing microspores. The nutritive cells constitute the **tapetum,** the innermost layer of the pollen sac wall (Figure 21–14a). The sporogenous cells become microsporocytes (pollen mother cells), which divide meiotically. Each diploid microsporocyte gives rise to a tetrad of haploid microspores. Microsporogenesis is completed with formation of the single-celled microspores.

During meiosis, each nuclear division may be followed immediately by cell wall formation, or the four microspore protoplasts may be walled off simultaneously after the second meiotic division. The first condition is common in monocots, the second in eudicots. Subsequently, the major features of the pollen grains are established (Figure 21–15). The pollen grains develop a resistant outer wall, the **exine,** and an inner wall, the **intine.** The exine is composed of the resistant substance

sporopollenin (page 406), which apparently is derived primarily from the tapetum. This polymer is present in the spore walls of all plants. The intine, which is composed of cellulose and pectin, is laid down by the microspore protoplasts.

Microgametogenesis in angiosperms is uniform and begins when the microspore divides mitotically, forming two cells within the original microspore wall. The division forms a large **tube cell** and a small **generative cell,** which moves to the interior of the pollen grain. This two-celled pollen grain is an immature microgametophyte. In about two-thirds of flowering plant species, the microgametophyte is in this two-celled stage at the time the pollen grains are liberated from the anther (Figure 21–16). In the remainder, the generative nucleus divides prior to the release of the pollen grains, giving rise to two male gametes, or sperm (Figure 21–17). Mature pollen grains may be packed with starch or oils, depending on the group, and are a nutritious source of food for animals.

Pollen grains, like the spores of seedless plants, vary considerably in size and shape, ranging from less than 20 micrometers to more than 250 micrometers in diameter. They also differ in the number and arrangement of the apertures through which the pollen tubes ultimately grow. These apertures can be elongate (furrows), circular (pores), or a combination of the two. Nearly all families, many genera, and a fair number of species of flowering plants can be identified solely by their pollen grains, on the basis of such characteristics as size, number, and type of apertures, and exine sculpturing. In contrast to larger pieces of plants—such as leaves, flowers, and fruits—pollen grains, because of their tough, highly resistant exine, are very widely represented in the fossil

(a)   10 μm

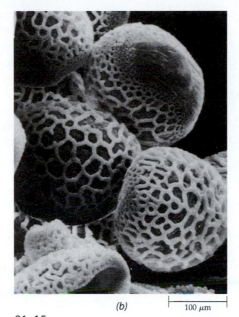

(b)   100 μm

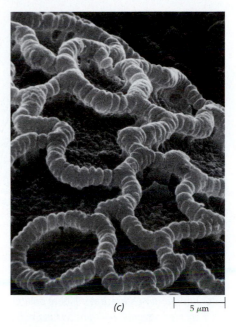

(c)   5 μm

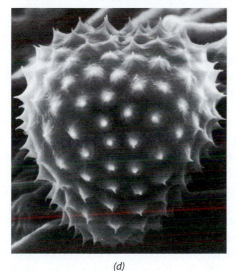

(d)

21–15

The wall of the pollen grain serves to protect the male gametophyte on its often hazardous journey from the anther to the stigma. The outer layer, or exine, is composed chiefly of a substance known as sporopollenin, which appears to be a polymer composed chiefly of carotenoids. The exine, which is remarkably tough and resistant, is often elaborately sculptured.

The sculpturing of pollen grain walls is distinctly different from one species to another, as revealed in these scanning electron micrographs of pollen grains. (a) Pollen grains of the horse chestnut (Aesculus hippocastanum). The pore of each grain through which a pollen tube may emerge is visible in the furrow. (b) Pollen grains of a lily (Lilium longiflorum). (c) Detail of the surface of a lily (L. longiflorum) pollen grain. (d) Pollen grain of the western ragweed (Ambrosia psilostachya). The pollen of ragweed is an important cause of hay fever (see essay on page 508). Spiny pollen grains such as these are common among members of the sunflower family, Asteraceae, of which ragweed is a member.

21–16

Mature pollen grain of Lilium, containing a two-celled male gametophyte. The spindle-shaped generative cell will divide mitotically after the germination of the pollen grain. The larger tube cell, which contains the generative cell, will form the pollen tube. The round structure above the generative cell is the tube cell nucleus.

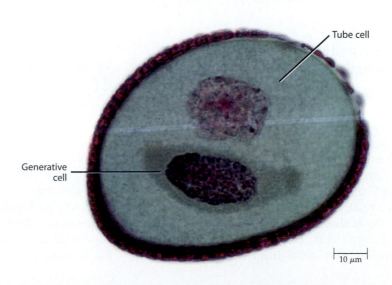

Tube cell

Generative cell

10 μm

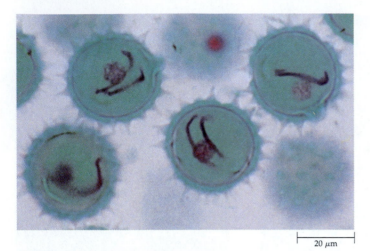

20 μm

21–17

Mature pollen grains—three-celled male gametophytes—of the telegraph plant, Silphium (family Asteraceae). Prior to pollination, each pollen grain contains two filamentous sperm cells, which are suspended in the cytoplasm of the larger tube cell. The pollen of Silphium is shed at the three-celled stage, whereas that of Lilium, shown in Figure 21–16, is shed at the two-celled stage.

record. Studies of fossil pollen can provide valuable insights into the kinds of plants and plant communities, and thus to the nature of the climates, that existed in the past.

In contrast to the spores of most seedless plants, which are also products of meiosis, pollen grains undergo mitosis within the pollen sac in which they are produced. Pollen grains, therefore, have two or three nuclei when shed, whereas most spores have only one. In addition, spores germinate through a characteristic Y-shaped suture on their surface, whereas pollen grains germinate through their apertures. The distinctive structure and arrangement of these apertures allows most kinds of pollen grains to be distinguished from spores morphologically.

Megasporogenesis and Megagametogenesis Culminate in Formation of an Egg and Polar Nuclei

Two distinct processes—megasporogenesis and megagametogenesis—lead to the formation of the megagametophyte. Megasporogenesis is the process of megaspore formation within the nucellus (megasporangium); it takes place inside the ovule. Megagametogenesis is the development of the megaspore into the megagametophyte (female gametophyte, or embryo sac).

The ovule is a relatively complex structure, consisting of a stalk—the **funiculus**—bearing a nucellus enclosed by one or two integuments. Depending on the species, one to many ovules may arise from the placentae, or

ovule-bearing regions of the ovary wall (Figure 21–9). Initially, the developing ovule is entirely nucellus, but soon it develops one or two enveloping layers, the integuments, which form a small opening, the **micropyle,** at one end of the ovule (Figure 21–18).

Early in the development of the ovule, a single megasporocyte arises in the nucellus (Figure 21–18a). The diploid megasporocyte divides meiotically (Figure 21–18b) to form four haploid megaspores, which are generally arranged in a linear tetrad. With this, megasporogenesis is completed. In most seed plants, three of the four megaspores disintegrate. The one farthest from the micropyle survives and develops into the megagametophyte.

The functional megaspore soon begins to enlarge at the expense of the nucellus, and the nucleus of the megaspore divides mitotically. Each of the resulting nuclei divides mitotically, followed by yet another mitotic division of the four resultant nuclei. At the end of the third mitotic division, the eight nuclei are arranged in two groups of four, one group near the micropylar end of the megagametophyte and the other at the opposite, or **chalazal,** end (Figure 21–18c). One nucleus from each group migrates into the center of the eight-nucleate cell; these two nuclei are then called **polar nuclei.** The three remaining nuclei at the micropylar end become organized as the **egg apparatus,** consisting of an **egg cell** and two short-lived cellular **synergids.** Cell wall formation also occurs around the three nuclei left at the chalazal end, forming the **antipodals.** The **central cell** contains the two polar nuclei. The eight-nucleate, seven-celled structure is the mature female gametophyte, or **embryo sac.**

The pattern of embryo sac development just described is the most common one. Other patterns occur in about a third of the flowering plants. In fact, one unusual pattern occurs in *Lilium,* which is the genus illustrated in Figure 21–18. In this case, there is no wall formation during megasporogenesis and all four megaspore nuclei participate in the formation of the embryo sac. Three of the nuclei move to the chalazal end of the embryo sac, while the remaining nucleus becomes situated at the micropylar end. This 3 + 1 arrangement of the nuclei represents the *first four-nucleate stage* in development of the embryo sac. What happens next is quite different at the two ends of the embryo sac. At the micropylar end of the sac, the single haploid nucleus undergoes mitosis, yielding two haploid nuclei. At the chalazal end, the mitotic spindles of the three sets of chromosomes unite and mitosis results in two nuclei that are $3n$ (triploid) in chromosome number. As a result of these events, a *second four-nucleate stage* is produced, with two haploid nuclei at the micropylar end of the embryo sac and two triploid nuclei at the chalazal end. Embryo sac development then proceeds in the manner described above for the more frequent kind of embryo sac formation, which has a single four-nucleate stage.

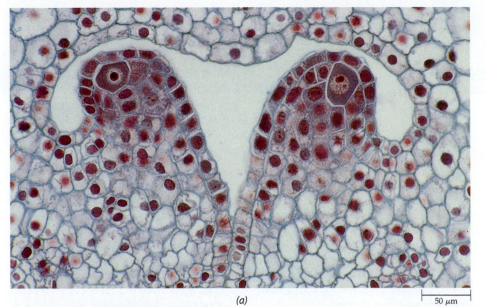

(a)

50 μm

21–18
Lilium. *Some stages in development of an ovule and embryo sac.* **(a)** *Two young ovules, each with a single, large megasporocyte surrounded by the nucellus. Integuments have not begun to develop.* **(b)** *Ovule has now developed integuments with a micropyle. The megasporocyte is in the first prophase of meiosis.* **(c)** *Ovule with eight-nucleate embryo sac (only six of the nuclei can be seen here, four at the micropylar end and two at the opposite, chalazal end). The polar nuclei have not yet migrated to the center of the sac. The funiculus is the stalk of the ovule.*

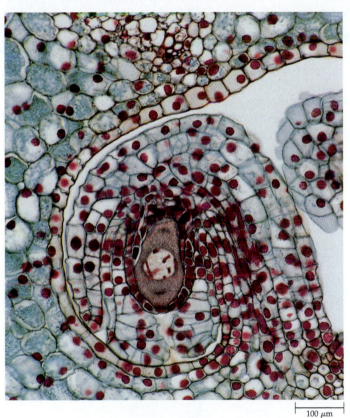

100 μm

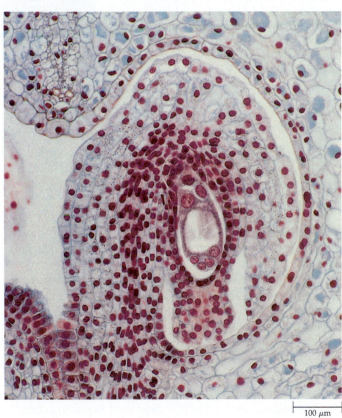

100 μm

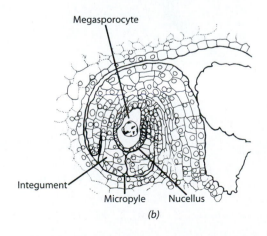

Megasporocyte

Integument

Micropyle   Nucellus

(b)

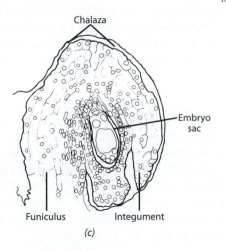

Chalaza

Embryo sac

Funiculus   Integument

(c)

Hay Fever

In the temperate areas of the Northern Hemisphere, it is estimated that between 10 and 18 percent of all people will suffer from hay fever, which can be highly debilitating, at some point in their lives. Some of the proteins that occur in hollow spaces in pollen grain walls, and which can be released immediately following contact with a moist surface, are generally the culprits. Among these proteins are some that can act as very powerful allergens and antigens, provoking strong reactions by the human immune system. These are probably also involved in genetic self-incompatibility. Proteins may also be released in tiny particles of the tapetum, smaller than pollen grains, which may become airborne as the anther splits open.

Wind-borne pollen, such as that of grasses, birch trees, and ragweed, is particularly important as an agent of hay fever, because it is shed in large amounts directly into the air and is thus more likely to reach susceptible victims than are the larger pollen grains of insect-pollinated plants. The quantity of pollen inhaled seems to be the most important factor in determining whether there will be an allergic response, but surprisingly, some wind-borne pollen that is shed in huge amounts, such as that of maize and pines, rarely causes any difficulty. The scent of certain flowers can also cause reactions that resemble hay fever, perhaps in part by increasing the sensitivity of the nasal membranes.

In temperate North America, the hay fever season can be divided into three parts. In spring, most hay fever is associated with tree pollen from such sources as oaks, elms, poplars, pecans, and birches. In summer, grass pollen predominates, with Bermuda grass, timothy, and orchard grass being important in different regions. By fall, ragweed and grasses different from those that predominate in summer become major irritants. The susceptibility of individual people to different kinds of plants varies greatly.

When new kinds of plants, such as rapeseed, become widely cultivated, they may become important new causes of hay fever. For example, in the arid southwestern United States, the irrigation of large areas for lawns and golf courses, and the introduction of many kinds of weeds to the area, have made hay fever common where it was once virtually unknown.

The incidence of hay fever has been rising rapidly for more than 50 years, even though the pollen count is actually falling in many areas. Part of the increase is related to better detection of the problem, but there is clearly a genuine increase in hay fever. Understanding why this has occurred will require better knowledge of the human immune system.

Pollination and Double Fertilization in Angiosperms Are Unique

With **dehiscence** of the anther—that is, shedding of its contents—the pollen grains are transferred to the stigmas in a variety of ways (see Chapter 22). The process whereby this transfer occurs is called **pollination.** Once in contact with the stigma, the pollen grains take up water from the cells of the stigma surface. Following this hydration, the pollen grain germinates, forming a pollen tube. If the generative cell has not already divided, it soon does, forming the two sperm. The germinated pollen grain, with its tube nucleus and two sperm, is the mature microgametophyte (Figure 21–19).

The stigma and style are modified both structurally and physiologically to facilitate the germination of the pollen grain and growth of the pollen tube. The surface of many stigmas is essentially glandular tissue, known as **stigmatic tissue,** which secretes a sugary solution. The stigmatic tissue is connected with the ovule by **transmitting tissue,** which serves as a path through the style for the growing pollen tubes. Some styles contain open canals lined with transmitting tissue. In such styles, the pollen tubes grow either along or among the cells of the lining. In most angiosperms, however, the styles are solid, with one or more strands of transmitting tissue extending from the stigma to the ovules. The pollen tubes grow either between the cells of the transmitting tissue or within their thick walls, depending on the kind of plant.

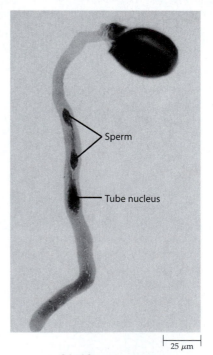

21–19

Microgametophyte, or mature male gametophyte, of Solomon's seal (Polygonatum). The two sperm and the tube nucleus can be seen in the pollen tube.

Commonly, the pollen tube enters the ovule through the micropyle and penetrates one of the two synergids next to the egg cell. The synergid begins to degenerate soon after pollination has taken place but before the pollen tube has reached the embryo sac. The two sperm and the tube nucleus are then released into the synergid through the pore that develops near the end of the pollen tube. Ultimately, one sperm nucleus enters the egg cell and the other enters the central cell, where it unites with the two polar nuclei (Figure 21–20). Recall that in most gymnosperms, only one of the two sperm is functional; one unites with the egg and the other degenerates. The involvement of both sperm in this process— the union of one with the egg and the other with the polar nuclei—is called **double fertilization.** It represents an unusual characteristic of the angiosperms, shared only with *Ephedra* and *Gnetum* (phylum *Gnetophyta*). In angiosperms with the most common type of embryo sac formation, the fusion of one of the sperm nuclei with the two polar nuclei, called **triple fusion,** results in a triploid (3n) **primary endosperm nucleus.** In *Lilium*, illustrated in Figures 21–18 and 21–20, in which one of the polar nuclei is triploid and the other haploid, triple fusion results in a pentaploid (5n) primary endosperm nucleus. Other situations occur in various groups of angiosperms. In any case, the tube nucleus degenerates during the process of double fertilization, and the remaining synergid and antipodals also degenerate near the time of fertilization or early in the course of differentiation of the embryo sac.

Significant progress has recently been made in our understanding of the origin of double fertilization and endosperm, features that have long been thought to be unique to flowering plants. These advances are based on the discovery of rudimentary double fertilization in *Ephedra* and *Gnetum*, two of the three genera of the phylum *Gnetophyta*, the closest living relatives of angiosperms (pages 490–492). The existence of a regular process of double fertilization in *Ephedra* and *Gnetum* as well as in the angiosperms suggests that a rudimentary type of double fertilization already existed in their common ancestor. Initially, however, the second fertilization product was diploid and yielded an extra embryo. Following the divergence of the angiosperm lineage from the common ancestor of this evolutionary line, the extra embryo apparently was modified to become a triploid embryo-nourishing endosperm. Thus endosperm may be regarded as a derivative of a second embryo, modified to perform a new function during the course of evolution.

The Ovule Develops into a Seed and the Ovary Develops into a Fruit

In double fertilization, several processes leading to development of the seed and fruit are initiated: the primary endosperm nucleus divides, forming the **endosperm;** the zygote develops into an embryo; the integuments develop into a seed coat; and the ovary wall and related structures develop into a fruit.

21–20

Double fertilization in Lilium. *Union of sperm and egg nuclei—"true" fertilization—can be seen in the lower half of the micrograph. Triple fusion of the other sperm nucleus and the two polar nuclei has taken place above. The three nuclei known as the antipodals can be seen at the chalazal end, opposite the micropyle, of the embryo sac.*

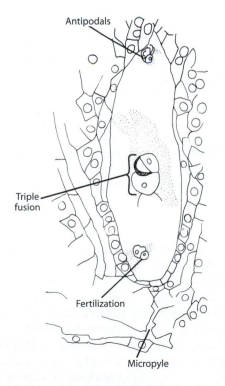

Antipodals

Triple fusion

Fertilization

Micropyle

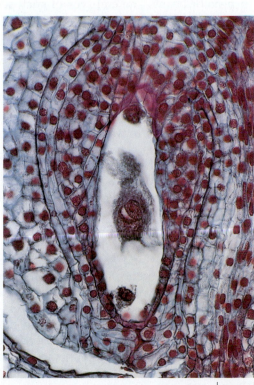

50 μm

In contrast to the embryogeny (embryo development) of the majority of gymnosperms, which begins with a free-nuclear stage, embryogeny in angiosperms resembles that of the seedless vascular plants in that the first nuclear division of the zygote is accompanied by cell wall formation. In the early stages of development, the embryos of monocots undergo sequences of cell division similar to those of other angiosperms, and the embryo becomes a spherical ball of cells. It is with the formation of the cotyledons that monocot embryos become distinctive, forming only one cotyledon. Other angiosperm embryos form two cotyledons. The details of angiosperm embryogeny are presented in Chapter 23.

Endosperm formation begins with the mitotic division of the primary endosperm nucleus and usually is initiated prior to the first division of the zygote. In some angiosperms, a variable number of free nuclear divisions precedes cell wall formation; this process is known as nuclear-type endosperm formation. In other species the initial and subsequent mitoses are followed by cytokinesis, which is known as cellular-type endosperm formation. Although endosperm development may occur in a variety of ways, the function of the resulting tissue remains the same: to provide essential food materials for the developing embryo and, in many cases, the young seedling as well. In the seeds of some groups of angiosperms, the nucellus proliferates into a food-storage tissue known as **perisperm.** Some seeds may contain both endosperm and perisperm, as do those of the beet *(Beta).* In many eudicots and some monocots, however, most or all of these storage tissues are absorbed by the developing embryo before the seed becomes dormant, as in peas or beans. The embryos of such seeds commonly develop fleshy, food-storing cotyledons. The principal food materials stored in seeds are carbohydrates, proteins, and lipids.

Angiosperm seeds differ from those of gymnosperms in the origin of their stored food. In four phyla of gymnosperms, the stored food is provided by the female gametophyte. In angiosperms, it is provided, at least initially, by endosperm, which is neither gametophytic nor sporophytic tissue. Another interesting difference is that the nutritious tissue is built up in angiosperms and *Gnetum* after fertilization occurs, whereas in other seed plants, the nutritious tissue is formed at least partly (in conifers) or entirely (other gymnosperms) *before* fertilization occurs.

Concomitantly with development of the ovule into a seed, the ovary (and sometimes other portions of the flower or inflorescence) develops into a fruit. As this occurs, the ovary wall, or **pericarp,** often thickens and becomes differentiated into distinct layers—the exocarp (outer layer), the mesocarp (middle layer), and the endocarp (inner layer), or exocarp and endocarp only. These layers are generally more conspicuous in fleshy fruits than in dry ones. Fruits are discussed in greater detail in Chapter 22.

An angiosperm life cycle is summarized in Figure 21–23 (pages 512 and 513).

A Variety of Conditions Promote Outcrossing in Angiosperms

Outcrossing is of critical importance for all eukaryotic organisms (pages 242–243). In plants, outcrossing is made possible by cross-pollination between individuals of the same species. In view of this relationship, it should therefore come as no surprise that angiosperms have evolved a variety of mechanisms that promote the transfer of pollen from one individual plant to another.

In dioecious plants, such as willows, the staminate and carpellate flowers occur on separate plants. In such plants, pollen must pass from one individual to another to achieve fertilization, and outcrossing is therefore inevitable. In monoecious plants, such as oaks, birches, and ragweed, there are separate staminate and carpellate flowers, but they occur together on the same individual. Their physical separation increases the chance that the wind, or an animal vector, will move pollen from one individual plant to another before depositing it on a receptive stigma. Maturation of the two kinds of flowers at different times further enhances the chance that outcrossing will occur.

Gymnosperms are also monoecious or dioecious, illustrating the importance of this condition in promoting outcrossing. Among them are *Ginkgo,* cycads, and junipers (conifers of the genus *Juniperus).* In all living gymnosperms, the ovule- and pollen-producing parts occur in different structures, as in different kinds of cones.

Another way in which angiosperms promote outcrossing is through **dichogamy,** a condition in which the stamens and carpels reach maturity at different times—even though they occur together in the same flowers. Those plants in which the stamens of an individual flower mature before the stigmas become receptive are said to be **protandrous** (see Figure 21–24a), and those in which the stigmas become receptive before the stamens mature are said to be **protogynous.** As a result of these conditions, at any given time a flower may be either effectively staminate or effectively carpellate. Dichogamy is widespread among flowering plants (Figure 21–21).

Another strategy by which plants achieve outcrossing is through physical separation of the stamens and stigma(s) within a flower (Figure 21–22). In this way, pollen derived from the stamens of a given flower will rarely reach the stigma(s) of that flower, even if the stamens and stigmas mature at the same time. Such a condition, then, greatly increases the likelihood of pollen being dispersed to another flower.

Genetic self-incompatibility also occurs widely in the angiosperms, with at least some, and often many, members of almost every plant family being genetically

(a)   *(b)*

21–21
These flowers of fireweed, Chamaenerion angustifolium, *illustrate dichogamy, in which the stamens and carpels on the same flower reach maturity at different times. (a) Flowers in the staminate phase, producing pollen. (b) Flowers in the carpellate phase. The flowers of fireweed open from the bottom of the long inflorescence upward. Soon after a flower opens, the anthers begin to shed pollen.*

About two days later, the style, which has been reflexed to one side, swings up into the center of the flower, the stigma opens, and the flower becomes carpellate. By this time, the anthers have completely shed their pollen. As a result, the lower flowers on a stem that has been blooming for more than a few days are always carpellate and the upper ones are

staminate. The lower flowers have more abundant nectar, and bees fly to the bottom of the inflorescence first when they arrive at a new plant, then move upward. Thus they carry pollen first to the carpellate flowers and ultimately, as they move up, pick up additional pollen from the staminate flowers before moving on.

self-incompatible. If a plant is genetically self-incompatible, it will be outcrossed even if its stamens and stigma come into contact regularly and mature at the same time. Two basic systems of genetic self-incompatibility occur in different groups of angiosperms.

In the more common system, *gametophytic self-incompatibility*, which is present in such economically important plant families as the grasses *(Poaceae)* and legumes *(Fabaceae)*, the behavior of the pollen is determined by its own (haploid) genotype. If the pollen grain's DNA carries an allele at its incompatibility locus that is identical to the allele that occurs at either of the two corresponding loci in the diploid stigma or style, then the entry of the pollen tube into the stigma, or its passage through the style, is barred. If the pollen grain carries at this locus an allele that is different from either of those present in the stigmatic tissue, then the pollen tube emerging from that grain is accepted.

In *sporophytic self-incompatibility*, present in the mustard family *(Brassicaceae)* and the sunflower family *(Asteraceae)*, the behavior of the pollen is determined by the genetics of the parent plant that produced the pollen, not by the alleles that occur in the pollen grain itself. In other words, the control system is based on the correspondence between the two kinds of diploid tissue. As in gametophytic self-incompatibility, the kind of "match" that occurs at the incompatibility locus determines acceptance or rejection.

21–22
This flower of Easter lily (Lilium longiflorum) *illustrates the wide separation of stigma and anthers that is characteristic of many plants.*

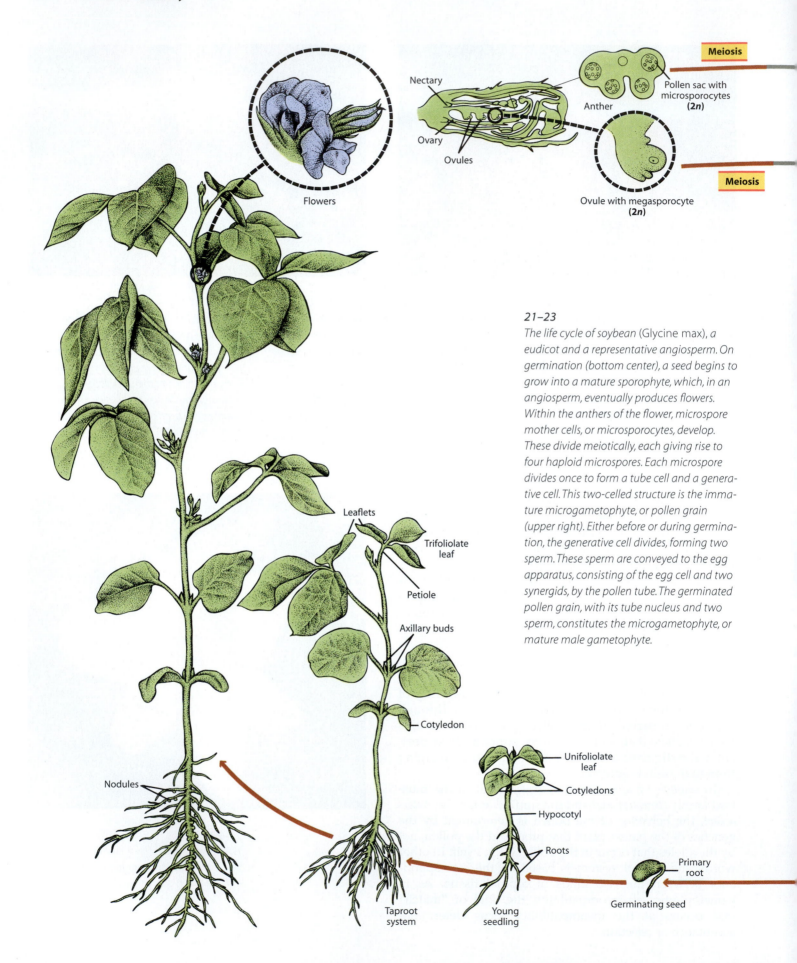

Meiosis

Nectary

Pollen sac with
microsporocytes
(2n)

Anther

Ovary

Ovules

Meiosis

Flowers

Ovule with megasporocyte
(2n)

21–23

*The life cycle of soybean (Glycine max), a
eudicot and a representative angiosperm. On
germination (bottom center), a seed begins to
grow into a mature sporophyte, which, in an
angiosperm, eventually produces flowers.
Within the anthers of the flower, microspore
mother cells, or microsporocytes, develop.
These divide meiotically, each giving rise to
four haploid microspores. Each microspore
divides once to form a tube cell and a generative cell. This two-celled structure is the immature microgametophyte, or pollen grain
(upper right). Either before or during germination, the generative cell divides, forming two
sperm. These sperm are conveyed to the egg
apparatus, consisting of the egg cell and two
synergids, by the pollen tube. The germinated
pollen grain, with its tube nucleus and two
sperm, constitutes the microgametophyte, or
mature male gametophyte.*

Leaflets

Trifoliolate
leaf

Petiole

Axillary buds

Cotyledon

Unifoliolate
leaf

Cotyledons

Hypocotyl

Nodules

Roots

Primary
root

Taproot
system

Young
seedling

Germinating seed

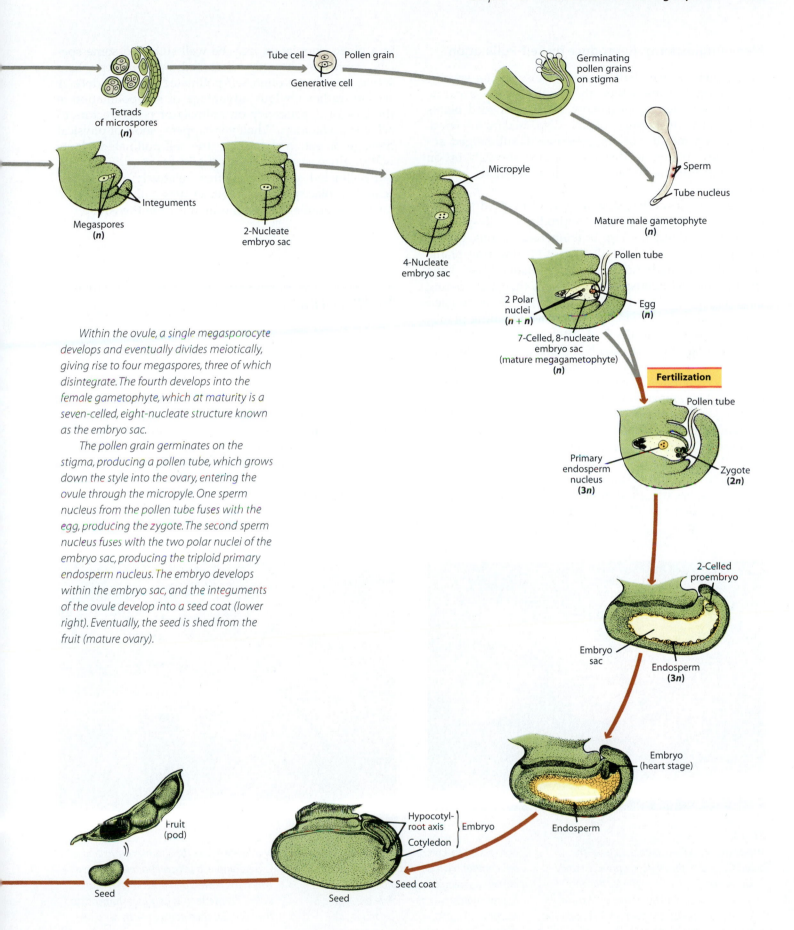

Tube cell
Pollen grain
Generative cell

Tetrads
of microspores
(*n*)

Germinating
pollen grains
on stigma

Sperm

Tube nucleus

Mature male gametophyte
(*n*)

Integuments

Micropyle

Megaspores
(*n*)

2-Nucleate
embryo sac

4-Nucleate
embryo sac

Pollen tube

2 Polar
nuclei
(*n* + *n*)

Egg
(*n*)

7-Celled, 8-nucleate
embryo sac
(mature megagametophyte)
(*n*)

Fertilization

Pollen tube

Primary
endosperm
nucleus
(3*n*)

Zygote
(2*n*)

Within the ovule, a single megasporocyte
develops and eventually divides meiotically,
giving rise to four megaspores, three of which
disintegrate. The fourth develops into the
female gametophyte, which at maturity is a
seven-celled, eight-nucleate structure known
as the embryo sac.

The pollen grain germinates on the
stigma, producing a pollen tube, which grows
down the style into the ovary, entering the
ovule through the micropyle. One sperm
nucleus from the pollen tube fuses with the
egg, producing the zygote. The second sperm
nucleus fuses with the two polar nuclei of the
embryo sac, producing the triploid primary
endosperm nucleus. The embryo develops
within the embryo sac, and the integuments
of the ovule develop into a seed coat (lower
right). Eventually, the seed is shed from the
fruit (mature ovary).

2-Celled
proembryo

Embryo
sac

Endosperm
(3*n*)

Embryo
(heart stage)

Hypocotyl-
root axis
Cotyledon

Embryo

Endosperm

Fruit
(pod)

Seed

Seed coat

Seed

Many Angiosperms Reproduce by Self-Pollination

In Chapter 22, we shall examine many of the specific plant–animal relationships that result in the accurate transfer of pollen between flowers. In most seed plants other than angiosperms, pollen is dispersed by the wind, as it is in a number of angiosperms. In submerged angiosperms, pollen is dispersed either through or on the surface of the water. All of these methods lead to outcrossing.

Many angiosperms, however, have adopted self-pollination as a regular mode of reproduction, despite the advantages of outcrossing. In temperate regions, for example, more than half of all species of flowering plants are self-pollinated. Self-pollinating plants often have smaller and less conspicuous flowers than outcrossing ones (Figure 21–24)—but then, the former have no need to attract animal visitors. In some self-pollinating plants, pollination occurs in the bud, after which the bud may or may not open. Often the bud simply falls off, leaving a ripening ovary behind. In others, pollination occurs only after the buds open—although pollen may also be dispersed by animals visiting such flowers.

The large numbers of angiosperms that are regularly self-pollinated clearly indicate that self-pollination is advantageous under certain circumstances. Populations of self-pollinating plants normally have higher proportions of genetically similar individuals than populations in which outcrossing predominates. Depending on their genotype, many or all of the self-pollinated individuals in a given population may be well suited to some specific habitat, such as the disturbed, open places where weeds, which are often self-pollinated, occur widely. A second, rather obvious advantage of self-pollination is the lack of dependency on animals or other vectors to achieve pollination. Whatever happens, short of physical damage or extreme drought, the self-pollinated plants will produce seed. This second advantage explains why self-pollinated plants are often relatively well represented in places where flower-visiting animals may be rare, for example on high mountains or in the Arctic.

Summary

Angiosperms, or Flowering Plants, Constitute the Phylum *Anthophyta*

The two largest classes of this phylum are the *Monocotyledones* (65,000 species) and the *Eudicotyledones* (165,000 species). Flowering plants differ from other seed plants in various distinctive characteristics, such as the presence of endosperm in their seeds; the fact that their ovules are enclosed within megasporophylls, the carpels; and their distinctive reproductive structure, the flower, which is characterized by the presence of carpels and distinctive microsporophylls, or stamens.

(a)

(b)

21–24

These two annual herbs are species of the genus Clarkia, *of the evening primrose family,* Onagraceae. *They grow in the foothills of California and are closely related, as shown by careful comparison of their nucleic acid components. (a)* Clarkia cylindrica *is an outcrossing species, although it is genetically self-compatible. It is protandrous, the eight anthers opening and shedding their pollen approximately two days before the stigma is receptive. In addition, as you can see, the stigma is widely separated from the anthers within the flower. (b) Sharply contrasting is* C. heterandra, *a self-compatible species in which the flowers are much smaller and paler. This species has only four stamens, and their anthers shed pollen directly onto the stigma.*

The Flower Is a Determinate Shoot That Bears Sporophylls

Individual flowers may have up to four whorls of appendages. From the outside in, the whorls are the sepals (collectively, the calyx); the petals (collectively, the corolla); the stamens (collectively, the androecium); and the carpels (collectively, the gynoecium). The sepals and petals are sterile, with the sepals often green and protective, covering the flower in bud, and the petals often colored and serving a function in attracting pollinators. Individual stamens are generally divided into a stalk, or filament, and an anther, containing four pollen sacs (two pairs). Carpels are usually differentiated into a swollen lower part, the ovary, and a slender upper part, the style, terminating in the receptive stigma. One or more of these whorls may be missing in the flowers of individual kinds of plants.

In Angiosperms, Pollination Is Followed by Double Fertilization

Pollination in angiosperms takes place by the transfer of pollen from anther to stigma. The male gametes, or sperm, of angiosperms are transmitted by means of a pollen grain, which is an immature male gametophyte. At the time of dispersal, such a gametophyte may contain either two or three cells. Initially, there is a tube cell and a generative cell, the latter dividing before or after dispersal to give rise to two sperm. The female gametophyte of an angiosperm is called an embryo sac. In many angiosperms, embryo sacs have eight nuclei at maturity, one of which is the egg (the number of cells varies in different groups). Both sperm function during angiosperm fertilization (double fertilization). One unites with the egg, producing a diploid zygote. The other unites with two polar nuclei, giving rise to the primary endosperm nucleus, which is usually triploid ($3n$). That nucleus divides, producing a unique kind of nutritive tissue, the endosperm, which may be absorbed by the embryo before the seed is mature or may persist in the mature seed. The angiosperms share double fertilization with the gnetophytes *Ephedra* and *Gnetum*, but in these gymnosperms, the process results in the formation of two embryos.

An Ovule Develops into a Seed, and an Ovary Develops into a Fruit

The ovaries (sometimes with some associated floral parts) develop into fruits, which enclose the seeds. Along with the flower from which it is derived, the fruit is a defining characteristic of the angiosperms.

A Variety of Conditions Promote Outcrossing in Angiosperms

Outcrossing in angiosperms is promoted by dioecism, in which staminate and carpellate flowers occur on different individual plants, and by monoecism, in which separate staminate and carpellate flowers occur on single individual plants. Outcrossing is also promoted by dichogamy, in which the stamens and carpels of a given flower mature at different times, or simply by the physical separation of these organs within an individual flower. Genetic self-incompatibility, which occurs widely in angiosperms, makes self-fertilization impossible even if the stamens and stigmas mature at the same time and come into contact with each other.

Many Angiosperms Reproduce by Self-Pollination

Self-pollination, resulting in the production of genetically uniform individuals, is characteristic of many angiosperms, including more than half of the species that occur in temperate regions. It clearly is favored over outcrossing under certain ecological conditions.

Selected Key Terms

anther p. 501	**pedicel** p. 499
antipodals p. 506	**peduncle** p. 499
carpel p. 499	**perigynous** p. 502
central cell p. 506	**perisperm** p. 510
chalazal p. 506	**petals** p. 501
dichogamy p. 510	**placenta** p. 501
dioecious p. 501	**placentation** p. 501
egg apparatus p. 506	**polar nuclei** p. 506
embryo sac p. 506	**pollen sac** p. 501
endosperm p. 509	**primary endosperm nucleus** p. 509
epigynous p. 502	**receptacle** p. 499
exine p. 504	**regular,** or **radially symmetrical** p. 502
filament p. 501	**sepals** p. 501
funiculus p. 506	**stamens** p. 501
generative cell p. 504	**stigmatic tissue** p. 508
hypogynous p. 502	**synergids** p. 506
inflorescence p. 499	**tapetum** p. 504
intine p. 504	**transmitting tissue** p. 508
irregular, or **bilaterally symmetrical** p. 502	**tube cell** p. 504
locule p. 501	
monoecious p. 501	

Questions

1. Distinguish among or between the following: calyx/corolla/perianth; stigma/style/ovary; complete/incomplete; perfect/imperfect; androecium/gynoecium.

2. Diagram and label as completely as possible a complete hypogynous flower, in which none of the floral parts are joined.

3. An imperfect flower is automatically incomplete, but not all incomplete flowers are imperfect. Explain.

4. Diagram and label completely a mature male gametophyte (germinated pollen grain) and a mature female gametophyte (embryo sac) of an angiosperm. Compare these gametophytes with their counterparts in pine.

5. Double fertilization followed by the formation of endosperm is unique to angiosperms. How does double fertilization in the gnetophytes *Ephedra* and *Gnetum* differ from that in angiosperms?

6. In some angiosperms, pollen production in the anthers occurs prior to or after the full development of the carpel of the same flower (dichogamy). What are the consequences of this shift in time frames?

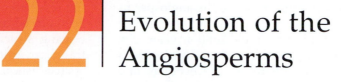

22 Evolution of the Angiosperms

22–1

The world's oldest flower. Imprint of Bevhalstia pebja in Early Cretaceous rocks (about 130 million years old) in Surrey, England. This plant was about 25 centimeters high and probably lived in water. It combines a fernlike structure with small flowerlike reproductive structures. The actual width of the flower is about 7 mm.

OVERVIEW

Reflect upon the evolutionary road we have traveled in the past few chapters of this book. One important trend concerns the mechanisms by which the sperm reach the eggs. In bryophytes and seedless vascular plants, sperm must swim through water to reach the eggs, whereas in gymnosperms the immature male gametophytes, or pollen grains, are largely carried by the wind to the nearby vicinity of the female gametophytes. There they germinate, producing pollen tubes and sperm. Hence, in the gymnosperms, water is no longer needed for the sperm to reach the eggs, and in two groups, the conifers and gnetophytes, the pollen tubes are true sperm conveyors, conveying nonmotile sperm more or less directly to the eggs. Nevertheless, gymnosperms are largely dependent upon wind for pollination. Angiosperms, by contrast, have evolved a set of features that attract a wide variety of pollinators—most notably insects— that assure a high degree of cross-pollination and evolutionary development.

In this chapter, we shall explore the angiosperms and learn why they have come to dominate the world's vegetation. It is a marvelous and unequaled success story. Let us begin with consideration of the origin of the angiosperms.

CHECKPOINTS

By the time you finish reading this chapter, you should be able to answer the following questions:

1. What are the current hypotheses on the origin of angiosperms, and what is the presumed relationship among the monocots, the eudicots, the woody magnoliids, and the paleoherbs?

2. What did the perianth (sepals and petals), stamens, and carpels of the earliest angiosperms look like? What are the four principal evolutionary trends among flowers?

3. What feature has evolved in angiosperms that has allowed them directed mobility in seeking a mate?

4. How do beetle-, bee-, moth-, bird-, and bat-pollinated flowers differ from one another?

5. What are some of the adaptations of fruits, in relation to their dispersal agents?

6. How, apparently, have secondary metabolites influenced angiosperm evolution?

517

In a letter to a friend, Charles Darwin once referred to the apparently sudden appearance of the angiosperms in the fossil record as "an abominable mystery." In the early fossil-bearing strata, about 400 million years old, one finds simple vascular plants, such as rhyniophytes and trimerophytes. Then there is a Devonian and Carboniferous proliferation of ferns, lycophytes, spheno-phytes, and progymnosperms, which were dominant until about 300 million years ago. The early seed plants first appeared in the Late Devonian period and led to the gymnosperm-dominated Mesozoic floras. Finally, early in the Cretaceous period, and at least 130 million years ago (Figure 22–1), angiosperms appear in the fossil record, gradually achieving worldwide dominance in the vegetation by about 90 million years ago. By about 75 million years ago, many modern families and some modern genera of this phylum already existed (Figure 22–2).

Despite their relatively late appearance in the fossil record, why did the angiosperms rise to world dominance and then continue to diversify to such a spectacular extent? In this chapter, we shall attempt to answer this question, centering our discussion on the relationships of the angiosperms, their origin and diversification, the evolution of the flower, the evolution of fruits, and the role of certain chemical substances in angiosperm evolution. All five topics will illustrate some of the reasons for the evolutionary success of the group.

Relationships of the Angiosperms

Since the time of Darwin, scientists have attempted to understand the ancestry of the angiosperms. One ap-proach has been to search for their possible ancestors in the fossil record. In this effort, particular emphasis has been placed on assessing the ease with which the ovule-bearing structures of various gymnosperms could be transformed into a carpel. Recently, phylogenetic analyses (cladistics) based on fossil, morphological, and molecular data have revitalized attempts to define the major natural groups of seed plants and to understand their interrelationships.

The most striking result from recent phylogenetic analyses is the support that they have provided for earlier ideas that the *Bennettitales* (page 472) and gneto-phytes (page 490) are the seed plants most closely related to angiosperms. The term "anthophytes" (not to be confused with the use of the term *Anthophyta* here to refer to angiosperms) has been proposed to refer collectively to the *Bennettitales*, gnetophytes, and angiosperms. It emphasizes the shared possession of flowerlike reproductive structures by these three groups of seed plants. Two contrasting hypotheses have been proposed for the phylogenetic relationships among anthophytes. One hypothesis proposes that the gnetophytes are monophyletic, and the derived similarities that *Gnetum* (Figure 20–41) and *Welwitschia* (Figure 20–43) share with angiosperms are interpreted as examples of convergent evolution (Figure 22–3a). The second hypothesis considers the gnetophytes to be paraphyletic, with *Gnetum* and *Welwitschia* as sister groups to angiosperms (Figure 22–3b). The latter hypothesis interprets the derived similarities of *Gnetum*, *Welwitschia*, and angiosperms to be homologous.

It is significant that the *Bennettitales* and gnetophytes first appear in the fossil record in the Triassic period, about 225 million years ago. This seems to place some constraints on the possible earliest date of the appear-

(a)

(b)

22–2

Fossils of angiosperms from Late Cretaceous deposits (about 70 million years ago) in Wyoming. (a) Leaf of a fan palm, Sabalites montana, *an extinct species distantly related to the palmettos of the southeastern United States. (b) Leaf of an extinct member of the family* Platanaceae, *commonly known as the sycamores.*

22–3

*Two proposed hypotheses for the phyloge-
netic relationships among the anthophytes.
(a) One hypothesis states that the Gnetales
are monophyletic, and the derived similarities
of Gnetum, Welwitschia, and angiosperms
are interpreted as convergent evolution.
This hypothesis is consistent with current
molecular data of both rbcL (the chloroplast-
encoded gene sequence for the large subunit
of Rubisco) and ribosomal RNA. (b) According to another hypothesis, the Gnetales are
paraphyletic, with Gnetum and Welwitschia
the sister group to the angiosperms. This
hypothesis interprets the derived similarities
of Gnetum, Welwitschia, and angiosperms as
being homologous.*

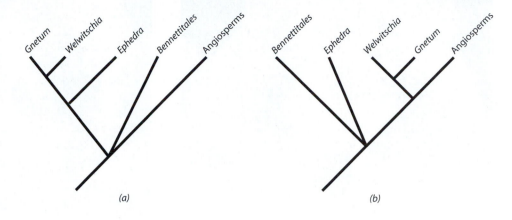

ance of the angiosperms—that is, on the possibility that
they could have arisen any earlier.

Origin and Diversification of the Angiosperms

The unique characteristics of the angiosperms include
flowers, closed carpels, double fertilization leading to
endosperm formation, a three-nucleate microgametophyte and an eight-nucleate megagametophyte, stamens
with two pairs of pollen sacs, and the presence of sieve
tubes and companion cells in the phloem (see Chapter
24). These similarities clearly indicate that the members
of this phylum were derived from a single common ancestor. This common ancestor of the angiosperms ultimately would have been derived from a seed plant that
lacked flowers, closed carpels, and fruits. The earliest
known, clearly identifiable fossils of angiosperms are
flowers and pollen grains up to 130 million years old,
from the Early Cretaceous period (Figure 22–1). There
are intriguing suggestions that much older fossils—up
to 200 million years old—may have had some, but perhaps not all, of the characteristic features of angiosperms. Currently the interpretation of these fossils is
enigmatic, and it appears most likely that the phylum
did in fact originate in the Early Cretaceous (or perhaps
uppermost Jurassic) period.

What were the earliest angiosperms like? Like gymnosperms, they clearly had pollen with a single aperture,
as found in monocots and a few other groups of angiosperms, as well as cycads, *Ginkgo,* and other groups. This
feature can therefore be considered an ancestral one that
has been retained in the course of evolution. Angiosperms that produce single-aperture pollen, therefore,
cannot be shown, on the basis of this character, to have
had a common ancestor distinct from other angiosperms.

The Magnoliids Are Ancestral to Both Monocots and Eudicots

In Chapter 21, we discussed the two largest classes of
angiosperms, the monocots and the eudicots, which between them comprise 97 percent of the members of the
phylum. The monocots clearly had a common ancestor,
as indicated by their single cotyledon and a number of
other features. The same is true of the eudicots, which
have a characteristic derived feature, their triaperturate
pollen (pollen with three slits or pores, and also pollen
types derived from the triaperturate group). The remaining 3 percent of living angiosperms, the **magnoliids,** include those with the most primitive features. These
ancestors of both monocots and eudicots have traditionally been grouped with the eudicots as "dicots," but this
is as illogical as grouping them with monocots. The evolutionary relationships of magnoliids are not well understood.

Although they have traditionally been regarded as
dicots, all magnoliids, like the monocots, have pollen
with a single aperture or some modification of this type.
One of their characteristic features is the possession of
oil cells with ethereal (ether-containing) oils, the basis of
the characteristic scents of nutmeg, pepper, and laurel
leaves. One group of magnoliids consists of plants
known informally as the **woody magnoliids.** These
plants have large, robust, bisexual flowers with many
free parts, arranged spirally on an elongate axis, as in
Magnolia (Figure 22–4). There are fewer than 20 families
of this group with living members. Among them, the
most familiar are the magnolia family *(Magnoliaceae),*
laurel family *(Lauraceae),* and spicebush family
(Calycanthaceae).

One of the numerous fossil woody magnoliids is
Archaeanthus linnenbergeri, which is probably very closely
related to living *Magnolia* and the tulip tree *(Liriodendron)*
(Figure 22–5). The stout flowers of *Archaeanthus* had an
elongate axis with 100 to 130 spirally arranged carpels,

(a)

(b)

(c)

22–4

Flowers and fruits of the southern magnolia (Magnolia grandiflora), *a woody magnoliid. The cone-shaped receptacle bears numerous spirally arranged carpels from which curved styles emerge. Below the styles in* **(a)** *and* **(b)** *are the cream-white stamens.* **(a)** *The anthers have not yet shed their pollen, whereas the stigmas are receptive. In other words, the species is protogynous.* **(b)** *The floral axis of a second-day flower, showing stigmas that are no longer receptive and stamens that are shedding pollen.* **(c)** *Fruit, showing carpels and bright red seeds, each protruding on a slender stalk.*

(a)

(b)

22–5

Archaeanthus linnenbergeri, an extinct angiosperm that generally resembles living magnolias. **(a)** *Fossil reproductive axis.* **(b)** *Fossil leaf.* **(c)** *Reconstruction of a flowering branch.* **(d)** *Reconstruction of a fruiting branch. The reconstructions are based on the studies of David Dilcher of the University of Florida and Peter Crane of the Field Museum of Natural History in Chicago and were drawn by Megan Rohn. Careful studies by Dilcher, Crane, and others are revealing a great deal about the nature of early angiosperms and their flowers.*

(c)

(d)

each eventually producing 10 to 18 seeds. There were three outer perianth parts and six to nine inner ones; the stamens were numerous and spirally arranged. *Archaeanthus*, which inhabited what were then the subtropical coastal plains of Kansas, was probably deciduous, and it seems to have been a small tree or shrub.

Apart from the woody magnoliids, the remaining magnoliids constitute a very diverse assemblage collectively called the **paleoherbs.** They are predominantly herbs with small flowers and often with relatively few flower parts; their flowers are often unisexual. Many paleoherbs, unlike the great majority of woody magnoliids, have carpels that are fused together. Among the living paleoherbs are families such as *Piperaceae*, the pepper family; *Aristolochiaceae*, the birthwort family (Figure 22–6); and *Nymphaeaceae*, the water lily family. A number of very early fossils, such as that shown in Figure 22–7, are also paleoherbs, although it is important to remember that this group basically includes all angiosperms that are not monocots, eudicots, or woody magnoliids. It is, therefore, a kind of evolutionary "grab bag" of angiosperms with primitive features.

The monocots arose from ancestors that may have been similar to the birthwort family, *Aristolochiaceae*, or to the pepper family, *Piperaceae*, judged from biochemical and morphological evidence. In any case, their ancestors were clearly not woody magnoliids. Although the fossil record of early monocots is rather poor, the first members of this class seem to have arisen as one of many diverse lines of paleoherbs, probably more than 120 million years ago. The major evolutionary lines of monocots were in existence at least 70 million years ago, before the close of the Cretaceous period.

22–6
Aristolochia grandiflora, *the Dutchman's pipe, belongs to the birthwort family* (Aristolochiaceae). Aristolochiaceae *are one of a few angiosperm families of living paleoherbs.*

22–7
A very early angiosperm, approximately 120 million years old, from fossil beds near Melbourne, Australia. This fossil, reported in 1990, probably represents an herbaceous plant that may have grown in marshy places. It has undivided leaves, and its flowers are subtended by bracts and are carpellate. This fossil paleoherb resembles members of the existing pepper family (Piperaceae).

5 mm

The nature of the ancestors of the eudicots is less clear but is under active investigation. The triaperturate pollen that marks the members of this class appears in the fossil record in the Lower Cretaceous period, about 127 million years ago or perhaps a little earlier. The group became quite diverse over the subsequent 30 to 40 million years, by the mid-Cretaceous. The features of the flower of *Magnolia* have often been used to demonstrate what the features of a primitive angiosperm flower might have been. It seems clear now, however, that the *Magnolia* flower, which is large, with numerous, spirally arranged parts, is in fact an early specialization. The common ancestor of the angiosperms, and perhaps the eudicots as well, seems likely to have had relatively small, simple, perhaps green, and rather unattractive flowers, with their petals and sepals not clearly differentiated. These features are characteristic of some Early Cretaceous fossils, and some living plants as well.

Angiosperms Spread Rapidly throughout the World

The appearance and rapid diversification of the eudicots and the monocots led to the increasing domination of angiosperms throughout the world during the 35 million years of the upper Cretaceous period (100–65 million years ago). By approximately 90 million years ago, several existing orders and families of angiosperms had appeared (Figures 22–2 and 22–8), and the flowering plants had achieved dominance throughout the Northern Hemisphere. During the subsequent 10 million years, they achieved dominance in the Southern Hemisphere as well.

The early angiosperms possessed many adaptive traits that made them particularly resistant to drought and cold. Among these were tough leaves, commonly reduced in size; vessel elements (efficient water-conducting cells); and a tough, resistant seed coat that provided protection against the young embryo drying out. These features are not found in all of the flowering plants, nor are they restricted to the angiosperms, but they certainly have played major roles in the success of the phylum. The early evolution of the deciduous habit (the seasonal loss of leaves), which allows woody plants to lose their leaves and to become relatively inactive physiologically during periods of drought, extreme heat, and cold, may also have contributed to the evolutionary success of the group—especially during the past 50 million years when world climates have been undergoing active change.

A number of other factors seem to have been important in the early and continuing success of the angiosperms, and we shall discuss some of them in more detail in this chapter. The evolution of sieve-tube elements presumably made possible the more efficient conduction of sugars throughout the plant in the phloem, just as vessel elements are more efficient than tracheids in the xylem. Perhaps of even greater importance, the precise systems of pollination and specialized mechanisms of seed dispersal that became characteristic of the more advanced flowering plants allowed them to exist as widely scattered individuals in many different kinds of habitats. The enormous chemical diversity of the angiosperms, which includes the many kinds of defenses against diseases and herbivores, has likewise been of

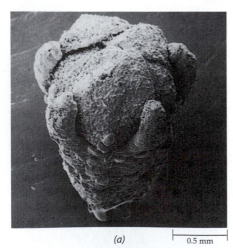

(a)   0.5 mm

(b)   0.2 mm

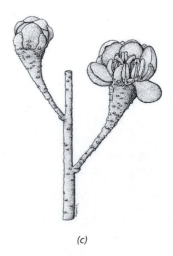

(c)

22–8
Silvianthemum suecicum, *a relatively specialized angiosperm that occurred in the Late Cretaceous period in southern Sweden, about 80 million years ago. Exquisitely preserved as fossil charcoal from ancient forest fires, these flowers are minute, perfect, and radially*

symmetrical, with five free sepals and five free petals, three fused carpels, disk-shaped nectaries, and minute, numerous seeds. Like two other genera that occurred with it, Silvianthemum *is broadly related to the saxifrage family, Saxifragaceae, and espe-*

cially to some of its woody relatives. (a), (b) Two views of the fossil flowers. (c) Reconstruction of a flower and bud. Else Marie Friis and her coworkers are making outstanding contributions to our knowledge of fossil angiosperms.

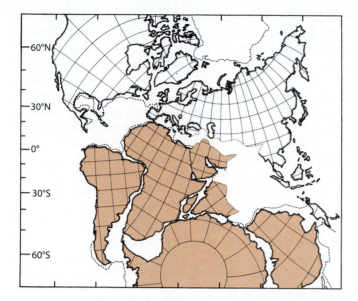

22-9

Relationship between the Earth's land areas at the time of the first appearance of the angiosperms in the fossil record (130 million years ago). In the middle of the Cretaceous period, about 100 million years ago, South America was still directly connected with Africa, Madagascar, and India and, via Antarctica, with Australia. These combined land masses, indicated here in color, formed the supercontinent of Gondwanaland.

forms. With the final separation of these two continents, at about the time the angiosperms became abundant in the fossil record worldwide, the world climate changed greatly. This was especially true in these equatorial regions, which became milder, with fewer extremes of temperature and humidity. Magnoliid angiosperms— those with the most archaic features—have survived most abundantly and are best represented today in regions with relatively uniform, warm climates, such as those of southeast Asia and the South Pacific.

22-10

*This forest of southern beech (*Nothofagus menziesii*), growing in Fiordland, South Island, New Zealand, is a relict from the cool temperate forest that extended from southern South America across Antarctica into Australia and New Zealand from about 80 million years ago until perhaps 30 million years ago. Increasingly large gaps have developed between the land masses throughout that entire period of time up to the present.*

great importance, as we shall see. These and other features bear directly on the fact that, compared with other plants, angiosperms are ecologically resilient and reproduce rapidly and efficiently.

About 130 million years ago, when the first definite angiosperm fossils appear in the fossil record, Africa and South America were directly connected with each other and also with Antarctica, India, and Australia in a great southern supercontinent called **Gondwanaland** (Figure 22–9). Africa and South America began to separate at about this time, forming the southern Atlantic Ocean, but they did not move completely apart in the tropical regions until about 90 million years ago. India began to move northward at about the same time, colliding with Asia starting about 65 million years ago and thrusting up the Himalayas in the process. Australia began to separate from Antarctica about 55 million years ago, but their separation did not become complete until more recently (Figure 22–10).

Within the central regions of West Gondwanaland, formed by what are now the continents of South America and Africa, habitats were arid to subhumid. Under these conditions, angiosperms and other kinds of organisms would have been challenged to produce new

Evolution of the Flower

What were the flowers of the earliest angiosperms like? Of course, we do not know this from direct observation, but we can deduce their nature from what we know of certain living plants and from the fossil record. In general, the flowers of these plants were diverse both in the numbers of floral parts and in the arrangement of these parts. Most modern families of angiosperms tend to have more fixed floral patterns that do not vary much in their basic structural features within a particular family. We shall discuss the derivation of these patterns over the course of evolution in the following sections, which deal with the different whorls of the flower from the outside in, moving from the perianth inward to the androecium and the carpel.

The Parts of the Flower Provide Clues to Angiosperm Evolution

The Perianth of Early Angiosperms Did Not Have Distinct Sepals and Petals   In the earliest angiosperms, the perianth, if present, was never sharply divided into calyx and corolla. Either the sepals and petals were identical, or there was a gradual transition in appearance between these whorls, as in modern magnolias and water lilies. In some angiosperms, including the water lilies, petals appear to have been derived from sepals. In other

words, the petals can be viewed as modified leaves that have become specialized for attracting pollinators. In most angiosperms, however, petals were probably derived originally from stamens that lost their sporangia—becoming "sterilized"—and then were specially modified for their new role. Most petals, like stamens, are supplied by just one vascular strand. In contrast, sepals are normally supplied by the same number of vascular strands as are the leaves of the same plant (often three or more). Within sepals and petals alike, the vascular strands usually branch so that the number of strands that enter them cannot be determined from the number of veins in the main body of the structures.

Petal fusion has occurred a number of times during the evolution of the angiosperms, resulting in the familiar tubular corolla that is characteristic of many families (Figure 22–11c). When a tubular corolla is present, the stamens often fuse with it and appear to arise from it. In a number of evolutionarily advanced families, the sepals are also fused into a tube.

The Stamens of Early Angiosperms Were Diverse in Structure and Function   The stamens of some families of woody magnoliids are broad, colored, and often scented, playing an obvious role in attracting floral visitors. In other archaic angiosperms, the stamens, although relatively small and often greenish, may also be fleshy. Many living angiosperms, in contrast, have stamens

(a)

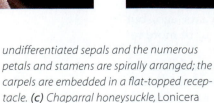

(b)

22–11

Examples of specialized flowers. (a) Wintergreen, Chimaphila umbellata. The sepals (not visible) and petals are reduced to five each, the stamens to ten, and the five carpels are fused into a compound gynoecium with a single stigma. (b) Lotus, Nelumbo lutea. The
undifferentiated sepals and the numerous petals and stamens are spirally arranged; the carpels are embedded in a flat-topped receptacle. (c) Chaparral honeysuckle, Lonicera hispidula. The ovary is inferior and has two or three locules; the sepals are reduced to small
teeth at its apex. The petals are fused into a corolla tube in the zygomorphic (bilaterally symmetrical) flower, and the five stamens, which protrude from the tube, are attached to its inner wall. The style is longer than the stamens, and the stigma is elevated above

with generally thin filaments and thick, terminal anthers (for example, see Figures 21–6 and 21–22). In general, the stamens of monocots and eudicots seem to be much less diverse in structure and function than the stamens of magnoliid angiosperms.

In some specialized flowers the stamens are fused together. Their fused filaments may then form columnar structures, as in the members of the pea, melon, mallow (Figure 22–11d), and sunflower families, or they may be fused with the corolla, as in the phlox, snapdragon, and mint families.

In certain plant families, some of the stamens have become secondarily sterile: they have lost their sporangia and become transformed into specialized structures, such as nectaries. Nectaries are glands that secrete **nectar,** a sugary fluid that attracts pollinators and provides food for them. Most nectaries are not modified stamens but arose instead in other ways. During the course of evolution of the angiosperm flower, the sterilization of stamens, as noted above, also played an important role in the evolution of petals.

The Carpels of Many Early Angiosperms Were Unspecialized A number of woody magnoliids have generalized and sometimes leaflike carpels, with no specialized areas for the entrapment of pollen grains comparable to the specialized stigmas of most living angiosperms. The carpels of many woody magnoliids

and other plants that retain archaic features are free from one another, instead of being fused together as in most contemporary angiosperms. In a few living woody magnoliids, the carpels are incompletely closed, although pollination is always indirect—the pollen does not contact the ovules directly. In the vast majority of living angiosperms, the carpels are closed (a condition that gave rise to the name of the phylum) and sharply differentiated into stigmas, styles, and ovaries. There is much variation in the arrangement of the ovules among contemporary groups of angiosperms, which often have fewer ovules than do some of the more generalized and archaic families of the phylum.

Four Evolutionary Trends among Flowers Are Evident

Insect pollination quite probably triggered the early evolution of angiosperms, both through the possibilities it provided for isolating small populations and with indirect pollination fostering competition between many pollen grains as they grew through the stigmatic tissue. The flowers of the earliest members of the phylum probably were bisexual, but unisexual flowers appeared very early in many different families. The undifferentiated perianth of early angiosperms soon gave rise to distinct petals and sepals. As the angiosperms continued to evolve, and relationships with specialized pollinators became more tightly linked, the number and arrangement

(c)

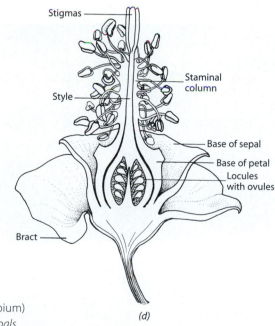

(d)

them. A pollinator visiting this flower would contact the stigma first, so that if it were carrying pollen from another flower, it would deposit that pollen on the stigma before reaching the anthers. Fruits of this species are shown in Figure 22–45c. **(d)** A diagram of a longitudinal section of a cotton (Gossypium) flower, of the mallow family, with the sepals and petals removed and showing the column of stamens fused around the style.

An Ambiguous Aquatic Plant

One of the great surprises revealed by contemporary studies of plant evolution concerns the relationship of the hornwort (Ceratophyllum), a unique aquatic angiosperm (a). Less than a decade ago, the perceptive studies of University of Connecticut botanist Donald Les, who contributed the material on which this essay is based, added a new dimension to the mystery of angiosperm evolution by offering a new hypothesis about the nature of Ceratophyllum. This small, odd genus, classified in a family of its own, had long thwarted the efforts of taxonomists to determine its relationships.

Ceratophyllum is a highly specialized plant that lacks roots entirely (even in the embryo), is drastically reduced in morphology and anatomy, and lives and reproduces entirely underwater (even pollination takes place below the water surface). Yet a number of its floral and vegetative features represent conditions that botanists regard as primitive for angiosperms. For many years, plant systematists have viewed Ceratophyllum as a highly modified descendant of water lilies, mainly because, like them, it is aquatic. However, a more critical evaluation of features indicated to Les that Ceratophyllum

was not closely related either to the water lilies (Nymphaeaceae) or water lotus (Nelumbonaceae). In fact, Les's investigations of the morphological and anatomical features of Ceratophyllum indicated that it could be a living fossil descended from one of the earliest of angiosperm evolutionary lines, and that it might not be closely related to any known group of living angiosperms.

At first glance it is not easy to envision how a reduced, herbaceous, aquatic plant with simple flowers and water pollination could possibly represent an ancient angiosperm. But what is easy to overlook is the possibility that Ceratophyllum is so ancient that these unusual features have accumulated over a very long time and may represent poorly what its ultimate ancestor looked like. Many of its features, such as unisexuality, lack of vessels, lack of a perianth, lack of a pollen exine, a single integument (seed coat) layer, and branching pollen tubes, might have been expected in an ancient angiosperm, but they could also be interpreted as adaptations to an aquatic habit. Ceratophyllum might have become aquatic long ago—even though we usually think of aquatic plants as recently derived from terrestrial ances-

tors. Actually, fossilized Ceratophyllum-like fruits from the Lower Cretaceous period, some 115 million years ago, are among the oldest known angiosperm reproductive remains.

Recently, the base sequences in the DNA of Ceratophyllum have been studied. In a surprise to many critics of Les's original hypothesis, several analyses of chloroplast-encoded rbcL gene sequences (coding for the large subunit of ribulose bisphosphate carboxylase/oxygenase, or Rubisco) have indicated that Ceratophyllum is likely to be one of the most basal groups of all angiosperms (b). In the light of these additional findings, it is also interesting that Ceratophyllum possesses the simple flowers that botanists believe were characteristic of early paleoherbs. Although the data are not simple to analyze, largely because of the unique nature of Ceratophyllum, other molecular studies are tending to confirm the rbcL results. Ceratophyllum clearly illustrates the point that modern living descendants of ancient evolutionary lineages almost always possess a mosaic of both archaic and derived features and, taken as a whole, they may no longer resemble the ancestral plants from which they were derived.

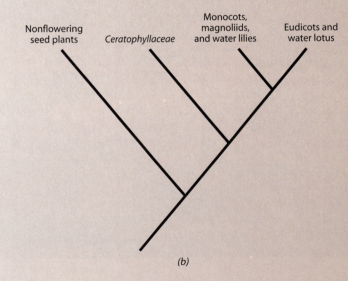

(a) A shoot of Ceratophyllum shows the simple, naked pistillate flower of this unique aquatic angiosperm. (b) A cladogram constructed from rbcL gene sequence data shows Ceratophyllum at the base of the flowering plants.

of floral patterns became more stereotyped. The following four trends are evident (Figure 22–11):

1. From flowers with few to many parts that are indefinite in number, flowers have evolved toward having few parts that are definite in number.

2. The number of floral whorls has been reduced from four in early flowers to three, two, or sometimes one in more advanced ones. The floral axis has become shortened so that the original spiral arrangement of parts is no longer evident. The floral parts often have become fused.

3. The ovary has become inferior rather than superior in position, and the perianth has become differentiated into a distinct calyx and corolla.

4. The radial symmetry (regularity), or actinomorphy, of early flowers has given way to bilateral symmetry (irregularity), or zygomorphy, in more advanced ones.

The *Asteraceae* and *Orchidaceae* Are Examples of Specialized Families

Among the most evolutionarily specialized of flowers are those of the family *Asteraceae* (*Compositae*), which are eudicots, and those of the family *Orchidaceae*, which are monocots. In number of species, these are the two largest families of angiosperms.

The Flowers of the *Asteraceae* Are Closely Bunched Together into a Head   In the *Asteraceae* (the composites), the epigynous flowers are relatively small and closely bunched together into a head. Each of the tiny flowers has an inferior ovary composed of two fused carpels with a single ovule in one locule (Figure 22–12).

In composite flowers, the stamens are reduced to five in number and are usually fused to one another (coalescent) and to the corolla (adnate). The petals, also five in number, are fused to one another and to the ovary, and

22–12

Composites (family Asteraceae*). (a) A diagram showing the organization of the head of a member of this family. The disk and ray flowers are subordinated to the overall display of the head, which functions as a single large flower in attracting pollinators. (b) Thistle,* Cirsium pastoris. *Members of the thistle tribe have only disk flowers. This particular species of thistle has bright red flowers and is regularly visited by hummingbirds, which are its primary agents of pollination. (c)* Agoseris, *a wild relative of the dandelion,* Taraxacum. *In the inflorescences of the chicory tribe (the group of composites to which dandelions and their relatives belong), there are no disk flowers. The marginal ray flowers, however, are often enlarged. (d) Sunflower,* Helianthus annuus.

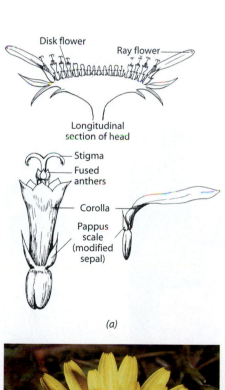

(a)

(b)

(c)

(d)

the sepals are absent or reduced to a series of bristles or scales known as the **pappus.** The pappus often serves as an aid to dispersal by wind, as it does in the familiar dandelion, a member of the *Asteraceae* (Figure 22–12c; see also Figures 22–41 and 22–42). In other members of this family, such as beggar-ticks *(Bidens),* the pappus may be barbed, serving to attach the fruit to a passing animal and thus to enhance its chances of being dispersed from place to place. In many members of the family *Asteraceae,* each head includes two types of flower: (1) disk flowers, which make up the central portion of the aggregate, and (2) ray flowers, which are arranged on the outer periphery. The ray flowers are often carpellate, but sometimes they are completely sterile. In some members of the *Asteraceae,* such as sunflowers, daisies, and black-eyed Susans, the fused bilaterally symmetrical (zygomorphic) corolla of each ray flower forms a long strap-shaped "petal."

In general, the composite head has the appearance of a single large flower. Unlike many single flowers, however, the head matures over a period of days, with the individual flowers opening serially in an inward-moving spiral pattern. As a consequence, the ovules in a given head may be fertilized by several different pollen donors. The success of this plan as an evolutionary strategy is attested to by the great abundance of the members of the *Asteraceae* and also their great diversity, which, with about 22,000 species, makes them the second largest family of flowering plants.

***Orchidaceae* Is the Largest Angiosperm Family**   Another successful flower plan is that of the orchids *(Orchidaceae),* which, unlike the composites, are monocots. There are probably at least 24,000 species of orchids, making them the largest family of flowering plants. In contrast to the composites, however, individual species of orchids are rarely very abundant. Most species of orchids are tropical, and only about 140 are native to the United States and Canada, for example. In the orchids, the three carpels are fused and, as in the composites, the ovary is inferior (Figure 22–13). Unlike the composites, however, each orchid ovary contains many thousands of minute ovules. Consequently, each pollination event may result in the production of a huge number of seeds. Usually only one stamen is present (in one subfamily, the lady-slipper orchids, there are two), and this stamen is characteristically fused with the style and stigma into a single complex structure—the **column.** The entire contents of an anther are held together and dispersed as a unit—the **pollinium** (see Figure 22–25b). The three petals of orchids are modified so that the two lateral ones form wings and the third forms a cuplike lip that is often very

(a)

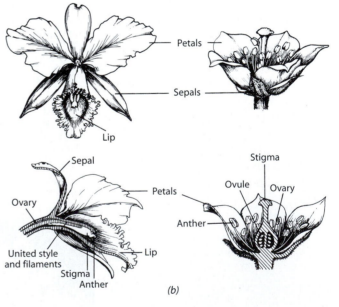

(b)

22–13
Orchids (family Orchidaceae*). (a) An orchid of the genus* Cattleya. *Orchids are one of the most specialized families of monocots. (b) A comparison of the parts of an orchid flower, shown on the left, with those of a radially symmetrical flower, shown on the right. The "lip" is a modified petal that serves as a landing platform for insects.*

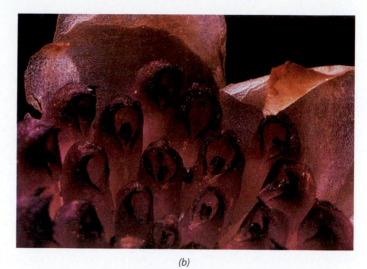

22–14

Rhizanthella, *a Western Australian orchid, is a saprophyte that grows entirely underground. Cracks in the soil form during the dry season, revealing the flowers, which never show* large and showy. *above the soil surface. (a) A view from above, with leaves and debris cleared away, of the parted bracts of* Rhizanthella *through which* *the plant's pollinators (flies) enter. (b) A dozen flowers of* Rhizanthella *are seen here surrounded by the protective bracts.*

large and showy. The sepals, also three in number, are often colored and similar to petals in appearance. The flower is always bilaterally symmetrical and often bizarre in appearance.

Among the orchids are some species with flowers the size of a pinhead and others with flowers more than 20 centimeters in diameter. Several genera contain saprophytic species. Two Australian species grow entirely underground, their flowers appearing in cracks in the ground, where they are pollinated by flies (Figure 22–14). In the commercial production of orchids, the plants are cloned by making divisions of meristematic tissue, and thousands of identical plants can be produced rapidly and efficiently (see Chapter 28). There are more than 60,000 registered hybrids of orchids, many of them involving two or more genera. The seed pods of orchids of the genus *Vanilla* are the natural source of the popular flavoring of the same name (Figure 22–15).

22–15

Vanilla, *an orchid that is the source of the flavoring of the same name. Originally used by the Aztecs in what is now Mexico, vanilla is cultivated primarily on Madagascar and other islands in the western Indian Ocean and elsewhere in the Old World. Vanilla is extracted from the dried, fermented seed pods of this orchid. Chocolate is a blend of an extract from the pods of the cacao plant and vanilla. Synthetic vanilla flavoring (vanillin) is now used as the source of about 95 percent of all vanilla consumed. (a) Flowers of the vanilla orchid (*Vanilla planifolia*). (b) Hand pollination of vanilla plants in Mexico. This procedure is carried out, even in wild plants, to ensure a good crop of the seed pods from which vanilla is extracted.*

Animals Serve As Agents of Floral Evolution

Plants, unlike most animals, cannot move from place to place to find food or shelter or to seek a mate. In general, plants must satisfy those needs by growth responses and by the structures that they produce. Many angiosperms, however, have evolved a set of features that, in effect, allows them directed mobility in seeking a mate. This set of features is embodied in the flower. By attracting insects and other animals with their flowers, and by directing the behavior of these animals so that cross-pollination (and therefore cross-fertilization) will occur at a high frequency, the angiosperms have transcended their rooted condition. In this one respect, they have become just as mobile as animals. How was this achieved?

Flowers and Insects Have Coevolved   The earliest seed-bearing plants were pollinated passively. Large amounts of pollen were blown about by the wind, reaching the vicinity of the ovules only by chance. The ovules, which were borne on the leaves or within cones, exuded sticky drops of sap from their micropyles. These drops served to catch the pollen grains and to draw them to the micropyle. As in most modern cycads (page 486) and gnetophytes, insects feeding on the pollen and other flower parts began returning to these new-found sources of food and thus transferred pollen from plant to plant. Such a system is more efficient than passive pollination by the wind. It allowed much more accurate pollination with many fewer pollen grains involved.

The more attractive the plants were to the insects (Figure 22–16), the more frequently they would be visited and the more seed they could produce. Any changes in the phenotype that made such visits more frequent or more efficient offered a selective advantage. Several important evolutionary developments followed. For example, plants that had flowers that provided special sources of food for their pollinators had a selective advantage. In addition to edible flower parts, pollen, and sticky fluid around the ovules, plants evolved floral nectaries. As previously mentioned, nectaries secrete the nutritious, sugary nectar that provides a source of energy for insects and other animals.

Attraction of insects to the naked ovules of these plants sometimes resulted in the loss of some of the ovules to the insects. The evolution of a closed carpel, therefore, gave certain seed plants—the ancestors of the angiosperms—a reproductive, and thus a selective, advantage. Further changes in the shape of the flower, such as the evolution of the inferior ovary, may have been additional means of protecting the ovules from being eaten by insects and other animals, thus providing a further reproductive advantage.

(a)

(b)

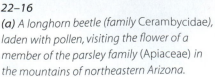

22–16

(a) A longhorn beetle (family Cerambycidae)*, laden with pollen, visiting the flower of a member of the parsley family* (Apiaceae) *in the mountains of northeastern Arizona.*

(b) Carapace of a beetle from the same strata as Archaeanthus, *an extinct angiosperm from 95 to 98 million years ago (Figure 22–5). The evolution of the flowering plants is, to a*   *large extent, the story of increasingly specialized relationships between flowers and their insect pollinators, in which beetles played an important early role.*

22–17

Flower-visiting beetles. (a) A pollen-eating beetle, Asclera ruficornis, *at the open, bowl-shaped flowers of round-leaved hepatica* (Hepatica americana) *in spring in the woods of eastern North America. All species of this family* (Oedemeridae) *are obligate pollen-feeders as adults. (b) A cetoniine scarabid beetle,* Eupoecila australasiae, *visiting the flowers of* Angophora woodsiana, *near Brisbane, Australia. The cetoniine scarabs have membranous mouthparts used for drawing up nectar.*

(a)

(b)

Another important evolutionary development was the appearance of the bisexual flower. The presence of both carpels and stamens in a single flower (in contrast, for instance, to the separate microsporangiate and megasporangiate cones of living conifers) offers a selective advantage by making each visit by a pollinator more effective. The pollinator can both pick up and deliver pollen at each stop.

In the early part of the Tertiary period, 40 to 60 million years ago, such specialized groups of flower-visiting insects as bees and butterflies—which had been evolving with the angiosperms for about 50 million years at that point—became even more abundant and diverse. The rise and diversification of these groups of insects were directly related to the increasing diversity of angiosperms. In turn, the insects profoundly influenced the evolutionary course of angiosperms and contributed greatly to their diversification.

If a given plant species is pollinated by only one or a few kinds of visitors, selection favors specializations related to the characteristics of these visitors. Many of the modifications that have evolved in flowers promoted constancy of a specific type of visitor to that particular kind of flower. Some of the special modifications of flowers that came about during the course of their evolution in response to specific pollinators will be described in the following pages.

Beetle-Pollinated Flowers Typically Have a Dull Color but a Strong Odor   Many modern species of angiosperms are pollinated solely or chiefly by beetles (Figure 22–17). The flowers of beetle-pollinated plants are either large and borne singly, such as those of magnolias, some lilies, California poppies, and wild roses, or small and aggregated in an inflorescence, such as those of dogwoods, elders, spiraeas, and many species of the parsley family (*Apiaceae*) (Figure 22–16a). Members of some 16 families of beetles are frequent visitors to flowers, although as a rule, these beetles derive most of their nourishment from other sources, such as sap, fruit, dung, and carrion. In beetles, the sense of smell is much more highly devel-

oped than the visual sense, and beetle-pollinated flowers are typically white or dull in color, with strong odors (Figure 22–18). These odors are usually fruity, spicy, or similar to the foul odors of fermentation. They are therefore distinct from the sweeter odors of flowers that are pollinated by bees, moths, and butterflies. Some beetle-pollinated flowers secrete nectar, which the beetles eat. In others, the beetles chew directly on the petals or on specialized food bodies (pads or clusters of cells on the surfaces of the various floral parts), and they also eat the pollen. Many beetle-pollinated flowers have inferior ovaries, with the ovules well buried in the floral tissues, out of reach of the chewing jaws of the beetles on which they depend for their pollination.

22–18

The foul-scented and often dark-colored flowers of many members of the milkweed family (Asclepiadaceae), *such as those of this African succulent plant,* Stapelia schinzii, *are pollinated by carrion flies.*

22–19

Bees have become as highly specialized as the flowers they have been associated with during the course of their evolution. Their mouthparts have become fused into a sucking tube containing a tongue. The first segment of each of the three pairs of legs has a patch of bristles on its inner surface. Those of the first and second pairs are pollen brushes that gather the pollen that sticks to the bee's hairy body. On the third pair of legs, the bristles form a pollen comb that collects pollen from these brushes and from the abdomen. From the comb, the pollen is forced up into pollen baskets, concave surfaces fringed with hairs on the upper segment of the third pair of legs. Shown here is a honeybee (Apis mellifera) foraging in a flower of rosemary (Rosemarinus officinalis). In the rosemary flower, the stamens and stigma arch upward out of the flower, and both come into contact with the hairy back of any visiting bee of the proper size. Here the anthers can be seen depositing white pollen grains on the bee.

Bee-Pollinated Flowers Are Usually Blue or Yellow with Distinctive Markings   Bees, the most important group of flower-visiting animals, are responsible for the pollination of more species of plants than the members of any other animal group. Modern families of bees have existed for at least 80 million years, and they subsequently became diverse along with evolutionary radiation of the angiosperms. Both male and female bees live on nectar, and the females also collect pollen to feed the larvae. Bees have mouthparts, body hairs, and other appendages with special adaptations that make them suitable for collecting and carrying nectar and pollen (Figure 22–19). As Karl von Frisch and other investigators of insect behavior have shown, bees can learn quickly to recognize colors, odors, and outlines. The portion of the light spectrum that is visible to most insects, including bees, is somewhat different from the portion visible to humans. Unlike human beings, bees perceive ultraviolet as a distinct color; however, they don't perceive red, which therefore tends to merge with the background.

Many kinds of bees—especially solitary bees, which constitute a majority of species of the group (Figure 22–20)—are highly constant in their visits to flowers, confining their visits to one or a few plant species. Such constancy increases the efficiency of the particular species of bee—or of an individual bee—while it is visiting the flowers of one plant species. In relation to this specialization, bee species with narrowly restricted foraging habits often feature conspicuous morphological and physiological adaptations, such as coarse bristles in their pollen-collecting apparatus (if they visit plants with large pollen grains) or elongated mouthparts (if they take nectar from plants with long-tubed flowers). When they are constant to this degree, and have morphological and behavioral factors such that they actually bring about pollination, bees exert a powerful evolutionary force for specialization of the plants they visit. The process by which two or more species act as selective forces on one another and each undergoes evolutionary change is known as **coevolution.** There are some 20,000 species of bees, the great majority of which visit flowers for food.

Bee flowers—that is, flowers that coevolved with bees—have showy, brightly colored petals that are usually blue or yellow. They often have distinctive patterns by which bees can efficiently recognize them. Such patterns may include "honey guides," special markings that indicate the position of the nectar (Figure 22–21). Bee flowers are never pure red, and, as special photographic techniques have shown, they often have distinctive markings that are normally invisible to humans (Figure 22–22).

In bee flowers, the nectary is characteristically situated at the base of the corolla tube, where it is accessible only to specialized organs, such as the mouthpiece of bees, and not, for example, to the chewing mouthparts of beetles. Bee flowers also characteristically have a "landing platform" of some sort (Figures 22–13 and 22–21).

22–20
A sweat bee (family Halictidae) *gathering pollen from the stamens of a cactus* (Echinocereus) *in Baja California, Mexico.*

The tall structures in the center of the flower are the stigmas.

22–21
"Honey guides" on the flowers of the foxglove (Digitalis purpurea) *serve as distinctive signals to insect visitors. The lower lip of the fused corolla serves as a landing platform of the kind that is commonly found in bee flowers.*

(a)

(b)

22–22
The color perception of most insects is somewhat different from that of human beings. To a bee, for example, ultraviolet light (which is invisible to humans) is seen as a distinct color. These photographs show a flower of marsh marigold (Caltha palustris) *(a) in natural light, showing the solid yellow color as the flower appears to humans, and (b) in ultraviolet light. The portions of the flower that appear light in (b) reflect both yellow and ultraviolet light, which combine to form a color known as "bee's purple," whereas the dark portions of the flower absorb ultraviolet and therefore appear pure yellow when viewed by a bee (see page 543).*

(a)

(b)

22–23

Bumblebees (Bombus). *These social bees are important pollinators of many genera of plants throughout the cooler parts of the Northern Hemisphere, and they have been introduced into regions where they are not native for the purpose of pollinating such plants as white clover* (Trifolium repens). *(a) A bumblebee gathering pollen from the flower of a California poppy* (Eschscholzia californica). *(b) Portion of the underground nest of a bumblebee colony, showing the cells in which the wormlike larvae complete their development. The bumblebees provision these cells with pollen and regurgitated nectar, which they obtain from flowers. Although a colony of bumblebees may visit a wide variety of flowers during the course of a season, an individual bee often visits only the flowers of one kind of plant on a single trip away from the nest.*

Bumblebees are among the most familiar flower visitors in the North Temperate zone (Figure 22–23). They are social bees that live in colonies. The queens (sexual females) overwinter and, when they emerge in the spring, lay their eggs to establish a new colony. Bumblebees cannot fly until their wing muscles reach a temperature of about 32°C. To maintain this temperature, they must forage constantly on flowers with a copious supply of nectar. Many plants of the cool parts of North America and Eurasia, including lupines, larkspurs, and fireweed (see Figure 21–21), are regularly pollinated by bumblebees throughout their ranges.

Some of the evolutionarily more advanced flowers, in particular the orchids, have developed complex passageways and traps that force the bees that visit them to follow a particular route into and out of the flower. This

22–24

The beelike flowers of the orchid Ophrys speculum *attract male bees, which are so deceived by the resemblance to female bees of their species that they attempt to copulate with the flowers. In doing so, they often pick up a pollen sac (pollinium) from a flower and may then carry it to another flower of the same species. This orchid was photographed in Sardinia.*

ensures that both anther and stigma come into contact with the bee's body at a particular point and in the proper sequence (Figure 22–19).

An even more bizarre pollination strategy has been adopted by orchids of the genus *Ophrys*. The flower resembles a female bee, wasp, or fly (Figure 22–24). The males of these insect species emerge early in the spring, before the females. The orchids bloom early in the spring as well, and the male insects attempt to copulate with the orchid flower. During the course of its "sexual" visit, a pollinium may be deposited on the insect's body. When the insect visits another flower of the same species, the pollinium may be caught in the appropriate grooves on the stigma and thus bring about the pollination of that flower.

An additional range of flowers with different charac-teristics is pollinated by flies of various kinds, including mosquitoes. These insects feed on nectar but do not gather pollen or store food for their larvae. Examples of flowers pollinated by mosquitoes and flies are shown in Figures 22–18 and 22–25.

Flowers Pollinated by Moths and Butterflies Often Have a Long Corolla Tube Flowers that coevolved with butter-flies and diurnal moths (those that are active during the day rather than at night) are similar in many respects to bee flowers. This is mainly because butterflies, moths, and bees are all guided to flowers by a combination of sight and smell (Figure 22–26). Some species of butter-flies, however, are able to perceive red as a distinct color, and some butterfly-pollinated flowers are red and orange.

(a)

(c)

(b)

22–25

Pollination by mosquitoes and other flies.
(a) Some small-flowered orchids, such as Habenaria elegans, in which the flowers are white or green and relatively inconspicuous, are visited and pollinated by mosquitoes in North Temperate and Arctic regions. The mosquitoes obtain nectar from the flowers.

(b) A female mosquito of the genus Aedes *with an orchid pollinium attached to its head. Other small-flowered orchids, such as those of the genus* Spiranthes, *are pollinated by bees.*
(c) A fly on a flower of a lily (Zigadenus fremontii). Notice the conspicuous yellow nectaries.

22–26

Copper butterfly (Lycaena gorgon) *sucking nectar from the flowers of a daisy. The long sucking mouthparts of moths and butterflies are coiled up at rest and extended when feeding. They vary in length from species to species. Only a few millimeters long in some of the smaller moths, they are 1 to 2 centimeters long in many butterflies, 2 to 8 centimeters long in some hawkmoths of the North Temperate zone, and as long as 25 centimeters in a few kinds of tropical hawkmoths.*

Most moths are nocturnal. The typical moth-pollinated flower—as seen, for example, in several species of tobacco (*Nicotiana*)—is white or pale in color and has a heavy fragrance, a sweet penetrating odor that often is emitted only after sunset. Well-known flowers that are pollinated by moths include the yellow-flowered species of evening primrose (*Oenothera*; see Figure 10–11b) and the pink-flowered amaryllis (*Amaryllis belladonna*).

The nectary of a moth or butterfly flower is often located at the base of a long, slender corolla tube or a spur and is usually accessible only to the long sucking mouthparts of moths and butterflies. Hawkmoths, for instance, do not usually enter flowers, as bees do, but hover above them, inserting their long mouthparts into the floral tube. Consequently, hawkmoth flowers do not have the landing platforms, traps, and elaborate internal structural modifications seen in some of the bee flowers. Most moth–flower relationships typically involve smaller moths that do not use nearly as much energy as the hawkmoths. The flowers over which the moths scramble are often smaller than those pollinated by hawkmoths and have short corolla tubes. One of the most specialized moth–flower relationships is shown in Figure 22–27.

22–27

Yucca moth (Tegeticula yucasella) *scraping pollen from a yucca flower. The female moth visits the creamy white flowers by night and gathers pollen, which she rolls into a tight little ball and carries in her specialized mouthparts to another flower. In the second flower, the moth pierces the ovary wall with her long ovipositor and lays a batch of eggs among the ovules. She then packs the sticky mass of pollen through the openings of the stigma. Moth larvae and seeds develop simultaneously, with the larvae feeding on the developing yucca seeds. When the larvae are fully developed, they gnaw their way through the ovary wall and lower themselves to the ground, where they pupate until the yuccas bloom again. It is estimated that only about 20 percent of the seeds are usually eaten.*

22–28
A male Anna's hummingbird (Calypte anna) *at a flower of the scarlet monkey-flower* (Mimulus cardinalis) *in southern California. Note the pollen on the bird's forehead, which is in contact with the stigma of the flower.*

Bird-Pollinated Flowers Produce Large Amounts of Nectar and Are Often Red and Odorless   Some birds regularly visit flowers to feed on nectar, floral parts, and flower-inhabiting insects. Many of these birds also serve as pollinators. In North and South America, the chief pollinators among the birds are hummingbirds (Figure 22–28). In other parts of the world, flowers are visited regularly by representatives of other specialized bird families (Figure 22–29).

Bird flowers have a copious, thin nectar (some actually drip with nectar when the pollen is mature) but usually have little odor because the sense of smell is poorly developed in birds. However, birds do have a keen color sense that is much like our own. It is not surprising, therefore, that most bird flowers are colorful, with red and yellow ones being the most common (Figure 22–12b). Bird-pollinated flowers include red columbine (Figure 22–30a), fuchsia, scarlet passion flower, eucalyptus, hibiscus, poinsettia (Figure 22–30b, c), and many members of the cactus, banana, and orchid families. Typically, these flowers are large or they form parts of large inflorescences, features that can be correlated with their importance as visual stimuli and their ability to hold large amounts of nectar.

22–29
A collared sunbird (Anthreptes collarii) *perching and feeding on a bird-of-paradise* (Strelitzia reginae) *flower in South Africa.*

(a)

(b)

(c)

22–30

*Examples of bird-pollinated flowers.
(a) Columbine (Aquilegia canadensis).
Alternating perianth segments are modified
into nectar-filled tubes. Hummingbirds visit
these hanging flowers, taking nectar from
them on the wing. The nectar of columbine
flowers is inaccessible to most other kinds of
animals. (b), (c) Poinsettias (Euphorbia
pulcherrima). In this familiar plant, a native
of Mexico, the flowers are small, greenish, and
clustered, but each cluster has a large, yellow
nectary from which abundant nectar flows.
Ants are seen feeding on the nectar in (b).
Modified upper leaves, bright red in color,
attract hummingbirds to the clusters of
flowers.*

Bird and other animal pollinators usually restrict
their visits to the flowers of a particular plant species, at
least for a short period of time. This is not the only factor
promoting outcrossing, however. The pollinator must
not confine its visits to a single flower or to the flowers
of a single plant. When flowers are visited regularly by
large animals with a high rate of energy expenditure—
such as birds, hawkmoths, or bats—the flowers must
produce large amounts of nectar to support the metabolic requirements of the animals and keep them coming
back. On the other hand, if an abundant supply of nectar
is available to animals with a lower rate of energy expenditure, such as small bees or beetles, they will tend to remain at a single flower. Being satisfied there, these
visitors will not move on to the flowers of other plants
where they might bring about outcrossing. Consequently, species that are regularly pollinated by animals with a high rate of energy consumption, such as
hummingbirds, have tended to evolve flowers with the
nectar held in tubes or otherwise unavailable to smaller
animals with lower rates of energy consumption.
Similarly, the color red is a signal to birds but not to
most insects. Birds, like ourselves, do not respond very
strongly to odor clues. Thus, odorless, red flowers, being
inconspicuous to insects, tend not to attract them. This
adaptation is advantageous in view of the copious production of nectar by such flowers.

**Bat-Pollinated Flowers Produce Copious Nectar and Have
Dull Colors and Strong Odors**   Flower-visiting bats are
found in tropical areas of both the Old World and the
New World, and more than 250 species of bats—about a
quarter of the total number of bat species—include at
least some nectar, fruit, or pollen in their diet. These
species of bats have slender, elongated muzzles and

long, extensible tongues, sometimes with a brushlike tip, and their front teeth are often reduced in size or are missing altogether.

Bat flowers are similar in many respects to bird flowers, being large, robust flowers that produce copious nectar (Figure 22–31). Because bats feed at night, bat flowers are typically dull in color, and many of them open only at night. Many bat-pollinated flowers are tubular or structurally modified to protect their nectar in other ways. Some bat-pollinated flowers—and fruits that are regularly dispersed by bats—hang down on long stalks below the foliage, where the bats can fly more easily. Other bat-pollinated flowers are borne on the trunks of trees. Bats are attracted to the flowers largely through their sense of smell. Bat-pollinated flowers therefore characteristically have either very strong fermenting or fruitlike odors, or musty scents like those produced by bats to attract one another. Bats fly from

22–31

By thrusting its face into the tubular corolla of a flower of an organ-pipe cactus (Steno-cereus thurberi), this bat, Leptonycteris curasoae, *is able to lap up nectar with its long, bristly tongue. Some of the pollen clinging to the bat's face and neck is transferred to the next flower it visits. This species of bat, which is one of the more specialized nectar-feeding bats, migrates from central and southern Mexico to the deserts of the south-western United States during late spring and early summer. Here it feeds on the nectar and pollen of organ-pipe and saguaro cacti and on the flowers of agaves.*

tree to tree, eating pollen and other flower parts, and carrying pollen from flower to flower on their fur. At least 130 genera of angiosperms are pollinated by bats, or have their seeds dispersed by bats, or both. Included among bat-pollinated angiosperms are such economically important plants as bananas, mangoes, kapok, and sisal.

Some bats derive a significant portion of their dietary protein from the pollen they consume. As yet another example of coevolution, the pollen of the flowers they visit has been found to contain significantly higher levels of protein than does the pollen of insect-pollinated flowers.

Wind-Pollinated Flowers Produce No Nectar, Have Dull Colors, and Are Relatively Odorless About a century ago, botanists considered wind-pollinated angiosperms to be the most primitive members of the phylum. They thought that all other kinds of angiosperms had evolved from them. The conifers—thought by some scientists at the time to have been the direct ancestors of the angiosperms—have small, drab-colored, odorless, unisexual cones and are pollinated by the wind. Similarly, the many examples of wind-pollinated flowers have dull colors, are relatively odorless, and do not produce nectar. The petals of these flowers are either small or absent, and the sexes are often separated on the same plant. However, studies of other characteristics of these wind-pollinated angiosperms—in particular, their specialized wood and pollen—have convinced most botanists that all wind-pollinated angiosperms evolved not from the conifers but from insect-pollinated angiosperms.

According to current interpretations of the evidence, wind-pollinated angiosperms originated independently from several different ancestral stocks. They are best represented in temperate regions and are relatively rare in the tropics. In temperate climates, many individual trees of the same species are often found close together, and the dispersal of pollen by wind can occur readily in early spring, when the trees are leafless. In the tropics, on the other hand, many more kinds of trees are found in a given area, and the distance to the nearest individual of the same species may be quite great. Furthermore, in many tropical communities the trees are evergreen, and the dispersal of pollen by wind is more difficult than in temperate deciduous forests when the trees are leafless. Under these circumstances, pollination by insects and other animals that have the ability to seek out other individuals of the same plant species, sometimes over relatively great distances (20 kilometers or more, in some cases), is much more efficient than wind pollination.

Because wind-pollinated angiosperms do not depend on insects to transport their pollen from place to place, they devote no energy to the production of nutritious rewards for insect visitors. Wind pollination is very inefficient, however, and it is successful only where a large

(a)

(b)

(c)

22–32

Unlike most angiosperms, the grasses have wind-pollinated flowers. Maize (Zea mays) has **(a)** *staminate inflorescences (tassels) at the top of the stem and* **(b)** *ovulate inflorescences, with long protruding stigmas (the "silk" on the ears of corn, or maize), lower on the stem.* **(c)** *Grasses characteristically have enlarged, feathery stigmas that efficiently*

catch the wind-blown pollen shed by the hanging anthers, as seen here in a grass of the genus Agropyron. **(d)** *Scanning electron micrograph of a pollen grain of maize, showing the smooth pollen wall found in most wind-pollinated plants and the single aperture characteristic of monocots.*

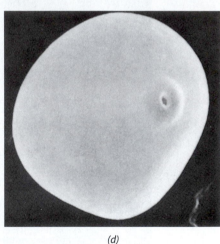

(d)

number of individuals of the same species grow fairly close together. Nearly all wind-borne pollen falls to the ground within 100 meters of the parent plant. Thus, if the individual plants are widely scattered, the chance that a pollen grain will reach a receptive stigma is very slim. Many wind-pollinated plants are either dioecious (having male and female flowers on separate plants), such as the willows; monoecious (having separate male and female flowers on the same plant), such as the oaks (see Figure 21–10); or genetically self-incompatible, such as many grasses (page 511). Even though their pollen moves about somewhat randomly, these plants have still evolved devices that foster a high degree of outcrossing.

Wind-pollinated flowers usually have well-exposed stamens that can easily lose their pollen to the wind. In some, the anthers are suspended from long filaments

hanging from the flower (Figures 22–32 and 22–33). The abundant pollen grains, which are generally smooth and small, do not adhere to one another as do the pollen grains of insect-pollinated species. The large stigmas are characteristically exposed, and they often have branches or feathery outgrowths adapted for intercepting wind-borne pollen grains. Most wind-pollinated plants have ovaries with single ovules (and hence single-seeded fruits) because each pollination event consists of the meeting of one pollen grain with one stigma and leads to the fertilization of one ovule for each flower. Thus, each oak flower produces only a single acorn and each grass flower produces only a single grain. To compensate, perhaps, plants with very small flowers tend to have them concentrated in large numbers into inflorescences (Figures 22–32 through 22–34).

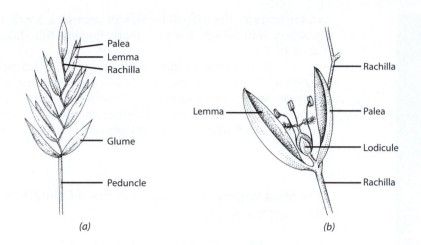

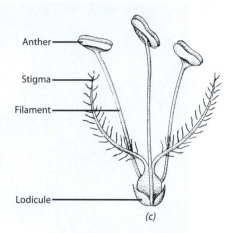

22–33
Grass flowers (florets) usually develop in clusters. (a) As a cluster matures, a single pair of dry, chaffy bracts—the glumes —separate a little, exposing the elongating spikelet, with from one to many florets (depending on the species of grass) attached to a central axis, or rachilla. (b) Each floret is surrounded by two distinctive bracts of its own, the palea and the lemma. These are forced apart, exposing the inner parts of the flower (c), by the swelling of the lodicules—small, rounded bodies at the base of the carpel—and are spread wide when the grass is in flower. The stamens, usually three in number, have slender filaments and long anthers, and the stigmas are typically long and feathery and so are efficient at intercepting the wind-borne pollen.

22–34
Most common species of trees in temperate regions are wind-pollinated. The staminate flowers of the paper birch (Betula papyrifera) hang down in catkins—flexible, thin tassels several centimeters long. These catkins are whipped by passing breezes, and the pollen, when mature, is scattered about by the wind.

Some Submerged Aquatic Angiosperms Are Pollinated via Water A few angiosperms—about 79 families and 380 genera—are submerged aquatic plants, living in marine and freshwater habitats. In 18 of these genera, the pollen is either transported underwater or floats from one plant to another across the surface of the water. In some of these plants, pollen grains are either threadlike (filiform), thus increasing their chances of coming into contact with receptive stigmas, or linked into chains, with similar effect (Figure 22–35a). The filiform pollen of some aquatic genera is unusual in lacking an exine. In other genera of submerged aquatic plants, the pollen is dispersed across the surface of the water. For example, in the freshwater eel grass *Vallisneria*, the entire staminate flower is released under water and floats to the surface where its three stamens become erect and function like a sail. Floating just below the tension zone at the

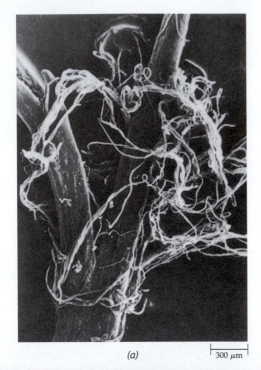

(a) |— 300 μm —|

(b)

22–35

Unique pollination systems in submerged aquatic angiosperms. (a) The slender, branching stigmas of the sea-nymph, Amphibolis, seen here in a scanning electron micrograph, have captured many grains of the threadlike pollen released by staminate plants of this species. (b) Tiny staminate flowers of eel grass, Vallisneria, float into the depression in the surface-tension layer of the water that surrounds the much larger carpellate flower. The individual pollen grains are visible on the staminate flowers.

water surface, the carpellate flower creates a small depression into which the staminate flowers fall (Figure 22–35b).

In contrast to these plants with very specialized pollination systems, which are clearly advanced in an evolutionary sense, most aquatic angiosperms are either wind-pollinated or insect-pollinated, like their terrestrial ancestors. Their flowers are held above the surface of the water.

The Most Important Pigments in Floral Coloration Are the Flavonoids

Color is one of the most conspicuous features of angiosperm flowers—a characteristic by which members of the phylum are easily recognized. The varied colors of different kinds of flowers evolved in relation to their pollination systems and, in general, are advertisements for particular kinds of animals, as we have just seen.

The pigments that are responsible for the colors of angiosperm flowers are generally common in vascular plants other than angiosperms. It is the way in which they are concentrated in angiosperm flowers, and particularly in their corollas, that is a special characteristic of the flowering plants. Surprisingly, all flower colors are produced by a small number of pigments. Many red, orange, or yellow flowers owe their color to the presence of carotenoid pigments similar to those that occur in leaves (and in all plants, in green algae, and in some other organisms, as well). The most important pigments in floral coloration, however, are **flavonoids,** which are compounds with two six-carbon rings linked by a three-carbon unit. Flavonoids probably occur in all angiosperms, and they are sporadically distributed among the members of other groups of plants. In leaves, flavonoids block far-ultraviolet radiation, which is destructive to nucleic acids and proteins. They usually selectively admit light of blue-green and red wavelengths, which are important for photosynthesis.

Pigments belonging to one major class of flavonoids, the **anthocyanins,** are major determinants of flower color (Figure 22–36). Most red and blue plant pigments are anthocyanins, which are water-soluble and are found in vacuoles. By contrast, the carotenoids are oil-soluble and are found in plastids. The color of an anthocyanin pigment depends on the acidity of the cell sap of the vacuole. Cyanidin, for example, is red in acid solution, violet in neutral solution, and blue in alkaline solution. In some plants, the flowers change color after pollination, usually because of the production of large amounts of anthocyanins, and then become less conspicuous to insects.

The **flavonols,** another group of flavonoids, are very commonly found in leaves and also in many flowers. A number of these compounds are colorless or nearly so,

Pelargonidin   Cyanidin   Delphinidin

22–36

Three anthocyanin pigments, the basic pigments on which flower colors in many angiosperms depend: pelargonidin (red), cyanidin (violet), and delphinidin (blue). Related compounds known as flavonols are

yellow or ivory, and the carotenoids are red, orange, or yellow. Betacyanins (betalains) are red pigments that occur in one group of eudicots. Mixtures of these different pigments, together with changes in cellular pH, produce

the entire range of flower color in the angiosperms. Changes in flower color provide "signals" to pollinators, telling them which flowers have opened recently and are more likely to provide food.

but they may contribute to the ivory or white hues of certain flowers.

For all flowering plants, different mixtures of flavonoids and carotenoids (as well as changes in cellular pH) and differences in the structural, and thus the reflective, properties of the flower parts produce the characteristic colors. The bright fall colors of leaves come about when large quantities of colorless flavonols are converted into anthocyanins as the chlorophyll breaks down. In the all-yellow flowers of the marsh marigold (*Caltha palustris*), the ultraviolet-reflective outer portion is colored by carotenoids, whereas the ultraviolet-absorbing inner portion is yellow to our eyes because of the presence of a yellow chalcone, one of the flavonoids. To a bee or other insect, the outer portion of the flower appears to be a mixture of yellow and ultraviolet, a color called "bee's purple," whereas the inner portion appears pure yellow (Figure 22–22). Most, but not all, ultraviolet reflectivity in flowers is related to the presence of carotenoids, and thus ultraviolet patterns are more common in yellow flowers than in others.

In the goosefoot, cactus, and portulaca families and in other members of the order *Chenopodiales* (*Centrospermae*), the reddish pigments are not anthocyanins or even flavonoids but a group of more complex aromatic compounds known as **betacyanins** (or betalains). The red flowers of *Bougainvillea* and the red color of beets are due to the presence of betacyanins. No anthocyanins occur in these plants, and the families characterized by betacyanins are closely related to one another.

Evolution of Fruits

Just as flowers have evolved in relation to their pollination by many different kinds of animals and other agents, so have fruits evolved for dispersal in many different ways. Fruit dispersal, like pollination, is a fundamental aspect of the evolutionary radiation of the

angiosperms. Before we consider this subject in more detail, however, we must present some basic information about fruit structure.

A fruit is a mature ovary, which may or may not include some additional flower parts. A fruit in which such additional parts are retained is known as an **accessory fruit.** Although fruits usually have seeds within them, some—**parthenocarpic fruits**—may develop without seed formation. The cultivated strains of bananas are familiar examples of this exceptional condition.

Fruits are generally classified as simple, multiple, or aggregate, depending on the arrangement of the carpels from which the fruit develops. **Simple fruits** develop from one carpel or from several united carpels. **Aggregate fruits,** such as those of magnolias, raspberries, and strawberries, consist of a number of separate carpels of one gynoecium. The individual parts of aggregate fruits are known as **fruitlets;** they can be seen, for example, in the magnolia fruit shown in Figure 22–4c. **Multiple fruits** consist of the gynoecia of more than one flower. The pineapple, for example, is a multiple fruit consisting of an inflorescence with many previously separate ovaries fused on the axis on which the flowers were borne (the other flower parts being squeezed between the expanding ovaries).

Simple fruits are by far the most diverse of the three groups. When ripe, they may be soft and fleshy, dry and woody, or papery. There are three main types of fleshy fruits—berries, drupes, and pomes. In **berries**—examples of which are tomatoes, dates, and grapes—there may be one to several carpels, each of which is typically many-seeded. The inner layer of the fruit wall is fleshy. In **drupes,** there may also be one to several carpels, but each carpel usually contains only a single seed. The inner layer of the fruit is stony and usually tightly adherent to the seed. Peaches, cherries, olives, and plums are familiar drupes. Coconuts are drupes whose outer layer is fibrous rather than fleshy, but in temperate regions we usually see only the coconut seed with the adherent stony inner layer of the fruit (Figure 22–37). **Pomes** are highly specialized fleshy fruits that are characteristic of one subfamily of the rose family. The pome is derived from a compound inferior ovary in which the fleshy portion comes largely from the enlarged base of the perianth. The inner portion, or endocarp, of a pome resembles a tough membrane, as you know from eating apples and pears, the two most familiar examples of this kind of fruit.

Dry simple fruits are classified as either dehiscent (Figures 22–38 and 22–39) or indehiscent (Figure 22–40). In **dehiscent fruits,** the tissues of the mature ovary wall (the pericarp) break open, freeing the seeds. In **indehiscent fruits,** on the other hand, the seeds remain in the fruit after the fruit has been shed from the parent plant.

There are several kinds of dehiscent simple dry fruits. The **follicle** is derived from a single carpel that splits down one side at maturity, as in the columbines and milkweeds (Figure 22–38a). Follicles were also characteristic of the extinct, Middle Cretaceous plant *Archaeanthus* (Figure 22–5), and they are also found in magnolias (Figure 22–4c). In the pea family *(Fabaceae)*, the characteristic fruit is a **legume.** Legumes resemble follicles, but they split along both sides (Figure 22–39). In the mustard family *(Brassicaceae)*, the fruit is called a **silique** and is formed of two fused carpels. At maturity, the two sides of the fruit split off, leaving the seeds attached to a persistent central portion (Figure 22–38c). The most common sort of dehiscent simple dry fruit is the **capsule**, which is formed from a compound ovary in plants with either a superior or an inferior ovary. Capsules shed their seeds in a variety of ways. In the poppy family *(Papaveraceae)*, the seeds are often shed when the capsule splits longitudinally, but in some members of this family they are shed through holes near the top of the capsule (Figure 22–38b).

Indehiscent simple dry fruits are found in many different plant families. Most common is the **achene,** a small, single-seeded fruit in which the seed lies free in

(a)

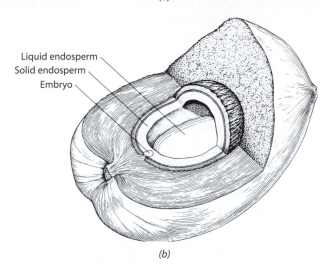

Liquid endosperm
Solid endosperm
Embryo

(b)

22–37

The coconut palm (Cocos nucifera) *has a very wide range on ocean shores throughout the world because its fruits are able to float for long periods and then germinate when they reach land.* **(a)** *A coconut germinating on the beach in Florida.* **(b)** *Diagram of a coconut fruit. The coconut milk is liquid endosperm. Cell walls form around the nuclei in the liquid endosperm, which becomes solid by the time of germination. When coconuts are transported commercially, their husks are usually removed first, so that in temperate countries people usually see only the stony inner shell of the fruit surrounding the seed.*

(a)

Capsule
(Papaver somniferum)

(b)

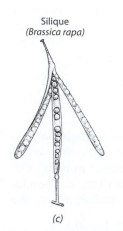

Silique
(Brassica rapa)

(c)

22–38

Dehiscent fruits. **(a)** *Bursting follicles of a milkweed* (Asclepias). **(b)** *In some members of the poppy family* (Papaveraceae), *such as poppies (genus* Papaver), *the capsule sheds its seeds through pores near the top of the fruit.* **(c)** *Plants of the mustard family* (Brassicaceae) *have a characteristic fruit known as a silique, in which the seeds arise from a central partition, and the two enclosing valves fall away at maturity.*

(a)   (b)

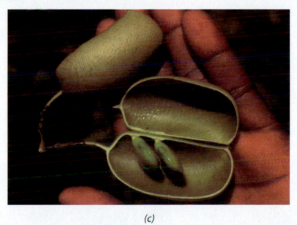

(c)

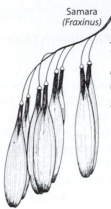

Samara
(Fraxinus)

22–40
The samara, a winged indehiscent fruit characteristic of ashes (Fraxinus) *and elms* (Ulmus), *retains its single seed at maturity. Samaras are dispersed by wind.*

the cavity except for its attachment by the funiculus. Achenes are characteristic of the buttercup family *(Ranunculaceae)* and the buckwheat family *(Polygonaceae).* Winged achenes, such as those found in elms and ashes, are commonly known as **samaras** (Figure 22–40). The achenelike fruit that occurs in grasses *(Poaceae)* is known as a **caryopsis,** or grain; in it, the seed coat is firmly united to the fruit wall. In the *Asteraceae,* the complex, achenelike fruit is derived from an inferior ovary; technically, it is called a **cypsela** (see Figure 22–42). Acorns and hazelnuts are examples of **nuts,** which resemble achenes but have a stony fruit wall and are derived from a compound ovary. Finally, in the parsley family *(Apiaceae)* and the maples *(Aceraceae),* as well as a number of other, unrelated groups, the fruit is a **schizocarp,** which splits at maturity into two or more one-seeded portions (Figure 22–41a).

22–39
The legume, a kind of fruit that is usually dehiscent, is the characteristic fruit of the pea family, Fabaceae *(also called* Leguminosae*). With about 18,000 species,* Fabaceae *is one of the largest families of flowering plants. Many members of the family are capable of nitrogen fixation because of the presence of nodule-forming bacteria of the genus* Rhizobium *on their roots (see Chapter 30). For this reason, these plants are often the first colonists on relatively infertile soils, as in the tropics, and they may grow rapidly there. The seeds of a number of plants of this family, such as peas, beans, and lentils, are important foods.*
(a) Legumes of the garden pea, Pisum sativum. *(b) Legumes of* Albizzia polyphylla, *growing in Madagascar. Each seed is in a separate compartment of the fruit.*
(c) Legume of Griffonia simplicifolia, *a West African tree. The two valves of the legume are split apart, revealing the two seeds inside.*

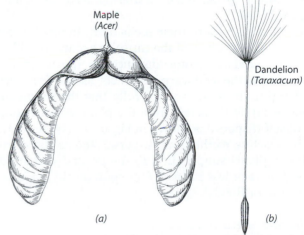

Maple
(Acer)

Dandelion
(Taraxacum)

(a)   (b)

22–41
Wind-dispersed fruits. (a) In maples (Acer), *each half of the schizocarp has a long wing. (b) The fruits of the dandelion* (Taraxacum) *and many other composites have a modified calyx, called the pappus, which is adherent to the mature cypsela and may form a plumelike structure that aids in wind dispersal.*

Fruits and Seeds Have Evolved in Relation to Their Dispersal Agents

Just as flowers have evolved according to the characteristics of the pollinators that visit them regularly, so have fruits evolved in relation to their dispersal agents. In both coevolutionary systems, there have, in general, been many changes in relation to different dispersal agents within individual families and a great deal of convergent evolution toward similar-appearing structures with similar functions. We shall review some of the adaptations of fruits here, in relation to their dispersal agents.

Many Plants Have Wind-Borne Fruits and Seeds   Some plants have light fruits or seeds that are dispersed by the wind (Figures 22–38a, 22–40, 22–41). The dustlike seeds of all members of the orchid family, for example, are wind-borne. Other fruits have wings, which are sometimes formed from perianth parts, that allow them to be blown from place to place. In the schizocarps of maples, for example, each carpel develops a long wing (Figure 22–41a). The two carpels separate and fall when mature. Many members of the *Asteraceae*—dandelions, for example—develop a plumelike pappus, which aids in keeping the light fruits aloft (Figures 22–41b and 22–42). In some plants, the seed itself, rather than the fruit, bears the wing or plume. The familiar butter-and-eggs *(Linaria vulgaris)* has a winged seed, and both fireweed *(Chamaenerion)* and milkweed *(Asclepias;* Figure 22–38a) have plumed seeds. In willows and poplars (family *Salicaceae)*, the seed coat is covered with woolly hairs. In tumbleweeds *(Salsola),* the whole plant (or a portion of it) is blown along by the wind, scattering seeds as it moves (Figure 22–43).

Other plants shoot their seeds aloft. In touch-me-not *(Impatiens),* the valves of the capsules separate suddenly, throwing seeds for some distance. In the witch hazel *(Hamamelis),* the endocarp contracts as the fruit dries, discharging the seeds so forcefully that they sometimes travel as far as 15 meters from the plant. Another example of self-dispersal is shown in Figure 22–44. In contrast to these active methods of dispersal, the seeds or fruits of many plants simply drop to the ground and are dispersed more or less passively (or sporadically, such as by rainwater or floods).

Fruits and Seeds Adapted for Floating Are Dispersed by Water   The fruits and seeds of many plants, especially those growing in or near water sources, are adapted for floating, either because air is trapped in some part of the fruit or because the fruit contains tissue that includes large air spaces. Some fruits are especially adapted for dispersal by ocean currents. Notable among these is the coconut (Figure 22–37), which is why almost every newly formed Pacific atoll quickly acquires its own co-

22–42
The familiar small, indehiscent fruits of dandelions, which are technically known as cypselas (but often loosely called achenes), have a plumelike, modified calyx (the pappus) and are spread by the wind. This photograph shows the fruiting heads of a plant of the genus Agoseris, which is closely related to the dandelions.

22–43
In tumbleweeds (Salsola), the whole plant breaks off and is blown across open country, scattering its seeds as it tumbles along. So many tumbleweeds blew into Mobridge, South Dakota, on November 8, 1989, that the town was besieged. The tumbleweeds are natives of Eurasia, but they are widely naturalized as weeds in North America and elsewhere.

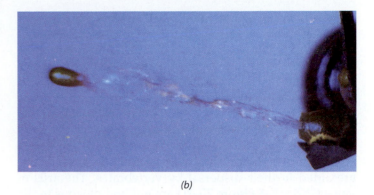

(a)

(b)

22–44

Dwarf mistletoe (Arceuthobium), a parasitic eudicot that is the most serious cause of loss of forest productivity in the western United States. (a) A plant growing on a pine branch in California. (b) Seed discharge. Very high hydrostatic pressure builds up in the fruit and shoots the seeds as much as 15 meters laterally. The seeds have an initial velocity of about 100 kilometers per hour. This is one of the ways in which the seeds are spread from tree to tree, although they are also sticky and can be carried from one tree to another over much longer distances by adhering to the feet or feathers of birds.

conut tree. Rain, also a common means of fruit and seed dispersal, is particularly important for plants that live on hillsides or mountain slopes.

Fruits and Seeds That Are Fleshy or Have Adaptations for Attachment Are Dispersed by Animals   The evolution of sweet and often highly colored, fleshy fruits was clearly involved in the coevolution of animals and flowering plants. The majority of fruits in which much of the pericarp is fleshy—bananas, cherries, raspberries, dogwoods, grapes—are eaten by vertebrates. When such fruits are eaten by birds or mammals, the seeds the fruits contain are spread by being passed unharmed through the digestive tract or, in birds, by being regurgitated at a distance from the place where they were ingested (Figure 22–45). Sometimes, partial digestion aids the germination of seeds by weakening their seed coats.

(a)

(b)

(c)

22–45

The seeds of fleshy fruits are usually dispersed by vertebrates that eat the fruits and either regurgitate the seeds or pass them as part of their feces. Examples of vertebrate-dispersed fruits are shown here. (a) Strawberries (Fragaria), an example of an aggregate fruit. The achenes are borne on the surface of a

fleshy receptacle. Immature strawberries, like the immature stages of many bird- or mammal-dispersed fruits, are green, but they become red when the seeds are mature and thus ready for dispersal. (b) The berries of many cacti, such as this prickly pear (Opuntia) growing in southern Mexico, are

conspicuous at maturity. (c) Berries of chaparral honeysuckle, Lonicera hispidula. These berries develop from inferior ovaries, and they therefore have fused portions of the outer floral whorls incorporated in them. A flower of this species is shown in Figure 22–11c.

(a)

When fleshy fruits ripen, they undergo a series of characteristic changes, mediated by the hormone ethylene, which will be discussed in Chapter 28. Among these changes are a rise in sugar content, a softening of the fruit caused by the breakdown of pectic substances, and often a change in color from inconspicuous, leaflike green to bright red (Figure 22–45a), yellow, blue, or black. The seeds of some plants, especially tropical ones, often have fleshy appendages, or arils, with the bright colors characteristic of fleshy fruits and, like these fruits, are aided in their dispersal by vertebrates. The arils of yews (*Taxus*; see Figure 20-31) are not homologous to the structures called arils in angiosperms, because yew arils are outgrowths of the seed, not of the fruit. Nevertheless, they play a similar role in seed dispersal.

Unripe fruits are often green or colored in such a way that they are inconspicuous among the plant's green leaves, and thus they are somewhat concealed from birds, mammals, and insects. They may also be disagreeable to the taste—such as unripe cherries (*Prunus*), which have a strong, acidic taste—thereby discouraging animals from eating them before the seeds are ripe. The changes in color that accompany ripening are the plant's "signal" that the fruit is ready to be eaten—the seeds are ripe and ready for dispersal (Figure 22–45). It is no coincidence that red is such a predominant color among ripe fruits. Red fruit is inconspicuous to insects, blending with the background of green leaves. Insects are too small to disperse the large seeds of fleshy fruits effectively, and such concealment is therefore advantageous to these plants. At the same time, the red fruits are very conspicuous to vertebrates, which then eat the fruits and help disperse their ripe seeds.

A number of angiosperms have fruits or seeds that are dispersed by adhering to fur or feathers (Figure 22–46). These fruits and seeds have hooks, barbs, spines, hairs, or sticky coverings that allow them to be transported, often for great distances, attached to the bodies of animals.

Other important agents of seed dispersal in some plants are ants (Figure 22–47). Such plants have evolved a special adaptation on the exterior of their seeds, called an elaiosome, a fleshy pigmented appendage that contains lipids, protein, starch, sugars, and vitamins. The ants usually carry such seeds back to their nests, where the elaiosomes are consumed by other workers or larvae and the seeds are left intact. The seeds readily germinate in this location, and seedlings often become established, protected from their predators and perhaps benefiting from nutrient enrichment as well. Up to a third of the species in some plant communities, such as the herb understory communities in the deciduous forests of the central and eastern United States, are dispersed by ants in this way. These plants include such familiar species as spring beauty (*Claytonia virginiana*), Dutchman's-

(b)

22–46

(a) The fruits of the African plant Harpagophytum, *a member of the sesame family* (Pedaliaceae), *are equipped with "grappling hooks," which catch in the fur on the legs of large mammals. In this way, the fruits are spread from place to place.* **(b)** *Mature inflorescences of cocklebur* (Xanthium), *which attach themselves to passing animals and are dispersed. In this case, the entire inflorescence is the dispersal unit, rather than the fruits alone, as in* Harpagophytum. Cocklebur *is a member of the family* Asteraceae.

22–47

An ant (Rhytidoponera metallica) *taking the seed of an acacia back to its nest in the Australian bushland. Note the large, yellow elaiosome.*

breeches *(Dicentra cucullaria)*, bloodroot *(Sanguinaria canadensis)*, many species of violets *(Viola)*, and *Trillium* (see Figure 1–6b).

Biochemical Coevolution

Also important in the evolution of angiosperms are the so-called *secondary metabolites*, or *products* (page 32). Once thought of as waste products, these include an array of chemically unrelated compounds, such as alkaloids, quinones, essential oils (including terpenoids), glycosides (including cyanogenic substances and saponins), flavonoids, and even raphides (needlelike crystals of calcium oxalate). The presence of certain of these compounds can characterize whole families, or groups of families, of flowering plants (Figure 22–48).

In nature, these chemicals appear to play a major role either in restricting the palatability of the plants in which they occur or in causing animals to avoid the plants altogether (Figure 22–49). When a given family of plants is characterized by a distinctive group of secondary plant products, those plants are apt to be eaten only by insects belonging to certain families. The mustard family *(Brassicaceae)*, for example, is characterized by the presence of mustard-oil glycosides and associated

Calactin

Sinigrin

Caffeine

Nicotine

Theobromine

22–48

Secondary metabolites: sinigrin, from black mustard, Brassica nigra; *calactin, a cardiac glycoside, from the milkweed* Asclepias curassavica; *nicotine, from tobacco,* Nicotiana tabacum, *a member of the nightshade family; caffeine, from* Coffea arabica *of the madder family; and theobromine, a prominent alkaloid in coffee, tea* (Thea sinensis), *and cocoa* (Theobroma cacao). *Nicotine, caffeine, and theobromine are alkaloids, members of a diverse class of nitrogen-containing ring compounds that are physiologically active in vertebrates.*

22–49

Poison ivy (Toxicodendron radicans) produces a secondary plant product, 3-penta-decanedienyl catechol, which causes an irritating rash on the skin of many people. The ability to produce this alcohol presumably evolved under the selective pressure exerted by herbivores. Fortunately, the plant is easily identifiable by its characteristic compound leaves with their three leaflets.

resent an unexploited food source for any group of insects that can tolerate or break down the poisons manufactured by the plant. The main evolutionary development of the butterfly group *Pierinae* probably occurred after its ancestors had acquired the ability to feed on plants of the mustard family by breaking down these toxic molecules.

Herbivorous insects that are narrowly restricted in their feeding habits to groups of plants with certain secondary plant products are often brightly colored. This coloration serves as a signal to their predators that they carry the noxious chemicals in their bodies and hence are unpalatable. For example, the assemblage of insects found feeding on a milkweed plant on a summer day might include bright green chrysomelid beetles, bright red cerambycid beetles and true bugs, and orange and black monarch butterflies, among others (see Figure 2–28). Milkweeds (*Asclepiadaceae*) are richly endowed with alkaloids and cardiac glycosides, heart poisons that have potent effects in vertebrates, the main potential predators of these insects. If a bird ingests a monarch butterfly, severe gastric distress and vomiting will follow, and orange-and-black patterns like that of the monarch will be avoided by the predator in the future. Other insects, such as the viceroy butterfly (which also has an orange-and-black pattern), have evolved similar coloration and markings. Thus, they escape predation by capitalizing on their resemblance to the poisonous monarch. This phenomenon, known as **mimicry,** is ultimately dependent upon the plant's chemical defenses. Various drugs and psychedelic chemicals, such as the active ingredients in marijuana (*Cannabis sativa*) and the opium poppy (*Papaver somniferum*), among others, are also secondary plant products that in nature presumably play a role in discouraging the attacks of herbivores (Figure 22–50).

Still more complex systems are known. When the leaves of potato or tomato plants are wounded, as by the Colorado potato beetle, the concentration of proteinase inhibitors, which interfere with the digestive enzymes in the guts of the beetle, rapidly increases in the wounded tissues. Other plants manufacture molecules that resemble the hormones of insects or other predators and thus interfere with the predators' normal growth and development. One of these natural products that resembles a human hormone is a complex molecule called diosgenin, which is obtained from wild yams. Diosgenin is only two simple chemical steps away from 16-dehydropregnenolone, or 16D, the main active ingredient in many oral contraceptives, and wild yams were once a major source for the manufacture of 16D. Unfortunately, these plants grow slowly, and the supply of wild yams was soon largely exhausted. Species of *Solanum* are being cultivated as alternative sources of molecules suitable for simple conversion to 16D.

enzymes that break down these glycosides to release the pungent odors associated with cabbage, horseradish, and mustard. Plant-eating insects of most groups ignore plants of the mustard family and will not feed on them even if they are starving. However, certain groups of true bugs and beetles, and the larvae of some groups of moths, feed only on the leaves of plants of the mustard family. The larvae of most of the members of the butterfly subfamily *Pierinae* (which includes the cabbage butterflies and orange-tips) also feed only on these plants. The same chemicals that act as deterrents to most groups of insect herbivores often act as feeding stimuli for these narrowly restricted feeders. For example, certain moth larvae that feed on cabbage will extrude their mouthparts and go through their characteristic feeding behavior when presented with agar or filter paper containing juices pressed from these plants.

Clearly the ability to manufacture these mustard-oil glycosides and to retain them in their tissues is an important evolutionary step that protects the plants of the mustard family from most herbivores. From the standpoint of herbivores in general, such protected plants rep-

(a) *(b)*

(c) *(d)*

22–50

*Some plants that produce hallucinogenic and medicinal compounds. **(a)** Mescaline, from the peyote cactus (Lophophora williamsii), is used ceremonially by many Native American groups of northern Mexico and the southwestern United States. **(b)** Tetrahydrocannabinol (THC) is the most important active molecule in marijuana*

*(Cannabis sativa). **(c)** Quinine, a valuable drug used in the treatment and prevention of malaria, is derived from tropical trees and shrubs of the genus Cinchona. **(d)** Cocaine, a drug that has recently been abused to an unparalleled extent, is derived from coca (Erythroxylum coca), a cultivated plant of*

northwestern South America. A Peruvian woman is shown here harvesting the leaves of cultivated coca. The secondary metabolites identified in these plants presumably protect them from the depredations of insects, but they are also physiologically active in vertebrates, including humans.

As mentioned previously, pollination and fruit-dispersal systems have developed particular coevolutionary patterns in which many of the possible variants have evolved not once but several times within a particular plant family or even genus. The resulting array of forms gives the angiosperms an extremely wide variety of pollination and fruit-dispersal mechanisms. In the case of biochemical relationships, however, the evolutionary steps appear to have been large and definitive, and whole families of plants can be characterized biochemically and associated with major groups of plant-eating insects. These biochemical relationships appear to have played a key role in the success of the angiosperms, which have a vastly more diverse array of secondary plant products than any other group of organisms.

Summary

The Angiosperms' Closest Relatives Are Thought to Be the *Bennettitales* and Gnetophytes

The ancestry of the angiosperms, long a subject of debate, has been investigated by phylogenetic analyses. Such methods have defined seed plants as one evolutionary line, or clade, and the *Bennettitales* and gnetophytes as the seed plants most closely related to angiosperms. All three groups possess flowerlike reproductive structures and collectively are referred to as "anthophytes." Only the angiosperms, however, belong to the phylum *Anthophyta*.

Several Factors Help Explain the Worldwide Success of Angiosperms

The earliest definite angiosperm remains are from the Early Cretaceous period, about 130 million years ago; they include both flowers and pollen. The flowering plants became dominant worldwide between 80 and 90 million years ago. Possible reasons for their success include various adaptations for drought resistance, including the evolution of the deciduous habit, as well as the evolution of efficient and often specialized mechanisms for pollination and seed dispersal. The primary radiation of angiosperms took place when the southern continents were united in the great land masses of Gondwanaland, which have separated progressively during the history of the group.

The Magnoliids, the Living Relatives of Early Angiosperms, Fall into Two Groups

The woody magnoliids have large, robust, bisexual flowers, often with many, spirally arranged floral parts. The remaining magnoliids, the paleoherbs, have relatively few flower parts and often unisexual flowers. Both of these groups, which together comprise about 3 percent of living angiosperm species, are characterized by pollen with a single pore (or furrow), as are the monocots, which include about 22 percent of living angiosperms. The eudicots, with pollen that has three apertures (pores or furrows), comprise about three-quarters of the species of angiosperms and are an enormously successful group.

The Four Whorls of Flower Parts Have Evolved in Different Ways

Typical angiosperm flowers consist of four whorls. The outermost whorl consists of sepals, which are special-ized leaves that protect the flower in bud. In contrast, the petals of most angiosperms have evolved from stamens that have lost their sporangia during the course of evolution. Stamens with anthers that comprise two pairs of pollen sacs are one of the diagnostic features of angiosperms. In the course of evolution, differentiation between the anther and the slender filament seems to have increased. Carpels are somewhat leaflike structures that have been transformed during the course of evolution to enclose the ovules. In most plants, the carpels have become specialized and differentiated into a swollen, basal ovary, a slender style, and a receptive terminal stigma. The loss of individual floral whorls and fusion within and between adjacent whorls have led to the evolution of many specialized floral types, which are often characteristic of particular families.

Angiosperms Are Pollinated by a Variety of Agents

Pollination by insects is basic in the angiosperms, and the first pollinating agents were probably beetles. The closing of the carpel, in an evolutionary sense, may have contributed to protection of the ovules from visiting insects. Pollination interactions with more specialized groups of insects seem to have evolved later in the history of the angiosperms, and wasps, flies, butterflies, and moths have each left their mark on the morphology of certain angiosperm flowers. The bees, however, are the most specialized and constant of flower-visiting insects and have probably had the greatest effect on the evolution of angiosperm flowers. Each group of flower-visiting animals is associated with a particular group of floral characteristics related to the animals' visual and olfactory senses. Some angiosperms have become wind-pollinated, shedding copious quantities of small, non-sticky pollen and having well-developed, often feathery stigmas that are efficient in collecting pollen from the air. Water-pollinated plants have either filamentous pollen grains that float to submerged flowers or various ways of transmitting pollen through or across the surface of the water.

Various Factors Affect the Relationship between Plant and Pollinator

Flowers that are regularly visited and pollinated by animals with high energy requirements, such as humming-birds, hawkmoths, and bats, must produce large amounts of nectar. These sources of nectar must then be protected and concealed from other potential visitors with lower energy requirements. Such visitors might satiate themselves with nectar from a single flower (or from the flowers of a single plant) and therefore fail to move on to another plant of the same species to effect cross-pollination. Wind pollination is most effective when individual plants grow together in large groups, whereas insects, birds, or bats can carry pollen great distances from plant to plant.

Flower Colors Are Determined Mainly by Carotenoids and Flavonoids

Carotenoids are yellow, oil-soluble pigments that occur in the chloroplasts and act as accessory pigments in photosynthesis. Flavonoids are water-soluble ring compounds present in the vacuole. Anthocyanins, which are blue or red pigments that constitute one major class of flavonoids, are especially important in determining the colors of flowers and other plant parts.

Fruits Are Basically Mature Ovaries

Fruits are just as diverse as the flowers from which they are derived, and they can be classified either morphologically, in terms of their structure and development, or functionally, in terms of their methods of dispersal. Fruits are basically mature ovaries, but if additional flower parts are retained in their mature structure, they are said to be accessory fruits. Simple fruits are derived from one carpel or from a group of united carpels, aggregate fruits from the free carpels of one flower, and multiple fruits from the fused carpels of several or many flowers. Dehiscent fruits split open to release the seeds, and indehiscent fruits do not.

Fruits and Seeds Are Dispersed by Wind, Water, or Animals

Wind-borne fruits or seeds are light and often have wings or tufts of hairs that aid in their dispersal. The fruits of some plants expel their seeds explosively. Some seeds or fruits are borne away by water, in which case they must be buoyant and have water-resistant coats. Others are disseminated by birds or mammals and frequently have fleshy coverings that are tasty or hooks, spines, or other devices that adhere to the coats of mammals or to feathers. Ants disperse the seeds and fruits of many plants; such dispersal units typically have an oily appendage, an elaiosome, which the ants consume.

Secondary Metabolites Are Important in the Evolution of Angiosperms

Biochemical coevolution has been an important aspect of the evolutionary success and diversification of the angiosperms. Certain groups of angiosperms have evolved various secondary products, or secondary metabolites, such as alkaloids, which protect them from most foraging herbivores. However, certain herbivores (normally those with narrow feeding habits) are able to feed on those plants and are regularly found associated with them. Potential competitors are excluded from the same plants because of their inability to handle the toxins. This pattern indicates that a stepwise pattern of coevolutionary interaction has occurred, and it appears likely that the early angiosperms also may have been protected by their ability to produce some chemicals that functioned as poisons for herbivores.

Selected Key Terms

accessory fruit p. 543	**legume** p. 544
achene p. 544	**magnoliids** p. 519
aggregate fruits p. 543	**mimicry** p. 550
anthocyanins p. 542	**multiple fruits** p. 543
berries p. 543	**nectar** p. 525
betacyanins p. 543	**nut** p. 545
capsule p. 544	**paleoherbs** p. 521
caryopsis p. 545	**pappus** p. 528
coevolution p. 532	**parthenocarpic fruits** p. 543
cypsela p. 545	**pollinium** p. 528
dehiscent fruits p. 544	**pomes** p. 544
drupes p. 543	**samara** p. 545
flavonoids p. 542	**schizocarp** p. 545
flavonols p. 542	**silique** p. 544
follicle p. 544	**simple fruits** p. 543
fruitlets p. 543	**woody magnoliids** p. 519
Gondwanaland p. 523	
indehiscent fruits p. 544	

Questions

1. What concept is embodied in the term "anthophyte" (not to be confused with *Anthophyta*)?

2. What unique characteristics of *Anthophyta* (angiosperms) indicate that the members of this phylum were derived from a single common ancestor?

3. Evolutionarily, petals apparently have been derived from two different sources. What are they?

4. Explain what is meant by coevolution, and provide two examples involving different insects and flowers.

5. Why are wind-pollinated angiosperms best represented in temperate regions and relatively rare in the tropics?

6. Distinguish among simple, aggregate, and multiple fruits, and give an example of each.

The Angiosperm Plant Body: Structure and Development

SECTION 5

Expanding shoot of the shagbark hickory (Carya ovata). Over winter, this shoot was greatly telescoped and existed as a terminal bud. As the bud expanded, the protective bud scales (below) separated and folded back. After the shoot is fully expanded, a new terminal bud will form and pass through a dormant period before it will be capable of expanding and repeating the cycle.

Chapter 23

Early Development of the Plant Body

OVERVIEW

Depending upon your point of view, when you watch a seedling emerge from the soil you are witnessing the beginning, middle, or end of a process. To a gardener, the emergence of the seedling is the first visible sign of life—a harbinger of the flower, fruit, or leaf that is the gardener's goal. But we know from Chapter 21 that the life of the new plant begins much earlier, with the fusion of sperm and egg to form a zygote within the embryo sac. Thus, to the botanist, the emergence of the seedling is merely an intermediate step in plant development. In the present chapter we take yet a third perspective: the emergence of the seedling is the culmination of embryonic development.

Embryonic development begins once the zygote divides; the two daughter cells constitute the first cells of the young embryo. In this chapter we follow the embryo as it progresses through a typical series of stages, eventually forming a structurally mature embryo consisting of a specific arrangement of embryonic tissues and organs within the seed.

But as gardeners know, the embryonic root and shoot will not emerge from the seed in the process of seed germination until the seed is exposed to the proper external conditions, including adequate moisture, temperature, and oxygen. Even if these conditions are met, some seeds still will not germinate because conditions within the seed are inadequate. Such seeds are said to be dormant. We discuss the factors affecting seed germination and dormancy and then describe the emerging young seedling.

CHECKPOINTS

By the time you finish reading this chapter, you should be able to answer the following questions:

1. How is polarity important in embryonic development of plants?

2. What are the three primary meristems of plants, and to which tissues do they give rise?

3. Through what sequence of stages of development do the embryos of eudicots progress? How does embryo development in monocots differ from that in eudicots?

4. How have mutations helped us understand embryo development?

5. What are the main parts of a mature eudicot or monocot embryo? What additional structures are present in grass embryos?

In the previous section, we traced the long evolutionary development of the angiosperms beginning with their presumed ancestor, a relatively complex, multicellular green alga. As we pointed out, the forked axes of early vascular plants were the forerunners of the shoots and roots of most modern vascular plants.

In this section, we shall be concerned with the structure and development of the angiosperm plant body, or sporophyte, which is a result of that long period of evolutionary specialization. This chapter begins with the formation of the embryo, a process known as **embryogenesis.** Embryogenesis establishes the body plan of the plant, consisting of two superimposed patterns: an **apical-basal pattern** along the main axis and a **radial pattern** of concentrically arranged tissue systems (Figure 23–1). Embryogenesis is accompanied by seed development. The seed, with its mature embryo, stored food, and protective seed coat, confers significant selective advantages over plants lacking seeds: the seed enhances the plant's ability to survive adverse environmental conditions and facilitates dispersal of the species.

As we follow the development of the angiosperm plant body, keep in mind the evolution of the vascular plants that we explored in the previous section. Developmental and evolutionary biologists are particularly excited over what can be learned about the evolution of developmental patterns. Great strides have been made by studying highly conserved genes—genes with similar DNA sequences in distantly related organisms—that regulate key developmental pathways. Much of what we know about this regulation comes from studying mutations that disrupt normal embryo development.

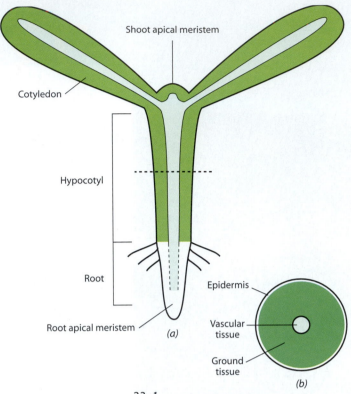

23–1

Diagrammatic representation of the body plan of an Arabidopsis *seedling.* **(a)** *The apical-basal pattern consists of an axis with a shoot tip at one end and a root tip at the other.* **(b)** *A transverse section through the hypocotyl reveals the radial pattern, consisting of the three tissue systems, as represented by the epidermis, ground tissue, and vascular tissue.*

Formation of the Embryo

The early stages of embryogenesis are essentially the same in magnoliids, eudicotyledons, and monocotyledons (Figures 23–2 and 23–3). Formation of the embryo begins with division of the zygote within the embryo sac of the ovule. In most flowering plants, the first division of the zygote is asymmetrical and transverse with regard to the long axis of the zygote (Figures 23–2a and 23–3a). With this division, the **polarity** of the embryo is established. The upper (chalazal) pole, consisting of a small *apical cell,* gives rise to most of the mature embryo. The lower (micropylar) pole, consisting of a large *basal cell,* produces a stalklike **suspensor** that anchors the embryo at the micropyle, the opening in the ovule through which the pollen tube enters.

Polarity is a key component of biological pattern formation. The term arises by analogy with a magnet, which has plus and minus poles. "Polarity" means simply that whatever is being discussed—a plant, an animal, an organ, a cell, or a molecule—has one end that is

different from the other. Polarity in plant stems is a familiar phenomenon. In plants that are propagated by stem cuttings, for example, roots will form at the lower end of the stem, with leaves and buds at the upper end.

The establishment of polarity is an essential first step in the development of all higher organisms, because it fixes the structural **axis** of the body, the "backbone" on which the lateral appendages will be arranged. In some angiosperms, polarity is already established in the egg cell and zygote, where the nucleus and most of the cytoplasmic organelles are located in the upper portion of the cell, and the lower portion is dominated by a large vacuole.

Through an orderly progression of divisions, the embryo eventually differentiates into a nearly spherical structure—the **embryo proper**—and the suspensor (Figures 23–2d and 23–3b, c). Before this stage is reached, the developing embryo is frequently referred to as the **proembryo.**

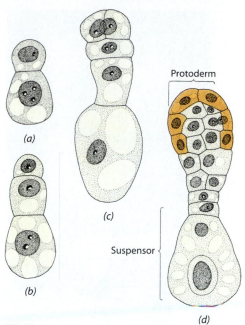

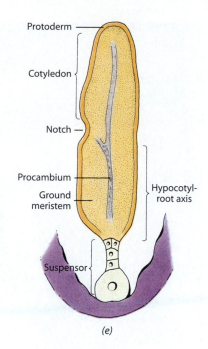

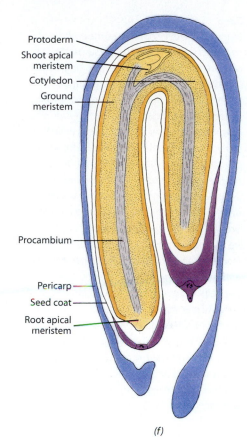

23–2

Stages in the development of the embryo of arrowhead (Sagittaria), a monocot. Early stages: (a) The two-celled stage, resulting from transverse division of the zygote. (b) The three-celled proembryo. (c) Disregarding the large basal cell, the proembryo is now at the four-celled stage. All four of these cells, through a series of divisions, contribute to formation of the embryo proper. (d) The protoderm has been initiated at the terminal *end of the embryo proper. At this stage, the suspensor consists of only two cells, one of which is the large basal cell. Late stages: (e) A depression, or notch (the site of the future apical meristem of the shoot), has formed at the base of the emerging cotyledon. (f) The cotyledon curves, and the embryo is approaching maturity. The suspensor has disappeared.*

The Protoderm, Procambium, and Ground Meristem Are the Primary Meristems

When first formed, the embryo proper consists of a mass of relatively undifferentiated cells. Soon, however, changes in the internal structure of the embryo proper result in the initial development of the tissue systems of the plant. The future epidermis, the **protoderm,** is formed by periclinal divisions—divisions parallel to the surface—in the outermost cells of the embryo proper (Figures 23–2d and 23–3c). Subsequently, vertical divisions within the embryo proper result in the initial distinction between **procambium** and **ground meristem** (Figure 23–3d, e). The ground meristem, the precursor of the **ground tissue,** surrounds the procambium, the precursor of the vascular tissues, xylem and phloem. The protoderm, ground meristem, and procambium—the so-called **primary meristems,** or primary meristematic tissues—extend into the other regions of the embryo as embryogenesis continues (Figures 23–2e, f and 23–3e, f).

Embryos Typically Progress through a Sequence of Developmental Stages

The stage of embryo development preceding cotyledon development—that is, when the embryo proper is spherical—is often referred to as the *globular stage.* Development of the cotyledons, the first leaves of the plant, may begin either during or after the time at which the procambium becomes discernible. As it develops, the globular embryo in eudicots gradually assumes a two-lobed form, or heart shape, and this stage is often called the *heart stage* (Figure 23–3d). The globular embryos of monocots, which form only one cotyledon, become cylindrical in shape (Figure 23–2e). It is during the transition between the globular stage of development and the emergence of the cotyledon(s) that the apical-basal pattern of the embryo becomes discernible, with the partitioning of the axis into shoot meristem, cotyledon(s), hypocotyl (the stemlike axis below the cotyledon(s)), embryonic root, and root meristem.

23–3

Stages in the development of the embryo of shepherd's purse (Capsella bursa-pastoris), *a eudicot.* **(a)** *Two-celled stage, resulting from transverse division of the zygote into an upper apical cell and a lower basal cell.* **(b)** *Six-celled proembryo. The suspensor is now distinct from the two terminal cells, which develop into the embryo proper. Endosperm provides food for the developing embryo.* **(c)** *The embryo proper is globular and has a protoderm, which will develop into the epidermis. The large cell near the bottom is the basal cell of the suspensor.* **(d)** *Embryo at the heart stage, when the cotyledons, the first leaves of the plant, begin to emerge.* **(e)** *The embryo seen here is at the torpedo stage. In* Capsella *the embryos curve. Ground meristem, the precursor of the ground tissue, surrounds the procambium, which will develop into the vascular tissues, xylem and phloem.* **(f)** *Mature embryo. The part of the embryo below the cotyledons is the hypocotyl. At the lower end of the hypocotyl is an embryonic root, or radicle.*

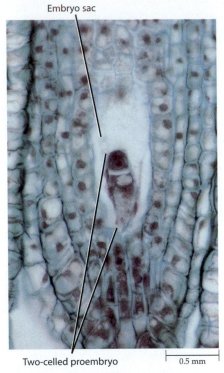

Embryo sac

Two-celled proembryo   0.5 mm

(a)

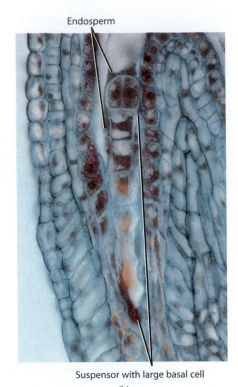

Endosperm

Suspensor with large basal cell

(b)

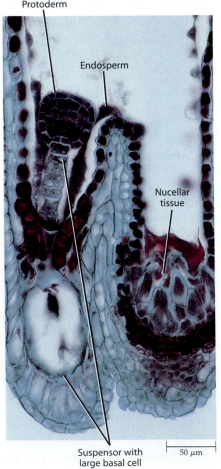

Protoderm

Endosperm

Nucellar tissue

Suspensor with large basal cell   50 µm

(c)

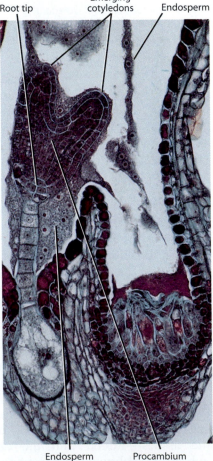

Root tip   Emerging cotyledons   Endosperm

Endosperm   Procambium

(d)

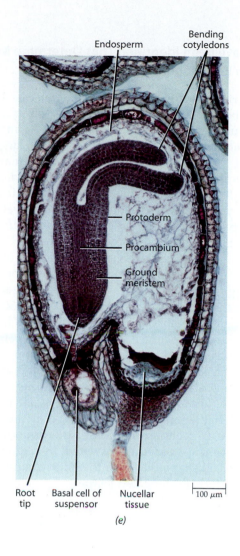

Endosperm · Bending cotyledons

— Protoderm

— Procambium

— Ground meristem

Root tip · Basal cell of suspensor · Nucellar tissue · 100 µm

(e)

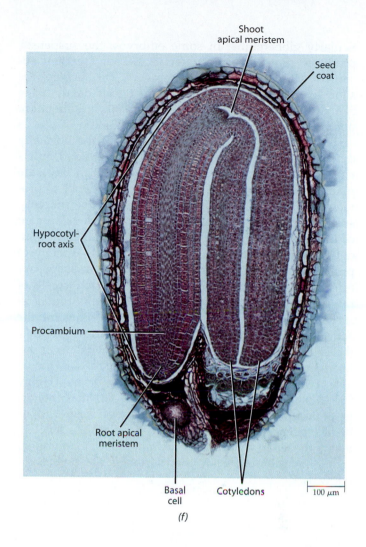

Shoot apical meristem · Seed coat

Hypocotyl-root axis

Procambium —

Root apical meristem

Basal cell · Cotyledons · 100 µm

(f)

As embryo development continues, the cotyledon(s) and axis elongate (the so-called *torpedo stage* of embryo development), and the primary meristems extend along with them (Figures 23–2f and 23–3e). During elongation, the embryo either remains straight or becomes curved. The single cotyledon of the monocot often becomes so large in comparison with the rest of the embryo that it is the dominating structure (see Figure 23–7c).

During the early stages of embryogenesis, cell division takes place throughout the young sporophyte. However, as the embryo develops, the addition of new cells gradually becomes restricted to the apical meristems of the shoot and root. As discussed previously (page 8), apical meristems are found at the tips of all shoots and roots and are composed of cells that are capable of repeated division. In magnoliid and eudicot embryos, the apical meristem of the shoot arises between the two cotyledons (Figure 23–3f). In monocot embryos, on the other hand, the shoot apical meristem arises on

one side of the cotyledon and is completely surrounded by a sheathlike extension from the base of the cotyledon (Figure 23–2f). The apical meristems of both shoot and root are of great importance, because these tissues are the source of virtually all of the new cells responsible for the development of the seedling and the adult plant.

The Suspensor Plays a Supporting Role in Development of the Embryo Proper

The suspensors associated with the embryos of *Selaginella*, a seedless vascular plant (page 442), and pine (page 482) function merely to push the developing embryos into nutritive tissues. In the past, it was believed that the suspensors of angiosperms played a similar role. We now know, however, that angiosperm suspensors are metabolically active. They play a role in supporting early development of the embryo proper by providing it with nutrients and growth regulators, particularly gibberellins. In some embryos, the suspensor produces proteinaceous materials, which are then absorbed by the rapidly growing embryo. The suspensor is short-lived, undergoing programmed cell death (page 576) at the torpedo stage of development. It is therefore not present in the mature seed (Figure 23–2f).

Several lines of evidence indicate that the normally developing embryo proper limits the growth and differentiation of the suspensor, the cells of which have the potential to generate an embryo. Such evidence is provided by several *Arabidopsis* embryo-defective mutants, such as *raspberry*1, *sus,* and *twn,* in which development of the embryo proper is disrupted. In all of these mutants, disrupted embryo development leads directly to proliferation of suspensor cells. Some of these suspensor cells acquire characteristics normally restricted to cells of the embryo proper, including the production of food reserves. The *twn* mutants are the most striking of the *Arabidopsis* embryo-defective mutants. In these mutants, the cells of the suspensor undergo embryogenic transformation, resulting in the formation of seeds with viable twin and occasionally triplet embryos (Figure 23–4). The results of these studies reveal that interactions occur between the suspensor and embryo proper. It has been suggested that during normal development, the embryo proper transmits specific inhibitory signals to the suspensor that suppress the embryonic pathway.

Genes Have Been Identified That Determine Major Events in Embryogenesis

The description of embryo formation tells us how the primary body plan of the plant develops, but it reveals little about the underlying mechanisms. As mentioned in the *Arabidopsis thaliana* essay on page 228, large populations of mutagen-treated *Arabidopsis* plants are being systematically screened for mutations that have an effect on plant development. Using this procedure, it is possible to identify the genes that govern plant development, a first step in determining how the genes function.

Using this method, very promising results have been obtained in identifying genes responsible for major events in *Arabidopsis* embryogenesis. (One approach to generating mutations that affect embryogenesis in *Arabidopsis* is outlined in Figure 23–5.) Over 50 distinct genes governing embryonic pattern formation have thus far been identified. Some of these regulatory genes affect the apical-basal pattern of the embryo and seedling. Mutations in these genes delete different regions of the apical-basal pattern (Figure 23–6). Another group of genes is involved in determining the radial pattern of tissue differentiation. Mutations in one of these genes, for example, prevents formation of the protoderm. Still another group of genes is involved in regulating the changes in cell shape that give the embryo and seedling their characteristic elongated shape.

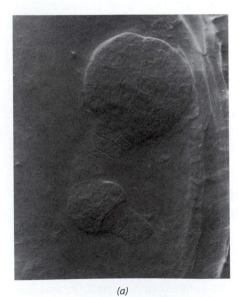

(a)

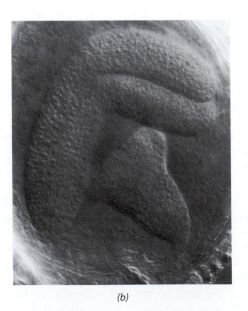

(b)

(c)

23–4

Twin embryo development in the twn *mutant of* Arabidopsis thaliana, *a eudicot.* **(a)** *A secondary embryo can be seen developing from the suspensor of the larger, primary*

embryo. Both embryos are at the globular stage. **(b)** *The cotyledons of the primary embryo are partly developed. Cotyledon development in the secondary embryo is at*

an early stage. **(c)** *Twin seedlings from a germinated seed. The seedling on the left resembles the wild-type (nonmutant) seedling. Its sibling has one large cotyledon.*

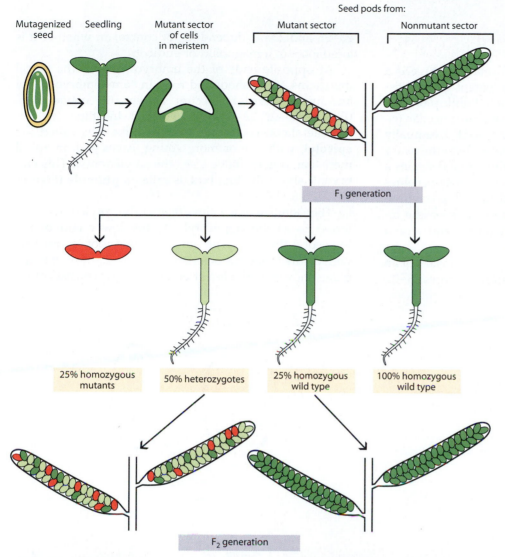

Mutagenized seed   Seedling   Mutant sector of cells in meristem

Seed pods from:
Mutant sector   Nonmutant sector

F₁ generation

25% homozygous mutants   50% heterozygotes   25% homozygous wild type   100% homozygous wild type

F₂ generation

23–5

A procedure for identifying mutants of Arabidopsis. A large population of seed is treated with the chemical mutagen ethyl methanosulfonate. Random mutations in cells of the shoot apical meristem of the embryo will give rise to mutant sectors on the adult plant. Flowers produced by these sectors will give rise to seed (F_1) that are heterozygous for a particular mutant gene. Self-fertilization of each F_1 plant will give rise to F_2 progeny that segregate 1:2:1 (1/4 homozygous wild type to 1/2 heterozygotes to 1/4 homozygous mutants). Phenotypic effects of recessive mutations will thus be observable in the F_2 generation of the mutagenized populations and can be selected for further study.

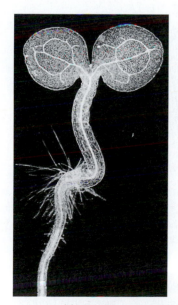

(a) wild type

(b) gurke

(c) fackel

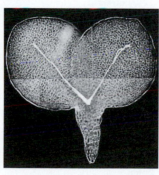

(d) monopteros

(e) gnom

23–6

Mutant seedlings of Arabidopsis. (a) A normal (wild-type) seedling for comparison with four types of mutants that lack major parts of the seedling structure. In the mutant seedlings shown here, (b) lacks a shoot apical meristem and cotyledons; (c) lacks a hypocotyl, so that the shoot meristem and cotyledons are attached directly to the root; (d) lacks a root; and (e) lacks both apical and basal portions—the remaining stem contains epidermis, ground tissue, and vascular tissue. The seedlings have been "cleared" to reveal the vascular tissue within them.

The Mature Embryo and Seed

Throughout the period of embryo formation, there is a continuous flow of nutrients from the parent plant to tissues of the ovule, resulting in a massive buildup of food reserves within the endosperm, perisperm (nucellar tissue), or cotyledons of the developing seed. Eventually the stalk, or **funiculus,** connecting the ovule to the ovary wall separates from the ovule, and the ovule becomes a nutritionally closed system. Finally, the seed becomes desiccated as it loses water to the surrounding environment, and the seed coat, which is derived from the integuments, hardens, encasing the embryo and stored food in "protective armor."

The mature embryo of flowering plants consists of an axis bearing either one or two cotyledons (Figures 23–2f,

23–3f, and 23–7), depending, of course, on whether it is a monocot or a magnoliid or eudicot.

At opposite ends of the embryo axis are the apical meristems of the shoot and root. In some embryos, only an apical meristem occurs above the cotyledon(s) (Figures 23–2f, 23–3f, and 23–7b, c). In others, an embryonic shoot, consisting of a stemlike axis called the **epicotyl,** with one or more young leaves and an apical meristem, occurs above (*epi-*) the cotyledon(s). This embryonic shoot, the first bud, is called a **plumule** (Figures 23–7a, d and 23–8).

The stemlike axis below (*hypo-*) the cotyledons is referred to as the **hypocotyl.** At the lower end of the hypocotyl there may be an embryonic root, or **radicle,** with distinct root characteristics (Figure 23–8b). In many plants, however, the lower end of the axis consists of lit-

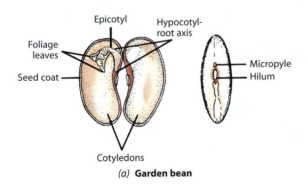

(a) **Garden bean**

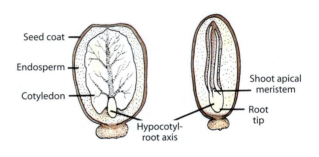

(b) **Castor bean**

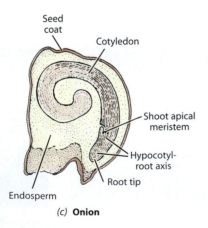

(c) **Onion**

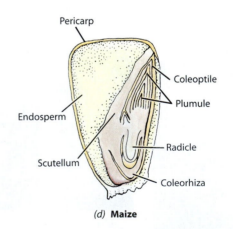

(d) **Maize**

23–7

Seeds of some common eudicotyledons and monocotyledons. *(a)* Seeds of the garden bean (Phaseolus vulgaris), *a eudicot, shown open and from an external edge view. The garden bean embryo has a plumule above the cotyledons, consisting of a short stem (epicotyl), two pairs of foliage leaves, and an apical meristem. The apical meristem is found between the foliage leaves and cannot be*

seen here. The fleshy cotyledons of the garden bean embryo contain the stored food. *(b)* A seed of the castor bean (Ricinus communis), *another eudicot, opened to show both flat and edge views of the embryo. The castor bean embryo has only an apical meristem above the cotyledons. The stored food occurs in the endosperm. The embryos of the monocots (c)* onion (Allium cepa) *and (d)* maize

(Zea mays) *are shown in longitudinal view. The shoot apical meristem of the onion embryo lies on one side and at the base of the cotyledon, which is much larger than the rest of the embryo. The maize embryo has a well-developed scutellum (cotyledon) and root, or radicle. The stored food in both seeds is in the endosperm.*

Wheat: Bread and Bran

Like all grasses, bread wheat (Triticum aestivum) is a monocotyledon, and its fruit—the grain or kernel—is one-seeded. The endosperm and the embryo are surrounded by covering layers composed of the pericarp and the remains of the seed coat. More than 80 percent of the volume of the wheat kernel is made up of starchy endosperm. The outermost layer of the endosperm, called the aleurone layer, contains lipid and protein reserves. The aleurone layer surrounds both the starchy endosperm and the embryo.

White flour is made from the starchy endosperm. In the milling of wheat, the bran—consisting of the covering layers and the aleurone layer—is removed. The bran constitutes about 14 percent of the kernel. Actually, the bran somewhat decreases the nutritional value of the wheat kernel. Because bran is mostly cellulosic, it cannot be digested by humans and tends to speed the passage of food through the intestinal tract, resulting in lower absorption. The embryo ("wheat germ"), which

represents about 3 percent of the kernel, is also removed during milling, because its high oil content would reduce the length of time that the flour can be stored. The bran and wheat germ, which contain most of the vitamins found in wheat, are increasingly being used for human consumption, as well as for livestock feed.

During the past few years, oat (Avena sativa) bran has become a popular food item among health-conscious people in many parts of the world. Research results indicate that oat bran, as part of a low-fat diet, can help reduce the amount of cholesterol in the blood. Apparently, the water-soluble fibers in the oat bran form a gel in the small intestine that traps cholesterol and prevents it from being reabsorbed into the bloodstream. Instead, the trapped cholesterol is excreted with other bodily wastes. Cholesterol is also reduced by the water-soluble fibers found in rice (Oryza sativa) and barley (Hordeum vulgare) bran.

tle more than an apical meristem covered by a rootcap. If a radicle cannot be distinguished in the embryo, the embryo axis below the cotyledons is called the **hypocotyl-root axis** (Figures 23–2e, 23–3f, and 23–7a, b, c).

In the discussion of the development of the angiosperm seed in Chapter 21, we noted that in many eudicots most or all of the food-storing endosperm and the perisperm, if present, is absorbed by the developing embryo (page 510). The embryos of such seeds develop large, fleshy, food-storing cotyledons that nourish the embryo as it resumes growth. Familiar examples of seeds with sizable cotyledons and no endosperm are sunflower, walnut, pea, and garden bean (Figure 23–7a). In eudicotyledons with large amounts of endosperm, such as the castor bean, the cotyledons are thin and membranous (Figure 23–7b). These cotyledons serve to absorb stored food from the endosperm itself during resumption of growth of the embryo.

In the monocotyledons, the single cotyledon, in addition to functioning as a food-storing or photosynthetic organ, also performs an absorptive function (Figure 23–7c, d). Embedded in endosperm, the cotyledon absorbs food digested from the endosperm by enzymatic activity. The digested food is then moved by way of the

cotyledon to the growing regions of the embryo. Among the most highly differentiated of monocot embryos are those of grasses (Figures 23–7d and 23–8). When fully formed, the grass embryo possesses a massive cotyledon, the **scutellum.** The scutellum, like the cotyledons of most monocots, functions in the absorption of food stored in the endosperm. The scutellum is attached to one side of the axis of the embryo, which has a radicle at its lower end and a plumule at its upper end. Both the radicle and the plumule are enclosed by sheathlike protective structures called the **coleorhiza** and the **coleoptile,** respectively (Figures 23–7d and 23–8).

All seeds are enclosed by a **seed coat,** which develops from the integument(s) of the ovule and provides protection for the enclosed embryo. The seed coat is usually much thinner than the integument(s) from which it is formed. The thin, dry seed coat may have a papery texture, but in many seeds it is very hard and highly impermeable to water.

The micropyle is often visible on the seed coat as a small pore. Commonly, the micropyle is associated with a scar, called the **hilum** (Figure 23–7a), which is left on the seed coat after the seed has separated from its stalk, or funiculus.

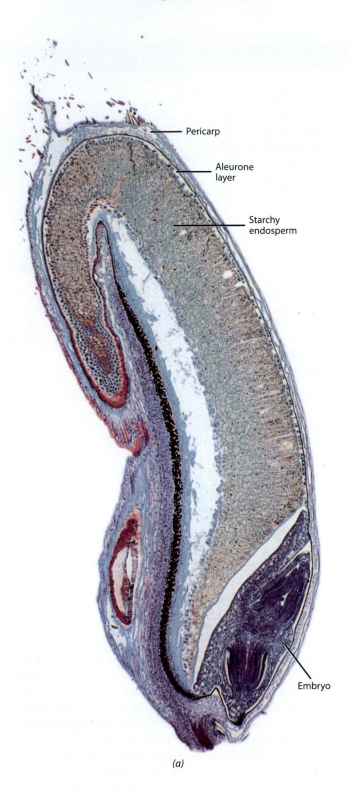

Pericarp

Aleurone layer

Starchy endosperm

Embryo

(a)

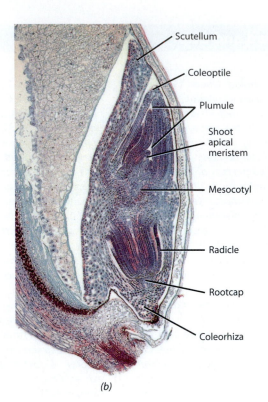

Scutellum

Coleoptile

Plumule

Shoot apical meristem

Mesocotyl

Radicle

Rootcap

Coleorhiza

(b)

23–8
Longitudinal section of mature grain, or kernel, of wheat (Triticum aestivum), *a monocot (**a**). The starchy endosperm is surrounded by a protein-containing aleurone layer. The covering layers of the wheat kernel consist largely of the pericarp, the mature ovary wall. The seed coat, which becomes fused with the pericarp, disintegrates during development of the kernel. (**b**) Detail of the mature wheat embryo, showing the large cotyledon known as the scutellum. The sheathlike coleorhiza and coleoptile enclose the radicle and plumule, respectively. The part of the embryo axis between the point of attachment of the scutellum and that of the coleoptile is known as the mesocotyl.*

Requirements for Seed Germination

The growth of the embryo is usually delayed while the seed matures and is dispersed. Resumption of growth of the embryo, or **germination** of the seed, is dependent upon many factors, both external and internal. Among the external, or environmental, factors, three are especially important: water, oxygen, and temperature. In ad-

dition, small seeds, such as those of lettuce (*Lactuca sativa*) and many weeds, commonly require exposure to light for germination (see page 713).

Most mature seeds are extremely dry, normally containing only 5 to 20 percent of their total weight as water. Thus, germination is not possible until the seed imbibes the water required for metabolic activities. Enzymes already present in the seed are activated, and new ones are synthesized for the digestion and utilization of the stored foods accumulated in the cells of the seed during the period of embryo formation (see page 681). The same cells that earlier had been synthesizing enormous amounts of reserve materials now completely reverse their metabolic processes and digest the stored food. Cell enlargement and cell division are initiated in the embryo and follow patterns characteristic of the species. Further growth requires a continuous supply of water and nutrients. As the seed imbibes water, it swells, and considerable pressure may develop within it (see the essay "Imbibition" on page 80).

During the early stages of germination, glucose breakdown may be entirely anaerobic, but as soon as the seed coat is ruptured, the seed switches to respiration, which requires oxygen. If the soil is waterlogged, the amount of oxygen available to the seed may be inadequate for aerobic respiration, and the seed will fail to grow into a seedling.

Although many seeds germinate over a fairly wide range of temperatures, they usually will not germinate below or above a certain temperature range specific for the species. The minimum temperature for many species is 0° to 5°C, the maximum is 45° to 48°C, and the optimum range is 25° to 30°C.

Dormant Seeds Will Not Germinate Even under Favorable External Conditions

Even when external conditions are favorable, some seeds will fail to germinate. Such seeds are said to be **dormant.** The most common causes of dormancy in seeds are the physiological immaturity of the embryo and the impermeability of the seed coat to water and, sometimes, to oxygen. Some physiologically immature seeds must undergo a complex series of enzymatic and biochemical changes, collectively called **after-ripening,** before they will germinate. In temperate regions, after-ripening is triggered by the low temperatures of winter. Thus, the necessity for a period of after-ripening helps prevent germination of the seed during the inclement period of winter, when the seedling would be unlikely to survive.

Dormancy is of great survival value to the plant. As in the example of after-ripening, it is a method of ensuring that conditions will be favorable for growth of the seedling when germination does occur. Some seeds must pass through the digestive tracts of birds or mammals before they will germinate, resulting in wider dispersal of the plant species. Some seeds of desert species will germinate only when inhibitors in their coats are leached away by rainfall; this adaptation ensures that the seed will germinate only during those rare intervals

23–9

Manzanita (Arctostaphylos viscida) *of the California chaparral community. The long-lived seeds of manzanita remain viable in the soil for years. Scarification, or breakdown of the seed coat, by fire or other means is necessary in order to break dormancy and induce germination.*

when desert rainfall provides sufficient water for the seedling to mature. Other seeds must be cracked mechanically, as by tumbling along in the rushing water of a gravelly streambed. Still other seeds lie dormant in cones or fruits until the heat of a fire releases the seeds. The Mediterranean-type-climate vegetation of the California chaparral community, dominated by manzanita (*Arctostaphylos*), is dependent upon fire for long-term persistence at any one site because fire induces the germination of manzanita seeds (Figure 23–9). Finally, the seeds of species that live in woodland clearings depend on either the death of a canopy tree or some other disturbance to form an opening in the canopy before they will germinate. Hence, the germination strategies of plants are closely linked to the ecological problems existing in their particular habitats. (See Chapter 29, page 713, for further discussion of dormancy in seeds.)

23–10

*Stages in the germination of some common eudicots. Seed germination in both **(a)** the garden bean* (Phaseolus vulgaris) *and **(b)** the castor bean* (Ricinus communis) *is epigeous, or above ground. During germination, the cotyledons are carried above ground by the elongating hypocotyl. Note that in both of these seedlings, the elongating hypocotyl forms a hook, which then straightens out, pulling the cotyledons and plumule or shoot apex above ground. **(c)** By contrast, seed germination in the pea* (Pisum sativum) *is hypogeous, or below ground. The cotyledons remain underground and the hypocotyl does not elongate. In hypogeous germination, typified by the pea seedling, it is the epicotyl that elongates and forms a hook, which pulls the plumule above ground as it straightens out.*

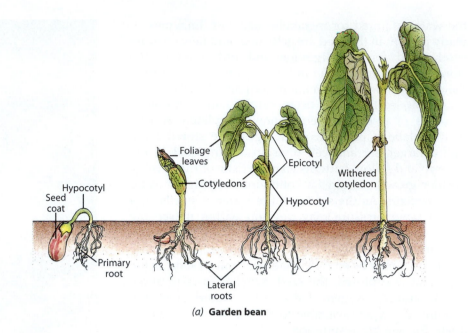

(a) **Garden bean**

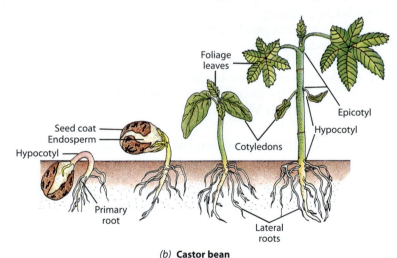

(b) **Castor bean**

From Embryo to Adult Plant

When germination occurs, the first structure to emerge from most seeds is the radicle, or embryonic root, which enables the developing seedling to become anchored in the soil and to absorb water. As this first root, called the **primary root,** or **taproot,** continues to grow, it develops branch roots, or **lateral roots.** These roots in turn may give rise to additional lateral roots. In this manner, a much-branched root system develops. Commonly, the primary root in monocots is short-lived, and the root system of the adult plant develops from **adventitious roots,** which arise at the nodes (the parts of the stem at which the leaves are attached) and then produce lateral roots. ("Adventitious" refers to structures arising not at their usual sites, as roots arising from stems or leaves rather than from other roots.)

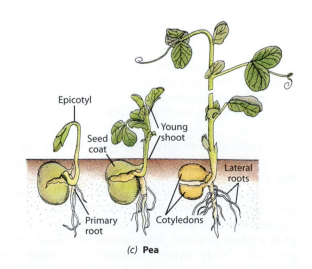

(c) **Pea**

Seed Germination May Be Epigeous or Hypogeous

The way in which the shoot emerges from the seed during germination varies from species to species. For example, after the root emerges from the seed of the garden bean (*Phaseolus vulgaris*), the hypocotyl elongates and becomes bent in the process (Figure 23–10a). Thus, the delicate shoot tip is protected from injury by being pulled rather than pushed through the soil. When the bend, or *hook,* as it is called, reaches the soil surface, it straightens out and pulls the cotyledons and plumule up into the air. This type of seed germination, in which the cotyledons are carried above ground level, is called **epigeous.** During germination and subsequent development of the seedling, the food stored in the cotyledons is digested and the products are transported to the growing parts of the young plant. The cotyledons gradually decrease in size, wither, and eventually drop off. By this time, the seedling has become *established;* that is, it is no longer dependent upon the stored food of the seed for its nourishment. The seedling is now a photosynthesizing, autotrophic organism.

Germination of the castor bean (*Ricinus communis*) (Figure 23–10b) is essentially similar to that of the garden bean, except that in the castor bean the stored food is found in the endosperm. As the hook straightens out, the endosperm and often the seed coat are carried upward, along with the cotyledons and plumule. During this period, the digested foods of the endosperm are absorbed by the cotyledons and are transported to the growing parts of the seedling. In both the garden bean and castor bean, the cotyledons become green upon exposure to light, but they do not have an important photosynthetic function. In some plants, such as squash (*Cucurbita maxima*), the cotyledons become important photosynthetic organs.

In the pea (*Pisum sativum*), the epicotyl is the structure that elongates and forms the hook. As the epicotyl straightens out, the plumule is raised above the soil surface. The cotyledons remain in the soil (Figure 23–10c), where they eventually decompose after the stored food has been mobilized for growth of the seedling. This type of seed germination, in which the cotyledons remain underground, is called **hypogeous.**

In the large majority of monocot seeds, the stored food is found in the endosperm (Figure 23–7c, d). In relatively simple monocot seeds, such as that of the onion (*Allium cepa*), it is the single tubular cotyledon that emerges from the seed and forms the hook (Figure 23–11a). When the cotyledon straightens, it carries the seed coat and enclosed endosperm upward. Throughout this period, and for some time afterward, the embryo obtains much of its nourishment from the endosperm by way of the cotyledon. Furthermore, the green cotyledon in onion functions as a photosynthetic leaf, contributing significantly to the food supply of the developing seed-

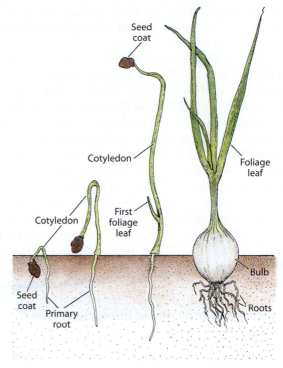

(a) **Onion**

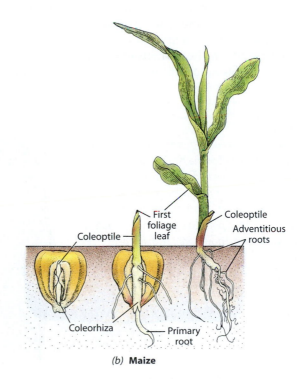

(b) **Maize**

23–11

Stages in the germination of two common monocotyledons: (a) onion (Allium cepa) and (b) maize (Zea mays). Seed germination in onion is epigeous; in maize, it is hypogeous.

ling. Soon the plumule, enclosed within the protective, sheathlike base of the cotyledon, elongates and emerges from the cotyledon.

Our last example of seedling development is provided by maize *(Zea mays)*, a monocot that has a highly differentiated embryo (Figure 23–12). The coleorhiza, which encloses the radicle, is the first structure to grow through the pericarp, the mature ovary wall, of the grain. (In maize, the integuments disintegrate during development of the seed and fruit, and hence the pericarp functions as a "seed coat.") The coleorhiza is then followed by the radicle, or primary root, which elongates very rapidly and quickly penetrates the coleorhiza (Figure 23–11b). After the primary root emerges, the coleoptile, which surrounds the plumule, is pushed upward by elongation of the first internode, called the *mesocotyl* (Figure 23–8b). (An internode is the part of the stem between two successive nodes; see Figure 1–10.) When the base of the coleoptile reaches the soil surface, its edges spread apart at the tip, and the first leaves of the plumule begin to emerge. In addition to the primary root, two or more seminal adventitious roots, which arise from the cotyledonary node (the part of the axis at which the cotyledons are attached), grow through the pericarp and then bend downward (Figure 23–12).

Regardless of the manner in which the shoot emerges from the seed, the activity of the apical meristem of the shoot results in the formation of an orderly sequence of leaves, nodes, and internodes. Apical meristems, which

develop in the leaf axils (the upper angles between leaves and stems), produce axillary shoots, and these, in turn, may form additional axillary shoots.

The period from germination to the time the seedling becomes established as an independent organism constitutes the most crucial phase in the life history of the plant. During this period, the plant is most susceptible to injury by a wide range of insect pests and parasitic fungi, and water stress can very rapidly prove fatal.

Summary

During Embryogenesis, the Body Plan of the Plant, Consisting of an Apical-Basal Pattern and a Radial Pattern, Is Established

Beginning with the zygote, the shoot and root of the young plant are initiated as one continuous structure. With the asymmetrical division of the zygote, the polarity of the embryo is established. Through an orderly progression of divisions, the embryo differentiates into a suspensor and an embryo proper, within which the primary meristems—the precursors of the epidermis, ground tissue, and vascular tissues—and radial pattern ultimately are formed. During the transition between the globular and heart stages, the apical-basal pattern of the embryo becomes discernible. Development of the cotyle-

23–12

Longitudinal section of mature grain, or kernel, of maize (Zea mays). Grain embryos commonly contain two or more seminal (seed) adventitious roots. One seminal adventitious root can be seen in this section (arrow). Although at first pointed upward, these roots become inclined downward with further growth.

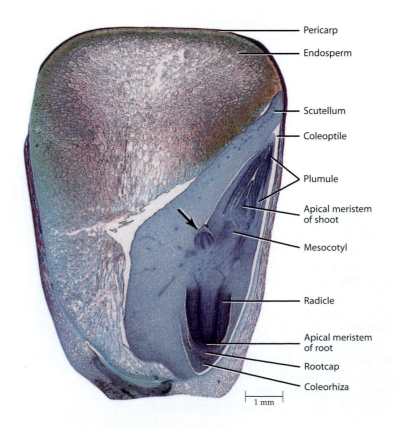

Pericarp
Endosperm
Scutellum
Coleoptile
Plumule
Apical meristem of shoot
Mesocotyl
Radicle
Apical meristem of root
Rootcap
Coleorhiza

1 mm

dons may begin either during or after the time at which the procambium becomes discernible. As the embryo develops, the addition of new cells is gradually restricted to the apical meristems.

Mutations Disrupt Normal Embryo Development

The suspensors of angiosperm embryos are metabolically active and play a role in supporting early development of the embryo proper. In some embryo-defective mutants of *Arabidopsis,* the cells of the suspensor occasionally form secondary embryos. Other mutations affect the apical-basal pattern of the embryo and seedling, resulting in the deletion of major portions of the plant.

The Mature Embryo Consists of a Hypocotyl-Root Axis and One or Two Cotyledons

The seeds of flowering plants consist of an embryo, a seed coat, and stored food. When fully formed, the embryo consists basically of a hypocotyl-root axis bearing either one or two cotyledons and an apical meristem at the shoot tip and at the root tip. The cotyledons of most eudicots are fleshy and contain the stored food of the seed. In other eudicots and in most monocots, food is stored in the endosperm, and the cotyledons function to absorb the simpler compounds resulting from the digestion of that food. The simpler compounds are then transported to growing regions of the embryo.

A Dormant Seed Will Not Germinate Even When External Conditions Are Favorable

Germination of the seed—resumption of growth of the embryo—depends upon environmental factors, including water, oxygen, and temperature. Many seeds must pass through a period of dormancy before they are able to germinate. Dormancy is of great survival value to the plant because it is a method of ensuring conditions will be favorable for growth of the seedling when germination occurs.

Following Emergence of the Root and Shoot, the Seedling Becomes Established

The root is the first structure to emerge from most germinating seeds, enabling the seedling to become anchored in the soil and to absorb water. The way in which the shoot emerges from the seed varies from species to species. Many magnoliids and eudicotyledons form a bend, or hook, in their hypocotyls or epicotyls. As the hook straightens out, the delicate shoot tip is pulled upward from the soil, thus preventing injury that might occur if the shoot tip were pushed through the soil.

Selected Key Terms

adventitious roots p. 566	**hypogeous** p. 567
apical-basal pattern p. 556	**lateral roots** p. 566
	plumule p. 562
coleoptile p. 563	**polarity** p. 556
coleorhiza p. 563	**primary meristems** p. 557
dormant p. 565	
embryo proper p. 556	**primary root** p. 566
epicotyl p. 562	**procambium** p. 557
epigeous p. 567	**proembryo** p. 556
funiculus p. 562	**protoderm** p. 557
germination p. 564	**radial pattern** p. 556
ground meristem p. 557	**radicle** p. 562
ground tissue p. 557	**scutellum** p. 563
hilum p. 563	**seed coat** p. 563
hypocotyl p. 562	**suspensor** p. 556
hypocotyl-root axis p. 563	**taproot** p. 566

Questions

1. Distinguish among or between the following: proembryo/embryo proper/suspensor; globular stage/heart stage/torpedo stage; epicotyl/plumule; hypocotyl-root axis/radicle; coleorhiza/coleoptile.

2. Explain what is meant by the apical-basal pattern and radial pattern of the plant.

3. What role is played by the suspensor in angiosperms, and what evidence indicates that the embryo proper suppresses the embryonic pathway in the suspensor?

4. Explain how mutagens can be used to generate mutations affecting embryogenesis.

5. Of what value, if any, is seed dormancy to a plant?

6. Suggest why the root is the first structure to emerge from a germinating seed.

7. How is seed dormancy defined, and what are some requirements for seed germination?

8. What are some of the ways in which the shoot emerges from the seed during germination? What is meant by epigeous and hypogeous germination?

Chapter 24

Cells and Tissues of the Plant Body

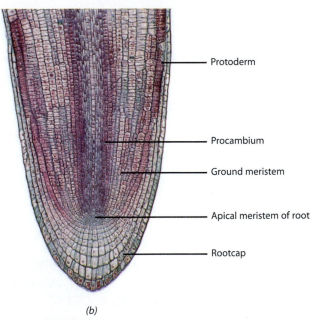

— Leaf primordium

— Apical meristem of shoot

— Axillary bud primordium

— Ground meristem

— Procambium

— Protoderm

(a)

— Protoderm

— Procambium

— Ground meristem

— Apical meristem of root

— Rootcap

(b)

24–1

Shoot and root apical meristems. (a) Longitudinal section of a shoot tip of lilac (Syringa vulgaris), *showing the shoot apical meristem and primordia of leaves and axillary buds. (b) Root tip of radish* (Raphanus sativus), *in longitudinal section, showing the root apical meristem covered by a rootcap. Note the files, or lineages, of cells behind the root apical meristem.*

Protoderm, procambium, and ground meristem are partly differentiated tissues known as primary meristems.

OVERVIEW

In the previous chapter we saw how the plant's basic body plan forms during embryonic development within the seed and how the young seedling emerges as a result of the activity of apical meristems. In this chapter, we continue this theme with an overview of the apical meristems and their role in the development of the seedling into an adult plant. You will find that apical meristems have two main functions. First, because they are principal sites of cell division, they are ultimately responsible for the extension growth of both the shoot and the root. Second, they produce cells that eventually form the mature tissues of the primary plant body.

To understand how apical meristems function you must consider changes that occur in both time and space. As cells in the meristem divide, they produce additional meristematic cells and also daughter cells that accumulate, increase in length, and push the meristematic cells forward. These daughter cells that are "left behind" then begin to change—they differentiate—first into partially differentiated tissues (the primary meristems) and later into the mature tissues of the plant.

The rest of this chapter describes these mature tissues and the types of cells of which they are composed. This information will be important in the next chapters to understand the internal structure of roots, stems, and leaves. These organs consist of many of the same basic tissues arranged in different ways.

CHECKPOINTS

By the time you finish reading this chapter, you should be able to answer the following questions:

1. What is the role of the apical meristems, and what is their composition?

2. What are the three overlapping processes of plant development, and how do they overlap?

3. What are the three tissue systems of the plant body? Of what tissues are they composed?

4. How do parenchyma, collenchyma, and sclerenchyma cells differ from one another? What are their respective functions?

5. What are the principal conducting cells in the xylem? In the phloem? What are the characteristic features of each cell type?

6. What roles are played by the epidermis?

As we discussed in the previous chapter, the process of embryogenesis establishes the apical-basal axis of the plant, with a shoot apical meristem at one end and a root apical meristem at the other. During embryogenesis, the radial pattern of the tissue systems within the axis is also determined. Embryogenesis is, however, only the beginning of development of the plant body. Most plant development occurs after embryogenesis through the activity of the **meristems.** These perpetually young tissues or populations of cells retain the potential to divide long after embryogenesis is over. With germination of the seed, the meristems known as the root and shoot apical meristems of the embryo generate cells that give rise to the roots, stems, leaves, and flowers of the adult plant.

Apical Meristems and Their Derivatives

Apical meristems are found at the tips of all roots and stems and are involved primarily with extension of the plant body (Figure 24–1). The term "apical meristem" refers to a population of cells composed of initials and their immediate derivatives. **Initials** are cells that divide in such a way that one of the sister cells remains in the meristem as an initial while the other becomes a new body cell or **derivative.** Derivatives usually divide near the root or shoot tip before undergoing differentiation. Divisions are not limited, however, to the initials and their immediate derivatives. For example, the so-called primary meristems—protoderm, procambium, and ground meristem—are initiated during embryogenesis and are extended throughout the plant body by the activity of the apical meristems. These primary meristems are partly differentiated tissues that remain meristematic for some time before they begin to differentiate into specific cell types in the primary tissues (Figure 24–2). Growth of this type, which involves extension of the plant body and formation of the primary tissues, is called **primary growth,** and the part of the plant body composed of these tissues is called the **primary plant body** (page 427). The apical meristems of root and shoot are discussed in greater detail in Chapters 25 and 26, respectively, and secondary growth, which involves thickening of the stem and root, is discussed in Chapter 27.

The existence of meristems, which add to the plant body throughout its life, underlies one of the principal differences between plants and animals. Birds and mammals, for example, stop growing when they reach maturity, although the cells of certain "turnover" tissues, such as skin or the lining of the intestine, continue to divide. Plants, however, continue to grow during their entire life span. Such unlimited or prolonged growth of the apical meristems is described as **indeterminate.**

Growth in plants is the counterpart, to some extent, of motility in animals. Plants "move" by extending their roots and shoots, both of which involve changes in size and form. As a result of these changes, a plant modifies its relationship with the environment, for example, by curving toward the light and extending its roots toward water. The sequence of growth stages in plants thus roughly corresponds to a whole series of motor acts in animals, especially those associated with obtaining food and water. In fact, growth in plants serves many of the functions that we group under the term "behavior" in animals.

Growth, Morphogenesis, and Differentiation

Development—the sum total of events that progressively form an organism's body—involves three overlapping processes: growth, morphogenesis, and differentiation. Development occurs in response to instructions contained in the genetic information an organism inherits from its parents. The specific developmental pathway followed, however, is determined in large part by environmental factors such as, in the case of plants, day length, light quality and quantity, and gravity (see Chapter 29).

Growth, an irreversible increase in size, is accomplished by a combination of cell division and cell enlargement (see Chapter 3). Cell division does not of itself constitute growth. It may simply increase cell number without increasing the overall volume of a structure. The addition of cells to the plant body by cell division increases the potential for growth by increasing the number of cells that can grow, but most plant growth is brought about by cell enlargement.

The shape of the plant and organization of its tissues are influenced greatly by both cell division and cell enlargement. The planes in which the cells divide and subsequently expand largely determine the shape and form of a plant part. For example, roots typically have a cylindrical form because most of the cells derived from the

24–2

Diagram summarizing the origin, from the apical meristem, of the primary meristems, which give rise to the tissue systems and tissues of the primary plant body.

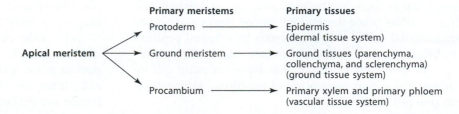

Primary meristems	Primary tissues
Protoderm ⟶	Epidermis (dermal tissue system)
Apical meristem → Ground meristem ⟶	Ground tissues (parenchyma, collenchyma, and sclerenchyma) (ground tissue system)
Procambium ⟶	Primary xylem and primary phloem (vascular tissue system)

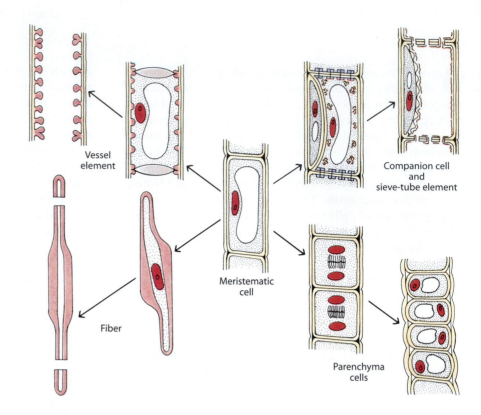

24–3

Diagram illustrating some of the cell types that may originate from a meristematic cell of the procambium or the vascular cambium. The meristematic cell depicted here (at the center), with a single large vacuole, is typical of the meristematic cells of the vascular cambium. Procambial cells typically contain several small vacuoles. Four different cell types are depicted here. The meristematic cells or precursors of these cells had identical genetic constitutions. The different cell types become distinct from one another because they express sets of genes not expressed by other cell types.

Vessel element

Companion cell and sieve-tube element

Fiber

Meristematic cell

Parenchyma cells

apical meristem divide transversely, in a plane perpendicular to the long axis of the root. This results in long files, or lineages, of cells (Figure 24–1b) that then expand mainly parallel to the long axis of the root. The acquisition of a particular shape or form is known as **morphogenesis.**

Differentiation—the process by which cells that have identical genetic constitutions become different from one another and from the meristematic cells from which they originated (Figure 24–3)—often begins while the cell is still enlarging. Cellular differentiation depends on the control of gene expression. Different types of cells and tissues synthesize different proteins because they express certain sets of genes not expressed by other types of cells and tissues. For example, both fibers and collenchyma cells are types of supportive cells, but the cell walls of fibers typically are rigid whereas those of collenchyma cells are flexible. During its development the fiber makes enzymes that produce lignin, which imparts the rigid property to its walls. The collenchyma cell, in contrast, makes enzymes that produce pectins, which impart plastic properties to its walls.

Although cellular differentiation depends on the control of gene expression, the fate of a plant cell—that is, what kind of cell it will become—is determined by its final position in the developing organ. Although cell lineages, such as those in the root, may be established, cell differentiation is not dependent on cell lineage. If an undifferentiated cell is displaced from its original position to a different one, it will differentiate into a cell type appropriate to its new position. One aspect of plant cell interaction is the communication of positional information from one cell to another.

Internal Organization of the Plant Body

Cells, the fundamental units of life, are associated in various ways with each other, forming coherent masses, or tissues. Moreover, the principal tissues of vascular plants are grouped together into larger units on the basis of their continuity throughout the plant body. These larger units, known as **tissue systems,** are readily recognized, often with the unaided eye. There are three tissue systems, and their presence in root, stem, and leaf reveals both the basic similarity of the plant organs and the continuity of the plant body. The three tissue systems are (1) the **dermal tissue system,** (2) the **vascular tissue system,** and (3) the **ground** (or fundamental) **tissue system.** As mentioned in Chapter 23, the tissue systems are initiated during the development of the embryo, where their precursors are represented by the primary meristems—protoderm, procambium, and ground meristem, respectively (Figure 24–2).

The ground tissue system consists of the three ground tissues—parenchyma, collenchyma, and sclerenchyma. Parenchyma is by far the most common of the ground tissues. The vascular tissue system consists of the two conducting tissues—xylem and phloem. The dermal tissue system is represented by the epidermis, which is the outer protective covering of the primary plant body, and later by the periderm in plant parts that undergo a secondary increase in thickness (page 427).

Within the plant body, the various tissues are distributed in characteristic patterns depending upon the plant part or plant taxon or both. The patterns are essentially alike from one plant part to another in that the vascular tissues are embedded within the ground tissue, with the

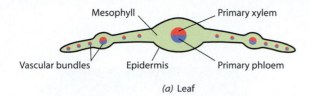

Mesophyll   Primary xylem

Vascular bundles   Epidermis   Primary phloem

(a) Leaf

parenchyma, collenchyma, and sclerenchyma—are simple tissues; xylem, phloem, and epidermis are complex tissues.

Ground Tissues

Parenchyma Tissue Is Involved in Photosynthesis, Storage, and Secretion

In the primary plant body, **parenchyma cells** commonly occur as continuous masses—as **parenchyma tissue**—in the cortex (Figure 24–5) and pith of stems and roots, in leaf mesophyll (see Figure 24–28), and in the flesh of fruits. In addition, parenchyma cells occur as vertical strands of cells in the primary and secondary vascular tissues and as horizontal strands called **rays** in the secondary vascular tissues (see Chapter 27).

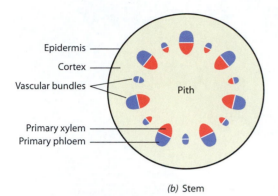

Epidermis

Cortex

Vascular bundles

Pith

Primary xylem
Primary phloem

(b) Stem

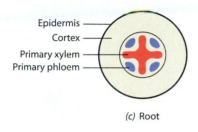

Epidermis

Cortex

Primary xylem
Primary phloem

(c) Root

24–4

*Diagrams showing distribution of vascular and ground tissues in leaves, stems, and roots of eudicotyledons. **(a)** In the leaf, the ground tissue is represented by the mesophyll, which is specialized for photosynthesis. Embedded in the mesophyll are the vascular bundles made up of primary xylem for water transport and primary phloem for sugar transport. **(b)** In the stem, the ground tissue is represented by the pith and cortex, and **(c)** in the root by cortex only. The epidermis forms the outer covering of all three plant parts.*

dermal tissue forming the outer covering. The principal differences in patterns depend largely on the relative distribution of the vascular and ground tissues (Figure 24–4). In the stem of a eudicotyledon, for example, the vascular system may form a system of interconnected strands embedded within the ground tissue. The region formed internal to the strands is called the pith, and the region external to them is called the cortex. In the root of the same plant, the vascular tissues may form a solid cylinder (vascular cylinder, or stele) surrounded by a cortex. In the leaf, the vascular system typically forms a system of vascular bundles (veins), embedded in photosynthetic ground tissue (mesophyll).

Tissues may be defined as groups of cells that are structurally and/or functionally distinct. Tissues composed of only one type of cell are called **simple tissues,** whereas those composed of two or more types of cells are called **complex tissues.** The ground tissues—

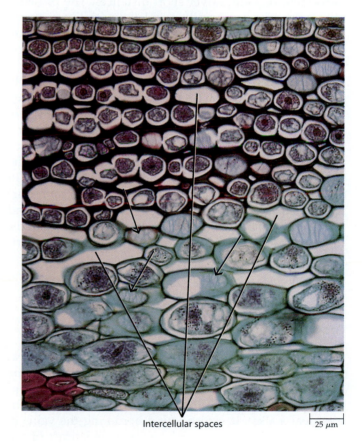

Intercellular spaces   25 μm

24–5

Parenchyma cells (below) and collenchyma cells with unevenly thickened walls (above), seen in a transverse section of the cortex of an elderberry (Sambucus canadensis) stem. In some of the parenchyma cells, a meshwork of lines is visible in the facing walls. The light areas within the meshwork are primary pit-fields (arrows), which are thin areas in the walls. The protoplasts of the collenchyma cells are plasmolyzed and therefore appear withdrawn from the walls. The clear areas between the cells are intercellular spaces.

Characteristically living at maturity, parenchyma cells are capable of cell division, and although their walls are commonly primary, some parenchyma cells also have secondary walls. Because they retain their meristematic ability, parenchyma cells having only primary walls play an important role in regeneration and wound healing. It is these cells that initiate adventitious structures, such as adventitious roots on stem cuttings (page 566). Parenchyma cells are involved in such activities as photosynthesis, storage, and secretion—activities dependent upon living protoplasts. In addition, parenchyma cells may play a role in the movement of water and the transport of food substances in plants.

Transfer Cells Are Parenchyma Cells with Wall Ingrowths

The cell wall ingrowths of **transfer cells** often greatly increase the surface area of the plasma membrane (Figure 24–6), and transfer cells are believed to facilitate the movement of solutes over short distances. The presence of transfer cells is generally correlated with the existence of intensive solute fluxes—in either inward (uptake) or outward (secretion) directions—across the plasma membrane.

Transfer cells are exceedingly common and probably serve a similar function throughout the plant body. They occur in association with the xylem and phloem of small, or minor, veins in cotyledons and in the leaves of many herbaceous eudicotyledons. Transfer cells are also associated with the xylem and phloem of leaf traces at the nodes in both eudicotyledons and monocotyledons. In addition, they are found in various tissues of reproductive structures (placentae, embryo sacs, endosperm) and in various glandular structures (nectaries, salt glands, the glands of carnivorous plants) where intensive short-distance solute transfer takes place.

Collenchyma Tissue Supports Young, Growing Organs

Collenchyma cells, like parenchyma cells, are living at maturity (Figures 24–5, 24–7, and 24–8). **Collenchyma tissue** commonly occurs in discrete strands or as continuous cylinders beneath the epidermis in stems and petioles (leaf stalks). It is also found bordering the veins in eudicot leaves. (The "strings" on the outer surface of celery stalks, or petioles, consist almost entirely of collenchyma.) Collenchyma cells typically are elongated. Their most distinctive feature is their unevenly thickened, nonlignified primary walls, which are soft and pliable and have a glistening appearance in fresh tissue. Because collenchyma cells are living at maturity, they can continue to develop thick, flexible walls while the organ is still elongating, making these cells especially well adapted for the support of young, growing organs.

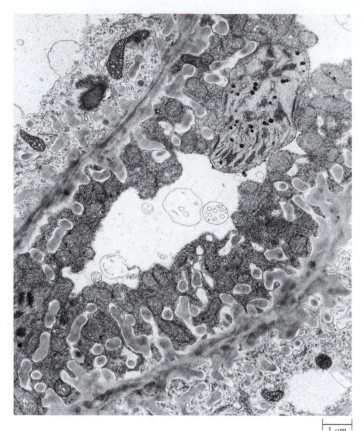

1 µm

24–6

Transverse section of a portion of the phloem from a small vein of a Sonchus *(sow thistle) leaf, showing transfer cells with their numerous wall ingrowths.*

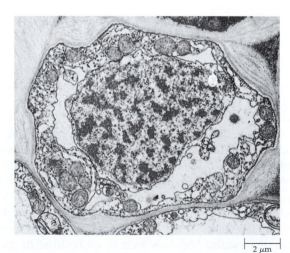

2 µm

24–7

Electron micrograph of a mature collenchyma cell from the filament of a wheat (Triticum aestivum) *stamen, in transverse section. Notice the protoplasmic contents and unevenly thickened wall of this cell.*

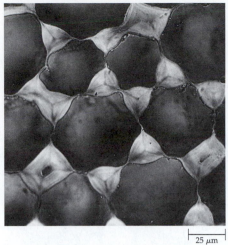

24–8

Transverse section of collenchyma tissue from a petiole in rhubarb (Rheum rhabarbarum). In fresh tissue like this, the unevenly thickened collenchyma cell walls have a glistening appearance.

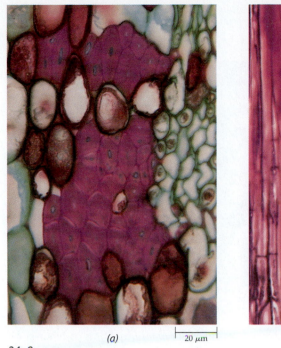

(a) 20 μm (b) 25 μm

24–9

Primary phloem fibers from the stem of a linden, or basswood (Tilia americana), seen here in both (a) cross-sectional and (b) longitudinal views. The secondary walls of these long, thick-walled fibers contain relatively inconspicuous pits. Only a portion of the length of these fibers can be seen in (b).

Sclerenchyma Tissues Strengthen and Support Plant Parts No Longer Elongating

Sclerenchyma cells may form continuous masses—as **sclerenchyma tissue**—or they may occur in small groups or individually among other cells. They may develop in any or all parts of the primary and secondary plant bodies and often lack protoplasts at maturity. The term "sclerenchyma" is derived from the Greek *sklēros*, meaning "hard," and the principal characteristic of sclerenchyma cells is their thick, often lignified secondary walls. Because of these walls, sclerenchyma cells are important strengthening and supporting elements in plant parts that have ceased elongating (see the description of the secondary wall on page 64).

Two types of sclerenchyma cells are recognized: **fibers** and **sclereids.** Fibers are generally long, slender cells that commonly occur in strands or bundles (Figure 24–9). The so-called bast fibers—such as hemp, jute, and flax—are derived from the stems of eudicots. Other economically important fibers, such as Manila hemp, are extracted from the leaves of monocots. Fibers vary in length, ranging from 0.8 to 6 millimeters in jute, 5 to 55 millimeters in hemp, and 9 to 70 millimeters in flax. Sclereids are variable in shape and are often branched (Figure 24–10), but compared with most fibers, sclereids are relatively short cells. Sclereids may occur singly or in

24–10

Branched sclereid from a leaf of the water lily (Nymphaea odorata), a magnoliid, as seen in (a) ordinary light and (b) polarized light. Numerous small, angular crystals of calcium oxalate are embedded in the wall of this sclereid.

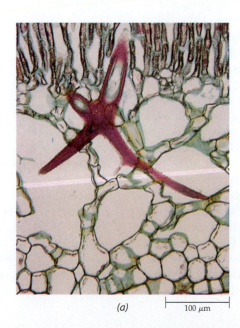

(a) 100 μm (b) 100 μm

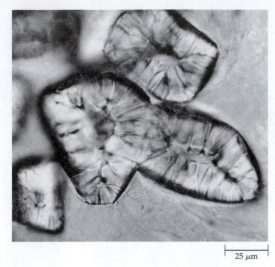

24–11

Sclereids (stone cells) from fresh tissue of the pear (Pyrus communis) fruit. The secondary walls contain conspicuous simple pits with many branches, known as ramiform pits. During formation of the clusters of stone cells in the flesh of the pear fruit, cell divisions occur concentrically around some of the sclereids formed earlier. The newly formed cells differentiate as stone cells, adding to the cluster.

aggregates throughout the ground tissue. They make up the seed coats of many seeds, the shells of nuts, and the stone (endocarp) of stone fruits, and they give pears their characteristic gritty texture (Figure 24–11).

Vascular Tissues

Xylem Is the Principal Water-Conducting Tissue in Vascular Plants

Xylem, the principal water-conducting tissue, is also involved in the conduction of minerals, in support, and in food storage. Together with the phloem, the xylem forms a continuous system of vascular tissue extending throughout the plant body (Figure 24–12). In the primary plant body, the xylem is derived from the procambium. During secondary growth, the xylem is derived from the vascular cambium (see Chapter 27).

The principal conducting cells of the xylem are the **tracheary elements,** of which there are two types, the **tracheids** and the **vessel elements,** or vessel members. Both are elongated cells that have secondary walls and lack protoplasts at maturity, and both types may have *pits* in their walls (Figure 24–13a through d; see page 66 for an introduction to pit structure). Unlike tracheids, vessel elements contain *perforations,* which are areas lacking both primary and secondary walls. Perforations are literally holes in the cell wall. The part of the wall bearing the perforation or perforations is called the **perforation plate** (Figure 24–14). Perforations generally occur

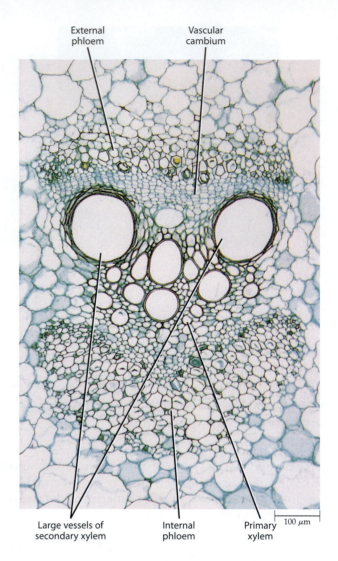

External phloem   Vascular cambium

Large vessels of secondary xylem   Internal phloem   Primary xylem   100 μm

24–12

Transverse section of a vascular bundle from the stem of a squash (Cucurbita maxima), a favorite species for the study of phloem. Phloem occurs both exterior to and interior to the xylem in squash vascular bundles. Typically, a vascular cambium develops between the external phloem and the xylem but not between the internal phloem and the xylem. Here the vascular cambium has produced some secondary phloem (two to three layers of cells) to the outside and some secondary xylem to the inside. The secondary xylem is largely represented by the two large vessels. All of the internal phloem is primary.

on the end walls, with the vessel elements joined end-to-end, forming long, continuous columns, or tubes, called **vessels** (Figure 24–15).

The tracheid is the only type of water-conducting cell found in most seedless vascular plants and gymnosperms. The xylem of the great majority of angiosperms, however, contains both vessel elements and tracheids. The tracheid, lacking perforations, is a less specialized type of cell than the vessel element, which is

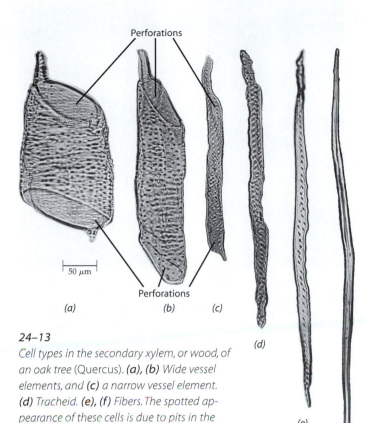

Perforations

Perforations

(a)   (b)   (c)

(d)

(e)

(f)

24–13

*Cell types in the secondary xylem, or wood, of an oak tree (Quercus). **(a)**, **(b)** Wide vessel elements, and **(c)** a narrow vessel element. **(d)** Tracheid. **(e)**, **(f)** Fibers. The spotted appearance of these cells is due to pits in the walls; pits are not discernible in **(f)**. Pits are areas lacking secondary walls. Only the vessel elements have perforations, which are areas lacking both primary and secondary cell walls (see Figure 24–14).*

the principal water-conducting cell in angiosperms. Vessel elements appear to have evolved independently in several groups of vascular plants.

Vessel elements are generally thought to be more efficient conductors of water than the tracheids, because water can flow relatively unimpeded from vessel element to vessel element through the perforations. Vessel elements, however, with their open system, are less safe for the plant than tracheids. Water flowing from tracheid to tracheid must pass through the pit membranes—the thin, modified primary cell walls—of the pit-pairs (page 66). Although the porous pit membranes offer relatively little resistance to the flow of water across them, they block even the smallest of air bubbles (see Chapter 31). Thus, air bubbles that form in a tracheid—for example, during the alternate freezing and thawing of xylem water in the spring—are restricted to that tracheid, and any resulting obstruction to water flow is also limited. On the other hand, air bubbles formed in a vessel element can potentially obstruct the flow of water for the entire length of the vessel. Wide vessels are more efficient at water conduction than narrow vessels, but wide vessels also tend to be longer and may not be as casualty-free as narrow vessels.

The tracheary elements of the primary xylem have a variety of secondary wall thickenings. For example, during the period of elongation or expansion of the roots, stems, and leaves, the secondary walls of many of the first-formed tracheary elements of the early-formed primary xylem (protoxylem; *proto-*, meaning "first") are deposited in the form of rings or spirals (Figure 24–16).

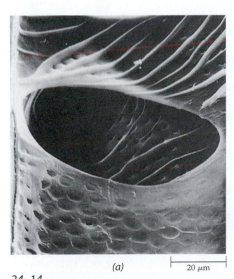

(a)   20 μm

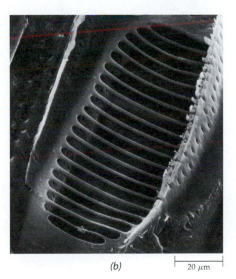

(b)   20 μm

100 μm

24–14

*Perforation plates. Scanning electron micrographs of the perforated end walls of vessel elements from secondary xylem. **(a)** A simple perforation plate, with its single large opening, seen here between two vessel elements in basswood, or linden (Tilia americana). **(b)** The ladderlike bars of a scalariform perforation plate between vessel elements of red alder (Alnus rubra). Pits can be seen in the wall below the perforation plate in **(a)** and in portions of the wall in **(b)**.*

24–15

Scanning electron micrograph showing parts of three vessel elements of a vessel in secondary xylem of red oak (Quercus rubra). Notice the rims (arrows) of the end walls between the vessel elements, which are arranged end-on-end.

These thickenings, which may be annular (ringlike) or helical (spiral), make it possible for such tracheary elements to be stretched or extended, although the cells are frequently destroyed during the overall elongation of the organ. In the primary xylem, the nature of the wall thickening is greatly influenced by the amount of elongation. If little elongation occurs, pitted elements rather than extensible elements appear. On the other hand, if much elongation takes place, many elements with annular and helical thickenings will develop. In the late-formed primary xylem (metaxylem; *meta-*, meaning "after") and secondary xylem, the secondary cell walls of the tracheids and vessel elements cover the entire primary walls, except at the pit membranes and at the perforations of the vessel elements (Figure 24–13a through d). Consequently, these walls are rigid and cannot be stretched.

Figure 24–17 illustrates some stages of differentiation of a vessel element with helical thickenings. Tracheary element differentiation is an example of **apoptosis** (from the Greek *apo*, meaning "away from," and *ptosis*, meaning "falling"), or programmed cell death. Apoptosis is the outcome of genetically programmed processes that result in cell death. In the case of the tracheary element, it results in total elimination of the protoplast. The cell walls are retained, except for the perforations of the ves-

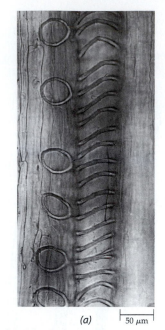

(a) 50 μm   (b) 50 μm

24–16

*Parts of tracheary elements from the first-formed primary xylem (protoxylem) of the castor bean (*Ricinus communis*). **(a)** Annular (the ringlike shapes at left) and helical wall thickenings in partly extended elements. **(b)** Double helical thickenings in elements that have been extended. The element on the left has been greatly extended, and the coils of the helices have been pulled far apart.*

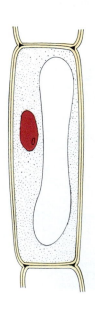
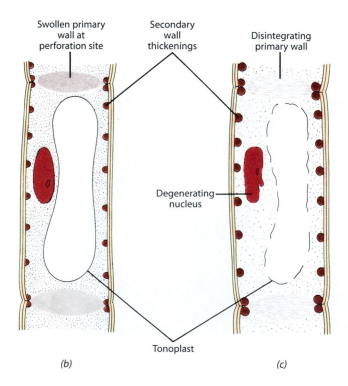

Swollen primary wall at perforation site — Secondary wall thickenings — Disintegrating primary wall — Perforation

Degenerating nucleus

Tonoplast

(a)   (b)   (c)   (d)

24–17

*Diagram illustrating development of a vessel element. **(a)** Young, highly vacuolated vessel element without a secondary wall. **(b)** The cell has expanded laterally, secondary wall deposition—in the form of a helix when seen in three dimensions—has begun, and the primary wall at the perforation site has increased in thickness. **(c)** Secondary wall deposition has been completed, and the cell is at the stage of lysis. The nucleus is degenerating, the tonoplast is ruptured, and the wall at the perforation site has partly disintegrated. **(d)** The cell is now mature; it lacks a protoplast and is open at both ends.*

TABLE 24–1	Cell Types of the Xylem and Phloem	
Cell Types	**Principal Function**
Xylem	
Tracheary elements	Conduction of water and
Tracheids	minerals
Vessel elements	
Fibers	Support; sometimes storage
Parenchyma	Storage
Phloem	
Sieve elements	Long-distance conduction of
Sieve cells	food materials
(with albuminous cells)	
Sieve-tube elements	
(with companion cells)	
Sclerenchyma	Support; sometimes storage
Fibers	
Sclereids	
Parenchyma	Storage

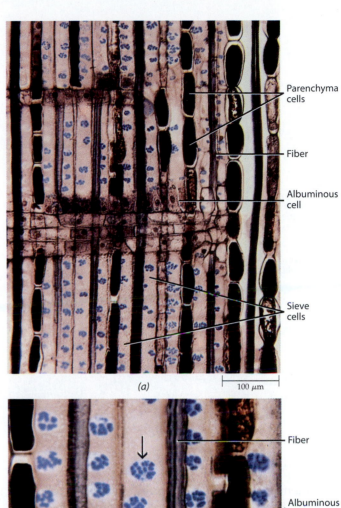

(a) 100 μm

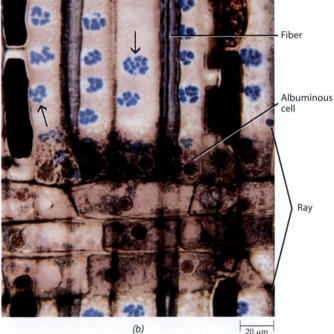

(b) 20 μm

sel elements, providing uninterrupted conduits for the transport of water and dissolved substances (see Chapter 31).

In addition to tracheids and vessel elements, the xylem tissue contains parenchyma cells that store various substances. Xylem parenchyma cells commonly occur in vertical strands. In the secondary xylem, they are also found in rays. Fibers also occur in xylem (Figure 24–13e, f). Many of these xylem fibers are living at maturity and serve a dual function of storage and support. Sclereids are sometimes present in xylem. The cell types of the xylem and their principal function are listed in Table 24–1.

Phloem Is the Principal Food-Conducting Tissue in Vascular Plants

The **phloem,** through which food is conducted to the various parts of the plant, may be primary or secondary in origin (Figure 24–12). As with primary xylem, the first-formed primary phloem (protophloem) is frequently stretched and destroyed during elongation of the organ.

The principal conducting cells of the phloem are the **sieve elements,** of which there are two types, the **sieve cells** (Figure 24–18) and the **sieve-tube elements,** or sieve-tube members (Figures 24–19 through 24–21). The term "sieve" refers to the clusters of pores, known as the **sieve areas,** through which the protoplasts of adjacent

24–18

Sieve cells. (a) Longitudinal (radial) view of secondary phloem of yew (Taxus canadensis), a conifer, showing vertically oriented sieve cells, strands of parenchyma cells, and fibers. Parts of two horizontally oriented rays can be seen traversing the vertical cells. Specialized parenchyma cells known as albuminous cells (see page 582) are characteristically associated with the sieve cells of gymnosperms. (b) Detail of portion of the secondary phloem of yew, showing sieve areas (arrows), with callose (stained blue) on the walls of the sieve cells, and albuminous cells, which here constitute the top row of cells in the ray. Note the lack of sieve plates on the sieve cells.

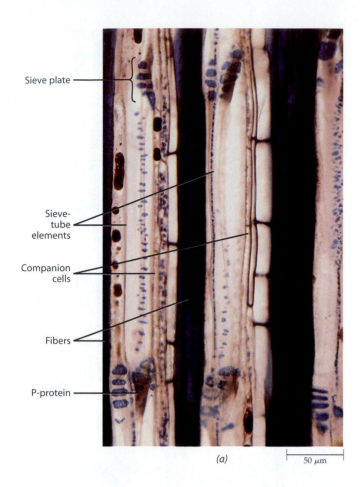

(a)   50 μm

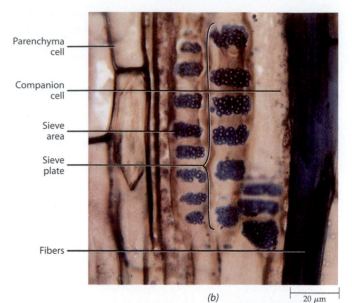

(b)   20 μm

24–19

Sieve-tube elements. ***(a)*** *Longitudinal (radial) view of secondary phloem of basswood (Tilia americana), showing sieve-tube elements, with sieve plates, and conspicuous groups of thick-walled fibers. Specialized parenchyma cells known as companion cells (see page 582) are characteristically associated with sieve-tube elements. The P-protein, a characteristic component of magnoliid and eudicot sieve-tube elements, has accumulated at the sieve plates in these elements.* ***(b)*** *Compound sieve plates of basswood sieve-tube elements. The sieve plates of basswood, seen here in detail, are known as compound sieve plates because they consist of two or more sieve areas. Each sieve area is composed of pores bordered by cylinders of callose, which are stained blue in this section.*

sieve elements are interconnected. In sieve cells, the pores are narrow and the sieve areas are rather uniform in structure on all walls. Most of the sieve areas are concentrated on the overlapping ends of the long, slender sieve cells (Figure 24–18a). In sieve-tube elements, however, the sieve areas on some walls have larger pores than those on other walls of the same cell. The part of the wall bearing the sieve areas with larger pores is called a **sieve plate** (Figures 24–19 through 24–21). Although sieve plates may occur on any wall, they generally are located on the end walls. Sieve-tube elements occur end-on-end in longitudinal series called **sieve tubes.** Thus, one of the principal distinctions between the two types of sieve elements is the presence of sieve plates in sieve-tube elements and their absence in sieve cells.

The sieve cell is a less specialized type of cell than the sieve-tube element. Sieve cells are the only type of food-conducting cell in gymnosperms, whereas only sieve-tube elements occur in angiosperms. The sieve elements of seedless vascular plants are variable in structure and are referred to simply as "sieve elements."

The walls of sieve elements are generally described as primary. In cut sections of phloem tissue, the pores of the sieve areas and sieve plates are generally blocked by or lined with a wall substance called **callose,** which is a polysaccharide composed of spirally wound chains of

glucose residues (Figures 24–18 and 24–19). Most, if not all, of the callose found in the pores of conducting sieve elements is deposited there in response to injury during preparation of the tissue for microscopy. This callose, and any callose resulting from wounding by natural occurrence, is referred to as "wound callose." Callose typically is also deposited at the sieve areas and sieve plates of senescing sieve elements; it is referred to as "definitive callose."

Unlike tracheary elements, sieve elements have living protoplasts at maturity (Figure 24–20). The protoplasts of mature sieve elements are unique, however, among the living cells of the plant. As the sieve element differentiates it undergoes profound changes, the major ones being the breakdown of the nucleus and of the tono-

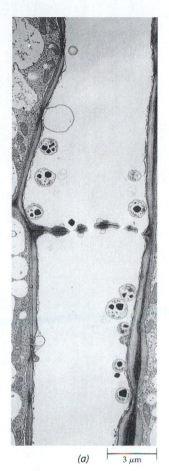

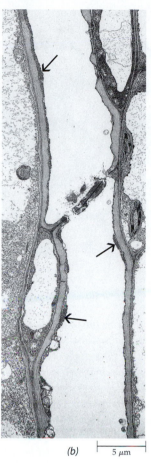

24–20
Electron micrographs of parts of mature sieve-tube elements from the phloem in the stems of maize and squash. (a) Longitudinal view of parts of two mature sieve-tube elements and a sieve plate of maize (Zea mays). The pores of the sieve plate are open. The numerous round organelles with dense inclusions composed of protein are plastids. The sieve-tube elements of maize, like those of many other monocots, lack P-protein. (b) Similar, longitudinal view of parts of two sieve-tube elements of squash (Cucurbita maxima). The sieve-tube elements of squash, like those of other eudicots, contain P-protein. In these sieve-tube elements, the P-protein is distributed along the wall (arrows) and the sieve-plate pores are open. At the lower left and upper right are companion cells. (c) Face view of part of a sieve plate between two mature sieve-tube elements of squash. As in (b), the sieve-plate pores are open, lined in part with P-protein (arrows).

(a) | 3 μm | *(b)* | 5 μm |

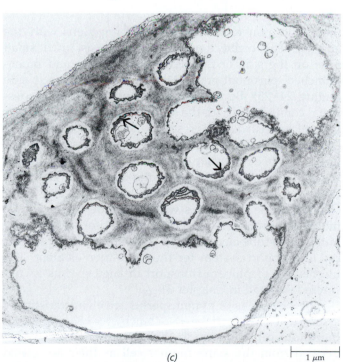

(c) | 1 μm |

plast, or vacuolar membrane. Sieve element differentiation also results in loss of the ribosomes, the Golgi complex, and the cytoskeleton. At maturity, all of the remaining components of the sieve-element protoplast are distributed along the wall. The components are the plasma membrane, a network of smooth endoplasmic reticulum—which is quite abundant in sieve cells, especially at the sieve areas—and some plastids and mitochondria. Thus, unlike the tracheary element protoplast, which undergoes a total breakdown during differentiation, that of the sieve element undergoes a selective breakdown. As we shall see, for the sieve element to perform its role as a food-conducting conduit, it must remain alive while at the same time providing an obstruction-free pathway for the movement of water and dissolved substances (see Chapter 31). A large number of viruses also find the sieve tube a suitable conduit for rapid movement throughout the plant (see Chapter 14).

The protoplasts of the sieve-tube elements of magnoliids and eudicotyledons (and some monocotyledons) are characterized by the presence of a proteinaceous substance once called "slime" and now known as **P-protein.** (The "P" in P-protein stands for phloem.) P-protein has its origin in the young sieve-tube element in the form of

Immature sieve-tube elements

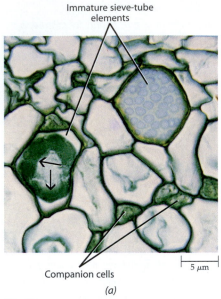

Companion cells

(a)

5 µm

Mature sieve-tube elements

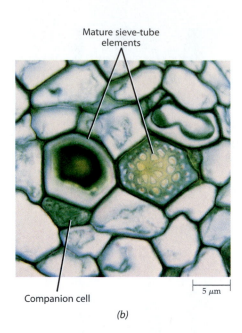

Companion cell

(b)

5 µm

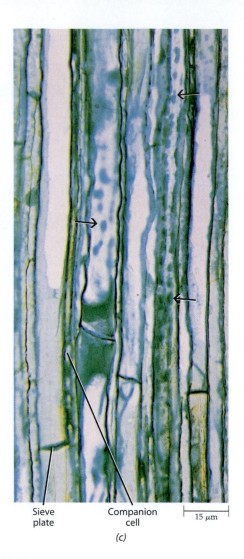

Sieve plate   Companion cell   15 µm

(c)

24–21

Immature and mature sieve-tube elements in the phloem in the stem of squash (Cucurbita maxima), as seen in photomicrographs. (a) Transverse section, showing two immature sieve-tube elements. P-protein bodies (arrows) can be seen in the sieve-tube element on the left, an immature sieve plate in the one to the right, above. The sieve plates of squash are simple sieve plates (one sieve area per plate).

The small, dense cells are companion cells. (b) Transverse section showing two mature sieve-tube elements. A slime plug can be seen in the sieve-tube element on the left; a mature sieve plate can be seen in the one on the right. The small, dense cells are companion cells. (c) Longitudinal section showing mature and immature sieve-tube elements. The arrows point to P-protein bodies in immature cells.

discrete bodies called P-protein bodies (Figure 24–21a, c). During the late stages of differentiation, the P-protein bodies elongate and disperse, and the P-protein, as well as other surviving components of the mature cell, becomes distributed along the walls. In cut sections of phloem tissues, "slime plugs" of P-protein are usually found near the sieve plates (Figure 24–21b). Slime plugs, which are not found in undisturbed cells, result from the surging of the contents of sieve tubes that are severed at the time the tissue is cut. In undisturbed, mature sieve-tube elements, the sieve-plate pores are lined with P-protein but are not plugged by it (Figure 24–20b, c). The function of P-protein has not been determined. However, some botanists believe that, together with wound callose, P-protein serves to seal the sieve-plate pores at the time of wounding, thus preventing the loss of contents from sieve tubes. Figure 24–22 illustrates some stages of differentiation of a sieve-tube element with P-protein.

Sieve-tube elements are characteristically associated with specialized parenchyma cells called **companion cells** (Figures 24–20b and 24–21), which contain all of the components commonly found in living plant cells, including a nucleus. The sieve-tube element and its associated companion cells are closely related developmentally (they are derived from the same mother cell), and

they have numerous cytoplasmic connections with one another. The connections typically consist of a small pore on the side of the sieve-tube element and much-branched plasmodesmata on the companion cell side (Figure 24–23). Because of their numerous plasmodesmatal connections with the sieve-tube elements and their general ultrastructural resemblance to secretory cells (high ribosome population and numerous mitochondria), it is believed that companion cells play a role in the delivery of substances to the sieve-tube elements. In the absence of a nucleus in the mature sieve-tube element, these substances would include informational molecules, proteins, and ATP necessary for maintenance of the sieve-tube element. The companion cell represents the life-support system for the sieve-tube element. The mechanism of phloem transport in angiosperms is considered in detail in Chapter 31.

The sieve cells of gymnosperms are characteristically associated with specialized parenchyma cells called **albuminous cells** (Figure 24–18b). Although generally not derived from the same mother cell as their associated sieve cells, albuminous cells are believed to perform the same roles as companion cells. Like the companion cell, the albuminous cell contains a nucleus, in addition to other cytoplasmic components characteristic of living cells. When the sieve elements die, their associated com-

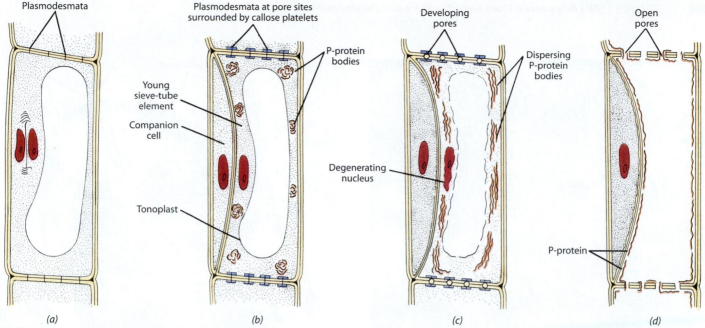

Plasmodesmata

Plasmodesmata at pore sites surrounded by callose platelets

P-protein bodies

Young sieve-tube element

Companion cell

Tonoplast

Developing pores

Dispersing P-protein bodies

Degenerating nucleus

Open pores

P-protein

(a)   (b)   (c)   (d)

24–22

*Differentiation of a sieve-tube element. **(a)** The mother cell of the sieve-tube element undergoing division. **(b)** Division has resulted in formation of a young sieve-tube element and a companion cell. After division, one or more P-protein bodies arise in the cytoplasm, which is separated from the vacuole by a tonoplast. The wall of the young sieve-tube element has thickened, and the sites of the*

*future sieve-plate pores are represented by plasmodesmata. Each plasmodesma is now surrounded by a platelet of callose on either side of the wall. **(c)** The nucleus is degenerating, the tonoplast is breaking down, and the P-protein bodies are dispersing in the cytoplasm lining the wall. At the same time, the plasmodesmata of the developing sieve plates are beginning to widen into pores.*

***(d)** At maturity, the sieve-tube element lacks a nucleus and a vacuole. All of the remaining protoplasmic components, including the P-protein, line the walls, and the sieve-plate pores are open. The callose platelets were removed as the pores widened. Not shown here but also present in the mature sieve-tube element are smooth endoplasmic reticulum, mitochondria, and plastids.*

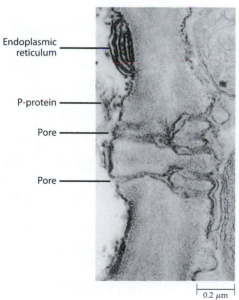

Endoplasmic reticulum

P-protein

Pore

Pore

0.2 μm

24–23

Electron micrograph showing pore–plasmodesmata connections between a companion cell (right) and a sieve-tube element (left) of the cottonwood (Populus deltoides) leaf. The branched plasmodesmata in the companion cell wall connect to a small pore in the wall of the sieve-tube element.

panion cells or albuminous cells also die, which is one more indication of the interdependence between sieve elements and their companion cells or albuminous cells.

Other parenchyma cells occur in the primary and secondary phloem (Figures 24–18 and 24–19). They are largely concerned with the storage of various substances. Fibers (Figures 24–18 and 24–19) and sclereids may also be present. The cell types of the phloem and their principal function are listed, along with those of the xylem, in Table 24–1.

Dermal Tissues

The Epidermis Is the Outermost Cell Layer of the Primary Plant Body

The **epidermis** constitutes the dermal tissue system of leaves, floral parts, fruits, and seeds and of stems and roots until they undergo considerable secondary growth. Epidermal cells are quite variable both functionally and structurally. In addition to the relatively unspecialized cells, which form the bulk of the epidermis, the epidermis may contain **guard cells** (Figures 24–24 through 24–27), many types of appendages, or **trichomes** (Figures 24–26, 24–28, and 24–29), and other kinds of specialized cells.

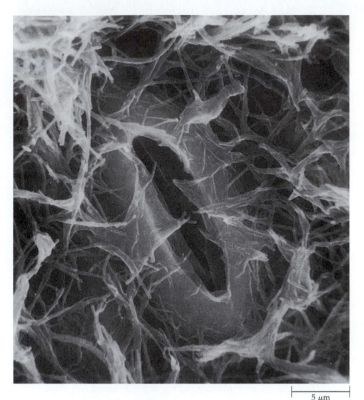

24–24

Surface view of lower epidermis of a
Eucalyptus globulus *leaf, taken with a scanning electron microscope. A single stoma—
flanked by two guard cells—and numerous
filaments of epicuticular wax deposits can be
seen here.*

The bulk of epidermal cells is compactly arranged
and gives considerable mechanical protection to the
plant part. The walls of the epidermal cells of the aerial
parts are covered with a cuticle, which minimizes water
loss. The cuticle consists mainly of cutin and wax (page
25). In many plants the wax is exuded over the surface
of the cuticle, either in smooth sheets or as rods or filaments extending upward from the surface, the so-called
epicuticular wax (Figure 24–24; see also Figure 2–12). It
is this wax that is responsible for the whitish or bluish
"bloom" on the surface of some leaves and fruits.

Interspersed among the flat, tightly packed epidermal cells, which typically lack chloroplasts, are the
chloroplast-containing guard cells (Figures 24–24
through 24–27), which regulate the small pores, or **stomata** (singular: stoma), in the aerial parts of the plant
and hence control the movement of gases, including
water vapor, into and out of those parts. (The term
"stoma" commonly is applied to the pore and the two
guard cells. The mechanism of stomatal opening and
closing is discussed in Chapter 31.) Although stomata
occur on all aerial parts, they are most abundant on
leaves. Guard cells are often associated with epidermal

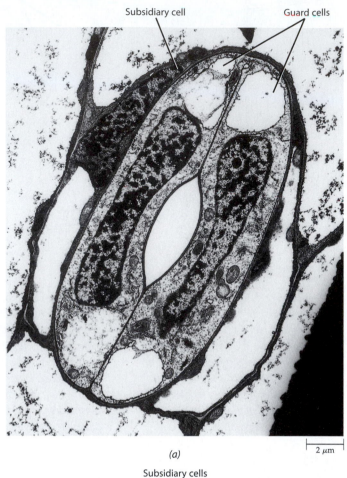

Subsidiary cell   Guard cells

(a)

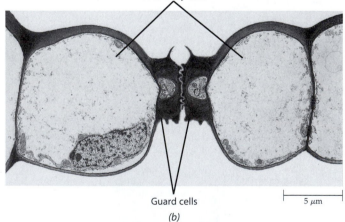

Subsidiary cells

Guard cells

(b)

24–25

*Electron micrographs of maize (Zea mays)
stomata.* **(a)** *Section taken parallel to the
surface of the leaf showing open pore between immature guard cells, whose walls
have not yet thickened, and two associated
subsidiary cells.* **(b)** *Transverse section
through closed stoma. Each thick-walled
guard cell is attached to a subsidiary cell. The
interior of the leaf is below.*

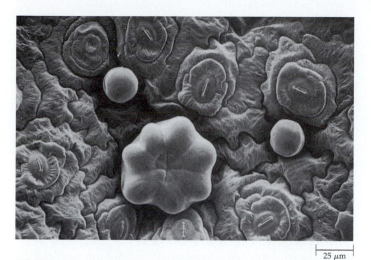

24–26

Scanning electron micrograph of surface view of the lower epidermis of a Swedish ivy (Plectranthus) leaf. Several stomata—pairs of guard cells—each flanked by two curved subsidiary cells can be seen. Also visible are two kinds of trichomes.

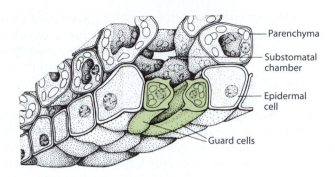

24–27

Diagram of a mature stoma, showing its relationship to the epidermis and underlying cells. The walls next to the pores of stomata are generally thicker than those adjacent to other epidermal cells. While the stomata are being formed, they may be raised above or recessed below the surface of the epidermis. Often there is a large air space, or substomatal chamber, just behind the stoma. Unlike other epidermal cells, guard cells contain chloroplasts.

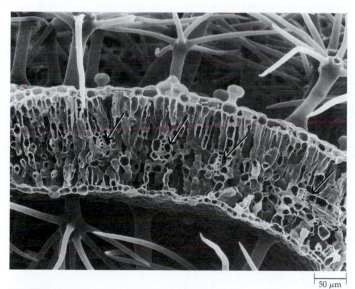

24–28

Trichomes. Transverse section of a common mullein (Verbascum thapsus) leaf, taken with a scanning electron microscope. The leaves and stems of the common mullein appear densely woolly because of the presence of great numbers of highly branched trichomes such as those seen here on both the upper and lower surfaces of the leaf. Short, unbranched glandular trichomes are also present. The ground tissue of the leaf is represented by the mesophyll, which is permeated by numerous vascular bundles, or veins (arrows).

cells that differ in shape from the ordinary epidermal cells. Such cells are termed *subsidiary cells* (Figures 24–25 and 24–26).

Trichomes have a variety of functions. Root hairs facilitate the absorption of water and minerals from the soil. Studies of plants from arid locations indicate that an increase in leaf pubescence (hairiness) results in increased reflectance of solar radiation, lower leaf temperatures, and lower rates of water loss. Many "air plants," such as epiphytic bromeliads, utilize leaf trichomes for the absorption of water and minerals. In contrast, in the saltbush (*Atriplex*), which grows in soil containing high levels of salt, the trichomes secrete salts from the leaf tissue, preventing an accumulation of these toxic substances in the plant. Trichomes may also provide a defense against insects. For example, in many species a positive correlation exists between hairiness and resistance to attack by insects. The hooked hairs of some species impale insects and their larvae, and the trichomes of carnivorous plants play an important role in trapping insect prey (see Chapter 30). Glandular (secretory) hairs may provide a chemical defense.

The development of trichomes on the epidermis of *Arabidopsis* has provided an excellent object for study of the control of cell fate and cell differentiation in plants (Figure 24–29). Some of the genes involved in trichome development on the shoot of *Arabidopsis* have been identified, and other development-altering mutants are available for analysis. Two genes, *GL1* (*GLABROUS1*) and

TTG (TRANSPARENT TESTA GLABRA), have been identified as essential for the initiation of trichome development. The *GL1* gene apparently acts locally rather than over long distances, indicating that an increase in the expression of *GL1* is an early event in commitment of a protodermal cell to trichome development. By contrast, evidence indicates that *TTG* may play a role in inhibiting neighboring protodermal cells from becoming trichomes. The *TTG* gene has also been implicated in cell fate in the root epidermis (see Chapter 25); *ttg* mutants lack trichomes on aerial plant parts but have extra root hairs. The mutant gene is indicated by lower case.

Periderm Is Secondary Protective Tissue

A **periderm** commonly replaces the epidermis in stems and roots having secondary growth. Although the cells of the periderm are generally arranged compactly, portions—the lenticels—are loosely arranged and provide

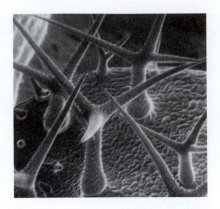

24–29

Scanning electron micrograph of branched trichomes on the leaf of Arabidopsis thaliana. *Genetic analysis has identified two genes that are essential for the initiation of trichome development in* Arabidopsis.

Summary TABLE   Tissues and Cell Types

TISSUES		CELL TYPE	CHARACTERISTICS	LOCATION	FUNCTION
Dermal tissues	Epidermis		Unspecialized cells; guard cells and cells forming trichomes; sclerenchyma cells	Outermost layer of cells of the primary plant body	Mechanical protection; minimizes water loss (cuticle); aeration of internal tissue via stomata
	Periderm		Cork cells; cork cambium cells; parenchyma cells of the phelloderm; sclereids	Initial periderm beneath epidermis; subsequently formed periderms deeper in bark	Replaces epidermis as protective tissue in roots and stems; aeration of internal tissue via lenticels
Ground tissues	Parenchyma tissue	Parenchyma	Shape: commonly polyhedral (many-sided); variable Cell wall: primary, or primary and secondary; may be lignified, suberized, or cutinized Living at maturity	Throughout the plant, as parenchyma tissue in cortex; as pith and pith rays; in xylem and phloem	Such metabolic processes as respiration, digestion, and photosynthesis; storage and conduction; wound healing and regeneration
	Collenchyma tissue	Collenchyma	Shape: elongated Cell wall: unevenly thickened, primary only—nonlignified Living at maturity	On the periphery (beneath the epidermis) in young elongating stems; often as a cylinder of tissue or only in patches; in ribs along veins in some leaves	Support in primary plant body
	Sclerenchyma tissue	Fiber	Shape: generally very long Cell wall: primary and thick secondary—often lignified Often (not always) dead at maturity	Sometimes in cortex of stems, most often associated with xylem and phloem; in leaves of monocotyledons	Support; storage
		Sclereid	Shape: variable; generally shorter than fibers Cell wall: primary and thick secondary—generally lignified May be living or dead at maturity	Throughout the plant	Mechanical; protective

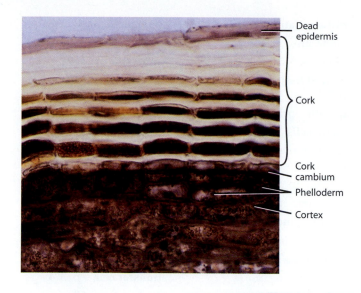

Dead epidermis

Cork

Cork cambium

Phelloderm

Cortex

24–30

Transverse section of the periderm from the stem of apple (Malus sylvestris). The periderm shown here consists largely of cork cells, which are laid down to the outside (above) in radial rows by the cells of the cork cambium. One or two layers of phelloderm cells lie below the cork cambium.

for aeration of the internal tissues of roots and stems. The periderm consists largely of protective **cork,** or *phellem,* which is nonliving and has walls that are heavily suberized at maturity; the periderm also includes **cork cambium,** or *phellogen,* and **phelloderm,** a living parenchyma tissue (Figure 24–30). The cork cambium forms cork tissue on its outer surface and phelloderm on its inner surface. The origin of the cork cambium is variable, depending on the species and plant part. The periderm is considered in detail in Chapter 27.

TISSUES		CELL TYPE	CHARACTERISTICS	LOCATION	FUNCTION
Vascular tissues	Xylem	Tracheid	Shape: elongated and tapering Cell wall: primary and secondary; lignified; contains pits but not perforations Dead at maturity	Xylem	Chief water-conducting element in gymnosperms and seedless vascular plants; also found in angiosperms
		Vessel element	Shape: elongated, generally not as long as tracheids; several vessel elements end-on-end constitute a vessel Cell wall: primary and secondary; lignified; contains pits and perforations Dead at maturity	Xylem	Chief water-conducting element in angiosperms
	Phloem	Sieve cell	Shape: elongated and tapering Cell wall: primary in most species; with sieve areas; callose often associated with wall and pores Living at maturity; either lacks or contains remnants of a nucleus at maturity; lacks distinction between vacuole and cytoplasm; contains large amounts of tubular endoplasmic reticulum; lacks P-protein	Phloem	Food-conducting element in gymnosperms
		Albuminous cell	Shape: generally elongated Cell wall: primary Living at maturity; associated with sieve cell, but generally not derived from same mother cell as sieve cell; has numerous plasmodesmatal connections with sieve cell	Phloem	Believed to play a role in the delivery of substances to the sieve cell, including informational molecules and ATP
		Sieve-tube element	Shape: elongated Cell wall: primary, with sieve areas; sieve areas on end wall with much larger pores than those on side walls—this wall part is termed a sieve plate; callose often associated with walls and pores Living at maturity; either lacks a nucleus at maturity or contains only remnants of nucleus; in magnoliids, eudicots, and some monocots, contains a proteinaceous substance known as P-protein; several sieve-tube elements in a vertical series constitute a sieve tube	Phloem	Food-conducting element in angiosperms
		Companion cell	Shape: variable, generally elongated Cell wall: primary Living at maturity; closely associated with sieve-tube elements; derived from same mother cell as sieve-tube element; has numerous plasmodesmatal connections with sieve-tube element	Phloem	Believed to play a role in the delivery of substances to the sieve-tube element, including informational molecules and ATP

Summary

Primary Growth Results from the Activity of Apical Meristems

After embryogenesis, most plant development occurs through the activity of meristems, which consist of initials and their immediate derivatives. The apical meristems are involved primarily with extension of roots and stems. Also called primary growth, this growth results in the formation of primary tissues, which constitute the primary body of the plant.

Development Involves Three Overlapping Processes: Growth, Morphogenesis, and Differentiation

Growth, an irreversible increase in size, is accomplished primarily by cell enlargement. Morphogenesis is the acquisition of a particular shape or form, and differentiation is the process by which cells with identical genetic constitutions become different from one another through differential gene expression.

Vascular Plants Are Composed of Three Tissue Systems

The tissue systems—dermal, vascular, and ground, which are present in root, stem, and leaf—reveal the basic similarity of the plant organs and the continuity of the plant body. A summary of plant tissues and their cell types is found in the Summary Table.

Selected Key Terms

albuminous cell, p. 582
apical meristem   p. 571
apoptosis   p. 578
collenchyma   p. 574
companion cell   p. 582

complex tissue   p. 573
derivatives   p. 571
dermal tissue system p. 572
development   p. 571

differentiation   p. 572
epidermis   p. 583
fiber   p. 575
ground tissue system p. 572
growth   p. 571
guard cells   p. 583
indeterminate growth p. 571
initials   p. 571
meristem   p. 571
morphogenesis   p. 572
parenchyma   p. 573
perforation plate   p. 576
periderm   p. 586
phloem   p. 579
P-protein   p. 581
primary growth   p. 571
primary plant body p. 571
sclereid   p. 575

sclerenchyma   p. 575
sieve area   p. 579
sieve cell   p. 579
sieve element   p. 579
sieve plate   p. 580
sieve tube   p. 580
sieve-tube element p. 579
simple tissue   p. 573
stomata   p. 584
tissue system   p. 572
tracheary element p. 576
tracheid   p. 576
transfer cell,   p. 574
trichome   p. 583
vascular tissue system p. 572
vessel   p. 576
vessel element   p. 576
xylem   p. 576

Questions

1. Distinguish between the following: collenchyma cell/sclerenchyma cell; tracheid/vessel element; perforation plate/pit; sieve cell/sieve-tube element; callose/P-protein.

2. What is growth?

3. The growth of apical meristems is described as indeterminate. Explain.

4. How do simple tissues differ from complex tissues? Give examples of each.

5. How do sclereids differ from fibers?

6. What is the relationship between a sieve-tube element and its companion cell?

25

The Root: Structure and Development

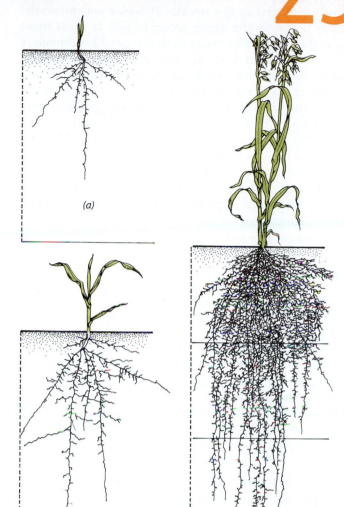

(a)

(b)   *(c)*

25–1

Diagrams of oat (Avena sativa) *plants, showing relative sizes of the root and shoot systems (a) 31, (b) 45, and (c) 80 days after planting. The oat plant, a monocot, has a fibrous root system. The roots are involved primarily with anchorage and absorption. Each of the vertical units depicted here represents 1 foot (approximately 30.5 centimeters).*

OVERVIEW

To most people, the beauty of a flowering plant lies not in its roots—but perhaps it should. Because most roots are underground and thus not readily visible, we tend either to ignore them or else take them for granted. However, as you will learn in this chapter, root systems not only play a variety of roles crucial to the survival of the plant but are usually more extensive than the aboveground portions of the plant.

This chapter begins with a discussion of the various types of root systems. The emphasis then switches to the patterns of growth observed in roots, including the structure and function of the root apical meristem, the role of the rootcap, and the presence of the three principal growth regions—all of which contribute to the formation of the mature root tissues.

These mature tissues are then examined in detail, and you will be able to draw on your knowledge of basic tissue systems from the previous chapter. As you will see, although the same tissue systems are present throughout the plant, those of roots have some structural features that are characteristic of roots.

Many roots exhibit secondary growth as well as primary growth. Thus, the latter part of this chapter describes the events associated with the activity of the vascular cambium and cork cambium and the ways in which they alter the primary body of the root. The chapter ends with a discussion of various root adaptations such as fleshy roots for storage and air roots for aeration.

CHECKPOINTS

By the time you finish reading this chapter, you should be able to answer the following questions:

1. What are the two principal types of root systems, and how do they differ from one another in both origin and structure?

2. What changes occur to the rootcap during elongation of the root, and what are some functions of the rootcap?

3. What are the two main types of apical organization, and how do they differ from one another?

4. What tissues are found in a root at the end of primary growth, and how are they arranged?

5. What effect does secondary growth have on the primary body of the root?

6. Why are lateral roots said to be endogenous?

The first structure to emerge from the germinating seed is the embryonic root, enabling the developing seedling to become anchored in the soil and to absorb water. This reflects the two primary functions of roots, namely **anchorage** and **absorption** (Figure 25–1). Two other functions associated with roots are **storage** and **conduction**. Most roots are important storage organs, and some, such as those of the carrot, sugarbeet, and sweet potato, are specifically adapted for the storage of food. Foods manufactured above ground, in photosynthesizing portions of the plant, move through the phloem to the storage tissues of the root. This food may eventually be used by the root itself, but more often the stored food is digested and the products transported back through the phloem to the aboveground parts. In biennial plants, which complete their life cycle over a two-year period, such as the sugarbeet and carrot, large food reserves accumulate in the storage regions of the root during the first year. These food reserves are then used during the second year to produce flowers, fruits, and seeds. Water and minerals, or inorganic ions, absorbed by the roots move through the xylem to the aerial parts of the plant.

Hormones (particularly cytokinins and gibberellins) synthesized in meristematic regions of the roots are transported upward in the xylem to the aerial parts, where they stimulate growth and development (see Chapter 28). Roots also synthesize a wide variety of secondary metabolites, such as nicotine, which ends up in the leaves of the tobacco plant (see Chapter 2).

Root Systems

The first root of the plant originates in the embryo and is usually called the **primary root.** In gymnosperms, magnoliids, and eudicotyledons, the primary root, termed the **taproot,** grows directly downward, giving rise to branch roots, or **lateral roots.** The older lateral roots are found nearer the base of the root (where the root and stem meet), and the younger ones nearer the root tip. This type of root system—that is, one formed from a strongly developed primary root and its branches—is called a **taproot system** (Figure 25–2a).

In monocotyledons, the primary root is usually short-lived. Instead, the main root system of the plant develops from adventitious roots, which arise from the stem. These stem-borne roots and their lateral roots give rise to a **fibrous root system,** in which no one root is more prominent than the others (Figures 25–1 and 25–2b). Taproot systems generally penetrate deeper into the soil than fibrous root systems. The shallowness of fibrous root systems and the tenacity with which they cling to soil particles make such plants especially well suited as ground cover for the prevention of soil erosion.

The extent of a root system—that is, the depth to which it penetrates the soil and the distance it spreads laterally—is dependent upon several factors, including the moisture, temperature, and composition of the soil. The bulk of most so-called "feeder roots" (the fine roots actively engaged in the uptake of water and minerals) usually occurs in the upper meter of soil. In most trees, feeder roots primarily occur in the upper 15 centimeters of soil, the part of the soil normally richest in organic matter. Some trees, such as spruces, beeches, and poplars, rarely produce deep taproots, whereas others, such as oaks and many pines, commonly produce relatively deep taproots, making such trees difficult to transplant. The record for depth of penetration by roots belongs to the desert shrub mesquite (*Prosopis juliflora*). Mesquite roots were found growing at a depth of 53.3 meters (nearly 175 feet) at an open-pit mine near Tucson, Arizona. During the digging of the Suez Canal in Egypt, roots of *Tamarix* and *Acacia* trees were found at a depth of 30 meters. The lateral spread of tree roots is usually greater—frequently four to seven times greater—than the spread of the crown of the tree. The root systems of maize plants (*Zea mays*) often reach a depth of about 1.5 meters, with a lateral spread of about a meter on all sides of the plant. The roots of alfalfa (*Medicago sativa*) may extend to depths of 6 meters or more.

One of the most detailed studies on the extent of the root and shoot systems of any one plant was conducted on a 4-month-old rye plant (*Secale cereale*). The total surface area of the root system, including root hairs, was 639 square meters, or 130 times the surface area of the shoot system. Even more amazing is the fact that the roots occupied only about 6 liters of soil.

The Plant Maintains a Balance between Its Shoot and Root Systems

In a growing plant, a balance is maintained between the total surface area available for the manufacture of food (the photosynthesizing surface) and the surface area available for the absorption of water and minerals. In seedlings, the total water- and mineral-absorbing surface usually far exceeds the photosynthesizing surface. However, the root-to-shoot ratio decreases gradually as the plant ages.

If damage to the root system seriously reduces its absorbing surfaces, shoot growth is reduced by lack of water, minerals, and root-produced hormones. Reduction in the size of the shoot system limits root growth by decreasing the availability of carbohydrates and shoot-produced hormones to the roots. In nature, injury and death of feeder roots are frequent and caused by a variety of agents, ranging from extremes of temperature and frost heaving to the activities of such members of the soil microfauna as nematodes and springtails that nibble away at these succulent roots. As injured roots die, new roots are rapidly formed; hence, the fluctuation

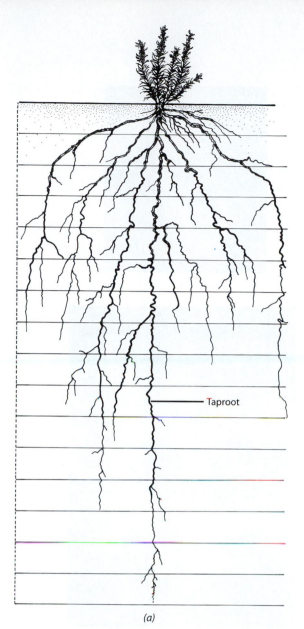

Taproot

(a)

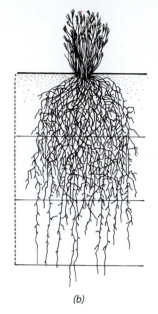

(b)

25–2
Two types of root systems, as represented by two prairie plants. (a) Taproot system of the blazing star (Liatris punctata), *a eudicot. (b) Fibrous root system of wire grass* (Aristida purpurea), *a monocot. Each of the vertical units depicted here represents 1 foot (approximately 30.5 centimeters). Taproot systems generally penetrate deeper into the soil than fibrous root systems.*

in the population and concentration of roots in the soil is as dynamic as that of the leaves and twigs in the air above. Even when plants are carefully transplanted, the balance between shoot and root is invariably disturbed. Most of the fine feeder roots are left behind when the plant is removed from the soil. Cutting back the shoot helps to reestablish a balance between the root system and the shoot system, as does repotting a plant that is root bound into a larger container.

Origin and Growth of Primary Tissues

The growth of many roots is apparently a continuous process that stops only under such adverse conditions as drought and low temperatures. During their growth through the soil, roots follow the path of least resistance and frequently follow spaces left by earlier roots that have died and rotted.

The Tip of the Root Is Covered by a Rootcap, Which Produces Mucigel

The **rootcap** is a thimblelike mass of living parenchyma cells that protects the apical meristem behind it and aids the root in its penetration of the soil (Figures 25–3 through 25–6). As the root grows longer and the rootcap is pushed forward, the cells on the periphery of the rootcap are sloughed. These sloughed cells and the growing root tip are covered by a slimy sheath, or **mucigel,** which lubricates the root during its passage through the soil (Figure 25–4). As rootcap cells are sloughed, new ones are added by the apical meristem. The longevity (from time of origin to sloughing) of rootcap cells ranges from four to nine days, depending upon the length of the cap and the species.

The slimy substance forming the mucigel is a highly hydrated polysaccharide, probably a pectin that is secreted by the outer rootcap cells. It accumulates in Golgi vesicles, which fuse with the plasma membrane and then release the slime into the cell wall. Eventually the slime passes to the outside.

In addition to protecting the apical meristem and aiding the root in its penetration of the soil, the rootcap plays an important role in controlling the response of the root to gravity (gravitropism). The site of perception of gravity in the rootcap is the central column of cells, the **columella,** in which starch-containing amyloplasts are found. Many botanists believe that the amyloplasts act as gravity sensors (see Chapter 29).

Apical Organization in Roots May Be Either Open or Closed

Apart from the rootcap, the most striking structural feature of the root apex is the arrangement of longitudinal files, or lineages, of cells that emanate from the apical

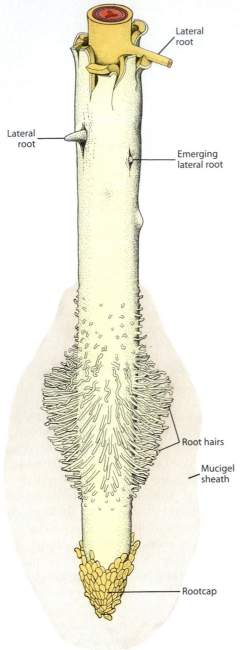

Lateral
root

Lateral
root

Emerging
lateral root

Root hairs

Mucigel
sheath

Rootcap

25–3

A portion of a eudicot root, showing the spatial relationship between the rootcap and region of root hairs, and (near the top) the sites of emergence of lateral roots, which arise from deep within the parent root. New root hairs arise just behind the region of elongation at about the same rate as the older hairs die off. The root tip is covered by a mucigel sheath, which lubricates the root during its passage through the soil.

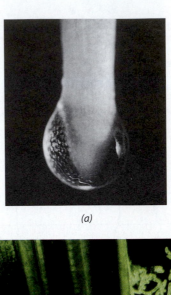

(a)

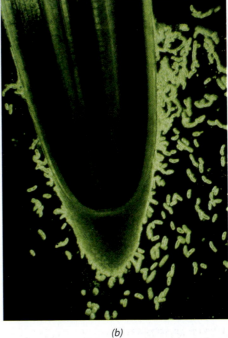

(b)

25–4

(a) Mucigel on the rootcap of a maize (Zea mays) root, containing sloughed rootcap cells. (b) A dark-field image of a living root showing a "cloud" of detached rootcap cells suspended in the mucigel sheath (not apparent).

meristem. The apical meristem is composed of relatively small, many-sided cells—the initials and their immediate derivatives (page 571)—with dense cytoplasm and large nuclei (Figures 25–5 and 25–6). The organization and number of initiating layers in the apical meristems of roots are both quite variable.

Two main types of apical organization are found in the roots of seed plants. In one, the rootcap, the vascular cylinder, and the cortex are traceable to independent layers of cells in the apical meristem, with the epidermis having a common origin with either the rootcap or the cortex (Figure 25–5). This first type of root apical organization is referred to as the "closed type," and each of the three regions—rootcap, vascular cylinder, and cortex—is interpreted as having its own initials. In the second type of root apical organization, all regions, or at least the cortex and the rootcap, converge on a common group of cells (Figure 25–6). This type of organization is called the "open type," with all regions having common initials.

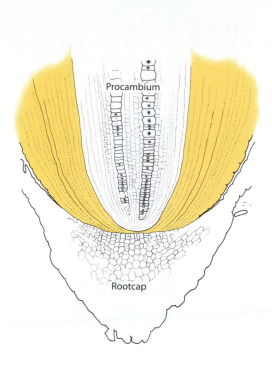

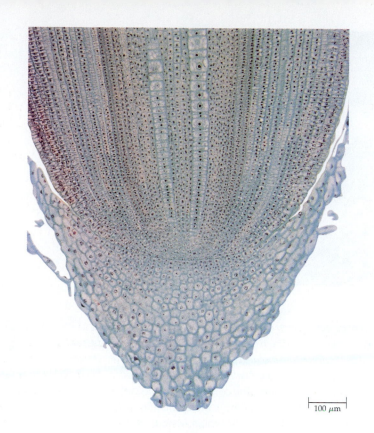

`100 µm`

25–5

Closed type of root apical organization, as seen in an apical meristem of a maize (Zea mays) root tip, in longitudinal section. Notice the three distinct layers of initials. The lower *layer gives rise to the rootcap, the middle layer to the protoderm and to the ground meristem, or cortex, and the upper layer to the procambium or vascular cylinder. Compare* *the organization of this apical meristem with that of the onion root tip shown in Figure 25–6b.*

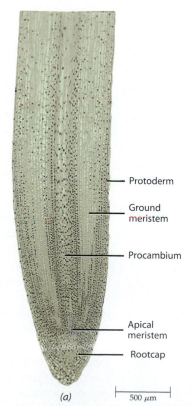

Protoderm

Ground meristem

Procambium

Apical meristem

Rootcap

(a) `500 µm`

(b) `200 µm`

25–6

Open type of root apical organization, as seen in longitudinal sections of onion (Allium cepa) root tip. (a) The primary meristems— *protoderm, procambium, and ground meristem—can be distinguished close to the apical meristem. (b) Detail of the apical* *meristem. Compare the organization of this apical meristem with that of the maize root tip shown in Figure 25–5.*

25–7

Quiescent center (dashed oval), as seen in a longitudinal section through the apical meristem of a maize (Zea mays) root tip. To prepare this autoradiograph, the root tip was supplied for one day with thymidine (a DNA precursor) labeled with the radioactive hydrogen isotope tritium (³H). In the rapidly dividing cells around the quiescent center, the radioactive material was quickly incorporated into the nuclear DNA. A photographic emulsion coating the section, having been exposed to the radiation from the labeled DNA, resulted in the formation of the dark grains on this autoradiograph.

Although the region of initials in the root apical meristems is mitotically active early in root development, divisions become infrequent in this region later in the growth of the root. Most cell division then occurs a short distance beyond the quiescent initials. The relatively inactive region of the apical meristem is known as the **quiescent center** (Figure 25–7).

As indicated by the term "relatively," the quiescent center is not totally devoid of divisions under normal conditions. Moreover, the quiescent center is able to repopulate the bordering meristematic regions when they are injured. In one study, for example, isolated quiescent centers of maize (*Zea mays*) grown in sterile culture were found to be able to form whole roots without first forming callus, or wound, tissue. In another study of maize roots, a striking correlation was found between the size of the quiescent center and the complexity of the primary vascular pattern of the root. These and other studies suggest that the quiescent center may play an essential role in organization and development of the root.

Growth in Length of Roots Occurs Near the Root Tip

The distance behind the apical meristem at which most cell division takes place varies from species to species and also within the same species, depending on the age of the root. The combination of the apical meristem and the nearby portion of root in which cell division does occur is called the **region of cell division** (Figure 25–8).

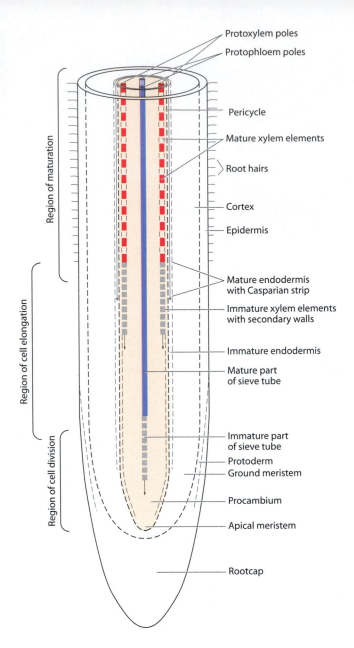

25–8

Diagram illustrating early stages in the primary development of a root tip. The region of cell division extends for a considerable distance behind the apical meristem. These cell divisions overlap with cell elongation and enlargement and also with cell maturation, or differentiation. At various distances from the apical meristem, cells enlarge and develop as specific cell types according to their position in the root. The three primary meristems— protoderm, ground meristem, and procambium—are delimited close to the apical meristem. Elements of the protophloem sieve tubes mature before (nearer to the apical meristem) the protoxylem elements. The endodermis matures (with Casparian strip) in advance of the protoxylem and the development of root hairs.

Behind the region of cell division, but not sharply delimited from it, is the **region of elongation,** which usually is only a few millimeters in length (Figure 25–8). The elongation of cells in this region results in most of the increase in length of the root. Above this region the root does not increase in length. Thus, the growth in length of the root occurs near the root tip and results in a very limited portion of the root constantly being pushed through the soil.

The region of elongation is followed by the **region of maturation,** or of differentiation, in which most of the cells of the primary tissues mature (Figure 25–8). Root hairs are also produced in this region, and sometimes this part of the root is called the root-hair zone (Figure 25–3). Obviously, if root hairs were to arise in the region of elongation, they would soon be sheared off due to the abrasive action as the root is pushed through the soil.

It is important to note that there is a gradual transition from one region of the root to another. The regions are not sharply delimited from one another. Some cells begin to elongate and differentiate in the region of cell division, whereas others reach maturity in the region of elongation. For example, the first-formed elements of the phloem and xylem mature in the region of elongation and are often stretched and destroyed during elongation of the root. As can be seen in Figure 25–8, the first-formed phloem elements (protophloem sieve tubes) reach maturity nearer the root tip than do the first-formed xylem elements (protoxylem elements), an indication of the need for food substances transported in the sieve tubes for root growth.

The protoderm, ground meristem, and procambium can be distinguished in very close proximity to the apical meristem (Figures 25–6 and 25–8). These are the primary meristems that differentiate into the epidermis, the cortex, and the primary vascular tissues, respectively (see Chapter 23).

Primary Structure

Compared with that of the stem, the internal structure of the root is usually relatively simple. This is due in large part to the absence of leaves and the corresponding absence of nodes and internodes (see Figure 1–10). Thus, the arrangement of the primary tissues shows very little difference from one level of the root to another.

The three tissue systems of the root in the primary stage of growth—the epidermis (dermal tissue system), the cortex (ground tissue system), and the vascular tissues (vascular tissue system)—can be readily distinguished from one another. In most roots, the vascular tissues form a solid cylinder, but in some monocotyledons they form a hollow cylinder around a pith (see Figures 25–10 and 25–11).

The Epidermis in Young Roots Absorbs Water and Minerals

The uptake of water and minerals by the root is facilitated by **root hairs**—tubular extensions of the epidermal cells—which greatly increase the absorptive surface of the root (Figure 25–9). In the study of a four-month-old rye plant, it was estimated that the plant contained approximately 14 billion root hairs, with an absorbing surface of 401 square meters. Placed end to end, these root hairs would extend well over 10,000 kilometers.

25–9

Root hairs. **(a)** *A radish* (Raphanus sativus) *seedling. Note the discarded seed coat, the cotyledons, curved hypocotyl, and primary root, with numerous root hairs. Most of the uptake of water and minerals occurs through the root hairs, which form just behind the growing tip of the root.* **(b)** *Root of a bent-grass* (Agrostis tenuis) *seedling. The root hairs may be as much as 1.3 centimeters long and may attain their full size within hours. Each hair is comparatively short-lived, but the formation of new root hairs and the death of old ones continue as long as the root is growing.*

(a)

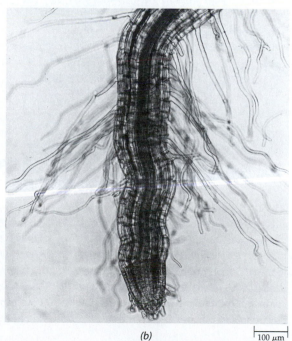

(b)

100 μm

Root hairs are relatively short-lived and are confined largely to the region of maturation. The production of new root hairs occurs just behind the region of elongation (Figure 25–8) and at a rate that nearly matches the rate at which older root hairs are dying off at the upper end of the root hair zone. As the tip of the root penetrates the soil, new root hairs develop behind it, providing the root with surfaces capable of absorbing new supplies of water and minerals, or inorganic ions. (See Chapter 31 for a discussion of absorption of water and inorganic ions by roots.) It is, of course, the new and growing roots—the feeder roots—that are primarily involved in the absorption of water and minerals. For this reason, great care must be taken by gardeners to move as much soil as possible along with the root system during transplantation. If the plant is simply "torn" from the soil, most of the feeder roots will be left behind and the plant probably will not survive.

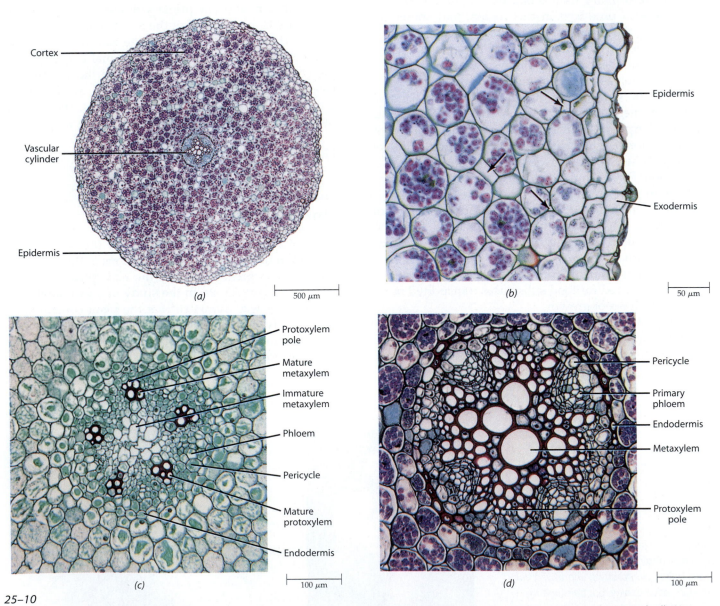

(a)   500 µm

(b)   50 µm

Cortex

Vascular cylinder

Epidermis

Epidermis

Exodermis

Protoxylem pole

Mature metaxylem

Immature metaxylem

Phloem

Pericycle

Mature protoxylem

Endodermis

Pericycle

Primary phloem

Endodermis

Metaxylem

Protoxylem pole

(c)   100 µm

(d)   100 µm

25–10
*Transverse sections of the root of a buttercup (Ranunculus). **(a)** Overall view of mature root. **(b)** Detail of outer portion of a mature root. In this root, the epidermis has died and the exodermis, the outer layer of cells of the cortex, has replaced the epidermis as the functional surface layer. Note intercellular spaces (arrows) among the cortical cells interior to the compactly arranged exodermis. The intercellular spaces are essential for aeration of the root cells. **(c)** Detail of immature vascular cylinder. Notice the intercellular spaces among the cortical cells. **(d)** Detail of mature vascular cylinder. Numerous starch grains are evident in the cortical cells. Note that in this section the root is tetrarch (with four ridges), whereas in **(c)** it is pentarch (with five ridges). Variation in the number of ridges of primary xylem is common along the axis of many roots.*

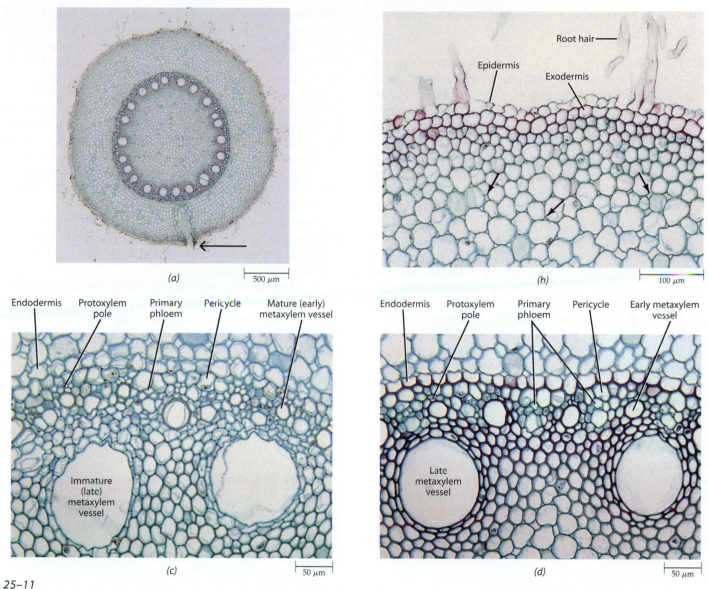

25–11

Transverse sections of a maize (Zea mays) root. (a) Overall view of mature root. Part of a lateral root is indicated by the arrow. The vascular cylinder, with its pith, is quite distinct.

(b) Detail of outer portion of a mature root, showing the epidermis with root hairs and part of the cortex. The outer layer of cortical cells is differentiated as a compactly arranged

exodermis. Note intercellular spaces (arrows) among the other cortical cells. (c) Detail of immature vascular cylinder. (d) Detail of mature vascular cylinder.

A thin cuticle has been identified on the epidermis in the absorbing part of some roots. In other roots, the epidermal cell walls appear to contain suberin. Nevertheless, the walls of the epidermal cells offer little resistance to the passage of water and minerals into the root. The mucigel covering the surface of roots enables the roots to establish a much more intimate contact with the particles of soil and may influence the availability of ions to the roots. The mucigel has been shown to provide an environment favorable to beneficial nitrogen-fixing bacteria (see Chapter 30). It may also provide the root with short-term protection from drying out

(desiccation). The layer of soil bound to the root by the mucigel and root hairs contains a variety of microorganisms and sloughed rootcap cells. This layer is called the **rhizosphere.**

Mutually beneficial symbiotic associations—mycorrhizae—occur between fungi and the roots of most vascular plants (see Chapter 15). The hyphal network of the fungus may extend far beyond the mycorrhizae, making it possible for the plant to obtain water and nutrients from a much larger volume of soil than made possible by root hairs. Typically, root hairs do not develop on ectomycorrhizae.

The Cortex Represents the Ground Tissue System in Most Roots

As seen in transverse section, the cortex occupies by far the greatest area of the primary body of most roots (Figure 25–10a). The plastids of the cortical cells commonly store starch but are usually devoid of chlorophyll. Roots that undergo considerable amounts of secondary growth—those of gymnosperms and most magnoliids and eudicots—shed their cortex early. In such roots, the cortical cells remain parenchymatous. In monocots, by contrast, the cortex is retained for the life of the root, and many of the cortical cells develop secondary walls that become lignified.

Regardless of the degree of differentiation, cortical tissue contains numerous intercellular spaces—air spaces essential for aeration of the cells of the root (Figures 25–10 and 25–11). The cortical cells have numerous contacts with one another, and their protoplasts are connected by plasmodesmata. Thus substances moving across the cortex may follow a symplastic pathway, moving from one protoplast to another by way of plasmodesmata, or an apoplastic pathway via the cell walls, or both. (The concept of symplast and apoplast is discussed on page 87, and the uptake of water and minerals by roots in Chapter 31.)

Unlike the rest of the cortex, the innermost layer is compactly arranged and lacks air spaces. This layer, the **endodermis** (Figures 25–10 and 25–11), is characterized by the presence of **Casparian strips** in its anticlinal walls (the radial and transverse walls, which are perpendicular to the surface of the root). The Casparian strip is not merely a wall thickening but an integral bandlike portion of the primary wall and middle lamella that is impregnated with suberin and sometimes lignin. The suberin and lignin infiltrate the spaces in the wall normally occupied by water, thus imparting a hydrophobic property to these specific wall regions. The plasma membranes of endodermal cells are quite firmly attached to the Casparian strips (Figures 25–12 and 25–13). Inasmuch as the endodermis is compact and the Casparian strips are impermeable to water and ions, apoplastic movement of water and solutes across the endodermis is blocked by the strips. Hence, all substances entering and leaving the vascular cylinder must pass through the protoplasts of the endodermal cells. This is accomplished either by crossing the plasma membranes of these cells or by moving symplastically via the numerous plasmodesmata connecting the endodermal cells with the protoplasts of neighboring cells of the cortex and vascular cylinder.

As mentioned previously, in roots that undergo secondary growth, the cortex and its innermost layer, the endodermis, are shed early. Although these older roots may still absorb water and minerals from the soil, they largely transport water and minerals absorbed by younger roots, with which they are connected. Roots in which growth has ceased, senesce, die, and decay. During vigorous growth, the roots are replaced at least as rapidly as they die back. In older roots in which the cortex is retained, a suberin lamella, consisting of alternating layers of suberin and wax, is eventually deposited internally over all wall surfaces of the endodermis. This is followed by the deposition of cellulose, which may become lignified (Figures 25–14 and 25–15). These changes in the endodermis begin opposite the phloem strands and spread toward the protoxylem (Figure 25–10d).

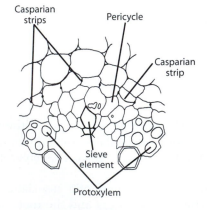

25–12

High-magnification view of a portion of an immature buttercup (Ranunculus) root, showing Casparian strips in the endodermal cells. Notice that the plasmolyzed protoplasts of the endodermal cells cling to the strips.

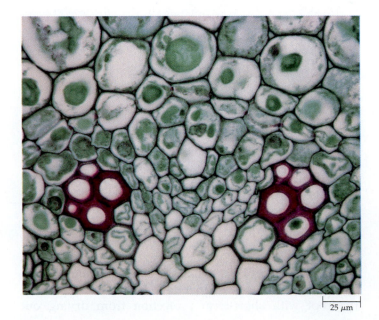

25 μm

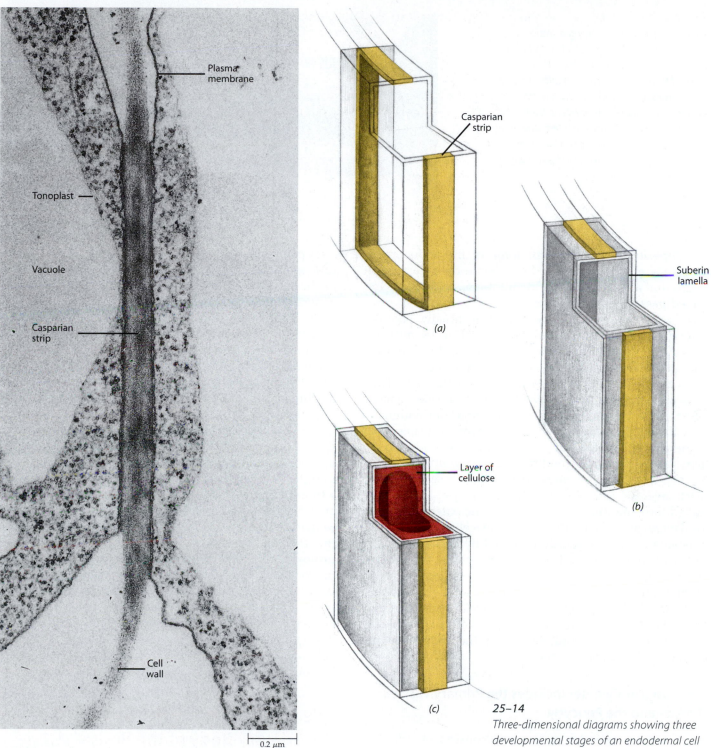

25–13

Transverse section of a radial anticlinal wall between two endodermal cells in the root of field bindweed (Convolvulus arvensis). The region of the Casparian strip is more intensely stained than the rest of the wall. Notice how firmly the plasma membrane (plasmalemma) adheres to the wall in the region of the Casparian strip in each of these plasmolyzed cells.

25–14

Three-dimensional diagrams showing three developmental stages of an endodermal cell in a root that remains in a primary state. (a) Initially, the endodermal cell is character-ized by the presence of a Casparian strip in its anticlinal walls. (b) Then a suberin lamella is deposited internally over all wall surfaces. (c) Finally, the suberin lamella is covered internally by a thick, often lignified, layer of cellulose. The outside of the root is to the left in all three diagrams.

25–15

Electron micrograph showing a section through the cell wall between two adjacent endodermal cells of a squash (Cucurbita pepo) root. In this late stage of differentiation of the endodermis, suberin lamellae are covered by cellulose wall layers on both sides of the primary wall. Note the alternating light and dense bands in the suberin lamellae, which are interpreted as consisting of wax and suberin, respectively.

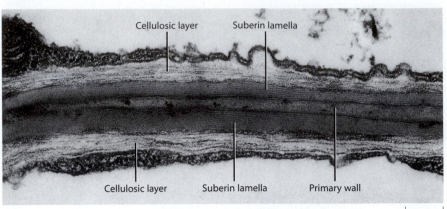

Cellulosic layer   Suberin lamella

Cellulosic layer   Suberin lamella   Primary wall

0.2 μm

Opposite the protoxylem, some of the endodermal cells may remain thin-walled and retain their Casparian strips for a prolonged period of time. Such cells are called **passage cells.** In some species they remain as passage cells, whereas in others they eventually become suberized and deposit additional cellulose. A misconception concerning the endodermis is that development of the suberin lamellae prevents the movement of substances across this innermost layer of cortical cells. This is not the case. So long as the endodermal cells remain alive, their plasmodesmata remain intact, providing a symplastic pathway for the movement of water and minerals. The uptake of water and minerals by roots may take place well behind the region of root hairs.

The roots of many angiosperms have a second compact layer of cells with Casparian strips. This layer, called the **exodermis,** develops from the outermost layer or layers of cells of the cortex. Development of the Casparian strips is quickly followed by deposition of a suberin lamella and, at least in some species, by a cellulosic layer as well. The suberized cell walls of the exodermis apparently reduce water loss from the root to the soil and provide a defense against attack by microorganisms. (See Chapter 31 for further discussion of the role of the exodermis and endodermis in the movement of water and solutes across the root.)

The Vascular Cylinder Includes the Primary Vascular Tissues and the Pericycle

The **vascular cylinder** of the root consists of primary vascular tissues and one or more layers of nonvascular cells, the **pericycle,** which completely surrounds the vascular tissues (Figures 25–10 and 25–11). The pericycle is considered part of the vascular cylinder because, like the vascular tissues, it originates from the procambium. In the young root, the pericycle is composed of parenchyma cells with primary walls, but as the root ages, the cells of the pericycle may develop secondary walls (Figure 25–11d).

The pericycle plays several important roles. In most

seed plants, lateral roots arise in the pericycle. In plants undergoing secondary growth, the pericycle contributes to the vascular cambium opposite the protoxylem and generally gives rise to the first cork cambium. Pericycle often proliferates—that is, gives rise to more pericycle.

The center of the vascular cylinder of most roots is occupied by a solid core of primary xylem from which ridgelike projections extend toward the pericycle (Figure 25–10). Nestled between the ridges of xylem are strands of primary phloem. Lacking a pith, the vascular cylinder in such roots is obviously a protostele (page 429).

The number of ridges of primary xylem varies from species to species, sometimes even varying along the axis of a given root. If two ridges are present, the root is said to be diarch; if three are present, triarch; four, tetrarch (Figure 25–10); and if many are present, polyarch (Figure 25–11). The first (*proto-*) xylem elements to mature in roots are located next to the pericycle, and the tips of the ridges are commonly referred to as **protoxylem poles** (Figures 25–10 and 25–11). The **metaxylem** (*meta-,* meaning "after") occupies the inner portions of the ridges and the center of the vascular cylinder and matures after the protoxylem. The roots of some monocots (for example, maize) have a pith (Figure 25–11), which some botanists regard as part of the vascular cylinder because they consider it to be of procambial origin.

Effect of Secondary Growth on the Primary Body of the Root

As we have mentioned, secondary growth in roots and stems consists of the formation of (1) secondary vascular tissues—secondary xylem and secondary phloem—from a vascular cambium and (2) periderm, composed mostly of cork tissue, from a cork cambium. Commonly, the roots of monocots lack secondary growth and, hence, consist entirely of primary tissues. In addition, the roots of many herbaceous eudicots undergo little or no secondary growth and remain largely primary in composition. (See Chapter 27 for a discussion of the vascular cambium.)

In roots that exhibit secondary growth, the vascular cambium is initiated by divisions of procambial cells that remain meristematic and are located between the primary xylem and primary phloem in portions of the root that are no longer elongating. Thus, depending on the number of phloem strands present in the root, two or more independent regions of cambial activity are initiated more or less simultaneously (Figure 25–16). Soon afterward, the pericycle cells opposite the protoxylem poles also divide, and the inner sister cells resulting from those divisions contribute to the vascular cambium. Now the cambium completely surrounds the core of xylem.

As soon as it is formed, the vascular cambium opposite the phloem strands begins to produce secondary xylem toward the inside, and in the process the strands of primary phloem are displaced outward from their positions between the ridges of primary xylem. By the time the cambium opposite the protoxylem poles is actively dividing, the cambium is circular in outline and the primary phloem has been separated from the primary xylem (Figure 25–16).

25–16

Root development in a woody eudicot.
(a) Early stage in primary development,
showing primary meristems. (b) At the
completion of primary growth, showing the
primary tissues and the meristematic pro-
cambium between the primary xylem and
primary phloem. (c) Origin of vascular cam-
bium. In the triarch root represented here,
cambial activity has been initiated in three
independent regions from procambium
between the three primary phloem strands
and the primary xylem. The pericycle cells
opposite the three protoxylem poles will also
contribute to the vascular cambium. Some
secondary xylem has already been produced
by the newly formed vascular cambium of
procambial origin. (d) After formation of
some secondary phloem and additional
secondary xylem, which further separate the
primary phloem from the primary xylem. A
periderm has not yet formed. (e) After forma-
tion of additional secondary xylem and
secondary phloem and of periderm. (f) At end
of first year's growth, showing the effect of
secondary growth—including periderm
formation—on the primary plant body. In (d)
through (f), the radiating lines represent rays.

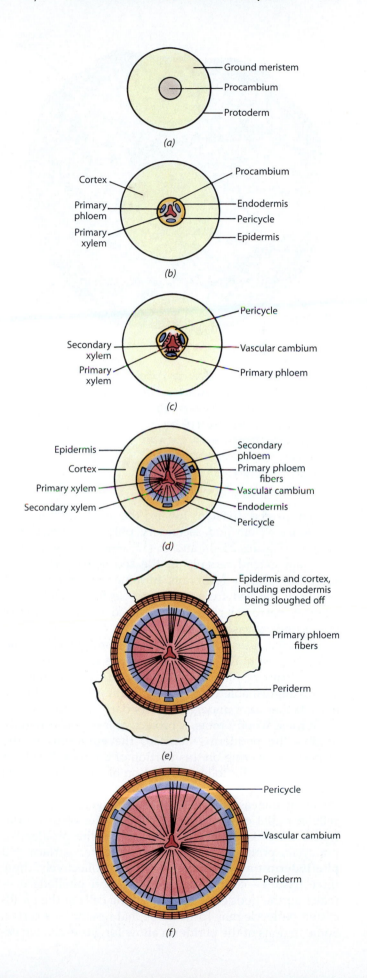

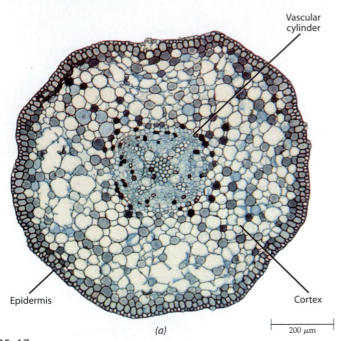

Vascular
cylinder

Epidermis

Cortex

(a)

200 μm

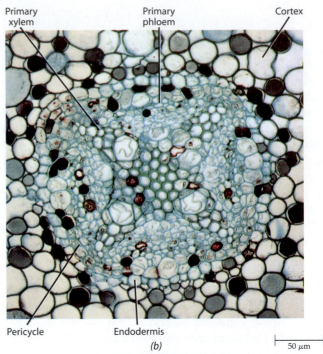

Primary
xylem

Primary
phloem

Cortex

Pericycle

Endodermis

(b)

50 μm

25–17

*Transverse sections of the willow (Salix) root,
which becomes woody. **(a)** Overall view of
root near completion of primary growth.
(b) Detail of primary vascular cylinder.
(c) Overall view of root at end of first year's
growth, showing the effect of secondary
growth on the primary plant body.*

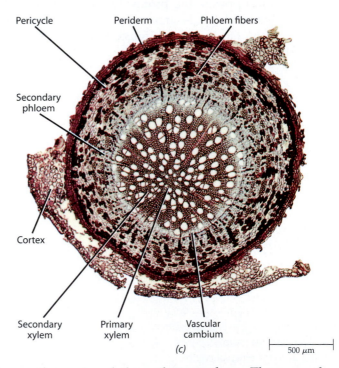

Pericycle

Periderm

Phloem fibers

Secondary
phloem

Cortex

Secondary
xylem

Primary
xylem

Vascular
cambium

(c)

500 μm

By repeated divisions toward the inside and outside, secondary xylem and secondary phloem are added to the root (Figures 25–16 and 25–17). Files of parenchyma cells that extend radially in the secondary xylem and secondary phloem form rays. In some roots, the vascular cambium derived from the pericycle forms wide rays, whereas narrower rays are produced in other parts of the secondary vascular tissues.

With increases in width of the secondary xylem and phloem, most of the primary phloem is crushed or obliterated. Primary phloem fibers may be the only remaining distinguishable components of the primary phloem in roots that have undergone secondary growth.

In most woody roots, a protective layer of secondary origin—the periderm—replaces the epidermis as the protective covering on this portion of the root. Periderm formation usually follows initiation of secondary xylem and phloem production. Divisions of pericycle cells cause an increase in the number of layers of pericycle cells in radial extent. A complete cylinder of **cork cambium**, which arises in the outer part of the proliferated pericycle, produces **cork** toward its outer surface and **phelloderm** toward its inner surface. Collectively, these three tissues—cork, cork cambium, and phelloderm—make up the periderm. The remaining cells of the proliferated pericycle may form tissue that resembles a cortex. Some regions of the periderm allow for gas exchange be-

tween the root and the soil atmosphere. These are the lenticels, spongy areas in the periderm with numerous intercellular spaces that permit the passage of air.

With the formation of the first periderm in the root, the cortex (including the endodermis) and the epidermis are isolated from the rest of the root, eventually die, and are sloughed. At the end of the first year's growth, the following tissues are present in a woody root (from outside to inside): possibly remnants of the epidermis and cortex, periderm, pericycle, primary phloem (fibers and crushed soft-walled cells), secondary phloem, vascular cambium, secondary xylem, and primary xylem (Figures 25–16f and 25–17c).

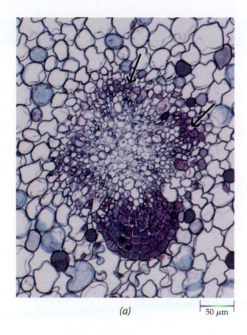

(a) | 50 μm

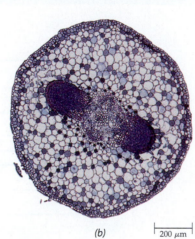

(b) | 200 μm

25–18

*Three stages in the origin of lateral roots in a willow (Salix). **(a)** One root primordium is present (below) and two others are being initiated in the region of the pericycle (arrows). The vascular cylinder is still very young.*

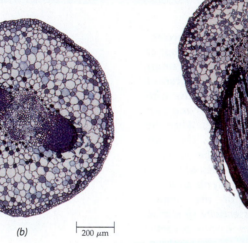

(c) | 250 μm

*(b) Two root primordia penetrating the cortex. **(c)** One lateral root has reached the outside, and the other is about to break through.*

Origin of Lateral Roots

In most seed plants, lateral roots (branch roots) originate in the pericycle. Because lateral roots originate deep from within the parent root, they are said to be endogenous, meaning "originating within" (Figures 25–3 and 25–18).

Divisions in the pericycle that initiate lateral roots occur some distance behind the region of elongation in partially or fully differentiated root tissues. In angiosperm roots, derivatives of both the pericycle and the endodermis commonly contribute to the new root primordium, although in many cases the derivatives of the endodermis are short-lived. As the young lateral root, or **root primordium,** increases in size, it pushes its way through the cortex (Figure 25–18), possibly secreting enzymes that digest some of the cortical cells lying in its path. While still very young, the root primordium develops a rootcap and apical meristem, and the primary meristems appear. Initially, the vascular cylinders of lateral root and parent root are not connected to one another. The two vascular cylinders are joined later, when derivatives of intervening parenchyma cells differentiate into xylem and phloem.

Aerial Roots and Air Roots

Aerial roots are adventitious roots produced from aboveground structures. The aerial roots of some plants serve as **prop roots** for support, as in maize (Figure 25–19).

When they come in contact with the soil, they branch and function also in the absorption of water and minerals. Prop roots are produced from the stems and branches of many tropical trees, such as the red mangrove (*Rhizophora mangle*), the banyan tree (*Ficus benghalensis*), and some palms. Other aerial roots, as in the ivy (*Hedera helix*), cling to the surface of objects such as walls and provide support for the climbing stem.

25–19

Prop roots of maize (Zea mays), a type of adventitious root.

Getting to the Root of Organ Development

Much recent research on organ development in plants has focused on the root of Arabidopsis thaliana. The two foremost reasons for this are the simple structure of the Arabidopsis root and the ease with which Arabidopsis seedlings can be grown on nutrient agar in Petri plates. When the plates are oriented vertically, the roots grow along the surface of the solidified agar media where abnormalities, which represent mutations, can be easily observed.

The organization of the mature Arabidopsis root is essentially similar to that of the embryonic root, which becomes distinct during the heart stage of embryogenesis (Figure 23–3d). The radial pattern is as follows: an outer region of epidermal cells, a middle region consisting of one layer each of parenchymatous cortical cells and endodermal cortical cells, and an innermost region (the vascular cylinder, or stele) consisting of pericycle and vascular tissue. The two cortical layers invariably consist of eight cells each. Two types of cells occur in the epidermis: those that produce root hairs and those that remain hairless. The root-hair cells are always located over the junction of the radial walls between two cortical cells, and the hairless cells always directly over cortical cells.

The apical meristem of the Arabidopsis root has a closed type of organization, with three layers of initials (Figure 25–5). The lower layer consists of columella rootcap initials and the initials of the lateral rootcap cells and epidermis. The middle layer consists of the initials of the cortex (parenchymatous and endodermal cortical cells) and the upper layer of initials of the vascular cylinder (pericycle and vascular tissue). At the center of the middle layer is a set of four cells that rarely divide. This set of cells constitutes the quiescent center of the root.

A great many root mutants of Arabidopsis have been identified. In particular, the control of cell fate in the epidermis has been examined in some detail. In one study, it was found that the TTG (TRANSPARENT TESTA GLABRA) gene is required to specify epidermal cell fate and cell patterns. The normal positional relationship—with root-hair cells located over radial walls between cortical cells and hairless cells positioned directly over cortical cells—was absent in ttg mutants. (Note

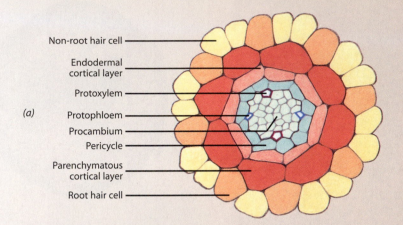

(a)

Non-root hair cell
Endodermal cortical layer
Protoxylem
Protophloem
Procambium
Pericycle
Parenchymatous cortical layer
Root hair cell

The root of Arabidopsis thaliana. (a) Transverse section of root prior to the development of root hairs, showing the outer epidermal cells, the middle cortical cells, and the innermost vascular cylinder. Note that the epidermal cells destined to form root hairs are located over radial walls of cortical cells, whereas those that will remain hairless are located directly over cortical cells.

that the mutant form of the gene is indicated in lowercase.) In such mutants, epidermal cells in all positions differentiate into root hairs, which are indistinguishable from normal, wild-type hairs. In transverse sections from the mature portion of the root, the number of epidermal and cortical cell files (lineages) is similar to that of the wild-type root. Overall, the results of this study indicate that the ttg mutation alters the positional control of root-cell hair differentiation but affects neither root-hair formation nor the structure of the mature root. Apparently, in wild-type roots, the TTG gene either provides or responds to positional signals that cause differentiating epidermal cells lying directly over cortical cells to remain hairless.

That positional control via cell-cell interaction, rather than cell lineage relationships (as determined by the organization of the apical meristem), plays the most important role in the determination of cell fate in the Arabidopsis root has clearly been demonstrated by laser ablation experiments. In these experiments, specific cells were ablated, or removed, with a laser and the effect on neighboring cells observed. For example, when the cells of the quiescent center were ablated, they were replaced and displaced toward the rootcap

(b) Longitudinal section of a root tip showing the relationship of the different layers or regions of the root to the tiers of initials in the apical meristem.

(c) Seedling roots of wild-type Arabidopsis thaliana, showing normal frequency and arrangement of root hairs. (d) Seedling roots of a ttg mutant with an excessive number of root hairs.

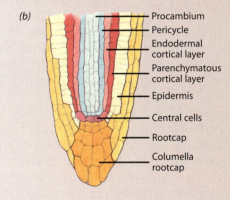

(b)

Procambium
Pericycle
Endodermal cortical layer
Parenchymatous cortical layer
Epidermis
Central cells
Rootcap
Columella rootcap

by cells derived from the vascular cylinder (procambium). These new cells subsequently acquired rootcap characteristics. Ablated cortical initials were replaced by pericycle cells, which then switched fate and behaved as cortical initials. Ablation of a single daughter cell of a cortical initial had no effect on subsequent divisions occurring in that initial, which was in contact with other cortical daughter cells of neighboring cortical initials. When all cortical daughter cells bordering a cortical initial

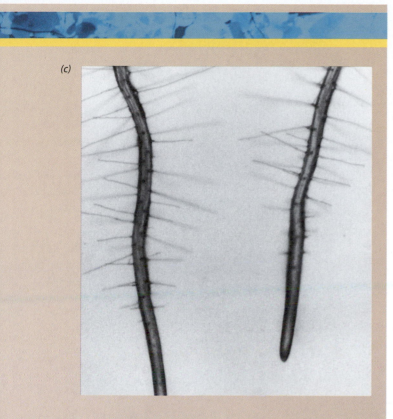

(c)

(d)

were ablated, however, that initial was unable to generate files of parenchymatous and endodermal cortical cells. Apparently, the cortical initials, and perhaps all initials, depend on positional information from more mature daughter cells within the same cell layer. In other words, the initials of the root apical meristem apparently lack intrinsic pattern-generating information. This is contrary to the traditional view of meristems as autonomous pattern-generating machines.

25–20
Pneumatophores (air roots) of the white mangrove (Laguncularia racemosa) *protruding from the mud near the base of a tree.*

Roots require oxygen for respiration, which is why most plants cannot live in soil that is inadequately drained and consequently lacks air spaces. In some trees that grow in swampy habitats, portions of the roots grow out of the water. Hence, the roots of such trees serve not only to anchor but also to aerate the root system. For example, the root systems of the black (*Avicennia germinans*) and white (*Laguncularia racemosa*) mangroves develop negatively gravitropic extensions called **air roots,** or **pneumatophores,** which grow upward out of the mud and so provide adequate aeration (Figure 25–20).

Many special adaptations of roots are found among epiphytes—plants that grow on other plants but are not parasitic on them. The root epidermis of epiphytic orchids, for example, is several layers thick (Figure 25–21) and, in some species, is the only photosynthetic tissue of the plant. This multiple epidermis, called velamen, provides mechanical protection for the cortex and reduces water loss. The velamen may also function in the absorption of water.

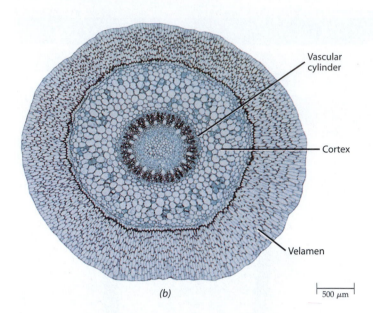

Vascular
cylinder

Cortex

Velamen

500 μm

(a)

(b)

25–21

(a) Aerial roots of an epiphytic orchid (Oncidium sphacelatum). (b) Transverse section of an orchid root, showing the multiple epidermis, or velamen.

Among epiphytes, *Dischidia rafflesiana,* the "flower pot plant," has a very unusual modification. Some of its leaves are flattened, succulent structures, but others form hollow containers—the "flower pots"—that collect debris and rainwater (Figure 25–22). Ant colonies may live in the pots and add to the nitrogen supply of the plant. Roots, formed at the node above the modified leaf, grow downward and into the pot, from which they absorb water and minerals.

(a)

(b)

25–22

Aerial roots of the epiphyte Dischidia rafflesiana, *or "flower pot plant." (a) A modified leaf, or "pot," which collects debris and rainwater. (b) Modified leaf cut open to show roots that have grown down into the pot.*

Adaptations for Food Storage: Fleshy Roots

Most roots are storage organs, and in some plants the roots are specialized for this function. Such roots are fleshy because of an abundance of storage parenchyma, which is permeated by vascular tissue. The development of some storage roots, such as that of the carrot *(Daucus carota)*, is essentially similar to that of nonfleshy roots, except for a predominance of parenchyma cells in the secondary xylem and phloem of the storage roots. The root of the sweet potato *(Ipomoea batatas)* develops in a manner similar to that of the carrot; however, in the sweet potato, additional vascular cambium cells develop within the secondary xylem around individual vessels or groups of vessels (Figure 25–23). These additional cambia (plural of cambium), while producing a few tracheary elements toward the vessels and a few sieve tubes away from them, mainly produce storage parenchyma cells in both directions. In the sugarbeet *(Beta vulgaris)*, most of the increase in thickness of the root results from the development of extra cambia (supernumerary cambia) around the original vascular cambium (Figure 25–24). These concentric layers of cambia, which superficially resemble growth rings in woody roots and stems, produce parenchyma-dominated xylem toward the inside and phloem toward the outside. The upper portion of most fleshy roots actually develops from the hypocotyl.

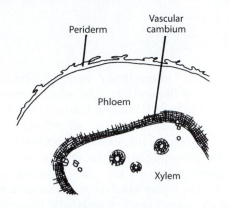

Periderm   Vascular cambium

Phloem

Xylem

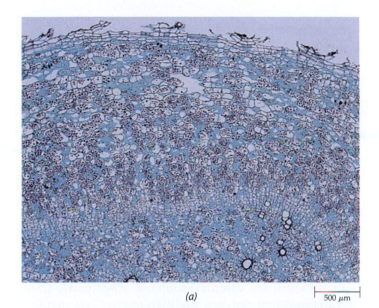

(a)

500 μm

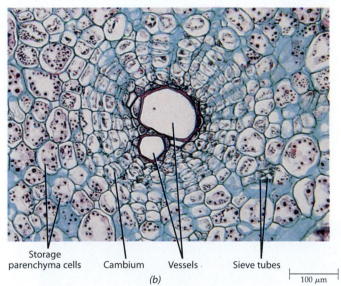

Storage
parenchyma cells   Cambium   Vessels   Sieve tubes

(b)

100 μm

25–23

Transverse sections of the root of a sweet potato (Ipomoea batatas). **(a)** *Overall view.* **(b)** *Detail of xylem, showing cambium around vessels. Most of the xylem and phloem is composed of storage parenchyma cells.*

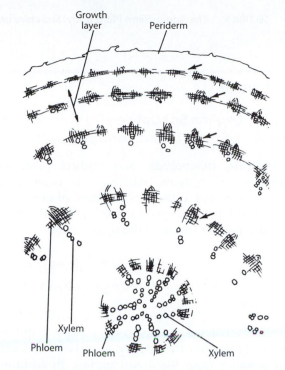

Growth layer   Periderm

Xylem

Phloem   Phloem   Xylem

500 μm

25–24

Transverse section of a sugarbeet (Beta vulgaris) root, with supernumerary cambia indicated by arrows on the diagram. The original vascular cambium produces relatively little secondary xylem and secondary phloem (in the center of the root).

607

Summary

Roots Are Organs Specialized for Anchorage, Absorption, Storage, and Conduction

Gymnosperms, magnoliids, and eudicots commonly produce taproot systems, whereas monocots usually produce fibrous root systems. The extent of the root system is dependent upon several factors, but the bulk of most feeder roots is found in the upper meter of the soil.

The Root Tip Can Be Roughly Divided into Regions of Cell Division, Elongation, and Maturation

The apical meristems of most roots contain a quiescent center; most meristematic activity, or cell division, occurs a short distance from the apical initials. In addition to this region of cell division, two other growth regions can be recognized in growing roots—the region of elongation and the region of maturation. During primary growth, the apical meristem gives rise to the three primary meristems—protoderm, ground meristem, and procambium—which differentiate into epidermis, cortex, and vascular cylinder, respectively. In addition, the apical meristem produces the rootcap, which serves to protect the meristem and aid the root in its penetration of the soil. Mucigel produced by the outer rootcap cells lubricates the root during its passage through the soil. It also enables the root to establish intimate contact with soil particles and provides an environment favorable to beneficial microorganisms.

The Root Epidermis and Cortex May Be Modified with Age

Many epidermal cells of the root develop root hairs, which greatly increase the absorbing surface of the root. With the exception of the endodermis, the cortex con-
tains numerous intercellular spaces. The compactly arranged cells of the endodermis, which forms the inner boundary of the cortex, contain Casparian strips on their anticlinal walls. Consequently, all substances moving between the cortex and vascular cylinder must pass through the protoplasts of the endodermal cells. The roots of many angiosperms also have an exodermis, which forms the outer boundary of the cortex and also consists of a compact layer of cells with Casparian strips.

The Vascular Cylinder Consists of the Primary Vascular Tissues and the Encircling Pericycle

The primary xylem usually occupies the center of the vascular cylinder and has radiating ridges that alternate with strands of primary phloem. Branch roots originate in the pericycle and push their way to the outside through the cortex and epidermis.

Secondary Growth in Roots Involves Both the Vascular Cambium and Cork Cambium

Secondary growth results in disruption of the primary body of the root as the strands of primary phloem are separated from the primary xylem through the formation of secondary vascular tissues by the vascular cambium. In roots, the vascular cambium arises partly from procambium that remains undifferentiated between the primary xylem and the primary phloem strands and partly from pericycle opposite the ridges of primary xylem. In most woody roots, the cork cambium of the first periderm originates from the pericycle. Consequently, formation of the periderm results in isolation and eventual separation of the cortex and epidermis from the rest of the root. Figure 25–25 presents a summary of root development of a woody eudicotyledon, beginning with the apical meristem and ending with the secondary tissues produced during the first year's growth.

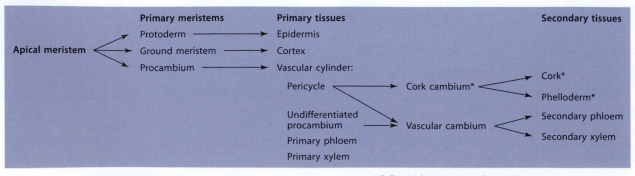

25–25

Summary of root development in a woody eudicot during the first year of growth.

Root Modifications Include Aerial Roots, Air Roots, and Fleshy Roots

Most roots are storage organs, and in some plants, such as the carrot, sweet potato, and sugarbeet, the roots are specialized for this function. The fleshy roots have an abundance of storage parenchyma permeated by vascular tissue.

Selected Key Terms

air roots   p. 605

Casparian strip   p. 598

columella   p. 591

cork   p. 602

cork cambium   p. 602

endodermis   p. 598

exodermis   p. 600

fibrous root system p. 590

lateral roots   p. 590

metaxylem   p. 600

mucigel   p. 591

passage cells   p. 600

pericycle   p. 600

phelloderm   p. 602

pneumatophores   p. 605

primary root   p. 590

primary structure   p. 595

prop roots   p. 603

protoxylem poles   p. 600

quiescent center   p. 594

region of cell division p. 594

region of elongation p. 595

region of maturation p. 595

rhizosphere   p. 597

rootcap   p. 591

root hairs   p. 595

root primordium   p. 603

root system   p. 590

taproot   p. 590

taproot system   p. 590

vascular cylinder   p. 600

Questions

1. Distinguish between the following: endodermal cells/passage cells; endodermis/exodermis; protoxylem/metaxylem; protophloem/metaphloem; aerial roots/air roots.

2. What are the principal functions of roots?

3. Discuss the need for a plant to maintain a balance between its shoot and root systems.

4. Of what "value" is mucigel to a root?

5. "During growth in length of a root, a very limited portion of the root is constantly being pushed through the soil." Explain.

6. How do the Casparian strips of endodermal cells affect the movement of water and solutes across the endodermis?

7. Structurally, what are the principal differences between the primary structure of monocot roots and that of other angiosperm roots?

8. What structural features do all fleshy storage roots have in common?

The Shoot: Primary Structure and Development

26–1

A portion of a Croton *shoot. The leaves of* Croton, *a eudicot, have a mottled appearance due to clonal variations in the ability of leaf cells to produce chlorophyll and are spirally arranged along the stem. At the apex the leaves are so close together that nodes and internodes are not distinguishable as separate regions of the stem. Growth in length of the stem between successive leaves, which are attached to the stem at the nodes, results in formation of the internodes.*

OVERVIEW

Before we begin our discussion of the shoot — the collective term for the stem and its leaves — it is useful to recall primitive vascular plants such as rhyniophytes that consist of stems alone. As we saw in Chapter 19, such stems are thought to have been the precursors of leaves and thus the forerunners of the shoot itself. This evolutionary viewpoint emphasizes the intimate relationship between the stem and the leaf and helps us understand the arrangement of tissues that we observe in the shoot of present-day plants.

As you will see, the primary tissues of the shoot arise from the shoot apical meristem. Accordingly, this chapter begins with a discussion of the structure and function of the shoot apex and the formation of the primary meristems. The primary structure of the stem is examined next, including variations in that structure as exemplified by selected species.

The emphasis then switches to the other component of the shoot — the leaf — beginning with a short section showing that the vascular tissues of the leaf are merely extensions of the vascular tissues of the stem. The anatomy of monocot and other angiosperm leaves is then compared, followed by a discussion of leaf development, an overview of the process by which leaves are shed from the plant, and a look at the development of floral parts — that is, the modified leaves we know as sepals, petals, stamens, and carpels. The chapter ends with an examination of some well-known leaf and stem modifications, including tubers, bulbs, and the succulent stems of cacti.

CHECKPOINTS

By the time you finish reading this chapter, you should be able to answer the following questions:

1. What is the structure of the shoot apical meristem of angiosperms, and what is the relationship between the zones of that meristem and the primary meristems of the shoot?

2. What three basic types of organization are found in the primary structure of the stems of seed plants?

3. What are leaf traces, and how are they indicative of the intimate relationship that exists between the stem and the leaf? What hypotheses have been proposed to explain the pattern of leaf arrangement on stems?

4. What structural differences exist between the leaves of monocots and those of other angiosperms?

The **shoot,** which consists of the stem and its leaves, is the aboveground portion of the plant that is familiar to us. It is initiated during development of the embryo, where it may be represented by a plumule, consisting of an epicotyl (the stem above the cotyledons), one or more young leaves, and an apical meristem, or by an apical meristem only. As we shall see, the shoot is structurally more complex than the root. In contrast to the root, the shoot has nodes and internodes, with one or more leaves attached at each node (Figures 26–1 and 26–2). Whereas the shoot apex produces leaves and axillary buds, which develop into lateral shoots, the root apex produces no lateral organs. (You will recall that lateral roots arise in the region of maturation.) At each node, one or more strands of the vascular cylinder of the stem turn outward and extend into the leaf, leaving one or more gaps in the vascular cylinder opposite the leaf. Gaps do not exist in the vascular cylinders (protosteles) of roots.

The two principal functions associated with stems are **support** and **conduction.** The leaves—the principal photosynthetic organs of the plant—are supported by the stems, which place the leaves in favorable positions for exposure to light. Substances manufactured in the leaves are transported through the stems by way of the phloem to sites of utilization, such as developing plant parts and storage tissues, including those of the stems. At the same time water and minerals are transported upward in the xylem from the roots and into the leaves via the stem.

Origin and Growth of the Primary Tissues of the Stem

The apical meristem of the shoot is a dynamic structure that, in addition to adding cells to the primary plant body, repetitively produces leaf primordia and bud primordia, resulting in a succession of repeated units called **phytomeres** (Figure 26–3). The **leaf primordia** develop into leaves and the **bud primordia** into lateral shoots.

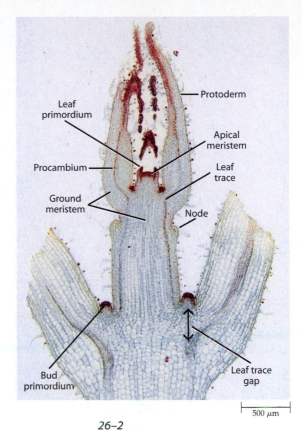

26–2
Longitudinal section of shoot tip of the common houseplant Coleus blumei, a eudicot. The leaves in Coleus are arranged opposite one another at the nodes, each successive pair at right angles to the previous pair (decussate phyllotaxy); thus the leaves of the labeled node are at right angles to the plane of section. (See pages 622–624 for a discussion of leaf traces, leaf trace gaps, and phyllotaxy.)

26–3
The apical meristem at the tip of the shoot is protected by young leaves that fold over it, as seen in this longitudinal section of a eudicot shoot. Activity of the apical meristem, which repetitively produces leaf and bud primordia, results in a succession of repeated units called phytomeres. Each phytomere consists of a node with its attached leaf, the internode below that leaf, and the bud at the base of the internode. The boundaries of the phytomeres are indicated by the dashed lines. Note that the internodes are of increasing length the farther they are from the apical meristem. Internodal elongation accounts for most of the increase in length of the stem.

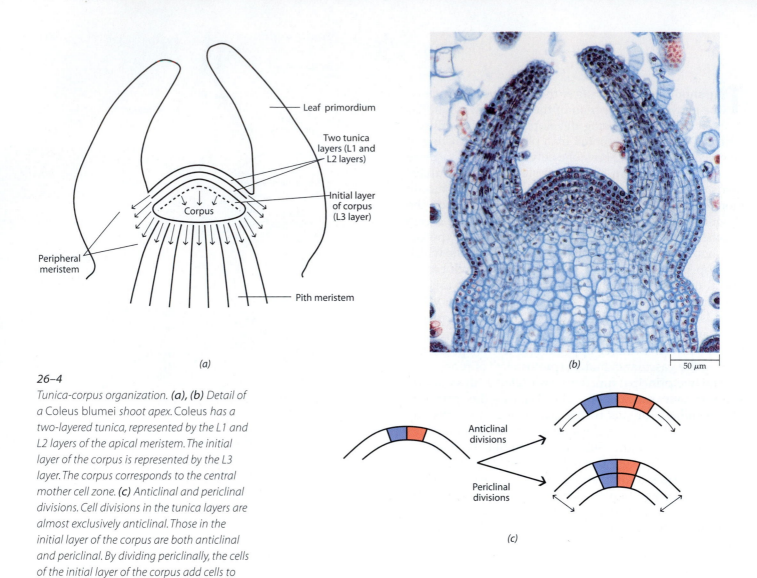

(a)

26–4

*Tunica-corpus organization. **(a), (b)** Detail of a* Coleus blumei *shoot apex. Coleus has a two-layered tunica, represented by the L1 and L2 layers of the apical meristem. The initial layer of the corpus is represented by the L3 layer. The corpus corresponds to the central mother cell zone. **(c)** Anticlinal and periclinal divisions. Cell divisions in the tunica layers are almost exclusively anticlinal. Those in the initial layer of the corpus are both anticlinal and periclinal. By dividing periclinally, the cells of the initial layer of the corpus add cells to the corpus.*

(b)

Anticlinal divisions

Periclinal divisions

(c)

50 μm

Unlike the root apical meristem, the vegetative shoot apical meristem lacks a specialized protective covering comparable to the rootcap. On the other hand, it commonly is surrounded by young leaves that fold over it and provide it with protection.

The vegetative shoot apex of most flowering plants has what is termed a **tunica-corpus** type of organization (Figure 26–4). The two regions—tunica and corpus—are distinguished by the planes of cell division that occur in them. The tunica consists of the outermost layer or layers of cells, which divide **anticlinally,** that is, in planes perpendicular to the surface of the meristem (Figure 26–4c). These divisions contribute to surface growth, without increasing the number of cell layers in the meristem. The corpus consists of a body of cells that lie beneath the tunica layers. In the corpus, the cells divide in various planes and add bulk to the developing shoot. The corpus and each layer of tunica have their own initials. The initials of the corpus occur beneath those of the tunica and add cells to the corpus by dividing **periclinally,** that is, parallel with the apical surface (Figure 26–4c). Hence, the number of layers of initials in

any given meristem is equal to the number of layers of tunica plus one.

The number of tunica layers varies from species to species. Most angiosperms have apices consisting of three superimposed layers of cells: two tunica layers and the initial layer of the corpus. These three cell layers commonly are designated L1, L2, and L3, with the outermost layer being the L1 and the innermost the L3 (Figure 26–4a). Although the L1 layer divides almost exclusively anticlinally, L1 cells occasionally divide periclinally. When this happens, the inner daughter cell is displaced into the L2 layer where it differentiates as though it had been derived from the L2 layer. Similar displacements may occur between derivatives of the L2 and L3 layers, with comparable results. Here again is evidence that cell differentiation is not dependent on cell lineage but rather on the final position of a cell in the developing organ (see the essay on page 604).

In the shoot apices of many angiosperms, the bulk of the corpus corresponds to an area of conspicuously vacuolated cells called the **central mother cell zone.** This zone of vacuolated cells is surrounded by the **peri-**

(a)

(b)

(c)

pheral meristem, or peripheral zone, which originates partly from the tunica (L1 and L2 layers) and partly from the corpus, or central mother cell zone, whose origin can be traced to the L3 layer (Figure 26–4). Three-dimensionally, the peripheral meristem forms a ring around the central mother cell zone. To the inside of this ring and just beneath the central mother cell zone is the **pith meristem.** Cell divisions are relatively infrequent in the central mother cell zone, and in this respect, the central mother cell zone is analogous to the quiescent center of the root apical meristem. By contrast, the peripheral zone is mitotically very active.

What is the relationship between these zones and the primary meristems of the shoot? The protoderm always originates from the outermost tunica (L1) layer, whereas the procambium and part of the ground meristem (the cortex and sometimes part of the pith) are derived from the peripheral meristem. The rest of the ground meristem (all or most of the pith) is formed by the pith meristem.

Although the primary tissues of the stem pass through periods of growth similar to those of the root, the stem cannot be divided along its axis into regions of cell division, elongation, and maturation as in the case of roots. When actively growing, the apical meristem of the shoot gives rise to leaf primordia in such rapid succession that nodes and internodes cannot at first be distinguished. As growth begins to occur between the levels of leaf attachment, the elongated parts of the stem take on the appearance of internodes, and the portions of the stem at which the leaves are attached become recognizable as nodes (Figures 26–1 and 26–5). Thus, increase in length of the stem occurs largely by internodal elongation, which may occur simultaneously over several internodes.

(d)

26–5

Stages in growth of the terminal bud and two lateral buds of the horse chestnut (Aesculus hippocastanum). (a) The young shoots are tightly packed in the buds and are protected by bud scales, which are highly modified leaves initiated late in the previous growing season. (b) The buds open to reveal the oldest rudimentary leaves. (c) Internodal elongation has separated the nodes from one another. The terminal bud of the horse chestnut is a mixed bud, containing both leaves and flowers, although the flowers are not visible here. The lateral buds produce only leaves. (d) Detail of the lower portions of the young shoots; the bud scales are separated and folded back.

The meristematic activity causing the elongation of the internode may be fairly uniform throughout the internode. In some species it occurs as a wave progressing from the base of the internode upward, while in others, such as grasses, it is restricted largely to the base of the internode. A localized meristematic region in the elongating internode is called an **intercalary meristem** (a meristematic region between two more highly differentiated regions). Certain elements of the primary xylem and primary phloem—specifically the protoxylem and protophloem—differentiate within the intercalary meristem and connect the more highly differentiated regions of the stem above and below the meristem.

Increase in stem thickness during primary growth involves both longitudinal (periclinal) divisions and cell enlargement. In plants with secondary growth, this primary thickening is moderate. Monocots usually lack secondary growth, but a few, such as palms, have massive primary growth. This growth occurs so close to the apical meristem that the shoot apex appears inserted on a shallow cone or even in a depression (Figure 26–6). Activity in the apex proper is not great, but immediately behind it cell division is intensive. The meristem responsible for the abrupt expansion of the apical region to a wide crown lies below the young leaf bases. Within this meristematic region, localized cell divisions result in the formation of procambial strands. This zone of procambium formation is known as the **meristematic cap.** The bulk of the meristem responsible for stem thickening is located below the cap, although the ground tissue between the procambial strands of the cap also contributes to the thickening.

The apical meristem of the shoot gives rise to the same primary meristems found in the root: protoderm, procambium, and ground meristem (Figure 26–2). These primary meristems in turn develop into the mature tissues of the primary plant body: epidermis, primary vascular tissues, and ground tissue, respectively.

Primary Structure of the Stem

Considerable variation exists in the primary structure of stems of seed plants, but three basic types of organization can be recognized: (1) In some conifers, magnoliids, and eudicots, the vascular system of the internode appears as a more or less continuous cylinder within the ground tissue (Figure 26–7a). (2) In others, the primary vascular tissues develop as a cylinder of discrete strands, or bundles, separated from one another by ground tissue (Figure 26–7b). (3) In the stems of most monocots and of some herbaceous (nonwoody) eudicots, the arrangement of the procambial strands and vascular bundles is more complex. As seen in transverse sections, the vascular bundles occur in more than one ring of bundles or ap-

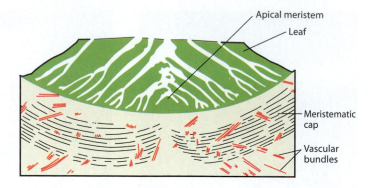

26–6

Diagrammatic representation of the anatomy of the top, or crown, of a thick-stemmed monocot without secondary growth, such as a palm tree. Increase in thickness is due to meristematic activity below the young leaf bases. The apical meristem and youngest leaf primordia are conventional in size, although they appear sunken below broad stem tissues. The zone of procambium formation is called the meristematic cap.

pear scattered throughout the ground tissue. In the latter instance, the ground tissue often cannot be distinguished as cortex and pith (Figure 26–7c).

In the discussion that follows, the stem of basswood, or linden *(Tilia americana)*, will be used to exemplify the first type of organization. The second type of organization will be exemplified by the elderberry *(Sambucus canadensis)*, alfalfa *(Medicago sativa)*, and buttercup *(Ranunculus)* stems, and the third type by the stem of maize *(Zea mays)*. The basswood and elderberry stems are also examples of stems that undergo much secondary growth. (They will be revisited during our discussion on secondary growth, in Chapter 27.) By contrast, the alfalfa stem undergoes relatively little secondary growth, and the stems of buttercup (a eudicot) and maize (a monocot) none at all.

The Primary Vascular Tissues of the *Tilia* Stem Form an Almost Continuous Vascular Cylinder

Figure 26–8 shows the *Tilia* stem, with what appears to be a continuous cylinder of primary vascular tissues. In fact, the vascular cylinder is composed of vascular bundles that are separated from one another by very narrow, inconspicuous regions of ground parenchyma. These parenchymatous regions, called **interfascicular parenchyma,** interconnect the cortex and pith. (Interfascicular means "between the bundles, or fascicles.")

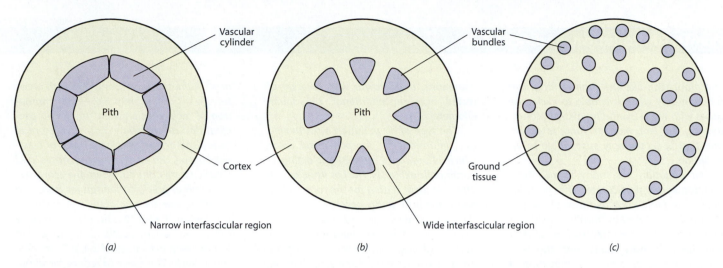

26–7

The three basic types of organization in primary structure of stems as seen in transverse section. (a) The vascular system appears as a continuous hollow cylinder around the pith. (b) Discrete vascular bundles form a single ring around the pith. (c) The vascular bundles appear scattered throughout the ground tissue.

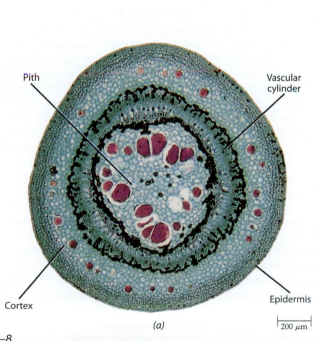

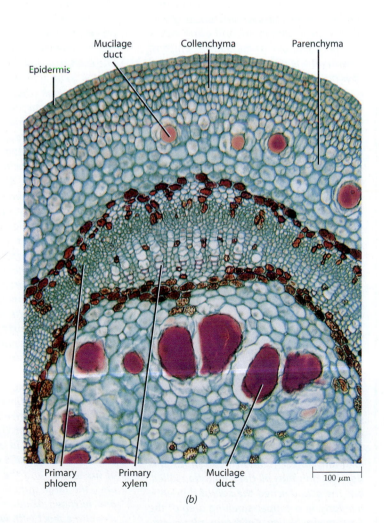

26–8

(a) Transverse section of basswood (Tilia americana) stem in a primary stage of growth. The vascular tissues appear as a continuous hollow cylinder that divides the ground tissue into pith and cortex. (b) Detail of a portion of the same basswood stem.

Plants, Air Pollution, and Acid Rain

The plant leaf, like the human lung, can function only when it is able to exchange gases with the surrounding air. As a consequence, the leaf, like the lung, is an organ that is exceedingly susceptible to air pollution.

Air pollution has many forms. Some pollutants are particulate. The particles may be organic, such as those present in the smoke produced by burning fossil fuels and garbage, or inorganic, such as cement kiln dusts, foundry dusts, and the lead compounds released in the combustion of leaded gasoline. As a major component of smog, these particles reduce the amount of sunlight reaching the Earth's surface. Such particles also have direct ill effects on plants. They may clog stomata and prevent them from functioning, or they—particularly metallic particles—may act as plant poisons.

Fluorides, which enter the air as waste products from the manufacture of phosphates, steel, aluminum, and other industrial products, act as cumulative poisons, entering the leaf through the stomata and causing collapse of leaf tissue, apparently by inhibiting enzymes concerned with cellulose synthesis. Thousands of acres of Florida citrus groves have been damaged by fluorides discharged from phosphate fertilizer factories.

When ores containing sulfur are processed, sulfur dioxide is produced:

$$2CuS + 3O_2 \longrightarrow 2CuO + 2SO_2$$

Sulfur dioxide has a disagreeable pungent-sweet taste; it is unusual among air pollutants in that it can be tasted at lower concentrations than it can be smelled. Sulfur oxides are also produced by the burning of fossil fuels containing sulfur. In moist air, sulfur oxides react with water to form droplets of sulfuric acid, H_2SO_4, a strong corrosive acid and a component of acid rain. In several parts of the United States, virtual deserts have been created by the emission of sulfur dioxide combined with metals. As long ago as 1905, air pollution controls were instituted in areas surrounding copper smelters in Tennessee, but the surrounding area, once covered by luxuriant forest, remains barren to this day. Not only was all the vegetation killed, but the acid leached the soil of nutrients. Similarly, in a copper-smelting area of the Sacramento Valley, California, all vegetation was killed over an area of 260 square kilometers, and growth was severely affected over an additional 320 square kilometers.

The type of air pollution most familiar to Californians is photochemical smog, which is produced by sunlight acting on automobile exhausts. The Los Angeles area offers an ideal setting for the formation of photochemical smog because the life of that area is heavily geared to the use of the automobile, and the mountains to the north and east form a "basin" that prevents reactants from dispersing. Not only are many species of plants unable to survive in the city itself (as is true in many other major cities of the world), but smog moving out of the Los Angeles basin is damaging agricultural crops and killing pine forests in mountains as far as 160 kilometers away.

One of the principal ingredients of photochemical smog is nitrogen dioxide, NO_2, which is produced by any combustion process that occurs in air (dry air is 77 percent nitrogen) and so is present in automobile exhausts. Under the influence of light, NO_2 is split into NO (nitric oxide) and atomic oxygen. The latter is extremely reactive and forms ozone, O_3, by reaction with molecular oxygen, O_2.

Similar reactions, powered by ultraviolet light, occur at the outer layers of the atmosphere, producing the ozone shield described on page 6. Ozone can also be produced by an electric spark and is the source of the "clean" smell after a lightning storm. Ozone is a highly toxic substance. In plants, it damages the thin-walled palisade cells, apparently affecting the permeability of the membranes of both the cells and their chloroplasts. Another component of photochemical smog is PAN (peroxyacetyl nitrate, $C_2H_3O_5N$). It is several times more toxic than ozone but is normally present in much lower concentrations. Photosynthesis is reduced 66 percent by a photochemical smog concentration of 0.25 part per million.

At the present time, there is much concern over the adverse effects of acid rain on the environment. Under normal conditions, rain arising in a nonpolluted area should have a pH of 5.6. With the burning of fossil fuels and the smelting of sulfide ores, the large quantities of sulfur oxides and nitrogen oxides emitted into the atmosphere react with water to form strong acids (primarily sulfuric acid and nitric acid). Rain and snow formed in such areas have a pH of less than 5.6. The phenomenon of "acid precipitation" (meaning precipitation of pH below 5.6) is widespread today and occurs over large areas of western Europe, the eastern United States, and southeastern Canada, where the yearly average pH of precipitation ranges from 4 to 4.5. Moreover, the rain of individual storms is often much more acidic. Acid rains have been recorded in Scotland, Norway, and Iceland with pH values of 2.4, 2.7, and 3.5, respectively. The move to reduce local pollution problems through the building of tall stacks on smelters and industrial plants has created regional problems. Pollutants emanating from tall stacks are transported over long distances through the atmosphere. For example, more than 75 percent of the sulfur in rain that falls in the Scandinavian countries is believed to originate in the British Isles and central Europe.

The effect of acid rain on vegetation is not well understood but is currently being investigated throughout the Northern Hemisphere. Acid rains have been shown to decrease growth of forest trees in Sweden. In addition, simulated acid rains have been shown to injure leaves and inhibit seed germination. In just a few years the observed occurrences of damage in the forests of western Germany have increased from just a few percent to more than 50 percent. This suggests that plants can tolerate the pollutants to a certain threshold value, but once that value is met, rapid changes may result from small increases in specific pollutants.

The clearest effect of acid rain has been on fish populations, which have been virtually eliminated in acidified lakes in some parts of the world. It has been suggested that much of the toxicity to fish is actually due to increased aluminum concentrations and is not directly attributable to acidic water. Aluminum, which makes up about 5 percent of the Earth's crust, is almost completely insoluble in neutral or alkaline water and thus has not been biologically available. As a result of acid rain, however, concentrations of dissolved aluminum ions in some lakes may increase to levels toxic to fish and other aquatic organisms. The solubility of other toxic metals, such as lead, cadmium, and mercury, also increases sharply with decreasing pH.

(a) Sulfur dioxide injury on a blackberry (Rubus) leaf is characterized by areas of injured leaf tissue surrounded by healthy tissue. *(b)* Ozone injury on a tobacco (Nicotiana tabacum) leaf appears as stipples or flecks of dead tissue on the upper surface of the leaf. In cases of severe ozone injury, the flecks coalesce into larger lesions that are visible on both leaf surfaces. *(c)* Acid rain arises when sulfur oxides and nitrogen oxides react with water in the atmosphere to form sulfuric acid and nitric acid.

(a)

(b)

(c)

As in most stems, the epidermis is a single layer of cells covered by a cuticle. The stem epidermis generally contains far fewer stomata than the leaf epidermis.

The cortex consists of collenchyma and parenchyma cells. The several layers of collenchyma cells, which provide support to the young stem, form a continuous cylinder beneath the epidermis. The rest of the cortex consists of parenchyma cells that will contain chloroplasts when mature. The innermost layer of cortical cells, which have deeply colored contents, sharply delimits the cortex from the cylinder of primary vascular tissues.

In the great majority of stems, including those of *Tilia*, the primary phloem develops from the outer cells of the procambium, and the primary xylem develops from the inner ones. However, not all of the procambial cells mature as primary tissues. A single layer of cells between the primary xylem and the primary phloem remains meristematic and becomes the vascular cambium. *Tilia* is also an example of a woody stem—a stem that produces much secondary xylem later in growth. After internodal elongation is completed in the *Tilia* stem, fibers develop in the primary phloem. These fibers are called **primary phloem fibers** (see Figure 27–9).

The inner boundary of the primary xylem in *Tilia* is sharply delimited by one or two layers of pith cells that have deeply colored contents. The pith is composed primarily of parenchyma cells and contains numerous large ducts, or canals, containing mucilage (a slimy carbohydrate). Similar ducts are formed in the cortex (Figure 26–8). As the cortical and pith cells increase in size, numerous intercellular spaces develop among them; these air spaces are essential for exchange of gases with the atmosphere. The cortical and pith parenchyma cells store various substances.

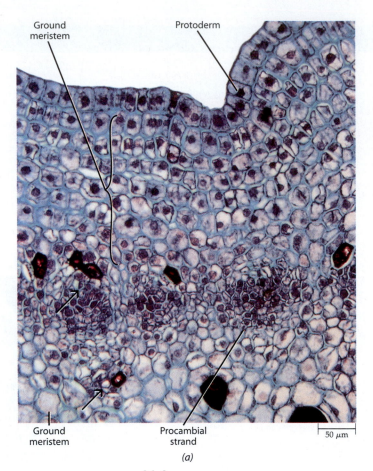

26–9

Transverse sections of the stem of the elderberry (Sambucus canadensis) *in a primary stage of growth. (a) A very young stem, showing protoderm, ground meristem, and three*

The Primary Vascular Tissues of the *Sambucus* Stem Form a System of Discrete Strands

In the stem of *Sambucus*, the interfascicular regions—also called **pith rays**—are relatively wide, and hence the procambial strands and primary vascular bundles form a system of discrete strands around the pith. The epidermis, cortex, and pith are similar in organization to those of *Tilia*. For this reason, the following discussion of the *Sambucus* stem will be used to explain in more detail the development of the primary vascular tissues of stems.

Figure 26–9a shows three procambial strands in which the primary vascular tissues have just begun to differentiate. The strand on the left is somewhat older than the two on the right and contains at least one mature sieve element and one mature tracheary element. Notice that the first mature sieve element appears in the outer part of the procambial strand (next to the cortex)

and that the first mature tracheary element appears in the inner part (next to the pith). Comparing Figures 26–9a and 26–9c, we see that the more recently formed sieve elements appear closer to the center of the stem and that the xylem differentiates in the opposite direction.

The first-formed primary xylem and primary phloem elements (protoxylem and protophloem, respectively) are stretched during elongation of the internode and are frequently destroyed. As in the *Tilia* stem, fibers develop in the primary phloem after internodal elongation is completed (see Figure 27–8).

Like the stems of *Tilia*, those of *Sambucus* become woody. In *Tilia*, almost all of the vascular cambium originates from procambial cells between the primary xylem and primary phloem because the interfascicular regions are very narrow. In *Sambucus*, with its relatively wide interfascicular regions, a substantial portion of the vas-

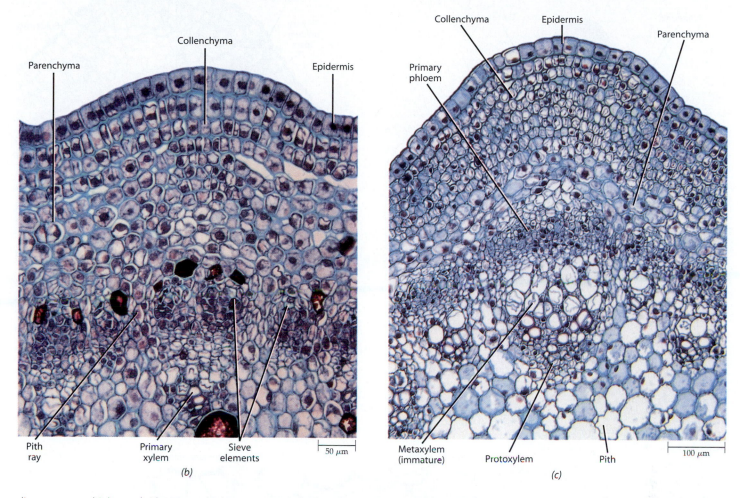

(b)

(c)

discrete procambial strands. The procambial strand on the left contains one mature sieve element (upper arrow) and one mature tracheary element (lower arrow). **(b)** *Primary* *tissues farther along in development.* **(c)** *Stem near completion of primary growth. Fascicular and interfascicular cambia are not* *yet formed. (For further stages in the growth of the elderberry stem, see Figures 27–7, 27–8, and 27–10.)*

cular cambium develops from the interfascicular parenchyma.

The Stems of *Medicago* and *Ranunculus* Are Herbaceous

The stems of many eudicots undergo little or no secondary growth and therefore are **herbaceous,** or nonwoody (see Chapter 27). Examples of herbaceous eudicot stems can be found in alfalfa (*Medicago sativa*) and in the buttercups (*Ranunculus*).

Medicago is an example of an herbaceous eudicot that exhibits some secondary growth (Figure 26–10). The structure and development of the primary tissues of the *Medicago* stem are similar to those of *Sambucus* and other woody angiosperms. The vascular bundles are separated by wide interfascicular regions and surround a large

pith. The vascular cambium is partly fascicular (procambial) and partly interfascicular (interfascicular parenchyma) in origin. During secondary growth, the secondary vascular tissues are formed mainly from the cambial cells derived from procambium. The interfascicular cambium generally produces only sclerenchyma cells on the xylem side.

The stem of *Ranunculus* is an extreme example of herbaceousness, and its vascular bundles resemble those of many monocots. The vascular bundles retain no procambium after the primary vascular tissues mature; hence, the bundles never develop a vascular cambium and lose their potential for further growth. Vascular bundles such as those of *Ranunculus* (Figure 26–11) and the monocots, in which all the procambial cells mature and the potential for further growth within the bundle is lost, are said to be **closed.** Closed vascular bundles are

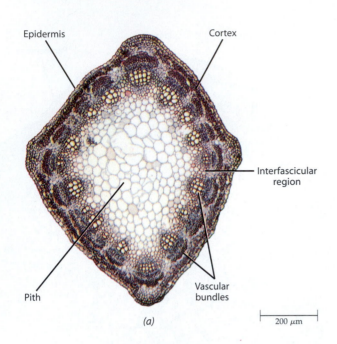

Epidermis   Cortex

Interfascicular region

Pith   Vascular bundles

(a)

200 μm

Primary phloem   Collenchyma

Parenchyma

Stoma

Primary phloem fibers   Epidermis

Interfascicular cambium   Primary xylem   Fascicular cambium

50 μm

(b)

26–10

(a) Transverse section of stem of alfalfa (Medicago sativa), *a eudicot with discrete vascular bundles.* **(b)** *Detail of a portion of the same alfalfa stem.*

Bundle sheath   Primary phloem

Primary xylem   50 μm

26–11

A closed vascular bundle. Transverse section of vascular bundle of the buttercup (Ranunculus), an herbaceous eudicot. The *vascular bundles of the buttercup are closed, that is, all of the procambial cells mature, precluding secondary growth. The primary phloem and primary xylem are surrounded by a bundle sheath of thick-walled sclerenchyma cells. Compare the vascular bundle shown here with the mature vascular bundle of maize shown in Figure 26–13c.*

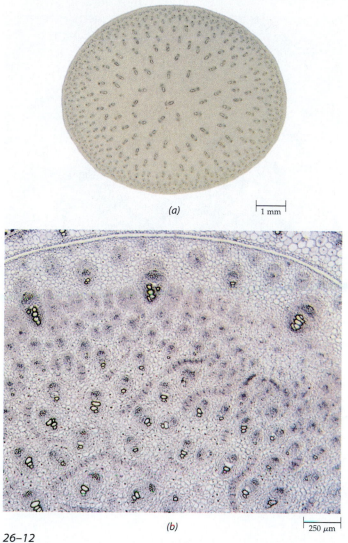

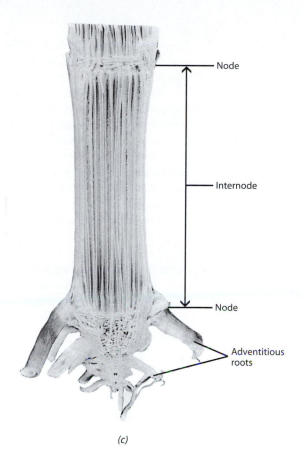

26–12

*Stem of maize (Zea mays). (**a**) Transverse section of the internodal region, showing numerous vascular bundles scattered throughout the ground tissue. (**b**) Transverse* *section of the nodal region of a young maize stem, showing horizontal procambial strands that interconnect the vertical bundles. (**c**) A* *mature stem split longitudinally; the ground tissue has been removed to expose the vascular system.*

usually entirely surrounded by a sheath composed of sclerenchyma cells. Vascular bundles that do give rise to a cambium are said to be **open**. In most eudicots, the vascular bundles are of the open type; they produce some secondary vascular tissues.

In the *Zea* Stem the Vascular Bundles Appear Scattered in Transverse Section

The herbaceous stem of maize (*Zea mays*) exemplifies the stems of monocots in which the vascular bundles appear scattered throughout the ground tissue in transverse section (Figure 26–12). As in other monocots, the vascular bundles of maize are closed.

Figure 26–13 shows three stages in the development of a maize vascular bundle. As in the bundles of eudicot stems, the phloem develops from the outer cells of the procambial strand, and the xylem develops from the inner cells. Also, as described previously, the phloem and the xylem differentiate in opposite directions. The first-formed phloem and xylem elements (protophloem and protoxylem) are stretched and destroyed during elongation of the internode. This results in the formation of a very large space, called a protoxylem lacuna, on the xylem side of the bundle (Figure 26–13c). The mature vascular bundle contains two large metaxylem vessels, and the phloem (metaphloem) is composed of sieve-tube elements and companion cells. The entire bundle is enclosed in a sheath of sclerenchyma cells.

26–13

*Three stages in the differentiation of the vascular bundles of maize (Zea mays), as seen in transverse sections of the stem. **(a)** The protophloem elements and two protoxylem elements are mature. **(b)** The protophloem sieve elements are now crushed, and much of the metaphloem is mature. Three protoxylem elements are now mature, and the two metaxylem vessel elements are almost fully expanded. **(c)** Mature vascular bundle surrounded by a sheath of thick-walled sclerenchyma cells. The metaphloem is composed entirely of sieve-tube elements and companion cells. The portion of the vascular bundle once occupied by the protoxylem elements is now a large space known as the protoxylem lacuna. Note the wall thickenings of destroyed protoxylem elements bordering the air space.*

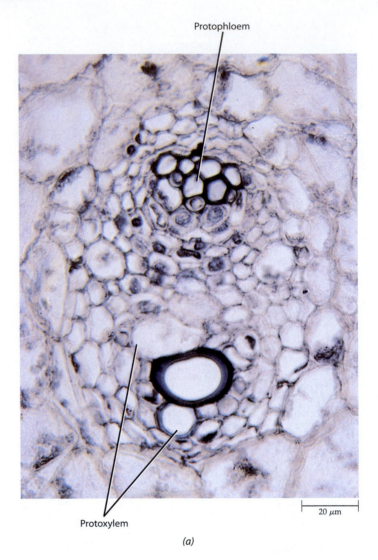

Protophloem

20 μm

Protoxylem

(a)

Relation between the Vascular Tissues of the Stem and the Leaf

The pattern formed by the vascular bundles in the stem reflects the close structural and developmental relationship between the stem and its leaves. The term "shoot" serves not only as a collective term for these two vegetative organs but also as an expression of their intimate physical and developmental association.

The procambial strands of the stem arise behind the apical meristem just below the developing leaf primordia and sometimes are present below the sites of future leaf primordia even before the primordia are discernible. As the leaf primordia increase in length, the procambial strands differentiate upward within them. From its inception, the procambial system of the leaf is continuous with that of the stem.

At each node, one or more vascular bundles diverge from the cylinder of strands in the stem, cross the cortex, and enter the leaf or leaves attached at that node (Figure 26–14). The extensions from the vascular system in the stem toward the leaves are called **leaf traces,** and the wide gaps or regions of ground tissue in the vascular cylinder located above the level where leaf traces di-

verge toward the leaves are called **leaf trace gaps.** A leaf trace extends from its connection with a bundle in the stem—a **stem bundle**—to the level at which it enters the leaf. A single leaf may have one or more leaf traces connecting its vascular system with that in the stem. The number of internodes that leaf traces traverse before they enter a leaf differs, so the traces vary in length.

If the stem bundles are followed either upward or downward in the stem, they will be found to be associated with several leaf traces. A stem bundle and its associated leaf traces are called a *sympodium* (plural: sympodia) (Figure 26–14). In some stems, some or all the sympodia are interconnected, whereas in others, all the sympodia are independent units of the vascular system. Regardless, the pattern of the vascular system in the stem is a reflection of the arrangement of the leaves on the stem. Buds commonly develop in the axils of leaves, and their vascular system is connected with that of the main stem by **branch traces.** Hence, at each node both leaf traces and branch traces (commonly two per bud) diverge outward from the main stem (Figure 26–15).

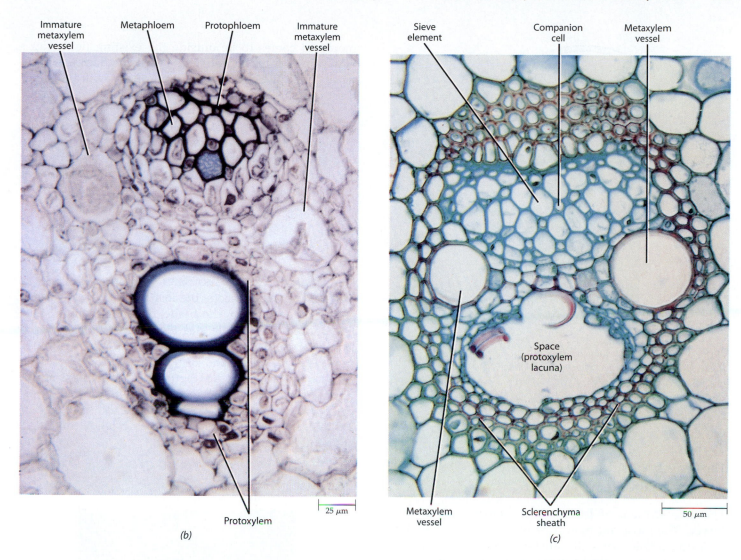

Immature metaxylem vessel · Metaphloem · Protophloem · Immature metaxylem vessel

Protoxylem

(b)

25 µm

Sieve element · Companion cell · Metaxylem vessel

Space (protoxylem lacuna)

Metaxylem vessel · Sclerenchyma sheath

50 µm

(c)

26–14

*Diagrams of the primary vascular system in the stem of an elm (Ulmus), a eudicot. **(a)** A transverse section of the stem showing the discrete vascular bundles encircling the pith. **(b)** Longitudinal view showing the vascular cylinder as though cut through leaf trace 5 in **(a)** and spread out in one plane. The transverse section in **(a)** corresponds with the topmost view in **(b)**. The numbers in both views indicate leaf traces. Three leaf traces—a median trace and two lateral traces—connect the vascular system of the stem with that of the leaf. A stem bundle and its associated leaf traces are called a sympodium.*

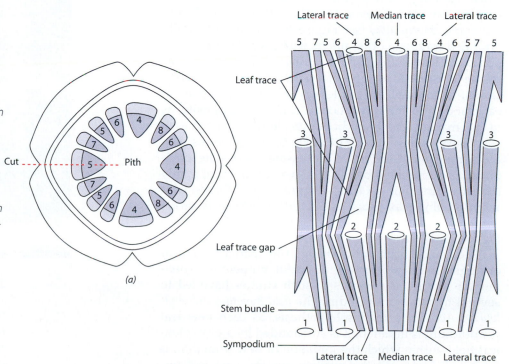

Cut · Pith

(a)

Lateral trace · Median trace · Lateral trace

5 7 5 6 4 8 6 4 6 8 4 6 5 7 5

Leaf trace

3 3 3 3

Leaf trace gap

2 2 2

Stem bundle

Sympodium

1 1 1 1

Lateral trace · Median trace · Lateral trace

(b)

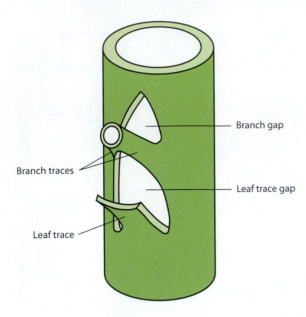

26–15

The relation between branch traces and a leaf trace to the vascular system in the main stem. Actually, the branch traces are leaf traces—the leaf traces of the first leaves of the bud or lateral branch. In magnoliids and eudicots, there are ordinarily two branch traces per bud.

The Leaves Are Arranged in Orderly Patterns on the Stem

The arrangement of leaves on a stem is called **phyllotaxy.** The most common type of phyllotaxy is **helical,** or spiral, phyllotaxy, with one leaf at each node and the leaves forming a helical pattern around the stem. For example, the oaks (*Quercus*), *Croton* (Figure 26–1), and mulberry (*Morus alba*) (see Figure 26–17a) have helically arranged leaves. In other plants with a single leaf at each node, the leaves are disposed in two opposite ranks, as in the grasses. This type of phyllotaxy is called **distichous.** In some plants the leaves are formed in pairs at each node and the phyllotaxy is said to be **opposite,** as in the maples (*Acer*) and honeysuckle (*Lonicera*). If each successive pair is at a right angle to the previous pair, the arrangement is termed **decussate.** Decussate phyllotaxy is exemplified by the members of the mint family (*Labiatae*), including *Coleus* (Figure 26–2). Plants with **whorled** phyllotaxy, such as Culver's-root (*Veronicastrum virginicum*), have three or more leaves at each node (see Figure 26–17b).

The mechanism underlying the pattern of leaf arrangement has been the object of numerous experimental studies. The results of such studies have led to several hypotheses, including the *field hypothesis of phyllotaxy.* According to this hypothesis, as each new leaf primordium is initiated, it is surrounded by a physiological field within which the initiation of new primordia is inhibited. Not until the position for the next leaf pri-

mordium comes to lie outside the existing fields can a new primordium be initiated. The fields decrease in intensity with the age of the primordium, with those farthest from the shoot tip being least intense.

A different explanation for the arrangement of leaves on a stem is the *first available space hypothesis.* According to this hypothesis, the major factor limiting the initiation of a new leaf primordium is space. A new primordium arises in the first space that becomes available—that is, when sufficient width and distance from the summit of the apex are attained.

Morphology of the Leaf

Leaves vary greatly in form and in internal structure. In magnoliids and eudicots, the leaf commonly consists of an expanded portion, the **blade,** or lamina, and a stalk-like portion, the **petiole** (see Figure 26–17). Scalelike or leaflike appendages called *stipules* develop at the base of some leaves (Figure 26–16). Many leaves lack petioles and are said to be *sessile* (Figure 26–18). In most monocots and certain eudicots, the base of the leaf is expanded into a **sheath,** which encircles the stem (Figure 26–18b). In some grasses, the sheath extends the length of an internode.

26–16

The pinnately compound leaf of the pea (Pisum sativum). Notice the stipules at the base of the leaf and the slender tendrils at the tip of the leaf. In the pea leaf, the stipules are often larger than the leaflets.

26–17

*Several examples of simple leaves. **(a)** Mulberry (Morus alba). **(b)** Culver's-root (Veronicastrum virginicum). **(c)** Sugar maple (Acer saccharum). **(d)** Silver maple (Acer saccharinum). **(e)** Red oak (Quercus rubra). Note the helical arrangement of the leaves in mulberry, and the whorled arrangement of those in Culver's-root. Leaf arrangement in the maples is opposite, and in oak it is helical, although only single leaves of these trees are shown here.*

26–18

*Sessile leaves (leaves without a petiole) are often found among eudicots, such as Moricandia, a member of the mustard family **(a)**, but are particularly characteristic of grasses and other monocots. **(b)** In maize (Zea mays), a monocot, the base of the leaf forms a sheath around the stem. The ligule, a small flap of tissue extending upward from the sheath, is visible. The parallel arrangement of the longitudinal veins is conspicuous in the portion of the blade shown here.*

(a) *(b)*

Leaf Dimorphism in Aquatic Plants

In the natural environment, the leaves of aquatic flowering plants may develop into two distinct forms (see Figure 10–16). Under water they develop into narrow and often highly dissected structures (water forms), and above the surface they develop into rather ordinary-looking leaves (land forms). It is possible to force immature leaves to develop into the form atypical for a given environment by applying any one of a wide range of treatments.

During a study on the aquatic plant Callitriche heterophylla, *it was found that the plant hormone gibberellic acid induced shoots growing in the air (emergent shoots) to produce water-form leaves. Abscisic acid, another plant hormone (see Chapter 28), led to the formation of land-form leaves on submerged shoots. Higher temperatures or the addition of mannitol, a sugar alcohol, to the water also caused submerged shoots to produce land-form leaves.*

In nature, the cellular turgor pressure of the submerged (water-form) leaves is relatively high, while that of the emergent (land-form) leaves is relatively low. The lower values in the emergent leaves may be due in part to transpirational loss of water vapor via the numerous stomata on the leaf surfaces. Associated with the higher turgor pressures in developing water-form leaves are long epidermal cells in the mature leaves.

Gibberellic acid caused the cells of emergent leaves to elongate, apparently by increasing cell wall plasticity, leading to increased water uptake and, hence, increased turgor pressure. At maturity these leaves had all the characteristics of typical water forms, including long epidermal cells. The limited cell expansion in submerged shoots exposed to abscisic acid or high temperatures apparently was not the result of low turgor; instead, the walls of the treated cells became less yielding so that high turgor failed to promote cell expansion. Growing submerged shoots in a mannitol solution resulted in turgor pressures similar to those of the emergent controls and the production of leaves with short epidermal cells.

The results of these experiments suggest that the relative magnitude of cellular turgor pressure determines the final leaf size and form in Callitriche heterophylla. *Thus, merely the presence or absence of water surrounding the developing leaf ensures that the leaf will be suitably adapted to life above or below the water.*

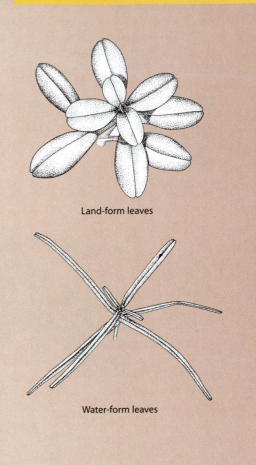

Land-form leaves

Water-form leaves

The leaves of magnoliids and eudicots are either simple or compound. In **simple leaves,** the blades are not divided into distinct parts, although they may be deeply lobed (Figure 26–17). The blades of **compound leaves** are divided into leaflets, each usually with its own small petiole (which is called a petiolule). Two types of compound leaves can be distinguished: pinnately compound and palmately compound leaves (Figure 26–19). In pinnately compound leaves, the leaflets arise from either side of an axis, the **rachis,** like the pinnae of a feather. (The rachis is an extension of the petiole.) The leaflets of a palmately compound leaf diverge from the tip of the petiole, and a rachis is lacking.

Since leaflets are similar in appearance to simple leaves, it is sometimes difficult to determine whether the structure is a leaflet or a leaf. Two criteria may be used to distinguish leaflets from leaves: (1) buds are found in the axils of leaves—both simple and compound—but not in the axils of leaflets, and (2) leaves extend from the stem in various planes, whereas the leaflets of a given leaf all lie in the same plane.

Structure of the Leaf

Variations in the structure of angiosperm leaves are to a great extent related to the habitat, and the availability of water is an especially important factor affecting their form and structure. On the basis of their water requirements or adaptations, plants are commonly characterized as **mesophytes** (plants that require an environment that is neither too wet nor too dry), **hydrophytes** (plants that require a large supply of water or grow wholly or partly submerged in water), and **xerophytes** (plants that are adapted to arid habitats). Such distinctions are not sharp, however, and leaves often exhibit a combination of features that are characteristic of different ecological types. Regardless of their varying forms, the foliage

26–19

Some examples of compound leaves. A palmately compound leaf is shown in (a); all the others are pinnately compound. (a) Red buckeye (Aesculus pavia). (b) Shagbark hickory (Carya ovata). (c) Green ash (Fraxinus pennsylvanica var. subintegerrima). (d) Black locust (Robinia pseudo-acacia). (e) Honey locust (Gleditsia triacanthos). In the honey locust, each leaflet is subdivided into smaller leaflets, and therefore it is twice, or bipinnately, compound. Two honey locust leaves are depicted here.

leaves of angiosperms are specialized as photosynthetic organs and, like roots and stems, consist of dermal, ground, and vascular tissue systems.

The Epidermis, with Its Compact Structure, Provides Strength to the Leaf

The mass of epidermal cells of the leaf, like those of the stem, are compactly arranged and covered with a cuticle that reduces water loss (page 25). Stomata may occur on both sides of the leaf or only on one side, either the upper or, more commonly, the lower side (Figure 26–20). In leaves of hydrophytes that float on the surface of the water, stomata may occur in the upper epidermis only (Figure 26–21); the submerged leaves of hydrophytes usually lack stomata entirely. The leaves of xerophytes generally contain greater numbers of stomata than those of other plants. Presumably these numerous stomata

permit a higher rate of gas exchange during the relatively rare periods of favorable water supply. In many xerophytes, the stomata are sunken in depressions on the lower surface of leaves (Figures 26–22 and 26–23). The depressions may also contain many epidermal hairs. Together these two features may serve to reduce water loss from the leaf. Epidermal hairs may occur on either or both surfaces of a leaf. Thick coats of epidermal hairs and the secreted resins of some hairs may also retard water loss from leaves.

In the leaves of magnoliids and eudicots, the stomata are usually scattered over the surface (Figure 26–24a); their development is mixed—that is, mature and immature stomata occur side by side in a partially developed leaf. In most monocots, the stomata are arranged in rows parallel with the long axis of the leaf (Figure 26–24b). Their formation begins at the tips of the leaves and progresses downward toward the base.

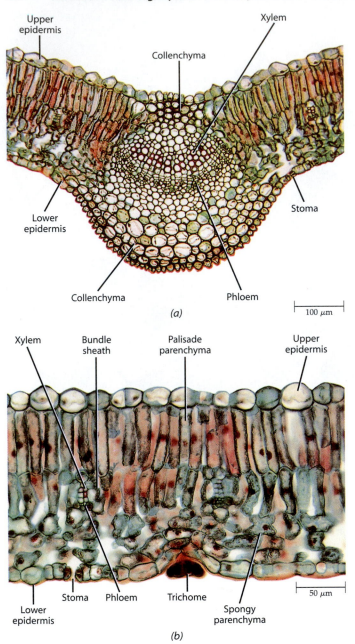

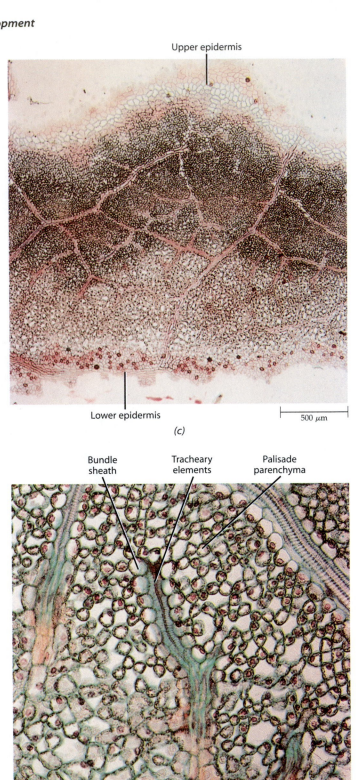

26–20

Sections of lilac (Syringa vulgaris) *leaf. (a) A transverse section through a midrib showing the midvein. (b) A transverse section through a portion of the blade. Two small veins (minor veins) can be seen in this view. (c) This is a "paradermal section" of the leaf. Strictly speaking, a paradermal section is one cut parallel to the epidermis. In practice, such sections are more or less oblique and extend from the upper to the lower epidermis. Thus, part of the upper epidermis can be seen in the light area at the top of this micrograph, and part of the lower epidermis can be seen at the bottom. Notice the greater number of stomata in the lower epidermis, as evidenced by the number of guard cells. (The darkly stained areas are trichomes.) The venation in lilac is netted. (d), (e) Enlargements of portions of (c). (d) Palisade parenchyma and spongy parenchyma with a vein ending, sectioned through some tracheary elements and surrounded by a bundle sheath. (e) A portion of the lower epidermis with two trichomes (epidermal hairs) and several stomata.*

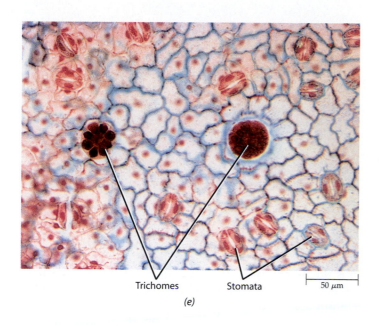

Trichomes   Stomata   50 μm

(e)

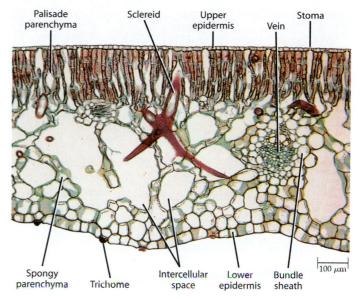

Palisade parenchyma   Sclereid   Upper epidermis   Vein   Stoma

Spongy parenchyma   Trichome   Intercellular space   Lower epidermis   Bundle sheath   100 μm

26–21

Transverse section of a leaf of the water lily (Nymphaea odorata), *a magnoliid, which floats on the surface of the water and has stomata in the upper epidermis only. As is typical of hydrophytes, the vascular tissue in the* Nymphaea *leaf is much reduced, especially the xylem. The palisade parenchyma consists of several layers of cells above the spongy parenchyma. Note the large intercellular (air) spaces, which add buoyancy to this floating leaf.*

26–22

Transverse section of oleander (Nerium oleander) *leaf. Oleander is a xerophyte, and this is reflected in the structure of the leaf. Note the very thick cuticle covering the multiple (several-layered) epidermis on the upper and lower surfaces of the leaf. The stomata and trichomes are restricted to invaginated portions of the lower epidermis called stomatal crypts.*

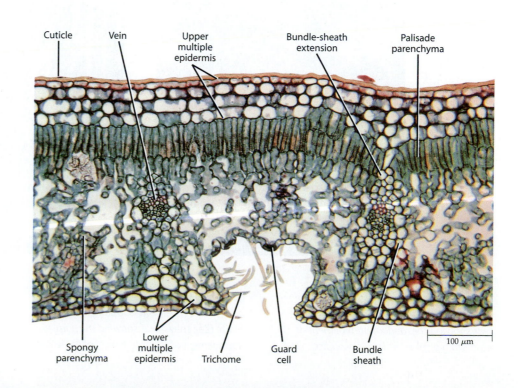

Cuticle   Vein   Upper multiple epidermis   Bundle-sheath extension   Palisade parenchyma

Spongy parenchyma   Lower multiple epidermis   Trichome   Guard cell   Bundle sheath   100 μm

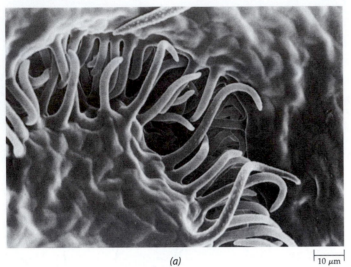

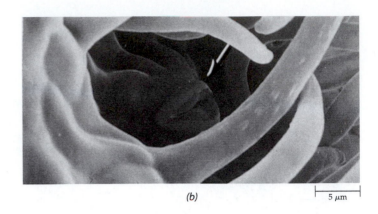

(b)   5 μm

26–23

Stomatal crypt. Scanning electron micrographs of a stomatal crypt in the lower epidermis of an oleander (Nerium oleander) *leaf. (a) Lower-magnification view of a crypt, showing numerous trichomes lining the crypt. (b) A higher magnification showing one of the stomata (arrow), which are restricted to crypts.*

(a)   10 μm

The Mesophyll Is Specialized for Photosynthesis

It is the **mesophyll**—the ground tissue of the leaf—with its large volume of intercellular spaces and numerous chloroplasts, that is particularly specialized for photosynthesis. The intercellular spaces are connected with the outer atmosphere through the stomata, which facilitate rapid gas exchange, an important factor in photosynthetic efficiency. In mesophytes, the mesophyll commonly is differentiated into **palisade parenchyma** and **spongy parenchyma.** The cells of the palisade tissue are columnar, with their long axes oriented at right angles to the epidermis, and the spongy parenchyma cells are irregular in shape (Figure 26–20b, d). Although the

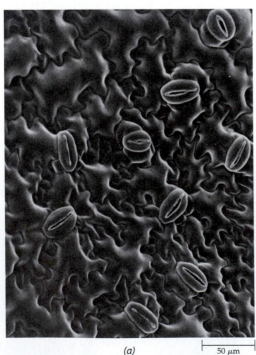

(a)   50 μm   *(b)*   25 μm

26–24

Stomata, shown in scanning electron micrographs. (a) Potato (Solanum tuberosum) *leaf, showing the random arrangement of stomata typical of the leaves of eudicots. The* *guard cells in potato are crescent-shaped and are not associated with subsidiary cells. (b) Maize* (Zea mays) *leaf, showing the parallel arrangement of stomata typical of the leaves* *of monocots. In maize each pair of narrow guard cells is associated with two subsidiary cells, one on each side of the stoma.*

palisade parenchyma appears more compact than the spongy parenchyma, most of the vertical walls of the palisade cells are exposed to intercellular spaces, and in some leaves the palisade surface may be two to four times greater than the spongy surface. Chloroplasts are also more numerous in palisade cells than in spongy cells. Most of the photosynthesis in the leaf, therefore, appears to take place within the palisade parenchyma.

The palisade parenchyma is usually located on the upper side of the leaf, and the spongy parenchyma is generally on the lower side (Figure 26–20). In certain plants, including many xerophytes, palisade parenchyma often occurs on both sides of the leaf. In some plants—for instance, maize (see Figure 7–25) and other grasses (see Figures 26–26 through 26–28)—the mesophyll cells are of more or less similar shape, and a distinction between spongy and palisade parenchyma does not exist.

Vascular Bundles Are Distributed throughout the Mesophyll

The mesophyll of the leaf is thoroughly permeated by numerous vascular bundles, or **veins,** which are continuous with the vascular system of the stem. In most magnoliids and eudicots, the veins are arranged in a branching pattern, with successively smaller veins branching from somewhat larger ones. This type of vein arrangement is known as **netted venation,** or reticulate venation (Figure 26–25). Often the largest vein extends along the long axis of the leaf as a midvein. The midvein, along with its associated ground tissue, makes up the so-called midrib of such leaves (Figure 26–20a). By contrast, most monocot leaves have many veins of fairly

similar size that extend along the long axis of the leaf. This vein arrangement is called **parallel venation,** or striate venation (Figure 26–18b). In parallel-veined leaves, the longitudinal veins are interconnected by smaller veins, forming a complex network.

The veins contain xylem and phloem, which generally are entirely primary in origin. The midvein and sometimes the coarser veins, however, undergo limited secondary growth in some magnoliid and eudicot leaves. In the vein endings of magnoliid and eudicot leaves, xylem elements often extend farther than the phloem elements, but in some plants both xylem and phloem elements extend to the ends of the vein. Commonly, the xylem occurs on the upper side of the vein, and the phloem on the lower side (Figure 26–20a, b).

The small veins of the leaf that are more or less completely embedded in mesophyll tissue are called **minor veins,** while the large veins associated with ribs (protrusions most commonly on the underside of the leaf) are called **major veins.** It is the minor veins that play the principal role in the collection of photosynthates (organic compounds produced by photosynthesis) from the mesophyll cells. With increasing vein size, the veins become less closely associated spatially with the mesophyll and increasingly embedded in nonphotosynthetic rib tissues. Hence, as the veins increase in size, their primary function changes from collection of photosynthates to transport of photosynthates out of the leaf.

The vascular tissues of the veins are rarely exposed to the intercellular spaces of the mesophyll. The large veins are surrounded by parenchyma cells that contain few chloroplasts, whereas the small veins usually are enclosed by one or more layers of compactly arranged cells

26–25

A cleared leaf (with the chlorophyll removed) of a cottonwood (Populus deltoides), at two successive magnifications. The cottonwood leaf has the netted venation characteristic of a eudicot. The small areas of mesophyll delimited by veins are called areoles. No mesophyll cell of the leaf is far from a vein. Water and dissolved minerals are carried to the leaf through the xylem; organic molecules produced by photosynthesis in the leaf are carried out of the leaf through the phloem.

10 mm

1 mm

that form a **bundle sheath** (Figures 26–20b, d through 26–22). The cells of the bundle sheath often resemble the mesophyll cells in which the small veins are located. The bundle sheaths extend to the ends of the veins, assuring that no part of the vascular tissue is exposed to air in the intercellular spaces and that all substances entering and leaving the vascular tissues must pass through the sheath (Figure 26–20d). Thus the bundle sheath is positioned similarly to the endodermis of the root and may similarly control the movement of substances to and from the vascular tissues.

In many leaves, the bundle sheaths are connected with either upper or lower epidermis or both by panels of cells resembling the bundle-sheath cells (Figure 26–22). These panels are called **bundle-sheath extensions.** Besides offering mechanical support to the leaf, in some leaves they apparently conduct water from the xylem to the epidermis.

The epidermis itself provides considerable strength to the leaf because of its compact structure and its cuticle. In addition, collenchyma or sclerenchyma may be present beneath the epidermis of the vein ribs of the larger veins in magnoliid and eudicot leaves, providing support to the leaf. In monocot leaves, the veins may be bordered by fibers. Collenchyma cells and fibers may also be found along the leaf margins of magnoliid and eudicot leaves and of monocot leaves, respectively.

Grass Leaves

After the discovery of the C_4 pathway of photosynthesis in sugarcane (page 144), a great many studies were devoted to the comparative anatomy of grass leaves in relation to photosynthetic pathways. It was discovered that the leaves of C_3 and C_4 grasses have rather consistent anatomical differences. For example, in the leaves of C_4 grasses, the mesophyll cells and bundle-sheath cells typically form two concentric layers around the vascular bundles, as seen in transverse sections (Figure 26–26). The compactly arranged bundle-sheath cells of the C_4 grasses are very large parenchyma cells that contain many large, conspicuous chloroplasts. This concentric arrangement of the mesophyll and bundle-sheath layers in C_4 plants is referred to as *Kranz* (German for "wreath") *anatomy.* The significance of the Kranz anatomy in relation to C_4 photosynthesis is discussed on pages 144 to 146.

In the leaves of C_3 grasses, by contrast, the mesophyll cells and bundle-sheath cells are not concentrically arranged. Moreover, the relatively small cells of the parenchymatous bundle sheaths in these plants have rather small chloroplasts, and at low magnifications the cells appear empty and clear. Commonly, an inner, more or less thick-walled sheath (the so-called mestome sheath) is also present in the C_3 grasses (Figure 26–27).

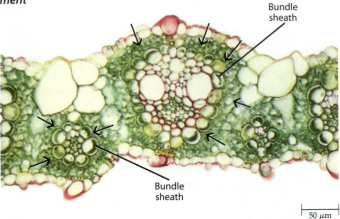

Bundle sheath

Bundle sheath

50 μm

26–26

Transverse section of sugarcane (Saccharum officinarum) leaf. As is typical of C_4 grasses, the mesophyll cells (arrows) are radially arranged around the bundle sheaths, which consist of large cells containing many large chloroplasts. (See also the maize leaf section in Figure 7–25.)

Another consistent structural difference between the leaves of C_3 and C_4 grasses is their interveinal distances, that is, the distances between laterally adjacent bundle sheaths. In C_4 grasses, only two to four mesophyll cells intervene between laterally adjacent bundle sheaths. In C_3 grasses, however, more than four (an average of 12 for the C_3 species in one study) mesophyll cells intervene between adjacent bundle sheaths.

The leaves of C_4 plants generally export photosynthates both more rapidly and more completely than leaves of C_3 plants. The reasons for these differences are unknown, but it has been suggested that the differences in physical distance between the mesophyll cells and the phloem of the vascular bundles may influence the rate and efficiency of loading of photosynthates into the sieve tubes.

The epidermis of grasses is made up of a variety of cell types. Most of the epidermal cells are narrow, elongate cells. Some especially large ones, termed **bulliform cells,** or motor cells, occur in longitudinal rows and are believed to participate in the mechanism of folding or rolling and unfolding or unrolling of the leaves, responses resulting from changes in water potential (Figure 26–28). During excessive loss of water, the bulliform cells become flaccid and the leaf folds or rolls. The epidermis also contains narrow, thick-walled guard cells, which are associated with subsidiary cells (Figure 26–24b; see also Figure 24–25).

Development of the Leaf

Much of our understanding of leaf development comes from **clonal analysis,** that is, from marking individual cells in a meristem and tracing their progeny, or clones,

26–27

Transverse section of wheat (Triticum aestivum) leaf. As is typical of C_3 grasses, the mesophyll cells are not radially arranged around the bundle sheaths. The bundle sheaths in the wheat leaf consist of two layers of cells: an outer sheath of relatively thin-walled parenchyma cells and an inner sheath, the mestome sheath, of thick-walled cells.

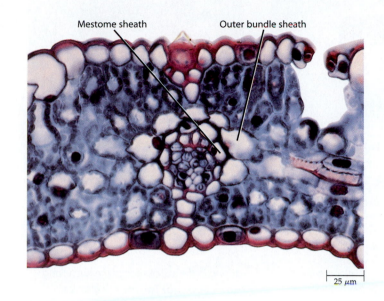

Mestome sheath   Outer bundle sheath

25 μm

26–28

Transverse sections of a leaf of annual bluegrass (Poa annua), a C_3 grass. Portions of (a) folded and (b) unfolded leaves including midvein. In the grass leaf, the mesophyll is not differentiated as palisade and spongy parenchyma. Strands of sclerenchyma cells commonly occur above and below the veins. The epidermis of the grass leaf contains bulliform cells—large epidermal cells thought to play a part in the folding and unfolding (rolling and unrolling) of grass leaves. In the Poa leaf shown in (a), the bulliform cells located in the upper epidermis are partly collapsed and the leaf is folded. An increase in turgor in the bulliform cells would presumably cause the leaf to unfold (b).

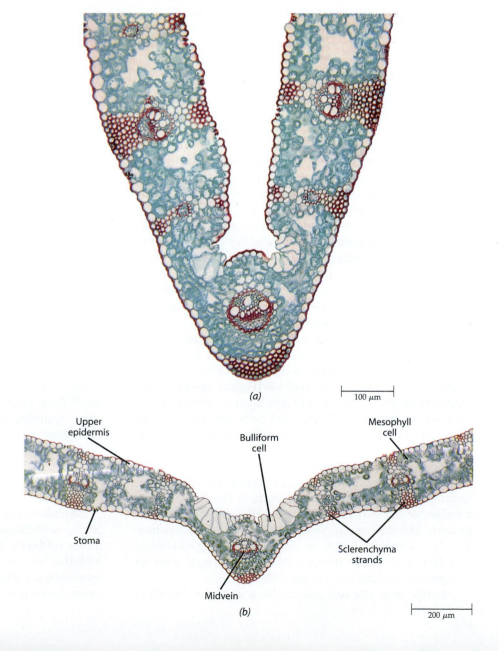

(a)

100 μm

Upper
epidermis

Bulliform
cell

Mesophyll
cell

Stoma

Sclerenchyma
strands

Midvein

(b)

200 μm

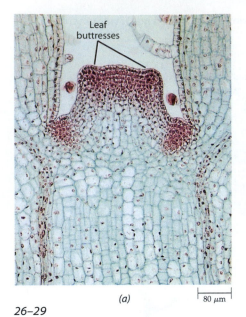

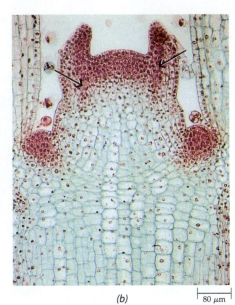

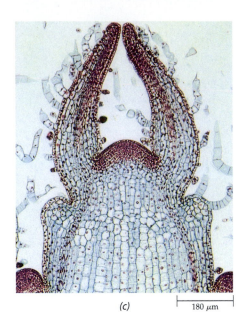

(a)   80 μm   (b)   80 μm   (c)   180 μm

26–29

Some early stages of leaf development in Coleus blumei, as seen in longitudinal sections of the shoot tip. The leaves in Coleus occur in pairs, opposite one another at the nodes (Figure 26–2). (a) Two small bulges, or leaf buttresses, can be seen opposite one another on the flanks of the apical meristem. In addition, a bud primordium can be seen arising in the axil of each of the two young leaves, below. *(b) Two erect, peglike leaf primordia have developed from the leaf buttresses. Notice the procambial strands (arrows) extending upward into the leaf primordia. The bud primordia, below, are farther along in development than those in (a). (c) As the leaf primordia elongate, the* procambial strands, which are continuous with the leaf-trace procambium in the stem, continue to develop into the leaves. Trichomes, or epidermal hairs, develop from certain protodermal cells very early, long before the protoderm matures to become the epidermis.

via cell lineages into the differentiated regions of the plant body. Such meristems are called *chimeral meristems*—meristems composed of layers or sectors of cells of different genetic composition. Genes involved in the production of anthocyanin and those involved in chlorophyll synthesis have commonly been used as genetic markers. Many variegated plants, such as crotons (Figure 26–1) and certain cultivars of ivy and geranium, are green-white chimeras. The white or yellow sectors are produced by mutant cells containing defective genes for chlorophyll synthesis. Although such mutations are known to occur spontaneously, they can be induced by exposure to X-rays or to chemical agents.

Clonal analysis has revealed that leaf primordia are initiated by groups of cells in the peripheral zone of the apical meristem. The group of cells spans all three layers of the meristem—L1, L2, and L3 (Figure 26–4a)—and ranges from about 5 to 10 cells per layer in *Arabidopsis* to somewhere between 50 and 100 cells per layer in tobacco, cotton, and maize. These cells are referred to as the **founder cells** of the leaf.

The earliest structural evidence of leaf initiation is a change in the orientation of cell division and expansion within the founder cells. This results in the formation of a bulge, or *leaf buttress* (Figure 26–29a). With continued growth, the buttress develops into the *leaf primordium,* which is generally flatter on its surface facing the apical meristem (the future upper surface of the leaf) than on its opposite surface (Figure 26–29b).

Shortly after the leaf primordium emerges from the

buttress, a distinctive, dense band of cells forms on opposite sides (along the margins) of the primordium. Formation of the blade is initiated in these narrow marginal bands, whereas the central region of the primordium differentiates into the midrib or rachis (Figure 26–30). Variation in the amount of expansion within the plane of the lamina accounts for much of the diversity of leaf shape.

In leaves with two tunica layers, the L1 layer gives rise to the epidermis, and the L2 and L3 layers contribute to the internal tissues. Expansion and increase in length of the leaf occur largely by *intercalary growth*, that is, by cell division and cell enlargement throughout the blade, with cell enlargement contributing the most. Differences in rates of cell division and cell enlargement in the various layers of the blade result in the formation of numerous intercellular spaces and produce the mesophyll form characteristic of the leaf. Typically, the leaf stops growing first at the tip and last at the base. Compared with growth of the stem, the growth of most leaves is of short duration. The restricted type of growth exhibited by the leaf and by floral apices is said to be *determinate*, as compared with the unlimited, or *indeterminate*, type of growth of the vegetative apical meristems.

Vascular development in a magnoliid or eudicot leaf begins with the differentiation of the procambium of the future midvein. This procambium differentiates upward into the primordium as an extension of the leaf-trace procambium (Figure 26–29c). All of the coarse, or major, veins develop upward and/or outward toward the mar-

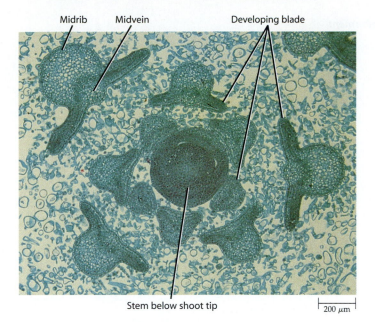

Midrib   Midvein   Developing blade

Stem below shoot tip

⊢ 200 μm ⊣

26–30

Transverse section of developing leaves of tobacco (Nicotiana tabacum) grouped around the shoot tip, sectioned below the apical meristem. The younger leaves are nearer the axis. A leaf primordium at first lacks differentiation into midrib and blade. Some early stages in development of blade and midrib can be seen here. (Portions of numerous trichomes surround the developing leaves.)

(b)   ⊢ 1 mm ⊣

26–31

Diagrams illustrating two stages in the development of the vascular system in the leaf of lettuce (Lactuca sativa). (a) Whereas the major, or coarse, veins develop upward in the blade, (b) the minor, or smaller, veins develop from the tip to the base of the blade. Thus, the tip of the leaf is the first region to have a complete system of veins.

(a)   ⊢ 1 mm ⊣

gins of the leaf in continuity with the procambium of the midvein (Figure 26–31a). The smaller, or minor, veins of the leaf are initiated at the tip of the leaf (Figure 26–31b) and develop from the tip to the base in continuity with the coarser veins. Thus, the tip of the leaf is the first part to have a complete system of veins. This course of development reflects the overall maturation of the leaf, which is from the tip to the base.

Development of the monocot leaf differs from that of the other angiosperm leaves in several respects. For example, in grasses such as maize and barley, growth activity quickly spreads laterally from the flanks of the developing leaf primordium and completely encircles the shoot apex. As the primordium increases in length, it gradually acquires a hoodlike shape (Figure 26–32). Further development of the blade proceeds in a linear manner, with new cells being added by the activity of a basal intercalary meristem. Elongation of the blade is restricted to a small zone above the region of cell division, and successively higher portions of the blade are more advanced in development. Growth of the sheath begins relatively late and lags behind the blade in development. The boundary between blade and sheath does not become distinct until the ligule, a thin projection from the top of the sheath, begins to develop (Figure 26–18b).

⊢ 50 μm ⊣

26–32

Scanning electron micrograph of developing leaf of barley (Hordeum vulgare), a monocot. At this stage, the developing lamina has a hoodlike form. The shoot apex can be seen through the slitlike opening in the hood.

Sun and Shade Leaves

Environmental factors, especially light, can have substantial developmental effects on the size and thickness of leaves. In many species, leaves grown under high light intensities—the so-called *sun leaves*—are smaller and thicker than the so-called *shade leaves* that develop under low light intensities (Figure 26–33). The increased thickness of sun leaves is due mainly to a greater development of the palisade parenchyma. The vascular system of sun leaves is more extensive, and the walls of epidermal cells are thicker than those of shade leaves. In addition, the ratio of the internal surface area of the mesophyll to the area of the leaf blade is much higher in sun leaves. One effect of these differences is that, although the two leaf types have similar photosynthetic rates at low light intensities, shade leaves are not adapted to high light intensities and consequently they have considerably lower maximum photosynthetic rates under these conditions.

Because light intensities vary greatly in different parts of the crowns of trees, extreme forms of sun and shade leaves can be found there. Sun and shade leaves also occur in shrubs and in herbaceous plants. They can be induced to form by growing the plants under high or low light intensities.

Leaf Abscission

In many plants, the normal separation of the leaf from the stem—the process of **abscission**—is preceded by certain structural and chemical changes near the base of the petiole. These changes result in the formation of an **abscission zone** (Figure 26–34). In woody angiosperms, two layers can be distinguished in the abscission zone: a separation (abscission) layer and a protective layer. The separation layer consists of relatively short cells with poorly developed wall thickenings that make it structurally weak. Prior to abscission, certain reusable ions and molecules are returned to the stem, among them magnesium ions, amino acids (derived from proteins), and sugars (some derived from starch). Then, enzymes break down the cell walls in the separation layer. The cell wall changes may include weakening of the middle lamella and hydrolysis of the cellulosic walls themselves. Cell division may precede the actual separation. If cell division occurs, the newly formed cell walls are the ones largely affected by the degradation phenomena. Beneath the separation layer, a protective layer composed of heavily suberized cells is formed, further isolating the leaf from the main body of the plant before the leaf drops. Tyloses may form in the tracheary elements before abscission (see page 665).

Eventually, the leaf is held to the plant by only a few strands of vascular tissue, which may be broken by the

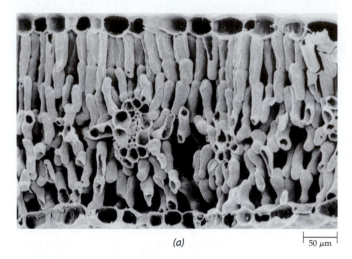

(a)   50 μm

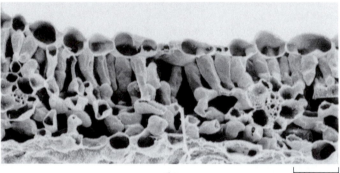

(b)   50 μm

26–33
*Scanning electron micrographs showing transverse veins of **(a)** sun and **(b)** shade leaves of* Thermopsis montana, *a member of the pea family* (Fabaceae). *Note that the sun leaf is considerably thicker than the shade leaf, a condition due mainly to a greater development of palisade parenchyma in the sun leaf.*

enlargement of parenchyma cells in the separation layer. After the leaf falls, the protective layer is recognized as a **leaf scar** on the stem (see Figure 27–17). Hormonal factors associated with abscission are discussed in Chapter 28.

Transition between Vascular Systems of the Root and the Shoot

As discussed in previous chapters, the distinction between plant organs is based primarily on the relative distribution of the vascular and ground tissues. For example, in eudicot roots the vascular tissues generally form a solid cylinder surrounded by the cortex. In addition, the strands of primary phloem alternate with the radiating ridges of primary xylem. By contrast, in the stem the vascular tissues often form a cylinder of discrete strands around a pith, with the phloem on the out-

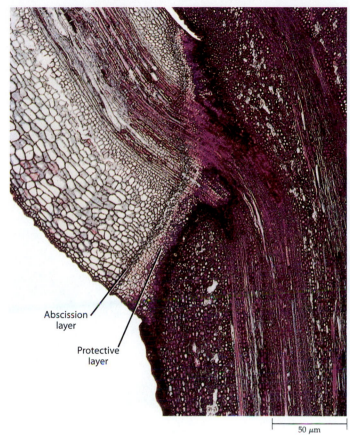

26–34

Abscission zone in maple (Acer) leaf, as seen in longitudinal section through the base of the petiole.

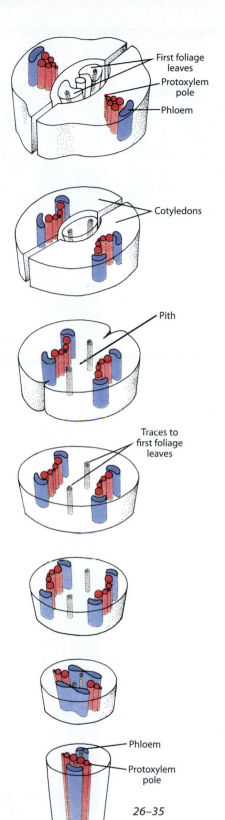

26–35

The transition region—the connection between root and cotyledons—in the seedling of a eudicotyledon with a diarch root, that is, a root with two protoxylem poles. In the root, the primary vascular system is represented by a single cylinder of vascular tissue. In the hypocotyl-root axis, the vascular system branches and diverges into the cotyledons, and the xylem and phloem become reoriented along the hypocotyl-root axis.

side of the vascular bundles and the xylem on the inside. Obviously, somewhere in the primary plant body, a change from the type of structure found in the root to that found in the shoot must take place. This change is a gradual one, and the region of the plant axis through which it occurs is called the **transition region.**

As described in Chapter 23, the shoot and root are initiated as a single continuous structure during development of the embryo. Consequently, vascular transition occurs in the axis of the embryo or young seedling. This transition is initiated during the appearance of the procambial system in the embryo and is completed with the differentiation of the variously distributed procambial tissues in the seedling. Vascular continuity between the root and shoot systems is maintained throughout the life of the plant.

The structure of the transition region can be very complex, and much variation exists in the transition regions of different kinds of plants. In most gymnosperms, magnoliids, and eudicots, vascular transition occurs within the vascular system connecting the root and the cotyledons. Figure 26–35 depicts a type of transition region commonly found in eudicots. Notice the diarch

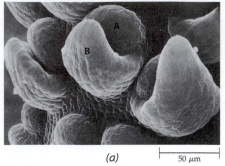

(a)   50 μm

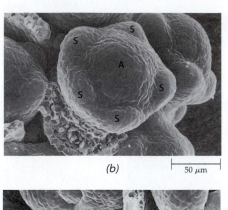

(b)   50 μm

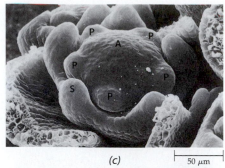

(c)   50 μm

26–36

Scanning electron micrographs showing some stages in the development of the perfect flower of Neptunia pubescens, *a legume with radial symmetry and whorled floral parts. (a) Floral apex (A) in axil of a bract (B). (b) Five sepal primordia (S) have been initiated around the floral apex. (c) Five petal primordia (P) have been initiated around the floral apex, alternating with the sepals (S). During their development, the sepals will form a calyx tube. (d) Five stamen primordia (two are indicated by arrows) have been initiated around the floral apex, alternating with the petals (P). (e) A second whorl of stamens (arrow) has been initiated, its members alternating with members of the first stamen whorl (ST$_1$). The carpel (C) has been initiated at the center of the floral apex. All floral parts are now present. (f) The carpel has now developed a cleft, which will form the locule of the ovary. The outer, or first, whorl of stamens (ST$_1$) are beginning to differentiate anthers and filaments. (ST$_2$ designates the inner, or second, whorl of stamens.) (g) The carpel is now beginning to differentiate into a style and ovary. (h) Older flower with both whorls of stamens in view. (i) Older flower with some stamens removed to reveal the carpel, with differentiated ovary (O), style, and stigma (arrow). Tips of the subtending bracts have been removed in (b) through (i). In (f) through (i) most sepals and petals have also been removed.*

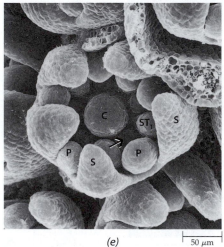

(e)   50 μm

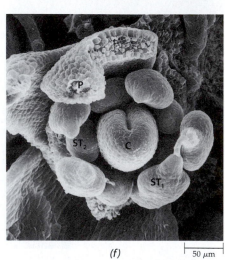

(f)   50 μm

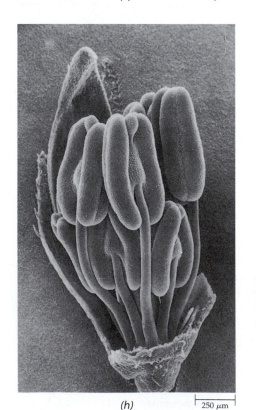

(h)   250 μm

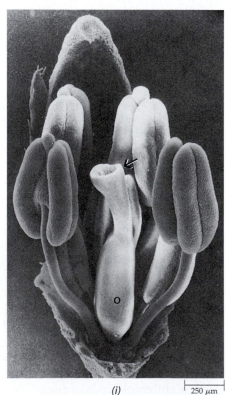

(i)   250 μm

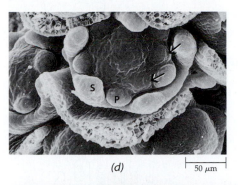

(d) |—— 50 μm ——|

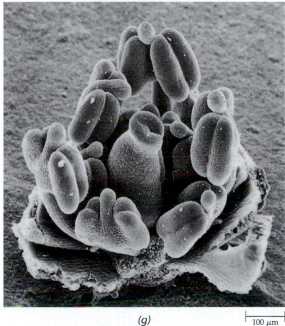

(g) |—— 100 μm ——|

(having two protoxylem poles) structure of the root; the branching and reorientation of the primary xylem and the primary phloem, which in the upper part of the axis result in the formation of a pith; and the traces of the first leaves of the epicotyl.

Development of the Flower

The development of the flower or inflorescence ends the meristematic activity of the vegetative shoot apex. During the transition to flowering, the vegetative shoot apex undergoes a sequence of physiological and structural changes and is transformed into a reproductive apex. Consequently, flowering may be considered as a stage in the development of the shoot apex and of the plant as a whole. Inasmuch as the reproductive apex exhibits a determinate growth pattern, flowering in annuals indicates that the plant is approaching completion of its life cycle. By contrast, flowering in perennials may be repeated annually or more frequently. Various environmental factors, including the length of day and the temperature, are known to be involved in the induction to flowering (see Chapter 29).

The transition from a vegetative to a floral apex is often preceded by an elongation of the internodes and the early development of lateral buds below the shoot apex. The apex itself undergoes a marked increase in mitotic activity, accompanied by changes in dimensions and organization: the relatively small apex with a tunica-corpus type of organization becomes broad and dome-like.

The initiation and early stages of development of the sepals, petals, stamens, and carpels are quite similar to those of leaves, their evolutionary precursors. Commonly, the initiation of the floral parts begins with the sepals, followed by the petals, then the stamens, and finally the carpels (Figure 26–36). This usual order of appearance of the floral parts may be modified in certain flowers, but the floral parts always have the same relative spatial relation to one another. The floral parts may remain separate during their development, or they may become united within whorls (connation) and between whorls (adnation).

The basic structure of the flower and some of its variations are discussed in Chapters 21 and 22.

A Small Set of Regulatory Genes Determines Organ Identity in the *Arabidopsis* Flower

In recent years our understanding of the genetic control of flower development has been greatly increased through the study of mutations that change floral part, or floral organ, identity. Such mutations, which result in the formation of the wrong organ in the wrong place, are called **homeotic mutations.** Horticulturists long have selected and propagated mutations that cause "double flowers," including the rose varieties grown in most gardens today. Whereas the wild rose contains only five petals, the rose doubles have 20 or more, the result of homeotic mutations that change the stamens of the wild-type rose into petals. Mutations of floral organ **homeotic genes**—genes affecting floral organ identity—have been studied most intensively in snapdragon (*Antirrhinum majus*) and *Arabidopsis thaliana* (Figure 26–37).

In *Arabidopsis*, the study of homeotic mutations has identified three classes of genes—designated A, B, and C—that are essential to the normal development and order of appearance of the floral organs produced in the flowers. The three classes are expressed in overlapping fields within the floral meristem; that is, each class acts in two adjacent whorls and is partly or wholly responsible for the identity of the organs in those two whorls (Figure 26–38). Class A genes function in the first and second whorls (sepals and petals), class B genes in the second and third whorls (petals and stamens), and class C genes in the third and fourth whorls (stamens and carpels).

(a)

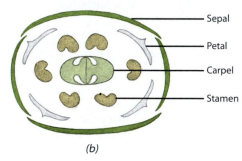

(b)

26–37
The wild-type, or normal, Arabidopsis *flower.*
*(a) Photograph. (b) Floral diagram showing
basic plan of flower. The flower contains four
sepals, four petals, six stamens, and a single
pistil (apocarpous gynoecium) composed of
two fused carpels.*

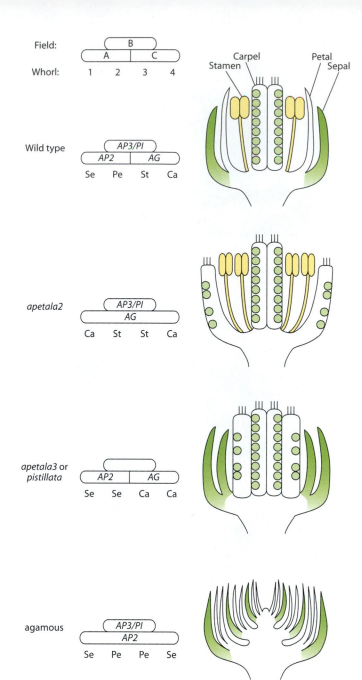

26–38
A model for the determination of floral organ identity in Arabidopsis *by
three classes of floral homeotic genes. Each class affects two whorls of
organs. In wild-type (normal)* Arabidopsis *flowers, the APETALA2 (AP2)
gene specifies field A, APETALA3 (AP3) and PISTILLATA (PI) genes specify
field B, and the AGAMOUS (AG) gene specifies field C, as indicated by
the boxes. The absence of letters in a box indicates the effect of a particular mutation (named on the left). For example, for sepals and petals to
be produced, both the AP3 and AP2 genes must be present, even though
the AP3 gene itself does not code for the production of sepals. Thus, if the
mutant ap2 gene is expressed, neither sepals nor petals will be produced,
even though the AP3 gene is present. The appearance, or phenotype, of
the wild-type and mutant flowers is depicted on the right. Se, sepal; Pe,
petal; St, stamen; Ca, carpel.*

The class A gene is exemplified by *APETALA2*, and
in the *apetala*2 mutant the sepals are converted into carpels and the petals into stamens. Two class B genes have
been identified—*APETALA3* and *PISTALLATA*—and, in
either mutant (*apetala*3 or *pistallata*), the petals are converted into sepals and the stamens into carpels. The class
C gene is exemplified by *AGAMOUS*. In *agamous* mutants the stamens are converted into petals and the
carpels are missing. At the site of the missing carpels the
cells behave as a floral meristem of sorts, generating additional sets of sepals and petals inside the first set. The
potential exists for the *agamous* pattern of development
to continue indefinitely. Figure 26–39 shows several
Arabidopsis flowers with homeotic mutations.

Similar types of homeotic mutations and regulatory
genes have been found in snapdragon, which is only dis-

(a)

(b)

(c)

26–39
Arabidopsis *flowers showing homeotic
mutations.* **(a)** *In* apetala2, *sepals are converted into carpels and petals into stamens.*
(b) *In* agamous, *stamens are converted into
petals and the site of the carpels into a floral
meristem.* **(c)** *In a triple mutant, with all three
classes of genes defective, all floral organs are
converted into leaves, the evolutionary precursors of the floral organs.*

tantly related to *Arabidopsis*. Apparently the basic mechanisms that control organ identity in developing flowers
are the same in both species, despite their remarkable
size difference.

Stem and Leaf Modifications

Stems and leaves may undergo modifications and perform functions quite different from those commonly associated with these two components of the shoot. One of
the most common modifications is the formation of **tendrils,** which aid in support. Some tendrils are modified
stems. In Boston ivy (*Parthenocissus tricuspidata*) and
Virginia creeper (*Parthenocissus quinquefolia*), the tendrils
form adhesive disks at their tips. The tendrils of the
grape *(Vitis)* (Figure 26–40) are modified stems that coil
around the supporting structure. In the grape, the tendrils sometimes produce small leaves or flowers. Most
tendrils, however, are leaf modifications. In legumes,
such as the garden pea *(Pisum sativum)*, the tendrils constitute the terminal part of the pinnately compound leaf
(Figure 26–16).

26–40
The tendrils of grape (Vitis) *are modified stems.*

26–41
The filmy branches of the common edible asparagus (Asparagus officinalis) *resemble leaves. Such modified stems are called cladophylls.*

Modified stems that assume the form of and closely resemble foliage leaves are called **cladophylls.** The leaflike branches of asparagus (*Asparagus officinalis*) are familiar examples of cladophylls (Figure 26–41). The thick and fleshy aerial shoots ("spears") of asparagus are the edible portion of the plant. The scales found on the spears are true leaves. If asparagus plants are allowed to continue growing, cladophylls develop in the axils of the minute, inconspicuous scales and then act as photosynthetic organs. In some cacti, the branches resemble leaves but are, in fact, cladophylls (Figure 26–42). As mentioned previously, true leaves typically have buds in their axils, while cladophylls do not. This characteristic may be used to distinguish between the two.

In some plants, the leaves are modified as spines, which are hard, dry, and nonphotosynthetic. The terms "spine" and "thorn" are frequently used interchangeably, but technically thorns are modified branches that arise in the axils of leaves (Figure 26–43). Another term commonly used interchangeably with thorn and spine is "prickle." A prickle, however, is neither a stem nor a leaf

but a small, more-or-less slender, sharp outgrowth from the cortex and epidermis. The so-called thorns on rose stems are prickles. All three structures—spines, thorns, and prickles—may serve as defensive structures, reducing predation by herbivores (plant-eating animals). In a special plant-herbivore interaction, the spines of the bull's-horn acacia provide shelter for ants that kill other insects attempting to feed on the acacia plant (see page 775).

Among the most spectacular of modified or specialized leaves are those of the carnivorous plants, such as the pitcher plant, the sundew, and the Venus flytrap, which capture insects and digest them with secreted enzymes. The available nutrients are then absorbed by the plant (see the essay on page 737).

Some Stems and Leaves Are Specialized for Food Storage

Stems, like roots, serve food-storage functions. Probably the most familiar type of specialized storage stem is the **tuber,** as exemplified by the Irish, or white, potato (*Solanum tuberosum*). In the white potato, tubers arise at the tips of **stolons** (slender stems growing along the surface of the ground) of plants grown from seed. However, when cuttings of tuber—the so-called "seed pieces"—are used for propagation, the tubers arise at the ends of long, thin **rhizomes,** or underground stems (Figure 26–44). Except for vascular tissue, almost the entire mass of the tuber inside the periderm ("skin") is storage parenchyma. The so-called "eyes" of the white potato

26–42
The branches of the spineless cactus Epiphyllum *resemble leaves but are actually modified stems called cladophylls.*

26–43

(a) Spines, as in this Ferocactus melocacti-formis, *are modified leaves. The spines origi-nate in the position of bud scales. (b) Thorns are modified branches, which, as this photo-graph of a hawthorn (*Crataegus*) shows, arise in the axils of the leaves.*

(b)

are nodal depressions containing groups of buds. Each seed piece cut from a potato tuber must include at least one eye. The depression is the axil of a scalelike leaf. The scalelike leaves are helically arranged on the tuber as are the leaves of the aboveground stems.

A **bulb** is a large bud consisting of a small, conical stem with numerous modified leaves attached to it. The leaves are scalelike and have thickened bases where food is stored. Adventitious roots arise from the bottom of the stem. Familiar examples of plants with bulbs are the onion (Figure 26–45a) and the lily.

Although superficially similar to bulbs, **corms** consist primarily of thickened, fleshy stem tissue. Their leaves commonly are thin and much smaller than those of bulbs; consequently, the stored food of the corm is found within the fleshy stem. Several well-known plants, such as gladiolus (Figure 26–45b), crocus, and cyclamen, pro-duce corms.

Kohlrabi (*Brassica oleracea* var. *caulorapa*) is one exam-ple of an edible plant with a fleshy storage stem. The short, thick stem stands above the ground and bears sev-eral leaves with very broad bases (Figure 26–45c). The common cabbage (*Brassica oleracea* var. *capitata*) is closely related to kohlrabi. The so-called "head" of cabbage con-sists of a short stem bearing numerous thick, overlap-ping leaves. In addition to a terminal bud, several well-developed axillary buds may be found within the head.

The leaf stalks, or petioles, of some plants become quite thick and fleshy. Celery (*Apium graveolens*) and rhubarb (*Rheum rhabarbarum*) are familiar examples.

26–44

*White potato (*Solanum tuberosum*), with tubers attached to a rhizome, or underground stem.*

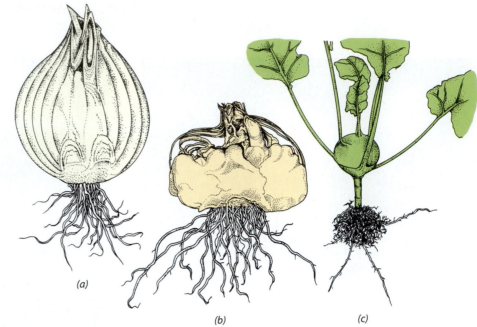

26–45
Examples of modified leaves or stems. (a) An onion (Allium cepa) bulb, which consists of a conical stem with scalelike, food-containing leaves attached. The leaves are the part of the onion we eat. (b) A gladiolus (Gladiolus grandiflorus) corm, which is a fleshy stem with small, thin leaves. (c) The fleshy storage stem of kohlrabi (Brassica oleracea var. caulorapa).

(a) (b) (c)

Some Stems and Leaves Are Specialized for Water Storage

Succulent plants are plants that have tissues specialized for the storage of water. Most of these plants—such as the cacti of the American deserts, euphorbias of similar appearance of the African deserts (see "Convergent Evolution" on page 266), and the century plant *(Agave)*—normally grow in arid regions, where the ability to store water is necessary for their survival. The green, fleshy stems of the cacti serve as both photosynthetic and storage organs. The water-storing tissue consists of large, thin-walled parenchyma cells that lack chloroplasts.

In the century plant, the leaves are succulent. As in succulent stems, nonphotosynthetic parenchyma cells of the ground tissue constitute the water-storing tissue. Other examples of plants with succulent leaves are the ice plant *(Mesembryanthemum crystallinum)*, the stonecrops *(Sedum)*, and certain species of *Peperomia*. In the ice plant, large epidermal cells with appendages (trichomes), called water vesicles, which superficially resemble beads of ice, serve a water-storage function. The water-storing cells of the *Peperomia* leaf are part of a multiple (several-layered) epidermis derived by periclinal divisions of the protoderm (Figure 26–46).

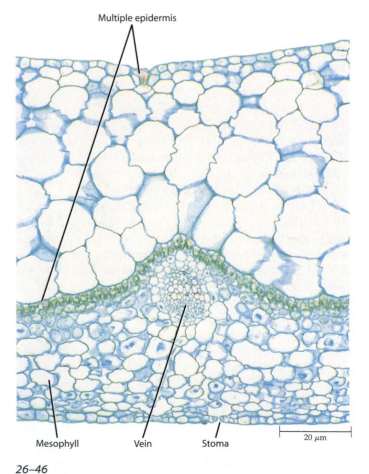

Multiple epidermis

Mesophyll Vein Stoma

20 µm

26–46
Transverse section of leaf blade of Peperomia. The very thick multiple epidermis, visible on its upper surface, presumably functions as water-storage tissue.

Summary

The Shoot Apical Meristem Produces Leaf Primordia, Bud Primordia, and Stem Primary Tissues

The vegetative shoot apices of most flowering plants have a tunica-corpus type of organization consisting of one or more peripheral layers of cells (the tunica) and an interior body of cells (the corpus). Most angiosperms have apices consisting of three superimposed layers of cells—two tunica layers and the initial layer of the corpus—which are designated L1, L2, and L3. Although the primary tissues of the stem pass through periods of growth similar to those of the root, the stem cannot be divided into regions of cell division, elongation, and maturation in the same manner as roots. The stem increases in length largely by internodal elongation.

Three Basic Types of Organization Exist in the Primary Structure of Stems

As in the root, the apical meristem of the shoot gives rise to protoderm, procambium, and ground meristem, which develop into the primary tissues. There are three basic patterns in the primary structure of stems, with regard to the relative distribution of ground and primary vascular tissues: the primary tissues may develop (1) as a more or less continuous cylinder within the ground tissue, (2) as a cylinder of discrete strands, or (3) as a system of strands that appear scattered throughout the ground tissue. Regardless of the type of organization, the phloem is commonly located outside the xylem.

Leaves and Stems Are Closely Related Physically and Developmentally

The term "shoot" serves not only as a collective term for the stem and its leaves but also as an expression of their intimate physical and developmental association. Leaf primordia are initiated by groups of cells, called founder cells, in the peripheral region of the shoot apex, and their position on the stem is reflected in the pattern of the vascular system in the stem. At each node, one or more leaf traces diverge from the stem and enter the leaf or leaves at that node. The procambial strands from which the leaf traces are formed develop just below the developing leaf primordia and are sometimes present below the sites of the future leaf primordia even before the primordia are discernible. Several hypotheses have been advanced to explain the mechanism underlying the pattern of leaf arrangement, or phyllotaxy, among them the field hypothesis and the first space available hypothesis.

Variations in Leaf Structure Are to a Great Extent Related to Habitat

In magnoliids and eudicotyledons, most leaves consist of a blade and petiole. The blades of some leaves (compound leaves) are divided into leaflets, whereas those of others (simple leaves) are not. Stomata are commonly more numerous on the lower than the upper surface of the leaf. The ground tissue, or mesophyll, of the leaf is specialized as a photosynthetic tissue and, in mesophytes, is differentiated into palisade parenchyma and spongy parenchyma. The mesophyll is thoroughly permeated by air spaces and by veins, which are composed of xylem and phloem surrounded by a parenchymatous bundle sheath. The xylem commonly occurs on the upper side of the vein, whereas the phloem occurs on the lower side.

Most C_3 and C_4 Grasses Can Be Distinguished on the Basis of Leaf Anatomy

In most monocots, including the grasses, the leaf consists of a blade and a sheath, which encircles the stem. The leaves of C_3 and C_4 grasses have rather distinct anatomical differences. Most notable among these differences is the presence in C_4 grasses and absence in C_3 grasses of a Kranz anatomy, that is, an arrangement in which the mesophyll cells and the bundle-sheath cells form two concentric layers around the vascular bundles.

Leaves Exhibit Determinate Growth, and Stems Exhibit Indeterminate Growth

Leaves are determinate in growth; that is, their development is of relatively short duration. However, the vegetative shoot apices that initiate leaves may exhibit unlimited, or indeterminate, growth. In many species, leaves grown under high light intensities are smaller and thicker than those grown under low light intensities. The former are called sun leaves, and the latter are called shade leaves.

The Separation of a Leaf from a Branch by Abscission Is a Complex Process

In many plants, leaf abscission is preceded by formation of an abscission zone, consisting of a separation layer and a protective layer, at the base of the petiole. After the leaf falls, the protective layer is recognized as a leaf scar on the stem.

The Transition Region Is Where the Root and Shoot Are United

The change in the distribution of the vascular and ground tissues found in the root to that seen in the shoot occurs in a region of the plant axis of the embryo and young seedling called the transition region. In most gymnosperms, magnoliids, and eudicots, vascular transition is located within the vascular system connecting the root and the cotyledons.

A Flower Is a Determinate Stem Tip Bearing Modified Leaves

At flowering, in many species, the vegetative shoot apex is directly transformed into a reproductive apex. Commonly, the initiation of the floral parts begins with the sepals, followed by the petals, then the stamens, and finally the carpels. Homeotic mutations may result in the formation of the wrong floral part in the wrong place.

Stems May Serve Food-Storage or Water-Storage Functions

Like roots, stems may be specialized for food storage. Examples of fleshy stems are tubers, bulbs, and corms.

Water-storing plants are known as succulents. The water-storage tissue of succulent plants is made up of large parenchyma cells. Stems or leaves or both may be succulent.

Selected Key Terms

abscission   p. 636

abscission zone   p. 636

blade   p. 624

branch trace   p. 622

bud primordium   p. 611

bulb   p. 643

bulliform cells   p. 632

bundle sheath   p. 632

bundle-sheath extension   p. 632

central mother cell zone   p. 612

cladophyll   p. 642

clonal analysis   p. 632

closed vascular bundle   p. 619

compound leaf   p. 626

corm   p. 643

founder cells   p. 634

herbaceous   p. 619

homeotic mutation   p. 639

hydrophyte   p. 626

intercalary meristem   p. 614

interfascicular parenchyma   p. 614

leaf primordium   p. 611

leaf scar   p. 636

leaf trace   p. 622

leaf trace gap   p. 622

major vein   p. 631

meristematic cap   p. 614

mesophyll   p. 630

mesophyte   p. 626

minor vein   p. 631

netted venation   p. 631

open vascular bundle   p. 621

palisade parenchyma   p. 630

parallel venation   p. 631

peripheral meristem   p. 613

petiole   p. 624

phyllotaxy   p. 624

phytomere   p. 611

pith meristem   p. 613

pith ray   p. 618

primary phloem fiber   p. 618

rachis   p. 626

rhizome   p. 642

sheath   p. 624

shoot   p. 611

simple leaf   p. 626

spongy parenchyma   p. 630

stem bundle   p. 622

stolon   p. 642

tendril   p. 641

transition region   p. 637

tuber   p. 642

tunica-corpus   p. 612

vein   p. 631

xerophyte   p. 626

Questions

1. Distinguish between the following: leaf primordium/bud primordium; leaf trace/leaf trace gap; simple leaf/compound leaf; separation layer/protective layer; closed vascular bundle/open vascular bundle.

2. By means of simple, labeled diagrams, compare the structure of a eudicot root with that of a eudicot stem at the end of primary growth. Assume that the root is triarch and the vascular cylinder of the stem consists of a system of discrete vascular bundles.

3. How might the distribution of stomata differ among the leaves of mesophytes, hydrophytes, and xerophytes?

4. Explain why the mesophyll tissue is particularly suited for photosynthesis.

5. What are the principal roles of the major and minor veins of leaves?

6. In what ways does the leaf anatomy of C_3 grasses differ from that of C_4 grasses?

7. Structurally, how do sun leaves differ from shade leaves?

8. In what ways are some stems and leaves adapted for storage of food and water?

9. What are the main events in the initiation and development of a leaf?

10. How have mutations contributed to our understanding of the genetic control of flower development?

27

Secondary Growth in Stems

27–1
A solitary shagbark hickory (Carya ovata) in the summer condition. Plants have been able to achieve such great stature because of the ability of their roots and stems to increase in girth, that is, to undergo secondary growth. Most of the tissue produced in this manner is secondary xylem, or wood, which not only conducts water and minerals to the far reaches of the shoot but also provides great strength to the roots and stems.

OVERVIEW

Mention wood and bark to most people and they might think of logs burning on a fire or names carved on a tree trunk. To a botanist, however, wood and bark are indicators of secondary growth—evidence that a plant has grown in diameter rather than height.

This chapter is basically about wood and bark and the meristems that produce them. In contrast to the primary growth produced by apical meristems, secondary growth results from the activity of lateral meristems: the vascular cambium and the cork cambium. As you will see, wood is one of the tissues produced by the vascular cambium, and it is largely the accumulation of woody tissue that causes the stem and root to expand laterally. In many woody plants, so much wood is produced that the epidermis stretches and cracks, providing potential entryways for invasion by insects, microorganisms, and other pathogens. But before this can happen, the cork cambium forms and produces impermeable cork cells that seal off the stem or root from the environment. Bark consists of all tissues outside the vascular cambium, and cork and cork cambium are normally important components of bark.

After an introductory section on annuals, biennials, and perennials, the first half of this chapter examines the structure and function of the vascular and cork cambia and the tissues they produce. The second half focuses on wood itself, including the different types of woods, growth rings, and some of wood's external features.

CHECKPOINTS

By the time you finish reading this chapter, you should be able to answer the following questions:

1. How do annuals, biennials, and perennials differ?

2. What types of cells make up the vascular cambium, and how do these cells function?

3. How does secondary growth affect the primary body of the stem?

4. What tissues are produced by the cork cambium, and what is the function of the periderm?

5. What is bark, and how does its composition change during the life of a woody plant?

6. What is wood, and how does conifer wood differ from angiosperm wood?

In many plants—that is, most monocots and certain herbaceous eudicots, such as buttercup (*Ranunculus*)—growth in a given part of the plant body ceases with maturation of the primary tissues. At the other extreme are the gymnosperms and woody magnoliids and woody eudicots, in which the roots and stems continue to increase in diameter in regions that are no longer elongating (Figure 27–1). This increase in thickness or girth of the plant body—termed secondary growth—results from activity of the two **lateral meristems:** the **vascular cambium** and the **cork cambium.**

Herbs, or herbaceous plants, are plants with shoots that undergo little or no secondary growth. In temperate regions either the shoot or the entire plant lives for only one season, depending on the species. Woody plants—trees and shrubs—can live for many years. At the start of each growing season, primary growth is resumed, and additional secondary tissues are added to the older plant parts through reactivation of the lateral meristems. Although most monocots lack secondary growth, some (such as the palms) may develop thick stems by primary growth alone (page 614). Some palms undergo a type of secondary growth called diffuse secondary growth, which takes place in older parts of the stem a considerable distance from the shoot apex. There the parenchyma cells of the ground tissue continue to divide and expand for a long period, accompanied by a proportional increase in the size of the intercellular spaces.

Annuals, Biennials, and Perennials

Plants are often classified according to their seasonal growth cycles as annuals, biennials, or perennials. In the **annuals**—which include many weeds, wildflowers, garden flowers, and vegetables—the entire cycle from seed to vegetative plant to flowering plant to seed again occurs within a single growing season, which may be only a few weeks in length. Only the dormant seed bridges the gap between one season of growth and the next.

In the **biennials,** two seasons are needed for the period from seed germination to seed formation. The first season of growth results in the formation of a root, a short stem, and a rosette of leaves near the soil surface. In the second growing season, flowering, fruiting, seed formation, and death occur, completing the life cycle. In temperate regions, annuals and biennials seldom become woody, although both their stems and their roots may undergo a limited amount of secondary growth.

Perennials are plants in which the vegetative structures live year after year. The herbaceous perennials pass unfavorable seasons as dormant underground roots, rhizomes, bulbs, or tubers. The woody perennials, which include vines, shrubs, and trees, survive above ground but usually stop growing during the unfavorable seasons. Woody perennials flower only when they become adult plants, which may take many years. For example, the horse chestnut, *Aesculus hippocastanum*, does not flower until it is about 25 years old. *Puya raimondii*, a large (up to 10 meters high) relative of the pineapple that is found in the Andes, takes about 150 years to flower. Many woody plants in temperate regions are deciduous, losing all their leaves at the same time and developing new leaves from buds when the season again becomes favorable for growth. In evergreen trees and shrubs, leaves are also lost and replaced, but not all leaves simultaneously.

The Vascular Cambium

Unlike the many-sided initials of the apical meristems, which contain dense cytoplasm and large nuclei, the meristematic cells of the vascular cambium are highly vacuolated. They exist in two forms: as vertically oriented **fusiform initials,** which are several to many times longer than they are wide, and as horizontally oriented **ray initials,** which are slightly elongated or squarish (Figures 27–2 and 27–3). The fusiform initials appear flattened or brick-shaped in transverse section.

Secondary xylem and secondary phloem are produced through periclinal divisions of the cambial initials and their immediate derivatives. In other words, the cell plate that forms between the dividing cambial initials is parallel to the surface of the root or stem (Figure 27–4a). If the derivative of a cambial initial is produced toward the outside of the root or stem, it eventually becomes a phloem cell; if it is produced toward the inside, it becomes a xylem cell. In this manner a long, continuous radial file, or row, of cells is formed, extending from the cambial initial outward into the phloem and inward into the xylem (Figure 27–5).

The xylem and phloem cells produced by the fusiform initials have their long axes oriented vertically and make up what is known as the **axial system** of the secondary vascular tissues. The ray initials produce horizontally oriented **ray cells,** which form the **vascular rays** or **radial system** (Figure 27–5). Composed largely of parenchyma cells, the vascular rays are variable in length. They serve as pathways for the movement of food substances from the secondary phloem to the secondary xylem and of water from the secondary xylem to the secondary phloem. Vascular rays also serve as storage centers for such substances as starch, proteins, and lipids, and may also synthesize some secondary metabolites.

In a restricted sense, the term "vascular cambium" is used to refer only to the cambial initials, of which there is one per radial file. However, it is often difficult, if not impossible, to distinguish between the initials and their immediate derivatives, which may remain meristematic for a considerable period of time before differentiating.

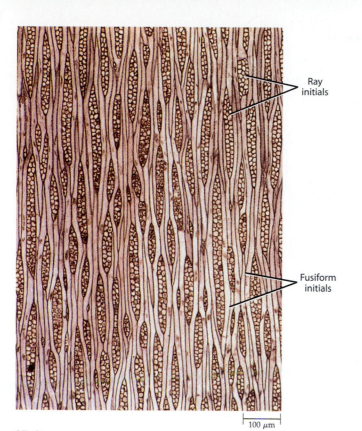

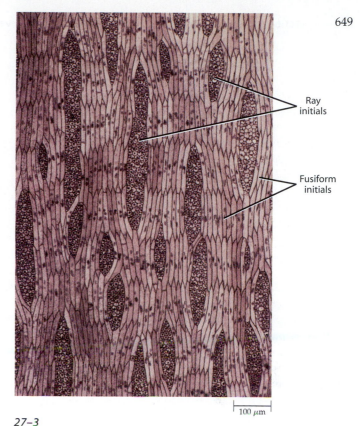

27–2

Vascular cambium of the apple (Malus sylvestris) tree, seen here in tangential view. Tangential sections are cut at right angles to the rays, so we see the rays here in transverse section (see Figure 27–12). A cambium such as this, in which the fusiform initials are not arranged in horizontal tiers on tangential surfaces, is said to be nonstoried. The fusiform initials average 0.53 millimeter in length in the apple.

27–3

Vascular cambium of the black locust (Robinia pseudo-acacia) tree, seen here in tangential view (see Figure 27–12). A cambium such as this, in which the fusiform initials are arranged in horizontal tiers on tangential surfaces, is said to be storied. The fusiform initials average 0.17 millimeter in length in black locust.

Even in the winter condition, when the cambium is dormant, or inactive, several layers of undifferentiated cells that are similar in appearance can be seen between the xylem and phloem. Consequently, some botanists use the term "vascular cambium" in a broader sense to refer to the initials and their immediate derivatives, which are indistinguishable from the initials. Other botanists refer to this region of initials and derivatives as the **cambial zone.**

As the vascular cambium adds cells to the secondary xylem and the core of xylem increases in diameter, the cambium is displaced outward, thus increasing in circumference. To accommodate the circumference increase, new cells are added to the vascular cambium by anticlinal divisions of the initials (Figure 27–4b). Along with an increase in the number of fusiform initials, new ray initials and rays are added, so that a fairly constant ratio of rays to fusiform cells is maintained in the secondary vascular tissues. Obviously, the developmental changes that occur in the cambium are exceedingly complex.

In temperate regions, the vascular cambium is dormant during winter and becomes reactivated in the spring. During reactivation, the cambial cells take up water, expand radially, and begin to divide periclinally.

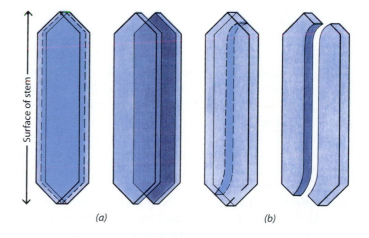

27–4

Periclinal and anticlinal divisions of fusiform initials. (a) Periclinal divisions are involved in the formation of secondary xylem and secondary phloem cells, and result in the formation of radial rows of cells (see Figure 27–5). When an initial divides periclinally, the two daughter cells appear one behind (or in front of) the other. (b) Anticlinal divisions are involved in the multiplication of fusiform initials. When an initial divides anticlinally, the two daughter cells appear side by side.

27–5

Diagram showing the relationship of the vascular cambium to its derivative tissues—secondary xylem and secondary phloem. The darker cells are the more recently derived ones. The vascular cambium is made up of two types of cells—fusiform initials and ray initials—which form the axial and radial systems, respectively. The arrangement of the cambial initials determines the organization of the secondary vascular tissues.

When the cambial initials produce secondary xylem and secondary phloem, they divide periclinally. Following division of an initial, one daughter cell (the initial) remains meristematic, and the other (the derivative of the initial) eventually develops into one or more cells of the vascular tissue. Cells produced toward the inner surface of the vascular cambium become xylem elements, and those produced toward the outer surface become phloem elements. The ray initials divide to form vascular rays, which lie at right angles to the derivatives of the fusiform initials. With the production of additional secondary xylem, the vascular cambium and secondary phloem are displaced in an outward direction. The diagrams (left to right) represent successively more mature stages.

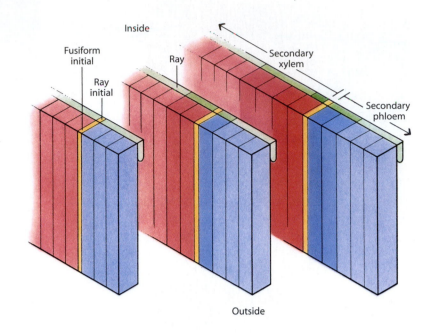

During expansion, the radial walls of the cambial cells become thinner, and as a result, the bark (all tissues to the outside of the vascular cambium) may be easily peeled or "slipped" off the stem. New growth layers, or increments, of secondary xylem and secondary phloem are laid down during the growing season. Reactivation of the vascular cambium is triggered by the expansion of the buds and resumption of their growth. The hormone auxin, produced by the developing shoots, moves downward in the stems and stimulates the resumption of cambial activity. Other factors are also involved in cambial reactivation and in continued normal growth of the cambium (see Chapter 28).

In some plants, the cambial cells divide more or less continuously, and the xylem and phloem elements undergo gradual differentiation. This type of cambial activity is found in plants growing in tropical regions. Not all tropical plants exhibit continuous cambial activity, however. About 75 percent of the trees growing in the rainforest of India exhibit continuous cambial activity. The percentage of such trees drops to 43 percent in the rainforest of the Amazon basin and to only 15 percent in that of Malaysia.

Effect of Secondary Growth on the Primary Body of the Stem

As mentioned in Chapter 26, the vascular cambium of the stem arises from the procambium that remains undifferentiated between the primary xylem and primary phloem, as well as from parenchyma of the interfascicular (between the bundles, or fascicles) regions. That portion of the cambium arising within the vascular bundles is known as **fascicular cambium,** and that arising in the interfascicular regions, or pith rays, is called **interfascicular cambium.** The vascular cambium of the stem, unlike that of the root, is essentially circular in outline from its inception (Figure 27–6).

In woody stems, the production of secondary xylem and secondary phloem results in the formation of a cylinder of secondary vascular tissues, with the rays extending radially through the cylinder (Figure 27–6). Commonly, much more secondary xylem than secondary phloem is produced in the stem in any given year; this situation is also true in the root. With secondary growth the primary phloem is pushed outward, and its thin-walled cells are destroyed. Only the thick-walled primary phloem fibers, if present, remain intact (see Figure 27–8). For a discussion of secondary growth in roots, see pages 600 to 602.

An elderberry (*Sambucus canadensis*) stem in two stages of secondary growth is shown in Figures 27–7 and 27–8. (Primary growth in the *Sambucus* stem is described on pages 618 and 619.) Only a small amount of secondary xylem and secondary phloem has been produced in the stem shown in Figure 27–7. In Figure 27–8, which shows a stem at the end of the first year's growth, considerably more secondary xylem than secondary phloem has been formed. The thick-walled cells outside the secondary phloem are primary phloem fibers. The soft-walled

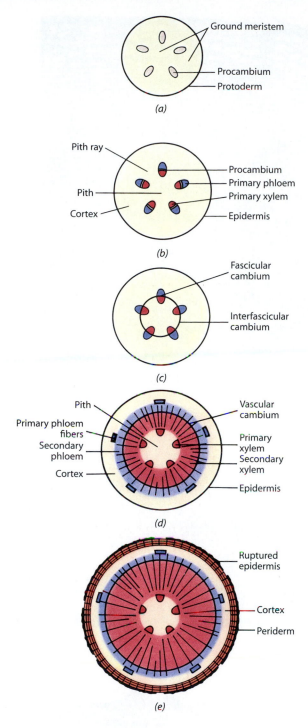

27–6

*Stem development in a woody angiosperm.
(a) Early stage in primary development,
showing primary meristems. (b) At completion of primary growth. (c) Origin of vascular
cambium. (d) After formation of some secondary xylem and secondary phloem. (e) At
end of first year's growth, showing the effect
of secondary growth—including periderm
formation—on the primary plant body. In (d)
and (e), the radiating lines represent rays.
(Compare this with root development as
depicted in Figure 25–16.)*

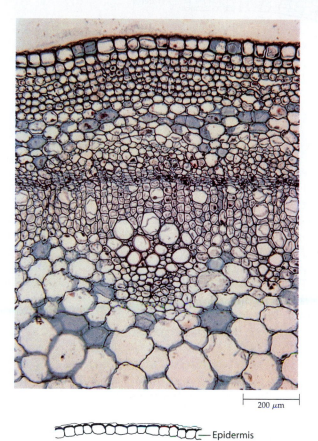

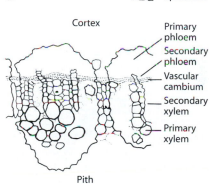

27–7

*Transverse section of an elderberry
(*Sambucus canadensis*) stem in which a
small amount of secondary growth has taken
place. A cork cambium has not yet formed.*

cells—sieve-tube elements and companion cells—are no
longer discernible. They were obliterated during development of the primary phloem fibers.

Figure 27–9 shows one-year-old, two-year-old, and
three-year-old stems of basswood (*Tilia americana*). In
Chapter 26, the stem of *Tilia* was given as an example of
one in which the primary vascular system appears as a
more or less continuous cylinder (see Figure 26–8), the
vascular bundles being separated from one another by
very narrow interfascicular regions, or pith rays. Thus
most of the vascular cambium in the *Tilia* stem is fascicular in origin. Some of the rays in the secondary phloem

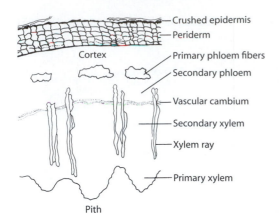

Crushed epidermis
Periderm
Cortex
Primary phloem fibers
Secondary phloem
Vascular cambium
Secondary xylem
Xylem ray
Primary xylem
Pith

27–8

Transverse section of an elderberry (Sambucus canadensis) *stem at the end of the first year's growth.*

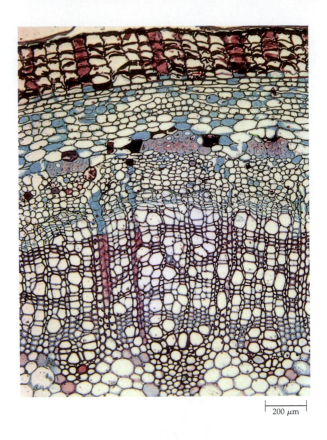

200 μm

of *Tilia* become dilated ("flaring") toward the periphery as the stem increases in girth (Figure 27–9a). This is one way in which the tissues outside the vascular cambium keep up with the increase in girth of the core of xylem.

The vascular cambia and secondary tissues of root and stem are continuous with one another. There is no transition region in the secondary plant body as there is in the primary plant body (page 637).

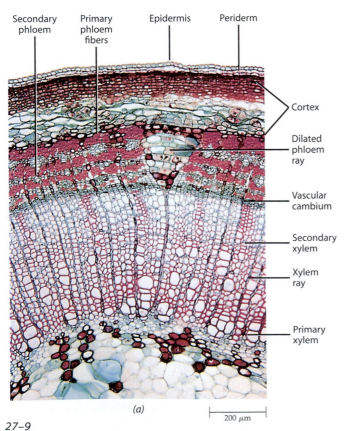

Secondary phloem
Primary phloem fibers
Epidermis
Periderm
Cortex
Dilated phloem ray
Vascular cambium
Secondary xylem
Xylem ray
Primary xylem

(a)

200 μm

27–9

Transverse sections of basswood (Tilia americana) *tree stems. (a) One-year-old stem. (b) Two-year-old stem. (c) Three-year-old stem. Numbers indicate growth increments in the secondary xylem. Note the difference in ring widths from year to year.*

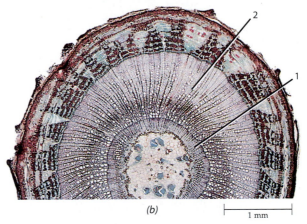

(b)

1 mm

2
1

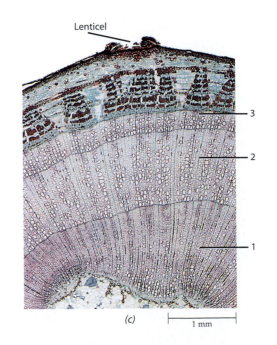

Lenticel

3
2
1

(c)

1 mm

652

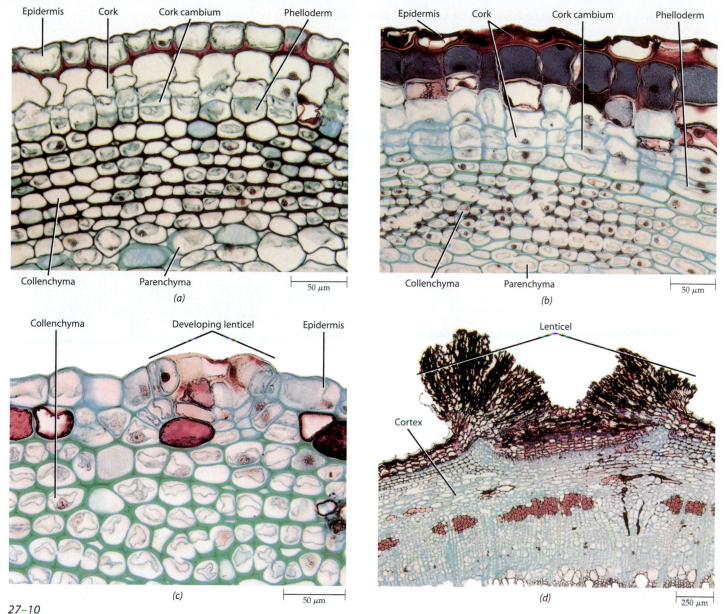

Epidermis **Cork** **Cork cambium** **Phelloderm**

Collenchyma **Parenchyma**
50 μm
(a)

Epidermis **Cork** **Cork cambium** **Phelloderm**

Collenchyma **Parenchyma**
50 μm
(b)

Collenchyma **Developing lenticel** **Epidermis**

(c) 50 μm

Lenticel

Cortex

(d) 250 μm

27–10
Some stages of periderm and lenticel devel-opment in elderberry (Sambucus canaden-sis), as seen in transverse sections. (a) Newly formed periderm, which consists of cork cambium, cork, and phelloderm, is seen beneath the epidermis. The epidermis has been separated from the cortex, consisting of collenchyma and parenchyma, by the peri-derm. (b) Periderm in more advanced stage of development, with increase in size of cork layer. Note that the epidermis is degenerating. *(c) Initiation of a lenticel. Collenchyma cells of the cortex can be seen beneath the develop-ing lenticel. (d) Well-developed lenticel. The phelloderm in Sambucus generally consists of a single layer of cells.*

The Periderm Is the Dermal Tissue System of the Secondary Plant Body

In most woody stems, as in most woody roots, periderm formation usually follows the initiation of secondary xylem and secondary phloem production. The **periderm** replaces the epidermis as the protective covering on those portions of the plant. Structurally, the periderm consists of three parts: the **cork cambium,** or phellogen, the meristem that produces the periderm; **cork,** or phellem, the protective tissue formed to the outside by the cork cambium; and the **phelloderm,** a living parenchyma tissue formed to the inside by the meristem (Figure 27–10).

In the stems of most woody plants, the first periderm usually appears during the first year of growth, most commonly originating in a layer of cortical cells immedi-ately below the epidermis, occasionally in the epidermis. In some species, the first periderm arises deeper in the stem, usually in the primary phloem.

Repeated divisions of the cork cambium result in the formation of radial rows of compactly arranged cells,

27–11
Lenticel of the stem of Dutchman's pipe
(Aristolochia), *as seen in transverse section.*
Unlike that of Sambucus, *the phelloderm of*
Aristolochia *consists of several layers of cells.*

Cuticle

Epidermis

Cork

Phelloderm

Cortex

200 µm

most of which are cork cells (Figures 27–10 and 27–11). During differentiation of the cork cells, their inner wall surfaces are lined by suberin lamellae, consisting of alternating layers of suberin and wax, which make the tissue highly impermeable to water and gases. The walls of the cork cells may also become lignified. At maturity, the cork cells are dead.

The cells of the phelloderm are living at maturity, lack suberin lamellae, and resemble cortical parenchyma cells. The phelloderm cells may be distinguished from cortical cells by their inner position in the radial rows of other periderm cells (Figure 27–11). Because the first periderm of the stem usually arises in the outer layer of cortical cells, the cortex of the stem is not sloughed during the first year as it is in woody roots (Figures 27–6 and 27–8; compare with Figures 25–16 and 25–17c), although the epidermis dries up and peels off.

At the end of the first year's growth, the following tissues are present in the stem: remnants of the epidermis, periderm, cortex, primary phloem (fibers and crushed soft-walled cells), secondary phloem, vascular cambium, secondary xylem, primary xylem, and pith (Figure 27–6). (Compare this list of tissues with that for the woody root at the end of the first year's growth on page 602).

The Lenticels Allow Gas Exchange through the Periderm

In the preceding discussion, we noted that the suberized cork cells are compactly arranged and, as a tissue, present an impermeable barrier to water and gases. However, the inner tissues of the stem, like all metabolically active tissues, need to exchange gases with the surrounding air, just as the inner tissues of the root need to

exchange gases with the surrounding air spaces between soil particles. In stems and roots containing periderms, this necessary gas exchange is accomplished by means of **lenticels** (Figures 27–10c, d and 27–11)—portions of the periderm with numerous intercellular spaces.

Lenticels begin to form during development of the first periderm (Figure 27–10) and, in the stem, generally appear below a stoma or group of stomata. On the surface of the stem or root, the lenticels appear as raised circular, oval, or elongated areas (see Figure 27–17). Lenticels are also formed on some fruits—for instance, the small dots on the surface of apples and pears are lenticels. As the roots and stems grow older, lenticels continue to develop at the bottom of cracks in the bark in newly formed periderms.

The Bark Includes All Tissues outside the Vascular Cambium

The terms "periderm," "cork," and "bark" are often unnecessarily confused with one another. As we've discussed, cork is one of three parts of the periderm; it is a secondary tissue that replaces the epidermis in most woody roots and stems. The term **bark** refers to all the tissues outside the vascular cambium, including the periderm or periderms when present (Figures 27–12 and 27–13). When the vascular cambium first appears, and secondary phloem has not yet been formed, the bark consists entirely of primary tissues. At the end of the first year's growth, the bark includes any primary tissues still present, the secondary phloem, the periderm, and any dead tissues remaining outside the periderm.

Each growing season, the vascular cambium adds secondary phloem to the bark, as well as secondary xylem, or wood, to the core of the stem or root. Usually

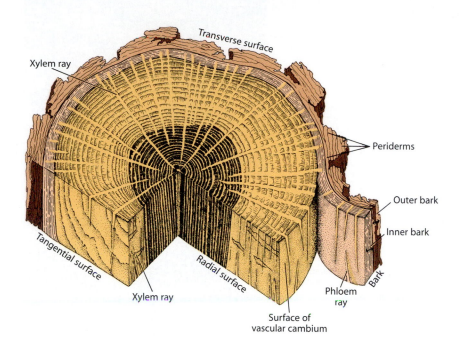

27–12
Diagram of part of a red oak (Quercus rubra) *stem, showing the transverse, tangential, and radial surfaces. The dark area in the center is heartwood. The lighter part of the wood is sapwood.*

the vascular cambium produces less secondary phloem than secondary xylem. In addition, the soft-walled cells (sieve elements and various kinds of parenchymatous elements) of the old secondary phloem are commonly crushed (Figures 27–14 through 27–16). Eventually, the old secondary phloem is separated from the rest of the phloem by newly formed periderms and is ultimately sloughed. As a result, considerably less secondary phloem accumulates in the stem or root than secondary xylem, which continues to accumulate year after year.

As the stem or root increases in girth, considerable stress is placed on the older tissues of the bark. In some plants, tearing of these tissues results in the formation of large air spaces. In many plants, the parenchyma cells of the axial system and rays divide and enlarge. In this manner, the old secondary phloem keeps up for a while with the increase in circumference of the stem or root. We noted earlier that, in the *Tilia* stem, certain rays, known as dilated rays, become very wide as the stem increases in girth (Figure 27–9a).

The first-formed periderm may keep up with the increase in girth of the root or stem for several years, with the cork cambium exhibiting periods of activity and inactivity that may or may not correspond to the periods of activity of the vascular cambium. In the stems of apple *(Malus sylvestris)* and pear *(Pyrus communis)* trees, the first cork cambium may remain active for up to 20 years. In most woody roots and stems, additional periderms are formed as the axis increases in circumference. After the first periderm, subsequently formed periderms originate deeper and deeper in the bark (Figures 27–12 and 27–13) from parenchyma cells of the phloem no longer actively engaged in the transport of food substances. These parenchyma cells become meristematic and form new cork cambia.

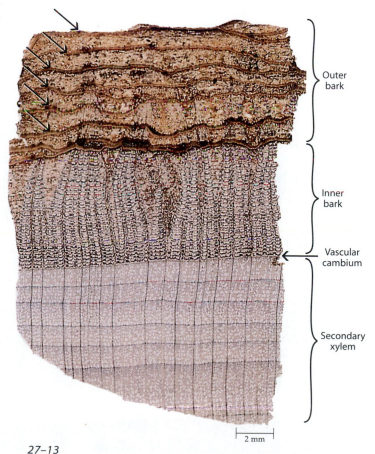

27–13
Transverse section of the bark and some secondary xylem from an old stem of basswood (Tilia americana). *Several periderms (arrows) can be seen traversing the mostly brownish outer bark in the upper third of the section. Below the outer bark is the inner bark, which is quite distinct in appearance from the more lightly stained secondary xylem in the lower third of the section.*

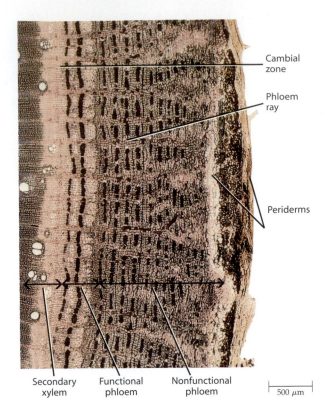

Cambial
zone

Phloem
ray

Periderms

Secondary
xylem

Functional
phloem

Nonfunctional
phloem

500 μm

27–14
Transverse section of the bark of a black locust (Robinia pseudo-acacia) *stem, consisting mostly of nonfunctional phloem.*

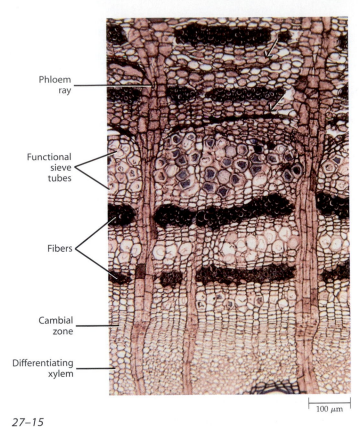

Phloem
ray

Functional
sieve
tubes

Fibers

Cambial
zone

Differentiating
xylem

100 μm

27–15
Transverse section of secondary phloem of the black locust (Robinia pseudo-acacia), *showing mostly functional phloem. Sieve tubes (indicated by arrows) of the nonfunctional phloem have collapsed.*

Cambial zone   Functional sieve tubes   Fibers   Cortex   Periderm

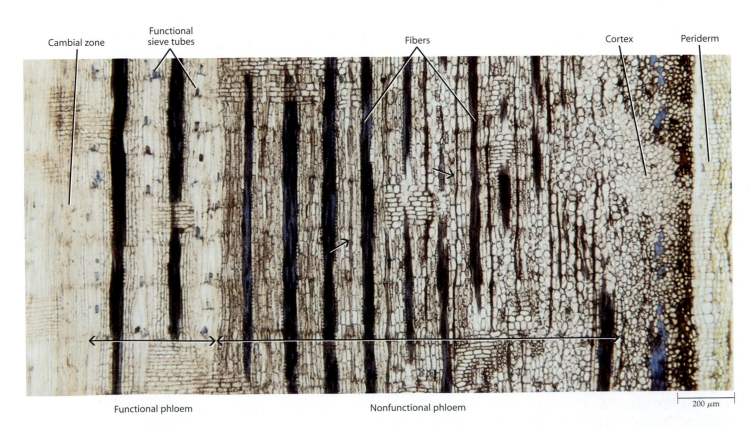

Functional phloem   Nonfunctional phloem

200 μm

27–16
Radial section of the bark of black locust (Robinia pseudo-acacia). *Most of the section consists of nonfunctional phloem, in which* the sieve tubes are collapsed (arrows). In black locust, the functional phloem consists only of the current season's growth increment, which *becomes nonfunctional in late autumn when its sieve tubes die and collapse.*

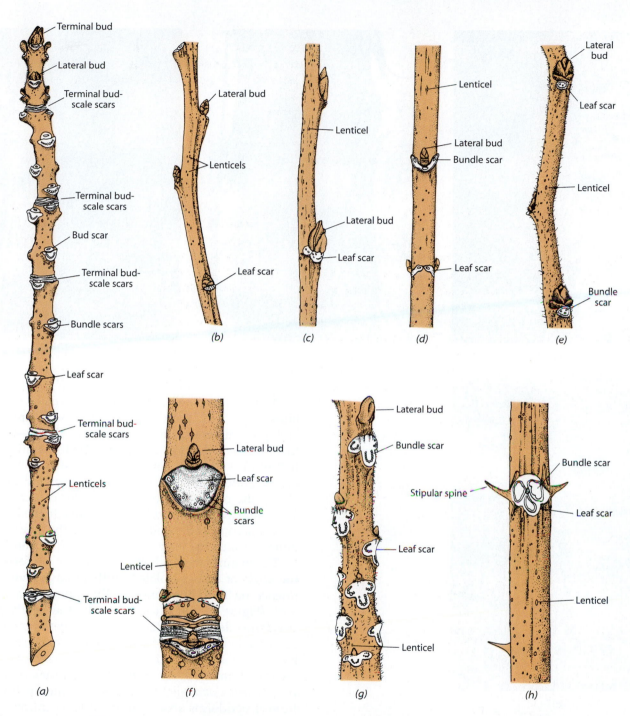

27–17

External features of woody stems. Examination of the twigs of deciduous woody plants reveals many important developmental and structural features of the stem. The most conspicuous structures on the twigs are the buds. Buds occur at the tips—the terminal buds—and in the axils of the leaves—the lateral, or axillary, buds of the twigs. In addition, accessory buds occur in some species. Commonly occurring in pairs, the accessory buds are located one on each side of a lateral bud. In some species, the accessory buds do not develop if their associated lateral bud

undergoes normal development. In others, the accessory buds form flowers and the lateral bud forms a leafy shoot.

After the leaves fall, leaf scars, with their bundle scars, can be seen beneath the lateral buds. The protective layer of the abscission zone produces the leaf scar. The bundle scars are the severed ends of vascular bundles that extended from the leaf traces into the petiole of the leaf, prior to abscission.

Groups of terminal bud-scale scars reveal the locations of previous terminal buds, and until they are obscured by secondary growth,

these groups of scars may be used to determine the age of portions of the stem. The portion of stem between two groups of such scars represents one year's growth. The lenticels appear as slightly raised areas on the stem. (a) Green ash (Fraxinus pennsylvanica var. subintegerrima). (b) White oak (Quercus alba). (c) Basswood (Tilia americana). (d) Box elder (Acer negundo). (e) American elm (Ulmus americana). (f) Horse chestnut (Aesculus hippocastanum). (g) Butternut (Juglans cinerea). (h) Black locust (Robinia pseudo-acacia).

(a)

(b)

(c)

(d)

27–18

*Bark of four species of trees. **(a)** Thin, peeling bark of the paper birch (Betula papyrifera). Horizontal lines on the surface of the bark are lenticels. **(b)** Shaggy bark of the shagbark hickory (Carya ovata). **(c)** Scaly bark of a sycamore, or buttonwood (Platanus occidentalis). **(d)** Deeply furrowed bark of the black oak (Quercus velutina).*

All of the tissues outside the innermost cork cambium—all of the periderms, together with any cortical and phloem tissues included among them—make up the *outer bark* (Figures 27–12 and 27–13). With maturation of the suberized cork cells, the tissues outside them are separated from the supply of water and nutrients. Hence the outer bark consists entirely of dead tissues. The living part of the bark, which is inside the innermost cork cambium and extends inward to the vascular cambium, is called the *inner bark* (Figures 27–12 and 27–13).

The manner in which new periderms are formed and the kinds of tissues isolated by them have a marked influence on the appearance of the outer surface of the bark (Figure 27–18). In some barks the newly formed periderms develop as discontinuous overlapping layers, resulting in formation of a scaly type of bark, called *scale bark* (Figures 27–12 and 27–13). Scale barks, for example, are found on relatively young stems of pine *(Pinus)* and pear *(Pyrus communis)* trees. Less commonly, the newly formed periderms arise as more or less continuous, concentric rings around the axis, resulting in formation of a *ring bark*. Grape *(Vitis)* and honeysuckle *(Lonicera)* are examples of plants with ring barks. The barks of many plants are intermediate between ring and scale barks.

Commercial cork is obtained from the bark of the cork oak, *Quercus suber*, which is native to the Mediterranean region (Figure 27–19). The first cork cambium of this tree has its origin in the epidermis, and the cork produced by it is of little commercial value. When the tree is about 20 years old, the original periderm is removed, and a new cork cambium is formed in the cortex, just a few millimeters below the site of the first one. The cork produced by the new cork cambium accumu-

27–19

Commercial cork being harvested from the trunk of the cork oak (Quercus suber). *The commercial value of cork arises from its imperviousness to liquids and gases and its strength, elasticity, and lightness.*

lates very rapidly and after about 10 years is thick enough to be stripped off the tree. Once again a new cork cambium arises beneath the previous one, and after about another 10 years the cork can be stripped again. After several strippings the new cork cambia are formed in the secondary phloem. This procedure may be repeated at about 10-year intervals until the tree is 150 or more years old. The spots and long dark streaks seen on the surfaces of commercial cork are lenticels.

In most woody roots and stems, very little secondary phloem is actually involved in the conduction of food. In most species, only the current year's growth increment, or growth ring, of secondary phloem is active in the long-distance transport of food through the stem. This is because the sieve elements are short-lived (see Chapter 24); most of them die by the end of the same year in which they are derived from the vascular cambium. In some plants, such as black locust (*Robinia pseudo-acacia*), the sieve elements collapse and are crushed relatively soon after they die (Figures 27–14 through 27–16).

The part of the inner bark actively engaged in the transport of food substances is called **functional phloem.** Although the sieve elements outside the functional phloem are dead, the phloem parenchyma cells

(axial parenchyma) and parenchyma cells of the rays may remain alive and continue to function as storage cells for many years. This part of the inner bark is known as **nonfunctional phloem.** Only the outer bark is composed entirely of dead tissue (Figures 27–14 and 27–16).

The Wood: Secondary Xylem

Apart from the use of various plant tissues as food for humans, no single plant tissue has played a more indispensable role in human survival throughout recorded history than **wood,** or secondary xylem (see Table 27–1). Commonly, woods are classified as either **hardwoods** or **softwoods.** The so-called hardwoods are angiosperm (magnoliid and eudicot) woods, and the softwoods are conifer woods. The two kinds of woods have basic structural differences, but the terms "hardwood" and "softwood" do not accurately express the relative density (weight per unit volume) or hardness of the wood. For example, one of the lightest and softest of woods is balsa (*Ochroma lagopus*), a tropical eudicot. By contrast, the woods of some conifers, such as slash pine (*Pinus elliottii*), are harder than some hardwoods. Although this chapter deals largely with secondary growth in angiosperms, for practical reasons we will consider conifer and angiosperm woods together.

Conifer Wood Lacks Vessels

The structure of conifer wood is relatively simple compared with that of most angiosperms. The principal features of conifer wood are its lack of vessels (see Chapter 24 for a discussion of tracheary elements) and its relatively small amount of axial, or wood, parenchyma. Long, tapering tracheids constitute the dominant cell type in the axial system. In certain genera, such as *Pinus*, the only parenchyma cells of the axial system are those associated with **resin ducts.** Resin ducts are relatively large intercellular spaces lined with thin-walled parenchyma cells that secrete resin into the duct. In *Pinus*, resin ducts occur in both the axial system and the rays (Figures 27–20 and 27–21). Wounding, pressure, and injuries by frost and wind can stimulate the formation of resin ducts in conifer wood, leading some investigators to suggest that all resin ducts result from trauma. Resin apparently protects the plant from attack by decay-producing fungi and bark beetles.

The tracheids of conifers are characterized by large, circular, bordered pits on their radial walls. Pits are most abundant on the ends of the cells, where they overlap with other tracheids (Figures 27–20 through 27–22). The pairs of pits, known as pit-pairs (page 66), between conifer tracheids are each characterized by the presence of a **torus** (plural: tori). The torus, a thickened central

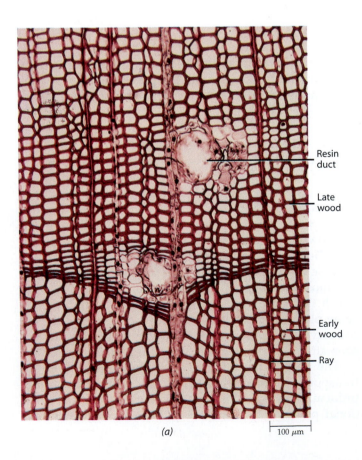

27–20

Block diagram of the secondary xylem of white pine (Pinus strobus), *a conifer. With the exception of the parenchyma cells associated with the resin ducts, the axial system consists entirely of tracheids. The rays are only one cell wide, except for those containing resin ducts. (Early and late wood are described on page 665.)*

27–21

Wood of white pine (Pinus strobus), *a conifer, in **(a)** transverse, **(b)** radial, and **(c)** tangential sections. (White pine wood has a specific gravity of 0.34; see page 668.)*

portion of the pit membrane (Figure 27–23), consists mainly of middle lamella and two primary walls. It is slightly larger than the openings, or apertures, in the pit-borders (Figure 27–22). The pit membrane is flexible, and under certain conditions, the torus may block one of the apertures and prevent the movement of water or gases through the pit-pair. Although long thought to occur only in certain genera of gymnosperms, tori have recently been reported in the bordered pit-pairs of tracheids and vessel elements in several genera of eudicots.

Figure 27–20 is a three-dimensional diagram of the wood of white pine *(Pinus strobus)* based on the three wood sections shown in Figure 27–21. In transverse sections, which are cut at right angles to the long axis of the root or stem, the tracheids appear angular or squarish, and the rays can be seen extending through the wood (Figure 27–21a). There are two kinds of longitudinal sections—radial and tangential. Radial sections are cut along a radius and parallel to the rays, and in such sections, the rays appear as sheets of cells oriented at right angles to the vertically elongated tracheids of the axial system (Figures 27–21b and 27–22d). Tangential sections are cut at right angles to the rays and reveal the width and height of the rays. In *Pinus* the rays are one cell wide, except for those containing resin ducts (Figure 27–21c), and one to 15 or more cells in height. Details of white pine wood are shown in Figure 27–22.

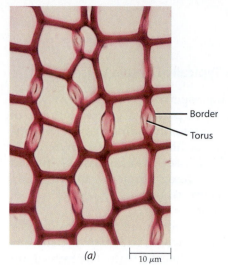

(a)

10 μm

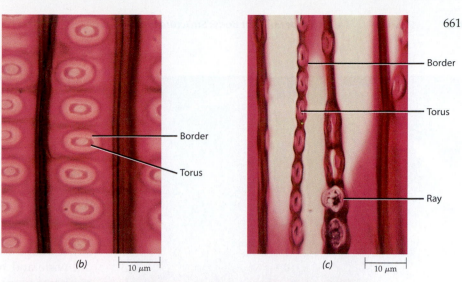

Border

Torus

(b)

10 μm

Border

Torus

Ray

(c)

10 μm

27–22

Details of white pine (Pinus strobus) *wood.* **(a)** *Transverse section, showing bordered pit-pairs on radial walls of tracheids.* **(b)** *Radial section, showing the face view of bordered pit-pairs in walls of tracheids.* **(c)** *Tangential section, showing bordered pit-pairs of tracheids.* **(d)** *Radial section, showing ray. The rays of pine and other conifers are composed of ray tracheids and ray parenchyma cells. Here ray tracheids occur at the top and bottom of the ray and ray parenchyma cells in the middle. Notice the bordered pits of ray tracheids. Above the ray parenchyma are two adjacent bordered pit-pairs.*

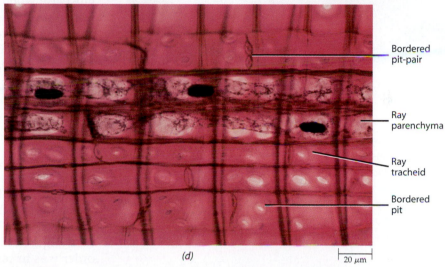

Bordered pit-pair

Ray parenchyma

Ray tracheid

Bordered pit

(d)

20 μm

Border

Torus

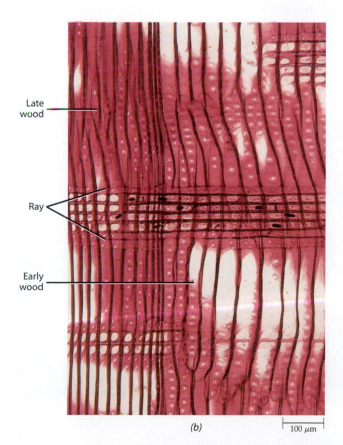

Late wood

Ray

Early wood

(b)

100 μm

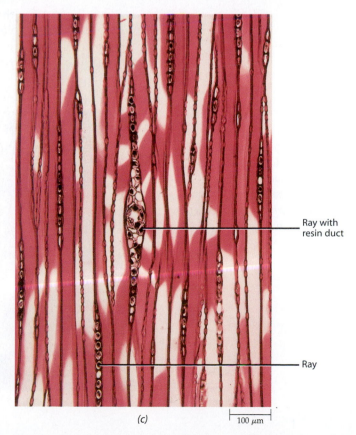

Ray with resin duct

Ray

(c)

100 μm

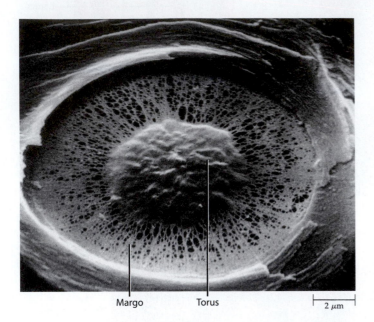

Margo   Torus   2 μm

27–23

*Scanning electron micrograph of the pit
membrane of a bordered pit-pair in a white
pine (Pinus strobus) tracheid. The thickened
part of the membrane is the torus, which is
impermeable to water. The part of the membrane surrounding the torus, called the
margo, is very porous, allowing the movement
of water and ions from tracheid to tracheid.*

Angiosperm Woods Typically Contain Vessels

Wood structure in angiosperms is much more varied
than in conifers, owing in part to a greater number of
cell types in the axial system, including vessel elements,
tracheids, several types of fibers, and parenchyma cells
(Figures 27–24 and 27–25; see also Figures 24–13 and
27–26). The presence of vessel elements, in particular,
distinguishes angiosperm woods from conifer wood,
with only a few exceptions.

The rays of angiosperm woods are often considerably
larger than those of conifer wood, which are usually one
cell wide and from one to 20 cells high. The rays of angiosperm woods range from one to many cells wide and
from one to several hundred cells high. In some angiosperm woods, such as oak, the large rays can be seen
with the unaided eye (Figure 27–12). The large rays of
the red oak (*Quercus rubra*) wood illustrated in Figure
27–25c are 12 to 30 cells wide and hundreds of cells high.
Besides the large rays, oak wood has numerous rays that
are only one cell wide. In red oak wood, the rays make
up, on average, about 21 percent of the volume of the
wood. Overall, the rays of hardwoods average about 17
percent of the volume of the wood, while the average for
conifer wood is about 8 percent.

As in conifer wood, transverse sections of angiosperm woods reveal radial files of cells of both the
axial and radial systems derived from the cambial initials (Figures 27–24 and 27–25a). The files may not be as
orderly as in conifer wood, however, for enlargement of
the vessels and elongation of fibers tend to push many of
the cells out of position. The displacement of rays by
vessel elements is particularly conspicuous in the transverse section of red oak wood, as shown by the wavy
rays to the left of the vessels in Figure 27–25a.

27–24

*Growth layers of wood, in transverse sections.
(a) Red oak (Quercus rubra). The large vessels
of ring-porous wood such as red oak are
found in the early wood. The dark vertical
lines are rays. (b) Tulip tree (Liriodendron
tulipifera), a diffuse-porous wood.*

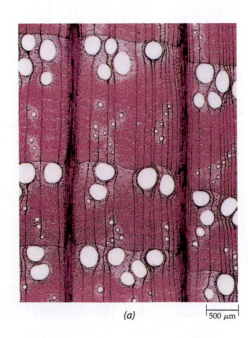

(a)   500 μm

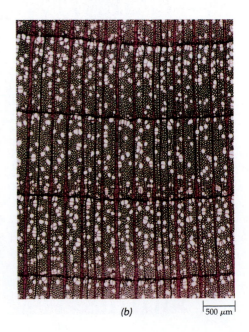

(b)   500 μm

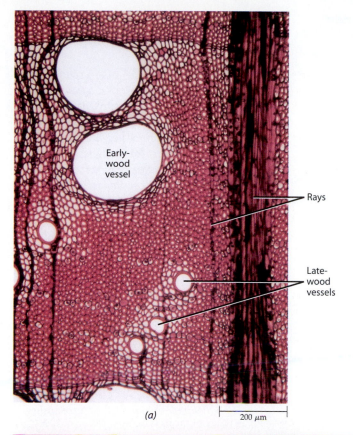

Early-wood vessel

Rays

Late-wood vessels

(a)

200 μm

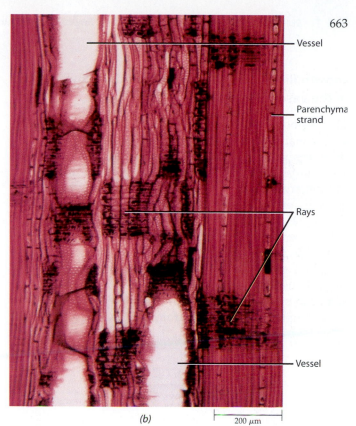

Vessel

Parenchyma strand

Rays

Vessel

(b)

200 μm

27–25

Wood of red oak (Quercus rubra) *in (a) transverse, (b) radial, and (c) tangential sections. (Red oak wood has a specific gravity of 0.57.)*

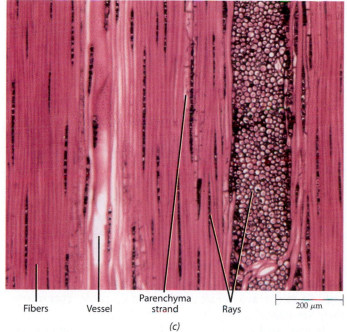

Fibers   Vessel   Parenchyma strand   Rays   200 μm

(c)

27–26

Scanning electron micrograph of block of American elm (Ulmus americana) *wood, showing the three faces, or surfaces, of the wood. By comparison with Figures 27–20, 27–24, and 27–25, you should be able to identify each surface. This is a semi-ring-porous wood, with the late-wood vessels arranged in wavy lines, a characteristic feature of the elms. Identify the early-wood and late-wood vessels and the rays in all three faces. The dense portion of the wood is composed largely of fibers. Axial parenchyma cells are also present but are not distinguishable at this magnification. (American elm wood has a specific gravity of 0.46.)*

2.5 mm

Growth Rings Result from the Periodic Activity of the Vascular Cambium

The periodic activity of the vascular cambium, which is a seasonally related phenomenon in temperate zones, produces growth increments, or **growth rings,** in both secondary xylem and secondary phloem (in the phloem the increments are not always readily discernible). If a growth layer represents one season's growth, it is called an **annual ring** (Figure 27–27). Abrupt changes in available water and other environmental factors may be responsible for the production of more than one growth ring in a given year; such rings are called *false annual rings.* Thus the age of a given portion of the old woody stem can be estimated by counting the growth rings, but the estimates may be inaccurate if false annual rings are

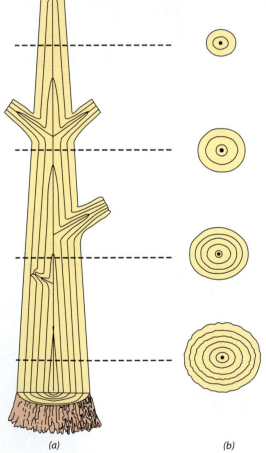

(a)   (b)

27–27

Annual rings in wood each represent one year's increment of growth. The number of rings varies with the distance above ground, the oldest part of the trunk occurring at ground level. (a) Diagram of a median longitudinal section of a tree trunk, and (b) transverse sections taken at four different levels. Once secondary growth has begun in a portion of stem (or root), that portion no longer increases in length.

27–28

(a) Bristlecone pine (Pinus longaeva) in the White Mountains of eastern California. These pines, which grow near the timberline, are the oldest living trees; one tree has reached an age of 4900 years. (b) Transverse section of wood from a bristlecone pine, showing the variation in width of annual rings. This section begins approximately 6260 years ago; the shaded band of rings represents the thirty years from 4240 B.C. to 4210 B.C. The overlapping patterns of rings in dead trees have made it possible to determine relative precipitation extending back some 8200 years.

Despite its advanced age, the 4900-year-old bristlecone pine may not be the oldest living thing on Earth. In fact, several years ago a ring of creosote bushes (Larrea divaricata), all apparently derived from one seed, was estimated to be about 12,000 years old. Growing in the Mojave Desert 150 miles northeast of Los Angeles, the ring of bushes has been dubbed the "King Clone" (see Figure 33–14). More recently, a group of Tasmanian botanists found a shrub known as king's holly (Lomatia tasmanica, family Proteaceae), which they believe to be more than 43,000 years old.

included. Trees that exhibit continuous cambial activity, such as those of tropical rainforests, may lack growth rings entirely. It is therefore difficult to judge the age of such trees.

The width of individual growth rings may vary greatly from year to year as a function of such environmental factors as light, temperature, rainfall, available soil water, and length of the growing season. The width of a growth ring is a fairly accurate index of the rainfall of a particular year. Under favorable conditions—that is, during periods of adequate or abundant rainfall—the growth rings are wide; under unfavorable conditions, they are narrow.

In semiarid regions, where there is very little rain, the tree is a sensitive rain gauge. An excellent example of this is the bristlecone pine *(Pinus longaeva)* of the western Great Basin (Figure 27–28). Each growth ring is different, and a study of the rings tells a story that dates back thousands of years. The oldest-known living specimen of bristlecone pine is 4900 years old. Dendrochronologists—scientists who conduct historical research through the growth rings of trees—have been able to match samples of wood from living and dead trees, and in this way they have built up a continuous series of rings dating back more than 8200 years. The widths of the growth rings of bristlecone pines at higher elevations

(a)

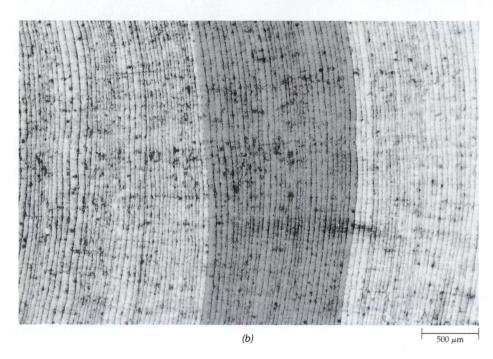

(b)   500 μm

(the upper tree line) have also been found to be closely related to temperature changes, which dramatically affect the length of the growing season in alpine climates. A record of average ring width in these trees provides a valuable guide to past temperatures and climatic conditions. For example, in the White Mountains of California, the summers were relatively warm from 3500 B.C. to 1300 B.C., and the tree line was about 150 meters above its present level. Summers were cool from 1300 B.C. to 200 B.C.

The structural basis for the visibility of growth layers in the wood is the difference in density of the wood produced early in the growing season and that produced later (Figures 27–21, 27–24, and 27–25). The **early wood** is less dense (with wider cells and proportionally thinner walls) than the **late wood** (with narrower cells and proportionally thicker walls). In a given growth layer, the change from early to late wood may be very gradual and almost imperceptible. However, where the late wood of one growth layer abuts on the early wood of a following growth layer, the change is abrupt and thus clearly discernible.

In some angiosperm woods, size differences of the vessels, or pores, in early and late woods are quite marked, the pores of the early wood being distinctly larger than those of the late wood. (The term "pore" is used by the wood anatomist for a vessel seen in cross section.) Such woods are called **ring-porous** woods (Figures 27–24a and 27–25a). In other angiosperm woods, the pores are fairly uniform in distribution and size throughout the growth layer. These woods are called **diffuse-porous** woods (Figure 27–24b). In ring-porous woods, almost all the water is conducted in the outermost growth layer, at speeds about ten times greater than in diffuse-porous woods.

Sapwood Conducts and Heartwood Does Not

As the wood becomes older, it gradually becomes nonfunctional in conduction and storage. Before this happens, however, the wood often undergoes visible changes, which involve the loss of reserve foods and the infiltration of the wood by various substances (such as oils, gums, resins, and tannins) that color it and sometimes make it aromatic. This often darker, nonconducting wood is called **heartwood,** while the generally lighter, conducting wood is called **sapwood** (Figure 27–12). Heartwood formation is believed to be a process that enables the plant to remove from regions of growth secondary metabolites that may be inhibitory or even toxic to living cells. The accumulation of these substances in the heartwood results in death of the living cells of the wood.

The proportion of sapwood to heartwood and the degree of visible difference between them vary greatly from species to species. Some trees, such as maple (*Acer*), birch (*Betula*), and ash (*Fraxinus*), have thick sapwoods; others, such as locust (*Robinia*), catalpa (*Catalpa*), and yew (*Taxus*), have thin sapwoods. Still other trees, such as the poplars (*Populus*), willows (*Salix*), and firs (*Abies*), have no clear distinction between sapwood and heartwood.

In many woods, tyloses are formed in the vessels when they become nonfunctional (Figure 27–29). **Tyloses** are balloonlike outgrowths from ray or axial parenchyma cells through pit cavities in the vessel wall. They may completely block the vessel. Tyloses are often induced to form prematurely or unnaturally by plant pathogens and may serve as a defensive mechanism by inhibiting spread of the pathogen throughout the plant via the xylem.

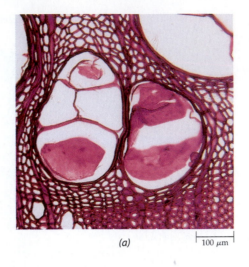

27–29

*Tyloses are balloonlike outgrowths of parenchyma cells that partially or completely block the lumen of a vessel. **(a)** Transverse and **(b)** longitudinal sections showing tyloses in vessels of white oak (Quercus alba), as seen with a light microscope.*

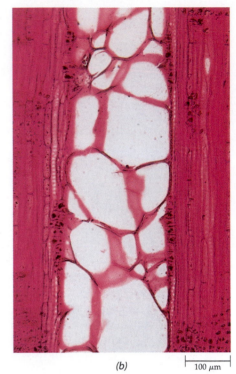

(a) 100 μm

(b) 100 μm

Reaction Wood Is a Type of Wood That Develops in Leaning Trunks and Limbs

The formation of **reaction wood** is assumed to result from the tendency of a branch or stem to counteract the force inducing the inclined position. In conifers, the reaction wood develops on the underside of the leaning part and is called *compression wood*. In angiosperms, it develops on the upper side and is called *tension wood*.

Compression wood is produced by the increased activity of the vascular cambium on the lower side of the bent stem and results in the formation of eccentric growth rings. Portions of growth rings located on the lower side are generally much wider than those on the upper side (Figure 27–30). Hence, compression wood causes straightening by expanding or pushing the trunk or limb upright. Compression wood has more lignin and less cellulose than normal wood, and its lengthwise shrinkage upon drying is often ten or more times as great as that of normal wood. (Normal wood usually shrinks lengthwise not more than 0.1 to 0.3 percent.) The difference in relative lengthwise shrinkage of normal and compression wood in a drying board often causes the board to twist and cup. Such wood is virtually useless except as fuel.

Tension wood is produced by the increased activity of the vascular cambium on the upper side of the stem and, as in the case of compression wood, is recognized by the presence of eccentric growth rings. To straighten the stem, the tension wood must exert a pull; hence the name "tension wood." Positive identification of tension wood requires microscopic examination of wood sections. Anatomically, the principal distinguishing feature is the presence of gelatinous fibers—fibers with little or no lignification, in which part of the secondary wall has a gelatinous appearance. Lengthwise shrinkage of tension wood rarely exceeds 1 percent, but boards containing it twist out of shape in drying. When such logs are sawed green, tension wood tears loose in bundles of fibers, imparting a wooly appearance to the boards.

27–30

Reaction wood in a conifer. Transverse section of the stem of a pine (Pinus sp.), showing compression wood with wider growth rings on the lower side.

The Gross Features of Wood Are Highly Variable

The appearance of wood depends on several factors, among them color, grain, texture, and figure. Not only are some of these characteristics helpful in identification of various kinds of woods, but they are also responsible for their decorative qualities.

Color in wood is variable both between different kinds of wood and within a species. The color of heartwood can be important in identifying a particular wood and in contributing to its aesthetic appeal. The rich chocolate or purplish brown heartwood of black walnut (*Juglans nigra*) and the red-brown heartwood of black cherry (*Prunus serotina*) are examples of longtime favorites for high-quality furniture.

Grain in wood refers to the direction of alignment of wood elements, or axial components—fibers, tracheids, parenchyma cells, and vessel elements—when considered *en masse*. The alignment of the axial components reflects the alignment of the fusiform initials that gave rise to them. When all the elements are oriented parallel to the longitudinal axis of a piece of wood, the grain is said to be *straight*. When the alignment does not coincide with the longitudinal axis of the piece, the wood is said to be *cross-grained*. *Spiral grain* is the term applied to a spiral arrangement of elements in a log or tree trunk, which has a twisted appearance after the bark has been removed (Figure 27–31). If the orientation of the spiral is reversed at more or less regular intervals along a single radius, the grain is said to be *interlocked*.

Texture of wood refers to the relative size and amount of size variation of elements within the growth increments. *Coarse* texture can result from the presence of wide bands of large vessels and broad rays, as in some ring-porous woods, and *fine* texture from the presence of small vessels and narrow rays. *Even* texture is the result of uniformity in cell dimensions—there is no perceptible difference between the early and late woods. The texture is *uneven* when distinct differences exist between early and late woods of a growth ring.

Figure refers to the patterns found on the longitudinal surfaces of wood. In a restricted sense, the term "figure" is used to refer to the more decorative woods, such as bird's-eye maple, prized in the furniture and cabinetmaking industries. Figure depends on grain and texture and on the orientation of the surface that results from sawing.

Boards are sawn from logs in two ways with reference to the radius of the log (Figure 27–32). In one, the wood is cut such that the broad faces of the board roughly parallel the tangential-longitudinal plane of the log. Tangentially cut boards are said to be **plainsawed**, or flat-cut. The growth rings in plainsawn boards appear as wavy bands with interestingly variable patterns. These are used to make decorative veneers but may warp if cut into thicker boards. In the other way of saw-

27–31

Trunk of a dead white oak (Quercus alba) *tree from which the bark has fallen, revealing the spiral grain of the wood.*

ing, a radial cut is made longitudinally through the center of the log. Radially cut boards are said to be **quartersawed**. In this type of stock the growth rings appear as parallel lines extending the length of the board, with the rays crossing the growth increments at right angles. Quartersawed boards are generally preferable for most uses because the radial surfaces have more uniform wearing and finishing properties, and the boards do not warp. Quartersawing is more time-consuming, however, and often more wasteful than plainsawing.

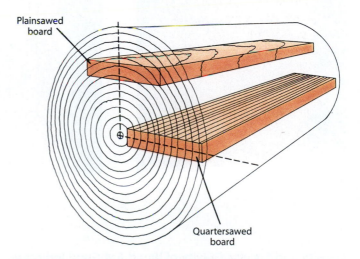

27–32

Plainsawed and quartersawed boards. In plainsawed boards the growth rings are roughly parallel to the broad faces. In quartersawed boards, the rings are more or less at right angles to the broad faces.

The Truth about Knots

Young trees typically have numerous branches growing from their trunks, and yet branches often are entirely lacking from the lower part of the trunks of older trees. What becomes of the branches found on the trunks of younger trees? Providing they remain on the tree, they are buried within the growing trunk, forming knots.

Branches have their origin from buds, and as long as they are living, they will undergo a periodic increase in length and thickness just like the trunk in which they are embedded. The vascular cambium of the living branch is continuous with that of the trunk. Thus, during periods of cambial activity, new wood is added as a continuous layer over branch and trunk, tightly fixing the knot in the wood of the trunk (a). Such knots are called tight knots (c). They will remain in place when the trunk is sawed for lumber.

When a branch dies, it stops growing and gradually is engulfed, bark and all, by the wood of the still growing trunk. Inasmuch as its cambium is no longer active, from that point onward the branch will lack physical continuity with the trunk (b). The dead branch may lose its bark, but if it remains on the trunk it will be engulfed by wood and continue as a knot. Such knots, called loose knots, may drop out of sawed lumber (d), (e). Sometimes changes in the wood of dead branches lead to the formation of substances that make the knots extremely hard and the wood very difficult to work with hand tools. On the other hand, woods with prominent knots, such as knotty pine, are valued for their decorative quality.

Radial sections of wood containing knots. (a) Trunk with tight knot. The cambium and growth rings are continuous between wood and knot. (b) View of trunk after branch has died. The dashed line on the left extends through the region of the tight knot. The dashed line on the right extends through the region of the loose knot.

Different portions of the same knot from white pine (Pinus strobus) wood: (c) a tight knot; (d) a loose knot; (e) the loose knot removed from the branch wood.

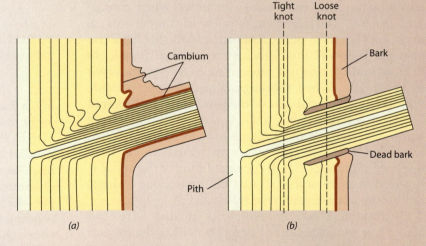

(a)   (b)

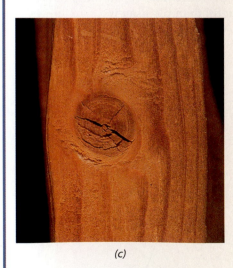

(c)

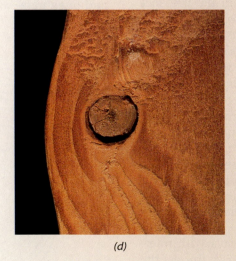

(d)

(e)

Density and Specific Gravity of Wood Are Good Indicators of the Strength of Wood

Density is the single most important indicator of the strength of wood and can be used to predict such characteristics as hardness, nailing resistance, and ease of machining. Dense woods generally shrink and swell more than light woods. In addition, the densest woods make the best fuel.

The *specific gravity* of a substance is the ratio of the weight of the substance to the weight of an equal volume of water. In the case of wood, the oven-dry weight

TABLE 27–1 Wood Uses of Some Common North American Trees

Angiosperm woods

Alder, red *(Alnus rubra)*—principal hardwood of Pacific Northwest. Used for furniture, especially chairs; sash, doors, and other millwork; plywood; charcoal; important source of pulpwood.

Ash, white *(Fraxinus americana)*—handles, especially long ones (shovel, spade, rake) because of its straightness of grain, strength, moderate weight, and other qualities; almost all baseball bats; oars, paddles, tennis-racket frames, hockey sticks; kitchen cabinets; toys and woodenware.

Basswood, or linden *(Tilia americana)*—veneer for plywood used as drawer panels and other concealed furniture parts; excelsior (wood shavings used in packing); sash, doors, and other millwork; piano keys; boxes and crates; caskets and coffins.

Beech, American *(Fagus americana)*—one of three important northern hardwoods, the others being yellow birch and sugar maple. Used for planing-mill products, especially flooring; veneer; fuel wood; hardwood distillation yielding acetic acid, methanol, and other chemicals; toys and woodenware.

Birch, yellow *(Betula alleghaniensis)*—veneer; furniture; toys and woodenware; musical instruments; toothpicks; boxes and baskets.

Cherry, black *(Prunus serotina)*—a superb cabinet wood; finished furniture; printers' blocks to which electrotypes are mounted; piano actions; interior trim; paneling; handles; toys and woodenware.

Cottonwood *(Populus deltoides)*—pulpwood for high-grade paper used in books and magazines; concealed parts of furniture; excelsior; tubs and pails for food products; boxes and crates.

Elm *(Ulmus* spp.*)*—has interlocked grain and, consequently, is difficult to split. Used for staves and hoops; boxes and crates; veneer for fruit and vegetable containers and round cheeseboxes; bent parts in furniture; interior trim.

Hickory, bitternut *(Carya cordiformis)*—tool handles, especially for axes, picks, and sledges; ladders; furniture; woodenware; prized for smoking meats; prime fuel wood.

Locust, black *(Robinia pseudo-acacia)*—mine timbers; railroad ties; fence posts; in construction where strength and durability are of great importance.

Maple, sugar *(Acer saccharum)*—veneer; railroad ties; fuel wood; furniture; flooring, especially bowling alleys and dance floors; toys and woodenware; musical instruments.

Oak, red and white *(Quercus rubra* and *Q. alba)*—railroad ties; veneer; flooring; sash, doors, and other millwork; firewood; ship- and boat-building; caskets and coffins; casks of white oak for aging wine and whiskey.

Persimmon *(Diospyros virginiana)*—golf-club heads; boxes and crates; handles.

Sycamore, American *(Platanus occidentalis)*—has interlocked grain. Used for veneer; boxes and crates; interior trim and paneling; flooring; butcher's blocks; concealed parts of furniture.

Walnut, black *(Juglans nigra)*—the finest cabinet wood native to the continental United States; veneer for plywood faces used in manufacture of furniture; goes directly into high-grade tables and chairs; leading wood for gunstocks; caskets and coffins.

Yellow poplar, or tulip tree *(Liriodendron tulipifera)*—veneer for plywood used for interior finish, furniture, and cabinetwork; pulpwood; boxes and crates; sash, doors, and other millwork.

Conifer woods

Cedar, western red *(Thuja plicata)*—premier wood for shingles; poles and posts; boat-building; greenhouse construction; exterior siding; sash and doors; millwork and interior finishing; caskets and coffins.

Douglas fir *(Pseudotsuga menziesii)*—the western forests of the United States are about 50 percent Douglas fir, which furnishes more timber than any other single species grown in the United States. Used for building construction, including veneer converted largely into plywood; railroad ties; mine timbers; pulpwood; boxes and crates; ship- and boat-building.

Hemlock, eastern *(Tsuga canadensis)*—pulpwood; general construction; boxes and crates; sash and doors; kitchen cabinets; tannins used in tanning leather.

Pine, loblolly *(Pinus taeda)*—interior finish; frame and sash; wainscoting, joists, and subflooring.

Pine, ponderosa *(Pinus ponderosa)*—boxes and crates; sash, doors, and other millwork; building construction; turned work (posts, balusters, porch columns); poles; toys; caskets and coffins.

Pine, slash *(Pinus elliottii)*—pulpwood; heavy timbers; railroad ties; veneer; turpentine and rosin; boxes; baskets and crates.

Pine, sugar *(Pinus lambertiana)*—boxes and crates; sash, doors, and other millwork; signs; piano keys and organ pipes.

Pine, western white *(Pinus monticola)*—matches; boxes and crates; building construction; sash, doors, and other millwork; core stock for plywood, especially tabletops.

Redwood *(Sequoia sempervirens)*—building construction of all sorts; ship- and boat-building; garden furniture; shingles and shakes; caskets and coffins.

Spruce, red *(Picea rubens)*—most important use is for pulpwood used in making pulp for paper; sounding boards for musical instruments; paddles and oars; ladder rails; ship- and boat-building; boxes and crates.

of the wood is used in figuring its specific gravity:

$$\text{Specific gravity} = \frac{\text{oven-dry weight of wood}}{\text{weight of the displaced volume of water}}$$

The specific gravity of dry solid wood substance (that is, dry cell wall material) of all plants is about 1.5. The differences in specific gravity of woods depend, therefore, upon the proportion of wall substance to lumen (the space bounded by the cell wall).

Fibers are especially important in determination of the specific gravity. If the fibers are thick-walled and narrow-lumened, the specific gravity tends to be high. Conversely, if the fibers are thin-walled and wide-lumened, it tends to be low. The presence of numerous thin-walled vessels also tends to lower the specific gravity.

Density is expressed as weight per unit volume, either as pounds per cubic foot (English) or as grams per cubic centimeter (metric). Water has a density of 62.4 lb/ft³, or 1 g/cm³. A wood weighing 31.2 lb/ft³, or 0.5 g/cm³, is therefore one-half as dense as water and has a specific gravity of 0.5. The *Guinness Book of World Records* lists black ironwood (*Olea capensis*) of South Africa as the heaviest (93 lb/ft³, or 1.49 g/cm³) wood and *Aeschynomene hispida* of Cuba as the lightest (2.75 lb/ft³, or 0.044 g/cm³) wood. Their respective specific gravities are 1.49 and 0.044. The specific gravities of most commercially useful woods are between 0.35 and 0.65.

See Table 27–1 for the uses of wood of some common North American trees.

27–33

Summary of stem development in a woody angiosperm during the first year of growth. (Compare this with root development, as summarized in Figure 25–25.)

Summary

Secondary Growth Causes an Increase in the Girth of Stems and Roots

Secondary growth (the increase in girth in regions that are no longer elongating) occurs in all gymnosperms and in most angiosperms other than monocots and involves the activity of the two lateral meristems—the vascular cambium and the cork cambium, or phellogen. Herbaceous plants may undergo little or no secondary growth, whereas woody plants—trees and shrubs—may continue to increase in thickness for many years. Figure 27–33 presents a summary of stem development in a woody plant, beginning with the apical meristem and ending with the secondary tissues produced during the first year's growth. Compare it with the summary of root development presented in Figure 25–25.

The Vascular Cambium Contains Two Types of Initials—Fusiform Initials and Ray Initials

Through periclinal divisions, the fusiform initials give rise to the components of the axial system, and the ray initials produce ray cells, which form the vascular rays, or radial system. Increase in circumference of the cambium is accomplished by anticlinal division of the initials.

The Cork Cambium Produces a Protective Covering on the Secondary Plant Body

The first cork cambium in most stems originates in a layer of cells immediately below the epidermis. The cork cambium produces cork toward the outside and phelloderm toward the inside. Together, the cork cambium, cork, and phelloderm constitute the periderm. Although most of the periderm consists of compactly arranged

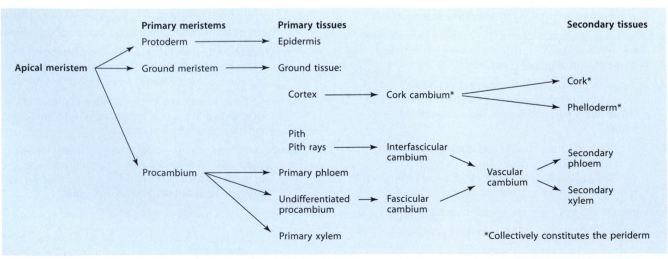

cells, isolated areas called lenticels have numerous intercellular spaces and play an important role in the exchange of gases through the periderm.

Bark Consists of All Tissues outside the Vascular Cambium

In old roots and stems, most of the phloem of the bark is nonfunctional. Sieve elements are short-lived, and generally only the present year's growth increment contains conducting, or functional, sieve elements. After the first periderm, subsequently formed periderms originate deeper and deeper in the bark from parenchyma cells of nonfunctional phloem.

Wood Is Secondary Xylem

Woods are classified as either softwoods or hardwoods. All so-called softwoods are conifers and all so-called hardwoods are angiosperms (woody magnoliids and woody eudicots). Conifer woods, which are simpler than angiosperm woods, consist of tracheids and parenchyma cells. Some contain resin ducts. Angiosperm woods may contain a combination of all of the following cell types: vessel elements, tracheids, several types of fibers, and parenchyma cells.

Growth Rings Result from the Periodic Activity of the Vascular Cambium

Growth layers that correspond to yearly increments of growth are called annual rings. The difference in density between the late wood of one growth increment and the early wood of the following increment makes it possible to distinguish the growth layers. Density and specific gravity are good indicators of the strength of wood. In many plants, the nonconducting heartwood is visibly distinct from the actively conducting sapwood.

Reaction Wood Develops in Response to the Force Inducing an Inclined Position of a Branch or Stem

Commonly, reaction wood develops on the underside of leaning trunks and limbs of conifers and on the upper side of similar parts in angiosperms; its formation causes straightening of the trunk or limb. Reaction wood is called compression wood in conifers and tension wood in angiosperms.

Selected Key Terms

annual   p. 648

annual ring   p. 664

axial system   p. 648

bark   p. 654

biennial   p. 648

cambial zone   p. 649

cork   p. 653

cork cambium   p. 653

diffuse-porous   p. 665

early wood   p. 665

fascicular cambium p. 650

functional phloem p. 659

fusiform initials   p. 648

growth ring   p. 664

hardwood   p. 659

heartwood   p. 665

interfascicular cambium p. 650

lateral meristem   p. 648

late wood   p. 665

lenticel   p. 664

nonfunctional phloem p. 659

perennial   p. 648

periderm   p. 653

phelloderm   p. 653

phloem ray Figure 27–12

radial system   p. 648

ray initials   p. 648

ray parenchyma cell p. 648

ray tracheid   Figure 27–22

reaction wood   p. 666

resin duct   p. 659

ring-porous   p. 665

sapwood   p. 665

softwood   p. 659

torus   p. 659

tyloses   p. 665

vascular cambium   p. 648

vascular ray   p. 648

wood   p. 659

xylem ray Figure 27–12

Questions

1. Distinguish between the following: axial system/radial system; fascicular cambium/interfascicular cambium; inner bark/outer bark; functional phloem/nonfunctional phloem.

2. By means of simple, labeled diagrams, compare the structure of a woody eudicot root with that of a woody eudicot stem at the end of the first year's growth. Assume that the root is triarch and the primary vascular system of the stem consisted of discrete vascular bundles.

3. If a nail were driven into a tree at a height of five feet from ground level and thereafter the tree grew, on average, two feet in height per year, approximately how high above ground level would the nail be ten years later? Explain your answer.

4. What structural feature of wood is responsible for the visibility of growth rings?

5. Of what importance are lenticels to the plant?

6. The terms "hardwood" and "softwood" do not accurately express the degree of density or hardness of wood. Explain.

7. The age of a woody stem cannot always be accurately estimated by counting the growth rings. Why?

8. Of what "value" is heartwood to a plant?

9. What are some gross features of wood that are of value in identification of various kinds of woods?

10. What are knots?

Physiology of Seed Plants

SECTION 6

The green, photosynthesizing leaves of the grapevine are its principal source of sugars, and its fruits, the grapes, one of the important destinations for those sugars. The sugar-filled juices of the grapes can be extracted, stored under anaerobic conditions with yeast cells, and converted into wine by converting glucose to ethanol. The tendrils of the grapevine are modified stems, which wrap around any object with which they come in contact, a phenomenon known as thigmotropism.

Chapter

Regulating Growth and Development: The Plant Hormones

OVERVIEW

In plant development, the ordinary becomes extraordinary and the mundane miraculous. We take for granted that stems grow tall, fruits ripen, leaves drop, and seeds germinate. But when we seek to understand the mechanisms underlying these developmental events we find processes so intricate, pervasive, and subtle that their details remain largely unknown. At the epicenter of plant growth and development are the hormones—small organic molecules that function as highly specific chemical signals between cells. Hormones are able to regulate growth and development in part because they produce amplified effects. That is, a single hormone molecule can trigger an increase in the concentration of many other molecules that in turn produce developmental changes within a cell. Hormones are involved in virtually every aspect of plant growth and development.

This chapter begins with an examination of the five major groups of plant hormones, including their discovery, principal effects, and commercial applications. This is followed by a discussion of the principal molecular mechanisms by which hormones are known to cause changes in growth and development. The chapter ends with a section on various laboratory techniques used to manipulate plant cells and tissues for practical purposes.

CHECKPOINTS

By the time you finish reading this chapter, you should be able to answer the following questions:

1. What are the five major groups of plant hormones? Why should hormones be considered regulators rather than stimulators?

2. What are the sites of biosynthesis of each of the major groups of plant hormones?

3. What are some of the effects caused by each of the major groups of plant hormones?

4. How was the discovery of cytokinins linked with the development of tissue culture?

5. What are some of the mechanisms by which plant hormones exert their effects at the molecular level?

6. What are some of the techniques used in plant biotechnology to manipulate the genetic potential of plants?

A plant, in order to grow, needs light from the sun, carbon dioxide from the air, and water and minerals, including nitrogen, from the soil. As discussed in Section 5, the plant does far more than simply increase its mass and volume as it grows. It differentiates, develops, and takes shape, forming a variety of cells, tissues, and organs. Many of the details of how these processes are regulated are not known. It has become clear, however, that normal development depends on the interplay of a number of internal and external factors. The principal internal factors that regulate plant growth and development are chemical, and they are the subject of this chapter. Some of the external factors—such as light, temperature, day length, and gravity—that affect plant growth are discussed in Chapter 29.

Plant hormones, or phytohormones, are organic substances that play a major role in regulating growth. Some hormones are produced in one tissue and transported to another tissue, where they produce specific physiological responses. Other hormones act within the same tissues where they are produced. In both cases, these chemical signals communicate information about the developmental or physiological state of cells, tissues, and, in some cases, widely separated organ systems. Hormones are active in very small quantities. In the shoot of a pineapple plant (*Ananas comosus*), for example, there are only 6 micrograms of indoleacetic acid, a common plant hormone, per kilogram of plant material. One enterprising plant physiologist calculated that the weight of the hormone in relation to that of the shoot is comparable to the weight of a needle in 20 metric tons of hay.

The word **hormone** comes from the Greek *horman*, meaning "to set in motion." It is now clear, however, that some hormones have inhibitory influences. Therefore,

TABLE 28–1   Plant Hormones: Their Nature, Occurrence, and Effects

Hormone(s)	Chemical Nature	Sites of Biosynthesis	Transport	Effects
Auxins	Indole-3-acetic acid is the principal naturally occurring auxin. It is synthesized primarily from tryptophan.	Primarily in leaf primordia and young leaves and in developing seeds.	IAA is transported from cell to cell, and transport is unidirectional (polar)	Apical dominance; tropic responses; vascular tissue differentiation; promotion of cambial activity; induction of adventitious roots on cuttings; inhibition of leaf and fruit abscission; stimulation of ethylene synthesis; inhibition or promotion (in pineapples) of flowering; stimulation of fruit development.
Cytokinins	N^6-adenine derivatives, phenyl urea compounds. Zeatin is the most common cytokinin in plants.	Primarily in root tips.	Cytokinins are transported via the xylem from roots to shoots.	Cell division; promotion of shoot formation in tissue culture; delay of leaf senescence; application of cytokinin can cause release of lateral buds from apical dominance.
Ethylene	The gas ethylene (C_2H_4) is synthesized from methionine. It is the only hydrocarbon with a pronounced effect on plants.	In most tissues in response to stress, especially in tissues undergoing senescence or ripening.	Being a gas, ethylene moves by diffusion from its site of synthesis.	Fruit ripening (especially in climacteric fruits, such as apples, bananas, and avocados); leaf and flower senescence; leaf and fruit abscission.
Abscisic acid	"Abscisic acid" is a misnomer for this compound, for it has little to do with abscission. ABA is synthesized from mevalonic acid.	In mature leaves, especially in response to water stress. May be synthesized in seeds.	ABA is exported from leaves in the phloem.	Stomatal closure; induction of photosynthate transport from leaves to developing seeds; induction of storage-protein synthesis in seeds; embryogenesis; may affect induction and maintenance of dormancy in seeds and buds of certain species.
Gibberellins	Gibberellic acid (GA_3), a fungal product, is the most widely available. GA_1 is probably the most important gibberellin in plants. GAs are synthesized from mevalonic acid.	In young tissues of the shoot and developing seeds. It is uncertain whether synthesis also occurs in roots.	GAs are probably transported in the xylem and phloem.	Hyperelongation of shoots by stimulating both cell division and cell elongation, producing tall, as opposed to dwarf, plants; induction of seed germination; stimulation of flowering in long-day plants and biennials; regulation of production of seed enzymes in cereals.

rather than thinking of hormones as stimulators, it is more useful to regard them as chemical regulators. But this term also needs qualification because the response to a particular regulator depends not only on its chemical structure but also on how it is "read" by the target tissue. The same hormone can elicit different responses in different tissues or at different times of development in the same tissue. Tissues may require different amounts of hormones. Such differences are referred to as differences in **sensitivity.** Thus, plant systems may vary the intensity of hormone signals by altering hormone concentrations or by changing the sensitivity to hormones that are already present.

Traditionally, five groups, or classes, of plant hormones have received the most attention: auxins, cytokinins, ethylene, abscisic acid, and gibberellins (Table 28–1). More recently, it has become clear that additional chemical signals are used by plants (Table 28–2). The brassinolides—complex organic molecules related to animal steroids—appear to be required for normal growth of most plant tissues. Salicylic acid, with a structure similar to aspirin (see Figure 2–31), has been implicated as a signal in defense responses to plant pathogens. The jasmonates—volatile compounds long recognized as components of floral fragrances—are now known to act as regulators of plant development. The recent discovery that a small peptide, systemin, is used in tomato (*Solanum lycopersicum*) as a long-distance signal to activate chemical defenses against herbivores suggests that plants have a much richer repertoire of internal chemical signals than previously thought.

We begin with auxin because it was the first substance to be identified as a plant hormone.

Auxins

Some of the first recorded experiments on growth-regulating substances were performed by Charles Darwin and his son Francis and were reported in *The Power of Movement in Plants*, published in 1881. The Darwins first made systematic observations of the bending or curving of plants toward light (phototropism; see Chapter 29), using seedlings of canary grass (*Phalaris canariensis*) and oats (*Avena sativa*). They found that bending did not occur if they covered the upper portion of the *coleoptile* (the sheathlike, protective structure covering the shoot of grass seedlings) with a cylinder of metal foil or a hollow tube of blackened glass (or a blackened quill) and illuminated the plant from the side (Figure 28–1). If, however, the tip was enclosed in a transparent glass tube (or quill), bending occurred. From these experiments, the Darwins concluded that "when seedlings are freely exposed to a lateral light some influence is transmitted from the upper to the lower part, causing the latter to bend." We now know that this bending is the result of differential cell elongation (see Chapter 29).

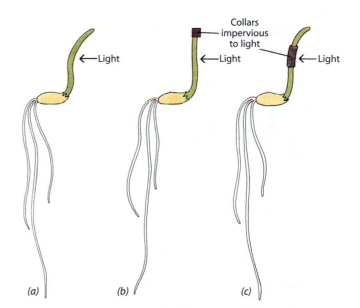

28–1
The Darwins' experiment. (a) Seedlings normally bend toward the light. (b) When the tip of a seedling was covered by a lightproof collar, this bending did not occur. (Bending did occur when the tip of a seedling was covered with a transparent collar.) (c) When a lightproof collar was placed below the tip, the characteristic light response took place. From these experiments, the Darwins concluded that, in response to light, an "influence" that causes bending is transmitted from the tip of the seedling to the area below the tip where bending normally occurs.

TABLE 28–2	More Recently Discovered Plant Growth Regulators	
Regulator(s)	Chemical Nature	Effects
Brassinolides	Steroids	Stimulate cell division and elongation; *Arabidopsis* mutations that block synthesis of brassinolides produce dwarf plants.
Salicylic acid	Phenolic compound	Activates pathogen defense genes.
Jasmonates	Volatile fatty acid derivatives	Regulate seed germination, root growth, storage-protein accumulation, and synthesis of defense proteins.
Systemin	Small peptide	Produced in wounded tissue systems; may induce defense genes in remote tissues.

Indole ring

Acetic acid side chain

CH₂ — COOH

N

H

(a) Indoleacetic acid (IAA)

O — CH₂ — COOH

Cl   Cl

(b) 2,4-Dichlorophenoxyacetic acid (2,4-D)

CH₂ — COOH

(c) 1-Naphthaleneacetic acid (NAA)

28–2

Auxins. (a) Indoleacetic acid (IAA) is the principal naturally occurring auxin. (b) Dichlorophenoxyacetic acid (2,4-D), a synthetic auxin, is widely used as an herbicide. (c) Naphthaleneacetic acid (NAA), another synthetic auxin, is commonly employed to induce the formation of adventitious roots in cuttings and to reduce fruit drop in commercial crops. The synthetic auxins, unlike IAA, are not readily broken down by natural plant enzymes and microbes and so are better suited than IAA for commercial purposes.

In 1926, the plant physiologist Frits W. Went succeeded in isolating this "influence" from the coleoptile tips of oat (*Avena*) seedlings. Went named this chemical substance **auxin,** from the Greek *auxein,* meaning "to increase."

As can be seen in Figure 28–2, the principal naturally occurring auxin, which is called indoleacetic acid (abbreviated IAA), closely resembles the amino acid tryptophan (see Figure 2–15c). Both tryptophan and IAA are synthesized from indole, and some evidence indicates that plants can produce IAA directly from tryptophan. However, recently discovered mutants in maize and *Arabidopsis* that are unable to synthesize tryptophan can still make IAA. Thus, plants are apparently capable of producing this essential plant growth regulator by a variety of pathways. Auxin is produced in the coleoptile

tips of grasses and in shoot tips. Although IAA has been found in root tips, most evidence indicates that it is not produced there but is transported there via the vascular cylinder. It is synthesized in leaf primordia and young leaves and is also found in flowers, fruits, and seeds.

Auxin Transport Is Polar

The movement of auxin in both shoots and roots is slow, only about one centimeter per hour. In addition, its transport is **polar,** or unidirectional (Figure 28–3): always toward the base **(basipetal)** in stems and leaves and toward the tip **(acropetal)** in roots. In contrast to the movement of sugars and other solutes, the primary route of auxin transport is not through the conduits (sieve tubes and vessels, respectively) of the phloem and xylem but through phloem parenchyma cells and parenchyma cells surrounding the vascular tissues. These parenchyma cells appear to have specialized carriers that move auxin molecules out of the cells at their basal ends only (Figure 28–3b). Since auxin can diffuse into cells passively, the activity of these carriers creates a net downward flow through the tissue. In plant parts capable of secondary growth, auxin transport also occurs through cells in the region of the vascular cambium.

Auxin Plays a Role in the Differentiation of Vascular Tissue

The gradient of auxin caused by basipetal transport influences the differentiation of the vascular tissue in the elongating shoot. When a stem of cucumber (*Cucumis sativus*) or some other herbaceous eudicotyledon is wounded in such a way as to sever and remove portions of the vascular bundles, new vascular tissues will form from cells in the pith and will connect the severed bundles. However, if the leaves and buds above the wound are removed, the formation of new cells is delayed. With the addition of IAA to the stem just above the wound, new vascular tissue begins to form (Figure 28–4). Auxin similarly plays an important role in the joining of vascular traces from developing leaves to the bundles in the stem.

More recent studies with cells isolated from the mesophyll of *Zinnia elegans* leaves showed a requirement for both auxin and cytokinin for the differentiation of xylem elements. Significantly, many of the isolated mesophyll cells differentiated directly into tracheary elements without proceeding through DNA synthesis or cytokinesis. Cell division was therefore not a prerequisite for differentiation. The results of all these studies illustrate the subtlety of growth regulator function. They also highlight the critical fact that growth regulators rarely act alone but instead operate in concert with other internal regulators of growth and development.

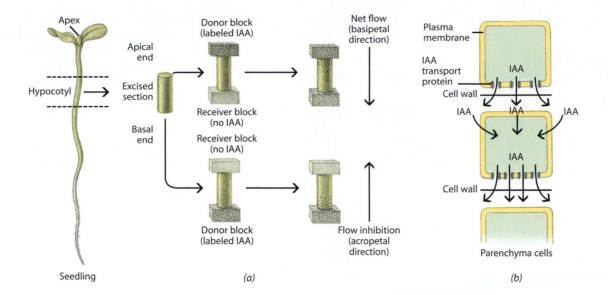

28–3

Auxin transport. (a) Experimental demonstration of polar auxin transport in stems, here represented by a segment of hypocotyl from a seedling. Hypocotyl segments are placed between agar blocks. The donor block contains radioactively labeled auxin. The rate of auxin transport is measured as radioactivity accumulating in the receiver block after a set time. The rate is much faster in the basipetal than in the acropetal direction. (b) Mechanism of polar auxin transport. Auxin moves into cells across the plasma membrane throughout the whole surface of the cell but is transported out of the cell by transport proteins, which are located only at the basal end of each parenchymatous cell.

28–4

A longitudinal view of vascular tissue (xylem) regeneration (arrow) around a seven-day-old wound in a young internode of cucumber (Cucumis sativus) from which the leaves and buds had been removed. IAA (0.1 percent in lanolin) was applied to the upper end of the internode immediately after wounding. The photograph shows the typical pattern of xylem differentiation induced by the basipetal polar movement of auxin.

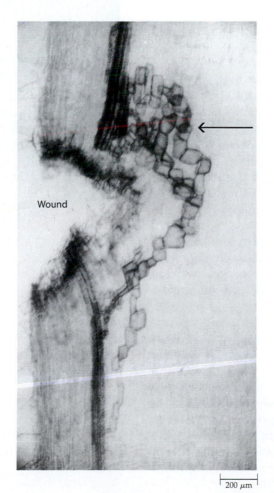

200 μm

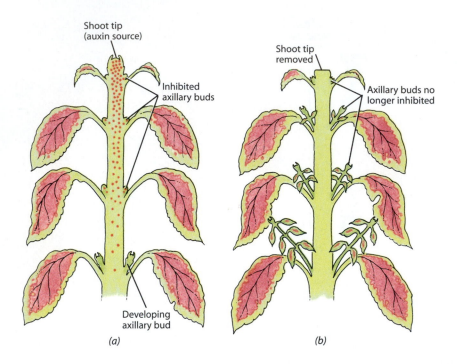

Shoot tip
(auxin source)

Inhibited
axillary buds

Developing
axillary bud

(a)

Shoot tip
removed

Axillary buds no
longer inhibited

(b)

28–5
Apical dominance in Coleus. **(a)** *Auxin produced in the shoot tip moves down the stem, inhibiting the growth of lateral shoots from the axillary buds. As the distance between the shoot tip and the axillary buds increases— and the concentration of auxin decreases— the buds are gradually freed from the inhibition.* **(b)** *If the shoot tip is removed, eliminating further production of auxin, the axillary buds are no longer inhibited and begin to grow vigorously.*

Auxin Provides Chemical Signals That Communicate Information over Long Distances

In many plant species, the basipetal flow of auxin from the growing apical bud inhibits the growth of axillary (lateral) buds (see Figure 26–3). If growth at the shoot tip is interrupted, the flow of auxin is diminished and the lateral shoots begin to develop (Figure 28–5). The inhibitory influence of an apical bud upon the lateral buds is referred to as **apical dominance.**

The role of auxin in apical dominance can be demonstrated experimentally. For instance, when the shoot tip of a bean plant *(Phaseolus vulgaris)* is removed, the lateral buds begin to grow. However, when auxin is applied to the cut surface, the growth of the lateral buds is inhibited.

In woody plants, auxin promotes activity of the vascular cambium. With expansion of buds and resumption of their growth in the spring, auxin moves downward in the stems and stimulates the cambial cells to divide, forming secondary vascular tissue.

Auxin Promotes the Formation of Adventitious Roots and the Growth of Many Fruits

The first practical application of auxin involved its promoting effect on the initiation of adventitious roots in cuttings (Figure 28–6). The practice of treating cuttings with auxin is commercially important, especially for the vegetative propagation of woody plants. (Application of a high concentration of auxin to already-growing roots, however, usually inhibits their growth.)

28–6
Adventitious root formation. The stalk of the African violet leaf on the left was placed in a solution containing the synthetic auxin naphthaleneacetic acid (NAA) for 10 days before the picture was taken. The stalk of the leaf on the right was placed in pure water. Note the growth of adventitious roots on the stalk of the hormone-treated leaf.

(a)

(b)

(c)

28–7

Auxin, produced by developing embryos, promotes maturation of the ovary wall and the development of fleshy fruits. (a) Normal

strawberry (Fragaria ananassa), (b) strawberry from which all seeds have been removed, and (c) strawberry from which a

horizontal band of seeds has been removed. If a paste containing auxin were applied to (b), the strawberry would grow normally.

Auxin is also involved with the formation of fruit. Ordinarily, if the flower is not pollinated and fertilization does not take place, the fruit will not develop. In some plants, fertilization of one egg cell is sufficient for normal fruit development, but in others, such as apples or melons, which have many seeds, several must be fertilized for the ovary wall to mature and become fleshy. By treating the female flower parts (carpels) of certain species with auxin, it is possible to produce parthenocarpic fruit (from the Greek *parthenos*, meaning "maiden, virgin"), which is fruit produced without fertilization— for example, seedless tomatoes, cucumbers, and eggplants. In many or most of these seedless fruits, however, immature ovules still exist in the fruit.

Developing seeds are a source of auxin. If during development of the aggregate fruit of the strawberry (*Fragaria ananassa*) all of the seeds are removed, the receptacle will stop growing altogether. (Actually, in the strawberry, it is small achenes—one-seeded fruits—that are removed. What is recognized as the strawberry fruit is the fleshy receptacle.) If a narrow ring of seeds is left, the fruit forms a bulging girdle of growth in the area of the seeds (Figure 28–7). If auxin is applied to the deseeded receptacle, growth proceeds normally.

Synthetic Auxins Are Used to Kill Weeds

Synthetic auxins such as 2,4-dichlorophenoxyacetic acid (2,4-D) (Figure 28–2) have been used extensively for the control of weeds on agricultural lands. In economic terms, this is the major practical use for plant growth

regulators. How these herbicides kill weeds is not completely known. 2,4-D is not broken down in plants as readily as natural auxins, and the resulting artificially high levels of auxinlike compounds certainly contribute to the lethal effects. The mechanism by which herbicides kill only *certain* weeds is also largely unknown. The selectivity of these compounds against broad-leaf weeds is due in part to the greater absorption and rates of transport of the herbicides by broad-leaf weeds than by grasses.

Considerable attention has been given to Agent Orange, the herbicide most commonly used as a defoliant during the Vietnam conflict. Agent Orange is a mixture of the *n*-butyl esters of 2,4-D and 2,4,5-trichlorophenoxyacetic acid (2,4,5-T), another synthetic auxin. It also contains the dioxin 2,3,7,8-TCDD, a contaminant of 2,4,5-T that has been demonstrated to be toxic to experimental animals and to humans. In humans, 2,3,7,8-TCDD causes chloracne, a severe skin lesion that occurs on the head and upper body, and, based on the positive evidence in animal studies, it is also probably carcinogenic in humans. The manufacture and use of 2,4,5-T have been banned in the United States.

Cytokinins

In 1941, Johannes van Overbeek found that coconut (*Cocos nucifera*) milk (which is liquid endosperm) contained a potent growth factor different from anything known at that time. This factor, or factors, greatly accelerated the development of plant embryos and promoted the growth of isolated tissues and cells in the test tube. Van Overbeek's discovery had two effects: it gave impetus to studies of isolated plant tissues, and it launched the search for another major group of plant growth regulators.

The basic medium used for tissue culture of plant cells contains sugar, vitamins, and various salts. In the early 1950s, Folke Skoog and his coworkers showed that a stem segment of the tobacco plant *(Nicotiana tabacum)* grew initially in such a culture medium, but that its growth soon slowed or stopped. Apparently, some growth stimulus originally present in the tobacco stem became exhausted. The addition of IAA had no effect. When coconut milk was added to the medium, however, the cells began to divide and growth of the tobacco stem resumed.

Skoog and his coworkers set out to identify the growth factor in the coconut milk. After many years of effort, they succeeded in producing a thousandfold purification of a growth factor, but they could not isolate it. So, changing course, they tested a variety of purine-containing substances—largely nucleic acids—in the hope of finding a new source of the growth factor. This led to the discovery by Carlos O. Miller that a breakdown product of DNA contained material that was highly active in promoting cell division.

Subsequently, Miller, Skoog, and their coworkers succeeded in isolating the growth factor from a DNA preparation and identifying its chemical nature. They called this substance *kinetin* and named the group of growth regulators to which it belongs the **cytokinins** because of the involvement of cytokinins in cytokinesis, or cell division. As shown in Figure 28–8, kinetin resembles the purine adenine, which was the clue that led to its discovery. Kinetin, which probably does not naturally occur in plants, has a relatively simple structure, and biochemists were soon able to synthesize a number of related compounds that behaved like cytokinins. Eventually, a natural cytokinin was isolated from kernels of maize *(Zea mays)*; called *zeatin*, it is the most active of the naturally occurring cytokinins.

Cytokinins have now been isolated from many different species of seed plants, where they are found primarily in actively dividing tissues, including seeds, fruits, and leaves, and in root tips. They have also been found in bleeding sap—the sap that drips out of pruning cuts, cracks, and other wounds in many types of plants. Cytokinins also have been identified in two seedless vascular plants, a horsetail *(Equisetum arvense)* and the fern *Dryopteris crassirhizoma*.

Although practical applications for cytokinins are not as extensive as those for auxin, the former have been important in plant development research. Cytokinins are central to tissue culture methods and are extremely important for biotechnology (see page 693). Treatment of lateral buds with cytokinin often causes the buds to grow, even in the presence of auxin, thus modifying apical dominance.

28–8

*Cytokinins. Note the similarities between the purine adenine **(a)** and these four cytokinins. **(b)** Kinetin and benzylamino purine (BAP) are commonly used synthetic cytokinins. **(c)** Zeatin and isopentenyl adenine (i^6 Ade) have been isolated from plants.*

Adenine

(a)

Kinetin

Zeatin

6-Benzylamino purine (BAP)

Isopentenyl adenine (i^6 Ade)

(b) Synthetic cytokinins

(c) Naturally occurring cytokinins

The Cytokinin/Auxin Ratio Regulates the Production of Roots and Shoots in Tissue Cultures

Apparently, the undifferentiated plant cell has two courses open to it: either it can enlarge, divide, enlarge, and divide again or, without undergoing cell division, it can elongate. The cell that divides repeatedly remains essentially undifferentiated, or meristematic, whereas the elongating cell will ultimately differentiate. In studies of tobacco stem tissues, the addition of IAA to the tissue culture produces rapid cell expansion, so that giant cells are formed. Kinetin alone has little or no effect, but IAA plus kinetin results in rapid cell division, so that large numbers of relatively small, undifferentiated cells are formed. In other words, cells remain meristematic in the presence of certain concentrations of both cytokinin and auxin.

In the presence of a high concentration of auxin, callus (undifferentiated) tissue frequently gives rise to organized roots. In tobacco pith callus, the relative concentrations of auxin and kinetin determine whether roots or buds form (Figure 28–9a). With higher concen-

trations of auxin, roots are formed, and with higher concentrations of kinetin, buds are formed. When both auxin and kinetin are present in roughly equal concentrations, the callus continues to produce undifferentiated cells (Figure 28–9b).

Cytokinins Delay Leaf Senescence

In most species of plants, leaves begin to turn yellow as soon as they are removed from the plant. This yellowing, which is due to a loss of chlorophyll, can be delayed by cytokinins. For example, when excised and floated on plain water, leaves of the cocklebur (*Xanthium strumarium*) turn yellow in about ten days. With kinetin (10 milligrams per liter) present in the water, much of the green color and fresh appearance of the leaf are maintained. If an excised leaf is spotted with kinetin-containing solutions, the spots remain green while the rest of the leaf yellows. Furthermore, if the cytokinin-spotted leaf contains radioactive amino acids, it can be shown that the amino acids migrate from other parts of the leaf to the cytokinin-treated areas. Another dramatic demon-

IAA concentration (mg/liter)

(a)

Kinetin concentration (mg/liter)

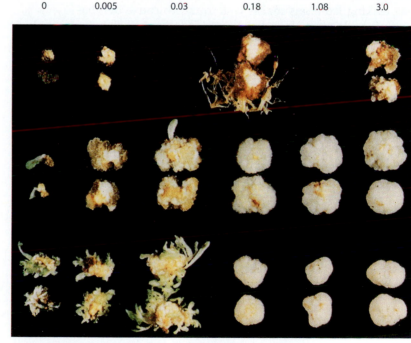

(b)

28–9

*Callus development. **(a)** Shoots and roots growing from undifferentiated tissue—a callus—of a carrot plant (Daucus carota) following treatment with both an auxin and a cytokinin. Depending on the relative proportions of auxin and cytokinin, callus from various types of plants will continue to grow*

as undifferentiated tissue, will produce roots, or will produce buds and shoots.

*   **(b)** *Effect of increasing IAA concentration at different kinetin levels on growth and organ formation of tobacco (Nicotiana tabacum) callus cultured on nutrient agar. Note that very little growth occurred without the addition of either IAA or kinetin (top left).*

Higher levels of IAA alone (top rows) promoted root formation, whereas they repressed bud formation when used alone or in combination with kinetin. The higher level of kinetin (lower rows) was more effective than the lower level (middle rows) in promoting bud development, but both kinetin levels were too high for promotion of root growth.

stration of the effect of cytokinins on leaf longevity is provided in Figure 28–30.

One interpretation of the senescence of detached leaves is that cytokinins are limiting in the detached leaf. This interpretation leads to an important and still unanswered question as to the site(s) of cytokinin production within plants. As mentioned previously, cytokinins are most abundant in actively dividing seeds, fruits, and leaves, and in root tips. This is not evidence, however, that cytokinins are synthesized in these organs because it is possible that cytokinins are transported there from some other sites. Based on several lines of evidence, root tips most certainly are involved in cytokinin synthesis. It is widely accepted that cytokinins synthesized in root tips are transported from them in the xylem to all other parts of the plant.

Ethylene

Ethylene was known to have effects on plants long before the discovery of auxin. The "botanical" history of **ethylene,** a simple hydrocarbon ($H_2C{=}CH_2$), goes back to the 1800s, when city streets were lighted with lamps that burned illuminating gas. In Germany, illuminating gas leaking from gas mains was found to cause defoliation of shade trees along the streets.

In 1901, Dimitry Neljubov demonstrated that ethylene was the active component of illuminating gas. He noticed that exposure of pea seedlings to illuminating gas caused stems to grow in a horizontal direction. When the gaseous components of illuminating gas were individually tested for effects, all were inactive except ethylene, which was active at concentrations as low as 0.06 part per million (ppm) in air. Neljubov's findings have led to the realization that ethylene exerts a major influence on many, if not all, aspects of growth and development in plants, including growth of most tissues, fruit maturation, fruit and leaf abscission, and senescence.

The biosynthesis of ethylene begins with the amino acid methionine, which reacts with ATP to form a compound known as *S*-adenosylmethionine, or SAM (Figure 28–10). Next, SAM is split into two different compounds, one of which is called ACC (1-aminocyclopropane-1-carboxylic acid). Enzymes on the tonoplast then convert ACC into ethylene, CO_2, and ammonium ion. Apparently, the formation of ACC is the step that is affected by those treatments (for example, high auxin concentrations, air pollution damage, wounding) that stimulate ethylene production.

Ethylene May Inhibit or Promote Cell Expansion

In most plant species, ethylene has an inhibitory effect on cell expansion. The triple response in pea seedlings is

a classic example. Ethylene treatment of dark-grown (etiolated) pea seedlings results in (1) decrease in longitudinal growth and (2) increase in radial expansion of the epicotyls and roots as well as (3) the horizontal orientation of epicotyls observed by Neljubov a century ago (Figure 28–11). On the other hand, ethylene elicits rapid stem growth in some semiaquatic species. In deep-water or floating varieties of rice, submergence of young rice plants during the monsoon season triggers an increase in ethylene biosynthesis. The resulting ethylene-induced internodal elongation provides a mechanism for the rice plants to keep pace with rising flood waters. In some cases the rice crop is actually harvested from boats. Another ethylene-mediated response to flooding in mesophytic plants is an increase in the development of air spaces in submerged tissues. The formation of these air spaces, or lacunae, results from the ethylene-mediated degeneration of cortical parenchyma tissues.

$$CH_3 - S - CH_2 - CH_2 - CH - COO^-$$
$$\underset{NH_3^+}{|}$$

Methionine

ATP

$PP_i + P_i$

$$CH_3 - \overset{+}{S} - CH_2 - CH_2 - CH - COO^-$$
Adenine-ribose
$$\underset{NH_3^+}{|}$$

S-adenosylmethionine (SAM)

Stimulated by high auxin concentrations, air pollution, wounding

$$\underset{H_2C}{\overset{H_2C}{>}} C \underset{COO^-}{\overset{NH_3^+}{<}}$$

1-Aminocyclopropane-1-carboxylic acid (ACC)

$CO_2 + NH_4^+$

$$CH_2 {=} CH_2$$
Ethylene

28–10

Biosynthesis of ethylene from methionine, which serves as the precursor of ethylene in all higher plant tissues. 1-Aminocyclopropane-1-carboxylic acid (ACC) is the immediate precursor of ethylene. The enzyme catalyzing conversion of SAM to ACC is ACC synthase.

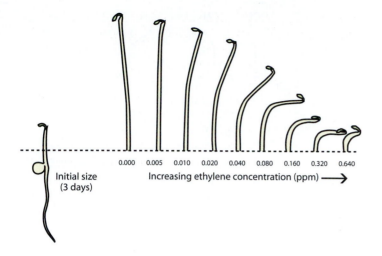

28–11

Triple response of pea seedlings to ethylene. The effects of increasing ethylene concentrations on the growth of dark-grown pea (Pisum sativum) *seedlings are shown. The so-called triple response includes a decrease in epicotyl elongation, a thickening of the shoot, and a change in orientation of growth from vertical to horizontal.*

Ethylene Plays a Role in Fruit Ripening

Ripening in fruit involves a number of changes. In fleshy fruits, the chlorophyll is degraded and other pigments may form, changing the fruit color. Simultaneously, the fleshy part of the fruit softens as a result of the enzymatic digestion of pectin, the principal component of the middle lamella of the cell wall. During this same period, starches and organic acids or, as in the case of the avocado (*Persea americana*), oils are metabolized into sugars. As a consequence of these changes, fruits become conspicuous and palatable and thus attractive to animals that eat the fruit and so scatter the seed.

During the ripening of many fruits—including tomatoes, avocados, and pome fruits, such as apples and pears—there is a large increase in cellular respiration evidenced by an increased uptake of oxygen. This phase is known as the **climacteric,** and such fruits are called climacteric fruits. (Fruits that show a steady decline, or gradual ripening, such as citrus fruits, grapes, and strawberries, are called nonclimacteric fruits.)

In climacteric fruits, increased ethylene synthesis precedes and is responsible for many of the ripening processes described above. The effect of ethylene on fruit ripening has agricultural importance. A major use is in promoting the ripening of tomatoes that are picked green and stored in the absence of ethylene until just before marketing. Ethylene is also used to hasten ripening of walnuts and grapes. Biotechnologists are now able to genetically alter both ethylene synthesis and sensitivity using gene transfer techniques (see Figure 28–30).

Whereas Ethylene Promotes Abscission, Auxin Prevents Abscission

Ethylene promotes **abscission,** or shedding, of leaves, flowers, and fruits in a variety of plant species. In leaves, ethylene presumably triggers the enzymes that cause the cell wall dissolution associated with abscission (page 636). Ethylene is used commercially to promote fruit loosening in cherries, blackberries, grapes, and blueberries, thus making mechanical harvesting possible. It is also used as a fruit-thinning agent in commercial orchards of prunes and peaches.

In many systems, abscission is controlled by an interaction of ethylene and auxin. Whereas ethylene triggers abscission, auxin appears to reduce the sensitivity of abscission zone cells to ethylene and thus prevents abscission. This auxin effect has also been used commercially. For instance, auxin treatment prevents preharvest drop of citrus fruits. In some cases, high concentrations of auxin actually stimulate abscission, however, this effect is thought to be due to an auxin stimulation of ethylene production.

Ethylene Apparently Plays a Role in Sex Expression in Cucurbits

Ethylene appears to play a major role in determining the sex of flowers in some monoecious plants (those plants having male and female flowers borne on the same individual). In cucurbits (family *Cucurbitaceae*; cucumber, squash), for example, high levels of gibberellins (see below) are associated with maleness, and treatment with ethylene changes the expression of sex to femaleness. In studies with cucumbers (*Cucumis sativus*), female buds produced greater quantities of ethylene than male buds. In addition, cucumbers grown under short light periods, or short-day conditions, which promote femaleness, produced more ethylene than those grown under long-day conditions (see Chapter 29). Hence, in cucurbits, ethylene apparently participates in the regulation of sex expression and is associated with the promotion of femaleness.

Abscisic Acid

At certain times, the survival of the plant depends on its ability to restrain its growth or its reproductive activities. In 1949, Paul F. Wareing discovered that the dormant buds of ash and potatoes contain large amounts of a growth inhibitor, which he called *dormin.* During the 1960s, Frederick T. Addicott reported the discovery in leaves and fruits of a substance capable of accelerating abscission, which he called *abscisin.* Soon abscisin and dormin were found to be identical chemically. The compound is now known as **abscisic acid,** or ABA (Figure 28–12). This is an unfortunate name choice since it now appears that this substance has no direct role in abscission.

Abscisic Acid Prevents Seed Germination and Induces Stomatal Closure

Abscisic acid levels increase during early seed development in many plant species. This increase in ABA stimulates the production of seed storage proteins and is also responsible for preventing premature germination. The breaking of dormancy in many seeds is correlated with declining ABA levels in the seed. In maize, there are single-gene mutants that either lack the ability to make ABA or exhibit a reduced sensitivity to the hormone. As a result, mutant embryos lack the ability to become dormant and germinate directly on the cob. Such mutants are called viviparous mutants (Figure 28–13).

Abscisic acid induces the closing of stomata in most plant species (see page 692 and Chapter 31). Since its synthesis is stimulated by water deficiency (water stress), ABA is most likely involved in the stomatal regulation of transpiration. In support of this idea, application of ABA to thin layers of epidermis peeled from the leaves of a variety of plants results in stomatal closure within a few minutes. In addition, mutant plants incapable of synthesizing ABA show a wilting phenotype; that is, they are capable of growing normally only in very humid environments.

28–12

Abscisic acid. Exogenous applications of abscisic acid may inhibit plant growth, but it also appears to act as a promoter (for instance, of storage-protein synthesis in seeds).

28–13

Viviparous-1 (vp1) *mutants of maize* (Zea mays). *The* vp1 *gene reduces the sensitivity of the mutant embryos to ABA. Hence, the mutant embryos fail to enter developmental arrest causing precocious germination of the seed (arrows) on the plant.*

Gibberellins

The early history of gibberellin research was an exclusive product of Japanese scientists. In 1926, the same year that Went isolated auxin from oat (*Avena*) coleoptile tips, E. Kurosawa of Japan was studying a disease of rice (*Oryza sativa*) called "foolish seedling disease," in which the plants grew rapidly; were spindly, pale-colored, and sickly; and tended to fall over. The cause of these symptoms, Kurosawa discovered, was a substance produced by a fungus, *Gibberella fujikuroi*, which was parasitic on seedlings.

Gibberellin was named and isolated in 1934 by the chemists T. Yabuta and Y. Sumiki. The discovery attracted little interest in the western world until after the Second World War. In 1956, J. MacMillan, in England, first successfully isolated gibberellin from a plant (the seed of the bean *Phaseolus vulgaris*). Since then, gibberellins have been identified from many species of plants, and it is now believed that they occur in all plants. They are present in various amounts in all parts of the plant, but the highest concentrations are found in immature seeds. More than 84 gibberellins now have been isolated and identified chemically. They vary slightly in structure (Figure 28–14), as well as in biological activity. The best-studied of the group is GA_3 (known as gibberellic acid), which is also produced by the fungus *Gibberella fujikuroi*.

The gibberellins have dramatic effects on stem and leaf elongation in intact plants by stimulating both cell division and cell elongation.

Application of Gibberellin Can Cause Dwarf Mutants to Grow Tall

The role of gibberellins in stem growth is most clearly demonstrated when this class of hormones is applied to many dwarf mutants (Figure 28–15). Under gibberellin treatment, such plants become indistinguishable from normal tall, nonmutant plants, indicating that these mu-

Gibberellic acid (GA₃)

GA₁₃

GA₄

28–14

Gibberellins. Shown here are three of the more than 84 gibberellins that have been isolated from natural sources. GA₃ (gibberellic acid) is *the most abundant in fungi and biologically active in many tests. The arrows indicate where minor structural differences occur that* *distinguish the other two examples of gibberellins, GA₄ and GA₁₃.*

28–15

Gibberellin effects on dwarf mutants. Contender beans, a dwarf cultivar of the common bean (Phaseolus vulgaris), were treated with gibberellin (right) to produce normal tall growth. The plant on the left was not treated and served as a control.

tants are unable to synthesize gibberellin and that tissue growth requires gibberellin. In maize, for example, four different types of dwarfs have been identified, each defective in a specific step of the gibberellin biosynthetic pathway. Studies of the hormone biochemistry in these mutant plants have led to a very important conclusion. Even though maize plants contain nine or more gibberellins, only the final compound in the pathway, GA₁, can cause effects directly. The other eight gibberellins must be further metabolized before a growth response is obtained. (See essay on page 228.)

Gibberellins Play Multiple Roles in Breaking Seed Dormancy and in Germination

The seeds of many plants require a period of dormancy before they will germinate. In certain plants, dormancy (see Chapter 29) usually cannot be broken except by exposure to cold or to light. In many species, including lettuce, tobacco, and wild oats, gibberellins will substitute for the dormancy-breaking cold or light requirement and promote the growth of the embryo and the emergence of the seedling. Specifically, the gibberellins enhance cell elongation, making it possible for the root to penetrate the growth-restricting seed coat or fruit wall. This effect of gibberellin has at least one practical application. Gibberellic acid hastens seed germination and thus ensures germination uniformity for the production of the barley malt used in brewing.

In barley (*Hordeum vulgare*) and other grass seeds, a specialized layer of endosperm cells, called the **aleurone** (see Figure 23–8), lies just inside the seed coat. The cells in the aleurone layer are rich in protein. When the seeds begin to germinate (triggered by the uptake of water), the embryo releases gibberellins, which diffuse to the aleurone cells and stimulate them to synthesize hydrolytic enzymes. One of these enzymes is α-amylase, which hydrolyzes starch. The enzymes digest the stored food reserves of the starchy endosperm. These food reserves, in the form of sugars, amino acids, and nucleic

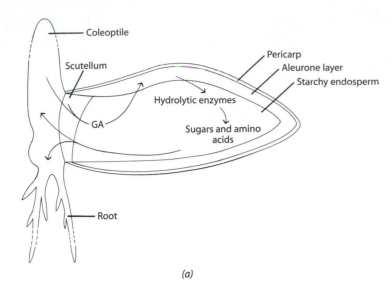

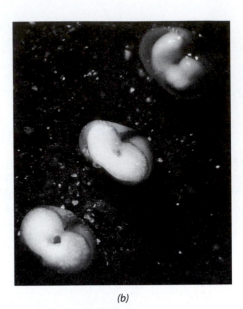

(a)

(b)

28–16

*Action of gibberellin in barley seeds. (**a**) Gibberellin (GA) produced by the embryo migrates into the aleurone layer, stimulating the synthesis of hydrolytic enzymes. These enzymes are released into the starchy endosperm, where they break down the endosperm reserves into sugars and amino acids, which are soluble and diffusible. The sugars and amino acids are then absorbed by the scutellum (cotyledon) and transported to the shoot and root for growth. (**b**) Each of these three seeds has been cut in half and the embryo removed. Forty-eight hours before the picture was taken, the seed at the lower left was treated with plain water, the seed in the center was treated with a solution of 1 part per billion of gibberellin, and the seed at the upper right was treated with 100 parts per billion of gibberellin. Digestion of the starchy storage tissue has begun to take place in the treated seeds.*

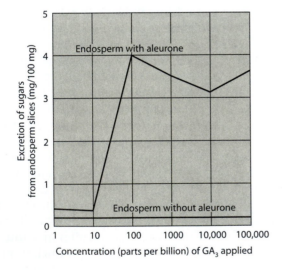

28–17

The release of sugar from endosperm can be induced by gibberellin (GA₃) treatment. These data show that sugars are produced only when the aleurone layer is present. It is, in fact, the aleurone layer that is the source of the enzyme α-amylase, which digests the starches stored in the endosperm.

acids, are absorbed by the scutellum and are then transported to the growing regions of the embryo (Figures 28–16 and 28–17).

Gibberellin Can Cause Bolting and Can Affect Fruit Development

Some plants, such as cabbages (*Brassica oleracea* var. *capitata*) and carrots (*Daucus carota*), form rosettes before flowering. (In a rosette, leaves develop but the internodes between them do not elongate.) In these plants, flowering can be induced by exposure to long days, to cold (as in the biennials), or to both. Following the appropriate exposure, the stems elongate—a phenomenon known as *bolting*—and the plants flower (Figure 28–18). Application of gibberellin to such plants causes bolting and flowering without appropriate cold or long-day exposure. Bolting is brought about by an increase both in cell number and in cell elongation. Gibberellin can thus be used for early seed production of biennial plants.

Gibberellins, like auxin, can cause the development of parthenocarpic fruits, including apples, currants, cucumbers, and eggplants. In some fruits, such as mandarin oranges, almonds, and peaches, the gibberellins have been effective in the promotion of fruit development where auxin has not. The major commercial application of gibberellins, however, is in the production of table grapes. In the United States, large amounts of gib-

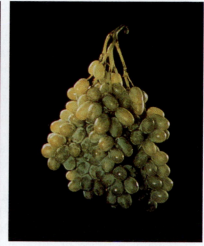

28–18

In this row of cabbages (Brassica oleracea var. capitata), *the plant in the middle has bolted and flowered naturally. Bolting and flowering can also be induced artificially by treatment with a gibberellin.*

28–19

The effect of GA$_3$ on the growth of Thompson Seedless grapes, a cultivar of Vitis vinifera. The bunch of grapes on the left was untreated, whereas that on the right was treated with GA$_3$. The result is looser clusters of larger grapes.

berellic acid are applied annually to the Thompson Seedless grapes, a cultivar of *Vitis vinifera.* Treatment causes larger fruit and much looser clusters (Figure 28–19).

The Molecular Basis of Hormone Action

Up to this point, we have discussed plant hormones in terms of the effects caused by external applications of hormones to a variety of plant systems. We now turn to some of the molecular mechanisms by which these chemical regulators influence growth, development, and rapid responses at the cellular level.

The development of organs (morphogenesis) can be described in terms of a coordinated series of cell divisions and subsequent cell enlargements. Specialization of cell types within an organ (differentiation) is the result of the selective expression of a particular set of genes within the genome of each individual cell. Clearly, in order to coordinate these cellular processes during development, individual cells must communicate with each other. This communication is due to plant hormones, which help to coordinate growth and development by acting as chemical messengers between cells. This concept is supported in part by numerous observable influences of plant hormones on the rate of cell division and on the rate and direction of cell expansion (Table 28–3). In addition, there is increasing evidence that the traditional as well as the newly discovered plant hormones can act either to stimulate or to repress specific genes within the nucleus. In fact, it appears that many observable hormone responses are the result of such differential gene expression.

TABLE 28–3	**Hormonal Influences on Basic Cellular Processes**			
Hormone(s)	Rate of Cell Division	Rate of Cell Expansion	Direction of Cell Expansion	Differentiation (Gene Expression)
Auxins	+	+	Longitudinal	+
Cytokinins	+	Little or no effect	None	+
Ethylene	+ or −	+ or −	Lateral	+
Abscisic acid	−	−	None	+
Gibberellins	+	+	Longitudinal	+

Key: + positive effect; − negative effect.

Hormones Control the Expression of Specific Genes

The **totipotency** of plant cells—that is, the potential of plant cells to give rise to entire plants—is clear evidence that all of the genes present in the zygote are also present in each living cell of the adult plant (see the essay on page 695). In any one cell, however, only selected genes are expressed and transcribed into mRNA and subsequently translated into proteins. The specific proteins that are produced determine the identity of the cell. It is proteins, specifically enzymes, that catalyze most of the cell's chemical reactions, and it is also proteins that form or produce most of the structural elements within and around the cell. Thus, a cortical cell in a root and a mesophyll cell in a leaf differ from each other structurally and functionally because of differences in gene expression during the course of their development.

The molecular mechanisms by which individual genes are switched on and off in the eukaryotic nucleus are not completely understood. A number of principles are beginning to emerge, however, that are common to both plants and animals (Figure 28–20). A eukaryotic gene is composed of a *coding sequence,* which specifies the amino acid sequence of the gene's protein product, as well as *regulatory sequences,* which are regions of DNA

bordering the coding sequence that play a regulatory role in gene transcription. Proteins called *regulatory transcription factors* can bind directly to specific DNA sequences within a regulatory sequence, activating (switching on) or repressing (switching off) that particular gene.

Plant molecular biologists are currently studying a number of plant genes that are either activated or repressed by such factors as light, environmental stress (see Chapter 29), and hormones. A classic example of how hormones control gene expression is provided by studies on the antagonistic effects of gibberellin (GA) and abscisic acid (ABA) on the synthesis of the starch-degrading enzyme α-amylase in the aleurone layer of barley seeds. If aleurone tissues are incubated in a medium containing radioactive amino acids and various hormones, the amino acids are incorporated into any new proteins synthesized during the treatment. These proteins can then be visualized and identified by separating them by size using gel electrophoresis (page 229) and detecting radioactive proteins using x-ray film. Treatment of aleurone tissues with GA causes an increased abundance of the α-amylase protein. This effect is counteracted by simultaneous application of ABA. Thus, ABA represses α-amylase gene expression while GA activates the gene.

28–20

*Regulation of gene expression. (**a**) Every gene in the nucleus that codes for protein is composed of an amino acid coding sequence and an associated regulatory sequence, which determines when the amino acid coding sequence will be expressed (that is, transcribed into mRNA). The regulatory sequence itself does not code for any protein. (**b**) A DNA binding protein called a transcription factor binds to the regulatory sequence. The transcription factor activates the gene; that is, RNA polymerase activity is stimulated and DNA is transcribed into mRNAs. The mRNAs are transported to the cytoplasm where they are translated into protein.*

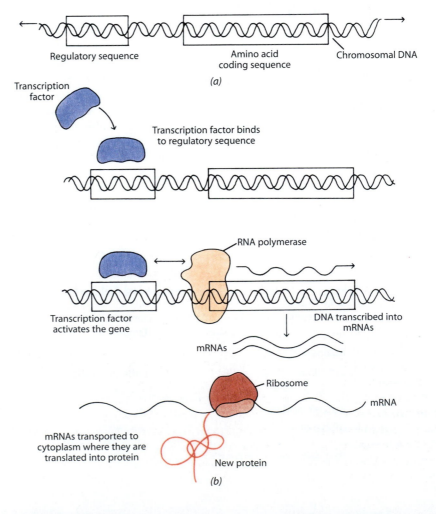

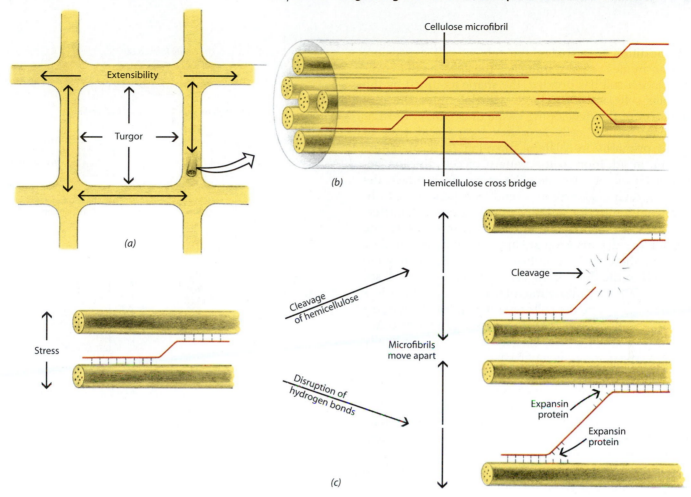

Cellulose microfibril

(b)

Hemicellulose cross bridge

Extensibility

Turgor

(a)

Stress

Cleavage of hemicellulose

Microfibrils move apart

Disruption of hydrogen bonds

Cleavage

Expansin protein

Expansin protein

(c)

28–21

*Hormonal regulation of the rate of cell expansion. **(a)** Turgor pressure within the cell pushes out on the cell walls. **(b)** Stretching of the cell wall is necessary for cell expansion but is limited by hemicellulose cross bridges between cellulose microfibrils. **(c)** Hormones may cause an increase in extensibility by stimulating a reversible cleavage of the hemicellulose cross bridges or through a disruption of hydrogen bonds between the cellulose microfibrils and the hemicellulose cross-bridges. Disruption of the hydrogen bonds is mediated by a cell wall protein called expansin. These modifications allow cellulose microfibrils to move apart from each other, resulting in the irreversible expansion of the wall.*

Hormones Can Regulate the Rate and Direction of Cell Expansion

The rate at which plant cells grow depends on several factors, including their position in the plant, the cell type, and a variety of environmental influences. The rate at which an individual cell expands is controlled by (1) the amount of turgor pressure inside the cell pushing against the cell wall (page 81) and (2) the extensibility of the cell wall (Figure 28–21; see also page 66). Extensibility, a physical property of the wall, is a measure of how much the wall will stretch permanently when a force is applied to it. As indicated in Table 28–3, the five traditional groups of hormones are capable of influencing the rate of cell expansion. In most cases examined, hormones affect the extensibility of the cell wall but have little direct influence on the turgor pressure. Auxin and GA stimulate plant growth by increasing the extensibility of cell walls, whereas ABA and ethylene inhibit plant growth by causing a decrease in extensibility.

The mechanisms by which hormones alter the extensibility of cell walls are not well understood. Two hypotheses are currently in favor. In the *acid growth hypothesis,* hormones—particularly auxin—activate a proton-pumping enzyme in the plasma membrane. Protons are pumped from the cytosol into the cell wall. The resulting drop in pH is thought to cause a loosening of the cell wall structure. This occurs either through the breakage and reformation of noncellulosic polysaccharides, which normally cross-link the cellulose microfibrils, or through the action of a new class of proteins called **expansins,** which disrupt hydrogen bonds between polysaccharides in the wall (Figure 28–21). An alternative hypothesis is based on recent discoveries that auxin activates the expression of specific genes within a few minutes of application. The products of these genes are thought to influence the delivery of new wall materials in such a way as to affect cell wall extensibility. These two hypotheses are not mutually exclusive, and both may be required to explain the influence of hormones on cell expansion.

In addition to affecting the rate of cell expansion, plant hormones can also influence the *direction* of expansion. Once a cell has divided, the shape assumed by the daughter cells as they enlarge will determine the ultimate form of the developing tissue or organ. For example, many of the cells in a developing leaf tend to expand primarily in a lateral direction. This lateral expansion, in addition to the pattern of cell division, results in formation of a platelike organ. In contrast, the cells in growing stem tissues tend to expand longitudinally, resulting in the "unidirectional" growth characteristic of an elongating stem. These differences in the direction of cell expansion are apparently determined by the orientation of the cellulose microfibrils as they are deposited in the developing cell wall (Figure 28–22; see also page 59). If cellulose microfibrils are deposited in a random orientation, the cells tend to expand in all directions. If the fibrils are laid down in a primarily transverse orientation, the cells tend to expand longitudinally (just as a coiled spring is much easier to stretch in the direction perpendicular to the orientation of the coils).

The orientation of cellulose microfibrils appears to be governed by the orientation of microtubules lying just inside the plasma membrane (page 66), and this arrangement of microtubules is influenced by hormones. Gibberellins, for example, promote a transverse arrangement of microtubules, resulting in greater longitudinal growth, or elongation. On the other hand, in stems, ethylene treatment causes some degree of reorientation of microtubules to the longitudinal direction, which promotes a more lateral (radial) expansion of the cells. This response to ethylene results in a stem that is shorter and thicker.

Hormones Interact with Specific Proteins Called Receptors, Which Then Activate Particular Response Pathways

In order for plant hormones to operate as chemical signals between cells, the target cells must have mechanisms for (1) identifying the specific hormone, (2) measuring the amount that is present, (3) transferring this information via biochemical pathways, and (4) converting the information into a complex set of developmental changes. Cells recognize plant hormones using proteins called hormone **receptors**. Each receptor protein contains a hormone-binding site that is specific for a particular hormone. Binding of the hormone to the receptor activates a particular response pathway in the cell.

Figure 28–23 illustrates a variety of ways in which a hormone bound to a receptor may activate response pathways. Binding of a hormone to its receptor results in a change in conformation (shape) of the receptor protein.

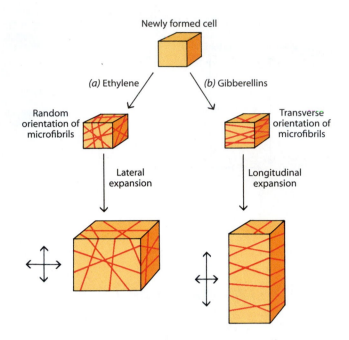

28–22

Ethylene and gibberellins have opposite effects on the orientation of cellulose microfibrils as they are laid down in the developing cell wall. The orientation of cellulose microfibrils apparently is governed by the orientation of microtubules just inside the plasma membrane. **(a)** *Ethylene causes some reorientation of microtubules from a transverse arrangement to a longitudinal direction. Thus, when the cell expands, expansion occurs more or less equally in all dimensions because of the random arrangement of the cellulose microfibrils.* **(b)** *Gibberellins, on the other hand, promote a transverse arrangement of microtubules. When the cell expands, it does so primarily longitudinally because the cellulose microfibrils are arranged in a transverse manner.*

This conformational change alters the receptor protein, allowing the receptor to interact with other components in the cell. For example, the activated receptor may interact directly with regulatory sequences of DNA to stimulate the transcription of specific hormone-activated genes. The steroid hormones in animals operate in this way. Other types of hormone receptors are located in the plasma membrane. Through their interaction with other membrane proteins, some of these receptors may activate ion pumps, such as the proton pump; others may act to open ion channels in the membrane (see Chapter 4).

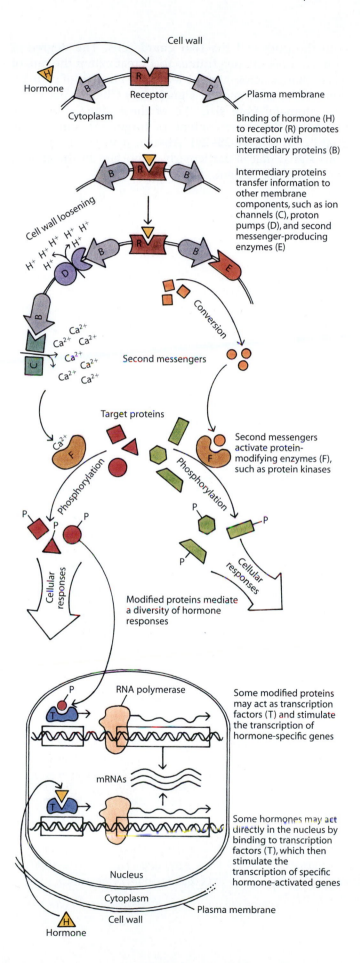

Cell wall

Hormone

Receptor

Cytoplasm

Plasma membrane

Cell wall loosening

Conversion

Second messengers

Target proteins

phosphorylation

Phosphorylation

Cellular responses

Cellular responses

Modified proteins mediate a diversity of hormone responses

RNA polymerase

mRNAs

Nucleus

Cytoplasm

Cell wall

Plasma membrane

Hormone

Binding of hormone (H) to receptor (R) promotes interaction with intermediary proteins (B)

Intermediary proteins transfer information to other membrane components, such as ion channels (C), proton pumps (D), and second messenger-producing enzymes (E)

Second messengers activate protein-modifying enzymes (F), such as protein kinases

Some modified proteins may act as transcription factors (T) and stimulate the transcription of hormone-specific genes

Some hormones may act directly in the nucleus by binding to transcription factors (T), which then stimulate the transcription of specific hormone-activated genes

28–23

Hormone response pathways. Hormones act by binding to proteins called receptors. The hormone signals are often amplified and passed down biochemical pathways by intermediary compounds called second messengers. Many responses to plant hormones are mediated through changes in gene expression within the nucleus. Specific genes are switched on or off as a result of hormone action. Hormones ultimately affect cell growth by causing changes in the architecture and chemical properties of the cell wall. See also Figure 28–20.

The calcium ion is of particular interest in hormone action. Generally, Ca^{2+} levels in the cytoplasm are very low. Hormonal stimulation of calcium ion channels results in a temporary elevation of Ca^{2+} levels. Binding of Ca^{2+} to the calcium-binding sites of certain proteins alters the activity of these proteins, much as hormones activate receptor proteins. *Protein kinases* are a class of enzymes that may be activated by Ca^{2+} or other "second messengers." The protein kinases may modify "target" proteins by transferring phosphate groups onto certain amino acids of a target protein, altering its activity.

Second Messengers Mediate Hormonal Responses Substances, such as calcium, that mediate hormonal responses are often referred to as **second messengers** (Figure 28–23). Second messengers perform two important functions: (1) they are involved in the transfer of information from the hormone-receptor complex to the target proteins, and (2) they amplify the signal produced by the hormone. Receptor activation of a single calcium ion channel may result in the release of hundreds of calcium ions into the cytosol. Each calcium ion can in turn activate a protein kinase molecule, and each protein kinase molecule can phosphorylate many molecules of target protein. Complex response pathways involving second messengers also contribute to the diversity of possible responses to a given hormone. Different cell types may have the same plasma membrane receptor but may respond quite differently to the same hormone if they carry a different complement of protein kinases and target proteins. (For a more detailed discussion of the signal transduction pathway and the role of second messengers, see page 86 and Figure 4–17.)

Stomatal Movement Involves a Specific Hormone Response Pathway

The regulation of stomatal movement by ABA is a hormone response pathway. Stomata (singular: stoma) are small openings in the epidermis, each surrounded by two guard cells, which change their shape to bring about the opening and closing of the pores. The term *stoma* (Greek for "mouth") is conventionally used to designate both the pore and the two guard cells. The degree of stomatal opening determines to a great extent the rate of gas exchange across the epidermis. A number of endogenous and environmental signals (see Chapter 31) influence stomatal pore size. All of these signals work by regulating the water content, or turgor pressure, of the guard cells (Figure 28–24). Abscisic acid is one endogenous signal that is particularly important in the control of stomatal movement.

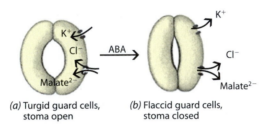

(a) Turgid guard cells, stoma open

(b) Flaccid guard cells, stoma closed

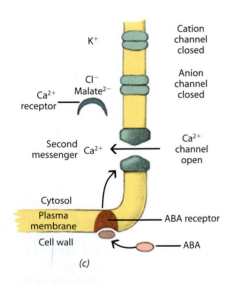

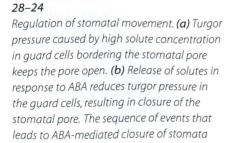

(c)

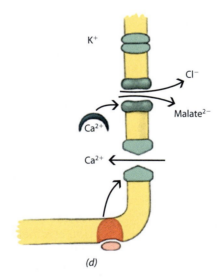

(d)

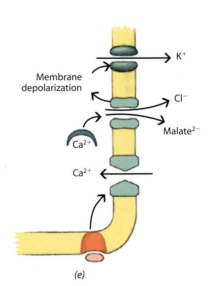

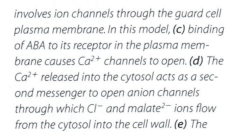

(e)

28–24

Regulation of stomatal movement. (a) Turgor pressure caused by high solute concentration in guard cells bordering the stomatal pore keeps the pore open. (b) Release of solutes in response to ABA reduces turgor pressure in the guard cells, resulting in closure of the stomatal pore. The sequence of events that leads to ABA-mediated closure of stomata involves ion channels through the guard cell plasma membrane. In this model, (c) binding of ABA to its receptor in the plasma membrane causes Ca^{2+} channels to open. (d) The Ca^{2+} released into the cytosol acts as a second messenger to open anion channels through which Cl^- and malate^{2-} ions flow from the cytosol into the cell wall. (e) The resulting drop in electrical potential (membrane depolarization) across the plasma membrane opens the K^+ channels and permits the release of K^+ into the cell wall. The flow of solutes from cytosol to cell wall results in decreased turgor pressure in the guard cells, causing closure of the stoma.

The receptor for ABA has not been identified, but details of cellular events that occur within minutes of ABA addition to isolated guard cell protoplasts indicate that rapid changes in osmotic potential of guard cells are mediated by ABA. The opening of a stoma is due to the uptake of solutes by its guard cells, which results in a more negative osmotic potential of the guard cell contents. (Recall that the osmotic potential, which is a negative quantity, is a function of the solute concentration; see page 79.) The important solutes that contribute to the osmotic potential of guard cells are Cl^- and K^+ ions, which are actively pumped into the cells, and $malate^{2-}$, a negatively charged (anionic) carbon compound synthesized by the guard cells.

Studies using the patch-clamp technique (see the essay on page 83) indicate that anion-specific channels in the plasma membrane of the guard cell open in response to ABA. Some experiments indicate that Ca^{2+} ions may act as a second messenger in this system. In this model (Figure 28–24), ABA activates Ca^{2+} channels in the plasma membrane, which results in an influx of Ca^{2+} from the cell wall to the cytosol. The Ca^{2+} then causes opening of anion channels in the plasma membrane by activating protein kinases. The opening of anion channels results in the rapid movement of anions, primarily Cl^- and $malate^{2-}$, from the cytosol to the cell wall. The subsequent depolarization of the plasma membrane—that is, the loss in electrical charge difference across the membrane—triggers the opening of K^+ channels. The result is a movement of K^+ from the cytosol to the cell wall. This rapid movement of Cl^-, $malate^{2-}$, and K^+ results in a less negative osmotic potential (higher water potential) of the cytosol and a more negative osmotic potential (lower water potential) of the wall. Water then moves down its water potential gradient from the cytosol to the cell wall, reducing the turgor of the guard cells and causing closure of the stomatal pore. When the ABA signal is removed, the guard cells slowly transport the potassium and chloride ions back into the cell using an electrochemical proton gradient generated by a proton pump (H^+-ATPase) in the plasma membrane. A more negative osmotic potential is reestablished within the guard cells, water flows into the cells by osmosis, and the resulting increase in turgor causes the stomatal pore to reopen.

Plant Biotechnology

The origins of **plant biotechnology,** the application of an array of techniques to manipulate the genetic potential of plants, can be traced back to the late 1850s and 1860s and the work of the German plant physiologists Julius von Sachs and W. Knop. It was they who demonstrated that many kinds of plants could be grown in water if provided with a few essential elements such as salts dissolved in the water—in other words, plants could be grown without placing their roots in soil—a technique we now call *hydroponics*. By the mid-1880s, it was known that at least ten chemical elements found in plants are necessary for normal growth. Today, 17 elements are generally considered to be essential for most plants (see Table 30–1).

The use of hydroponics and the quest to understand the mineral nutrition of plants provided the impetus for studies on the growth of excised plant parts in nutrient solutions. Gradually, substances such as sucrose, various vitamins, and other organic substances were added to the nutrient solutions in an attempt to sustain growth. It was not until the discovery of plant hormones and an understanding of their roles in the control of plant growth and development, however, that organ and tissue culture truly became feasible.

Plant Tissue Culture Can Be Used in Clonal Propagation

Plant **tissue culture** can be broadly defined as a collection of methods of growing large numbers of cells in a sterile and controlled environment (Figure 28–25). At present, the greatest impact of tissue culture is in the area of plant multiplication, referred to as **micropropagation,** or **clonal propagation,** since the individuals produced from single cells are genetically identical. The goal is to induce individual cells to express their totipotency.

Somatic Hybrids Have Been Produced by Protoplast Fusion The most elegant use of tissue culture for biotechnology has involved plant regeneration from protoplasts, that is, from cells that have had their walls removed by enzymatic digestion. In one procedure, known as **protoplast fusion,** protoplasts from two different plants are made to fuse together, producing a **somatic hybrid** cell. Most frequently, mesophyll cells have been used, although any living cell lacking a secondary wall can be utilized. Shortly after the walls have been removed and before new walls are formed, the naked protoplasts are made to fuse by the addition of an appropriate agent, such as polyethylene glycol, or by use of electric shock (electroporation; see page 699). The hybrid cells must then be grown on solid growth medium so that callus can develop. Provided all goes well, the callus will form somatic embryos, and then plantlets can be nurtured to grow into adult, fertile plants. Haploid hybrid plants, which are obtained through the use of pollen or of portions (explants) of anthers, are unable to reproduce sexually and hence must be propagated vegetatively.

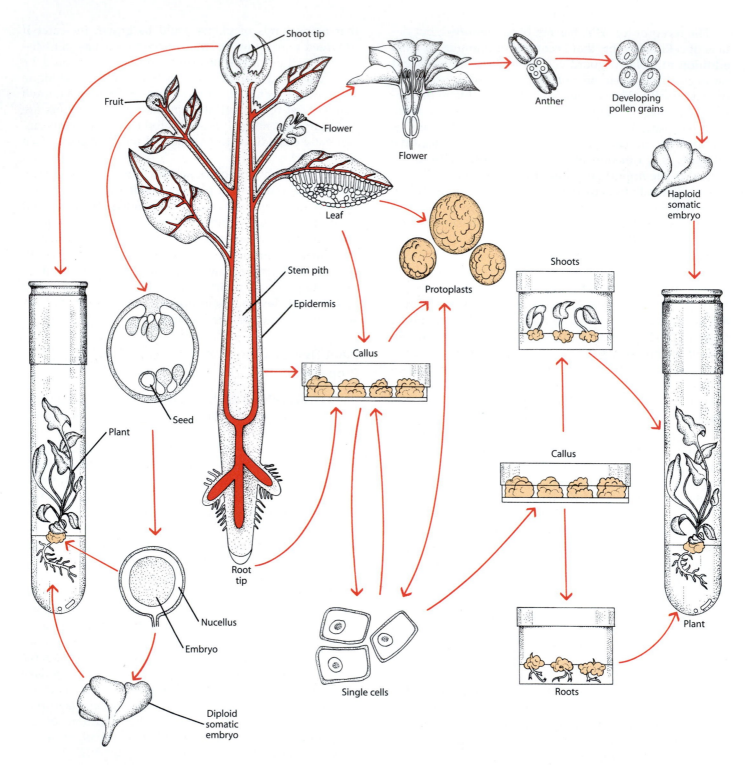

Shoot tip

Fruit

Flower

Flower

Anther

Developing pollen grains

Haploid somatic embryo

Leaf

Stem pith

Epidermis

Protoplasts

Shoots

Callus

Seed

Plant

Callus

Nucellus

Embryo

Root tip

Plant

Single cells

Roots

Diploid somatic embryo

28–25

Plant tissue culture. Theoretically, any plant cell—except those lacking a nucleus or enclosed by a rigid, secondary wall—is potentially capable of regenerating the organism from which it was derived. Such a cell is said to be totipotent. Collections of similar cells form tissues, tissues and tissue systems are organized into organs, and the specific spatial arrangement of organs constitutes the organism. Plants can be regenerated in vitro (in an artificial environment) from organ explants (root and shoot tips, buds, leaf primordia, developing embryos, bud scales, and so on), tissue explants (pith, cortex, epidermis, phloem, nucellus), cells (parenchyma, collenchyma, uninucleate or binucleate pollen grains), and protoplasts. The scheme shown here illustrates some of the pathways by which micropropagation can be achieved.

Totipotency

The fact that either roots or shoots can be generated from the same undifferentiated callus cells has important implications for plant genetics. As early as 1902, the German botanist Gottlieb Haberlandt suggested that all living plant cells are totipotent—that is, that each cell possesses the potential to develop into an entire plant— but he was never able to demonstrate this. In fact, more than half a century passed before his hypothesis was proved essentially correct. Haberlandt did not know which substances to provide to the cells since plant hormones had not yet been discovered.

In the late 1950s, F. C. Steward isolated small bits of phloem tissue from carrot (Daucus carota) root and placed them in liquid growth medium in a rotating flask. (Such pieces of tissue are called explants.) The medium contained sucrose and the in-

organic nutrients necessary for plant growth (see Chapter 30), certain vitamins, and coconut milk, which Steward knew to be rich in plant growth compounds—although the nature of these compounds was not then understood.

In the rotating flask, individual cells continuously broke away from the growing cell mass and floated free in the medium. These individual cells were able to grow and divide. Before long, Steward observed that roots had developed in many of these new cell clumps. If left in the swirling medium, the cell clumps did not continue to differentiate, but if they were transferred to a solid medium—which was agar in these experiments—some of the clumps developed shoots. If the clumps were then transplanted to soil, the little plants leafed out, flowered, and produced seed. Similar results were obtained a few

years later by V. Vasil and A. C. Hildebrandt, who used explants of tobacco (Nicotiana) pith from a fresh stem of a hybrid. Rather than coconut milk, the medium used by Vasil and Hildebrandt contained IAA and kinetin.

The results indicated that at least some of the cells of the mature carrot phloem and tobacco pith contained all the genetic potential for full plant development, although this potential was not expressed by these cells in the living plant. These experiments also showed that such differentiated cells can express portions of their previously unexpressed genetic potential in order to trigger those particular developmental patterns. By achieving these results, Steward and Vasil and Hildebrandt confirmed Haberlandt's hypothesis concerning totipotency.

Several interspecific (between species) somatic hybrids have been produced by protoplast fusion within genera that include tobacco, petunia, potato, and carrot (Figure 28–26). Most intergeneric hybrids have been sterile. Fertile intergeneric hybrid plants have been obtained by protoplast fusion between potato and tomato, both of which are members of the *Solanaceae*, or nightshade, family.

Pathogen-Free Plants Have Been Produced by Meristem Culture Apart from providing a means to produce identical copies (**clones**) of a plant, micropropagation provides a way to circumvent many plant diseases. This is due, in part, to the decontamination of the explants and the sterile conditions practiced in micropropagation, but primarily it is due to the use of meristem and shoot-tip culture techniques. Here, only very small explants of meristem and shoot tips lacking differentiated vascular tissues are cultured. Such explants are often virus-free because virus particles, which may be present in mature vascular elements below the meristems, can reach the meristematic regions of the apices only very slowly by cell-to-cell movement. The production of virus-free plants by meristem culture has greatly increased the yields of several crop plants, including potatoes and rhubarb.

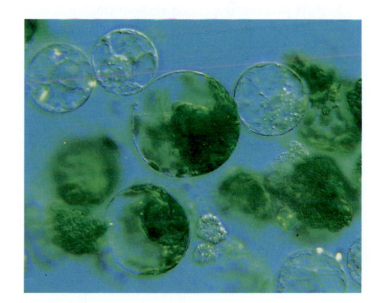

28–26

Protoplasts of a wild tobacco known as tree tobacco (Nicotiana glauca) (green) and commercial tobacco (N. tabacum) (clear). Some have fused, giving rise to hybrid cells. With appropriate treatment, the hybrid cells will regenerate cell walls, multiply, and give rise to whole plants.

Genetic Engineering Allows the Manipulation of Genetic Material for Practical Purposes

Genetic engineering, the application of recombinant DNA technology, is providing one of the most important means by which crop plants will be improved in the future. It has a distinct advantage over recombination of plant genetic material brought about by natural and somatic hybridization because it allows individual genes to be inserted into organisms in a way that is both precise and simple. In addition, the species involved in the gene transfer do not have to be capable of hybridizing with one another.

Restriction enzymes (page 224) allow research scientists to break up large genomes into defined fragments that can be maintained and propagated in bacterial plasmids. (See Chapter 11 for a discussion of recombinant DNA technology.)

Agrobacterium tumefaciens Is a Natural Genetic Engineer

The most popular organism for the transfer of foreign genes into plants is *Agrobacterium tumefaciens*, a soil-dwelling bacterium that infects a wide range of eudicots, typically gaining entry through wounds. *Agrobacterium tumefaciens* induces the formation of tumors, called crown-gall tumors (Figure 28–27), on the plant by transferring a specific region, the **T-region,** or **T-DNA,** of a tumor-inducing (Ti) plasmid to the host's nuclear DNA.

Each **Ti plasmid** is a closed circle of DNA that consists of about 100 genes (Figure 28–28). The T-DNA consists of approximately 20,000 base pairs of DNA bounded by 25-base-pair repeats at each end (Figure 28–28b). The T-DNA carries a number of genes, including one (gene *O*) that codes for an opine-synthesizing enzyme. In addition, *onc* is a group of three genes, two of which code for enzymes involved in the synthesis of auxin, and one of which codes for an enzyme that cat-

28–27
Crown galls growing on a Nicotiana glauca *stem.*

alyzes the synthesis of a cytokinin. The presence of the genes that code for the enzymes involved in hormone biosynthesis confers on host cells the ability to grow and divide uncontrollably. The opines, unique amino acid derivatives, are used by the bacteria as sources of carbon and nitrogen. Another region of the Ti plasmid, the *vir* region, is essential for the transfer process but is not incorporated into the host's DNA. Thus, *Agrobacterium* is a natural genetic engineer. It reprograms plant cells by transferring new genetic information into the host's genome.

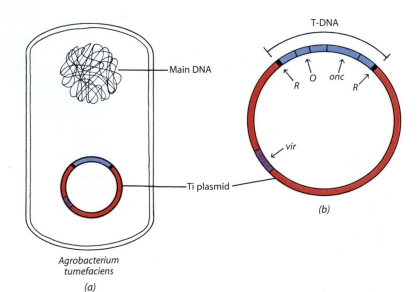

(a)

Agrobacterium tumefaciens

(b)

28–28
*Ti plasmid. **(a)** Diagrammatic representation of an* Agrobacterium tumefaciens *cell, showing the main DNA (the bacterial "chromosome") and the Ti plasmid. **(b)** Detail of the Ti plasmid. O is the gene that codes for an opine-synthesizing enzyme; onc is a group of three genes coding for enzymes that are involved in the biosynthesis of plant hormones; and R represents sequences of 25 base pairs. Only the DNA between the two R regions is transferred into the plant's genome. The group of genes known as vir controls the transfer of the T-region, or T-DNA, to the host (plant) chromosome.*

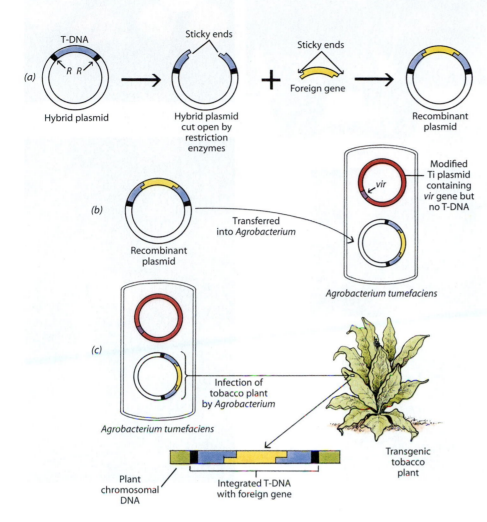

Procedure for using plasmids of Agrobacterium tumefaciens as vectors in DNA, or gene, transfer. (a) A hybrid plasmid that carries only the T-DNA from a Ti plasmid is cut open with restriction enzymes and a foreign gene is inserted, creating a recombinant plasmid. (b) The recombinant plasmid is transferred into an A. tumefaciens cell that contains a Ti plasmid with its T-DNA removed, creating an engineered plasmid. (c) The A. tumefaciens containing the engineered plasmid is used to infect a plant. The vir region of the Ti plasmid without T-DNA controls the transfer of the foreign gene from the recombinant plasmid into the plant's chromosomes.

Agrobacterium tumefaciens, with its Ti plasmid, is a powerful tool for genetic engineering of eudicots. The tumor-promoting genes on the T-DNA can be removed and replaced by foreign genes (Figure 28–29). Infection of a plant with *Agrobacterium* containing these engineered plasmids will result in transfer of the foreign genes into the plant's genome. **Transgenic plants** (plants containing foreign genes) obtained through the use of plasmids (page 282) will transmit the foreign genes to their progeny in a Mendelian fashion (Chapter 10).

Considerable progress has been made in applying this method to the transfer of agriculturally useful genes. These include genes for resistance to insects, herbicides, and viruses. In some cases, genes that mediate pathogen resistance in one species of plant are being identified and transferred into disease-susceptible species. In an extreme case, a gene from the bacterium *Bacillus thuringiensis* (the *BT* gene) has been transferred into plants. The *BT* gene codes for a protein toxin that specifically kills lepidopteran larvae—that is, the larvae of butterflies and moths. As shown in Figure 28–30a, plants expressing this gene are resistant to damage from caterpillars.

Plants have been engineered for resistance to several herbicides, including glyphosate (marketed by Monsanto under the name Roundup), which works by blocking a single enzyme essential to plants for the production of aromatic amino acids. Although extremely effective and nontoxic to animals, glyphosate kills all plants, including crops. One successful approach to genetically engineering resistance into crop plants resulted from the identification of a mutant form of the target enzyme from the bacterium *Salmonella* that is not blocked by glyphosate. Transfer of this mutant gene into crop plants, using the Ti plasmid, resulted in plants that were resistant to the herbicide.

Genetic Engineering Is Being Used to Manipulate Hormone Production As genes involved in hormone biosynthesis and perception are identified, genetic engineering is also being used to alter growth and development in plants. For example, genetically altered forms of ethylene biosynthetic enzymes have been transferred into tomato plants, resulting in diminished ethylene production and a significant delay in fruit ripening (Figure 28–30b). Such fruits can be induced to ripen by application of ethylene after they have arrived at market.

28–30

Examples of how gene transfer technology allows genetic engineers to alter growth, development, and disease resistance in plants. (a) The BT gene from the bacterium Bacillus thuringiensis codes for a protein that is toxic to lepidopteran insects. When the BT gene is transferred into plants, the plant cells produce the toxin and become resistant to caterpillar damage, as seen here in transformed (left) and untreated (right) tomato (Solanum lycopersicum) plants 4 days after exposure to caterpillars. (b) Transfer of a mutant form of the ethylene receptor gene (etr1-1) from Arabidopsis into tomatoes renders the fruits insensitive to ethylene. On the left are transgenic tomatoes that remained the same golden color 100 days after picking, whereas untreated fruits, on the right, turned deep red and began to rot after the same length of time. (c, d) Transfer of the (etr1-1) gene into petunia (Petunia hybrida cv. Mitchell) plants renders flowers insensitive to ethylene, resulting in an increased longevity of flowers. The flower from a transgenic plant in (c) is viable 8 days after pollination, whereas the flower from an untreated plant in (d) is withered 3 days after pollination. (e) When tobacco (Nicotiana tabacum) plants are transformed with a gene that codes for a cytokinin biosynthetic gene that is expressed only in older leaves, a dramatic delay in leaf senescence results, as seen here 20 weeks after transplanting transgenic (left) and untreated (right) seedlings into soil.

Mutant forms of the ethylene receptor gene from *Arabidopsis* have been used to delay senescence or wilting of flowers (Figure 28–30c). This achievement is of great interest to the cut-flower industry.

Genetic engineering has also been used to manipulate cytokinin production. The coding sequence of an *Agrobacterium* gene for an enzyme that synthesizes a cytokinin was fused to the promoter sequence of an *Arabidopsis* gene that is expressed only in senescent leaves. When this chimeric gene was transferred into tobacco plants, the production of cytokinin in aging leaves dramatically delayed leaf senescence (Figure 28–30d). If this delayed senescence effect could be made to operate in desirable crop plants, a significant impact on yield might result.

Genetic Engineering Is Also Being Used to Improve Food Quality Alterations in fatty acid biosynthetic pathways have been used to reduce the proportion of saturated fats in soybean (*Glycine max*) and canola (*Brassica napus*). Alterations in the starch content of potatoes (*Solanum tuberosum*) and the nutritional quality of protein in maize

kernels are being developed. Gene transfer is also being used to produce desirable products that the plant does not normally make. The possibilities range from the production of pharmaceutical proteins such as vaccines to the production of polyhydroxybutanol, a compound used to make plastic.

Reporter Genes Are Used for Gene Monitoring One of the technical problems encountered in attempts at gene transfer is knowing whether a particular gene has actually been introduced into a new host cell and, if transferred, whether it is directing the synthesis of protein. To overcome this problem, **reporter genes** have been developed, which can be transferred to the plant cell with the help of *Agrobacterium*. For gene monitoring, the reporter gene must be linked to a promoter. When the promoter is switched on, the reporter gene is activated. Three reporter genes currently in use are *GUS*, the *Escherichia coli* β-glucuronidase gene; the luciferase gene, which is obtained either from fireflies (*Photinus pyralis*) or the marine bacterium *Vibrio harveyi*; and the gene for a green fluorescent protein from jellyfish.

In a notable experiment, investigators at the University of California at San Diego spliced the gene for luciferase from fireflies into the DNA of tobacco cells, and plants were then regenerated from these cells. The final result was a tobacco plant that glows when it is provided with luciferin, the substrate of luciferase (Figure 28–31). Whenever the promoter is switched on, the luciferase gene is activated and luciferase is produced; in the presence of oxygen, luciferin plus ATP produces bioluminescence, as seen in the flash of a firefly.

There Are Limitations to the *Agrobacterium* Gene-Transfer System Despite its great usefulness for the transfer of foreign genes into many plant species, *Agrobacterium* cannot under natural conditions transform monocots. The latter include, of course, the world's major cereal crops: wheat, rice, maize, barley, and sorghum. Moreover, the tissue-culture systems and micropropagation techniques that have served so well for eudicots do not work well for monocots. Considerable effort is being made to develop suitable procedures for the regeneration of plants from monocot protoplasts.

Fortunately, two new methods of gene transfer that are applicable to both monocots and eudicots are providing promise. One of these, which was mentioned briefly earlier, is called **electroporation.** In this method, a high-voltage electric pulse is applied to a solution containing protoplasts and DNA. The electric shock temporarily opens pores in the plasma membrane through which the DNA passes.

In the second method, called **particle bombardment,** high-velocity microprojectiles (small tungsten particles) are used to deliver RNA or DNA into cells. The RNA or DNA is adsorbed to the surface of the microprojectiles. This approach has an advantage over other procedures in that it can be used on intact tissues. In addition, the process does not require either cell culture or pretreatment of the recipient tissue.

Genetic Engineering Has Both Benefits and Risks
Genetic engineering technology is providing biologists with an opportunity never before available—to transfer genetic traits between very different organisms. For the plant biologist this means potentially increasing crop production by introducing genes that increase the crop's resistance to various pathogens or herbicides and enhance its tolerance to various stresses. Examples of the former have already been given for the control of lepidopteran larvae and resistance to the herbicide glyphosate. A modified "ice-minus" strain of the bacterium *Pseudomonas syringa* has already been developed to reduce the susceptibility of certain crops to frost and, therefore, allow early planting. And attempts are being made to increase nitrogen availability, a limiting factor in crop production, by transferring genes responsible for nitrogen fixation into crops such as wheat and maize, which do not fix nitrogen. In addition to benefits derived

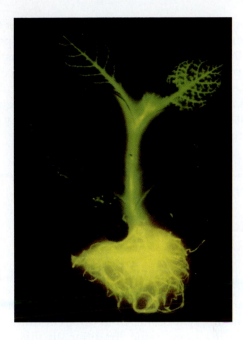

28–31

Genes for the production of the enzyme luciferase were inserted into cells isolated from a normal tobacco plant (Nicotiana tabacum), using the Ti plasmid of Agrobacterium tumefaciens as a vector. After the undifferentiated callus cells developed into a whole plant, the cells that incorporated the luciferase gene into their DNA became luminescent in the presence of luciferin, ATP, and oxygen.

from increased crop production, billions of dollars could be saved in fertilizer, insecticide, and herbicide costs.

We are reminded, however, that the benefits of genetic engineering do not come without some risks. Some agriculturists argue that genetic engineering and biotechnology will only extend what they regard as the "invade and conquer" style of farm production and pest management at a time when more emphasis should be placed on **sustainable agriculture,** which minimizes chemical use and "focuses on working with nature, rather than trying to conquer it." It is argued that increasing the number of herbicide-resistant crops may actually encourage wider herbicide use rather than limit it and thus contribute to environmental problems. The potential also exists for genetic transfer between engineered organisms and wild organisms in the environment. The impact of such transfer would depend on the nature of the engineered trait. If, for example, it were a gene for pest resistance, it could make a hybrid weed superior to its pure wild types. Because of risks such as these, many people feel that genetic engineering technology must be carefully managed. In order to avoid serious environmental, economic, and social problems, genetically engineered crops should be used as one element in promoting systems of sustainable agriculture. In addition, biological control and improved cultivation practices are also important elements in putting together such systems, which have a critical role to play in feeding a hungry world.

Summary

Plant Hormones Play a Major Role in Regulating Growth

Plant hormones are chemical regulators that produce physiological responses and are active in extremely small amounts. The same hormone can elicit different responses in different tissues or at different times of development in the same tissue. Traditionally, five groups of phytohormones (plant hormones) have received the most attention—auxins, cytokinins, ethylene, abscisic acid, and gibberellins—but it has become increasingly clear that additional chemical signals are used by plants.

Auxins Such As Indoleacetic Acid (IAA) Are Transported in a Polar Manner

Auxin is produced in the apical meristems of shoots and the tips of coleoptiles. It is transported unidirectionally, always toward the base of the plant, where it controls the lengthening of the shoot, chiefly by promoting cell elongation. Auxin also plays a role in differentiation of vascular tissue and initiates cell division in the vascular cambium. It often inhibits growth in lateral buds, thus maintaining apical dominance. The same quantity of auxin that promotes growth in the stem inhibits growth in the main root system. Auxin promotes the formation of adventitious roots in cuttings and retards abscission in leaves, flowers, and fruits. In fruits, auxin produced by seeds stimulates growth of the ovary wall.

Cytokinins Are Involved in Cytokinesis

The cytokinins are chemically related to certain components of nucleic acids. They are most abundant in actively dividing tissues such as seeds, fruits, and leaves, but may be transported there via the xylem from root tips. Cytokinins act in concert with auxin to cause cell division in plant tissue culture. In tobacco pith cultures, a high concentration of auxin promotes root formation, whereas a high concentration of cytokinins promotes bud formation. In intact plants, cytokinins promote the growth of lateral buds, acting in opposition to the effects of auxin. Cytokinins prevent senescence in leaves by stimulating protein synthesis.

Ethylene Is the Only Gaseous Plant Hormone

Ethylene, a simple hydrocarbon, exerts a major influence on many, if not all, aspects of growth and development in plants, including growth of most tissues, fruit and leaf abscission, and fruit ripening. In most plant species, ethylene has an inhibitory effect on cell expansion. In some semiaquatic species, however, ethylene elicits rapid stem growth. Ethylene apparently plays a major role in determining the sex of flowers in some monoecious plants.

Abscisic Acid Induces Dormancy and Stomatal Closure

Abscisic acid (ABA) is a growth-inhibiting hormone found in dormant buds and fruits. ABA stimulates the production of seed proteins and the closing of stomata. Studies using the patch clamp technique indicate that Ca^{2+} channels in the guard cell plasma membrane open in response to ABA, triggering a sequence of events that lowers the turgor of the guard cells and causes stomatal closure.

Gibberellins Stimulate the Growth of Stems and the Germination of Seeds

The gibberellins control shoot elongation, as is especially evident in dwarf plants. When such plants are treated with gibberellin, normal growth is restored. Studies of dwarf maize plants indicate that even though maize plants normally contain nine or more gibberellins, only the final compound in the pathway affects growth directly. In many species, gibberellins can substitute for the dormancy-breaking cold or light required for their seeds to germinate. In barley and other grass seeds, the embryo releases gibberellins that cause the aleurone layer of the endosperm to produce α-amylase. This enzyme breaks down the starch stored in the endosperm, releasing sugar that nourishes the embryo and promotes germination. Whereas gibberellins activate α-amylase gene expression, abscisic acid represses the gene.

Hormones Bind Specific Receptors Which Then Activate Particular Response Pathways

At the molecular level, hormones influence developmental processes through their interaction with receptors in the plant cell. The plant hormones mediate many processes by activating or repressing sets of genes in the cell's nucleus. At the cellular level, hormones influence the rate and direction of cell expansion and the rate of cell division. These effects are mediated through complex biochemical response pathways in the cell, often involving second messengers.

Plant Biotechnology Involves the Use of Tissue Culture

Advances in hormone research and DNA biochemistry have made it possible to manipulate the genetics of plants in specific ways. Among the most important procedures in biotechnology is tissue culture. In ideal cases, tissue culture is used to obtain complete plants starting with single, genetically altered cells. The potential for single cells to develop into entire plants is called totipotency.

Genetic Engineering Involves the Manipulation of Genes for Practical Purposes

Genetic engineering (recombinant DNA) technology is based on the ability to cut DNA molecules precisely into specific pieces and to combine those pieces to produce new combinations. The Ti plasmid of *Agrobacterium tumefaciens*, which induces the formation of crown-gall tumors, is being used as a vector for introducing foreign genes into a plant's genes. The transgenic plants then transmit the foreign genes to their progeny in a Mendelian fashion. Through the use of reporter genes, it is possible to determine visually whether genes carried by the plasmid have been successfully transferred to the plant cells and are being expressed.

Questions

1. Home gardeners commonly pinch off the shoot tips of certain plants in order to promote fuller, bushier growth. Explain why removal of the shoot tip should promote such growth.

2. What is meant by parthenocarpic fruit? Which two plant hormones are used to produce such fruit?

3. Compare and/or contrast auxins and cytokinins in each of the following: principal sites of biosynthesis; polarity of transport; cell types or tissues involved in transport; effect on cell division; effect on the production of roots and shoots in tissue cultures.

4. In what ways is ethylene a unique plant hormone?

5. What is the relationship between abscisic acid levels and viviparous mutants?

6. In what way are gibberellins of value to the brewing industry?

7. Explain how abscisic acid regulates the opening and closing of stomata.

8. Explain how the Ti plasmid of *Agrobacterium tumefaciens* is used in the production of transgenic plants.

Selected Key Terms

abscisic acid (ABA) p. 683

abscission p. 683

acropetal p. 676

aleurone p. 685

apical dominance p. 678

auxin p. 676

basipetal p. 676

climacteric p. 683

clonal propagation p. 693

clones p. 695

cytokinins p. 680

electroporation p. 699

ethylene p. 682

expansins p. 689

genetic engineering p. 696

gibberellin p. 684

hormone p. 674

micropropagation p. 693

particle bombardment p. 699

plant biotechnology p. 693

polar transport p. 676

protoplast fusion p. 693

receptors p. 690

reporter genes p. 698

restriction enzymes p. 696

second messengers p. 691

sensitivity p. 675

somatic hybrids p. 693

sustainable agriculture p. 699

Ti plasmid p. 696

tissue culture p. 693

totipotency p. 688

transgenic plants p. 697

T-region, or T-DNA p. 696

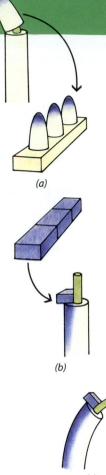

(a)

(b)

(c)

29–1

*Went's experiment. **(a)** He cut the coleoptile tips from oat (Avena sativa) seedlings and placed the tips on a slice of agar for about an hour. **(b)** He then cut the agar into small blocks and placed a block on one side of each decapitated shoot. **(c)** The seedlings, which were kept in the dark during the entire experiment, subsequently were observed to bend away from the side on which the agar block was placed. From this result, Went concluded that the "influence" that caused the seedling to bend was chemical and that it accumulated on the side away from the light. In Went's experiment, auxin molecules (blue) produced in the coleoptile tip were first transferred to the agar and then to one side of the seedling by means of an agar block.*

OVERVIEW

In some ways plants behave like animals—they can tell time, plan ahead for harsh weather, and respond to environmental changes. From an evolutionary perspective this should come as no surprise. For example, it is evolutionarily advantageous for plants in temperate regions to detect changes in the seasons because they can then flower in time for the resulting embryo to become encased by the protective seed coat before snow arrives. And it is similarly advantageous for seeds not to germinate and buds not to open prematurely—for example during an uncharacteristically mild day in the midst of winter.

This chapter deals with mechanisms by which plants respond to the external environment. Some of these mechanisms deal with everyday phenomena. Why, for example, do roots grow downward and stems grow upward? What causes tendrils to twine around fenceposts, and leaves of the Venus flytrap to close upon its prey? Other aspects may be less familiar, such as why plants grown in a greenhouse are often taller and thinner than the same type of plant grown outside. In some instances, the underlying mechanisms are mediated by the hormones discussed in the previous chapter. In most instances, the mechanisms are not completely understood.

Because of its importance in the life cycle, particular emphasis in this chapter is placed on flowering, a process that involves the interaction of many different phenomena including circadian rhythms and the biological clock, photoperiodism and the measurement of daylength, and interactions among hormones.

CHECKPOINTS

By the time you finish reading this chapter, you should be able to answer the following questions:

1. What is a tropism? By what mechanisms do plants respond to light? To gravity?

2. Why is it important that plants be able to "tell time"? What are some characteristics of the biological clock in plants?

3. What effect does daylength have on flowering?

4. What is phytochrome, and how is it involved in flowering, seed germination, and stem growth?

5. What is dormancy, and what environmental cues may be necessary to break the dormancy of seeds and buds?

L iving things must regulate their activities in accordance with the world around them. Many animals, being mobile, can change their circumstances to some extent—foraging for food, courting a mate, and seeking or even making shelters in bad weather. A plant, by contrast, is immobilized once it sends down its first root. However, plants do have the ability to respond and make adjustments to a wide range of changes in their external environment. This ability is manifested chiefly in changing patterns of growth. The environmental inputs into developmental programs in plants are responsible for the large variation in form that is often observed between genetically identical individuals.

The Tropisms

A growth response involving bending, or curving, of a plant part toward or away from an external stimulus that determines the direction of movement is called a **tropism.** A response toward the stimulus is said to be *positive;* a response away from the stimulus is *negative.*

Phototropism Is Growth in Response to Directional Light

Perhaps the most familiar interaction between plants and the external world is the curving of growing shoot tips toward light (see Figure 28–1). This growth response, now known as **phototropism,** is caused by elongation—under the influence of the hormone auxin—of the cells on the shaded side of the tip. What role does the light play in the phototropic response? Three possibilities are: (1) light decreases the auxin sensitivity of the cells on the lighted side, (2) light destroys auxin, or (3) light drives auxin to the shaded side of the growing tip.

To test these hypotheses, Winslow Briggs and his coworkers carried out a series of experiments based on earlier work by Frits Went (Figure 29–1). These investigators first showed that the same total amount of auxin is obtained from the tips of coleoptiles whether they remain in the dark or the light. However, following exposure to light on one side, the amount of auxin obtained from the shaded side is greater than the amount from the lighted side. If the tip is split and a barrier, such as a thin piece of glass, is placed between the two halves (the light and dark sides), this differential distribution of auxin is no longer observed (Figure 29–2). In other words, Briggs demonstrated that auxin migrates from the light side to the dark side. Experiments using ^{14}C-labeled auxin (indoleacetic acid, or IAA) have shown that it is migration of auxin, not destruction of auxin, that accounts for the different amounts obtained from the light and dark sides. These experiments are consistent with the hypothesis that auxin redistribution is responsible for phototropic curvature.

The redistribution of auxin in response to light is thought to be mediated by a photoreceptor, specifically, a pigment-containing protein that absorbs light and converts the signal into a biochemical response (for a discussion of receptors and signal processing, see page 690).

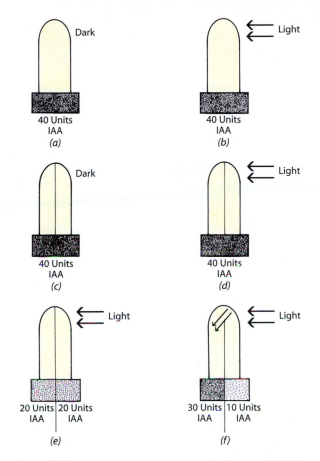

29–2

Experiments by Briggs and coworkers showing lateral displacement of IAA in unilaterally illuminated maize coleoptile tips. The diffusible auxin was collected in agar blocks. The units indicate the relative amounts of IAA present in each block. (a), (b) The same amount of IAA is collected from intact coleoptiles whether they remain in the dark or are illuminated. (c), (d) If the coleoptiles are split in half by a thin piece of glass but remain intact, the same amount of IAA is collected as in (a) and (b). (e) If the two halves and the agar block below are completely separated, the same amounts of auxin are collected from the dark and illuminated sides. (f) When the coleoptile is not split entirely to the top, more auxin diffuses from the darkened half than from the illuminated half. These results support the idea that unilateral illumination induces a lateral movement of IAA from the illuminated to the darkened half of the coleoptile.

The phototropic response is triggered by blue wavelengths of light (400 to 500 nanometers). Recently, mutants of *Arabidopsis* that are insensitive to blue light have been used to clone the genes that code for photoreceptors. These photoreceptor proteins are related to other well-studied proteins that bind pigments called flavins. Flavins absorb light primarily in the blue wavelengths of the spectrum, providing an explanation for why blue light is most effective in phototropic responses.

Gravitropism Is Growth in Response to Gravity

Another familiar tropism is **gravitropism,** a response to gravity (Figure 29–3). If a seedling is placed on its side, its root will grow downward (positive gravitropism) and its shoot will grow upward (negative gravitropism). The original explanation for this mechanism involved the redistribution of auxin from the upper to the lower side. In shoots, the higher auxin concentration on the lower side stimulates more rapid cell expansion on that side of the stem, resulting in an upward curving of the stem. For the root, however, which is more sensitive to auxin, the increased auxin concentration on the lower side actually inhibits cell expansion, resulting in a downward bending of the root as cells on the upper side expand more rapidly than those on the lower side.

Experimental evidence in support of this hypothesis of redistribution of auxin has been obtained largely for the shoot. Auxin-inducible genes have been identified and cloned from a number of species. Auxin-activated transcription of some of these genes occurs only on the side of the stem showing increased growth. It is still not clear whether this transcriptional activation is due to an increase in auxin concentration or to an increase in sensitivity of the tissue to auxin already present.

In many cases, particularly with eudicots, bulk redistribution of auxin has not been observed. One possible explanation is that auxin redistribution occurs between such closely associated tissues that it is difficult to measure. If, as thought, the growth rate of stems is largely determined by the extensibility of the epidermal cell layers, regulation of stem growth could involve lateral movement of auxin over the distance of only a few cell layers from the cortex to the epidermal layer. Such minor changes in auxin distribution would be difficult to detect but could have dramatic effects on growth. The *AUX*1 gene, the product of which mediates the hormonal control of root gravitropism in *Arabidopsis*, has recently been isolated and characterized. Mutations within the *AUX*1 gene result in a phenotype in which root growth is resistant to auxin. The result is a root that does not exhibit gravitropic curvature. The gene was expressed predominantly within epidermal cells of the root.

Recent studies indicate that calcium plays an important role in the gravitropic response of both shoots and roots, and that this response is mediated by the calcium-binding protein calmodulin (page 87). Calcium has been shown to move toward the upper surface of gravistimulated shoots before the shoots curve upward and toward the lower surface of roots before the roots curve downward. In intact maize (*Zea mays*) roots, the influence of calcium may be at the epidermis.

The perception of gravity is correlated with the sedimentation of amyloplasts (starch-containing plastids) within specific cells of the shoot and root. Such cells are often found contiguous to or surrounding the vascular bundles in shoots. In roots, however, these cells are localized in the rootcap, particularly in the central column (columella) of the rootcap (Figure 29–4). When a root is placed in a horizontal position, the amyloplasts, which

29–3

Gravitropic responses in the shoot of a young tomato plant (Solanum lycopersicum). *The plant in* **(a)** *was placed on its side and kept in a stationary position; the plant in* **(b)** *was placed upside down and held in position in a ring stand. Although originally straight, the stem bent and grew upward in both cases. If a plant is held horizontally and slowly rotated around its horizontal axis, no curvature (gravitropic response) will occur; that is, the plant will continue to grow horizontally. Can you explain the differences in the growth of a horizontally rotated plant and the plants shown here?*

(a)   (b)

29–4

Amyloplasts. (a) Photomicrograph of a median longitudinal section of the rootcap of the primary root of the common bean, Phaseolus vulgaris. Arrows point to amyloplasts (starch-containing plastids) sedimented near the transverse walls in the central column of cells (columella) of the rootcap. (b) Electron micrograph of columella cells from a rootcap similar to that shown in (a). The amyloplasts (arrows) are visible at the bottom of each cell near the transverse walls. The starch grains are clearly evident within the amyloplasts.

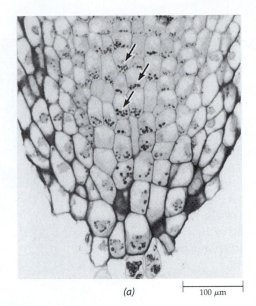

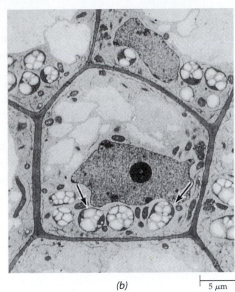

(a) `|—— 100 μm ——|`

(b) `|— 5 μm —|`

were sedimented near the transverse walls of the vertically growing roots, slide downward and come to rest near what were previously vertically oriented walls (Figure 29–5). After several hours, the root curves downward and the amyloplasts return to their previous position along the transverse walls. According to the *starch-statolith hypothesis,* the sedimentable amyloplasts play the role of statoliths, or gravity sensors. How the movement of the starch-containing amyloplasts is translated into differential growth rates has yet to be explained.

An alternative to the starch-statolith hypothesis for gravity sensing by plant cells is the *hydrostatic pressure hypothesis,* which proposes that plant cells detect gravity changes by sensing the hydrostatic pressure exerted by the protoplast on its cell wall. The sensing mechanism presumably is capable of detecting pulling and pushing forces applied by the mass of the protoplast on the top and bottom sides of the cell. A third hypothesis, the *plas-*

malemma (plasma membrane) central control hypothesis, incorporates the hydrostatic pressure hypothesis. It proposes that Ca^{2+}-selective cation channels, clustered around attachment centers linking the plasma membrane to the cytoskeleton and the cell wall, open in response to tensions at the plasma membrane. Such tensions could be caused by changes in the distribution of forces (including those due to gravity) exerted by the protoplast, the cytoskeleton, or the cell wall at the attachment sites. Opening of the calcium channels would cause a transient increase in the cytosolic Ca^{2+} level, triggering local activation of Ca^{2+}-dependent regulatory proteins, such as calmodulin and protein kinases, involved in the signal-transduction pathways responsible for the final response. In addition, local increases in cytosolic Ca^{2+} would trigger local auxin release from the stimulated cells, resulting in an auxin gradient across the gravity-stimulated organ.

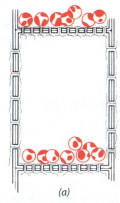

(a)

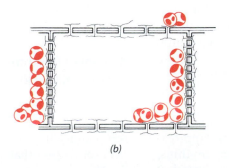

(b)

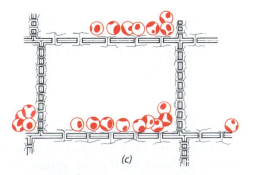

(c)

29–5

The response to gravity of amyloplasts in the columella cells of a rootcap. (a) The amyloplasts are normally sedimented near the transverse walls in the rootcap of a root

growing vertically downward. When the root is placed on its side, (b) and (c), the amyloplasts slide toward the normally vertical walls that are now parallel with the soil surface.

This movement of amyloplasts may play an important role in the perception of gravity by roots.

Hydrotropism Is Growth in Response to a Moisture Gradient

Roots also respond to differences in soil moisture content by growing toward regions of greater water potential. This tropism is called **hydrotropism.** Hydrotropism has been difficult to study, in part because the response of roots to gravity is much stronger than their response to a moisture gradient. This problem was resolved with the discovery of a mutant of pea *(Pisum sativum)* called *ageotropum,* which is neither gravitropic nor phototropic. Results of studies on *ageotropum* indicate that the sensing cells for a moisture gradient inducing positive hydrotropism reside in the rootcap and that calcium plays a role in the induction of root hydrotropism.

Thigmotropism Is Growth in Response to Touch

Another common tropism is **thigmotropism** (from the Greek *thigma,* meaning "touch"), a response to contact with a solid object. One of the most common examples of thigmotropism is seen in tendrils, which are modified leaves in some species and modified stems in others (page 641). The tendrils wrap around any object with which they come in contact (Figure 29–6) and so enable the plant to cling and climb. The response can be rapid; a tendril may wrap around a support one or more times in less than one hour. Cells touching the support shorten slightly, and those on the other side elongate.

Studies by M. J. Jaffe have shown that tendrils can store the "memory" of tactile stimulation. For example, if pea *(Pisum sativum)* tendrils are kept in the dark for three days and then rubbed, they will not coil, perhaps because of the requirement for ATP. If, however, they are illuminated within two hours of the rubbing, they will show the coiling response.

29–6

Thigmotropism. Tendrils of bur cucumber (Cucumis anguria). Twisting is caused by different growth rates on the inside and outside of the tendril.

Circadian Rhythms

It is a common observation that the flowers of some plants open in the morning and close at dusk. Also, many leaves unfold in the sunlight and refold at night (Figure 29–7). As long ago as 1729, the French scientist Jean-Jacques de Mairan noticed that these diurnal (daily) movements continue even when the plants are kept in dim light (Figure 29–8). More recent studies have shown that photosynthesis, auxin production, and the rate of cell division also have regular daily rhythms and that these rhythms continue even when all environmental conditions are kept constant. These regular, approximately 24-hour cycles are called **circadian rhythms,** from the Latin words *circa,* meaning "approximately," and *dies,* meaning "a day." Examples of circadian rhythms are universal among eukaryotes. Although long thought to be absent in bacteria, the presence of circadian rhythms has recently been demonstrated in cyanobacteria.

Circadian Rhythms Are Controlled by Biological Clocks

Are these rhythms actually internal—that is, caused by factors entirely within the organism—or is the organism keeping itself in time with some external stimulus? Most workers agree that the rhythms are endogenous—that is, they are controlled internally. The internal timing mechanism is often referred to as the organism's **biological clock.**

(a)

(b)

29–7

Diurnal movements. Leaves of the wood sorrel (Oxalis) during the day (a) and at night (b). One hypothesis for the function of such sleep movements is that they prevent the leaves from absorbing moonlight on bright nights, thus protecting the photoperiodic phenomena discussed later in this chapter. Another hypothesis, proposed by Charles Darwin more than a century ago, is that the folding reduces heat loss from the leaves at night.

Biological Clocks Are Synchronized by the Environment

Under constant environmental conditions, such as can be maintained only in a laboratory, the period of the circadian rhythm is **free-running**—that is, its natural period (usually between 21 and 27 hours) does not have to be reset at each cycle. It behaves as a self-sustained oscillator. In the natural world, the environment acts as a synchronizing agent (or *Zeitgeber*, German for "time giver"). In fact, the environment is responsible for keeping a circadian rhythm in step with the daily 24-hour light-dark cycle. If a circadian rhythm of a plant were greater or less than 24 hours, the rhythm would soon get out of step with the 24-hour light-dark cycle. A phenomenon such as flowering, which generally takes place in the light period, would occur at a different time each day, including the dark period. Thus the plant must become resynchronized—that is, **entrained**—to the 24-hour day.

Entrainment is the process by which a periodic repetition of light and dark (or some other external cycle)

29–8

Circadian rhythms. In many plants, the leaves move outward, to a position perpendicular to the stem and the sun's rays, during the day and upward, toward the stem, at night. These sleep movements can be recorded on a revolving drum using a delicately balanced pen-and-lever system attached to a leaf by a fine thread (a). Many plants, such as the common bean (Phaseolus vulgaris) shown here, will continue to exhibit these movements for several days even when kept in continuous dim light. (b) A recording of this circadian rhythm, showing its persistence under a condition of constant dim light.

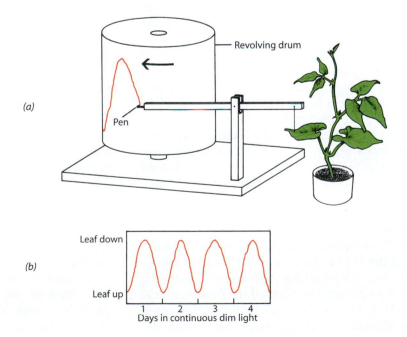

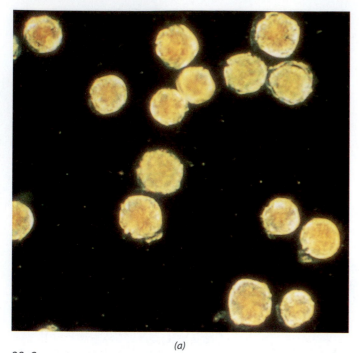

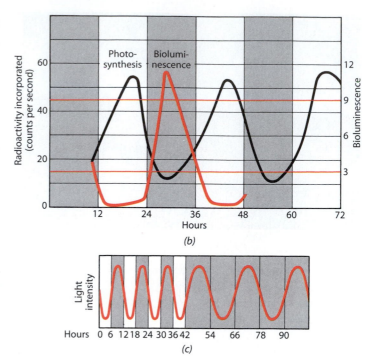

29–9

*Entrainment. **(a)** Photomicrograph of the dinoflagellate* Gonyaulax polyedra, *a single-celled marine alga. **(b)** In G. polyedra, three different functions follow separate circadian rhythms: bioluminescence, which reaches a peak in the middle of the night (colored curve); photosynthesis, which reaches a peak in the middle of the day (black curve); and cell* division *(not shown), which is restricted to the hours just before dawn. If* Gonyaulax *is kept in continuous dim light, these three functions continue to occur with the same rhythm for days and even weeks, long after a number of cell divisions have taken place. **(c)** The rhythm of bioluminescence in* Gonyaulax, *like most circadian rhythms, can be altered by modify-* ing the cycles of illumination. For example, if cultures of the alga are exposed to alternating light and dark periods, each 6 hours in duration, the rhythmic function will become entrained to this imposed cycle (left). If the cultures are then placed in continuous dim light, the organisms will return to their original rhythm of about 24 hours.

causes a circadian rhythm to remain synchronized with the same cycle as the entraining factor. Light-dark cycles and temperature cycles are the principal factors in entrainment (Figure 29–9).

Another feature of interest is that these rhythms do not automatically speed up as the temperature rises, even though biochemical activities—and an internal biological clock must have a biochemical basis—take place more rapidly at high temperatures than at low ones. Some clocks run slightly faster as the temperature rises, but others go more slowly, and many are almost unchanged. Thus, the biological clock must contain within its workings a compensatory mechanism—a feedback system that allows it to adjust to temperature changes. Such a temperature-compensating feature would be very important in plants.

Genetic analyses in *Neurospora* have led to the discovery of some of the molecular components of its biological clock, including the *frequency, white collar-1,* and *white collar-2* genes. The protein products of the *frequency* gene have been shown to feed back and inhibit their own transcription. (Similar feedback loops have been described in other organisms.) *White collar-1* and *white col-* lar-2, both of which encode DNA binding proteins, apparently act as transcriptional activators. Both are required for *Neurospora*'s responses to light.

An excellent plant example of the clock's control of the expression of specific genes is found in the genes that encode chlorophyll *a/b* binding (Cab) proteins of the light-harvesting complexes. This was elegantly demonstrated by fusing a fragment of the *Arabidopsis CAB2* promoter and the firefly luciferase gene *(Luc)* coding region, providing a bioluminescent reporter of *CAB* transcriptional activity (page 698). When this so-called reporter gene was transferred into *Arabidopsis* plants, the influence of the clock on the expression of the chlorophyll *a/b*-binding protein *(CAB)* genes could be monitored by detecting the luminescence resulting from the activity of luciferase (Figure 29–10). This reporter system has been used to screen for mutants that affect the clock mechanism in *Arabidopsis.*

In addition to coordinating daily events, the primary usefulness of the biological clock is that it enables the plant or animal to respond to the changing seasons of the year by accurately measuring changing daylength. Even at tropical latitudes, many plants are responsive to

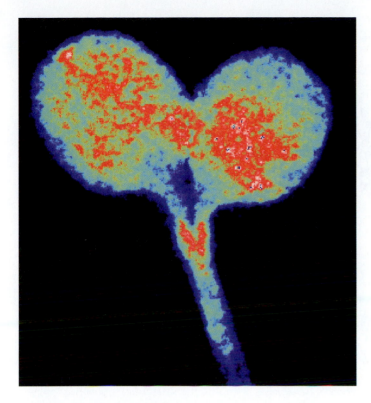

29–10

Control of gene expression by the biological clock. Shown here is an Arabidopsis seedling in which the firefly luciferase gene was fused to the promoter region of the CAB2 gene. Expression of the CAB2 gene, which is regulated by light and the biological clock, is greatest (red) in the cotyledons and absent (blue) in the root.

daylength and may use this signal to synchronize flowering or other activities with such seasonal events as wet or dry periods. In this way, changes in the environment trigger responses that result in adjustments of growth, reproduction, and other activities of the organism.

Photoperiodism

The effect of daylength on flowering was discovered some seventy years ago by two investigators from the U.S. Department of Agriculture, W. W. Garner and H. A. Allard, who found that neither the Maryland Mammoth variety of tobacco (*Nicotiana tabacum*) nor the Biloxi variety of soybean (*Glycine max*) would flower unless the daylength was shorter than a critical number of hours. Garner and Allard called this phenomenon **photoperiodism.** Plants that flower only under certain daylength conditions are said to be *photoperiodic*. Photoperiodism is a biological response to a change in the proportions of light and dark in a 24-hour daily cycle. Although the concept of photoperiodism began with studies of plants, it has now been demonstrated in various fields of biology, including the mating behavior of such diverse animals as codling moths, spruce budworms, aphids, potato worms, fish, birds, and mammals.

Daylength Is a Major Determinant of Flowering Time

Garner and Allard went on to test and confirm their discovery with many other species of plants. They found that plants are of three general types, which they called **short-day, long-day,** and **day-neutral.** Short-day plants flower in early spring or fall; they must have a light period *shorter* than a critical length. For instance, the common cocklebur (*Xanthium strumarium*) is induced to flower by 16 hours or less of light (Figures 29–11 and 29–12). Other short-day plants are some chrysanthemums (Figure 29–13a), poinsettias, strawberries, and primroses.

29–11

The relative length of day and night determines when plants flower. The four curves depict the annual change of daylength in four North American cities at four different latitudes. The horizontal green lines indicate the effective photoperiod of three different short-day plants. The cocklebur, for instance, requires 16 hours or less of light. In Miami, San Francisco, and Chicago, it can flower as soon as it matures, but in Winnipeg, the buds do not appear until early August, so late that the frost will probably kill the plants before the seed is set.

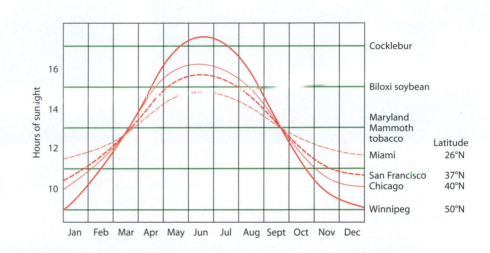

29–12
Short-day plants flower only when the pho-toperiod is less than some critical value. The common cocklebur (Xanthium strumarium), a short-day plant, requires less than 16 hours of light to flower. Long-day plants flower only when the photoperiod is longer than some critical value. The long-day plant henbane (Hyoscyamus niger) requires about 10 hours (depending on temperature) or more to flower. If the dark period is interrupted by a flash of light, Hyoscyamus will flower even when the daily light period is shorter than 10 hours. However, a pulse of light during the dark period has the opposite effect in a short-day plant—it prevents flowering. The bars at the top indicate the duration of light and dark periods in a 24-hour day.

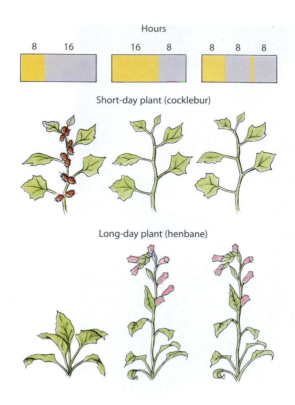

Hours

Short-day plant (cocklebur)

Long-day plant (henbane)

Long-day plants, which flower chiefly in the summer, will flower only if the light periods are *longer* than a critical length. Spinach, some potatoes, some wheat varieties, lettuce, and henbane *(Hyoscyamus niger)* are examples of long-day plants (Figures 29–12 and 29–13b).

Cocklebur and spinach will both bloom if exposed to 14 hours of daylight, yet only one is designated as a long-day plant. The important factor is not the absolute length of the photoperiod but whether it is longer or shorter than some critical interval. Day-neutral plants flower without respect to daylength. Some examples of day-neutral plants are cucumber, sunflower, rice, corn, and garden pea.

Within individual species of plants that cover a large north-south range, different photoperiodic ecotypes (locally adapted variants of an organism) have often been observed. Thus, in many species of prairie grasses, which range from southern Canada to Texas, northern ecotypes flower before southern ones when they are grown together in a common environment. Different

(a)

(b)

29–13
*Representative short-day (**a**) and long-day (**b**) plants, each grown under short-day (on left) and long-day (on right) conditions. (**a**)*
*Chrysanthemum. (**b**) Spinach (Spinacia oleracea). Note that the plants exposed to long days have longer stems than those*
exposed to short days, regardless of whether or not they flower.

populations are precisely adjusted to the demands of the local day-night regimen.

The photoperiodic response can be remarkably precise. At 22.5°C, the long-day plant henbane will flower when exposed to photoperiods of 10 hours and 20 minutes (Figure 29–12), but not with a photoperiod of 10 hours. Environmental conditions also affect photoperiodic behavior. For instance, at 28.5°C henbane requires 11.5 hours of light, whereas at 15.5°C it requires only 8.5 hours.

The response varies with different species. Some plants require only a single exposure to the critical day-night cycle, whereas others, such as spinach, require several weeks of exposure. In many plants, there is a correlation between the number of induction cycles and the rapidity of flowering or the number of flowers formed. Some plants have to reach a certain degree of maturity before they will flower, whereas others will respond to the appropriate photoperiod when they are seedlings. Some plants, when they get older, will eventually flower even if not exposed to the appropriate photoperiod, although they will flower much earlier with the proper exposure.

Plants Keep Track of Daylength by Measuring the Length of Darkness

In 1938, Karl C. Hamner and James Bonner began a study of photoperiodism, using the short-day plant cocklebur, which requires 16 hours or less of light per 24-hour cycle to flower. This plant is particularly useful for experimental purposes because a single exposure under laboratory conditions to a short-day cycle will induce flowering two weeks later, even if the plant is immediately returned to long-day conditions. Hamner and Bonner showed that it is the leaf blade of the cocklebur that perceives the photoperiod. A completely defoliated plant cannot be induced to flower. But if as little as one-eighth of a fully expanded leaf is left on the stem, the single short-day exposure induces flowering.

In the course of these studies, in which they tested a variety of experimental conditions, Hamner and Bonner made a crucial and totally unexpected discovery. If the period of darkness is interrupted by as little as a one-minute exposure to light from a 25-watt bulb, flowering does not occur. Interruption of the light period by darkness has absolutely no effect on flowering. Subsequent experiments with other short-day plants showed that they, too, required periods of uninterrupted darkness rather than uninterrupted light.

The most sensitive part of the dark period with respect to light interruption is in the middle of the period. If a short-day plant, such as cocklebur, is given an 8-hour light period and then an extended period of darkness, the plant passes through a stage of increasing sensitivity to light interruption lasting about 8 hours, followed by a period when light interruption has a diminishing effect. In fact, a one-minute light exposure after 16 hours of darkness stimulates flowering.

On the basis of the findings of Garner and Allard, described earlier, commercial growers of chrysanthemums had found that they could hold back blooming in short-day plants by extending the daylight with artificial light. On the basis of the new experiments by Hamner and Bonner, growers were able to delay flowering simply by switching on the light for a short period in the middle of the night.

What about long-day plants? They also measure darkness. A long-day plant that will flower if kept under light for 16 hours and dark for 8 hours will also flower on 8 hours of light and 16 hours of dark if the dark is interrupted by a one-hour exposure to light.

Chemical Basis of Photoperiodism

An important clue to the mechanism of plants' response to the relative proportions of light and darkness came from a team of research workers at the U.S. Department of Agriculture Research Station in Beltsville, Maryland. The clue came from an earlier study performed with lettuce (*Lactuca sativa*) seeds, which germinated only if they were exposed to light. This requirement is true of many small seeds, which must germinate in loose soil and near the surface in order for the seedling to emerge. In studying the light requirement of germinating lettuce seeds, earlier workers had shown that red light stimulates germination and that light of a slightly longer wavelength (far-red) inhibits germination even more effectively than does no illumination at all.

Hamner and Bonner had shown that when the dark period is interrupted by a single flash of light from an ordinary light bulb, the cocklebur will not flower. The Beltsville group, following this lead, began to experiment with light of different wavelengths, varying the intensity and duration of the flash. They found that red light with a wavelength of about 660 nanometers was most effective in preventing flowering in the cocklebur and other short-day plants and in promoting flowering in long-day plants.

Using lettuce seeds, the Beltsville group found that when a flash of red light was followed by a flash of far-red light, the seeds did not germinate. The red light most effective in inducing germination in the lettuce seeds was of the same wavelength as that involved in the flowering response—about 660 nanometers. Furthermore, the light most effective in inhibiting the effect produced by red light had a wavelength of 730 nanometers, in the far-red region. The sequence of red and far-red flashes could be repeated over and over; the number of flashes did not matter, but the nature of the final one

29–14

*Light and the germination of lettuce seeds.
(a) Seeds exposed briefly to red light; (b) seeds
exposed to red light followed by far-red light;
(c) seeds exposed to a sequence of red, far-
red, red; (d) seeds exposed to a sequence of
red, far-red, red, far-red. Whether or not the
seeds germinate depends on the final wavelength in the series of exposures, with red light
promoting and far-red light inhibiting germination.*

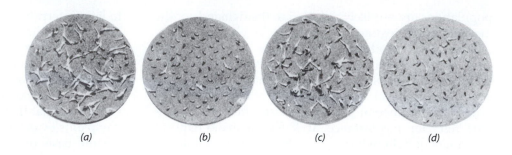

(a)   (b)   (c)   (d)

did. If the sequence ended with a red flash, the majority of the seeds germinated. If it ended with a far-red flash, the majority did not germinate (Figure 29–14).

Phytochrome Is Involved in Photoperiodism

The photoreceptors involved both in flowering and in lettuce-seed germination exist in two different interconvertible forms: P_r, which absorbs red light, and P_{fr}, which absorbs far-red light. When a molecule of P_r absorbs a photon of 660-nanometer light, the molecule is converted to P_{fr} in a matter of milliseconds; when a molecule of P_{fr} absorbs a photon of 730-nanometer light, it is quickly reconverted to the P_r form. These are called *photoconversion reactions*. The P_{fr} form is biologically active (that is, it will trigger a response such as seed germination), whereas P_r is inactive. The photoreceptors can thus function as a biological switch, turning responses on or off.

The lettuce-seed germination experiments are easily understood in these terms. Since P_r absorbs red light most efficiently (Figure 29–15), this light will convert a high proportion of the molecules to the P_{fr} form, thereby inducing germination. Subsequent far-red light absorbed

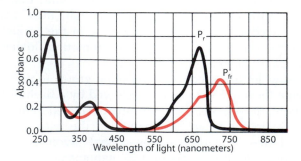

29–15

Absorption spectra of the two forms of phytochrome, P_r and P_{fr}. This difference in the absorption spectra made it possible to isolate the photoreceptor.

by P_{fr} will convert essentially all of the molecules back to P_r, canceling the effect of the prior red light.

What about flowering under natural day-night cycles? Since white light contains both red and far-red wavelengths, both forms of the photoreceptor are exposed simultaneously to photons that are efficient in promoting their photoconversion. After a few minutes in the light, then, a photoequilibrium is established in which the rates of conversion of P_r to P_{fr} and P_{fr} to P_r are equal, and the proportion of each type of photoreceptor molecule is constant (about 60 percent P_{fr} in noontime sunlight).

When plants are switched to darkness, the level of P_{fr} steadily declines over a period of several hours. If a high level of P_{fr} is regenerated by pulse irradiation with red light in the middle of the dark period (Figure 29–12), it will inhibit flowering in short-day (that is, "long-night") plants that otherwise would have flowered and promote flowering in long-day (that is, "short-night") plants that otherwise would not have flowered. In either case, the effect of the red pulse can be canceled by an immediate far-red pulse, which reconverts the P_{fr} to P_r.

In 1959, Harry A. Borthwick and his coworkers at Beltsville named this photoreceptor **phytochrome** and presented conclusive physical evidence for its existence. The principal characteristics of the receptor, as they are presently understood, are summarized schematically in Figure 29–16.

Phytochrome Can Be Detected in Tissues with a Spectrophotometer   Phytochrome is present in plants in much smaller amounts than pigments such as chlorophyll. To detect phytochrome, one needs to use a spectrophotometer that is sensitive to extremely small changes in light absorbance. Such an instrument was not available until some seven years after the existence of phytochrome was proposed; in fact, the first use of this spectrophotometer was to detect and isolate phytochrome.

In order to avoid interference from chlorophyll, which also absorbs light of about 660 nanometers, dark-grown seedlings (in which chlorophyll has not yet devel-

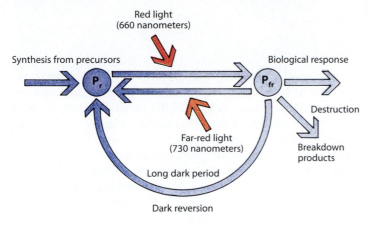

29–16

Phytochrome is continuously synthesized in the P$_r$ form from its precursors, amino acids, and accumulates in this form in dark-grown plants. P$_r$ changes to P$_{fr}$ when exposed to red light, which is present in sunlight. P$_{fr}$ is the active form that induces a biological response. P$_{fr}$ is converted back to P$_r$ by photo-conversion when exposed to far-red light. In darkness, P$_{fr}$ reverts to P$_r$ (dark reversion) or is lost through a process termed "destruction" that occurs over several hours and probably involves hydrolysis by a protease. All three alternative pathways for P$_{fr}$ removal provide the potential for reversing induced responses. It should be noted, however, that dark reversion has been detected only in eudicots, not in monocots.

29–17

The phytochrome chromophore in its P$_r$ form, showing its attachment to the protein part of the molecule.

oped) were chosen as the source of phytochrome. The pigment associated with the phytochrome receptor proved to be blue in color (why might you expect this color?), and it showed the characteristic P$_r$ to P$_{fr}$ conversion in the test tube by reversibly changing color slightly in response to red or far-red light.

The phytochrome molecule contains two distinct parts: a light-absorbing portion (the chromophore) and a large protein portion (Figure 29–17). The chromophore is much like the phycobilins that serve as accessory pigments in cyanobacteria and red algae. Genes for the protein portion have now been isolated from several species, and the amino acid sequence has been deduced from the nucleotide sequence. Most plants probably have several different phytochromes encoded by a family of divergent genes. In *Arabidopsis*, five such genes are known.

The specific mechanism by which phytochrome works has not yet been established. However, it is clear that phytochrome-regulated morphogenesis results from changes in gene transcription. A number of phytochrome-activated genes have been identified and isolated. The regulatory regions of these genes are quite complex, with several different DNA sequences interacting with a number of DNA binding proteins. As yet there is no evidence that phytochrome interacts directly with these genes.

Phytochrome Is Involved in a Wide Variety of Plant Responses The germination of many seeds occurs in the dark. In the seedlings, the stem elongates rapidly, pushing the shoot (or, in grasses, the coleoptile) up through the dark soil. During this early stage of growth, there is essentially no enlargement of the leaves, which would interfere with the passage of the shoot through the soil. Soil is not necessary for this growth pattern; any seedling grown in the dark will be elongated and spindly and will have small leaves. It will also be yellow to colorless because plastids do not turn green until exposed to light. Such a seedling is said to be **etiolated** (Figure 29–18).

29–18

Dark-grown seedlings, such as the bean plants on the left, are thin and pale; they have longer internodes and smaller leaves than light-grown seedlings, such as those on the right. The group of physical characteristics exhibited by these dark-grown seedlings, known as etiolation, has survival value for the seedlings because it increases their chances of reaching light before their stored energy supplies are exhausted.

When the seedling tip emerges into the light, the etiolated growth pattern gives way to normal plant growth. In eudicots, the hook unbends, the growth rate of the stem may slow down somewhat, and leaf growth begins (see Figure 23–10). In grasses, growth of the mesocotyl (the part of the embryo axis between scutellum and coleoptile) stops, the stem elongates, and the leaves open (see Figure 23–11). Such light-triggered growth and developmental responses are called **photomorphogenic responses.**

A dark-grown bean seedling that receives five minutes of red light a day, for instance, will show these light effects beginning on the fourth day. If the exposure to red light is followed by a five-minute exposure to far-red, none of the changes usually produced by the red light will appear (Figure 29–19). Similarly, in the seedlings of grains, termination of mesocotyl growth is triggered by exposure to red light, and the effect of red light is canceled by far-red.

Studies of the *long hypocotyl* mutants of *Arabidopsis* have provided information on the roles of specific photoreceptor genes in photomorphogenesis. In these mutants, light does not arrest hypocotyl elongation, indicating that the wild-type *HY* genes are involved in light perception or response. In the *hy1* and *hy2* mutants, the biosynthesis of the phytochrome chromophore is blocked. Two additional mutants, *hy8* and *hy3*, are known to have mutations in two different genes, designated *PhyA* and *PhyB*, respectively. The wild-type *PHYA* and *PHYB* genes code for the protein portions of two different phytochromes, phytochrome A and phytochrome B. The *hy8* mutant, which lacks phytochrome A, is reduced in its response to red light, whereas the *hy3* mutant, which lacks phytochrome B, has a reduced response to far-red light, indicating different roles for different forms of the phytochrome receptors.

In another mutant, the *hy4* mutant, the ability to respond to blue light is blocked. Cloning of the gene responsible for the *hy4* mutant revealed that it is a blue-light receptor gene. This finding illustrates the complexity of responses to light that involve the contributions of several specific photoreceptors, each responsible for a different aspect of photomorphogenesis.

An important feature of phytochrome in plants growing in a natural environment is the detection of shading by other plants. Radiation of wavelengths below 700 nanometers is almost completely reflected or absorbed by vegetation, whereas that between 700 and 800 nanometers (in the far-red range) is largely transmitted. Thus plants grown in the shade receive more far-red light than red light. This causes a dramatic upward shift in the equilibrium ratio of P_r to P_{fr} (that is, more P_{fr} is converted to P_r) in shaded plants and results in a rapid increase in the rate of internodal elongation. Given the competition for light under the forest canopy, a plant's ability to sense the level of light and adjust its growth accordingly has obvious adaptive significance.

Reversible red/far-red reactions are also involved in formation of anthocyanin, the red or blue pigment found in apples, turnips, and cabbage. These reactions also participate in changes in chloroplasts and other plastids, and in a tremendous variety of other plant responses in all phases of the plant life cycle.

When the existence of phytochrome was first demonstrated, its discoverers hypothesized that its behavior might explain the phenomenon of photoperiodism—that is, the reversible red/far-red reactions might be involved in the time-measuring mechanism, or biological clock. It is now generally agreed, however, that the time-measuring phenomenon of photoperiodism is not controlled by the interconversion of P_r and P_{fr} alone. A more complex explanation must be sought.

29–19

Photomorphogenic response. All three bean plants received eight hours of light each day. The center plant was exposed to five minutes of far-red light, which promotes elongation, at the start of each dark period. The plant on the right received the same five-minute exposure to far-red light followed by a five-minute exposure to red light, which counteracted the effect of the far-red light. The plant on the left served as the control; its phytochrome is predominantly in the P_{fr} form as it enters the dark period because of its exposure to daylight.

Hormonal Control of Flowering

Hamner and Bonner, in their early cocklebur experiments, showed that the leaf "perceived" the light, which caused the bud to flower. Apparently, some flowering-promoting substance is transmitted from leaf to bud. This hypothetical substance has been termed the flowering hormone, or *floral stimulus*.

Identification of the Hypothetical Flowering Hormone Remains Elusive

The early experiments on the floral stimulus were carried out independently in several laboratories in the 1930s. The plant physiologist M. Kh. Chailakhyan carried out some of the earliest experiments just a few years before the first cocklebur studies. Using the short-day *Chrysanthemum indicum*, he showed that if the upper portion of the plant is defoliated and the leaves on the lower part are exposed to a short-day induction period, the plant will flower. If, however, the upper, defoliated part is kept on short days and the lower, leafy part on long days, no flowering occurs. He interpreted these results as indicating that the leaves form a hormone that moves to the stem apex and initiates flowering. Chailakhyan named this hypothetical hormone *florigen*, the "flower maker."

Further experiments showed that the flowering response will not take place if the leaf is removed immediately after photoinduction. But if left on the plant for a few hours after the induction cycle is complete, the leaf can then be removed without affecting flowering. The flowering hormone can pass through a graft from a photoinduced plant to a noninduced plant. Unlike auxin, however, which can pass through agar or nonliving tissue, florigen can move from one plant tissue to another only if they are connected by living tissue. If a branch is girdled, that is, if a circular strip of bark is removed, florigen movement ceases. On the basis of these data, it was concluded that florigen moves via the phloem, the pathway followed by most organic substances in plants.

Gibberellins Can Induce Some Plants to Flower

Subsequently, Anton Lang showed that several long-day plants and biennials, such as celery and cabbage, could be made to flower by treatment with gibberellin, even when the plants were grown under the wrong (that is, noninducing) photoperiod. This finding led Chailakhyan to modify his florigen hypothesis and to speculate that florigen actually consists of two classes of hormones, gibberellin and the as-yet-unidentified *anthesin*. According to this modified hypothesis, long-day plants produce anthesin but not gibberellin during noninducing

photoperiods. Gibberellin treatments at this time can cause flowering. On the other hand, short-day plants produce gibberellin but fail to make anthesin when they are grown under noninducing conditions. The concept of florigen as a combination of gibberellin and anthesin has stimulated considerable research, but it cannot account for a critical observation: short-day plants grown under noninductive conditions (and thus, in theory, producing gibberellin) will not cause flowering of grafted long-day plants that are also under noninductive conditions (and, in theory, producing anthesin).

Both Inhibitors and Promoters May Be Involved in the Control of Flowering

In some plants—the Biloxi soybean (*Glycine max*) is an example—leaves must be removed from the grafted noninduced plant or it will not flower. This observation suggests that the leaves of noninduced plants may produce an inhibitor. In fact, some investigators have concluded that there is no substance that initiates flowering but rather a substance that inhibits flowering until it is removed. Strong evidence now suggests that, at least in certain plants, both inhibitors and promoters are involved in the control of flowering.

The most convincing evidence for the existence of both flower-inducing and flower-inhibiting substances in the same plant is provided by the experimental studies conducted by Lang, Chailakhyan, and I. A. Frolova. These researchers chose three kinds of tobacco plants for their studies: the day-neutral *Nicotiana tabacum* cultivar Trabezond, the short-day tobacco cultivar Maryland Mammoth, and the long-day *Nicotiana silvestris*. They found that flower formation in the day-neutral tobacco was accelerated (compared with grafted day-neutral controls) by a graft union with the long-day plant when the grafts were kept on long days, and by a graft union with the short-day tobacco when the grafts were kept on short days. When long-day plants grafted to day-neutral plants were exposed to short days, flowering in the day-neutral receptor was greatly inhibited (Figure 29–20). By contrast, when short-day plants grafted to day-neutral plants were exposed to long days, there was little or no delay in flowering.

These results indicate that the leaves of the long-day plants are capable of producing flower-inducing substances on long days and flower-inhibiting substances on short days, and that both substances can be transferred through a graft union, an indication that both substances are transported through the plant. In the case of the short-day tobacco, the leaves apparently produce little or no flower-inhibiting substance when exposed to long days. If such substances are produced by the short-day plant, they are much less effective in retarding flowering than those produced by the long-day plants.

29–20
Evidence for flower-inducing and flower-inhibiting substances using grafts of Nicotiana silvestris *(a long-day plant) onto the* Nicotiana tabacum *cultivar Trabezond (a day-neutral plant). Under long days, flower formation in the day-neutral tobacco was accelerated by the graft (left). Under short days, the day-neutral tobacco did not flower for the duration (90 to 94 days) of the experiment (right).*

All attempts to isolate either florigen or the flower inhibitor have thus far been unsuccessful.

Genetic Control of Flowering

As discussed earlier in this chapter, Garner and Allard's study of Maryland Mammoth tobacco led to the discovery of photoperiodism. The Maryland Mammoth variety was the result of a spontaneous mutation that converted a day-neutral variety of tobacco into a short-day plant. This mutant was rather conspicuous in the field where it first appeared because it never flowered during the long days of summer and therefore grew quite large. The mutant plant was moved from the field to a greenhouse where it flowered during the winter.

Garner and Allard tested many hypotheses—such as poor nutrition and low light intensity—as to why this mutant flowered in the winter before concluding that the only variable remaining was daylength. Their test of the daylength hypothesis, which involved covering plants in the field with a light-proof, ventilated shed each late afternoon to produce short days, led to the discovery of photoperiodism.

The analysis of mutations that affect flowering time continues to advance our understanding of the flowering process. One species in which flowering control is being genetically studied is *Arabidopsis thaliana* (pages 639 to 641). *Arabidopsis* is a long-day plant, and several mutations have been identified that affect the promotion of flowering by long-day photoperiods (Figure 29–21).

Some of these mutations result in plants that exhibit delayed flowering. These mutants behave as if they are unable to sense that they are growing under long-day (inductive) photoperiods (Figure 29–21). Thus, the likely role of the normal (wild-type) protein product of these genes is to promote flowering during long days. In the mutants that lack this protein product, promotion of flowering does not occur. Mutations in other genes result in plants that flower rapidly, regardless of photoperiod. The mutations in these rapidly flowering plants identify genes whose wild-type role is to inhibit flowering during short-day (noninductive) photoperiods. It is not surprising to find genes that promote flowering and other genes that delay it. Many biological regulatory systems are controlled by promoters and inhibitors. Indeed, it is tempting to speculate that some of the genes that affect the photoperiodic response may be involved in the production of promoters and inhibitors of flowering similar to those revealed by the grafting studies described above.

Other mutations in *Arabidopsis* have identified genes that are involved in the conversion, or transition, from vegetative to floral apices. Two such genes are *LEAFY* and *APETALA*. In plants with mutations in these genes, flowers are replaced by leafy-type shoots. The systems that plants use to regulate flowering time, such as photoperiod responsiveness, must somehow lead to the activation of genes such as *LEAFY*. Moreover, when certain plant species are genetically engineered to make the *LEAFY* protein all of the time, flowering can occur very early in development, presumably because the requirement for the normal signals to activate *LEAFY* are bypassed.

29–21

Arabidopsis thaliana *wild-type plants and flowering-time mutants. (a) Plants grown under long-day, inductive conditions. On the right is a wild-type plant and on the left is a plant containing a mutation that delays flowering. The mutant plant has greatly increased in size vegetatively because of the delay in flowering. (b) Plants grown under short-day, noninductive conditions. On the right is a wild type and on the left is a plant containing a mutation that causes rapid flowering independent of photoperiod.*

(a)

(b)

Dormancy

Plants do not grow at the same rate at all times. During unfavorable seasons, they limit their growth or cease to grow altogether. This ability enables plants to survive periods of water scarcity or low temperature.

Dormancy is a special condition of arrested growth. After periods of ordinary rest, growth resumes when the temperature becomes milder or when water or any other limiting factor becomes available again. A dormant bud or embryo, however, can be "activated" only by certain, often quite precise, environmental cues. This adaptation is of great survival importance to the plant. For example, the buds of plants expand, flowers form, and seeds germinate in the spring—but how do they recognize spring? If warm weather alone were enough, in many years all the plants would flower and all the seedlings would start to grow during warm autumn weather, only to be destroyed by the winter frost. The same could be said for any one of the warm spells that often punctuate the winter season. The dormant seed or bud does not respond to these apparently favorable conditions because of endogenous inhibitors, which must first be removed or neutralized before the period of dormancy can be terminated. In contrast to this "reluctance" to grow too rapidly, commercial seeds are artificially selected for their readiness to germinate promptly when they are exposed to favorable conditions, a trait that would be a great hazard for wild seeds.

Seeds Require Specific Environmental Cues to Break Dormancy

The seeds of almost all plants growing in areas with marked seasonal temperature variations require a period of cold prior to germination. This requirement is normally satisfied by winter temperatures. The seeds of many ornamental plants have a similar cold require-

ment. If moist seed is exposed to a low temperature for many days (average optimum temperature and time, 5°C for 100 days), the dormancy may be broken and the seed germinated. This horticultural procedure is termed **stratification.** Many seeds require drying before they germinate (although some may be nondormant before they dry out); this prevents their germination within the moist fruit of the parent plant. Some seeds, such as lettuce seeds, require exposure to light, but others are inhibited by light.

Some seeds will not germinate in nature until they have become abraded, as by soil action. Such abrasion wears away the seed coat, permitting water or oxygen to enter the seed and, in some cases, removing the source of inhibitors. Hard seed coats that interfere with water absorption and embryo enlargement are common among legumes.

The seeds of some desert species germinate only when sufficient rain has fallen to leach away inhibitory chemicals present in the seed coat. The amount of rainfall necessary to wash off these germination inhibitors is directly related to the supply of water that the desert plant needs to become established as a seedling. Mechanical abrasion or breaking of the seed coat (**scarification**) with a knife, file, or sandpaper may allow the "hard seed" condition or inhibitor to be removed, or metabolic activity requisite to germination to be initiated. Germination may also be induced by soaking the seeds in alcohol or some other fat solvent (to dissolve waxy substances that impede entry of water) or in concentrated acids. These procedures are widely used by horticulturalists. (See also pages 564 and 565.)

Some seeds may remain viable for a long time in the dormant condition, enabling them to exist for many years, decades, and even centuries under favorable conditions. The oldest, directly dated seed known to have germinated is that of a sacred lotus (*Nelumbo nucifera*) found in an ancient lake bed at Pulantien in Liaoning Province, China. Radiocarbon dating indicated that the seed was 1288 years old. Although this demonstration of endurance is impressive, even this record may have been broken by seeds of the arctic lupine (*Lupinus arcticus*). Some of these seeds, which were found in a frozen lemming burrow in the Yukon with animal remains estimated by carbon dating to be at least 10,000 years old, germinated within 48 hours. The age of the seeds themselves, however, was not determined.

Plant scientists have become increasingly interested in the factors involved in maintaining seed viability. Various enzyme systems may progressively fail in the stored seed, eventually leading to a complete loss of viability. Under what circumstances might viability be prolonged? Such questions are relevant to the worldwide interest in developing **seed banks,** with the goal of preserving the genetic characteristics of the wild and early cultivated varieties of various crop plants for use in future breeding programs. The need for such seed banks arises from the progressive replacement of older varieties with newer ones and the elimination of the replaced varieties, chiefly through the destruction of their habitats. In addition, many wild species of plants are in danger of becoming extinct, and the seeds of these plants should be preserved if possible.

The Dormant Condition in Buds Is Preceded by Acclimation

Dormancy in buds is essential to the survival of temperate herbaceous and woody perennials that are exposed to low temperatures during winter. Although dormant

buds do not elongate—that is, do not exhibit observable growth—they may undergo meristematic activity during various phases of dormancy.

Bud dormancy in many trees is initiated in midsummer, long before leaf fall in autumn. The dormant bud is an embryonic shoot consisting of an apical meristem, nodes and internodes (not yet extended), and small rudimentary leaves or leaf primordia with buds or bud primordia in their axils, all enclosed by *bud scales* (Figure 29–22). The bud scales are very important because they help prevent desiccation, restrict the movement of oxygen into the bud, and insulate the bud from heat loss. Growth inhibitors are known to accumulate in bud scales as well as in bud axes and leaves within buds. In many respects, therefore, the roles of the bud scales parallel those of the seed coat.

During and following the cessation of growth leading to the dormant condition, the plant tissues begin to undergo numerous physical and physiological changes to prepare the plant for winter, a process known as **acclimation.** Decreasing daylength is the primary factor involved in the induction of dormancy in buds (Figure

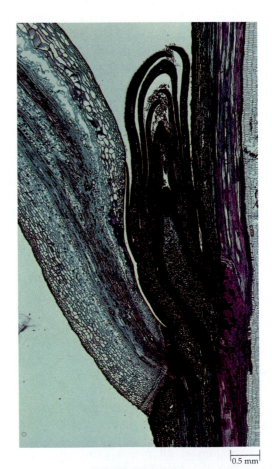

0.5 mm

29–22

Longitudinal section of a dormant axillary bud of a maple (Acer). The bud consists of an embryonic shoot enclosed by bud scales.

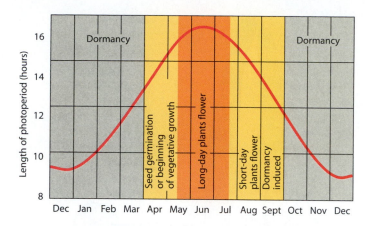

29–23

Relationship between daylength and the developmental cycle of plants in the North Temperate Zone.

(labels within figure: Length of photoperiod (hours); Dormancy; Dormancy; Seed germination or beginning of vegetative growth; Long-day plants flower; Short-day plants flower; Dormancy induced; months Dec Jan Feb Mar Apr May Jun Jul Aug Sept Oct Nov Dec)

29–23). Generally, more inhibitors are found in leaves and buds under short-day conditions than under long-day conditions. Acclimation to cold leads to **cold hardiness**—the ability of the plant to survive the extreme cold and drying effects of winter weather.

As with seeds, the buds of many plant species require cold to break dormancy. If branches of flowering trees and shrubs are cut and brought inside in autumn they do not flower; but if the same branches are left outside until late winter or early spring, they will bloom in the warmer indoor temperatures. Deciduous fruit trees, such as the apple, chestnut, and peach, cannot be grown in climates where the winters are not cold. Similarly, bulbs such as those of tulips, hyacinths, and narcissus can be "forced," that is, made to bloom indoors in the winter, but only if they have previously been outside or in a cold place. As discussed in Chapter 26, such bulbs are actually large buds in which the leaves are modified for storage.

Cold is not required to break dormancy in all cases. In the potato, for instance, in which the "eyes" are dormant buds, at least two months of dry storage are the chief requirement; temperature is not a factor. In many plants, particularly trees, the photoperiodic response breaks winter dormancy, with the dormant buds being the receptor organs.

Application of gibberellins sometimes breaks dormancy. For instance, gibberellin treatment of a peach bud may induce development after the bud has been kept for 164 hours below 8°C. Does this mean that under normal conditions, increase in gibberellin terminates the dormancy? Not necessarily. Dormancy may be a state of balance between growth inhibitors and growth stimulators. Addition of any growth stimulator (or removal of inhibitors, such as abscisic acid) may alter the balance so that growth begins.

Cold and the Flowering Response

Cold may affect the flowering response. For example, if winter rye (*Secale cereale*) is planted in the autumn, it germinates during the winter and flowers the following summer, 7 weeks after growth resumes. If it is planted in the spring, it does not flower for 14 weeks. In 1915, the plant physiologist Gustav Gassner discovered that he could influence the flowering of winter rye and other cereal plants by controlling the temperature of the germinating seeds. He found that if the seeds of the winter strain are kept at near-freezing (1°C) temperatures during germination, the winter rye, even when planted in late spring, will flower the same summer it is planted. This procedure, which came to be known as **vernalization** (from the Latin *vernus*, meaning "of the spring"), is now a common practice in agriculture.

Even after vernalization, the plant must be subjected to a suitable photoperiod, usually long days. The vernalized winter rye behaves like a typical long-day plant, flowering in response to the long days of summer. A similar example is seen in a biennial strain of henbane (*Hyoscyamus niger*). The vegetative rosette, which culminates the first year's growth, flowers only if it is exposed to cold. After cold exposure, it becomes a typical long-day plant, with the same photoperiodic response as that of the annual strain.

As the above examples indicate, in some plants the cold treatment affects the photoperiodic response. Spinach, ordinarily a long-day plant, does not usually flower until the days are 14 hours in length. If the spinach seeds are cold treated, however, they will flower when the days are only 8 hours long. Similarly, cold treatment of the clover *Trifolium subterraneum* can completely remove its dependency on daylength for flowering.

In the biennial henbane and most other biennial long-day plants that form rosettes, gibberellin treatment can substitute for the cold requirement. If gibberellin is applied, such plants will elongate rapidly and then flower. Application of gibberellin to short-day plants or to non-rosette long-day plants has little effect on (or inhibits) the flowering response. However, if gibberellin synthesis is inhibited when the plant is being exposed to the appropriate inductive cycle, the plant will not flower unless gibberellin is added to it.

Nastic Movements

Nastic movements are plant movements that occur in response to a stimulus but whose direction of movement is independent of the position of the origin of the stimulus. Probably the most widely occurring nastic movements are the sleep movements. Known technically as **nyctinastic movements** (meaning "night closure," from the Greek words *nyx*, "night," and *nastos*, "close-pressed"), they constitute the up and down movements of leaves in response to the daily rhythms of light and darkness. The leaves are oriented vertically in darkness and horizontally in the light. These movements are especially common in leguminous plants.

Most nyctinastic leaf movements result from changes in the size of parenchyma cells in the jointlike thickenings at the base of each leaf (or, if the leaf is compound, at the base of each leaflet also). This thickening (**pulvinus**) is a flexible cylinder with the vascular system concentrated in the center. Anatomically all pulvini consist of a core of vascular tissue surrounded by a bulky cortex of thin-walled parenchyma cells (Figure 29–24). Pulvinar movement is associated with changes in turgor and with the concomitant contractions and expansions of the ground parenchyma on opposite sides of the pulvinus. Turgor changes in the contracting and expanding cells are brought about by mechanisms similar to those associated with the closing and opening of guard cells (page 692) and are under the control of the biological clock and phytochrome.

Thigmonastic (Seismonastic) Movements Are Nastic Movements Resulting from Mechanical Stimulation

Thigmonastic movements are exemplified by the familiar sensitive plant *(Mimosa pudica)*, whose leaflets and sometimes entire leaves droop suddenly in response to touch, shaking, or electrical or thermal stimulation (Figure 29–25). As with sleep movements (also exhibited by *M. pudica)*, this response is a result of a sudden

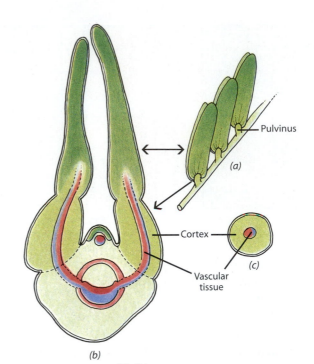

29–24
*Diagrammatic representation of pulvini in Mimosa pudica. **(a)** Portion of rachis showing three leaflets, each with a pulvinus at its base. **(b)** Transverse section through a rachis with two leaflets in the closed condition, showing longitudinal view of the pulvini. **(c)** Transverse section through a pulvinus showing the core of vascular tissue surrounded by a cortex, which consists largely of thin-walled parenchyma cells.*

29–25
*Thigmonasty in the sensitive plant (Mimosa pudica). **(a)** Normal position of leaves and leaflets. Responses to touch **(b)** result from changes in turgor pressure in certain cells of jointlike thickenings (pulvini) at the base of the leaflets. Only a single leaflet need be stimulated for the response in **(b)** to occur.*

(a)

(b)

(a) (b)

29–26

Touch response in the Venus flytrap (Dionaea muscipula). Here, an unwary fly, attracted by nectar secreted on the leaf surface, can be seen on a leaf (a) before and (b) after its closure. Each leaf half is equipped with three sensitive hairs. When an insect walks on one of the leaves, it brushes against the hairs, triggering the traplike closing of the leaf. The toothed edges mesh, the leaf halves gradually squeeze shut, and the insect is pressed against digestive glands on the inner surface of the trap.

The trapping mechanism is so specialized that it can distinguish between living prey and inanimate objects, such as pebbles and small sticks, that fall on the leaf by chance: the leaf will not close unless two of its hairs are touched in succession or one hair is touched twice.

change in turgor pressure in specific cells of the pulvinus at the base of each leaflet and leaf. The loss of water from these cells follows the efflux of potassium ions from the cells into the apoplast. The accumulation of ions in the apoplast seems to be initiated by a decrease of water potential triggered by an apoplastic accumulation of sucrose unloaded from the phloem. Only a single leaflet needs to be stimulated; the stimulus then moves to other parts of the leaf and throughout the plant. Two distinct mechanisms, one electrical and the other chemical, appear to be involved with the spread of the stimulus in the sensitive plant.

The mechanism responsible for the rapid closure (within approximately 0.5 second) of the leaves of the carnivorous Venus flytrap (*Dionaea muscipula*) has generally been explained as being due to either a sudden loss of turgor pressure in the upper epidermis, which becomes flexible and causes the trap lobes to bend inward, or a sudden acid-induced wall loosening of the motor cells and their subsequent expansion. Recent studies cast doubt on both of these explanations and indicate that trap movement may be driven by turgor pressure in the layer of mesophyll cells underlying the upper epidermis. When the trap is open, the mesophyll cells are compressed and the tissue is under tension. The turgor-driven extension of the mesophyll cells, which results in trap closure, does not require a wall-loosening mechanism. The biochemical processes associated with trap closure remain obscure. They are accompanied, however, by a marked decrease in ATP levels. The touch response exhibited in the trapping mechanism of the Venus flytrap is highly specialized (Figure 29–26).

Generalized Effects of Mechanical Stimuli on Plant Growth and Development: Thigmomorphogenesis

In addition to the specialized responses of some plants—such as the sensitive plant and Venus flytrap—to touch and other mechanical stimuli, plants also respond to mechanical stimuli by altering their growth patterns, a phenomenon known as **thigmomorphogenesis.** Although botanists have long known that plants grown in a greenhouse tend to be taller and more spindly than plants grown outside, it was not until the 1970s that systematic studies undertaken by M. J. Jaffe revealed that regular rubbing or bending of stems inhibits their elongation and stimulates their radial expansion, resulting in shorter, stockier plants. Plants in a natural environment are, of course, subjected to similar stimuli, in the form of wind, raindrops, and rubbing by passing animals and machines. The swaying of plants by wind is a powerful causal factor in thigmomorphogenesis.

As we saw in Chapter 28, such growth responses are caused by changes in gene expression. Studies on *Arabidopsis thaliana* show that some of the genes whose expression is induced by touch encode proteins related to the calcium-binding protein calmodulin, suggesting a role for Ca^{2+} in mediating the growth responses. (Ca^{2+} has been implicated in the regulation of a number of plant processes, including the regulation of mitosis, polarized cell growth, cytoplasmic streaming, and both tropic and nastic movements.) *Arabidopsis* plants stimu-

29–27

Arabidopsis thaliana *plants six weeks of age. The plant on the left was touched twice daily; the plant on the right is the untreated control. The inhibition of growth by touch or other noninjurious mechanical stimulation resulting in reduced length and increased width is referred to as thigmomorphogenesis.*

lated with touch were conspicuously shorter than the untreated plants (Figure 29–27).

Solar Tracking

The leaves and flowers of many plants have the ability to move diurnally, orienting themselves either perpendicular or parallel to the sun's direct rays. Commonly known as **solar tracking** (Figure 29–28), this phenomenon is technically called **heliotropism** (from the Greek *hēlios,* meaning "sun"). Unlike stem phototropism, the leaf movement of heliotropic plants is not the result of asymmetrical growth. In most cases the movements involve pulvini at the bases of leaves and/or leaflets, and apparently involve mechanisms similar to those associated with stomatal and nyctinastic movements. Some petioles appear to have pulvinal characteristics along most or all of their length. Some common plants exhibiting heliotropic leaf movements are cotton, soybeans, cowpeas, lupines, and sunflowers.

There are two types of heliotropism. In one (diaheliotropism), the movement of the leaves is such that the broad surfaces of the blades remain perpendicular to the sun's direct rays throughout the day. In such leaves, more photons are available for photosynthesis, and photosynthetic rates appear to be higher throughout the day than in nontracking leaves or leaves exhibiting the second type of heliotropism (Figure 29–29). In this second type, plants actively avoid direct sunlight during periods of drought by orienting their leaf blades parallel to

29–28

(a) Leaves of a lupine (Lupinus arizonicus), *orienting to track the course of the sun. This phenomenon is commonly known as solar tracking. (b) Solar tracking by a field of sunflowers* (Helianthus annuus).

(a)

(b)

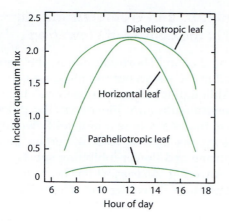

29–29

Two types of heliotropism. Comparison of the photosynthetically useful solar radiation between 400 and 700 nanometers incident on a diaheliotropic leaf (kept perpendicular to the sun's rays), a nontracking horizontal leaf, and a paraheliotropic leaf (kept parallel to the sun's rays) over the course of a day.

the sun's rays (paraheliotropism). This orientation minimizes absorption of solar radiation rather than maximizing it (Figure 29–29), decreasing leaf temperature and transpirational water loss and enhancing survival during drought periods.

Summary

Plants Possess a Variety of Adaptations That Enable Them to Detect and Respond to Alterations in Their Environment

Phototropism is the curving of a growing shoot toward light. The differential growth of the seedling is caused by lateral migration of the growth hormone auxin under the influence of light. The photoreceptor for this response is a pigment-containing protein that absorbs blue light and converts the signal into a biochemical response. Gravitropism is the response of a shoot or a root

Summary TABLE   The Major Types of Plant Movements or Growth Responses to External Stimuli

TYPE OF MOVEMENT	DESCRIPTIONS AND EXAMPLES	MECHANISMS AND/OR OTHER FEATURES
Tropism: directional growth in response to an external stimulus	Phototropism: growth of shoot, coleoptile, or petiole of leaf toward the light	May be caused by a light-induced lateral redistribution of auxin to the shaded side of the structure. Auxin then stimulates cell elongation on the shaded side. Redistribution of auxin is believed to be mediated by a flavoprotein photoreceptor.
	Gravitropism: downward growth of roots and upward growth of shoots	Apparently caused by the gravity-induced vertical redistribution of auxin to the lower side of the stem and root. Higher auxin concentration in stems stimulates cell elongation; in roots, it inhibits cell elongation. Three hypotheses to explain the perception of gravity: 1. Starch-statolith hypothesis 2. Hydrostatic pressure hypothesis 3. Plasmalemma central control hypothesis
	Hydrotropism: growth of roots toward regions of greater water potential	Sensing cells are in the rootcap. Calcium ions play a role in the induction of hydrotropism.
	Thigmotropism: response to contact with a solid object	Responsible for the coiling of tendrils around a support.
Nastic movement: movement in response to an external stimulus, with direction of movement unrelated to the direction of the stimulus	Nyctinasty: sleep movements of leaves	Results from turgor changes in cells of the pulvinus. Under the control of the biological clock and phytochrome.
	Thigmonasty: movement resulting from mechanical stimulation, such as closing of leaves of sensitive plant and Venus flytrap	Results from turgor changes in cells of the pulvinus. Initiated by apoplastic accumulation of sucrose, which decreases the water potential of the apoplast.
Thigmomorphogenesis: growth response unrelated to the direction of the external stimulus	Mechanical stimulus such as rubbing or bending of stems inhibits their elongation and stimulates their radial expansion	Involves changes in expression of genes encoding proteins related to the calcium-binding protein calmodulin. Thus calcium ions are implicated in the mechanism.
Heliotropism: solar tracking	Leaves and flowers orient themselves with respect to the sun's rays during the day	Results from turgor changes in cells of the pulvinus.

to gravity. The movement of auxin to the lower surface of a horizontally oriented shoot or root may play a role in the upward curvature of the shoot and the downward curvature of the root. Calcium apparently plays an important role in the gravitropic response of both shoots and roots. The site of perception of gravity in roots appears to be in the central column of cells of the rootcap. Hydrotropism is growth in response to a moisture gradient. The sensing cells for a moisture gradient apparently reside in the rootcap. Thigmotropism is a response to contact with a solid object, such as the winding of tendrils.

Living Organisms Exhibit Various Rhythms

Circadian rhythms are cycles of activity that recur at intervals of about 24 hours in an organism under constant environmental conditions. These rhythms are endogenous, caused not by environmental factors but by an internal timing mechanism. This timing mechanism, called the biological clock, controls the expression of specific genes.

Photoperiodism Is the Response of Organisms to Changing 24-Hour Cycles of Light and Darkness

Photoperiodic responses control the onset of flowering in many plants. Some plants, known as long-day plants, will flower only when the periods of light exceed a critical length. Other plants, short-day plants, flower only when the periods of light are less than some critical length. Day-neutral plants flower regardless of photoperiods. Experiments have shown that the dark period rather than the light period is the critical factor.

Phytochrome Is Involved in Photoperiodism

Phytochrome, a photoreceptor commonly present in the tissues of plants, is the molecule that detects transitions between light and darkness. This photoreceptor can exist in two forms, P_r and P_{fr}. P_r absorbs red light and is thereby converted to P_{fr}. The P_{fr} absorbs far-red light and is converted to P_r. P_{fr} is the active form of the photoreceptor; it promotes flowering in long-day plants, inhibits flowering in short-day plants, promotes germination in lettuce seeds, and promotes normal growth in seedlings. The phytochrome molecule contains two distinct parts: a light-absorbing portion (the chromophore) and a large protein portion.

Both Inhibitors and Promoters May Be Involved in the Control of Flowering

In both long-day and short-day plants, the photoperiod is perceived in the leaves but the response takes place in the bud. Although it has not yet been isolated or identified, a chemical substance moves via the phloem from the leaves to the bud, where it induces flowering. Strong evidence suggests that, at least in certain plants, both flower-inducing and flower-inhibiting substances are involved in flowering. *Arabidopsis* has been shown to have genes that promote flowering and others that delay it.

Dormancy Permits the Plant to Survive Water Shortages and Extremes of Hot or Cold

Dormancy is a special condition of arrested growth in which entire plants, or such structures as seeds or buds, do not renew growth without special environmental cues. The requirement for such cues, which include cold exposure, dryness, and a suitable photoperiod, prevents the tissues from breaking dormancy during superficially favorable conditions. Decreasing daylength is the primary factor involved in the induction of dormancy in buds. Acclimation to cold leads to cold hardiness, the ability of the plant to survive the extreme cold of winter weather. Vernalization refers to the promotion of flowering in winter strains by keeping seeds at low temperatures. Hormones, cold, and light interact to modify plant responses.

Many Plants Exhibit Nastic Movements

Plant movements that occur in response to a stimulus, but whose direction is independent of the direction of the stimulus, are called nastic movements. Among these are the widely occurring sleep, or nyctinastic, movements—the up and down movements of leaves in response to the daily rhythm of light and darkness. Nastic movements resulting from touch (thigmonastic movements) include the triggered closing of the carnivorous Venus flytrap.

Thigmomorphogenesis and Solar Tracking Are Familiar Phenomena

Plants may respond to mechanical stimuli by altering their growth patterns, a phenomenon known as thigmomorphogenesis. Such plants typically are shorter and stockier than nonstimulated plants. In addition, the leaves and flowers of many plants follow or track the sun during the course of the day, either maximizing or minimizing absorption of solar radiation.

Selected Key Terms

acclimation p. 718

biological clock p. 706

circadian rhythms p. 706

cold hardiness p. 719

day-neutral plant p. 709

dormancy p. 717

entrainment p. 707

etiolation p. 713

free-running period p. 707

gravitropism p. 704

heliotropism p. 722

hydrotropism p. 706

long-day plant p. 709

nastic movements p. 719

nyctinastic movements p. 719

photomorphogenic responses p. 714

photoperiodism p. 709

phototropism p. 703

phytochrome p. 712

pulvinus p. 720

scarification p. 718

seed banks p. 718

short-day plant p. 709

solar tracking p. 722

stratification p. 717

thigmomorphogenesis p. 721

thigmonastic movements p. 720

thigmotropism p. 706

tropism p. 703

vernalization p. 719

Questions

1. How do the starch-statolith hypothesis, hydrostatic pressure hypothesis, and plasmalemma central control hypothesis differ in their explanations of gravity sensing by plant cells?

2. Distinguish between the following: circadian rhythm/biological clock; phototropism/photoperiodism; thigmotropism/thigmonastic movement.

3. What evidence suggests that hormones—both stimulatory and inhibitory—are involved in flowering?

4. "Plants keep track of daylength by measuring the length of darkness." Explain.

5. Explain how it is possible for a long-day plant and a short-day plant growing at the same location to flower on the same day of the year.

6. Suppose you were given a chrysanthemum plant, in bloom, one autumn and you decided to keep it indoors as a house plant. What precautions would you need to take to ensure that it would bloom again?

7. The flowers of some plants open in the morning and close at dusk. Design and describe an experiment that could be used to determine whether these daily rhythms are controlled by the plant's biological clock or by the presence or absence of light.

8. Of what advantage is solar tracking to a plant?

9. How do nastic movements differ from tropisms? What mechanisms are responsible for sleep movements and for the closure of leaves of carnivorous plants?

30–1

Nitrogen-fixing nodules on the roots of a soybean (Glycine max) plant, a legume. These nodules are the result of a symbiotic relationship between the soil bacterium Bradyrhizobium japonicum *and the cortical cells of the root.*

Chapter

30

Plant Nutrition and Soils

OVERVIEW

To many people, soil is merely dirt. From a plant's perspective, however, soil is crucial for survival because it provides support, water, and a variety of elements essential for growth. This chapter begins by discussing these essential elements, focusing on their roles in the cell and the symptoms that result when they are not present in sufficient quantities. Next, we examine the origin and composition of soils and the processes that affect the utilization of nutrients and water. For example, the weathering of rock not only provides elements the plant requires, but the size and electric charge of the soil particles produced by weathering affect the availability of ions, water, and oxygen.

Other important soil components are bacteria and fungi. As we have seen in previous chapters, these microorganisms decompose dead plants and animals, thus recycling their chemical constituents into the environment. The remainder of the chapter deals with this recycling process, emphasizing the flow of nitrogen and phosphorus among plants, other organisms, and the physical environment. Particular emphasis is placed on nitrogen fixation—the process by which atmospheric nitrogen (N_2) is converted to chemical forms that can be used by plants and other organisms.

The chapter ends with a discussion of ways in which humans have disrupted nutrient cycles and the techniques being used to solve the problems caused by these disturbances.

CHECKPOINTS

By the time you finish reading this chapter, you should be able to answer the following questions:

1. Which elements are essential for plant growth, and what are their functions?

2. What are some of the common symptoms associated with nutrient deficiencies, and how does the mobility of a nutrient affect the symptom associated with its deficiency?

3. What are the sources of the inorganic nutrients utilized by plants?

4. Why are nutrient cycles so important for plants? What are the main components of the nitrogen and phosphorus cycles?

5. In what ways have humans disrupted nutrient cycles? How is plant nutrition research contributing to solutions for problems associated with agriculture and horticulture?

Plants must obtain from the environment the specific raw materials required in the complex biochemical reactions necessary for the maintenance of their cells and for growth. In addition to light, plants require water and certain chemical elements for metabolism and growth. Much of the evolutionary development of plants has involved structural and functional specialization necessary for the efficient uptake of these raw materials and their distribution to living cells throughout the plant.

In contrast to those of animals, the nutritional demands of plants are relatively simple. Under favorable environmental conditions, most green plants can use light energy to transform CO_2 and H_2O into organic compounds for their energy source. They can also synthesize all of their required amino acids and vitamins, using inorganic nutrients drawn from the environment.

Plant nutrition involves the uptake from the environment of all the raw materials required for essential biochemical processes, the distribution of these materials within the plant, and their utilization in metabolism and growth.

More than 60 chemical elements have been identified in plants, including gold, silver, lead, mercury, arsenic, and uranium. Obviously, not all of the elements present in plants are essential. Their presence is, to a certain extent, a reflection of the composition of the soil in which the plants are growing. Most of the chemical elements found in plants are absorbed as inorganic ions from the soil solution.

The roles played by root hairs and by the fungal hyphae of mycorrhizae in the absorption of inorganic ions are discussed in Chapters 31 and 15, respectively. In Chapter 31, we consider the mechanism of ion absorption by roots and the pathway followed by the ions from the soil solution to the tracheary elements in the vascular cylinder of the root.

Essential Elements

As early as 1800, chemists and plant biologists had analyzed plants and demonstrated that certain chemical elements were absorbed from the environment. Opinions differed, however, on whether the absorbed elements were impurities or constituents required for essential functions. By the mid-1880s it had been established that at least ten of the chemical elements present in plants were necessary for normal growth. In the absence of any one of these elements, plants displayed characteristic abnormalities of growth, or deficiency symptoms, and often such plants did not reproduce normally. These ten elements—carbon, hydrogen, oxygen, potassium, calcium, magnesium, nitrogen, phosphorus, sulfur, and iron—were designated as **essential elements** for plant growth. They are also referred to as essential minerals and essential inorganic nutrients.

In the early 1900s, manganese was also found to be an essential element. During the next 50 years, with the aid of improved techniques for removing impurities from nutrient cultures, five additional elements—zinc, copper, chlorine, boron, and molybdenum—were determined to be essential, bringing the total to 16 elements. A seventeenth element, nickel, has also been added to the list by many plant nutritionists.

There are two primary criteria by which an element is judged to be essential to a plant: (1) if it is needed for the plant to complete its life cycle (that is, to produce viable seed) and/or (2) if it is part of any molecule or constituent of the plant that is itself essential to the plant, such as the magnesium in the chlorophyll molecule or the nitrogen in proteins. A third criterion also used by many plant nutritionists is whether deficiency symptoms appear in the absence of the element, even though the plant may be capable of producing viable seed.

The Essential Elements Can Be Divided into Micronutrients and Macronutrients

Chemical analyses for essential elements are a useful means of determining the relative amounts of various elements required for the normal growth of different plant species. One way to determine which elements are essential and in what concentrations is to chemically analyze healthy plants. The freshly harvested plants or plant parts are heated in an oven to drive off the water, then the remaining material, or **dry matter,** is analyzed.

Table 30–1 lists the 17 elements currently believed to be essential to all vascular plants and the approximate internal concentrations considered adequate for normal plant growth and development. Note that the concentrations vary over a wide range. The first eight elements are called **micronutrients** or trace elements because they are required in very small, or trace, amounts (concentrations equal to or less than 100 mg/kg of dry matter). The last nine elements are called **macronutrients** because they are required in large amounts (concentrations of 1000 mg/kg of dry matter or greater). Many ions are accumulated inside cells at concentrations much higher than those in the soil solution, and often at concentrations much higher than those minimally required. As discussed in Chapter 4, plant cells must use energy to accumulate solutes, that is, to move solutes against an electrochemical gradient.

Certain plant species and taxonomic groups are characterized by unusually high or low amounts of specific elements (Figure 30–2). As a result, plants growing in the same nutrient medium may differ markedly in nutrient content. Eudicots generally require greater amounts of calcium and boron than do monocots. Analyses of this sort are particularly valuable in agriculture as a guide to the nutritional well-being of plants and the need for ap-

TABLE 30–1   Essential Elements for Most Vascular Plants and Internal Concentrations Considered Adequate

Element	Chemical Symbol	Form Available to Plants	Atomic Weight	Adequate Concentration in Dry Tissue		Number of Atoms Relative to Molybdenum
				mg/kg	%	
Micronutrients						
Molybdenum	Mo	MoO_4^{2-}	95.95	0.1	0.00001	1
Nickel	Ni	Ni^{2+}	58.71	?	?	?
Copper	Cu	Cu^+, Cu^{2+}	63.54	6	0.0006	100
Zinc	Zn	Zn^{2+}	65.38	20	0.0020	300
Manganese	Mn	Mn^{2+}	54.94	50	0.0050	1000
Boron	B	H_3BO_3	10.82	20	0.002	2000
Iron	Fe	Fe^{3+}, Fe^{2+}	55.85	100	0.010	2000
Chlorine	Cl	Cl^-	35.46	100	0.010	3000
Macronutrients						
Sulfur	S	SO_4^{2-}	32.07	1000	0.1	30,000
Phosphorus	P	$H_2PO_4^-, HPO_4^{2-}$	30.98	2000	0.2	60,000
Magnesium	Mg	Mg^{2+}	24.32	2000	0.2	80,000
Calcium	Ca	Ca^{2+}	40.08	5000	0.5	125,000
Potassium	K	K^+	39.10	10,000	1.0	250,000
Nitrogen	N	NO_3^-, NH_4^+	14.01	15,000	1.5	1,000,000
Oxygen	O	O_2, H_2O, CO_2	16.00	450,000	45	30,000,000
Carbon	C	CO_2	12.01	450,000	45	35,000,000
Hydrogen	H	H_2O	1.01	60,000	6	60,000,000

After P. R. Stout, *Proceedings of the Ninth Annual California Fertilizer Conference*, pages 21–23 (1961).

30–2

Some plants contain large amounts of specific elements. **(a)** *Plants of the mustard family, such as wintercress (Barbarea vulgaris), use sulfur in the synthesis of the mustard oils that give the plants their sharp taste. These mustard plants were growing in Ithaca, New York.* **(b)** *Horsetails incorporate silicon into their cell walls, making the plants indigestible to most herbivores but useful, at least in colonial America, for scouring pots and pans. These bushy vegetative shoots, as well as several stalk-like fertile shoots, of the horsetail Equisetum telemateia were photographed in California.*

(a)

(b)

plications of fertilizer. Potential nutritional deficiencies in livestock that consume specific plants can also be predicted by inorganic analyses.

Nutritional studies have established that some elements are essential for only limited groups of plants or for plants grown under specific environmental conditions. Legumes, such as alfalfa *(Medicago sativa)*, benefit from the addition of cobalt to the culture medium. It is not the alfalfa, however, that requires cobalt but the symbiotic nitrogen-fixing bacteria growing in association with the roots of the alfalfa. Sodium is present in many plants in relatively high concentrations. It has been recognized for many years that sodium can partially replace the requirement for potassium in some species. Studies have shown that in other species sodium is an essential element in its own right. For example, sodium appears to be required by C_4 plants and by certain halophytes (see the essay on page 744).

Functions of Essential Elements

Although the essential elements are usually classified as micronutrients and macronutrients, they have sometimes been classified functionally into two groups: those that have a role in the structure of an important compound and those that have an enzyme-activating role. There is, however, no sharp distinction between these two functions. For example, nitrogen and sulfur are major components of both proteins and coenzymes, and magnesium, in addition to being part of the chlorophyll molecule, is an activator of many enzymes. All nutrients have well-known specific functions (Table 30–2) that are impaired when the supply of the nutrient is inadequate. Because the nutrients fill such basic needs and are involved in such fundamental processes, the deficiencies affect a wide variety of structures and functions in the plant body.

Nutrient Deficiency Symptoms Depend on the Function(s) and Mobility of the Essential Element

Because they are easily observed, most well-described nutrient deficiency symptoms are associated with the shoot. They include such symptoms as stunted growth of stems and leaves, localized death of tissues (necrosis), and yellowing of the leaves due to loss or reduced development of chlorophyll (chlorosis) (Figure 30–3). The deficiency symptoms for any essential element depend not only on the element's role in the plant but also upon its mobility within the plant, that is, the relative ease with which it is transported in the phloem from older to younger plant parts, particularly leaves.

Take, for example, magnesium, which as we know is an essential part of the chlorophyll molecule. Without magnesium, chlorophyll cannot be formed, resulting in chlorosis. Chlorosis of the older leaves of magnesium-deficient plants typically becomes more severe than that of the younger leaves. The reason for this is the ability of younger leaves to withdraw nutrients from older ones. The withdrawal of the magnesium from the older leaves to the younger leaves also depends upon magnesium's mobility in the phloem. Elements that readily move through the phloem are said to be phloem-mobile. In addition to magnesium, phosphorus, potassium, and nitrogen are phloem-mobile. Other elements, such as boron, iron, and calcium, are relatively immobile, whereas copper, manganese, molybdenum, sulfur, and zinc are usually intermediate in mobility. Deficiency symptoms of phloem-mobile elements appear earliest and are most pronounced in older leaves, whereas those of phloem-immobile elements appear first in younger leaves.

30–3

Chlorosis, the loss or reduced development of chlorophyll resulting from mineral deficiency. **(a)** *Deficiency of magnesium, a phloem-mobile element, in maize (Zea mays). The older leaves are more affected than the younger leaves, which are able to translocate magnesium from the older leaves.* **(b)** *Deficiency of iron, a phloem-immobile element, results in symptoms of chlorosis in younger leaves, as seen here is sorghum (Sorghum bicolor).*

(a)

(b)

TABLE 30–2   Essential Elements: Functions and Deficiency Symptoms

Element	Functions	Deficiency Symptoms
Micronutrients		
Molybdenum	Required for nitrogen fixation and nitrate reduction	Interveinal (between the veins) chlorosis appearing first on older leaves and then progressing to youngest leaves; chlorosis followed by gradual necrosis of interveinal areas and then of remaining tissues.
Nickel	Essential part of enzyme functioning in nitrogen metabolism	Necrotic spots in leaf tips
Copper	Activator or component of some enzymes involved in oxidation and reduction	Young leaves dark green, twisted, misshapen, and often with necrotic spots
Zinc	Activator or component of many enzymes	Reduction in leaf size and length of internodes; leaf margins often distorted; interveinal chlorosis; mostly older leaves affected
Manganese	Activator of some enzymes; required for integrity of chloroplast membrane and for oxygen release in photosynthesis	Initially, interveinal chlorosis on younger or older leaves, depending on species, followed by or associated with interveinal necrotic spots; disorganization of thylakoid membranes of chloroplasts
Boron	Influences Ca^{2+} utilization, nucleic acid synthesis, and membrane integrity	Earliest symptom is failure of root tips to elongate; young leaves light green at bases; leaves become twisted and shoot dies back at terminal bud
Iron	Required for chlorophyll synthesis; component of cytochromes and nitrogenase	Interveinal chlorosis of young leaves; stems short and slender
Chlorine	Involved in osmosis and ionic balance; probably essential in photosynthetic reactions that produce oxygen	Wilted leaves with chlorotic and necrotic spots; leaves often become bronze color; roots stunted in length and thickened near tips
Macronutrients		
Sulfur	Component of some amino acids and proteins and of coenzyme A	Young leaves with light green veins and interveinal areas
Phosphorus	Component of energy-carrying phosphate compounds (ATP and ADP), nucleic acids, several coenzymes, phospholipids	Plants dark green, often accumulating anthocyanins and becoming red or purple; in later stages of growth, stems stunted; oldest leaves become dark brown and die
Magnesium	Component of the chlorophyll molecule; activator of many enzymes	Mottled or chlorotic leaves; may redden; sometimes with necrotic spots; leaf tips and margins turned upward; mostly older leaves affected; stems slender
Calcium	Component of cell walls; enzyme cofactor; involved in cellular membrane permeability; component of calmodulin, a regulator of membrane and enzyme activities	Shoot and root tips die; young leaves at first hooked, then die back at tips and margins, developing cut-out appearance at these sites
Potassium	Involved in osmosis and ionic balance and in opening and closing of stomata; activator of many enzymes	Mottled or chlorotic leaves with small spots of necrotic tissue at tips and margins; weak, narrow stems; mostly older leaves affected
Nitrogen	Component of amino acids, proteins, nucleotides, nucleic acids, chlorophylls, and coenzymes	General chlorosis, especially in older leaves; in severe cases, leaves become completely yellow and then become tan as they die; some plants exhibit purple coloration due to accumulation of anthocyanins

Some of the more common nutrient deficiency symptoms are given in Table 30–2. The table does not include the macronutrients carbon, oxygen, and hydrogen, which are acquired primarily from CO_2 and H_2O during photosynthesis. These elements are the major components of the plant's organic compounds.

The Soil

Soil is the primary nutrient medium for plants. Soils must provide plants not only with physical support but also with adequate inorganic nutrients at all times, as well as with adequate water and a suitable gaseous environment for the root systems. Understanding the origins of soils and their chemical and physical properties in relation to plant growth requirements is critical in planning for the nutrition of field crops.

The Weathering of Rocks Produces the Inorganic Nutrients Utilized by Plants

The inorganic nutrients utilized by plants are derived from the atmosphere and from the weathering of rocks in the crust of the Earth. The Earth is composed of 92 naturally occurring elements, which are often found in the form of minerals. **Minerals** are naturally occurring inorganic compounds that are usually composed of two or more elements in definite proportions by weight (see Appendix A). Quartz (SiO_2), calcite ($CaCO_3$), and kaolinite ($Al_4Si_4O_{10}(OH)_8$) are examples of minerals.

Most rocks consist of several different minerals and are divided into three groups—igneous, sedimentary, and metamorphic—based on origin and formation. *Igneous* rocks, such as granite, are derived directly from molten material and, for the most part, were originally formed when the Earth cooled and solidified. As a result of weathering, igneous rocks and other kinds of rocks may be broken down into soluble and insoluble components. After being transported by water, wind, or glaciers, these components form new deposits—usually in water—that in time become cemented and solidified into *sedimentary* rocks, such as shale, sandstone, and limestone. Limestone is rich in carbonates from the shells of lake or ocean organisms. Although sedimentary rocks form only about 5 percent of the Earth's crust, they are of great importance because they occur extensively at or near the surface. Under the extreme heat and pressure deep within the Earth, sedimentary and igneous rocks may be transformed into a third type of rock—*metamorphic*. In this way, quartzite is formed from sandstone, slate from shale, and marble from limestone.

Weathering processes, involving the physical disintegration and chemical decomposition of minerals and rocks at or near the Earth's surface, produce the inorganic materials from which soils are formed. Weathering

30–4

The fibrous root systems of grasses bind and anchor prairie soil in place.

involves freezing and thawing or heating and cooling, which cause substances in the rocks to expand and contract, splitting rocks apart. Water and wind often carry the rock fragments great distances, exerting a scouring action that breaks and wears the fragmented rock into even smaller particles. Water enters between the particles, and soluble materials dissolve in the water. Water combines with carbon dioxide and air impurities, such as sulfur dioxide and oxides of nitrogen, to form dilute acids that help to dissolve materials that are less soluble in pure water. Soil formation may occur at the site of weathering, or the parent materials may be transported elsewhere by gravity, wind, water, or glaciers.

Soils also contain organic materials. If light and temperature conditions permit, bacteria, fungi, algae, lichens, and bryophytes and small vascular plants gain a foothold on or among the weathered rocks and minerals. Growing roots also split the rocks, and the disintegrating bodies of plants and those of the animals associated with them add to the accumulating organic material. Finally, larger plants move in, anchoring the soil in place with their root systems (Figure 30–4), and a new community begins.

Soils Consist of Layers Called Horizons

Examining a vertical section of soil, one can see variations in the color, the amount of living and dead organic matter, the porosity, the structure, and the extent of

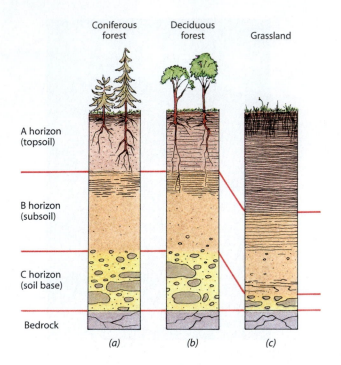

Coniferous forest   Deciduous forest   Grassland

A horizon (topsoil)

B horizon (subsoil)

C horizon (soil base)

Bedrock

(a)   (b)   (c)

30–5

Diagrams of soil layers of three major soil types. (a) The litter of the northern coniferous forest is acidic and slow to decay, and the soil has little accumulation of humus, is very acidic, and is leached of minerals. (b) In the cool, temperate deciduous forest, decay is somewhat more rapid, leaching less extensive, and the soil more fertile. Such soils have been widely used for agriculture, but they need to be prepared by adding lime (to reduce acidity) and fertilizer. (c) In the grasslands, almost all of the plant material above the ground dies each year, as do many of the roots, and thus large amounts of organic matter are constantly returned to the soil. In addition, the finely divided roots penetrate the soil extensively. The result is highly fertile soil, often black in color, with a topsoil sometimes more than a meter in depth. Natural soils, particularly forest soils, often contain a layer of decomposing litter on top of the A horizon. This layer is called the O horizon.

weathering. These variations generally result in a succession of rather distinct layers that soil scientists refer to as **horizons.** At least three horizons—designated A, B, and C—are recognized (Figure 30–5).

The A horizon (sometimes called the "topsoil") is the upper region, that of the greatest physical, chemical, and biological activity. The A horizon contains the greatest portion of the soil's organic material, both living and dead. It is the horizon where **humus**—the dark-colored mixture of colloidal organic decay products—accumu-

lates. It is alive with populations of roots, insects and other small arthropods, earthworms, protists, nematodes, and decomposer organisms (Figure 30–6).

The B horizon (sometimes called the "subsoil") is a region of deposition. Iron oxide, clay particles, and small amounts of organic matter are among the materials leached from the A horizon to the B horizon by water percolating, or moving down, through the soil. The B horizon contains much less organic material and is less weathered than the A horizon above it. Human activity has often mixed together the A and B horizons through tillage operations, forming an Ap ("p" for "plow") horizon, which blends into the B horizon.

The C horizon, or soil base, is composed of the broken-down and weathered rocks and minerals from which the true soil in the upper horizons is formed.

Soils Are Composed of Solid Matter and Pore Space

The pore space is the space around the soil particles. Different proportions of air and water occupy the pore space, depending on prevailing moisture conditions. The soil water is present primarily as a film on the surfaces of soil particles. The fragments of rocks and minerals in the soil vary in size from sand grains, which can be seen easily with the naked eye, to clay particles too small to be seen even under the low power of a light microscope. The following classification is one scheme for categorizing soil particles according to size:

Particles	Diameter (in micrometers)
Coarse sand	200–2000
Fine sand	20–200
Silt	2–20
Clay	Less than 2

Soils contain a mixture of particles of different sizes and are divided into textural classes according to the proportions of different particles present in the mixture. For example, soils that contain 35 percent or less clay and 45 percent or more sand are sandy; those containing 40 percent or less clay and 40 percent or more silt are silty. **Loam soils** contain sand, silt, and clay in proportions that result in ideal agricultural soils. The coarser soil particles of loam aid drainage, while the finer soil particles have high nutrient-retention capabilities.

The solid matter of soils consists of both inorganic and organic materials, with the proportions varying greatly in different soils. The organic component includes the remains of organisms in various stages of decomposition, a highly decomposed fraction (humus), and a wide range of living plants and animals. Structures as large as tree roots may be included, but the living phase is dominated by fungi, bacteria, and other microorganisms.

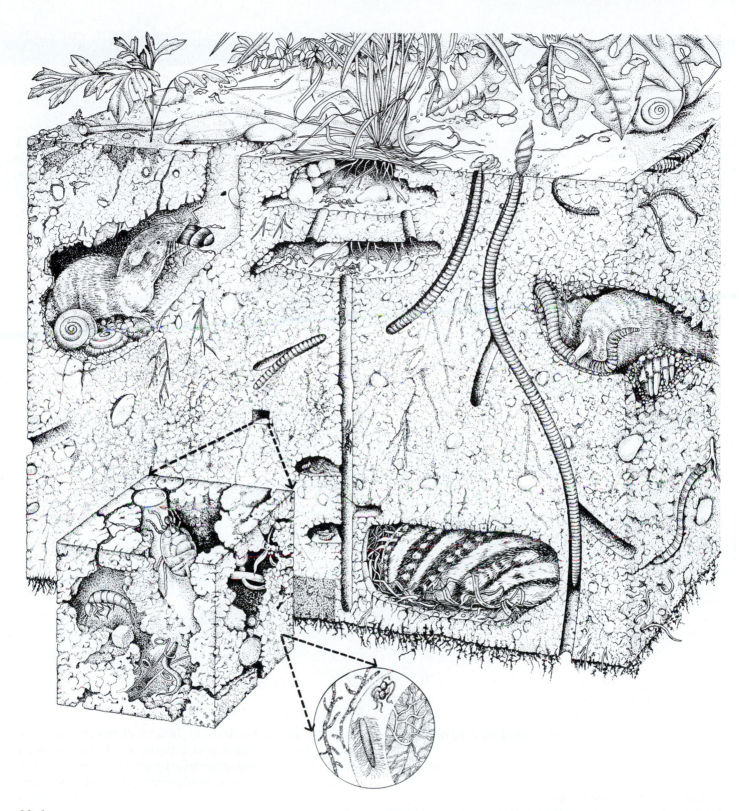

30–6

Plants share the soil with a vast number of living organisms, ranging from microbes to small mammals such as moles, shrews, and ground squirrels. Multitudes of burrowing creatures—most notably ants and earthworms—aerate the soil and improve its ability to absorb water. Called by Aristotle "the intestines of the Earth," earthworms refine the soil by processing it through their gut. The refined soil is then deposited on the soil surface in the form of castings. In a single year, the combined activities of earthworms may produce as much as 500 metric tons of castings per hectare. The castings are very fertile, containing five times the nitrogen content of the surrounding soil, seven times the phosphorus, 11 times the potassium, three times the magnesium, and twice the calcium. Bacteria and fungi are the principal decomposers of the organic matter in soils.

The Water Cycle

The Earth's supply of water is stable and is used over and over again. Most of the water (98 percent) is present in oceans, lakes, and streams. Of the remaining 2 percent, some is frozen in polar ice and glaciers, some is found in the soil, some is in the atmosphere as water vapor, and some is in the bodies of living organisms.

Sunshine evaporates water from the oceans, lakes, and streams, from the moist soil surfaces, and from the bodies of living organisms, drawing the water back up into the atmosphere, from which it falls again as rain. Evaporation exceeds precipitation over the oceans, resulting in a net movement of water vapor, carried by wind, from the ocean to the land. Over 90 percent of the water lost on the land is by plant transpiration (evaporation of water from the soil plus transpiration from plants is called evapotranspiration). This constant movement of water from the Earth into the atmosphere and back again is known as the water cycle. The water cycle is driven by solar energy.

Some of the water that falls on the land percolates down through the soil until it reaches a zone of saturation. In the zone of saturation, all holes and cracks in the rock are filled with water. Below the zone of saturation is solid rock through which the water cannot penetrate. The upper surface of this zone of saturation is known as the water table.

The Pore Space of Soils Is Occupied by Air and Water

Approximately 50 percent of the total soil volume is represented by pore space, which is occupied by varying proportions of air and water, depending on moisture conditions. When no more than half the pore space is occupied by water, adequate oxygen is available for root growth and other biological activity.

Following a heavy rain or irrigation, soils retain a certain amount of water and remain moist even after gravity has removed loosely bound water. If the fragments that make up a soil are large, the pores and spaces between them will be large; water will drain through the soil rapidly, and relatively little will be available for plant growth in the A and B horizons. Because of their finer pores and the attractive forces that exist between water molecules and colloidal-size clay particles (particles small enough to remain in suspension but larger than true solute particles), clay soils are able to hold a much greater amount of water against the action of gravity. Clay soils may retain three to six times more water than a comparable volume of sand; hence, soils with more clay can hold more water that is available to plants. The percentage of water that a soil can hold against the action of gravity is called its **field capacity.**

If a plant is allowed to grow indefinitely in a sample

of soil and no water is added, the plant will eventually be unable to absorb water rapidly enough to meet its needs, and it will droop and wilt. When wilting is severe, plants fail to recover even when placed in a humid chamber. The percentage of water remaining in a soil when such irreversible wilting occurs is called the **permanent wilting percentage** of that soil.

Figure 30–7 shows the relationship between the soil water content and the potential at which water is held by sandy, loam, and clay soils. The forces that retain water in the soil can be expressed in the same terms (in this case, water potential) as the forces for water uptake in cells and tissues (see pages 76–78). The soil water potential decreases gradually with a decrease in the soil moisture below field capacity. Soil scientists have agreed to consider soil with a water potential of −1.5 megapascals to be at its permanent wilting percentage, or point.

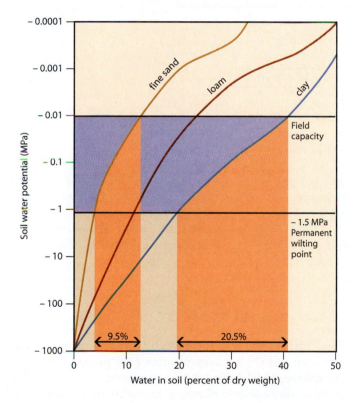

30–7

Relationship between soil water content and soil water potential in sandy, loam, and clay soils. (The curves are plotted on a logarithmic scale.) Note that whereas water available to plants in a fine sand soil is only about 9.5 percent of the dry weight of the sand, that available to plants in a clay soil is considerably greater, about 20.5 percent of the dry weight of clay. Soil with a water potential of −1.5 megapascals (MPa) is considered to be at its permanent wilting percentage.

Cation Exchange Is Important Because Exchangeable Ions Are Available to Plants and Are Not Leached Out by Water

The inorganic nutrients taken in through the roots of plants are present in the soil solution as ions. Most metals form positively charged ions, that is, cations, such as Ca^{2+}, K^+, Na^+, and Mg^{2+}. Both clay particles and humus may have an excess of negative charges on their colloidal surfaces where cations can be bound and thus held against the leaching action of percolating soil water.

These weakly bound cations can be replaced by other cations and thus released into the soil solution, where they become available for plant growth. This process is called **cation exchange.** For example, when CO_2 is released by respiring roots, it dissolves in the soil solution to become carbonic acid (H_2CO_3). The carbonic acid then ionizes to produce bicarbonate (HCO_3^-) and hydrogen (H^+) ions. The H^+ produced in this way may *exchange* for the nutrient cations on the clay and humus.

The principal negatively charged ions, or anions, found in soil are NO_3^-, SO_4^{2-}, HCO_3^-, and OH^-. Anions are leached out of the soil more rapidly than cations because anions do not attach to clay particles. An exception is phosphate, which is retained against leaching because it forms insoluble precipitates. Phosphate is specifically adsorbed by, or held on the surface of, compounds containing iron, aluminum, and calcium.

The acidity or alkalinity of soil is related to the availability of inorganic nutrients for plant growth. Soils vary widely in pH, and many plants have a narrow range of tolerance on this scale. In alkaline soils, some cations are precipitated, and such elements as iron, manganese, copper, and zinc may thereby become unavailable to plants. Mycorrhizae (see Chapter 15) are especially important in the absorption and transfer of phosphorus in most plants, but these structures have also been implicated in the increased absorption of manganese, copper, and zinc by plants.

Nutrient Cycles

As we now know, virtually all vascular plants require 17 essential elements, or inorganic nutrients, for normal growth and development. Inasmuch as the Earth is essentially a closed system, these elements are available only in limited supply. Life on Earth depends, therefore, on the recycling of these elements. Both macronutrients and micronutrients are recycled through plant and animal bodies, returned to the soil, broken down, and taken up into plants again. Each element has a different cycle, involving many different organisms and different enzyme systems. Some cycles, such as those of carbon, oxygen, sulfur, and nitrogen, which exist in gaseous forms (as elements or compounds) in the atmosphere, are es-

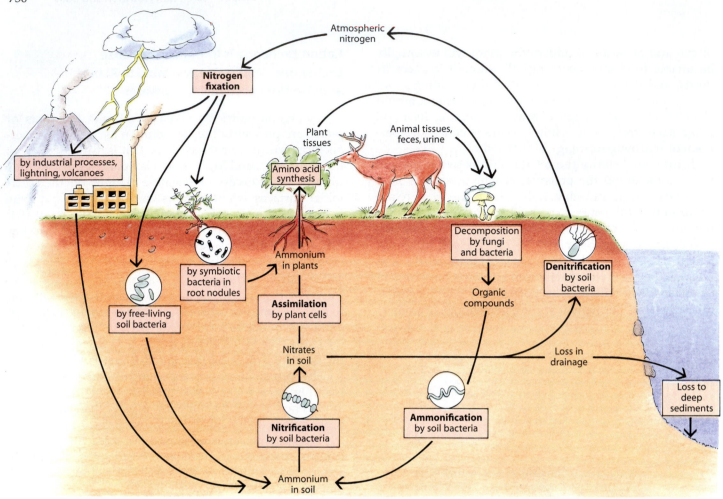

30–8

The nitrogen cycle in a terrestrial ecosystem. The primary reservoir of nitrogen is the atmosphere, where nitrogen makes up 78 percent of dry air. Only a few microorganisms, some symbiotic and others free-living, are capable of fixing nitrogen gas into inorganic compounds that can be used by plants in the synthesis of amino acids and other organic nitrogenous compounds.

The nitrogen cycle in aquatic ecosystems is similar to the terrestrial cycle. Although the specific organisms differ, the processes are essentially the same, with specific groups of bacteria carrying out the various chemical reactions on which the cycle depends.

sentially global in nature. Other cycles, such as those of phosphorus, calcium, potassium, and the micronutrients, which are not found in a gaseous state, are generally more localized. Because nutrient cycles involve both living organisms and the physical environment, they are also called **biogeochemical cycles.**

Nutrient cycles are said to be "leaky" because not all of the nutrients returned to the soil become available for plant use. Some are lost from the system. Soil erosion, for instance, removes the nutrient-rich (especially phosphorus and nitrogen) topsoil, which is carried away in streams and rivers and eventually ends up in the ocean. The removal of nutrients (especially nitrogen and potassium) with harvesting and the losses of nitrogen- and sulfur-containing gases to the atmosphere when plants are burned also contribute to nutrient loss. In addition, leaching accounts for losses of all types of soluble nutrients, most notably potassium, nitrate, and sulfate.

Nitrogen and the Nitrogen Cycle

The chief reservoir of nitrogen is the atmosphere; in fact, nitrogen gas (N_2) makes up about 78 percent of the atmosphere. Most living things, however, cannot use elemental atmospheric nitrogen to make amino acids and other nitrogen-containing compounds and therefore are dependent on the more reactive nitrogenous compounds, such as ammonium and nitrate, present in soil. Unfortunately, these compounds are not nearly as abundant as nitrogen gas. So, despite the abundance of nitrogen in the atmosphere, a shortage of nitrogen in the soil is often the major limiting factor in plant growth.

The process by which this limited amount of nitrogen is circulated and recirculated throughout the world of living organisms is known as the nitrogen cycle (Figure 30–8). The three principal stages of this cycle are (1) ammonification, (2) nitrification, and (3) assimilation.

Carnivorous Plants

A few species of plants are able to use animal proteins directly as a nitrogen source. These carnivorous plants have special adaptations that are used to lure and trap insects and other very small animals. The plants digest the trapped organisms, absorbing the nitrogenous compounds they contain as well as other organic compounds and minerals, such as potassium and phosphate. Most of the carnivores of the plant world are found in bogs, a habitat that is usually quite acidic and thus not favorable for the growth of nitrifying bacteria. Carnivorous plants entrap their prey in a variety of ways. The common bladderwort (Utricularia vulgaris) *(a)* is a freefloating aquatic plant with traps that are tiny, flattened, pear-shaped bladders. Each

bladder has a mouth guarded by a hanging door. The tripping mechanism consists of four stiff bristles near the lower free edge of the door. When a small animal brushes against these bristles, the hairs distort the lower edge of the door, causing it to spring open. Water then rushes into the bladder, carrying the animal inside, and the door snaps shut behind it. A range of enzymes secreted by the internal wall of the bladder and by the residential bacterial population digest the animal. Released minerals and organic compounds are taken up through the cell walls of the trap, while the undigested exoskeleton remains within the bladder.

The sundew (Drosera rotundifolia) *(b)* is a tiny plant, often only a few centimeters

across, with club-shaped hairs on the upper surface of its leaves. The tips of these glandular hairs secrete a clear, sticky liquid, or mucilage, that attracts insects such as the damselfly shown here. When an insect is caught in the mucilage, the hairs bend inward until the leaf curves around the insect. The hairs secrete at least six enzymes, which, together with bacteria-produced enzymes, especially chitinase, digest the insect. The nutrients released from the prey mix with the mucilage which is then resorbed by the glands that secreted the digestive enzymes. Similar secretion mechanisms are found in other carnivorous plants, including the Venus flytrap (Dionaea muscipula) *and the butterwort* (Pinguicula grandiflora).

(a)   (b)

(a) The common bladderwort (Utricularia vulgaris). (b) A sundew (Drosera rotundifolia).

Ammonium Is Released As Organic Material Decays

Much of the soil nitrogen is derived from dead organic materials in the form of complex organic compounds such as proteins, amino acids, nucleic acids, and nucleotides. These nitrogenous compounds are usually rapidly decomposed into simpler compounds by soil-dwelling saprophytic bacteria and various fungi. These organisms incorporate the nitrogen into amino acids and proteins and release excess nitrogen in the form of ammonium ions (NH_4^+) by a process known as **ammonification** or nitrogen mineralization. In alkaline media, the

nitrogen may be converted to ammonia gas (NH_3). This conversion usually occurs, however, only during the decomposition of large amounts of nitrogen-rich material, as in a manure pile or a compost heap that has contact with the atmosphere. Within soil, the ammonia produced by ammonification dissolves in the soil water, where it combines with protons to form the ammonium ion. In some ecosystems, the ammonium ion is not rapidly oxidized but remains in the soil. Plants growing in these soils are able to take up NH_4^+ and use it in the synthesis of plant protein.

In Some Soils Nitrifying Bacteria Convert Ammonium into Nitrite and Then into Nitrate

Several species of bacteria common in soils are able to oxidize ammonia or ammonium ions. The oxidation of ammonium, or **nitrification,** is an energy-yielding process, and the energy released in the process is used by these bacteria to reduce carbon dioxide in much the same way that photosynthetic autotrophs use light energy in the reduction of carbon dioxide. Such organisms are known as chemosynthetic autotrophs (as distinct from photosynthetic autotrophs). The chemosynthetic nitrifying bacterium *Nitrosomonas* is primarily responsible for oxidation of ammonium to nitrite ions (NO_2^-):

$$2NH_4^+ + 3O_2 \longrightarrow 2NO_2^- + 4H^+ + 2H_2O$$

Nitrite is toxic to plants, but it rarely accumulates in the soil. *Nitrobacter,* another genus of bacteria, oxidizes the nitrite to form nitrate ions (NO_3^-), again with a release of energy:

$$2NO_2^- + O_2 \longrightarrow 2NO_3^-$$

Because of nitrification, nitrate is the form in which almost all nitrogen is absorbed by most crop plants grown on dry land where nitrification is strongly favored by the oxidizing tillage practices of agriculture. Most nitrogen fertilizer used commercially contains either ammonium ions (NH_4^+) or urea, which breaks down into NH_4^+ in soils. The NH_4^+ is converted to NO_3^- by nitrification.

Besides Cycling Nitrogen within Itself, the Soil–Plant System Also Loses Nitrogen

The major loss of nitrogen from the soil–plant system occurs by **denitrification,** an anaerobic process in which nitrate is reduced to volatile forms of nitrogen, such as nitrogen gas (N_2) and nitrous oxide (N_2O), which then return to the atmosphere. This process is carried out by numerous microorganisms. The low oxygen conditions necessary for denitrification have long been perceived as characteristic of waterlogged soils and such habitats as swamps and marshes. Scientists now recognize that these conditions commonly exist within soil aggregates even in the absence of excessive water. Consequently, denitrification is virtually a universal process in soils. A fresh supply of readily decomposable organic matter provides the energy source required by denitrifying bacteria and, if other conditions are appropriate, promotes denitrification. Lack of an energy source allows nitrate concentrations to build up to high levels in groundwater.

Nitrogen is also lost from an ecosystem by the removal (harvesting) of plants from the soil, by soil erosion, by burning of plants, and by leaching. Nitrates and nitrites, both of which are anions, are particularly susceptible to being washed from the root zone by water percolating through the soil.

The Replenishment of Nitrogen Occurs Primarily by Nitrogen Fixation

If the nitrogen that is removed from the soil were not steadily replaced, virtually all life on this planet would slowly disappear. Nitrogen is replenished in the soil primarily by nitrogen fixation. Much lesser amounts are added via precipitation and the weathering of rocks.

Nitrogen fixation is the process by which atmospheric N_2 is reduced to NH_4^+ and made available for transfer to carbon-containing compounds to produce amino acids and other nitrogen-containing organic compounds. Nitrogen fixation, which can be carried out only by certain bacteria, is a process on which all living organisms are dependent, just as most organisms are ultimately dependent on photosynthesis as the source of their energy.

The enzyme that catalyzes the fixation of nitrogen is called **nitrogenase.** It is similar in all organisms from which it has been isolated. Nitrogenase contains molybdenum, iron, and sulfide prosthetic groups, and therefore these elements are essential for nitrogen fixation. Nitrogenase also uses large amounts of ATP as an energy source, making nitrogen fixation an expensive metabolic process.

Nitrogen-fixing bacteria may be classified according to their mode of nutrition: those that are free-living (nonsymbiotic) and those that live in symbiotic association with certain vascular plants.

The Most Effective Nitrogen-Fixing Bacteria Form Symbiotic Relationships with Plants

Of the two classes of nitrogen-fixing organisms, the symbiotic bacteria are by far the most important in terms of total amounts of nitrogen fixed. The most common of the nitrogen-fixing bacteria are *Rhizobium* and *Bradyrhizobium,* both of which invade the roots of legumes, such as alfalfa (*Medicago sativa*), clovers (*Trifolium*), peas (*Pisum sativum*), soybean (*Glycine max*), and beans (*Phaseolus*). In the symbiotic association between bacteria and legumes, the bacteria provide the plant with a form of nitrogen that it can use to make protein. The plant, in turn, provides the bacteria with both an energy source for their nitrogen-fixing activity and the carbon-containing molecules necessary for the production of nitrogenous compounds.

The beneficial effects on the soil of growing leguminous plants have been recognized for centuries. Theophrastus, who lived in the third century B.C., wrote that the Greeks used crops of broad beans (*Vicia faba*) to enrich the soils. In modern agriculture, it is common practice to rotate a nonleguminous crop, such as maize (*Zea mays*), with a leguminous crop, such as alfalfa, or maize with soybeans and then wheat. The leguminous plants are then either harvested for hay, leaving behind the nitrogen-rich roots, or, better still, simply plowed under. A crop of alfalfa that is plowed back into the soil may add as much as 300 to 350 kilograms of nitrogen per hectare of soil. As a conservative estimate, 150 to 200 million metric tons of fixed nitrogen are added to the Earth's surface each year by such biological systems.

Nodules Are Produced by the Host Plant Root upon Infection by Bacteria *Rhizobium* and *Bradyrhizobium* bacteria, commonly called **rhizobia,** enter the root hairs of leguminous plants when the plants are still seedlings. Establishment of the nitrogen-fixing symbiosis between *Bradyrhizobium japonicum* and soybean *(Glycine max)* begins with the attachment of rhizobia to emerging root hairs (Figure 30–9a). The root hairs typically develop into tightly curled structures, entrapping the rhizobia (Figure 30–9b). Invasion of the root hairs and underlying cortical cells by the rhizobia occurs via **infection threads,** tubular structures formed by progressive inward growth of the root-hair cell walls from the sites of

penetration (Figure 30–9b, c). A single root hair may be penetrated by several rhizobia and, hence, may contain several infection threads (Figure 30–9b). The bacterial symbiont also induces cell division in localized regions of the cortex, which it then enters through growth and branching of the infection thread. Rhizobia from the infection thread are released into envelopes derived from the host-cell plasma membrane. The rhizobia develop into **bacteroids,** the name given the now-enlarged nitrogen-fixing rhizobia (Figure 30–9d). Proliferation of the membrane-enclosed bacteroids and cortical cells of the root result in the formation of tumorlike growths known as **nodules** (Figure 30–1). The manner of infection and nodule formation in the roots of other legumes is apparently similar to that for soybean.

The root nodules of legumes consist of a relatively narrow cortex, which surrounds a large central zone containing both bacteroid-infected and uninfected cells (Figure 30–10). Vascular bundles, which radiate from the point of attachment of the nodule to the root, occur in

30–9

Events in the infection of soybean by Bradyrhizobium japonicum. *(a) Scanning electron micrograph showing rhizobia (arrows) attached to recently emerged root hair. (b) Differential-interference contrast photomicrograph showing a short, curled root hair containing multiple infection threads (arrows). (c) Electron micrograph of an infection thread containing rhizobia. The plasma membrane and cell wall of the root cell are continuous with the infection thread. Each rhizobium is surrounded by a halo of capsular polysaccharides. (d) Electron micrograph of groups of bacteroids, each surrounded by a membrane derived from the infected root nodule cell. Note the uninfected cell adjacent (above) to the infected cell.*

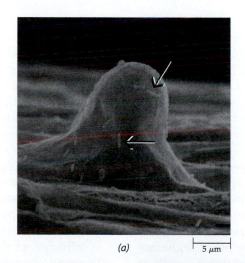
(a)   5 μm

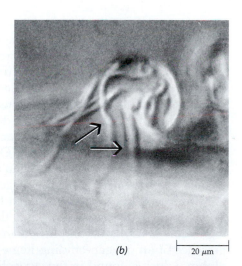
(b)   20 μm

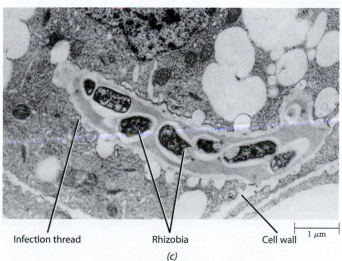
Infection thread   Rhizobia   Cell wall   1 μm
(c)

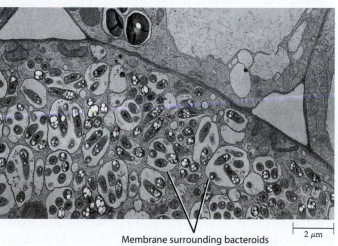
Membrane surrounding bacteroids   2 μm
(d)

30–10
Section through a mature soybean root nodule. Vacuolate uninfected cells can be seen among the darkly stained infected cells in the central zone of the nodule. The nodule cortex, containing vascular bundles (arrows) and a layer of darkly stained sclerenchyma cells, surrounds the central zone.

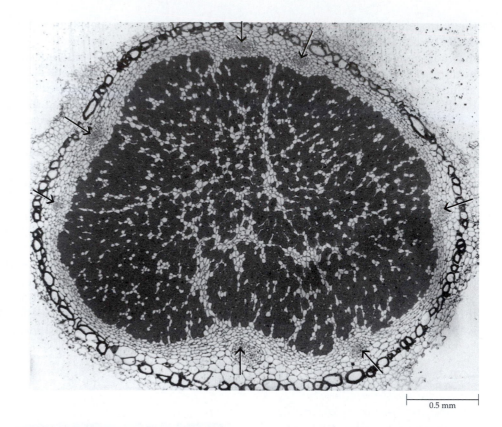

0.5 mm

the inner cortex. In soybean, both infected and uninfected cell types play a role in the production of ureides (derivatives of urea), the end product of nitrogen fixation that is exported from the nodule to the plant. Other legumes produce other nitrogenous compounds, the amino acid asparagine being one of the most common.

The O_2 concentration in the bacteria-infected cells must be carefully regulated since O_2 is a potent irreversible inhibitor of nitrogenase and yet is required for aerobic respiration to meet the large ATP demands of this enzyme. Oxygen is also needed for other metabolic activities in both the bacteria and the plant cells. The regulation of O_2 is accomplished in large part by the presence of an oxygen-binding heme protein, **leghemoglobin,** which is found in the cytosol of infected cells at rather high concentrations. This protein, which imparts a pink color to the central region of the nodule, is produced partly by the bacteroid (the heme portion) and partly by the plant (the globin portion). The leghemoglobin is thought to buffer the O_2 concentration within the nodule, allowing respiration without inhibiting the nitrogenase. Leghemoglobin also acts as an O_2 carrier, facilitating the diffusion of O_2 to the bacteroids.

The Association between the Bacterium and the Plant Is Highly Specific The symbiosis between a species of *Rhizobium* or *Bradyrhizobium* and a legume is quite specific; for example, bacteria that invade and induce nodule formation in clover (*Trifolium*) roots will not induce nodules on the roots of soybeans (*Glycine*). The interac-

tion between the symbiotic partners involves an intricate exchange of molecular signals that regulate the expression of genes essential for infection and nodule formation. Two groups of bacterial genes are necessary for the formation of nitrogen-fixing nodules. One group, the *nod* genes, are involved in the host-specific response and in formation of the nodule, and the other group, the *nif* genes, are involved in nitrogen fixation. Generally, the genes for symbiotic interactions are located on a large plasmid in *Rhizobium* species and on the chromosome in *Bradyrhizobium* strains. The complete nucleotide sequence and gene complement of the plasmid of a *Rhizobium* species (NGR234) has recently been determined.

Initiation of the symbiotic relationship begins when flavonoids secreted by the legume bind to and activate the bacterial *nod* gene known as *nodD*. The *nodD* gene product then induces the expression of other bacterial *nod* genes, whose products are required for such processes as root-hair curling, cell wall degradation, and formation of infection threads. In addition, other bacterial *nod* gene products activate the plant genes called *Nod* genes. The *Nod* genes encode plant cell proteins called *nodulins*, which are essential for cortical cell division and for growth and function of the nodule.

The mechanism of the initial attachment of the bacteria to the root-hair surface is poorly understood. It has been suggested that sugar-binding proteins called **lectins,** which are secreted by legume roots, interact with the bacteria and facilitate their binding to the root-hair cell walls.

(a)

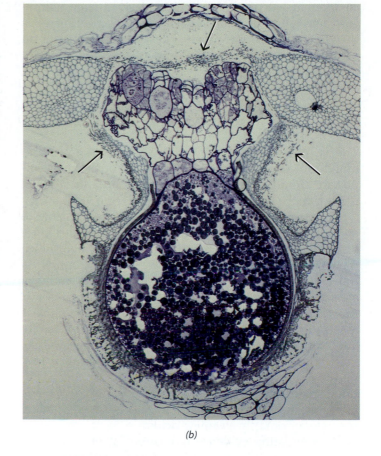

(b)

30–11

Azolla-Anabaena *symbiosis.* **(a)** Azolla filiculonides, *a water fern that grows in symbiotic association with the cyanobacterium* Anabaena. **(b)** *Filaments of* Anabaena *can be seen (arrows) associated with a female gametophyte (megagametophyte) that has developed from a germinated megaspore of Azolla.*

There Are Other Nitrogen-Fixing Symbioses That Do Not Involve Legumes The legumes are by far the largest group of plants that enters into a nitrogen-fixing partnership with symbiotic bacteria. There are, however, a few nitrogen-fixing symbioses that involve plants other than legumes. Alder trees *(Alnus),* for example, form nodules that are induced by and contain nitrogen-fixing actinomycetes, rather than *Rhizobium* or *Bradyrhizobium.* Sweet gale *(Myrica gale),* sweet fern *(Comptonia),* and mountain lilacs *(Ceanothus)* also form symbiotic associations with actinomycetes (see Chapter 14).

Another symbiotic relationship is of considerable practical interest in certain parts of the world. *Azolla* is a small floating water fern, and *Anabaena* is a nitrogen-fixing cyanobacterium that lives in the cavities of the *Azolla* fronds (Figure 30–11). The *Azolla-Anabaena* symbiosis is unique among nitrogen-fixing symbioses in that the relationship is sustained throughout the life cycle of the host. *Azolla* infected with *Anabaena* may contribute as much as 50 kilograms of nitrogen per hectare. In the Far East, for example, heavy growths of *Azolla-Anabaena* are permitted to develop on rice paddies (see also Figure 14–14). The rice plants eventually shade out the *Azolla,* and as the fern dies, nitrogen is released for use by the rice plants.

Nonsymbiotic Nitrogen-Fixing Bacteria Are Found Free-Living in the Soil

Nonsymbiotic bacteria of the genera *Azotobacter,* *Azotococcus, Beijerinckia,* and *Clostridium* are able to fix nitrogen. The first three genera are aerobic, while *Clostridium* is anaerobic. All four genera are common saprophytic soil bacteria, which rely on the oxidation of organic matter in the soil to provide energy for the fixation process. It is estimated that they probably add about 7 kilograms of nitrogen to a hectare of soil per year. Another important group includes many photosynthetic bacteria, such as cyanobacteria.

Industrial Nitrogen Fixation Has High Energy Costs

Since the process was first developed in 1914, the industrial, or commercial, production of fixed nitrogen has increased steadily to the current level of approximately 50 million metric tons per year. Most of this nitrogen is used in agricultural fertilizers. Industrial fixation, unfortunately, is accomplished at high energy cost in terms of fossil fuels. In most such processes, N_2 is reacted with H_2 at high temperature and pressure, in the presence of metal catalysts, to form ammonia. The energy-expensive component is hydrogen, which is derived from natural gas, petroleum, or coal. In such developed countries as the United States, this process, despite its high cost, can account for as much as one-third of the annual newly fixed nitrogen.

Assimilation of Nitrogen

The assimilation of inorganic nitrogen (nitrates and ammonia) into organic compounds is one of the most important processes in the biosphere, on an almost equal footing with photosynthesis and respiration.

The principal source of nitrogen that is available to crop plants grown under field conditions is nitrate. Once nitrate enters a cell, it is reduced to ammonia, which then is rapidly incorporated into organic compounds via the glutamine synthetase–glutamate synthase pathway shown in Figure 30–12. In most herbaceous plants this process takes place mainly in leaf chloroplasts, in close association with photosynthesis. (Many plant physiologists consider it to be an extension of photosynthesis.) When the amount of nitrate supplied to the roots is low, nitrate reduction in many plants takes place primarily in the plastids of the roots. The organic nitrogen made available by root-metabolized nitrate is transported in the xylem primarily as amino acids.

Accumulating evidence indicates that organic nitrogen is a major and direct source of nitrogen for plants in nitrogen-limited ecosystems. For example, Arctic plants readily absorb amino acids and grow more rapidly on organic than on inorganic sources of nitrogen. In addition, ectomycorrhizae and the mycorrhizae of heather (see Chapter 15), which are common in infertile soils, directly break down proteins in the organic matter found in the soil. They then absorb and transfer amino acids to the host plant directly, without nitrogen being mineralized to nitrate and ammonia.

The Phosphorus Cycle

The phosphorus cycle (Figure 30–13) seems simpler than the nitrogen cycle because there are fewer steps, and because particular steps are not dependent on specific groups of microorganisms. The phosphorus cycle also differs from the nitrogen cycle in that the Earth's crust, rather than the atmosphere, is the primary reservoir for replenishing phosphorus. There are no significant phosphorus-containing gases. Because humus and soil particles bind phosphate (PO_4^{3-})—the only important inorganic form of phosphorus—phosphorus recycling tends to be quite localized. As previously discussed, the weathering of rocks and minerals over long periods of time is the source of most of the phosphorus in the soil solution.

Compared with the amount of nitrogen, the amount of phosphorus required by plants is relatively small (Table 30–1). Nevertheless, of all the elements for which the Earth's crust is the primary reservoir, phosphorus is the most likely to limit plant growth. In Australia, for example, where the soils are extremely weathered and deficient in phosphorus, the distribution and limits of native plant communities are often determined by the available soil phosphate.

Phosphorus circulates from plants to animals and is returned to the soil in organic forms in residues and wastes. These organic forms of phosphorus are converted to inorganic phosphate through the activities of microorganisms, and thus the phosphorus again becomes available to plants (Figure 30–13).

Some phosphorus is lost from the terrestrial ecosystem by leaching and erosion, but in most natural ecosystems the weathering of rock can keep pace with this loss. The lost phosphorus eventually reaches the oceans, where it is deposited in sediments as precipitates and in the remains of organisms. In the past, the use of guano (deposits of seabird feces) as agricultural fertilizer re-

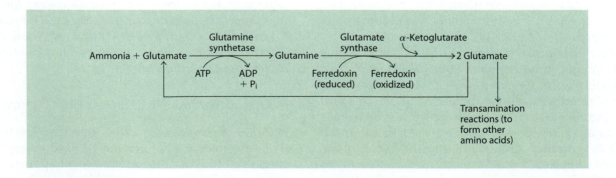

30–12

The glutamine synthetase–glutamate synthase pathway (abbreviated) in leaves. The assimilation of nitrogen into organic compounds requires ATP and reduced ferredoxin, both of which are readily available in photosynthesizing cells. Of the two glutamate molecules produced, one recycles to bind with ammonia, perpetuating the pathway, whereas the other glutamate is transaminated to form other amino acids. In roots, either NADH or NADPH replaces the ferredoxin in the pathway.

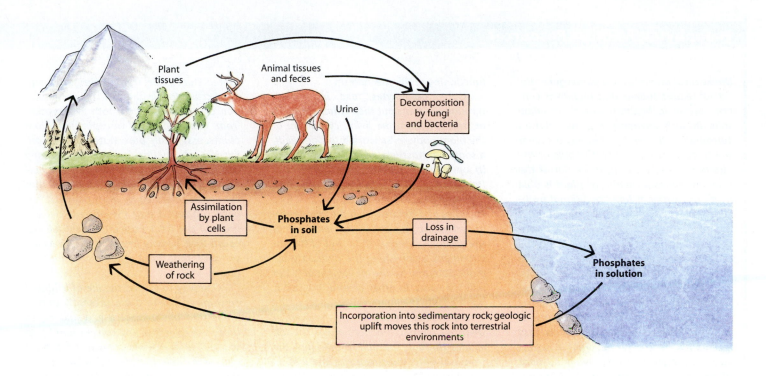

30–13

The phosphorus cycle in a terrestrial ecosystem. Phosphorus is essential to all living organisms as a component of energy-carrier molecules, such as ATP, and of the nucleotides of DNA and RNA. Like other minerals, it is released from dead tissues by the activities of decomposers, taken up from the soil by

plants, and cycled through the ecosystem.

The phosphorus cycle in an aquatic ecosystem involves different organisms but is, in most respects, similar to the terrestrial cycle shown here. However, in aquatic ecosystems a considerable amount of phosphorus is incorporated into the shells and skeletons of

aquatic organisms. This phosphorus, along with phosphates that precipitate out of the water, is subsequently incorporated into sedimentary rock. Such rock, returned to the land surface as a result of geologic uplift, is the primary terrestrial reservoir of phosphorus.

turned some of the ocean phosphorus to terrestrial ecosystems. However, most of the phosphorus in deep-sea sediments will become available only as a result of major geologic uplifts. Phosphate fertilizers are used for agricultural soils that have inadequate levels of phosphorus available to plants. Deposits of phosphate rock are being mined on a large scale for use as agricultural fertilizer.

Human Impact on Nutrient Cycles and Effects of Pollution

Normal functioning of the phosphorus, nitrogen, and other nutrient cycles requires the orderly transfer of elements between steps in a cycle in order to prevent buildup or depletion of nutrients at any one stage. Over millions of years, organisms have been provided with needed quantities of essential inorganic nutrients as a result of the normal functioning of such cycles. However, the need to adequately feed an exponentially growing human population has had drastic effects on some cycles, and in time this may lead to harmful accumulations

and depletions of nutrients at specific stages of one or more nutrient cycles. For example, crop removal and increased soil erosion have accelerated phosphorus loss from soils. The addition of phosphorus to aquatic habitats in the form of sewage and drainage from fertilized agricultural fields has resulted in massive growths of algae and flowering plants, thus seriously reducing the recreational value of the affected areas. Many states have banned phosphorus-containing detergents for this reason.

Normal functioning of the nitrogen cycle, which has a relatively high degree of efficiency, involves a balance between fixation processes, which remove nitrogen from the atmospheric reservoir, and denitrification processes, which return nitrogen to the atmosphere. Unfortunately, with the increasing introduction of fixed nitrogen (nitrates) into the environment through the extensive use of commercial fertilizers, the efficiency of the cycle declines. In addition, marshes and wetlands, the primary sites of denitrification, are being destroyed at an alarming rate through their conversion to building sites, agricultural land, and dump sites.

Both terrestrial and, particularly, aquatic ecosystems

Halophytes: A Future Resource?

Unlike most animals, most plants generally do not require sodium and, moreover, cannot survive in brackish waters or saline soils. In such environments, the solution surrounding the roots often has a higher solute concentration than the cells of the plants, causing water to move out of the roots by osmosis. Even if the plant is able to absorb water, it faces additional problems from the high level of sodium ions. If the plant takes up water and excludes sodium ions, the solution surrounding the roots becomes even saltier, increasing the likelihood of water loss through the roots. The salt may eventually become so concentrated that it forms a crust around the roots, effectively blocking the supply of water to them. Another problem is that sodium ions may enter the plant in preference to potassium ions, depriving the plant of an essential nutrient as well as inhibiting some enzyme systems.

Some plants—known as halophytes—can grow in saline environments such as deserts, salt marshes, and coastal areas. All of these plants have evolved mechanisms for dealing with high sodium concentrations, and for some of them sodium appears to be a required nutrient. The adaptations of halophytes vary. In many halophytes, a sodium-potassium pump seems to play a major role in maintaining a low sodium concentration within the cells while simultaneously ensuring that a sufficient supply of potassium ions enters the plant. In some species, the pump operates primarily in the root cells, pumping sodium back to the environment and potassium into the root. The presence of calcium ions (Ca^{2+}) in the soil solution is thought to be essential for the effective functioning of this mechanism.

Other halophytes take in sodium through the roots but then either secrete it or isolate it from the living cytoplasm of the plant body. In Salicornia (pickleweed), a sodium-potassium pump (or a variant of it) operates in the vacuolar membranes (tonoplasts) of leaf cells. Sodium ions enter the cells but are immediately pumped into vacuoles and isolated from the cytoplasm. In such plants, the solute concentration of the vacuoles is higher than that of the environment, establishing the necessary osmotic potential for the movement of water into the roots. In other genera, the salt is pumped into the intercellular spaces of the leaves and then secreted from the plant. In Distichlis palmeri (Palmer's grass), the salt exudes through specialized cells (not the stomata) onto the surface of the leaf. In Atriplex (saltbush), it is concentrated by special salt glands and pumped into bladders. The bladders expand as the salt accumulates, finally bursting. Rain or the passing tide washes the salt away.

Halophytes are of current interest not only because of the light they may shed on the osmoregulatory mechanisms of plants but also because of their potential as crop plants. In a world with an ever-increasing need for food, vast areas are unsuitable for agricultural purposes because of the salinity of the soil. For example, there are over 30,000 kilometers of desert coastline and about 400 million hectares of desert with potential water supplies that are too salty for crop plants. Moreover, each year about 200,000 hectares of irrigated crop land becomes so salty that further agriculture is impossible. When arid land is heavily irrigated, as in large areas of the western United States, salts from the irrigation waters accumulate in the soil. This accumulation occurs because, in both evaporation from the soil and transpiration from plants, essentially pure water is given off, leaving all solutes behind. Over many years, the salt concentration of the soil increases, eventually reaching levels that cannot be tolerated by most plants. It has been suggested that the ancient civilizations of the Near East ultimately fell because their heavily irrigated land became so salty that food could no longer be grown on it.

One way to extend the life of irrigated crop lands and to bring presently barren areas into agricultural use would be to breed salt tolerance into conventional crop plants. Thus far, however, such efforts have met with little success. Scientists at the University of Arizona's Environmental Research Laboratory are taking what appears to be a more promising approach. They have gathered halophytes from all over the world and are engaged in an extensive research program to determine optimal growing conditions, potential yields, and the nutritional value and palatability of the seeds and vegetative parts of the various species. Their results suggest that a number of halophyte species have great potential for use in livestock feed and, quite possibly, for human consumption as well.

(a) Atriplex *(saltbush) is one of several halophytes being evaluated as potential crop plants.*
(b) The surface of a leaf of Atriplex. *Salt is pumped from the leaf tissues through narrow stalk cells into the large, expandable bladder cells.*

are subject to widespread environmental damage because of acid rain. Acid rain results from the interaction of sulfur dioxide and oxides of nitrogen—derived principally from the combustion of fossil fuels—with atmospheric moisture to form sulfuric and nitric acids. These acids impart a high degree of acidity to the rainfall. Acid rain can have adverse effects on plants, the weathering of rocks and minerals, the solubilities of potentially toxic metals in the environment, and even human health. (See the essay on pages 616 and 617.)

Soils and Agriculture

In natural situations, the elements present in the soil recirculate and so become available again for plant growth. As discussed previously, negatively charged clay particles and organic matter are able to bind such positively charged ions as Ca^{2+}, Na^+, K^+, and Mg^{2+}. These ions are then displaced from the particles by other cations (cation exchange) and absorbed by roots from the soil solution. In general, the cations that are required

(a)

(b)   100 µm

soils that are transferred from natural ecosystems to agricultural systems do not have enough plant-available nutrients to support crops for commercial harvest, although the quantities are sufficient for the native plant communities.

Programs for supplementing nutrient supplies for agricultural and horticultural crops should be based on soil testing, which is used to diagnose nutrient deficiency and to predict a likely response to the addition of fertilizer at a recommended amount. Nitrogen, phosphorus, and potassium are the three elements that are commonly included in commercial fertilizers. Fertilizers are usually labeled with a formula that indicates the percentage of each of these elements. A 10-5-5 fertilizer, for example, is one that contains 10 percent nitrogen (N), 5 percent phosphorus (reported as phosphorus pentoxide, P_2O_5), and 5 percent potassium (reported as potassium oxide, K_2O). This method of reporting phosphorus and potassium contents of fertilizer is a historical relic dating from the days when analytical chemists reported all of their analyses as oxides.

Other essential inorganic nutrients, although required in very small amounts, can sometimes become limiting factors in soils on which crops are grown. Experience has shown that the most common deficiencies are those for iron, sulfur, magnesium, zinc, and boron.

Plant Nutrition Research

Research on the inorganic nutrients essential for crop plants—particularly on the quantities of nutrients required for optimal crop yields and on the capacities of various soils to provide the nutrients—has been of great practical value in agriculture and horticulture. Because of the steady increase in worldwide food needs, this type of research undoubtedly will continue to be essential.

Ways Are Being Sought to Overcome Soil Deficiencies and Toxicities

Modification and manipulation of soils by adding nutrients as fertilizers, by raising the pH with lime, or by removing excess salts by leaching with water may not be the only means of improving and maintaining crop production in below-optimum soils. By using the knowledge and techniques of plant breeding and plant nutrition, it may be possible to select and develop cultivars of crop species that are better adapted for growth in nutrient-deficient environments. The validity of this research is confirmed by the occurrence of wild plants in nutritional environments that are very different from the average soil environments in which crop plants are grown. Examples of such less optimal environments are

by plants are present in large amounts in fertile soils, and the amounts removed by a single crop are small. However, when a series of crops is grown on a particular field and the nutrients are continuously removed from their cycles as the crops are harvested, some of these cations may no longer be present in sufficient plant-available forms for the intended purpose. For example, in almost all soils, most of the potassium is present in forms that are not exchangeable and not plant-available. The same is true for phosphorus and nitrogen. Often

Compost

Composting, a practice as old as agriculture itself, has attracted increased interest as a means of utilizing organic wastes by converting them to fertilizer. The starting material is any collection of organic matter—leaves, kitchen garbage, animal manure, straw, lawn clippings, sewage sludge, sawdust—and the population of bacteria and other microorganisms normally present. The only other requirements are oxygen and moisture. Grinding of the organic matter is not essential, but it provides a greater surface area for microbial attack and thus speeds the process.

In a compost heap, microbial growth accelerates quite rapidly, generating heat, much of which is retained because the outer layers of organic matter act as insulation. In a large heap (2 meters × 2 meters × 1.5 meters, for instance), the interior temperature rises to about 70°C; in small heaps, it usually reaches 40°C. As the temperature rises, the population of decomposers changes, with thermophilic and thermotolerant forms replacing the organisms previously present. As the original forms die, their organic matter also becomes part of the product. A useful side effect of the temperature increase is that most of the common pathogenic bacteria that may have been present, for example, in sewage sludge, are destroyed, as are cysts, eggs, and other immature forms of plant and animal parasites.

With the passage of time, changes in pH also occur in a compost heap. The initial pH value is usually slightly acidic, about 6.0, which is comparable to the pH of most plant fluids. During the early stages of decomposition, the production of organic acids causes a further acidification, with the pH decreasing to about 4.5 to 5.0. However, as the temperature rises, the pH also increases; the composted material eventually levels off at slightly alkaline values (7.5 to 8.5).

An important factor in composting (as in any biological growth process) is the ratio of carbon to nitrogen (C/N). A C/N ratio of about 30 to 1 (by weight) is optimal. If the C/N ratio is too high, microbial growth slows; too low, and some nitrogen escapes as ammonia. If the compost materials are quite acidic, limestone (calcium carbonate) may be added to balance the pH; however, if too much is added, it will increase the nitrogen loss.

Studies involving municipal compost piles in Berkeley, California, demonstrated that if large piles were kept moist and aerated, composting could be completed in as little as two weeks. Generally, though, three months or more during the winter is needed to complete the process. If compost is added to the soil before the composting process is complete, it may temporarily deplete the soil of soluble nitrogen.

Because it greatly reduces the bulk of plant wastes, composting can be a very useful means of waste disposal. In Scarsdale, New York, for example, leaves composted in a municipal site were reduced to one-fifth their original volume. At the same time, they formed a useful soil conditioner, improving both the aeration and water-holding capacity. Chemical analyses indicate, however, that a rich compost commonly contains, in dry weight, only about 1.5 to 3.5 percent nitrogen, 0.5 to 1.0 percent phosphorus, and 1.0 to 2.0 percent potassium, far less than commercial fertilizers. Yet, unlike commercial fertilizers, compost can be a source of nearly all the elements known to be needed by plants. Compost provides a continuous balance of nutrients, releasing them gradually as it continues to decompose in the soil.

Today, the main driving force behind composting is the increasing cost of waste disposal and the increasing difficulty of finding suitable disposal sites—sites that will not negatively affect the environment or pollute our waters. Unfortunately, composting is a long way from becoming economically viable as a substitute for manufacturing fertilizers for commercial agriculture.

acidic *Sphagnum* bogs, in which the pH may be less than 4.0, and mine tailings, which often contain high concentrations of potentially toxic metals.

Plants that tolerate high concentrations of potentially toxic metals, such as zinc, nickel, chromium, and lead, are being investigated as part of phytoremediation strategies to restore the soils around old smelters, metal finishers, and nuclear weapons plants. Of special interest are the plants known as **hyperaccumulators,** which concentrate trace elements, heavy metals, or radionuclides to 100 times the normal level, or even higher levels. Such plants apparently accumulate such substances to produce toxic foliage, which helps them to evade predators, including caterpillars, fungi, and bacteria. Among the hundreds of plants that have been screened, the Indian mustard plant (Brassica juncea), which is grown throughout the world for its oilseed, and alpine penny cress (Thlaspi caerulescens) have shown the greatest promise.

A quite different problem is to find a means of making plants tolerant to certain metals. The most common metal in soils, aluminum, causes problems for agriculture on 30 to 40 percent of the world's arable land, most commonly in the tropics, where the soils are acidic. In nonacidic soils aluminum is locked up in insoluble compounds, but in acidic soils, it becomes soluble, is taken up by roots, and inhibits growth. Aluminum has a direct effect on phosphate availability and apparently inhibits the absorption of iron, as well as having direct toxic effects on plant metabolism. Tobacco and papaya plants have now been genetically engineered that secrete five to six times as much citric acid from their roots as control plants. This was accomplished by transferring the bacterial gene for the enzyme citrate synthase into the two plant species. The secreted citric acid binds to the aluminum and prevents it from entering the root. The citric acid–producing plants were able to grow well in aluminum concentrations tenfold higher than those tolerated by control plants. The next step is to transfer the citrate synthase gene into major food crops such as rice and maize.

Biological Nitrogen Fixation Is Being Manipulated to Improve Efficiency

Manipulation of biological nitrogen fixation also offers tremendous potential for improved efficiency in nitrogen utilization. One aspect of research in this area is concerned with improving the efficiency of the association of legumes with either Rhizobium or Bradyrhizobium, for example, through the genetic screening of both legumes and bacteria. Such screening could identify combinations that would result in increased nitrogen fixation in specific environments. This could result from a greater photosynthetic efficiency in legumes, so that more carbohydrate is available for bacterial nitrogen fixation and growth. However, nitrogen fixation requires considerable energy, and so any increase in fixation might be at the expense of shoot productivity.

A second research approach is to develop additional and more effective associations of free-living nitrogen-fixing bacteria and higher plants. In Brazil in the early 1970s, several types of nitrogen-fixing bacteria were found growing in the rhizosphere (see Chapter 25) in association with the roots of certain tropical grasses. Similar associations with some of the world's major food crops, such as maize (Zea mays) and sugarcane (Saccharum), have also been reported. More recently, three newly discovered endophytic, nitrogen-fixing bacteria,

Acetobacter diazotrophicus, Herbaspirillum seropedicae, and Herbaspirillum rubrisubalbicans, were found in stems and leaves of a sugarcane, which was able to obtain more than 60 percent of its nitrogen needs from biological nitrogen fixation.

Probably the most exciting research approach is to be found in genetic engineering: one example being the genetic modifications of the *nif* genes necessary for nitrogen fixation and their transfer from one organism to another (see Chapter 28).

Summary

Plants Require Micronutrients and Macronutrients for Growth and Development

A total of 17 inorganic nutrients are required by most plants for normal growth. Of these, carbon, hydrogen, and oxygen are derived from air and water. The rest are absorbed by roots in the form of ions. These 17 elements are categorized as either micronutrients or macronutrients, depending on the amounts in which they are required. The micronutrients are molybdenum, nickel, copper, zinc, manganese, boron, iron, and chlorine. The macronutrients are sulfur, phosphorus, magnesium, calcium, potassium, nitrogen, hydrogen, carbon, and oxygen. Some inorganic nutrients, such as sodium and cobalt, are essential only for specific organisms.

Inorganic Nutrients Perform a Number of Important Roles in Cells

Inorganic nutrients regulate osmosis and affect cell permeability. Some also serve as structural components of cells, as components of critical metabolic compounds, and as activators and components of enzymes. These functions are impaired when the supply of the essential nutrient is inadequate, and this results in nutrient deficiency symptoms such as stunted growth of stems and leaves, localized death of tissues (necrosis), and yellowing of leaves (chlorosis).

Soils Provide Both a Chemical and a Physical Environment for Plant Growth

The chemical and physical properties of soils are critical in determining their ability to provide the inorganic nutrients, water, and other conditions necessary for maximum crop plant production. The weathering of rocks and minerals supplies the inorganic component of soils. In addition to inorganic nutrients, soils contain organic matter and pore space occupied by varying proportions of water and gases. At least three layers, or horizons—designated A, B, and C—exist in all soils. The A horizon contains the greatest portion of the soil's organic matter, including humus and large numbers of living organisms. Loam soils contain sand, silt, and clay in proportions that result in ideal agricultural soils. Under agricultural conditions, nitrogen, phosphorus, and potassium are the nutrients most often limiting to plant growth and most frequently added to soils in fertilizers.

The Elements Required by Plants Are Recycled Locally and Globally

Each essential inorganic nutrient is circulated, in a complex cycle, among organisms and between organisms and the environment. Because nutrient cycles involve both living organisms and the physical environment, they are also called biogeochemical cycles. Nutrient cycles are leaky: not all of the nutrients returned to the soil become available for plant use.

Ammonification, Nitrification, and Denitrification Are Reactions in the Nitrogen Cycle Carried Out by Soil Bacteria

The circulation of nitrogen through the soil, through the bodies of plants and animals, and back to the soil again is known as the nitrogen cycle. Most soil nitrogen is derived from dead organic materials of plant and animal origin. These substances are decomposed by soil organisms. Ammonification—the release of ammonium ions (NH_4^+) from nitrogen-containing compounds—is carried out by soil bacteria and fungi. Nitrification is the oxidation of ammonia or ammonium ions to form nitrites and nitrates. One type of bacterium is responsible for the oxidation of ammonia to nitrite, and another for the oxidation of nitrite to nitrate. Nitrogen enters plants almost entirely in the form of nitrates. The major loss of nitrogen from the soil is by denitrification. Nitrogen is also lost by crop removal, soil erosion, fire, and leaching.

Nitrogen Fixation Is a Crucial Aspect of the Nitrogen Cycle

Nitrogen is replenished in the soil primarily by nitrogen fixation, the process by which N_2 is reduced to ammonium and made available for assimilation into amino acids and other organic nitrogenous compounds. Biological nitrogen fixation is carried out only by bacteria, including those (*Rhizobium* and *Bradyrhizobium*) that are symbionts of leguminous plants, free-living bacteria, and actinomycetes in symbiotic relationships with a few genera of nonleguminous plants. The most efficient nitrogen-fixing bacteria are those that form symbiotic relationships with plants, which produce root nodules upon infection. In agriculture, plants are removed from the soil, and nitrogen and other elements are not recycled, as they are in natural ecosystems; these elements must be replenished in either organic or inorganic form.

Phosphorus Recycling Is Quite Localized

The phosphorus cycle differs from the nitrogen cycle in part because the Earth's crust, rather than the atmosphere, is the primary replenishment reservoir of phosphorus. The weathering of rocks and minerals over long periods of time is the source of most phosphorus in the soil solution. Phosphorus circulates from plants to animals and is returned to the soil in organic forms, which then are converted to inorganic forms by microorganisms and thus made available to plants. Although some phosphorus is lost from the terrestrial ecosystem by leaching and erosion, weathering of rock usually keeps pace with this loss.

Human Activities Have Had Drastic Effects on Some Nutrient Cycles

Much of the damage to nutrient cycles is related to the need to adequately feed an exponentially growing human population. Crop removal and increased soil erosion have accelerated phosphorus loss from soils; marshes and wetlands, the primary sites of denitrification, are being destroyed; and aquatic environments are subject to widespread damage because of acid rain. Soils that are transferred from natural ecosystems to agricultural systems often do not have enough plant-available nutrients to support crops for commercial harvest, although nutrient quantities are sufficient for native plant communities.

Plant Nutrition Research Is of Great Practical Value to Agriculture

The knowledge and techniques of plant breeding and plant nutrition are being used to select and develop cultivars of crop species that are better adapted for growth in nutrient-deficient environments. Hyperaccumulators, which concentrate trace elements, heavy metals, or radionuclides at levels much higher than normal, are being identified and used to restore the soils around old smelters, metal finishers, and nuclear weapons plants. Both legumes and bacteria are being genetically screened for combinations that would result in increased nitrogen fixation in specific environments. Genetic engineering offers the possibility of transferring the genes necessary for nitrogen fixation from one organism to another.

Selected Key Terms

ammonification   p. 737

bacteroids   p. 739

biogeochemical cycles
p. 736

cation exchange   p. 735

denitrification   p. 738

dry matter   p. 727

essential elements   p. 727

field capacity   p. 734

horizons   p. 732

humus   p. 732

hyperaccumulators
p. 746

infection threads   p. 739

lectins   p. 740

leghemoglobin   p. 740

loam soils   p. 732

macronutrients   p. 727

micronutrients   p. 727

minerals   p. 731

nitrification   p. 738

nitrogen fixation   p. 738

nitrogenase   p. 738

nodules   p. 739

nutrient cycles   p. 735

permanent wilting
percentage   p. 735

rhizobia   p. 739

Questions

1. Distinguish among or between the following: micronutrients/macronutrients; A horizon/B horizon/C horizon; necrosis/chlorosis; igneous rocks/sedimentary rocks/metamorphic rocks; ammonification/nitrification/denitrification; symbiotic nitrogen-fixing bacteria/nonsymbiotic nitrogen-fixing bacteria.

2. What attributes of loam soils help make them ideal agricultural soils?

3. How does the size of the spaces around soil particles influence the amount of water that is available to plants?

4. Why is cation exchange important to plants?

5. Explain the sequence of events leading to nodule formation on legume roots.

6. What role does leghemoglobin play in the root nodules of legumes?

The Movement of Water and Solutes in Plants

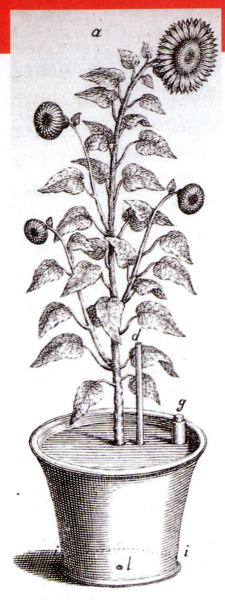

31–1

Diagram of the sunflower plant from Stephen Hales's description of his experiments in the early 1700s on the movement of water through the plant body. Hales found that most of the water "imbibed" by the plant was lost by "perspiration."

OVERVIEW

The long-distance transport of water and minerals in plants takes place in the conduits of the xylem, which extend from root to leaf. Water enters the conduits from cells in the root and exits them in the leaf, where it evaporates from the surfaces of the mesophyll cells into the intercellular spaces. When stomata are open, water vapor diffuses from the saturated intercellular spaces into the atmosphere, a process called transpiration. The water lost by transpiration is replaced by water that is pulled upward through the xylem from the root. A key event in water transport is the opening and closing of the stomata. Hence, this chapter focuses on the mechanism of stomatal movement and how stomata are influenced by environmental factors.

The transport of sugars takes place in the conduits of the phloem. Sugars are always transported from source to sink—that is, from regions of supply such as photosynthesizing leaves to regions of metabolism or storage such as apical meristems or roots, respectively. At the source, the entry of sugars into the conduits causes water to follow by osmosis. With the removal of sugar—followed by water—at the sink, the sugars are thus transported along a gradient of turgor pressure from source to sink. The chapter ends with a discussion of the evidence for this osmotically driven "pressure-flow mechanism" and the metabolic activity necessary for sugar transport to occur.

CHECKPOINTS

By the time you finish reading this chapter, you should be able to answer the following questions:

1. What is transpiration, and why is it dubbed an "unavoidable evil"?

2. What is the role of turgor and the orientation of cellulose microfibrils in the guard cell walls in the opening and closing of stomata?

3. How does the cohesion-tension theory account for the movement of water to the top of tall trees?

4. How does the osmotically generated pressure-flow mechanism account for the movement of sugars from source to sink?

5. By what mechanism do apoplastic phloem-loading plants secrete sugars into the sieve tube–companion cell complexes?

Both organic and inorganic nutrients and water are transported throughout the plant body. The capacity to do so is critical in determining the ultimate structure and function of the plant's component parts, as well as its development and overall form. The tissues involved with the long-distance transport of substances in the plant are the xylem and the phloem. As discussed in Chapters 24 through 26, these two tissues form a continuous vascular system that penetrates practically every part of the plant. Xylem and phloem are closely associated both spatially and functionally. Although we often think of the xylem as *the* water-conducting tissue and the phloem as *the* food-conducting tissue, their functions overlap. The phloem, for example, moves large volumes of water throughout the plant. The phloem, in fact, is the principal source of water for many developing plant parts, such as the developing fruits of a wide variety of crops. In addition, substances are transferred from the phloem to the xylem and recirculated throughout the plant.

The first investigators of "circulation" in plants were seventeenth-century physicians, who searched for pathways and pumping mechanisms analogous to the circulation of blood in animals. Great strides were made during the eighteenth century toward understanding the movement of water and dissolved minerals in xylem. By the end of the nineteenth century, a plausible mechanism was proposed for the rise of water in tall plants. It was not until the 1920s and 1930s, however, that the role of phloem as a food-conducting tissue was widely accepted. We have come a long way since then in our understanding of phloem transport.

Movement of Water and Inorganic Nutrients through the Plant Body

Plants Lose Large Quantities of Water to Transpiration

In the early eighteenth century, Stephen Hales, an English physician, noted that plants "imbibe" a much greater amount of water than animals do. He calculated that one sunflower plant, bulk for bulk, "imbibes" and "perspires" 17 times more water than a human every 24 hours (Figure 31–1). Indeed, the total quantity of water absorbed by any plant is enormous—far greater than that used by any animal of comparable weight. An animal uses less water because much of its water is recirculated through its body over and over again, in the form (in vertebrates) of blood plasma and other fluids. In plants, nearly 99 percent of the water taken in by the roots is released into the air as water vapor. Table 31–1 shows the amount of water released by several crop plants in a single growing season. These amounts pale, however, when compared with the amount of water lost

in a single day—200 to 400 liters (50 to 100 gallons)—by a single tree growing in a deciduous forest of southwestern North Carolina. This loss of water vapor from plants, known as **transpiration,** may involve any aboveground part of the plant body, but leaves are by far the principal organs of transpiration.

Why do plants lose such large quantities of water to transpiration? This question can be answered by considering the requirements for the chief function of the leaf, photosynthesis—the source of all the food for the entire plant body. The energy necessary for photosynthesis comes from sunlight. Therefore, for maximum photosynthesis, a plant must spread a maximum surface to the sunlight, creating a large transpiring surface at the same time. But sunlight is only one of the requirements for photosynthesis; the chloroplasts also need carbon dioxide. Under most circumstances, carbon dioxide is readily available in the air surrounding the plant. However, in order for carbon dioxide to enter a plant cell, which it does by diffusion, it must go into solution because the plasma membrane is nearly impervious to the gaseous form of carbon dioxide. The gas must therefore come into contact with a moist cell surface, but wherever water is exposed to unsaturated air, evaporation occurs. In other words, the uptake of carbon dioxide for photosynthesis and the loss of water by transpiration are inextricably bound together in the life of the green plant.

Water Vapor Diffuses from the Leaf to the Atmosphere through Stomata

Transpiration—sometimes called an "unavoidable evil"—can be extremely injurious to a plant. Excessive transpiration (water loss exceeding water uptake) retards the growth of many plants and kills many others by dehydration. Despite their long evolutionary history, plants have not developed a structure that is both favorable to the entrance of the carbon dioxide essential in photosynthesis and unfavorable to the loss of water vapor by transpiration. However, a number of special adaptations minimize water loss while promoting carbon dioxide gain.

TABLE 31–1	Water Loss by Transpiration in One Plant in a Single Growing Season
Plant	Water Loss (liters)
Cowpea (*Vigna sinensis*)	49
Potato (*Solanum tuberosum*)	95
Wheat (*Triticum aestivum*)	95
Tomato (*Solanum lycopersicum*)	125
Maize (*Zea mays*)	206

After J. F. Ferry, *Fundamentals of Plant Physiology* (New York: Macmillan Publishing Company, 1959).

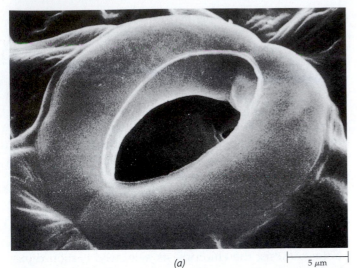

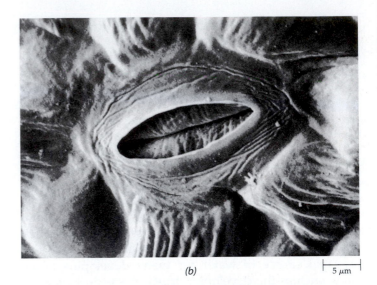

(a)   5 μm   (b)   5 μm

31–2

Scanning electron micrographs showing **(a)** *open stoma in epidermis of cucumber* (Cucumis sativus) *leaf and* **(b)** *closed stoma in epidermis of parsley* (Apium petro-

selinum) leaf. The stomata lead into a honeycomb of air spaces that surround the thin-walled, photosynthetic mesophyll cells within

the leaf. The air spaces are saturated with water vapor that has evaporated from the surfaces of the mesophyll cells.

The Cuticle Serves as an Effective Barrier to Water Loss
Leaves are covered by a **cuticle** that makes their surfaces largely impervious to both water and carbon dioxide (page 25). Only a small fraction of the water transpired by plants is lost through this protective outer coating, and another small fraction is lost through the lenticels in the bark (see Figures 27–10d and 27–11). By far the largest amount of water transpired by a vascular plant is lost through the stomata (Figure 31–2). Stomatal transpiration involves two steps: (1) evaporation of water from cell wall surfaces bordering the intercellular spaces (air spaces) of the leaf, and (2) diffusion of the resultant water vapor from the intercellular spaces into the atmosphere by way of the stomata (see Figure 31–20).

The Opening and Closing of Stomata Controls the Exchange of Gases across the Leaf Surfaces   As discussed previously (page 692), changes in shape of the guard cells of the stomata bring about the opening and closing of the stomatal pores. Although stomata occur on all aerial parts of the primary plant body, they are far more abundant in leaves. The number of stomata may be quite large; for example, there are approximately 12,000 stomata per square centimeter of leaf surface in tobacco (*Nicotiana tabacum*) leaves. The stomata lead into a honeycomb of air spaces that surround the thin-walled mesophyll cells within the leaf. These spaces, which make up 15 to 40 percent of the total volume of the leaf, contain air saturated with water vapor that has evaporated from the damp surfaces of the mesophyll cells. Although the stomatal pores account for only about 1 percent of the total leaf surface, more than 90 percent of the water tran-

spired by the plant is lost through the stomata. The rest is lost through the cuticle.

Closing of stomata not only prevents loss of water vapor from the leaf but, as mentioned, also prevents entry of carbon dioxide into the leaf. A certain amount of carbon dioxide, however, is produced by the plant during respiration, and as long as light is available, this carbon dioxide can be used to sustain a very low level of photosynthesis even when the stomata are closed.

Stomatal Movements Result from Changes in Turgor Pressure within the Guard Cells   Stomatal opening occurs when solutes are actively accumulated in the guard cells. The accumulation of solutes (and resultant decrease in guard cell water potential) causes osmotic movement of water into the guard cells and a buildup of turgor pressure in excess of that in the surrounding epidermal cells. Stomatal closing is brought about by the reverse process: with a decline in guard cell solutes (and resultant increase in guard cell water potential), water moves out of the guard cells and the turgor pressure decreases. Thus, turgor is maintained or lost due to the passive osmotic movement of water into or out of the cells along a gradient of water potential that itself is created by active solute transport. The important solutes responsible for these gradients in water potential and their involvement in the mechanism of stomatal movement are discussed in Chapter 28.

The Radial Orientation of Cellulose Microfibrils in the Guard Cell Walls Is Required for Pore Opening   The structure of the guard cell walls plays a crucial role in sto-

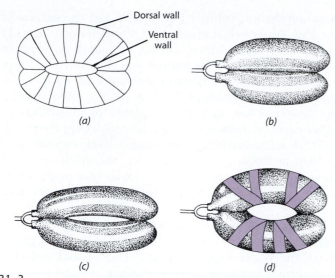

31–3

Radial micellation in guard cells. (a) A pair of guard cells, with lines indicating the radial arrangement of cellulose microfibrils in the guard cell walls. (b) Two partially inflated balloons have been glued together near their ends in order to model the effect of radial micellation on the opening of stomata. (c) The same balloons at higher pressure, that is, fully inflated. A narrow slit is visible. (d) A pair of fully inflated balloons after bands of tape have been added to simulate radial micellation. The opening is much greater than in (c).

matal movements. During expansion of the paired guard cells, two physical constraints cause the cells to bend and thus to open the pore. One of these constraints is the radial orientation of the cellulose microfibrils in the guard cell walls (Figure 31–3a). This **radial micellation** allows the guard cells to lengthen while preventing them from expanding laterally. The second constraint is found at the ends of the guard cells, where they are attached to one another. This common wall remains almost constant in length during opening and closing of the stoma. Consequently, increase in turgor pressure causes the outer (dorsal) walls of the guard cells to move outward relative to their common walls. As this happens, the radial micellation transmits this movement to the wall bordering the pore (the ventral wall), and the pore opens. Figure 31–3b through d depicts the results of some experiments with balloons that have been used in support of the role of radial micellation in stomatal movement.

Water Loss Is the Major Influence Affecting Stomatal Closing When the turgor of a leaf drops below a certain critical point, which varies with different species, the stomatal opening becomes smaller. The effect of water loss overrides other factors affecting the stomata, but stomatal changes can occur independently of overall water gain or loss by the plant. The most conspicuous example is found in the many species in which the stomata regularly open in the morning and close in the evening, even though there may be no changes in the amount of water available to the plant.

In many plants, there is a marked increase in the level of abscisic acid (ABA) during periods of water stress. When "fed" or applied to leaves, ABA causes stomatal closure within a few minutes. Moreover, the effect of ABA on stomatal movement is readily reversible (page 693). Solute loss from guard cells begins when ABA produced by the mesophyll arrives at the stomata, signaling the stomata that the mesophyll cells are experiencing water stress. Stomatal closure can also be caused by ABA produced in roots, without any changes in the water status or turgor of the leaf. Upon exposure to drying soil, extra ABA is synthesized by the root and transported to the leaves in the xylem. The xylem ABA concentration and, therefore, the stomatal behavior reflect the roots' access to soil water.

Mutants—such as *flacca*, a wilty mutant of tomato (*Solanum lycopersicum*)—that are unable to synthesize ABA remain permanently wilted. When ABA is applied to the leaves of such mutants, their stomata close and the plants become turgid.

Environmental Factors or Signals Also Affect Stomatal Movement Environmental factors affecting stomatal movement include *carbon dioxide concentration, light,* and *temperature.* In most species, an increase in CO_2 concentration causes the stomata to close. The magnitude of this response to CO_2 varies greatly from species to species and with the degree of water stress a given plant has undergone or is undergoing. In maize (*Zea mays*), the stomata may respond to changes in CO_2 in a matter of seconds. The site for sensing the level of CO_2 is located within the guard cells.

In most species, the stomata open in the light and close in the dark. This can be explained in part by the photosynthetic utilization of CO_2, which brings about a reduction in the CO_2 level within the leaf. Light may, however, have a more direct effect on stomata. Blue light has long been known to stimulate stomatal opening independently of CO_2. For example, guard cell protoplasts of onion (*Allium cepa*) swell in the presence of potassium (K^+) when illuminated with blue light. The blue-absorbing pigment (a flavin or a carotenoid located in the tonoplast and possibly the plasma membrane) promotes K^+ uptake by the guard cells. The blue light response is involved in stomatal opening in the early morning and in stomatal responses to sunflecks, spots of light that penetrate the leaf canopy and range in duration from a few seconds to minutes.

Evidence also exists for the presence of red-light-stimulated stomatal opening mediated by the guard cell chloroplasts. Apparently the chloroplasts supply the ATP that fuels proton pumping at the guard cell plasma membrane, the site of active K^+ uptake. It is believed that the guard cell chloroplasts are involved with stomatal adaptation to sun, shade, and temperature.

Within normal ranges (10° to 25°C), changes in temperature have little effect on stomatal behavior, but temperatures higher than 30° to 35°C can lead to stomatal closure. The closing can be prevented, however, by holding the plant in air without carbon dioxide, which suggests that temperature changes work primarily by affecting the concentration of carbon dioxide in the leaf. An increase in temperature results in an increase in respiration and a concomitant increase in the concentration of intercellular carbon dioxide, which may actually be the cause of stomatal closure in response to heat. Many plants in hot climates close their stomata regularly at midday, apparently because carbon dioxide has accumulated in the leaf and because the leaves are dehydrated as water loss by transpiration exceeds water uptake by absorption.

Stomata not only respond to environmental factors but also exhibit daily rhythms of opening and closing that appear to be controlled from within the plant—that is, they also exhibit circadian rhythms (page 706).

A wide variety of succulents—including cacti, the pineapple *(Ananas comosus)*, and members of the stonecrop family *(Crassulaceae)*—open their stomata at night, when conditions are least favorable to transpiration. The crassulacean acid metabolism (CAM) characteristic of such plants has a pathway for carbon flow not substantially different from that of C_4 plants (page 147). At night, when their stomata are open, CAM plants take in carbon dioxide and convert it to organic acids. During the day, when their stomata are closed, the carbon dioxide is released from these organic acids for use in photosynthesis.

Environmental Factors Affect the Rate of Transpiration

Although stomatal opening and closing are the major plant factors affecting the rate of transpiration, a number of other factors both in the environment and in the plant itself influence transpiration. One of the most important of these is *temperature*. The rate of water evaporation doubles for every 10°C rise in temperature. However, because evaporation cools the leaf surface, its temperature does not rise as rapidly as that of the surrounding air. As noted previously, stomata close when temperatures exceed 30° to 35°C.

Humidity is also important. This is because the rate of transpiration is proportional to the vapor pressure difference, which is the difference in water vapor pressure between the intercellular (air) spaces and the surface of the leaf. Water is lost much more slowly into air already laden with water vapor. Leaves of plants growing in shady forests, where the humidity is generally high, typically spread large, luxuriant leaf surfaces because for these plants the main problem is getting enough light, not losing water. In contrast, plants of grasslands or other exposed areas often have narrow leaves characterized by relatively little leaf surface, thick cuticles, and sunken stomata. Grassland plants obtain all the light they can use but are constantly in danger of excessive water loss because of the low humidity in the surrounding air.

Air currents also affect the rate of transpiration. A breeze cools your skin on a hot day because it blows away the water vapor that has accumulated near the skin surface and so accelerates the rate of evaporation of water from your body. Similarly, wind blows away the water vapor from leaf surfaces, affecting the vapor pressure difference across the surface. Sometimes, if the air is very humid, wind may decrease transpiration by cooling the leaf, but a dry breeze will greatly increase evaporation.

Water Is Conducted through the Vessels and Tracheids of the Xylem

Water enters the plant by the roots and is given off, in large quantities, by the leaves. How does the water get from one place to another, often over large vertical distances? This question has intrigued many generations of botanists.

The general pathway that water follows in its ascent has been clearly identified. One can trace this pathway simply by placing a cut stem in water that is colored with any harmless dye (preferably, the stem should be cut under the water to prevent air from entering the conducting elements of the xylem) and then tracing the path of the liquid into the leaves. The stain quite clearly delineates the conducting elements of the xylem. Experiments using radioactive isotopes confirm that the isotope and, presumably, the water do indeed travel by way of vessels (or tracheids) in the xylem. In the experiment shown in Figure 31–4, care had to be taken to separate the xylem from the phloem. Earlier experiments in which this separation was not made produced ambiguous results because there is a great deal of lateral movement from the xylem into the phloem. This lateral movement, however, as the experiment shows, is not necessary for the overall movement of water and minerals from soil to leaf.

Water Is Pulled to the Top of Tall Trees: The Cohesion-Tension Theory Now that we know the path water takes, the next question we ask is, "How does the water move?" Logic suggests two possibilities: it can be

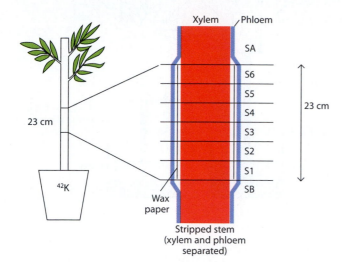

	Stem Segment	^{42}K in Xylem (ppm)	^{42}K in Phloem (ppm)
Above strip	SA	47	53
Stripped section	S6	119	11.6
	S5	122	0.9
	S4	112	0.7
	S3	98	0.3
	S2	108	0.3
	S1	113	20
Below strip	SB	58	84

31–4

Radioactive potassium (^{42}K) added to the soil water shows that the xylem is the channel for the upward movement of both water and inorganic ions. Wax paper was inserted be- *tween the xylem and the phloem to prevent lateral transport of the isotope. The relative amounts of ^{42}K detected in each segment of the stem are given in the table. Note the* *reduced amounts of ^{42}K in the phloem of the segments from the stripped section of the stem.*

pushed from the bottom or pulled from the top. As we shall see, the first of these possibilities is not the answer. Briefly, root pressure (see page 760) does not exist in all plants, and in those plants in which it is present, it is not sufficient to push water to the top of a tall tree. Moreover, the simple experiment just described (that involving the cut stem) rules out root pressure as a crucial factor. So we are left with the hypothesis that water is pulled up through the plant body, and this hypothesis is correct according to all present evidence.

When water evaporates from the cell wall surfaces bordering the intercellular spaces in the interior of a leaf during transpiration, it is replaced by water from within the cell. This water diffuses across the plasma membrane, which is freely permeable to water but not to the solutes of the cell. As a result, the concentration of solutes within the cell increases and the water potential of the cell decreases. A gradient of water potential then becomes established between this cell and adjacent, more saturated cells. These cells, in turn, gain water from other cells until eventually this chain of events reaches a vein and exerts a "pull," or tension, on the water of the xylem. Because of the extraordinary cohesiveness among water molecules, this tension is transmitted all the way down the stem to the roots. As a result, water is withdrawn from the roots, pulled up the xylem, and distributed to the cells that are losing water vapor to the atmosphere (Figure 31–5). This water loss makes the water potential of the roots more negative and increases their capacity to extract water from the soil. Hence, the lowered water potential in the leaves,

brought about by transpiration and/or by the use of water in the leaves, results in a gradient of water potential from the leaves to the soil solution at the surface of the roots. This gradient of water potential provides the driving force for the movement of water along the soil-plant-atmosphere continuum.

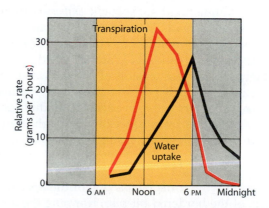

31–5

Measurements of water movement in ash (Fraxinus) trees show that a rise in water uptake follows a rise in transpiration. These data suggest that the loss of water generates the force needed for its uptake.

31–6

Demonstration of the cohesion-tension theory. (a) A simple physical system. A porous clay pot is filled with water and attached to the end of a long, narrow glass tube that is also filled with water. The water-filled tube is placed with its lower end below the surface of a volume of mercury in a beaker. Water evaporates from the pores in the pot and is replaced by water "pulled up" through the tube in a continuous column. As the water rises, mercury is pulled up into the tube to replace it. (b) Transpiration from leaves results in sufficient water loss to create a similar negative pressure.

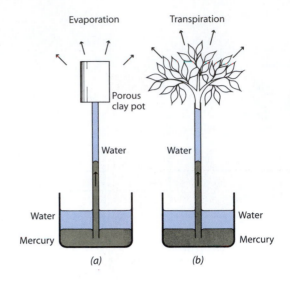

This theory of water movement is called the **cohesion-tension theory** because it depends on the cohesiveness of water, the property of water that permits it to withstand tension (Figure 31–6). However, the theory might also be called the *cohesion-adhesion-tension theory* because adhesion of the water molecules to the walls of the tracheids and vessels of the xylem and to the cell walls of the leaf and root cells is just as important for the rise of water as are cohesion and tension. The cell walls along which the water moves have evolved as a very effective water-attracting surface. This surface is capable of holding water with a force great enough to support a column of water several kilometers high against gravity.

Air Bubbles Can Break the Continuity of Water in the Xylem   The cohesion of water in the xylem is increased by the filtering effect of the roots, which remove fine particles that can act as nuclei for bubble formation. Water cohesion is also enhanced by the small diameter of the xylem conduits—vessels and tracheids—through which the water moves. Despite the filtering provided by the roots, bubble formation does occur, and it is a normal occurrence in many trees. *Cavitation* (rupture of the water columns) and subsequent *embolism* (filling of vessels and/or tracheids with air or water vapor) are the bane of the cohesion-tension mechanism (Figure 31–7). Embolized xylem conduits cannot conduct water. Fortunately, the surface tension of the air-water meniscus spanning the small pores in the pit membranes of the bordered pit-pairs between adjacent conduits usually prevents air bubbles from squeezing through the pores, helping to isolate them in a single vessel or tracheid (Figure 31–8). In conifer tracheids, the passage of air is prevented by the lateral displacement of the pit membrane such that the torus blocks one of the apertures, or openings, of the bordered pit-pair, trapping the air bubble. Thus, the pit membranes are very important to the safety of water transport. (See also the discussion on pages 576 and 577.)

Most embolisms are triggered by air sucked into the vessel or tracheid via a pore in the wall or pit membrane adjacent to an already embolized conduit. This process,

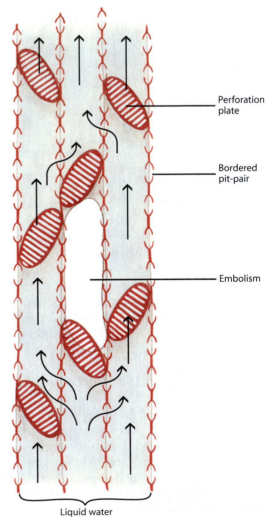

31–7

Detours around an embolized vessel element. An embolism consisting of water vapor has blocked the movement of water through a single vessel element. However, water is able to detour around the embolized element via the bordered pit-pairs between adjacent vessels. The vessel elements shown here are characterized by ladderlike perforation plates.

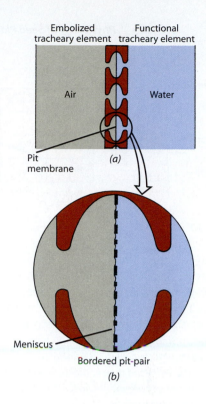

Air   Water

Pit
membrane   (a)

Meniscus

Bordered pit-pair

(b)

31–8

Embolized tracheary element. (a) Diagram showing bordered pit-pairs between tracheary elements, one of which is embolized (filled with air) and thus nonfunctional. (b) Detail of a pit membrane. When a tracheary element is embolized, air is prevented from spreading to the adjacent functional tracheary element by the surface tension of the air-water meniscus spanning the pores in the pit membrane. The bordered pit-pairs diagrammed here do not have a torus, the impermeable central thickening typical of the pit membranes of conifer tracheids. Tori are also found in the pit membranes between vessel members of some angiosperms.

known as *air seeding,* occurs only when the pressure difference across the wall or pit membrane exceeds the surface tension at the air-water meniscus spanning the pores (Figure 31–8b). The largest pores are the most vulnerable to the penetration of air. A plant is susceptible to this mode of embolism any time even one of its vessels or tracheids becomes air-filled by physical damage (for example, by an insect bite or broken branch). Freezing can also induce embolisms because air is not soluble in ice, and the xylem sap—that is, the fluid contents of the xylem—contains dissolved air. In addition, it is now becoming clear that xylem dysfunction induced by drought is a serious problem to plants.

The Cohesion-Tension Theory Holds up to the Test   There is no doubt that the tensile strength of water is great enough to prevent the pulling apart of adjacent water molecules under the tension required to move water up the xylem of tall trees, despite some accounts to the contrary. For example, it has been demonstrated that a column of water in a fine capillary tube is capable of withstanding a tension of -26.4 megapascals (MPa); the estimated tension required to move water to the top of a tall redwood (*Sequoia sempervirens*) is only about -2.0 MPa.

How can the cohesion-tension theory be tested? One way to test this theory directly is by measuring the water potential of large pieces of tissues, such as whole shoots, with a *pressure chamber* (also called a pressure bomb). The pressure chamber measures the negative hydrostatic pressure, or tension, that exists in the whole plant part. The assumption is made that the water potential of the xylem is fairly close to that of the whole plant part. For example, when a twig is cut from a transpiring tree, the columns of water, which were under tension in the vessels, abruptly recede below the cut surface. At this time, the cut surface appears dry. To make a measurement, the twig is mounted in a pressure chamber, as shown in Figure 31–9. The chamber is then pressurized with gas

Magnifier

Gas
pressure

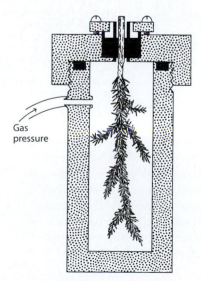

31–9

Measuring plant water potential in a cut branch. When the branch is cut, some of the xylem sap—which was under negative pressure, or tension, before the branch was cut—recedes into the xylem below the cut surface. It is assumed that the water potential of the xylem is fairly close to that of the whole branch. The branch is placed in a pressure chamber, where the pressure is increased until sap emerges from the cut end of the stem. If one assumes that equal pressure is required to force the sap in either direction, the positive pressure needed to force out the sap is, ideally, equal to the tension that existed in the branch before it was severed.

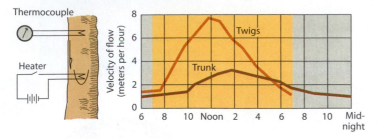

31–10

Method for measuring velocity of sap flow. A small heating element inserted into the xylem heats the ascending sap for a few seconds. A thermocouple above the heating element records the passing wave of heat. The experimenter times the interval between these two events. As shown in the graph, in the morning the sap begins to increase its velocity of flow first in the twigs and then in the trunk. In the evening, velocity diminishes first in the twigs and then in the trunk.

(air or nitrogen) until the curved upper surfaces of the water columns just appear (when magnified) at the cut surface of the twig. The magnitude of the pressure required to return the water to the cut surface is called the balance pressure and is equal in magnitude (but opposite in sign) to the negative pressure, or tension, that existed in the xylem before the twig was excised. Results obtained by this method are entirely consistent with the

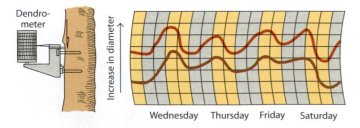

31–11

A dendrometer (left) records small daily fluctuations in the diameter of a tree trunk at two different heights. As shown in the graph, in the morning, shrinkage occurs in the upper trunk slightly before it occurs in the lower trunk. These data suggest that transpiration from the leaves "pulls" water out of the trunk before it can be replenished from the roots. The shaded strips signify nighttime.

predictions of the cohesion-tension theory. Although the validity of pressure-chamber estimates of xylem tension has recently been questioned, other studies apparently have confirmed the existence of large tensions in the xylem.

A second set of data that is in accord with the cohesion-tension theory indicates that the movement of water begins at the top of the tree. The velocity of sap flow in various parts of the tree has been measured by an ingenious method involving a small heating element to warm the xylem contents for a few seconds and a sensitive thermocouple to detect the moment at which the heated xylem sap moves past a specific point (Figure 31–10). As shown in the graph, in the morning the sap begins to flow first in the twigs, as tension arises close to the leaves, and later in the trunk. In the evening, the flow diminishes first in the twigs, as water loss from the leaves diminishes, and later in the trunk. Trees with wide vessels (200 to 400 micrometers in diameter) showed midday peak velocities of 16 to 45 meters per hour (measured at breast height), while those with narrow vessels (50 to 150 micrometers in diameter) had slower midday peak velocities, ranging from 1 to 6 meters per hour.

A third set of supporting data comes from measurements of minute changes in the diameter of the tree trunk (Figure 31–11). The shrinking of the trunk is interpreted by some investigators as occurring because of the negative pressures in the water passages of the xylem. Presumably, the water molecules adhering to the sides of the vessels pull the sides inward. When transpiration begins in the morning, first the upper part of the trunk shrinks, as water is pulled out of the xylem before it can be replenished from the roots; then the lower part shrinks. Later in the day, as the transpiration rate decreases, the upper trunk expands before the lower trunk does. There is some evidence that changes in trunk diameter are due in part or largely to shrinkage and expansion of bark tissues as water moves laterally to and from the xylem in response to changing pressures in the xylem.

Note that the energy for evaporation of water molecules—and thus for movement of water and inorganic nutrients through the plant body—is supplied not by the plant but directly by the sun. Note also that movement is possible because of the extraordinary cohesive and adhesive properties of water, to which the plant is so exquisitely adapted.

The cohesion-tension theory is sometimes called the "transpiration-pull theory." This is unfortunate because "transpiration-pull" implies that transpiration is essential for water to move to leaves. Although transpiration may increase the rate at which water moves, any use of water in leaves produces forces causing the movement of water into them.

Water Absorption by Roots Is Facilitated by Root Hairs

The root system serves to anchor the plant in the soil and, above all, to meet the tremendous water requirements of the leaves resulting from transpiration. Most of the water that a plant takes from the soil enters through the younger parts of the root (see Chapter 25). Absorption takes place directly through the epidermis of younger roots. The root hairs, located several millimeters above the root tip, provide an enormous surface area for absorption (Figure 31–12; Table 31–2). From the root hairs, the water moves through the cortex (the outer layer(s) of which may be differentiated as an exodermis, a subepidermal layer of cells with Casparian strips). From there, the water progresses through the endodermis (the innermost layer of cortical cells) and into the vascular cylinder. Once in the conducting elements of the xylem, the water moves upward through the root and stem and into the leaves, from which most of it is lost to the atmosphere by transpiration. Hence, the soil-plant-atmosphere pathway can be viewed as a continuum of water movement.

Water May Follow One or More of Three Possible Pathways across the Root The pathway followed by water across the root depends in large part on the degree of differentiation of the various tissues that make up the root. In

TABLE 31–2	Density of Root Hairs on the Root Surface in Three Plant Species
Plant	Root Hair Density (per square centimeter)
Loblolly pine (*Pinus taeda*)	217
Black locust (*Robinia pseudo-acacia*)	520
Rye (*Secale cereale*)	2500

After J. F. Ferry, *Fundamentals of Plant Physiology* (New York: Macmillan Publishing Company, 1959).

each of the tissues, water may follow one or more of three possible pathways: (1) **apoplastic** (via the cell walls), (2) **symplastic** (from protoplast to protoplast via plasmodesmata), and/or (3) **transcellular** (from cell to cell, passing from vacuole to vacuole) (Figure 31–13). For example, in a root without an exodermis, water can move apoplastically as far as the endodermis. At the endodermis, however, water is forced to traverse the plasma membranes and protoplasts of the tightly packed endodermal cells because of the presence of the water-impermeable Casparian strips in the radial and transverse walls of these cells (page 598). In roots having an exodermis, by contrast, the Casparian strips in the radial and transverse walls of the compactly

31–12

*Root hairs. (a) Primary root of a radish (*Raphanus sativus*) seedling, showing its numerous root hairs. (b) Root hairs surrounded by soil particles; each particle is covered by a layer of adhered water.*

(a)

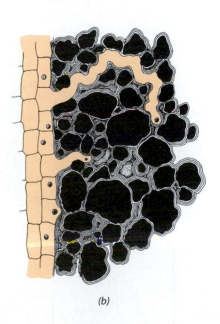

(b)

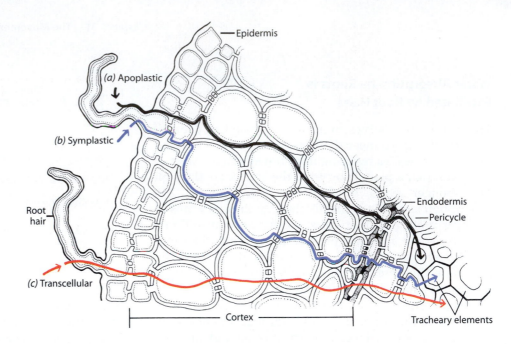

Epidermis

(a) Apoplastic

(b) Symplastic

Root hair

(c) Transcellular

Endodermis

Pericycle

Cortex

Tracheary elements

31–13

*Possible pathways for the movement of water from the soil, across the epidermis and cortex, and into the tracheary elements, or water-conducting elements, of the root: **(a)** apoplastic movement (black line) occurs via cell walls; **(b)** symplastic movement (blue line) is from protoplast to protoplast via plasmodesmata; and **(c)** transcellular movement (red line) occurs from cell to cell, the water passing from vacuole to vacuole. The root depicted here lacks an exodermis (a subepidermal layer of cells with Casparian strips).*

Note that water following an apoplastic pathway is forced by the Casparian strips of the endodermal cells to cross the plasma membranes and protoplasts of these cells on its way to the xylem. Having crossed the plasma membrane on the inner surface of the endodermis, the water may once again enter the apoplastic pathway and make its way into the lumina of the tracheary elements. Inorganic ions are actively absorbed by the epidermal cells and then follow a symplastic pathway across the cortex and into parenchyma cells, from which they may be secreted into the tracheary elements.

arranged exodermal cells prevent apoplastic movement of water across that cell layer. Water could follow either a symplastic or transcellular pathway across such cells. If, however, the outer tangential walls of the exodermal cells possess suberin lamellae, movement across that surface may be limited to the symplast. Having traversed the exodermis, subsequent movement of the water across the cortex could be by one or more of the three possible pathways listed above.

In the Absence of Transpiration, Roots Can Generate Positive Pressure The driving force for movement of water across the root is the difference in water potential between the soil solution at the surface of the root and the xylem sap. When transpiration is very slow or absent, as it is at night, the gradient of water potential is brought about by the secretion of ions into the xylem. Because the vascular tissue of the root is surrounded by the endodermis, ions do not tend to leak back out of the xylem. Therefore, the water potential of the xylem becomes more negative, and water moves into the xylem by osmosis through the surrounding cells. In this manner, a positive pressure, called **root pressure,** is created, and it forces both water and dissolved ions up the xylem (Figure 31–14).

Dewlike droplets of water at the tips of grass and

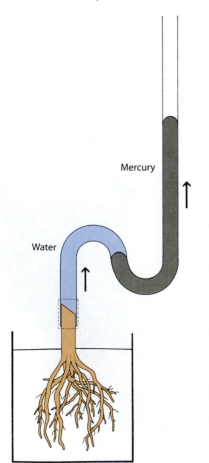

Mercury

Water

31–14

Demonstration of root pressure in the cut stump of a plant. Uptake of water by the plant's roots causes the mercury to rise in the column. Pressures of 0.3 to 0.5 megapascal have been demonstrated by this method.

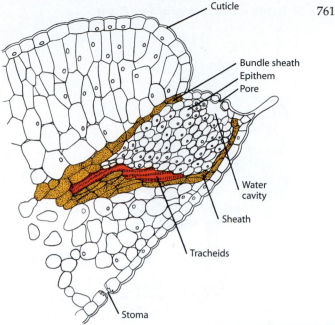

31–15

Guttation droplets at the margins of a leaf of lady's mantle (Achemilla vulgaris) also demonstrate the presence of root pressure. These droplets are not condensation from water vapor in the surrounding air; rather, they are forced out of the leaf through openings in structures called hydathodes at the tips along the leaf margins (see Figure 31–16).

31–16

Longitudinal view of a hydathode of the leaf of saxifrage (Saxifraga lingulata). The hydathode consists of the terminal tracheids of a bundle end, thin-walled parenchyma (the epithem) with numerous intercellular spaces, and an epidermal pore. The tracheids are in direct contact with epithem, which empties into a water cavity behind the pore. The epidermal pores commonly are stomata that lack the capacity to close and open.

other leaves in the early morning demonstrate the effects of root pressure (Figure 31–15). These droplets are not dew—which is water that has condensed from the air—but come from within the leaf by a process known as **guttation** (from the Latin *gutta*, meaning "drop"). They exude through openings—commonly stomata that lack the capacity to close and open—in special structures called **hydathodes,** which occur at the tips and margins of leaves (Figure 31–16). The water of guttation is literally forced out of the leaves by root pressure.

Root pressure is least evident during the day, when the movement of water through the plant is the fastest, and the pressure never becomes high enough to force water to the top of a tall tree. Moreover, many plants, including conifers such as pine, do not develop root pressure. Thus, root pressure can perhaps be regarded as a by-product of the mechanism of pumping ions into the xylem and only indirectly as a mechanism of moving water into the shoot under special conditions.

Many Plants Redistribute Soil Water by Hydraulic Lift
Recent studies indicate that in many plants, water taken up at night by deep roots located in moist soil regions is transferred to dry soil regions via shallower roots. This phenomenon is known as **hydraulic lift.** Although the process is believed to be passive—driven by root and soil water potential gradients—an active role by the

roots has not been ruled out. Hydraulic lift not only improves the soil water status of the plant exhibiting this phenomenon, but neighboring plants can also use a significant proportion of this water source (Figure 31–17).

Hydraulic lift apparently is widespread and has been found both in plants adapted to arid conditions, such as sagebrush and the oaks of the California chaparral, and in plants of wetter habitats, such as alfalfa, barley, and maize. Even the sugar maple, which grows in wet environments, exhibits hydraulic lift.

Todd Dawson, a plant ecologist at Cornell University, estimated that a 40-foot-tall sugar maple was able to deliver between 40 and 60 gallons of water to the upper soil levels every night. Distinguishing between water stored deep in the ground and water of summer rainfalls on the basis of their hydrogen isotope content (rain tends to have a higher concentration of the heavier hydrogen isotope deuterium, ^{2}H), Dawson was able to determine the source of water absorbed by nearby plants. He found that many plants growing near the sugar maples were using water released by the trees (Figure 31–17).

Water Absorption by Roots of Transpiring Plants May Be Passive During periods of high transpiration rates, ions accumulated in the xylem of the root are swept away in the **transpiration stream,** or flow of water, and the amount of osmotic movement across the endodermis de-

Species	Distance from Base of Tree				
	0.5 m	1.0 m	1.5 m	2.5 m	5.0 m
May apple (*Podophyllum peltatum*)	61%	53%	50%	9%	0%
False Solomon's seal (*Smilacina racemosa*)	60	58	5	0	0
Wild strawberry (*Fragaria virginia*)	58	54	50	13	1
Meadow rue (*Thalictrum dioicum*)	55	50	11	2	0
Wild ginger (*Asarum canadense*)	31	21	8	0	0
Large-flowered trillium (*Trillium grandiflorum*)	25	18	0	0	0
Goldenrod (*Solidago flexicaulis*)	20	19	6	0	0
Lowbush blueberry (*Vaccinium vacillans*)	19	10	5	1	0
Velvet grass (*Holcus lanatus*)	21	7	0	0	0
Spicebush (*Lindera benzoin*)	11	6	0	0	0
Linden tree (*Tilia heterophylla*)	0	1	0	1	0
Beech tree (*Fagus grandifolia*)	1	0	0	0	0

31–17

Effects of hydraulic lift on nearby plants. The table shows mean percentages of total xylem water that was hydraulically lifted in each type of plant growing in the vicinity of a sugar maple (Acer saccharum) *tree.*

The Uptake of Inorganic Nutrients by Roots Is an Energy-Requiring Process

The uptake, or absorption, of inorganic ions takes place through the epidermis of the younger roots. Current evidence suggests that the major pathway followed by ions from the epidermis to the endodermis of the root is symplastic—that is, from protoplast to protoplast via plasmodesmata. Ion uptake by the symplastic route begins at the plasma membrane of the epidermal cells. The ions then move from the epidermal cell protoplasts to the first layer of cortical cells (possibly an exodermis) through the plasmodesmata in the epidermal-cortical cell walls (Figure 31–13). Movement of ions into the root continues in the cortical symplast—again, from protoplast to protoplast via plasmodesmata—through the endodermis and into the parenchyma cells of the vascular cylinder by diffusion. This movement is possibly aided by cytoplasmic streaming within the individual cells.

Nutrient uptake from soil by most seed plants is greatly enhanced by naturally occurring mycorrhizal fungi associated with the root systems of the plants (page 340). Mycorrhizae are especially important in the absorption and transfer of phosphorus, but the increased absorption of zinc, manganese, and copper has also been demonstrated. These nutrients are relatively immobile in soil, and depletion zones for them quickly develop around the root and root hairs. The hyphal network of mycorrhizae extends several centimeters out from colonized roots, thus exploiting a large volume of soil more efficiently.

How the ions enter the mature vessels (or tracheids) of the xylem from the parenchyma cells of the vascular

creases. At such times, the roots become passive absorbing surfaces through which water is pulled by bulk flow generated in the transpiring shoots. Some investigators believe that almost all of the absorption of water by the roots of transpiring plants occurs in this passive manner.

During periods of rapid transpiration, water may be removed from around the root hairs so quickly that the local soil becomes depleted of water. Water will then move from some distance away toward the root hairs through fine pores in the soil. In general, however, the roots come into contact with additional water by growing, although roots will not grow in dry soil. For example, under normal conditions, roots of apple trees grow an average of about 3 to 9 millimeters a day, roots of prairie grasses may grow more than 13 millimeters a day, and the main roots of maize plants average 52 to 63 millimeters a day. The results of such rapid growth can be remarkable: a four-month-old rye (*Secale cereale*) plant has over 10,000 kilometers of roots and many billions of root hairs.

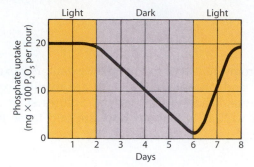

31–18

Energy requirement for nutrient uptake. The rate of phosphate absorption (measured as P_2O_5; see page 745) by maize (Zea mays) plants fell to near zero after four days of continuous darkness. The rate began to rise again when light was restored. These and other data indicate that mineral ion uptake in plants is a process that requires energy.

cylinder has been the subject of considerable debate. At one time it was suggested that the ions leak passively from the parenchyma cells into the vessels, but substantial evidence now suggests that the ions are secreted either directly into the vessels from the parenchyma cells or indirectly via apoplastic (cell wall) channels that extend from the parenchyma cells to the vessels.

The mineral composition of root cells is far different from that of the medium in which the plant grows. For example, in one study, cells of pea (*Pisum sativum*) roots had a K^+ ion concentration 75 times greater than that of the nutrient solution. Another study showed that the vacuoles of rutabaga (*Brassica napus* var. *napobrassica*) cells contained 10,000 times more K^+ ions than the external solution.

Since substances do not diffuse against a concentration gradient, it is clear that minerals are absorbed by **active transport** (page 83). Indeed, the uptake of minerals is known to be an energy-requiring process. For instance, if roots are deprived of oxygen, or poisoned so that respiration is curtailed, mineral uptake is drastically decreased. Also, if a plant is deprived of light, it will cease to absorb salts after its carbohydrate reserves are depleted (Figure 31–18) and will finally release minerals back into the soil solution. Hence, ion transport from the soil to the vessels of the xylem requires two active, carrier-mediated membrane events: (1) uptake at the plasma membrane of the epidermal cells and (2) secretion into the vessels at the plasma membrane of the parenchyma cells bordering the vessels.

Inorganic Nutrients Are Exchanged between Transpiration and Assimilate Streams

Once secreted into the xylem vessels (or tracheids), the inorganic ions are rapidly transported upward and throughout the plant in the transpiration stream. Some ions move laterally from the xylem into surrounding tissues of the roots and stems, while others are transported into the leaves (Figure 31–19).

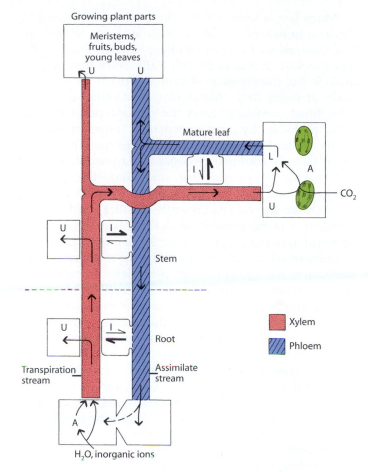

31–19

Diagram showing the circulation of water, inorganic ions, and assimilates in the plant. Water and inorganic ions taken up by the root move upward in the xylem in the transpiration stream. Some water and ions move laterally into tissues of the root and stem, while others are transported to growing plant parts and mature leaves. In the leaves, substantial amounts of water and inorganic ions are transferred to the phloem and are exported with sucrose in the assimilate stream. The growing plant parts, which are relatively ineffective in acquiring water through the transpiration stream, receive much of their nutrients and water via the phloem. Both water and solutes entering the roots in the phloem may be transferred to the xylem and recirculated in the transpiration stream. The symbol A designates sites specialized for the absorption and assimilation of raw materials from the environment. L and U designate sites of loading and unloading, respectively, and I designates principal points of interchange between the xylem and phloem.

Much less is known about the pathways followed by the ions in leaves than about those in roots. Within the leaf, the ions are transported along with the water in the leaf apoplast, that is, in the cell walls. Some ions may remain in the transpiration stream and reach the main regions of water loss—the stomata and other epidermal cells. Most eventually enter the protoplasts of the leaf cells, probably by carrier-mediated transport mechanisms similar to those in roots. The ions may then move symplastically (through protoplasts via plasmodesmata) to other parts of the leaf, including the phloem. Inorganic ions may also be absorbed in small amounts through the leaves; consequently, fertilization by direct application of micronutrients to the foliage has become standard agricultural practice for some crop plants.

Substantial amounts of the inorganic ions that are imported into the leaves through the xylem are exchanged with the phloem of the leaf veins and exported from the leaf together with sucrose in the assimilate stream of the phloem (Figure 31–19; see also the discussion of assimilate transport that follows). For instance, in a study of the annual white lupine (*Lupinus albus*), transport in the phloem accounted for more than 80 percent of the fruit's vascular intake of nitrogen and sulfur, and 70 to 80 percent of its phosphorus, potassium, magnesium, and zinc intake. The uptake of such inorganic ions by developing fruits is undoubtedly coupled to the flow of sucrose in the phloem.

Recycling may occur in the plant as nutrients reaching the roots in the descending assimilate stream of the phloem are transferred to the ascending transpiration stream of the xylem (Figure 31–19). Only those ions that can move in the phloem—*phloem-mobile ions*—can be ex-

ported from the leaves to any great extent. For example, K^+, Cl^-, and phosphate (HPO_4^{2-}) are readily exported from leaves, whereas Ca^{2+} is not. Solutes such as calcium, as well as boron and iron, are said to be *phloem-immobile*.

Assimilate Transport: Movement of Substances through the Phloem

Whereas water and inorganic solutes ascend the plant in the transpiration stream of the xylem, the sugars manufactured during photosynthesis move out of the leaf in the **assimilate stream** of the phloem (Figure 31–20). The sugars are transported not only to sites where they are used, such as growing shoot and root tips, but also to sites of storage, such as fruits, seeds, and the storage parenchyma of stems and roots (Figure 31–19).

Assimilate movement is said to follow a source-to-sink pattern. The principal **sources,** or exporters, of assimilate solutes are the photosynthesizing leaves, but storage tissues may also serve as important sources. All plant parts unable to meet their own nutritional needs may act as **sinks,** that is, importers of assimilates. Thus, storage tissues act as sinks when they are importing assimilates and as sources when they are exporting assimilates.

Source-sink relations may be relatively simple and direct, as in some young seedlings, where cotyledons containing reserve food often represent the major source and growing roots represent the major sink. In older plants, the upper, most recently formed mature leaves commonly export assimilates primarily toward the shoot

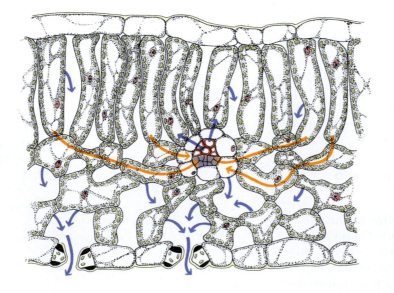

31–20

Diagram of leaf showing the pathways followed by water molecules of the transpiration stream as they move from the xylem of a minor vein to the mesophyll cells, evaporate from the surface of the mesophyll cell walls, and then diffuse out of the leaf through an open stoma (blue arrows).

Also shown are the pathways followed by sugar molecules manufactured during photosynthesis as they move from mesophyll cells to the phloem of the same vein and enter the assimilate stream. The sugar molecules manufactured in the palisade parenchyma cells are believed to move to the spongy parenchyma cells and then laterally to the phloem via the spongy cells (gold arrows).

(a)   (b)

31–21
Diagrams of assimilate transport in a plant in
(a) the vegetative stage and (b) the fruiting
stage. The arrows indicate the direction of
assimilate transport in each stage.

tip, the lower leaves export assimilates primarily to the roots, and those in between export assimilates in both directions (Figure 31–21a). This pattern of assimilate distribution is markedly altered during the change from vegetative to reproductive growth. Developing fruits are highly competitive sinks that monopolize assimilates from the nearest leaves and frequently from distant ones as well, often causing a marked decline of vegetative growth (Figure 31–21b).

Experiments with Radioactive Tracers Provide Evidence of Sugar Transport in Sieve Tubes

Early evidence supporting the role of the phloem in assimilate transport came from observations of trees from which a complete ring of bark had been removed. As noted in Chapter 27, the bark in older stems is composed largely of phloem. When a photosynthesizing tree is ringed, or "girdled," in this manner, the bark above the ring becomes swollen, indicating the formation of new wood and bark tissues stimulated by an accumulation of assimilates moving downward in the phloem from the photosynthesizing leaves (Figure 31–22).

Much convincing evidence of the role of the phloem in assimilate transport has been obtained with radioactive tracers. Experiments with radioactive assimilates (such as ^{14}C-labeled sucrose) have not only confirmed

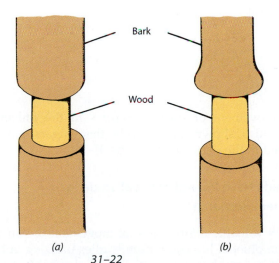

(a)   (b)

31–22
As early as the seventeenth century, Marcello Malpighi of Italy noticed that (a) when a ring of bark was removed from a stem (b) the tissues above the ring became swollen. He correctly interpreted this phenomenon as new growth of wood and bark tissues stimulated by an accumulation of food moving down from the leaves and intercepted at the ring. Malpighi studied the effect of ringing at different times of the year and found that no swelling occurred during the winter months.

31–23

*Sugar transport in sieve tubes. Microauto-radiographs of **(a)** transverse and **(b)** longitu-dinal sections of a vascular bundle from a stem of the broad bean (Vicia faba). A leaf of this stem was exposed to $^{14}CO_2$ for 35 min-utes. During that time, $^{14}CO_2$ was incorpo-rated into sugars, which were then transported to other parts of the plant. The leaf was sectioned, and the sections were placed in contact with autoradiographic film for 32 days. When the film was developed and compared with the underlying tissue sections, it was apparent that the radioactivity (visible as dark grains on the film) was confined almost entirely to the sieve tubes.*

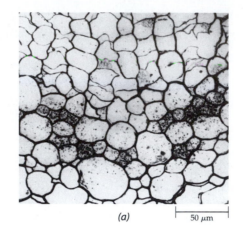

(a)   50 μm

(b)   50 μm

the movement of such substances in the phloem but have also shown conclusively that sugars are trans-ported in the sieve tubes (Figure 31–23).

Aphids Have Proved of Great Value in Phloem Research

Much valuable information on movement of substances in the phloem has come from studies utilizing aphids—small insects that suck the juices of plants. Most species of aphids are phloem-feeders. These aphids insert their modified mouthparts, or stylets, into a stem or leaf, ex-tending them until the tips of the stylets puncture a con-ducting sieve tube (Figure 31–24). The turgor pressure of the sieve tube then forces the sieve-tube sap through the aphid's digestive tract and out through its posterior end as droplets of "honeydew." If feeding aphids are anes-thetized (to prevent them from withdrawing their stylets from the sieve tube) and severed from their stylets, exu-dation of sieve-tube sap often continues from the cut stylets for many hours. The exudate can be collected

with a micropipette. Analyses of exudates obtained in this manner reveal that sieve-tube sap contains 10 to 25 percent dry matter, 90 percent or more of which is sugar—*mainly sucrose* in most plants. Low concentra-tions (less than 1 percent) of amino acids and other nitro-gen-containing substances are also present.

Data obtained from studies utilizing aphids and ra-dioactive tracers indicate that the rates of longitudinal movement of assimilates in the phloem are remarkably fast. For example, sieve-tube sap moved at a rate of about 100 centimeters per hour at the sites of the stylet tips.

Phloem Transport Is Driven by an Osmotically Generated Pressure Flow

Several mechanisms have been proposed to explain as-similate transport in the sieve tubes of the phloem. Probably the earliest explanation was that of diffusion, followed by that of cytoplasmic streaming. Normal dif-fusion and cytoplasmic streaming of the kind found in

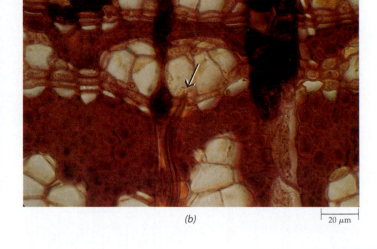

(a)

(b)   20 μm

31–24

(a) *Aphid* (Longistigma caryae) *feeding on a basswood* (Tilia ameri-cana) *stem. A droplet of "honeydew" can be seen emerging from the aphid.* **(b)** *A photomicrograph showing part of the modified mouthparts (stylets) of the aphid in a sieve tube of the secondary phloem of the basswood stem. An arrow points to the tips of the stylets.*

plant cells were largely abandoned as possible transport mechanisms when it became known that the velocities of assimilate transport (typically 50 to 100 centimeters per hour) are far too great for either of these phenomena to account for long-distance transport via sieve tubes.

Alternative hypotheses have been advanced to explain the mechanism of phloem transport, but only one, the pressure-flow hypothesis, satisfactorily accounts for practically all of the data obtained in experimental and structural studies on phloem.

Originally proposed in 1927 by the German plant physiologist Ernst Münch, and since modified, the pressure-flow hypothesis is the simplest and the most widely accepted explanation for long-distance assimilate transport in sieve tubes. It is the simplest explanation because it depends only on osmosis as the driving force for assimilate transport.

Briefly stated, the **pressure-flow hypothesis** asserts that assimilates are transported from sources to sinks along a gradient of turgor pressure developed osmotically. For example, sucrose produced in mesophyll cells by photosynthesizing leaves is secreted into the sieve tubes of the minor veins, resulting in higher sucrose concentrations in the sieve tubes than in the mesophyll cells (Figure 31–25). This active process, called **phloem loading,** decreases the water potential in the sieve tube and causes water entering the leaf in the transpiration stream to move into the sieve tube by osmosis. With the movement of water into the sieve tube at this source, the sucrose is carried passively by the water to a sink, such as

growing tissue or a storage root, where the sucrose is unloaded, or removed, from the sieve tube. The removal of sucrose results in an increased water potential in the sieve tube at the sink and subsequent movement of water out of the sieve tube there. The sucrose may be either used in growth or respiration or stored at the sink, but most of the water returns to the xylem and is recirculated in the transpiration stream.

Note that the pressure-flow hypothesis casts the sieve tubes in a passive role in the movement of the sugar solution through them. Active transport is also involved in the pressure-flow mechanism, but not directly in the long-distance transport through the sieve tubes. Instead, active transport is involved with loading and possibly unloading of sugars and other substances into and out of the sieve tubes at the sources and sinks.

Phloem Loading Can Be Apoplastic or Symplastic   The pathway followed by sucrose from the mesophyll cells to the sieve tube–companion cell complexes of the minor veins might be entirely symplastic—that is, from cell to cell via plasmodesmata—or, alternatively, the sucrose might enter the apoplast (cell walls) prior to being actively loaded into the sieve tube–companion cell complexes. The sieve tube–companion cell complexes of apoplastic loaders have virtually no plasmodesmatal connections with other cell types of the leaf (Figure 31–26a), whereas the companion cells of symplastic loaders are connected with the mesophyll symplast via numerous plasmodesmata.

Considerable evidence indicates that apoplastic loading of sucrose is driven by a proton gradient, which is generated by a proton pump at the expense of ATP (Figure 31–26b). The pump is a plasma-membrane-bound H^+-ATPase that utilizes the energy from the hydrolysis of ATP to transport protons (H^+) across the membrane. The proton gradient established by this primary active transport then is used by specific carrier molecules in the plasma membrane to couple the movement of protons back into the symplast to the transport of sucrose. This type of secondary active transport is known as **sucrose-proton cotransport,** or symport (page 83).

The metabolic energy required for apoplastic loading may be expended entirely by companion cells, rather than by the sieve tubes. It was assumed at one time that loading occurred across the plasma membrane of the companion cell, which then transferred the sugar to its associated sieve tube via the many plasmodesmatal connections in their common wall. It now appears, however, that some sieve tubes are capable of loading themselves, the site of active transport being their own plasma membranes. Whatever the case, the mature sieve tube probably is dependent upon its companion cell for much or most of its energy needs.

The mechanism of symplastic phloem loading is still a matter of debate. Interestingly, families of flowering plants exhibiting symplastic loading dominate in the tropics and subtropics, and apoplastic-loading families in the temperate and boreal zones. There are also strong indications that symplastic phloem loading is evolutionarily older than apoplastic loading.

Phloem loading is a selective process. As mentioned previously, sucrose is by far the most common sugar transported; in addition, all of the sugars found in sieve-tube sap are nonreducing sugars. Certain amino acids and ions also are selectively loaded into the phloem.

Phloem Unloading and Transport into Recipient Sink Cells Can Be Apoplastic or Symplastic   The process by which sugars and other assimilates transported in the phloem exit the sieve tubes of sink tissues is called **phloem unloading.** The transport events that immediately follow the unloading of assimilates are referred to as *post-*

31–25

Diagram of osmotically generated pressure-flow mechanism. Yellow dots represent sugar molecules. Sugar is actively loaded into the sieve tube at the source. With the increased concentration of sugar, the water potential is decreased, and water from the xylem enters the sieve tube by osmosis. Sugar is removed (unloaded) at the sink, and the sugar concentration falls; as a result, the water potential is increased, and water leaves the sieve tube. With the movement of water into the sieve tube at the source and out of it at the sink, the sugar molecules are carried passively by the water along the concentration gradient between source and sink. Note that the sieve tube between source and sink is bounded by a selectively permeable membrane, the plasma membrane. Consequently, water enters and leaves the sieve tube not only at the source and sink, but all along the pathway. Evidence indicates that few if any of the original water molecules entering the sieve tube at the source end up in the sink, because they are exchanged with other water molecules that enter the sieve tube from the phloem apoplast along the pathway.

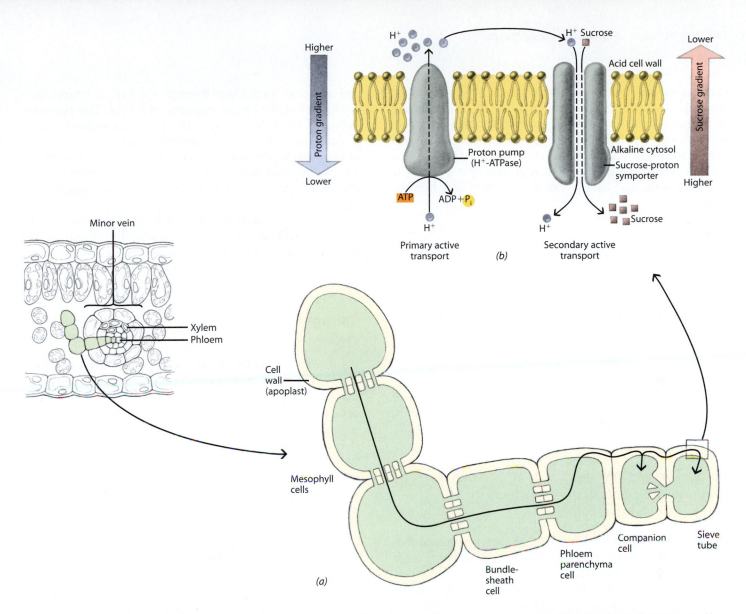

31–26

Phloem loading from the apoplast, or cell wall. (a) In apoplastic phloem-loading species, sucrose manufactured in the mesophyll cells of source leaves follows a symplastic pathway through the protoplasts via plasmodesmata to the near vicinity of the sieve tube–companion cell complexes. There

it exits the symplastic pathway and enters the apoplastic pathway. The sucrose then is actively loaded from the apoplast into the sieve tube–companion cell complexes, which have few or no plasmodesmatal connections with other cell types of the leaf. (b) The driving force for the transport of sucrose into the sieve tube–companion cell complex is

provided by a proton gradient generated by proton pumps (H$^+$-ATPase) in the plasma membrane. The proton gradient then is used by a specific carrier (sucrose-proton symporter) in the plasma membrane that couples sucrose transport to the diffusion of H$^+$ back into the cell.

phloem or *post-sieve-tube transport.*

In growing vegetative sinks, such as young leaves and roots, unloading and transport into sink cells are usually symplastic. In other sinks, unloading is apoplastic. Although the actual unloading process is probably passive, transport into sink tissues depends on metabolic activity. For example, in symplastic unloaders, energy is needed to maintain the concentration gradient between the sieve tubes and sink cells. In storage organs, such as sugarbeet roots and sugarcane stems, where apoplastic unloading takes place, energy is needed to accumulate sugars to high concentrations in the sink cells.

Summary

Most of the Water Transpired by a Vascular Plant Is Lost through the Stomata

Most of the water taken up by plant roots is lost to the air as water vapor. This process, called transpiration, is inextricably linked to the uptake by the leaf of CO_2, which is essential in photosynthesis.

A pair of guard cells can change their shape to bring about the opening and closing of the stoma, or pore.

Closing of stomata prevents the loss of water vapor from the leaf. Stomatal movements result from changes in turgor pressure within the guard cells, and the structure (radial micellation) of the guard cell walls. The turgor changes are closely correlated with changes in the solute level in the guard cells. A stoma opens when its guard cells become turgid and closes when they become flaccid.

A Number of Factors Affect Stomatal Movements

Water loss is the major influence affecting stomatal closure. During periods of water stress the abscisic acid (ABA) level increases in many plants, causing stomatal closure. Environmental factors also affect stomatal movement, including carbon dioxide concentration, light, and temperature. Temperature, humidity, and air currents also affect the rate of transpiration. In most species, stomata open in the light and close in the dark. In CAM plants the reverse is true.

Water Moves from the Roots to the Leaves through the Conduits—Vessels and Tracheids—of the Xylem

The current and widely accepted theory for the mechanism of water movement to the top of tall plants through the xylem is the cohesion-tension theory. According to this theory, water is pulled up through the plant body. This pull, or tension, is brought about by transpiration and/or by the use of water in the leaves, which results in a gradient of water potential from the leaves to the soil solution at the surface of the roots. It is the cohesiveness of water that permits it to withstand tension. Embolisms—the presence of air or water vapor—in xylem conduits are the bane of the cohesion-tension mechanism. Fortunately, the pit membranes of the bordered pit-pairs between adjacent tracheary elements usually prevent the passage of air from an embolized conduit into a functional one.

Uptake of Water by Roots Takes Place Largely through the Root Hairs

The root hairs provide an enormous surface area for water uptake. In some plants, the uptake of water from the soil results in the buildup of positive pressure, or root pressure, when transpiration is very slow or absent. This osmotic uptake depends on transport of inorganic ions from the soil into the xylem by the living cells of the root and can result in guttation, a process in which liq-

uid water is forced out through special structures (hydathodes) in the tips or margins of leaves. The pathway followed by water across roots may be apoplastic, symplastic, or transcellular; however, apoplastic movement is blocked at the endodermis by the Casparian strips. The water must pass through the plasma membrane and protoplast of the endodermal cells on its way to the xylem.

Many plants redistribute soil water by hydraulic lift. Hydraulic lift is a nighttime process by which plants transfer water taken up by deep roots located in moist soil regions to dry soil regions via shallower roots.

Inorganic Nutrients Become Available to Plants in Soil Solution in the Form of Ions

Plants employ metabolic energy to concentrate the ions they require. Ion transport from the soil to the vessels of the xylem requires two active, carrier-mediated events: uptake at the plasma membrane of epidermal cells and secretion into the vessels at the plasma membrane of parenchyma cells bordering the vessels. Inorganic ions follow a mostly symplastic pathway from the epidermis to the xylem. Substantial amounts of the inorganic ions that are imported into the leaves through the xylem are exchanged with the phloem of the leaf veins and exported from the leaf in the assimilate stream. Nutrient uptake from the soil by most seed plants is greatly enhanced by mycorrhizal fungi.

Assimilate Movement in the Phloem Is from Source to Sink

Research on movement of substances in the phloem has been greatly aided by the use of aphids and radioactive tracers. Sieve-tube sap contains sugar (mainly sucrose), small quantities of nitrogenous substances, and phloem-mobile ions. Rates of longitudinal movement of substances in the phloem greatly exceed the normal rate of diffusion of sucrose in water—the rates typically range from 50 to 100 centimeters per hour.

According to the pressure-flow hypothesis, assimilates move from sources to sinks along turgor pressure gradients developed osmotically. Sugars are actively loaded into the sieve tube–companion cell complexes at a source. This decreases the water potential in the sieve tube and causes water to move into the sieve tube by osmosis. Meanwhile, the removal of sugar at the sink increases the water potential of the sieve tube there. With the movement of water into the sieve tube at the source and out of it at the sink, the sugar molecules are carried passively by water along the concentration gradient from source to sink.

Selected Key Terms

apoplastic loading
p. 767

apoplastic pathway
p. 759

assimilate stream   p. 764

cohesion-tension theory
p. 756

cuticle   p. 752

guttation   p. 761

hydathode   p. 761

hydraulic lift   p. 761

phloem loading   p. 767

phloem unloading
p. 768

pressure-flow hypothesis
p. 767

radial micellation   p. 753

root pressure   p. 760

sinks   p. 764

sources   p. 764

**sucrose-proton cotransport
(symport)**   p. 768

symplastic loading
p. 767

symplastic pathway
p. 759

transcellular pathway
p. 759

transpiration   p. 751

transpiration stream
p. 761

Questions

1. Explain how each of the following factors or signals affects stomatal movement: carbon dioxide concentration, light, temperature.

2. Explain how each of the following factors affects the rate of transpiration: temperature, humidity, air currents.

3. By means of a simple diagram, outline the pathway followed by a water molecule in the transpiration stream, beginning with a root hair and ending in the atmosphere outside a leaf. Label all pertinent tissues and cell layers along the way.

4. Pit membranes are very important to the safety of water transport. Explain.

5. What evidence supports the cohesion-tension theory?

6. Explain the relationship between root pressure and guttation.

7. How does hydraulic lift benefit plants growing in the neighborhood of a plant exhibiting this phenomenon?

8. Distinguish between the symplast and the apoplast.

9. The uptake of inorganic nutrients is an energy-requiring process. Explain.

10. What evidence supports the role of sieve tubes as the food-conducting conduits of the phloem?

Ecology

SECTION 7

Rose hips, the fruit of the rose, are rich in vitamin C and have been used since ancient times to prepare medicinal distillations. Pliny the Elder (A.D. 23–79) wrote "Rose juice is used for the ears, sores in the mouth, the gums, as a gargle for the tonsils, for the stomach, uterus, rectal trouble, headache."

The use of rose hips appears to go back even farther in our history. Recent excavations of cave dwellings dated at over 300,000 years ago have revealed that Peking man, consisting of races of the extinct species closest to modern humans, gathered rose hips, as well as walnuts, hazelnuts, and pine nuts.

Chapter 32

The Dynamics of Communities and Ecosystems

OVERVIEW

Ecology—the study of the interactions of organisms with each other and with their physical environment—deals with some of the most familiar, yet most mysterious and unpredictable aspects of the world around us. For example, when we see a flower in a garden or a tree in a park we are largely unaware of the myriad ways in which the plant's growth is influenced by a variety of organisms such as bacteria, fungi, small and large animals, and other plants. Plants are affected not only by familiar environmental conditions such as sunlight, temperature, and rainfall, but also by subtle factors such as soil pH and the levels of macronutrients and micronutrients. To complicate matters, organisms in the plant's environment may exert their influence indirectly by interacting with nonliving components in ways that affect plant growth. For example, bacteria may alter the soil pH, which in turn influences the plant.

One goal of ecology is to describe these complex interactions in sufficient detail that we can predict what would happen if humans altered a particular environment such as by cutting or burning a forest, introducing new species, or adding pesticides. Making accurate predictions has been extremely difficult in the past because subtle interactions are often overlooked.

This chapter provides information about the basic structure and function of ecological systems—ecosystems, for short. Topics discussed are the types of organisms in ecosystems and the ways in which they interact, nutrient cycling, the transfer of energy through food chains, and the changes in ecosystems over time.

CHECKPOINTS

By the time you finish reading this chapter, you should be able to answer the following questions:

1. What does the science of ecology encompass, and what is the difference between a population, a community, and an ecosystem?

2. What is mutualism?

3. What do ecologists mean by the term "competition," and how do plants compete with one another?

4. How do plants defend themselves against herbivores and pathogens?

5. What is a food chain, what types of organisms are found in each link of the chain, and how does energy flow through it?

Ecology, by definition, is the study of organisms and their interactions with each other and with all the components of their environment. The questions ecologists ask are among the most fundamental in biology. Why, for example, can particular plants and animals be found living together in one area and not in others? Why are there so many of one kind of organism and so few of another? And, how do **ecosystems**—that is, all of the organisms that occur together at a particular place, as well as the environment with which they interact—respond to human disturbance, and how might we minimize our impact on ecosystems and ultimately all life on Earth?

At the individual level, ecologists ask how the individual organism functions in and interacts with its environment. Individuals exist in **populations,** groups of individuals, usually of one species, that occur together at a given place. At the population level, ecologists ask what determines the abundance and distribution of individuals of a particular species.

Communities, which consist of populations, include the groups of plants, animals, and other organisms that live in a particular area. Although we can speak of an herb community or a community of vertebrate animals, the word "community" when unmodified is generally taken to mean *all* of the organisms that occur together at a particular place. An ecosystem includes not only the diverse organisms within the community, but also their physical environment and their interactions, both with other organisms and with the environment.

As we have seen throughout this text, the thread that links together the enormous diversity of the living world is evolution, a process of change through time. Evolution is inextricably intertwined with ecology in what G. E. Hutchinson of Yale University aptly called "the ecological theater and the evolutionary play." In previous chapters, we have considered the fundamental points of evolutionary theory and the major categories of evidence, as well as the mechanisms by which the plot of the evolutionary play moves forward. Here and in the following chapters, we shall place ourselves in the ecological theater to observe the actors—that is, all living things—in the variety of interactions with each other and the physical environment that constitute the play.

Interactions between Organisms

Interaction is central to the field of ecology. No organism living in a community—whether that community be a patch of woodland, a pasture, a pond, or a coral reef—exists in isolation. Each organism participates in a number of interactions, both with other organisms and with the nonliving components of the environment. We shall organize our discussion around three major kinds of interactions between species of organisms: mutualism, competition, and plant-herbivore (and plant-pathogen) interactions.

Mutualism Is an Interaction in Which Both Species Benefit

Mutualism is a biological interaction in which the growth, survival, and/or reproduction of both interacting species are enhanced. In many mutualisms, neither partner can survive without the other, particularly when competition from other plants and predation are taken into account. We have already come across several examples of mutualism in earlier chapters. Lichens are an outstanding example (see Chapter 15). Others are the relationship between legumes and the nitrogen-fixing bacteria that live in nodules on their roots (see Chapter 30) and some of the close pollination relationships discussed in Chapter 22, such as that between the yucca moth and the yucca plant. Two additional examples of mutualism are mycorrhizae and ant-acacia interactions.

Mycorrhizae Are Associations between Roots and Fungi
One of the most interesting and ecologically significant examples of mutualism concerns the interaction between fungi and plants. As discussed in Chapters 15 and 30, the roots of most vascular plants are associated with fungi, forming compound structures known as mycorrhizae. These fungi play a vital role in the absorption of phosphorus and other essential nutrients required by the plants. Without the fungi, the normal growth of the plants would be impossible. Mycorrhizae also appear to have played a crucial role in the establishment of early plants on land.

In many vascular plants, nonmycorrhizal individuals are rarely encountered under natural conditions, even though growth may be possible without fungi if nutrients are abundant. Most vascular plants are dual organisms in the same sense that lichens are dual organisms, although the relationship is not obvious above ground. As University of Wisconsin soil scientist S. A. Wilde stated, "A tree removed from the soil is only a part of the whole plant, a part surgically separated from its…absorptive and digestive organ."

The fungi that form mycorrhizal associations in most plants are zygomycetes. As discussed in Chapter 15, the associations are called endomycorrhizae, and they are characteristic of a majority of species of herbs, shrubs, and trees. In some groups of conifers and eudicots—mainly trees—the associations are mostly with basidiomycetes but also with certain ascomycetes; such associations are called ectomycorrhizae. Some of these associations are highly specific, with one species of fungus forming ectomycorrhizal associations only with a particular species or a group of related species of vascular plants. For example, the pore fungus *Boletus elegans* is known to associate only with larch *(Larix)*, a conifer. Other fungi, such as *Cenococcum geophilum,* have been discovered living in ectomycorrhizal association with forest trees of more than a dozen genera. Ecto-mycorrhizae are particularly characteristic of relatively pure stands of trees growing at high latitudes in the Northern Hemisphere or at high elevations, two places where slow decomposition rates may make soil nutrients particularly difficult to obtain.

Acacia Trees and Ants Interact in a Mutually Beneficial Way

The most intricate examples of mutualism occur in the tropics, where many more kinds of organisms are found than in temperate regions. For example, trees and shrubs of the genus *Acacia* occur widely in tropical and subtropical regions. The association between certain species of *Acacia* in the lowlands of Mexico and Central America and the ants that inhabit their thorns (actually, stipular "thorns," or spines) is a remarkably complex plant-animal interaction. In particular, the relationship between the bull's-horn acacias *(Acacia cornigera)* and ants of the genus *Pseudomyrmex* has been well documented (Figure 32–1).

The bull's-horn acacias have a pair of greatly swollen thorns more than 2 centimeters long at the base of each leaf. Nectaries occur on the petioles, and small nutritive structures known as Beltian bodies are located at the tip of each leaflet. The ants live inside the hollow thorns, obtaining sugars by drinking nectar and fats and proteins by eating the Beltian bodies. The acacias grow extremely rapidly and are particularly common in disturbed areas, where the competition between rapidly growing colonizing plants is often intense.

Thomas Belt first described the relationship between *Pseudomyrmex* and the bull's-horn acacias in his book, *The Naturalist in Nicaragua* (1874). Following his observations, there was a prolonged controversy about whether the presence of the ants actually benefited the acacia plants. This question was finally and definitively solved in 1964 by Daniel Janzen. Janzen found that the worker ants, which swarm over the surface of the plant, bite and sting animals of all sizes that contact the plant, thus protecting it from the activities of herbivores and ensuring a home for themselves. Moreover, whenever the branches of another plant touch an inhabited acacia tree, the ants girdle the other plant's bark, destroying the invading branches and producing a tunnel to the light through the dense surrounding tropical vegetation.

When Janzen removed the ants from a plant by poisoning them or clipping off the portions of the plant that contained ants, the plant grew very slowly and usually died within a few months as a result of insect damage and shading by other plants. On the other hand, plants inhabited by ants grew very rapidly, soon reaching 6 meters or more in height and overtopping the other second-growth vegetation. Ants of the genus *Pseudomyrmex*

(a)

(b)

32–1

Ants and acacias. (a) A worker ant (Pseudomyrmex ferruginea) *drinking from a nectary of a bull's-horn acacia* (Acacia cornigera). *At the right is an entrance hole into a thorn cut out by the queen ant. After hollowing out the thorn, the queen raises her brood within it. (b) Worker ants collecting Beltian bodies from the tips of acacia leaflets. Rich in protein and oils, the Beltian bodies are an important food source for both adult and larval ants. The ants kill other insects that attempt to feed on the acacia and girdle plants that come into contact with it.*

make their nests only in these particular acacias and are completely dependent on the acacias' nectaries and Beltian bodies for food. Thus the ant-acacia system is as much a dual biological entity as, for example, a lichen. One element usually cannot survive without the other in the community in which it occurs.

Competition Results When Organisms Require the Same Limited Resource

Competition is defined as interaction between members of the same population or of two or more populations in order to obtain a mutually required resource available in limited supply. In competition, one species may interfere with another species enough to keep the second species from gaining access to the resource.

Growth Rate Is an Important Factor Affecting Competition among Plants   Unlike animals, which ingest food, green plants are dependent on a single process, photosynthesis, to obtain their energy. Competition in plants is therefore manifested largely in terms of a "struggle for light." Plants whose photosynthetic rate, growth form, and pattern of allocation of energy to leaves, roots, and stems result in the highest rate of growth in a particular environment will often have an advantage in this struggle. Plants with a lower growth rate must either be able to photosynthesize at low light intensities or have a high growth rate relative to competitors elsewhere. If not, these slow growers will die. The evolution of plants with C_4 and CAM photosynthesis (pages 147–149) is a case in point. The physiological properties of such species enable them to dominate climatic areas where their C_3 ancestors would have either perished or else been competitively excluded by the more rapidly growing C_4 or CAM species.

Differences in height, leaf arrangement, crown shape, and allocation of energy to roots versus leaves affect a plant's growth rate relative to potential competitors. Different combinations of these traits are likely to maximize growth in different environments; no single combination can produce the best competitor in all environments. As a consequence, vegetation in different climatic areas is dominated by plants with different growth forms. Within a given area—indeed, within a single community, such as a temperate forest or prairie—differences in growth form, photosynthetic physiology, and energy allocation may permit species to coexist by granting each a competitive advantage in different microenvironments.

The ways in which individual plants are able to enhance their overall growth and thus compete for light, water, and mineral nutrients largely determine their success in different habitats. An understanding of such factors is therefore the key to increasing yield in extreme environments, such as those that are relatively poor in nutrients or water or are heavily shaded. These relationships are therefore important in agriculture. Knowledge of these factors will also provide the information necessary to predict the performance of individual plant species and communities in a world that is rapidly changing due to global warming, which is resulting from increased levels of CO_2 and other atmospheric gases produced largely by human activities.

The Principle of Competitive Exclusion Provides a Baseline for Studying Competition   Competition between plant species can be visualized more easily under experimental conditions than in nature. From observations of the growth of organisms in simple environments, it has been deduced that two species with similar environmental requirements cannot coexist indefinitely in the same habitat. This is a simplified version of what Garrett Hardin has called the **principle of competitive exclusion**. A classic study of competition was conducted on two species of duckweed, *Lemna polyrhiza* and *Lemna gibba*. When grown in individual pure cultures, *L. gibba* always grew more slowly than *L. polyrhiza*, yet when the two species were grown together *L. polyrhiza* was always replaced by *L. gibba*. The plant bodies of *L. gibba* have air-filled sacs that enable them to form a floating mass over

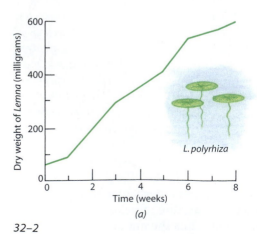

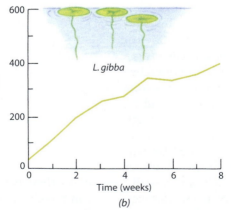

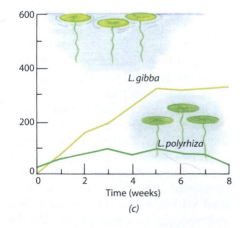

32–2

An experiment with two species of floating duckweed, tiny angiosperms found in ponds and lakes. One species, Lemna polyrhiza (a),   *grows more rapidly in pure culture than the other species, Lemna gibba (b). But L. gibba has tiny air-filled sacs that float it on the*   *surface. When the two species are grown together, L. gibba shades L. polyrhiza and is the victor in the competition for light (c).*

the other species, cutting off the light. As a consequence, in mixed cultures the shaded *L. polyrhiza* died out (Figure 32–2).

If the environment is complex, as it often is in nature, various organisms may relate to it in different ways, in effect subdividing the environment and achieving competitive success in different microenvironments. These various organisms may then continue to coexist indefinitely.

For example, Engelmann spruce *(Picea engelmannii)* and subalpine fir *(Abies lasiocarpa)* coexist and form the dominant tree community in the subalpine zone of the central and northern Rocky Mountains. Careful field measurements suggest that these two tree species are maintained as codominants in this area. The greater longevity and size of the spruce are balanced by the faster growth in height of firs and more flexible requirements for establishing fir seedlings. Spruce seedlings are found primarily in forest gaps or associated with a canopy of firs, whereas fir seedlings are found more commonly in the forest. Fir seedlings outcompete spruce in shaded understories by the simple fact that they survive. Spruce seedlings, however, outcompete fir in sunnier sites by virtue of higher growth rates there and lower sensitivity to drought. Constant disturbance from storms, flooding, and avalanches, among other factors, and the relatively short life span of fir prevent it from taking over the entire landscape. Patterns of this sort are frequent in different plant communities, but they are often not so obvious. The different requirements of the species involved remove them from direct competition and allow them to coexist indefinitely.

If populations of coexisting species are kept at low levels, competitive exclusion may be avoided even though the species do, in principle, compete for the same limiting resources. Such a situation occurred in the chalk grasslands of England when the grasses were kept closely cropped by rabbits. As a result, many kinds of flowering plants were able to grow in this habitat. Diversity is enhanced in such situations when competitive dominant species—grasses and other tall herbs in this case—are attacked preferentially. This allowed many other kinds of shorter grasses and other herbs to coexist. However, the situation changed drastically earlier in this century, when a severe epidemic of myxomatosis, a disease caused by a virus, drastically reduced the population of rabbits. After this decline, the grass cover of the chalk soils became deeper and more dense, and many of the formerly abundant species of flowering plants became rare. Similar effects are often seen when comparing grazed and ungrazed pastures or grasslands, and they also arise in areas where natural disasters, such as hurricanes, are common. Thus, as we might expect, the diversity of species is greater along a wave-lashed coast, where disturbance is continuous, than in a more stable environment.

Clonal Reproduction May Complicate Studies of Competition Reproduction of genetically identical individuals, or clonal reproduction, which is important in many plants, sometimes makes it difficult to define the limits of individual plants in nature. This, in turn, may make it difficult to evaluate the competitive interactions of specific genotypes within a species. Genetically identical individuals may be widely dispersed but still occur in widely separated but similar environments within a single complex habitat. This is characteristic not only of rhizomatous perennials, such as many grasses and sedges, but also of dandelions and other plants in which the seeds are produced asexually and contain embryos that are genetically identical to their parents (Figure 32–3). It is characteristic also of those plants, such as white clover *(Trifolium repens)*, in which an original plant simply grows apart into distinct individuals.

32–3

One of the most widely distributed of all the flowering plants is the dandelion. Its flowers produce enormous numbers of seeds asexually by the process of apomixis (page 255). The seeds are enclosed in fruits bearing a plumelike structure that is caught by the wind, thereby aiding dispersal. As a result, genetically identical individuals may occur in widely separated environments, making it difficult to define the limits of individual plants. There are far fewer genetically distinct individuals in the population than if dandelions reproduced sexually.

Competing for Light

Forest herbs show a remarkable diversity in leaf height and arrangement. Some species, such as rattlesnake plantain (a), bear foliage at or near ground level. Others, such as trillium (see Figures 1–6b and 33–19a) and bloodroot, hold their leaves 20 to 50 centimeters above the ground. Yet others, such as jewelweed, spread their crowns a meter or more above the ground. Most forest herbs have umbrella-like canopies, but some have several layers of leaves along an erect stem, and a few have tightly packed foliage along an arching stem. What is the significance of these variations, and how are they related to the distribution of individual species?

Recent studies by Thomas Givnish, of the University of Wisconsin at Madison, have shed light on some of these issues. According to his analysis, leaf height is likely to affect an herb's growth relative to its competitors—and hence its competitive ability—primarily through its effect on the way the herb captures light and allocates energy. Taller herbs, which capture more light than shorter ones, spend more of their annual energy income from photosynthesis to build unproductive stems at the expense of productive leaves. Herbs should evolve taller and taller canopies until the growth advantage each obtains from increased height is balanced by the growth loss caused by reduced energy allocation to leaves. In sparsely covered areas, there is little advantage to growing tall and "wasting" a large amount of energy on stems, and so short plants are favored because they are better able to capture light. Indeed, there is evidence to support these predictions. Tall species—such as jewelweed, stinging nettles, and many asters and goldenrods (b)—are usually found in dense stands under partially open canopies, particularly on moist, fertile floodplains. Short species—such as rattlesnake plantain and striped wintergreen—are found in areas of sparse herb cover, especially in dry, infertile oak forests.

An intriguing variation among forest herbs involves an arching growth form. Species with leaves packed along arching stems are less efficient than umbrella-like herbs in that they require more stem tissue to hold the same amount of leaf mass a given height above level ground. On steep slopes, however, such plants gain an advantage because they orient downhill; as a result, erect plants rooted downslope require longer stems to hold their leaves in the same position. Again, as predicted, in virgin forests, dominance by arching herbs increases dramatically on slopes steeper than approximately 20 degrees.

These findings illustrate the general principle that different environments favor different growth forms and patterns of energy allocation: adaptation and competitive success are strongly dependent on the conditions in which the plant grows.

(a)

(b)

Some Organisms Produce Chemicals That Inhibit the Growth of Others   Depletion of a shared resource, such as light or water, is not the only mechanism by which plants may compete. In some competitive interactions, one (or both) of the competing organisms produces chemical substances that inhibit either the growth of members of its own species, resulting in increased spacing of like individuals, or the growth of other species. For example, the fungus *Penicillium chrysogenum*, which grows on organic substrates such as seeds, produces significant quantities of penicillin in nature. Penicillin inhibits the growth of gram-positive bacteria, which might otherwise compete directly with the fungus for the same nutrients. However, bacteria that produce penicillinases (enzymes that break down penicillin), such as *Bacillus cereus*, often replace the *Penicillium*.

Analogous relationships among plants are grouped under the general heading of **allelopathy.** An excellent example of this phenomenon is provided by the effects of black walnuts (*Juglans nigra*) on other plants, which are very sparse under the walnut trees. Tomatoes (*Solanum lycopersicum*) and alfalfa (*Medicago sativa*) wilt when grown near black walnuts, and their seedlings die if their roots contact walnut roots. Similarly, white pine (*Pinus strobus*) and black locust (*Robinia pseudoacacia*) are often killed by black walnuts growing in their vicinity. This is especially true in poorly drained soils, where the toxic substances leached from the walnuts apparently accumulate. The bare zones around sage (*Salvia leucophylla*) shrubs in California (Figure 32–4) may also occur in part because of allelopathic effects, although the birds and rodents that the shrubs shelter exert an important influence on the patterns observed.

Allelopathic effects are likewise being applied in agriculture. For example, a strip planted with sorghum will have two to four times fewer weeds the following year than other strips. The sorghum plants evidently leave behind allelopathic compounds in the soil that depress the growth of weeds.

Plant-Herbivore and Plant-Pathogen Interactions Involve a Variety of Defense Mechanisms

We have already considered one important area of plant-herbivore interactions involving flowering plants—the relationships between flowers and their visitors, and between fruits and their dispersers (see Chapter 22). These specialized interactions arose during the course of evolution from the more general relationships between plants and the animals that consume them. Relationships between plants and pathogenic organisms, especially fungi and bacteria, are similar in their effects.

Knowledge of plant-herbivore interactions may have important applications in determining the structure of natural communities. For example, vast areas of Australia were at one time covered with spiny clumps of prickly-pear cactus (*Opuntia*), a plant that was introduced from Latin America. Fertile lands became useless for grazing, and the economy of great stretches of the interior was severely threatened. Today, the cactus has been nearly eliminated by a cactus moth (*Cactoblastis cactorum*) discovered in South America and deliberately introduced into Australia to control the cactus. The larvae of this moth destroy the cactus plants by eating them. The moth, once abundant in Australia, can scarcely be found today, even by a careful inspection of the few remaining cactus clumps; yet there is no doubt that it continues to exert a controlling influence over the populations of this plant (Figure 32–5).

Plants Produce Toxic Chemicals in Response to Herbivores
The effects of herbivores on plants are profound, both in the short term and in the long term. Herbivores control the reproductive potential of plants by destroying their photosynthetic surfaces, their food-storage organs, or their reproductive structures. As discussed in Chapter 22, these interactions have led, over the course of time, to the evolution by plants of a wide variety of chemical defenses—in the form of molecules commonly referred

32–4

Purple sage (Salvia leucophylla) *shrubs produce terpenes that evaporate and spread through the air, ultimately reaching the soil, where they inhibit the growth of other plants. In this aerial view, the bare zone around the individual Salvia shrubs and colonies can be seen plainly. Immediately around the shrubs is a completely bare zone, and then a zone of inhibited grassland with a few stunted annual herbs.*

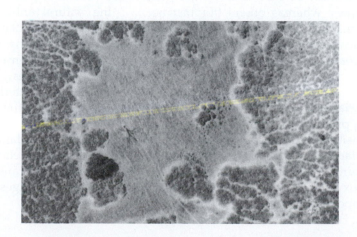

32–5
A plant-herbivore interaction. (a) Dense prickly-pear cactus (Opuntia inermis) *growing on a pasture in Queensland, Australia, in November 1926; (b) the same pasture in October 1929, after the cacti were destroyed by the deliberately introduced South American moth* Cactoblastis cactorum. *First introduced in May 1925, the larvae of this moth destroyed the cacti on more than 120 million hectares of rangeland.*

(a)   (b)

to as secondary metabolites. The ability of plants to produce toxic chemicals and to retain them in their tissues gives the plants a tremendous competitive advantage. Indeed, these chemicals are apparently the most important factors in controlling herbivorous insects in nature. This advantage is analogous to the advantage achieved by the production of thorns or tough, leathery leaves, which obviously protect plants from grazing. Scientists working to improve the resistance of crops to herbivores are focusing much of their effort on these chemicals.

In the sea, many seaweeds have evolved comparable defenses. These marine algae are consumed by many different kinds of herbivores, including fishes, sea urchins, mollusks, and other animals; nearly all of the biomass is ultimately consumed in some habitats. To escape these herbivores, some seaweeds grow in cracks and holes or in other habitats that the herbivores do not visit. Palatable seaweeds may gain protection by growing intermixed with chemically protected ones. Many seaweeds produce chemical defenses that render them distasteful to herbivores, whereas others (for instance, coralline red algae) may be too tough to consume. In other words, the array of defenses against herbivores that is employed by seaweeds is virtually as extensive as that which occurs among flowering plants on land.

Plant-Herbivore and Plant-Pathogen Interactions May Be Quite Complex

Pea plants (*Pisum sativum*) are largely protected from parasitic fungi by a substance called pisatin, which the plants produce. Many strains of the important parasitic fungus *Fusarium*, however, have enzymes called monooxygenases, which convert pisatin into a less toxic compound. These fungi then have the ability to attack peas. Humans also utilize monooxygen-

ases to detoxify certain chemicals that would otherwise be harmful to the body. In such ways, "chemical warfare" between plants and their herbivores is continuously being waged.

The protective chemicals that plants produce are often not only distasteful but may display still other features that deter herbivores. Chromenes, for example, can interfere with insect juvenile hormone (essential to an insect's life cycle) and thus can act as true insecticides. A Mexican sneezeweed (*Helenium* sp.) produces helanalin, which functions as a powerful insect repellent. Pyrethrum is another natural insecticide, which is produced commercially from a species of *Chrysanthemum*. Even the waxy surfaces of leaves, which are difficult to digest, may be important in retarding attacks by insects and fungi (see Figure 2–12).

Phytoalexins Are Molecules Produced in Response to Invasion by Microorganisms

When infected with fungi or bacteria, plants often defend themselves by producing natural antibiotics called **phytoalexins** (page 32). These are lipidlike compounds whose synthesis can also be stimulated by leaf damage. They appear to be produced in response to the presence of specific carbohydrate molecules, called **elicitors,** that are present in fungal and bacterial cell walls. The elicitors, which are released from the fungal or bacterial cell walls by enzymes present in the plants being attacked, diffuse through the plant cells more or less like hormones. Ultimately the elicitors bind to specific receptors on the plasma membranes of the plant cells, bringing about metabolic changes that result in the production of the phytoalexins.

Pesticides and Ecosystems

Approximately half a billion metric tons of pesticides and herbicides are produced annually for application to crops in the United States alone. Of this enormous total, it has been estimated that only approximately 1 percent actually reaches the target organisms. Most of the remainder either reaches soil, water, or nontarget organisms in the same ecosystem or spreads into neighboring ecosystems, where it may have important detrimental effects. This residue may result in the elimination of certain species, for example, and the loss of those species may affect the functioning of the ecosystem as a whole or lead indirectly to the elimination of other species. The abundance of important decomposers, such as earthworms and other soil organisms, may be greatly reduced by pesticides and herbicides, so that the ecosystem as a whole ceases to function normally. The intensity of such effects depends on the toxicity of the chemicals and on their persistence in the environment.

One of the problems associated with some chemical pollutants is that they tend to be concentrated as they pass up through food chains, reaching their highest concentrations in the top predators (a). For example, chlorinated hydrocarbons, such as DDT (now outlawed in the United States for use but not for manufacture and export, unfortunately, and not outlawed in many other industrialized countries), become concentrated in the tissues of predatory birds and cause the shells of their eggs to be abnormally thin. The shell of such an egg is likely to break before the chick is ready to hatch, causing its death (b). In addition, many strains of insects, bacteria, and fungi have evolved that are resistant to the pesticides designed to control them. Of the estimated 2000 species of major insect pests, about a quarter have already evolved strains that are resistant to one or more insecticides. Similarly, a number of species of weeds have evolved resistance to herbicides.

Other effects are less direct. For example, predator species that naturally control populations of pest species may be eliminated by pesticide poisoning, leading to outbreaks of the very pest organisms that the chemicals were employed to control in the first place. As with most human activities, the net effect of the application of pesticides and herbicides is to lessen the diversity of the ecosystem affected. Yet productive modern agriculture depends, to a large extent, on the application of these useful substances.

Because the effects of pesticides and herbicides are often so drastic, however, scientists are actively searching for less damaging methods of improving agricultural yields. These include selective breeding to produce crops that are resistant to pests (the methods of genetic engineering described in Chapter 28 will be especially useful in achieving this goal), increased research to develop pesticides and herbicides that are less toxic and less persistent than those currently used, and integrated pest-management systems, involving combinations of control measures, including the encouragement of predators and diseases of pest species as well as the judicious application of pesticides.

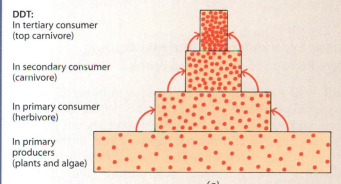

DDT:
In tertiary consumer
(top carnivore)

In secondary consumer
(carnivore)

In primary consumer
(herbivore)

In primary producers
(plants and algae)

(a)

(a) Concentration of DDT residues being passed along a simple food chain. The concentration of DDT increases as the material passes up the chain, and high concentrations occur in the carnivores.

(b)

(b) This dead peregrine falcon embryo, its development almost completed, was found with two broken, infertile eggs in a nest in southwestern Scotland in May of 1971. It is not known if the developing falcon died as a direct consequence of the high levels of DDT residue in its body or as a result of the collapse of its shell. Birds at the top of food chains, such as the peregrine falcon, the osprey, and the bald eagle, were the principal victims. With the banning of DDT in many countries, dramatic recoveries have been observed in populations of all three species.

In principle, it should be possible to spray elicitors onto crops before the crops become infected and thereby protect them from fungal and bacterial pests. Such a process would be analogous to vaccination in humans and domestic animals. One possible problem, however, is that the energy cost to the plant of producing large quantities of phytoalexins might lower the plant's ultimate yield more than would the fungal or bacterial infection. A natural advantage of the phytoalexins as defensive substances is that the plant does not need to expend the energy necessary to produce them unless it is actually attacked. Nevertheless, understanding phytoalexin production in plants is of considerable importance for crop protection, and several synthetic elicitors have already been produced and tested. Manipulating the genetic basis of resistance, now possible through the methods of genetic engineering (described in Chapter 28), also offers new possibilities for enhancing crop resistance that do not carry a high energy cost.

Tannins Provide a Static Chemical Defense   Just as some plants produce phytoalexins, others produce tannins and other phenolic compounds (pages 34–37), and these compounds seem to play a similar role in nature. Tannins are generally static defenses, always present in the plant parts where they occur. In some instances, however, they may be marshaled by the plant when it is attacked. For example, when gypsy moths (*Lymantria dispar*) attack and defoliate oak trees (*Quercus* spp.), the trees produce new leaves that are much higher in tannins and other phenolic compounds than normal. The new leaves produced under such conditions are also tougher, and they contain less water than those they replace. Indeed, the differences are great enough that larvae feeding on the new leaves experience reduced growth, and further outbreaks of gypsy moths are diminished in intensity. The tannins apparently interfere with digestion in the insects by combining with plant proteins, making them indigestible. Similar effects may be common in other plants as well. For example, when snowshoe hares heavily browse some trees and shrubs, such as paper birch (*Betula papyrifera*), these plants produce new shoots that are much richer in resins and phenolic compounds than the earlier shoots.

The secondary metabolites ingested by herbivores may, in turn, play a role in the animals' ecological relationships with other animals. For example, some insects, such as the monarch butterfly, store these poisons within their tissues and are thereby protected from their predators (see Figure 2–28). In addition, some sex attractants in insects are derived from the plants on which they feed.

Viewed as a whole, the relationships within a community are incredibly complex. Organisms that coexist within a community often have evolved together. Within the community, they affect one another in an endless variety of ways, a few of which scientists are just beginning to understand.

Nutrient Cycling

Ecosystems have the property of regulating the flow of energy, originally derived from the sun, and regulating the cycling of nutrients. Aspects of these properties will be explored in the remaining pages of this chapter.

In terms of its nutrient supply, an ecosystem is more or less self-sustaining. One of the more important reasons for this autonomy is the continuous cycling of chemical elements between organisms and the environment. The pathways of some of these essential elements, known as nutrient cycles, were discussed in Chapter 30. Ideally, nothing is lost, so that the pool of nutrients is continually renewed and continually available for the growth of organisms. The rate of flow from the nonliving pool to the organisms and back again, the amount of available material in the nonliving pool, and the form of this pool differ among nutrients and from habitat to habitat.

By considering the interaction of organisms, it is easy to see that competition for nutrients must often occur. In some cases, this competition has become so intense that organisms benefit from a mutualistic interaction, such as the formation of mycorrhizae, in which one partner assumes the responsibility of providing nutrients for the other.

Classic Experiments on Nutrient Recycling Were Performed at Hubbard Brook

Studies of a deciduous forest ecosystem in the Hubbard Brook Experimental Forest of the White Mountain National Forest of New Hampshire have shown that the plants of this community play a major role in the retention of nutrient elements. The investigators first established a procedure for determining the mineral budget—input and output, or "gain" and "loss"—of different areas in the forest. By analyzing the nutrient content of rain and snow, they were able to estimate atmospheric input, and by constructing concrete weirs that channeled the water flowing out of selected areas, they were able to calculate output (Figure 32–6). A particular advantage of this site was that a granitic bedrock is present just below the soil surface, so that very little material leaches downward; that is, the soil water percolates only a short distance. In addition, few nutrients are added by dissolution of the highly resistant bedrock.

The investigators discovered that the natural forest was extremely efficient in conserving its mineral elements. For example, annual net loss of calcium from the ecosystem was 9.2 kilograms per hectare. This represents only about 0.3 percent of the calcium in the system. In

32–6

Weir in the Hubbard Brook Experimental Forest in New Hampshire. Water from each of six experimental ecosystems was channeled through weirs, built where the water leaves the watershed, and was analyzed for chemical elements. The trees and shrubs in the watershed behind this weir have been cut down. The experiments showed that deforestation disrupted the tight cycling of nutrients by various living components of the ecosystem and greatly increased the loss of nutrient elements from that system.

of dead plants and animals continued to be decomposed to ammonia or ammonium ions, which were then acted upon by nitrifying bacteria to produce nitrates, the form in which nitrogen is usually assimilated by plants. However, no plants were present, and the nitrate ions were not held in the soil. The net loss of nitrogen averaged 120 kilograms per hectare per year from 1966 to 1968. As a side effect, the stream that drained the area, polluted with nitrates, supported an algal bloom. The nitrate concentration of the stream increased to levels above those established by the U.S. Public Health Service as safe for drinking water.

Nitrogen is not always lost so readily from disturbed forest ecosystems. In the Hubbard Brook experiment, vegetation regrowth was prevented by herbicides so no uptake was possible. These results therefore represent the maximum possible loss. The actual local rates of nitrogen loss depend on the details of the nitrogen cycle in the particular forest before the disturbance takes place. For example, if the microorganisms in the ecosystem have a high demand for the nitrogen released after forest clear-cutting, forest disturbance would not result in losses of the magnitude measured at Hubbard Brook.

Trophic Levels

In addition to its *physical* (or nonliving) components, each ecosystem includes two classes of *biotic* (or living) components—autotrophs and heterotrophs. Autotrophs are mainly photosynthetic organisms, which are able to use light energy to manufacture their own food; they are known as **primary producers.** Autotrophs consist of green plants, algae, and autotrophic bacteria. Because heterotrophs cannot manufacture their own food, they must use the organic molecules made by the autotrophs. Several feeding levels, or **trophic levels,** occur among the heterotrophs: **primary consumers,** or **herbivores**—animals that feed on living plants; **secondary consumers,** or parasites and carnivores—animals that feed on animals; and **decomposers**—fungi, bacteria, and various small animals that break down organic matter stored in the bodies of other organisms. All three levels are present in most ecosystems.

In a given ecosystem, organisms from each of the trophic levels make up what is called a **food chain.** The relationships between the organisms in a food chain regulate the flow of energy and nutrients through the ecosystem. The length and complexity of such food chains vary a great deal. Usually an organism has more than one source of food and is itself preyed on by more than one kind of organism. Under most circumstances, it is more nearly correct to speak of a **food web** (Figure 32–7). The complexity of trophic relationships has a number of important implications for the overall properties of that ecosystem.

the case of nitrogen, the ecosystem was actually accumulating this element at a rate of about 2 kilograms per hectare per year. There was a similar, though somewhat smaller, net gain of potassium in the system.

At Hubbard Brook, the biological regulation of element cycling was tested by the following experiment. In the winter of 1965–1966, all trees, saplings, and shrubs in one 15.6-hectare area in one small watershed of the forest were cut down. No organic materials were removed, however, and the soil was undisturbed. During the following spring, the area was sprayed with a herbicide to inhibit regrowth. During the four months from June through September 1966, the runoff of water from the area was four times greater than in previous years. Net loss of calcium and potassium were both about 20 times higher than in the undisturbed forest. The most severe disturbance was seen in the nitrogen cycle. The tissues

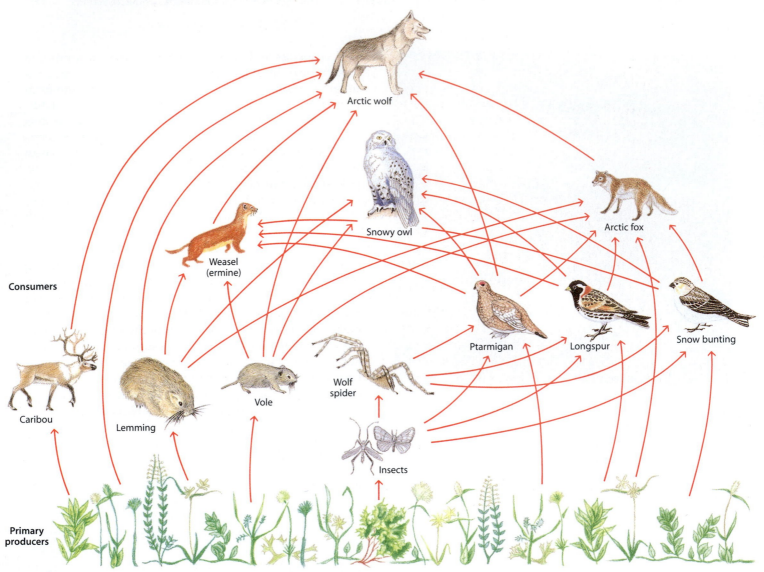

32–7

Diagram of a food web in the Arctic tundra during the spring and summer. The arrows point in the direction of energy flow. This food web is simplified considerably. In reality, many more species of both plants and animals are involved. The fungi, bacteria, and other small animals that function as decomposers (not shown here) also play an important role in food webs.

The Flow of Energy through Ecosystems Is Linear, Not Cyclical

In an ecosystem, the flow of energy typically begins with solar energy that is captured in photosynthesis and used to make carbohydrate molecules. Energy does not flow in a cycle in ecosystems; rather, it flows through autotrophs (usually photosynthetic organisms, that is, plants, algae, and some bacteria) to consumers (animals and heterotrophic protists) and then to decomposers. At each step, the great majority of the energy is dissipated in the form of heat energy, which eventually is returned to space as infrared radiation.

A very large amount of **biomass** is produced on Earth each year. ("Biomass" is a convenient shorthand term for organic matter.) It is currently estimated that the Earth's annual production of biomass amounts to about 200 billion metric tons. Despite this enormous figure, photosynthetic organisms are not very efficient in converting the sun's energy into organic compounds. Generally, less than 1 percent of the light that falls on a plant is utilized in photosynthesis (see Chapter 5). However, particularly productive stands of vegetation, and some aquatic systems, may convert up to 3 percent of the annual incident solar radiation into chemical energy.

Only a Small Fraction of the Available Energy Is Transferred from One Trophic Level to the Next When the organic material produced by plants is consumed by an herbivore, energy is released. Most of this energy is lost as heat, and a fraction of the organic material consumed is converted to animal tissue. In general, less—and often far less—than 20 percent of the usable energy (potential energy) of the plant is added to the mass of the herbivore; the remainder is lost in respiration. A similar gain-loss relationship is found at each succeeding level. Thus, assuming that an average of 1500 kilocalories of light energy per square meter of land surface is utilized by plants each day, about 1 percent (as noted above), or 15 kilocalories, is converted to plant material. Of this amount, perhaps 10 percent, or 1.5 kilocalories, is incorporated into the bodies of the herbivores that eat the plants, and 10 percent of that, or about 0.15 kilocalorie, is incorporated into the bodies of the carnivores that prey on the herbivores.

To give a concrete example, Lamont Cole of Cornell University, in his studies of Lake Cayuga, near the Cornell campus in New York State, calculated that, for every 1000 kilocalories of light energy utilized by algae in the lake, about 150 kilocalories are reconstituted as small aquatic animals. Of these 150 kilocalories, 30 kilocalories are reconstituted as smelt (a small fish). If we were to eat these smelt, we would gain about 6 kilocalories from the original 1000 kilocalories used by the algae. But if trout eat the smelt and we then eat the trout, we gain only about 1.2 kilocalories from the original 1000 kilocalories. Smelt are much more abundant and constitute a much larger biomass in Lake Cayuga than the trout. Thus, more of the original energy is available to us if we eat smelt rather than the trout that feed on smelt; yet trout are considered a delicacy, smelt a much less desirable food for humans. Under conditions of starvation, people cannot afford the tenfold loss of energy that occurs when plants are fed to animals; they must then become herbivores to obtain as much food as possible.

Food chains are generally limited to three or four links; the amount of energy remaining at the end of a longer food chain is so small that few organisms can be supported by it. Body size also plays a role in the structure of food chains. For example, an animal constituting one link generally has to be large enough to capture prey from the previous, lower link on the food chain (though small insects do feed on large trees!). In the end, most of the biomass in an ecosystem is used by decomposers, such as fungi and bacteria.

In General, Ecosystems Can Be Described by Pyramids of Energy, Biomass, and Numbers

Owing to the relationships just discussed, the total energy at successively higher trophic levels in an ecosystem generally decreases sharply, setting up the sort of relationship described by the expression "pyramid of energy" (Figure 32–8). This relationship may not hold if there is rapid turnover among the primary producers, such as algae in a lake. The turnover rate then becomes the controlling factor, and the total energy present at that level may be relatively small at any one time.

A similar relationship is characteristic of mass, leading to the existence of "pyramids of biomass." Most pyramids of biomass take the form of an upright pyramid, as shown in Figure 32–9a, whether the producers are large or small. Pyramids of biomass are inverted only when the producers have very high reproduction rates. For example, in the ocean, the standing crop of phyto-

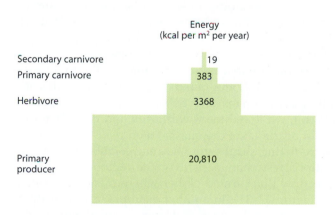

32–8
Pyramid of energy flow for a river ecosystem in Florida. A relatively small proportion of the energy in the system is transferred at each trophic level. Much of the energy is used metabolically and is measured as heat (in kilocalories) lost in respiration.

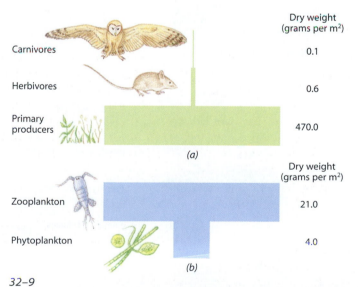

32–9
Pyramids of biomass for (a) the plants and animals in a field in Georgia and (b) plankton in the English Channel. Such pyramids reflect the mass present at any one time. The seemingly paradoxical relationship between the biomass of phytoplankton and zooplankton exists because the high reproduction rate of the smaller phytoplankton population is sufficient to support a large zooplankton population.

plankton may be smaller than the biomass of the zooplankton that feeds upon it (Figure 32–9b). Because the reproduction rate of the phytoplankton population equals or exceeds the rate of grazing by the zooplankton population, a small biomass of phytoplankton can supply food for a large biomass of zooplankton.

In general, there are also far more individuals at the lower trophic levels than at the higher levels, which leads to a "pyramid of numbers." Figure 32–10a, for example, shows a pyramid of numbers for a grassland ecosystem. In this type of ecosystem, the primary producers (the individual grass plants) are small, and so a large number of them is required to support the primary consumers (the herbivores). In an ecosystem in which the primary producers are large (for instance, trees), one primary producer may support many herbivores, as indicated in Figure 32–10b.

Like pyramids of biomass, pyramids of numbers indicate only the quantity of organic material present at one time. They do not give information on the total amount of material produced or, as do pyramids of energy, the rate at which it is produced.

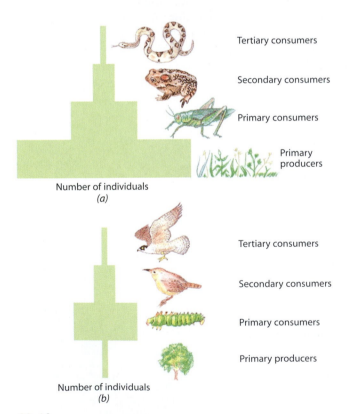

32–10

*Pyramids of numbers for **(a)** a grassland ecosystem, in which the number of primary producers (grass plants) is large, and **(b)** a temperate forest, in which a single primary producer, a tree, can support a large number of herbivores.*

A practical aspect of the flow of energy through ecosystems concerns human efforts to develop renewable sources of energy from plants, a process known as biological energy conversion. Plantings of fast-growing trees and other plants may provide a major, renewable energy source in the future, and they may constitute one of the most environmentally benign ways of efficiently capturing solar energy. Even now, scientists estimate that the waste materials available each year after the harvest of crop and forest products could provide an amount of energy equivalent to 1 percent of the gasoline consumed annually in the United States, or 4 percent of the nation's annual consumption of electrical energy. The potential is limited, however, by the energy costs of harvesting the material. In developing countries, more than 2 billion people rely on biomass for virtually all their cooking, heating, and lighting, so that the ability of plants to convert energy from the sun to a form useful to humans is a matter of central importance on a global scale.

Development of Communities and Ecosystems

Succession Is the Change in a Community over Time

Some plant communities appear to remain the same year after year, whereas others change rapidly. In the latter case, there is a progression of changes in the community called **succession.** A cleared woodlot is rapidly colonized by the remaining trees in the vicinity. In a similar fashion, a pasture eventually gives way to a forest. Analogous series of events occur in naturally disturbed areas, such as lakes, floodplain forests, or steep hillsides. Natural disturbances (for instance, floods, windstorms, earthquakes, landslides, fires) are a pervasive feature of all ecosystems, even "pristine" wilderness areas that are little disturbed by humans. The process of succession is both continuous and worldwide in scope.

Succession occurs at a variable rate in all temporarily disturbed areas. Some ponds, for example, fill with aquatic plant remains and debris; emergent vegetation builds soil, while sediment washes in and contributes to filling in of the lake; the site is taken over by meadow; moisture-loving shrubs may become established; and finally the forest characteristic of the region develops in the meadow that formed earlier where there had been a pond (Figure 32–11). In another example, rocks weather and break down because of freezing and thawing and other physical factors. The process is sometimes expedited by the action of lichens, which secrete chemicals that erode the rocks directly, and by mosses, which expand when wet, continually breaking off little flakes of rock (Figure 32–12). Soil accumulates around the bases of the lichens and mosses, and seedlings of flowering plants eventually may become established. Their roots

(a)

(b)

(c)

(d)

32–11

Succession. (a) Emerging vegetation grows along the edge of a pond. (b) Aquatic plants with floating leaves, such as water lily (Nymphaea odorata), *grow across the surface of a pond and eventually choke out bottom-dwelling plants. (c) Water hyacinths* (Eichhornia crassipes) *play a similar role in warmer climates. (d) Marsh grasses, sedges, and cattails* (Typha spp.) *growing on an old pond bed continue the process of succession.*

32–12

An early stage of succession. Lichens have begun to erode the rocks, while ferns and bryophytes are accumulating soil in a small crevice.

32–13
Seedling trees of balsam fir (Abies balsamea) *growing under and replacing quaking aspen* (Populus tremuloides) *in northern Arizona— a stage in forest succession leading to a climax community of white spruce* (Picea glauca) *and balsam fir.*

penetrate cracks, breaking the rocks down further. Eventually, perhaps after many centuries, the rock may be completely reduced to a component of soil, which also includes organic matter from the generations of organisms that have grown in it. The soil will ultimately be occupied by forests or other types of vegetation characteristic of the region. In the early stages of succession, plants with symbionts that have the ability to fix nitrogen may be prominent, at least in certain areas. Another example of succession is shown in Figure 32–13.

Succession Is Often Not Completely Unidirectional, Particularly in Its Later Stages   It is generally simple to observe the early stages of succession and very difficult to interpret the later ones. In more mature communities, reversals in the expected direction of succession often occur, and the ultimate outcome may be heavily influenced by the nature of the adjoining communities.

The creation and refilling of **gaps** created by natural disturbances are processes that play a key role in succession and in the maintenance of species diversity in various plant communities. The gaps that occur when trees fall in a forest, for example, provide opportunities for many plant species with relatively high light requirements to flourish. In temperate North America, a number of species that characteristically occur in gaps, such as black cherries and blackberries, have fleshy fruits eaten by birds. The birds digest the fleshy parts of these

fruits and drop the seeds in new gaps, which are thus colonized efficiently. Such pioneer species often have light, nondurable wood and are characterized by multi-layered, diffuse crowns and high rates of growth under sunny conditions. The dominant, late successional trees in the same forest often have very different characteristics, such as heavy but durable wood, more tightly packed crowns, and slow rates of growth under sunny conditions. These trees can, however, grow in the shade and live longer than the pioneers. Such differences between tree species play a major role in determining their local success and thus the structure of the mature forest.

One of the most significant forms of natural disturbance affecting plant communities is fire (Figure 32–14). For example, when European settlers first arrived in California, they found a magnificent forest of sugar pine *(Pinus lambertiana)* along much of the length of the Sierra Nevada. Although conservationists tried to preserve some of this forest in national parks and forests, many of the stands of pines were eventually replaced by other trees, such as white fir *(Abies concolor)* and incense cedar *(Calocedrus decurrens)*. Why did this change take place?

Sugar pine was a member of a successional stage in the forests of this region that was maintained by periodic fires. These fires were greatly reduced in number and scope after the influx of settlers to the area. Without lightning-set fires of low intensity periodically racing through the groves, a thick growth of brush and smaller

(a)

(b)

32–14
(a) When fire sweeps through a forest, recolonization—with regeneration from nearby unburned stands of vegetation—is initiated. Some plants produce sprouts from the base, others seed abundantly on the burned area. In one group of pines, the closed-cone (serotinous) pines, the cones do not open to release their seeds until they have been exposed to fire. (b) Sugar pines (Pinus lambertiana) *in Yosemite National Park in California. With the prevention of forest fires by human beings, sugar pines are being replaced by other trees, such as white fir* (Abies concolor), *the trees seen here growing at the base of the sugar pines.*

trees grew up, evidently creating conditions so shady that sugar pine seedlings could not compete effectively. Only a policy of letting the occasional fires that occur burn, or one of controlled burning, can preserve the remaining groves of sugar pine in the open form that most people find so attractive (Figure 32–14b). Similar relationships are found in all vegetation types in which fires periodically burn, either naturally or through human activity.

Following natural disasters, recolonization produces similar successional changes. For example, in August 1883, a violent volcanic eruption destroyed half of the island of Krakatau, in the Java Straits about 40 kilometers from Java, Indonesia. The remaining half of the island was covered by a layer of pumice and ash more than 31 meters thick. Neighboring islands were also buried, and the entire assemblage of plants and animals on these islands was wiped out. Soon afterward, however, the recolonization of Krakatau began, and the expected number (based on the number originally occupying the area) of about 30 species of land and freshwater birds was reached within about 30 years. Recolonization by plants also proceeded rapidly, with a total of more than 270 species being recorded for the island of Krakatau by 1934.

In the state of Washington, the violent eruption of Mount St. Helens on May 18, 1980, sent a massive avalanche of volcanic debris from the top and north side of the mountain into the North Toutle River Valley.

The Great Yellowstone Fire

It was an exceptionally dry summer in 1988, with the hottest, driest conditions since the Dust Bowl of the 1930s. But the park managers of Yellowstone National Park were confident that major fires could not happen. Yet in July of that year, Yellowstone broke into flames. Soon new fires were breaking out in every section of the park; they continued to spread throughout August and September. Park concessions were endangered. Tourists had to be evacuated. A holocaust had begun. What had gone wrong?

Yellowstone Park was established in 1872 in order to preserve its natural beauty, and that included protecting it from fires. Eventually, however, scientists and managers realized that there was a gradual decrease in certain species, as seen in photographs taken through the years. This evidence, along with the existence of charred stumps, strongly suggested that fire was a normal part of the Yellowstone ecosystem. In 1972, park offi-

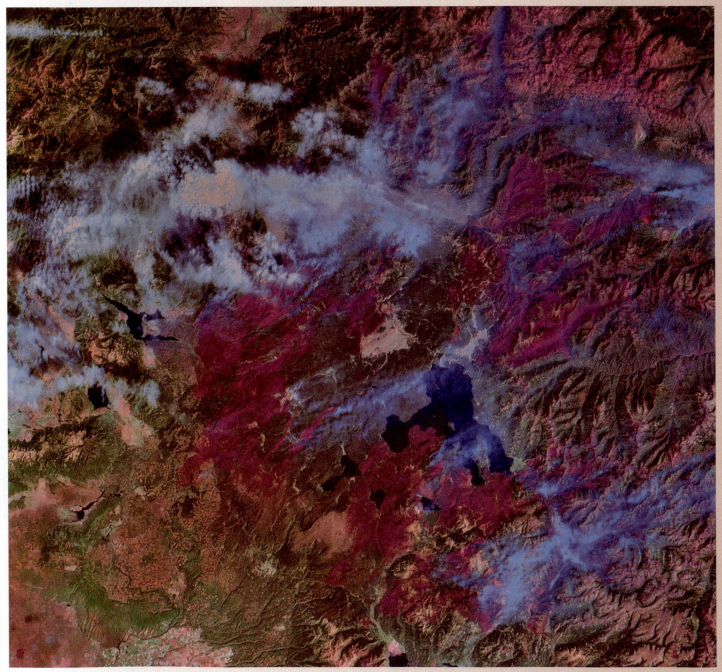

(a)

cials initiated a program in which they allowed natural fires in remote areas to burn without interference. The program was subsequently extended throughout the park to all but 5 percent of the total area—places where people were living. This policy was widely viewed as a success.

But by October 1988, 4300 square kilometers lay charred, and nowhere in the park could you get a vantage point where there was no visible fire damage. Dry winds, gusting to more than 100 kilometers per hour, had frustrated the efforts of more than 9000 fire fighters to control the spread of the flames. Was this havoc the result of buildup of abnormal amounts of biomass in the trees and shrubs? Could it have been prevented by controlled burning?

Fire scars among the annual growth rings of lodgepole pines (Pinus contorta) indicated that any one spot in Yellowstone had burned naturally about every 20 to 25 years, with a wide range of variation. Lodgepole pine, like other pines, is adapted to fire. Its cones open only in intense heat, its seeds regenerate best following fires, and its seedlings benefit when the light-blocking mature trees are killed and thrive on the nutrients released from burned pine needles. The beautiful, even-aged lodgepole pine stands that are such a characteristic feature of Yellowstone National Park are a direct consequence and reflection of periodic catastrophic fires.

At higher elevations, with cooler climates, mostly small fires have burned almost every decade. In areas with no significant fires, the lodgepole pines are eventually replaced by other tree species, such as subalpine fir (Abies lasiocarpa) and Engelmann spruce (Picea engelmannii). These species usually succeed lodgepole pine, starting about 150 years after a fire, and become dominant during the following century. As the original lodgepole pines die, however, the forest canopy becomes uneven, with many gaps, and is highly flammable. Fires that occur at early stages of succession are usually small, but as the forest becomes more mature, fires are extensive and highly destructive.

In Yellowstone, fire usually intervenes and the climax fir-spruce communities are very rare. The extensive fires of 1988 seem to have been similar to a series of fires that occurred around 1700 and perhaps should also be viewed as a natural feature of this area. Major efforts to suppress fires in the Yellowstone area may merely have delayed this natural cycle for a few years.

In any event, the great fire of 1988 appears to have had no major lasting effect on the ecosystem. Lush meadows feed healthy buffalo and deer. Colorful lupines help to replace the soil's depleted nitrogen.

Pine seedlings thrive on the sunny flats and slopes. Lodgepole pines grow slowly in Yellowstone. In 30 years, these seedlings will be 2 to 4 meters tall; perhaps in 80 or 100 years, the forests will begin to resemble those that burned in 1988. In two or three centuries, the stage may be set for a repeat of the events that occurred in the late summer of that year. Meanwhile, curiosity about the recovery of the Yellowstone ecosystem from the great fire has proved a great stimulus to tourism, and people have been visiting the area in record numbers.

In management terms, the continuation of a program in which lightning-set fires are allowed to burn is clearly an appropriate strategy for maintaining the natural ecosystems of Yellowstone, but the total area—like that of all protected areas—is so limited that fires cannot be allowed to burn completely freely. They affect too many interests both within the protected areas and beyond their boundaries to be treated with complete passivity. Scientists and managers currently are studying the possibility of burning areas near human habitation periodically to reduce the amount of natural fuel. The struggle between the interests of humans and the continuance of nature goes on. In this context, fire is clearly seen as an example of a natural, pervasive influence affecting an apparently pristine ecosystem.

(a) Satellite image of Yellowstone area at the height of the fire activity, September 8, 1988. Red denotes the area within the burn perimeter, and green identifies unaffected vegetation. Yellowstone Lake is the dark blue shape on the right, just below the center. *(b)* Regeneration near Elk Park in Yellowstone, a year after the fire.

(b)

Within 15 minutes, more than 61,000 hectares of forest and recreation land were devastated by the lateral blast, which blew down forests on about 21,000 hectares and killed trees and other plants but left them standing on another 9700 hectares. In addition, the nine-hour eruption covered the whole area with up to 0.5 meter of ash, pumice, and rock pulverized by the blast.

Life began to reappear on the slopes affected by the eruption almost immediately, however, with plants sprouting up through the volcanic debris in the following spring (Figure 32–15). Much of this debris soon eroded off, and wind-dispersed seeds and fruits blew back into the area. Such dispersal was especially important in areas that had been buried by debris avalanches, which were so deep that they killed the plants buried beneath them, and on the volcanic flows. Many small animals also survived, both underground and in lakes and streams, and terrestrial vertebrates soon moved back into the area. The eruptive periods of Mount St. Helens have been separated by only 100 to 500 years in the last 35 centuries, and both the blast itself and the recovery from its effects that scientists were able to observe in the early 1980s were typical of natural phenomena that periodically affect living systems in volcanic areas.

Theoretically, Succession Ends with a Climax Community

A general property of successional processes is that, if climatic factors remain constant, the process of succession should ultimately slow and nearly stop, and terminate in a so-called **climax community.** Such a community is self-perpetuating; its characteristics relate to the specific climatic conditions under which it is produced. A climax community, however, is an ideal concept that mainly serves as a reference point against which to measure community change. In reality, climatic conditions often change, natural disturbances such as hurricanes and landslides occur, and animals modify the nature of the changing communities. The communities that develop at a particular place reflect a balance of many different environmental factors.

As human beings become more and more numerous, and their impact on ecosystems correspondingly profound, the science of **restoration ecology,** which attempts to understand the process of succession better and use its principles to reestablish natural communities, will become more and more important. It is often not a simple matter to re-create natural communities once they have been destroyed; yet the process is one of great significance to an increasingly overcrowded world.

A well-known example of restoration ecology has taken place at the University of Wisconsin Arboretum in Madison (Figure 32–16). Begun in 1934 on damaged agricultural land, several distinct natural communities have been developed, among them a tallgrass prairie, a dry prairie, and several types of pine and maple forests. While restoration can be a costly and uncertain process, it provides a way of going on the offensive in the struggle to ensure the survival of classic ecosystems and the plants and animals within them. The Arboretum's Curtis Prairie, for example, supports more than 200 species of native plants, many of them now rare in the area. Early work on the Curtis Prairie provided new insights into the importance of fire in the prairie ecosystem. Fire plays a key role, not only by eliminating most species of trees, but also by influencing the nutrient economy of the prairie ecosystem and in other ways as well. In fact, a prairie may be regarded as a highly flammable ecosystem that "uses" fire to maintain itself against encroachment by nonprairie species. In many parts of North America, the persistence of prairie depends on recurrent fires—set by Native Americans or lightning before the

32–15

When Mount St. Helens in the state of Washington exploded on May 18, 1980, shock waves leveled all of the trees in an area of about 21,000 hectares, and a deep layer of ash was deposited. Fireweed (Chamaerion augustifolium) and grasses, as seen here four months after the eruption, were among the first plants to recolonize the area.

(a)

(b)

32-16

A restored prairie at the University of Wisconsin Arboretum in Madison. (a) Late summer on the prairie, with purple blazing

star (Liatris pycnostachya), *white flowering spurge* (Euphorbia corollata), *and yellow prairie-coneflower* (Ratibida pinnata).

(b) A burn carried out in late fall plays an important role in maintaining the prairie ecosystem.

arrival of Europeans. If these fires do not occur, the prairie may quickly be colonized by trees and accompanying, shade-tolerant species. This, indeed, is what happened to many of the tallgrass prairies in the American Midwest following European settlement and the elimination of fire from the landscape. Though fire played an important role in the management of vegetation in many areas even after settlement, the ecological significance of this was not understood until relatively recently. Attempts to restore prairies such as the Curtis Prairie at the University of Wisconsin Arboretum have played a key part in clarifying this role.

Although volcanic eruption provides a very dramatic example of natural disturbance and the ensuing early stages of succession, these phenomena are typical of all communities and occur throughout the world. Disturbance and succession are two important factors that account for the full extent of the diversity of life on Earth.

Summary

An Ecosystem Consists of a Community and Its Environment

Ecosystems are self-sustaining systems that include living organisms as well as the nonliving (physical) elements of the environment with which they interact. Communities consist of all the organisms that live in a particular area.

Some Relationships Found in Communities Are Mutualism, Competition, and Plant-Herbivore/Plant-Pathogen Interactions

Some of the relationships that occur in communities can be grouped under three headings: mutualism, competition, and plant-herbivore (and plant-pathogen) relationships. In mutualism, two populations interact to the benefit of both. Examples include lichens, mycorrhizal associations between fungi and the roots of plants, and the relationships between flowering plants and their pollinators and fruit and seed dispersers. In the bull's-horn acacias of Latin America, the thorns are inhabited by specialized ants that obtain their food from the plants and protect them from most herbivores and from competition with other plants.

Competition Results When Organisms Require the Same Limited Resource

Competitive interactions occur between most kinds of plants that grow in close proximity and between most individual plants also. The principle of competitive exclusion states that when two kinds of organisms occurring together compete for the same limiting resources, ultimately only one of them will survive in that area. One of the most important kinds of competition is competition for light. Frequently, plants with the highest growth rate relative to other species in a particular environment will be the most successful competitors there. In the course of evolution plants have also developed chemical weapons with which to compete aggressively with nearby plants, and such allelopathic relationships can also affect community composition. Flowering plants often compete for the services of pollinators and seed dispersers, a situation that can have a profound impact on the reproduction and long-term abundance of different species.

Plants Have a Variety of Physical and Chemical Defense Mechanisms

Plants counter the effects of herbivores, which limit the reproductive potential of the plants, through the evolution of spines, tough leaves, and similar structures or structural alterations, and, most important, chemical defenses. An insect or other herbivore that has overcome a plant's chemical defenses not only has a new and often largely untapped food resource at its disposal but may also utilize the toxic substances produced by the plant to gain a degree of protection from its own predators.

The Biotic Components of an Ecosystem Are the Primary Producers, Consumers, and Decomposers

An ecosystem consists of nonliving elements and two different kinds of living elements—autotrophs (primary producers) and heterotrophs (consumers). Among the heterotrophs are the primary consumers, or herbivores; the secondary consumers, or carnivores and parasites; and the decomposers. The organisms found at these levels are members of food chains or food webs.

Hubbard Brook Has Provided an Outdoor Laboratory for Studying Nutrient Cycling

The properties of ecosystems have been studied experimentally at Hubbard Brook, in New Hampshire, where it has been shown that undisturbed natural communities control the cycling of nutrients but that the control tends to be lost when the ecosystem is disturbed.

The Flow of Energy through an Ecosystem Affects the Mass and Number of Its Component Organisms

Energy flows through ecosystems, with 1 percent or less of the incident solar energy converted into chemical energy by green plants. When these plants are consumed, less than 20 percent of their potential energy is stored at the next trophic level; a similar degree of efficiency characterizes transfers farther up the food chain. The amounts of energy remaining after several transfers are so small that food chains are rarely more than three or four links long. In most ecosystems, more energy, biomass, and individuals occur at lower trophic levels, giving rise to the phenomena known as pyramids of energy, biomass, and numbers.

Succession Is the Change in a Community over Time

Succession occurs in naturally open areas, such as lakes, ponds, or meadows in a forested region, and after an area has been denuded by artificial or natural means. In the course of succession, the kinds of plants and animals in the area change continuously, some being characteristic only of the early stages of succession. The creation and refilling of gaps created by natural disturbances play a key role in the process of succession and in the maintenance of species diversity in various forest communities. The pioneer species that arise in the gaps grow rapidly under sunny conditions and have other characteristics different from those of the trees that dominate the mature forest. Eventually succession may result in the production of a climax community, which reproduces itself indefinitely unless there are major environmental changes.

Fire plays a very important role in the dynamics of many ecosystems, as in the maintenance of sugar pine forests in the Sierra Nevada of California and of prairie ecosystems. Succession after volcanic eruption, such as that of Krakatau in Java in 1883 or Mount St. Helens in Washington State in 1980, provides a spectacular example of the process, and these areas have been studied extensively.

Selected Key Terms

allelopathy p. 779
biomass p. 784
climax community p. 792
communities p. 774
competition p. 776
decomposers p. 783
ecology p. 774
ecosystems p. 774
elicitors p. 780
food chain p. 783
food web p. 783
gaps p. 788
herbivores p. 783
mutualism p. 774

nutrient cycling p. 782
phytoalexins p. 780
populations p. 774
primary consumers p. 783
primary producers p. 783
principle of competitive exclusion p. 776
restoration ecology p. 792
secondary consumers p. 783
succession p. 786
trophic levels p. 783

Questions

1. In what way does the ant-acacia system resemble a lichen?

2. Explain the role of growth rate in the competition among plants.

3. According to the principle of competitive exclusion, two species with similar environmental requirements cannot coexist indefinitely in the same habitat. How might competitive exclusion be avoided?

4. The diversity of species is greater in an environment where disturbance is continuous than in a more stable environment. Why?

5. Explain the role of phytoalexins and tannins in the defense of plants against microorganisms and herbivores, respectively.

6. Comment on the importance of plants in the retention of nutrients in forest ecosystems.

7. Why are food chains generally limited to three or four links?

8. In general, ecosystems can be described by pyramids of energy, biomass, and numbers. Explain.

9. What role do gaps play in succession?

10. Disturbance and succession are two important factors that account for the full extent of the diversity of life on Earth. Explain.

11. How does a plant community change over time?

Global Ecology

(a)

(b)

33-1

*Regions with mild winters and long, dry summers are often dominated by spiny shrubs with broad, thick evergreen leaves. This vegetation, known formally as Mediterranean scrub, has been given a variety of local names. (**a**) In North America it is known as chaparral, as seen here in southern California. (**b**) In Mediterranean regions it is known as maquis, as seen here on the Greek island of Corfu. Although the plants of these two areas are mostly unrelated, they closely resemble one another in their growth patterns and characteristic appearance.*

OVERVIEW

Imagine driving across the United States from New York City to Los Angeles. As an astute botanist, you would notice how the plant communities change along the way. For example, you would observe deciduous trees in the middle Atlantic states, grasses in the Great Plains, and cacti in the southwestern deserts. These large-scale communities are called biomes, and they are the subject of this chapter.

One advantage of learning about biomes is that you can feel at home—in a botanical sense—wherever you travel in the world. Biomes are determined largely by temperature and precipitation, and thus the same biome might very well be found on several different continents where the climate is similar. For example, you would encounter the temperate deciduous forest biome in France, China, and Russia, as well as in the United States.

On your cross-country drive you would also notice that there are no sharp boundaries between biomes. The deciduous forest slowly intergrades into grassland, and grassland intergrades into desert. If your route took you across mountains, you would find that communities also change as altitude increases. Such changes are similar to those you would encounter driving either north or south from the equator.

This chapter describes the world's major biomes—ranging from the Arctic tundra to the tropical rainforest. Special emphasis is placed on the impact of humans on these communities.

CHECKPOINTS

By the time you finish reading this chapter, you should be able to answer the following questions:

1. What is a biome, and what factors affect the distribution of biomes on Earth?

2. Given that tropical rainforests contain such a diversity of species, why are tropical soils unsuitable for agriculture?

3. How is an African savanna different from a North American grassland? What role does fire play in grasslands?

4. Which best characterizes a desert—high temperature or low precipitation? How have plants adapted to desert living?

5. How does the appearance of the temperate deciduous forest biome change from one season to the next? What accounts for these changes?

Y ou may already be familiar with many of the vast biomes of the Earth—the tundra, with its herds of caribou; the temperate deciduous forests, leafless in winter but bursting into life in the spring; and the deserts, with their cacti and other succulent and spiny plants, for example. Defined in ecological terms, a **biome** is a terrestrial, climatically controlled set of ecosystems that are characterized by distinctive vegetation and among which there is an exchange of water, nutrients, gases, and biological components, including people. The plants and animals that occur in particular biomes have characteristic growth forms and other adaptations that have evolved in relation to particular climates. It is because of these common growth forms that we can recognize biomes such as savannas or deserts wherever they occur, even though the individual organisms from a biome in one region are often completely different from those of the same biome in another region, having achieved similar characteristics as a result of parallel evolution. A single biome occupies large areas of land surface and usually occurs on more than one continent (Figure 33–1). Biomes can be classified in a number of different ways, but the categories that are discussed in this book (see Figure 33–3) provide a basis for an overall account of the vegetation of the world.

Life on the Land

Land plants face a variety of problems. In most regions, they are subjected to periodic drought and to rapid diurnal and seasonal changes in temperature. They must survive during seasons unfavorable to their growth, they must often grow on substrates of unfavorable mineral composition, and they are subject to the action of gravity (which affects organisms on land much more strongly than it does aquatic organisms). Living on land has some advantages, however, because oxygen is more uniformly distributed in air than it is in water, and carbon dioxide is more readily available.

The distribution of biomes results from three kinds of physical factors: (1) the distribution of heat from the sun and the relative seasonality of different portions of the Earth; (2) global patterns of air circulation (Figure 33–2), particularly the directions in which the prevailing moisture-laden winds blow; and (3) such geologic factors as the distribution of mountains and their height and orientation. All of these factors interact to produce the varying patterns of vegetation over the face of the Earth and underlie differences in productivity in different geographic regions (Figure 33–4).

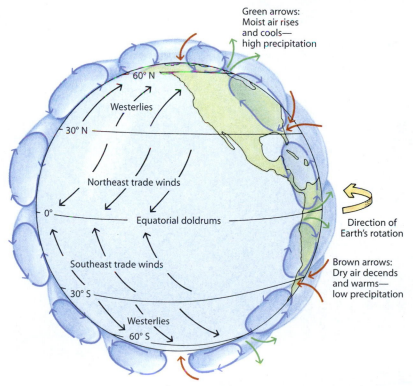

33–2

The Earth's surface is covered by belts of air currents, which determine the major patterns of wind and rainfall. In this diagram, the blue arrows indicate the direction of movement of the air within the belts. The green arrows indicate regions of rising air, which are char- *acterized by high precipitation, and the brown arrows indicate regions of descending air, which are characterized by low precipitation. The dry air descending at latitudes of 30° north and south is responsible for the great deserts of the world.* *The prevailing winds on the Earth's surface, indicated by the black arrows, are produced by the twisting effect of the Earth's rotation on the air currents within the individual belts.*

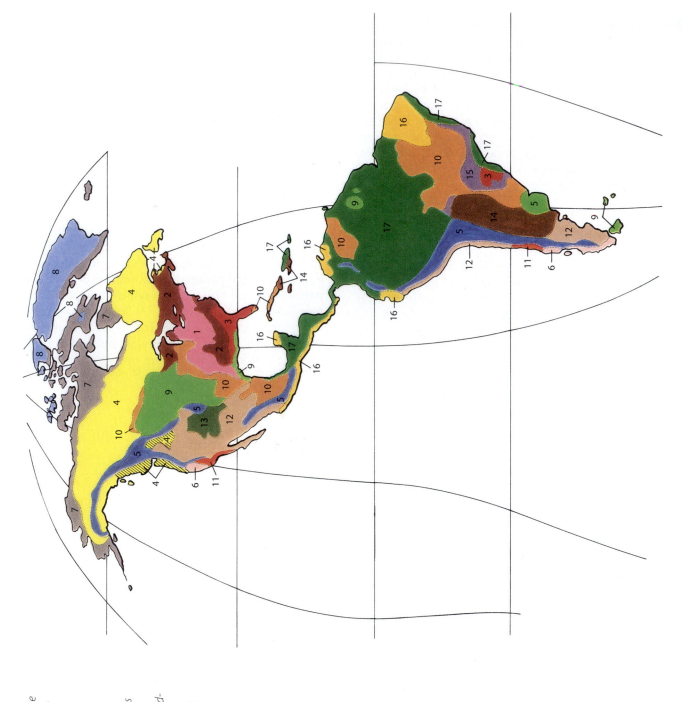

33–3

Biomes of the world. The information in these maps and the accompanying key was originally supplied by A. W. Küchler of the University of Kansas, one of the leading authorities on the distribution of biomes.

Because of the global coverage of the maps, the scale is relatively small and the content is generalized. Any given biome is not always uniform, and all of the biomes include considerable variations in vegetation. The boundaries between the biomes may be sharp, but more often they are blurred, consisting of broad zones of transition from one type of vegetation to another.

1	Temperate deciduous forests
2	Temperate mixed forests
3	Subtropical mixed forests
4	Taiga
4	Northwestern coniferous forest
5	Alpine tundra and mountain forests
6	Mixed west-coast forests
7	Arctic tundra
8	Ice desert
9	Grasslands
10	Savannas
11	Mediterranean scrub
12	Deserts and semideserts
13	Juniper savanna
14	Southern woodland and scrub
15	Tropical mixed forests
16	Monsoon forests
17	Rainforests

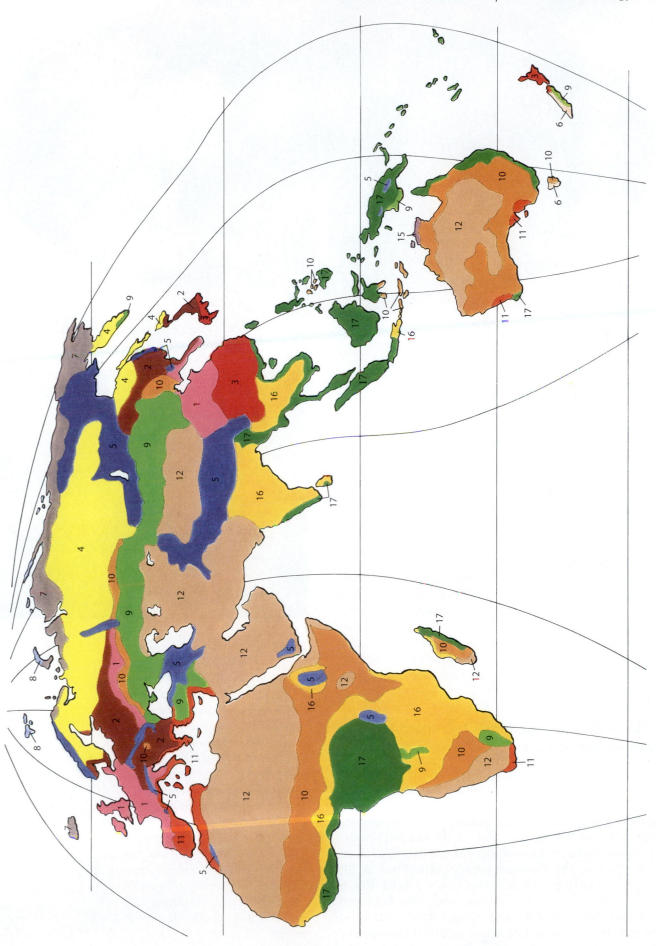

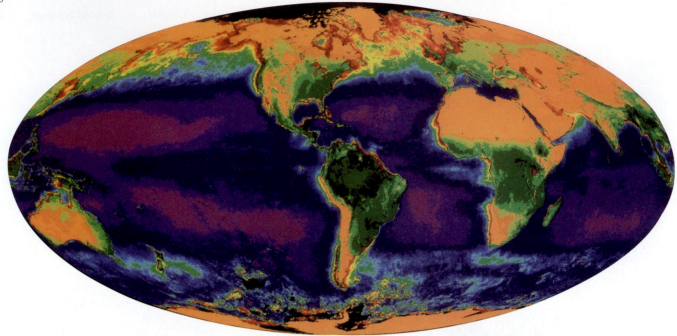

33–4

The Earth's biological productivity, as reconstructed from three years of satellite data by NASA. Rainforests and other highly productive areas are dark green, deserts yellow-orange. More than two-thirds of all biological productivity occurs on land, the remainder in the sea. The concentration of phytoplankton, which consists of photosynthetic protists, is represented by a scale that runs from red (highest productivity) to orange, yellow, green, blue, and purple (lowest productivity).

The Earth's land areas are discontinuous, and this separation of land masses has an important effect on the distribution of organisms. In addition, there are often sharp differences in precipitation, substrate, climate, and other factors between different places on land. Such differences mean that the distribution of any particular kind of land organism is likely to be much more limited than that of an organism of similar size and motility in the sea (Figure 33–5).

The Distribution of Organisms Is Influenced by Latitude and Altitude

Plants and plant communities vary with the latitude of their habitat. For instance, the mean atmospheric temperature decreases with increasing latitude. In terms of annual averages, the warmest atmospheric temperature is found not at the equator but at 10° north latitude. Only during January is the equator the warmest circle of latitude. In July, the warmest parallel of latitude is 20° north. The Northern Hemisphere as a whole has a higher annual mean temperature than the Southern Hemisphere because of the much larger land surfaces in the north. The annual mean temperature at the equator

33–5

This fern, the spleenwort Asplenium viride, *is growing in rock crevices at Sierra Buttes, Sierra County, northern California. It occurs in similar habitats all around the Northern Hemisphere—across both North America and Eurasia—but in North America the closest sites to Sierra Buttes where other A. viride plants are found are central Washington and northeastern Nevada, both nearly 1000 kilometers away. Throughout its range, A. viride grows on cliffs at scattered localities, reaching new places by means of its wind-dispersed spores. Sierra Buttes, however, is far from the edge of the fern's more continuous range, and its occurrence there provides an excellent example of the discontinuous and dispersed nature of the ranges of plants and animals in general.*

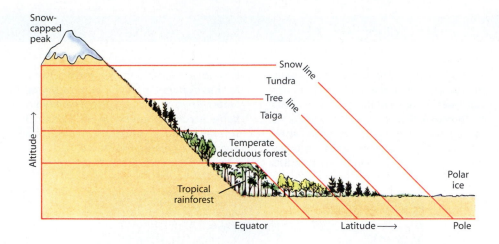

33–6

In the Northern Hemisphere, we can experience a similar sequence of dominant plant life and its associated animal life by either traveling north for hundreds of kilometers or ascending a mountain. This relationship between altitude and latitude was first pointed out by Alexander von Humboldt. To experience a similar sequence of vegetation in the Southern Hemisphere, we could ascend a mountain. However, by simply traveling south, we would never encounter vegetation corresponding to the taiga and the tundra of the Northern Hemisphere. Can you explain why?

is 26.3°C, while at 40° north latitude it is 14°C, and at 40° south latitude it is 12.4°C.

Just as one encounters lower temperatures in traveling north or south from the equator, one encounters them also when ascending to higher elevations. In general, a change in mean atmospheric temperature corresponding to an increase in latitude of 1° occurs with each increase in elevation of approximately 100 meters. This relationship has important consequences for the distribution of land organisms. For example, plants and animals characteristic of Arctic regions may be found at high elevations near or even at the equator, particularly in mountain ranges trending north and south (Figure 33–6).

There are, however, important differences between high-latitude habitats and high-altitude habitats. In the mountains, the air is clearer and the solar radiation is more intense. Most of the water vapor in the atmosphere—which plays a major role in preventing heat from radiating away from the Earth at night—occurs below 2000 meters. Consequently, nights are often much cooler in the mountains than at lower elevations at the same latitude. Moreover, in the Arctic and Antarctic, daylength varies considerably, from 24 hours of light in the summer to 24 hours of darkness in the winter in the extreme north and south, whereas at the equator daylength varies little from 12 hours per day throughout the year. Temperature variation is also much greater at Arctic and Antarctic locations than at high-altitude locations at the equator. Those "Arctic" or "Antarctic" organisms that extend their ranges toward the equator in the mountains must make physiological adjustments for such differences between high-latitude (Arctic and Antarctic) and high-altitude (alpine) environments.

There are often pronounced temperature variations from slope to slope on a particular mountain. In the middle latitudes of the Northern Hemisphere, for example, the southern and western slopes of mountains are often warmer than the northern and eastern ones, because the northern slopes receive no direct sunlight and the eastern slopes receive sunlight only in the mornings, when it is cooler and the solar radiation is less intense. Clouds that form in the mornings often dissipate by the afternoon, enhancing this effect.

Local precipitation is also affected by the presence of mountains. For example, along the West Coast of the United States, where the prevailing winds are from the west, the western slopes of the Sierra Nevada have abundant rainfall, while the eastern slopes are dry and desertlike. As air from the ocean hits the western slope, it rises, is cooled, and releases its water. Then, after passing the crest of the mountain range, the air descends again. As it becomes warmer, its water-holding capacity increases, producing a "rain shadow" on the eastern slope (Figure 33–7).

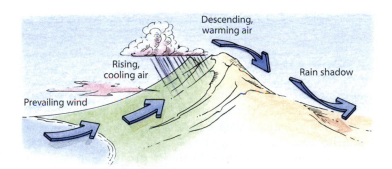

33–7

The effect of coastal mountains on patterns of precipitation in the Northern Hemisphere. As winds come off the water, the air is forced upward by the contour of the land, cools, and releases its moisture in the form of rain or snow. As the air descends on the far side of the mountains and becomes warmer, its capacity to hold water increases, producing a rain shadow.

Alexander von Humboldt

Alexander von Humboldt (1769–1859) was perhaps the greatest scientific traveler who ever lived and was certainly one of the greatest writers and scientists of his era. A native of Germany, Humboldt ranged widely across the interior of Latin America from 1799 to 1804 and climbed some of its highest mountains. Exploring the region between Ecuador and central Mexico, Humboldt was the first to recognize the incredible diversity of tropical life and, consequently, the first to realize just how vast a number of species of plants and animals there must be in the world.

In his travels, Humboldt was impressed with the fact that plants tend to occur in repeatable groups, or communities, and that wherever there are similar conditions—relating to climate, soil, or biological interactions—similar groupings of plants appear. He also discovered a second major principle—the relationship between altitude and latitude. He found that climbing a mountain in the tropics is analogous to traveling farther north (or south) from the equator. Humboldt illustrated this point with his well-known diagram of the zones of vegetation on Mount Chimborazo in Ecuador, which he climbed. On this mountain, he reached the highest elevation on record attained by any human being up to that date.

On leaving Latin America in 1804, Humboldt visited the United States for eight weeks. He spent three of these weeks as Thomas Jefferson's guest at Monticello, talking over many matters of mutual interest. It is thought that Humboldt's enthusiasm for learning about new lands encouraged Jefferson's own great scheme for the exploration of the western United States. Thus, it is fitting that Humboldt's name is commemorated in the names of several counties, mountain ranges, and rivers in the American West.

Because of the relationships just described, the driest slope of a mountain in the Northern Hemisphere is usually that oriented toward the southwest. (Which would you expect to be the driest slope of a mountain in the Southern Hemisphere?) For the same reasons, mosses and lichens often grow, or are better developed, on the northeastern side of trees in the Northern Hemisphere, leading to a well-known folk saying about how to identify the points of the compass when trying to find one's way in the woods. The relationship is valid, however, only in sufficiently moist and well-lighted situations.

As a result of the complex interactions between varying local conditions, the distribution of biomes is not what we would expect if the Earth were a perfect sphere. In addition, major shifts in vegetation types have occurred in the past, profoundly affecting the sort of vegetation that we see today. In the following pages, some examples are provided of the ways in which climate, topography, and soil type interact to control the distributions of the biomes. We shall discuss the 17 biomes shown on the map in Figure 33–3, grouped into nine major categories: rainforests; savannas and deciduous tropical forests (including savannas, subtropical mixed forests, monsoon forests, tropical mixed forests, and southern woodland and scrub); deserts (including deserts and semideserts, and juniper savanna); grasslands; temperate deciduous forests; temperate mixed and coniferous forests (including mixed west-coast forests, temperate mixed forests, and alpine tundra and mountain forests); Mediterranean scrub; taiga (including taiga and northwestern coniferous forest); and Arctic tundra (including Arctic tundra and ice desert).

(a)

33–8
Tropical rainforest. (a) The interior of a rainforest in Costa Rica. The broad-leaved plants with red flowers are Heliconia irrasa. *(b) The diversity of the trees in the forest, which may reach several hundred species per hectare, is revealed when individual trees burst into bloom, like these trees in the coastal rainforest of Brazil. (c) A buttressed tree in the rainforest of Ecuador. Note the woody vines known as lianas on the trunk.*

(b)

(c)

Rainforests

More species of plants and animals live in **tropical rainforests** than in all other biomes of the world combined. Neither water nor temperature is a limiting factor during any part of the year. Although there are many species, there are usually few individuals per species. A species of tree, for example, may be represented by only one individual per hectare. Not only are the species in tropical rainforests extremely numerous, but their interrelationships are more complex than those of plants and animals in any other biome (Figure 33–8).

Tropical rainforests are dominated by broad-leaved evergreen trees. Little light penetrates to the floor of these forests, and the rainfall is generally between 200 and 400 centimeters per year. There is little accumulation of organic debris because decomposers rapidly break down leaves, stems, and the bodies of animals in the constantly warm and moist climate. Nutrients released from this breakdown are quickly absorbed by mycorrhizal roots or are leached from the soil by the rain. Although there may be a notable variation in precipitation from month to month, there is no pronounced dry season.

Generally, plants in the tropical rainforest have not evolved particular mechanisms that would permit them to survive unfavorable seasons of drought or cold. Nearly all the plants are woody, and woody vines, **lianas,** are abundant. There is a large flora of **epiphytes,** which grow on the branches of other plants in the illuminated zone far above the forest floor. Epiphytes, which include orchids, ferns, and bromeliads (Figure 33–9), have no direct contact with the forest floor. They obtain water from the humid air of the canopy, as well as directly from rain, and they obtain minerals from dust and the surfaces of the plants on which they grow. Along with epiphytes and lianas, many kinds of animals live in the treetops; this is the area of the tropical rainforest in which animal life is most abundant and diverse.

Because so little light reaches the forest floor, very few herbaceous plants grow there, and those that do are found mostly in gaps in the forest canopy created by treefalls. Many plants in the tropical rainforest are trees that are usually larger than those found in temperate forests, often reaching 40 to 60 meters in height. Furthermore, the tree species are diverse, averaging more than 40 species per hectare. This is in sharp contrast with temperate zone forests, where there are rarely more than a few tree species per hectare. Nevertheless, the trees of the tropical rainforest are remarkably homogeneous in appearance. Generally, they branch only near the crown. Because their roots are usually shallow, they often have buttresses at the base of the trunk that provide a firm, broad anchorage. Their leaves are medium-sized, leathery, untoothed and unlobed, and dark green. The bark of these trees is thin and smooth, and their flowers are generally inconspicuous and greenish or whitish in color. Such forests often form several layers of foliage, the lower layers consisting of seedlings of the taller species together with a relatively small number of lower-growing species.

Tropical rainforests are well developed in three major areas of the world (Figure 33–3). The largest is in the Amazon Basin of South America, with extensions into coastal Brazil, Central America, and eastern Mexico, as well as some of the islands in the West Indies. In Africa, there is a large area of rainforest in the Congo (Zaire) Basin, with an extension westward along the coast of Liberia. The third area of rainforest extends from Sri Lanka and western India to Thailand, Cambodia, Vietnam, the Philippines, Indonesia, the large islands of Malaysia, and a narrow strip along the northeastern and eastern coast of Australia. In the North American region, tropical rainforest exists only in southern Mexico and on Puerto Rico and the northern side of Hispaniola in the West Indies.

Temperate rainforests are much simpler in structure than tropical rainforests, with less species diversity. They are dominated by broad-leaved evergreen trees like those in the tropics, and branches of trees are covered by a thick layer of epiphytic plants. Temperate rainforests occur along the eastern side of Australia, in Tasmania, and in parts of New Zealand.

33–9

Epiphytes collect and store water and nutrients from the surrounding air, rain, and dust, creating little patches of soil from accumulated debris. Bromeliads, seen here as leafy tufts, are among the most common of the epiphytes. The leaves of bromeliads merge at their bases to form watertight tanks that, in the larger species, can hold as much as 45 liters of rainwater. These pools of water are microcosms of bacteria, protozoa, larvae, insects, and insect-eaters. Many rainforest mosquitoes breed exclusively in bromeliad tanks. The bromeliads absorb water from their built-in reservoirs and are also supplied with nutrients from the debris. Lianas can also be seen in this photograph.

Rainforests Are Rapidly Being Destroyed

The tropical rainforest now forms about half of the forested area of the Earth, but it is in the process of being destroyed by human activities. The rapidly expanding human population in the tropics, coupled with traditional patterns of land ownership, has made the usual practices of tropical agriculture—clear-cutting of the forest, followed by short-term cultivation—immensely destructive when carried out on such a wide scale. Many kinds of plants, animals, and microorganisms will become extinct in the course of this destruction, as we shall discuss in the next chapter.

One reason for the rapidity with which tropical forests disintegrate under human pressure has to do with the nature of tropical soils. Many of these soils undergo extreme weathering due to the high and constant temperatures and abundant rainfall and as a consequence are relatively infertile. A majority of the many different mineral nutrients are locked up in the plants themselves, not, as in many temperate communities, in the soil. More than half of the soils of the tropics are acidic and deficient in calcium, phosphorus, potassium, magnesium, and other nutrients. Furthermore, the phosphorus in such soils tends to combine with iron or aluminum to form insoluble compounds that are not available for plant growth, and the soils often contain toxic levels of aluminum ions. Plant roots tend to spread out in a thin layer no more than a few centimeters in depth, and they rapidly transfer the nutrients released by the decomposition of fallen leaves and branches back into the living plants. Below this thin layer of topsoil, which is easily destroyed during the process of clearing away the trees, there is virtually no organic matter. For all of these reasons, the cultivation of tropical soils presents formidable problems, even though nutrients may be available in abundance immediately after the soils have been cleared and crops may be good for two or three years. Nevertheless, the tropical forests of the world are being cut and burned at an ever-increasing rate, mainly to produce fields that become completely useless to agriculture within a few years (Figure 33–10). It is estimated that, by early in the next century, most of the tropical rainforests will have disappeared, with the exception of those in the western Amazon Basin and central Africa.

33–10

Of the native vegetation that once covered Madagascar, an island in the Indian Ocean 420 kilometers east of the African mainland, only 10 percent remains, and that is fast disappearing. On Madagascar, as in other tropical rainforests, after the land is cleared, the brick-red soil turns as hard as concrete. Rainwater runs off the barren landscape, choking the rivers with silt and causing floods and mudslides. The sea around Madagascar is rust-red from the island's red soil. The rice paddies are being destroyed by the floods, and people are close to starving.

Savannas and Deciduous Tropical Forests

Savannas consist of grassland with scattered broad-leaved deciduous and evergreen trees, which may occur singly or in groves. Some savannas are dominated by trees, often thickly so, and others are dominated by shrubs. Savannas occupy the region between the evergreen tropical rainforest biome and the desert biome (Figure 33–11). Savanna trees are generally deciduous and lose their leaves in the dry season. Savannas cover large areas of East Africa and are also found on all continents on the margins of rainforests, where the rainfall is seasonal and limiting (Figure 33–3). Savannas also occur in the region between the prairies and temperate deciduous forests and in the region between prairies and taiga in North America and over much of eastern Mexico, Cuba, and southernmost Florida.

Savannas usually have much lower annual rainfall than tropical rainforests—frequently in the range of 90 to 150 centimeters a year. There is also a wider range in average monthly temperatures, owing to the seasonal drought and the sparse covering of vegetation. The thorn forest (cerrado) that covers wide areas in Brazil belongs to the savanna biome. Savannas and similar seasonally deciduous tropical and subtropical plant communities grade into tropical rainforests as the rainfall becomes higher and its seasonal distribution more

33–11

A savanna in East Africa with giraffes surrounded by a herd of impala. The transitional nature of this biome, relative to the character- *istic vegetation of the tropical rainforest biome and the desert biome, is evident in the* *grasses, shrubs, and short trees seen here. The trees are acacias.*

even, and corridors of tropical rainforest (called gallery forest) extend out into the savannas along rivers. At their poleward margins, savannas and related plant communities grade into deserts.

In savannas, because of the scattered distribution of trees, the ground is generally well illuminated, and perennial herbs (mostly grasses) are common. Bulbous plants, which are able to withstand periodic burning, are abundant. Because of the dense cover of perennial herbs made possible by the abundant seasonal rainfall, there are few annual herbs. Epiphytes are also rare.

Savanna trees often have thick bark. They are well branched, but they are seldom more than 15 meters tall. Nearly all are deciduous; they lose their leaves at the start of the dry season and flower when they are leafless. Their leaves are generally smaller than those of the evergreen trees of the rainforest, and so they lose less water by transpiration.

In central and southern Africa, the Indian subcontinent, and Southeast Asia, there are extensive areas of **monsoon forest** (Figure 33–3), which resembles savanna in that it is seasonally dry. In the monsoon forest, however, the density of trees is much higher than in savannas, and the trees and shrubs are mainly broad-leaved

and deciduous. Monsoon forests are one particularly widespread kind of seasonally deciduous forest, and a number of others are found in different parts of the tropics. Monsoon forests are also found in the northern Yucatán Peninsula in Mexico and in eastern Brazil. In the areas where such forests occur, there is high precipitation during portions of the year, when the moisture-bearing monsoon wind blows steadily off the ocean, but there is also a well-defined dry season during which the trees lose their leaves.

Other kinds of deciduous and partly deciduous forests occur in the tropics and subtropics, in areas where there is a pronounced dry season. Most of Florida and extensive areas throughout the southeastern United States are covered by subtropical mixed forests. In these forests, pines (Figure 33–12) and other evergreen trees are mixed with deciduous trees; the precipitation occurs mainly in the summer. Tropical mixed forests, in which evergreen trees and shrubs are better represented than in subtropical mixed forests, occur locally in eastern and southern Brazil and northern Australia (Figure 33–3). Related communities—subtropical and tropical mixed forests and southern woodland and scrub—are found in Argentina.

33–12
Slash pine (Pinus elliottii) *is one of the widespread evergreen tree species found in the subtropical mixed forests of the southeastern United States.*

The existence of the savannas and monsoon forests of Africa and other regions depends to a large extent on periodic burning. The winter is long and hot, and people frequently burn the plain to encourage the growth of young grass to support domestic herds of both grazing animals and game animals.

Deserts

The great **deserts** of the world are all located in the zones of atmospheric high pressure that flank the tropics at about 30° north latitude and 30° south latitude, and they extend poleward in the interior of the large continents (Figure 33–3). Many deserts receive less than 20 centimeters of rainfall per year. In the Atacama Desert of coastal Peru and northern Chile, the average rainfall is less than 2 centimeters per year. Extensive deserts are located in North Africa and in the southern part of the African continent, where the Namib Desert is inhabited by some of the world's most extraordinary organisms, including *Welwitschia* (see Figure 20–43). Other deserts occur in the Near East, in southeast Mongolia and northern China, in western North America and western and southern South America, and in Australia. The Sahara, which extends all the way from the Atlantic coast of Africa to the Arabian peninsula, is the largest desert in the world. Seventy percent of the entire continent of Australia is covered in semiarid or arid areas. Less than 5 percent of North America is desert.

The temperatures in many deserts—the *hot deserts*—are very high; summer temperatures of more than 36°C

are common in some deserts. Other deserts are cooler. An example of a *cold desert* is the Great Basin Desert of western North America, which lies between the Sierra Nevada–Cascade Mountain system and the Rocky Mountains. Here there are only a few weeks of high temperatures each year. In general, however, the water vapor in desert air, even when relatively abundant, fails to condense because of the high temperatures. Because the cover of vegetation is generally sparse, heat normally radiates from deserts rapidly at night, creating large differences in temperature in the course of a 24-hour period.

The annual distribution of rainfall in deserts generally reflects that of the adjacent areas. On the equatorial side, it rains in the summer; on the poleward side, it rains in the winter. Between the two, as in the lowlands of Arizona, there may be two annual peaks of precipitation. As a result, such an area generally has two periods of active plant growth—one in the winter and one in the summer—and different plants are active in each period. In general, the patterns of activity of desert plants reflect the origins of the plants. Characteristically, plant species that have migrated from areas where they grew actively in the winter continue doing so in the desert. Similarly, plant species that have migrated into the deserts from areas where they grew during periods of summer rainfall continue to grow actively in the summer in the desert, wherever a sufficient amount of summer rainfall makes this possible.

Desert Plants Are Adapted to Low Precipitation and Extremes of Temperature

Annual plants are better represented, both in number and in kind, in the deserts and semiarid regions of the world than anywhere else. Because of the erratic supply of water, perennial herbs do not succeed well in deserts and semideserts, and there is no dense covering of perennials to inhibit the growth of annuals. Annuals, with their very active growth, can germinate and complete their life cycles in the open areas during the limited periods when water is available. The seeds of these plants can survive in the soil during the long periods of drought, which sometimes extend over many years. Then, when sufficient water is available to induce their germination and support their growth to flowering, the seeds can germinate rapidly.

The relatively few perennial herbs that grow in the desert are often bulbous and dormant for much of the year. Most of the taller plants either are succulents—such as the cacti, euphorbias, and other characteristic desert plants (Figure 33–13)—or else have small leaves that either are leathery or are shed during unfavorable seasons (Figure 33–14). Usually the leaves have a thicker cuticle and fewer stomata than those of plants in less

33–13

Desert. Shown here are some representative plants of the principal deserts of North America. (a) The Sonoran Desert stretches from southern California to western Arizona and down into Mexico. A dominant plant, the giant saguaro cactus (Carnegiea gigantea) is often as much as 15 meters high, with a widespreading network of shallow roots. Water is stored in a thickened stem, which expands, accordionlike, after a rainfall. (b) To the east of the Sonoran is the Chihuahuan Desert, one of whose principal plants is the agave, or century plant (Agave shawii), a monocot. (c) North of the Sonoran is the Mojave Desert, whose characteristic plant is the Joshua tree (Yucca brevifolia). This plant was named by early Mormon colonists who thought that its form resembled that of a bearded patriarch gesticulating in prayer. The Mojave contains Death Valley, the lowest point on the continent (90 meters below sea level), only 130 kilometers from Mount Whitney, with an elevation of more than 4000 meters. (d) The Mojave blends into the Great Basin Desert, a cold desert bounded by the Sierra Nevada to the west and the Rockies to the east. It is the largest and bleakest of the American deserts. The dominant plant is sagebrush (Artemisia tridentata), shown here with the snow-covered Sierra Nevada mountains in the background.

(a)

(b)

(c)

(d)

33–14

One of the most characteristic plants of the Mojave, Sonoran, and Chihuahuan deserts of North America is the creosote bush (Larrea divaricata), which has small, leathery water-conserving leaves. Creosote bushes in the Mojave Desert may form circular or elliptical clones because of new branch production at the periphery of stem crowns and the segmentation and death of older stem segments, resulting in a ring of satellite bushes around a central bare area. The latter usually accumulates a mound of sand, which may reach a depth of about half a meter and may function as a water-storage medium, thus contributing to the survival of creosote bushes in periods of drought. Some clones may attain extreme ages: the King Clone, shown here, is estimated to be nearly 12,000 years old. Such ancient clones started from seeds that germinated near the end of the last glacial expansion.

How Does a Cactus Function?

The barrel cactus *Ferocactus acanthodes*, which grows along the northwestern edge of the Sonoran Desert in southern California, gets its name because it is shaped like a barrel. It looks as if its stem has been folded like a fan, and it is covered with spines like a porcupine. Why should this desert plant present such an unusual and formidable appearance?

When the barrel cactus is full of water, the folds swell and are barely visible, but when the plant dries, the folds are deep and the stem is able to contract without crushing the cells. This ridge-and-valley folding of the stem has other advantages, too. Deep inside the valleys between the folds are the stomata. The spines help to break up the wind currents, and thus the valleys serve as protected retreats where dry desert winds cannot easily reach to carry moist air away from the vicinity of the stomatal chambers. These formidable spines also help to protect the cactus from the rodents and birds that are in constant search of water, even going so far as to steal it from a succulent stem. In a study conducted by Park Nobel, of the University of California, Los Angeles, barrel cacti stored enough water in their succulent stems to permit them to open their stomata for about 40 days after the soil became too dry to furnish them with any additional water. Under such conditions, many of the fine roots are sloughed, to prevent water loss to the soil. After seven months of drought, stomatal activity ceased, and the osmotic potential of the stem was more than double the value it had been during wet periods, despite the ability of the stem to fold and shrink. When rain finally returned, the shallow root systems (with a mean depth of only about 8 centimeters) took in water so rapidly that the stomata were fully functional again within 24 hours of rainfall.

But these are not the only mechanisms the barrel cactus uses for conserving water. Like many other desert plants, the barrel cactus exhibits crassulacean acid metabolism (CAM). It opens its stomata only at night and so undergoes gas exchange with cooler air, which can hold less water than warmer air. Consequently, the plant loses less water to the atmosphere through transpiration. Its ratio of mass of water transpired to mass of CO_2 fixed is only about 70:1 for the entire year. This is considerably lower than for a typical C_3 plant, which requires larger quantities of water to fix an equivalent amount of carbon but has a higher maximum rate of photosynthesis.

Since seedlings of the barrel cactus, unlike their parents, cannot tolerate extremely high temperatures and prolonged drought, they survive only in certain years and in protected microhabitats. By the age of 26 years, these plants have usually grown to only about 34 centimeters tall, adding about 10 percent to the mass of their stems each year.

Barrel cactus. (a) In a hydrated state, and (b) in a dehydrated state.

(a)   (b)

arid regions. Many desert plants have green stems, rich in chlorophyll, which may contribute large amounts of photosynthate to the plants; such plants often lack leaves, as do most species of cacti. Many of the succulent plants have adopted CAM photosynthesis (page 147) and absorb carbon dioxide only at night; their stomata remain closed during the day. C_4 photosynthesis is likewise more common among the plants of deserts and other seasonally dry, warm habitats than it is elsewhere (page 146).

(a)

(b)

33–15

(a) *Juniper* (Juniperus osteosperma) *savanna in the Great Basin near Wellington, Utah.* **(b)** *Cold winters are characteristic of the pinyon-juniper savannas and woodlands of the Great Basin, as shown here south of Moab, Utah.*

The temperatures at which the maximum photosynthetic rate occurs in desert plants are often much higher than the corresponding temperatures for the plants of other biomes. For example, in *Tidestromia oblongifolia*, a C_4 perennial herb common in the Death Valley region of California and Nevada, the maximum photosynthetic rate is achieved at temperatures between 45° and 50°C in full sunlight in midsummer. In addition, the leaves of many desert plants are small, the reduction in surface area being another adaptation to conserving water. The leaves may also be oriented in such a way as to minimize heat absorption, or they may be covered with a dense coat of hairs (trichomes) that reflect solar radiation. The trees and shrubs of deserts either have wide-ranging roots that effectively absorb the periodic rainfall or are restricted to washes and arroyos, along which underground water is concentrated.

The juniper savannas that are such a familiar sight throughout the interior of the western United States (Figure 33–15) occur in areas of cold desert, but generally at higher elevations. During the pluvial periods of the Pleistocene epoch, at the times of maximum expansion of the continental glaciers, such plant communities moved down into lowland areas that have subsequently become, for the most part, deserts. Because the preservation of plant materials is so good in the dry desert air, the shifting patterns of vegetation in a particular area can be traced in the dried leaves and stems in old nests of pack rats.

Grasslands

Grasslands form the zone that lies between deserts and temperate deciduous forests, occurring where the amount of rainfall is intermediate between the amounts characteristic of those two biomes. Included in grasslands are a wide variety of plant communities, some of which intergrade with savannas, others with deserts, and still others with temperate deciduous forests. As the amount of precipitation decreases, the major grasslands of the world often grade into deserts toward the equator. The more fertile grasslands receive as much as 100 centimeters of precipitation per year; they grade into temperate deciduous forests, where the moisture supply is even more abundant. Grasslands are characterized by cold winters and, unlike savannas, generally lack trees, except along streams.

Grasslands have traditionally been heavily exploited for agriculture, both as pasture and, when plowed up, for cultivation. Much effort has been devoted to learning how to convert other biomes into grassland and, once the conversion has been made, how to maintain them in this condition. The most productive soils for temperate agriculture are found in areas formerly occupied by tall-grass prairies (Figure 30–5).

Grasslands generally occur over large areas in the interior portions of the continents, most notably in North America and across Eurasia (Figure 33–3). In North America, there is a transition from the more desertlike, western, shortgrass prairie (the Great Plains), through the moister, richer, tallgrass prairie (the Corn Belt), to the eastern temperate deciduous forest (Figure 33–16). Grasslands become progressively drier with increasing distance from the Atlantic Ocean and the Gulf of Mexico, which are the major sources of moisture-bearing

(a)

(b)

33–16

Grasslands. The grasslands of North America include large regions of shortgrass and tall-grass prairie. (a) A female bison nursing her calf in the shortgrass prairie of Custer State Park, South Dakota. *(b) A June day on a tallgrass prairie in North Dakota. The cotton* wood grove by the prairie creek is characteristic of this biome.

33–17

Coreopsis dominating a glade in temperate deciduous forest near St. Louis, Missouri. Such openings in the forest, which usually occur in areas of shallow soil, are often dominated by prairie plants that form continuous stands beyond the borders of the forest.

winds in the eastern half of the North American continent. Farther north, grasslands become moister again as evaporation decreases at relatively cool temperatures. "Typical" shortgrass prairie occurs under near-drought conditions. In years with greater moisture, taller grasses (but not the very tall ones of the tallgrass prairie) tend to predominate. In many forested areas in the central and eastern United States, there are open areas, called glades, that are typically dominated by prairie plants (Figure 33–17).

Perennial bunchgrasses and sod-forming grasses are dominant, but other perennial herbs are common. Although the growth of grassland plants is seasonal, there is little room for the development of annual herbs, and these are essentially absent from this biome. Occasionally, annuals and introduced weeds become established in disturbed areas, such as around the burrows of animals, near buildings, and along roadsides and railroads.

Fires play a critical role in the development of grasslands. Toward the deserts, there is too little biomass to sustain the spread of fires. Toward moister areas, such as temperate deciduous forests, trees often grow rapidly enough to escape the influence of ground fires, at least on fertile sites. Fire selects against plants with permanent aboveground parts and favors grasses, sedges, and

33–18

The prairie soils were once so bound together with the roots of grasses that they could not be cultivated until adequate plows were developed. But once the plants were removed by overgrazing or careless cultivation, prairie soils rapidly deteriorated and were blown away by the wind. This photograph of a badly eroded farmyard, taken in Oklahoma in 1937, vividly recalls the "dust bowl" conditions that led many thousands of people to migrate away from the central United States. John Steinbeck's novel The Grapes of Wrath *was based on the experiences of these migrants.*

other herbs that have basal meristems that are not easily destroyed by fire or by grazers. When disturbed, grasslands have often changed into either forests or deserts. Thus, the grasslands that once stretched from southern Arizona to western Texas, which were encountered by early European settlers, were largely overgrazed and converted to deserts during the past century. The failure of farmers and graziers to manage grasslands properly led to the "dust bowl" disasters of the central United States in the 1930s (Figure 33–18).

All of the great grasslands of the world were once inhabited by herds of grazing mammals associated with large predators. Such herds were widespread in the periods of maximum expansion of the glaciers during the Pleistocene epoch, when they were widely hunted by our ancestors. Many of these grazing mammals, such as the American bison, have since been hunted almost to extinction. They survive today mainly in refuges, having given way to herds of domestic animals and cultivated fields.

Temperate Deciduous Forests

Temperate deciduous forests are almost absent in the Southern Hemisphere, but they are represented on all the major land masses of the north (Figure 33–3). This type of forest reaches its best development in areas with warm summers and relatively cool winters (Figure 33–19). Annual precipitation generally ranges from about 75 to 250 centimeters and is either distributed

evenly throughout the year or concentrated somewhat during the summer months. The deciduous (leaf-dropping) habit may be related to the unavailability of water during much of the winter, which is a consequence of soil temperatures below the freezing point. These conditions delay the movement of soil water to the roots. In the colder regions occupied by the temperate deciduous forest, the soils are frozen for many months. In effect, the winters in regions occupied by temperate deciduous forest are analogous to the hot, dry seasons that occur in savannas and deciduous tropical forests. The unavailability of water in both biomes is associated with the seasonal loss of leaves in trees and shrubs.

The ecology of temperate deciduous forests differs from that of evergreen forests in the nature of the annual cycle of growth (Figure 33–20). In winter, the trees are leafless and their metabolic activity is greatly reduced. In spring, a variety of herbaceous plants bursts forth in profusion on the well-illuminated forest floor (Figure 33–19a). Some species (the **spring ephemerals**) have leaves that emerge fully formed from bulbs or rhizomes and complete their period of activity in sunlight before the trees leaf out overhead. Others (for example, **early** and **late summer species,** and **evergreen species**) emerge more slowly and sustain photosynthetic activity longer and later during the shady conditions prevailing in summer. Summer-active species generally have leaves that are broader but thinner in cross section than those active only in spring and usually have smaller storage organs than species with shorter growing periods. Forest herbs vary in height from a few centimeters to more

33–19
Temperate deciduous forest. Representative plants of the temperate deciduous forests of North America. (a) A beech and maple forest in Michigan, photographed in the spring. The forest floor is carpeted with large-flowered trillium (Trillium grandiflorum). (b) In the fall, as seen here in a forest in the southern Appalachian mountains, the leaves of the maples and eastern sourwoods (Oxydendrum) turn a beautiful scarlet.

(a)

(b)

than one meter. Variation in leaf height appears to be related to the typical density of the competing foliage that different species encounter in their microenvironments (see the essay on page 778). Most of the species that ripen their seeds in spring are dispersed by ants (pages 548 and 549), which are active when few other dispersers are present. Most of the species that ripen seeds in the fall, however, are dispersed by birds, at a season that coincides with the height of fall migration. Very few annual plants occur in deciduous forests, probably because the dense plant cover in the moist understory puts seedlings at a strong competitive disadvantage.

The soils in regions occupied by temperate deciduous forest are usually acidic, and they tend to contain relatively low amounts of nutrients. For these reasons, such soils are easily depleted of nutrients after the forests are cleared, and they may become highly infertile following years of intensive cultivation. Prairie soils are far more fertile and have characteristics much more favorable to sustained cultivation.

One of the outstanding characteristics of the temperate deciduous forest is that the plants found in each of its three main regions in the Northern Hemisphere are similar to one another, often representing related species of particular genera. For example, the herbaceous plants of the deciduous forests of east Asia generally resemble those of eastern North America more closely than either group resembles those of western North America, where temperate deciduous forests mostly disappeared millions of years ago.

33–20

The four seasons in a temperate deciduous forest in Illinois. The trees leaf out early in spring and begin to manufacture food; they lose their leaves in autumn and enter an essentially dormant state, in which they pass the unfavorable growing conditions of winter. Many herbs grow under the trees (see Figure 33–19a), and a number of them flower very early in spring, before the tree leaves reach full size and shade the forest floor. In spring, most of the trees produce abundant pollen, which is carried by the wind.

Jobs versus Owls

Some of the world's most magnificent forests grow along the western side of the Cascade Mountains in Washington, Oregon, and northern California. In the Olympic Peninsula of Washington, these forests are truly temperate rainforests, with more than 400 centimeters of rain per year; other forests, more distant from the coast and farther south, are drier. In ancient forests, also known as old-growth forests, the huge trees are more than 250 years old; some of them reach heights of nearly 100 meters and diameters of approximately 2 meters. These forests are biologically diverse, representing a late successional stage with well-developed understories of herbs, ferns, mosses, and other plants (see Chapter 32).

Originally, such ancient forests probably covered more than 76,000 square kilometers in the Pacific states. Estimates of the area now remaining vary widely because of disagreements about what constitutes an ancient forest, but it is not more than one-third of the original area and may be as little as one-sixth. Furthermore, the forests that are left tend to be highly fragmented, with the penetration of sun, wind, and other factors that do not affect undisturbed forest causing the loss of species and leading to diminished genetic variability.

An enormous controversy erupted in Oregon and Washington in 1989 when the U.S. Forest Service proposed taking more than 1500 square kilometers of national forest out of production in order to protect the northern spotted owl, which had recently been declared an endangered species by the U.S. Fish and Wildlife Service. This set-aside would reduce the available timber supply by about 5 percent and result in the loss of about 3000 jobs; it is projected to be of sufficient size to protect about 1300 pairs of owls.

In 1993, the Clinton administration proposed a compromise that angered both sides in this controversy. The U.S. Congress subsequently passed salvage legislation, which allowed loggers to remove standing dead and diseased trees from these forests. Loggers have also been confronted by fishermen, who have petitioned for endangered species status of the sockeye and coho salmon. The silt from clear-cutting may clog the egg-laying sites of salmon in forest streams. The arguments continue.

In a broader sense, the controversy in the Pacific Northwest does not concern owls versus jobs at all; it concerns regional values and the way in which we choose to treat the remnants of ancient forest, centuries old and fundamentally important to the appearance and sound ecological functioning of the region. Economically, tourism is even more important than lumbering in the Pacific Northwest, but the benefits of tourism accrue more to city-dwellers than to the inhabitants of rural towns, who may be directly dependent on the timber industry for their livelihoods. If all the ancient trees are cut—the only remaining ones, for all practical purposes, are on federal lands—the jobs will still run out in a few years, and the transition to sustainable forestry, or to other forms of employment, must then be made in any case. Although relatively few species of plants and animals seem to be absolutely restricted to ancient forests, these forests do protect many kinds of organisms besides the owls.

As is true for ecological problems in general, people must find new ways to determine their overall priorities and not necessarily chop down every tree in an effort to "protect" jobs over the short term. A country that has the will to do so can find better ways to take care of its people and encourage the development of jobs than to use irreplaceable natural resources. Meanwhile, the high prices paid by Japan for unsawn logs from the Pacific Northwest will continue to ensure the rapid destruction of the ancient forests, and the owl will remain as a symbol of the deeper regional values that are being permanently lost.

Northern spotted owl, found in old-growth forests from northern California to southern British Columbia, Canada

Temperate Mixed and Coniferous Forests

Bordering the deciduous forests on the north are mixed forests, in which conifers form an important element along with the deciduous trees. Such temperate mixed forests (Figure 33–21) are characteristic of the Great Lakes–Saint Lawrence River region, much of the southeastern United States, eastern Europe, the northern and eastern border regions of Manchuria (in northeast China) and adjacent Siberia, eastern Korea, and northern Japan (Figure 33–3). Temperate mixed forests sometimes occur in areas having colder winters than those in areas of temperate deciduous forest. Temperate mixed forests are also partly characteristic of regions with more nutrient-poor soils or with less seasonal environments. **Temperate mixed forests** represent a broad transitional

33–21
Temperate mixed forest. In this forest in southern Ontario, the evergreen conifers appear dark green and the deciduous trees are showing their fall colors.

33–22
Redwoods (Sequoia sempervirens) are a prominent feature of the mixed west-coast forests of California. Watered by the frequent fogs of the region during the dry summers and protected from freezing temperatures by their proximity to the ocean, redwoods often form spectacular groves, many of which, like the one shown here in Muir Woods near San Francisco, are now protected in parks and reserves.

zone between temperate deciduous forests to the south and taiga to the north.

Most of the deciduous trees and associated herbs were eliminated in western North America during the latter half of the Cenozoic era as the amount of summer rainfall was greatly reduced. In their place now grow the mountain **coniferous forests** and mixed west-coast forests of western North America, which contain such trees as the coast redwood (*Sequoia sempervirens;* Figure 33–22), the big tree or giant sequoia (*Sequoiadendron giganteum*), the Douglas fir (*Pseudotsuga menziesii*), and the sugar pine (*Pinus lambertiana;* see Figure 32–14b), all of which were much more widely distributed in the past (see "Jobs versus Owls" on page 815). Similar kinds of vegetation are found in areas with similar climatic characteristics in Scandinavia, central Europe, the Pyrenees, the Caucasus, the Urals, southern Tibet, and the Himalayas northward to eastern Siberia. Vegetation types that resemble these are also found in areas in western South America, central New Guinea, southwestern New Zealand, the southern Arabian peninsula, Ethiopia,

and the mountains of Central Africa (Figure 33–3). At higher elevations in these regions are found the open grassland communities called alpine tundra (Figure 33–23), which in North America are intermixed with mountain forests from the Brooks Range in southern Alaska southward into the Rocky Mountains, the Cascades, and the Sierra Nevada.

Mediterranean Scrub

Highly distinctive **scrub communities** have evolved from mixed deciduous–evergreen forests in areas with Mediterranean climates, which are characterized by cool, moist winters and hot, dry summers (Figure 33–24). Such climates are found along the shores of the Mediterranean Sea, over a large part of California (and for a short distance to the north and south of that state), in central Chile, on the southern coast of Africa, and along portions of the coast of southern and southwestern Australia (Figure 33–3). The plants in these areas—

33–23

Alpine tundra on the Olympic Peninsula in Washington State. Such alpine tundra is comparable in many respects to the Arctic tundra found hundreds of miles to the north. Here, however, forested slopes are found within a hundred meters or so of the alpine meadows.

often evergreen or summer-deciduous trees and shrubs—have relatively short growing seasons that are restricted to the cool part of the year, when moisture is relatively abundant. They may lock up nutrients efficiently in their evergreen leaves. In Mediterranean climates, the luxuriant growth of spring is followed by drought and dormancy during the summer.

Fire is a prominent ecological factor in Mediterranean-type vegetation, and fires apparently occurred regularly in such plant communities long before they were invaded by human beings. Today, fire can be a serious problem in such areas as southern California, where homes extend far up into the chaparral. **Chaparral** is the word used in California and elsewhere in western North America for evergreen, often spiny shrubs that typically form dense thickets. Similar vegetation types have evolved elsewhere, particularly in areas characterized by Mediterranean climates, and have been given different names locally. The equivalent vegetation around the Mediterranean Sea is called *maquis* (Figure 33–1), in Chile it is known as *matorral*, and in South Africa, *fynbos*.

As is true of the deserts of the world, each area of Mediterranean climate is isolated from the others, and each has its own distinctive assemblage of plants and animals. The degree of ecological convergence is high, however, and the chaparral of California is quite similar in appearance to the matorral of Chile or the maquis of the Mediterranean, even though the plants are mostly unrelated. Seasonal drought enhances the importance of edaphic (soil-related) and biotic variation, and small dif-

33–24

Mediterranean scrub. Chaparral in the mountains near Los Angeles, California. This sort of plant community, consisting of broad-leaved, drought-resistant, and often spiny evergreen shrubs, occurs only in limited areas of the world, but it has evolved independently on five continents in areas having Mediterranean (summer-dry) climates. (See also Figure 33–1.)

ferences in precipitation often have profound effects on the vegetation and animal life present in the area. Hence, these areas often have high proportions of extremely local species of plants and animals, many of them now in great danger of extinction. In their modern form, these areas have already been profoundly changed by people. Much of their vegetation is now in a very different condition from that occurring before people occupied the areas with their grazing animals—for example, today's areas might be inhabited by more shrubs and fewer trees than formerly, or by more spiny and poisonous plants.

Taiga

This northern coniferous forest, or boreal forest, is characterized by a persistent cover of snow in the winter and by a severe climate toward the north. In the southern reaches of the taiga, the trees are taller and more luxuriant, often reaching 75 meters or more in height (Figure 33–25). In its main, northern area, however, the trees are shorter, and thousands of square kilometers are covered by this uniform forest, with relatively few species of plants and animals (Figure 33–26). **Taiga** is the Russian word for vegetation of the sort found in this biome, which extends over much of Russia, Scandinavia, and northern North America (Figure 33–2). It is evergreen almost everywhere except in large areas of northeastern Siberia, where larches *(Larix)*, which are deciduous conifers, are dominant.

Taiga occurs in the interior of large continental masses at the appropriate latitudes and in North America extends along the Pacific coast south to California. In such regions, extreme temperatures range from −50° to 35°C. Taiga is flanked on the south by montane forests (as in western North America), deciduous forests, savannas, or grassland, depending on the amount of precipitation in the region. Because continental masses do not occur at the appropriate latitudes in the Southern Hemisphere, taiga is absent there. Owing to the influence of the prevailing westerlies blowing over relatively warm ocean currents between 40° and 50° north latitude, the western portions of North America and Eurasia are characterized by milder climates than their eastern portions. Consequently, taiga is found somewhat farther north toward the Pacific coast than it is along the Atlantic coast in North America, and the same is true of the individual distributions of many kinds of plants and animals that inhabit the biome. The northern limits of taiga are ultimately determined by the severity of the Arctic climate, reaching a limit where the maximum monthly temperature is approximately 10°C.

In the extensive northern reaches of the taiga, most of the precipitation falls in the summer; the cold winter air in these regions has a very low moisture content. The an-

33–25
Taiga. The dense, tall forest that grows on the rainy slopes and flats of the Olympic Peninsula in Washington is the southern expression of the taiga, or northern coniferous forest. Epiphytic mosses, liverworts, Selaginella, and lichens often grow luxuriantly on the trees, as seen here.

33–26
The northern taiga, which covers hundreds of thousands of square kilometers in the cooler part of the North Temperate Zone, is dominated by white spruce (Picea glauca) and larch (Larix). This photograph was taken in northern Manitoba, Canada. In the more northern part of their range, the trees are smaller than shown here.

nual total precipitation usually amounts to less than 30 centimeters. The rate of evaporation is low, however, and lakes, bogs, and marshes are common (Figure 33–26). More than three-quarters of the northern reaches of the taiga is underlain by permanent ice, or **permafrost,** usually within less than a meter of the surface; the permafrost tends to trap surface water and form lakes. Fires are common in the taiga, and they result in generally warmer, more productive sites for at least 10 to 20 years afterward, owing to the local melting of the permafrost. In general, taiga soils are highly acidic, very low in nutrients, and poorly suited for agriculture.

Species of a few genera of trees are common in the northern taiga, including spruce *(Picea)*, larch *(Larix)*, fir *(Abies)*, and poplar *(Populus)*. Among the more common shrubs are dewberries *(Rubus)*, Labrador tea *(Rhododendron)*, willows *(Salix)*, birches *(Betula)*, and alders *(Alnus)*. Occasionally, in warm, dry areas, groves of pine *(Pinus)* are found. The members of all these genera of trees and shrubs are ectomycorrhizal, and they occur in dense stands consisting of only one or a very few species. Perennial herbs are common, and mosses and lichens are especially prevalent, often forming luxuriant masses. Annual plants are essentially absent.

The massive evergreen coniferous forests of the Pacific Northwest of the United States and Canada (Figure 33–3) are adaptations to the winter-wet/summer-dry environment of that region. Because photosynthesis is limited by lack of moisture during the warm season, deciduous trees are at a disadvantage and are usually found only along stream banks. The evergreen conifers, however, can synthesize carbohydrates all year round and, because of their massive size, can store water and nutrients for use during the dry season. Their thick barks and high crowns protect them from the fires characteristic of the region.

At its northern limits, taiga grades unevenly into tundra. In both these biomes, the days are long during the relatively short growing season; north of the Arctic Circle, the sun does not set during at least part of the summer. Because of the seasonally abundant light and favorable temperatures, cool-season cultivated plants, such as cabbages *(Brassica oleracea* var. *capitata)*, may grow rapidly in cleared areas in the taiga, attaining large sizes in a remarkably short period of time. Yet the infertile, highly leached soils of the taiga do not allow most forms of agriculture.

Arctic Tundra

Arctic tundra is a treeless biome that extends to the farthest northern limits of plant growth (Figure 33–27). It occupies an enormous area: fully one-fifth of the Earth's land surface (Figure 33–3). The majority of tundra can be found in the Arctic, mostly above the Arctic Circle, al-

(a)

(b)

33–27

Arctic tundra. (a) Arctic wet coastal tundra near Prudhoe Bay, Alaska, in late summer. The reddish brown plants are the grass Arctophila fulva, *the green ones are the sedge* Carex aquatilis. *Note that the water table is above* the surface because of the presence of permafrost; such conditions are characteristic of the Arctic tundra. **(b)** Arctic tundra at Barrow, Alaska. A clone of the sedge Eriophrum angustifolium is growing out into a solid patch of another species of the same genus, E. scheuchzeri. Vegetative reproduction, like that in E. angustifolium illustrated here, is characteristic of many tundra plants.

though it extends farther south along the eastern sides of the continents than along the western sides. The Arctic tundra essentially constitutes one huge band across Eurasia and North America, with alpine tundra, more closely related to the adjacent mountain forests, extending southward in the mountains (Figure 33–23; see also Figure 33–6). Some species of plants that occur in the Arctic tundra have wide circumpolar ranges.

The entire region occupied by Arctic tundra is underlain by permafrost. The soils are acidic to neutral, low in nutrients, and generally poor for agriculture. Even though precipitation is usually less than 25 centimeters per year, much of it is held near the surface by underlying permafrost, and the ground is usually wet. Fixed nitrogen is generally in very short supply, and there are only a few species of legumes and other plants with symbiotic bacteria that fix atmospheric nitrogen. The evaporation rate is low because of the relatively high moisture content of the air and low temperatures. In contrast, some tundra areas are so dry as to constitute true polar deserts. Most of the land north of 75° north latitude is a desert or semidesert, with few plants taller

than 5 centimeters. This is a result of the very low precipitation in both summer and winter and the very cold winter temperatures.

For plant growth to occur at all, the mean temperature must be above freezing for at least one month of the year. The growing season (the time from one killing frost to the next) in many areas of Arctic tundra is less than two months. Several genera of low shrubs, including birch (*Betula*), willow (*Salix*), blueberry (*Vaccinium*), and Labrador tea (*Rhododendron*), are common in the tundra, and there are a number of genera of perennial herbs but only one native annual species (*Koenigia islandica*). Many of the plants that grow in the Arctic tundra, including a number of the grasses and sedges, are evergreen, thus enabling them to initiate photosynthesis soon after adequate light, moisture, and temperature become available. Plant height is controlled primarily by snow depth in winter; large woody plants are absent because of their relatively high energy allocation to unproductive stem tissue, which cannot be supported given the short, cool growing season. A number of tundra plants have relatively large, showy flowers, the production of which

Summary TABLE   Some Characteristics of the Earth's Principal Biomes

BIOME	TEMPERATURE AND PRECIPITATION	CHARACTERISTIC PLANTS	MISCELLANEOUS FEATURES
Rainforests	High temperature and high rainfall year round.	Broad-leaved evergreen trees, epiphytes, and lianas.	The biome with the greatest diversity of species. Infertile soils.
Savannas and deciduous tropical forests	High temperature and seasonal drought.	Grasslands with scattered broad-leaved deciduous or evergreen trees or shrubs.	Periodic burning is an important aspect.
Deserts	Precipitation generally very low except for occasional peaks; maximum temperature varies with the type of desert.	Succulents such as cacti; annual herbs.	Adaptations include small leaves, thick cuticles, and photosynthetic rates with high maximum temperatures.
Grasslands	Moderately low precipitation; cold winters and warm summers.	Perennial bunchgrasses and sod-forming grasses.	Heavily exploited for agriculture.
Temperate deciduous forests	Moderate precipitation evenly distributed; cool winters and warm summers.	Deciduous trees and many perennial herbs.	The dominant herbaceous plants vary with the seasons.
Temperate mixed and coniferous forests	Moderately low precipitation and moderately cold winters.	Mixtures of deciduous trees and conifers.	Occur as a transition zone north of the deciduous forest. Also found in areas with nutrient-poor soils or with less seasonal environments.
Mediterranean scrub	Cool, moist winters and hot, dry summers.	Evergreen or summer-deciduous, drought-resistant trees and shrubs in dense thickets.	Called chaparral in California and maquis around the Mediterranean Sea.
Taiga	Moderately low precipitation and cold winters, although in the Pacific Northwest the winters are very wet.	Forest of evergreen trees.	Soils are highly acidic and very low in nutrients. Permafrost may be present.
Arctic tundra	Very low precipitation in both summer and winter; very cold winters.	Low shrubs, grasses, sedges, and lichens.	Permafrost present throughout. Much of the biomass is underground.

costs the plants considerable quantities of energy. Such flowers hold energy-rich rewards for their pollinators—necessary rewards, given the low temperatures that prevail at high latitudes. Vegetative propagation is characteristic of many of the perennials, and this may be correlated with the uncertainties of setting seed during the brief Arctic summer. Much of the biomass of tundra plants—from half to as much as 98 percent—is underground, consisting not only of roots but also of rhizomes and other kinds of underground stems.

North of the Arctic tundra is ice desert, where physical conditions are even more extreme and vegetation is absent. Ice desert is characteristic of the interior of Greenland; Svalbard, a small group of islands off Norway; and Novaya Zemlya, two islands off the north coast of Siberia. Much of Antarctica, not mapped in Figure 33–3, is also covered by ice.

Summary

Biomes are terrestrial ecosystems characterized by distinctive vegetation. The distribution of biomes is a result of complex interactions between the distribution of heat from the sun, air-circulation patterns, and geologic features. These factors cause wide differences in temperature and precipitation from place to place and season to season. In addition to climate, differences in the surfaces of the continents, such as soil composition and altitude, affect the kinds of plant and animal life found in the various biomes on Earth.

Tropical Rainforests Have a Great Diversity of Species, with Few Individuals per Species

Tropical rainforest, in which neither water nor low temperature is a limiting factor, is by far the richest biome in terms of number of species. The trees are evergreen and characterized by medium-sized leathery leaves. A poorly developed layer of herbs grows on the forest floor, but there are many vines and epiphytes at higher levels. Tropical soils are often acidic and very poor in nutrients; such soils lose their fertility rapidly when the forest is cleared.

Savannas and Deciduous Tropical Forests Occur Where Rainfall Is Seasonal

Mostly tropical and subtropical communities that are characterized by a seasonal drought are termed savannas, subtropical mixed forests, monsoon forests, tropical mixed forests, and southern woodland and scrub. The trees and shrubs of these communities are wholly or partly deciduous, shedding their leaves during times of drought. Herbaceous perennials are common. Savannas also occur between the prairies and temperate deciduous forests and between the prairies and the taiga in North America. Subtropical mixed forests cover most of Florida and the Coastal Plain of the southeastern United States. In these forests evergreen trees, such as pines, grow intermixed with deciduous trees.

Desert Plants Are Adapted to Low Precipitation and Extremes of Temperature

Away from the equator, tropical and subtropical communities grade into deserts and semideserts, which are characterized by low precipitation and often by high daytime temperatures during at least part of the year. Succulent plants and annual herbs are common in deserts.

Grasslands Occur Where Precipitation and Fire Interact to Support Grasses but Not Trees

Grasslands, which intergrade with savannas, deserts, and temperate forests, are characterized by a general lack of trees, except along streams. The most productive soils for temperate agriculture are grassland soils.

Temperate Deciduous Forests Are Made up of Leaf-Shedding Trees and Many Types of Perennial Herbs

In the temperate deciduous forests, most of the trees lose their leaves during the cold (usually snowy) winters, when moisture may be unavailable for growth. Many genera are common to the temperate deciduous forests of eastern North America and eastern Asia. Temperate deciduous forests are bordered by temperate mixed and coniferous forests to the north, in which conifers play an important role.

Mediterranean Scrub Is Characterized by Evergreen, Drought-Resistant Shrubs or Trees That Form Thickets

Distinctive scrub communities, called chaparral in North America and maquis in the Mediterranean region, have evolved in the five widely separated areas of the world with a Mediterranean climate—a dry summer and a cool, rainy winter growing season. Such communities occur in western North and South America, around the Mediterranean, in the Cape Region of South Africa, and in southwestern Australia.

The Taiga Is Characterized by Forests of Evergreen Trees

The taiga is a vast northern coniferous forest that extends in unbroken bands across Eurasia and North America and down the Pacific coast to northern California. In its southern reaches, the taiga is dominated by tall trees with a lush growth of bryophytes and lichens; northward, it consists of vast monotonous stretches of forest with very few tree species.

The Arctic Tundra Has Low-Lying Shrubs and Grasses but No Trees

North of the taiga is the tundra, a treeless region that also extends around the Northern Hemisphere, mostly above the Arctic Circle, in a band that is broken only by bodies of water. Both the northern reaches of the taiga and all of the tundra are underlain by permafrost. Because of this permafrost, and especially because of the low rates of evapotranspiration, tundra and taiga soils are relatively moist and highly leached of nutrients.

Questions

1. Describe the influence of latitude and altitude on the distribution of organisms on Earth.

2. Describe the effect of mountains on local precipitation.

3. Compare tropical and temperate forests in terms of the numbers of species found in each, and in the size and appearance of the trees.

4. Explain why annual plants are better represented, both in number and in kind, in the deserts and semiarid regions of the world than anywhere else.

5. Compare the relative amounts of nutrients found in forest and grassland soils.

6. How are the evergreen conifers of the Pacific Northwest of the United States and Canada adapted to the winter-wet/ summer-dry environment of that region?

7. What are the principal differences between taiga and tundra? What role does permafrost play in these biomes?

Selected Key Terms

Arctic tundra p. 819

biome p. 797

chaparral p. 817

coniferous forest p. 815

deciduous tropical forest p. 805

desert p. 807

early summer species p. 812

epiphytes p. 804

evergreen species p. 812

grassland p. 810

late summer species p. 812

lianas p. 804

Mediterranean scrub p. 816

monsoon forest p. 806

permafrost p. 819

rainforest p. 803

savanna p. 805

scrub community p. 816

spring ephemerals p. 812

taiga p. 818

temperate deciduous forest p. 812

temperate mixed forest p. 815

tropical rainforest p. 803

34

The Human Prospect

34–1

A Kalahari San—a hunter-gatherer—with a guinea hen he has snared. At his waist are a bow and quiver containing highly poisonous arrows. The San once populated most of southern and east central Africa. Today they live in the region of the Kalahari Desert in Botswana.

OVERVIEW

As we saw in Chapter 32, all of us—regardless of our dietary preferences—are dependent either directly or indirectly on plants for our source of food. This chapter continues that theme and focuses on the types of plants that comprise the food supply for the world's human population. But food has a long history, and you will be taken on a journey that begins 5 million years ago and continues to the present day and into the future.

After a brief description of the early development of Homo sapiens, the chapter turns to the development of agriculture—first in the Near East and other parts of the Old World and later in the Americas. At every step you will encounter foods that today are familiar but once were found only in localized areas of the world. For example, barley and wheat were first domesticated in the Near East, rice and soybeans in China, bananas in tropical Asia, coffee in Africa, maize and beans in Mexico, and potatoes in South America. Spices such as cinnamon, black pepper, and cloves were first cultivated in the tropics of Asia, and herbs such as mint, thyme, and basil were domesticated in Europe. Today, humans are fed by a relatively small number of crops, although there is great potential for the cultivation of wild species.

Next, we examine the rapid growth of the human population and the way this growth is threatening to outpace the ability of agriculture to sustain it. The chapter ends with a description of some of the promising developments in agriculture that are potential solutions to the problem of world hunger.

CHECKPOINTS

By the time you finish reading this chapter, you should be able to answer the following questions:

1. When and where did agriculture begin? What plants were particularly important as early crops?

2. What plants were important in New World agriculture? How do these plants differ from those first cultivated in the Old World?

3. What is the difference between a spice and an herb? Where did spices and herbs originate?

4. What are the world's principal crop plants today?

5. How has the growth of the human population changed since the early decades of the Industrial Revolution? What problems have arisen because of this growth?

Our own species, *Homo sapiens,* has been in existence for at least 500,000 years and has been abundant for approximately 150,000 years. Like all other living organisms, we represent the product of at least 3.5 billion years of evolution. Our immediate antecedents, members of the genus *Australopithecus,* first appeared no less than 5 million years ago. At about that time, they apparently diverged within Africa from the evolutionary line that gave rise to the chimpanzees and gorillas, our closest living relatives. The members of the genus *Australopithecus* were relatively small apes that often walked on the ground on two legs.

Larger, tool-using human beings—members of the genus *Homo*—first appeared about 2 million years ago. They almost certainly evolved from members of the genus *Australopithecus,* but they had much larger brains, apparently associated with their ability to use tools and reinforced by it. Early members of the genus *Homo* probably subsisted mainly by gathering food (picking fruits, seeds, and nuts; harvesting edible shoots and leaves; and digging roots), scavenging dead animals, and occasionally hunting. They learned how to use fire no less than 1.4 million years ago. Their hunter-gatherer methods for obtaining food and shelter seem to have resembled that of some contemporary groups of humans (Figure 34–1). Our species, *Homo sapiens,* appeared in Africa by about 500,000 years ago and in Eurasia by about 250,000 years ago.

About 34,000 years ago, the powerfully built, short, stocky Neanderthal people, who had been abundant in Europe and western Asia, disappeared completely. They were replaced by human beings who were essentially just like us. From that time onward, our ancestors made increasingly complex tools of stone and also of bone, ivory, and antler, materials that had not been used earlier. These people were excellent hunters, preying on the herds of large animals with which they shared their environment. They also began to create often magnificent ritual paintings on the walls of caves. The foundations of modern society had been laid.

The Agricultural Revolution

The Beginnings of Agriculture Involved the Deliberate Planting of Wild Seeds

The modern human beings who replaced the Neanderthals soon migrated over the entire surface of the globe. They colonized Siberia soon after they appeared in Europe and western Asia, and they reached North America by 12,000 to 13,000 years ago. Their migration eastward took place during one of the cold periods of the Pleistocene epoch, when savannas, with their large herds of grazing mammals, were widespread. As these people migrated, they seem to have been responsi-

(a)

(b)

34–2
(a) Harvesting and (b) winnowing of wheat (Triticum) in Tunisia, North Africa. Similar small-scale cultivation of wheat has been taking place around the Mediterranean basin for more than 10,000 years.

ble for the extinction of many species of these animals. At any rate, extensive hunting by humans, together with major changes in climate, occurred at the same time that these animals were disappearing from many parts of the world.

About 18,000 years ago, the glaciers began to retreat, just as they had done 18 or 20 times before during the preceding 2 million years. Forests migrated northward across Eurasia and North America, while grasslands became less extensive and the large animals associated with them dwindled in number. Probably no more than 5 million humans existed throughout the world, and they gradually began to utilize new sources of food. Some of them lived along the seacoasts, where animals that could be used as sources of food were locally abundant; others, however, began to cultivate plants, thus gaining a new, relatively secure source of food.

The first deliberate planting of seeds was probably the logical consequence of a simple series of events. For example, the wild cereals (grain-producing members of the grass family, *Poaceae*) are weeds, ecologically speaking; that is, they grow readily on open or disturbed areas, patches of bare land where there are few other plants to compete with them. People who gathered these grains regularly might have spilled some of them accidentally near their campsites, or planted them deliberately, and thus created a more dependable source of food. When this sequence of events was initiated, cultivation began (Figure 34–2). In places where wild grains and legumes were abundant and readily gathered, humans would have remained for long periods of time, eventually learning how to increase their yields by saving and planting seeds, by protecting their crops from mice, birds, and other pests, and by watering and fertilizing them.

Through humans' gradual selection of particular genetic variants of these plants, the characteristics of the domesticated crops would have changed gradually, with more seeds selected from plants with specific characteristics that made the plants easier to gather, store, or use. For example, the stalk (rachis) breaks readily in the wild wheats and their relatives, scattering the ripe seeds. In the cultivated species of wheat, the rachis is tough and holds the seeds until they are harvested. Seeds held in this way would not be dispersed well in nature, but they can be gathered easily by humans for food and replanting. As this selection process is continued, a crop plant steadily becomes more and more dependent on the humans who cultivate it, just as the humans become more and more dependent on the plant.

Agriculture in the Old World Began in the Fertile Crescent

The domestication of plants and animals began about 11,000 years ago in the area known as the **Fertile Crescent** of the eastern Mediterranean, in lands that extend through parts of what are now Lebanon, Syria, Turkey, Iraq, Iran, Jordan, and Israel. In this region, barley (*Hordeum vulgare*) and wheat (*Triticum*) were apparently the first plants to be brought into cultivation, with lentils (*Lens culinaris*) and peas (*Pisum sativum*) soon following (Figure 34–3). Additional plants that were domesticated early in this area were chickpeas, or garbanzos (*Cicer arietinum*), vetch (*Vicia* spp.), olives (*Olea europaea*), dates (*Phoenix dactylifera*), pomegranates (*Punica granatum*), and grapes (*Vitis vinifera*). Wine made from the grapes and beer brewed from the grains were used from very early times. Flax (*Linum usitatissimum*) was also cultivated very early, probably both as a source of food (the seeds are still eaten today in Ethiopia) and as a source of fiber for weaving cloth.

Among the first cultivated plants, the cereals provided a rich source of carbohydrates, and the legumes provided an abundant source of proteins. The seeds of legumes are among the richest in proteins of all plant parts, and their proteins, in turn, often are rich in the particular amino acids that are poorly represented in the cereals. It is not surprising, therefore, that legumes have been cultivated, along with cereals, from the very beginnings of agriculture in various parts of the world. Of all the protein consumed by human beings worldwide, plants contribute about 70 percent, animals about 30 percent. Only 18 percent of the total plant protein comes from legumes; about 70 percent comes from cereals, even though they contain smaller percentages of protein. Nevertheless, legume proteins are very important in human diets, and there seem to be great opportunities for increasing the quality—in terms of their amino acid composition—of the legumes that we consume.

The Domestication of Plants Influenced Other Aspects of Culture The cultivation of plants became more and more organized with the passage of time. For example, specialized implements associated with the harvesting and processing of grains, including flint sickle blades, grinding stones, and stone mortars and pestles, were in

(a)

(b)

34–3

Two of the first plants to be cultivated in the Near East were (a) barley (Hordeum vulgare) and (b) peas (Pisum sativum).

use more than 10,000 years ago. About 8000 years ago, humans began to make pottery vessels for storing grain. At the same time that plants were first being cultivated in the Near East, various animals—including dogs (which may have been the first animals kept regularly by humans), goats, sheep, cattle, and pigs—were also domesticated. Horses were domesticated later in south-western Europe, cats in Egypt, and chickens in Southeast Asia, but all of these animals spread rapidly throughout the world.

Everywhere that they were kept, grazing animals ate the plants that were available to them, whether culti-vated or not; produced wool, hides, milk, cheese, and eggs; and could themselves be eaten by their owners. As human beings increased in number, they built up herds of grazing animals so large that they began to destroy their own pastures, causing widespread ecological de-struction (Figure 34–4). Much of the Near East and other arid areas around the Mediterranean Sea are still badly overgrazed, and deserts continue to spread, as they have from the initial formation of large herds of domestic animals.

At any rate, humans now had steady sources of food, in the form of domesticated plants and animals, and they were able to organize villages, starting about 10,000 years ago, and eventually towns, about 4000 years later. Land that was productive and on which humans could live permanently could be owned, accumulated, and passed on to their descendants.

Evidence Exists for Early Domestication of Food Plants in China, Tropical Asia, and Africa

The agriculture that had originated in the Near East spread northwestward, ex-tending over much of Europe and reaching Britain by about 4000 B.C. At the same time, agriculture was devel-oped independently in other parts of the world. There is evidence that agriculture may have been practiced in the subtropical Yellow River area of China nearly as long ago as it was in the Near East. Several genera of ce-real grasses known as millets were cultivated for their grain, and eventually rice (*Oryza sativa*), now one of the most valuable cereals in the world, was added to them. Later, rice displaced the millets as a crop throughout much of their earlier range. Soybeans (*Glycine max*) have been cultivated in China for at least 3100 years (Figure 34–5).

34–4

Herds of domesticated animals, like these karakul sheep in Afghanistan, devastated large areas of the eastern Mediterranean region as their numbers increased. In many areas, only spiny or poisonous plants survived, while previously fertile fields were converted to desert.

34–5

Soybeans (Glycine max) have been a major crop in the United States for only about 50 years. They are one of the richest sources of nutrients among food plants, the seeds con-sisting of 40 to 45 percent protein and 18 percent fats and oils. In the countries of east-ern Asia, soybeans are used to make bean curd and soy sauce, among other products. Soybeans can be grown most successfully in temperate regions, and the United States produces more than half of the world's crop. Like many legumes, they harbor nitrogen-fixing bacteria in nodules on their roots, by means of which they are able to obtain nitrogen for their own growth and to enrich the soil (see Chapter 30). Soybeans are often rotated with maize in the United States, primarily to interrupt the life cycles of the important nematode and insect pests that attack these two major crops.

(a)   *(b)*

34–6

Rice (Oryza sativa) provides half of the food consumed by some 1.6 billion people and more than a quarter of the food consumed by another 400 million. It has been cultivated for at least 6000 years, and it is currently grown on 145 million hectares, about 11 percent of *the world's arable land. Rice is the source of several kinds of alcoholic beverages, including sake, a wine that is a traditional drink in Japan. When rice is grown wet, in paddy fields, fish are often farmed in the flooded fields and harvested along with the rice.* *Since its establishment in 1962, the International Rice Research Institute in the Philippines has made important contributions to the improvement of this crop. Seen here are (a) rice terraces and (b) water buffalo being used to cultivate rice in Bali, Indonesia.*

In other parts of subtropical Asia, an agriculture based on rice and different legumes and root crops was developed. There is some archaeological evidence of the cultivation of rice in Thailand about 10,000 years ago, but much further research is needed before these dates can be established with certainty. The humid, rainy conditions of the tropics tend to destroy most of the evidence with which an archaeologist might trace these ancient events. Such animals as water buffaloes, camels, and chickens were domesticated in Asia long ago, and they became important elements in the systems of cultivation that were practiced there (Figure 34–6).

Eventually, plants such as the mango (*Mangifera indica*) and the various kinds of citrus (*Citrus* spp.) were brought into cultivation in tropical Asia, and others, such as rice and soybeans, began to be cultivated farther north. Taro (*Colocasia esculenta*) is a very important food plant in tropical Asia, where it is grown for its starchy tubers; *Xanthosoma* is a related food plant of the New World tropics. *Colocasia* and other similar genera, including *Xanthosoma*, are the source of poi, a staple starch food of the Pacific Islands, including Hawaii, where these plants were brought by Polynesian settlers about 1500 years ago.

Bananas (*Musa* × *paradisiaca*) are among the most important domesticated plants to come from tropical Asia; their fruits are a staple food throughout the tropical parts of the world. Starchy varieties of bananas, known as plantains, are much more important as a source of food in tropical countries, where two-thirds of the total banana crop is consumed, than are the sweet varieties, which are more familiar in temperate regions. Wild bananas have large, hard seeds, but the cultivated varieties, like many kinds of cultivated citrus fruits, are seedless. The banana reached Africa about 2000 years ago, and it was brought to the New World soon after the voyages of Columbus.

There was also early domestication of food plants in Africa, but, again, we have little direct evidence of the timing of its origins. At any event, there is a gap of at least 5000 years between the origins of agriculture in the Fertile Crescent and its appearance at the southern tip of the African continent. Such grains as sorghum (*Sorghum* spp.) and various kinds of millet (*Pennisetum* spp. and *Panicum* spp.); a number of kinds of vegetables, including okra (*Hibiscus esculentus*); several kinds of root crops, but notably yams (*Dioscorea* spp.); and a species of cotton (*Gossypium*)—all were initially brought into cultivation in Africa. The various species of cotton are widespread as wild plants, mainly in seasonally arid but mild climates, and they are obviously useful; the long hairs on their seeds can easily be woven into cloth (Figure 34–7). Fragments of cotton cloth 4500 years old have been found in India. Cotton seeds are also used as a source of oil, and the seed meal from which the oil has been expressed is used for feeding animals. Coffee (*Coffea arabica*) is another crop of African origin. It was brought into cultivation much later than the other plants men-

34–7

Cotton (Gossypium), one of the most useful of plants cultivated for fiber, seems to have been domesticated independently in India and China (the same species of cotton is cultivated in both areas), Africa (another species), Mexico (another species), and western South America (yet a fourth species). Cotton has been cultivated for thousands of years for cloth and has also become an important source of edible oil during the past century. The cotton that is grown widely throughout the world today is a polyploid from the New World; the diploid species of the Old World are more locally cultivated.

34–8

Maize, or corn (Zea mays), the most important crop plant in the United States. Originally used mainly to feed people, maize is now one of the world's chief foods for domestic animals. In the United States, some 80 percent of the crop is consumed by animals. At the time of Columbus, maize was cultivated from southern Canada to southern South America. Five main types are recognized: popcorn, flint corns, flour corns, dent corns, and sweet corn. Dent corns, in which there is a dent in each kernel, are primarily responsible for the productivity of the U.S. corn belt and are mainly used as animal food. Dent corn is increasingly important as a source of high-fructose corn syrup, as used in canned soft drinks, and of ethanol.

tioned here, but it is now a very important commercial crop in the tropics.

Agriculture in the New World Utilized Many New Species

A parallel development of agriculture took place in North and South America. No domesticated plants appear to have been brought by human beings from the Old World to the New World prior to 1492. Dogs were certainly brought by people migrating to North America across the Bering Straits but were the only kind of domestic animal that these humans brought with them. This attests to the great value of dogs for protection, hunting, herding, and, commonly, as a source of meat. It may also be related to the very early domestication of dogs.

The plants that were brought into cultivation in the New World were different from those that were first cultivated in the Old World. In place of wheat, barley, and

rice, there was maize (*Zea mays*; Figure 34–8), and in place of lentils, peas, and chickpeas, the inhabitants of the New World cultivated kidney beans (*Phaseolus vulgaris*), lima beans (*Phaseolus lunatus*), and peanuts (*Arachis hypogaea*), among other legumes. Among the other important crop plants of Mexico were cotton (*Gossypium* spp.), chili peppers (*Capsicum* spp.), tomatoes (*Solanum* (*Lycopersicon*) spp.), tobacco (*Nicotiana tabacum*), cacao (*Theobroma cacao*, which yields cocoa, the major ingredient in chocolate), pineapple (*Ananas comosus*), pumpkins and squashes (*Cucurbita* spp.), and avocados (*Persea americana*). Cotton was domesticated independently in the New and in the Old World, with different species involved in the different centers of domestication. Its cultivation by humans in Mexico dates back at least 4000 years, and cotton has been cultivated even longer in Peru. The New World cottons are polyploid, and they are the source of nearly all the cotton that is cultivated throughout the world today; in contrast, the Old World cottons are diploid. After the diverse culti-

The Origin of Maize

Ears of maize, or corn, differ so greatly from those of its ancestor that for many years the wild progenitor species was not recognized. We now think, however, that maize (Zea mays subspecies mays) is simply the domesticated form of a large wild grass, the southern Mexican annual teosinte (Zea mays subspecies parviglumis). Teosinte plants have hundreds of small, narrow, two-rowed ears, each made up of 5 to 12 fruit cases (each permanently enclosing one grain) that disarticulate and fall free at maturity. Because the fruit cases are woody and not separable from the grains, the flour (meal) that results from grinding them is unpalatable. The several species of teosinte grow from southern Chihuahua, Mexico, to Nicaragua. They are capable of forming hybrids with maize and may do so spontaneously when maize and teosinte grow together (the maíz de coyote, or wild maize, as the hybrids have been called).

Maize is known only as a cultivated plant and could not survive on its own in the wild. Not only are its large naked grains vulnerable, but they are permanently attached to a central axis (the "corn cob") and permanently protected by many overlapping leaf sheaths (the "husks") that would not permit dissemination even if the grains were free. In short, the maize ear is a well-packaged, high-yielding, and easily harvested agricultural artifact, with all its principal characteristics selected by the early farmers.

The domestication of maize, which began in southern Mexico more than 7000 years ago, obviously involved making the teosinte grains accessible as food, as well as increasing grain and ear size and harvestability. As a result of human selection for these characteristics, maize came to differ from teosinte in much the same way as the unbranched and monocephalic (single-headed) cultivated sunflower differs from its highly branched and many-headed wild ancestor. Both maize and the cultivated sunflower have all the reproductive resources of a single branch concentrated in one, gigantic, many-seeded, terminal structure, the maize ear or the sunflower, which results in easier harvesting and higher yield.

An exciting new development in the story of maize evolution has been the discovery of a new species of wild perennial teosinte, Z. diploperennis. This rare plant was first found in 1977 in southwestern Mexico by Rafael Guzmán, then an undergraduate student at the University of Guadalajara, and was recognized in 1978 as a new species by Hugh Iltis, John Doebley, and their associates at the University of Wisconsin in Madison. Interfertile with annual maize, it carries the genes for resistance to seven of the nine major viruses that infect maize in the United States; for five of these, no other source of resistance is known. The economic implications are obvious when one considers the worldwide gross value of

maize—nearly $60 billion in 1991 alone.

Zea diploperennis occurs naturally in only three small areas, totaling a little more than two hundred hectares, in the cloud forests of the Sierra de Manantlán, a mountain range between Guadalajara and Puerto Vallarta. There it might easily have been destroyed by spreading cultivation and cattle ranching without ever having been made known to science. To protect this one species, as well as the associated rich flora and fauna (which include ocelots, jaguars, and mountain lions), a large biosphere nature reserve and research laboratory has been established by the University of Guadalajara.

Above are Zea mays *subspecies* parviglumis *(left) and* Zea mays *subspecies* mays *(right).*

vated plants of the New World were discovered by Europeans following the voyages of Columbus, many of them were brought into cultivation in Europe, spreading from there to the rest of the world. Not only were some of these plants totally new to Europeans, but others, such as the New World species of cotton, were better than the forms already in cultivation in Eurasia and soon replaced them.

The Early Domestication of Plants Occurred in Both Central and South America The earliest evidence of the existence of domesticated plants in Mexico dates from about 9000 years ago, but extensive agriculture appears not to have become well established until considerably

later. The evidence thus suggests that the domestication of plants began later in the New World than in Eurasia. Many of the crops that were originally domesticated in Mexico eventually became widespread from Central America and Mexico northward into Canada. Similar crops were also widely grown throughout the lowlands and middle elevations of South America. In fact, it is likely that agriculture was developed independently in Mexico and in Peru, although the question cannot be settled with the evidence that is now available. The earliest evidence of agriculture in Peru is almost as ancient as that from Mexico, and some domestic plants, such as peanuts, may well have been brought from South America to Mexico by humans.

34–9

A distinctive form of agriculture was developed at high elevations in the Andes of South America. (a) Cultivated fields in the mountains of northwestern Argentina. (b) A potato (Solanum tuberosum) field in Ecuador. Only a limited selection of the available genetic diversity of potatoes has been used for the improvement of the cultivated crop, one of the most important in the world. (c) Three of the four major root crops that are cultivated

in the Andes, shown here for sale in the market at Tarma, Peru—potatoes (S. tuberosum), añu (Tropaeolum tuberosum), and ullucu (Ullucus tuberosus). Ullucu, which can be grown at higher elevations than potatoes, forms large, nutritious tubers; it might well become a useful crop elsewhere in the world. The fourth root crop that is common in the Andes is oca (Oxalis tuberosa), which is also cultivated to a limited extent in New Zealand

and elsewhere. The tubers of all four of these plants are naturally freeze-dried by Andean farmers, after which they can be easily stored for later consumption. (d) Quinoa (Chenopodium quinoa), a grain of the pigweed family (Chenopodiaceae), is an important crop in the Andes—here shown in cultivation in northern Chile—and is now being tested for more widespread cultivation.

(a)

(b)

(c)

(d)

In the south central Andes of South America, a distinctive kind of agriculture was developed (Figure 34–9), based on tuberous crops such as potatoes (*Solanum tuberosum* and related species) and seed crops such as quinoa *(Chenopodium quinoa)* and lupines (*Lupinus* spp., of the family *Fabaceae*). Potatoes were cultivated throughout the highlands of South America at the time of Columbus, but they did not reach Central America or Mexico until they were taken there by the Spaniards.

They have been one of the most important foods in Europe for two centuries, yielding more than twice as many calories per hectare as wheat.

A Variety of Plants and Animals Were Domesticated in Different Regions of the New World Other crops were first brought into cultivation outside these main centers. For example, the sunflower (*Helianthus annuus*) was domesticated by Native Americans living in what is now the United States (Figure 34–10). Another very important crop plant of the New World, manioc (*Manihot* spp.), also called cassava or yuca, was domesticated in the drier areas of South America but is now cultivated on a major scale throughout the tropics (Figure 34–11).

Although the familiar white, or Irish, potato is a member of the nightshade family (*Solanaceae*), the sweet potato (*Ipomoea batatas*), as you might know if you have seen a field of it in flower, is a member of the morning-glory family (*Convolvulaceae*) and is therefore not closely related to the white potato. At the time of the voyages of Columbus, the sweet potato was widely cultivated in Central and South America, but it was also widespread on some Pacific islands, as far as New Zealand and Hawaii, to which it was apparently carried by human beings in the course of their early voyages. Subsequent to the time of Columbus, the sweet potato became a very important crop throughout most of Africa and tropical Asia.

Very few animals were domesticated in the New World. The so-called Muscovy duck (which, despite its name, does not come from Moscow), turkeys, guinea pigs, llamas, and alpacas are among the few to have originated there. Towns and eventually large cities were organized in the New World, just as they had been earlier throughout Eurasia, wherever the agricultural sys-

34–10
Sunflowers (Helianthus annuus), *an important crop because of the oil that is obtained from their seeds, were first domesticated at least 3000 years ago in what is now the central United States.*

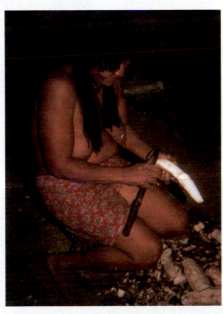

(a)　　　　　　　(b)

34–11
Manioc (Manihot esculenta), *one of the most important root crops throughout the tropics. Tapioca is produced from the starch extracted from the roots of this plant. Some of the cultivated strains, the bitter maniocs, contain poisonous cyanide compounds, which must be removed before the tuber is eaten.* **(a)** *Manioc cultivated in a forest clearing in southern Venezuela.* **(b)** *A Tirió woman in southern Surinam peeling a manioc root to prepare it for cooking.*

tems were well enough developed to support them. Crops were extensively cultivated around such centers, but there were no large herds of domestic animals comparable to those that had become so prominent throughout Europe and Asia.

When Europeans colonized the Western Hemisphere, however, they brought their herds with them. In time, these herds brought about the same sort of widespread ecological devastation in parts of the New World as that which had taken place millennia earlier in the Near East and other parts of Eurasia. Many natural plant communities cannot readily be converted to pastures. For example, the widespread clearing of moist tropical forests to make pastures has been enormously destructive everywhere that it has been attempted (see Chapter 33). For the most part, pastures in such areas are productive only in the short term, until the nutrients in the soil have been exhausted, and they must often be abandoned after no more than 10 or 15 years.

Spices and Herbs Are Plants Prized for Their Flavors

The substances that are produced by plants primarily to defend themselves from insects and other herbivores were discussed in Chapters 2 and 22. These substances contribute to the flavors, odors, and tastes of many plants and thus add features that have been utilized by human beings since prehistoric times. Some of these substances make plants poisonous to humans and other animals, but others add characteristics that humans find desirable.

Spices—strongly flavored parts of plants, usually rich in essential oils—may be derived from the roots, bark, seeds, fruits, or buds. **Herbs,** on the other hand, are usually the leaves of nonwoody plants, although laurel, or bay, leaf and a few other condiments derived from trees or shrubs are considered herbs also. In practice, herbs and spices intergrade completely. Both have traditionally been used by humans to flavor food, especially if that food has become stale or somewhat spoiled.

Historically, the Most Important Spices Were Cultivated in Tropical Asia   Spices and herbs have been used widely in cooking for as long as we have records; the search for spices played a major role in the great Portuguese, Dutch, and English voyages that began in the thirteenth century and eventually led to Europeans' discovery of lands in the Western Hemisphere. The most important spices came from the tropics of Asia; they were responsible for the voyages and for a great deal of warfare as well. By the third century B.C., caravans of camels—often requiring two years for the trip—were carrying spices from tropical Asia back to the civilizations of the Mediterranean region. Included among these spices were cinnamon (the bark of *Cinnamomum zeylanicum*), black pepper (the dried, ground fruits of *Piper nigrum*;

34–12
Black pepper (Piper nigrum) *has been known for thousands of years as an important spice. On a worldwide basis, the use and consumption of black pepper is approximately equal to that of all other spices combined.*

Figure 34–12), cloves (the dried flower buds of *Eugenia aromatica*), cardamom (the seeds of *Elettaria cardamomum*), ginger (the rhizomes of *Zingiber officinale*), and nutmeg and mace (the seeds and the dried outer seed coverings, respectively, of *Myristica fragrans*; Figure 34–13). When the Romans learned that by taking advan-

34–13
Nutmeg (Myristica fragrans) *is one of the most important traditional spices from tropical Asia. The spice nutmeg is derived from the ground seeds, whereas the spice mace comes from the fleshy aril, seen here as bands of red tissue. Birds act as natural seed distributors by removing the seed and aril to consume the aril for its high protein content. The seed is soon dropped and thus planted on the forest floor away from the mother tree of this dioecious species.*

tage of the seasonal shifts of the monsoon winds they could reach India by sea from Aden, they shortened the trip to about a year—but it was still a highly dangerous and uncertain enterprise. A smaller number of additional spices, including vanilla (the dried, fermented seedpods of the orchid *Vanilla planifolia*; see Figure 22–15), red peppers (*Capsicum* spp.), and allspice (the dried, unripe berries of *Pimenta officinalis*), so-called because it was considered to combine the flavors of cinnamon, clove, and nutmeg, came from the New World tropics after the voyages of Columbus.

Herbs Originated in Many Parts of the World In Europe and the Mediterranean region generally, there were many different kinds of native herbs, ones that were more familiar locally and, perhaps for that reason, not as highly prized as some of the spices that were available only from distant lands. Especially prominent among these herbs are members of the mint family (*Lamiaceae*). These include thyme (*Thymus* spp.), mint (*Mentha* spp.), basil (*Ocimum vulgare*), oregano (*Origanum vulgare*), and sage (*Salvia* spp.). Also important were members of the parsley family (*Apiaceae*), including parsley (*Petroselinum crispum*), dill (*Anethum graveolens*), caraway (*Carum carvi*), fennel (*Foeniculum vulgare*), coriander (*Coriandrum sativum*), and anise (*Anethum graveolens*). Some members of this family (parsley, for example) are grown primarily for their leaves, and some (such as caraway) are grown for their seeds, but many (such as dill and coriander) are valued for both.

Tarragon (*Artemisia dracunculus*) is an herb that consists of the leaves of a plant belonging to the same genus as wormwood and the sagebrush of the western United States and Canada. Mustard (*Brassica nigra*) seed, which can be ground into the spice we call mustard, likewise comes from a plant native to Eurasia. Bay leaves, traditionally taken from the tree *Laurus nobilis* of the Mediterranean region but now also often taken from *Umbellularia californica* of California and Oregon, are another herb, one that is derived from temperate members of the largely tropical laurel family (*Lauraceae*). Saffron, popular in the Near East and adjacent regions, consists of the dried stigmas of *Crocus sativus*, a small, bulbous plant of the iris family (*Iridaceae*). The stigmas are gathered laboriously by hand, which accounts for the extremely high price of saffron and the fact that it is so prized—as much for its color as for its taste.

Coffee (Figure 34–14) and tea (*Camellia sinensis*) provide the two most important beverages in the world; both are consumed primarily because of the stimulating alkaloid, caffeine, that they contain. Coffee is made from seeds of the coffee plant that have been dried, roasted, and ground, whereas tea is prepared from the dried leafy shoots of the tea plant. Coffee, as mentioned earlier, was domesticated in the mountains of northeastern

Africa, whereas tea was first cultivated in the mountains of subtropical Asia; both are now widespread crops throughout the warm regions of the world. Coffee now provides the livelihood for some 25 million people, and it provides a major source of income for the 50 tropical nations that export it. A third of the world's supply of coffee comes from Brazil.

Agriculture Is a Global Phenomenon

For the last 500 years, the important crops have been carried throughout the world and cultivated wherever they grow best. The major grains—wheat, rice, and maize—are grown everywhere that the climate will permit. Plants unknown in Europe before the voyages of Columbus, including maize, tomatoes, and *Capsicum* peppers, are now cultivated throughout the world. Sunflowers, first domesticated in the area that is now the United States, now produce more than half of their total worldwide yield in Russia. Sunflower seeds are widely

34–14

Coffee (Coffea arabica) is an important cash crop throughout the tropics. It is a member of the madder family (Rubiaceae), as is cinchona (Cinchona), which produces the medically important alkaloid quinine. Rubiaceae is one of the larger families of flowering plants, with about 6000 species, mainly tropical.

consumed by humans as a snack, and sunflower meal is important for animal feed. Sunflowers are also displacing the traditional olives as a source of oil in many parts of Spain and elsewhere in the Mediterranean region. Throughout the world, the sunflower is being extensively grown for its oil, and it is second only to the soybean among plants grown for this purpose.

Some tropical crops have also become widespread. For example, rubber (derived from several species of trees of the genus *Hevea*, of the family *Euphorbiaceae*) was brought into cultivation on a commercial scale about 150 years ago. The main area of rubber production is in tropical Asia (see Figure 2–27). For rubber, as for many other crops, cultivation away from the areas where the plants are native seems to be favored. The plants are then often free of the pests and diseases that attack them in their native lands, unless, of course, the pest has been transported too—for example, in seed stores. The need for strict quarantines to avoid transferring pests and diseases of plants between countries or continents is often overlooked, however, in the eagerness to establish trade links between nations.

Oil palms *(Elaeis guineensis)* are native to West Africa but are grown now in all tropical regions. Although they have been grown on a commercial scale for only about 75 years, oil palms are among the most important cash crops of the tropics today. Among the others are coffee and bananas, both widespread. Cacao, which was first semidomesticated in tropical Mexico and Central America, is now most important as a crop in West Africa (Figure 34–15). Sugarcane *(Saccharum officinale)* was domesticated in New Guinea and adjacent regions, whereas sugarbeets were developed from other cultivated members of their species in Europe. Yams *(Dioscorea* spp.) are an important tropical root crop. A number of species of yams are cultivated throughout the tropics, some of them from West Africa, others from Southeast Asia, and a few—less important—are cultivated in Latin America. The better yams have now spread throughout the tropics, and they provide a staple food over wide areas. Manioc is important as an industrial crop in addition to being one of the most important human food crops (Figure 34–11). As a result of large-scale plantings, manioc has become a significant source of industrial starch and animal feed. Tons of processed, dried, and pelletized manioc are exported from Southeast Asia, particularly Thailand, to Europe to be used as a major food supplement in the swine and dairy cattle industries. From small plantings to large machine-operated farms in Central and South America as well as in Africa and Asia, this tuberous crop plant provides one of the most important food sources for the expanding populations of the tropics.

Another of the most important cultivated plants of the tropics is the coconut palm *(Cocos nucifera)*, which seems to have originated in the western Pacific–tropical Asian region but was widespread in the western and central Pacific Ocean area before the European voyages of exploration. Natural stands of coconuts are rare in the eastern Pacific, with a few known from Central America. The vast range of the coconut may be the result of natural dispersal of the fruits floating in the sea, rather than human intervention. Each tree produces about 50 to 100 fruits (drupes) each year, and they are a rich source of protein, oils, and carbohydrates. Coconut shells, leaves, husk fibers, and trunks are used to make many useful items, including clothing, buildings, and utensils; it is the solid and liquid endosperm that we eat.

The World's Food Supply Is Based Mainly on Fourteen Kinds of Crop Plants   Modern agriculture has become highly mechanized in temperate regions and also in some parts of the tropics. It has also become highly specialized, with just six kinds of plants—wheat, rice, maize, potatoes, sweet potatoes, and manioc—directly or indirectly (that is, after having been fed to animals) providing more than 80 percent of the total calories consumed by human beings. These plants are rich in carbohydrates, but they do not provide a balanced diet for humans. They are usually eaten with legumes, such as common beans, peas, lentils, peanuts, or soybeans, which are rich in protein, and with leafy vegetables,

34–15
Cacao (Theobroma cacao), *the source of chocolate and cocoa. The fruits, shown here, contain several large seeds, or "beans." Cacao was first domesticated in Mexico, where chocolate was a prized drink among the Aztecs; at times, the beans were used as currency.*

34–16

Sugarbeet (Beta vulgaris). The sugarbeet is simply a variety of the ordinary beet—selected from strains grown earlier for fodder, not those cultivated as a root crop—in which the sucrose content has been increased by selection from about 2 percent to more than 20 percent. Beets were domesticated in Europe, where their leaves have long been used for food; Swiss chard is another variety of beet, one that is grown for its edible leaves. For about 300 years, beets have also been used as a source of sugar that has been competitive with sugarcane, which must be grown in the tropics and imported to the countries of the developed world. In the United States, production of raw sugar from beets amounts to about a third of the total sugar consumed domestically.

such as lettuce, cabbage, spinach *(Spinacia oleracea),* and chard, which are abundant sources of vitamins and minerals. Such plants as sunflowers and olives provide fats, which are also necessary in the human diet.

In addition to the six major food crops, there are eight others of considerable importance to human beings: sugarcane, sugarbeet (Figure 34–16), common beans, soybeans, barley, sorghum, coconuts, and bananas. Taken together, these 14 kinds of plants constitute the great majority of crops widely cultivated as sources of food.

There are great regional differences in the human diet. For example, rice (Figure 34–6) provides more than three-quarters of the diet in many parts of Asia, and wheat (Figure 34–17) is equally dominant in parts of North America and Europe. In regions with scarce rainfall, maize can be grown successfully only with supplementary irrigation, which is usually not necessary for wheat. Extending the productive ranges of these major grains and finding additional crop species are tasks of the greatest importance to the human species, as we shall see.

The Growth of Human Populations

The roughly 5 million humans who lived 11,000 years ago were already the most widely distributed large land mammal in the world. Subsequently, however, with the development of agriculture, this number has grown at an accelerating pace.

Various mechanisms come into play to limit the numbers of individuals in groups of humans who make their living by hunting. A woman on the move cannot carry more than one infant along with her household baggage, minimal though that baggage may be. When simple means of birth control—often simply abstention from sex—are not effective, such a woman may resort to abortion or, more commonly, infanticide. In addition, there is a high natural mortality in these populations, particularly among the very young, the old, the ill, the disabled, and women in childbirth. As a result of these factors, populations that are dependent on hunting tend to re-

34–17

Bread wheat (Triticum aestivum), cultivated with modern techniques. First domesticated in the Near East, wheat has become the most widely grown crop in the world today. Along with barley, which is now largely used for animal food and as the source of malt for making beer, wheat was probably one of the first two plants to be cultivated. Because of the special properties of some of its proteins, wheat is used more extensively than any other cereal for making bread. Wheat proteins form a sticky substance called gluten that facilitates the handling of the dough and helps to hold the bread together.

main small. In addition, there is not much incentive for specialization of knowledge or skills; those basic skills on which individual survival depends are of the greatest importance.

The Development of Agriculture Dramatically Affected Population Growth

Once most groups of humans became sedentary, there was no longer the same urgent need to limit the number of births, and children may have become more of an asset to their families than previously, helping with agricultural and other chores. People could live much more densely on the land than ever before. For hunting and food-gathering economies, an area of 5 square kilometers, on the average, is required to provide enough food for one family to eat. In crop-based economies, only a small fraction of that amount of land is required. In the cities and towns that agricultural productivity made possible, human knowledge became increasingly specialized. Because the efforts of a few people could produce enough food for everyone, patterns of life became more and more diversified. People became merchants, artisans, bankers, scholars, poets—all the rich mixture of which a modern community is composed. The development of agriculture set human society on its modern course.

As a consequence of the development of agriculture, the numbers of human beings grew to about 130 million, distributed all over the Earth, by the beginning of the Christian era. Over a period of about 8000 years, the human population had increased about 25-fold. By 1650, the world population had reached 500 million, with many people living in urban centers (Figure 34–18). The development of science and technology had begun, as had the process of industrialization, bringing about further profound changes in the lives of human beings and their relationships with the natural world. The human birth rate has remained essentially constant throughout the world from the seventeenth century onward, but the death rate has decreased dramatically on a regional basis, with the result that the population overall has grown at an unprecedented rate. In the twentieth century, the birth rate itself has dropped in developed countries.

How Will the World's Rapidly Growing Population Be Fed?

As the twentieth century comes to an end, there are just under 6 billion humans on our planet; this compares with 2.5 billion in 1950. The global population doubled in less than 40 years. In 1997, about 21 percent of the people lived in the developed world (the United States, Canada, Europe, Russia, Australia, and New Zealand), some 21 percent in China, and the remainder in the de-

34–18

London was one of the thriving cities of seventeenth-century Europe, its influence reaching out to distant parts of the world as the global population increased. The world population reached a level of 1 billion people by 1850.

veloping world, mostly consisting of countries that are tropical or subtropical, at least in part.

For the planet as a whole, the population is growing at about 1.5 percent per year. This means that about 165 humans are being added to the world population every minute—more than 240,000 each day, or about 88 million every year—about equal to the total population of Germany or Mexico. A high proportion—usually about 35 to 48 percent—of the humans living in the developing countries are under 15 years of age. The comparable percentage for developed countries is about 12 to 18 percent in Europe and Japan, 20 percent in Canada, and 22 percent in the United States. Such young people have not yet reached the age at which humans usually bear children. Consequently, population growth in developing countries cannot soon be brought under control, even though government policy and individual choice often favor such a trend. With 6 billion people in the world in the year 2000, nearly 1 billion—equal to the population of the entire globe 200 years ago in the early decades of the Industrial Revolution—will have been added during the 1990s alone. More than 95 percent of this growth will take place in developing countries.

If worldwide efforts to promote family planning continue to be pursued strongly, the Earth's population may stabilize at about 8–8.5 billion people; otherwise, the ul-

34–19

These impoverished people, living on the outskirts of Tegucigalpa, Honduras, represent the condition of a majority of the world's population. Their prospects for the future depend directly on limiting population growth, incorporating the poor into the global economy, and finding new and improved methods for productive agriculture in the tropics and subtropics.

timate total will be higher, perhaps in the 10–12 billion range—but it all depends on future actions. In the context of the next 100 years, the next few decades are likely to be one of the most difficult periods ever faced by the human race (Figure 34–19). In 1996, the World Bank estimated that 1.4 billion people—nearly one out of every four of us—were living in absolute poverty. These people are unable to obtain food, shelter, or clothing dependably. Some 500 million people—nearly one out of every ten—were receiving less than 80 percent of the daily intake of food calories recommended by the United Nations; their minds and bodies are literally wasting away.

Humans are presently estimated to be consuming, wasting, or diverting more than 40 percent of the total net photosynthetic productivity on land, and that productivity has been greatly reduced by our burning and clearing in the past. Although we have achieved a 2.6-fold increase in world grain production since 1950, this increase has been accomplished at the expense of no less than 25 percent of the topsoil and more than 15 percent of the land that is under cultivation. As we saw in Chapter 33, we lack the agricultural technology to convert most of the lands in the tropics to sustainable productivity. Most of the world's lands that can be cultivated using available techniques are already in cultivation. Despite this, the rapidly growing majority of the world's population that lives in the tropics must somehow be fed. The job cannot be done by exporting surpluses from the more productive lands of the developed world. The solution to the problem of feeding the world's population must be found in the regions where most people live—the tropics and subtropics.

Substantial food increases are needed if the world's population is to be fed adequately, especially given the greatly increased demand for protein-rich foods in the developing world. Realistically, there appears to be little hope of attaining this. In some tropical areas, such as Africa south of the Sahara, food production per capita has actually been falling. Recently, even the *total* food production has declined in this vast area, whose population is well over 500 million and rapidly growing. For U.S. citizens, who spend on the average less than a fifth of their personal income on food, the increasing cost of food already occasions serious concern. For those in developing nations, who may spend 80 to 90 percent of their income on food, it can be a death sentence. Indeed, in such countries as Bangladesh and Haiti and in such regions as East Africa, humans are dying in increasing numbers because of the lack of food. How can the situation be improved?

Agriculture in the Future

Advances in Agriculture Have Brought Problems As Well As Benefits

The first significant advance that led to a massive increase in agricultural productivity was the development of irrigation (Figure 34–20). The necessity of providing water to crops has always been so evident that irrigation was practiced in the Near East as early as 7000 years ago, and it was developed independently in Mexico apparently about 5000 years ago. During the past two cen-

34–20

This irrigated cotton field in Texas is representative of modern, intensive agriculture. Irrigation may pose severe environmental problems over the long run, however, especially if it is coupled with the intensive use of pesticides and herbicides. More pesticides are used on cotton than on any other crop in the world.

turies, increasingly specialized and efficient machinery has been developed for agriculture, and great increases in productivity have been achieved. Fertilizers have been widely used, with their production drawing heavily on the use of fossil fuels. One of the great problems worldwide is how to utilize the gains in productivity that are made possible by increased mechanization, irrigation, and fertilization without simultaneously displacing millions of workers. Throughout much of the developing world, well over three-quarters of the people are directly engaged in food production, compared with fewer than 3 percent in the United States in the late twentieth century.

Research has already done a great deal to improve agriculture. In the United States, the land-grant college system and the associated state agricultural experiment stations have made major contributions in this area. Nevertheless, many problems remain. The energy cost of producing crops in the United States and other developed countries is very high. Modern agriculture likewise depends on an elaborate distribution system, which is expensive in terms of energy and is easily disrupted. A significant proportion of each crop, the exact amount depending on the region and the year, is lost to insects and other pests. In many regions, an additional significant proportion is lost after harvest, either to spoilage or to insects, rats, mice, and other pests. Water is increasingly expensive in many areas, and the quality of local water supplies is often degraded by runoff from fields to which fertilizers and pesticides have been applied. Soil erosion is a problem everywhere, and it increases with the intensity of the agriculture (Figure 34–21). Many efforts are under way to achieve improvements in crop productivity, in the protection of crops from pests, and in the efficiency with which crops use water. Each of these will be discussed in this chapter.

Improving the Quality of Existing Crops Is an Important Goal

The most promising approach to alleviating the world's food problem seems to lie in the further development of existing crops grown on land that is already in cultivation. Most land suitable for plant agriculture is already being cultivated, and increasing the supplies of water, fertilizers, and other chemicals for crops is not economically feasible in many parts of the world. Thus the genetic improvement of existing crops is of exceptional importance. Such development entails improving not only the yield of these crops but also the quantity of proteins and other nutrients that they contain. The *quality* of the protein in food plants is also of the greatest importance to human nutrition: animals, including human beings, must be able to obtain from food the right amounts of all the essential amino acids—the ones that they cannot manufacture themselves. Nine of the 20 amino acids

34–21

Use of the no-tillage cropping system, which combines ancient and modern agricultural practices, has been increasing rapidly. By the year 2000, as much as 65 percent of the acreage of crops grown in the United States may be grown under no-tillage conditions. Soil erosion is virtually eliminated in this system, as shown in the maize planted without tillage into a killed cover crop of clover (right). Compare this with the evident erosion in maize planted conventionally (left). This photograph was taken soon after a spring rainstorm. When no-tillage methods are used, the energy input into maize and soybean production is reduced by 7 and 18 percent, respectively, and the crop yields are as high as or higher than those obtained by conventional plowing and disking methods.

required by human adults must be obtained from food (page 26); the other 11 can be manufactured in the human body. Plants that have been selected for an improved protein content, however, inevitably have higher requirements for nitrogen and other nutrients than do their less modified ancestors. For this reason, improved crops cannot always be grown on the marginal lands where they would be especially useful.

The quality of crops can also be improved in many respects other than yield and protein composition and quantity. Newly developed crop varieties may be more resistant to disease, as when they contain secondary metabolites that their herbivores find distasteful; more interesting in form, shape, or color (such as apples of a brighter red); more amenable to storage and shipping (such as tomatoes that stack better in boxes); or improved in other features that are important for the crop in question.

For decades, the plant breeder, patiently sifting the available genetic diversity and producing hybrids and varieties of better promise, has produced thousands of improved strains of our major crops (Figure 34–22). Typically, thousands of hybrids are made and evaluated

34–22

Norman Borlaug, who was awarded the Nobel Peace Prize in 1970. Borlaug was the leader of a research project sponsored by the Rockefeller Foundation under which new strains of wheat were developed at the International Center for the Improvement of Maize and Wheat (CIMMYT) in Mexico. Widely planted, these new strains changed the status of Mexico from that of a wheat importer when the program began in 1944 to that of an exporter by 1964. Borlaug also played a vital role in introducing higher-yielding wheat varieties to India and Pakistan.

in order to find a few that represent genuine improvements on those already in wide cultivation. For example, the production of maize per hectare in the United States was increased about eightfold between the 1930s and the 1980s, even though relatively little of the genetic diversity of this remarkable plant was utilized in the process.

Hybrid Maize Plants Produce High Yields   The increase in the production of maize was made possible largely because of the introduction of **hybrid maize** seed. Inbred lines of maize (themselves originally of hybrid origin) are used as parents. When they are crossed, the result is seed that produces very vigorous hybrid plants. The strains to be crossed are grown in alternating rows, with the tassels (staminate inflorescences) being removed from one row of plants by machine or by hand, so that all of the seeds set on those plants will be of hybrid origin. Through a careful selection of the best inbred lines, vigorous strains of hybrid maize that are suitable for cultivation at any given locality can be produced. Because of the uniform characteristics of the hybrid plants, they are easier to harvest, and they uniformly produce much higher yields than nonhybrid individuals. Less than 1 percent of the maize grown in the United States in 1935 was hybrid maize, but now virtually all of it is. Far less water, fertilizer, pesticides, and labor are now needed to produce much greater yields per hectare.

Research at International Agricultural Research Centers Made Possible the Green Revolution   Extensive efforts have been made during the past few decades to increase the yields of wheat and other grains, particularly in warmer regions. These efforts have led to dramatic improvements in crop yields, largely at international crop-improvement centers located in subtropical areas. When the improved strains of wheat, maize, and rice produced at these centers were grown in such countries as Mexico, India, and Pakistan, they made possible the pattern of improved agricultural productivity that has been called the Green Revolution. The techniques of breeding, fertilization, and irrigation that have been developed as a part of the Green Revolution have been applied in many countries of the developing world.

Any grain requires excellent conditions to produce high yields, and fertilization, mechanization, and irrigation have all been necessary components of the successes of the Green Revolution. Only relatively wealthy landowners have been able to cultivate the new crop strains, and in many areas the net effect has been to accelerate the consolidation of farmlands into a few large holdings by the very wealthy. Such consolidation has not necessarily provided either jobs or food for the majority of the local peoples.

Triticale Is a Promising Hybrid of Wheat and Rye   The traditional methods of the plant breeder can sometimes lead to surprising results. For example, hybrids of wheat (*Triticum*) with rye (*Secale*), which are called **triticale** (*Triticosecale* is the scientific name), have become a crop of increasing importance in several areas and hold much promise for the future (Figure 34–23). Triticale hy-

34–23

Triticale (Triticosecale), a modern polyploid hybrid between wheat and rye, which combines the high-quality yield of wheat with the ruggedness of rye.

brids arose following the spontaneous doubling of the chromosome number in a sterile hybrid of wheat and rye (see Chapter 12) and were later produced artificially. In the mid-1950s J. G. O'Mara of Iowa State University produced such hybrids using the chemical colchicine, which arrests the formation of the cell plate and thus allows the chromosome number of the treated plant cells to double.

Unfortunately, triticale is only partly fertile, and its endosperm often develops poorly. It combines the high yield and quality of wheat with the winter hardiness and disease resistance of rye, but the best combinations of these features are still being sought by plant breeders. Improved strains of triticale are gaining in popularity in the 1990s and are already grown on more than 1 million hectares in over 30 countries around the world. Most triticale is used for animal feed, but it is being consumed in increasing amounts by humans also.

To Provide Protection against Pathogens, the Genetic Diversity of Crop Plants Must Be Preserved and Utilized

Intensive programs of breeding and selection have tended to narrow the genetic variability of crop plants. Much artificial selection, for obvious reasons, has been concerned with yield, and sometimes disease resistance has been lost among the highly uniform members of a progeny that has been selected strongly for increased yield. Overall, individual crop plants have tended to become more and more uniform, as particular traits have been stressed more strongly than others, and these plants have become more vulnerable to attack by diseases and pests. In 1970, for example, the fungus that causes southern leaf blight of maize, *Cochliobolus heterostrophus*, destroyed approximately 15 percent of the U.S. maize crop—a loss of approximately $1 billion (Figure 34–24). These losses were apparently related to the appearance of a new race of the fungus that was highly destructive to some of the major strains of maize that were used extensively in the production of hybrid maize seed. The non-nuclear genetic factors in many of the commercially important strains of maize were identical, having been introduced through the methods used to produce hybrid maize when identical carpellate parents were employed repeatedly.

In order to help guard against such losses, it is necessary to locate and to preserve distinct strains of our important crops, because these strains—even though their overall characteristics may not be attractive economically—may contain genes useful in the continuing fight against pests and diseases (Figure 34–25). The germ plasm (hereditary material) in reserve in our seed and clonal banks may also provide genes for elevating yields, for adaptation, or for certain high-value traits, such as special oils. Since the beginnings of agriculture, huge reserves of variability have accumulated in all crop plants by the processes of mutation, hybridization, artificial se-

34–24
Southern leaf blight of corn, a disease that is caused by the fungus Cochliobolus heterostrophus.

(a)

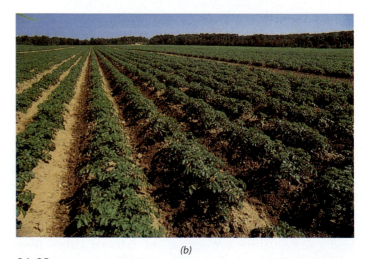

(b)

34–25
Preservation of crop strains. (a) U.S. Department of Agriculture seed bank at Fort Collins, Colorado, showing seeds being sorted out and sealed for long-term storage. Approximately 200,000 germ plasm lines are stored at this facility. (b) The seed potato farm at Three Lakes, Wisconsin, is the national repository for strains of potatoes.

lection, and adaptation to a wide range of conditions. For such crops as wheat, potatoes, and maize, there are literally thousands of known strains. In addition, even more genetic variability exists among the wild relatives of the cultivated crops, often, however, in areas where it is being lost to advancing civilization. The problem lies in finding, preserving, and using the genetic variability of cultivated plants and their wild relatives before they are lost.

As an example of the role of genetic diversity in the history of, and the prospects for, a single crop plant, we shall consider the potato. There are more than 60 species of potatoes, most of them never cultivated, and thousands of different strains of cultivated potatoes are known (Figure 34–9). Despite this, most of our cultivated potatoes are descended from a very few strains that were brought to Europe in the late sixteenth century. This genetic uniformity led directly to the Irish potato famine of 1846 and 1847, in which the potato crop was nearly wiped out by a blight caused by the mold *Phytophthora infestans* (see Chapter 17). Within three years, the population of Ireland fell from 8.5 million to 6.5 million, with about one in ten people either starving to death or dying of diseases associated with starvation and an additional one in five emigrating. The subsequent breeding of strains of potatoes resistant to the blight restored the status of the plant as a crop in Ireland and elsewhere. Looking to the future, the potential for further developing potatoes as a crop through the use of additional cultivated and wild strains is enormous.

Striking examples of the potential of obtaining additional genetic material from the wild have been achieved by tomato breeders. The collection of strains of tomatoes, mostly carried out in recent years by Charles Rick and his associates at the University of California, Davis, has led to effective control of many of the important diseases of tomatoes, such as the rots caused by the deuteromycetes *Fusarium* and *Verticillium* and by several viral diseases. The nutritive value of tomatoes has been greatly enhanced, and their ability to tolerate saline or other basically unfavorable conditions has been increased greatly through the systematic collection, analysis, and use of strains of wild tomatoes in breeding programs.

Various Wild Plants Have Great Potential for Becoming Important Crops

In addition to the plant species already widely cultivated, there are many wild plants, or plants that are cultivated only locally, that could, if brought into more general cultivation, make important contributions to the world economy. For example, as mentioned earlier, we still derive more than 80 percent of our calories from only six kinds of angiosperms out of the approximately 235,000 species that exist. Only about 3000 angiosperm species have ever been cultivated for food, and the great majority of those either are no longer used or are used very locally. Only about 150 plant species have ever been cultivated widely.

Many plants other than this limited number, however—especially those that have been used before but have been completely abandoned or seen as having less importance—may prove to be very useful. Some of these species are still in cultivation in different areas of the world. Although we are accustomed to thinking of plants primarily as important sources of food, they also produce oils, drugs, pesticides, perfumes, and many other products that are important to our modern industrial society. In fact, about 25 percent of the prescription drugs commonly available are derived directly from plant products. We have come to regard the production of such substances from plants, however, as a rather archaic method that has now been fully supplanted by chemical synthesis in commercial laboratories. Their production by plants, however, requires no energy other than that of the sun: it occurs naturally. As our sources of nonrenewable energy near exhaustion and prices increase, it has become increasingly important to find ways to produce complex chemical molecules less expensively. Further, the vast majority of plants have never been examined or tested to determine their usefulness.

A few examples of crops that have recently been developed will demonstrate the great potential that exists in nature. Jojoba (pronounced "ho-*ho*-ba"), *Simmondsia chinensis,* despite its Latin name, is a shrub native to the deserts of northwestern Mexico and the adjacent United States (Figure 34–26). The large seeds of jojoba contain

34–26
Jojoba (Simmondsia chinensis), *an important crop that is increasingly being planted throughout the arid areas of the world. Jojoba is a source of waxes with special lubricating characteristics and other useful properties.*

about 50 percent liquid wax, a substance that has impressive industrial potential. Wax of this sort is indispensable as a lubricant where extreme pressures are developed, as in the gears of heavy machinery and in automobile transmissions. It is difficult to produce synthetic liquid wax commercially, and the endangered sperm whale is the only alternative natural source. Jojoba wax is also used in cosmetics and as a food additive, and other applications are still being discovered for this unusual substance. The plant thrives in hot deserts that are not suitable for the cultivation of most other crops, and some varieties are even highly salt-tolerant. Crops of jojoba are being produced in arid areas throughout the world, and they may make an important contribution to the economic development of these areas, especially by helping to provide employment for local peoples in the cultivation of the crop.

Jojoba might also help to stop the spread of deserts, since it can grow in sandy soil when the rainfall is as little as 7.5 centimeters per year. In some areas, plantings of jojoba might make a useful contribution to stabilizing the soil. Such considerations are of prime importance in view of the fact that the Sahara, for example, is spreading southward at the rate of about 5 kilometers per year and thus limiting the potential for food production of areas that are not now desert. At present, Saudi Arabia, Kuwait, Egypt, Morocco, Ecuador, and Nigeria are experimenting with jojoba for this purpose.

Another interesting new crop is guayule, *Parthenium argentatum* (Figure 34–27). A member of the daisy, or sunflower, family *(Asteraceae)*, guayule is closely related to ragweed *(Ambrosia* spp.). It is a low shrub native to northern Mexico and the southwestern United States. As much as 20 percent of the living plant body may consist of rubber.

Synthetic rubber has replaced natural rubber for many purposes and now constitutes about two-thirds of the world supply. It is manufactured from petroleum and is therefore not renewable, as is rubber produced by plants. Almost all natural rubber is now obtained from Para rubber trees, members of the genus *Hevea* of the spurge family *(Euphorbiaceae)*. According to a World Bank report, by the beginning of this coming century, while synthetic rubbers will continue to constitute approximately two-thirds of our requirements, a total of 30 million metric tons of natural rubber will be needed for products ranging from heavy-duty tires to condoms and surgical gloves (the human immunodeficiency virus (HIV) will pass through anything else). Although they are native to the Amazon Basin of South America, Para rubber trees are cultivated most successfully in tropical Asia. While natural rubber had been extracted from jungle plants since the middle of the nineteenth century, 1899 was the first official year of record for production from plantation plantings in Southeast Asia. In that year, 4 tons of rubber were harvested from a total of 4000 acres of the Para rubber tree, the tree having only recently been transferred to Southeast Asia from Brazil via Kew Gardens, near London.

Guayule provides a promising alternative to *Hevea* species. It can be cultivated in the desert and has the potential of greatly increasing the world rubber supply. Wild plants of guayule have been used as a source of rubber for nearly a century, and cultivated fields in the United States produced more than 1.3 million kilograms of rubber during World War II, when the rubber supply was cut off by the Japanese. Although guayule received relatively little attention during the four decades following the end of the war, it is under active investigation again. Guayule rubber is hypoallergenic and hence can substitute for *Hevea* rubber in condoms and other health-related materials.

A third crop that might well be grown more widely is the grain amaranths (various species of *Amaranthus*), plants that have been grown for food for thousands of years in Latin America, but mainly on a relatively small scale. The grain amaranths have recently emerged as an excellent example of a potential new food crop. Weedy varieties of *Amaranthus* are called "pigweed" in English-speaking countries, but in pre-Columbian times the seeds of these plants were one of the basic foods of the New World—almost as important as maize and beans. Some 20,000 metric tons of seeds were sent annually to Tenochtitlán (present-day Mexico City) in tribute to the chief ruler of the Aztecs. The Spanish conquistadors forbade the Mexican people to use amaranth because of its association with pagan rituals and human sacrifice, and it survived as a crop only in very small, local areas. The protein content of amaranth seeds is as high as that of

34–27

Guayule (Parthenium argentatum), *a desert shrub that produces natural rubber. First harvested from natural populations, guayule is now being cultivated widely.*

the cereals, and, furthermore, amaranth protein contains large amounts of the amino acid lysine, making it nutritionally complementary to the cereals. (Many cereals are low in lysine, which is an essential nutrient for human beings.) The leaves of some races of amaranth can also be used as nutritious vegetables. In view of all these virtues, it is not surprising that the cultivation of amaranth is spreading throughout the world.

One important area of investigation in the search for valuable new crops focuses on salt-tolerant species. Increasingly, intensive agriculture is being practiced in the arid and semiarid areas of the world, primarily in response to the demands of the ever-increasing human population. Such practices have placed enormous demands on very limited local water supplies, which have become increasingly brackish as the water is used, reused, and polluted with fertilizer from adjacent cultivated areas. In addition, there are many parts of the world, especially near seashores, where the soil and the local water supplies are naturally saline. Unproductive when farmed using traditional agricultural methods, such areas might be brought under cultivation if the right kinds of plants could be found (Figure 34–28).

Finding suitable plants for saline areas entails not only seeking out entirely new crop plants but also breeding salt-tolerance into traditional crops (see the essay on pages 744 and 745). In tomatoes, for example, a wild species that grows on the sea-cliffs of the Galápagos Islands and is therefore very salt-resistant, *Solanum cheesmaniae*, has been used as a source of salt-tolerance in new hybrids with the commonly cultivated tomato, *Solanum lycopersicum*. The selection of these hybrids was carried out in a medium having half the salinity of seawater. Researchers have succeeded in selecting hybrid individuals that will complete their development in water of that salinity. Salt-tolerant strains of barley have been developed using similar methods.

Plants Continue to Be Important Sources of Drugs   In addition to their other uses, plants are important sources of medicinal drugs. In fact, about a quarter of the prescriptions written in the United States contain at least one product that has been derived from a plant. For millennia, humans have been using plants for medicinal purposes. Botany, in fact, was traditionally regarded as a branch of medicine, and only in the past 150 years or so have there been professional botanists, as distinct from physicians. There has not, however, been any comprehensive effort to identify and bring into use previously unexploited secondary metabolites of the sort discussed in Chapters 2 and 22.

One of the reasons that plants will continue to be of importance as sources of drugs, despite the ease with which many drugs can be synthesized in the laboratory, is that plants manufacture the drugs inexpensively, without requiring additional energy. Furthermore, the structure

(a)

(b)

34–28
The development of new crops, or new strains of presently cultivated crops, that can succeed at relatively high salt concentrations is important in many areas of the world, particularly those that are arid or semiarid. **(a)** Eggplants (Solanum melongena) *cultivated in the Arava Valley of Israel using drip irrigation (water reaches the plants through plastic tubes and is greatly conserved in this way)* with highly saline water (1800 parts per million). Fifteen years ago, this amount of salinity was thought to be incompatible with commercial crop production. **(b)** At an experimental site on the Mediterranean coast of Israel, a highly nutritious fodder shrub, Atriplex nummularia, *is grown with 100 percent seawater. Yields are similar to those achieved with alfalfa, but the leaves and* stems of Atriplex *are highly salty and thus not as valuable for fodder as alfalfa. Research being conducted by D. Pasternak and his coworkers at the Ben-Gurion University of the Negev aims at overcoming the problems of seawater irrigation, and this type of irrigation may one day make possible the cultivation of huge areas that are now coastal deserts.*

A New Millennium: The Transition to Sustainability

"We cannot afford another century like this one," warned conservationist George Schaller at a national conference on biodiversity and ecosystems held in Washington, D.C., in the fall of 1997. But with evidence for progress all around—cleaner air, safer supplies of food and water, and more attention to the preservation of wild habitats everywhere—what could this statement mean to a citizen of the United States, the most prosperous nation on Earth?

Simply put, our system is unsustainable. It was made possible by use of energy and raw materials that were both abundant and cheap, and by ignoring the environmental costs of their large-scale exploitation. The evidence of those costs is now alarmingly clear: dramatic population growth, resource depletion, reduction in biodiversity, and global atmospheric changes whose effects are staggering in their scope and consequences for all forms of life. Developing sustainable growth strategies, ones that do not depend on such environmental degradation, is the challenge we face if we are to avoid "another century like this one."

Sustainability begins with the recognition that several simple, unalterable facts shape our relationship to the environment. Most important, the Earth is a single planet, functioning in the way that it does as a result of the activities of its living systems. For all practical purposes, the only thing that reaches the Earth's surface is energy from the sun, and nothing leaves it—nothing at all—except some of the energy, which is radiated back into space. Collectively, we must face up to the fact that we can't throw anything "away," because there is no away!

We must also recognize that the twentieth century has been a unique period in history. Just during the second half of this century, world population has grown from 2.5 billion to over 6 billion. This population growth has had profound ecological consequences. During this period, we have wasted about a quarter of the Earth's topsoil and nearly a fifth of its agricultural land through erosion, due to a combination of bad farming practice and abuse of marginal lands. We have substantially changed the composition of the atmosphere. We have cut without replacing about a third of the forests that existed in 1950. Almost 50 percent of the total net photosynthetic productivity on land—the production of nearly half of all the land plants in the world—is directly or indirectly consumed, wasted, or diverted (as when a city replaces a forest) by humans. All the rest of the estimated 5 to 7 million species of eukaryotic organisms have available only about the same amount of energy as we use for ourselves. It is little wonder so many of them are becoming extinct.

And they are becoming extinct at an alarming rate. Human activities have led to a thousandfold increase in the worldwide rate of extinction, so that for each new species that evolves, more than 1,000 disappear. Even after recent increased recognition of the problem, this rate continues to climb steeply. By the end of the twenty-first century, unless there are huge changes in our behavior well beforehand, as many as three-quarters of all species living on land may either be extinct or on the way to extinction. There have been only 5 great extinction events in the more than 3.5-billion-year history of life on Earth, each apparently caused by environmental cataclysms such as asteroid collisions. The twentieth century has witnessed the beginning of the sixth such event, but this one is caused by the activities of a single species, ourselves.

Even if there were not moral, ethical, and religious reasons to preserve the species of the world, there are quite practical reasons to do so. We believe that the twenty-first century will be the age of biology, in the sense that genes, individual organisms, and whole ecosystems will be used to produce goods and repair environmental damage. It is clearly not in our best interests to allow so many organisms to become extinct, especially since we have catalogued only an estimated quarter of the species that exist on Earth. We have never even seen, let alone analyzed the potential of, most of the species we are losing.

This loss of biodiversity may accelerate even more due to global atmospheric changes brought about by the twin threats of chlorofluorocarbons and greenhouse gases.

Over some 3.5 billion years, the photosynthetic activity of living organisms has changed our atmosphere from a reducing one, rich in hydrogen, to an oxidizing one, with about 21 percent oxygen. This oxygen (O_2) is in equilibrium with ozone (O_3). The ozone exists primarily in the stratosphere, hundreds of kilometers above the Earth's surface, where it plays a major role in blocking solar UV-B radiation, which is highly destructive to proteins and nucleic acids—the fundamental building blocks of life. Formation of this protective ozone layer marked a critical turning point in the evolution of life. Chlorofluorocarbons (CFCs), a class of long-lived gases first manufactured about 75 years ago and widely used as refrigerants, destroy ozone. Percolating up into the stratosphere, they have been responsible for breaking down about 6 to 8 percent of the ozone so far, increasing the incidence of malignant skin cancer by 20 to 25 percent in mid-latitudes, and even more nearer the poles. This trend is so serious that the world's nations have agreed to stop it, and CFC production is currently being phased out in the light of international agreements. Such international agreements signal an important, if belated, recognition that global sustainability requires global solutions.

A separate round of international negotiations took place in Kyoto, Japan, at the end of 1997, to address the proliferation of greenhouse gases, most notably carbon dioxide. The atmospheric level of carbon dioxide has risen rapidly and steadily since the beginning of the Industrial Revolution in the mid-eighteenth century. From the 190 parts per million at the end of the last Ice Age (approximately 18,000 years ago), to perhaps 250 parts per million early in the Industrial Revolution in the late eighteenth century, the carbon dioxide level has now climbed to about 362 parts per million. It continues to rise rapidly. Excess carbon dioxide is entering the atmosphere primarily as a result of burning "fossil fuels," formed from ancient living organisms and preserved underground as oil, coal, and natural gas. In addition, about a quarter of the total carbon dioxide being added to the atmosphere comes from cutting forests, where much carbon is naturally stored.

The reason that the climbing level of carbon dioxide poses a problem is that carbon dioxide absorbs some of the infrared radiation by which heat escapes into space. Along with water vapor, methane, and CFCs, carbon dioxide thus acts like the glass in a greenhouse, causing the Earth to be warmer than it would be if these gases were not present. Planets that have atmospheres rich in carbon dioxide are very warm, while those that lack it are very cold. Without any carbon dioxide in the atmosphere, the average year-round temperature of the Earth would be about −15°C (−5°F); with it, the temperature is about 12° C (59°F). Throughout the history of our planet, unusually cold conditions have prevailed when carbon dioxide levels were low, and conditions have been unusually warm when levels were high.

Over the course of the 1990s, it has become evident both that global warming has begun, with an increase of 0.3 to 0.6°C already documented, and that that it is already proceeding more rapidly than it did at the end of the Ice Ages. The International Panel on Climate Change, a United Nations-sponsored body that includes many of the leading experts in climate modeling and other relevant fields, has calculated an increase in global temperatures by 1.5 to 5°C by the end of the next century unless present patterns of activity are altered radically.

Many challenging conditions are likely to arise as a result of the projected warming. For example, sea level has already risen 8 to 18 centimeters, and it is projected to rise 60 centimeters or more during the next century, putting hundreds of millions of people at risk of flooding or displacement throughout the world. Major shifts in ocean currents such as the Gulf Stream are possible, perhaps altering rainfall and other aspects of climate, making agriculture difficult or impossible in many of the areas that are rich farmlands today. Radical shifts in patterns of precipitation are likely to increase the number and intensity of violent storms such as hurricanes. These climatic changes, as well as the atmospheric changes that caused them, may accelerate the rate of extinction, further eroding biodiversity.

The fossil fuels driving these climatic changes also drive much of the world's economy. It is difficult to see how their levels can be controlled, but it is obvious they must be. Arguments about precisely how fast the climate is actually warming cannot change the fundamental fact that we must stop pumping huge quantities of carbon dioxide into the atmosphere. Very few people believe that much of our civiliza-

Clearcut old-growth forest in Gifford Pinchot National Forest in southern Washington state. Logging continues in the nation's remaining 5 percent of old-growth forests, many of which provide critical habitat to threatened wildlife species and are also located in important watershed areas.

tion could survive a doubling of the present carbon dioxide level, and yet to hold the increase to, say, 550 parts per million, will require drastic changes to be carried out starting right now. Since making these changes will in some cases prove disruptive to economic development, many people are reluctant to begin. Others, like John Browne, the CEO of British Petroleum, have wisely recognized that the inertia in our massive global economy means that we must begin now—and that the businesses and nations that do so, instead of engaging in wishful thinking or self-deception, will be the winners of the future.

Sustainability requires an end to decisions made as though the world and its productive systems have no limits. An alternative decision-making system has been proposed by Karl-Henrik Robert, a distinguished Swedish physician, and his colleagues. Called "The Natural Step," this system is important because of the simplicity of its principles, and because it focuses on what is commonly agreed about the environment, not on the areas on which people disagree. Its basic "system conditions" by which we must organize our activities are simply stated:

Neither manufactured nor mined substances, nor their waste products, can be allowed to increase in the biosphere indefinitely

We must preserve the plants, algae, and bacteria that carry out the photosynthesis on which all life on Earth, including our own, ultimately depends

Social justice must prevail to the extent that there are not insupportable inequities between different groups of people.

This set of principles must be the broad outline on which we build a sustainable world. As former U.S. Undersecretary of State Tim Wirth put it, "The economy is a wholly owned subsidiary of the environment." The world will prosper in the future because of the farsighted actions of those people who recognize the simple, unalterable facts about our relationship to the environment. As one of those people, you have a vital role to play—as a pioneer of achieving global sustainability for the new millennium. No challenge you will ever face is more important, and none will ever so profoundly affect, and continue to affect, every aspect of your life in the future.

of some molecules—such as the steroids, including cortisone and the hormones used in birth-control pills (Figure 34–29)—is so complex that, although they can be synthesized chemically, the production methods make them prohibitively expensive. For this reason, antifertility pills and cortisone were manufactured in the past principally from substances extracted from the roots of wild yams (*Dioscorea*), obtained largely from Mexico. When these sources were virtually exhausted, other plants, including *Solanum aviculare*, a member of the same genus as the nightshades and potatoes, were developed into crops for the same purpose.

In addition to considerations of cost, the marvelous variety of substances that plants produce is important: a source of new products that is seemingly inexhaustible (Figure 34–30). One way in which scientists have learned about potential new drugs is by studying the medical uses to which plants are put by rural or indigenous peoples (Figure 34–31). For example, the contraceptive properties of Mexican yams were "discovered" in this way.

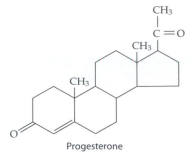

Progesterone

34–29

The steroid progesterone is the precursor of human male and female sex hormones. It is closely related in structure to cholesterol (see Figure 2–14), another abundant human steroid, and to cortisol, which is marketed as cortisone and used to treat inflammation. Other steroids are found in plants, and they are abundant in yams (Dioscorea), from which they are extracted and used as the starting material for the synthesis of the active ingredient in birth-control pills. Steroids are extremely active physiologically in vertebrates.

34–30

Periwinkle (Catharanthus roseus), *the natural source of the drugs vinblastine and vincristine. These drugs, which were developed in the 1960s, are highly effective against certain cancers. Vinblastine is typically used to treat Hodgkin's disease, a form of lymphoma, and vincristine is used in cases of acute leukemia. Before the development of vinblastine, a person with Hodgkin's disease had a one in five chance of survival; now, the odds have been increased to nine in ten. This periwinkle is widespread throughout the warmer regions of the Earth, but it is native only to Madagascar, an island where only a small proportion of the natural vegetation remains undisturbed.*

34–31

Hortense Robinson, a traditional healer from Belize, explains the medicinal uses of breadfruit leaves to Michael Balick, Director of the Institute of Economic Botany at The New York Botanical Garden. Boiling the leaves in water results in an extract that is used to treat diabetes and high blood pressure. Although the study of medicinal uses of forest plants has led to the discovery of important drugs, such as D-tubocurarine chloride (used as a muscle relaxant in open-heart surgery) and ipecac (employed in the treatment of amoebic dysentery), the opportunities for gaining such knowledge are rapidly being lost as indigenous cultures disappear. Whole groups of people lose their traditional life styles, and forests as well as other wilderness areas that contain medicinal plants are destroyed. Valuable knowledge that has accumulated through thousands of years of trial-and-error experimentation is being forgotten in a very short period of time. Much of this information has been transmitted orally because written records do not exist.

As we expand our search for useful plants, we should keep in mind the rapid loss of plant species that is related to (1) the rapid growth of the human population; (2) poverty, especially in the tropics, where about two-thirds of the world's species of plants occur; and (3) our incomplete understanding of how to construct productive agricultural systems in the tropics. With the complete destruction of undisturbed tropical forests, which seems almost certain to occur within a century, many kinds of plants, animals, and microorganisms will become extinct. Because our knowledge of plants, especially those of the tropics, is so rudimentary, we are faced with the prospect of losing many plant species before we even learn of their existence, much less have the opportunity to find out whether they might be of some use to us. The examination of wild plants for potential human use must be accelerated, and promising species must be preserved in seed banks, in cultivation, or, preferably, in natural reserves.

Genetic Engineering Will Play an Increasingly Important Role in Crop Improvement

One of the most important ways in which crops will be improved is by the methods of genetic engineering. Molecular biologists have learned how to transfer foreign genes into plant cells. As we saw in Chapter 12, natural hybridization, which brings about the recombination of genetic material, plays an important role in the ongoing process of plant evolution. Similarly, the plant breeder has utilized the techniques of hybridization to recombine genetic material with the aim of developing crop plants with improved characteristics. What is new about the methods of genetic engineering is that they allow individual genes to be inserted into organisms in a way that is both precise and simple. Characteristics of interest can therefore be enhanced directly, with far less necessity for backcrossing and selection of the progeny than has been necessary in the past. Genetic engineering is discussed further in Chapter 28.

Solving the Problem of World Hunger Requires an Integrated Approach

The problem of how to ease the burdens of hunger and extreme poverty that are borne by at least a fifth of the world's people remains severe. The Green Revolution must of course go forward, but, at the same time, we must recognize that the broader solutions to this problem are social, political, and ethical. They involve not only the growth of food and its distribution but also the creation of jobs whereby the poor can earn the means to buy food. They involve not only limiting population growth but also raising the living standards of the poor to tolerable levels. No matter how striking the advances of agricultural science might be, they will never be adequate to eliminate hunger in a world in which the population is growing rapidly. In countries with large numbers of people living in poverty, there must be an organizational system that facilitates the introduction of new agricultural practices. There must be sources of fertilizers, pesticides, equipment, credit, and water that are available to all; farmers must be able to sell their products, and they must be able to get these products to market in order to do so. Genetic engineering, coupled with an increased knowledge of plant physiology, will certainly produce superior new crops, but will the billion poor farmers of the Third World have access to them?

Plants also must be developed more fully as sources of nonfood products, such as medicines, other chemicals, and energy. Our major crop plants were mainly brought into widespread cultivation thousands of years ago, and many additional ones could be of great use if we were to identify them, work out their requirements for cultivation, and put them into production. However, many observers believe that, as vegetation is being destroyed rapidly throughout the world, 15 to 20 percent of the total number of plant species may become extinct during the next 30 years. Such a loss (of up to 50,000 plant species) would tragically and unnecessarily limit our options, and it must be reduced as much as possible. Certainly, the prospect of a loss of this magnitude makes the search for useful plants and their conservation seem especially urgent.

A broad knowledge of plant biology has become increasingly relevant to achieving solutions to some of society's most crucial problems. The stabilization of the human population must come first, but then our attention must turn both to the amelioration of poverty and malnutrition and to finding ways in which to provide food for the people of all countries. All the principles of plant growth and development must be brought into play in improving the practice of agriculture throughout the world, in making it possible for our planet to support unprecedented numbers of people in dignity and relative prosperity. Doing this will tax human imagination and ingenuity to the limit, but the task is of immense importance. We must make every effort to accomplish it.

Summary

Humans Evolved Relatively Late in the Earth's History

The human race originated in Africa. The genus *Australopithecus* existed there at least 5 million years ago. Our genus, *Homo*, apparently evolved from *Australopithecus* about 2 million years ago, and our species, *Homo sapiens*, has existed for at least 500,000 years.

Agriculture Began in the Fertile Crescent and Spread to Other Areas of the Old World

Starting about 11,000 years ago, in the Fertile Crescent—an area that extends from Lebanon and Syria through Iraq to Iran—humans began to cultivate such plants as barley, lentils, wheat, and peas. In cultivating and caring for these crops, the early farmers changed the characteristics of the plants so that they became more nutritious and easier to harvest and differed in other ways from their wild relatives. Agriculture spread from this center across Europe, reaching Britain by about 6000 years ago. It also seems to have spread southward through Africa, although agriculture may have originated there independently in one or more centers. Many crops were first brought into domestication in Africa, including yams, okra, coffee, and cotton, which was also domesticated independently in the New World and perhaps in Asia. In Asia, agriculture based on staples such as rice and soybeans—and, farther south, citrus, mangoes, taro, bananas, and other crops—was developed.

The Domestication of Animals Accompanied the Domestication of Plants

Domestic animals, starting with the dog, were an important feature of agriculture in the Old World from the earliest times. Herds of grazing animals were ecologically destructive to many of the semiarid areas of the Old World, especially as the animals increased in number, but they were also important sources of food. When these grazing animals were introduced into Latin America following the voyages of Columbus, they proved to be enormously destructive in many habitats, including tropical forests.

Agriculture in the New World Utilized Many New Species

Agriculture was developed independently in the New World. It began as much as 9000 years ago in Mexico and Peru. Dogs were brought to the New World by people migrating from Asia, but apparently no other domestic animals or plants were introduced in this way. Columbus and those who followed him found a virtual cornucopia of new crops to bring back to the Old World. These crops included maize, kidney beans, lima beans, tomatoes, tobacco, chili peppers, potatoes, sweet potatoes, manioc, pumpkins and squashes, avocados, cacao, and the major cultivated species of cotton.

Spices and Herbs Are Plants Prized for Their Flavors and Odors

Spices, which are strongly flavored plant parts usually rich in essential oils, may be derived from the roots, bark, seeds, fruits, or buds, whereas herbs are usually the leaves of nonwoody plants. Cinnamon, black pepper, and cloves are examples of spices; mint, dill, and tarragon are herbs.

The World's Food Supply Is Based Mainly on a Relatively Small Number of Crop Plants

For the last 500 years, the important crops have been cultivated throughout the world. Wheat, rice, and maize, which provide most of the calories we consume, are cultivated wherever they will grow, and a limited number of other plants have achieved worldwide commercial status. Six crops—wheat, rice, maize, potatoes, sweet potatoes, and manioc—provide over 80 percent of the calories consumed by the world's human population.

Rapid Population Growth Has Caused Numerous Problems

The human population has grown from an estimated 5 million people at the time when agriculture was first developed to 6 billion people by the end of the twentieth century. The world population is growing very rapidly, and more than 95 percent of the growth is taking place in developing countries. Nearly 25 percent of the world population lives in absolute poverty. As a result of this growth and of widespread poverty, and because relatively little has been done to develop agricultural practices suitable for tropical regions, the tropics are being devastated ecologically, with up to 20 percent of the world's species likely to be lost over the next 30 years.

Solving Today's Problems Requires an Integrated Approach

The world food supply can be improved by the traditional methods of plant breeding and selection, by the cultivation of new crops, and by the methods of genetic engineering. Genetic diversity is necessary to allow crops to be modified to meet the different requirements of growing successfully in different regions; modern commercial agriculture tends to reduce genetic diversity. Among the more promising of recently developed new

crops are such food plants as triticale and the grain amaranths and various salt-tolerant species. Also promising
are such industrial crops as jojoba, valuable as a source
of liquid wax for lubrication, and guayule, a source of
rubber. Many medicinal drugs are also derived from
plants, and without doubt numerous additional ones
await discovery. The widespread destruction of habitats
in the tropics, however, is threatening the existence of
many potentially useful plant species before they can be
identified and exploited.

Selected Key Terms

Agricultural Revolution   p. 824

Fertile Crescent   p. 825

herbs   p. 832

hybrid maize   p. 839

spices   p. 832

triticale   p. 839

Questions

1. Among the first cultivated plants were cereals and
   legumes. What food value does each of these plant groups
   contribute to the human diet?

2. Explain how the development of agriculture affected population growth.

3. What is meant by the Green Revolution?

4. Explain the importance of preserving the genetic diversity
   of crop plants as a safeguard against pathogens.

5. Comment on the importance of breeding salt-tolerance into
   traditional crop plants.

6. Although many medicinal drugs can be synthesized in the
   laboratory, plants will continue to be an important source
   of such drugs. Why?

7. What developments in agriculture have the potential for
   helping solve the problem of world hunger?

Suggestions for Further Reading

Chapter 1

Allègre, C.J., and S.H. Schneider: "The Evolution of the Earth," *Scientific American*, October 1994, 66–71, 74–75.

Balik, Michael J., and Paul A. Cox: *Plants, People, and Culture: The Science of Ethnobotany,* Scientific American Library, W.H. Freeman, New York, 1996.

Balter, Michael: "Looking for Clues to the Mystery of Life on Earth," *Science*, 273:870–872, 1996.

Baskin, Yvonne: *The Work of Nature: How the Diversity of Life Sustains Us,* Island Press, Washington, DC, 1997.

Benton, M.J.: "Diversification and Extinction in the History of Life," *Science*, 268:52–58, 1995.

Berner, Robert A.: "The Rise of Plants and Their Effect on Weathering and Atmospheric CO_2," *Science*, 276:544–545, 1997.

Botanical Society of America, "Botany for the Next Millennium," June 1995, Department of Botany, Ohio State University, 1735 Neil Avenue, Columbus, OH 43210.

Botanical Society of America, "Careers in Botany, A Guide to Working with Plants," Department of Botany, Ohio State University, 1735 Neil Avenue, Columbus, OH 43210.

Bowring, Samuel A., and Todd Housh: "The Earth's Early Evolution," *Science*, 269: 1535–1540, 1995.

Daily, Gretchen C., and Paul R. Ehrlich: "Population, Sustainability, and Earth's Carrying Capacity," *BioScience*, 42:761–771, 1992.

de Duve, Christian: "The Beginnings of Life on Earth," *American Scientist*, 83:428–437, 1995.

Gordon, Malcolm S., and Everett C. Olson: *Invasions of the Land: The Transitions of Organisms from Aquatic to Terrestrial Life,* Columbia University Press, New York, 1994.

Holland, Heinrich D.: "Evidence for Life on Earth More Than 3850 Million Years Ago," *Science*, 275:38–39, 1997.

Monastersky, Richard: "The Rise of Life on Earth," *National Geographic*, March 1998, 54–81.

Smith, Bruce D.: *The Emergence of Agriculture,* Scientific American Library, W.H. Freeman, New York, 1995.

Section 1
Chapters 2–4

Agre, Peter, Dennis Brown, and Soren Nielsen: "Aquaporin Water Channels: Unanswered Questions and Unresolved Controversies," *Current Opinion in Cell Biology*, 7:472–483, 1995.

Alberts, Bruce, Dennis Bray, Julian Lewis, Martin Raff, Keith Roberts, and James D. Watson: *Molecular Biology of the Cell*, 3rd ed., Garland, New York, 1994.

Bartley, Glenn E., and Pablo A Scolnik: "Plant Carotenoids: Pigments for Photoprotection, Visual Attraction, and Human Health," *The Plant Cell*, 7:1027–1038, 1995.

Becker, Wayne M., Jane B. Reece, and Martin F. Poenie: *The World of the Cell*, 3rd ed., Benjamin/Cummings, Menlo Park, CA, 1996.

Bednarek, Sebastian Y., and Natasha V. Raikhel: "Intracellular Trafficking of Secretory Proteins," *Plant Molecular Biology*, 20:133–150, 1992.

Bolwell, G. Paul: "Cyclic AMP, the Reluctant Messenger in Plants," *Trends in Biochemical Sciences*, 20:492–495, 1995.

Brett, C., and K. Waldron (eds.): *Physiology and Biochemistry of Plant Cell Walls,* 2nd ed., Chapman & Hall, London, 1996.

Carpita, Nick, Maureen McCann, and Lawrence R. Griffing: "The Plant Extracellular Matrix: News from the Cell's Frontier," *The Plant Cell*, 8:1451–1463, 1996.

Cyr, Richard J.: "Microtubules in Plant Morphogenesis: Role of the Cortical Array," *Annual Review of Cell Biology*, 10:153–180, 1994.

Dakora, Felix D.: "Plant Flavonoids: Biological Molecules for Useful Exploitation," *Australian Journal Plant Physiology*, 22:87–99, 1995.

Delmer, Deborah P., and Yehudit Amor: "Cellulose Biosynthesis," *The Plant Cell*, 7:987–1000, 1995.

Driouich, Azeddine, L. Faye, and L. Andrew Staehelin: "The Plant Golgi Apparatus: A Factory for Complex Polysaccharides and Glycoproteins," *Trends in Biochemical Sciences*, 18:210–214, 1993.

Epel, Bernard L: "Plasmodesmata: Composition, Structure and Trafficking," *Plant Molecular Biology*, 26:1343–1356, 1994.

Fosket, Donald E: *Plant Growth and Development, A Molecular Approach,* Academic Press, San Diego, CA, 1994.

Gibeaut, David M., and Nicholas C. Carpita: "Biosynthesis of Plant Cell Wall Polysaccharides," *FASEB Journal*, 8:906–915, 1994.

Goddard, Russell H., Susan M. Wick, Carolyn D. Silflow, and **D. Peter Snustad:** "Microtubule Components of the Plant Cell Cytoskeleton," *Plant Physiology,* 104:1–6, 1994.

Goldberg, Martin W., and **Terence D. Allen:** "Structural and Functional Organization of the Nuclear Envelope," *Current Opinion in Cell Biology,* 7:301–309, 1995.

Hepler, Peter K., and **Julia M. Hush:** "Behavior of Microtubules in Living Plant Cells," *Plant Physiology,* 112:455–461, 1996.

Hyams, Jeremy S., and **Clive W. Lloyd** (eds.): *Microtubules,* John Wiley & Sons, New York, 1994.

Kerr, Richard A.: "Martian 'Microbes' Cover Their Tracks," *Science,* 276:30–31, 1997.

King, John: *Reaching for the Sun: How Plants Work,* Cambridge University Press, New York, 1997.

Kragler, Friedrich, William J. Lucas, and **Jan Monzer:** "Plasmodesmata: Dynamics, Domains and Patterning," *Annals of Botany,* 81:1–10, 1998.

Kuroiwa, Tsuneyoshi, and **Hidenobu Uchida:** "Organelle Division and Cytoplasmic Inheritance,": *BioScience,* 46:827–835, 1996.

Kutchan, Toni M.: "Alkaloid Biosynthesis—The Basis for Metabolic Engineering of Medicinal Plants," *The Plant Cell,* 7:1059–1070, 1995.

Lackey, Robert T.: "The Future of the Earth," *BioScience,* 45: 295–296, 1995.

Lambert, Anne-Marie: "Microtubule-Organizing Centers in Higher Plants: Evolving Concepts," *Botanica Acta,* 108:535–537, 1995.

Lehninger, Albert L., David L. Nelson, and **Michael M. Cox:** *Principles of Biochemistry,* 2nd ed., Worth, New York, 1992.

Leigh, R.A., D. Sanders, and **J.A. Callow** (eds.): *"The Plant Vacuole,"* *Advances in Botanical Research,* 25:1–463, 1997.

Lloyd, Clive W. (ed.): *The Cytoskeletal Basis of Plant Growth and Form,* Academic Press, New York, 1991.

Marano, Maria R., Esteban C. Serra, Elena G. Orellano, and **Nestor Carrillo:** "The Path of Chromoplast Development in Fruits and Flowers," *Plant Science,* 94:1–17, 1993.

Martin, Cathie, and **Alison M.:** "Starch Biosynthesis," *The Plant Cell,* 7:971–985, 1995.

Martin, Ronald E.: "Late Permian Extinctions," *Science,* 274:1549–1550, 1996.

Mazza, G., and **E. Miniati:** *Anthocyanins in Fruits, Vegetables, and Grains,* CRC Press, Boca Raton, FL 1993.

McGarvey, Douglas J., and **Rodney Croteau:** "Terpenoid Metabolism," *The Plant Cell,* 7:1015–1026, 1995.

McKay, David S., et al.: "Search for Past Life on Mars: Possible Relic Biogenic Activity in Martian Meteorite ALH84001," *Science,* 273:924–930, 1996.

Mohnen, Debra, and **Michael G. Hahn:** "Cell Wall Carbohydrates as Signals in Plants," *Cell Biology,* 4:93–102, 1993.

Mojzsis, S.J., et al.: "Evidence for Life on Earth Before 3,800 Million Years Ago," *Nature,* 384:55–59, 1996.

Monastersky, Richard: "Impact Wars: Debate Rages Over Killer Comets and Dinosaur Deaths," *Science News,* 145:156–157, 1994.

Moore, Peter D.: "Life in the Slow Lane," *Nature,* 374:26, 1995.

Niklas, Karl J.: *The Evolutionary Biology of Plants,* University of Chicago Press, Chicago, 1997.

Oparka, Karl J.: "Signaling Via Plasmodesmata—The Neglected Pathway," *Cell Biology,* 4:131–138, 1993.

Saap, Jan: "Cell Evolution and Organelle Origins: Metascience to Science," *Archiv Protistenkunde,* 145:263–275, 1995.

Sackheim, G.: *Introduction to Chemistry for Biology Students,* 5th ed., Benjamin/Cummings, Redwood City, CA, 1996.

Sharon, Nathan, and **Halina Lis:** "Carbohydrates in Cell Recognition," *Scientific American,* January 1993, 82–89.

Silverstein, Samuel C.: "Phagocytosis of Microbes: Insights and Prospects," *Trends in Cell Biology,* 5:141–142, 1995.

Smith, C. J.: "Accumulation of Phytoalexins: Defense Mechanism and Stimulus Response System," *New Phytologist,* 132: 1–45, 1996.

Verma, D.P.S.: *Signal Transduction in Plant Growth and Development,* Elsevier Science, Ireland, 1997.

Waterman, Peter G.: "Roles for Secondary Metabolites in Plants," in *Secondary Metabolites: Their Function and Evolution,* pp. 255–269, John Wiley & Sons, New York, 1992.

Whatley, Jean M.: "The Endosymbiotic Origin of Chloroplasts," *International Review of Cytology,* 144:259–299, 1993.

Wolf, S., and **W.J. Lucas:** "Virus Movement Proteins and Other Molecular Probes of Plasmodesmal Function," *Plant, Cell and Environment,* 17:573–585, 1994.

Wyatt, Sarah E., and **Nicholas C. Carpita:** "The Plant Cytoskeleton–Cell Wall Continuum," *Trends in Cell Biology,* 3:413–417, 1993.

Young, Allen M.: *The Chocolate Tree: A Natural History of Cacao,* Smithsonian Institution Press, Washington, DC, 1994.

Section 2
Chapters 5–7

Alberts, Bruce, Dennis Bray, Julian Lewis, Martin Raff, Keith Roberts, and **James D. Watson:** *Molecular Biology of the Cell,* 3rd ed., Garland, New York, 1994.

Apel, P.: "Evolution of the C_4 Photosynthetic Pathway: A Physiologist's Point of View," *Photosynthetica,* 30:495–502, 1994.

Becker, Wayne M., Jane B. Reece, and **Martin F. Poenie:** *The World of the Cell,* 3rd ed., Benjamin/Cummings, Menlo Park, CA, 1996

Birky, C. William, Jr.: "Uniparental Inheritance of Mitochondrial and Chloroplast Genes: Mechanisms and Evolution," *Proceedings of the National Academy of Sciences USA,* 92:11331–11338, 1995.

Culotta, Elizabeth: "Will Plants Profit from High CO_2?", *Science,* 268: 654–656, 1995.

Dai, Ziyu, S.B. Ku, and **Gerald E. Edwards:** "C_4 Photosynthesis: The CO_2-Concentrating Mechanism and Photorespiration," *Plant Physiology,* 103:83–90, 1993.

Egginston, S., and **H.F. Ross** (eds.): *Oxygen Transport in Biological Systems: Modelling of Pathways from Environment to Cell,* Cambridge University Press, New York, 1992.

Ehleringer, James R., and **Russell K. Monson:** "Evolutionary and Ecological Aspects of Photosynthetic Pathway Variation," *Annual Review of Ecology and Systematics,* 24:411–439, 1993.

Falkowski, Paul G., and **John A. Raven:** *Aquatic Photosynthesis,* Blackwell Science, Malden, MA, 1997.

Fosket, Donald E.: *Plant Growth and Development, A Molecular Approach,* Academic Press, San Diego, CA, 1994.

Furbank, Robert T., and **William C. Taylor:** "Regulation of Photosynthesis in C_3 and C_4 Plants: A Molecular Approach," *The Plant Cell,* 7:797–807, 1995.

Garby, Lars, and **Poul S. Larsen:** *Bioenergetics: Its Thermodynamic Foundations,* Cambridge University Press, New York, 1995.

Gibson, Arthur C.: *Structure-Function Relations of Warm Desert Plants,* Springer-Verlag, New York, 1996.

Hall, D.O., and **K.K. Rao:** *Photosynthesis,* 5th ed., Cambridge University Press, Cambridge, 1994.

Leegood, R.C.: "The Regulation of C_4 Photosynthesis," *Advances in Botanical Research,* 26:251–316, 1997.

Lehninger, Albert L., David L. Nelson, and **Michael M. Cox:** *Principles of Biochemistry,* 2nd ed., Worth, New York, 1992.

MacKenzie, Debora: "Where Has All the Carbon Gone?" *New Scientist,* January 8, 1994, 30–33.

MacKenzie, Sally, Shichuan He, and **Anna Lyznik:** "The Elusive Plant Mitochondrion as a Genetic System," *Plant Physiology,* 105:775–780, 1994.

Murray, David R.: *Carbon Dioxide and Plant Responses,* Research Studies Press, UK, 1997.

Nisbet, Euan G., Johnson R. Cann, and **Cindy Lee Van Dover:** "Origins of Photosynthesis," *Nature,* 373:479–480, 1995.

Rögner, Matthias, Egbert J. Boekema, and **Jim Baker:** "How Does Photosystem 2 Split Water: The Structural Basis of Efficient Energy Conversion," *Trends in Biochemical Sciences,* 21:44–49, 1996.

Salisbury, Frank B., and **Cleon W. Ross:** *Plant Physiology,* 4th ed., Wadsworth, Belmont, CA, 1992.

Siedow, James N., and **Ann L. Umbach:** "Plant Mitochondrial Electron Transfer and Molecular Biology," *The Plant Cell,* 7:821–831, 1995.

Sitte, P.: "Facts and Concepts in Cell Compartmentation," *Progress in Botany,* 59: 3–36, 1998.

Taiz, Lincoln, and **Eduardo Zeiger:** *Plant Physiology,* Benjamin/Cummings, Redwood City, CA, 1991.

Winter, K., and **J. A. C. Smith** (eds.): *Crassulacean Acid Metabolism: Biochemistry, Ecophysiology and Evolution,* Springer-Verlag, New York, 1996.

Zelitch, Israel: "Control of Plant Productivity by Regulation of Photorespiration," *BioScience,* 42:510–516, 1992.

Section 3
Chapter 8

Barinaga, Marcia: "A New Twist to the Cell Cycle," *Science,* 269:631–632, 1995.

Doerner, Peter W.: "Cell Cycle Regulation in Plants," *Plant Physiology,* 106:823–827, 1994.

Doree, Marcel, and **Simon Galas:** "The Cyclin-Dependent Protein Kinases and the Control of Cell Division," *Journal FASEB,* 8:1114–1121, 1994.

Forer, A., and **P.J. Wilson:** "A Model for Chromosome Movement During Mitosis," *Protoplasma,* 179:95–105, 1994.

Fuller, Margaret T.: "Riding the Polar Winds: Chromosomes Motor Down East," *Cell,* 81:5–8, 1995.

Khodjakov, Alexey, Richard W. Cole, Andrew S. Bajer, and **Conly L. Rieder:** "The Force for Poleward Chromosome Motion in *Haemanthus* Cells Acts Along the Length of the Chromosome During Metaphase but Only at the Kinetochore During Anaphase," *Journal of Cell Biology,* 132:1093–1104, 1996.

Murray, Andrew: "Cyclin Ubiquitination: The Destructive End of Mitosis," *Cell,* 81:149–152, 1995.

Shaul, Orit, Marc van Montagu, and **Dirk Inze:** "Regulation of Cell Division in *Arabidopsis,*" *Critical Reviews in Plant Sciences,* 15:97–112, 1996.

Smirnova, Elena A., and **Andrew S. Bajer:** "Microtubule Converging Centers and Reorganization of the Interphase Cytoskeleton and the Mitotic Spindle in Higher Plant *Haemanthus,*" *Cell Motility and the Cytoskeleton,* 27:219–233, 1994.

Staiger, Chris, and **John Doonan:** "Cell Division in Plants," *Current Opinion in Cell Biology,* 5:226–231, 1993.

Vernos, Isabelle, and **Eric Karsenti:** "Chromosomes Take the Lead in Spindle Assembly," *Trends in Cell Biology,* 5:297–301, 1995.

Chapter 9

Asker, Sven E., and **Lenn Jerling:** *Apomixis in Plants,* CRC Press, Boca Raton, FL, 1992.

Bickel, Sharon E., and **Terry L. Orr-Weaver:** "Holding Chromatids Together to Ensure They Go Their Separate Ways," *BioEssays,* 18:293–300, 1996.

Calzada, Jean-Philippe Vielle, Charles F. Crane, and **David M. Stelly:** "Apomixis: The Asexual Revolution," *Science,* 274:1322–1323, 1996.

Carpenter, Adelaide T.C.: "Chiasma Function," *Cell,* 77:959–962, 1994.

Cresti, M., and **A. Tiezzi** (eds.): *Sexual Plant Reproduction,* Springer-Verlag, New York, 1992.

Kenrick, Paul: "Alternation of Generations in Land Plants: New Phylogenetic and Palaeobotanical Evidence," *Biological Review,* 69:293–330, 1994.

Koltunow, Anna M., Ross A. Bicknell, and **Abdul M. Chaudhury:** "Apomixis: Molecular Strategies for the Generation of Genetically Identical Seeds Without Fertilization," *Plant Physiology*, 108:1345–1352, 1995.

McKim, Kim S., and **R. Scott Hawley:** "Chromosomal Control of Meiotic Cell Division," *Science*, 270:1595–1600, 1995.

Mogie, Michael: *The Evolution of Asexual Reproduction in Plants*, Chapman & Hall, New York, 1992.

Nilsson, Nils-Otto, Torbjorn Sall, and **Bengt O. Bengtsson:** "Chiasma and Recombination Data in Plants: Are They Compatible?" *Trends in Genetics*, 9:344–348, 1993.

Petes, Thomas D., and **Patricia J. Pukkila:** "Meiotic Sister Chromatid Recombination," *Advances in Genetics*, 33:41–62, 1995.

Richards, A. I.: *Plant Breeding Systems*, 2nd ed., Chapman & Hall, New York, 1997.

Schmekel, Karin, and **Bertil Daneholt:** "The Central Region of the Synaptonemal Complex Revealed in Three Dimensions," *Trends in Cell Biology*, 5:239-242, 1995.

Valero, Myriam, Sophie Richerd, Veronique Perrot, and **Christophe Destombe:** "Evolution of Alternation of Haploid and Diploid Phases in Life Cycles," *Trends in Ecology and Evolution*, 7:25–29, 1992.

Chapter 10

Blackburn, Elizabeth H.: "Telomerases," *Annual Review of Biochemistry*, 61:113–129, 1992.

Blackburn, Elizabeth H.: "Telomeres: No End in Sight," *Cell*, 77:621–623, 1994.

Coghlan, Andy: "Engineering the Therapies of Tomorrow," *New Scientist*, 138(1870):26–31, 1993.

Corcos, A.F., and **F.V. Monaghan:** *Gregor Mendel's Experiments on Plant Hybrids: A Guided Study*, Rutgers University Press, New Brunswick, NJ, 1993.

Fuchs, J., A. Brandes, and **I. Schubert:** "Telomere Sequence Localization and Karyotype Evolution in Higher Plants," *Plant Systematics and Evolution*, 196:227–241, 1995.

Griffiths, Anthony J.F., Jeffrey H. Miller, David T. Suzuki, Richard C. Lewontin, and **William M. Gelbart:** *An Introduction to Genetic Analysis*, 6th ed., W.H. Freeman, New York, 1996.

Klug, W.S., and **M.R. Cummings:** *Concepts of Genetics*, 5th ed., Prentice Hall, Upper Saddle River, NJ, 1997.

Marx, Jean: "Chromosome Ends Catch Fire," *Science*, 265:1656–1658, 1994.

Marx, Jean: "How DNA Replication Originates," *Science*, 270:1585–1587, 1995.

Osborne, Brian I., and **Barbara Baker:** "Movers and Shakers: Maize Transposons as Tools for Analyzing Other Plant Genomes," *Current Opinion in Cell Biology*, 7:406–413, 1995.

Stent, Gunther: "DNA's Stroke of Genius," *New Scientist*, 138 (1870):21–25, 1993.

Zakian, Virginia A.: "Telomeres: Beginning to Understand the End," *Science*, 270:1601–1606, 1995.

Chapter 11

Alberts, Bruce, Dennis Bray, Julian Lewis, Martin Raff, Keith Roberts, and **James D. Watson:** *Molecular Biology of the Cell*, 3rd ed., Garland, New York, 1994.

Becker, Wayne M., Jane B. Reece, and **Martin F. Poenie:** *The World of the Cell*, 3rd ed., Benjamin/Cummings, Menlo Park, CA, 1996.

Charlesworth, Brian, Paul Sniegowski, and **Wolfgang Stephan:** "The Evolutionary Dynamics of Repetitive DNA in Eukaryotes," *Nature*, 371:215–220, 1994.

Griffiths, Anthony J.F., Jeffrey H. Miller, David T. Suzuki, Richard C. Lewontin, and **William M. Gelbart:** *An Introduction to Genetic Analysis*, 6th ed., W.H. Freeman, New York, 1996.

McElfresh, Kevin C., Debbie Vining-Forde, and **Ivan Balazs:** "DNA-based Identity Testing in Forensic Science," *BioScience*, 43:149–157, 1993.

Melese, Teri, and **Zhixiong Xue:** "The Nucleolus: An Organelle Formed By the Act of Building a Ribosome," *Current Opinion in Cell Biology*, 7:319–324, 1995.

Meyerowitz, Elliot M., and **Chris R. Somerville** (eds.): *Arabidopsis*, Cold Spring Harbor Laboratory Press, Plainview, NY, 1994.

Wang, Thomas A., and **Joachim J. Li:** "Eukaryotic DNA Replication," *Current Opinion in Cell Biology*, 7:414–420, 1995.

Weising, Kurt, Hilde Nybom, Kirsten Wolff, and **Wieland Meyer:** *DNA Fingerprinting in Plants and Fungi*, CRC Press, Boca Raton, FL, 1994.

Chapter 12

Baskin, Yvonne: "California's Ephemeral Vernal Pools May Be a Good Model for Speciation," *BioScience*, 44:384–388, 1994.

Bell, Graham: *Selection: The Mechanism of Evolution*, Chapman & Hall, New York, 1997.

Gibbons, Ann: "On the Many Origins of Species," *Science*, 273:1496–1499, 1996.

Givnish, Thomas J.: "Adaptive Plant Evolution on Islands: Classical Patterns, Molecular Data, New Insights," in P.R. Grant (ed.), *Evolution on Islands*, Oxford University Press, London, pp. 281–304, 1998.

Gould, Stephen Jay, and **Niles Eldredge:** "Punctuated Equilibrium Comes of Age," *Nature*, 366:223–227, 1993.

Hecht, Max K., Ross J. MacIntyre, and **Michael T. Clegg** (eds.): *Evolutionary Biology*, Plenum Press, New York, 1996.

Kerr, Richard A.: "Did Darwin Get It All Right?," *Science*, 267:1421–1422, 1995.

Landman, Otto E.: "Inheritance of Acquired Characteristics Revisited," *BioScience*, 43:696–705, 1993.

McEvey, Shane F.: *"Hugh E.H. Paterson: Evolution and the Recognition Concept of Species: Collected Writings,"* Johns Hopkins University Press, Baltimore, 1993.

Mlot, Christine: "Microbes Hint at a Mechanism Behind Punctuated Evolution," *Science*, 272:1741, 1996.

Morell, Virginia: "Starting Species with Third Parties and Sex Wars," *Science*, 273:1499–1502, 1996.

Rosenzweig, Michael L.: *Species Diversity in Space and Time*, Cambridge University Press, New York, 1995.

Thompson, John N.: *The Coevolutionary Process*, University of Chicago Press, Chicago, 1994.

Zeyl, Clifford, and **Graham Bell:** "Symbiotic DNA in Eukaryotic Genomes," *Trends in Ecology and Evolution*, 11:10–15, 1996.

Section 4
Chapter 13

Barinaga, Marcia: "Archaea and Eukaryotes Grow Closer," *Science*, 264:1251, 1994.

Brooks, Daniel R., Deborah A. McLennan, James M. Carpenter, Stephen G. Weller, and **Jonathan A. Coddington:** "Systematics, Ecology, and Behavior," *BioScience*, 45:687–695, 1995.

Davis, George M.: "Systematics and Public Health," *BioScience*, 45:705–715, 1995.

Donoghue, Michael J., James A. Doyle, Jacques Gauthier, Arnold G. Kluge, and **Timothy Rowe:** "The Importance of Fossils in Phylogeny Reconstruction," *Annual Review of Ecology and Systematics*, 20:431–460, 1989.

Doolittle, Russell F., Da-Fei Feng, Simon Tsang, Glen Cho, and **Elizabeth Little:** "Determining Divergence Times of the Major Kingdoms of Living Organisms with a Protein Clock," *Science*, 271:470–477, 1996.

Doyle, Jeff J.: "DNA, Phylogeny, and the Flowering of Plant Systematics," *BioScience*, 43:380–389, 1993.

Gogarten, J. Peter: "The Early Evolution of Cellular Life," *Trends in Ecology and Systematics*, 10:147–151, 1995.

Hoch, Peter C., and **A.G. Stephenson** (eds.): *Experimental and Molecular Approaches to Plant Biosystematics*, Monographs in Systematic Botany from the Missouri Botanical Garden, vol. 53, Missouri Botanical Garden, St. Louis, MO, 1995.

Huston, Michael A.: *Biological Diversity: The Coexistence of Species on Changing Landscapes*, Cambridge University Press, New York, 1994.

Keeling, Patrick J., and **W. Ford Doolittle:** "Archaea: Narrowing the Gap Between Prokaryotes and Eukaryotes," *Proceedings of the National Academy of Sciences USA*, 92:5761–5764, 1995.

Lauder, George V., Ray B. Huey, Russell K. Monson, and **Richard J. Jensen:** "Systematics and the Study of Organismal Form and Function," *BioScience*, 45:696–704, 1995.

Margulis, Lynn: *Symbiosis in Cell Evolution*, 2nd ed., W.H. Freeman, New York, 1993.

Margulis, Lynn, and **Karlene V. Schwartz:** *Five Kingdoms, An Illustrated Guide to the Phyla of Life on Earth*, 3rd ed., W.H. Freeman, New York, 1998.

Miller, Douglass R., and **Amy Y. Rossman:** "Systematics, Biodiversity, and Agriculture," *BioScience*, 45:680–686, 1995.

Mishler, Brent D.: "Cladistic Analysis of Molecular and Morphological Data," *American Journal of Physical Anthropology*, 94:143–156, 1994.

Morell, Virginia: "Proteins 'Clock' the Origins of All Creatures—Great and Small," *Science*, 271:448, 1996.

Pearson, Lorentz C.: *The Diversity and Evolution of Plants*, CRC Press, Inc., Boca Raton, FL, 1995.

Sapp, Jan: *Symbiosis Evolving*, Oxford University Press, New York, 1994.

Savage, Jay M.: "Systematics and the Biodiversity Crisis", *BioScience*, 45:673–680, 1995.

Schiebinger, Londa: "The Loves of the Plants," *Scientific American*, February 1996, 110–115.

Schopf, J. William: "Microfossils of the Early Archean Apex Chert: New Evidence of the Antiquity of Life," *Science*, 260:640–646, 1993.

Service, Robert F.: "Microbiologists Explore Life's Rich, Hidden Kingdoms," *Science*, 275:1740–1742, 1997.

Simpson, Beryl B., and **Joel Cracraft:** "Systematics: The Science of Biodiversity." *BioScience*, 45:670–672, 1995.

Soltis, Pamela S., and **Douglas E. Soltis:** "Plant Molecular Systematics," *Evolutionary Biology*, 28:139–194, 1995.

Taylor, Thomas N., and **Edith L. Taylor:** *The Biology and Evolution of Fossil Plants*, Prentice Hall, Upper Saddle River, NJ, 1993.

Williams, David M., and **T. Martin Embley:** "Microbial Diversity: Domains and Kingdoms," *Annual Review of Ecology and Systematics*, 27:569–595, 1996.

Woodland, Dennis W.: *Contemporary Plant Systematics*, 2nd ed., Andrews University Press, Berrien Springs, MI, 1997.

Chapter 14

Agrios, George N.: *Plant Pathology*, 3rd ed. Academic Press, New York, 1988.

Atlas, Ronald M.: *Microbiology: Fundamentals and Applications*, 2nd ed., Macmillan, New York, 1988.

Barinaga, Marcia: "Archaea and Eukaryotes Grow Closer," *Science*, 264:1251, 1994.

Baumann, Peter, Sohail A. Qureshi, and **Stephen P. Jackson:** "Transcription: New Insights from Studies on Archaea," *Trends in Genetics*, 11:279–283, 1995.

Carmichael, Wayne W.: "The Toxins of Cyanobacteria," *Scientific American*, January 1994, 78–85.

Dangl, J.L.: *Bacterial Pathogenesis of Plants and Animals*, Springer-Verlag, New York, 1994.

Day, Lucille: "Microbial Phylogeny," *Mosaic*, 22:47–56, 1991.

DeLong, Edward F., Ke Ying Wu, Barbara B. Prezelin, and **Raffael V.M. Jovine:** "High Abundance of Archaea in Antarctic Marine Picoplankton," *Nature*, 371:695–697, 1994.

Forterre, Patrick: "Thermoreduction, a Hypothesis for the Origin of Prokaryotes," *C.RC. Acad. Sci. Paris, Serie III, Life Sciences,* 318:415–422, 1995.

Gianinazzi-Pearson, Vivienne, and **Jean Denarie:** "Red Carpet Genetic Programmes for Root Endosymbioses," *Trends in Plant Science,* 2:371–372, 1997.

Hecht, Jeff: "'Rare' Bug Dominates the Oceans," *New Scientist,* 144:21, 1994.

Hoffman, Michelle: "Researchers Find Organism They Can Really Relate To," *Science,* 257:32, 1992.

Holm, Constance: "Genes Confirm Archaea's Uniqueness," *Science,* 271:1061, 1996.

Koiwa, Hisashi, Ray A. Bressan, and **Paul M. Hasegawa:** "Regulation of Protease Inhibitors and Plant Defense," *Trends in Plant Science,* 2:379–384, 1997.

Levin, Morris A., Ramon J. Seidler, and **Marvin Rogul** (eds.): *Microbial Ecology: Principles, Methods and Applications,* McGraw-Hill, New York, 1992.

Losick, Richard: "Differentiation and Cell Fate in a Simple Organism," *BioScience* 45(6):400–405, 1995.

Madigan, Michael T., and **Barry L. Marrs:** "Extremophiles," *Scientific American,* April 1997, 82–87.

Madigan, Michael T., John M. Martinko, and **Jack Parker:** *Biology of Microorganisms,* 8th ed., Prentice Hall, Upper Saddle River, NJ, 1997.

Matthews, R.E.F.: *Plant Virology,* 3rd ed., Academic Press, New York, 1991.

Matthews, R.E.F.: *Diagnosis of Plant Virus Diseases,* CRC Press, Boca Raton, FL, 1993.

Meeks, John C.: "Symbiosis between Nitrogen-fixing Cyanobacteria and Plants," *BioScience,* 48:266–276, 1998.

Miller, Virginia L., James B. Kaper, Daniel A. Protnoy, and **Ralph R. Isberg** (eds.): *Molecular Genetics of Bacterial Pathogenesis,* ASM Press, Washington, DC, 1994.

Mohan, S., C. Dow, and **J.A. Cole** (eds.): *Prokaryotic Structure and Function: A New Perspective,* Cambridge University Press, New York, 1992.

Nisbet, E.G., and **C.M.R. Fowler:** "Some Like It Hot," *Nature,* 382:404, 1996.

Shapiro, James A., and **Martin Dworkin** (eds.): *Bacteria as Multicellular Organisms,* Oxford University Press, New York, 1997.

Walter, Malcolm: "Old Fossils Could Be Fractal Frauds," *Nature,* 383:385–386, 1996.

Wolf, S., and **W.J. Lucas:** "Virus Movement Proteins and Other Molecular Probes of Plasmodesmal Function," *Plant, Cell and Environment,* 17:573–585, 1994.

Chapter 15

Ahmadjian, Vernon: *The Lichen Symbiosis,* John Wiley & Sons, New York, 1993.

Alexopoulous, C.J., C.W. Mims, and **M. Blackwell:** *Introductory Mycology,* 4th ed., John Wiley & Sons, New York, 1996.

Benjamin, Denis R.: *Mushrooms: Poisons and Panaceas—A Handbook for Naturalists, Mycologists, and Physicians,* W.H. Freeman, San Francisco, 1995.

Castello, John D., Donald J. Leopold, and **Peter J. Smallidge:** "Pathogens, Patterns and Processes in Forest Ecosystems," *BioScience,* 45:16–24, 1995.

Chen, Mei-Yu, Robert H. Insall, and **Peter N. Devreotes:** "Signaling Through Chemoattractant Receptors in *Dictyostelium*," *Trends in Genetics,* 12:52–57, 1996.

Chiu, Sui-Wai, and **David Moore** (eds.): *Patterns in Fungal Development,* Cambridge University Press, New York, 1996.

Crittenden, P.D., J.C. David, D.L. Hawksworth, and **F.S. Campbell:** "Attempted Isolation and Success in the Culturing of a Broad Spectrum of Lichen-forming and Lichenicolous Fungi," *New Phytologist,* 130:267–297, 1995.

Deacon, J.W.: *Modern Mycology,* 3rd ed., Blackwell Science, Malden, MA, 1997.

Dix, Neville J., and **John Webster:** *Fungal Ecology,* Chapman & Hall, New York, 1995.

Esser, Karl: "Fungal Genetics: From Fundamental Research to Biotechnology," *Progress in Botany,* 58:3–38, 1997.

Frankland, J.C., N. Magan, and **G.M. Gadd:** *Fungi and Environmental Change,* Cambridge University Press, New York, 1996.

Fry, William E., and **Stephen B. Goodwin:** "Resurgence of the Irish Potato Famine Fungus," *BioScience,* 47:363–371, 1997.

Gargas, Andrea, Paula T. DePriest, Martin Grube, and **Anders Tehler:** "Multiple Origins of Lichen Symbioses in Fungi Suggested by SSU rDNA Phylogeny," *Science,* 268:1492–1495, 1995.

Hagiwara, Hiromitsu: *Magic of the Myxomycetes,* National Science Museum, Tokyo, 1997.

Hawksworth, D.L.: "The Recent Evolution of Lichenology: A Science for Our Times," *Cryptogamic Botany,* 4:117–129, 1994.

Hawksworth, D.L., P.M. Kirk, B.C. Sutton, and **D.N. Pegler:** *Ainsworth and Bisby's Dictionary of the Fungi,* 8th ed., Oxford University Press, New York, 1996.

Holden, Constance (ed.): "A Fungus on Our Family Tree," *Science,* 260:295, 1993.

Jennings, D.H.: *The Physiology of Fungal Nutrition,* Cambridge University Press, New York, 1995.

Mittler, Ron, and **Eric Lam:** "Sacrifice in the Face of Foes: Pathogen-induced Programmed Cell Death in Plants," *Trends in Microbology,* 4:10–15, 1996.

Mukerji, K.G.: *Concepts in Mycorrhizal Research,* Kluwer Academic Publishers, Boston, 1996.

Retallack, G.J.: "Were Early Animals Really Lichens?" *Science,* 267:967, 1995.

Smith, F.A., and **S.E. Smith:** "Mutualism and Parasitism: Diversity in Function and Structure in the 'Arbuscular' (VA) Mycorrhizal Symbiosis," *Advances in Botanical Research,* 22:1–43, 1996.

Smith, Myron L., Johann N. Bruhn, and **James B. Anderson:** "The Fungus *Armillaria bulbosa* is Among the Largest and Oldest Living Organisms, *Nature,* 356:428–431, 1992.

Stephenson, Steven L., and **Henry Stempen:** *Myxomycetes: A Handbook of Slime Molds,* Timber Press, Portland, OR, 1994.

Strange, Richard N.: *Plant Disease Control: Towards Environmentally Acceptable Methods,* Chapman & Hall, London, 1993.

Sutton, Brian: *A Century of Mycology,* Cambridge University Press, New York, 1996.

Taylor, T.N., H. Hass, W. Remy, and **H. Kerp:** "The Oldest Fossil Lichen," *Nature,* 378:244, 1995.

Taylor, T.N., and **Jeffrey M. Osborn:** "The Importance of Fungi in Shaping the Paleoecosystem," *Review of Palaeobotany and Palynology,* 90:249–262, 1996.

Yoon, Carol Kaesuk: "Pariahs of the Fungal World, Lichens Finally Get Some Respect," *The New York Times,* June 13, 1995.

Chapters 16 and 17

Bhattacharya, Debashis (ed.): *Origins of Algae and Their Plastids,* Plant Systematics and Evolution, Supplement 11, pp. 1–287, Springer-Verlag, Wien, 1997.

Bhattacharya, Debashish, and **Linda Medlin:** "Algal Phylogeny and the Origin of Land Plants," *Plant Physiology,* 116:9–15, 1998.

Buchheim, Mark A., Melinda A. McAuley, Elizabeth A. Zimmer, Edward C. Theriot, and **Russell L. Chapman:** "Multiple Origins of Colonial Green Flagellates from Unicells: Evidence from Molecular and Organismal Characters," *Molecular Phylogenetics and Evolution,* 3:322–343, 1994.

Bold, H.C., and **Michael J. Wynne:** *Introduction to the Algae Structure and Reproduction,* 2nd ed., Prentice Hall, Englewood Cliffs, NJ, 1985.

Canter-Lund, Hilda, and **John W.G. Lund:** *Freshwater Algae: Their Microscopic World Explored,* Biopress, Bristol, England, 1995.

Cornillon, Sophie, et al.: "Programmed Cell Death in *Dictyostelium,*" *Journal of Cell Science,* 107:2691–2704, 1994.

Dawes, Clinton J.: *Marine Botany,* 2nd ed., John Wiley & Sons, New York, 1998.

Emslie, Kerry R., Martin B. Slade, and **Keith L. Williams:** "From Virus to Vaccine: Developments Using the Simple Eukaryote, *Dictyostelium discoideum,*" *Trends in Microbiology,* 12:476–479, 1995.

McCourt, Richard M.: "Green Algal Phylogeny," *Trends in Ecology and Evolution,* 10:159–163, 1995.

McFadden, Geoff, and **Paul Gilson:** "Something Borrowed, Something Green: Lateral Transfer of Chloroplasts by Secondary Endosymbiosis," *Trends in Ecology and Evolution,* 10:12–17, 1995.

Richardson, Laurie L.: "Remote Sensing of Algal Bloom Dynamics," *BioScience* 46:492–501, 1996.

Siegert, Florian, and **Cornelis J. Weijer:** "Spiral and Concentric Waves Organize Multicellular *Dictyostelium* Mounds," *Current Biology,* 5:937–943, 1995.

Smayda, Theodore J., and **Yuzuru Shimizu:** *Toxic Phytoplankton Blooms in the Sea: Proceedings of the Fifth International Conference on Toxic Marine Phytoplankton,* Elsevier, New York, 1993.

Stiller, John W., and **Benjamin D. Hall:** "The Origin of Red Algae: Implications for Plastid Evolution," *Proceedings of the National Academy of Sciences USA,* 94:4520–4525, 1997.

van den Hoek, C., D.G. Mann, and **H.M. Jahns:** *Algae: An Introduction to Phycology,* Cambridge University Press, New York, 1995.

Chapters 18–22

Agosta, William: *Bombardier Beetles and Fever Trees: A Close-up Look at Chemical Warfare and Signals in Animals and Plants,* Addison-Wesley, Reading, MA, 1996.

Barth, Friedrich G.: *Insects and Flowers,* Princeton University Press, Princeton, NJ, 1985.

Basile, Dominick V., and **Margaret R. Basile:** "The Role and Control of the Place-Dependent Suppression of Cell Division in Plant Morphogenesis and Phylogeny," *Memoirs of the Torrey Botanical Club,* 25: 63–84, 1993.

Bell, Peter R.: *Green Plants: Their Origin and Diversity,* Dioscorides Press, Portland, OR, 1992.

Brown, Roy C., and **Betty E. Lemmon:** "Diversity of Cell Division in Simple Land Plants Holds Clues to Evolution of the Mitotic and Cytokinetic Apparatus in Higher Plants," *Memoirs of the Torrey Botanical Club,* 25:45–62, 1993.

Buckles, Mary Parker: *The Flowers Around Us: A Photographic Essay on Their Reproductive Structures,* University of Missouri Press, Columbia, MO, 1985.

Coen, Enrico S., and **Jacqueline M. Nugent:** "Evolution of Flowers and Inflorescences," *Development,* Supplement: 107–116, 1994.

Cox, Paul Alan: "Water-Pollinated Plants," *Scientific American,* October 1993, 68–74.

Crane, Peter R., Else Marie Friis, and **Kaj Raunsgaard Pedersen:** "The Origin and Early Diversification of Angiosperms," *Nature,* 374:27–33, 1995.

Cronquist, Arthur: *The Evolution and Classification of Flowering Plants,* 2nd ed., The New York Botanical Garden, Bronx, NY, 1988.

Cullen, J.: *The Identification of Flowering Plant Families, Including a Key to Those Native and Cultivated in North Temperate Regions,* Cambridge University Press, New York, 1997.

Doust, Jon Lovett, and **Lesley Lovett Doust** (eds.): *Plant Reproductive Ecology: Patterns and Strategies,* Oxford University Press, New York, 1988.

Doyle, Jeff J.: "DNA, Phylogeny, and the Flowering of Plant Systematics," *BioScience* 43:380–389, June 1993.

Endress, Peter K.: "Floral Structure and Evolution of Primitive Angio-sperms: Recent Advances," *Plant Systematics and Evolution,* 192:79–97, 1994.

Fleming, Theodore H.: "Plant-Visiting Bats," *American Scientist,* 81:460–467, 1993.

Frahm, Jan-Peter, "Systematics of the Bryophytes," *Progress in Botany,* 58:455–469, 1997.

Friedman, William E.: "The Evolutionary History of the Seed Plant Male Gametophyte," *Trends in Ecology and Evolution,* 8:15–21, 1993.

Friedman, William E.: *Biology and Evolution of the Gnetales,* University of Chicago Press, Chicago, 1996.

Friedman, William E., and **Jeffrey S. Carmichael:** "Double Fertilization in Gnetales: Implications for Understanding Reproductive Diversification Among Seed Plants," *International Journal of Plant Sciences,* 157 (6 Supplement): S77–S94, 1996.

Friis, Else Marie, and **Peter K. Endress:** "Flower Evolution," *Progress in Botany,* 57:253–280, 1996.

Gamlin, Linda: "The Big Sneeze," *New Scientist,* 37–41, June 2, 1990.

Gifford, Ernest M., and **Adriance S. Foster:** *Morphology and Evolution of Vascular Plants,* 3rd ed., W.H. Freeman, New York, 1989.

Graham, Linda E.: *Origin of Land Plants,* John Wiley & Sons, New York, 1993.

Hughes, Norman F.: *The Enigma of Angiosperm Origins,* Cambridge University Press, New York, 1994.

Iwatsuki, K., and **P.H. Raven** (eds.): *Evolution and Diversification of Land Plants,* Springer-Verlag, New York, 1997.

Kearns, Carol Ann, and **David William Inouye:** "Pollinators, Flowering Plants, and Conservation Biology," *BioScience,* 47:297–307, 1997.

Kenrick, Paul, and **Peter R. Crane:** "Water-Conducting Cells in Early Fossil Land Plants: Implications for the Early Evolution of Tracheophytes, *Botanical Gazette,* 152:335–356, 1991.

Kenrick, Paul, and **Else Marie Friis:** "Paleobotany of Land Plants," *Progress in Botany,* 56:372–395, 1995.

Kenrick, Paul, and **Peter R. Crane:** *The Origin and Early Diversification of Land Plants: A Cladistic Study,* Smithsonian Institution Press, Washington, DC, 1997.

King-Hele, Desmond: "Chronicle of the Lustful Plants," *New Scientist,* pp. 57–61, April 22, 1989.

Kroken, Scott B., Linda E. Graham, and **Martha E. Cook:** "Occurrence and Evolutionary Significance of Resistant Cell Walls in Charophytes and Bryophytes," *American Journal of Botany,* 83:1241–1254, 1996.

Lanner, Ronald M.: *Made for Each Other: A Symbiosis of Birds and Pines,* Oxford University Press, New York, 1996.

Lawrence, Susan V.: "Recent Advances in Hay Fever Research Are Nothing to Sneeze At," *Smithsonian,* September 1984, 100–111.

Lloyd, David G., and **Spencer C.H. Barrett:** *Floral Biology: Studies on Floral Evolution in Animal-Pollinated Plants,* Chapman & Hall, New York, 1996.

Mabberley, D.J.: *The Plant-Book: A Portable Dictionary of the Vascular Plants,* 2nd ed., Cambridge University Press, New York, 1997.

Niklas, Karl J.: "The Aerodynamics of Wind Pollination," *The Botanical Review,* 51:328–386, 1985.

Niklas, Karl J., *Plant Allometry: The Scaling of Form and Process,* University of Chicago Press, Chicago, IL, 1994.

Niklas, Karl J.: *The Evolutionary Biology of Plants,* University of Chicago Press, Chicago, 1997.

Norstog, Knut J., and **Trevor J. Nicholls:** *The Biology of the Cycads,* Cornell University Press, New York, 1997.

Pearson, Lorentz C.: *The Diversity and Evolution of Plants,* CRC Press, New York, 1995.

Philbrick, C. Thomas, and **Donald H. Les:** "Evolution of Aquatic Angiosperm Reproductive Systems," *BioScience,* 46:813–826, 1996.

Press, Malcolm C., and **Jonathan D. Graves** (eds.): *Parasitic Plants,* Chapman & Hall, New York, 1995.

Reski, R.: "Development, Genetics, and Molecular Biology of Mosses," *Botanica Acta,* 111:1–15, 1998.

Richards, A.J.: *Plant Breeding Systems,* Allen and Unwin, Winchester, MA, 1986.

Robacker, David C., Bastiaan J.D. Meeuse, and **Eric H. Erickson:** "Floral Aroma," *BioScience,* 38:390–396, 1988.

Schiebinger, Londa: "The Loves of the Plants," *Scientific American,* February 1996, 110–115.

Silvertown, Jonathan, Miguel Franco, and **John L. Harper:** *Plant Life Histories: Ecology, Phylogeny and Evolution,* Cambridge University Press, New York, 1997.

Stewart, Wilson N., and **Gar W. Rothwell:** *Paleobotany and the Evolution of Plants,* 2nd ed., Cambridge University Press, New York, 1993.

Stuessy, Tod F.: *Plant Taxonomy: The Systematic Evaluation of Comparative Data,* Columbia University Press, New York, 1990.

Taylor, David Winship, and **Leo J. Hickey:** *Flowering Plant Origin, Evolution and Phylogeny,* Chapman & Hall, New York, 1996.

Taylor, Thomas N., and **Edith L. Taylor:** *The Biology and Evolution of Fossil Plants,* Prentice Hall, Englewood Cliffs, NJ, 1993.

Thomas, Barry A., and **R.A. Spicer:** *The Evolution and Paleobotany of Land Plants,* Croom Helm, London, 1987.

White, Mary E.: *The Flowering of Gondwana,* Princeton University Press, Princeton, NJ, 1990.

Winship, D.W., and **L.J. Hickey** (eds.): *Flowering Plant Origin, Evolution, and Phylogeny,* Chapman & Hall, New York, 1996.

Young, James A.: "Tumbleweed," *Scientific American,* March 1991, 82–87.

Section 5
Chapters 23–27

Benfey, Philip, and John W. Schiefelbein: "Getting to the Root of Plant Development: The Genetics of *Arabidopsis* Root Formation," *Trends in Genetics*, 10:84–88, 1994.

Bowman, John (ed.): *Arabidopsis, An Atlas of Morphology and Development*, Springer-Verlag, New York, 1993.

Core, Harold A., Wilfred A. Côté, and Arnold C. Day: *Wood Structure and Identification*, 2nd ed., Syracuse University Press, Syracuse, NY, 1979.

Esau, Katherine: *Anatomy of Seed Plants*, 2nd ed., John Wiley & Sons, New York, 1977.

Fahn, Abraham: *Plant Anatomy*, 4th ed., Pergamon Press, Elmsford, NY, 1990.

Gartner, B.L. (ed.): *Plant Stems: Physiology and Functional Morphology*, Chapman & Hall, New York, 1995.

Goldberg, Robert B., Genaro de Paiva, and Ramin Yadegari: "Plant Embryogenesis: Zygote to Seed," *Science*, 266:605–614, 1994.

Greenberg, Jean T.: "Programmed Cell Death: A Way of Life for Plants," *Proceedings of the National Academy of Sciences USA*, 93:12094–12097, 1996.

Hara, Noboru: "Developmental Anatomy of the Three-Dimensional Structure of the Vegetative Shoot Apex," *Journal Plant Research*, 108:115–125, 1995.

Hart, Stephen: "The Drama of Cellular Death, *BioScience*, 44:451–455, 1994.

Haughn, George W., Elizabeth A Schultz, and Jose M. Martinez-Zapater: "The Regulation of Flowering in *Arabidopsis thaliana*: Meristems, Morphogenesis, and Mutants," *Canadian Journal of Botany*, 73:959–981, 1995.

Havel, L., and D.J. Durzan: "Apoptosis in Plants," *Botanica Acta*, 109:268–277, 1996.

Hoadley, R.B.: *Understanding Wood*, The Taunton Press, Newtown, CT, 1980.

Kalthoff, Klaus: *Analysis of Biological Development*, McGraw-Hill, New York, 1996.

Kolek, J., and V. Konzinka (eds.): *Physiology of the Plant Root System*, Kluwer Academic Publishers, Boston, 1992.

Kozlowski, Theodore T., Paul J. Kramer, and Stephen G. Pallardy: *The Physiological Ecology of Woody Plants*, Academic Press, San Diego, CA, 1991.

Kozlowski, Theodore T., and Stephen G. Pallardy: *Growth Control in Woody Plants*, Academic Press, San Diego, CA, 1997.

Laux, Thomas, and Gerd Jurgens: "Establishing the Body Plan of the *Arabidopsis* Embryo," *Acta Botanica Neerlandica*, 43:247–260, 1994.

Metcalfe, C.R., and L. Chalk: *Anatomy of the Dicotyledons*, vol. 1, 2nd ed., Clarendon Press, Oxford, 1979.

Metcalfe, C.R., and L. Chalk: *Anatomy of the Dicotyledons*, vol. 2, 2nd ed., Clarendon Press, Oxford, 1983.

Meyerowitz, Elliot M.: "The Genetics of Flower Development," *Scientific American*, May, 1994, 56–65.

Panshin, A.J., and C. De Zeeuw: *Textbook of Wood Technology*, vol. 1, 4th ed., McGraw-Hill, New York, 1980.

Raghavan, V.: *Molecular Embryology of Flowering Plants*, Cambridge University Press, New York, 1997.

Sinha, Neelima: "Simple and Compound Leaves: Reduction or Multiplication," *Trends in Plant Science*, 2:396–402, 1997.

Smith, Laurie G., and Sarah Hake: "The Initiation and Determination of Leaves," *The Plant Cell*, 4:1017–1027, 1992.

Smith, William K., Thomas C. Vogelmann, Evan H. DeLucia, David T. Bell, and Kelly A. Shepherd: "Leaf Form and Photosynthesis," *BioScience*, 47:785–793, 1997.

Steeves, Taylor A., and Ian M. Sussex: *Patterns in Plant Development*, 2nd ed., Prentice Hall, Englewood Cliffs, NJ, 1989.

Vincent, Carol A., Rosemary Carpenter, and Enrico S. Coen: "Cell Lineage Patterns and Homeotic Gene Activity During *Antirrhinum* Flower Development," *Current Biology*, 5:1449–1458, 1995.

Waisel, Yoav, Amram Eshel, and Uzi Kafkafi: *Plant Roots: The Hidden Half*, Marcel Dekker, New York, 1991.

Weigel, Detlef: "Patterning the *Arabidopsis* Embryo," *Current Biology*, 3:443–445, 1993.

Yeung, Edward C., and David W. Meinke: "Embryogenesis in Angiosperms: Development of the Suspensor," *The Plant Cell*, 5:1371–1381, 1993.

Zimmermann, Martin H.: *Xylem Structure and the Ascent of Sap*, Springer-Verlag, New York, 1983.

Section 6
Chapters 28 and 29

Barendse, Gerard W.M., and Ton J.M. Peeers: "Multiple Hormonal Control in Plants," *Acta Botanica Neerlandica*, 44:3–17, 1995.

Bevan, M.W., B.D. Harrison, and C.J. Leaver (eds.): *The Production and Uses of Genetically Transformed Plants*, Chapman & Hall, New York, 1994.

Bowles, Dianna J.: "Signal Transduction in Plants," *Trends in Cell Biology*, 5:404–408, 1995.

Chamovitz, Daniel A., and Xing-Wang Deng: "Light Signaling in Plants," *Critical Reviews in Plant Sciences*, 15:455–478, 1996.

Cohen, Jon: "The Genomics Gamble," *Science*, 275:767–772, 1997.

Cohen, Jon: "Developing Prescriptions with a Personal Touch," *Science*, 275:776, 1997.

Davies, Eric: "Intercellular and Intracellular Signals and Their Transduction Via the Plasma Membrane-Cytoskeleton Interface," *Cell Biology*, 4:139–147, 1993.

Davies, Peter J. (ed.): *Plant Hormones and Their Role in Plant Growth and Development*, Martinus Nijhoff Publishers, Boston, MA, 1987.

Davies, Peter J. (ed.): *Plant Hormones: Physiology, Biochemistry and Molecular Biology*, Kuwer Academic Publishers, Dordrech, 1995.

Fluhr, Robert, and **Autar K. Mattoo:** "Ethylene—Biosynthesis and Perception," *Critical Reviews in Plant Sciences*, 15:479–523, 1996.

Fosket, Daniel E.: *Plant Growth and Development, A Molecular Approach*, Academic Press, San Diego, 1994.

Frederick, Robert J., and **Margaret Egan:** "Environmentally Compatible Applications of Biotechnology," *BioScience*, 44:529–535, 1994.

Galston, Arthur W.: *Life Processes of Plants*, Scientific American Library, W.H. Freeman, New York, 1994.

Giampietro, Mario: "Sustainability and Technological Development in Agriculture," *BioScience*, 44:677–689, 1994.

Irving, M.S., Sigalit Ritter, A.D. Tomos, and **D. Koller:** "Phototropic Response of the Bean Pulvinus: Movement of Water and Ions," *Botanica Acta*, 110:118–126, 1997.

Jean, Roger V.: *Phyllotaxis: A Systemic Study in Plant Morphogenesis*, Cambridge University Press, New York, 1994.

Jenkins, Gareth I., John M. Christie, Geeta Fuglevand, Joanne C. Long, and **Jennie A. Jackson:** "Plant Responses to UV and Blue Light: Biochemical and Genetic Approaches," *Plant Science*, 112:117–138, 1995.

Kieliszewski, Marcia J., and **Derek T.A. Lamport:** "Entensin: Repetitive Motifs, Functional Sites, Post-Translational Codes, and Phylogeny," *The Plant Journal*, 5:167–172, 1994.

King, John: *Reaching for the Sun: How Plants Work*, Cambridge University Press, New York, 1997.

Koller, D.: "Light-Driven Leaf Movements," *Plant, Cell and Environment*, 13:615–632, 1990.

Kozlowski, Theodore T., and **Stephen G. Pallardy:** *Growth Control in Woody Plants*, Academic Press, San Diego, 1997.

Lang, G.A. (ed.): *Plant Dormancy: Physiology, Biochemistry and Molecular Biology*, CAB International, Wallingford, Oxon, UK, 1996.

Marion-Poll, Annie: "ABA and Seed Development," *Trends in Plant Science*, 2:447–487, 1997.

Marshall, Eliot: "Gene Tests Get Tested," *Science*, 275:782, 1997.

Mohr, Hans, and **Peter Schopfer:** *Plant Physiology*, Springer-Verlag, New York, 1995.

Parks, Brian M., and **Roger P. Hangarter:** "Blue Light Sensory Systems in Plants," *Cell Biology*, 5:347–353, 1994.

Quail, Peter H.: "Photosensory Perception and Signal Transduction in Plants," *Current Opinion in Genetics and Development*, 4:652–771, 1994.

Rissler, Jane, and **Margaret Mellon:** *Perils Amidst the Promise, Ecological Risks of Transgenic Crops in a Global Market*, Union of Concerned Scientists, Cambridge, MA, 1993.

Salisbury, F.B., and **C.W. Ross:** *Plant Physiology*, 4th ed., Wadsworth, Belmont, CA, 1992.

Smith, Harry: "Phytochrome Transgenics: Functional, Ecological and Biotechnological Applications," *Cell Biology*, 5:315–325, 1994.

Stearns, Tim: "The Green Revolution," *Current Biology*, 5:262–264, 1995.

Taiz, L., and **E. Zeiger:** *Plant Physiology*, Benjamin/Cummings, Redwood City, CA, 1991.

Takahashi, Hideyuki: "Hydrotropism: The Current State of Our Knowledge," *Journal Plant Research*, 110:163–169, 1997.

Vasil, Indra K.: "Molecular Improvement of Cereals," *Plant Molecular Biology*, 25:925–937, 1994.

Vernooij, Bernard, Scott Uknes, Eric Ward, and **John Ryals:** "Salicylic Acid as a Signal Molecule in Plant-Pathogen Interactions," *Current Opinion in Cell Biology*, 6:275–279, 1994.

Ward, John M., Zhen-Ming Pei, and **Julian I. Schroeder:** "Roles of Ion Channels in Initiation of Signal Transduction in Higher Plants," *The Plant Cell*, 7:833–844, 1995.

Wayne, Randy: "Excitability in Plant Cells," *American Scientist*, 81:140–151, 1993.

Wilkes, Garrison: "Gene Banks," *Encyclopedia of Environmental Biology*, 2:181–190, 1995.

Wilkes, Garrison: "Germplasm Conservation," *Encyclopedia of Environmental Biology*, 2:191–201, 1995.

Chapters 30 and 31

Allen, Michael F.: *The Ecology of Mycorrhizae*, Cambridge University Press, New York, 1991.

Asner, Gregory P., Timothy R. Seastedt, and **Alan R. Townsend:** "The Decoupling of Terrestrial Carbon and Nitrogen Cycles," *BioScience*, 47:226–234, 1997.

Baker, David A., and **John L. Hall** (eds.): *Transport of Photoassimilates*, Longman Scientific & Technical, Harlow, Essex, England, and John Wiley & Sons, New York, 1989.

Brown, Kathryn S.: "The Green Clean," *BioScience*, 45:579–582, 1995.

Campbell, Stu: *The Gardener's Guide to Composting*, revised ed., Storey Publishing, Pownal, VT, 1990.

Campbell, Stu: *The Mulch Book: A Complete Guide for Gardeners*, Storey Publishing, Pownal, VT, 1991.

Canny, Martin J.: "What Becomes of the Transpiration Stream," *New Phytologist*, 114:341–368, 1990.

Canny, Martin J.: "Apoplastic Water and Solute Movement: New Rules for an Old Space," *Annual Review of Plant Physiology and Plant Molecular Biology*, 46:215–236, 1995.

Canny, Martin J.: "Transporting Water in Plants," *American Scientist*, 86:152–159, 1998.

Chernicoff, Stanley, and **Ramesh Venkatakrishnan:** *Geology*, Worth, New York, 1995.

Coleman, David C., and **D.A. Crossley, Jr.:** *Fundamentals of Soil Ecology*, Academic Press, San Diego, CA, 1996.

Crawford, Nigel M.: "Nitrate: Nutrient and Signal for Growth," *The Plant Cell*, 7:859–868, 1995.

Dawson, Todd E.: "Hydraulic Lift and Water Use by Plants: Implications for Water Balance, Performance, and Plant-Plant Interactions," *Oecologia*, 95:565–574, 1993.

Geurts, René, and Henk Franssen: "Signal Transduction in *Rhizobium*-induced Nodule Formation," *Plant Physiology*, 112:447–453, 1996.

Juniper, B.E., R.J. Robins, and D.M. Joel: *Carnivorous Plants*, Academic Press, London, 1998.

Knoblauch, Michael, and Aart J.E. van Bel: "Sieve Tubes in Action," *The Plant Cell*, 10:35–50, 1998.

Kozlowski, Theodore T., and Stephen G. Pallardy: *Physiology of Woody Plants*, Academic Press, San Diego, CA, 1997.

Kramer, Paul J., and John S. Boyer: *Water Relations of Plants and Soils*, Academic Press, San Diego, CA, 1995.

Larcher, Walter: *Physiological Plant Ecology: Ecophysiology and Stress Physiology of Functional Groups*, 3rd ed., Springer-Verlag, Berlin, 1995.

Lösch, Rainer: "Plant-Water Relations," *Progress in Botany*, 56:56–96, 1995.

Madore, Monica A., and William J. Lucas (eds.): *Carbon Partitioning and Source-Sink Interactions in Plants*, American Society of Plant Physiologists, Baltimore, 1995.

Marschner, H.: *Mineral Nutrition of Higher Plants*, 2nd ed., Academic Press, San Diego, CA, 1995.

Marschner, H., E.A. Kirby, and C. Engels: "Importance of Cycling and Recycling of Mineral Nutrients within Plants for Growth and Development," *Botanica Acta*, 110:265–273, 1997.

Merson, John: "Mining with Microbes," *New Scientist*, January 4, 1992, 17–19.

Mylona, Panagiota, Katharina Pawlowski, and Ton Bisseling: "Symbiotic Nitrogen Fixation," *The Plant Cell*, 7:869–885, 1995.

Plucknett, Donald L.: "International Agricultural Research for the Next Century," *BioScience*, 43:432–440, 1993.

Richter, Daniel D., and Daniel Markewitz: "How Deep is Soil?" *BioScience*, 45:600–609, 1995.

Salisbury, Frank B., and Cleon W. Ross: *Plant Physiology*, 4th ed., Wadsworth, Belmont, CA, 1992.

Schulz, Alexander: "Phloem. Structure Related to Function," *Progress in Botany*, 59:429–475, 1998.

Singer, Michael J., and Donald N. Munns: *Soils, An Introduction*, 3rd ed., Prentice Hall, Upper Saddle River, NJ, 1996.

Taiz, Lincoln, and Eduardo Zeiger: *Plant Physiology*, Benjamin/Cummings, Redwood City, CA, 1991.

Zimmermann, Martin H., and Claud L. Brown: *Trees: Structure and Function*, Springer-Verlag, New York, 1975.

Zimmermann, Martin H.: *Xylem Structure and the Ascent of Sap*, Springer-Verlag, New York, 1983.

Section 7
Chapters 32—34

Bailey, Robert G.: *Ecosystem Geography*, Springer-Verlag, New York, 1996.

Balick, Michael J., and Paul Alan Cox: *Plants, People, and Culture: The Science of Ethnobotany*, Scientific American Library, W.H. Freeman, New York, 1996.

Balick, Michael J., Elaine Elisabetsky, and Sarah A. Laird: *Medicinal Resources of the Tropical Forest: Biodiversity and Its Importance to Human Health*, Columbia University Press, New York, 1996.

Bartley, Glenn E., and Pablo A. Scolnik: "Plant Carotenoids: Pigments for Photoprotection, Visual Attraction, and Human Health," *The Plant Cell*, 7:1027–1038, 1995.

Bazzaz, Fakhri A., and Eric D. Fajer: "Plant Life in a CO_2-Rich World," *Scientific American*, January 1992, 68–74.

Behrensmeyer, Anna K., John D. Damuth, William A. DiMichele, Richard Potts, Hans-Dieter Sues, and Scott L. Wing (eds.): *Terrestrial Ecosystems through Time, Evolutionary Paleoecology of Terrestrial Plants and Animals*, The University of Chicago Press, Chicago, IL, 1992.

Bisset, Norman G. (ed.): *Herbal Drugs and Phytopharmaceuticals*, CRC Press, Boca Raton, FL, 1994.

Bohnert, Hans J., Donald E. Nelson, and Richard G. Jensen: "Adaptations to Environmental Stresses," *The Plant Cell*, 7:1099–1111, 1995.

Buhner, Stephen H.: *Sacred Plant Medicine*, Roberts Rinehard, Boulder, CO, 1996.

Bullock, Stephen H., Harold A. Mooney, and Ernesto Medina (eds.): *Seasonally Dry Tropical Forests*, Cambridge University Press, New York, 1995.

Cairns, John, Jr., and John R. Heckman: "Restoration Ecology: The State of an Emerging Field," *Annual Review of Energy Environment*, 21:167–189, 1996.

Chadwick, Derek J., and Joan Marsh (eds.): *Ethnobotany and the Search for New Drugs*, Ciba Foundation Symposium 185, John Wiley & Sons, New York, 1994.

Chrispeels, Maarten J., and David E. Sadava: *Plants, Genes and Agriculture*, Jones and Bartlett, Boston, MA, 1994.

Committee on Sustainable Agriculture and the Environment in the Humid Tropics: *Sustainable Agriculture and the Environment in the Humid Tropics*, National Academy Press, Washington, DC, 1993.

Cowan, C. Wesley, and Patty Jo Watson (eds.): *The Origins of Agriculture*, Smithsonian Institution Press, Washington, DC, 1992.

Cox, Paul A., and Michael J. Balick: "The Ethnobotanical Approach to Drug Discovery," *Scientific American*, June 1994, 82–87.

Ellis, Gerry, and Karen Kane: *America's Rain Forest*, North Word Press, Minocqua, WI, 1991.

Forsyth, A.; *Portraits of the Rainforest*, Camden House, Camden East, Ontario, 1990.

Frankel, Otto H., Anthony H.D. Brown, and **Jeremy J. Burdon:** *The Conservation of Plant Biodiversity*, Cambridge University Press, New York, 1995.

Givnish, T. J., and **K. J. Sytsma** (eds.): *Molecular Evolution and Adaptive Radiation*, Cambridge University Press, New York, 1997.

Goulding, Michael: "Flooded Forests of the Amazon," *Scientific American*, March 1993, 114–120.

Guertin, David, S., William E. Easterling, and **James R. Brandle:** "Climate Change and Forests in the Great Plains," *BioScience*, 47:287–294, 1997.

Holloway, Marguerite: "Nurturing Nature," *Scientific American*, April 1994, 98–108.

Holloway, Marguerite: "Sustaining the Amazon," *Scientific American*, June 1993, 90–99.

Homer-Dixon, Thomas F., Jeffrey H. Boutwell, and **George W. Rathjeus:** "Environmental Change and Violet Conflict," *Scientific American*, February 1993, 38–45.

Jeffries, Michael J.: *Biodervisity and Conservation*, Routledge, New York, 1997.

Kareiva, Peter M., Joel G. Kingsolver, and **Raymond B. Huey,** (eds.): *Biotic Interactions and Global Change*, Sinauer Associates, Sunderland, MA, 1993.

Kusler, Jon A., William J. Mitsch, and **Joseph S. Larson:** "Wetlands," *Scientific American*, January 1994, 64–70.

Larson, Douglas: "The Recovery of Spirit Lake," *American Scientist*, 81:166–177, 1993.

Martin, Gary J.: *Ethnobotany—A Methods Manual*, Chapman & Hall, New York, 1995.

Mitsch, W. J. (ed.): *Global Wetlands: Old World and New*, Elsevier Science, Amsterdam, The Netherlands 1994.

National Research Council: *Lost Crops of the Incas*, National Academy Press, Washington, DC, 1990.

New York Botanical Garden: *Advances in Economic Botany*, a series of monographs, The New York Botanical Garden, Bronx, NY.

Norse, Elliott A.: *Ancient Forests of the Pacific Northwest*, The Island Press, Washington, DC, 1990.

Peterken, George F.: *Natural Woodland, Ecology and Conservation in Northern Temperate Regions*, Cambridge University Press, New York, 1996.

Pickett, T. A., Jurek Kolasa, and **Clive G. Jones:** *Ecological Understanding: The Nature of Theory and the Theory of Nature*, Academic Press, San Diego, CA, 1994.

Polis, Gary A., and **Kirk O. Winemiller:** *Food Webs, Integration of Patterns and Dynamics*, Chapman & Hall, New York, 1996.

Prance, G.T., J. Chadwick, and **J. Marsh** (eds.): *Ethnobotany and the Search for New Drugs*, John Wiley & Sons, New York, 1994.

Primack, Richard B.: *Essentials of Conservation Biology*, Sinauer Associates, Sunderland, MA, 1993.

Raven, Peter H., Linda R. Berg, and **George B. Johnson:** *Environment*, 2nd ed., Saunders College Publishing, Fort Worth, NY, 1995.

Reagan, Douglas P., and **Robert B. Waide:** *The Food Web of a Tropical Rain Forest*, University of Chicago Press, Chicago, IL, 1996.

Redford, Kent, and **Christine Padoch** (eds.): *Conservation of Neotropical Forests—Working from Traditional Resource Use*, Columbia University Press, New York, 1992.

Reid, Walter V., et al.: *Biodiversity Prospecting—Using Genetic Resources for Sustainable Development*, World Resources Institute, Washington, DC, 1993.

Richards, P. W.: *The Tropical Rain Forest*, An Ecological Study, 2nd ed., Cambridge University Press, New York, 1996.

Sandlund, O. T., et al. (eds.): *Conservation of Biodiversity for Sustainable Development*, Scandinavian University Press; outside Scandinavia, distributed by Oxford University Press, London, 1992.

Schulter, Richard Evans, and **Siri von Reis:** *Ethnobotany—Evolution of a Discipline*, Timber Press, Portland, OR, 1995.

Simpson, Beryl B., and **Molly Conner Ogorzaly:** *Economic Botany—Plants in Our World*, 2nd ed., McGraw-Hill, New York, 1995.

Smith, Bruce D.: *The Emergence of Agriculture*, Scientific American Library, New York, 1995.

Smith, C. I.: "Accumulation of Phytoalexins: Defense Mechanism and Stimulus Response System," *New Phytologist*, 132:1–45, 1996.

Solbrig, O. T., H. M. van Emden, and **P. G.W. J. van Oordt:** *Biodiversity and Global Change*, CAB International, Wallingford, UK, 1994.

Vitousek, Peter M., et al.: "Human Alteration of the Global Nitrogen Cycle: Sources and Consequences," *Ecological Applications*, 7:737–750, 1997.

Weiner, Jonathan: *The Next One Hundred Years: Shaping the Future of Our Living Earth*, Bantam Books, New York, 1990.

Wilson, Edward O.: *The Diversity of Life*, The Belknap Press of Harvard University Press, Cambridge, MA, 1992.

Young, James, A.: "Tumbleweed," *Scientific American*, March 1991, 82–87.

Appendix A

Fundamentals of Chemistry

Atoms

All matter is made up of combinations of **elements.** There are 92 naturally occurring elements on Earth. Elements are, by definition, substances that cannot be broken down into other substances by ordinary chemical means. The smallest particle of an element that has the properties of that element is an **atom** (from the Greek *atomos*, meaning "indivisible"). Each element has a unique atomic structure.

Every atom has a core, or **nucleus,** that contains one or more positively charged particles called **protons.** The number of protons distinguishes the atoms of different elements from one another. For example, an atom of hydrogen, the simplest element, has 1 proton in its nucleus; an atom of carbon has 6 protons. For any element, the number of protons in the nucleus of its atoms is called its **atomic number.** Thus the atomic number of hydrogen is 1 and the atomic number of carbon is 6.

Outside the atomic nucleus are negatively charged particles, the **electrons,** which are attracted by the positive charge of the protons. The number of electrons in an atom is equal to the number of protons. The number and arrangement of electrons determine whether an atom will react with other atoms, and how it will do so.

Atomic nuclei also contain **neutrons,** uncharged particles of about the same weight as protons. The weight of an atom is essentially made up of the weight of the protons and neutrons in its nucleus; electrons are so light that their weight is usually disregarded. The **atomic weight** of an element is defined as the weight of an atom relative to the weight of a carbon atom having 6 protons and 6 neutrons and a designated atomic weight of 12 (^{12}C; see Isotopes below). Because these atomic weights are relative values, they are expressed without units of weight. Similarly, the **atomic mass** of an element is the mass of an atom relative to that of a carbon atom with a designated atomic mass of 12 (^{12}C). Atomic mass is expressed in units called *atomic mass units* or *daltons,* one dalton being one-twelfth the mass of an atom of carbon (^{12}C). Table A–1 shows the atomic structures of some elements.

Isotopes

Although all atoms of a particular element have the same number of protons in their nuclei, different atoms

TABLE A-1 · Atomic Structures of Some Familiar Elements

Element	Symbol	Nucleus		Number of Electrons	Atomic Weight*
		Number of Protons (Atomic Number)	Number of Neutrons*		
Hydrogen	H	1	0	1	1
Helium	He	2	2	2	4
Carbon	C	6	6	6	12
Nitrogen	N	7	7	7	14
Oxygen	O	8	8	8	16
Sodium	Na	11	12	11	23
Phosphorus	P	15	16	15	31
Sulfur	S	16	16	16	32
Chlorine	Cl	17	18	17	35
Potassium	K	19	20	19	39
Calcium	Ca	20	20	20	40

*For the most common isotope

of the element may contain different numbers of neutrons. These different forms of the same element, differing in atomic weight (number of protons plus number of neutrons) but not in atomic number, are known as **isotopes.** The atomic weight of an element having two or more isotopes is an average value calculated for the naturally occurring mixture of isotopes.

Most elements have several isotopic forms. Hydrogen, for example, exists in three isotopic forms (Table A–2). The common form of hydrogen has 1 proton and an atomic weight of 1; it is symbolized as ^1H, or simply H. (Atomic weight, by convention, is written as a superscript number to the left of the element's symbol.) A second isotope, called deuterium, contains 1 proton and 1 neutron and has an atomic weight of 2; this isotope is symbolized as ^2H. A third, very rare, isotope is tritium, ^3H, with 1 proton and 2 neutrons and an atomic weight of 3. All three isotopes have essentially the same chemical properties: all three have the same number of electrons, and it is the electrons that determine chemical properties. (Note that the atomic weight of naturally occurring hydrogen, a mixture of the three isotopes, is 1.008.)

Some isotopes (tritium, ^3H, is one example) are *radioactive.* This means that the nucleus is unstable and emits energy as it changes to a more stable form. The energy (particles or rays) released by **radioactive isotopes**, also called **radioisotopes**, can be detected by various means, such as with a Geiger counter or on photographic film.

A sample of a radioisotope emits energy at a rate proportional to the number of atoms in the sample. The isotope is said to undergo *radioactive decay* as its atoms change to another isotope or another element. The rate of decay is measured in terms of half-life: the *half-life* of a radioisotope is defined as the time in which half the atoms in a sample lose their radioactivity and become stable. Because the half-life of an isotope is constant, it is possible to calculate the fraction of decay that will occur for a given isotope over a given period of time. Half-lives vary widely. The radioactive nitrogen isotope ^{13}N has a half-life of 10 minutes; tritium (^3H) has a half-life of 12.25 years. The most common isotope of uranium (^{238}U) has a half-life of 4.5 billion years.

Radioisotopes have a number of important uses in biological research. They can be used to determine the age of, or to *date,* fossils or the rocks in which fossils are found. For example, the uranium isotope ^{238}U undergoes a series of decays, eventually being transformed to an isotope of lead, ^{206}Pb. Thus the ratio of ^{238}U to ^{206}Pb in a given rock sample is a good indication of how long ago that rock was formed. Five different isotopes are now employed in the dating of rocks. Radioactive isotopes are also used as *tracers.* Because isotopes of an element have the same chemical properties, a radioisotope behaves in an organism in the same way as the more common, nonradioactive isotope. As a result, biologists are able to use isotopes of a number of elements, especially carbon, nitrogen, and oxygen, to trace the course of many essential chemical reactions in living systems. For example, use of radioactive carbon dioxide ($^{14}CO_2$) played an important role in elucidating the pathways of photosynthesis (see Chapter 7). A third use of isotopes is in *autoradiography,* a technique in which a sample of material containing a radioisotope is placed on a

Isotope	Symbol	Number of Protons (Atomic Number)	Number of Neutrons	Number of Electrons	Atomic Weight
Hydrogen	^1H	1	0	1	1
Deuterium	^2H	1	1	1	2
Tritium	^3H	1	2	1	3

TABLE A–2   Isotopes of Hydrogen

sheet of photographic film. Energy emitted from the isotope leaves traces on the film and so reveals the exact location of the isotope within the specimen (see Figure 31–23).

Electrons and Energy

Atoms, the indivisible units of the elements, are the fundamental building blocks of all matter. Yet they are mostly empty space. As electrons move around the nucleus at almost the speed of light, the distance from electron to nucleus is, on average, about 1000 times the diameter of the nucleus. Because the electrons are so exceedingly small, most of the space within the atom is almost entirely empty.

The distance between an electron and the nucleus around which it moves is determined by the electron's *potential energy,* or energy of position (pages 95–96). The more energy an electron has, the farther it will be from the nucleus. Thus an electron with a relatively small amount of energy is found close to the nucleus and is said to be at a low **energy level.** An electron with more energy is farther from the nucleus, at a higher energy level. With a sufficient input of energy, an electron can move from a lower to a higher energy level—but not to an energy state somewhere in between. An atom that has absorbed sufficient energy to boost an electron to a higher energy level is said to be in an *excited state.* As long as the electron remains at the higher level it possesses the added energy. When it returns to the lower energy level, as it tends to do, the added energy is released (Figure A–1).

The Arrangement of Electrons: Models of Atomic Structure

The concept of the atom as the indivisible unit of elements is almost 200 years old. However, our ideas about atomic structure have undergone many changes over the years. One of the most useful models of atomic structure is derived from the work of the physicist Niels Bohr, who discovered that different electrons have different amounts of energy and are at different distances from the nucleus (Figure A–2). In the Bohr model, the energy levels are depicted as concentric rings surrounding the

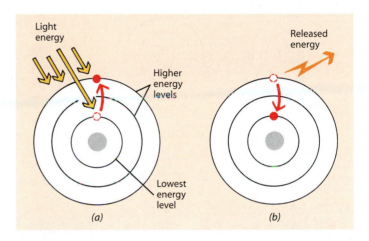

A–1

(a) *When an atom, such as the hydrogen atom in this diagram, receives an input of energy, such as light energy, an electron (shown here in red) may be boosted to a higher energy level. The electron thus gains potential energy. The gray area in the center of the diagram represents the nucleus.* **(b)** *When the electron returns to its previous energy level, this energy is released in the form of heat or light.*

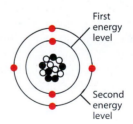

A–2

A Bohr model of the carbon atom.

nucleus. This model is not a true "picture" of the atom and has been superseded by another model (described below), but schematic diagrams derived from Bohr's model (such as those in Figure A–3) can help us visual-

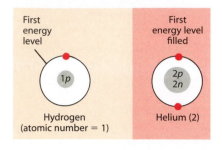

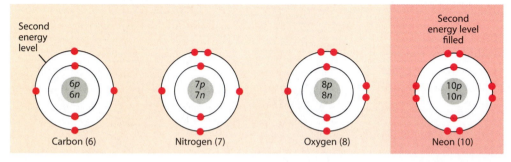

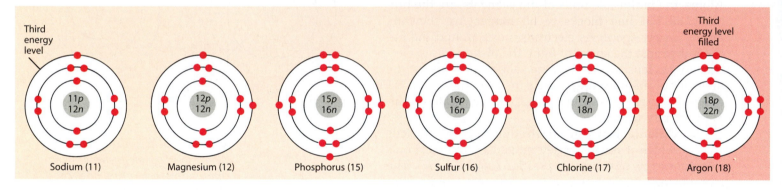

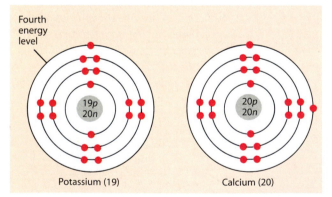

A–3

Schematic diagrams of the electron arrangements in some common elements. In the nuclei, p represents protons and n represents neutrons. In each diagram, the electrons are at their lowest possible energy levels. Note how the first three energy levels are filled in sequence as atomic number increases. As you can see, potassium and calcium have one and two electrons, respectively, in the fourth energy level. In atoms of higher atomic number, additional electrons fill the third energy level (up to 10 electrons) before more electrons are added to the fourth level.

ize the energy levels of an atom, keep track of the number of electrons in each energy level, and see how atoms are likely to interact.

An atom is most stable when all of its electrons are at their lowest possible energy levels. Therefore the electrons of an atom fill the energy levels in order: the first is filled before the second, the second before the third, and so on. The first (lowest) energy level, closest to the nu-

cleus, can hold up to two electrons. Thus the single electron of hydrogen moves around the nucleus in the first energy level, as do the two electrons of helium (Figure A–3). All other elements have atoms with more than two electrons. Since the first energy level is filled by two electrons, the additional electrons must occupy higher energy levels, farther from the nucleus. The second energy level can hold up to eight electrons, and so can the third energy level of elements through atomic number 20 (calcium).

The most recent model of atomic structure provides a more accurate picture of the atom (Figure A–4). An electron is so small and moves so rapidly that it is impossible to determine, at any given moment, both its precise location and the exact amount of energy it possesses. The model therefore describes the *pattern* of an electron's motion rather than its position. The volume of space in which an electron will be found 90 percent of the time is defined as its **orbital**. The one or two electrons at the lowest (first) energy level occupy a single spherical orbital. The second energy level, which can contain up to eight electrons, consists of four orbitals. The third energy level of elements through atomic number 20 also consists of four orbitals. For elements of higher atomic number the pattern becomes more complex.

The electrons in each orbital traverse a particular region of space, and, taken together, these patterns of movement give the atom as a whole a particular three-dimensional shape (Figure A–5). The atom, however, does not have rigid boundaries. Instead it is defined by regions of charge.

The Basis of Chemical Reactivity

The way in which an atom reacts chemically is determined by the number and arrangement of its electrons. An atom is most stable when all of its electrons are at their lowest possible energy levels. Moreover, an element having atoms with a completely filled outermost energy level is more stable than an element having atoms with a partially filled outer energy level. For example, helium (atomic number 2), neon (atomic number 10), and argon (atomic number 18), as shown in Figure A–3, have completely filled outer energy levels and so tend to be unreactive.

In the atoms of most elements, however, the outer energy level is only partially filled. These atoms tend to interact with other atoms in such a way that, after reaction, both atoms have completely filled outer energy levels. Some atoms lose electrons. Some gain electrons. And in many of the most important chemical reactions in living systems, atoms share their electrons with each other.

Electronegativity

The atomic nuclei of different elements have different degrees of attraction for electrons. The affinity of an element for electrons is known as its **electronegativity.** Electronegativity is expressed on a scale of 0 to 4, with helium and the other noble gases having electronegativities of 0 and fluorine having an electronegativity of 4. Electronegativity values for some elements are given in Table A–3. Differences in electronegativity have important consequences for the properties of molecules, as we shall see.

TABLE A–3	Electronegativities of Some Common Elements
Fluorine (F)	4.0
Oxygen (O)	3.5
Nitrogen (N)	3.0
Chlorine (Cl)	3.0
Carbon (C)	2.5
Sulfur (S)	2.5
Hydrogen (H)	2.1
Phosphorus (P)	2.1
Sodium (Na)	0.9
Helium (He)	0.0

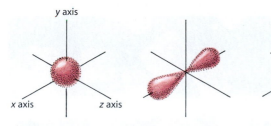

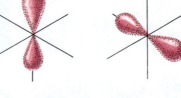

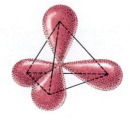

A–4
The four orbitals of the second energy level. Each orbital can hold up to two electrons. One orbital is spherical and encloses the spherical orbital of the first energy level, which, in turn, encloses the nucleus, located at the intersection of the axes. The other three

orbitals are dumbbell-shaped, with their axes perpendicular to one another. For clarity, the orbitals are shown individually. In reality, the orbitals influence one another and determine the overall shape of the atom.

A–5
An orbital diagram of the carbon atom. Each orbital of the second energy level holds one electron. Each electron moves in a pattern that provides maximum separation from the other electrons. The resulting shape resembles four teardrops, extending out from the nucleus.

Ions, Molecules, and Chemical Bonds

As noted above, atoms can interact to achieve completely filled outer energy levels in different ways: by losing electrons, by gaining electrons, or by sharing electrons with each other. The gain or loss of electrons produces charged atoms, called **ions.** The sharing of electrons produces new, larger particles called **molecules.**

When different elements combine in a definite and constant proportion and are held together by chemical bonds, the product is a **chemical compound.** Examples of chemical compounds are water (H_2O), sodium chloride (NaCl), carbon dioxide (CO_2), and glucose ($C_6H_{12}O_6$).

Ions and Ionic Interactions

For many atoms, the simplest way to attain a completely filled outer energy level is to gain or lose one or two electrons. For example, a chlorine atom (atomic number 17) needs one electron to complete its outer energy level; sodium (atomic number 11) has a single electron in its outer level (Figure A–3). This outer electron of sodium is strongly attracted by the chlorine atom (which is highly electronegative), and it jumps from the sodium to the chlorine atom (see Figure 5–5).

As a result of the transfer, the sodium and the chlorine have outer energy levels that are completely filled. And, each atom now has unequal numbers of electrons and protons: the sodium has 11 protons and 10 electrons (an extra positive charge); the chlorine has 17 protons and 18 electrons (an extra negative charge). Such electrically charged atoms are known as ions. Positively charged ions, such as Na^+, are called **cations,** and negatively charged ions, such as Cl^-, are called **anions.**

Oppositely charged ions attract one another, and the resulting substance in the case of Na^+ and Cl^- is sodium chloride (NaCl), ordinary table salt (Figure A–6). Interactions that involve the mutual attraction of oppositely charged ions are called **ionic interactions,** or ionic bonds. Potassium (atomic number 19) also has a single electron in its outermost energy level and reacts with chlorine to form potassium chloride (KCl).

The calcium ion (Ca^{2+}) is formed by loss of two electrons; it can attract and hold two Cl^- ions, forming calcium chloride, $CaCl_2$ (the subscript 2 indicates that two atoms of chlorine are present for each atom of calcium). Similarly, magnesium forms magnesium chloride, $MgCl_2$.

Small ions such as Na^+ and Cl^- make up less than 1 percent of the weight of most living matter, but they play crucial roles. For instance, K^+ is the principal positively charged ion in most organisms, and many essential biological reactions occur only in the presence of K^+. Both Na^+ and K^+ are involved in the active transport of sugars and amino acids across plasma membranes; Ca^{2+} has a direct effect on the physical properties of the membrane and has numerous other roles in cells; Mg^{2+} is an integral part of chlorophyll, the molecule in green plants and algae that traps radiant energy from the sun.

Molecules and Covalent Bonds

Another way for atoms to complete their outer energy levels is by sharing electrons with each other. Chemical bonds formed by sharing one or more pairs of electrons are known as **covalent bonds.** In a covalent bond, each electron spends part of its time around one nucleus and part of its time around the other. Thus the sharing of electrons completes the outer energy level of each atom and neutralizes the positive charge of each nucleus.

Atoms that need to gain electrons to achieve filled and thus stable outer energy levels have a strong ten-

A–6

*Ionic interactions. **(a)** Oppositely charged ions, such as sodium and chloride, depicted here as spheres, attract one another. Table salt is crystalline NaCl, a latticework of alternating Na^+ and Cl^- ions held together by their opposite charges. Such bonds between oppositely charged ions are known as ionic interactions. **(b)** The regularity of the latticework is reflected in the structure of salt crystals, magnified here about 30 times.*

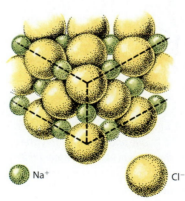

Na$^+$ Cl$^-$

(a)

(b)

dency to form covalent bonds. The simplest example is the covalent bond formed between two hydrogen atoms in a hydrogen molecule, H_2 (Figure A–7).

Of extraordinary importance in living systems is the capacity of carbon to form covalent bonds. Carbon (atomic number 6) has four electrons in its outer energy level (Figure A–3). It can share each of these electrons with another atom, forming covalent bonds with as many as four other atoms, thus producing a stable, filled outer energy

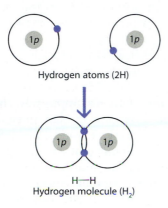

A–7

Covalent bonds. (a) When a molecule of hydrogen is formed, each atom shares its single electron with the other atom. As a result, both atoms effectively have a filled first energy level containing two electrons—a highly stable arrangement. This type of bond, in which electrons are shared, is called a covalent bond. (b) A representation of the pattern of movement of the shared electrons around the hydrogen nuclei. Because the electrons move so rapidly, the charges of both nuclei are effectively neutralized at all times.

level (8 electrons) (Figure A–8). The covalent bonds may be with different elements—most often hydrogen, oxygen, or nitrogen—or with other carbon atoms.

Single and Double Bonds   There are various ways in which atoms can form covalent bonds and fill their outer energy levels. Oxygen, for example, has six electrons in its outer energy level (Figure A–3); four are grouped into pairs and are generally unavailable for bonding, but each of the two unpaired electrons can be shared with another atom. In the water molecule (H_2O) one of these electrons participates in a covalent bond with one hydrogen atom and the other in a covalent bond with a different hydrogen atom (Figure A–9a). Two **single bonds** are formed, and all three atoms have filled outer energy levels.

The bonding situation is different in another familiar substance, carbon dioxide (CO_2). In this molecule, each oxygen atom is joined to the carbon atom by *two pairs* of electrons (four electrons). Such bonds are called **double bonds** (Figure A–9b). Carbon atoms can form double bonds with each other as well as with other elements, as in ethylene, $H_2C\!\!=\!\!CH_2$, and the components of some fats and oils (see Figure 2–9). Carbon atoms can also form triple bonds (in which three pairs of electrons are shared), but these are rare in living systems.

Single bonds are flexible, allowing the bonded atoms to rotate in relation to one another. Double bonds are much more rigid and restrict the relative movement of the bonded atoms. The presence of double bonds in a molecule significantly affects its properties, as in fats and oils (pages 23–24).

Polar Covalent Bonds   As noted previously, elements differ in electronegativity—that is, in their affinity for electrons. In covalent bonds between atoms of different elements, the electrons are not shared equally between the two atoms: the shared electrons tend to spend more

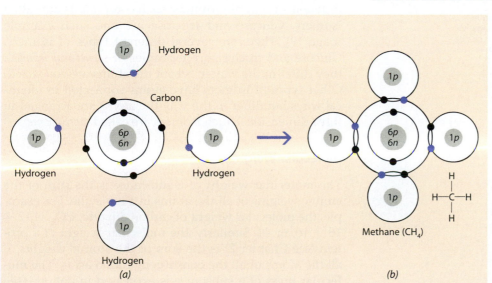

A–8

A carbon atom, with four electrons in its outer energy level, can form covalent bonds with as many as four other atoms. (a) When a carbon atom reacts with four hydrogen atoms, each of the electrons in its outer energy level forms a covalent bond with the single electron of one hydrogen atom, producing (b) a molecule of methane, CH_4.

A–9

Single and double bonds. (a) Diagram of a water molecule, H₂O. Each single covalent bond consists of one electron contributed by the oxygen and one electron contributed by the hydrogen. The covalent bond in the hydrogen molecule (Figure A–7) and the four covalent bonds in the methane molecule (Figure A–8) are also single bonds. (b) The carbon dioxide molecule, CO₂. The carbon atom participates in two double bonds, one with each oxygen atom. Each double bond consists of two pairs of electrons. A double bond is represented by two parallel lines, ═.

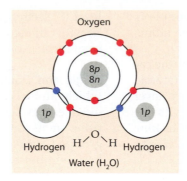

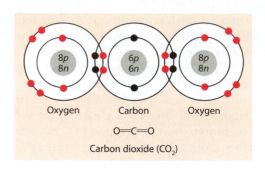

time around the nucleus of the more electronegative atom. As a consequence, this atom has a slightly negative charge, and the less electronegative atom has a slightly positive charge (Figure A–10).

Covalent bonds in which electrons are shared unequally are known as **polar covalent bonds**. Such bonds often involve oxygen, which is highly electronegative. In molecules that are perfectly symmetrical, such as carbon dioxide (Figure A–9b), the unequal charges cancel out and the molecule as a whole is nonpolar. However, in asymmetrical molecules such as water (Figure A–9a), the molecule as a whole is polar, with regions of partial negative charge and regions of partial positive charge (see Figure A–12). Many of the special properties of water, upon which life depends, derive largely from its polar nature.

Ionic interactions, polar covalent bonds, and nonpolar covalent bonds may be considered as chemical bonds that differ along a spectrum of degree of difference in electronegativity between combining atoms. In ionic interactions, there is no electron sharing but rather an elec-

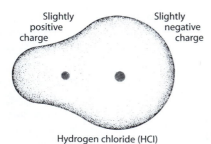

Hydrogen chloride (HCl)

A–10

In a polar molecule, such as hydrogen chloride (HCl), the shared electrons tend to spend more time around the nucleus of the more electronegative atom—in this case, the chlorine atom. As a result, the chlorine atom has a slightly negative charge, and the less electronegative atom (hydrogen) has a slightly positive charge.

trostatic attraction between oppositely charged ions—a case of extreme difference in electronegativity (e.g., Na and Cl). In polar covalent bonds, electrons are shared, but, because of a difference in electronegativity between bonding atoms (e.g., H and O), they are shared unequally. In wholly nonpolar covalent bonds, electrons are shared equally; such bonds can exist only between identical atoms, as in H_2, Cl_2, O_2, and N_2.

Chemical Formulas

The properties of molecules depend on their three-dimensional structure—the shape and volume of space occupied by electrons in their outermost energy levels (orbitals). However, chemists have developed methods for representing molecules on paper—in two dimensions—that allow them to keep track of all the atoms and bonds. **Molecular formulas** indicate the number and types of atoms in a molecule; **structural formulas** show the way in which the atoms are bonded to one another (see the essay on page 19). Sometimes two or more compounds can have the same molecular formula but different structural formulas; such compounds are called **isomers.** Glucose and fructose are one such example (page 19). There are several different types of isomers. Glucose and fructose are examples of *structural isomers*: they differ in the way in which the atoms are connected. Other types of isomers have atoms connected in identical ways but differ in the way the atoms are arranged in three dimensions. Such isomers are called *stereoisomers*.

Molecular Weights

The **molecular weight** of a substance is the sum of the atomic weights of all the atoms in a molecule. For example, the molecular weight of carbon dioxide, CO_2, is 12 + 16 + 16, or 44. Similarly, the molecular weight of a protein (see Chapter 2) is the sum of the atomic weights of all the atoms of all the constituent amino acids. The **molecular mass** of a substance is expressed in daltons (see page 863).

Molecules, as well as atoms, are measured in amounts called **moles**. One mole of any substance contains the same number of particles (atoms, ions, or molecules) as 1 mole of any other substance. This number, 6.022×10^{23}, is known as *Avogadro's number*. Thus 1 mole of sodium contains 6.022×10^{23} atoms of sodium; 1 mole of chloride ions contains 6.022×10^{23} Cl^- ions; and 1 mole of water contains 6.022×10^{23} molecules of water. One mole of a substance weighs an amount, in grams, that is numerically equal to its atomic weight or molecular weight. Thus 1 mole of sodium weighs 23 grams, and 1 mole of water (H_2O) weighs 18 grams. The mole is useful for defining quantities involved in chemical reactions. In order to form water, for example, one would combine 2 moles of hydrogen (H_2) (4 grams) and 1 mole of oxygen (O_2) (32 grams); that is, four hydrogen atoms for every two oxygen atoms. This would give 2 moles of water, each weighing 18 grams.

Functional Groups

Sometimes clusters of atoms joined by covalent bonds tend to react together as a group, known as a **functional group.** The specific chemical properties of organic molecules are determined primarily by their functional groups.

One example is the hydroxyl group, —OH (not to be confused with OH^-, the hydroxide ion). This, for example, is the functional group of methanol (CH_3OH) and ethanol (C_2H_5OH). Glycerol, $C_3H_5(OH)_3$, a component of many lipids, has three hydroxyl groups (see Figure 2–9). Glucose ($C_6H_{12}O_6$), a sugar, has five hydroxyl groups (page 19). Table A–4 shows a number of biologically important functional groups.

Chemical Reactions

Chemical reactions are exchanges of electrons between atoms. This often involves the breaking of bonds and the formation of new bonds. Reactions are described by **chemical equations.** For example, the equation for the formation of water from oxygen gas and hydrogen gas is

$$2H_2 + O_2 \longrightarrow 2H_2O$$

Reactants   Product

The arrow in the equation means "forms" and shows the direction of the chemical change. Like algebraic equations, chemical equations balance: the number and kinds of atoms in the product(s) must equal the number and kinds of atoms in the original reactant(s). The equation above tells us that two molecules of hydrogen react with one molecule of oxygen to yield two molecules of water.

The numerous different chemical reactions occurring in living systems can be classified into a few general types, catalyzed by different classes of enzymes (biological catalysts; see Chapter 5). Some reactions common in

Group		Name
—OH		Hydroxyl
—NH₂	—NH₃⁺	Amino*
(carbonyl structure)		Carbonyl
(carboxyl —COOH)	(carboxyl —COO⁻)	Carboxyl*
(phosphate —O—P(O)(OH)—OH)	(phosphate —O—P(O)(O⁻)—O⁻)	Phosphate*
—SH		Sulfhydryl (thiol)

TABLE A–4   Some Functional Groups of Biological Molecules

*For these groups, the ionized form (shown on the right) predominates at biological pH.

biology are condensation, or dehydration, reactions (page 20); hydrolysis reactions (page 20); oxidation-reduction reactions (pages 98–99); phosphorylation reactions (page 106); rearrangements of molecules, such as those catalyzed by isomerase and mutase enzymes in glycolysis (pages 111–112); and carboxylation reactions, such as the photosynthetic reaction catalyzed by Rubisco, the most abundant enzyme in the living world (page 140).

Chemical Equilibrium

Some chemical reactions can go in either direction. When net change ceases, the reaction is said to be at **equilibrium.** Consider the following imaginary reaction:

$$A + B \rightleftharpoons C + D$$

Equilibrium is reached when as many molecules of C and D are being converted to molecules of A and B as molecules of A and B are being converted to molecules of C and D. At equilibrium, the concentration of reactants does not have to equal the concentration of products; only the *rates* of the forward and reverse reactions must be the same. In the equation above, the different lengths of the arrows indicate that, at equilibrium, more C + D is present than A + B.

Suppose the reaction is set up so that only A and B molecules are present initially. At first the reaction goes to the right, with A reacting with B to yield C and D. Figure A–11 shows the relative changes in concentration as the reaction continues. As C and D accumulate, the rate of the reverse reaction increases; at the same time, the rate of the forward reaction decreases because the concentrations of A and B are decreasing. At some point (after six microseconds in the example in Figure A–11), the rates of the forward and reverse reactions equalize and no further changes in concentration take place. The proportions of reactants (A + B) and products (C + D) will remain the same, but there will always be more C and D than A and B.

The Structure and Properties of Water

As noted above, the water molecule is polar. Because of the high electronegativity of oxygen, the paired electrons of the two covalent bonds spend more time around the oxygen nucleus than around the hydrogen nuclei. Consequently, the region near each hydrogen nucleus has a weak positive charge. Moreover, the oxygen atom has four other electrons in its outer energy level. These electrons, not involved in covalent bonding to hydrogen, are paired in two orbitals, each having a weak negative charge. Thus, the water molecule, in terms of its polarity, is four-cornered, with two positively charged corners and two negatively charged ones (Figure A–12a).

A–12

The polarity of the water molecule and its consequences. (a) This orbital model shows the four-cornered shape of the water molecule. The two orbitals formed by the shared pairs of electrons bonding the hydrogen atoms to the oxygen have a slightly positive charge because their electrons spend more time around the oxygen nucleus than around the hydrogen nuclei. The other two orbitals, with nonbonding electrons, have a slightly negative charge. (b) Each water molecule can form hydrogen bonds (dashed lines) with up to four other water molecules. In liquid water, the hydrogen bonds constantly break and re-form in a shifting pattern. (c) When water freezes it forms a crystalline structure. Each water molecule is hydrogen-bonded to four other water molecules in a three-dimensional latticework. The hexagonal arrangement shown here is repeated throughout the crystal. The water molecules are farther apart in ice than in liquid water, which is why ice is lighter in weight (less dense) than liquid water.

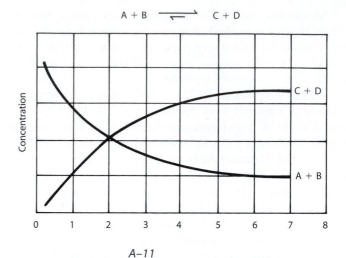

A + B ⇌ C + D

A–11

A graph of the changes in concentration of products and reactants in a reversible reaction. At first, only A and B are present. The reaction begins when A and B react to form C and D. At the end of two microseconds (2×10^{-6} seconds), the concentrations of A + B and C + D are equal. As the reaction proceeds, the concentration of C + D continues to increase until equilibrium is reached (at about six microseconds) and thereafter remains greater than the concentration of A + B. At equilibrium, the rates of forward and reverse reactions are the same.

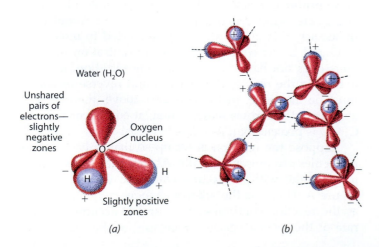

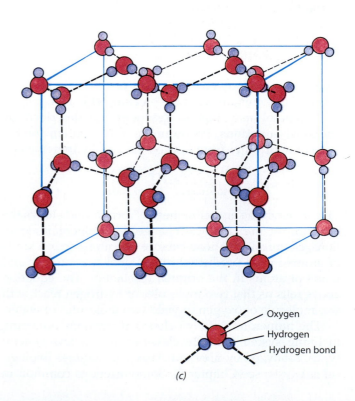

TABLE A–5	Some Biological Molecules in Which Hydrogen Bonding Is Important
Type of Molecule	Hydrogen Bonding
DNA	Between base pairs of the double helix (pages 199–200) Between DNA and RNA during transcription (page 210) Between DNA and regulatory proteins (page 221)
RNA	Between portions of the molecule in transfer RNA (page 211) Between DNA and RNA during transcription (page 210) Between messenger RNA and transfer RNA during translation (page 212)
Proteins	Between amino acid residues in secondary structure (alpha helix, beta pleated sheet); in tertiary structure (three-dimensional folding of the polypeptide chain); in quaternary structure (between polypeptide chains) (pages 28–30)
Polysaccharides	Between glucose residues in starch, cellulose (page 21)
Water	Among water molecules Between water molecules and other polar molecules or polar regions of molecules

Hydrogen Bonds

When one of the charged regions of a water molecule comes close to the oppositely charged region of another water molecule, the force of attraction forms a noncovalent bond between them, which is known as a **hydrogen bond**. A hydrogen bond forms between the negative "corner" of one water molecule and the positive "corner" of another. Every water molecule can establish hydrogen bonds with as many as four other water molecules. The degree of hydrogen bonding varies with water temperature (Figure A–12b).

Any single hydrogen bond is significantly weaker than a covalent bond or ionic interaction. Moreover, it has an exceedingly short lifetime; on average, each hydrogen bond lasts about 1/100,000,000,000th of a second (10^{-12} second). But, as one bond is broken, another is formed. All together, hydrogen bonds have considerable strength.

A hydrogen bond can form between any hydrogen atom that is covalently bonded to an electronegative atom—usually oxygen or nitrogen—and the electronegative atom of another molecule. Hydrogen bonds also occur between various regions of many large biological molecules, where they help maintain structural stability. Hydrogen bonding is of crucial importance in the structure of many biological molecules (Table A–5).

Water As a Solvent

Many substances within living systems are found in aqueous solution. A **solution** is a uniform mixture of the molecules or ions of two or more substances. The substance present in the greatest amount—usually a liquid—is called the **solvent**; the substances present in lesser amounts are called **solutes**.

The polarity of water molecules is responsible for water's capacity as a solvent for polar or charged substances. The polar water molecules tend to separate substances such as NaCl into their constituent ions. As shown in Figure A–13, the water molecules cluster around and segregate the charged ions.

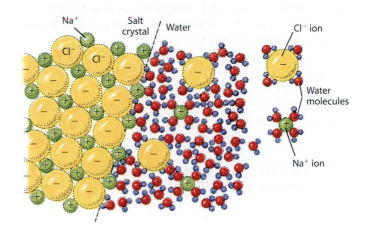

A–13

Because of the polarity of water molecules, water can serve as a solvent for polar molecules and ions. This diagram shows sodium chloride (NaCl) dissolving in water. The polar water molecules cluster around the individual ions, separating them from one other. Note the different arrangements of the water molecules around the Na^+ ions and around the Cl^- ions.

Many molecules important in living systems are polar or have polar regions. Because of their polarity, these molecules attract water molecules. Small polar molecules readily dissolve in water—the sugars glucose and sucrose, for example. Larger molecules with polar regions interact with water at these regions. Polar molecules or polar regions are said to be **hydrophilic** ("water-loving").

Molecules that lack polar regions tend to be very insoluble in water. The hydrogen bonding between water molecules acts as a force to exclude nonpolar molecules. As a result of this exclusion, nonpolar molecules tend to cluster together in water—this is why droplets of fat coalesce on the surface of chicken soup. Such molecules are said to be **hydrophobic** ("water-fearing"), and the clustering together is known as **hydrophobic interactions.** Large molecules such as proteins and some types of lipids have both hydrophilic and hydrophobic regions. Their overall structure and the way in which they interact with the aqueous solutions of cells depend on the distribution of these hydrophilic and hydrophobic regions within the molecule.

Acids and Bases

In liquid water, there is a slight tendency for the nucleus of one of the hydrogen atoms of a water molecule to leave the oxygen atom to which it is covalently bonded and jump to the oxygen atom to which it is hydrogen-bonded. In this reaction, two ions are produced: the hydronium ion, H_3O^+, and the hydroxide ion, OH^- (Figure A–14). For simplicity, the H_3O^+ ion is usually represented in chemical equations as a hydrogen ion, H^+. A hydrogen ion, consisting of the hydrogen nucleus only, is also called a **proton.**

In any given volume of pure water, a small but constant number of water molecules are ionized in this way. The number is constant because the tendency of water to ionize is offset by the tendency of the ions to reunite; that is, the dissociation of water is in equilibrium:

$$H_2O \rightleftharpoons H^+ + OH^-$$

The longer arrow indicates that, at equilibrium, only a small fraction exists in the ionized form.

In pure water, the number of H^+ ions exactly equals the number of OH^- ions. This is necessarily the case, because neither ion can be formed without the other when only H_2O molecules are present. The concentration of H^+ ions in pure water at 25°C is known to be 10^{-7} mole per liter. Thus at 25°C, $[H^+] = [OH^-] = 1 \times 10^{-7}$ mole per liter. (Chemists use square brackets as shorthand for "concentration of.")

When an ionic or polar substance is dissolved in water, it may change the relative numbers of H^+ and OH^- ions such that $[H^+]$ no longer equals $[OH^-]$. For example, when hydrogen chloride (HCl) dissolves in water, it is almost completely ionized into H^+ and Cl^- ions. As a result, in an HCl solution (hydrochloric acid) $[H^+]$ exceeds $[OH^-]$. Conversely, when sodium hydroxide (NaOH) dissolves in water, it ionizes into Na^+ and OH^- ions, and $[OH^-]$ exceeds $[H^+]$ in the sodium hydroxide solution.

A solution is *acidic* when $[H^+]$ is greater than $[OH^-]$; conversely, a solution is *basic,* or *alkaline,* when $[OH^-]$ is greater than $[H^+]$. There is always an inverse relationship between $[H^+]$ and $[OH^-]$ because the product of their concentrations is a constant. For all aqueous solutions, at 25°C,

$$[H^+][OH^-] = 1 \times 10^{-14} \text{ (moles per liter)}^2$$

Thus, we define an **acid** as a substance that causes an increase in the relative number of H^+ ions, or protons, in a solution; acids are also called *proton donors.* And we define a **base** as a substance that causes a decrease in the relative number of H^+ ions—that is, an increase in the relative number of OH^- ions—in a solution. Bases are *proton acceptors.*

Strong and Weak Acids and Bases   Strong acids and bases are substances, like HCl and NaOH, that ionize almost completely in solution, resulting in relatively large increases in $[H^+]$ and $[OH^-]$, respectively. Weak acids and bases, by contrast, are substances that ionize only slightly, resulting in relatively small increase in $[H^+]$ or $[OH^-]$. The functional groups of many biologically important molecules are weak acids or bases.

For example, the carboxyl group (—COOH) acts as

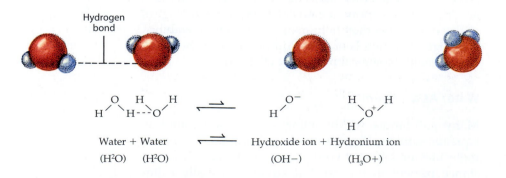

A–14

When water ionizes, a hydrogen nucleus (a proton) shifts from the oxygen atom to which it is covalently bonded to the oxygen atom to which it is hydrogen-bonded. The resulting ions are the hydroxide ion, OH⁻, and the hydronium ion, H₃O⁺.

Water + Water   Hydroxide ion + Hydronium ion

(H²O)   (H²O)   (OH−)   (H₃O+)

a weak acid, because in solution it ionizes to release a proton:

$$R-COOH \rightleftharpoons R-COO^- + H^+$$

(R represents any chemical structure to which the carboxyl group is attached.) The amino group ($-NH_2$) acts as a weak base, accepting a proton:

$$R-NH_2 + H^+ \rightleftharpoons R-NH_3^+$$

Both of these functional groups occur in all amino acids (see Figure 2–15) and in proteins.

The pH Scale   Chemists express degrees of acidity by means of the **pH scale.** (The symbol "pH" is derived from the German *potenz Hydrogen*, "power of hydrogen.") As noted above, a liter of pure water contains 10^{-7} mole of H^+ ions; on the pH scale this is simply referred to as pH 7 (Table A–6). At pH 7, the concentrations of free H^+ and OH^- ions are exactly the same, $[H^+] = [OH^-] = 10^{-7}$ mole per liter. A solution of pH 7 is said to be *neutral*; pure water is also neutral. Any pH below 7 is acidic, and any pH above 7 is basic. The lower the pH number, the higher the concentration of hydrogen ions and the more acidic the solution. A solution of pH 2 contains 10^{-2} mole of H^+ per liter of solution and is highly acidic; a solution of pH 10 contains 10^{-10} mole of H^+ per liter and is strongly basic. A difference of one pH unit represents a tenfold difference in $[H^+]$.

Many chemical reactions in living systems take place at about pH 7. Some biological fluids are quite acidic: lemon juice has a pH of about 2, as do the stomach contents of humans and other animals. Orange juice has a pH of about 3. Human blood has a pH of 7.4. The best soil pH for most plants is about 6.4, but alkaline soils have a pH between 7 and 9, and peat bogs may have a pH as low as 3.

The Energy Factor in Chemical Reactions

Molecules react with each other only when they collide with sufficient energy to (1) overcome the repulsive forces between their negatively charged electron orbitals and (2) break existing chemical bonds. The energy required, called the **energy of activation** (pages 99–100), varies with the nature of the molecules: the more stable the substance, the more forceful the collision must be for a reaction to occur. Each chemical bond has a characteristic energy content, or **bond energy**; the higher the bond energy, the stronger the chemical bond and the greater the energy required to break it. The total bond energy of any molecule is the amount of energy required to break it into its constituent atoms.

In any given sample of molecules, some are moving with sufficient energy (kinetic energy) for a reaction to

TABLE A–6	The pH Scale			
	Concentration of H^+ Ions (mole per liter)		**pH**	**Concentration of OH^- Ions (mole per liter)**
	1.0	10^0	0	10^{-14}
	0.1	10^{-1}	1	10^{-13}
	0.01	10^{-2}	2	10^{-12}
Acidic	0.001	10^{-3}	3	10^{-11}
	0.0001	10^{-4}	4	10^{-10}
	0.00001	10^{-5}	5	10^{-9}
	0.000001	10^{-6}	6	10^{-8}
Neutral	0.0000001	10^{-7}	7	10^{-7}
		10^{-8}	8	10^{-6} 0.000001
		10^{-9}	9	10^{-5} 0.00001
		10^{-10}	10	10^{-4} 0.0001
Basic		10^{-11}	11	10^{-3} 0.001
		10^{-12}	12	10^{-2} 0.01
		10^{-13}	13	10^{-1} 0.1
		10^{-14}	14	10^0 1.0

occur. But at normal temperatures and pressures, the proportion with this energy may be so small that, for all practical purposes, the reaction does not take place. Reaction rates can be increased by increasing the likelihood of sufficiently forceful collisions between molecules. This can be achieved by raising the temperature, thereby increasing the average velocity at which the molecules move, increasing their energy, and increasing the likelihood of their colliding with sufficient force to react. The use of heat to drive a chemical reaction is common in chemistry laboratories and in industry. High pressures are also used to increase the rate of reactions. Other ways used by chemists to increase the rates of reactions are to increase the concentration of reacting molecules and to use catalysts. *Catalysts* lower the energy of activation of a reaction.

Living systems cannot use increased temperatures and pressures to increase reactions rates, and the concentrations of reacting substances are often very low. Catalysts called *enzymes* are used by all organisms to lower activation energies and increase reaction rates, as described in Chapter 5.

The energy changes that occur in chemical reactions can be measured, and the relationships among various forms of energy are the subject of the science of **thermo-**

dynamics. Consider the complete oxidation (combustion) of glucose, represented by the following equation:

$$C_6H_{12}O_6 + 6O_2 \longrightarrow 6CO_2 + 6H_2O$$

This reaction releases energy to the surroundings. The reaction can be set in motion by electrical ignition, and energy is released in the form of heat, which can be measured quite precisely in a calorimeter (Figure A–15). The release of heat in the combustion of glucose can be expressed as

$$\Delta H = -673 \text{ kilocalories per mole of glucose}$$

where ΔH is the change in *heat content* (the Greek letter delta (Δ) represents "change," and H represents "heat content"). The negative value indicates that heat is released. A **calorie** is defined as the amount of heat necessary to raise the temperature of 1 gram of water by 1°C; 1000 calories = 1 kilocalorie. A reaction that liberates heat, such as glucose combustion, is said to be *exothermic*.

ΔH is one of two important factors in determining whether a reaction is energy-releasing (*exergonic*) or energy-consuming (*endergonic*), and whether the reaction will occur under specified conditions. The other factor is ΔS, the change in *entropy*, or randomness. The overall energy change in a reaction is determined by the relationship

$$\Delta G = \Delta H - T\Delta S$$

where ΔG is known as the change in *free energy*, and T is the absolute temperature. **Absolute temperature** is the temperature measured with respect to absolute zero, the temperature at which all molecular motion ceases and there is no heat; this is equivalent to about −273.16°C. Absolute temperatures are expressed in units called kelvins (K); 25°C is equivalent to about 298 K. Free energy, entropy, and endergonic and exergonic reactions are discussed in Chapter 5.

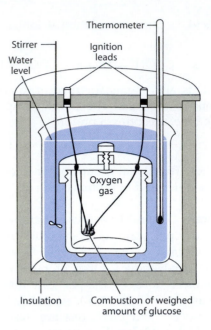

A–15

A calorimeter. A known quantity of glucose or some other material is ignited electrically and completely burned (oxidized) to form carbon dioxide and water. The increase in temperature of a known amount of water is measured. Based on the specific heat of water (the heat in calories required to raise the temperature of one gram of water by 1°C), the number of calories released by the burning of the sample can be calculated. In the case of glucose, combustion releases 673 kilocalories per mole, or 3.74 kilocalories per gram.

Appendix B

The Hardy–Weinberg Equation

As we noted in the text (page 240), Hardy and Weinberg demonstrated the constancy of allele and genotype frequencies in an idealized population in which five conditions hold:

1. No mutations.
2. Isolation from other populations.
3. Large population size.
4. Random mating.
5. No natural selection.

To understand how Hardy and Weinberg arrived at their equation expressing the equilibrium of allele and genotype frequencies in a population meeting these five conditions, let us look at just one gene. For simplicity, we select one that has only two alleles, which we will call A and a. We are interested in the relative proportions—that is, the frequencies—of A and a from one generation to the next. By convention, the letter p is used to designate the frequency of one allele and the letter q to designate the frequency of the other. When there are only two alleles, p and q together must equal one: $p + q = 1$.

These proportions—p and q—could be expressed in terms of fractions, as done by Mendel, but since the proportions of the two alleles will probably not be equal, as they were in Mendel's carefully controlled experiments, it is more convenient to express the numbers as decimals. For example, suppose that in a particular population 80 percent of the alleles of the gene under study are allele A. The frequency of A is 0.8, or $p = 0.8$. Given that there are only two alleles, we then know that the frequency of allele a is 0.2. In other words, $q = 1 - p$.

Let us assume that the relative frequencies of A and a are the same in both males and females (as they are for most alleles in natural populations). Now suppose that males and females mate at random with respect to alleles A and a. We can calculate the frequencies of the resulting genotypes by drawing a Punnett square. As you can see from the illustration, the genotypes in the population produced by this random mating would consist of 64 percent AA, 32 percent Aa, and 4 percent aa.

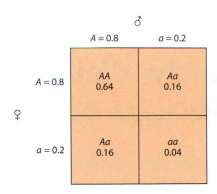

Results of random mating in a population in which the frequency (p) of allele A equals 0.8 and the frequency (q) of allele a equals 0.2. In setting up this Punnett square, we are assuming that the alleles appear in the same frequencies in males as in females.

Instead of drawing a Punnett square, we can do the same thing algebraically. Because $p + q = 1$, it follows that:

$$(p + q)(p + q) = 1 \times 1 = 1$$

Or, as you probably remember from algebra:

$$p^2 + 2pq + q^2 = 1$$

This algebraic expression of the genotype frequencies is the Hardy–Weinberg equation.

Let us apply this equation to the random mating that just occurred in our population. Taking the initial values for the frequencies of the two alleles, we obtain the following results:

$$p^2 = 0.8 \times 0.8$$
$$= 0.64 \text{ (the frequency of } AA \text{ genotypes)}$$
$$2pq = 2 \times 0.8 \times 0.2$$
$$= 0.32 \text{ (the frequency of } Aa \text{ genotypes)}$$
$$q^2 = 0.2 \times 0.2$$
$$= 0.04 \text{ (the frequency of } aa \text{ genotypes)}$$

What has happened to the frequencies of the two alleles in the gene pool as a result of this round of mating? We know from our calculations that the frequency of *AA* is 0.64. In addition, half of the alleles in the heterozygotes (*Aa*) are *A*, so the total frequency of allele *A* is 0.64 plus one-half of 0.32—that is, 0.64 plus 0.16, to give a total of 0.8. The frequency of allele *A* (*p*) has not changed. Similarly, the total frequency of allele *a* is 0.04 (in the homozygotes) plus 0.16 (half the alleles in the heterozygotes), or 0.2. The frequency of allele *a* (*q*) has also remained the same.

If another round of mating occurs, the proportion of *AA*, of *Aa*, and of *aa* genotypes in our population will again be 64 percent, 32 percent, and 4 percent, respectively. Again the frequency of allele *A* will be 0.8 and of allele *a* 0.2. And so on, and so on, generation after generation. In an ideal population in which the five conditions are met, neither the allele frequencies nor the genotype frequencies change from generation to generation.

The Hardy–Weinberg equilibrium also applies to situations in which there are more than two alleles of the same gene. However, the equation representing the equilibrium is more complex. For example, the genotype frequencies of three alleles are expressed by the algebraic expansion of $(p + q + r)^2 = 1$, with *r* representing the frequency in the gene pool of the third allele.

An Application of the Hardy–Weinberg Equation

As we have noted, the Hardy–Weinberg equilibrium holds only in populations in which the five stated conditions are met. What happens if one of these conditions is not met? Let us imagine another pair of alleles in which allele *a* has a damaging effect in the homozygous state, reducing the likelihood that a homozygous *aa* individual will survive to reproduce. We can estimate the number of *aa* genotypes in the population, perhaps by screening tests on newborn infants. Suppose, for instance, that the condition shows up in 1 in 10,000 infants. In other words, $q^2 = 1/10,000$, or 0.0001. Thus, $q = \sqrt{0.0001}$, or 0.01. If $q = 0.01$, then $p = 0.99$ and $2pq = 0.0198$, or almost 0.02. Thus about 2 percent of the population—one person in every 50—can be estimated to be a heterozygous carrier of this allele.

Now suppose that we do the same screening tests five years later and find that *q* is not 0.01 but 0.009, and then we repeat it a few years later and find that it has once again decreased very slightly, perhaps to 0.008. In other words, evolution is occurring. One allele is decreasing in frequency while the other is increasing. Using the Hardy–Weinberg equilibrium as our yardstick, we know not only that change has taken place and its direction but also that there must be some reason for it. We can then look for the factors causing the change.

Appendix C

Metric Table

	Fundamental Unit	Quantity	Numerical Value	Symbol	English Equivalent
Area		hectare	10,000 m^2	ha	2.471 acres
Length	meter			m	39.37 inches
		kilometer	1000 (10^3) m	km	0.62137 mile
		centimeter	0.01 (10^{-2}) m	cm	0.3937 inch
		millimeter	0.001 (10^{-3}) m	mm	
		micrometer	0.000001 (10^{-6}) m	μm	
		nanometer	0.000000001 (10^{-9}) m	nm	
		angstrom	0.0000000001 (10^{-10}) m	Å	
Mass	gram			g	0.03527 ounce
		kilogram	1000 g	kg	2.2 pounds
		milligram	0.001 g	mg	
		microgram	0.000001 g	μg	
Time	second			sec	
		millisecond	0.001 sec	msec	
		microsecond	0.000001 sec	μsec	
Volume (solids)	cubic meter			m^3	35.314 cubic feet
		cubic centimeter	0.000001 m^3	cm^3	0.061 cubic inch
		cubic millimeter	0.000000001 m^3	mm^3	
Volume (liquids)	liter			l	1.06 quarts
		milliliter	0.001 liter	ml	
		microliter	0.000001 liter	μl	

Temperature Conversion Scale

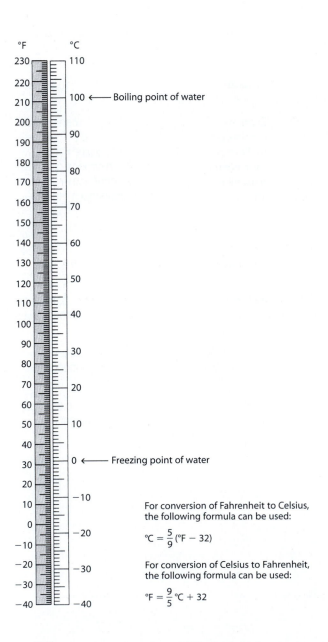

°F   °C

← Boiling point of water

← Freezing point of water

For conversion of Fahrenheit to Celsius, the following formula can be used:

$$°C = \frac{5}{9}(°F - 32)$$

For conversion of Celsius to Fahrenheit, the following formula can be used:

$$°F = \frac{9}{5}°C + 32$$

Appendix D

Classification of Organisms

There are several different ways to classify organisms. The one presented here follows the overall scheme described in Chapter 13, in which organisms are divided into three domains: *Archaea, Bacteria,* and *Eukarya. Archaea* and *Bacteria* are distinct lineages of prokaryotic organisms. *Eukarya,* which consists entirely of eukaryotic organisms, includes four kingdoms: *Protista, Animalia, Fungi,* and *Plantae.* The chief taxonomic categories are domain, kingdom, phylum, class, order, family, genus, species.

The classification that follows includes the phyla of *Protista,* except those considered *Protozoa,* as well as the *Fungi* and *Plantae.* Certain classes given prominence in this book are also included, but the listings are far from complete. The number of species given for each group is the estimated number of living species that have been described and named. Only groups that include living species are described. Viruses are not included in this appendix, but are discussed in Chapter 14.

Domain *Archaea*

Archaea are prokaryotic cells. They lack a nuclear envelope, plastids, mitochondria, and other membrane-bounded organelles, and 9-plus-2 flagella. They are unicellular but sometimes aggregate into filaments or other superficially multicellular bodies. Their predominant mode of nutrition is absorption, but one group of genera obtain their energy by metabolizing sulfur, and another genus, *Halobacterium,* does so through the operation of a proton pump. Many archaea are methanogens, generators of methane. Others are among the most "salt-loving" (extreme halophiles) and "heat-loving" (extreme thermophiles) of all known prokaryotes. Reproduction is asexual, by fission; genetic recombination has not been observed. They are diverse morphologically, being motile flagellated or nonmotile rods, cocci, and spirilla. *Archaea* differ fundamentally from *Bacteria* in the base sequences of their ribosomal RNAs and lipid composition of their plasma membranes. They also differ from *Bacteria* in lacking peptidoglycans in their cell walls. There are fewer than 100 named species.

Domain *Bacteria*

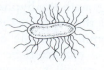

Like *Archaea, Bacteria* lack a nuclear envelope, plastids, mitochondria, and other membrane-bounded organelles, and 9-plus-2 flagella. They are unicellular, but many form aggregates. Their predominant mode of nutrition is absorption, but some groups are photosynthetic or chemosynthetic. Reproduction is predominantly asexual, by fission or budding, but portions of DNA molecules may also be exchanged between cells under certain circumstances. They are motile by simple flagella or by gliding, or they may be nonmotile.

About 2600 species of bacteria are recognized at present, but this is probably only a small fraction of the actual number. The recognition of species is not comparable with that in eukaryotes and is based largely upon metabolic features. One group, the class *Rickettsiae*—very small bacteria—occurs widely as parasites in arthropods and may consist of tens of thousands of species, depending upon the classification criteria used; they are not included in the estimate given here.

The *Bacteria* can be divided into twelve major lineages, or kingdoms. Among these, cyanobacteria are an ancient group that is abundant and important ecologically. Formerly and misleadingly called "blue-green algae," cyanobacteria have a type of photosynthesis that is based on chlorophyll *a*. Cyanobacteria, like the red algae, also have accessory pigments called phycobilins. Many cyanobacteria can fix atmospheric nitrogen, often in specialized cells called heterocysts. Some cyanobacteria form complex filaments or other colonies. Although some 7500 species of cyanobacteria have been described, a more reasonable estimate puts the number of these specialized bacteria at about 200 distinct nonsymbiotic species.

Domain *Eukarya*

Kingdom *Fungi*

Eukaryotic multicellular or rarely unicellular organisms in which the nuclei occur in a basically continuous mycelium; this mycelium becomes septate in certain groups and at certain stages of the life cycle. Fungi are heterotrophic; they obtain their nutrition by absorption. Members of all but one phylum *(Chytridiomycota)* form important symbiotic relationships with the roots of plants, called mycorrhizae. Reproductive cycles typically include both sexual and asexual phases. There are over 70,000 valid species of fungi to which names have been given, and many more will eventually be found. Some have been named two or more times; this is particularly so for fungi that may be classified both as ascomycetes and as members of the deuteromycetes. The major characteristics of the phyla of *Fungi* are provided in Table 15–1.

Phylum *Chytridiomycota*:   Chytrids. Predominantly aquatic heterotrophic organisms with motile cells characteristic of certain stages in their life cycle. The motile cells of most have a single, posterior, whiplash flagellum. Their cell walls are composed of chitin, but other polymers may also be present, and they store their food as glycogen. There are about 790 species.

Phylum *Zygomycota*:   Terrestrial fungi with the hyphae septate only during the formation of reproductive bodies; chitin is predominant in the cell walls. Zygomycetes can usually be recognized by their profuse, rapidly growing hyphae. The class includes about 1060 described species, some of which occur as components of the endomycorrhizae that are found in about 80 percent of all vascular plants.

Phylum *Ascomycota*:   Terrestrial and aquatic fungi with the hyphae septate but the septa perforated; complete septa cut off the reproductive bodies, such as spores or gametangia. Chitin is predominant in the cell walls. Sexual reproduction involves the formation of a characteristic cell—the ascus—in which meiosis takes place and within which ascospores are formed. The hyphae in many ascomycetes are packed together into complex "bodies" known as ascomata. There are about 32,300 species of ascomycetes.

Phylum *Basidiomycota*: Terrestrial fungi with the hyphae septate but the septa perforated; complete septa cut off reproductive bodies, such as spores. Chitin is predominant in the cell walls. Sexual reproduction involves formation of basidia, in which meiosis takes place and on which the basidiospores are borne. *Basidiomycota* are dikaryotic during most of their life cycle, and there is often complex differentiation of "tissues" within their basidiomata. They are the fungal components of most ectomycorrhizae. There are some 22,300 described species.

Class *Basidiomycetes*: Includes the hymenomycetes and gasteromycetes. The hymenomycetes produce basidiospores in a hymenium exposed on a basidioma; the mushrooms, coral fungi, and shelf, or bracket, fungi. The gasteromycetes produce basidiospores inside basidiomata, where they are completely enclosed for at least part of their development; puffballs, earthstars, stinkhorns, and their relatives. The basidia of most *Basidiomycetes* are aseptate (internally undivided).

Class *Teliomycetes*: Consists of fungi commonly referred to as rusts. Unlike the *Basidiomycetes*, the rusts do not form basidiomata, and they have septate basidia.

Class *Ustomycetes*: Commonly referred to as smuts. Like the *Teliomycetes*, they do not form basidiomata, and they form septate basidia.

Yeasts: A yeast, by definition, is simply a unicellular fungus that reproduces primarily by budding. The yeasts are not a formal taxonomic group. The yeast growth form is exhibited by a broad range of unrelated fungi encompassing the *Zygomycota*, *Ascomycota*, and *Basidiomycota*. Most yeasts are ascomycetes, but at least a quarter of the genera are *Basidiomycota*.

Deuteromycetes: The deuteromycetes are an artificial assemblage of about 15,000 distinct species of fungi for which only the asexually reproductive state is known or in which the sexual reproductive features are not used as the basis of classification. They are often referred to as the "Fungi Imperfecti."

Lichens: A lichen is a mutualistic symbiotic association between a fungal partner and a certain genus of green algae or of cyanobacteria. The fungal component of the lichen is called the mycobiont, and the photosynthetic component is called the photobiont. About 98 percent of the mycobionts belong to the *Ascomycota*, the remainder to the *Basidiomycota*. About 13,250 species of lichen-forming fungi have been described.

Kingdom *Protista*

Eukaryotic unicellular or multicellular organisms. Their modes of nutrition include ingestion, photosynthesis, and absorption. True sexuality is present in most phyla. They move by means of 9-plus-2 flagella or are nonmotile. Fungi, plants, and animals are specialized multicellular groups derived from *Protista*. The phyla treated in this book are categorized as heterotrophic protists (slime molds and water molds; the first three phyla listed below) and photosynthetic protists (the algae). The characteristics of the phyla of *Protista* are outlined in Tables 16–1 and 17–1.

Phylum *Myxomycota:* Plasmodial slime molds. Heterotrophic amoeboid organisms that form a multinucleate plasmodium that creeps along as a mass and eventually differentiates into sporangia, each of which is multinucleate and eventually gives rise to many spores. Sexual reproduction is occasionally observed. The predominant mode of nutrition is by ingestion. There are about 700 species.

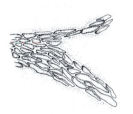

Phylum *Dictyosteliomycota:* Cellular slime molds, or dictyostelids. Heterotrophic organisms that exist as separate amoebas (called myxamoebas). Eventually, the myxamoebas swarm together to form a pseudoplasmodium, within which they retain their individual identities. Ultimately, the pseudoplasmodium differentiates into a fruiting body. Sexual reproduction involves structures known as macrocysts. Pairs of amoebas first fuse, forming zygotes. Subsequently, these zygotes attract and then engulf nearby amoebas. The principal mode of nutrition is by ingestion. There are about 50 known species in four genera.

Phylum *Oomycota:* Water molds and related organisms. Aquatic or terrestrial organisms with motile cells characteristic of certain stages of their life cycle. The flagella are two in number—one tinsel and one whiplash as is characteristic of heterokonts. Their cell walls are composed of cellulose or celluloselike polymers, and they store their food as glycogen. There are about 694 species.

Phylum *Euglenophyta:* Euglenoids. About a third of the approximately 40 genera of euglenoids have chloroplasts, with chlorophylls *a* and *b* and carotenoids; the others are heterotrophic and essentially resemble members of the phylum *Zoomastigina,* within which they would probably best be included. They store food as paramylon, an unusual carbohydrate. Euglenoids usually have two apical flagella and a contractile vacuole. The flexible pellicle is rich in proteins. Sexual reproduction is unknown. There are some 900 species, most of which occur in fresh water.

Phylum *Cryptophyta:* Cryptomonads. Photosynthetic organisms that possess chlorophylls *a* and *c* and carotenoids; in addition some cryptomonads contain a phycobilin, either phycocyanin or phycoerythrin. They are rich in polyunsaturated fatty acids. In addition to a regular nucleus the cryptomonads contain a reduced nucleus called a nucleomorph. There are 200 known species.

Phylum *Rhodophyta:* Red algae. Primarily marine algae characterized by the presence of chlorophyll *a* and phycobilins. Particularly abundant in tropical and warm waters. Their carbohydrate food reserve is floridean starch, and the cell walls are composed of cellulose or pectins, with calcium carbonate in many. No motile cells are present at any stage in the complex life cycle. The vegetative body is built up of closely packed filaments in a gelatinous matrix and is not differentiated into roots, leaves, and stems. It lacks specialized conducting cells. There are some 4000 to 6000 known species.

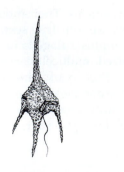

Phylum *Dinophyta:* Dinoflagellates. Autotrophic organisms, about half of which possess chlorophylls *a* and *c* and carotenoids; the other half lack a photosynthetic apparatus and hence obtain their nutrition either by ingesting solid food particles or by absorbing dissolved organic compounds. Food is stored as starch. A layer of vesicles—often containing cellulose—lies beneath the plasma membrane. This phylum contains some 2000 to

4000 known species, mostly biflagellated organisms. These all have lateral flagella, one of which beats in a groove that encircles the organism. Sexual reproduction is generally isogamous, but anisogamy is also present. The mitosis of dinoflagellates is unique. Many—in a form called zooxanthellae—are symbiotic in marine animals, and they make important contributions to the productivity of coral reefs.

Phylum _Haptophyta:_   Haptophytes. Mostly photosynthetic organisms that contain chlorophyll _a_ and some variation of chlorophyll _c._ Some have the accessory pigment fucoxanthin. The most distinctive feature of the haptophytes is the haptonema, a threadlike structure that extends from the cell along with two flagella. There are about 300 known species of haptophytes.

Phylum _Chrysophyta:_   Chrysophytes. Primarily unicellular or colonial organisms that possess chlorophylls _a_ and _c_ and carotenoids, mainly fucoxanthin. Food is stored as the water-soluble carbohydrate chrysolaminarin. The cell walls are absent or consist of cellulose that may be impregnated with minerals; some are covered with silica scales. There are about 1000 known living species.

Phylum _Bacillariophyta:_   Diatoms. Unicellular or colonial organisms with two-part siliceous cell walls, the two halves of which fit together like a Petri dish. Diatoms have chlorophylls _a_ and _c,_ as well as fucoxanthin. Reserve storage materials include lipids and chrysolaminarin. Diatoms lack flagella, except on some male gametes. It is estimated that there are at least 100,000 living species, plus thousands of extinct ones.

Phylum _Phaeophyta:_   Brown algae. Multicellular, nearly entirely marine algae characterized by the presence of chlorophylls _a_ and _c_ and fucoxanthin. The carbohydrate food reserve is laminarin, and the cell walls have a cellulose matrix containing algin. Motile cells are biflagellated, with one forward flagellum of the tinsel type and one trailing flagellum of the whiplash type. A considerable amount of differentiation is found in some of the kelps (some of the large brown algae of the order _Laminariales_), with specialized conducting cells for transporting the products of photosynthesis to the regions of the body that receive little light. There is, however, no differentiation into roots, leaves, and stems, as in the vascular plants. Although there are only about 1500 species, the brown algae dominate rocky shores throughout the cooler regions of the world.

Phylum _Chlorophyta:_   Green algae. Unicellular or multicellular photosynthetic organisms characterized by the presence of chlorophylls _a_ and _b_ and various carotenoids. The carbohydrate food reserve is starch; only green algae and plants, which are clearly descended from the green algae, store their reserve food inside their plastids. The cell walls of green algae are formed of polysaccharides, sometimes cellulose. Motile cells generally have two apical whiplash flagella. True multicellular genera do not exhibit complex patterns of differentiation. Multicellularity has arisen at least twice. There are about 17,000 known species.

Class _Chlorophyceae:_   Green algae in which the unique mode of cell division involves a phycoplast—a system of microtubules parallel to the plane of cell division. The nuclear envelope persists throughout mitosis, and chromosome division occurs within it. Motile cells, if present, are symmetrical and possess two, four, or many flagella that are apical and directed forward.

Sexual reproduction always involves the formation of a dormant zygote and zygotic meiosis. These algae predominantly occur in fresh water.

Class *Ulvophyceae:* Green algae with a closed mitosis in which the nuclear envelope persists; the spindle is persistent through cytokinesis. Motile cells, if present, are symmetrical and possess two, four, or many flagella that are apical and directed forward. Sexual reproduction often involves alternation of generations and sporic meiosis, and dormant zygotes are rare. These are predominantly marine algae.

Class *Charophyceae:* Unicellular, few-celled, filamentous, or parenchymatous green algae in which cell division involves a phragmoplast—a system of microtubules perpendicular to the plane of cell division. The nuclear envelope breaks down during the course of mitosis. Motile cells, if present, are asymmetrical and possess two flagella that are subapical and extend laterally at right angles from the cell. Sexual reproduction always involves the formation of a dormant zygote and zygotic meiosis. Certain members of this class resemble plants more closely than do any other organisms. These algae predominantly occur in fresh water.

Kingdom *Plantae*

The plants are autotrophic (some are derived heterotrophs), multicellular organisms possessing advanced tissue differentiation. All plants have an alternation of generations, in which the diploid phase (sporophyte) includes an embryo and the haploid phase (gametophyte) produces gametes by mitosis. Their photosynthetic pigments and food reserves are similar to those of the green algae. Plants are primarily terrestrial. Summaries of some of the characteristics of the various phyla are found in the Summary Tables of Chapters 18 to 20 and in Table 21–1.

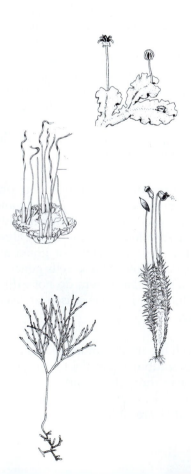

Phylum *Hepatophyta:* Liverworts. *Hepatophyta* and the two following phyla, all of which constitute the bryophytes, have multicellular gametangia with a sterile jacket layer; their sperm are biflagellated. In all three phyla, most photosynthesis is carried out in the gametophyte, upon which the sporophyte is dependent. Liverworts lack specialized conducting tissue (with possibly a few exceptions) and stomata; they are the simplest of all living plants. The gametophytes are thallose or leafy, and the rhizoids are single-celled. There are about 6000 species.

Phylum *Anthocerophyta:* Hornworts. Bryophytes with thallose gametophytes; the sporophyte grows from a basal intercalary meristem for as long as conditions are favorable. Stomata are present on the sporophyte; there is no specialized conducting tissue. There are about 100 species.

Phylum *Bryophyta:* Mosses. Bryophytes with leafy gametophytes; the sporophytes have complex patterns of dehiscence. Specialized conducting tissue is present in both gametophytes and sporophytes of some species. Rhizoids are multicellular. Stomata are present on the sporophytes. There are about 9500 species.

Phylum *Psilotophyta:* Psilotophytes. The *Psilotophyta* and the three following phyla constitute the living phyla of seedless vascular plants. Psilotophytes are homosporous. There are two genera, one of which has leaflike appendages on the stem; both genera have extremely simple sporophytes, with no differentiation between root and shoot. The sperm are motile. There are several species.

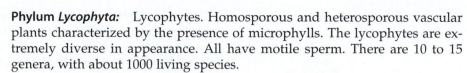

Phylum *Lycophyta:* Lycophytes. Homosporous and heterosporous vascular plants characterized by the presence of microphylls. The lycophytes are extremely diverse in appearance. All have motile sperm. There are 10 to 15 genera, with about 1000 living species.

Phylum *Sphenophyta:* Horsetails. A single genus of homosporous vascular plants, *Equisetum,* with jointed stems marked by conspicuous nodes and elevated siliceous ribs. Sporangia are borne in a strobilus at the apex of the stem. Leaves are scalelike. Sperm are motile. There are 15 living species of horsetails.

Phylum *Pterophyta:* Ferns. Mostly homosporous, although some are heterosporous. All possess a megaphyll. The gametophyte is more or less free-living and usually photosynthetic. Multicellular gametangia and free-swimming sperm are present. There are about 11,000 species.

Phylum *Cycadophyta:* Cycads. This and the following three phyla make up the gymnosperms. Cycads have sluggish cambial growth and pinnately compound, palmlike or fernlike leaves; ovules and seeds are exposed. The sperm are flagellated and motile but are carried to the vicinity of the ovule in a pollen tube. There are 11 genera with about 140 species.

Phylum *Ginkgophyta:* *Ginkgo.* Gymnosperm with considerable cambial growth and fan-shaped leaves with open dichotomous venation; ovules and seeds exposed; seed coats fleshy. Sperm are carried to the vicinity of the ovule in a pollen tube but are flagellated and motile. There is only one species.

Phylum *Coniferophyta:* Conifers. Gymnosperms with active cambial growth and simple leaves; ovules and seeds exposed; sperm nonflagellated. The most familiar group of gymnosperms. There are some 50 genera with about 550 species.

Phylum *Gnetophyta:* Gnetophytes. Gymnosperms with many angiospermlike features, such as vessels; the gnetophytes are the only gymnosperms in which vessels occur. They are the group of gymnosperms most closely related to angiosperms. Motile sperm are absent. There are three very distinctive genera with about 70 species.

Phylum *Anthophyta:* Flowering plants; angiosperms. Seed plants in which ovules are enclosed in a carpel and seeds are borne within fruits. The angiosperms are extremely diverse vegetatively but are characterized by the flower, which is basically insect-pollinated. Other modes of pollination, such as wind pollination, have been derived in a number of different lines. The gametophytes are much reduced, with the female gametophyte often consisting of only seven cells at maturity. Double fertilization involving the two sperm of the mature microgametophyte gives rise to the zygote (sperm and egg) and to the primary endosperm nucleus (sperm and polar nuclei); the former becomes the embryo and the latter becomes a special nutritive tissue called the endosperm. There are about 235,000 species.

Class *Monocotyledones:* Monocots. Flower parts usually in threes; leaf venation usually parallel; primary vascular bundles in the stem are scattered; true secondary growth is not present; one cotyledon. There are about 65,000 species.

Class *Eudicotyledones*:   Eudicots. Flower parts usually in fours or fives; leaf venation usually netlike; primary vascular bundles in the stem are in a ring; many with a vascular cambium and true secondary growth; two cotyledons. There are about 165,000 species.

Together, the monocots and eudicots represent about 97 percent of angiosperms. The remaining 3 percent of living angiosperms are the magnoliids, those angiosperms with the most primitive features and the ancestors of both monocots and eudicots.

Some other common terms used to describe major groups of plants deserve mention here. In systems in which the algae and fungi are regarded as plants, they are often grouped as a subkingdom, *Thallophyta,* the thallophytes: organisms with no highly differentiated tissues, such as root, stem, or leaf, and no vascular tissues (xylem and phloem). The bryophytes and vascular plants are then grouped into a second subkingdom, *Embryophyta,* in which the zygote develops into a multicellular embryo still encased in an archegonium or an embryo sac. All embryophytes are marked by an alternation of heteromorphic generations. The term "embryophyte" commonly is used as a synonym for plants.

Although they are no longer used in formal schemes of classification, terms such as "algae," "thallophytes," "vascular plants," and "gymnosperms" are still sometimes useful in an informal sense. An even earlier scheme divided all plants into "phanerogams," those with flowers, and "cryptogams," those lacking flowers; these terms are still occasionally seen.

Glossary

A

Å: *See* ångstrom.

a- [Gk. *a-*, not, without]: Prefix that negates the succeeding part of the word; "an-" before vowels and "h".

abscisic acid [L. *abscissus*, to cut off]: A plant hormone that brings about dormancy in buds, maintains dormancy in seeds, and brings about stomatal closing, among other effects.

abscission (ăb•sizh´ŭn): The dropping off of leaves, flowers, fruits, or other plant parts, usually following the formation of an abscission zone.

abscission zone: The area at the base of a leaf, flower, or fruit, or other plant part containing tissues that play a role in the separation of a plant part from the plant body.

absorption spectrum: The spectrum of light waves absorbed by a particular pigment.

accessory bud: A bud generally located above or on either side of the main axillary bud.

accessory cell: *See* subsidiary cell.

accessory fruit: A fruit, or assemblage of fruits, whose fleshy parts are derived largely or entirely from tissues other than the ovary. An example is the strawberry, whose receptacle is fleshy and whose fruits (achenes) are embedded in its surface.

accessory pigment: A pigment that captures light energy and transfers it to chlorophyll *a*.

acclimation: The process by which numerous physical and physiological processes prepare a plant for winter.

achene: A simple, dry, one-seeded indehiscent fruit in which the seed coat is not adherent to the pericarp.

acid: A substance that dissociates in water, releasing hydrogen ions (H^+) and thus causing a relative increase in the concentration of these ions; having a pH in solution of less than 7; a proton donor; the opposite of "base."

acid growth hypothesis: The hypothesis that acidification of the cell wall leads to hydrolysis of restraining bonds within the wall and, consequently, to cell elongation driven by the turgor pressure of the wall.

actin filament: A helical protein filament, 5 to 7 nanometers thick, composed of globular actin molecules; a major constituent of the cytoskeleton of all eukaryotic cells; also called microfilament.

actinomorphic [Gk. *aktis*, ray of light, + *morphē*, form]: Pertaining to a type of flower that can be divided into two equal halves in more than one longitudinal plane; also called radially symmetrical or regular; *see also* zygomorphic.

action spectrum: The spectrum of light waves that elicits a particular reaction.

active site: The region of an enzyme surface that binds the substrate during the reaction catalyzed by the enzyme.

active transport: Energy-requiring transport of a solute across a membrane in the direction of increasing concentration (against the concentration gradient).

ad- [L. *ad-*, toward, to]: Prefix meaning "toward" or "to."

adaptation [L. *adaptare*, to fit]: A peculiarity of structure, physiology, or behavior that aids in fitting an organism to its environment.

adaptive radiation: The evolution from one kind of organism to several divergent forms, each specialized to fit a distinct and diverse way of life.

adenine (ăd´e•nēn): A purine base present in DNA, RNA, and nucleotide derivatives, such as ADP and ATP.

adenosine triphosphate (ATP): A nucleotide consisting of adenine, ribose sugar, and three phosphate groups; the major source of usable chemical energy in metabolism. On hydrolysis, ATP loses one phosphate to become adenosine diphosphate (ADP), releasing usable energy.

adhesion [L. *adhaerere*, to stick to]: The sticking together of unlike objects or materials.

adnate [L. *adnatus*, grown together]: Said of fused unlike parts, as stamens and petals; *see also* connate.

ADP: *See* adenosine triphosphate.

adsorption [L. *ad-*, to, + *sorbere*, to suck in]: The adhesion of a liquid, gaseous, or dissolved substance to a solid, resulting in a higher concentration of the substance.

adventitious [L. *adventicius*, not properly belonging to]: Referring to a structure arising from an unusual place, such as buds at other places than leaf axils, or roots growing from stems or leaves.

aeciospore (ē´sĭ•o•spor) [Gk. *aikia*, injury, + *spora*, seed]: A binucleate spore of rust fungi; produced in an aecium.

aecium, *pl.* **aecia:** In rust fungi, a cuplike structure in which aeciospores are produced.

aerobic [Gk. *aer*, air, + *bios*, life]: Requiring free oxygen.

aerobic respiration: *See* respiration.

after-ripening: Term applied to the metabolic changes that must occur in some dormant seeds before germination can occur.

agar: A gelatinous substance derived from certain red algae; used as a solidifying agent in the preparation of nutrient media for growing microorganisms.

aggregate fruit: A fruit developing from the several separate carpels of a single flower.

akinete: A vegetative cell that is transformed into a thick-walled resistant spore in cyanobacteria.

albuminous cell: Certain ray and axial parenchyma cells in gymnosperm phloem that are spatially and functionally associated with the sieve cells; also called Strasburger cells.

aleurone [Gk. *aleuron*, flour]: A proteinaceous material, usually in the form of small granules, occurring in the outermost cell layer of the endosperm of wheat and other grains.

alga, *pl.* **algae** (ăl´ga, ăl´je): Traditional term for a series of unrelated groups of photosynthetic eukaryotic organisms lacking multicellular sex organs (except for the charophytes); the "blue-green algae," or cyanobacteria, are one of the groups of photosynthetic bacteria.

algin: An important polysaccharide component of brown algal cell walls; used as a stabilizer and emulsifier for some foods and for paint.

alkali [Arabic *algili*, the ashes of the plant saltwort]: A substance with marked basic properties.

alkaline: Pertaining to substances that release hydroxyl ions (OH⁻) in water; having a pH greater than 7.

alkaloids: Bitter-tasting nitrogenous compounds that are basic (alkaline) in their chemical properties; includes morphine, cocaine, caffeine, nicotine, and atropine.

allele (ă•lēl') [Gk. *allēlōn*, of one another, + *morphē*, form]: One of the two or more alternative forms of a gene.

allelopathy [Gk. *allelon*, of one another, + *pathos*, suffering]: The inhibition of one species of plant by chemicals produced by another plant.

allopatric speciation [Gk. *allos*, other, + *patra*, fatherland, country]: Speciation that occurs as the result of the geographic separation of a population of organisms.

allopolyploid: A polyploid formed from the union of two separate chromosome sets and their subsequent doubling.

allosteric interaction [Gk. *allos*, other, + *steros*, shape]: A change in the shape of a protein resulting from the binding to the protein of a non-substrate molecule; in its new shape, the protein typically has different properties.

alternate phyllotaxy: Leaf arrangement in which there is one bud or one leaf at a node.

alternation of generations: A reproductive cycle in which a haploid (*n*) phase, the gametophyte, produces gametes, which, after fusion in pairs to form a zygote, germinate, producing a diploid (2*n*) phase, the sporophyte. Spores produced by meiotic division from the sporophyte give rise to new gametophytes, completing the cycle.

amino acids [Gk. *Ammon*, referring to the Egyptian sun god, near whose temple ammonium salts were first prepared from camel dung]: Nitrogen-containing organic acids, the units, or "building blocks," from which protein molecules are built.

ammonification: Decomposition of amino acids and other nitrogen-containing organic compounds, resulting in the production of ammonia (NH_3) and ammonium ions (NH_4^+).

amoeboid [Gk. *amoibe*, change]: Moving or eating by means of pseudopodia (temporary cytoplasmic protrusions from the cell body).

amphi- [Gk. *amphi-*, on both sides]: Prefix meaning "on both sides," "both," or "of both kinds."

amylase (ăm'ĭ•lās): An enzyme that breaks down starch into smaller units.

amyloplast: A leucoplast (colorless plastid) that forms starch grains.

an- [Gk. *an-*, not, without]: Prefix equivalent to "a-," meaning "not" or "without"; used before vowels and "h."

anabolism [Gk. *ana-*, up, + *-bolism* (as in metabolism)]: The constructive part of metabolism; the total chemical reactions involved in biosynthesis.

anaerobic [Gk. *an-*, without, + *aer*, air, + *bios*, life]: Referring to any process that can occur without oxygen, or to the metabolism of an organism that can live without oxygen; strict anaerobes cannot survive in the presence of oxygen.

analogous [Gk. *analogos*, proportionate]: Applied to structures similar in function but different in evolutionary origin, such as the phyllodes of an Australian *Acacia* and the leaves of an oak.

anaphase [Gk. *ana*, away, + *phasis*, form]: A stage in mitosis in which the chromatids of each chromosome separate and move to opposite poles; similar stages in meiosis in which chromatids or paired chromosomes move apart.

anatomy: The study of the internal structure of organisms; morphology is the study of their external structure.

andro- [Gk. *andros*, man]: Prefix meaning "male."

androecium [Gk. *andros*, man, + *oikos*, house]: (1) The floral whorl that comprises the stamens; (2) in leafy liverworts, a packetlike swelling containing the antheridia.

aneuploid: A chromosomal aberration in which the chromosome number differs from the normal chromosome number for the species by a small number.

angiosperm [Gk. *angion*, a vessel, + *sperma*, a seed]: Literally, a seed borne in a vessel (carpel); thus one of a group of plants whose seeds are borne within a mature ovary (fruit).

ångstrom [after A. J. Ångstrom, a Swedish physicist, 1814–74]: A unit of length equal to 10^{-10} meter; abbreviated Å.

anion [Gk. *anienae*, to go up]: A negatively charged ion.

anisogamy [Gk. *aniso*, unequal, + *gamos*, marriage]: The condition of having dissimilar motile gametes.

annual [L. *annulus*, year]: A plant whose life cycle is completed in a single growing season.

annual ring: In wood, the growth layer formed during a single year; *see also* growth layer.

annulus [L. *anus*, ring]: In ferns, a row of specialized cells in a sporangium; in gill fungi, the remnant of the inner veil forming a ring on the stalk.

antenna complex: The portion of a photosystem that consists of pigment molecules (antenna pigments) that gather light and "funnel" it to the reaction center.

anterior: Situated before or toward the front.

anther [Gk. *anthos*, flower]: The pollen-bearing portion of a stamen.

antheridiophore [Gk. *anthos*, flower, + *phoros*, bearing]: In some liverworts, a stalk that bears antheridia.

antheridium: A sperm-producing structure that may be multicellular or unicellular.

anthocyanin [Gk. *anthos*, flower, + *kyanos*, dark blue]: A water-soluble blue or red pigment found in the cell sap.

Anthophyta: The phylum of angiosperms, or flowering plants.

anthophytes: Collectively the *Bennettitales*, gnetophytes, and angiosperms, all with a shared possession of flowerlike reproductive structures; not to be confused with the *Anthophyta*, the angiosperm phylum.

antibiotic [Gk. *anti*, against or opposite, + *biotikos*, pertaining to life]: Natural organic substances that retard or prevent the growth of organisms; generally used to designate substances formed by microorganisms that prevent growth of other microorganisms.

anticlinal: Perpendicular to the surface.

anticodon: In a tRNA molecule, the three-nucleotide sequence that base-pairs with the mRNA codon for the amino acid carried by that particular tRNA; the anticodon is complementary to the mRNA codon.

antipodals: Three (sometimes more) cells of the mature embryo sac, located at the end opposite the micropyle.

apical dominance: The influence exerted by a terminal bud in suppressing the growth of lateral, or axillary, buds.

apical meristem: The meristem at the tip of the root or shoot in a vascular plant.

apomixis [Gk. *apo*, separate, away from, + *mixis*, mingling]: Reproduction without meiosis or fertilization; vegetative reproduction.

apoplast [Gk. *apo*, away from, + *plastos*, molded]: The cell wall continuum of a plant or organ; the movement of substances via the cell walls is called apoplastic movement or transport.

apoptosis (ā-po-to-sis): Programmed cell death.

apothecium [Gk. *apotheke*, storehouse]: A cup-shaped or saucer-shaped open ascoma.

arch-, archeo- [Gk. *arche*, archos, beginning]: Prefix meaning "first," "main," or "earliest."

Archaea: A phylogenetic domain of prokaryotes consisting of the methanogens, most extreme halophiles and hyperthermophiles, and *Thermoplasma*.

archegoniophore [Gk. *archegonos*, the first of a race, + *phoros*, bearing]: In some liverworts, a stalk that bears archegonia.

archegonium, *pl.* **archegonia:** A multicellular structure in which a single egg is produced; found in the bryophytes and some vascular plants.

aril (ăr´ĭl) [L. *arillus*, grape, seed]: An accessory seed covering, often formed by an outgrowth at the base of the ovule; often brightly colored, which may aid in dispersal by attracting animals that eat it and, in the process, carry the seed away from the parent plant.

artifact [L. *ars*, art, + *facere*, to make]: A product that exists because of an extraneous, especially human, agency and does not occur in nature.

artificial selection: The breeding of selected organisms to produce strains with desired characteristics.

ascogenous hyphae [Gk. *askos*, bladder, + *genous*, producing]: Hyphae containing paired haploid male and female nuclei; they develop from an ascogonium and eventually give rise to asci.

ascogonium: The oogonium or female gametangium of the ascomycetes.

ascoma, *pl.* **ascomata:** A multicellular structure in ascomycetes lined with specialized cells called asci, in which nuclear fusion and meiosis occur. Ascomata may be open or closed. Also called an ascocarp.

ascospore: A spore produced within an ascus; found in ascomycetes.

ascus, *pl.* **asci:** A specialized cell, characteristic of the ascomycetes, in which two haploid nuclei fuse to produce a zygote that immediately divides by meiosis; at maturity, an ascus contains ascospores.

aseptate [Gk. *a-*, not, + L. *septum*, fence]: Nonseptate; lacking cross walls.

asexual reproduction: Any reproductive process, such as fission or budding, that does not involve the union of gametes.

assimilate stream: The flow of assimilates, or food materials, in the phloem; moves from source to sink.

atom [Gk. *atomos*, indivisible]: The smallest unit into which a chemical element can be divided and still retain its characteristic properties.

atomic nucleus: The central core of an atom, containing protons and neutrons, around which electrons orbit.

atomic number: The number of protons in the nucleus of an atom.

atomic weight: The weight of a representative atom of an element relative to the weight of an atom of carbon ^{12}C, which has been assigned the value 12.

ATP: *See* adenosine triphosphate.

ATP synthase: An enzyme complex that forms ATP from ADP and phosphate during oxidative phosphorylation in the inner mitochondrial membrane.

auto- [Gk. *autos*, self, same]: Prefix meaning "same" or "self-same."

autoecious [Gk. *autos*, self, + *oikia*, dwelling]: In some rust fungi, completing the life cycle on a single species of host plant.

autopolyploid: A polyploid formed from the doubling of a single genome.

autoradiograph: A photographic print made by a radioactive substance acting upon a sensitive photographic film.

autotroph [Gk. *autos*, self, + *trophos*, feeder]: An organism that is able to synthesize the nutritive substances it requires from inorganic substances in its environment; *see also* heterotroph.

auxin [Gk. *auxein*, to increase]: A class of plant hormones that control cell elongation, among other effects.

axial system: In secondary xylem and secondary phloem; the term applied collectively to cells derived from fusiform cambial initials. The long axes of these cells are oriented parallel with the main axis of the root or stem. Also called longitudinal system and vertical system.

axil [Gk. *axilla*, armpit]: The upper angle between a twig or leaf and the stem from which it grows.

axillary: Term applied to buds or branches occurring in the axil of a leaf.

B

bacillus, *pl.* **bacilli** (ba•sĭl´ŭs) [L. *baculum*, rod]: A rod-shaped bacterium.

backcross: The crossing of a hybrid with one of its parents or with a genetically equivalent organism; a cross between an individual whose genes are to be tested and one that is homozygous for all of the recessive genes involved in the experiment.

Bacteria: The phylogenetic domain consisting of all prokaryotes that are not members of the domain *Archaea*.

bacteriophage [Gk. *bakterion*, little rod, + *phagein*, to eat]: A virus that parasitizes bacterial cells.

bacterium, *pl.* **bacteria:** A prokaryotic organism. *See also Bacteria.*

bacteroid: An enlarged, deformed *Rhizobium* or *Bradyrhizobium* cell found in root nodules; capable of nitrogen fixation.

bark: A nontechnical term applied to all tissues outside the vascular cambium in a woody stem; *see also* inner bark *and* outer bark.

basal body: A self-reproducing, cylinder-shaped cytoplasmic organelle from which cilia or flagella arise; identical in structure to the centriole, which is involved in mitosis and meiosis in most animals and protists.

base: A substance that dissociates in water, causing a decrease in the concentration of hydrogen ions (H$^+$), often by releasing hydroxyl ions (OH$^-$); bases have a pH in solution of more than 7; the opposite of "acid."

basidioma, *pl.* **basidiomata:** A multicellular structure, characteristic of the basidiomycetes, within which basidia are formed.

basidiospore: A spore of the basidiomycota, produced within and borne on a basidium following nuclear fusion and meiosis. Also called a basidiocarp.

basidium, *pl.* **basidia:** A specialized reproductive cell of the *Basidiomycota*, often club-shaped, in which nuclear fusion and meiosis occur.

berry: A simple fleshy fruit that includes a fleshy ovary wall and one or more carpels and seeds; examples are the fruits of grapes, tomatoes, and bananas.

bi- [L. *bis*, double, two]: Prefix meaning "two," "twice," or "having two points."

biennial: A plant that normally requires two growing seasons to complete its life cycle, flowering and fruiting in its second year.

bilaterally symmetrical: *See* zygomorphic.

biological clock [Gk. *bios*, life, + *logos*, discourse]: The internal timing mechanism that governs the innate biological rhythms of organisms.

biomass: Total dry weight of all organisms in a particular population, sample, or area.

biome: A complex of terrestrial communities of very wide extent, characterized by its climate and soil; the largest ecological unit.

biosphere: The zone of air, land, and water at the surface of the Earth that is occupied by organisms.

biotechnology: The practical application of advances in hormone research and DNA biochemistry to manipulate the genetics of plants.

biotic: Relating to life.

bisexual flower: A flower that has at least one functional stamen and one functional carpel.

bivalent [L. *bis*, double, + *valere*, to be strong]: A pair of synapsed homologous chromosomes. Also called a tetrad.

blade: The broad, expanded part of a leaf; the lamina.

body cell: Vegetative or somatic cell; *see also* spermatogenous cell.

bordered pit: A pit in which the secondary wall arches over the pit membrane.

bract: A modified, usually reduced leaflike structure.

branch root: *See* lateral root.

bryophytes (brī′o•fĭts): The members of the phyla of nonvascular plants; the mosses, hornworts, and liverworts.

bud: (1) An embryonic shoot, often protected by young leaves; (2) a vegetative outgrowth of yeasts and some bacteria as a means of asexual reproduction.

bulb: A short underground stem covered by enlarged and fleshy leaf bases containing stored food.

bulk flow: The overall movement of water or some other liquid induced by gravity, pressure, or an interplay of both.

bulliform cell: A large epidermal cell present, with other such cells, in longitudinal rows in grass leaves; also called motor cell. Believed to be involved with the mechanism of rolling and unrolling of the leaves.

bundle scar: Scar or mark left on leaf scar by vascular bundles broken at the time of leaf fall, or abscission.

bundle sheath: Layer or layers of cells surrounding a vascular bundle; may consist of parenchyma or sclerenchyma cells, or both.

bundle-sheath extension: A group of cells extending from a bundle sheath of a vein in the leaf mesophyll to either upper or lower epidermis or both; may consist of parenchyma, collenchyma, or sclerenchyma.

C

C_3 pathway: *See* Calvin cycle.

C_3 plants: Plants that employ only the Calvin cycle, or C_3 pathway, in the fixation of CO_2; the first stable product is the three-carbon compound 3-phosphoglycerate.

C_4 pathway: The set of reactions through which carbon dioxide is fixed to a compound known as phosphoenolpyruvate (PEP) to yield oxaloacetate, a four-carbon compound.

C_4 plants: Plants in which the first product of CO_2 fixation is a four-carbon compound (oxaloacetate); both the Calvin cycle (C_3 pathway) and C_4 pathway are employed by C_4 plants.

callose: A complex branched carbohydrate that is a common wall constituent associated with the sieve areas of sieve elements; may develop in reaction to injury in sieve elements and parenchyma cells.

callus [L. *callos*, hard skin]: Undifferentiated tissue; a term used in tissue culture, grafting, and wound healing.

calorie [L. *calor*, heat]: The amount of energy in the form of heat required to raise the temperature of one gram of water 1°C. In making metabolic measurements, the kilocalorie (kcal), or Calorie—the amount of heat required to raise the temperature of one kilogram of water 1°C—is generally used.

Calvin cycle: The series of enzymatically mediated photosynthetic reactions during which carbon dioxide is reduced to 3-phosphoglyceraldehyde and the carbon dioxide acceptor, ribulose 1,5-bisphosphate, is regenerated. For every three molecules of carbon dioxide entering the cycle, a net gain of one molecule of glyceraldehyde 3-phosphate results.

calyptra [Gk. *kalyptra*, covering for the head]: The hood or cap that partially or entirely covers the capsule of some species of mosses; it is formed from the expanded archegonial wall.

calyx (kā′lĭks) [Gk. *kalyx*, a husk, cup]: The sepals collectively; the outermost flower whorl.

CAM: *See* crassulacean acid metabolism.

cambial zone: A region of thin-walled, undifferentiated meristematic cells between the secondary xylem and secondary phloem; consists of cambial initials and their recent derivatives.

cambium [L. *cambiare*, to exchange]: A meristem that gives rise to parallel rows of cells; commonly applied to the vascular cambium and the cork cambium, or phellogen.

capsid: The protein coat of a virus particle.

capsule: (1) In angiosperms, a dehiscent, dry fruit that develops from two or more carpels; (2) a slimy layer around the cells of certain bacteria; (3) the sporangium of bryophytes.

carbohydrate [L. *carbo*, ember, + *hydro*, water]: An organic compound consisting of a chain of carbon atoms to which hydrogen and oxygen are attached in a 2:1 ratio; examples are sugars, starch, glycogen, and cellulose.

carbon cycle: Worldwide circulation and utilization of carbon atoms.

carbon fixation: The conversion of CO_2 into organic compounds during photosynthesis.

carbon-fixation reactions: In photosynthetic cells, the light-independent enzymatic reactions concerned with the synthesis of glucose from CO_2, ATP, and NADPH; also called light-independent reactions and dark reactions.

carnivorous: Feeding upon animals, as opposed to feeding upon plants (herbivorous); also refers to plants that are able to utilize proteins obtained from trapped animals, chiefly insects.

carotene (kăr′o•tēn) [L. *carota*, carrot]: A yellow or orange pigment belonging to the carotenoid group.

carotenoids (kă•rŏt′e•noids): A class of fat-soluble pigments that includes the carotenes (yellow and orange pigments) and the xanthophylls (yellow pigments); found in chloroplasts and chromoplasts of plants. Carotenoids act as accessory pigments in photosynthesis.

carpel [Gk. *karpos*, fruit]: One of the members of the gynoecium, or inner floral whorl; each carpel encloses one or more ovules. One or more carpels form a gynoecium.

carpellate: Pertaining to a flower with one or more carpels but no functional stamens; also called pistillate.

carpogonium [Gk. *karpos*, fruit, + *gonos*, offspring]: In red algae, the female gametangium.

carposporangium [Gk. *karpos*, fruit, + *spora*, seed, + *angeion*, vessel]: In red algae, a carpospore-containing cell.

carpospore: In red algae, the single diploid protoplast found within a carposporangium.

carriers: Transport proteins that bind specific solutes and undergo conformational change in order to transport the solute across the membrane.

caryopsis [Gk. *karyon*, a nut, + *opsis*, appearance]: Simple, dry, one-seeded indehiscent fruit with the pericarp firmly united all around the seed coat; a grain characteristic of the grasses (family *Poaceae*).

Casparian strip [after Robert Caspary, German botanist]: A bandlike region of primary wall containing suberin and lignin; found in anticlinal—radial and transverse—walls of endodermal and exodermal cells.

catabolism [Gk. *katabolē*, throwing down]: Collectively, the chemical reactions resulting in the breakdown of complex materials and involving the release of energy.

catalyst [Gk. *katalysis*, dissolution]: A substance that accelerates the rate of a chemical reaction but is not used up in the reaction; enzymes are catalysts.

category [Gk. *katēgoria*, category]: In a hierarchical classification system, the level at which a particular group is ranked.

cation [Gk. *katienai*, to go down]: A positively charged ion.

catkin: A spikelike inflorescence of unisexual flowers; found only in woody plants.

cDNA: *See* complementary DNA.

cell [L. *cella*, small room]: The structural unit of organisms; in plants, cells consist of the cell wall and the protoplast.

cell division: The division of a cell and its contents, usually into two roughly equal parts.

cell plate: The structure that forms at the equator of the spindle in the dividing cells of plants and a few green algae during early telophase.

cell sap: The fluid contents of the vacuole.

cellular respiration: *See* respiration.

cellulase: An enzyme that hydrolyzes cellulose.

cellulose: A carbohydrate; the chief component of the cell wall in plants and some protists; an insoluble complex carbohydrate formed of microfibrils of glucose molecules attached end to end.

cell wall: The rigid outermost layer of the cells found in plants, some protists, and most prokaryotes.

central mother cells: Relatively large vacuolate cells in a subsurface position in apical meristems of shoots.

centriole [Gk. *kentron*, center, + L. *-olus*, little one]: A cytoplasmic organelle found outside the nuclear envelope, and identical in structure to a basal body; centrioles are found in the cells of most eukaryotes other than fungi, red algae, and the nonflagellated cells of plants. Centrioles divide and organize spindle fibers during mitosis and meiosis.

centromere [Gk. *kentron*, center, + *meros*, a part]: Region of constriction of chromosome that holds sister chromatids together.

chalaza [Gk. *chalaza*, small tubercle]: The region of an ovule or seed where the funiculus unites with the integuments and the nucellus.

channel proteins: Transport proteins that form water-filled pores that extend across cellular membranes; when open, the channel proteins allow specific solutes to pass through them.

chemical potential: The activity or free energy of a substance; it is dependent upon the rate of motion of the average molecule and the concentration of the molecules.

chemical reaction: The making or breaking of chemical bonds between atoms or molecules.

chemiosmotic coupling: Coupling of ATP synthesis to electron transport via an electrochemical H+ gradient across a membrane.

chemoautotrophic: Refers to prokaryotes that are able to manufacture their own basic foods by using the energy released by specific inorganic reactions; *see also* autotroph.

chiasma (kī•ăz′ma) [Gk. *chiasma*, a cross]: The X-shaped figure formed by the meeting of two nonsister chromatids of homologous chromosomes; the site of crossing-over.

chitin (kī′tĭn) [Gk. *chiton*, tunic]: A tough, resistant, nitrogen-containing polysaccharide forming the cell walls of certain fungi, the exoskeleton of arthropods, and the epidermal cuticle of other surface structures of certain protists and animals.

chlor- [Gk. *chloros*, green]: Prefix meaning "green."

chlorenchyma: Parenchyma cells that contain chloroplasts.

chlorophyll [Gk. *chloros*, green, + *phyllon*, leaf]: The green pigment of plant cells, which is the receptor of light energy in photosynthesis; also found in algae and photosynthetic bacteria.

chloroplast: A plastid in which chlorophylls are contained; the site of photosynthesis. Chloroplasts occur in plants and algae.

chlorosis: Loss or reduced development of chlorophyll.

chroma- [Gk. *chroma*, color]: Prefix meaning "color."

chromatid [Gk. *chroma*, color, + L. *-id*, daughters of]: One of the two daughter strands of a duplicated chromosome, which are joined at the centromere.

chromatin: The deeply staining complex of DNA and proteins that forms eukaryotic chromosomes.

chromatophore [Gk. *chroma*, color, + *phorus*, a bearer]: In some bacteria, a discrete vesicle delimited by a single membrane and containing photosynthetic pigments.

chromoplast: A plastid containing pigments other than chlorophyll, usually yellow and orange carotenoid pigments.

chromosome [Gk. *chroma*, color, + *soma*, body]: The structure that carries the genes. Eukaryotic chromosomes are visualized as threads or rods of chromatin, which appear in contracted form during mitosis and meiosis and otherwise are enclosed in a nucleus; each eukaryotic chromosome contains a linear DNA molecule; prokaryotes typically have a single chromosome consisting of a circular DNA molecule.

chrysolaminarin: The storage product of the chrysophytes and diatoms.

cilium, *pl.* **cilia** (sĭl′ē•ŭm) [L. *cilium*, eyelash]: A short, hairlike flagellum, usually numerous and arranged in rows.

circadian rhythms [L. *circa*, about, + *dies*, a day]: Regular rhythms of growth and activity that occur on an approximately 24-hour basis.

circinate vernation [L. *circinare*, to make round, + *vernare*, to flourish]: As in ferns, the coiled arrangement of leaves and leaflets in the bud; such an arrangement uncoils gradually as the leaf develops further.

cisterna, *pl.* **cisternae** [L. *cistern*, a reservoir]: A flattened or saclike portion of the endoplasmic reticulum or a Golgi body (dictyosome).

clade: One evolutionary line of organisms.

cladistics: System of arranging organisms following an analysis of their primitive and advanced features so that their phylogenetic relationships will be reflected accurately.

cladogram: A line diagram that branches repeatedly, suggesting phylogenetic relationships among organisms.

cladophyll [Gk. *klados*, shoot, + *phyllon*, leaf]: A branch resembling a foliage leaf.

clamp connection: In the *Basidiomycota*, a lateral connection between adjacent cells of a dikaryotic hypha; ensures that each cell of the hypha will contain two dissimilar nuclei.

class: A taxonomic category between phylum and order in rank. A class contains one or more orders, and belongs to a particular phylum.

cleistothecium [Gk. *kleistos*, closed, + *thekion*, small receptacle]: A closed, spherical ascoma.

climax community: The final stage in a successional series; its nature is determined largely by the climate and soil of the region.

cline: A graded series of changes in some characteristics within a species, often correlated with a gradual change in climate or another geographical factor.

clone [Gk. *klon*, twig]: A population of cells or individuals derived by asexual division from a single cell or individual; one of the members of such a population.

cloning: Producing a cell line or culture all of whose members are characterized by a specific DNA sequence; a key element in genetic engineering.

closed vascular bundle: A vascular bundle in which a cambium does not develop.

coalescence [L. *coalescere*, to grow together]: The union of floral parts of the same whorl, as petals to petals.

coccus, *pl.* **cocci** (kŏk′ŭs) [Gk. *kokkos*, a berry]: A spherical bacterium.

codon (kō′dŏn): Sequence of three adjacent nucleotides in a molecule of DNA or mRNA that form the code for a single amino acid, or for the termination of a polypeptide chain.

coenocytic (se•nō•sĭ′tic) [Gk. *koinos*, shared in common, + *kytos*, a hollow vessel]: A term used to describe an organism or part of an organism that is multinucleate, the nuclei not separated by walls or membranes; also called siphonaceous, siphonous, or syncytial.

coenzyme: An organic molecule, or nonprotein organic cofactor, that plays an accessory role in enzyme-catalyzed processes, often by acting as a donor or acceptor of electrons; NAD+ and FAD are common coenzymes.

coevolution [L. *co*, together, + *e-*, out, + *volvere*, to roll]: The simultaneous evolution of adaptations in two or more populations that interact so closely that each is a strong selective force on the other.

cofactor: One or more nonprotein components required by enzymes in order to function; many cofactors are metal ions, while others are called coenzymes.

cohesion [L. *cohaerere*, to stick together]: The mutual attraction of molecules of the same substance.

cold hardiness: The ability of a plant to survive the extreme cold and drying effects of winter weather.

coleoptile (kō′lē•op′till) [Gk. *koleos,* sheath, + *ptilon,* feather]: The sheath enclosing the apical meristem and leaf primordia of the grass embryo; often interpreted as the first leaf.

coleorhiza (kō′lē•o•rī′za) [Gk. *koleos,* sheath, + *rhiza,* root]: The sheath enclosing the radicle in the grass embryo.

collenchyma [Gk. *kolla,* glue]: A supporting tissue composed of collenchyma cells; common in regions of primary growth in stems and in some leaves.

collenchyma cell: Elongated living cell with unevenly thickened nonlignified primary cell wall.

colloid (kŏl′oid): A permanent suspension of fine particles.

community: All the organisms inhabiting a common environment and interacting with one another.

companion cell: A specialized parenchyma cell associated with a sieve-tube element in angiosperm phloem and arising from the same mother cell as the sieve-tube element.

competition: Interaction between members of the same population or of two or more populations to obtain a resource that both require and that is available in limited supply.

complementary DNA (cDNA): A single-stranded molecule of DNA that has been synthesized from an mRNA template by reverse transcription.

complete flower: A flower having four whorls of floral parts—sepals, petals, stamens, and carpels.

complex tissue: A tissue consisting of two or more cell types; epidermis, periderm, xylem, and phloem are complex tissues.

compound: A combination of atoms in a definite ratio, held together by chemical bonds.

compound leaf: A leaf whose blade is divided into several distinct leaflets.

compression wood: The reaction wood of conifers; develops on the lower sides of leaning trunks or limbs.

concentration gradient: The concentration difference of a substance per unit distance.

cone: *See* strobilus.

conidiophore: Hypha on which one or more conidia are produced.

conidium, *pl.* **conidia** [Gk. *konis,* dust]: An asexual fungal spore not contained within a sporangium; it may be produced singly or in chains; most conidia are multinucleate.

conifer: A cone-bearing tree.

conjugation: The temporary fusion of pairs of bacteria, protozoa, and certain algae and fungi during which genetic material is transferred between the two individuals.

conjugation tube: A tube formed during the process of conjugation to facilitate the transfer of genetic material.

connate (kŏn′āt): Said of similar parts that are united or fused, as petals fused in a corolla tube; *see also* adnate.

consumer: In ecology, an organism that derives its food from another organism.

continuous variation: Variation in traits to which a number of different genes contribute; the variation often exhibits a "normal" or bell-shaped distribution.

contractile vacuole: A clear, fluid-filled vacuole in some groups of protists that takes up water within the cell and then contracts, expelling its contents from the cell.

convergent evolution [L. *convergere,* to turn together]: The independent development of similar structures in organisms that are not directly related; often found in organisms living in similar environments.

cork: A secondary tissue produced by a cork cambium; made up of polygonal cells, nonliving at maturity, with suberized cell walls, which are resistant to the passage of gases and water vapor; the outer part of the periderm. Also called phellem.

cork cambium: The lateral meristem that forms the periderm, producing cork (phellem) toward the surface (outside) of the plant and phelloderm toward the inside; common in stems and roots of gymnosperms and woody angiosperms. Also called phellogen.

corm: A thickened underground stem, upright in position, in which food is accumulated, usually in the form of starch.

corolla [L. *corona,* crown]: The petals collectively; usually the conspicuously colored flower whorl.

corolla tube: A tubelike structure resulting from the fusion of the petals along their edges.

cortex: Ground-tissue region of a stem or root bounded externally by the epidermis and internally by the vascular system; a primary-tissue region; also used to refer to the peripheral region of a cell protoplast.

cotransport: Membrane transport in which the transfer of one solute depends on the simultaneous or sequential transfer of a second solute.

cotyledon (kŏt′i•lē′dŭn) [Gk. *kotyledon,* cup-shaped hollow]: Seed leaf; generally absorbs food in monocotyledons and stores food in other angiosperms.

coupled reactions: Reactions in which energy-requiring chemical reactions are linked to energy-releasing reactions.

covalent bond: A chemical bond formed between atoms as a result of the sharing of two electrons.

crassulacean acid metabolism, or CAM: A variant of the C_4 pathway; phosphoenolpyruvate fixes CO_2 in C_4 compounds at night and then, during the daytime, the fixed CO_2 is transferred to the ribulose bisphosphate of the Calvin cycle within the same cell. Characteristic of most succulent plants, such as cacti.

cristae, *sing.* **crista:** The enfoldings of the inner mitochondrial membrane, which form a series of crests or ridges containing the electron-transport chains involved in ATP formation.

crop rotation: The practice of growing different crops in regular succession to aid in the control of insects and diseases, to increase soil fertility, and to decrease erosion.

cross-fertilization: The fusion of gametes formed by different individuals; the opposite of self-fertilization.

crossing-over: The exchange of corresponding segments of genetic material between the chromatids of homologous chromosomes at meiosis.

cross-pollination: The transfer of pollen from the anther of one plant to the stigma of a flower of another plant.

cross section: *See* transverse section.

cryptogam: An archaic term for all organisms except the flowering plants (phanerogams), animals, and heterotrophic protists.

cultivar: A variety of plant found only under cultivation.

cuticle: Waxy or fatty layer on outer wall of epidermal cells, formed of cutin and wax.

cutin [L. *cutis,* skin]: Fatty substance deposited in many plant cell walls and on outer surface of epidermal cell walls, where it forms a layer known as the cuticle.

cyclic electron flow: In chloroplasts, the light-induced flow of electrons originating from and returning to photosystem I.

cyclosis (sī•klō′sis) [Gk. *kyklosis,* circulation]: The streaming of cytoplasm within a cell.

-cyte, cyto- [Gk. *kytos,* hollow vessel, container]: Suffix or prefix meaning "pertaining to the cell."

cytochrome [Gk. *kytos,* container, + *chroma,* color]: Heme proteins serving as electron carriers in respiration and photosynthesis.

cytokinesis [Gk. *kytos,* hollow vessel, + *kinesis,* motion]: Division of the cytoplasm of a cell following nuclear division.

cytokinin [Gk. *kytos,* hollow vessel, + *kinesis,* motion]: A class of plant hormones that promotes cell division, among other effects.

cytology: The study of cell structure and function.

cytoplasm: The living matter of a cell, exclusive of the nucleus; the protoplasm.

cytoplasmic ground substance: *See* cytosol.

cytosine: One of the four pyrimidine bases found in the nucleic acids DNA and RNA.

cytoskeleton: The flexible network within cells, composed of microtubules and actin filaments, or microfilaments.

cytosol: The cytoplasmic matrix of the cytoplasm in which the nucleus, various organelles, and membrane systems are suspended.

D

day-neutral plants: Plants that flower without regard to daylength.

de- [L. *de-*, away from, down, off]: Prefix meaning "away from," "down," or "off"; for example, dehydration means removal of water.

deciduous [L. *decidere*, to fall off]: Shedding leaves at a certain season.

decomposers: Organisms (bacteria, fungi, heterotrophic protists) in an ecosystem that break down organic material into smaller molecules that are then recirculated.

dehiscence [L. *de*, down, + *hiscere*, split open]: The opening of an anther, fruit, or other structure, which permits the escape of reproductive bodies contained within.

dehydration synthesis: The synthesis of a compound or molecule involving the removal of water; also called a condensation reaction.

denitrification: The conversion of nitrate to gaseous nitrogen; carried out by a few genera of free-living soil bacteria.

deoxyribonucleic acid (DNA): Carrier of genetic information in cells; composed of chains of phosphate, sugar molecules (deoxyribose), and purines and pyrimidines; capable of self-replication as well as determining RNA synthesis.

deoxyribose [L. *deoxy*, loss of oxygen, + *ribose*, a kind of sugar]: A five-carbon sugar with one less atom of oxygen than ribose; a component of deoxyribonucleic acid.

dermal tissue system: The outer covering tissue of the plant; the epidermis or the periderm.

desmotubule [Gk. *desmos*, to bind, + L. *tubulus*, small tube]: The tubule traversing a plasmodesmatal canal and uniting the endoplasmic reticulum of the two adjacent cells.

determinate growth: Growth of limited duration, characteristic of floral meristems and of leaves.

deuterium: Heavy hydrogen; a hydrogen atom, the nucleus of which contains one proton and one neutron. (The nucleus of most hydrogen atoms consists of only a proton.)

dichotomy: The division or forking of an axis into two branches.

dicotyledon: Obsolete term used to refer to all angiosperms other than monocotyledons; characterized by having two cotyledons; *see also* eudicotyledons *and* magnoliids.

dictyosome: *See* Golgi body.

differentiation: A developmental process by which a relatively unspecialized cell undergoes a progressive change to a more specialized cell; the specialization of cells and tissues for particular functions during development.

diffuse-porous wood: A wood in which the pores, or vessels, are fairly uniformly distributed throughout the growth layers or in which the size of pores changes only slightly from early wood to late wood.

diffusion [L. *diffundere*, to pour out]: The net movement of suspended or dissolved particles from a more concentrated region to a less concentrated region as a result of the random movement of individual molecules; the process tends to distribute such particles uniformly throughout a medium.

digestion: The conversion of complex, usually insoluble foods into simple, usually soluble forms by means of enzymatic action.

dikaryon (Gk. *di*, two, + *karyon*, a nut]: In fungi, mycelium with paired nuclei, each usually derived from a different parent.

dikaryotic: In fungi, having pairs of nuclei within cells or compartments.

dimorphism [Gk. *di*, two, + *morphē*, form]: The condition of having two distinct forms, such as sterile and fertile leaves in ferns, or sterile and fertile shoots in horsetails.

dioecious [Gk. *di*, two, + *oikos*, house]: Unisexual; having the male and female (or staminate and ovulate) elements on different individuals of the same species.

diploid: Having two sets of chromosomes; the 2*n* (diploid) chromosome number is characteristic of the sporophyte generation.

disaccharide [Gk. *di*, two, + *sakcaron*, sugar]: A carbohydrate formed of two simple sugar molecules linked by a covalent bond; sucrose is an example.

disk flowers: The actinomorphic, tubular flowers in *Asteraceae*; contrasted with flattened, zygomorphic ray flowers. In many *Asteraceae*, the disk flowers occur in the center of the inflorescence, the ray flowers around the margins.

distal: Situated away from or far from the point of reference (usually the main part of body); opposite of proximal.

DNA: *See* deoxyribonucleic acid.

DNA sequencing: Determination of the order of nucleotides in a DNA molecule.

domain: The taxonomic category above the kingdom level; the three domains are *Archaea, Bacteria,* and *Eukarya.*

dominant allele: One allele is said to be dominant with respect to an alternative allele if the homozygote for the dominant allele is indistinguishable phenotypically from the heterozygote; the other allele is said to be recessive.

dormancy [L. *dormire*, to sleep]: A special condition of arrested growth in which the plant and such plant parts as buds and seeds do not begin to grow without special environmental cues. The requirement for such cues, which include cold exposure and a suitable photoperiod, prevents the breaking of dormancy during superficially favorable growing conditions.

double fertilization: The fusion of the egg and sperm (resulting in a 2*n* fertilized egg, the zygote) and the simultaneous fusion of the second male gamete with the polar nuclei (typically resulting in a 3*n* primary endosperm nucleus); a unique characteristic of all angiosperms.

doubling rate: The length of time required for a population of a given size to double in number.

drupe [Gk. *dryppa*, overripe olive]: A simple, fleshy fruit, derived from a single carpel, usually one-seeded, in which the inner fruit coat is hard and may adhere to the seed.

druse: A compound, more or less spherical crystal with many component crystals projecting from its surface; composed of calcium oxalate.

E

early wood: The first-formed wood of a growth increment; it contains larger cells and is less dense than the subsequently formed late wood; replaces the term "spring wood."

eco- [Gk. *oikos*, house]: Prefix meaning "house" or "home."

ecology: The study of the interactions of organisms with their physical environment and with one other.

ecosystem: A major interacting system that involves both living organisms and their physical environment.

ecotype [Gk. *oikos*, house, + L. *typus*, image]: A locally adapted variant of an organism, differing genetically from other ecotypes.

edaphic [Gk. *edaphos*, ground, soil]: Pertaining to the soil.

egg: A nonmotile female gamete, usually larger than a male gamete of the same species.

egg apparatus: The egg cell and synergids located at the micropylar end of the female gametophyte, or embryo sac, of angiosperms.

elater [Gk. *elater*, driver]: (1) An elongated, spindle-shaped, sterile cell in the sporangium of a liverwort sporophyte (aids in spore dispersal); (2) clubbed, hygroscopic bands attached to the spores of the horsetails.

electrochemical gradient: The driving force that causes an ion to move across a membrane due to the difference in the electric charge across the membrane in combination with the difference in the ion's concentration on the two sides of the membrane.

electrolyte: A substance that dissociates into ions in aqueous solution and so makes possible the conduction of an electric current through the solution.

electromagnetic spectrum: The entire spectrum of radiation, which ranges in wavelength from less than a nanometer to more than a kilometer.

electron: A subatomic particle with a negative electric charge equal in magnitude to the positive charge of the proton, but with a mass of 1/1837 of that of the proton. Electrons orbit the atom's positively charged nucleus and determine the atom's chemical properties.

electron-dense: In electron microscopy, not permitting the passage of electrons and so appearing dark.

electron transport: The movement of electrons down a series of electron-carrier molecules that hold electrons at slightly different energy levels; as electrons move down the chain, the energy released is used to form ATP from ADP and phosphate. Electron transport plays an essential role in the final stage of cellular respiration and in the light-dependent reactions of photosynthesis.

element: A substance composed of only one kind of atom; one of more than 100 distinct natural or synthetic types of matter that, singly or in combination, compose virtually all materials of the universe.

embryo [Gk. *en*, in, + *bryein*, to swell]: A young sporophytic plant, before the start of a period of rapid growth (germination in seed plants).

embryogenesis: Development of an embryo from a fertilized egg, or zygote; also called embryogeny.

embryophytes: The bryophytes and vascular plants, both of which produce embryos; a synonym for plants.

embryo sac: The female gametophyte of angiosperms, generally an eight-nucleate, seven-celled structure; the seven cells are the egg cell, two synergids and three antipodals (each with a single nucleus), and the central cell (with two nuclei).

endergonic: Describing a chemical reaction that requires energy to proceed; opposite of exergonic.

endo- [Gk. *endo*, within]: Prefix meaning "within."

endocarp [Gk. *endo*, within, + *karpos*, fruit]: The innermost layer of the mature ovary wall, or pericarp.

endocytosis [Gk. *endon*, within, + *kytos*, hollow vessel]: The uptake of material into cells by means of invagination of the plasma membrane; if solid material is involved, the process is called phagocytosis; if dissolved material is involved, it is called pinocytosis.

endodermis [Gk. *endon*, within, + *derma*, skin]: A single layer of cells forming a sheath around the vascular region in roots and some stems; the endodermal cells are characterized by a Casparian strip within radial and transverse walls. In roots and stems of seed plants, the endodermis is the innermost layer of the cortex.

endogenous [Gk. *endon*, within, + *genos*, race, kind]: Arising from deep-seated tissues, as in the case of lateral roots.

endomembrane system: Collectively, the cellular membranes that form a continuum (plasma membrane, tonoplast, endoplasmic reticulum, Golgi bodies, and nuclear envelope).

endoplasmic reticulum: A complex, three-dimensional membrane system of indefinite extent present in eukaryotic cells, dividing the cytoplasm into compartments and channels. Those portions that are densely coated with ribosomes are called rough endoplasmic reticulum, and other portions with fewer or no ribosomes are called smooth endoplasmic reticulum.

endosperm [Gk. *endon*, within, + *sperma*, seed]: A tissue, containing stored food, that develops from the union of a male nucleus and the polar nuclei of the central cell; it is digested by the growing sporophyte either before or after the maturation of the seed; found only in angiosperms.

energy: The capacity to do work.

energy of activation: The energy that must be possessed by atoms or molecules in order to react.

energy-transduction reactions: *See* light reactions.

entrainment: The process by which a periodic repetition of light and dark, or some other external cycle, causes a circadian rhythm to remain synchronized with the same cycle as the modifying, or entraining, factor.

entropy: A measure of the randomness or disorder of a system.

enzyme: A protein that is capable of speeding up specific chemical reactions by lowering the required activation energy, but is unaltered itself in the process; a biological catalyst.

epi- [Gk. *epi*, upon]: Prefix meaning "upon" or "above."

epicotyl: The upper portion of the axis of an embryo or seedling, above the cotyledons (seed leaves) and below the next leaf or leaves.

epidermis: The outermost layer of cells of the leaf and of young stems and roots; primary in origin.

epigeous [Gk. *epi*, upon, + *ge*, the Earth]: Type of seed germination in which the cotyledons are carried above ground level.

epigyny [Gk. *epi*, upon, + *gyne*, woman]: A pattern of floral organization in which the sepals, petals, and stamens apparently grow from the top of the ovary; *see also* of hypogyny.

epiphyte (ĕp′ĭ•fīt): An organism that grows upon another organism but is not parasitic on it.

epistatic [Gk. *epistasis*, a stopping]: Term used to describe a gene the action of which modifies the phenotypic expression of a gene at another locus.

essential elements: Chemical elements essential for normal plant growth and development; also referred to as essential minerals and essential inorganic nutrients.

ethylene: A simple hydrocarbon that is a plant hormone involved in the ripening of fruit; $H_2C = CH_2$.

etiolation (e′tĭ•o•lā′shŭn) [Fr. *etioler*, to blanch]: A condition involving increased stem elongation, poor leaf development, and lack of chlorophyll; found in plants growing in the dark or with a greatly reduced amount of light.

etioplast: Plastid of a plant grown in the dark and containing a prolamellar body.

eudicotyledons: One of two major classes of angiosperms, *Eudicotyledones*; formerly grouped with the magnoliids, a diverse group of archaic flowering plants, as "dicots"; abbreviated as eudicot.

Eukarya: The phylogenetic domain containing all eukaryotic organisms.

eukaryote [Gk. *eu*, good, + *karyon*, kernel]: A cell that has a membrane-bounded nucleus, membrane-bounded organelles, and chromosomes in which the DNA is associated with proteins; an organism composed of such cells. Plants, animals, fungi, and protists are the four kingdoms of eukaryotes.

eusporangium: A sporangium that arises from several initial cells and, before maturation, forms a wall of more than one layer of cells.

eustele [Gk. *eu-*, good, + *stele*, pillar]: A stele in which the primary vascular tissues are arranged in discrete strands around a pith: typical of gymnosperms and angiosperms.

evolution: The derivation of progressively more complex forms of life from simple ancestors; Darwin proposed that natural selection is the principal mechanism by which evolution takes place.

exergonic [L. *ex*, out, + Gk. *ergon*, work]: Energy-yielding, as in a chemical reaction; applied to a "downhill" process.

exine: The outer wall layer of a spore or pollen grain.

exocarp [Gk. *exo*, without, + *karpos*, fruit]: The outermost layer of the mature ovary wall, or pericarp.

exocytosis [Gk. *ex*, out of, + *kytos*, vessel]: A cellular process in which particulate matter or dissolved substances are enclosed in a vesicle and transported to the cell surface; there, the membrane of the vesicle fuses with the plasma membrane, expelling the vesicle's contents to the outside.

exodermis: The outer layer, one or more cells in depth, of the cortex in some roots; these cells are characterized by Casparian strips within the radial and transverse walls. Following development of Casparian strips, a suberin lamella is deposited on all walls of the exodermis.

exon [Gk. *exo*, outside]: A segment of DNA that is both transcribed into RNA and translated into protein; exons are characteristic of eukaryotes. *See also* intron.

eyespot: A small, pigmented structure in flagellate unicellular organisms that is sensitive to light; also called a stigma.

F

F₁: First filial generation. The offspring resulting from a cross. F_2 and F_3 are the second and third generations resulting from such a cross.

facilitated diffusion: Passive transport with the assistance of carrier proteins.

family: A taxonomic group between order and genus in rank; the ending of family names in animals and heterotrophic protists is *-idae*; in all other organisms it is *-aceae*. A family contains one or more genera, and each family belongs to an order.

fascicle (făs′ĭ•kŭl) [L. *fasciculus*, a small bundle]: A bundle of pine leaves or other needlelike leaves of gymnosperms; an obsolete term for a vascular bundle.

fascicular cambium: The vascular cambium originating within a vascular bundle, or fascicle.

fat: A molecule composed of glycerol and three fatty acid molecules; the proportion of oxygen to carbon is much less in fats than it is in carbohydrates. Fats in the liquid state are called oils.

feedback inhibition: Control mechanism whereby an increase in the concentration of some molecule inhibits the further synthesis of that molecule.

fermentation: The extraction of energy from organic compounds without the involvement of oxygen.

ferredoxin: Electron-transferring proteins of high iron content; some are involved in photosynthesis.

fertilization: The fusion of two gamete nuclei to form a diploid zygote.

fiber: An elongated, tapering, generally thick-walled sclerenchyma cell of vascular plants; its walls may or may not be lignified; it may or may not have a living protoplast at maturity.

fibril: Submicroscopic threads composed of cellulose molecules, which constitute the form in which cellulose occurs in the cell wall.

field capacity: The percentage of water a particular soil will hold against the action of gravity; also called field moisture capacity.

filament: (1) The stalk of a stamen; (2) a term used to describe the threadlike bodies or segments of certain algae or fungi.

fission: Asexual reproduction involving the division of a single-celled individual into two new single-celled individuals of equal size. Also applied to the division of plastids.

fitness: The genetic contribution of an organism to future generations, relative to the contributions of organisms living in the same environment that have different genotypes.

flagellum, *pl.* **flagella** [L. *flagellum*, whip]: A long threadlike organelle that protrudes from the surface of a cell. The flagella of bacteria are capable of rotary motion and consist of a single protein fiber each; eukaryotic flagella, which are used in locomotion and feeding, consist of an array of microtubules with a characteristic internal 9 + 2 microtubule structure and are capable of a vibratory, but not rotary, motion. A cilium is a small eukaryotic flagellum.

flavonoids: Phenolic compounds; water-soluble pigments present in the vacuoles of plant cells; those found in red wines and grape juice have been reported to lower cholesterol levels in the blood.

flavoprotein: A dehydrogenase that contains a flavin and often a metal and plays a major role in oxidation; abbreviated FP.

floral tube: A cup or tube formed by the fusion of the basal parts of the sepals, petals, and stamens; floral tubes are often found in plants that have a superior ovary.

floret: One of the small flowers that make up the composite inflorescence or the spike of the grasses.

florigen [L. *flor-*, flower, + Gk. *-genes*, producer]: A hypothetical plant hormone that promotes flowering.

flower: The reproductive structure of angiosperms; a complete flower includes calyx, corolla, androecium (stamens), and gynoecium (carpels), but all flowers contain at least one stamen or one carpel.

fluid-mosaic model: Model of membrane structure, with the membrane composed of a lipid bilayer in which globular proteins are embedded.

follicle [L. *folliculus*, small ball]: A dry, dehiscent simple fruit derived from a single carpel and opening along one side.

food chain, food web: A chain of organisms existing in any natural community such that each link in the chain feeds on the one below and is eaten by the one above; there are seldom more than six links in a chain, with autotrophs on the bottom and the largest carnivores at the top.

fossil [L. *fossils*, dug up]: The remains, impressions, or traces of an organism that has been preserved in rocks found in the Earth's crust.

fossil fuels: The altered remains of once-living organisms that are burned to release energy; oil, gas, and coal.

founder cells: The group of cells in the peripheral zone of the apical meristem involved with the initiation of a leaf primordium.

founder effect: Type of genetic drift that occurs as the result of the founding of a population by a small number of individuals.

FP: *See* flavoprotein.

free energy: Energy available to do work.

frond: The leaf of a fern; any large, divided leaf.

fruit: In angiosperms, a mature, ripened ovary (or group of ovaries), containing the seeds, together with any adjacent parts that may be fused with it at maturity; sometimes applied informally, and misleadingly, as in "fruiting body," to the reproductive structures of other kinds of organisms.

fucoxanthin (fū′kŏ•zăn′thĭn) [Gk. *phykos*, seaweed, + *xanthos*, yellowish-brown]: A brownish carotenoid found in brown algae and chrysophytes.

fundamental tissue system: *See* ground tissue system.

funiculus [L. *funiculus*, small rope or cord]: The stalk of the ovule.

fusiform initials [L. *fusus*, spindle]: The vertically elongated cells in the vascular cambium that give rise to the cells of the axial system in the secondary xylem and secondary phloem.

G

gametangium, *pl.* **gametangia** [Gk. *gamein,* to marry, + L. *tangere,* to touch]: A cell or multicellular structure in which gametes are formed.

gamete [Gk. *gamete,* wife]: A haploid reproductive cell; gametes fuse in pairs, forming zygotes, which are diploid.

gametic meiosis: Meiosis resulting in the formation of haploid gametes from a diploid individual; the gametes fuse to form a diploid zygote that divides to form another diploid individual.

gametophore [Gk. *gamein,* to marry, + *phoros,* bearing]: In the bryophytes, a fertile stalk that bears gametangia.

gametophyte: In plants, which have an alternation of generations, the haploid (*n*), gamete-producing generation, or phase.

gel: A mixture of substances having a semisolid or solid constitution.

gemma, *pl.* **gemmae** (jĕm′ă) [L. *gemma,* bud]: A small mass of vegetative tissue; an outgrowth of the thallus, for example, in liverworts or certain fungi; it can develop into an entire new plant.

gene: A unit of heredity; a sequence of DNA nucleotides that codes for a protein, tRNA, or rRNA molecule, or regulates the transcription of such a sequence.

gene flow: The movement of alleles into and out of a population.

gene frequency: The relative occurrence of a particular allele in a population.

gene pool: All the alleles of all the genes of all the individuals in a population.

generative cell: (1) In many gymnosperms, the cell of the male gametophyte that divides to form the sterile and spermatogenous cells; (2) in angiosperms, the cell of the male gametophyte that divides to form two sperm.

genetic code: The system of nucleotide triplets (codons) in DNA and RNA that dictates the amino acid sequence in proteins; except for three "stop" signals, each codon specifies one of 20 amino acids.

genetic drift: Evolution (change in allele frequencies) owing to chance processes.

genetic engineering: The manipulation of genetic material for practical purposes; also referred to as recombinant DNA technology.

genetic recombination: The occurrence of gene combinations in the progeny that are different from the combinations present in the parents.

genome: The totality of genetic information contained in the nucleus, plastid, or mitochondrion.

genomic library: A library encompassing an entire genome either in the nucleus or in the nucleoid of an organelle (mitochondrion, plastid) in eukaryotes, or in the nucleoid of a prokaryote.

genotype: The genetic constitution, latent or expressed, of an organism, as contrasted with the phenotype; the sum total of all the genes present in an individual.

genus, *pl.* **genera:** The taxonomic group between family and species in rank; genera include one or more species.

geotropism: *See* gravitropism.

germination [L. *germinare,* to sprout]: The beginning or resumption of growth by a spore, seed, bud, or other structure.

gibberellins (jĭb•ĕ•rĕ′lĭns) [*Gibberella,* a genus of fungi]: A class of plant hormones, the best known effect of which is to increase the elongation of plant stems.

gill: The strips of tissue on the underside of the cap in *Basidiomycetes.*

girdling: The removal from a woody stem of a ring of bark extending inward to the cambium; also called ringing.

glucose: A common six-carbon sugar ($C_6H_{12}O_6$); the most common monosaccharide in most organisms.

glycerol: A three-carbon molecule with three hydroxyl groups attached; glycerol molecules combine with fatty acids to form fats or oils.

glycogen [Gk. *glykys,* sweet, + *gen,* of a kind]: A carbohydrate similar to starch that serves as the reserve food in bacteria, fungi, and most organisms other than plants.

glycolysis: The anaerobic breakdown of glucose to form two molecules of pyruvate, resulting in the net liberation of two molecules of ATP; catalyzed by enzymes in the cytosol.

glyoxylate cycle: A variant of the Krebs cycle; present in bacteria and some plant cells, for net conversion of acetate into succinate and, eventually, new carbohydrate.

glyoxysome: A peroxisome containing enzymes necessary for the conversion of fats into carbohydrates; glyoxysomes play an important role during the germination of seeds.

Golgi body (gôl′jē): In eukaryotes, a group of flat, disk-shaped sacs that are often branched into tubules at their margins; serve as collecting and packaging centers for the cell and concerned with secretory activities; also called dictyosomes. The term "Golgi complex" is used to refer collectively to all the Golgi bodies, or dictyosomes, of a given cell.

grafting: A union of different individuals in which a portion, called the scion, of one individual is inserted into a root or stem, called the stock, of the other individual.

grain: *See* caryopsis; also a term used to refer to the alignment of wood elements.

grana, *sing.* **granum:** Structures within chloroplasts, seen as green granules with a light microscope and as a series of stacked thylakoids with an electron microscope; the grana contain the chlorophylls and carotenoids and are the sites of the light reactions of photosynthesis.

gravitropism [L. *gravis,* heavy, + Gk. *tropes,* turning]: The response of a shoot or root to the pull of the Earth's gravity; also called geotropism.

ground meristem [Gk. *meristos,* divisible]: The primary meristem, or meristematic tissue, that gives rise to the ground tissues.

ground tissue: Tissues other than the vascular tissues, the epidermis, and the periderm; also called fundamental tissue.

ground tissue system: All tissues other than the epidermis (or periderm) and the vascular tissues; also called fundamental tissue system.

growth layer: A layer of growth in the secondary xylem or secondary phloem; *see also* annual ring.

growth ring: A growth layer in the secondary xylem or secondary phloem, as seen in transverse section; may be called a growth increment, especially where seen in other than transverse section.

guanine [Sp. from Quechua, *huanu,* dung]: A purine base found in DNA and RNA. Its name is based on the fact that guanine is abundant in the form of a white crystalline base in guano and other kinds of animal excrement.

guard cells: Pairs of specialized epidermal cells surrounding a pore, or stoma; changes in the turgor of a pair of guard cells cause opening and closing of the pore.

guttation [L. *gutta,* a drop]: The exudation of liquid water from leaves; caused by root pressure.

gymnosperm [Gk. *gymnos,* naked, + *sperma,* seed]: A seed plant with seeds not enclosed in an ovary; the conifers are the most familiar group; not a monophyletic group.

gynoecium [Gk. *gyne,* woman, + *oikos,* house]: The aggregate of carpels in the flower of a seed plant.

H

habit [L. *habitus,* condition, character]: Characteristic form or appearance of an organism.

habitat [L. *habitare,* to inhabit]: The environment of an organism; the place where it is usually found.

hadrom: The central strand of water-conducting cells found in the axes of some moss gametophytes and sporophytes.

haploid [Gk. *haploos*, single]: Having only one set of chromosomes (*n*), in contrast to diploid (2*n*).

hardwood: A name commonly applied to the wood of a magnoliid or eudicot tree.

Hardy–Weinberg law: The mathematical expression of the relationship between the relative frequencies of two or more alleles in a population. It demonstrates that the frequencies of alleles and genotypes will remain constant in a random-mating population in the absence of inbreeding, selection, or other evolutionary forces.

haustorium, *pl.* **haustoria** [L. *haustus*, from *haurire*, to drink, draw]: A projection of a fungal hypha that functions as a penetrating and absorbing organ; in parasitic angiosperms, a modified root capable of penetrating and absorbing materials from host tissues.

heartwood: Nonliving and commonly dark-colored wood in which no water transport occurs; it is surrounded by sapwood.

helical phyllotaxy: Leaf arrangement in which there is one leaf at a node, forming a helical pattern around the stem; also called alternate phyllotaxy.

heliotropism [Gk. *helios*, sun]: *See* solar tracking.

hemicellulose (hĕm′ē•sĕl′ū•lōs): A polysaccharide resembling cellulose but more soluble and less ordered; found particularly in cell walls.

herb [L. *herba*, grass]: A nonwoody seed plant with a relatively short-lived aerial portion.

herbaceous: An adjective referring to nonwoody plants.

herbarium: A collection of dried and pressed plant specimens.

herbivorous: Feeding upon plants.

heredity [L. *heredis*, heir]: The transmission of characteristics from parent to offspring through the gametes.

hermaphrodite [Gk. for Hermes and Aphrodite]: An organism possessing both male and female reproductive organs.

hetero- [Gk. *heteros*, different]: Prefix meaning "other" or "different."

heterocyst [Gk. *heteros*, different, + *cystis*, a bag]: A transparent, thick-walled, nitrogen-fixing cell that forms in the filaments of certain cyanobacteria.

heteroecious (hĕt′er•ē′shŭs) [Gk. *heteros*, different, + *oikos*, house]: As in some rust fungi, requiring two different host species to complete the life cycle.

heterogamy [Gk. *heteros*, other, + *gamos*, union or reproduction]: Reproduction involving two types of gametes.

heterokaryotic [Gk. *heteros*, other, + *karyon*, kernel]: In fungi, having two or more genetically distinct types of nuclei within the same mycelium.

heterokonts: Organisms with one long, ornamented (tinsel) flagellum and one shorter and smooth (whiplash) flagellum; includes oomycetes, chrysophytes, diatoms, brown algae, and certain other groups.

heteromorphic [Gk. *heteros*, different, + *morphē*, form]: A term used to describe a life history in which the haploid and diploid generations are dissimilar in form.

heterosis [Gk. *heterosis*, alteration]: Hybrid vigor, the superiority of the hybrid over either parent in any measurable character.

heterosporous: Having two kinds of spores, designated as microspores and megaspores.

heterothallic [Gk. *heteros*, different, + *thallus*, sprout]: A term used to describe a species, the haploid individuals of which are self-sterile or self-incompatible; two compatible strains or individuals are required for sexual reproduction to take place.

heterotroph [Gk. *heteros*, other, + *trophos*, feeder]: An organism that cannot manufacture organic compounds and so must feed on organic materials that have originated in other plants and animals; *see also* autotroph.

heterozygous: Having two different alleles at the same locus on homologous chromosomes.

Hill reaction: The oxygen evolution and photoreduction of an artificial electron acceptor by a chloroplast preparation in the absence of carbon dioxide.

hilum [L. *hilum*, a trifle]: (1) Scar left on seed after separation of seed from funiculus; (2) the part of a starch grain around which the starch is laid down in more or less concentric layers.

histone: The group of five basic proteins associated with the chromosomes of all eukaryotic cells.

holdfast: (1) Basal part of a multicellular alga that attaches it to a solid object; may be unicellular or composed of a mass of tissue; (2) cuplike structure at the tips of some tendrils, by means of which they become attached.

homeo-, homo- [Gk. *homos*, same, similar]: Prefix meaning "similar" or "same."

homeostasis (hŏ′mē•ō•stā′sĭs) [Gk. *homos*, similar, + *stasis*, standing]: The maintaining of a relatively stable internal physiological environment within an organism, or a steady-state equilibrium in a population or ecosystem. Homeostasis usually involves feedback mechanisms.

homeotic mutation: A mutation that changes organ identity so that the wrong structures appear in the wrong place or at the wrong time.

homokaryotic [Gk. *homos*, same, + *karyon*, kernel]: In fungi, having nuclei with the same genetic makeup within a mycelium.

homologous chromosomes: Chromosomes that associate in pairs in the first stage of meiosis; each member of the pair is derived from a different parent. Homologous chromosomes are also called homologs.

homology [Gk. *homologia*, agreement]: A condition indicative of the same phylogenetic, or evolutionary, origin, but not necessarily the same in present structure and/or function.

homosporous: Having only one kind of spore.

homothallic [Gk. *homos*, same, + *thallus*, sprout]: A term used to describe a species in which the individuals are self-fertile.

homozygous: Having identical alleles at the same locus on homologous chromosomes.

hormogonium, *pl.* **hormogonia:** A portion of a filament of a cyanobacterium that becomes detached and grows into a new filament.

hormone [Gk. *hormaein*, to excite]: An organic substance produced usually in minute amounts in one part of an organism, from which it is transported to another part of that organism on which it has a specific effect; hormones function as highly specific chemical signals between cells.

host: An organism on or in which a parasite lives.

humus: Decomposing organic matter in the soil.

hybrid: Offspring of two parents that differ in one or more heritable characteristics; offspring of two different varieties or of two different species.

hybridization: The formation of offspring between unlike parents.

hybrid vigor: *See* heterosis.

hydrocarbon [Gk. *hydro*, water, + L. *carbo*, charcoal]: An organic compound that consists only of hydrogen and carbon atoms.

hydrogen bond: A weak bond between a hydrogen atom attached to one oxygen or nitrogen atom and another oxygen or nitrogen atom.

hydroids: The water-conducting cells of the moss hadrom; they resemble the tracheary elements of vascular plants except that they lack specialized wall thickenings.

hydrolysis [Gk. *hydro*, water, + *lysis*, loosening]: Splitting of one molecule into two by addition of the H^+ and OH^- ions of water.

hydrophyte [Gk. *hydro*, water, + *phyton*, a plant]: A plant that depends on an abundant supply of moisture or that grows wholly or partly submerged in water.

hydrotropism: The growth of roots toward regions of greater water potential, or in response to a moisture gradient.

hydroxyl group: An OH⁻ group; a negatively charged ion formed by the dissociation of a water molecule.

hymenium [Gk. *hymen,* a membrane]: The layer of asci on an ascoma, or of basidia on a basidioma, together with any associated sterile hyphae.

hyper- [Gk. *hyper,* above, over]: Prefix meaning "above" or "over."

hypertonic: Refers to a solution that has a concentration of solute particles high enough to gain water across a selectively permeable membrane from another solution.

hypha, *pl.* **hyphae** [Gk. *hyphe,* web]: A single tubular filament of a fungus, oomycete, or chytrid; the hyphae together comprise the mycelium.

hypo- [Gk. *hypo,* less than]: Prefix meaning "under" or "less."

hypocotyl: The portion of an embryo or seedling situated between the cotyledons and the radicle.

hypocotyl-root axis: The embryo axis below the cotyledon or cotyledons, consisting of the hypocotyl and the apical meristem of the root or the radicle.

hypodermis [Gk. *hypo,* under, + *derma,* skin]: One or more layers of cells beneath the epidermis, which are distinct from the underlying cortical or mesophyll cells.

hypogeous [Gk. *hypo,* under, + *ge,* the Earth]: Type of seed germination in which the cotyledons remain underground.

hypogyny [Gk. *hypo,* under, + *gyne,* woman]: Floral organization in which the sepals, petals, and stamens are attached to the receptacle below the ovary; *see also* epigyny.

hypothesis [Gk. *hypo,* under, + *tithenai,* to put]: A temporary working explanation or supposition based on accumulated facts and suggesting some general principle or relation of cause and effect; a postulated solution to a scientific problem that must be tested by experimentation and, if disproved or shown to be unlikely, is discarded.

hypotonic: Refers to a solution that has a concentration of solutes low enough to lose water across a selectively permeable membrane to another solution.

I

IAA: *See* indoleacetic acid.

imbibition (ĭm·bĭ·bĭsh'ŭn): Adsorption of water and swelling of colloidal materials because of the adsorption of water molecules onto the internal surfaces of the materials.

imperfect flower: A flower lacking either stamens or carpels.

imperfect fungi: The deuteromycetes, or conidial fungi, which reproduce only asexually, or in which the sexual cycle has not been observed; most deuteromycetes are ascomycetes.

inbreeding: The breeding of closely related plants or animals; in plants, it is usually brought about by repeated self-pollination.

incomplete flower: A flower lacking one or more of the four kinds of floral parts, that is, lacking sepals, petals, stamens, or carpels.

indehiscent (ĭn'de·hĭs'ĕnt): Remaining closed at maturity, as are many fruits (samaras, for example).

independent assortment: *See* Mendel's second law.

indeterminate growth: Unrestricted or unlimited growth, as with a vegetative apical meristem that produces an unrestricted number of lateral organs indefinitely.

indoleacetic acid (IAA): Naturally occurring auxin, a kind of plant hormone.

indusium, *pl.* **indusia** (ĭn·dū'zĭ·ŭm) [L. *indusium,* a woman's undergarment]: Membranous growth of the epidermis of a fern leaf that covers a sorus.

inferior ovary: An ovary that is completely or partially attached to the calyx; the other floral whorls appear to arise from the ovary's top.

inflorescence: A flower cluster, with a definite arrangement of flowers.

initial: (1) In the meristem, a cell that remains within the meristem indefinitely and at the same time, by division, adds cells to the plant body; (2) a meristematic cell that eventually differentiates into a mature, more specialized cell or element.

inner bark: In older trees, the living part of the bark; the bark inside the innermost periderm.

integral proteins: Transmembrane proteins and other proteins that are tightly bound to the membrane.

integument: The outermost layer or layers of tissue enveloping the nucellus of an ovule; develops into the seed coat.

inter- [L. *inter,* between]: Prefix meaning "between," or "in the midst of."

intercalary [L. *intercalare,* to insert]: Descriptive of meristematic tissue or growth not restricted to the apex of an organ; that is, growth in the regions of the nodes.

interfascicular cambium: The vascular cambium arising between the fascicles, or vascular bundles, from interfascicular parenchyma.

interfascicular region: Tissue region between vascular bundles in a stem; also called pith ray.

internode: The region of a stem between two successive nodes.

interphase: The period between two mitotic or meiotic cycles; the cell grows and its DNA replicates during interphase.

intine: The inner wall layer of a spore or pollen grain.

intra- [L. *intra,* within]: Prefix meaning "within."

intron [L. *intra,* within]: A portion of mRNA as transcribed from eukaryotic DNA that is removed by enzymes before the mRNA is translated into protein. *See also* exon.

ion: An atom or molecule that has lost or gained one or more electrons, and thus is positively or negatively charged.

irregular flower: A flower in which one or more members of at least one whorl differ in form from other members of the same whorl.

iso- [Gk. *isos,* equal]: Prefix meaning "similar," "alike." Is- before a vowel.

isogamy: A type of sexual reproduction in which the gametes (or gametangia) are alike in size; found in some algae and fungi.

isomer [Gk. *isos,* equal, + *meros,* part]: One of a group of compounds identical in atomic composition but differing in structural arrangement; for example, glucose and fructose.

isomorphic [Gk. *isos,* equal, + *morphē,* form]: Identical in form.

isotonic: Having the same osmotic concentration.

isotope: One of several possible forms of a chemical element that differ from other forms in the number of neutrons in the atomic nucleus, but not in chemical properties.

K

karyogamy [Gk. *karyon,* kernel, + *gamos,* marriage]: The union of two nuclei following fertilization, or plasmogamy.

kelp: A common name for any of the larger members of the order *Laminariales* of the brown algae.

kinetin [Gk. *kinetikos,* causing motion]: A purine that probably does not occur in nature but that acts as a cytokinin in plants.

kinetochore [Gk. *kinetikos,* putting in motion, + *chorus,* chorus]: Specialized protein complexes, which develop on each centromere, and to which spindle fibers are attached during mitosis or meiosis.

kingdom: One of the seven chief taxonomic categories; for example, *Fungi* or *Plantae.*

Kranz anatomy [Ger. *Kranz,* wreath]: The wreathlike arrangement of mesophyll cells around a layer of large bundle-sheath cells, forming two concentric layers around the vascular bundle; typically found in the leaves of C_4 plants.

Krebs cycle: The series of reactions that results in the oxidation of pyruvate to hydrogen atoms, electrons, and carbon dioxide. The electrons, passed along electron-carrier molecules, then go through the oxidative phosphorylation and terminal oxidation processes. Also called the tricarboxylic acid cycle or TCA cycle.

L

L₁, L₂, L₃ layers: The outer cell layers of angiosperm apical meristems with a tunica-corpus organization.

lamella (la•měl′a) [L. *lamella,* thin metal plate]: Layer of cellular membranes, particularly photosynthetic, chlorophyll-containing membranes; *see also* middle lamella.

lamina: The blade of a leaf.

laminarin: One of the principal storage products of the brown algae; a polymer of glucose.

lateral meristems: Meristems that give rise to secondary tissue; the vascular cambium and cork cambium.

lateral root: A root that arises from another, older root; also called a branch root, or secondary root, if the older root is the primary root.

late wood: The last part of the growth increment formed in the growing season; it contains smaller cells and is denser than the early wood; replaces the term "summer wood."

leaching: The downward movement and drainage of minerals, or inorganic ions, from the soil by percolating water.

leaf: The principal lateral appendage of the stem; highly variable in both structure and function; the foliage leaf is specialized as a photosynthetic organ.

leaf buttress: A lateral protrusion below the apical meristem; represents the initial stage in the development of a leaf primordium.

leaf gap: Region of parenchyma tissue in the primary vascular cylinder above the point of departure of the leaf trace or traces in ferns.

leaflet: One of the parts of a compound leaf.

leaf primordium [L. *primordium,* beginning]: A lateral outgrowth from the apical meristem that will eventually become a leaf.

leaf scar: A scar left on a twig when a leaf falls.

leaf trace: That part of a vascular bundle extending from the base of the leaf to its connection with a vascular bundle in the stem.

leaf trace gap: Region of parenchyma tissue in the primary vascular cylinder of a stem above the point of departure of the leaf trace or traces in seed plants.

legume [L. *legumem,* leguminous plant]: (1) A member of the *Fabaceae,* the pea or bean family; (2) a type of dry simple fruit that is derived from one carpel and opens along both sides.

lenticels (lĕn′tĭ•sĕls) [L. *lenticella,* a small window]: Spongy areas in the cork surfaces of stem, roots, and other plant parts that allow interchange of gases between internal tissues and the atmosphere through the periderm; occur in vascular plants.

leptoids: The food-conducting cells associated with the hydroids of some moss gametophytes and sporophytes; they resemble the sieve elements of some seedless vascular plants.

leptom: The food-conducting tissue consisting of leptoids; surrounds the hadrom in the axes of some moss gametophytes and sporophytes.

leptosporangium: A sporangium that arises from a single initial cell and whose wall is composed of a single layer of cells.

leucoplast (lū′kō•plăst) [Gk. *leuko,* white, + *plasein,* to form]: A colorless plastid; leucoplasts are commonly centers of starch formation.

liana [F. *liane,* from *lier,* to bind]: A large, woody vine that climbs on other plants.

life cycle: The entire sequence of phases in the growth and development of any organism from time of zygote formation to gamete formation.

ligase: An enzyme that joins together (ligates) two molecules in an energy-dependent process; DNA ligase, for example, is essential for DNA replication, catalyzing the covalent bonding of the 3′ end of a new DNA fragment to the 5′ end of a growing chain.

light reactions: The reactions of photosynthesis that require light and cannot occur in the dark; also called light-dependent reactions and energy-transduction reactions.

lignin: One of the most important constituents of the secondary wall of vascular plants, although not all secondary walls contain lignin; after cellulose, lignin is the most abundant plant polymer.

ligule [L. *ligula,* small tongue]: A minute outgrowth or appendage at the base of the leaves of grasses and those of certain lycophytes.

linkage: The tendency for certain genes to be inherited together owing to the fact that they are located on the same chromosome.

lipid [Gk. *lipos,* fat]: One of a large variety of nonpolar organic molecules that are insoluble in water (which is polar) but dissolve readily in nonpolar organic solvents; lipids include fats, oils, steroids, phospholipids, and carotenoids.

loam soils: Soils containing sand, silt, and clay in proportions that result in ideal agricultural soils.

locule (lŏk′ūl) [L. *loculus,* small chamber]: A cavity within a sporangium or a cavity of the ovary in which ovules occur.

locus, *pl.* **loci:** The position on a chromosome occupied by a particular gene.

long-day plants: Plants that must be exposed to light periods longer than some critical length for flowering to occur; they flower in spring or summer.

lumen [L. *lumen,* light, an opening for light]: The space bounded by the plant cell wall; the thylakoid space in chloroplasts; the narrow, transparent space of endoplasmic reticulum.

lysis [Gk. *lysis,* a loosening]: A process of disintegration or cell destruction.

lysogenic bacteria: Bacteria-carrying viruses (phages) that eventually break loose from the bacterial chromosome and set up an active cycle of infection, producing lysis in their bacterial hosts.

lysosome [Gk. *lysis,* loosening, + *soma,* body]: An organelle, bounded by a single membrane, and containing hydrolytic enzymes that are released when the organelle ruptures and are capable of breaking down proteins and other complex macromolecules.

M

macrocyst: A flattened, irregular structure, encircled by a thin membrane, in which zygotes are formed during the life cycle of cellular slime molds.

macroevolution: Evolutionary change on a grand scale, involving major evolutionary trends.

macrofibril: An aggregation of microfibrils, visible with the light microscope.

macromolecule [Gk. *makros,* large]: A molecule of very high molecular weight; refers specifically to proteins, nucleic acids, polysaccharides, and complexes of these.

macronutrients [Gk. *makros,* large, + L. *nutrire,* to nourish]: Inorganic chemical elements required in large amounts for plant growth, such as nitrogen, potassium, calcium, phosphorus, magnesium, and sulfur.

magnoliids: The 3 percent of living angiosperms with the most primitive features; the ancestors of both monocots and eudicots.

major veins: The larger leaf vascular bundles, which are associated with ribs; they are largely involved with the transport of substances into and out of the leaf.

maltase: An enzyme that hydrolyzes maltose to glucose.

mannitol: One of the storage molecules of the brown algae; an alcohol.

mating type: A particular genetically defined strain of an organism that is incapable of sexual reproduction with another member of the same strain but capable of such reproduction with members of other strains of the same organism.

matrotrophy: Pertaining to a form of nutrition provided by the maternal gametophyte as, for example, in the case of a moss gametophyte providing nutrients to the zygote and developing sporophyte.

mega- [Gk. *megas,* large]: Prefix meaning "large."

megagametophyte [Gk. *megas,* large, + *gamos,* marriage, + *phyton,* plant]: In heterosporous plants, the female gametophyte; located within the ovule of seed plants.

megaphyll [Gk. *megas,* large, + *phyllon,* leaf]: A generally large leaf with several to many veins; its leaf trace or traces is/are associated with a leaf gap (in ferns) and a leaf trace gap (in seed plants); in contrast to microphyll.

megasporangium, *pl.* **megasporangia:** A sporangium in which megaspores are produced; *see also* nucellus.

megaspore: In heterosporous plants, a haploid (*n*) spore that develops into a female gametophyte; in most groups, megaspores are larger than microspores.

megaspore mother cell: A diploid cell in which meiosis will occur, resulting in the production of four megaspores; also called a megasporocyte.

megasporocyte: *See* megaspore mother cell.

megasporophyll: A leaf or leaflike structure bearing a megasporangium.

meiosis (mī•ō′sĭs) [Gk. *meioun,* to make smaller]: The two successive nuclear divisions in which the chromosome number is reduced from diploid (2*n*) to haploid (*n*) and segregation of the genes occurs; gametes or spores (in organisms with an alternation of generations) may be produced as a result of meiosis.

meiospores: Spores that arise through meiosis and are therefore haploid.

membrane potential: The voltage difference across a membrane due to the differential distribution of ions.

Mendel's first law: The factors for a pair of alternate characteristics are separate, and only one may be carried in a particular gamete (genetic segregation).

Mendel's second law: The inheritance of one pair of characteristics is independent of the simultaneous inheritance of other traits, such characteristics assorting independently as though there were no others present (later modified by the discovery of linkage).

meristem [Gk. *merizein,* to divide]: The undifferentiated, perpetually young plant tissue from which new cells arise.

meso- [Gk. *mesos,* middle]: Prefix meaning "middle."

mesocarp [Gk. *mesos,* middle, + *karpos,* fruit]: The middle layer of the mature ovary wall, or pericarp, between the exocarp and endocarp.

mesocotyl: The internode between the scutellar node and the coleoptile in the embryo and seedling of grasses (*Poaceae*).

mesophyll: The ground tissue (parenchyma) of a leaf, located between the layers of epidermis; mesophyll cells generally contain chloroplasts.

mesophyte [Gk. *mesos,* middle, + *phyton,* a plant]: A plant that requires an environment that is neither too wet nor too dry.

messenger RNA (mRNA): The class of RNA that carries genetic information from the gene to the ribosomes, where it is translated into protein.

metabolism [Gk. *metabolē,* change]: The sum of all chemical processes occurring within a living cell or organism.

metaphase: The stage of mitosis or meiosis during which the chromosomes lie in the equatorial plane of the spindle.

metaxylem [Gk. *meta,* after, + *xylon,* wood]: The part of the primary xylem that differentiates after the protoxylem; the metaxylem reaches maturity after the portion of the plant part in which it is located has finished elongating.

micro- [Gk. *mikros,* small]: Prefix meaning "small."

microbody: *See* peroxisome.

microevolution: Evolutionary change within a population over a succession of generations.

microfibril: A threadlike component of the cell wall, composed of cellulose molecules, visible only with the electron microscope.

microfilament: *See* actin filament.

microgametophyte [Gk. *mikros,* small, + *gamos,* marriage, + *phyton,* plant]: In heterosporous plants, the male gametophyte.

micrometer: A unit of microscopic measurement convenient for describing cellular dimensions; 1/1000 of a millimeter; its symbol is μm.

micronutrients [Gk. *mikros,* small, + L. *nutrire,* to nourish]: Inorganic chemical elements required only in very small, or trace, amounts for plant growth, such as iron, chlorine, copper, manganese, zinc, molybdenum, nickel, and boron.

microphyll [Gk. *mikros,* small, + *phyllon,* leaf]: A small leaf with one vein and one leaf trace not associated with either a leaf gap or leaf trace gap; in contrast to megaphyll. Microphylls are characteristic of the lycophytes.

micropyle: In the ovules of seed plants, the opening in the integuments through which the pollen tube usually enters.

microsporangium: A sporangium within which microspores are formed.

microspore: In heterosporous plants, a spore that develops into a male gametophyte.

microspore mother cell: A cell in which meiosis will occur, resulting in four microspores; in seed plants, often called a pollen mother cell; also called a microsporocyte.

microsporocyte: *See* microspore mother cell.

microsporophyll: A leaflike organ bearing one or more microsporangia.

microtubule [Gk. *mikros,* small, + L. *tubulus,* little pipe]: Narrow (about 25 nanometers in diameter), elongate, nonmembranous tubule of indefinite length; microtubules occur in the cells of eukaryotes. Microtubules move the chromosomes in cell division and provide the internal structure of cilia and flagella.

middle lamella: The layer of intercellular material, rich in pectic compounds, cementing together the primary walls of adjacent cells.

mimicry [Gk. *mimos,* mime]: The superficial resemblance in form, color, or behavior of certain organisms (mimics) to other more powerful or more protected ones (models), resulting in protection, concealment, or some other advantage for the mimic.

mineral: A naturally occurring chemical element or inorganic compound.

minor veins: The small leaf vascular bundles, which are located in the mesophyll and enclosed by a bundle sheath; they are involved with distribution of the transpiration stream and the uptake of the products of photosynthesis.

mitochondrion, *pl.* **mitochondria** [Gk. *mitos,* thread, + *chondrion,* small grain]: A double-membrane-bounded organelle found in eukaryotic cells; contains the enzymes of the Krebs cycle and the electron-transport chain; the major source of ATP in nonphotosynthetic cells.

mitosis (mī•tō′sĭs) [Gk. *mitos,* thread]: A process during which the duplicated chromosomes divide longitudinally and the daughter chromosomes then separate to form two genetically identical daughter nuclei; usually accompanied by cytokinesis.

mitotic spindle: The array of microtubules that forms between the opposite poles of a eukaryotic cell during mitosis.

mole: The name for a gram molecule. The number of particles in 1 mole of any substance; always equal to Avogadro's number: 6.022×10^{23}.

molecular weight: The relative weight of a molecule when the weight of the most frequent kind of carbon atom is taken as 12; the sum of the relative weights of the atoms in a molecule.

molecule: Smallest possible unit of a compound, consisting of two or more atoms.

mono- [Gk. *monos*, single]: Prefix meaning "one" or "single."

monocotyledon: A plant whose embryo has one cotyledon; one of the two great classes of angiosperms, *Monocotyledones;* often abbreviated as monocot.

monoecious (mō·nē'shŭs) [Gk. *monos*, single, + *oikos*, house]: Having the anthers and carpels produced in separate flowers on the same individual.

monokaryotic [Gk. *monos*, one, + *karyon*, kernel]: In fungi, having a single haploid nucleus within one cell or compartment.

monomers [Gk. *monos*, single, + *meros*, part]: Small, repeating units that can be linked together to form polymers.

monophyletic: Pertaining to a taxon descended from a single ancestor.

monosaccharide [Gk. *monos*, single, + *sakcaron*, sugar]: A simple sugar, such as five-carbon and six-carbon sugars, that cannot be dissociated into smaller sugar particles.

-morph, morph- [Gk. *morphē*, form]: Suffix or prefix meaning "form."

morphogenesis: The development of form.

morphology [Gk. *morphē*, form, + *logos*, discourse]: The study of form and its development.

mRNA: *See* messenger RNA.

mucigel: A slime sheath covering the surface of many roots.

multigene family: A collection of related genes on a chromosome; most eukaryotic genes appear to be members of multigene families.

multiple epidermis: A tissue composed of several cell layers derived from the protoderm; only the outer layer assumes characteristics of a typical epidermis.

multiple fruit: A cluster of mature ovaries produced by a cluster of flowers, as in the pineapple.

mutagen [L. *mutare*, to change, + Gk. *genaio*, to produce]: An agent that increases the mutation rate.

mutant: A mutated gene or an organism carrying a gene that has undergone a mutation.

mutation: Any change in the hereditary state of an organism; such changes may occur at the level of the gene (gene mutations, or point mutations) or at the level of the chromosome (chromosome mutations).

mutualism: The living together of two or more organisms in an association that is mutually advantageous.

myc-, myco- [Gk. *mykes*, fungus]: Prefix meaning "pertaining to fungi."

mycelium [Gk. *mykes*, fungus]: The mass of hyphae forming the body of a fungus, oomycete, or chytrid.

mycology: The study of fungi.

mycorrhiza, *pl.* **mycorrhizae:** A symbiotic association between certain fungi and plant roots; characteristic of most vascular plants.

N

NAD+: *See* nicotinamide adenine dinucleotide.

NADP+: *See* nicotinamide adenine dinucleotide phosphate.

nanoplankton (năn'ō·plăngk'tŏn) [Gk. *nanos*, dwarf, + *planktos*, wandering]: Plankton with dimensions of less than 70 to 75 micrometers.

nastic movement: A plant movement that occurs in response to a stimulus, but whose direction is independent of the direction of the stimulus.

natural selection: The differential reproduction of genotypes based on their genetic constitution.

nectary [Gk. *nektar*, the drink of the gods]: In angiosperms, a gland that secretes nectar, a sugary fluid that attracts animals to plants.

netted venation: The arrangement of veins in the leaf blade that resembles a net; characteristic of the leaves of magnoliids and eudicots; also called reticulate venation.

neutron [L. *neuter*, neither]: An uncharged particle with a mass slightly greater than that of a proton, found in the atomic nucleus of all elements except hydrogen, in which the nucleus consists of a single proton.

niche: The role played by a particular species in its ecosystem.

nicotinamide adenine dinucleotide (NAD+): A coenzyme that functions as an electron acceptor in many of the oxidation reactions of respiration.

nicotinamide adenine dinucleotide phosphate (NADP+): A coenzyme that functions as an electron acceptor in many of the reduction reactions of biosynthesis; similar in structure to NAD+ except that it contains an extra phosphate group.

nitrification: The oxidation of ammonium ions or ammonia to nitrate; a process carried out by a specific free-living soil bacterium.

nitrogen fixation: The incorporation of atmospheric nitrogen into nitrogen compounds; carried out by certain free-living and symbiotic bacteria.

nitrogen-fixing bacteria: Soil bacteria that convert atmospheric nitrogen into nitrogen compounds.

nitrogenous base: A nitrogen-containing molecule having basic properties (tendency to acquire an H atom); a purine or pyrimidine; one of the building blocks of nucleic acids.

node [L. *nodus*, knot]: The part of a stem where one or more leaves are attached; *see also* internode.

nodules: Enlargements or swellings on the roots of legumes and certain other plants inhabited by symbiotic nitrogen-fixing bacteria.

noncyclic electron flow: The light-induced flow of electrons from water to NADP+ in oxygen-evolving photosynthesis; it involves both photosystems I and II.

nonseptate: *See* aseptate.

nucellus (noo·sĕl'ŭs) [L. *nucella*, a small nut]: Tissue composing the chief part of the young ovule, in which the embryo sac develops; equivalent to a megasporangium.

nuclear envelope: The double membrane surrounding the nucleus of a cell.

nucleic acid: An organic acid consisting of joined nucleotide complexes; the two types are deoxyribonucleic acid (DNA) and ribonucleic acid (RNA).

nucleoid: A region of DNA in prokaryotic cells, mitochondria, and chloroplasts.

nucleolar organizer region: A special area on a certain chromosome associated with the formation of the nucleolus.

nucleolus, *pl.* **nucleoli** (noo·klē'ō·lŭs) [L. *nucleolus*, a small nucleus]: A small, spherical body found in the nucleus of eukaryotic cells, which is composed chiefly of rRNA in the process of being transcribed from copies of rRNA genes; the site of production of ribosomal subunits.

nucleoplasm: The ground substance of a nucleus.

nucleosome: A complex of DNA and histone proteins that forms the fundamental packaging unit of eukaryotic DNA; its structure resembles a bead on a string.

nucleotide: A single unit of nucleic acid, composed of a phosphate, a five-carbon sugar (either ribose or deoxyribose), and a purine or a pyrimidine.

nucleus, *pl.* **nuclei:** (1) A specialized body within the eukaryotic cell bounded by a double membrane and containing the chromosomes; (2) the central part of an atom of a chemical element.

nut: A dry, indehiscent, hard, one-seeded simple fruit, generally produced from a gynoecium of more than one fused carpel.

O

obligate anaerobe: *See* strict anaerobe.

-oid [Gk. *oid*, like, resembling]: Suffix meaning "like" or "similar to."

Okazaki fragments [after R. Okazaki, Japanese geneticist]: In DNA replication, the discontinuous segments in which the 3′ to 5′ strand (the lagging strand) of the DNA double helix is synthesized; typically 1000 to 2000 nucleotides long in prokaryotes, and 100 to 200 nucleotides long in eukaryotes.

ontogeny [Gk. *on*, being, + *genesis*, origin]: The development, or life history, of all or part of an individual organism.

oo- [Gk. *oion*, egg]: Prefix meaning "egg."

oogamy: Sexual reproduction in which one of the gametes (the egg) is large and nonmotile, and the other gamete (the sperm) is smaller and motile.

oogonium (ō′o•gō′nē•ŭm): A unicellular female sex organ that contains one or several eggs.

oospore: The thick-walled zygote characteristic of the oomycetes.

open vascular bundle: A vascular bundle in which a vascular cambium develops.

operator: A segment of DNA that interacts with a repressor protein to regulate the transcription of the structural genes of an operon.

operculum (o•pûr′ku•lŭm) [L. *operculum*, lid]: In mosses, the lid of the sporangium.

operon [L. *opus, operis*, work]: In the bacterial chromosome, a segment of DNA consisting of a promoter, an operator, and a group of adjacent structural genes; the structural genes, which code for products related to a particular biochemical pathway, are transcribed onto a single mRNA molecule, and their transcription is regulated by a single repressor protein.

opposite phyllotaxy: Term applied to leaves occurring in pairs at a node.

order: A category of classification between the rank of class and family; classes contain one or more orders, and orders in turn are composed of one or more families.

organ: A structure composed of different tissues, such as root, stem, leaf, or flower parts.

organelle (or′găn•el) [Gk. *organella*, a small tool]: A specialized, membrane-bounded part of a cell.

organic: Pertaining to living organisms in general, to compounds formed by living organisms, and to the chemistry of compounds containing carbon.

organism: Any individual living creature, either unicellular or multicellular.

osmosis (ŏs•mō′sĭs) [Gk. *osmos*, impulse or thrust]: The diffusion of water, or any solvent, across a selectively permeable membrane; in the absence of other forces, the movement of water during osmosis will always be from a region of greater water potential to one of lesser water potential.

osmotic potential: The change in free energy or chemical potential of water produced by solutes; carries a negative (minus) sign; also called solute potential.

osmotic pressure: The potential pressure that can be developed by a solution separated from pure water by a selectively permeable membrane; in the absence of other forces, the movement of water during osmosis will always be from a region of greater water potential to one of lesser water potential.

osmotic potential: The change in free energy or chemical potential of water produced by solu permeable or semipermeable membrane; it is an index of the solute concentration of the solution.

outcrossing: Cross-pollination between individuals of the same species.

outer bark: In older trees, the dead part of the bark; the innermost periderm and all tissues outside it; also called rhytidome.

ovary [L. *ovum*, an egg]: The enlarged basal portion of a carpel or of a gynoecium composed of fused carpels; a mature ovary, sometimes with other adherent parts, is a fruit.

ovule (ō•vūl) [L. *ovulum*, a little egg]: A structure in seed plants containing the female gametophyte with egg cell, all being surrounded by the nucellus and one or two integuments; when mature, an ovule becomes a seed.

ovuliferous scale: In certain conifers, the appendage or scalelike shoot to which the ovule is attached.

oxidation: The loss of an electron by an atom or molecule. Oxidation and reduction (gain of an electron) take place simultaneously, because an electron that is lost by one atom is accepted by another. Oxidation-reduction reactions are an important means of energy transfer in living systems.

oxidative phosphorylation: The formation of ATP from ADP and inorganic phosphate; oxidative phosphorylation takes place in the electron-transport chain of the mitochondrion.

P

pairing of chromosomes: Side-by-side association of homologous chromosomes.

paleobotany [Gk. *palaios*, old]: The study of fossil plants.

paleoherbs: Apart from the woody magnoliids, collectively the remaining magnoliids; constitute a very diverse assemblage; include all angiosperms that are not woody magnoliids, eudicots, or monocots.

palisade parenchyma: A leaf tissue composed of columnar chloroplast-bearing parenchyma cells with their long axes at right angles to the leaf surface.

panicle (păn′ĭ•kûl) [L. *panicula*, tuft]: An inflorescence, the main axis of which is branched, and whose branches bear loose flower clusters.

para- [Gk. *para*, beside]: Prefix meaning "beside."

paradermal section [Gk. *para*, beside, + *derma*, skin]: Section cut parallel to the surface of a flat structure, such as a leaf.

parallel evolution: The development of similar structures having similar functions in two or more evolutionary lines as a result of the same kinds of selective pressures.

parallel venation: The pattern of venation in which the principal veins of the leaf are parallel or nearly so; characteristic of monocots.

paramylon: The storage molecule of euglenoids.

paraphyletic: Pertaining to a taxon that excludes species that share a common ancestor with species included in the taxon.

paraphysis, *pl.* **paraphyses** [Gk. *para*, beside, + *physis*, growth]: In certain fungi, a sterile filament growing among the reproductive cells in the fruiting body; also applied to the sterile filaments growing among the gametangia and sporangia of certain brown algae and among the antheridia and archegonia of mosses.

parasexual cycle: The fusion and segregation of heterokaryotic haploid nuclei in certain fungi to produce recombinant nuclei.

parasite: An organism that lives on or in an organism of a different species and derives nutrients from it; the association is beneficial to the parasite and harmful to the host.

parenchyma (pa•rĕng′kĭ•ma) [Gk. *para*, beside, + *en*, in, + *chein*, to pour]: A tissue composed of parenchyma cells.

parenchyma cell: Living, generally thin-walled cell of variable size and form; the most abundant kind of cell in plants.

parthenocarpy [Gk. *parthenos*, virgin, + *karpos*, fruit]: The development of fruit without fertilization; parthenocarpic fruits are usually seedless.

passage cell: Endodermal cell of root that retains a thin wall and Casparian strip when other associated endodermal cells develop thick secondary walls.

passive transport: Non-energy-requiring transport of a solute across a membrane down the concentration or electrochemical gradient by either simple diffusion or facilitated diffusion.

pathogen [Gk. *pathos*, suffering, + *genesis*, beginning]: An organism that causes a disease.

pathogenic: Disease-causing.

pathology: The study of plant or animal diseases, their effects on the organism, and their treatment.

PCR: See polymerase chain reaction.

pectin: A highly hydrophilic polysaccharide present in the intercellular layer and primary wall of plant cell walls; the basis of fruit jellies.

pedicel (ped'ĭ•sĕl): The stalk of an individual flower in an inflorescence.

peduncle (pe•dŭng'kûl): The stalk of an inflorescence or of a solitary flower.

pentose phosphate cycle: The pathway of oxidation of glucose 6-phosphate to yield pentose phosphates.

peptide: Two or more amino acids linked by peptide bonds.

peptide bond: The type of bond formed when two amino acid units are joined end to end by the removal of a molecule of water; the bonds always form between the carboxyl (−COOH) group of one amino acid and the amino (−NH₂) group of the next amino acid.

perennial [L. *per*, through, + *annuus*, a year]: A plant in which the vegetative structures live year after year.

perfect flower: A flower having both stamens and carpels; hermaphroditic flower.

perfect stage: That phase of the life history of a fungus that includes sexual fusion and the spores associated with such fusions.

perforation plate: Part of the wall of a vessel element that is perforated.

peri- [Gk. *peri*, around]: Prefix meaning "around" or "about."

perianth (pĕr'ĭ•anth) [Gk. *peri*, around, + *anthos*, flower]: (1) The petals and sepals taken together; (2) in leafy liverworts, a tubular sheath surrounding an archegonium, and later, the developing sporophyte.

pericarp [Gk. *peri*, around, + *karpos*, fruit]: The fruit wall, which develops from the mature ovary wall.

periclinal: Parallel to the surface.

pericycle [Gk. *peri*, around, + *kykos*, circle]: A tissue characteristic of roots that is bounded externally by the endodermis and internally by the phloem.

periderm [Gk. *peri*, around, + *derma*, skin]: Outer protective tissue that replaces epidermis when it is destroyed during secondary growth; includes cork, cork cambium, and phelloderm.

perigyny [Gk. *peri*, around, + *gyne*, female]: A form of floral organization in which the sepals, petals, and stamens are attached to the margin of a cup-shaped extension of the receptacle; superficially, the sepals, petals, and stamens appear to be attached to the ovary.

perisperm [Gk. *peri*, around, + *sperma*, seed]: Food-storing tissue derived from the nucellus that occurs in the seeds of some flowering plants.

peristome (pĕr'ĭ•stōm) [Gk. *peri*, around, + *stoma*, a mouth]: In mosses, a fringe of teeth around the opening of the sporangium.

perithecium: A spherical or flask-shaped ascoma.

permanent wilting percentage: The percentage of water remaining in a soil when a plant fails to recover from wilting even if placed in a humid chamber.

permeable [L. *permeare*, to pass through]: Usually applied to membranes through which liquid substances may diffuse.

peroxisome: A spherical, single membrane-bounded organelle, ranging in diameter from 0.5 to 1.5 micrometers; some peroxisomes are involved in photorespiration, and others (called glyoxysomes) with the conversion of fats to sugars during seed germination; also called microbodies.

petal: A flower part, usually conspicuously colored; one of the units of the corolla.

petiole: The stalk of a leaf.

pH: A symbol denoting the relative concentration of hydrogen ions in a solution; pH values run from 0 to 14, and the lower the value the more acidic a solution, that is, the more hydrogen ions it contains; pH 7 is neutral, less than 7 is acidic, and more than 7 is alkaline.

phage: See bacteriophage.

phagocytosis: See endocytosis.

phellem: See cork.

phelloderm (fĕl'o•durm) [Gk. *phellos*, cork, + *derma*, skin]: A tissue formed inwardly by the cork cambium, opposite the cork; inner part of the periderm.

phellogen (fĕl'o•jĕn): See cork cambium.

phenolics: A broad range of compounds, all of which have a hydroxyl group (−OH) attached to an aromatic ring (a ring of six carbons containing three double bonds); includes flavonoids, tannins, lignins, and salicylic acid.

phenotype: The physical appearance of an organism; the phenotype results from the interaction between the genetic constitution (genotype) of the organism and its environment.

phloem (flō'ĕm) [Gk. *phloos*, bark]: The food-conducting tissue of vascular plants, which is composed of sieve elements, various kinds of parenchyma cells, fibers, and sclereids.

phloem loading: The process by which substances (primarily sugars) are actively secreted into the sieve tubes.

phosphate: A compound formed from phosphoric acid by replacement of one or more hydrogen atoms.

phospholipids: A phosphorylated lipid; similar in structure to a fat, but with only two fatty acids attached to the glycerol backbone, with the third space occupied by a phosphorus-containing molecule; important components of cellular membranes.

phosphorylation (fŏs'fo•rĭl•a'shŭn) [Gk. *phosphoros*, bringing light]: A reaction in which phosphate is added to a compound; for example, the formation of ATP from ADP and inorganic phosphate.

photo-, -photic [Gk. *photos*, light]: Prefix or suffix meaning "light."

photobiont [Gk. *photos*, light, + *bios*, life]: The photosynthetic component of a lichen.

photolysis: The light-dependent oxidative splitting of water molecules that takes place in photosystem II of the light reactions of photosynthesis.

photon [Gk. *photos*, light]: The elementary particle of light.

photoperiodism: Response to duration and timing of day and night; a mechanism evolved by organisms for measuring seasonal time.

photophosphorylation [Gk. *photos*, light, + *phosphoros*, bringing light]: The formation of ATP in the chloroplast during photosynthesis.

photorespiration: The oxygenase activity of Rubisco combined with the salvage pathway, consuming O_2 and releasing CO_2; occurs when Rubisco binds O_2 instead of CO_2.

photosynthesis [Gk. *photos*, light, + *syn*, together + *tithenai*, to place]: The conversion of light energy to chemical energy; the production of carbohydrates from carbon dioxide and water in the presence of chlorophyll by using light energy.

photosystem: A discrete unit of organization of chlorophyll and other pigment molecules embedded in the thylakoids of chloroplasts and involved with the light-dependent reactions of photosynthesis.

phototropism [Gk. *photos*, light, + *trope*, turning]: Growth in which the direction of the light is the determining factor, as the growth of a plant toward a light source; turning or bending in response to light.

phragmoplast: A spindle-shaped system of fibrils, which arises between two daughter nuclei at telophase and within which the cell plate is formed during cell division, or cytokinesis. The fibrils of the phragmoplast are composed of microtubules. Phragmoplasts are found in all green algae except the members of the class *Chlorophyceae* and in plants.

phragmosome: The layer of cytoplasm that forms across the cell where the nucleus becomes located and divides.

phycobilins: A group of water-soluble accessory pigments, including phycocyanins and phycoerythrins, which occur in the red algae and cyanobacteria.

phycology [Gk. *phykos*, seaweed]: The study of algae.

phycoplast: A system of microtubules that develops between the two daughter nuclei parallel to the plane of cell division. Phycoplasts occur only in green algae of the class *Chlorophyceae*.

phyllo-, phyll- [Gk. *phyllon*, leaf]: Prefix meaning "leaf."

phyllode (fĭl′ōd): A flat, expanded, photosynthetic petiole or stem; phyllodes occur in certain genera of vascular plants, where they replace leaf blades in photosynthetic function.

phyllotaxy: The arrangement of the leaves on a stem; also called phyllotaxis.

phylogeny [Gk. *phylon*, race, tribe]: Evolutionary relationships among organisms; the developmental history of a group of organisms.

phylum, *pl.* **phyla** [Gk. *phylon*, race, tribe]: A taxonomic category of related, similar classes; a higher-level category below kingdom and above class; until recently called division by botanists.

physiology: The study of the activities and processes of living organisms.

phyto-, -phyte [Gk. *phyton*, plant]: Prefix or suffix meaning "plant."

phytoalexin: [Gk. *phyton*, plant, + *alexein*, to ward off, protect]: A chemical substance produced by a plant to combat infection by a pathogen (fungi or bacteria).

phytochrome: A phycobilinlike pigment found in the cytoplasm of plants and a few green algae that is associated with the absorption of light; photoreceptor for red and far-red light; involved in a number of timing processes, such as flowering, dormancy, leaf formation, and seed germination.

phytoplankton [Gk. *phyton*, + *planktos*, wandering]: Aquatic, free-floating, microscopic, photosynthetic organisms.

pigment: A substance that absorbs light, often selectively.

pileus [L. *pileus*, a cap]: The caplike part of mushroom basidiomata and of certain ascomata.

pinna, *pl.* **pinnae** (pĭn′a) [L. *pinna*, feather]: A primary division, or leaflet, of a compound leaf or frond; may be divided into pinnules.

pinocytosis: *See* endocytosis.

pistil [L. *pistillum*, pestle]: A term sometimes used to refer to an individual carpel or a group of fused carpels.

pistillate: *See* carpellate.

pit: A recessed cavity in a cell wall where the secondary wall does not form.

pit membrane: The middle lamella and two primary cell walls between two pits.

pit-pair: Two opposite pits plus the pit membrane.

pith: The ground tissue occupying the center of the stem or root within the vascular cylinder; usually consists of parenchyma.

pith ray: *See* interfascicular region.

placenta, *pl.* **placentae** [L. *placenta*, cake]: The part of the ovary wall to which the ovules or seeds are attached.

placentation: The manner of ovule attachment within the ovary.

plankton [Gk. *planktos*, wandering]: Free-floating, mostly microscopic, aquatic organisms.

plaque: Clear area in a sheet of cells resulting from the killing or lysis of contiguous cells by viruses.

-plasma, plasmo-, -plast [Gk. *plasma*, form, mold]: Prefix or suffix meaning "formed" or "molded"; examples are protoplasm, "first-molded" (living matter), and chloroplast, "green-formed."

plasma membrane or **plasmalemma:** Outer boundary of the cytoplasm, next to the cell wall; consists of a single membrane; also called cell membrane, or ectoplast.

plasmid: A relatively small fragment of DNA that can exist free in the cytoplasm of a bacterium and can be integrated into and then replicated with a chromosome. Plasmids make up about 5 percent of the DNA of many bacteria, but are rare in eukaryotes.

plasmodesma, *pl.* **plasmodesmata** [Gk. *plasma*, form, + *desma*, bond]: The minute cytoplasmic threads that extend through openings in cell walls and connect the protoplasts of adjacent living cells.

plasmodium (plăz·mō′dĭ·ŭm): Stage in life cycle of myxomycetes (plasmodial slime molds); a multinucleate mass of protoplasm surrounded by a membrane.

plasmogamy [Gk. *plasma*, form, + *gamos*, marriage]: Union of the protoplasts of gametes that is not accompanied by union of their nuclei.

plasmolysis (plăz·mŏl′ĭ·sĭs) [Gk. *plasma*, form, + *lysis*, a loosening]: The separation of the protoplast from the cell wall because of the removal of water from the protoplast by osmosis.

plastid (plăs′tĭd): Organelle in the cells of certain groups of eukaryotes that is the site of such activities as food manufacture and storage; plastids are bounded by two membranes.

pleiotropy [Gk. *pleros*, more, + *trope*, a turning]: The capacity of a gene to affect more than one phenotypic characteristic.

plumule [L. *plumula*, a small feather]: The first bud of an embryo; the portion of the young shoot above the cotyledons.

pneumatophores [Gk. *pneuma*, breath, + *-phoros*, carrying]: Negatively gravitropic extensions of the root systems of some trees growing in swampy habitats; they grow upward and out of the water and probably function to ensure adequate aeration.

point mutation: An alteration in one of the nucleotides in a chromosomal DNA molecule; an allele of a gene changes, becoming a different allele; also called gene mutation.

polar molecule: A molecule with positively and negatively charged ends.

polar nuclei: Two nuclei (usually), one derived from each end (pole) of the embryo sac, which become centrally located; they fuse with a male nucleus to form the primary (typically $3n$) endosperm nucleus.

pollen [L. *pollen*, fine dust]: A collective term for pollen grains.

pollen grain: A microspore containing a mature or immature microgametophyte (male gametophyte); pollen grains occur in seed plants.

pollen mother cell: *See* microspore mother cell.

pollen sac: A cavity in the anther that contains the pollen grains.

pollen tube: A tube formed after germination of the pollen grain; carries the male gametes into the ovule.

pollination: In angiosperms, the transfer of pollen from an anther to a stigma. In gymnosperms, the transfer of pollen from a pollen-producing cone directly to an ovule.

poly- [Gk. *polys*, many]: Prefix meaning "many."

polyembryony: Having more than one embryo within the developing seed.

polygenic inheritance: The inheritance of quantitative characteristics determined by the combined effects of multiple genes.

polymer (pŏl′ĭ·mer): A large molecule composed of many similar molecular subunits.

polymerase chain reaction (PCR): A technique for amplifying specific regions of DNA by multiple cycles of DNA polymerization, utilizing special primers, DNA polymerase molecules, and nucleotides; each cycle is followed by a brief heat treatment to separate complementary strands.

polymerization: The chemical union of monomers such as glucose or nucleotides to form polymers such as starch or nucleic acid.

polynucleotide: A single-stranded DNA or RNA molecule.

polypeptide: A molecule composed of amino acids linked together by peptide bonds, not as complex as a protein.

polyphyletic: Pertaining to a taxon whose members are derived from two or more ancestors not common to all members of the taxon.

polyploid (pŏl′ĭ•ploid): Referring to an organism, tissue, or cell with more than two complete sets of chromosomes.

polysaccharide: A polymer composed of many monosaccharide units joined in a long chain, such as glycogen, starch, and cellulose.

polysome or **polyribosome:** An aggregation of ribosomes actively involved in the translation of the same RNA molecule, one after another.

pome (pōm) [Fr. *pomme*, apple]: A simple fleshy fruit, the outer portion of which is formed by the floral parts that surround the ovary and expand with the growing fruit; found only in one subfamily of the *Rosaceae* (apples, pears, quince, pyracantha, and so on).

population: Any group of individuals, usually of a single species, occupying a given area at the same time.

P-protein: Phloem-protein; a proteinaceous substance found in cells of angiosperm phloem, especially in sieve-tube elements; also called slime.

preprophase band: A ringlike band of microtubules, found just beneath the plasma membrane, that delimits the equatorial plane of the future mitotic spindle of a cell preparing to divide.

primary endosperm nucleus: The result of the fusion of a sperm nucleus and the two polar nuclei.

primary growth: In plants, growth originating in the apical meristems of shoots and roots, as contrasted with secondary growth.

primary meristem or **primary meristematic tissue** (mĕr′ĭ•ste•măt′ĭk): A tissue derived from the apical meristem; of three kinds: protoderm, procambium, and ground meristem.

primary metabolites: Molecules that are found in all plant cells and are necessary for the life of the plant; examples are simple sugars, amino acids, proteins, and nucleic acids.

primary pit-field: Thin area in a primary cell wall through which plasmodesmata pass, although plasmodesmata may occur elsewhere in the wall as well.

primary plant body: The part of the plant body arising from the apical meristems and their derivative meristematic tissues; composed entirely of primary tissues.

primary root: The first root of the plant, developing in continuation of the root tip or radicle of the embryo; the taproot.

primary tissues: Cells derived from the apical meristems and primary meristematic tissues of root and shoot; as opposed to secondary tissues derived from a cambium; primary growth results in an increase in length.

primary cell wall: The wall layer deposited during the period of cell expansion.

primordium, *pl.* **primordia** [L. *primus*, first, + *ordiri*, to begin to weave]: A cell or organ in its earliest stage of differentiation.

pro- [Gk. *pro*, before]: Prefix meaning "before" or "prior to."

procambium (prō•kăm′bē•ŭm) [L. *pro*, before, + *cambiare*, to exchange]: A primary meristematic tissue that gives rise to primary vascular tissues.

proembryo: Embryo in early stages of development, before the embryo proper and suspensor become distinct.

prokaryote [Gk. *pro*, before, + *karyon*, kernel]: A cell lacking a membrane-bounded nucleus and membrane-bounded organelles; *Bacteria* and *Archaea*.

prolamellar body: Semicrystalline body found in plastids arrested in development by the absence of light.

promoter: A specific segment of DNA to which RNA polymerase attaches to initiate transcription of mRNA from an operon.

prophase [Gk. *pro*, before, + *phasis*, form]: An early stage in nuclear division, characterized by the shortening and thickening of the chromosomes and their movement to the metaphase plate.

proplastid: A minute self-reproducing body in the cytoplasm from which a plastid develops.

prop roots: Adventitious roots arising from the stem above soil level and helping to support the plant: common in many monocots, for example, maize (*Zea mays*).

prosthetic group: A heat-stable metal ion or an inorganic group (other than an amino acid) that is bound to a protein and serves as its active group.

protease (prō′tē•ās): An enzyme that digests protein by the hydrolysis of peptide bonds; proteases are also called peptidases.

protein [Gk. *proteios*, primary]: A complex organic compound composed of many (100 or more) amino acids joined by peptide bonds.

prothallial cell [Gk. *pro*, before, + *thallos*, sprout]: The sterile cell or cells found in the male gametophytes, or microgametophytes, of vascular plants other than angiosperms; believed to be remnants of the vegetative tissue of the male gametophyte.

prothallus (prō•thăl′ŭs): In homosporous vascular plants, such as ferns, the more or less independent, photosynthetic gametophyte; also called the prothallium.

protist: A member of the kingdom *Protista*.

proto- [Gk. *protos*, first]: Prefix meaning "first"; for example, *Protozoa*, "first animals."

protoderm [Gk. *protos*, first, + *derma*, skin]: Primary meristematic tissue that gives rise to epidermis.

proton: A subatomic, or elementary, particle, with a single positive charge equal in magnitude to the charge of an electron and a mass of 1; the basic component of every atomic nucleus.

protonema, *pl.* **protonemata** [Gk. *protos*, first, + *nema*, a thread]: The first stage in development of the gametophyte of mosses and certain liverworts; protonemata may be filamentous or platelike.

protoplasm: A general term for the living substance of all cells.

protoplast: The protoplasm of an individual cell; in plants, the unit of protoplasm inside the cell wall.

protostele [Gk. *protos*, first, + *stele*, pillar]: The simplest type of stele, consisting of a solid column of vascular tissue.

protoxylem: The first part of the primary xylem, which matures during elongation of the plant part in which it is found.

protracheophyte: An organism with branched axes and multiple sporangia but with water-conducting cells similar to the hydroids of modern mosses rather than to the tracheary elements of vascular plants; an intermediate stage in the evolution of vascular plants, or tracheophytes.

proximal (prŏk′sĭ•măl) [L. *proximus*, near]: Situated near the point of reference, usually the main part of a body or the point of attachment; opposite of distal.

pseudo- [Gk. *pseudes*, false]: Prefix meaning "false."

pseudoplasmodium: A multicellular mass of individual amoeboid cells, representing the aggregate phase in the cellular slime molds.

pulvinus, *pl.* **pulvini:** Jointlike thickening at the base of the petiole of a leaf or petiolule of a leaflet, and having a role in the movements of the leaf or leaflet.

pumps: Transport proteins driven by either chemical energy (ATP) or light energy; in plant and fungal cells, they typically are proton pumps.

punctuated equilibrium: A model of evolutionary change that proposes that there are long periods of little or no change punctuated with brief intervals of rapid change.

purine (pū′rēn): The larger of the two kinds of nucleotide bases found in DNA and RNA; a nitrogenous base with a double-ring structure, such as adenine or guanine.

pyramid of energy: Energy relationships among various feeding levels involved in a particular food chain; autotrophs (at the base of the pyramid) represent the greatest amount of available energy; herbivores are next; then primary carnivores, secondary carnivores, and so forth. Similar pyramids of mass, size, and number also occur in natural communities.

pyrenoid [Gk. *pyren*, the stone of a fruit, + *oides*, like]: Differentiated regions of the chloroplast that are centers of starch formation in green algae and hornworts.

pyrimidine: The smaller of the two kinds of nucleotide bases found in DNA and RNA: a nitrogenous base with a single-ring structure, such as cytosine, thymine, or uracil.

Q

quantasome [L. *quantus*, how much, + Gk. *soma*, body]: Granules located on the inner surfaces of chloroplast lamellae; believed to be involved in the light-dependent reactions of photosynthesis.

quantum: The ultimate unit of light energy.

quiescent center: The relatively inactive initial region in the apical meristem of a root.

R

raceme [L. *racemus*, bunch of grapes]: An indeterminate inflorescence in which the main axis is elongated but the flowers are borne on pedicels that are about equal in length.

rachis (rā′kĭs) [Gk. *rachis*, a backbone]: Main axis of a spike; the axis of a fern leaf (frond), from which the pinnae arise; in compound leaves, the extension of the petiole corresponding to the midrib of an entire leaf.

radially symmetrical: *See* actinomorphic.

radial micellation: The radial orientation of cellulose microfibrils in guard cell walls; plays a role in the movement of guard cells.

radial section: A longitudinal section cut parallel to the radius of a cylindrical body, such as a root or stem; in the case of secondary xylem, or wood, and secondary phloem, parallel to the rays.

radial system: In secondary xylem and secondary phloem, the term applied to all the rays, the cells of which are derived from ray initials; also called the horizontal system, or ray system.

radicle [L. *radix*, root]: The embryonic root.

radioisotope: An unstable isotope of an element that decays or disintegrates spontaneously, emitting radiation; also called a radioactive isotope.

raphe [Gk. *raphe*, seam]: (1) Ridge on seeds, formed by the stalk of the ovule, in those seeds in which the stalk is sharply bent at the base of the ovule; (2) groove on the frustule of a diatom.

raphides (răf′ĭ•dēz) [Gk. *rhaphis*, a needle]: Fine, sharp, needlelike crystals of calcium oxalate found in the vacuoles of many plant cells.

ray flowers: *See* disk flowers.

ray initial: An initial in the vascular cambium that gives rise to the ray cells of secondary xylem and secondary phloem.

reaction center: The complex of proteins and chlorophyll molecules of a photosystem capable of converting light energy to chemical energy in the photochemical reaction.

reaction wood: Abnormal wood that develops in leaning trunks and limbs; *see also* compression wood *and* tension wood.

receptacle: That part of the axis of a flower stalk that bears the floral organs.

recessive: Describing a gene whose phenotypic expression is masked in the heterozygote by a dominant allele; heterozygotes are phenotypically indistinguishable from dominant homozygotes.

recombinant DNA: DNA formed either naturally or in the laboratory by the joining of segments of DNA from different sources.

reduction [L. *reductio*, a bringing back; originally "bringing back" a metal from its oxide]: Gain of an electron by an atom; reduction takes place simultaneously with oxidation (the loss of an electron by an atom), because an electron that is lost by one atom is accepted by another.

regular: *See* actinomorphic.

regulator gene: A gene that prevents or represses the activity of the structural genes in an operon.

replicate: Produce a facsimile or a very close copy; used to indicate the production of a second molecule of DNA exactly like the first molecule or the production of a sister chromatid.

replication fork: In DNA synthesis, the Y-shaped structure formed at the point where the two strands of the original molecule are being separated and the complementary strands are being synthesized.

repressor: A protein that regulates DNA transcription; this occurs because RNA polymerase is prevented from attaching to the promoter and transcribing the gene. *See also* operator.

resin duct: A tubelike intercellular space lined with resin-secreting cells (epithelial cells) and containing resin.

resonance energy transfer: The transfer of light energy from an excited chlorophyll molecule to a neighboring chlorophyll molecule, exciting the second molecule and allowing the first one to return to its ground state.

respiration: An intracellular process in which molecules, particularly pyruvate in the Krebs cycle, are oxidized with the release of energy. The complete breakdown of sugar or other organic compounds to carbon dioxide and water is termed aerobic respiration, although the first steps of this process are anaerobic.

restriction enzymes: Enzymes that cleave the DNA double helix at specific nucleotide sequences.

reticulate venation: *See* netted venation.

reverse transcription: The process by which an RNA molecule is used as a template to make a single-stranded copy of DNA.

rhizobia [Gk. *rhiza*, root, + *bios*, life]: Bacteria of the genera *Rhizobium* or *Bradyrhizobium*, which may be involved with leguminous plants in a symbiotic relationship that results in nitrogen fixation.

rhizoids [Gk. *rhiza*, root]: (1) Branched rootlike extensions of fungi and algae that absorb water, food, and nutrients; (2) root-hairlike structures in liverworts, mosses, and some vascular plants, which occur on free-living gametophytes.

rhizome: A more or less horizontal underground stem.

ribonucleic acid (RNA): Type of nucleic acid formed on chromosomal DNA and involved in protein synthesis; composed of chains of phosphate, sugar molecules (ribose), and purines and pyrimidines; RNA is the genetic material of many kinds of viruses.

ribose: A five-carbon sugar; a component of RNA.

ribosomal RNA: Any of a number of specific molecules that form part of the structure of a ribosome and participate in the synthesis of proteins.

ribosome: A small particle composed of protein and RNA; the site of protein synthesis.

ring-porous wood: A wood in which the pores, or vessels, of the early wood are distinctly larger than those of the late wood, forming a well-defined ring in cross sections of the wood.

RNA: *See* ribonucleic acid.

root: The usually descending axis of a plant, normally below ground, which serves to anchor the plant and to absorb and conduct water and minerals into it.

rootcap: A thimblelike mass of cells that covers and protects the growing tip of a root.

root hairs: Tubular outgrowths of epidermal cells of the root; greatly increase the absorbing surface of the root.

root pressure: The pressure developed in roots as the result of osmosis, which causes guttation of water from leaves and exudation from cut stumps.

rRNA: *See* ribosomal RNA.

Rubisco: RuBP carboxylase/oxygenase, the enzyme that catalyzes initial reaction of the Calvin cycle, involving the fixation of carbon dioxide to ribulose 1,5-bisphosphate (RuBP).

runner: *See* stolon.

S

samara: Simple, dry, one-seeded or two-seeded indehiscent fruit with pericarp-bearing, winglike outgrowths.

sap: (1) A name applied to the fluid contents of the xylem or the sieve elements of the phloem; (2) the fluid contents of the vacuole are called cell sap.

saprophyte [Gk. *sapros*, rotten, + *phyton*, plant]: An organism that secures its food directly from nonliving organic matter; also called a saprobe.

sapwood: Outer part of the wood of stem or trunk, usually distinguished from the heartwood by its lighter color, in which conduction of water takes place.

satellite DNA: A short nucleotide sequence repeated in tandem fashion many thousands of times; this region of the chromosome has a distinctive base composition and is not transcribed.

savanna: Grassland containing scattered trees.

scarification: The process of cutting or softening a seed coat to hasten germination.

schizo- [Gk. *schizein*, to split]: Prefix meaning "split."

schizocarp (skĭz′ō•karp): Dry simple fruit with two or more united carpels that split apart at maturity.

sclereid [Gk. *skleros*, hard]: A sclerenchyma cell with a thick, lignified secondary wall having many pits. Sclereids are variable in form but typically not very long; they may or may not be living at maturity.

sclerenchyma (skle•rĕng′kĭ•ma) [Gk. *skleros*, hard, + L. *enchyma*, infusion]: A supporting tissue composed of sclerenchyma cells, including fibers and sclereids.

sclerenchyma cell: Cell of variable form and size with more or less thick, often lignified, secondary walls; may or may not be living at maturity; includes fibers and sclereids.

scutellum (sku•tĕl′ŭm) [L. *scutella*, a small shield]: The single cotyledon of a grass embryo, specialized for absorption of the endosperm.

secondary growth: In plants, growth derived from secondary or lateral meristems, the vascular and cork cambiums; secondary growth results in an increase in girth, and is contrasted with primary growth, which results in an increase in length.

secondary plant body: The part of the plant body produced by the vascular cambium and the cork cambium; consists of secondary xylem, secondary phloem, and periderm.

secondary root: *See* lateral root.

secondary tissues: Tissues produced by the vascular cambium and cork cambium.

secondary cell wall: Innermost layer of the cell wall, formed in certain cells after cell elongation has ceased; secondary walls have a highly organized microfibrillar structure.

secondary metabolites: Molecules that are restricted in their distribution, both within the plant and among different plants; important for the survival and propagation of the plants that produce them; there are three major classes—alkaloids, terpenoids, and phenolics. Also called secondary products.

second messenger: A small molecule that is formed in or released into the cytosol in response to an external signal; it relays the signal to the interior of the cell; examples are calcium ions and cyclic AMP.

seed: A structure formed by the maturation of the ovule of seed plants following fertilization.

seed coat: The outer layer of the seed, developed from the integuments of the ovule.

seedling: A young sporophyte, which develops from a germinating seed.

segregation: The separation of the chromosomes (and genes) from different parents at meiosis; *see* Mendel's first law.

selectively permeable [L. *seligere*, to gather apart, + *permeare*, to go through]: Applied to membranes that permit passage of water and some solutes but block the passage of others; semipermeable membranes permit the passage of water but not solutes.

sepal (sē′păl) [L. *sepalum*, a covering]: One of the outermost flower structures, a unit of the calyx; sepals usually enclose the other flower parts in the bud.

septate [L. *septum*, fence]: Divided by cross walls into cells or compartments.

septum, *pl.* **septa:** A partition or cross wall.

sessile (sĕs′ĭl) [L. *sessilis*, of or fit for sitting, low, dwarfed]: Attached directly by the base; referring to a leaf lacking a petiole or to a flower or fruit lacking a pedicel.

seta, *pl.* **setae** (sē′ta) [L. *seta*, bristle]: In bryophytes, the stalk that supports the capsule, if present; part of the sporophyte.

sexual reproduction: The fusion of gametes followed by meiosis and recombination at some point in the life cycle.

sheath: (1) The base of a leaf that wraps around the stem, as in grasses; (2) a tissue layer surrounding another tissue, such as a bundle sheath.

shoot: The above-ground portions, such as the stem and leaves, of a vascular plant.

short-day plants: Plants that must be exposed to light periods shorter than some critical length for flowering to occur; they usually flower in autumn.

shrub: A perennial woody plant of relatively low stature, typically with several stems arising from or near the ground.

sieve area: A portion of the sieve-element wall containing clusters of pores through which the protoplasts of adjacent sieve elements are interconnected.

sieve cell: A long, slender sieve element with relatively unspecialized sieve areas and with tapering end walls that lack sieve plates; found in the phloem of gymnosperms.

sieve element: The cell of the phloem that is involved in the long-distance transport of food substances; sieve elements are further classified into sieve cells and sieve-tube elements.

sieve plate: The part of the wall of sieve-tube elements bearing one or more highly differentiated sieve areas.

sieve tube: A series of sieve-tube elements arranged end-to-end and interconnected by sieve plates.

sieve-tube element: One of the component cells of a sieve tube; found primarily in flowering plants and typically associated with a companion cell; also called sieve-tube member.

signal transduction: The process by which a cell converts an extracellular signal into a response.

silique [L. *siliqua*, pod]: The fruit characteristic of the mustard family; two-celled, the valves splitting from the bottom and leaving the placentae with the false partition stretched between. A small, compressed silique is called a silicle.

simple fruit: A fruit derived from one carpel or several united carpels.

simple leaf: An undivided leaf; as opposed to a compound leaf.

simple pit: A pit not surrounded by an overarching border of secondary wall; as opposed to a bordered pit.

simple tissue: A tissue composed of a single cell type; parenchyma, collenchyma, and sclerenchyma are simple tissues.

siphonaceous [Gk. *siphon*, a tube, pipe]: In algae, multinucleate cells without cross walls; coenocytic; also called siphonous.

siphonostele [Gk. *siphon*, pipe, + *stele*, pillar]: A type of stele containing a hollow cylinder of vascular tissue surrounding a pith.

slime: *See* P-protein.

softwood: A name commonly applied to the wood of a conifer.

solar tracking: The ability of the leaves and flowers of many plants to move diurnally, orienting themselves either perpendicular or parallel to the sun's direct rays; also called heliotropism.

solute: A molecule dissolved in a solution.

solute potential: *See* osmotic potential.

solution: Usually a liquid, in which the molecules of the dissolved substance, the solute (for example, sugar), are dispersed between the molecules of the solvent (for example, water).

somatic cells [Gk. *soma*, body]: All cells except the gametes and the cells from which the gametes develop.

soredium, *pl.* **soredia** [Gk. *soros*, heap]: A specialized reproductive unit of lichens, consisting of a few cyanobacterial or green algal cells surrounded by fungal hyphae.

sorus, *pl.* **sori** (sō′rŭs) [Gk. *soros*, heap]: A group or cluster of sporangia or spores.

specialized: (1) Of organisms, having special adaptations to a particular habitat or mode of life; (2) of cells, having particular functions.

speciation: The origin of new species in evolution.

species, *pl.* **species** [L. *species*, kind, sort]: A kind of organism; species are designated by binomial names written in italics.

specific epithet: The second part of a species name; for example, *mays* of *Zea mays*, which is maize.

specificity: Uniqueness, as in proteins in given organisms or enzymes in given reactions.

sperm: A mature male gamete, usually motile and smaller than the female gamete.

spermatangium, *pl.* **spermatangia** (spûr′mă•tăn′jĭ•ŭm) [Gk. *sperma*, sperm, + L. *tangere*, to touch]: In the red algae, the structure that produces spermatia.

spermatium, *pl.* **spermatia** [Gk. *sperma*, sperm]: In the red algae and some fungi, a minute, nonmotile male gamete.

spermatogenous cell: The cell of the male gametophyte, or pollen grain, of gymnosperms, which divides mitotically to form two sperm.

spermatophyte [Gk. *sperma*, seed, + *phyton*, plant]: A seed plant.

spermogonium, *pl.* **spermogonia** (spûr′mă•gŏn′ĭ•ŭm) [Gk. *sperma*, sperm, + *gonos*, offspring]: In the rust fungi, the structure that produces spermatia.

spike [L. *spica*, head of grain]: An indeterminate inflorescence in which the main axis is elongated and the flowers are sessile.

spikelet: The unit of inflorescence in grasses; a small group of grass flowers.

spindle fibers: Bundles of microtubules, some of which extend from the kinetochores of the chromosomes to the poles of the spindle.

spine: A hard, sharp-pointed structure; usually a modified leaf, or part of a leaf.

spirillum, *pl.* **spirilli** [L. *spira*, coil]: A long coiled or spiral bacterium.

spongy parenchyma: A leaf tissue composed of loosely arranged, chloroplast-bearing cells.

sporangiophore (spo•răn′jĭ•o•fōr′) [Gk. *spora*, seed, + *pherein*, to carry]: A branch bearing one or more sporangia.

sporangium, *pl.* **sporangia** (spo•răn′jĭ•ŭm) [Gk. *spora*, seed, + *angeion*, a vessel]: A hollow unicellular or multicellular structure in which spores are produced.

spore: A reproductive cell, usually unicellular, capable of developing into an adult without fusion with another cell.

spore mother cell: A diploid (2*n*) cell that undergoes meiosis and produces (usually) four haploid cells (spores) or four haploid nuclei.

sporic meiosis: Meiosis resulting in the formation of haploid spores by a diploid individual, or sporophyte; the spores give rise to haploid individuals, or gametophytes, which eventually produce gametes that fuse to form diploid zygotes; the zygotes, in turn, develop into sporophytes; this kind of life cycle is known as an alternation of generations.

sporophyll (spŏ′rŏ•fĭl): A modified leaf or leaflike organ that bears sporangia; applied to the stamens and carpels of angiosperms, fertile fronds of ferns, and other similar structures.

sporophyte (spŏ′rŏ•fīt): The spore-producing, diploid (2*n*) phase in a life cycle characterized by alternation of generations.

sporopollenin: The tough substance of which the exine, or outer wall, of spores and pollen grains is composed; a cyclic alcohol highly resistant to decay.

stalk cell: *See* sterile cell.

stamen (stā′mĕn) [L. *stamen*, thread]: The part of the flower producing the pollen, composed (usually) of anther and filament; collectively, the stamens make up the androecium.

staminate (stăm′ĭ•nat): Pertaining to a flower having stamens but no functional carpels.

starch [M.E. *sterchen*, to stiffen]: A complex insoluble carbohydrate; the chief food storage substance of plants; composed of a thousand or more glucose units.

statoliths [Gk. *statos*, stationary, + *lithos*, stone]: Gravity sensors; starch-containing plastids (amyloplasts) or other bodies in the cytoplasm.

stele (stē′le) [Gk. *stele*, a pillar]: The central cylinder, inside the cortex, of roots and stems of vascular plants.

stem: The part of the axis of vascular plants that is above ground, as well as anatomically similar portions below ground, such as rhizomes or corms.

stem bundle: Vascular bundle belonging to the stem.

sterigma, *pl.* **sterigmata** [Gk. *sterigma*, a prop]: A small, slender protuberance of a basidium, bearing a basidiospore.

sterile cell: One of two cells produced by division of the generative cell in developing pollen grains of gymnosperms; it is not a gamete, and eventually degenerates.

stigma: (1) The region of a carpel that serves as a receptive surface for pollen grains and on which they germinate; (2) a light-sensitive, pigmented structure; see eyespot.

stigmatic tissue: The pollen-receptive tissue of the stigma.

stipe: A supporting stalk, such as the stalk of a gill fungus or the leaf stalk of a fern.

stipule (stĭp′yūl): An appendage, often leaflike, that occurs on either side of the basal part of a leaf, or encircles the stem, in many kinds of flowering plants.

stolon (stō′lŏn) [L. *stolo*, shoot]: A stem that grows horizontally along the ground surface and may form adventitious roots, such as the runners of a strawberry plant.

stoma, *pl.* **stomata** (stō′ma) [Gk. *stoma*, mouth]: A minute opening bordered by guard cells in the epidermis of leaves and stems through which gases pass; also used to refer to the entire stomatal apparatus—the guard cells plus their included pore.

stratification: The process of exposing seeds to low temperatures for an extended period before attempting to germinate them at warm temperatures.

strict anaerobe: an anaerobe that is killed by oxygen; can live only in the absence of oxygen.

strobilus, *pl.* **strobili** (strŏb′ĭ•lŭs) [Gk. *strobilos*, a cone]: A reproductive structure consisting of a number of modified leaves (sporophylls) or ovule-bearing scales grouped terminally on a stem; a cone. Strobili occur in many kinds of gymnosperms, lycophytes, and sphenophytes.

stroma [Gk. *stroma,* anything spread out]: The ground substance of plastids.

structural gene: Any gene that codes for a protein; in distinction to regulatory genes.

style [Gk. *stylos,* column]: A slender column of tissue that arises from the top of the ovary and through which the pollen tube grows.

sub- [L. *sub,* under, below]: Prefix meaning "under" or "below"; for example, subepidermal, "underneath the epidermis."

suberin (sū'ber•ĭn) [L. *suber,* the cork oak]: Fatty material found in the cell walls of cork tissue and in the Casparian strip of the endodermis.

subsidiary cell: An epidermal cell morphologically distinct from other epidermal cells and associated with a pair of guard cells; also called an accessory cell.

subspecies: The primary taxonomic subdivision of a species. Varieties are used as equivalent to subspecies by some botanists, or subspecies may be divided into varieties.

substrate [L. *substratus,* strewn under]: The foundation to which an organism is attached; the substance acted on by an enzyme.

substrate phosphorylation: Phosphorylation—the formation of ATP from ADP and inorganic phosphate—that takes place during glycolysis.

succession: In ecology, the orderly progression of changes in community composition that occurs during the development of vegetation in any area, from initial colonization to the attainment of the climax typical of a particular geographic area.

succulent: A plant with fleshy, water-storing stems or leaves.

sucker: A sprout produced by the roots of some plants that gives rise to a new plant; an erect sprout that arises from the base of stems.

sucrase (sū'krās): An enzyme that hydrolyzes sucrose into glucose and fructose; also called invertase.

sucrose (sū'krōs): A disaccharide (glucose plus fructose) found in many plants; the primary form in which sugar produced by photosynthesis is translocated.

superior ovary: An ovary that is free and separate from the calyx.

suspension: A heterogeneous dispersion in which the dispersed phase consists of solid particles sufficiently large that they will settle out of the fluid dispersion medium under the influence of gravity.

suspensor: A structure at the base of the embryo in many vascular plants. In some plants, it pushes the embryo into nutrient-rich tissue of the female gametophyte.

symbiosis (sĭm'bi•ō'sĭs) [Gk. *syn,* together with, + *bios,* life]: The living together in close association of two or more dissimilar organisms; includes parasitism (in which the association is harmful to one of the organisms) and mutualism (in which the association is advantageous to both).

sympatric speciation [Gk. *syn,* together with, + *patra,* fatherland, country]: Speciation that occurs without geographic isolation of a population of organisms; usually occurs as the result of hybridization accompanied by polyploidy; may occur in some cases as a result of disruptive selection.

symplast [Gk. *syn,* together with, + *plastos,* molded]: The interconnected protoplasts and their plasmodesmata; the movement of substances in the symplast is called symplastic movement, or symplastic transport.

sympodium, *pl.* **sympodia:** A stem bundle and its associated leaf traces.

syn-, sym- [Gk. *syn,* together with]: Prefix meaning "together."

synapsis [Gk. *synapsis,* a contract or union]: The pairing of homologous chromosomes that occurs prior to the first meiotic division; crossing-over occurs during synapsis.

synergids (sĭ•nûr'jĭds): Two short-lived cells lying close to the egg in the mature embryo sac of the ovule of flowering plants.

syngamy [Gk. *syn,* together with, + *gamos,* marriage]: *See* fertilization.

synthesis: The formation of a more complex substance from simpler ones.

systematics: Scientific study of the kinds and diversity of organisms and of the relationships between them.

T

taiga: The northern coniferous forest.

tangential section: A longitudinal section cut at right angles to the radius of a cylindrical structure, such as a root or stem; in the case of secondary xylem, or wood, and secondary phloem, at right angles to the rays.

tapetum (ta•pē'tŭm) [Gk. *tapes,* a carpet]: Nutritive tissue in the sporangium, particularly an anther.

taproot: The primary root of a plant formed in direct continuation with the root tip or radicle of the embryo; forms a stout, tapering main root from which arise smaller, lateral roots.

taxon: General term for any one of the taxonomic categories, such as species, class, order, or phylum.

taxonomy [Gk. *taxis,* arrangement, + *nomos,* law]: The science of the classification of organisms.

teliospore (tē'lĭ•ō•spōr): In the rust fungi, a thick-walled spore in which karyogamy and meiosis occur and from which basidia develop.

telium, *pl.* **telia:** In the rust fungi, the structure that produces teliospores.

telomere: The end of a chromosome; has repetitive DNA sequences that help counteract the tendency of the chromosome otherwise to shorten with each round of replication.

telophase: The last stage in mitosis and meiosis, during which the chromosomes become reorganized into two new nuclei.

template: A pattern or mold guiding the formation of a negative or complement; a term applied especially to DNA duplication, which is explained in terms of a template hypothesis.

tendril [L. *tendere,* to extend]: A modified leaf or part of a leaf or stem modified into a slender coiling structure that aids in support of the stems; tendrils occur only in some angiosperms.

tension wood: The reaction wood of magnoliids and eudicots; develops on the upper side of leaning trunks and limbs.

tepal: One of the units of a perianth that is not differentiated into sepals and petals.

test cross: A cross of a dominant with a homozygous recessive; used to determine whether the dominant is homozygous or heterozygous.

tetrad (tĕt'răd): A group of four spores formed from a spore mother cell by meiosis; *see also* bivalent.

tetraploid (tĕt'ra•ploid) [Gk. *tetra,* four, + *ploos,* fold]: Twice the usual, or diploid (2*n*), number of chromosomes (that is, 4*n*).

tetrasporangium, *pl.* **tetrasporangia** [Gk. *tetra,* four, + *spora,* seed, + *angeion,* vessel]: In certain red algae, a sporangium in which meiosis occurs, resulting in the production of tetraspores.

tetraspore [Gk. *tetra,* four, + *spora,* seed]: In certain red algae, the four spores formed by meiotic division in the tetrasporangium of a spore mother cell.

tetrasporophyte [Gk. *tetra,* four, + *spora,* seed, + *phyton,* plant]: In certain red algae, a diploid individual that produces tetrasporangia.

texture: Of wood, refers to the relative size and amount of variation in size of elements within the growth rings.

thallophyte: A term previously used to designate fungi and algae collectively, now largely abandoned.

thallus (thăl'ŭs) [Gk. *thallos,* a sprout]: A type of body that is undifferentiated into root, stem, or leaf; the word *thallus* was used commonly when fungi and algae were considered to be plants, to distinguish their simple construction, and that of certain gametophytes, from the differentiated bodies of plant sporophytes and the elaborate gametophytes of the bryophytes.

theory [Gk. *theorein*, to look at]: A well-tested hypothesis; one unlikely to be rejected by further evidence.

thermodynamics [Gk. *therme*, heat, + *dynamis*, power]: The study of energy exchanges, using heat as the most convenient form of measurement of energy. The first law of thermodynamics states that in all processes, the total energy of the universe remains constant. The second law of thermodynamics states that the entropy, or degree of randomness, tends to decrease.

thermophile: An organism with an optimal growth temperature between 45 and 80°C.

thigmomorphogenesis: The alteration of plant growth patterns in response to mechanical stimuli.

thigmotropism [Gk. *thigma*, touch]: A response to contact with a solid object.

thorn: A hard, woody, pointed branch.

thylakoid [Gk. *thylakos*, sac, + *oides*, like]: A saclike membranous structure in cyanobacteria and the chloroplasts of eukaryotic organisms; in chloroplasts, stacks of thylakoids form the grana; chlorophylls are found within the thylakoids.

thymine: A pyrimidine occurring in DNA but not in RNA; *see also* uracil.

Ti plasmid: A circular plasmid of *Agrobacterium tumifaciens* that enables the bacterium to infect plant cells and produce a tumor (crown gall tumor); a powerful tool in the transfer of foreign genes into plant genomes.

tissue: A group of similar cells organized into a structural and functional unit.

tissue culture: A technique for maintaining fragments of plant or animal tissue alive in a medium after removal from the organism.

tissue system: A tissue or group of tissues organized into a structural and functional unit in a plant or plant organ. There are three tissue systems: dermal, vascular, and ground, or fundamental.

tonoplast [Gk. *tonos*, stretching, tension, + *plastos*, formed, molded]: The cytoplasmic membrane surrounding the vacuole in plant cells; also called vacuolar membrane.

torus, *pl.* **tori:** The central thickened part of the pit-membrane in the bordered pits of conifers and some other gymnosperms.

totipotent: The potential of a plant cell to develop into an entire plant.

tracheary element: The general term for a water-conducting cell in vascular plants; tracheids and vessel elements.

tracheid (trā′kē•ĭd): An elongated, thick-walled conducting and supporting cell of xylem. It has tapering ends and pitted walls without perforations, as contrasted with a vessel element. Found in nearly all vascular plants.

tracheophyte: A vascular plant.

transcription: The enzyme-catalyzed assembly of an RNA molecule complementary to a strand of DNA.

transduction: The transfer of genes from one organism to another by a virus.

transfer cell: Specialized parenchyma cell with wall ingrowths that increase the surface area of the plasma membrane; apparently functions in the short-distance transfer of solutes.

transfer RNA (tRNA): Low-molecular-weight RNA that becomes attached to an amino acid and guides it to the correct position on the ribosome for protein synthesis; there is at least one tRNA molecule for each amino acid.

transformation: The transfer of naked DNA from one organism to another; also called "gene transfer." Transposons are often used as vectors in transformation when it is carried out in the laboratory.

transgenic organism: An organism whose genome contains DNA—from the same or a different species—that has been modified by the methods of genetic engineering.

transition region: The region in the primary plant body showing transitional characteristics between structures of root and shoot.

translation: The assembly of a protein on the ribosomes; mRNA is used to direct the order of the amino acids.

translocation: (1) In plants, the long-distance transport of water, minerals, or food; most often used to refer to food transport; (2) in genetics, the interchange of chromosome segments between nonhomologous chromosomes.

transmembrane proteins: Globular proteins that traverse the lipid bilayer of cellular membranes. Some extend across the lipid bilayer as a single alpha helix and others as multiple alpha helices.

transmitting tissue: A tissue similar to stigmatic tissue and serving as a path for the pollen tube in the style.

transpiration [Fr. *transpirer*, to perspire]: The loss of water vapor by plant parts; most transpiration occurs through stomata.

transpiration stream: The flow of water through the xylem, from the roots to the leaves.

transport protein: A specific membrane protein responsible for transferring solutes across membranes; grouped into three broad classes: pumps, carriers, and channels.

transposon [L. *transponere*, to change the position of something]: A DNA sequence that carries one or more genes and is flanked by sequences of bases that confer the ability to move from one DNA molecule to another; an element capable of transposition, which is the changing of a chromosomal location.

transverse section: A section cut perpendicular, or at right angles, to the longitudinal axis of a plant part.

tree: A perennial woody plant generally with a single stem (trunk).

tricarboxylic acid cycle or **TCA cycle:** *See* Krebs cycle.

trichogyne [Gk. *trichos*, a hair, + *gyne*, female]: In the red algae and certain ascomycetes and *Basidiomycota*, a receptive protuberance of the female gametangium for the conveyance of spermatia.

trichome [Gk. *trichos*, hair]: An outgrowth of the epidermis, such as a hair, scale, or water vesicle.

triglyceride: Glycerol ester of fatty acids; the main constituent of fats and oils.

triose [Gk. *tries*, three, + *ose*, suffix indicating a carbohydrate]: Any three-carbon sugar.

triple fusion: In angiosperms, the fusion of the second male gamete, or sperm, with the polar nuclei, resulting in formation of a primary endosperm nucleus, which is triploid ($3n$) in most groups.

triploid [Gk. *triploos*, triple]: Having three complete chromosome sets per cell ($3n$).

tritium: A radioactive isotope of hydrogen, ^{3}H. The nucleus of a tritium atom contains one proton and two neutrons, whereas the more common hydrogen nucleus consists only of a proton.

tRNA: *See* transfer RNA.

-troph, tropho- [Gk. *trophos*, feeder]: Suffix or prefix meaning "feeder," "feeding," or "nourishing"; for example, autotrophic, "self-nourishing."

trophic level: A step in the movement of energy through an ecosystem, represented by a particular set of organisms.

tropism [Gk. *trope*, a turning]: A response to an external stimulus in which the direction of movement is usually determined by the direction from which the most intense stimulus comes.

tube cell: In male gametophytes, or pollen grains, of seed plants, the cell that develops into the pollen tube.

tuber [L. *tuber*, swelling]: An enlarged, short, fleshy underground stem, such as that of the potato.

tundra: A treeless circumpolar region, best developed in the Northern Hemisphere and most found north of the Arctic Circle.

tunica-corpus: The organization of the shoot apex of most angiosperms and a few gymnosperms, consisting of one or more peripheral layers of cells (the tunica layers) and an interior (the corpus). The tunica layers undergo surface growth (by anticlinal divisions), and the corpus undergoes volume growth (by divisions in all planes).

turgid (tûr'jĭd) [L. *turgidus,* to be swollen]: Swollen, distended; referring to a cell that is firm due to water uptake.

turgor pressure [L. *turgor,* a swelling]: The pressure within the cell resulting from the movement of water into the cell.

tylose [Gk. *tylos,* a lump]: A balloonlike outgrowth from a ray or axial parenchyma cell through the pit in a vessel wall and into the lumen of the vessel.

type specimen: Usually a dried plant specimen housed in an herbarium; selected by a taxonomist to serve as the basis for comparison with other specimens in determining whether they are members of the same species or not.

U

umbel (ŭm'bĕl) [L. *umbella,* sunshade]: An inflorescence, the individual pedicels of which all arise from the apex of the peduncle.

unicellular: Composed of a single cell.

unisexual: Usually applied to a flower lacking either stamens or carpels; a perianth may be present or absent.

unit membrane: A visually definable, three-layered membrane, consisting of two dark layers separated by a lighter layer, as seen with an electron microscope.

uracil (ū'ra•sĭl): A pyrimidine found in RNA but not in DNA; *see also* thymine.

uredinium, *pl.* **uredinia** [L. *uredo,* a blight]: In rust fungi, the structure that produces urediniospores.

urediniospore [L. *uredo,* a blight, + *spora,* spore]: In rust fungi, a reddish, binucleate spore produced in summer.

V

vacuolar membrane: *See* tonoplast.

vacuole [L. *vacuus,* empty]: A space or cavity within the cytoplasm filled with a watery fluid, the cell sap; a lysosomal compartment.

variation: The differences that occur within the offspring of a particular species.

variety: A group of plants or animals of less than species rank. Some botanists view varieties as equivalent to subspecies, and others consider them divisions of subspecies.

vascular [L. *vasculum,* a small vessel]: Pertains to any plant tissue or region consisting of or giving rise to conducting tissue; for example, xylem, phloem, vascular cambium.

vascular bundle: A strand of tissue containing primary xylem and primary phloem (and procambium if still present) and frequently enclosed by a bundle sheath of parenchyma or fibers.

vascular cambium: A cylindrical sheath of meristematic cells, the division of which produces secondary phloem and secondary xylem.

vascular plant: A plant that has xylem and phloem; also called tracheophyte.

vascular rays: Ribbonlike sheets of parenchyma that extend radially through the wood, across the cambium, and into the secondary phloem; they are always produced by the vascular cambium.

vascular system: All the vascular tissues in their specific arrangement in a plant or plant organ.

vector [L., a bearer, carrier, from *vehere,* to carry]: (1) A pathogen that carries a disease from one organism to another; (2) in genetics, any virus or plasmid DNA into which a gene is integrated and subsequently transferred into a cell.

vegetative: Of, relating to, or involving propagation by asexual processes; also referring to nonreproductive plant parts.

vegetative reproduction: (1) In seed plants, reproduction by means other than by seeds; apomixis; (2) in other organisms, reproduction by vegetative spores, fragmentation, or division of the somatic body. Unless a mutation occurs, each daughter cell or individual is genetically identical to its parent.

vein: A vascular bundle forming a part of the framework of the conducting and supporting tissue of a leaf or other expanded organ.

velamen [L. *velumen,* fleece]: A multiple epidermis covering the aerial roots of some orchids and aroids; also occurs on some terrestrial roots.

venation: Arrangement of veins in the leaf blade.

venter [L. *venter,* belly]: The enlarged basal portion of an archegonium containing the egg.

vernalization [L. *vernalis,* spring]: The induction of flowering by cold treatment.

vessel [L. *vasculum,* a small vessel]: A tubelike structure of the xylem composed of elongate cells (vessel elements) placed end to end and connected by perforations. Its function is to conduct water and minerals through the plant body. Found in nearly all angiosperms and a few other vascular plants (for example, gnetophytes).

vessel element: One of the cells composing a vessel; also called vessel member.

viable [L., *vita,* life]: Able to live.

volva [L. *volva,* a wrapper]: A cuplike structure at the base of the stalk of certain mushrooms.

W

wall pressure: The pressure of the cell wall exerted against the turgid protoplast; opposite and equal to the turgor pressure.

water potential: The algebraic sum of the solute potential and the pressure potential, or wall pressure; the potential energy of water.

water vesicle: An enlarged epidermal cell in which water is stored; a type of trichome.

weed [O.E. *weod,* used at least since the year 888 in its present meaning]: Generally an herbaceous plant not valued for use or beauty, growing wild, and regarded as using ground or hindering the growth of useful vegetation.

whorled phyllotaxy: Arrangement of three or more leaves or floral parts in a circle at a node.

wild type: In genetics, the phenotype or genotype that is characteristic of the majority of individuals of a species in a natural environment.

wood: Secondary xylem.

woody magnoliids: Magnoliids with large, robust, bisexual flowers with many free parts, arranged spirally on an elongated axis, or receptacle.

X

xanthophyll: (zǎn'thō•fĭl) [Gk. *xanthos,* yellowish-brown, + *phyllon,* leaf]: A yellow chloroplast pigment; a member of the carotenoid class.

xerophyte [Gk. *xeros,* dry, + *phyton,* a plant]: A plant that has adapted to arid habitats.

xylem [Gk. *xylon,* wood]: A complex vascular tissue through which most of the water and minerals of a plant are conducted; characterized by the presence of tracheary elements.

Z

zeatin (zē′ă•tĭn): Plant hormone; a natural cytokinin isolated from maize.

zooplankton [Gk. *zoe*, life, + *plankton*, wanderer]: A collective term for the nonphotosynthetic organisms present in plankton.

zoosporangium: A sporangium bearing zoospores.

zoospore (zō′o•spōr): A motile spore, found among algae, oomycetes, and chytrids.

zygomorphic [Gk. *zygo*, pair, + *morphē*, form]: A type of flower capable of being divided into two symmetrical halves only by a single longitu-dinal plane passing through the axis; also called bilaterally symmetrical.

zygosporangium, *pl.* **zygosporangia:** A sporangium containing one or more zygospores.

zygospore: A thick-walled, resistant spore that develops from a zygote, resulting from the fusion of isogametes.

zygote (zī′gōt) [Gk. *zygotos*, paired together]: The diploid (2*n*) cell resulting from the fusion of male and female gametes.

zygotic meiosis: Meiosis by a zygote to form four haploid cells, which divide by mitosis to produce either more haploid cells or a multicellular individual that eventually gives rise to gametes.

Illustration Credits

All photographs not credited herein are by Ray F. Evert. All section openers are by Rhonda Nass.

Chapter 1

Opener, p. xvi, Rhonda Nass; **1.1** Tokai University Research Center; **1.2** Biological Photo Service; **1.3** Richard E. Dickerson, "Chemical Evolution and the Origin of Life," *Scientific American,* vol. 239(3), pages 70–86, 1978; **1.4** Michael Durham/Ellis Nature photography; **1.5** Sidney W. Fox; **1.6(a, b)** Gary Breckon; **1.7** Harlen P. Banks; **1.8** Anne Wertheim/Animals Animals; **1.9** Dr. Jeremy Burgess/Science Photo Library/Photo Researchers Inc.; **1.10** After W. Troll, *Vergleichende Morphologie der Hoheren Pflanzen,* vol. 1, pt. 1, Verlage von Gebruder Borntraeger, Berlin, 1937; **1.11(a)** Dr. Anne La Bastille/Photo Researchers, Inc.; **1.11(b)** B. C. Alexander/Photo Researchers, Inc.; **1.11(c)** Martin Harvey/The Wildlife Collection; **1.11(d)** Jack Swenson/The Wildlife Collection; **1.11(e)** Fred Hirschmann; **1.11(f)** Stephen P. Parker/Photo Researchers, Inc.; **1.12** From H. Curtis and N. Sue Barnes, *Invitation to Biology,* 5th ed., Worth Publishers, New York, 1994; **1.13** Line art adapted from: P. Ehrlich, A. Ehrlich, and Holdren, *Ecoscience: Population, Resources, Environment,* W. H. Freeman and Company, New York, 1977

Chapter 2

2.4(c) L. M. Biedler; **2.8** M. Kruatrachue and R. Evert, *American Journal of Botany,* vol. 64, pages 310–325, 1977; **2.9** From H. Curtis and N. Sue Barnes, *Invitation to Biology,* 5th ed., Worth Publishers, New York, 1994; **2.10** From Curtis and Barnes, *op. cit.,* 1994; **2.12** B. E. Juniper; **2.17–2.19** From H. Curtis and N. Sue Barnes, *Invitation to Biology,* 5th ed., Worth Publishers, New York, 1994; **2.25(a)** Dr. Jeremy Burgess/Photo Researchers, Inc.; **2.25(b)** Dr. Morley Read /Photo Researchers, Inc.; **2.25(c)** Gerry Ellis/Ellis Nature Photography; **2.25(d)** Bill Storde; **2.26(b)** Marcel Isy-Schwart/Image Bank; **2.27** Gerry Ellis/Ellis Nature Photography; **2.28(a)** Frans Lanting/Minden Pictures; **2.28(b)** Dwight Kuhn/Bruce Coleman, Inc.; **2.29** Albert F. W. Vick, Jr.; **2.30** Katherine Esau; **2.31(b)** Steve Solum/Bruce Coleman, Inc.

Chapter 3

3.1(a) Dr. Jeremy Burgess/Science Photo Library/Photo Researchers, Inc.; **3.1(b)** Courtesy of the National Library of Medicine; **3.2 A. Ryter; 3.3(photo)** N. J. Lang, *Journal of Psychology,* vol. 1, pages 127–134, 1965; **3.4(photo)** George Palade; **3.5(photo)** Michael A. Walsh; **3.11** Myron C. Ledbetter; **3.12** D.S. Neuberger; **3.13** R. R. Dute; **3.15** David Stetler; **3.16(photo)** K. Esau; **3.17** K. Esau; **3.18(a)** Mary Alice Webb; **3.19** M. Kruatrachue and R. Evert, *American Journal of Botany,* vol. 64, pages 310–325, 1977; **3.21** Dr. Peter K. Hepler; **3.22(a, b)** R. R. Dute; **3.27(b)** Dr. M. V. Parthasarathy; **3.28** R. R. Powers; **3.29(b)** Lewis Tilney; **3.30(a)** Brian Wells and Keith Roberts in Alberts et al., *Molecular Biology of the Cell,* 2nd ed., Garland Publishing, Inc., New York, 1989; **3.32** After K. Esau, *Anatomy of Seed Plants,* 2nd ed., John Wiley and Sons, Inc., New York, 1977; **3.34(a)** H. A. Core, W. A. Côté, and A. C. Day, *Wood Structure and Identification,* 2nd ed., Syracuse University Press, Syracuse, NY, 1979; **3.34b** After R. D. Preston, in A. W. Robards (ed.), *Dynamic Aspects of Plant Ultrastructure,* McGraw-Hill Book Company, New York, 1974; **Page 47** After Richard E. Williamson, Plant Physiology, vol. 82, pp. 631-634, 1986, and B. Alberts, D. Bray, J. Lewis, M. Raff, K. Roberts, and J. D. Watson, *Molecular Biology of the Cell,* 2nd ed., Garland Publishing, Inc., New York, 1989

Chapter 4
4.4–4.7 From H. Curtis and N. Sue Barnes, *Invitation to Biology,* 5th ed., Worth Publishers, New York, 1994; **4.8** After A. L. Lehninger, *Biochemistry,* 2nd ed., Worth Publishers, New York, 1975; **Page 80,** Doug Wechsler/Earth Scenes; **4.14** Birgit Satir; **4.16(a–c)** David G. Robinson; **4.19(a–c)** E. B. Tucker, *Protoplasma,* vol. 113, pages 193–201, 1982

Chapter 5
5.8(a, b) Thomas A. Steitz

Chapter 6
6.2 From H. Curtis and N. Sue Barnes, *Invitation to Biology,* 5th ed., Worth Publishers, New York, 1994; **6-15(b)** John N. Telfold; **6-18(b)** The Metropolitan Museum of Art; **Page 122,** M. P. Price/Bruce Coleman, Inc.; **6.20** From H. Curtis and N. Sue Barnes, *Invitation to Biology,* 5th ed., Worth Publishers, New York, 1994

Chapter 7
7.2 Paul W. Johnson/Biological Photo Service; **7.3** Colin Milkins/Oxford Scientific Films; **7.4** From Peter Gray, *Psychology,* Worth Publishers, New York, 1991; **7.5** From H. Curtis and N. Sue Barnes, *Invitation to Biology,* 5th ed., Worth Publishers, New York, 1994; **7.6** Prepared by Govindjee; **7.7** L. E. Graham; **7.12** D. Branton; **Page 137,** After Alberts et al., *Molecular Biology of the Cell,* 2nd ed., Figure 9–52, Garland Publishing, Inc., New York, 1989; **7.15** Adapted from B. Alberts et al., *Molecular Biology of the Cell,* 2nd ed., Garland Publishing, Inc., New York, 1989, page 378; **7-17** Dr. Jeremy Burgess/Photo Researchers, Inc.; **7.28(a)** Leonard L. Rue/Bruce Coleman, Inc.; **7.28(b)** Phil Degginger/Bruce Coleman, Inc.

Chapter 8
8.1 From H. Curtis and N. Sue Barnes, *Invitation to Biology,* 5th ed., Worth Publishers, New York, 1994; **Page 159 (a–d)** Susan Wick; **8.6(a–f)** W. T. Jackson; **8.10** J. Cronshaw; **8.11(a–c)** P. K. Hepler, *Protoplasma,* vol. 111, pages 121–133, 1982

Chapter 9
9.5 B. John; **9.7** W. Tai; **9.8(a, b)** P. B. Moens; **9.9** G. Ostergren; **Page 180, (a)** G. I. Bennard/Oxford Scientific Films; **(b, c)** Heather Angel/Biofotos

Chapter 10
10.1 Moravian Museum, Brno, Czechoslovakia; **10.2** From H. Curtis and N. Sue Barnes, *Invitation to Biology,* 5th ed., Worth Publishers, New York, 1994; **10.3** Adapted from K. von Frisch, *Biology,* translated by Jane Oppenheimer, Harper and Row Publishers, Inc., New York, 1964; **10.4–10.9** From H. Curtis and N. Sue Barnes, *Invitation to Biology,* 5th ed., Worth Publishers, New York, 1994; **10.10** After Figure 5-15 in: Anthony J. F. Griffiths, Jeffrey H. Miller, David T. Suzuki, Richard C. Lewontin, William M. Gelbart, *An Introduction to Genetic Analysis,* 6th ed., W. H. Freeman and Company, New York, 1996; **10.11(a)** C. G. G. J. van Steenis; **10.11(b)** Warren L. Wagner, Smithsonian, Washington, D.C.; **10.12** Nik Kleinberg; **10.13** From H. Curtis and N. Sue Barnes, *Invitation to Biology,* 5th ed., Worth Publishers, New York, 1994; **10.15** Dr. Max–B. Schroder and Hannelore Oldenburg; **10.16** Heather Angel; **10.17** From H. Curtis and N. Sue Barnes, *Invitation to Biology,* 5th ed., Worth Publishers, New York, 1994; **10.18(a)** From J. D. Watson, *The Double Helix,* Antheneum Publishers, New York, 1968; **10.18(b)** Will and Deni Mcintyre/Photo Researchers, Inc.; **10.19–10.21** From H. Curtis and N. Sue Barnes, *Invitation to Biology,* 5th ed., Worth Publishers, New York, 1994; **10.22** A. B. Blumenthal, H. J. Kreigstein, and D. S. Hognes, *Cold Spring Harbor Symposium on Quantitative Biology,* vol. 38, page 205, 1973; **10.23** From Curtis and Barnes, **Biology,** 5th ed., Worth Publishers, New York, 1989; **Essay, page 000** From Curtis and Barnes, *Invitation to Biology,* 5th ed., Worth Publishers, New York, 1994.

Chapter 11
11.1 Computer Graphics Laboratory, University of California at San Francisco; **11.2** From H. Curtis and N. Sue Barnes, *Invitation to Biology,* 5th ed., Worth Publishers, New York, 1994; **11.9** Hans Ris; **11.10** After W. M. Becker, J. B. Reece, M. F. Poenie, *The World of the Cell,* 3rd ed., Benjamin/Cummings Publishing Company, Inc., Menlo Park, CA, 1996, page 586; **11.11–11.14** From H. Curtis and N. Sue Barnes, *Invitation to Biology,* 5th ed., Worth Publishers, New York, 1994; **11.15(a)** John Bova/Photo Researchers, Inc.; **11.15(b)** Arnold Sparrow **11.16(a)** Victoria Foe; **11.17** Adapted from B. Alberts et al., *Molecular Biology of the Cell,* 2nd ed., Garland Publishing, Inc., New York, 1989; **11.18(a)** James German; **11.18** Line art adapted from P. Chambon, "Split Genes," *Scientific American,* May 1981, pages 60–71; **11.19** Adapted from R. Lewin, Science, vol. 212, pages 28–32, 1981; **11.20** Adapted from B. Alberts et al., *Molecular Biology of the Cell,* Garland Publishing, Inc., New York, 1983; **11.21** From H. Curtis and N. Sue Barnes, *Invitation to Biology,* 5th ed., Worth Publishers, New York, 1994; **11.22** Adapted from Neil A. Campbell, *Biology,* 4th ed., Benjamin/Cummings Publishing Company, Inc., Menlo Park, CA, 1996; **Page 228 (a–d)** T. Bleecker; (drawing) S. Ross-Craig, *Drawings of British Plants,* Part III, *Cruciferae,* Bell, London, 1949; **11.23** Adapted from a photo by John T. Fiddles and Howard M. Goodman; **11.24** From Curtis and Barnes, op. cit., 1994; **11.26** Vysis; **11.27** Redrawn from Curtis and Barnes, *Biology,* 5th ed., Worth Publishers, New York, 1989

Chapter 12
12.1 By permission of Mr. G. P. Darwin, courtesy of The Royal College of Surgeons of England; **12.2** From Curtis and Barnes, *Invitation to Biology,* 5th ed., Worth Publishers, New York, 1994; **12.3** After D. Lack, *Darwin's Finches,* Harper and Row, Publishers, Inc., New York, 1961; **12.4(a)** Mervin W. Larson/Bruce Coleman; **12.4(b)** Frans Lanting/Minden Pictures; **12.6(a)** E. R. Degginger/Bruce Coleman, Inc.; **12.6(b)** P. Hereneen; **12.6(c)** Dr David Grimaldi, American Museum of Natural History; **12.7** A. J. Griffiths; **12.10(a)** M&C Photography/Peter Arnold, Inc.; **12.10(b)** Kent and Donna Dannen/Photo Researchers, Inc.; **12.11(a, b)** Heather Angel/Biofotos; **12.12** Robert Ornduff, University of California, Berkeley; **12.15** Adapted from E. O. Wilson and W. H. Burrert, *A Primer of Population Biology,* Sinauer Associates, Sunderland, MA, 1971; **Page 250,** Gerald D. Carr; **12.16 and 12.17** From H. Curtis and N. Sue Barnes, *Invitation to Biology,* 5th ed., Worth Publishers, New York, 1994; **12.18** In Marion Ownbey, "Natural hybridization and amphiploidy in the genus *Tragopogon,*" *American Journal of Botany,* 37 (7), 487–499, 1950; **12.19(a, b)** Heather Angel/Biofotos; **12.19(c–e)** C. J. Marchant; **12.21(a)** Thase Daniel/Bruce Coleman, Inc.

Chapter 13
13.1 Corbis/Bettmann; **13.2(a)** Larry West; **13.2(b)** L. Campbell/NHPA; **13.2(c)** Imagery; **Table 13.1(top)** G. R. Roberts; **Table 13.1(bottom)** David Thomas/Oxford Scientific Films; Essay, page 266**(a–c)** E. S. Ross; **13.5** From Curtis and Barnes, *Biology,* 5th ed., Worth Publishers, New York, 1989; **13.7** From Curtis and Barnes, *Invitation to Biology,* 5th ed., Worth Publishers, New York, 1994; **13.9** Jones/Mayer; **13.10(a)** L. V. Leak, *Journal of Ultrastructural Research,* vol. 21, pages 61–74, 1967; **13.10(b)** Helmut Koing and Karl Stettler; **13.10(c)** K. Esau; **13.12(a)** R. M. Meadows/Peter Arnold, Inc.; **13.12(b)** L. E. Graham; **13.13(a)** R. M. Meadows/Peter Arnold, Inc.; **13.13(b)** L. E. Graham; **13.13(c)** Kim Taylor/Bruce Coleman, Inc.; **13.13(d)** G. I. Bernard/Oxford Scientific Films/Animals Animals; **13.13(e)** E. V. Gravé; **13.14(a)** Photo Researchers, Inc.; **13.14(b)** L. West; **13.14(c)** K. B. Sandved; **13.14(d)** John A. Lynch; **13.15(a)** L. West/Bruce Coleman, Inc.; **13.15(b)** J. W. Perry; **13.15(c)** E. S. Ross; **13.15(d)** Robert Carr; **13.15(e)** John Shaw/Bruce Coleman, Inc.; **13.15(f–h)** J. Dermid; **13.15(i)** Steve Solum 1985/Bruce Coleman, Inc.; **13.15(j)** E. Beals

Chapter 14
14.1 MSU Instructional Media Center/R. Hammerschmidt; **14.2** Paul Chesley/Photographers Aspen; **14.3(a, b)** M. T. Madigan; **14.4** D. A. Cuppels and A. Kelman, *Phytopathology,* vol. 70, pages 1110–1115, 1980; **14.5** C. C. Brinton, Jr., and John Carnahan; **14.6(a)** USDA; **14.6(b)** David Phillips/Visuals Unlimited; **14.6(c)** Richard Blakemore; **14.7** Hans Reichenbach; **14.8** P. Gerhardt; **14.9** E. J. Ordal; **14.10(photo)** M. Jost; **14.11(a)** E. V. Gravé; **14.11(c)** Winton Patnode/Photo Researchers, Inc.; **14.12**(top) Fred Bavendam/Peter Arnold, Inc.; **14.12(bottom)** After M. R. Walter, *American Scientist,* vol. 65, pages 563–571, 1977; **14.13(a)** Robert D. Warmbrodt; **14.13(b)** Paul W. Johnson/Biological Photo Service; **14.14** Robert and Linda Mitchell; **14.15** Germaine Cohen-Bazire; **14.16** T. D. Pugh and E. H. Newcomb; **14.17** J. F. Worley/USDA; **14.18(a)** M. V. Parthasarathy; **14.18(b)** Henry

Donselman; **14.19** After G. N. Agrios, *Plant Pathology*, 2nd ed., Academic Press, Inc., New York, 1978; **14.20** USGS Eros Data Center; **14.21** R. Robinson/Visuals Unlimited; **14.22** Leonard Lessin/Peter Arnold, Inc.; **14.24(b)** Jean-Yves Sgro; **14.25(a)–(d)** Jean-Yves Sgro; **14.26(a)** G. Gaard and R. W. Fulton; **14.26(b), (c)** G. Gaard and G. A. deZoeten; **14.27** M. Dollet and R. G. Milne; **14.29** D. Maxwell; **14.30(a)** K. Maramorosch **14.30(b)** E. Shikata **14.30(c)** E. Shjikata and K. Maramorosch; **14.31** T. White; **14.32** Th. Koller and J. M. Sogo, Swiss Federal Institute of Technology, Zurich.

Chapter 15
15.1 R. M. Meadows/Peter Arnold Inc.; **15.2** Charles M. Fitch/Taurus Photos; **15.3(a)** M. Powell; **15.3(c)** Thomas Volk; **15.3(d)** E. S. Ross; **15.4** R. J. Howard; **15.5** M. D. Coffey, B. A. Palevitz, and P. J. Allen, *The Canadian Journal of Botany*, vol. 50, pages 231–240, 1972; **15.6** E. C. Swann and C. W. Mims; **15.8** M. Powell; **15.9** John W. Taylor; **Page 315, (b)** John Hogdin, from A. H. R. Buller, *Researchers on Fungi*, vol. 6, Longman, Inc., New York; **15.13(a)** Thomas Volk; **15.13(b)** G. J. Breckon; **15.15(a, b)** J. C. Pendland and D. G. Boucias; **15.16(a)** C. Bracker; **15.16(b)** D. S. Neuberger; **15.16(c)** Bryce Kendrick; **15.19(a)** David J. McLaughlin, Alan Beckett, and Kwon S. Yoon, *Botanical Journal of the Linnaean Society*, vol. 91, page 253, 1985; **15.19(b)** D. J. McLaughlin and A. Beckett; **15.20** Haisheng Lü and D. J. McLaughlin, *Mycologia*, vol. 83, page 320, 1991; **15.21(b)** Haisheng Lü and D. J. McLaughlin, *Mycologia*, vol. 83, page 320, 1991; **15.22(a)** C. W. Perkins/Earth Scenes; **15.22(b)** Thomas Volk; **15.22(c)** Peter Katsaros/Photo Researchers, Inc.; **15.22(d)** J. W. Perry; **15.24** E. S. Ross; **15.25** Walter Hodge/Peter Arnold, Inc.; **15.26(a, b)** R. Gordon Wasson, Botanical Museum of Harvard University; **15.27(a)** M. Fogden; **15.27(b)** Thomas Volk; **15.27(c), (d)** Jeff Lopore/Photo Researchers Inc.; **15.28** E. S. Ross; **15.31(a, b)** Stephen J. Kron; **15.32** Grant Heilman Photography; **15.33(a)** Andrew McClenaghan/Photo Researchers Inc.; **15.33(b)** John Durham/Photo Researchers Inc.; **15.34(a, b)** G. L. Barron; **Page 333, (a, b)** G. L. Barron, University of Guelph; **(c)** N. Allin and G. L. Barron, University of Guelph; **Page 355,** John Webster, University of Exeter; **15.35(a, b)** E. Imre Friedmann; **15.36(a)** E. S. Ross; **15.36(b)** Meredith Blackwell/Louisiana State University; **15.36(c)** Robert A. Ross; **15.37(a, b)** E. S. Ross; **15.37(c)** Stephen Sharnoff/National Geographic Society; **15.37(d)** Larry West; **15.39(a–c)** V. Ahmadjian and J. B. Jacobs; **15.40** S. A. Wilde; **15.41(a, b)** Bryce Kendrick; **15.42** William J. Yawney and Richard C. Schultz; **15.43** D. J. Read; **15.44** B. Zak, U. S. Forest Service; **15.45(a)** R. D. Warmbrodt; **15.45(b)** R. L. Peterson and M. L. Farquhar; **15.46** Thomas N. Taylor, Ohio State University

Chapter 16
16.1 G. Vidal and T. D. Ford, *Precambrian Research*, 1985; **16.2** D. P. Wilson/Science Source/Photo Researchers, Inc.; **16.3(a)** Biophoto Associates/Photo Researchers, Inc.; **16.4** Linda Graham; **16.5** D. S. Neuberger; **16.7(a)** Victor Duran; **16.7(b)** Ed Reschke/Peter Arnold, Inc.; **16.8(a, b)** K. B. Raper **16.8(c),** London Scientific Films/Oxford Scientific Films; **16.8(e–g)** Robert Kay; **16.9** K. B. Raper; **16.10** Linda Graham; **16.11** Geoff McFadden and Paul Gilson; **16.12** Kim Taylor/Bruce Coleman, Inc.; **16.13(a)** D. P. Wilson/Eric and David Hosking Photography; **16.13(b)** E. S. Ross; **16.13(c)** R. C. Carpenter; **16.13(d)** Jean Baxter/Photo NATS; **16.14(a, b)** M. Littler and D. Litter, Smithsonian Institution; **16.15(a)** Linda Graham; **16.15(b)** J. Waaland; **16.16** Ronald Hoham, Colgate University; **16.17** (photo) C. Pueschel and K. M. Cole, *American Journal of Botany*, vol. 69, pages 703–720, 1982. **16.20(a, b)** D. P. Wilson/Science Source/Photo Researchers, Inc.; **16.20(c)** Florida Department of Natural Resources; **Page 364(a)** Erik Freeland/Matrix; **(b)** Florida Department of Natural Resources; **16.21** Robert F. Sisson/National Geographic Society; **16.22(a–c)** J. Burkholder, North Carolina State University; **16.23** J. Burkholder, North Carolina State University; **16.24(a)** Elizabeth Venrick, Scripps Institution of Oceanography, University of California, San Diego; **16.24(b)** Peggy Hughes/Institute of Marine Sciences

Chapter 17
17.1 Doug Wechsler/Animals Animals; **17.3(a)** A. W. Barksdale; **17.3(b)** A.W. Barksdale, *Mycologia*, vol. 55, pages 493–501, 1963; **17.5** After J. H. Niederhauser and W. C. Cobb, *Scientific American*, vol. 200, pages 100–112, 1959; **17.6(a)** M. I. Walker/Photo Researchers, Inc.; **17.6(b)** F. Rossi; **17.6(c)** Biophoto Associates/Science Source/Photo

Researchers, Inc.; **17.6(d)** Dr. Ann Smith/SPL/Photo Researchers, Inc.; **17.8(a, b)** C. Sandgren; **17.9(a)** G. R. Roberts; **17.9(b), (c)** D. P. Wilson/Eric and David Hosking Photography; **Page 380 (a)** Bob Evans/Peter Arnold, Inc.; **(b)** W. H. Hodge/Peter Arnold, Inc.; **(c)** Kelco Communications; **17.10(b)** Oxford Scientific Films; **17.11** C. J. O'Kelly; **17.12(a, b)** R. Evert and John West; **17.15(a–f)** Ronald Hohman; **17.17** After K. R. Mattox and K. D Stewart, in *Systematics of the Green Algae*, D. E. G. Irvine and D. M. John (eds.), 1984; **17.18** L. E. Graham; **17.19** W. L. Dentler/Biological Photo Service; **17.21(a–c)** L. Graham; **17.22** David L. Kirk, *Science*, vol. 231, page 51, 1986; **17.23** J. Robert Waaland/Biological Photo Resources; **17.24(a)** L. E. Graham; **17.24(b)** M. I. Walker/Photo Researchers, Inc.; **17.25** L. E. Graham; **17.26(a, b)** Larry Hoffman; **17.27** Gary Floyd; **17.28** After R. T. Skagel, R. J. Bandoni, G. E. Rouse, W. B. Schofield, J. R. Stein, and T. M. C. Taylor, *An Evolutionary Survey of the Plant Kingdom*, Wadsworth Publishing Company, Inc., Belmont, CA, 1966; **17.29(a)** James Graham; **17.29(b, c)** K. Esser, *Cryptograms*, Cambridge University Press, Cambridge, 1982; **17.29(d)** E. S. Ross; **17.30** D. P. Wilson/Eric and David Hosking Photography; **17.32(a)** Robert A. Ross; **17.32(b)** Grant Heilman Photography; **17.32(c)** L. R. Hoffman; **17.32(d)** V. Paul; **17.33(a, b)** Richard W. Greene; **17.34(a-d)** M. I. Walker/Science Source/Photo Researchers, Inc.; **17.35(a–d)** Lee W. Wilcox; **17.36(a, b)** L. E. Graham; **17.37(a)** William H. Amos/Bruce Coleman, Inc.

Chapter 18
18.1 T. S. Elias; **18.2** Brent Mishler; **18.3** D. R. Given; **18.5(b)** R. E. Magill, Botanical Research Institute, Pretoria; **18.6** L. Graham; **18.7(a, b)** D. S. Neuberger; **18.10** Karen Renzaglia; **18.11** D. K. Smith; **18.12(a)** John Wheeler; **18.13(a)** Field Museum of Natural History; **18.13(b)** Dr. G. J. Chafaris/Dr. E. R. Degginger; **18.15 (a, b)** Harold Taylor AB IPP/Oxford Scientific Films; **18.16(a,b)** J. J. Engel, *Fieldiana: Botany* (New Series), vol. 3, pages 1–229, 1980; **18.16(c)** J. J. Engel; **18.18** K. B. Sandved; **18.19(a)** Andrew Drinnan; **18.19(d)** D. S. Neuberger; **18.21** Larry West; **18.22** Martha Cook; **18.23(a)** M. C. F. Proctor; **18.24** C. Hebant; **18.25(a–c)** C. Hebant, *Journal of the Hattori Botanical Laboratory*, vol. 39, pages 235–254, 1975; **18.26(a)** D. S. Neuberger; **18.28** Rod Planck/Photo Researchers, Inc.; **18.29** Fred D. Sack; **18.30(b)** R. E. Magill, Missouri Botanical Garden, St. Louis; **18.31(a)** A. E. Staffen; **18.31(b)** E. S. Ross

Chapter 19
19.1 Diane Edwards; **19.2** M. K. Rasmussen and Stuart A. Naquin; **19.3** After A. S. Foster and E. M. Gifford, Jr, *Comparative Morphology of Vascular Plants*, 2nd ed., W. H. Freeman & Company, New York, 1974; **19.4** After Katherine Esau, *Plant Anatomy*, 2nd ed., John Wiley & Sons, New York, 1965; **19.5(a, d)** After K. K. Namboodiri and C. Beck, *American Journal of Botany*, vol. 55, pages 464–472, 1968; **19.5(b, c)** After K. Esau, *Plant Anatomy*, 2nd ed., John Wiley & Sons, Inc., New York, 1965; **19.10(b)** After J. Walton, *Phytomorphology*, vol. 14, pages 155–160, 1964; **19.10(c)** After F. M. Heuber, *International Symposium on the Devonian System*, vol. 2, D. H. Oswald (ed.), Alberta Society of Petroleum Geologists, Calgary, Alberta, Canada, 1968; **19.11** Field Museum of Natural History; **19.13** Specimen provided by Ripon Microslides, Ripon, WI; **19.15(a)** D. S. Neuberger; **19.16(a)** David Johnson, Big Bend National Park; **19.16(b, c)** D. S. Neuberger; **19.16(d)** Fletcher and Baylis/Photo Researchers, Inc.; **19.19** W. H. Wagner; **19.20** After Kristine Rasmussen and Stuart Naquin; **Page 456, (a)** After M. Hirmer, *Handbuch der Palaobotanik*, vol. 2, Druck and Verlag von R. Oldenbourg, Munich and Berlin, 1927; **Page 457, (b)** After W. N. Stewart and T. Delevoryas, *Botanical Review*, vol. 22, pages 45–80, 1956; **19.22(a)** R. L. Peterson, M. J. Howarth, and D. P. Whittier, *Canadian Journal of Botany*, vol. 59, pages 711–720, 1981; **19.22(b)** Heather Angel; **19.23(a)** D. Cameron; **19.23(b)** R. Schmid; **19.25(a)** R. Carr; **19.25(b)** E. S. Ross; **19.27** Dr. Jeremy Burgess/Science Photo Library/Photo Researchers, Inc.; **1930(a, c, e)** W. H. Wagner; **1930(b)** James L. Castner; **19.30(d)** Nancy A. Murray; **19.30(f)** David Johnson; **19.32(a)** W. H. Wagner; **19.32(b)** John D. Cunningham/Visuals Unlimited; **19.34** Bill Ivy/Tony Stone Images, Inc.; **19.35(a–d)** C. Neidorf; **19.37(a–c)** D. Farrar; **19.39(a)** D. S. Neuberger; **19.39(b, c)** W. H. Wagner

Chapter 20
20.1 M. K. Rasmussen and Stuart A. Naquin; **20.4(a, b)** After J. M. Pettitt and C. B. Beck, *Contributions from the Museum of Paleontology*, University of Michigan Press, vol. 2, pages 139–54, 1968; **20.4(c)** J. M.

Pettitt and C. B. Beck, *Science*, vol. 156, pages 1727–1729, 1967; **20.6** Charles B. Beck; **20.7** After S. E. Scheckler, *American Journal of Botany*, vol. 63, pages 923–934, 1975; **20.8, 20.9** M. K. Rasmussen and Stuart A. Naquin: **20.11(a)** Field Museum of Natural History; **20.13, 20.14(a)** J. Dermid; **20.15** E. S. Ross; **20.16** J. Kummerow; **20.19** B. Haley; **20.21(c)** Gary J. Breckon; **20.25** N. Fox-Davies/Bruce Coleman Ltd.; **20.28(a)** W. H. Hodge/Peter Arnold, Inc.; **20.28(b)** E. S. Ross; **20.29** H. H. Iltis; **20.30** Grant Heilman Photography; **20.31(a)** Larry West; **20.31(b)** J. Burton/Bruce Coleman, Inc.; **20.32** Geoff Bryant/Photo Researchers, Inc.; **20.33 (photo)** Carolina Biological Supply Company; **20.34** Mark Wetter; **20.35** Gene Ahrens/Bruce Coleman, Inc.; **20.36** Sichuan Institute of Biology; **20.37** D. A. Steingraeber; **20.38(a)** Knut Norstog; **20.38(b)** D. T. Hendricks and E. S. Ross; **20.39** Knut Norstog; **Page 484 (a, b)** W. Jones; **(c–e)** J. Plazz; **20.40(a)** J. W. Perry; **20.40(b)** Runk and Schoenberger/Grant Heilman Photography; **20.41(b, c)** G. Davidse; **20.42(a)** E. S. Ross; **20.42(b, d)** J. W. Perry; **20.42(c)** K. J. Niklas; **20.43(a)** C. H. Bornman; **20.43(b, c)** E. S. Ross

Chapter 21
21.1 E. S. Ross; **21.2(a–c)** W. P. Armstrong; **21.3(a, b)** G. J. Breckon; **21.3(c)** E. S. Ross; **21.4(a)** E. S. Ross; **21.4(b)** E. R. Degginger/Earth Scenes; **21.4(c)** T. Davis/Photo Researchers, Inc.; **21.5(a, b)** E. S. Ross; **21.5(c)** R. Carr; **21.8(a, c)** Larry West; **21.8(b, e)** J. H. Gerard; **21.8(d)** Grant Heilman Photography; **21.10** E. S. Ross; **21.12(a)** Larry West; **21.12(b)** Specimen provided by Rudolf Schmid; **21.13(a)** Runk and Schoenberger/Grant Heilman Photography; **21.15(a)** p. Echlin; **21.15(b, c)** J. Heslop-Harrison and Y. Heslop-Harrison; **21.15(d)** J. Mais; **21.21(a, b)** P. Hoch; **21.22** James L. Castner; **21.24(a, b)** C. S. Webber

Chapter 22
22-1 Chris R. Hill, courtesy of E. A. Jarzembowski, Maidstone Museum and Art Gallery; **22.2** E. Dorf; **22.4(a–c)** James L. Castner; **22.5(a–d)** David L Dilcher, (Reconstructions by Megan Rohn in consultation with D. Dilcher); **22.6** Geoff Bryant/Photo Researchers, Inc.; **22.7 (photo)** D. W. Taylor and L. J. Hickey, *Science*, vol. 247, page 702, 1990; **22.8(a, b)** E. M. Friis; **22.10** Mike Andrews/Animals Animals/Earth Scenes; **22.11(a, c)** E. S. Ross; **22.11(b)** J. W. Perry; **Page 526(a)** Donald H. Les; **22.12(b, c)** E. S. Ross; **22.12(d)** A. Sabarese; **22.13(a)** E. S. Ross; **22.14(a, b)** J. A. L. Cooke; **22.15(a, b)** W. H. Hodge; **22.16(a)** E. S. Ross; **22.16(b)** D. L. Dilcher; **22.17(a)** Larry West; **22.17(b)** T. J. Hawkeswood; **22.18, 22.19, 22.20** E. S. Ross; **22.21** Larry West; **22.22(a, b)** T. Eisner; **22.23, 22.24, 22.25(a, c)** E. S. Ross; **22.25(b)** L. B. Thein; **22.26** E. S. Ross; **22.27** M. P. L. Fogden/Bruce Coleman, Inc.; **22.28** R. A. Tyrell; **22.29** Oxford Scientific Films; **22.30(a–c)** E. S. Ross; **22.31** D. J. Howell; **22.32(a, b)** T. Hovland/ Grant Heilman Photography; **22.32(c)** E. S. Ross; **22.32(d)** J. F. Skvarla, University of Oklahoma; **22.33(b,c)** After D. B. Swingle, *A Textbook of Systematic Botany*, McGraw-Hill Book Company, New York, 1946; **22.28(c)** After L. Berson, *Plant Classification*, D. C. Heath and Company, Boston,1957; **22.34** D. S. Neuberger; **22.35(a)** John M. Pettitt and R. Bruce Knox, *Scientific American*, vol. 244, pages 134–144, 1981; **22.35(b)** Sean Morris/Oxford Scientific Films; **22.35(b), 22.38(b), 22.40(a)** After R. T. Skagel, R. J. Bandoni, G. E. Rouse, W. B. Schofield, J. R. Stein, and T. M. C. Taylor, *An Evolutionary Survey of the Plant Kingdom*, Wadsworth Publishing Company, Inc., Belmont, CA, 1966; **22.37(a)** J. L. Castner; **22.38(a)** E. S. Ross; **22.39(a)** E. S. Ross; **22.39(b ,c)** K. B. Sandved; **22.42** E. S. Ross; **22.43** Larry Atkinson/Mobridge Tribune; **22.44(a)** E. S. Ross; **22.44(b)** U.S. Forest Service; **22.45(a–c)** E. S. Ross; **22.46(a, b)** E. S. Ross; **22.47** J. Kendrick; **22.49** Robert and Linda Mitchell; **22.50(a, c, d)** T. Plowman; **22.50(b)** E. S. Ross

Chapter 23
23.2 After A. S. Foster and E. M. Gifford, Jr., *Comparative Morphology of Vascular Plants*, 2nd ed., W. H. Freeman & Company, New York, 1974; **23.4(a–c)** Daniel M. Vernon and David W. Meinke; **23.6(a–e)** Gerd Juergens, Lehrstuhl fuer Entwicklungsgenetik, Universitaet Tuebingen; Essay, page 563, Werner H. Muller/Peter Arnold, Inc.; **23.9** Tom McHugh/Photo Researchers, Inc.

Chapter 24
24.7 M. C. Ledbetter and K. B. Porter, *Introduction to the Fine Structure of Plant Cells*, Springer-Verlag, Inc., New York, 1970; **24.14(a, b)** H. A. Core, W. A. Cote, and A. C. Day, *Wood: Structure and Identification*, 2nd ed., Syracuse University Press, Syracuse, NY, 1979; **24.15** I. B. Sachs,

Forest Products Laboratory, U.S.D.A.; **24.17, 24.22, 24.27** After K. Esau, *Anatomy of Seed Plants*, 2nd ed., John Wiley & Sons, New York, 1977; **24.20(a)** M. A. Walsh; **4.24** J. S. Pereira; **24.26** Randall Brand; **24.28** T. Vogelman and G. Martin; **24.29** M. David Marks and Kenneth A. Feldman

Chapter 25
25.1, 25.2 After J. E. Weaver, *Root Development of Field Crops*, McGraw-Hill Book Company, New York, 1926; **25.4(a, b)** F. C. Guinel and M. E. McCully; **25.7** F. A. L. Clowes; **25.8** After K. Esau, Plant Anatomy, 2nd ed., John Wiley & Sons, Inc., New York, 1965; **25.9(a)** Robert Mitchell/Earth Scenes; **25.13** H. T. Bonnett, Jr., *Journal of Cell Biology*, vol. 37, pages 199-205, 1968; **25.14** After W. Braune, A. Leman, and H. Taubert, Pflanzenanatomisches Praktikum, VEB Gustav Fischer Verlag, Jena, 1967; **25.15** Robert D. Warmbrodt. *New Phytologist*, vol. 102, pages 175–192, 1986; **25.19** E. R. Degginger/Bruce Coleman, Inc.; **Page 604 (c, d)** M. E. Galway, J. D. Masucci, A. M. Lloyd, V. Walbot, R. W. Davis, and J. W. Sciefelbein. *Developmental Biology*, vol. 166, pages 740–754, 1994; **25.20** Robert and Linda Mitchell; **25.22(a, b)** D. A. Steingraeber

Chapter 26
26.5(a–d) Mary Ellen Gerloff; **Page 617(a)** H. C. Jones, Tennessee Valley Authority; **(b)** J. S. Jacobson and A. C. Hill (eds.), *Recognition of Air Pollution Injury to Vegetation: A Pictorial Atlas*, Air Pollution Control Association, Pittsburgh, PA, 1970; **(c)** After T. H. Maugh, II, *Science*, vol. 226, pages 1408–1410, 1984; **26.12(c)** W. Eschrich; **26.14, 26.15** After K. Esau, *Anatomy of Seed Plants*, 2nd ed., John Wiley & Sons, New York, 1977; **Page 626,** After P. A. Deschamp and T. J. Cooke, *Science*, vol. 219, pages 505–507, 1983; **26.17, 26.19** Rhonda Nass/Ampersand; **26.18(a)** James W. Perry; **26.24(a)** Michele McCauley; **26.25** William A. Russin; **26.31** Daniel J. Barta; **26.32** Joanne M. Dannenhoffer; **26.33(a, b)** Raymon Donahue and Greg Martin; **26.36(a–i)** Shirley C. Tucker; **26.37(a)** Leslie Sieburth; **26.39(a, b)** Leslie Sieburth; **6.39(c)** Mark Running; **26.40** James W. Perry; **26.41** E. R. Degginger/Earth Scenes; **26.42** David A. Steingraeber; **26.43(a, b)** James W. Perry; **26.44** G. R. Roberts

Chapter 27
27.19 S. Gutierrez/Photo Researchers, Inc.; **27.23** I. B. Sachs, Forest Products Laboratory, U.S.D.A.; **27.26** H.A. Core, W. A. Côté, and A. C. Day, *Wood Structure and Identification*, 2nd ed., Syracuse University Press, Syracuse, NY, 1979; **27.28(a)** Galen Rowell 1985/Peter Arnold, Inc.; **27.28(b)** C. W. Ferguson, Laboratory of Tree-Ring Research, University of Arizona; **27.30** Regis Miller; **Page 668 (c–e)** R. Evert

Chapter 28
28.4 Roni Aloni, Tel Aviv University; **28.6** Runk and Schoenberger/ Grant Heilman Photography; **27.7(a–c)** Bruce Iverson; **28.9(a)** E. Webber/Visuals Unlimited; **28.9(b)** F. Skoog and C. O. Miller, *Symposia of the Society for Experimental Biology*, vol. 11, pages 118–131, 1957; **28.11** J. D. Goeschl; **28.13** D. R. McCarty; **28.15** S. W. Wittwer; **28.16(b)** J. E. Varner; **28.17** J. van Overbeck, *Science*, vol. 152, pages 721–731, 1966; **28.18** Carolina Biological Supply Co.; **28.19** Abbott Laboratories; **28.25** After illustration by Hilleshög, Laboratory for Cell and Tissue Culture, Research Division, Landskvona, Sweden; **28.26** Phillip A. Harrington/Fran Heyl Associates; **28.27** Eugene W. Nester; **28.30** Bleecker; **28.31** Keith Wood, University of California, San Diego

Chapter 29
29.3(a, b) Department of Botany, University of Wisconsin; **29.4(a, b)** J. S. Ranson and R. Moore, *American Journal of Botany*, vol. 70, pages 1048–1056, 1983; **29.5** After B. E. Juniper, *Annual Review of Plant Physiology*, vol. 27, pages 385–406, 1976; **29.6** Stephen A. Parker/Photo Researchers, Inc.; **29.7(a, b)** Jack Dermid; **29.8** After A. W. Galston, *The Green Plants*, Prentice Hall, Inc., Upper Saddle River, NJ, 1968; **29.9(a)** Biophoto Associates/Science Source/Photo Researchers, Inc.; **29.9(b, c)** After B. Sweeney, *Rhythmic Phenomena in Plants*, Academic Press, Inc., New York, 1969; **29.10** Steve A. Kay; **29.11** After A. W. Naylor, *Scientific American*, vol. 286, pages 49–56, 1952; **29.12** After P. M. Ray, *The Living Plant*, Holt, Rinehart, & Winston, Inc., New York, 1963; **29.13(a, b)** Department of Botany, University of Wisconsin; **29.14(a–d)** U.S. Department of Agriculture; **29.18** Breck P. Kent/Earth Scenes; **29.19** U.S. Department of Agriculture; **29.20** A. Lang, M. Kh. Chailakhyan,

and I. A. Frolova, *Proceedings of the National Academy of Sciences*, vol. 74, pages 2412–2416, 1977; **29.21** R. Amasino; **29.23** After A. W. Naylor, *Scientific American*, vol. 286, pages 49–56, 1952; **29.25(a, b)** Robert L. Dunne/Bruce Coleman, Inc.; **29.26(a, b)** Runk and Schoenberger/Grant Heilman Photography; **29.27** Janet Braam; **29.28(a)** J. Ehleringer and I. Forseth, University of Utah; **29.28(b)** Gene Ahrens/Bruce Coleman, Inc.; **29.29** After J. Ehleringer and I. Forseth, *Science*, vol. 210, pages 10-94–10-98, 1980

Chapter 30
30.1 Lipha Tech; **30.2(a)** Donald Specker/Earth Scenes; **30.2(b)** Biological Photo Service; **30.3(a, b)** University of Wisconsin, Madison, Department of Soil Science; **30.4** E. Crichton/Bruce Coleman, Ltd.; **Page 737 (a)** Dwight Kuhn/Bruce Coleman, Inc.; **(b)** L. West, Bruce Coleman, Inc.; **30.6** After B. Gibbons, *National Geographic*, vol. 166, pages 3500–388, 1984 (Ned M. Seidler, artist); **30.7** After F. B. Salisbury and C. W. Ross, *Plant Physiology*, 4th ed., Figure 5–12b, Wadsworth Publishing Company, Belmont, CA, 1992; **30.8** From H. Curtis and N. Sue Barnes, *Invitation to Biology*, 5th ed., Worth Publishers, Inc., New York, 1994; **30.9(a, b)** B. F. Turgeon and W. D. Bauer, *Canadian Journal of Botany*, vol. 60, pages 152–161, 1982; **30–9(c)** Ann Hirsch, University of California, Los Angeles; **30-9(d)** E. H. Newcomb; **30.10** J. M. L. Selker and E. H. Newcomb, *Planta*, volume 165, pages 446–454, 1985; **30.11(b)** H. E. Calvert; **30.13** From H. Curtis and N. Sue Barnes, *Invitation to Biology*, 5th ed., Worth Publishers, Inc., New York, 1994; **Page 745(a)** Grant Heilman Photography; **(b)** J. H. Troughton and L. Donaldson, *Probing Plant Structures*, McGraw-Hill Book Company, New York, 1972; **Page 746** Foster/Bruce Coleman, Inc.

Chapter 31
31.1 Stephen Hale; **31.2(a, b)** J. H. Troughton; **31.3** After D. E. Aylor, J. Y. Parlange, and A. D. Krikorian, *American Journal of Botany*, vol. 60, pages 163–171, 1973; **31.4** After M. Richardson, *Translocation in Plants*, Edward Arnold Publishers, Ltd., London, 1968; **31.15** After A. C. Leopold, *Growth and Development*, McGraw-Hill Book Company, New York, 1964; **31.6** After M. Richardson, *Translocation in Plants*, Edward Arnold Publishers, Ltd., London, 1968; **31.7** L. Taiz and E. Zeiger, *Plant Physiology*, The Benjamin/Cummings Publishing Company, Inc., Redwood City, CA; **31.9** P. F. Scholander, H. T. Hammel, E. D. Bradstreet, and E. A. Hemmingsen, *Science*, vol. 148, 1965; **31.10, 31.11** After M. H. Zimmermann, *Scientific American*, vol. 208, pages 132–142, 1963; **31.12(a)** G. R. Roberts; **31.14** After M. Richardson, *Translocation in Plants*, Edward Arnold Publishers, Ltd., London, 1968; **31.15** H. Reinhard/Bruce Coleman, Inc.; **31.16** After E. Hausermann and A. Frey-Wyssling, *Protoplasma*, vol. 57, pages 37–80, 1963; **31.17** After Todd E. Dawson, "Hydraulic Lift and Water Use by Plants: Implications for Water Balance, Performance, and Plant-Plant Interactions," *Oecologia*, 95:565–574; **31.19** After J. S. Pate, *Transport in Plants*, I., *Phloem Transport*, M. H. Zimmermann and J. A. Milburn (eds.), Springer-Verlag, Berlin, 1975; **31.22** After Malpighii, *Opera Posthuma*, London, 1675; **31.23(a, b)** E. Fritz; **31.24(a, b)** M. H. Zimmermann; **31.26** After Neil A. Campbell, *Biology*, 4th ed., The Benjamin/Cummings Publishing Company, Inc., Menlo Park, CA, 1996

Chapter 32
32.1(a) N. H. Cheatham/DRK Photo; **32–1(b)** Michael Fogden/DRK Photo; **32.3** Harry Taylor/Oxford Scientific Films; **Page 778(a)** Kevin Byron/Bruce Coleman, Inc.; **(b)** John Shaw/Bruce Coleman, Inc.; **32.4** C. H. Muller; **32.5(a, b)** Australian Department of Lands; **Page 781(b)** R. T. Smith/Ardea; **32.6** G. E. Likens; **32.11(a)** Fred Bauendam/Peter Arnold, Inc. **32.11(b)** Wendell Metzen/Peter Arnold, Inc.; **32.11(c)** James H. Carmichael/Bruce Coleman, Inc.; **32.11(d)** L. West/Bruce Coleman, Inc.; **32.12** Jane Burton/Bruce Coleman, Inc.; **32.13** Bruce Coleman, Inc.; **32.14(a)** J. Dermid; **32.14(b)** E. S. Ross; **Page 790, (a)** U.S. Geological Survey; **Page 791, (b)** Jeff Henry/Peter Arnold, Inc.; **32.15** John Marshall; **32.16(a)** Keith Wendt/UW Arboretum; **32.16(b)** W. R. Jordan/UW Arboretum

Chapter 33
33.1(a) Jack Wilburn/Earth Scenes; **33.1(b)** Ardea Photographics; **33.3** After A. W. Kiichler; **33.4** Dr. Gene Feldman/NASA/Goddard Space Flight Center; **33.5** C. D. MacNeill, Jepson Herbarium, University of California, Berkeley; **Page 802,** F. G. Weitsch; **33.8(a)** Michael Fogden/DRK photo; **33.8(b)** Martin Wendler/NHPA; **33.8(c)** E. S. Ross; **33.9** Tom Bean/Allstock; **33.11** Peter Ward/Bruce Coleman, Inc.; **33.12** J. Dermid; **33.13(a)** Martin Wendler/NHPA; **33.13(b)** Max Thompson/NAS/Photo Researchers, Inc.; **33.13(c)** Ric Ergenbright; **33.13(d)** Greg Vaughn/Tom Stack and Associates; **33.14** F. C. Vasck; **Page 809, (a, b)** Ronald F. Thomas; **33.15(a, b)** J. Reveal; **33.16(a)** Jeff Foott; **33.16(b)** Pat Caulfield; **33.18** Soil Conservation Service; **33.19(a)** Rod Planck/Tom Stack and Associates; **33.19(b)** P. White; **33.20** J. H. Gerard; **Page 815,** Tom and Pat Leeson; **33.21** R. Burda/Taurus Photos, Inc.; **33.22** E. S. Ross; **33.23** E. Beals; **33.24** D. Brokaw; **33.25** E. S. Ross; **33.26** J. Bartlett and D. Bartlett/Bruce Coleman, Inc.; **33.27(a, b)** W. D. Bellings

Chapter 34
34.1 Irven de Vore/Anthrophoto; **34.2(a, b)** E. S. Ross; **34.3(a, b)** E. S. Ross; **34.4** E. S. Ross; **34.5** W. H. Hodge/Peter Arnold, Inc.; **34.6(a)** James P. Blair, © 1983, National Geographic Society; **34.6(b)** Susan Pierres/Peter Arnold, Inc.; **34.7** E. S. Ross; **34.8** G. R. Roberts; **Page 829,** John Doebley; **34.9(a)** E. Zardini; **34.9(b)** C. B. Heiser, Jr.; **34.9(c)** A. Gentry; **34.9(d)** M. K. Arroyo; **34.10** E. S. Ross; **34.11(a)** C. F. Jordan; **34.11(b)** M. J. Plotkin; **34.12** M. J. Plotkin; **34.13** K. B. Sandved; **34.14** W. H. Hodge/Peter Arnold, Inc.; **34.15** E. S. Ross; **34.16** W. H. Hodge/Peter Arnold, Inc.; **34.17** Harvey Lloyd/Peter Arnold, Inc.; **34.20** R. Abernathy; **34.21** C. A. Black; **34.22** AP/Wide World Photos; **34.23** W. H. Hodge/Peter Arnold, Inc.; **34.24** Agricultural Research Service, U.S.D.A.; **34.25(a)** Agricultural Research Service, U.S.D.A; **34.25(b)** University of Wisconsin; **34.26** Robert and Linda Mitchell; **34.27** 1982 Angelina Lax/Photo Researchers, Inc.; **34.28(a, b)** J. Aronson; **Page 845,** Gary Braasch/Tony Stone Images; **34.30** M. J. Plotkin; **34.31** Michael J. Balick

Index

Numbers in **boldface** indicate figures and tables.

A

ABA (*see* abscisic acid)
Abies, 483 (*see also* fir)
  balsamea, **483**, **788**
  concolor, 788, **789**
  lasiocarpa, 777, **791**
abscisic acid (ABA), **674t**, 683–84, **684**
  cell division and, **687t**
  cell wall extensibility and, 689
  dormancy and, 683–84
  molecular basis of action of, 688
  seed development and, 684
  stomatal movement and, **692**, 692–93, 753
  water relations and, 684
abscisin, 683
abscission, 636
  auxin and, 683
  ethylene and, 683
abscission zone, 636, **637, 657**
absolute temperature, 864
absorption
  of food by cotyledons, 563, 569
  by haustoria, 310, **311**
  by roots, 590, 595–600, 603, 605–6, 608
absorption spectrum, 131, **131**
Acacia (acacia), **10, 239, 549,** 590
  bull's-horn, 642
  cornigera, 775, **775**
  mutualism between ants and, 775, **775**
ACC (1-aminocyclopropane-1-carboxylic acid), 682, **682**
accessory buds, **657**
accessory fruit, 543
accessory pigments, 133 (*see also* carotenoids, chlorophylls, phycobilins)
acclimation, dormant condition in buds and, 718–19, **719**
ACC synthase, **682, 698**
Acer, 545, 624, 637, 718
  negundo, **657**
  saccharinum, **625**
  saccharum, **341, 625, 669t,** 762
Aceraceae, 545
Acetabularia, 395, **395**
Acetobacter diazotrophicus, 747
acetyl CoA, 114, **114,** 122
  in Krebs cycle, **115, 115**
acetyl group (CH₃CO), 114, **114,** 115
acetylsalicylic acid (aspirin), **36**
Achemilla vulgaris, 761
achenes, 544–45
Achlya, 373, **373**
  ambisexualis, **373**
  bisexualis, **26**
Achnanthes, 376
acid growth hypothesis, 689
acid rain, 616, **617,** 744
acids, 862–63
  fatty, 22–24, **23, 24,** 122, **124**
acorns, 545
acritarchs, **347,** 348
acropetal transport of auxin, 676
actin, 60
  myosin and, 47
actin filaments, **59,** 60, **60, 71t**
  cell division and, 158, 163, 166

actinomorphy, 502, 527
actinomycetes, **281,** 284–85
  nitrogen-fixing, 741
action spectrum, 131, **131**
active site, 101, **101,** 104
active transport, **82,** 83–84, **84**
adaptation, as result of natural selection, 245–47
adaptive radiation, 249, 250
adder's tongue, 454
Addicott, Frederick T., 683
adenine, 30, **30, 31,** 198, **198,** 199, **200**
  in ATP, **31,** 204
  in DNA, **30,** 198, 199, **200,** 201, 204
  in NAD, 102
  in RNA, **209t,** 223, **223**
adenosine diphosphate (*see* ADP)
adenosine triphosphate (*see* ATP)
Adiantum, **455**
adnation, 502
ADP (adenosine diphosphate), 31, 106
  in glycolysis, 111, **111**
  structure of, **105**
adventitious roots, 566, **568**
  auxin and, 678
aecia, aeciospores, 327, **329**
Aedes, **535**
aerial roots, 603
aerobes, 287
aerobic, 6
aerobic pathway in respiration, 113–22
Aeschynomene hispida, 670
Aesculus
  hippocastanum, **505, 613,** 648, **657**
  pavia, **627**
aethalia, 352, **353**
aflatoxins, 334
Africa, 523, 827
after-ripening, 565
agar, 358, **358,** 380
Agaricaceae, **265t**
Agaricales, **265t**
Agaricus
  bisporus, **265t,** 325
  campestris, 325
Agathis, 483, 488, 489
Agave (agave; century plant), 644, 808
  shawii, **808**
Agent Orange, 679
ageotropum (mutant of *Pisum sativum*), 706
aggregate fruits, 543
Aglaophyton major, 312, 434, **434**
Agoseris, **527, 546**
agricultural revolution, 824–35
agriculture, 11, 823, 824–35
  beginnings of, 824–25, 840
  domestication of plants, **825, 826, 827–31,** 832, **833–35**
  in Fertile Crescent, 825
  future of, 837–47
    genetic engineering, 847
    genetic variability of crop plants, 840–41
    improvement of existing crops, 838–43, **843**
    medicinal drugs, **833,** 843, 846, **846**
    new crops, 841–43, 846–47
    research efforts, 839
    sustainability, 844–45
  as global phenomenon, 833–35

  in the New World, 828–32
  population growth and, 835–37
  soils and, 744–45
  sustainable, 699
Agrobacterium, 294, **294**
  tumefaciens, **696,** 696–98, **698**
Agropyron, **540**
Agrostis tenuis, 146, 244–45, **245,** 595
A horizon (topsoil), 732, **732**
AIDS, 296, 308
air bubbles
  in tracheids and vessel elements, 577
  water transport in the xylem and, **756,** 756–58, **757**
air currents, 797
  transpiration and, 754
air plants, 585
air pollution, 616
  lichens and, 340
air roots, 605, **605**
air seeding, 757
air spaces (*see* intercellular spaces)
akinetes, **289, 290,** 290–91
alanine, **28**
Albizzia phylla, **545**
albuminous cells, 579, 582–83, **587t**
alcohol fermentation, 123, **123**
alcohol(s), 883
  ethyl (ethanol), 123, **123,** 308
  mannitol, 382
alder, 741
  red, 577, **669t**
Alectoria sarmentosa, **337**
aleurone layer, 563, **564**
  gibberellins and, 685, **686, 688**
alfalfa, 614, 619, **620,** 729, 738, 739
algae, 11, **275t,** 348, **349t, 372t** (*see also by name of group*)
  accessory pigments of, **349t,** 365, **372t,** 381, 383
  bioluminescence of, **363,** 366
  blooms of, 350, **364,** 365, 367, 378–79, 391
  blue-green (*see* cyanobacteria)
  brown (*see* brown algae)
  carbon cycle and, 350
  cell walls of, **349t, 372t**
  chlorophylls of, **349t,** 365, 367, **372t,** 378, 383
  chrysophytes (*Chrysophyta*) (*see* chryso-phytes)
  classification of, **275t**
  diatoms (*see* diatoms)
  ecology of, 348–50
  economic importance of, 350, 361, **364,** 366, **367,** 373–75, 379, 381–82
  evolution of, 348, 352, 360, 383
  green (*Chlorophyta*), (*see* green algae)
  edible, 350, 361, 380
  food reserves of, **349t, 372t**
  habitats of, **349t, 372t**
  haptophyte (*Haptophyta*) (*see* haptophyte algae)
  red (*Rhodophyta*) (*see* red algae)
  life cycle of, **171, 362, 374, 377, 384, 385, 386, 394**
  snow, 383, **386**
  symbiotic, 351, 356–57, **357,** 358, **365,** 378, 383
algin, 382
alginates, 380

alkaloids, **32,** 32–33, 549, **549,** 550
  fungal endophytes as producers of, 335
Allard, H. A., 709, 716
alleles
  crossing-over and, 191, **191**
  defined, 186
  dominant, 185, 186
  evolution and, 239
    gene flow, 241
    genetic drift, 241
  frequency of, 239, 240
  heterozygous/homozygous, 186
  interactions among
    of different genes, 195
    that affect the phenotype, 194
  multiple, 194
  principle of independent assortment and, 189, **189,** 190
  principle of segregation and, 186, **186**
  recessive
    diploidy and, 243, **243t**
    genetic variability and, 243
allelopathy, 33, 779
Allium, 162
  cepa, **562,** 567, **567, 593,** 644
  porrum, 341
allolactose, 217, 218
Allomyces, **308t**
  arbusculus, **314**
  life cycle of, **314**
  reproduction in, 313
allopatric speciation, 249
  punctuated equilibrium model and, 257
allopolyploidy (allopolyploids), **252,** 252–53
allosteric enzymes, 104
allspice, 833
Alnus, 741
  rubra, 577, **669t**
alpha-galacturonic acid, **22**
alpha-glucose, **19, 20,** 22
alpha helix, 28, **28, 29**
alpine penny cress, 747
alpine tundra, 798–99, **817**
alternation of generations, **171,** 172
  in algae, **171,** 360–61, 383, **384,** 393, **394**
  in *Allomyces*, 313, **314**
  in bryophytes, 402, **410–11**
  in chytrids, 313
  in vascular plants, 430
  isomorphic/heteromorphic, 172
altitude, distribution of organisms and, 801, **801,** 802
aluminum (Al)
  acid rain and, 616
  agriculture and, 747
Amanita, 325
  muscaria, **323**
  virosa, 325
amaranths, grain, 842–43
Amaranthus, 842
  retroflexus, **126**
Amaryllis (amaryllis), 536
  belladonna, **536**
Ambrosia psilostachya, **505**
American elm, **657,** 663
amino acids
  active site of enzymes and, 101, **101**
  coding sequences, 688, **688**

essential, 26
in food plants, 26
genetic code and, 209t
molecular structure of, 26–29, **28**, **29**
molecular systematics and, **269**, 269–70
nitrogen and, 26
proteinoid microspheres and, **4**
protein synthesis and, 208, 209, **209**, 210
sequence of, 28
in transfer RNA (tRNA), 210
aminoacyl (A) site, 211–13
aminoacyl-tRNA, 210
aminoacyl-tRNA synthetases, 210, 211
1-aminocyclopropane-1-carboxylic acid
(ACC), 682
ammonia (NH₃), 3, **3**, 26
ammonification, 736, **736**, 737
ammonium (NH₄⁺), nitrogen cycle and, **736**,
737–38
ammonium ions, 737, 738
amoebas, 365, 872 (see also myxamoebas)
of cellular slime molds, **354**
of plasmodial slime molds, 353
Amorphophallus titanum, 192
AMP (adenosine monophosphate), **31**, 106
structure of, **105**
Amphibolis, **542**
∝-amylase, **685**, 686, **688**
amylopectin, 20, **21**
amyloplasts, **23**, 51, **51**, 591 (see also plastids)
gravitropism and, 704–5, **705**
amylose, 20, **21**
Anabaena, **272**, 290, **290**, 741, **741**
azollae, **44**, 462
cylindrica, **288**
anabolism, 123–24, **124**
anaerobes, 287
facultative, 287
strict, 287
anaerobic pathways, in respiration, 122–23
anaerobic processes, 6
analogy (analogous features), systematics
and, 267
anaphase
meiotic, **174**, **175**, 176, 177
mitotic, **160**, **161**, 163
anatomy of plants, 11
Andreaea, 416
rothii, **415**
Andreaeidae, 412, 416
Andreaeobryum, 416
androecium, 501
of liverworts, **409**, 412
Anethum graveolens, 833
aneuploidy, 193
Aneurophyton-type progymnosperms, 471, **471**
angiosperms (flowering plants), **275t**, 470,
495–516, 875 (see also Anthophyta;
flowers)
ancestors of, **519–23**
aquatic, 541–42, **542**
cytoplasmic inheritance in, 196
diversity of, 496–98
embryogeny, **556–61**
evolution of, 428, 431, 432, 496, **517–53** (see
also flowers, evolution of; fruits,
evolution of)
biochemical coevolution, 549–51
hypotheses of relationships among
anthophytes, 518–19, **519**
magnoliids, 519, **520**, 521
origin and diversification, 519–23
resistance to drought and cold, 522
single common ancestor, 519
fossil, 517–18, 520, 522
gametophytes of, 431
life cycle of, 503–14, **512–13**
development of the seed and fruit,
509–10
double fertilization, 509, **509**
megasporogenesis and megagametogenesis, 506
microsporogenesis and microgametogenesis, 504–6
outcrossing, 510–11

pollination, 508–9, **530–42**
self-pollination, 514
parasitic, **498**, 547
phylogenetic relationships with other
embryophytes, 472
saprophytic, **498**
specialized families of, **527–29**
vessel elements of, 428
Angophora woodsiana, 531
angstrom (Å), 42
Animalia (animal kingdom), 276
animals (see also birds; insects; and specific
animals)
dispersal of fruits and seeds by, 547, 547–49
distribution of (see specific biomes)
domestication of, 826, **826**, 831
as fruit and seed distributors, **547–49**
germination and, 565
secondary plant metabolites and, 549
anions, 735, 856
anise, 833
anisogamy, **372**
Anna's hummingbird, **537**
annual rings, 664
annuals, 648
annular scars, in *Oedogonium*, 392, **392**
annulus, 454
Antarctica, 801, 821
lichens in, 821
antenna complex, 135, **136**
antenna pigments, 135
antheridia, 372
of algae, **385**, 392, **392**
of ancestor of plants, 398, **402**
of ascomycetes, 320
of bryophytes, **402**, 404, 405
of club mosses (Lycopodiaceae), **437**, 438
of Equisetum, **449**, **451**
evolution of vascular plants and, 430, 431
of ferns, 455, 458–59, **460–61**
of hornworts, 412
of Marchantia, **405**, **408**, **411**
of liverworts, 408, **408**, **409**
of Lycopodium, **437**
of mosses, 414, **417**, **419**, 420, **420**
of oomycetes, **374**
of Psilotum, 444, **447**
of seedless vascular plants, 431
of Selaginella, **441**, 442
of Takakia, **407**
antheridiol, 26
antheridiophores, 408, **408**
anthers, **184**, 499, 501, 504
dehiscence of, **504**, 508
of lily, **504**
of orchids, 528
outcrossing and, **511**
of wind-pollinated flowers, 540, **541**
anthesin, 715
Anthocerophyta, 874 (see also bryophytes;
hornworts)
Anthoceros, 412, **413**
anthocyanins, 35, **35**, 54–55, 542, 543, **543**
Anthophyta, **265t**, 470, 875 (see also
angiosperms)
diversity in, 496–98
anthophytes, 518, **519**
anthracnose diseases, 334
anthrax, 287
Anthreptes collarii, **537**
antibiotics, 287
deuteromycetes and, 334
anticlinal divisions, 612, **612**
of fusiform initials, **649**
anticodons, 210, **211–13**
antiparallel strands, 199, **200**
antipodals, 506, 509, **509**
antiporters, 83, **84**
Antirrhinum majus, 639
ants, **538**, 606, 642
mutualism between acacia trees and, 775,
775
seed dispersal by, 548
añu, 830
A₀, A₁, 138

apex (apices)
apical-basal pattern of, 556, **556**, 557, 560
floral, **638**, 639
reproductive, 639
root, **479**, 591, 611
shoot, **436**, **440**, **446**, **450**, **460**, **479**, **566**,
610–12, **612**, 614, 635, **635**
transition to flowering, 639
aphids, 298, 766, **767**
Ap horizon, 732
Apiaceae, 264, **502**, 530, 531, 545, 833
apical cell, 556, **558**
apical dominance, **678**, 678
apical meristems, 8, 157, **157**, 425, 426, 559,
562, 571, **571**
of embryo, **559**, **561**, 564
functions of, 570
of pines, 474, **475**, 482
of roots, 8, **559**, 562, **570**, 591–92, 604, 605
quiescent center, 594
types of organization of, 592, **593**
of shoots, 8, **556**, 557, **557**, **558**, 559, **561**, 562,
562, 568, **570**, 611–14
apical placentation, 501, **501**
Apis mellifera, 532
Apium
graveolens, **643**
petroselinum, **752**
Apocynales, 34
apomixis, 180, 255–56
apoplast, 87
apoplastic pathway, 759, **760**
apoplastic transport, 87
apoptosis, 355, 578
apple, 241, 587, 649, 655
flowers, 503
appressoria, 339
aquaporins, 83
aquatic plants, 397, 401, 407, 526 (see also
algae; water molds; and specific plants)
angiosperms, 541–42, **542**
leaf dimorphism in, 626, **626**
Aquilegia canadensis, **538**
Arabidopsis, **157**, **556**, 682, 698, **704**
thaliana, 47, 228, **228**, 560, **560**, **561**, 586, 692,
708, 714
embryo-defective mutants of, 560, **560**,
561
floral organ identity in, 639–40, **640**, 641
flowering, 716, **717**
roots of, 604, **604**
thigmomorphogenesis, 721–22, **722**
trichomes on the epidermis of, 585
Araceae, 37
Araucaria, 483, 488, 489
araucana, 483
heterophylla, 483, 484
Araucariaceae, 483, 488
arbuscules, 341, **341**, 344
Arceuthobium, 547
archaea, 6, 270, **270**, 282, 294–96
extreme halophilic, 295
extreme thermophilic, 296, **296**
methane-producing (methanogens), 295,
295, 296
Archaea (domain), **272**, 294–96, 869
Archaeanthus, 521, **530**, 544
linnenbergeri, 519, **520**
Archaeopteris, **471**
macilenta, **471**
Archaeopteris-type progymnosperms, 471
Archaeosperma arnoldii, 468, **469**
archegonia
of ancestors of plants, 398, **402**
of bryophytes, **402**, 404–9
of club mosses (Lycopodiaceae), **437**, 438
of Equisetum, **449**, **451**
evolution of vascular plants and, 431
of ferns, 455, 458–59, **460**
of hornworts, 412
of liverworts, 408, **408**, **409**
of Lycopodium, **437**
of Marchantia, 449
of mosses, 414, **417**, **418**, 420
of pines, 481, 482

of *Psilotum*, 444, **447**
of seedless vascular plants, 431
of *Selaginella*, **441**, 442
archegoniophores, 408, **408**
Arctic, 801
Arctic tundra, **798–99**, 819, 819–21
Arctophila fulva, **819**
Arctostaphylos, 565
viscida, **565**
Arcyria nutans, 353
areoles, 631
arginine, 27
Argyroxiphium sandwicense (silversword), 250,
250, 251
arils, 483, **484**
seed dispersal and, 548
Aristida purpurea, 591
Aristolochia, **428**, 654
grandiflora, **521**
Aristolochiaceae, 521, **521**
tridentata, 808
Aristotle, 733
Armillaria
gallica, 307
ostoyae, 307
arrowhead, **558–59**
Artemisia dracunculus, 833
tridentata, 808
Arthrobotrys anchonia, 333
Arthuriomyces peckianus, **311**
artificial selection, 238, **238**
Asarum canadense, 762
asci, 318, 318–19, **319**
in yeasts, 330, **331**
Asclepiadaceae, 266, 531, 550 (see also milkweed)
Asclepias, 34, **544**, 546
curassavica, **549**
Asclera ruficornis, 531
ascocarp (see ascomata)
Ascodesmis nigricans, **319**
ascogenous hyphae, 320
ascogonia, 320
ascomata, 319, 338
ascomycetes (Ascomycota), 275, **308t**, 309, **309**,
310, **310**, 312, **317**, 317–20, 870
life cycle of, 318, **318**, 320
ascospores, 318, **318**, 319, 319, 320, 330, 331,
331
aseptate hyphae, 310
asexual reproduction (vegetative reproduction), 166, 170, 179, **179**, 180
in algae, 352, 360, 371, 373, **374**, 377, 379,
389–90, 391–92, 396
in ascomycetes, 318, **318**
in bryophytes, 404
in *Chlamydomonas*, 388
in diatoms, 376, **376**, 377, 378
in euglenoids, 352
in fungi, 311–12
genetic variability and, 242
in mosses, 414, 421
in slime molds, 355–56
in vascular plants, 255, **256**
in yeasts, 330
ash, 545, 755
green, **627**
white, **669t**
Asia, 826–27
A (aminoacyl) site, 211, **212**, **213**
asparagine, 27, 740
asparagus, 642
Asparagus officinalis, 642
aspartate (aspartic acid), 27, 144, **144**
aspen, quaking, **788**
Aspergillus, 332, 333
flavus, 334
fumigatus, **332**
oryzae, 334
parasiticus, 334
soyae, 334
aspirin, 36
Asplenium
rhizophyllum, **180**
septentrionale, **453**
viride, **800**

assimilate stream of the phloem, 764, **764**
  nutrients exchanged between transpiration stream and, **763**, 763–64
assimilate transport, 764–69
  and aphids, 766, **767**
  phloem loading, 767–68, **769**
  phloem unloading, 768–69
  pressure-flow hypothesis, 766–69, **768**
  radioactive tracer experiments, 765–66
  in sieve tubes, 765–67
Asteraceae (Compositae), 256, 264, **505**, **506**, 511, **527**, 527–28, 545
aster yellows, 293
athlete's foot, 334
atmosphere, 844
atmospheric gases, in early Earth, 3–4
atom, 863 (*see also specific elements*)
atomic mass, 863
atomic mass units (daltons), 863
atomic number, 863
atomic structure
  of familiar elements, **864t**
  models of, **853**, 865–67, **866**, **867**
atomic weight, 863
  chemical reactivity of, 867
ATP (adenosine triphosphate), 31, **31**, 93
  alcohol fermentation and, 123
  chemiosmotic coupling and, 120, **121**
  in electron transport chain, 118
  energy supplied (released) by, **31**, 31–32, 105, **105**, 106, 120–22
  in glycolysis, 111, **111**, 112, **112**, 113, **113**
  in oxidative phosphorylation, 119, 120
  in photosynthesis, **134**, 135, **136**, 137
    cyclic photophosphorylation, 139, **139**
  in respiration, 108, **108**, 109, **110**
  role of, 105–6
  structure of, **105**
ATPases, 47, 105, **105**
ATP synthase, 119, 120, **120**, **121**, 137, **138**
Atriplex, 585, 744, **744**
Atriplex nummularia, **843**
atropine, 33
Aureococcus, 379
aureomycin, 287
Auricularia auricula, **322**
autoecious parasites, 327
autopolyploidy (autopolyploids), 252, **252**
autoradiography, 864–65
autotrophic cells, 124
autotrophs, 286, 287, 783
  chemosynthetic, 287
  photosynthetic, 5, 287
autumn crocus plant (*Colchicum autumnale*), 193
auxiliary cell, 361
auxins, 66, **674t**, 675–79, **676**, **679**, **687t**, 696
  (*see also* indoleacetic acid (IAA); naphthaleneacetic acid (NAA))
  abscission and, 683
  Agent Orange and, 679
  apical dominance and, **674t**, 678, **678**
  cambial reactivation and, 650
  cell differentiation and, 676, **677**
  cell division and, 678, 681, **687t**
  cell elongation and, 66, **675**, **687t**
  cell wall extensibility and, 689
  circinate vernation and, 458
  differentiation of vascular tissue and, 676
  ethylene and, **674t**, 682–83
  fruit growth and, **674t**, 678, **679**
  gibberellins and, 689
  gravitropism and, 704–5, **705**
  leaf development and, **674t**
  phototropism and, **702**, 703
  ratio of cytokinin to, 681
  root growth and, **674t**, 678
  synthetic, 679
  in tissue culture, 680, **681**
  transport of, 676, 677
  vascular cambium and, 678
  in Went's experiment, 676, **702**
auxospores, **376**, **377**
Avena sativa, 146, 563, **589**, 675
Avicennia germinans, 605

avocado, 683
Avogadro's number, 871
awns, wheat with, 196
axial core, 175, **176**
axial system, 648
axil, leaf, **8**, 611, 626
axile placentation, 501, **501**
axis
  of plant body, 556
  stemlike (*see* epicotyl; hypocotyl-root axis)
Azolla, 290, 462, 741, **741**
  filiculonides, **741**
Azotobacter, 741
Azotococcus, 741

B

Bacillariophyta (diatoms), **275t**, 349, 371, **372t**, **375**, 375–78, 873 (*see also* algae)
  heterotrophic, 378
  pennate, **276**
  resting stages of, 376, 378
bacilli (rods), 284, **285**
Bacillus
  cereus, **286**, 779
  thuringiensis, 697
bacteria, 287–94 (*see also* prokaryotes *and by name*)
  cell walls of, 283, **283**
  commercial uses of, 287
  diseases caused by, 287, 293–94
  endospores of, 286, **286**
  evolution of mitochondria and chloroplasts from, 53
  gram-negative, 283, **283**
  gram-positive, 283, **283**
  green, 291
  nitrifying, 738
  photosynthetic, 127–28, 133, 291 (*see also* cyanobacteria; green bacteria; prochlorophytes; purple bacteria)
  as prokaryotes, 6
  purple, 282, 283
    photosynthesis in, 291
  purple nonsulfur, 291, **291**
  purple sulfur, **127**, 128, 291
  thermophilic, **282**
Bacteria (domain), 270, **270**, **272**, 282, 287–94, 869–70
  major lineages (kingdoms) of, 287
bacterial chromosome, 156, **156**
bacteriochlorophyll, 133, 291
bacteriology, 12
bacteriophages, 286, 297, **297**
bacteroids, 739, **739**
badnaviruses, 300
Baker, Kathleen Drew, 361
bald cypresses, 483
balsa, 659
balsam fir, **483**, **788**
banana, **497**, 827
banyan tree, 603
bar (unit of pressure), 77
Barbarea vulgaris, **728**
barberry (*Berberis*), 327, **328**
bark, 647, 650, 654–59, **658**
  definition of, 654
  inner, **655**, 658, 659
  outer, **655**, 658
  ring, 658
  scale, 658, **658**
barley, **635**, 685, **686**, 688, 825, **825**
  bran, 563
  foxtail, **278**
barley malt, 685
barrel cactus, 809, **809**
basal bodies, 61, **387**
basal cell, 556, **557**, **558**
basal placentation, 501, **501**
bases, 862–63
basidia, **321**, 324, **324**
basidiocarp (*see* basidiomata)
basidiomata, 320, **320**, 322, 325
Basidiomycetes, **265t**, **275t**, 309, 322, **323**, 323–26, 871
  life cycle of, **320–21**

Basidiomycota, **265t**, 275, **308t**, 309, 311, 312, 320–30, 871 (*see also* fungi)
  classes of, 322
  life cycle of, **328–29**
  mycelium of, 322
basidiospores, 320, **320–321**, 322–24, **324**, 325, 326, **326**, 327, 328
  in *Ustomycetes* (smuts), 330
basil, 833
basipetal transport of auxin, 676
Bassham, James A., 140
basswood (linden; *Tilia americana*), **575**, **577**, **580**, **657**, **669t**
  bark of, 655
  stems of, 614, 615, 618, 651, 652, **652**
bast fibers, 575
Bateson, William, *190*
Batrachospermum
  moniliforme, 357
  sirodotia, 360
bats, pollination by, 538–39, **539**
Bauhin, Caspar, 262
bay leaves, 833
Beagle, HMS, 236, **236**
bean golden mosaic, 300
beans, 738
  broad, **8**, 738, **766**
  castor, **562**, 563, **566**, 567, **578**
  contender, **685**
  garden, **562**, 563, **566**, 567
beech, 762, 813
  American, **669t**
  family, 342
bees, pollination by, **532**, 532–35, **533**, **534**
bee's purple, **533**, 543
beetles, 550
  Colorado potato, 550
  longhorn, **530**
  pollination by, **530**, 531, **531**
  pollination of cycads and, 487
beets, 835
  sugar-, **272**, 606, **607**, 834, 835, **835**
beet yellows virus, 300, **300**
beggar-ticks, 528
Beijerinckia, 741
Belt, Thomas, 775
Beltian bodies, 775, **775**
Bennettitales (cycadeoids), 472, **473**, 518
Benson, Andrew A., 140
bent grass (*Agrostis tenuis*), 244–45, **245**, 595
benzylamino purine (BAP), **680**
Berberis, 327
berries, 543
beta-carotene, 133
betacyanins (betalains), 543, **543**
beta-fructose, 20
beta-glucose, **19**, 22
beta oxidation, 122
beta pleated sheet, 28, **29**
Beta vulgaris, 606, **607**, 835
Betula
  alleghaniensis, **343**, **669t**
  papyrifera, **541**, **658**
Betulaceae, 342
Bevhalstia pebja, **517**
B horizon (subsoil), 732, **732**
bicarbonate ion (HCO_3^-), 144, **144**, **149**
Bidens, 528
biennials, 648
bifacial vascular cambium, 470
binary fission, 156, 285
bindweed, field, 599
binomial names, 262–64
binomial system of nomenclature, 262
biochemical coevolution, 549–51
biodiversity, 844, 845 (*see also* genetic variability)
biogeochemical cycles, 736
biological clocks, **12**, 709
  circadian rhythms and, 706–9
biological species concept, 248
bioluminescence, 122, 699, 708
  of dinoflagellates, 366
biomass, 784
  pyramids of, **785**, 785

biomes, 9, 796, **820t** (*see also* forests; *specific types of biomes*)
  definition of, 797
  distribution of, 797, **798–99**, 802
  map, **798–99**
  patterns of wind and rainfall and, **801**, 802
  relationship between altitude and latitude and, 800, **801**
biosphere, 5
biotechnology, 231, 693–99 (*see also* genetic engineering; recombinant DNA)
birches, 342
  ectomycorrhiza of, **343**
  paper, **10**, 541
  yellow, **669t**
bird-of-paradise, 537
birds
  DDT and, **781**
  in fruit and seed dispersal, **547**, 548
  germination and, 565
  pollination by, 537–38, **538**
bird's-nest fungi, 323, 326, **326**
birth-control pills, 846, **846**
birthwort family, 521, **521**
bisexual gametophytes, 430, 431, **436**, 438, 444, **444**, 449, 449, **450**
bivalents (tetrads), 172, **173**
black ironwood, 670
black locust, **627**, **656**, **657**, 659
Blackman, F. F., 128
black oak, 256, **257**, **658**
black pepper, 832
black stem rust of wheat, 327
  life cycle of, **328–29**
black walnut, **669t**
black wart disease, 313
bladderwort, 737
blazing star, 591
blights, 293, **294**
blister rust, **327**
bloodroot, 549
blooms
  of algae, 350, **363**, **364**, 365, **367**, 378–79, 391
  of chrysophytes, 379
  of cyanobacteria, 289, 290
  of dinoflagellates, 364
  of extreme halophilic archaea, **295**
  of haptophytes, 367
bluebells, **500**
blueberry, lowbush, 762
bluegrass
  annual, **633**
  Kentucky (*Poa pratensis*), 146, 244
    apomixis in, 255–56
blue-green algae (*see* cyanobacteria)
body cell (*see* spermatogenous cell)
body plan, 556, **556**
Bohr, Niels, 865
bolting, gibberellin and, 686, **687**
bond energy, 875
bonds, chemical, 105, 106, 868–70
Bonnemaisonia hamifera, 358
Bonner, James, 711, 715
Bordeaux mixture, 373
bordered pits (pit-pairs), **64**, 659, 660, **661**, 662
boreal forests (taiga), 150, **818**, 818–19
Borlaug, Norman, **839**
boron (B), as essential element, 727, **730t**
Borthwick, Harry A., 712
Boston ivy, 641
botany (plant biology), 2, 11
  areas of study in, 11–12
  importance of knowledge of, 12–14
Botrychium, 453, 454–55
  parallelum, **454**
  virginianum, **59**
bottleneck effect, 242
botulism, 287
Bougainvillea, 543
box elder, **657**
Brachythecium, 421
bracken fern, **458**
bracket fungi (*see* shelf fungi)
Bradyrhizobium, 738–40
  japonicum, 739, **739**

bran, 563
branches of lycophyte trees, 456
branching in progymnosperms, 471, **471**
branch traces, 622, **624**
Brassica
   juncea, 747
   napus, 698
     var. *napobrassica*, 763
   nigra, **549**, 833
   oleracea, 47, **238**
     var. *capitata*, 643, 686
     var. *caulorapa*, 643, **644**
   rapa, **544**
Brassicaceae, **228**, 340, 511, 544, **544**, 549
brassinolides, 675, **675t**
brassins, 26
Brazil, **803**
Briggs, Winslow, 703, **703**
bristlecone pine, **664**, 664–65
British soldier lichens, **337**, 339
broad beans (*Vicia faba*), **8**, 738, **766**
broccoli, 238
Bromeliaceae, 149, **337**
bromeliads, 804, **804**
brown algae (*Phaeophyta*), **7**, 171, **275t**, 370,
    371, **372t**, 379–83, 873 (*see also* algae)
   conducting tissues in, **382**
   edible, 380
   fucoxanthin in, 381
   life cycles of, 383
   thalli, 381
Browne, John, 845
brown rot of stone fruits, 317
brown spot of corn, 313
brown tides, 379
brussels sprouts, 238
Bryidae ("true mosses"), 412, 416–22 (*see also*
    bryophytes; mosses)
   "cushiony" or "feathery" growth patterns
    in, 421, **422**
   sexual reproduction in, 416, **417–18**, 418–21
   specialized tissues for water and food
    conduction, 416, **416**, **417**
Bryophyta (*see* bryophytes; mosses)
bryophytes, 172, **275t**, 277, 400–423 (*see also*
    hornworts; liverworts; mosses)
   comparative structure of, **403**, 403–4
   comparative summary of characteristics of,
    422t
   life cycles in, **410–11**, **418–19**
   matrotrophy in, 404–6
   phyla of, 407
   phylogenetic relationships with other
    embryophytes, 472
   relationships to other groups, 401–4, **402**
   reproduction in, 404, **405**, 406
   sporopollenin-walled spores of, 402,
    406–7
   symbiosis and, 403
   vascular plants compared to, 425
BT gene, 697, **698**
buckeye, red, **627**
buckwheat family, 545
budding, of yeasts, 330, 331, **331**
bud primordia, 611, **611**, 634
buds, **8**, 611, 613, 622
   accessory, 657
   auxin-kinetin ratio and, 681, **681**
   dormancy of, 717, **718**, 718–19, **719**
   hormones and, 678
   lateral (axillary), **657**, 678, **678**
   in mosses, **416**
   terminal, **657**
   on woody stems, **657**
bud scale(s), **613**
   dormancy and, 718, **718**
   scars, 657
bugs (*see also* insects)
   true, 550
bulbs, 643, **644**
   forcing, 719
bulk flow, 77
bulliform cells, 632, **633**
bull kelp, **379**
bull's-horn acacia, 642, 775, **775**

bumblebees, 534, **534**
bundle scars, **657**
bundle-sheath cells, 144, **145**, 146
   of grasses, 632, **633**
bundle-sheath extensions, 632
bundle sheaths, **628**, **629**, 632
   in grasses, 632, **632**, 633
butter-and-eggs, **500**, 546
buttercup, **596**, **598**, 614, 619, **620**
   family, **502**, 545
   water, **197**
butterflies, 550
   monarch, 34, **35**, 550
   pollination by, 535–36, **536**
   viceroy, 550
butternut, **657**
butterwort, 737
buttonwood, **658**

C
C_3–C_4 intermediates, 147
C_3 pathway (*see* Calvin cycle)
C_4 pathway (C_4 photosynthesis; Hatch-Slack
    pathway), 144, **144**, **145**, 146, 147, **148**,
    149
   in grasses, 632, **633**
C_3 plants, 144
   global warming and, 151
C_4 plants, 144, **144**, **145**, 146, 147, **147**, **148**, 149
   global warming and, 151
cabbages, **238**, 550, 643
   bolting and flowering, 686, **687**
   family, **73**
cacao, 834, **834**
Cactaceae, **266**
cacti, 266, **266**, **497**, 543
   barrel, 809, **809**
   cladophylls of, 642, **642**
   dispersal of, **547**
   family, 266
   fishhook, **278**
   organ-pipe, **539**
   peyote, **551**
   prickly-pear, 779, **780**
   saguaro, **10**, 539, **808**
   water storage in, 644
Cactoblastis cactorum, 779, **780**
caffeine, 32–33, **33**, 549, 833
calactin, 549
calamites (giant horsetails), 445, 456–57, **466**
calcification
   algae and, 350
   of red algae, 358–59
calcite, 731
calcium (Ca), **76**
   cycling, 782–83
   deposits (stromatolites), **289**
   as essential element, 727, **730t**
   gravitropism and, 704, **705**
calcium carbonate (calcite), 731
calcium ions (Ca^{2+})
   halophytes and, 744
   hormone action and, 691, **692**
   as second messengers, 87
calcium oxalate, **575**
   in vacuoles, 54, **54**
California, chaparral of, 817, **817**
California poppy, **497**, **534**
Callitriche heterophylla, 626
Callixylon, 471
   newberryi, **470**
callose, 579, 580, **580**, **583**
   definitive, 580
   wound, 580
callus, 681, **681**
calmodulin, 87, 704
Calocedrus decurrens, **278**, 788
calorie, 864
Calostoma cinnabarina, **326**
Calothrix, **289**
Caltha palustris, **533**, 543
Calvin, Melvin, 139
Calvin cycle (C_3 pathway), **134**, 139–43, **140**,
    141, 147, **148**, 149
   C_4 pathway and, 144

photorespiration and, 143
Calycanthaceae, 519
Calypte anna, **537**
calyptra, 406, **410**, **411**, 419, 421 (*see also*
    venter)
calyx, 501, **502**, 527
calyx tube, **638**
cambial initials (*see* fusiform initials; ray
    initials)
cambial zone, 649
cambium, 443
   cork (phellogen), **8**, 427, 587, **587**, 600, 602,
    647, 648, 653–55, **655**, 658, **659**
   fascicular, **619**, 650–51
   interfascicular, **619**, 650–51
   of *Isoetes*, 443
   reactivation of, 649–50, 678
   storied/nonstoried, **649**
   supernumerary, 606, **607**
   vascular (*see also* vascular cambium), **8**, 427,
    576, **576**, 647–50, **649**, **650**
Cambrian period, 877
   Lower, 312
Camellia sinensis, 833
cAMP (cyclic adenosine monophosphate;
    cyclic AMP), 354
   cellular slime molds and, 355
CAM photosynthesis, 147, **148**, 149
CAM plants, 147, **148**
   stomatal movements and, 754
Campylopus, **401**
canary grass, 675
Candida albicans, 331
Candolle, Augustin-Pyramus de, 264
Cannabis sativa, 550, **551**
canola, 698
cap (pileus), 324
Capparidaceae, 263
Capsella bursa-pastoris, 557
Capsicum, 833
   frutescens, **299**
capsids, 298, **299**
capsules (*see also* sporangia)
   of angiosperms, **544**
   of bacteria, 284
   of bryophytes, **405**, 406, 408, **408**, **409**, 412,
    414, **414**, 416, **418**, 419, 420, 421, **421**,
    422
   of mosses, 414, 416, **418**, **419**, 420, 421, **421**,
    422
caraway, 833
carbohydrates, 18, **19–22**, **38t** (*see also by name*)
   in carbon cycle, 150
   formation of, **20–22**
   molecular structure of, 18–22
   oxidation of, 109–24
   in plasma membrane, 74–75
carbon (C), 18 (*see also* carbon cycle)
   covalent bonds of, 869
   as essential element, 727
   fixation of (*see* carbon-fixation reactions)
carbon balance, global (*see also* greenhouse
    effect)
   forests and, 150
   prokaryotes and, 287
carbon cycle, 150–51, **151**
   algae and, 150, 350
   peatlands and, 415
carbon dioxide (CO_2) (*see also* carbon cycle)
   in atmosphere, 3, 150–51, 845
   diffusion of, 78
   from fermentation, **110**, 122, **123**
   from glucose oxidation, 109
   Krebs cycle and, 109, **115**, 116, **121**, **124**
   in oceans, 150
   in photosynthesis, 751 (*see also* carbon-
    fixation reactions)
   from pyruvate oxidation and decarboxyla-
    tion, 114
   stomatal movement and concentration of,
    753
carbon-fixation reactions, 133, 134, **134**, 135,
    139–51
   advantages and disadvantages of, 149
   Calvin cycle (C_3 pathway), 139–43

crassulacean acid metabolism (CAM), 147,
    148, **149**, 149
   four-carbon pathway (C4 pathway; Hatch-
    Slack pathway), 144, **144**, **145**, 146, 147,
    148, 149
   in grasses, 632, **633**
Carboniferous period, 431, 435, 445, **456**, 877
   plants of, 456–57, **456–57**
   swamps of, **433**
carbon to nitrogen (C/N) ratio, 746
carbonyl groups (—C=O), 18
carboxyl groups (—COOH)
   in amino acids, 26, **27**
   in fatty acids, **23**
cardamom, 832
cardiac glycosides, 34, 550
Carex aquatilis, **819**
carinal canals, 445, **448**
carnauba wax, 25
Carnegiea gigantea, **497**, **808**
carnivorous plants, 642, 737, **737**
carotene, 133, **134**
carotenoids (carotenoid pigments), 49–51,
    133, **134**, 135, 542, 543, **543**
carpels, 499, **499**, 501, 521, 528, **638**
   development of, **638**, 639
   of early angiosperms, **524**, **525**, 527, 530
   fruits and, 543
   insect pollination and, 530, 531
   outcrossing and, 510, **511**
carpogonia, 359, 360
carposporangia, **362**
carpospores, 359, 360, **362**
carposporophytes, 361
carrageenan, 358, 380
carrier proteins, **82**, 83
carrion flies, 531
carrot, **62**, 606, 681, 686, 695
Carum carvi, 833
Carya
   cordiformis, **669t**
   ovata, **554**, 627, 647, 658
caryopsis, 545
Casparian strips, **594**, 598, **598**, **599**, 600
cassava, 831
castings, 733
castor bean, **562**, 563, **566**, 567, 578
catabolism, 123–24, **124**
catalysts, 863
   enzymes as, 30, 100
categories, taxonomic, 264
caterpillars
   of the monarch butterfly, 34, **35**
   resistance to damage from, 697, **698**
Catharanthus roseus, 846
cation exchange, 735
cations, 856
catkins, **499**, 541
Cattleya, **528**
cauliflower, **238**
caulimoviruses, 300
cavitation, 756
cDNA (complementary DNA) libraries, 226
Ceanothus, 741
cedar
   incense, **278**, 788
   western red, **669t**
celery, 643
cell cycle, 155, 157, **157**
   controls that regulate, **157**, 158
   phases of, 157, **157**
cell death, programmed, 355, 578
cell differentiation, 214, 572
   hormones and, 687, **687t**
cell division, 155, 156, **157**, 159 (*see also* cell
    cycle; cytokinesis; meiosis; mitosis)
   in algae, **376**, 376–78, **377**, 386–88
   embryogenesis and, 556, 557, **557**, **558–59**,
    559
   in eukaryotes, 156
   growth and, 571
   hormones and, **674t**, 687, **687t**
    auxins, 676
    cytokinins, 680, 681
    gibberellins, 684

in plants, 158, 160–66
in prokaryotes, 156, **156**
region of, 594, **594**
reproduction of the organism and, 166
cell elongation, hormones and, **674t**, 675, **675t**, 685, 686
cell enlargement, 571
cell expansion, 59, **66**, 166, 634
direction of, 690
gravitropism and, 704
hormones and, 681, 682, 687, **687**, **689**, 689–90
cell membrane (*see* plasma membrane)
cell plate, 163, **164**, 165, **165**, 166
cells (*see also* specific types and parts of cells)
components of, 41, 43, 45–46, **46t**, **70–71t**
diffusion and, 78–79
eukaryotic (*see* eukaryotic cells)
evolution of, 4–6
expansion of (*see* cell expansion)
files (lineages) of, **570**, 572, 591, 604
first use of the term, 40
forerunners of the first, 4–5
fossils, 2, **2**, 43
hormonal influences on basic processes of, **687t**
prokaryotic (*see* prokaryotes)
structure of, 40–71
totipotency and, 688, 695
cell sap, 54
cell theory, 41, 65
cell-to-cell communication, 86–89
plasmodesmata and, 87
signal transduction and, 86–87, **87**
cellular differentiation (*see* cell differentiation)
cellular membranes (*see* membranes, cellular)
cellular respiration (*see* respiration)
cellular slime molds (*Dictyosteliomycota;* dictyostelids), 275, 348, **349t**, 354–56 (*see also* plasmodial slime molds)
cellular-type endosperm formation, 510
cellulose, 21, **21**, 22, **38t**
in cell walls, **62**, 62–63, **63**, 689, **689**, 690, **690** (*see also* cells walls)
cellulose microfibrils (*see* microfibrils)
cellulose plates of dinoflagellates, 363, **363**
cellulose synthase, 66, **67**
cell walls, 21, 22, 43, **44**, 45, **45**, **63**, **70t**
of angiosperms, 504, 506, 510
cell division and, 156, 158, 163, **165**, 166
cellulose in, **62**, 62–63, **63**
of chytrids, 312
of diatoms (frustules), 376, **376**, **377**, 378
direction of cell expansion and, 690, **690**
extensibility of, 63, 689
of fungi, 310
growth of, 66
of guard cells, 752, **753**
lignin in, 36, 63
noncellulosic molecules in, 63
oomycetes and, 371, **372**
overview of, 61–62, **62**
of pines, 481
plasmolysis and, 81
primary, **62**, 64
of prokaryotes, 283–84
of red algae, 358
secondary, 64–65, **65**
pits in, 66
of slime molds, 355, **356**
suberized, 25
turgor pressure and, 81
Cenozoic era, 877
central cell, 506
central dogma of molecular biology, 209
central mother cell zone, 612, 613
centric diatoms, **375**, 376, 378
centrioles, 158
in fungi, 311
centripetal differentiation, 435
centromeres, 160, **160**, **162**
in meiosis, 176
centrosome, 158
century plant (*Agave*), 644, 808
shawii, **808**

Cerambycidae, 530
Ceratium, **363**
tripos, 363
Ceratophyllum, 526, **526**
Cercis canadensis, **54**
cereals (grains), 825, 843
cetoniine scarabid beetle, **531**
CFCs, 844
Chaetoceros, **375**
Chailakhyan, M. Kh., 715
chalazal, 506, **507**, 556
Chamaenerion, 546
angustifolium, **511**, 792
Chaney, Ralph, 486
channel proteins, **82**, 83
chaparral, 565, **565**, **796**, 817, **817**
chaparral honeysuckle, **524**
dispersal of, **547**
Chara, 47, **387**, 398
Charales, 397, 398, 402
Chargaff, Erwin, 198
Charophyceae (charophytes), 387, **387**, 396–98, 874 (*see also* algae; green algae)
bryophytes compared to, 401–2, 407
checkpoints, 157, 158
cheese, deuteromycetes and, 334
chemical compounds, 868
chemical equations, 871
chemical equilibrium, 871–72
chemical formulas, 870
chemical reactions, 871
energy and, 875–76
chemical reactivity, basis of, 867
chemiosmotic coupling, 120
in chloroplasts and mitochondria, 137, 138
in photophosphorylation, **138**
chemoorganotrophs, 295
chemosynthetic autotrophs, 287, 738
chemotaxis, 355
Chenopodiales (*Centrospermae*), 543
Chenopodium quinoa, 830, **830**
cherries, **502**
cherry
black, **669t**
flowers, **503**
chestnut, horse, **505**, **613**
chestnut blight, 317
chiasmata, 172–73, **173**, 176
chicken pox, 296
chicory tribe, 527
Chihuahuan Desert, **808**
Chimaphila umbellata, **524**
chimeral meristems, 634
China, 826
chitin, 22, **38t**, 310
Chlamydia pneumoniae, 287
Chlamydomonas, 44, **44**, 45, **61**, **171**, 383, **387**, **388**, 388–89, **389**
life cycle of, **389**
nivalis, 386
Chlorella, **274**
chloride (Cl⁻), 76
chloride ions, stomatal movement and, **692**, 693
chlorine (Cl)
as essential element, 727, **730t**
chlorobium chlorophyll, 133
Chlorococcum, 390–91, **391**
echinozygotum, **391**
chlorofluorocarbons (CFCs), 844
Chloromonas, 383
brevispina, **386**
granulosa, **386**
Chlorophyceae (chlorophytes), 386, **386**, **387**, 388–92, 873–74 (*see also* algae, green algae)
filamentous and parenchymatous, 392
life cycle of, 388, **389**, **390**
motile colonial, 389–90, **390**
motile unicellular, 388–89
nonmotile colonial, **391**
nonmotile unicellular, 390–**91**
chlorophyll *a*, 133 (*see also* P₆₈₀, P₇₀₀ chlorophyll molecules)
in light reactions, 135, 136

in prochlorophytes, 292
chlorophyll *a/b*-binding protein (CAB), 708, **709**
chlorophyll *b*, 133
in prochlorophytes, 292
chlorophyll *c*, 133
chlorophylls, 2, 49, 131, 133
absorption spectrum of, 131
magnesium and, 729
molecular structure of, **133**
in photosystems, 135
synthesis of, 634
Chlorophyta, 873 (*see also* algae; green algae)
chloroplast endoplasmic reticulum, 357, 367
chloroplasts, **44–45**, 48, **49**, **50**, **70t**, 132, **132**
of bryophytes, 401–2, 421
chemiosmotic coupling in, 137, 138
of cryptomonads, 357
cytoplasmic inheritance involving, 196
of euglenoids, 351
evolutionary origin of, 53, 272, **273**
genome of, **273**, 274
of green algae, 351
in photosynthesis, 49–50, **126**, **143**, 144, 146, **146**
of red algae, 358
structure of, 49
chlorosis, 729, **729**, **730t**
Choanephora, 317
choanoflagellates, 277, 312, **312**
cholera, 287
cholesterol, 25, **25**
bran and, 563
Chondromyces crocatus, **285**
Chondrus crispus, 358
C horizon (soil base), 732, **732**
chromatids
in meiosis, 172, **173**
in mitosis, 160, **160–162**, 163
chromatin, 47, **48**, 219, **220**
chromophore, 713
chromoplasts, 48, 50–51, **70t** (*see also* plastids)
chromosome mutations, 192–94
chromosomes, 41 (*see also* DNA)
bacterial, 156, **156**
cell division and, 156, **156**, 157, **157**, 158
prophase, 160, **160**
changes in number of (aneuploidy), 193
chemical structure of, 197
daughter, **161**
duplication of, 193
eukaryotic, 218–22 (*see also* DNA (deoxyribonucleic acid), of eukaryotic chromosomes)
homologous (homologs), **170**, 170–77, **173**
crossing-over and, 172, 191, **191**
linkage of genes and, 190, 191, **191**
in meiosis I, 174–77
in meiosis II, 177
inversions of, 193
linkage of genes and, 190, 191, **191**
in meiosis, 169
Mendel's laws and, 190, **190**
in mitosis, 160, **160**, 161, **161**, **162**, 163
number of, 170 (*see also* diploid; haploid)
prokaryotic, 216–18
translocations of, 193
chrysanthemum, **710**
Chrysanthemum, 780
indicum, 715
chrysanthemum stunt viroid (CSVd), 303
Chrysochromulina, 367
chrysolaminarin, 378
Chrysophyta (chrysophytes), **275t**, 349, 371, **372t**, 378, 378–79, 873 (*see also* algae)
Chytridiomycota (chytrids), 275, **308t**, 870
Chytridium confervae, **313**
chytrids (*Chytridiomycota*), 309, **309**, 312–13, **313**, **314**
evolutionary origin of, 312
Cicuta maculata, **500**
ciguatera, 364
cilia, 60 (*see also* flagella)
Cinchona, **551**, **833**
Cinnamomum zeylanicum, 832

cinnamon, 832
circadian rhythms, 706–9, **707**
circinate vernation, 458
Cirsium pastoris, **527**
cisternae, 56
citrate (citric acid), 115
citrate synthase, 747
citric acid cycle (*see* Krebs cycle)
citrus stubborn disease, 292, **292**
clades, 267
cladistics (phylogenetic analysis), 267
cladograms, 267–68, **268**, **432**, **472**, **519**, **526**
Cladonia
cristatella, 337, **339**
subtenuis, 337
Cladophora, 393, **393**
cladophylls, 642, **642**
Cladosporium herbarum, 307
clamp connections, 322, **322**, 333
Clarkia, 514
cylindrica, **514**
heterandra, **514**
Clasmatocolea humilis, **409**
classes, taxonomic, 264
classification (*see* systematics; taxonomy)
clathrin, 58, **58**, 86
Clausen, Jens, 246
Clavariaceae, **277**
Clavibacter, 294, **294**
Claviceps purpurea, 335, **335**
clay soils, 732, 734, 735, **735**
Claytonia virginiana, 548
cleavage furrow, 386
cleistothecium, 319, **319**
climacteric, 683
climatic changes (*see* global warming)
climax community, 792
clines, 246
clonal analysis, 632, 634
clonal propagation (micropropagation), 693–95, **694**
clonal reproduction, competition and, 777
clones, 695
cloning, DNA (gene cloning), 225, 226, **226** (*see also* recombinant DNA)
Clostridium, 741
botulinum, **285**, 286
clover, 719, 738, 740
white, 244, **534**
cloves, 832
club mosses (*Lycopodiaceae*), 429, **435**, 435–38, **436** (*see also* seedless vascular plants)
gametophytes of, 431, **436**, 437, 438
life cycle of, 435, **436–37**, 438
suspensor in, 442
coal age plants, **456–57**, **466–67**
coated pits, 86, **86**
coated vesicles, 58, **58**, 86, **86**
cobalt (Co), 729
coca, 551
cocaine, 32, **32**, 551
cocci, 284, **285**
coccolithophorids, 367, **367**
coccoliths, 367
Cochliobolus heterostrophus, 840, **840**
Cochrane, Theodore S., 263
cocklebur, 548, 681, 709, **709**, 710, **710**, 711
cocoa, 832
coconut, 543–44
dispersal by ocean currents, 546
fruit, **544**
lethal yellowing of, 293, **293**
palm, 497, **544**, 834
coconut cadang-cadang viroid (CCCVd), 303
coconut milk, **544**, 695
growth factor in, 679–80
Cocos nucifera, **293**, 497, **544**, 679, 834
coding sequences, 688, **688**
Codium, 394, **395**
fragile, 394
magnum, 383
codons, 209, **209**
initiation, 212
stop, 214
Coelomyces, **308t**

coenocytic hyphae, 310
coenzyme A (CoA), 114, **114**
coenzyme Q (CoQ; ubiquinone), **117**, 118, **118**
coenzymes, 102
coevolution, 247, 532
  of animals and flowering plants, 547
  biochemical, 549–51
cofactors, 102–3
Coffea arabica, **549**, 827–28, **833**
coffee, **549**, 827, 833, **833**
cohesion-tension theory of water transport, 754–58, **756**, **757**
  air bubbles and, **756**, 756–57, **757**
  tensile strength of water and, 757
  test of, **757**, 757–58
Colacium, 351
colchicine, 193, 840
cold (see also temperature)
  dormancy and, 717, 719
  flowering response and, 719
cold hardiness, 719
colds (viral), 296
Cole, Lamont, 785
Coleochaetales, 397, 402
Coleochaete, **387**, 397, **397**
coleoptile, **562**, 563, **564**, 568, 675, 676, **686**, **702**
  phototropism and, **703**, 703
coleorhiza, **562**, 563, **564**, 568
Coleus, 624, **678**
  blumei, **611**, **612**, 634
collared sunbird, **537**
collenchyma (cells or tissue), **571**, 572, **573**, 574, **574–575**
  of leaves, **628**, 632
  of stems, 618
Colocasia esculenta, 827
Colorado potato beetle, 550
color perception, in bees, 532, **533**
colors (see also pigments)
  of flowers
    bat-pollinated flowers, 539
    bee-pollinated flowers, 532
    beetle-pollinated flowers, 531
    pigments responsible for, 542–43, **543**
    wind-pollinated flowers, 539
  of wood, 667
columbine, **538**
columella, 591, 604
  gravitropism and, 704, **705**
column, 528
comets, 4, **4**
Commelina communis, 35
Commelinales, **265t**
commelinin, **35**
communication, cell-to-cell, 86–89
  plasmodesmata and, 87
  signal transduction and, 86–87, **87**
communities, 9, **10**, 11, 786 (see also biomes)
  climax, 792
  definition of, 774
companion cells, **580–582**, 582–83, **583**, **587t**, 651
competition, 776–79
  clonal reproduction and, 777
  defined, 776
  growth rate and, 776
  by inhibiting the growth of others, 779
competitive exclusion, principle of, 776–77
complementary DNA (cDNA) libraries, 226
complementary strands, 199
complex tissues, 573
Compositae, 264 (see Asteraceae)
composites, 256, **527**, 527–28
compost, 746
compression wood, 666, **666**
Comptonia, 741
concentration gradient, diffusion and, 78
conceptacles, 385
Conchocelis, 361
conchospores, **359**
conducting tissue (see also phloem; xylem)
  in algae, 382
  of *Anthoceros*, **413**
  bryophytes and vascular plants compared, 402
  of mosses, 416, **417**
cones (see also strobili)
cone scales (see ovuliferous scales)
conidia, 311, 318, **318**, **319**, 320, 332
conidial fungi (deuteromycetes), 332, **332**, 871
  predaceous, **333**
conidiogenous cells, 311
conidiophores, 318
Coniferophyta (see conifers)
coniferous forests, 802, 816, 819 (see also taiga)
  northern (taiga), **818**, 818–19
conifers (*Coniferophyta*), **275t**, **278**, 470, 472–86, 875 (see also pines)
  of Carboniferous period, 457
  commercial uses of, **669t**
  compression wood of, 666
  wood of, 659–61, **660**, **661**
Coniochaeta, 319
conjugation, **284**, 285–86
conjugation tubes, 396, **396**
connation, 502
contender beans, **685**
continuous variation, 195, **195**
contraceptive, oral, 78
contractile vacuoles, 80, 351
convergent evolution, 266, **266**, 267, 428, 518, 519
Convolvulaceae, 498
Convolvulus arvensis, **599**
Cooksonia, **6**, 424, **426**, 434
  pertoni, **434**
copepods, 366
co-pigmentation, 35, **35**
copper (Cu), as essential element, 727, **730t**
copper butterfly, 536
Coprinus, 324
  cinereus, **321**
CoQ (coenzyme Q), **117–19**, 121
coral fungi, 323
coralline algae (*Corallinaceae*), **358**, 359, **359**, 361 (see also algae; red algae)
coral reefs, **358**, 359
  dinoflagellates and, 365, **365**
Cordaites, **456**, 457, **457**
Coreopsis, 811
corepressors, 217, **217**, 218
coriander, 833
Coriandrum sativum, 833
cork (phellem), 7, **40**, 587, 602, 647, 653
  commercial, 658–59, **659**
cork cambium (phellogen), 8, 427, 587, **587**, 600, 602, 647, 648, 653–55
cork oak, 658, **659**
cork tissue, 600
corms, 442, **443**, 643
corn (see maize)
corn smut, 330, **330**
corn streak, 300
corn stunt disease, **292**
Cornus florida, 334
corolla, 501, 502, 527, 528
corolla tube, 524, **524**, 532, 535–36, **539**
corpus, 612, **612**, 613
Correns, Carl, 189
cortex, **8**, 573, **573**
  of lichens, 338
  of roots, 592, **593**, **597**, 598–600, 602
  of stems, 613, 614, **615**, 618
cortisone, 846
corymb, **499**
Costa Rica, **803**
cotranslational import, 214, **214**, 215
cotransport systems, carrier proteins as, 83
cotton, 827, 828, **837**
  flowers, **525**
cottonwood, 68, 342, **583**, 631, **669t**
cotyledons (seed leaves), **8**, 482, **562**, 563 (see also eudicots; monocots; scutellum)
  of angiosperms, 510
  development of, 557, **557**, 559, **560**
  germination and, **566**, 567–68
  of gymnosperms, 478–79, **482**
  notch at the base of, **558**
coupled reactions, 105
covalent bonds, 868–70, **869**
  polar, 869–70, **870**
crabgrass (*Digitaria sanguinalis*), 146–47
Crane, Peter, **520**
Crassulaceae, 147
crassulacean acid metabolism (CAM), 147, **148**, **149**, 149
Crataegus, **643**
creosote bush, **664**, 808
Cretaceous period, 470, 488, inside front cover
  Early, **517**, 519
  Lower, 522
  upper, 522
Crick, Francis, 197–99, **199**, 209
cristae, 52, **52**, 114
cross-grained wood, 667
crossing-over, 172, **173**, 176, 191, **191**, 193
cross-pollination, **184** (see also outcrossing)
Croton (crotons), **610**, 624, 634
crown gall, 294, **294**
crown-gall tumors, 696, **696**
crown wart of alfalfa, 313
crozier, 320
Crucibulum laeve, **326**
crustose coralline red algae, **358**, 359, **359**
crustose lichens, **336**, 338
Cryphonectria parasitica, 317
cryptomonads (*Cryptophyta*), **275t**, 348, **349t**, **356**, 356–57, **357**, 872 (see also algae)
crystal violet, 283
cucumber, 676, **677**, **752**
  ethylene and sex expression in, 683
Cucumis sativus, 676, **677**, 683, **752**
Cucurbita
  maxima, 567, **576**, **581**, **582**
  pepo, **600**
cucurbits (*Cucurbitaceae*), ethylene and sex expression in, 683
Culver's-root, 624, **625**
Cupressus, 483
  goveniana, 483
cupules, **469**
Curculionidae, 487
Curtis Prairie, 792–93, **793**
Cuscuta, 498
  salina, **498**
cuticles, 7, 25, **25**, 597
  as barrier to water loss, 752
  of bryophytes, 403
  of epidermal cells, 584, 597
cuticular wax, 25
cutin, 24, 25, 63
cyanidin, 542, **543**
Cyanidium, 357
cyanobacteria (blue-green algae), 43, **44**, 282, **288**, 288–91, **289**, 403, 870 (see also bacteria)
  akinetes of, **289**, **290**, 290–91
  chloroplast evolution, and 290
  environmental conditions and, 288
  gas vesicles of, 289
  glycogen and, 288
  heterocysts of, 290, **290**
  importance of, 288
  in lichens, 340
  marine, 289
  membranes of, 288
  motile, 288
  mucilaginous sheaths of, 288
  nitrogen-fixing, 290
  photosynthesis and, 288, 290
  pigments, 288
  planktonic, 289–90
  in stromatolites, 289, **289**
  structure of, **288–90**
  symbiotic, 290 (see also lichen)
Cyathea, 453, **453**
  australis, **445**
cycadeoids (*Bennettitales*), 472, **473**, 518
cycads (*Cycadophyta*), 239, **275t**, 470, 472, 473, 486–87, **487**, 875
Cycas siamensis, **487**
cycles
  nutrient (see nutrient cycles)
seasonal growth, 648
  water, **734**
cyclic adenosine monophosphate (see cAMP)
cyclic AMP (see cAMP)
cyclic electron flow, 139, **139**
cyclic photophosphorylation, 139, **139**
cyclosis (cytoplasmic streaming), 46
  in giant algal cells, 47, **47**
cyclosporin, 308
Cyclotella meneghiniana, **375**
cylinders (see steles)
Cymbidium orchids, **278**
Cyperaceae, 340
cypresses, 483, **483**
  bald, 483, **485**
  Gowen, **483**
Cypripedium calceolus, **260**
cypsela, 545, **545**, **546**
Cyrtomium falcatum, **459**
cysteine, **27**, 29
Cystopteris bulbifera, **278**
cysts, resting, 365
cytochrome c, 269, **269**
cytochromes, 116, **117**
cytokinesis, 156, 157, **157**, 158, 163, **164**, **165**
  in *Chlorophyceae*, **386**, **387**
cytokinins, 590, **674t**, 676, 679–82, **680**, 696
  genetic engineering and, 698, **698**
  leaf senescence delayed by, 681–82
cytology, 11, 190
cytoplasm, 41, **44**, 45, **70–71t**
  division of (see cytokinesis)
  of prokaryotes, 282
cytoplasmic inheritance, 196, **196**
cytoplasmic male sterility, 196
cytoplasmic sleeve, 87, **88**
cytoplasmic streaming (cyclosis), 46
  in giant algal cells, 47, **47**
cytosine, 198, **198**, 199, **200**
cytoskeleton, 43, 58, **71t**, **273**
cytosol, 46

D

Dactylis glomerata, 244
daisy, **536**
daltons, 863
damping-off diseases, 374
dandelion (*Taraxacum officinale*), **132**, 180, **278**, 528, **545**, **777**
  fruits of (cypselas), **546**
darkness, flowering and, 711
dark reactions (see light-independent reactions)
dark reversion, **713**
Darwin, Charles, 3, 35, **235**, 236–38, 248, 257, 518, 675, **707**
Darwin, Francis, 675
Dasycladus, **395**
dATP (deoxyadenosine triphosphate), 204 (see also ATP)
Daucus carota, **62**, 606, **681**, 686, 695
daughter cells, 156
daughter chromosomes, **161**, 163
Dawsonia superba, **417**
daylength
  dormancy in buds and, 718–19, **719**
  flowering and, **709**, 709–14 (see also photoperiodism)
day-neutral plants, 710
DDT, 781, **781**
Death Valley, **808**, 810
deciduous forests
  temperate, 812–14, **813**, **814**
  tropical, 806
deciduous trees, 522
decomposers, 287, 783
  fungi as, 307
decussate phyllotaxy, **611**, 624
deficiencies, nutrient, 727, 729, **730t**, 731
definitive callose, 580
dehiscence
  of anthers, **504**, 508
  of spores, 412, **414–15**, 416
dehiscent fruits, 544, **544**
dehydration synthesis, 20, **20**, 22
  amino acids and, 28

16-dehydropregnenolone (16D), 550
deletions, 192–93
delphinidin, **543**
denaturation, 104
denaturation of proteins, 30
dendrochronologists, 664–65
dendrometer, **758**
denitrification, **736**, 738
Dennstaedtia punctilobula, **458**
density of wood, 668, 670
deoxyribose, 30, **30**
deoxyribonucleic acid (*see* DNA)
derivatives, 571
dermal tissue system, 426, **427**, **571**, 572,
    583–86, **586t**
dermatophytes, 334
desert plants, **10**
    convergent evolution of, 266
    germination of seeds of, 565
deserts, **10**, **798–99**, 807–10, **808**
    cold (ice), 807, 821
desmids, 397, **397**
desmotubules, 68, **68**, 87, **88**
deuteromycetes (conidial fungi), 332, **332**, 871
    predaceous, **333**
development, 571 (*see also* cell differentiation;
    growth; morphogenesis)
    of flowers, **638**, 639, **640**, 641
    of fruits, 509–10
        gibberellins and, 686
    hormones and (*see* hormones)
developmental plasticity, 246
Devonian period, 425, **426**, 427, 428, 431, 435,
    443, 471, 877
    Early (Lower), 312, 339, **426**, 434
    Late (Upper), 312, 431, 434, 445, **456**, 468,
        469, 470
    Middle, **426**, 427
de Vries, Hugo, 189, 192, **192**, 252
diabetes, 846
diatomaceous earth, 378
diatoms (*Bacillariophyta*), **275t**, 349, 371, **372t**,
    375, 375–78, 873 (*see also* algae)
    cell wall of, 376
    centric, 376
    heterotrophic, 378
    life cycle of, **376–377**
    pennate, **276**, 376
    resting stages of, 376, **378**
Dicentra cucullaria, 549
2,4-dichlorophenoxyacetic acid (2,4-D), **676**,
    679
dichogamy, 510, **511**
Dicksonia, **455**
dicots, 496 (*see also* eudicots; magnoliids)
Dictyophora duplicata, **326**
dictyosomes (*see* Golgi complex)
Dictyosteliomycota (dictyostelids), **884** (*see also*
    cellular slime molds; *Dictyostelium*)
Dictyostelium
    discoideum, 354, **354**, 355
        life cycle of, **354**
    mucoroides, **356**
Diener, Theodor O., 302
differentiation, cell (*see* cell differentiation)
diffuse-porous woods, **662**, 665
diffusion, **77**, 77–79, **78**
    concentration gradient and, 78
    facilitated, **82**, 83
    osmosis as special case of, 78–79
    passive transport and, 82–83
    simple, **82**, 82–83
digger pine, **480**
Digitalis, 34
    purpurea, **533**
Digitaria, **300**
digitoxin, 34
digoxin, 34
dihybrid crosses, 185
dihydroxyacetone phosphate, 111, **111**, 142
dikaryon, 312
dikaryotic cells of ascomycetes, 320
dikaryotic mycelium (or hyphae), 322, **322**
dilated rays, 652, **652**, 655
Dilcher, David, **520**

dill, 833
Dinobryon, 378, **378**
dinoflagellates (*Dinophyta*) **275t**, 348, 349,
    349t, 350, 361–66, **363**, 872–73 (*see also*
    algae)
    feeding by, 365
    red tides and toxic blooms caused by, 364
    resting cysts formed by, 365
    as symbionts, 365
    toxic or bioluminescent compounds
        produced by, 366
Dinophyta, 872–73 (*see also* dinoflagellates)
dinucleotides, 102
dioecious plants, 501
Dionaea muscipula, 721, **721**, 737
Dioscorea, 834, 846, **846**
diosgenin, 550
Diospyros, **67**
    virginiana, **669t**
dioxin, 679
Diphasiastrum, 438
    complanatum, **435**
    digitatum, **278**
diptheria, 287
diploid number, 47, 170
diploidy, 171, 172
    preservation of variability and, 243
disaccharides, 18, **38t**
Dischidia rafflesiana, 606, **606**
Discula destructiva, 334
diseases (*see also* specific diseases and types of
    diseases)
    bacterial, 293–94
    fungal, 307–8
        ascomycetes, 317
        chytrids, 313
    mycoplasmalike organisms (phytoplasmas)
        as cause of, 292–93
    oomycetes as cause of, 373–74
    viral, 296–97, 301–2
    viroid, 302–3
disk flowers, **527**, 528
dispersal agents, evolution of fruits and,
    546–49
    biochemical coevolution and, 549–51
    vertebrates, **547**, 547–49
    water, 546–47
    wind, 546
Distichlis palmeri, 744
distichous phyllotaxy, 624
disulfide bridges, 29
diterpenoids, 33, 34
diurnal movements, 706, **707**
divisions, taxonomic, 264
DNA (deoxyribonucleic acid), **30**, 31, **31**, 41
    amount of, in selected bacterial and
        eukaryotic genomes, **221t**
    antiparallel strand of, 199
    base pairing of, 199, **200**
    cell division and, 156, **157**, 158
    of *Ceratophyllum*, 526
    chloroplasts and, 50
    in chromatin, 47
    composition of, in various species, **198**
    double helix, **198**
    of eukaryotic chromosomes, 218–22
        condensation of DNA, 220–21
        histones and nucleosomes, 219–20
        introns and exons, 222, **222**
        repetition of nucleotide sequences,
            221–22
        single-copy DNA, 222
        specific binding proteins, 221
    in genetic engineering, 224–231
    information-carrying capacity of, 204
    interspersed repeated, 222
    lagging strands of, 202, 203
    leading strands of, 202, 203
    of mitochondria, 53
    mutations and, 192
    in nucleolus, 48
    nucleotides in (*see* nucleotides)
    Okazaki fragments, 202, **203**
    polymerases, 201, **203**
    of prokaryotes, 216–18, 282

recombinant (*see* recombinant DNA)
replication of, **201**, 201–4, 208–9, **209**
    as bidirectional, 201–2
    energetics of, 203–4
    errors in, 203
    initiation of (origin of replication), 201,
        202
    overview of, **202**
    summary of, **203**
    telomeres, 203
    templates, 201
    RNA compared to, 208, **208**
    RNA synthesis and, 208
    simple-sequence repeated, 222
    single-copy, 222
    spacer, **219**, 222
    structure of, 197–200, **198–200**, 207
        complementary nature of strands, 199
    tandemly repeated, 221–22
    of viruses, 297, 300
    Watson-Crick model, 197, **198–199**
DNA cloning (gene cloning), 225, 226, **226–27**
    (*see also* recombinant DNA)
DNA ligase, 202, **203**, 224
DNA polymerases, 201, 202, **203**
    polymerase chain reaction (PCR) and, 226
    replication errors and, 203
DNA sequencing, **229**, 229–30
    molecular systematics and, 270
dodder, 498
Dodecatheon meadia, **500**
Doebley, John, 829
dogbane, 34
dogs, 826, 828
dolipore septa, 322, **322**
domains, looped, 220
domains, symplastic, 88
domains, taxonomic, 881–82
    major distinguishing features of, 271, **271t**
    nucleotide sequences as evidence for three,
        270, **270**
    representatives of, **272**
dominance (dominant characteristics), 185
    incomplete, 194
donor organism, 224
dormancy, 717–19
    acclimation and, 718–19
    breaking of, 684
    of buds, 717, **718**, 718–19, **719**
    gibberellin and, 685
    of seeds, 565, 685, 717–18
dormin, 683
double bonds, 857
double fertilization, 492, 509, **509**
Douglas fir, 483, **669t**
downy mildew in grapes, 373
Drepanophycus, **426**
Drew Baker, Kathleen 361
Drosera rotundifolia, **737**
drought, in Mediterranean climates, 817
drugs, plants as sources of, 843, 846
drupes, 543
druses, **54**
Dryopteris
    crassirhizoma, 680
    marginalis, **458**
Dubautia, 250
    reticulata, **250**
    scabra, 251
duckweeds, **496**
duplication of chromosomes (polyploidy), 193
    (*see also* allopolyploidy; autopoly-
    ploidy)
    speciation and, 249, 252–54
duplications, 193
DuPraw, E. J., 219
Durvillea antarctica, **379**
"dust bowl" conditions, 812, **812**
Dutch elm disease, 317
Dutchman's breeches, 548–49
Dutchman's pipe, **428**, **521**, 654
dwarf mutants, gibberellin and, 684–85, **685**
dyes, lichen, 338
dynamic instability, 59

E
early wood, 665
Earth, **2**
    biological productivity of, **800**
earthstars, 323, **326**
earthworms, **733**
Easter lily, **511**
Ebola virus, 296
Echinocereus, **266**, 533
ecology, 11, 773 (*see also* ecosystems; interac-
    tions between organisms)
    definition of, 774
    global, 796–822 (*see also* biomes)
    restoration, 792–93, **793**
economic botany, 11
*Eco*RI, **225**
ecosystems (ecological systems), 9, 773, 774
    development of communities and, 786–89,
        792–93
    flow of energy through, 784–86
    nutrient recycling and, 782–83
    pesticides and, 781
    trophic levels and, 783–86
ecotypes, 246–47, **247**
Ectocarpus, 382, **382**, 383
    siliculosus, **382**
ectomycorrhizae, 341, 342, **342**, 343, 774–75
Ecuador, 803
eel grass, 541, **542**
effector site, 104
egg apparatus, 506
egg cell, 506 (*see also* gametes)
    of algae, **362**, 371, **372**, **374**, 377, 383, **384–85**,
        392, 398
    of angiosperms, 506, 509, **513**
    of bryophytes, **405**, **410**, **419**
    of gymnosperms, 472, **473**, 474, **479**, **481**,
        482, **487**, 488
    of oomycetes, 371, **374**
    of seedless vascular plants, 442, 445, 449
eggplants, 843
Eichhornia crassipes, **787**
Einstein, Albert, 130
Elaeis guineensis, 834
elaiosome, 548, **549**
Elaphoglossum, **453**
elaters
    of *Equisetum*, 448–49, **449**, **450**
    of *Marchantia*, 408, **409**, 449
elderberry, **573**, 614, **618–19**, 650, **651–653**
Eldredge, Niles, 257
electrical energy, **2**
electrical gradient, 82
electric potential (voltage), membranes and, 74
electrochemical gradient, 82, 119, **121**
electromagnetic spectrum, 129
electronegativity, 867
electron carriers, 116, **117–19**, **121**, 122, 137,
    138–39
electron(s), 863
    energy level of, 865, **865**
    in oxydation-reduction, 98–99
    in photosynthesis, 131, 132, 134, 135–38,
        136, 138 (*see also* cyclic electron flow;
        noncyclic electron flow)
electron transport chain, 108, 109, **110**, 116,
    116, 117, 118, **119**, 122
    in photosynthesis, 135, 137
electrophoresis, 229, **229**
electroporation, 699
elements, 18, 851 (*see also* specific types)
    essential, 727, **728t**, 729, **730t**, 731
Elettaria cardamomum, 832
elevation (*see* altitude)
elicitors, 780, 782
Elkinsia polymorpha, 468, **469**
elm, 545, 623, **669t**
    American, **657**, **663**
Elodea, 81, **81**, 128
elongation
    cell
        gibberellins and, 685, 686
        hormones and, **674t**, 675, **675t**
    internodal, **611**, 613, **613**, 614, 618, 682
    region of, **594**, 595

embolism, water transport and, **756**, 756–57, **757**

embryogenesis, 556–61, 571
 in angiosperms, 509, 510
 genes and, 560, **561**
 primary meristems and, 557
 stages in, 557, **557**, **558–59**, 559
 suspensors and, 559–60

Embryophyta (embryophytes), 876
 phylogenetic relationships between major
  groups of, 472
 as synonym for plants, **402**, 406

embryos (young sporophytes), 9, 468, 470 (*see also* embryogenesis)
 of angiosperms, 509, 510, **513**, 555–68
 asexually produced, **180**, 255–56
 in bryophytes, **402**, 404, **405**, 406, **410**, **418**
 of Capsella, **558–59**
 dormancy of, 565
 eudicot, 557–59, **558–64**, 565, **566–68**
 evolution of seed and, 510, 562
 formation of, 556–561, **557**, **588–61**
 of gnetophytes, 492
 of gymnosperms, 472
 maize, 562, **567**, **568**
 mature, 562
 monocot, 557–59, **562–63**, **566**, **567**, **568**
 mutant of *Arabidopsis*, 560, **560**, **561**
 of pines, 482, **482**, 559
 proper, 556, 557, **557**
 of *Sagittaria*, **557**
 of seedless vascular plants, 438, 442, 449, 559
 vascular transition in, 636, **637**
 wheat, **564**

embryo sac, 506, **507**, **513**

Emiliania huxleyi, 367, **367**

enations, **430**

Encephalartos ferox, **487**

endangered species, 13, 805, **815**, 818, 844

endergonic processes (reactions), 96, **96**, 876
 ATP and, 105

endocarp, 510

endocytosis, 85, **85**, 86
 receptor-mediated, 86

endodermis, 448, 475, **598**, **599**, 600, 603

endomembrane system, 57, **71t** (*see also*
  plasma membrane)
 evolutionary origin of, 272, **273**
 mobility of cellular membranes exemplified
  by, 58

endomycorrhizae, **312**, **341**, **341**, **344**, 774 (*see
  also* mycorrhizae)

endophytes, 308, 335

endoplasmic reticulum (ER), 47, **48**, 53, **55**, **56**,
  57, **71t**
 cisternal, **55**, **56**
 cortical, 56, **56**
 plasmodesmata and, **67**, 68
 polypeptide targeting and sorting and, 214,
  214, **215**
 ribosomes and 55, 56, **57**
 rough, 55, 56, **57**, 58
 smooth, 56, **57**
 tubular, 56, **56**

endosperm, 509, 510, **557**, **558**, 562, **562**, 563
 gibberellins and, 685–86, **686**
 liquid, **544**

endospores, **285**, 286, **286**

endosymbionts, 272, **273**, 274, 290, 351,
  356–57, **357**, 358, **365**, 378, 383
 in *Vorticella*, 274, **274**

energy, 93–107 (*see also* ATP; respiration;
  photosynthesis)
 of activation, 99–100, **100**, 105, 863
 bond, 863
 concept of, 95
 ecosystems and, 9
 electrical, 2
 first law of thermodynamics, 95
 flow of, **94**
  through ecosystems, 784–86
 free, 864
 living organisms' need for a steady input
  of, 97–98

in nucleotides, 31–32
 from organic molecules, 4–5
 photosynthesis and, 2
 potential
  of electrons, 95, 853
  osmotic (solute), 79, **79**
  of water, 76–77
 pyramid of, 785, **785**
 second law of thermodynamics, 95–97
 solar (radiant), 2, 5, 94
 storage as fats and oils, 22, 24

energy level of electrons, 865, **865**

energy-transduction reactions, 133, **134**

Engelmann, T. W., 131

Engelmann spruce, 777, 791

Entogonia, **375**

Entomophthorales, 317

Entomophthora muscae, **277**

entrainment, 707, 708, **708**

entropy, 97

environment (environmental factors) (*see also*
  specific factors)
 clines and ecotypes as adaptations to,
  246–47, **247**
 lichens used to monitor the, 340
 stomatal movement and, 753
 transpiration and, 754

enzymes (enzymatic reactions), 30, 93, 99–106,
  100, 863
 active site of, 101, **101**
 allosteric, 104
 as catalysts, 100
 cofactors and, 102–3
 metabolic pathways and, 103, **103**
 pH of the surrounding solution and,
  104
 regulation of activity of, 103–5
 regulatory, 104
 restriction (restriction endonucleases),
  224–25, **225**, 696
  DNA sequencing techniques and, 229
 substrates and, 100–101
 temperature and, 103–4, **104**

Eocene epoch, inside front cover

Ephedra, 490, **491**, 492
 double fertilization in, 492, 509
 trifurca, **491**
 viridis, **491**

epicotyl, 562, **562**, **566**, **567**

epicuticular wax, 25, 584, **584**

epidermis, 7, **8**, **571**, 572, **573**, 583–86, **585**
 of grasses, 632, **633**
 of leaves, 583–86, **584–86**, 627, **628**, **629**, 630,
  632, **633**, **634**
 of roots, 595–97, **596**, **597**, 602
 of stems, 618

epigeous germination, **566**, **567**, 567

epigynous flowers (epigyny), 502, **502**, **503**,
  527

Epiphyllum, **642**

epiphytes, 421, 435, 445, 449, **453**
 roots of, 605, 606, **606**
 in tropical rainforests, 804, **804**

epistasis, 195

epithem, **761**

equatorial plane, **160–61**, 162

equilibrium, chemical, 871–72

Equisetites, 445

Equisetum, 445, **448**, 456, **463t** (*see also* horsetails)
 arvense, **448–9**, 680
 ferrissii, 255, **255**, 256
 hyemale, 57, **256**
 laevigatum, 256
 life cycle of, 449, **450–51**
 sylvaticum, 278
 telmateia, **728**

ergosterol, **25**

ergot, 335, **335**

ergotism, 335

Ericaceae, 343

Eriophrum
 angustifolium, **819**
 scheuchzeri, **819**

erosion, 736, **838**

Erwinia, 293, **294**

amylovora, **285**, 293
 tracheiphila, 294

Erysiphe aggregata, **319**

Erythroxylon coca, 551

Escherichia coli (E. coli), 26, **43**, 225, **226**, 229
 conjugating, **284**
 DNA of, 209, 216
 operons in, 218
 in protein synthesis studies, 210
 ribosomes in, **211**

Eschscholzia californica, **497**, 534

E (exit) site, 212, **212**, 213

essential elements (essential minerals;
  essential inorganic nutrients), 727–31,
  728t
 criteria for, 727
 functions of, 729–31, **730t**

essential oils, 34, 549

estrogen, 26

ethylene, 548, **674t**, **682**, 682–83
 abscission and, 683
 cell expansion and, 682
 cell wall extensibility and, 689
 fruit ripening and, 683
 genetic engineering and, 697–98, **698**
 orientation of cellulose microfibrils and,
  690, **690**
 sex expression in cucurbits and, 683

ethyl methanosulfonate, **561**

etiolation, 713–14

etioplasts, 51–52, **52**

Euastrum affine, **397**

Eucalyptus (eucalyptus), 342
 cloeziana, 25
 giant (red tingle), **495**
 globulus, **584**
 jacksonii, **495**

Eucheuma, 380

euchromatin, 220, 221

eudicots (eudicotyledons; *Eudicotyledones*),
  497, 504 (*see also* magnoliids)
 embryos of, 557, **557**, 559, **562**, 563
 evolution of, 519, **522**
 major features of, **498t**
 parasitic, 498
 seeds of, **562**
 types of flowers in, **502**

Eudicotyledones, 496, 876 (*see* eudicots)

Eudorina, **390**

Eugenia aromatica, 832

Euglena, 80, **351**, **351**

Euglenophyta (euglenoids), **275t**, 348–52, **349t**,
  872 (*see also* algae)

Eukarya (domain), 270, **270**, **272**, 274
 kingdoms in, 275–79

eukaryotes, **44**, 53 (*see also* *Eukarya*)
 chromosomes of, 218–20 (*see also* DNA
  (deoxyribonucleic acid), of eukaryotic
  chromosomes)
 DNA replication in, 202, **202**, 203
 evolutionary origin of, **272–73**, 272–74, 348
 mRNA in, 222–23, **223**
 regulation of gene expression in, 220–21
 transcription in, 221, 223, **223**

eukaryotic cells, 41–68 (*see also* eukaryotes)
 first appearance of, 6
 prokaryotic cells compared to, 41, 43, **44**,
  45t, 272

Euphorbia (euphorbias), **266**, 644
 corollata, **793**
 pulcherrima, **538**

Euphorbiaceae, 34, **266**, 834

Eupoecila australasiae, **531**

Europa, 3

European larch, **483**

Eurystoma angulare, **469**

eusporangia, **453**, 455

eusporangiate ferns, **453**, 453–55

eusteles, 428, **428**, 429, **463t**, 471

evapotranspiration, 734

evening primrose, 192, **192**, 252, 536
 family, **514**
 Oenothera glazioviana, 192, **192**
 Oenothera lamarckiana, 252

evergreen species, 648, 812

evergreen wood fern, **458**

evolution, 235–58 (*see also* individual organisms; natural selection; phylogenetic
  trees; speciation; systematics)
 agents of change in, 240–42
  gene flow, 241
  genetic drift, 241–42
  mutations, 193–94, 240–41
  nonrandom mating (inbreeding), 242
 clockface of, **12**
 coevolution, 247, 532
  of animals and flowering plants, 547
  biochemical, 549–51
 convergent, 266, **266**, 267
 of C4 pathway, 147
 Darwin's theory of, 236–38
 of eukaryotes, **272–73**, 272–74
 gradualism model of, 257, 258
 Hardy-Weinberg equilibrium and, 240
 of human beings, 11–12, 824
 of major groups of organisms, 256–58
 photorespiration and, 143
 of plants, 2–11 (*see also* angiosperms
  (flowering plants), evolution of;
  bryophytes; vascular plants, evolution
  of)
  aggregations of molecules, 4–5
  chemical precursors, 3–4
  communities of plants, 9, **10**, 11
  of Devonian period, 425, **426**, 427
  on Earth, 2–3
  photosynthesis, 5–7
  transition to land, 7–11
 punctuated equilibrium model of, 257–58

excited state, 131, 853

exergonic processes (reactions), **96**, 96–97, 864
 ATP and, 105

exine, 504, **505**

exocarp, 510

exocytosis, 58, 85, **85**

exodermis, **597**, 600
 of roots, 596

exons, 222, **222**, 223, **223**

explants, 695

extensins, 63, 689

extinction of plant species, present and
  future, 844, 847

extinct plants
 gymnosperms, 472
 seedless vascular plants, 425

extreme halophilic archaea, 295

extreme thermophilic archaea, 296, **296**

Exuviaella, 363

eyelash cup, 317

eyespot (*see* stigmas (eyespot))

F

F_1, F_2 (filial) generations, **185**, **187**, **189**, **190**

Fabaceae, 264, 511, 544, **545** (*see also* legumes)

facilitated diffusion, 82, 83

FAD (flavin adenine dinucleotide), 116, **116**, **119**

FADH2, 116, **117**, **119**, 122

Fagaceae, 342, **502**

Fagus
 americana, **669t**
 grandifolia, **762**

fairy rings, 325, **325**

false annual rings, 664

false Solomon's seal, **762**

families, taxonomic, 264, 266

fan palm, 518

fasciation, **294**

fascicles, 474, **475**

fascicular cambium, 650

fats, 22, **23**, **38t**
 respiration and, 122
 saturated, 24
 unsaturated, 24

fatty acids, 22, **23**, **38t**
 respiration and, 122

feedback inhibition, 104, **104**

Fd (*see* ferredoxin)

feeder roots, 590, 596

fennel, 833

fermentation, 109, **110**

alcohol, 123, **123**
lactate, 122, 123
ferns, 55, 59, 108, **275t**, 424, 431, 449, **452**, **453**, 453–55, 457–63, 804 (see also *Pterophyta*)
bracken, **458**
Carboniferous period, 457
cladogram of, 268, **268**
in cladograms showing phylogenetic relationships, **268**
eusporangiate, **453**, 453–55
evergreen wood, **458**
Filicales, **453**, 454, 455, 458–62
filmy, **453**
grape, 454
heterosporous, 454, 462 (see also water ferns)
homosporous, 430, **459** (see also *Filicales*)
leptosporangiate, 453, **453**, 454, 455, 460, 462–63
life cycle of, **460–61**
maidenhair, 455, **458**
ostrich, **458**
seed, **456**, 457, **457**
species of, 449
sweet, 741
tree, 453, **455**, 457
walking, **180**
water, **462**, 462–63
whisk, 443
Ferocactus
acanthodes, 809, **809**
melocactiformis, **643**
ferredoxin (Fd), **136**, 138, **139**
Fertile Crescent, 825
fertilization, **169**, 170, **170**, 171, **171**, 180, **184**, 312 (see also pollination; sexual reproduction; and specific plants)
in algae, 360, 372, 396, **396**
in angiosperms, 509
in bryophytes, 400, 402, 406, **410**, 414, **419**
in chytrids, **313**, **314**
in cycads and *Ginkgo*, 473, 490
double, 492, 509, **509**
in gnetophytes, 492
in fruit development, 679
in fungi, 311–12
in *Lycopodiaceae*, **437**, 438
in oomycetes, 373, **374**
in pines, **481**, 482
in seedless vascular plants, 438, 442, 449
fertilization tubes, **373**, **374**
fertilizers, 745, 838
nitrogen, 738, 741–42, 745
fescue, tall, 335
Festuca arundinacea, 335
F$_1$, F$_2$ generations, 185
fibers, 572, 575, **575**, **577**
bast, 575
gelatinous, 666
phloem, 618, 650, 651
xylem, 579
fiber tracheid, 576–77
fibrous proteins, 29
fibrous root system, 590, **591**
Ficus benghalensis, 603
fiddleheads, 455, **458** (see also leaves, of ferns)
uncoiling of, 458
field capacity of soils, 734
field hypothesis of phyllotaxy, 624
figure, of wood, 667
filaments, (angiosperm) **499**, 501, **503**
fused, 525
of wind-pollinated flowers, 540, **541**
filaments (fungal), 310
Filicales, **453**, 454, 455, 458–62
life cycle of, 459, **460–61**
filmy ferns, **453**
fimbriae, 284, **284**
fir, 483
balsam, **483**, 788
Douglas, 483, **669t**
subalpine, 777, 791
white, 788, **789**
fire blight, 293

fireflies, 698
fires
dormancy of seeds and, 565, **565**
forest, 788–91, **789–791**
grasslands and, 811–12
Mediterranean-type vegetation and, 817
in taigas, 819
fireweed, **511**, 546, **792**
fir mosses, 438
first available space hypothesis, 624
fish, acid rain and, 616
Fissidens, **400**
fitness of an individual, 239
flagella, 44, 52, 60, 60–61, **61**
of algae, 60, **349t**, **372t**, 383, 387–93, 396, 398
of chytrids, 312, 313
of cryptomonads, 356
of cycad sperm, 488
of diatoms, 376
of dinoflagellates, 361
emergent, **351**
of euglenoids, 351, **351**
of eukaryotes, 60–61
of haptophytes, 366, 367
of heterokonts, 371
nonemergent, **351**
of oomycetes, 371
of prokaryotes, 284
of slime molds, **353**, 354
structure of, 60, **61**
tinsel, 60, **349t**, 371, **372t**, 376
whiplash, 60, 313, **349t**, 354, **371**, **372t**,
flagellar roots, 386, 387, **387**
flavin adenine dinucleotide (FAD), 116, **116**, 119
flavin mononucleotide (FMN), **116**, **117**, 118, 119
flavins, 704
flavones, 35, **35**
flavonoids, 34–35, 542, 543, 549
flavonols, 35, 542–43, **543**
flax, 575, 825
Fleming, Sir Alexander, 334
flies, 529, **529**, **531**, 535, **535**
floral apex, **638**, 639
floral stimulus (flowering hormone), 715
florets, **541**
Florey, Howard, 334
floridean starch, 358
florigen, 715, 716
flour, 563
flour corns, **828**
flowering
cold and, 719
daylength and (photoperiodism), **709**, 709–14
genetic control of, 716
hormonal control of, 715–16
inhibitors and promoters of, 715, **716**
phytochrome and, 712
flowering hormone (floral stimulus), 715
flowering plants (see angiosperms)
flowers, 495, **498–502** (see also inflorescences; pollination)
bee, 532
bisexual, 531
bolting and, 686, **687**
carpellate (or pistillate), 501, **511**
colors of, 50, 531, 532, 539, 542–43, **543**
complete, 502
definition of, 498
development of, **638**, 639–41
floral organ identity, 639–40, **640**, 641
dioecious, 501
disk, **527**, 528
double, 639
epigynous, 502, **502**, 527
of eudicots, **497**, **498t**, **502**
evolution of, 524–43 (see also angiosperms, evolution of)
animals as agents of, 530–39
carpels, 525, 527
evolutionary trends, 527
most specialized flowers, 527–29
perianth, 524

pollination mechanisms, 525, 530–40
sepals and petals, 524
stamens, 524–25
hormones and, 715–16
hypogynous, **499**, 502, **502**
imperfect (unisexual), 501
incomplete, 502
irregular (bilateral; zygomorphic), 502
of magnoliids, 519, **520**, 521, 522
of monocots, **497**, **498t**
monoecious, 501, **502**
odors of, 531, 532, 535, 537–39
parts of, 499, **499**, 501–2
perfect (bisexual), 501
perigynous, 502, **502**
placentation of, 501, **501**
regular (actinomorphic; radial), 502, 527
senescence or wilting of, genetic engineering and, 698, **698**
staminate, 501, **511**
symmetry in, 502
terminology, 499–502
underground, 529, **529**
unisexual, 501
vasculature of, 524
flow of energy through ecosystems, 784–86
fluid-mosaic model of membranes, 74, **75**
fluorescence, 131
fluorides, as air pollutant, 616
fly agaric, **323**
fMet (see methionine)
FMN (flavin mononucleotide), **116**, **117**, 118, **119**
Foeniculum vulgare, 833
foliose lichens, **336**, 338
follicles, 544
food (see nutrients, inorganic; nutrition, plant)
food bodies, 531
food chains, 783, 785
food-conducting cells, tissues, 8, 382, 402, 403, 425, 579–583 (see also leptoids; phloem)
food reserves, 20, 22, **349t**, **372t** (see also by type)
food storage
in fleshy roots, 606, **607**
in leaves, 643
in seeds, 470, 482, 510, 563, **564**
in stems, 642–43
food transport (see also assimilate transport; leptoids; phloem)
in bryophytes, 402, 403
in vascular plants, 425, 764–69
food web, 783, **784**
foolish seedling disease, 684
foot
of bryophytes, **405**, 406, **410**, 413
of seedless vascular plants, 445
Foraminifera, 378
foreign genes, 696–99
forests
carbon cycle and, 150
coniferous, 816, 819
fires in, 482, 788–91, **789–791**
monsoon, 806, 807
old-growth, 815
rainforests, 489, **798–800**, 803–5
temperate, 804
tropical, **10**, **803**, 803–5
subtropical mixed, 806, **807**
taiga, **818**, 818–19
temperate
deciduous, **10**, 812–14, **813**, **814**
mixed, 815–16, **816**
pyramid of numbers for, **786**
tropical mixed, 806
forestry, sustainable, 815
formulas, chemical, 870
Forsythia, **50**
fossil fuels, 150, 845
fossils, 238, **239**
of algae, 348, 359
of angiosperms, 470, **517**, 518, **518**, 519, 521, **521**, 522
of bryophytes, 403
earliest known, 2, **3**

eukaryotic, 348
of flowers, **517**, 519, **520–22**
of fungi, 312
"gaps" in fossil record, 257
of gymnosperms, 457, 472, 488–89, **489**
of lichens, 339
of megaspore, **469**
of *Metasequoia*, **485**
of multicellular organisms, 6, **6**
of mycorrhizae, 343, **344**
of photosynthetic organisms, 5
of pollen, 504, 506, 519
of prokaryotes, 2, **3**, 282
radiocarbon dating of, 852
of seeds, 468
of vascular plants, **424**, 427, 429, 432, 434
of wood, 470
founder cells, 634
founder effect, 241–42, **242**
Fox, Sidney W., 4
foxglove, 34, **533**
Fragaria, 547
ananassa, 679, **679**
virginia, 762
Franklin, Rosalind, 198, **198**
Fraxinus, **545**, **755**
americana, **669t**
pennsylvanica var. *subintegerrima*, **627**, **657**
free central placentation, 501, **501**
free energy, 864
free-energy change, 97
Friis, Else Marie, 522
Frisch, Karl von, 532
Fritschiella, **387**, 392, **392**
Frolova, I. A., 715
fronds, 455, 457, 458 (see also leaves, of ferns)
frost, reducing susceptibility to, 699
fructans, 20
fructose, 19, **19**
fructose 1,6-bisphosphate, **111**, 113
fructose 6-phosphate, 111, **111**, 113
fruit drop, 683
fruitlets, 543
fruits, 503, 543
accessory, 543
aggregate, 543
auxin and formation of, 679
climacteria, 683
colors of, 50–51, 683
definition of, 543
dehiscent, 544, **544**
development and growth of, 509–10, 686
dispersal of 546–51 (see also seeds, dispersal of)
dry, 544–45
evolution of, 543–49
fleshy, 543, 547–48, 683
indehiscent, 544–45
lenticels on, 654
multiple, 543
parthenocarpic, 543
ripening of, 683, 697, **698**
genetic engineering and, 697, **698**
simple, 543, 544
Frullania, **409**, 412
frustules, 376, **376**, **377**, 378
fruticose lichens, 338
Fucales, 379, 381, **381**
fuchsia, 537
fucoxanthin, 365, 367, 378, 381–82 (see also carotenoids; xanthophylls)
Fucus, 171, 383, **385** (see also brown algae)
life cycle of, **171**, **385**
vesiculosus, **379**
Funaria hygrometrica, **56**, **420**
functional groups (chemical), 871, **871t**
functional phloem, 659
fundamental tissue system (see ground tissue system)
fungi, 5, **5**, **11**, 306–46 (see also mushrooms; specific types of fungi)
cell walls of, 310
classification of, 277, 309
coenocytic, 310, 313
conidial, 332–34

as decomposers, 307
destructiveness of, 307–8
diseases caused by, 307–8, 313
endophytic, 335
evolutionary origin of, 277, 312
fossil, 312
as heterotrophic absorbers, 310
hyphae of, 310
importance of, 307–9
life cycles of, **314**, **316**, **318**, **321**, **328–329**
mitosis and meiosis in, 310–12
mycorrhizal (*see* mycorrhizae)
parasitic, 310, 317, 334, 339
pathogenic, 307–8, 313, 317
phototropism in, 315, **315**
predaceous, 333, **333**
reproduction in, 311–12
saprophytic, 310, 312, 313
spores of, 308, 311–13, 315, **315**, **316**, 326
symbiotic relationships of, 308, 334–44 (*see also* lichens; mycorrhizae)
Fungi (kingdom), **265t**, 275, 277, **277**, 882
Fungi Imperfecti (*see* deuteromycetes)
funiculus, **506**, **507**, 562
furrowing, of cell, 386–87, **387**
Fusarium, 780
fusiform initials, **648**, **649**, **649**, **650**, 667
fynbos, 817

G

G_0, G_1, G_2 phases, 157, **157**, 158
GA_3 (gibberellic acid) 626, **674t**, 684, 689
galactans, **349t**
β-galactosidase, **218**
α-galacturonic acid, **22**
Galápagos Islands, 236, 237, **237**
Galápagos tortoises, 236, **237**
gametangia, 312, **316**, 317 (*see also* antheridia; archegonia)
of algae, 383, **384**, **385**
of bryophytes, **402**, **405**
of chytrids, **314**
of mosses, 416, **417**, 420
plurilocular, 383
of vascular plants, 402
gametes, 169, 170, **170**, **171** 372 (*see also* egg cell; sperm)
of algae, 360, 383, **384**, **385**, 388–89, **389**, 394
allopolyploidy and, **252**
of angiosperms, 504, 506, **506**
autopolyploidy and, **252**
of bryophytes, 400
of chytrids, **314**
defined, 170
evolution of, 8
of fungi, 312, **313**, **314**
Mendel's experiments with peas and, 184, 186, **186**, **187**, 188, **189**, 190, **190**
nonmotile, **372**
number of chromosomes in, 47
of oomycetes, 371, 372
plasmodium formation and, 354
segregation of alleles in formation of, 186, **186**
of slime molds, 352, **353**
gametic meiosis, 171, **171**
gametophores, 408
gametophytes, **171**, 172 (*see also* embryo sac; megagametophytes; microgametophytes; pollen)
of algae, 359, 360, 361, 383, **384**
of angiosperms, 431, 503 (*see also* pollen)
bisexual, 430, 431, **436**, 438, 444, **444**, 449, 449, **450**
of bryophytes, 403, 404, 425
of chytrids, **314**
of club mosses, 431, **436**, **437**, 438
of cycads and *Ginkgo*, 473, **473**, 474
of *Equisetum*, 449, **449**, **450**
evolution of, 430, 431
of ferns, 455, 458, 459, **459–461**, 462
of gymnosperms, 472–73
of heterosporous and homosporous plants, 430, 431
of hornworts, 412, **413**
of liverworts, 407, 408, **408–411**

of *Lycopodium*, **436–37**
of *Marchantia*, **410–11**
of mosses, 414, **414**, 416, **416**, **418**, **419**, 420, **420**, 421, **422**
and mycorrhizae, 438, 445
of pines, 472, 477, **477**, **478**
of *Polypodium*, 460–61
of *Psilotum*, 444, **444**, 445, **446**, **447**
of rhyniophytes, 434
of *Selaginella*, **440**, **441**, 442
of vascular plants, 430–31
gametophytic self-incompatibility, 511
gamma rays, **129**
Ganoderma applanatum, 323
gap junctions, 87
garbanzos, 825
garden bean, **562**, 563, **566**, 567
Garner, 716
Garner, W. W., 709
gas exchange, lenticels and, 654
Gassner, Gustav, 719
gasteromycetes, 323, 325, 326, **326**
gas vesicles, 289
gating, 83
Geastrum saccatum, **326**
gel electrophoresis, 688
geminiviruses, 300, **300**
gemma cups, 408, **409**
gemmae, 404, 408, **409**
gene cloning (*see* DNA cloning)
gene expression
biological clock and, 708, **709**
chromosome condensation and, 220–21
in eukaryotes, 220–21
hormones and, 687, **687t**, 688, **688**
regulation of (*see* gene regulation)
RNA and, 208
specific binding proteins and, 221
temperature and, 197
gene flow, 241
gene monitoring, 698
gene mutations (point mutations), 192
gene pool, 239–40
General Sherman sequoia, **485**
generation, filial (F_1, F_2), 185–89
generative cells, 196
of angiosperms, 504, **505**, 508, **512**
of pines, 477, **477**, **479**, 481
gene regulation
in eukaryotes, 220–21
in prokaryotes, 216
genes, 31, 184, 197 (*see also* alleles; chromosomes; DNA)
cell differentiation and, 214, 572
coding sequences of, 222, 688, **688**
dominant/recessive, 185–87
embryogenesis and, 560, **561**
epistasis of, 195
eukaryotic, 218, 221
expression (*see* gene expression)
foreign, 696–99
frequency, 239–40
homeotic, 639
interactions among alleles of different, 195
linkage of, 190–91, **191**
movable (transposons), 193
noncoding sequences, 221, 222
pleiotropic, 195
prokaryotic, 210, 216–18
regulator, 216, **217**
regulatory sequences of, 688, **688**
reporter, 698–99
sequencing, 229–30
structural, 216, **217**
traits controlled by several, 195
transfer of, 696–99
genetic code, **209**, 209–10
as nearly identical in all organisms, 210
genetic drift, 241–42
genetic engineering, 13, 696–99 (*see also* recombinant DNA)
Agrobacterium tumefaciens and, **696**, 696–97
benefits and risks of, 699
food quality improved by, 698

hormone production and, 697–98
nitrogen-fixing bacteria and, 747
genetic isolation (reproductive isolation), 248, 256
genetic linkage maps (*see* linkage maps)
genetic recombination, 172
genetics (heredity), 11, 183–204 (*see also* alleles; DNA; genes; genotypes; phenotypes)
chemical basis of, 197
continuous variation, 195, **195**
Darwin's theory of evolution and, 238
epistasis, 195
incomplete dominance, 194, **194**
linkage of genes and, 190–91, **191**
Mendel's experiments in (*see* Mendelian genetics)
mutations, 192–94
polygenic inheritance, 195
genetic self-incompatibility, 510–11
genetic variability (genetic diversity), 239
diploidy and, 243
meiosis and, 170, 177–79
natural selection and, 243–45
preservation and promotion of, 242–43
as protection against pathogens, 840–41
sexual reproduction and, 242
genome, 53, 196, 297
genomic libraries, 226
Genomosperma
kidstonii, **469**
latens, **469**
genotypes, 186, 197
genus (genera), 262
geologic eras, inside front cover
geotropism (*see* gravitropism)
germination, 54
of coconuts, **544**
cold treatment and, 717
dormancy and, 565, 717–18
epigeous, **566**, 567, **567**
in eudicots, 566
of fungal spores, 317, 330
hormones and, 684, **685**
hypogeous, **566**, 567, **567**
imbibition and, 80, **80**
light and, 564, 711–12, **712**, 717
photoperiodism (phytochrome) and, 711–12, **712**, **713**
requirements for, 564–65, 717–18
of soybeans, **512**
of sporocarps, 462
temperature and, 564–65, 717
water and, 564–65
giant cell, 356, **356**
Gibberella
acuminata, 310
fujikuroi, 684
gibberellic acid (GA_3), 626, **674t**, 684, 689
gibberellins, 560, **674t**, 684–87, **685**, **686**
bolting and, 686, **687**
commercial uses of, 686–87
dormancy and, 685–86, **686**, 719
dwarf mutants and, 684–85, **685**
flowering and, 715, 719
fruit development and, 686, **687**
molecular basis of action of, 688, 690
orientation of cellulose microfibrils and, 690, **690**
in roots, 590
in seeds, 685
in sex expression, 683
structure of, 685
Gibbs, Josiah Willard, 97
gill fungi, 325
gills of mushrooms, 324, **324**
ginger, **762**, 832
Ginkgo biloba, 60, **473**, 488, **488**, **490**, 490
Ginkgophyta, 60, **275t**, 470, 472, 473, **473**, 488, **488**, **490**, 490, 887
Givnish, Thomas, 778
glades, 811, **811**
gladiolus, 643, **644**
Gladiolus grandiflorus, **644**
gleba, 326

Gleditsia triacanthos, **627**
GL1 (*GLABROUS1*) gene, 585–86
global ecology, 796–822 (*see also* biomes)
global warming, 150–51, 415, 845
peatland carbon and, 415
globular proteins, 29
globular stage, 557, **560**
Glomus, **308t**, 317
etunicatum, **341**
versiforme, **341**
glucose, 19, **19**, 20, 22
fermentation of, 123
oxidation of, 99, 109, **110**, **121** (*see also* glycolysis)
as product of photosynthesis, 128, 142
glucose 6-phosphate, 111, **111**, 113
glumes, **541**
glutamate (glutamic acid), **27**, **28**
glutamine, **27**
glutamine synthetase–glutamate synthase pathway, 742, **742**
gluten, 835
glyceraldehyde, **19**
glyceraldehyde 3-phosphate (3-phosphoglyceraldehyde; PGAL), 111–13, **111–13**, 140–42, **140**, **141**
glycerol, **38t**
glycine, **28**
Glycine max, **51**, 698, 715, 738, 739, **740**, 826, **826**
life cycle, **512–13**
glycocalyx, 284
glycogen, 20, **22**, **38t**, 284
glycolipids, **38t**, 75
glycolysis, 108, 109–13, **110**, 122
anaerobic, 122, 123, **123**
steps of, 111–12, **111–12**
summary of, 112–13, **113**
glycoproteins, 58, 63, 74
glycosides, **549**, 549–50
glycosidic bond, 20
glyoxysomes, 54
glyphosate, resistance to, 697
Gnetales, **519**
Gnetophyta (gnetophytes), **275t**, 470, 472–74, 490–92, **491**, 492, **492**, 518, 887
relationship to angiosperms, 492, 509
Gnetum, 490, **491**, 492, 518, **519**
double fertilization in, 492, 509
goat's beard (*Tragopogon*), 253, **253**
goldenrods, **762**, 778, **778**
goldeye lichen, 337
Golgi complex (Golgi bodies), 57, 57–58, **71t**
Gondwanaland, 488, 523, **523**
Gonium, **390**
gonorrhea, 287
Gonyaulax
polyedra, **363**, 708
tamarensis, 364
goosefoot, 543
Gossypium, **525**, 828, **828**
Gould, Stephen Jay, 257
Gowen cypress, 483
gradualism model of evolution, 257, 258
grain of wood, 667
grains (cereal), 195, **564**, 568
Gram, Hans Christian, 283
gram-negative and gram-positive bacteria, 283, **283**
grana (grana thylakoids), 49, **50**, **126**, **132**
granite mosses (*Andreaeidae*), 412, 416
grape ferns, 454
grapes, 373, 825
gibberellic acid and, 686–87, **687**
tendrils of, 641, **642**
grapevine, **672**
grasses, 253, 256, 511 (*see also* names of specific grasses)
Agrostis tenuis, 244–45, **245**
C_3 and C_4, 146, **147**
caryopsis of, 545
eel, 541, **542**
embryos of, 563
epidermis of, 632, **633**
fungal endophytes in, 335

Kentucky bluegrass (*Poa pratensis*), 146, 244, 255–56
leaves of, 632, **633**
orchard, 244
wind-pollinated flowers of, **540**, **541**
grasslands, 798–99, 810–12, **811**, **812**
biome map of, 798–99
pyramid of numbers for, **786**
gravitropism, 591, 704–5, **705**, 723t
grazing mammals, 812, 826
Great Basin Desert, 807, **808**
green algae (*Chlorophyta*), 171, 275t, 348, 349, 371, **372t**, 383–98, **386**, 402, 873–74 (*see also* algae; *Charophyceae* (charophytes); *Chlorophyceae* (chlorophytes); *Ulvophyceae* (ulvophytes))
cell division in, 386–87, 387
classification of, 383, 386
cytoplasmic streaming in, 47, **47**
economic uses of, 380
as endosymbionts, 274, **274**
food reserves in, 372t, 383
life cycles of, **389**, **394**
relationship to plants, 383, 396–98, 401–2
sitosterol in, 25, **25**
green ash, **627**, 657
green bacteria, 291
greenhouse effect, 13, 150 (*see also* global warming)
greenhouse gases, 844–45
Green Revolution, 839, 847
green snow, 386
Griffonia simplicifolia, **545**
ground meristem, 557, **557**, 570, 571 (*see also* primary meristems)
of shoots, **611**, 613, 614
ground tissue system, 426, **427**, 557, **557**, 571, 572–73, **573**, 573–76, 586t
growth, 571 (*see also* meristems)
of bud, **611**, 612, **613**, 681, 718–19
deficiency symptoms of, **730t**
of embryo, 556–60
hormones and (*see* hormones)
increments (rings), 664–66
indeterminate, 571
intercalary, 634
nutritional requirements for (*see* nutrition, plant)
phytochrome and, 712–14
primary, 8, 426–27, 571
in stems, 611–14
regulators of, 675t (*see also* by name)
in roots, 591–95
secondary (*see* secondary growth)
in stems, 611–14
thigmomorphogenesis and, 721–22
growth rings, 664–66, **666**
Grypania, 348
guanine, 198, **198**, 199, **200**
guanosine triphosphate (GTP), 212
guard cells, 583, **584**, 584–85, **585**
radial orientation of cellulose microfibrils in, 752
stomatal movement and, 692, **692**, 693
turgor pressure in, 752
guayule, 842, **842**
GUS gene, 698
guttation, 761, **761**
Guzmán, Rafael, 829
Gymnodinium
breve, 364, **364**
catenella, 364
costatum, **363**
neglectum, **363**
gymnosperms, 467–94 (*see also* conifers; cycads; *Ephedra*; *Ginkgophyta*; *Gnetophyta*; *Welwitschia*; pines)
evolution of, 431
extinct, 472
living, 472–74
main features of, **493t**
phylogenetic relationships with other embryophytes, **472**
gynoecium, **499**, 501
gypsy moths, 782

H

Habenaria elegans, **535**
Haberlandt, Gottlieb, 695
hadrom, 416
Haeckel, Ernst, **264**
Haemanthus katherinae, **161**
Hagemann, Wolfgang, 65
hairs (*see also* trichomes)
glandular (secretory), 585
leaf, 34, **140**, 250, **585**, 585–86, **586**, 627, **628**, **634**
root (*see* root hairs)
Hale-Bopp comet, 4, **4**
Hales, Stephen, 751
half-life of radioisotopes, 864
Halimeda, 395, **395**
hallucinogenic compounds, 551
Halobacterium halobium, 295
halophilic archaea, extreme, 295
halophytes, 744
Hamamelis, 546
Hamner, Karl C., 711, 715
Hantaan virus, 296
haploid cells, **171**, 172, 173, 175, **175**, 177, 179, 331, 332 (*see also* meiosis)
haploid generation, 172, 279 (*see also* gametophytes; sporophytes)
haploidization, in deuteromycetes, 333
haploid number, 47, **169**, 170 (*see also* meiosis)
haploid organisms, 171, 243
haploid phase of life cycle, **171**
haploid spores, 313, 314, 316, 318, **324**, 329, 330, 354
haplophase, 353
Haplopappus gracilis, 47
haptonema, 366, **367**
haptophyte algae (*Haptophyta*), 275t, 348, 349, 349t, 366–67, **367**, 885 (*see also* algae)
hardwoods, 659 (*see also* by name)
Hardy, G. H., 239, 240
Hardy-Weinberg law (equilibrium), 240–41, 877–78
hares, snowshoe, 782
Harpagophytum, **548**
Hartig net, 342, **343**
Hatch, M. D., 144
Hatch-Slack pathway (C4 pathway), 144, **144**, **145**, 146, 147, **148**, 149
in grasses, 632, **633**
haustoria, 310, **311**, 339, **339**, 498
Hawaiian Islands, 250–51, **250–51**
hawkmoths, 536, **536**
Hawksworth, David L., 339
hawthorn, **643**
hay fever, 508
hazelnuts, 545
heads (flower), **499**
heart disease, 287
heart stage, 557, **557**
heartwood, **655**, 665, 667
heat content, 864
heather family, mycorrhizae in, 343
Hedera helix, 603
helanalin, 780
Helenium, 780
Helianthus, 255, **255**
annuus, 255, **255**, 527, 722, 831, **831**
anomalus, 255, **255**
petiolaris, 255, **255**
helical phyllotaxy, 624
helicases, 201
Helicobacter pylori, 287
Heliconia irrasa, **803**
heliotropism, **722**, 722–23, **723t**
heliotropism, **723t**
helix
alpha, 28, **28**, 29
double, **198–200**, 201–2
heme, **117**
hemicellulose cross bridges, **689**
hemicelluloses, 63, **62**, 66
Hemitrichia serpula, **353**
hemlock, 483
eastern, 669t
water, **500**

western, **342**
hemorrhagic fevers, 296
hemp, 575
henbane, 710, **710**, 711, 719
hepatica, **497**, 531
Hepatica americana, **497**, **531**
hepatics (*see* liverworts)
Hepatophyta (*see* liverworts)
hepatitis, 296
herbaceous plants, 619–21, 648, 812–13, 814
Herbaspirillum
rubrisubalbicans, 747
seropedicae, 747
herbicides, 679, 697, 781
herbivores, 783 (*see also* animals; mammals)
interactions between plants and, 779–82, **780**
herbs, 832, 833
heredity (*see* genetics; inheritance)
Hericium coralloides, **323**
herpes, 296
Herpothallon sanguineum, **277**
heterochromatin, 220–21
heterocysts, 290, **290**
heteroecious, 327
heterokaryosis, 333
heterokonts, 371
heteromorphic generations, alternation of, 172
Heterosigma, 379
heterospory, 431, 442, 443, 468
in angiosperms, 503
in ferns, 454, 462
in gymnosperms, 503
in *Isoetes*, 443
in progymnosperms, 471
in *Selaginella*, 442
heterothallic organisms, 317, 373
heterotrophs, 5, 124, 286, 783
heterozygotes, 242–43
heterozygous, 186, **187**
Hevea, 834, 842
brasiliensis, 34, **34**
hexokinase, **101**, 111, **111**
hexoses, 18
hickory
bitternut, 669t
shagbark, **554**, 627, **647**
hierarchical taxonomic categories, 264
Hiesey, William, 246
Hildebrandt, A. C., 695
Hill, Robin, 128
Hill reaction, 128
hilum, 563
*Hind*III, **225**
histidine, 26, **27**
histones, 41, 158, 219–21
Hodgkin's disease, 846
Holcus lanatus, 762
holdfast, 382
homeotic genes, 639
homeotic mutations, 639, 640, **641**
homologous chromosomes (homologs), **170**, 170–77, **173**
crossing-over and, 172, 191, **191**
linkage of genes and, 190, 191, **191**
in meiosis I, 174–77
in meiosis II, 177
homology (homologous features), 266
Homo sapiens, 824
homospory, 430
in *Equisetum*, 448
in ferns, 454, 458
in *Lycopodium*, 438
in *Psilotum*, 444
homothallic organisms, 317, 373
homozygotes, 242–43
homozygous, 186, 187, **189**, 190, **190**
honey guides, 532, **533**
honey locust, **627**
honeysuckle, 624
chaparral, **524**, **547**
Hoodia, **266**
hook
fungal, 320

seedling, 567
Hooke, Robert, **40**, 41
Hordeum
jubatum, **278**
vulgare, 563, **635**, 685, 825, **825**
horizons, soil, 731–32, **732**
hormogonia, 288
hormones, 86, 673–701, **674t** (*see also specific hormones*)
cellular processes and, **687t**
defined, 674
flowering and, 715–16
genetic engineering and production of, 697–98
groups of, 675
molecular basis of action of, 687–93
cell expansion, **689**, 689–90
gene expression, 687, 688, **688**
receptors, 690–91, **691**
second messengers, 691, **691**, **692**
stomatal movement, **692**, 692–93
sensitivity to, 675
in tissue culture, 680–81, **681**, 693
hornworts (*Anthocerophyta*), 275t, 400, 407, 412, 526, **526**, 886 (*see also* bryophytes)
comparative structure of, 403, **404**
horse chestnut, **505**, **613**, 648, **657**
horsetails, 57, 255, **256**, 275t, 278, 680, **728**, 875 (*see also* *Equisetum*; seedless vascular plants; *Sphenophyta*)
giant (calamites), 456–57, **466**
host cell, ancestral, 272
Hubbard Brook Experimental Forest, 782–83, **783**
human(s)
and agriculture, 824–25
culture and society, 824, 826, 836–37
evolution of, 11–12, 824
impact on environment, 824, 826, 832, 837, 844–45
impact on nutrient cycles, 743–45
Industrial Revolution and, 836
populations, 835–37
Humboldt, Alexander von, 802, **802**
humidity, transpiration and, 754
hummingbirds, 537, **537**, **538**
humus, 732, **732**
hunger and poverty, 847
hunter-gatherers, 823, 824–25
Huperzia, 438
lucidula, **438**
Hutchinson, G. E., 774
hyacinths, water, 787
hybridization, 243, 252, 253, 838–39, 847
recombination speciation and, 255
somatic, 693, 695
hybridization, nucleic acid, 230
hybrids, 243
fertile, 253
of maize, 243, 829, 839
sterile, 252, 255–56
of wheat, 254
hybrid vigor, 243
hydathodes, 761, **761**
hydraulic lift, 761, **762**
Hydrodictyon, 391, **391**
reticulatum, 391
hydrogen (H), 3, **3**, 18
as essential element, 727
isotopes of, 864, **865t**
hydrogen bonds, 873, **873t**, 874
hydrogen sulfide (H_2S), **127**, 128
hydroids, 416, **417**, 420
hydrolysis, 20, **20**, 871
hydrolytic enzymes, gibberellins and, 685, **686**
hydrophilic groups, molecules, 18, 874
hydrophobic interactions, 874
hydrophobic molecules, 22, 874
hydrophytes, leaves of, 626, 627, **629**
hydroponics, 693
hydrostatic pressure, 77, 80 (*see also* turgor pressure)
hydrostatic pressure hypothesis, 705
hydrotropism, 706, **723t**
hydroxyl groups (—OH), 18, 859

Hygrocybe aurantiosplendens, **309**
hymenium (hymenial layer), 319, **321**, 323, **323**, 324
hymenomycetes, **320**, 323, **323**, 324 (see also *Basidiomycota*)
Hymenophyllum tunbrigense, **453**
Hyoscyamus niger, 710, **710**, 719
hypanthium, 502, **502**, 503
hyperaccumulators, 746–47
hypertonic solutions, 79
hyphae, 310 (see also conidiophores; haustoria; rhizoids)
　of ascomycetes, 318, 320
　coenocytic, 310, 313, 315
　dikaryotic, monokaryotic, **322**
　of predaceous fungi, 333, **333**
　receptive, 327
　septate, 310, 318, 322
hypocotyl, **157**, 556, 557, 566
hypocotyl-root axis, 482, **482**, 513, 562, 563, **637**
hypodermis of pine leaves (needles), 474
hypogeous germination, 566, 567, **567**
hypogynous flowers, 502, **502**
hypotonic solutions, 79

I

IAA (indoleacetic acid), 676, **676**, 681, **681**, 695, 703
i⁶Ade (isopentenyladenine), **680**
ice desert, **798–99**, 821
　biome map of, **798–99**
ice plant, 644
igneous rocks, 731
Iltis, Hugh, 829
imbibition (hydration), 80, 81, 565
immunofluorescence microscopy, 159
Impatiens, 546
inbreeding, 242
incense cedar, 788
incomplete dominance, 194, **194**
indehiscent fruits, 544–45
independent assortment, principle of, 188–89, **189**, 190
indeterminate growth, 571
Indian pipe, 498, **498**
indoleacetic acid (see IAA)
induced-fit hypothesis, 101, **101**
inducers, 217
induction in signal transduction, 87, **87**
indusia, 458, **458**, **459**
Industrial Revolution, 836
infanticide, **835**
infection (see diseases)
infection threads, 739
inflorescences (flower clusters), **192**, 253, 293, 491, 499, **499**, 501 (see also flowers)
　of wind-pollinated flowers, 540
influenza, 296
infared radiation, **129**
Ingenhousz, Jan, 127
inheritance, 195–96 (see also genetics)
initials, 157, 453, 571 (see also apical meristems; cambium)
　defined, 157
　fusiform, 648, 649, **649**, **650**, 667
　ray, 648, 649, **650**
initiation codons, 212
initiation stage of translation, 212, **212**, 213
inky cap, **321**
inorganic nutrients (see essential elements)
insects (see also pesticides; names of specific insects)
　biochemical coevolution and, 549–51
　carnivorous plants and, 642, **737**
　juvenile hormones and, 780
　pollination by, **530**, 530–35
　　bees, **532**, 532–35, **533**, **534**
　　beetles, 531, **531**
　　in bisexual flowers, 531
　　in cycads, 487
　　evolution of angiosperms and, 525
　　in gnetophytes, 492, **492**
　　mosquitoes and flies, 535, **535**
　　moths and butterflies, 535–36, **536**

nectar, 530
　wind pollination and, 539
　resistance to insecticides, 781
　secondary plant products and, 549, 550
　symbiotic relationships between fungi and, 308
　trichomes as defense against, 585
　viruses transmitted by, 298
integral proteins, 74
integuments, 470, 507
　of angiosperms, 506, **507**, 509, **513**
　evolution of, 468, **469**
　of pines, 481, 482
interactions between organisms, 774–82
　competition, 776–79
　mutualism, 774–75
　phytoalexins and, 780, 782
　plant-herbivore and plant-pathogen, 779–82, **780**
　tannins and, 782
intercalary growth, 634
intercalary meristems, 614
intercellular spaces, **573**
　ethylene and, 682
　in leaves, **629**, 630, 631, 632, 634
　in root cortex, 598
　transpiration and, 752
interfascicular cambium, 650
interfascicular parenchyma, 614, 619
interlocked grain, 667
intermediate filament, 58
International Code of Botanical Nomenclature, 262, 264
International Panel on Climate Change, 845
International Rice Research Institute, **827**
internodal elongation, 613, **613**, 614, 618, 682
internodes, 8, 445, **450**, 568, **610**, 611, **611**, 613
interphase
　meiotic, 172
　mitotic, 157, **157**, 158, **165**
interspersed repeated DNA, 222
intine, 504
introns, 222–23, **222**, **223**
inversions, chromosomal, 193
ionic interactions (ionic bonds), 868, **868**
ion(s), 868 (see also specific elements)
　active transport of, 82–84
　channels, 83
　as cofactors, 102–3
　functions of, 729, **730t**
　phloem-immobile, phloem mobile, 729
　transport in plants, 729, 762–63, **763**, 764
ipecac, 846
Ipomoea batatas, 606, **607**, 831
Iridaceae (irises), 833
Irish moss, **358**, 412
Irish potato famine (1846–1847), 373, 841
iron (Fe)
　as essential element, 727, **730t**
　in electron carriers, 116
　oxide, 732
iron-sulfur proteins, 116, **117**
ironwood, black, 670
irrigation, 837, **837**, 843
isidia, 338
islands, speciation and, 249, **249**
isocitrate (isocitric acid), 115, **115**
Isoetes (quillworts), 23, 55, 149, 442–43, **443**
　muricata, 23, **55**
　storkii, **442**
Isoetaceae, 442–43, **463t**
isogametes (isogamy), 312, **372**, **394**
isolation, reproductive (genetic), 248, 256
isoleucine, 26, **27**
isomers, 858
isomorphic generations, alternation of, 172
isopentenyl adenine (i⁶Ade), **680**
isoprene, 33, **33**
isotonic solutions, 79
isotopes, 851–53
isozymes, 103
ivy, 603
　poison, **550**
　Swedish, **585**

J

jacket layer (sterile), 402, **405**, **461**
Jacob, François, 216
Jaffe, M. J., 706, 721
Janzen, Daniel, 775
jasmonates, 675, **675t**
Jefferson, Thomas, 802
Jeffrey pines, **244**
jelly fungi, 323, **324**
jojoba, **841**, 841–42
Joshua tree, 808
Juglans
　cinerea, **657**
　nigra, **669t**, 779
junipers (*Juniperus*), 483
　communis (common), 484, **484**
　osteosperma, 810
juniper savannas, **798–99**, 810, **810**
Jurassic period, 483, 488, 877
jute, 575

K

Kalanchoë daigremontiana, **180**
kale, **238**
kaolinite, 731
kapok, 539
Kaplan, Donald, 65
karyogamy, 312, **318**, **321**, 353, 389
　in *Chlamydomonas,* 389, **389**
　in mushrooms, **321**
　in plasmodial slime molds, 353
　in wheat rust, 328–29
Keck, David, 246
kelps, 7, **370**, 371, **379**, 380, **382–83** (see also brown algae)
　commercial uses of, 380, 382
　internal structure of, 382
kelvins (K), 876
Kentucky bluegrass (*Poa pratensis*), 146, 244
　apomixis in, 255–56
kernel (grain), 195, **564**, 568
α-ketoglutarate (α-ketoglutaric acid), **115**
kinases, 106
　protein, 691
kinetin, 680, **680**, 681, **681**, 695
kinetochore microtubules, 161, 162, **162**, 163
　meiotic, 176
kinetochores, **160**, 161, 162
King Clone, **664**, 808
kingdoms, taxonomic, 264, 271
　in the domain *Eukarya,* 275t, 275–79
Klebsormidium, 387
Knop, W., 693
knots in wood, 668, **668**
kohlrabi, **238**, 643, **644**
kombu, 380
Krakatau, 789
Kranz anatomy, 144, 632
Krebs, Sir Hans, 115
Krebs cycle, 108, 109, **110**, 114, 115, **115**, 122, 124
Küchler, A. W., 798
Kuehneola uredinis, 327
Kurosawa, E., 684

L

Labiatae, 624
Labrador tea, 819
lactate (lactic acid), 122–123
lactate fermentation, 122, 123
lactose, 217, 218
lactose (lac) operon, 210, **210**
Lactuca sativa, **164**, 635, 711
lacZ gene, 226
lady's mantle, 761
lady's slipper, yellow, **260**
lagging strands, 202, 203
Laguncularia racemosa, 605, **605**
Lamarck, Jean Baptiste de, 236
Lamiaceae, 833
lamina, 429, 624
Laminaria, 379, **379**, 381, 382 (see also kelps)
　life cycle of, **384**
laminarin, **372t**, 381
land, transition of plants to, 7–11

land-form leaves, 626, **626**
landing platform, on bee flowers, 532, **533**, 536
Lang, Anton, 715
larches, 483, **818**
　European, 483
Larix, 483, **818**
　decidua, 483
Larrea divaricata, **664**, 808
late blight of potatoes, 373, **375**
lateral (axillary) buds, **657**, 678, **678**
lateral meristems, 8, 648 (see also cork cambium; vascular cambium)
lateral roots (branch roots), 590, **592**, 603, **603**
late wood, 665
latex, 34
Lathyrus odoratus, 190–91
Latin names, 262–63
latitude, distribution of organisms and, 800–801, **801**, 802
Lauraceae (laurel family), 519, 833
Laurus nobilis, 833
LDA (lysergic acid amide), 335
leaching, 735, 738
lead (P⁶), 244–45, 616
leading strands, 202, 203
leaf axil, 8, 611, 626, 642
leaf blade, 624, **625**, 626, **628**, 635, **635**
leaf buttress, 634, **634**
leaf gaps, **428**, 429, **429**
leafhoppers, 298
leaflets, 8, 626
leaf primordia, 611, **611**, 614, 622, 634, **634**, 635, **635**
　phyllotaxy and, 624
leaf scars, 636, **657**
leaf sheath, 624, **625**
leaf spots, **294**
leaf stalk (see petiole)
leaf-stem relation, 622–24
leaf trace gaps, 429, 622
leaf traces, 429, 622, **623**, 624
leaves, 7, 426 (see also cotyledons; shoots; specific plants)
　abscission of, 636, 683
　air pollution and, 616
　of angiosperms, 624–36
　of aquatic plants, 626, **626**
　of bryophytes, **414**, 415, 416
　of CAM plants, 149
　coloration, 543
　composting, 746
　compound, 626, **627**
　of cordaites, 457
　of C₃, C₄ plants, 144, **145**, 146, 632
　of cycads, 486
　of desert plants, **266**, 644
　development of, 632–35, **634**, **635**
　dimorphic, **626**
　of *Ephedra,* 491
　epidermis of, 583–86, **584–86**, 627, **628**, 629, 630, 632, **633**, **634**
　of *Equisetum,* 448, **448**
　of eudicots, 624, **625**, 626, 627, 631
　evolution of, **428**, 429–30
　of ferns, 454, **454**, 455, 458, 462, **462**, 463 (see also fronds)
　food storage in, 643
　function of, 611, 630
　gas exchange by, 144, 627, 692, 752 (see also stomata)
　of *Gnetum,* **491**
　of *Ginkgo,* 488, 490, **490**
　of grasses, 632, **633**
　hairs (trichomes) of, 34, **140**, 250, 585, 585–86, **586**, 627, **628**, 634
　heliotropism of, 722–23
　of hydrophytes, 626, **627**
　of *Isoetes,* 442, **442–43**
　of larch, **483**
　of *Lycophyta,* 438, **438**
　of Magnoliids, 624, 626, 627, **629**, 631
　of mesophytes, 626, 630
　modifications of, 641–44, **644**

of monocots, 624, **625**, 631, 635
morphology of, 624–26
movements of, 707, 719–23
phyllotaxy of (arrangement on stem), 624
of pines (needles), 474, **475**, 476
scalelike, 643
seed (*see* cotyledons)
of *Selaginella*, 439, **439**
senescence of, 681–82, 698
sessile, 624, **625**
simple, **625**, 626
stomata on, **584**, 584–85, **585**, 626, 627, **628**, **629**, 630, **630**
structure of, 626–32
succulent, 644, 807
sun and shade, 636, **636**
vascular tissues of stems and, 622–24
veins of (*see* veins)
water storage in, 644
of *Welwitschia*, **492**
of *Wollemia nobilis*, 488, **489**
of xerophytes, 626, 627, **629**, 631
of yews, **484**
lectins, 75, 740
leghemoglobin, 740
Legionnaires' disease, 287
legumes, 511, 544, **545**, 641, 825
cultivation of, 825–26, **826**, 827, 828
nitrogen-fixing bacteria and, 738–41
Leguminosae, 264
lemma, **541**
Lemna
gibba, **496**, 776, **776**
polyrhiza, 776, **776**
Lemnaceae, **496**
lenticels, 586, 602, **653**, 654, **654**, 657, **658**
Lentinula edodes, 325
Lepidodendron, **456**, 466
leptoids, 416, **417**, 420
leptom, 416
Leptonycteris curasoae, **539**
leptosporangia (*see* sporangia, of ferns)
leptosporangiate ferns, 453, **453**, 454, 455, 460, 462–63
Les, Donald, 526
lethal yellowing of coconut, 293, **293**
lettuce, 164, 635, 711–12, **712**
leucine, **27**
leucoplasts, 48, 51, **51**, 70t (*see also* plastids)
leukemia, 846
lianas, 803, 804, **804**
Liatris
punctata, **591**
pycnostachya, **793**
lichen acids, 339–40
lichens, 334–40, 883
algal or cyanobacterial component of, 340, 334–35, **338**, 339
commercial uses of, 340
crustose, **336**, 338
ecological importance of, 339–40
evolution of, 334
foliose, **336**, 338
fruticose, **337**, 338
fungal component of, 334, 338, 339
marine, 336
mycobiont-photobiont relationship in, 339, **339**
pollution and, 340
reproduction in, 338
structure of, **338**, 338
survival of, 338–39
water content of, 338–39
weathering of rocks and, 786
Licmophora flagellata, **375**
lid (*see* operculum)
life cycles, 172 (*see also specific groups or organisms*)
types of, 171, **171**
light (*see also* energy; phototropism)
absorption of, 130–33 (*see also* pigments)
competition for, 776–78
flowering and, **709**, 709–14
leaf development and, 636
nature of, 128–30, **129**, **130**

in photosynthesis, 2, 130–39
plant responses to, 709–14
plastid formation and, 51–52, **52**
red, far-red, **129**, 711–14
in seed germination, 564, 711–12, **712**, 717
stomatal movement and, 753–54
ultraviolet (ultraviolet radiation), 6, 12, 129, 192, 203
bees' perception of, 532, **533**, 543
carotenoids and, 543
flavonoids and, 35, 542
visible, 129, **129**, 130
wavelengths of, 129, **129**
light-independent reactions, 134 (*see also* carbon-fixation reactions)
light reactions (light-dependent reactions), 133–39
lignification, 36
lignins, 36–37, 63, 428, 572
evolution of plants and, 425
ligule, 439, **440**, 443, **625**
lilac, 570, **628**
Liliaceae, 196
lilies
African blood, **161**
Easter, 511
parts of, **499**
pollen grains of, **505**
voodoo, 37
water, 521, 575, **629**, 787
Lilium, **176**, 504, **505**, 506, **506**
double fertilization in, 509, **509**
henryi, **499**
longiflorum, **505**, 511
limber pine, 482
limestone (calcium carbonate), 150, 731
Linaria vulgaris, **500**, 546
linden, 575, 577, 762 (*see also* basswood)
Lindera benzoin, 762
linkage maps, 191, **191**
linked genes, 190–91, **191**
Linnaeus, Carl, 262, **262**
linolenic acid, **23**
Linum usitatissimum, 825
lip cells, 454
lip of flower, 528, **528**, 529
lipid bilayer, 24, **24**, 74, **75**, 76
lipids, 18, 22–26, **38t** (*see also* fats; glycolipids; oils; phospholipids; steroids; *and specific types of lipids*)
lipopolysaccharide, 283
Liriodendron, 519
tulipifera, **662**, 669t
Lithocarpus densiflora, **502**
liverworts (*Hepatophyta*), 275t, 400, 407–12, 886 (*see also* bryophytes)
comparative structure of, 403
leafy, **409**, 409–12, **412**
life cycle of, **410–11**
sporophytes of, 406
thalloid, 407–8, **408**
loam soils, 732, 735, **735**
Lobaria verrucosa, **338**
locules, 501
locus (on chromosome), 186, 190, **191**
locust
black, **627**, 669t
honey, **627**
lodgepole pine, 481, 482, 791
lodicules, **541**
London plane tree, 248
long-day plants, 710, **710**, 711
Longistigma caryae, 767
long shoot, **483**, 488
Lonicera, 624
hispidula, **524**, 547
looped domains, 220
loose smut of oats, 330
Lophophora williamsii, **551**
lorica, 351
Los Angeles, smog in, 616
lotus, **524**
sacred, 718
luciferase, 698, **699**, 708, **709**
luciferin, 698–99

lumen of endoplasmic reticulum, 56
lupine, **500**, 722, 830
arctic, 718
white, 764
Lupinus
albus, 764
arcticus, 718
arizonicus, **722**
diffusus, **500**
Lycaena gorgon, **536**
Lycophyta (lycophytes), **275t**, **426**, 429–31, **432**, 435–43, **436t**, 887 (*see also* club mosses)
club mosses (*Lycopodiaceae*), 435–38
Isoetaceae, 442–43
Isoetes, 442, **442**, 443
life cycles, 436–37, **440–41**
Lycopodium, 436–37, 438, **438**
Selaginellaceae, 439–42
lycophyte trees, 435, 456, **456**, 457, 466
Lycopodiaceae (*see* club mosses)
Lycopodiella, 438
Lycopodium, 435, 438 (*see also* club mosses; *Lycophyta*)
lagopus, **427**, **436**, 438, **438**
Lyell, Charles, 238
Lygodium, 453, **453**
Lymantria dispar, 782
lymphoma, 846
lysergic acid amide (LDA), 335
lysine, 26, **27**
lysosomes, 55

M

McClintock, Barbara, 193, 203
mace, 832
MacMillan, J., 684
Macrocystis, 380, 382
integrifolia, 382
pyrifera, 380
macrocysts, 356, **356**
macroevolution, 257
macrofibrils, 63
macromolecules, 18
macronutrients, 727, **728t**, 730t
Madagascar, **805**
madder family, 833
Madia, 251
magnesium (Mg/Mg^{2+}), 76
in chlorophyll molecule, 133
as cofactor, 102
as essential element, 727, 729, **730t**
Magnolia, 519, 522
grandiflora, **520**
Magnoliaceae, 519
magnolias, 544
southern, **520**
magnoliids, 519, **520**, 521, 523, 525
embryos of, 559
maidenhair ferns, **455**, **458**
maidenhair tree (*Ginkgo biloba*), 60, 275t, 470, 472, 473, **473**, 488, **488**, 490, **490**, 888
Mairan, Jean-Jacques de, 706
maize (*Zea mays*), 45, **45**, **46**, 50, 145, 829
actin filaments in, 60
aflatoxins in, 334
Black Mexican, 195
classification of, 265
continuous variation in, 195
cultivation of, 828, **828**, 829, **829**, 838, 839
domestication of, 829, **829**
embryo of, **562**
endocytosis in, 86
flowers, **540**
germination in, **567**, 568, **568**
hybrid, 243, 829, 839
leaf stomata, **584**
leaves of, **625**
photosynthesis in, **134**, **145**, 146, **146**
pollen grain, **540**
roots, 590, **592**, **593**, 594, **594**, 597, 600, 603, **603**
sieve-tube elements, **581**
southern leaf blight of, 840, **840**
stems, 614, 621, **621**
viviparous, 684, **684**

lumen of endoplasmic reticulum
maize streak, 300
malaria, **551**
malate (malic acid), 144, **144**, 145, 147, 149, 693
male sterility, cytoplasmic, 196
Malloch, D. W., 343
Malpighi, Marcello, **765**
Malthus, Thomas, 237
Malus sylvestris, 503, 587, 649, 655
mammals (*see also* animals *and by name*)
soil dwelling, 733
Mammillaria microcarpa, **278**
manganese (Mn/Mn^{2+}), 137, **730t**
Mangifera indica (mango), 827
mangroves, 603, **605**
Manihot esculenta, 831, **831**
Manila hemp, 575
manioc, 831, **831**, 834
mannitol, 382
mantle of ectomycorrhizae, 342, **342**
manzanita, 565, **565**
maples, 545, **545**, 624, **637**, 718, 813
red, **10**
schizocarps of, 546
sugar, **625**, 669t
map units, 191
maquis, **796**, 817
Marasmius oreades, 325
Marattiales, 454–55, **463t**
marble, 731
Marchantia, **278**, 403, **403**, 405, 408, 409 (*see also* bryophytes)
life cycle of, **410–11**
margo, 662
mariculture, 350
marijuana, 550, **551**
Mars, 3
Marsilea, 462
polycarpa, **462**
Marsileales, 454, 462, **463t**
maternal inheritance, 196
maternity plant (*Kalanchoë daigremontiana*), **180**
mating strains, types, **316**, **321**, **328**, 388, **389**, **394**, **396**
matorral, 817
matrix of mitochondrion, 114, **114**, 118, **119**, 120, **120**, **121**, 122
matrotrophy, in bryophytes, 404–6
Matteuccia struthiopteris, **458**
Matthaei, Heinrich, 209
maturation, region of, **594**, 595, 596
Maxwell, James Clerk, 128–29
May apple, 762
meadow rue, 762
measles, 296
measurements, in microscopy, **42t**
mechanical stimulation, plant responses to, 721–22
Medicago sativa, 614, 619, **620**, 729, 738
medicinal compounds, plants, 550, **551**, 843, 846, **846**
medicine and botany, 11, 843, 846
Mediterranean climates, **10**, 816, 817
Mediterranean scrub, **796**, 798–99, 817
Medullosa noei, **456**, 457
megagametogenesis, 506
megagametophytes (female gametophytes), 431
of angiosperms, 506
evolution of seeds and, 468
of gymnosperms, 470
of pines, **478**, **479**, 481, 482
of *Selaginella*, **440**, 441, 442
megapascal (MPa), 77
megaphylls, **429**, 470
evolution of, 429, 430, **430**
of ferns, 455
megasporangia, 431
evolution of seeds and, 468
of *Isoetes*, 443, **443**
in ovules, **468**
of pines and other conifers, 476
of *Selaginella*, **440**, 441, 442
megaspore mother cells (megasporocytes), **480**, 481, 506, **513**

megaspores, 431, **513**
  of angiosperms, 506
  evolution of seeds and, 468
  in ovules, 468, **469**
  of pines, 481
  of *Selaginella*, **440, 441,** 442
megasporocytes (megaspore mother cells),
    480, 481, 506, **513**
megasporogenesis, 506
megasporophylls, 442, 443, **443,** 487, 501
meiocytes, 173
meiosis, 169, 169–78 (*see also* sexual reproduc-
    tion)
  in algae, 362, 383, 384, 385, 389, 394
  in angiosperms, 504
  in *Chlamydomonas*, 389, **389**
  in chytrids, 312
  in *Lycopodium*, **436**
  in diatoms, **376, 377**
  in *Equisetum*, **450**
  in *Fucus*, **385**
  in fungi, 310–12
  gametic, 171, **171**
  genetic variability and, 170, 177–79
  in gymnosperms, **478,** 481
  haploid, diploid numbers and, 170–71
  in hymenomycetes, 324
  life cycles and, 171, **171**
  in liverworts, **410**
  mitosis compared and, **178,** 179
  in mosses, **418, 419**
  in oomycetes, **374**
  phases of, 173–78
  in plasmodial slime molds, **353**
  in *Psilotum*, **446**
  sporic, 171, 172
  in *Ulvophyceae*, 393, **394**
  in vascular plants, **430**
  in wheat rust, **328**
  in yeasts, 330
  zygotic, 171, **171,** 379, 389, 391, 392, 397
meiospores, 172
Melampsora lini, **311**
Melilotus alba, **301**
melon, 525, 679
membrane potential, 82
membrane proteins (*see* integral proteins;
    transmembrane proteins)
membranes, cellular, **73** (*see also* endomem-
    brane system; plasma membrane; *and
    specific membranes*)
  fluid-mosaic model of, 74, **75**
  internal, 74
  solute transport across, 82–84
  structure of, 74–76
  unit membrane model of, 74
  water transport across (*see* water transport
    (movement))
Mendel, Gregor, **183,** 184–90, 195
Mendel's first law, 185–87
Mendel's second law, 188–89
Mendelian genetics, 184–90, **185, 185t,** 242
  chromosomal basis of Mendel's laws,
    189–90, **190**
  laws of probability, 188
  method of, 184–85
  pleiotropy, 195
  principle of independent assortment,
    188–89, **189,** 190
  principle of segregation, 185–87
Mentha, 833
meristematic cap, 614, **614**
meristematic cells, 572, **572**
meristem culture, pathogen-free plants
    produced by, 695
meristems, 8, 571 (*see also* cambium)
  apical (*see* apical meristems)
  cell cycle and, 157, 166
  chimeral, 634
  ground, 557, **557, 570, 571, 611,** 613, 614 (*see
    also* primary meristems)
  of hornworts, 412
  intercalary, 614
  lateral, 8, 427
  peripheral, 613

pith, 613
  primary, 557, 559, 570, **570, 571, 611,** 614
    (*see also* ground meristem; procam-
    bium; protoderm)
Mertensia virginica, **500**
mescaline, 551
Mesembryanthemum crystallinum, 644
mesocarp, 510
mesocotyl, 564, 568
mesophyll, 8, **132, 573, 585**
  of grass leaves, 632, **632, 633**
  photosynthesis and, 630–31
  of pine leaves (needles), 474, 475
mesophyll cells, 49, **126,** 144, **144, 145,** 146,
    146
  phloem loading and, 767
  somatic hybrids and, 693
  transpiration and, 752, **752, 764**
  of Venus flytrap, 721
mesophytes, 626
Mesozoic era, 431, 472, 486, 877
mesquite, 590
messenger RNA (*see* mRNA)
mestome sheath, 632, **633**
metabolic pathways, 103, **103**
  feedback inhibition of, **104**
metabolism, 123–24
metabolites (secondary plant products), **549,**
    549–51, **551**
metal ions, as cofactors, 102
metamorphic rocks, 731
metaphase
  meiotic, **174, 175,** 176, 177
  mitotic, 159, **160,** 161, **161, 162, 162,** 163, **165**
  plate, **161,** 162
metaphloem, 621 (*see also* primary phloem)
Metasequoia (dawn redwood), **485,** 486
  glyptostroboides, **486**
metaxylem, 600, **619,** 621, **622, 623** (*see also*
    primary xylem)
methane, 3, 3
Methanococcus jannaschii, 230, 270, **271**
methanogens, 295, **295,** 296
Methanothermus fervidus, **272**
methionine (fMet), 26, **27,** 212, 682, **682**
metric table, 879
Mexico, agriculture in, 828–30
micelles, 62, **63**
Micrasterias
  radiosa, 397
  thomasiana, 397
microbodies (*see* peroxisomes)
Micrococcus luteus, **285**
microcysts, 354
microenvironment, 777
microevolution, 240
microfibrils, 46, **59,** 62, **62, 63, 67, 165** (*see also*
    cell walls)
  growth of cell wall and, 66, **66**
  hormones and, 689, **689,** 690, **690**
  orientation of, **66,** 159, **165,** 690, **690**
microfilaments (*see* actin filaments)
microgametogenesis, 504
microgametophytes, 431, 508, **512** (*see also*
    pollen)
  of angiosperms, 504
  of *Selaginella*, **440, 441,** 442
microhabitat isolation, **257**
micronutrients (trace elements), 727, **728t,**
    730t
microphylls, **429, 442**
  of club mosses, 435, **438**
  of *Equisetum*, 445, **448**
  evolution of, 429, **430**
  fertile (*see* sporophylls)
  of lycophytes, 435, 436, **438,** 439, **442–43**
micropropagation (clonal propagation),
    693–95, **694**
micropyle, 468, **468, 469,** 470, **473, 479,** 481,
    506, **507,** 509, **509,** 556, **562,** 563
microscopes, 40, **42–43,** 159
microspheres, proteinoid, 4
microsporangia, 431, 442, **443** (*see also* pollen
    sacs)
  of pines and other conifers, 476, 477

of *Selaginella*, **440**
  of yews, **484**
microsporangiate cones
  of cycads, 487
  of *Ephedra*, 491
  of pines, 476, 477, **477, 478,** 481, 481
microspore mother cells (*see* microsporocytes)
microspores, 431, 468, **473**
  of angiosperms, 504, **512**
  of pines, 477, **479**
  of *Selaginella*, **440, 441,** 442
microsporogenesis, 504
microsporocytes (microspore or pollen
    mother cells)
  of angiosperms, 504
  of pines, 477, **478**
microsporophylls, 442, 443, **443,** 501
  of pines, 477, **477**
microtubule organizing centers, 59, 163, 311
    (*see also* polar rings)
microtubules, 29, **59, 71t**
  cell cycle and cell division and, 158–62,
    161, 162, 165, **165**
  cell plate formation and, 165, **165,** 166
  cell wall formation and, 66, **67, 165**
  in flagella and cilia, 61, **61,** 105
  functions of, 59
  hormones and orientation of, 690
  kinetochore, 161, 162, **162,** 163
  phragmoplast, 165
  polar, 163
  spindle, 161
  structure of, 58–59
middle lamella, **22,** 64, 166
midrib, **628,** 631, 635
midvein, 628, 631, **633,** 635
Miki, Shigeru, 486
mildews, 317, 373 (*see also* molds)
milkweed, 34, **35, 544,** 546, **549,** 550
  family, 266, **266, 531**
Miller, Carlos O., 680
Miller, Stanley L., 3, **3**
millets, 826
mimicry, 550
Mimosa pudica, **36,** 720, **720**
Mimulus cardinalis, **537**
minerals, 6
  absorption by roots, 762–63
  cycles (*see* nutrient cycles)
  as nutrients, 731 (*see also* essential elements)
  transport in plants, 729, 762–64
mint family, 34, **245,** 624, 833
miso, 334
Mississippian period, 471
mistletoe, 498, **547**
Mitchell, Peter, 118, 120
mitochondria, 44, **45, 52,** 52–53, **71t**
  cell cycle and, 158
  chemiosmotic coupling in, 137, 138
  cytoplasmic inheritance and, 196
  DNA of, 53
  electron transport chain in, 116, **117,** 118
  evolution of, 53, **272–73,** 272–74
  functions of, 52, **108,** 113–14, 143
  matrix of, 114, **114,** 118, **119,** 120, **120, 121,**
    122
  structure of, 114, **114**
mitosis, 156, 157, **157,** 158, **160, 161,** 166
  in algae, 360, 386–87, **388–98**
  in chytrids, 312
  in dinoflagellates, 363
  duration of, 163
  in euglenoids, 352
  in fungi, 310–11
  meiosis compared and, **178,** 179
  in oomycetes, 371, **374**
  phases of, 160–63, **162**
  in prokaryotes, 285–86
  in slime molds, 352, 355
mitotic spindle, 158–60, **160,** 161, **162,** 163, **165**
mixotrophs, 348
Mnium, **417**
moa plant, **444**
Mojave Desert, **808**
molds, 315, **316,** 317

slime (*see* cellular slime molds; plasmodial
    slime molds)
  water, **171,** 275, 373, 872
molecular chaperones, 11
molecular chaperones, 29
molecular clock, 270
molecular formulas, 19, 870
molecular genetics, 197
molecular mass, 870
molecular systematics, 268–71
molecular weights, 870–71
molecules, 18, 868–71
  biologically important, **38t**
  diffusion of, 76–78
  energy and, 97–99, 105–6, 875–76
  excited, 131–32, 135, 865
  organic, 3–4, 18, 869–70 (*see also* specific
    types)
  representations of, 19, **19**
moles (units), 97, 871
molybdenum (Mo/MoO$_4^{2-}$), as essential
    element, 727, **730t**
monarch butterfly, 34, **35,** 550
Monilinia fructicola, 317
monkeyflower, scarlet, **537**
monkey-puzzle tree, **483**
monocots (monocotyledons), **497, 498, 504,** 887
  cotyledon of, 559, 563
  embryos of, 557, **558–59,** 559, 563
  eudicots compared and, **498t**
  evolution of, 519, 521
  genetic engineering and, 699
  leaf development in, 635
  major features of, **498t**
  parasitic, 498
  seeds of, 562
Monocotyledones, **265t,** 496, 887
Monod, Jacques, 216
monoecious plants, 501, **502**
monohybrid crosses, 185
monokaryotic (uninucleate) cells, 322
monomers, 18
monooxygenases, 780
monophyletic taxa, 265
monosaccharides, 18, **19**
monospores, 360
monoterpenoids, 33, 34
Monotropa uniflora, **498**
monsoon forests, **798–99,** 806, 807
Monterey pines, **475, 477**
Morchella, **308t, 320,** 342
morels, **309,** 317, **320,** 342
Moricandia, **625**
  arvensis, **73**
morphine, 32, **32**
morphogenesis, 166, 572
  hormones and, 687
morphological species concept, 248
Morus alba, 624, **625**
mosaics (mosaic diseases), 301
mosquitoes, pollination by, 535, **535**
moss, Irish, 358
mosses (*Bryophyta*), **56, 275t,** 400, 407, 412–22,
    432, 886 (*see also* bryophytes; *and
    specific names of mosses*)
  asexual reproduction in, 414, 421
  cladogram of, 267–68, **268, 402**
  classification of, 412
  club (*see* club mosses)
  comparative structure of, 403
  comparative summary of characteristics of,
    422t
  economic uses of, 415
  fir, 438
  granite (*Andreaeidae*), 412, 416
  habitats of, 401, **401**
  Irish (algae), 412
  life cycles of, 414, 416, **418–19**
  liverworts compared and, 412
  peat (*Sphagnidae*), 412, **414,** 414–15
  protonemata of, 407
  reindeer (lichens), 412
  scale (liverworts), 412
  sea (algae), 412
  Spanish, 412

sporophytes of, 406
"true," (Bryidae), 412, 416–22
conducting tissues of, 416, **416**, **417**
"cushiony," "feathery," 421, **422**
sexual reproduction in, 416, **417–18**, 418–21
weathering of rocks and, 786
moths, 550, 782
pollination by, 535–36, **536**
motility
of algae, 388–90, 393
of chytrids, 313
of diatoms, 376
of dinoflagellates, 361, 363–65
of euglenoids, 351
of oomycetes, 371, 373
of prokaryotes, 284, 288–89, 292–93
of slime molds, 352, **354–55**
motor proteins, 163
mountain forests, **798–99**
mountain lilacs, 741
mountains, precipitation and, 801, **801**, 802
Mount St. Helens, 789, 792, **792**
movement proteins, 300, 302
M phase, 157, **157**
mRNA (messenger RNA), 208, 209, **688**
in eukaryotes, 222–23, **223**
gene regulation in prokaryotes and, 216, **216**
protein synthesis and, 210–14
mRNA-binding sites, 211
mucigel, 591, **592**, 597
mucilage, 618
mucilaginous sheaths, 288, 397
mulberry, 624, **625**
mullein, **585**
Müller, Hermann J., 181
multicellular organisms, evolution of, **6**, **7**, 11, **12**, 14, 45
multilayered structure, 387, **387**
Multinational Coordinated *Arabidopsis thaliana* Genome Research Project, 228
multipass transmembrane proteins, 75, **75**, 82
multiple fruits, 543
mumps, 296
Münch, Ernst, 767
Musa × paradisiaca, **497**, 827
mushrooms, **277**, **308t**, **309**, 323 (see also *Basidiomycetes*)
commercially cultivated, 325
edible, classification of, 265
fairy rings, 325, **325**
hallucinogenic, 325, **325**
inky cap, **321**
life cycle of, 320, **320–21**
oyster, 333, **333**
parts of, 324–25
poisonous, 325
reproduction in, 324
mustard, 833
black, **549**
family, **228**, **238**, 340, 511, 544, **544**, 549, 550, **728**
Indian, 747
mustard-oil glycosides, 549–50
mutagens, 192
mutants
dwarf, gibberellin and, 684–85, **685**
viviparous, 684, **684**
mutations, 172, 192–94, 240, **241**
chromosome, 192–94
embryogenesis and, 560, **560**, **561**
evolutionary change and, 193–94, 240–41
gene (point), 192
homeotic, 639, 640, **641**
of prokaryotes, 285–86
mutualism (mutualistic relationships), 334, 774–75 (see also lichens; mycorrhizae; symbiosis)
of mycobiont and photobiont, 339, **339**
mycelia, **306**, 310, **320**, 325
of ascomycetes, 318, 320
of *Basidiomycota,* 322
dikaryotic, monokaryotic, 322
heterokaryotic, 333

of mushrooms, 324–25
septate, 310, 318, 322
Mycena
haematopus, **309**
lux-coeli, **122**
mycobionts, 334
photobionts' relationship with, 339, **339**
mycologists, 348
mycology, 12, 310
mycoplasmalike organisms (MLOs; phytoplasmas), 292–93, **293**
mycoplasmas, 292
mycorrhizae, 308, 312, 340–44, 431, 597, 762
of club mosses (*Lycopodiaceae*), 431, **436**, **437**, 438
evolution of, 343
fossil, 343, **344**
mutualism and, 774–75
tree nutrition and, 340, **340**, 341
vesicular-arbuscular (V/A), 341
mycorrhizal fungus, 340–44, 762, 774
of angiosperms, 342–43
of ferns, 455
of *Lycopodiaceae,* 431, 436
of *Psilotum,* 431, 444, 445
mycotoxins, 307, 334
myosin, actin and, 47
Myrica gale, 741
Myristica fragrans, 832
myxamoebas (amoeba-like cells), **354**, 354–56, **356**
myxobacteria, 285, **285**
myxomatosis, 777
Myxomycota (myxomycetes), 884 (see also plasmodial slime molds)
life cycle of, **353–55**
myxospores, 285, **285**

N

NAA (naphthalenacetic acid), **676**, 678
NAD⁺/NADH, **102**, 102–3, **117**
in alcohol fermentation, 123
in electron transport chain, 118
in glycolysis, 109, 112, **112**, 113, **113**, 122
in Krebs cycle, 114, **114**, 116
NADP⁺/NADPH, 134, **134**, 135, **136**, 138, 141
naphthaleneacetic acid (NAA), **676**, 678
nastic movements, 719–21, **723t**
natural disasters, 789
natural gas, 295
Naturalist in Nicaragua, The, 775
natural selection, 236, 237–38, 240
adaptation as result of, 245–47
genetic variability and, 243–45
rapid changes in natural populations and, 244–45
neck canal cells, 404, **405**
necrosis, 729
nectar, 525
of bat-pollinated flowers, 538, 539, **539**
of bird flowers, 537, **538**
insect pollination and, 530, 531, **531**
nectaries, **522**, 525, 530
insect pollination and, 532, **535**, 536
bee flowers, 532
needles, pine (see leaves, of pines)
Neljubov, Dimitry, 682
Nelumbo
lutea, **524**
nucifera, 718
nematodes
mycorrhizae and, 340
predaceous fungi and, **333**
neomycin, 287
Nepeta cataria, 262
Neptunia pubescens, **638**
Nereocystis, 382
Nerium oleander, **629**
netted venation, 631, **631**
nettle, stinging, 778
neurological illness, 364
Neurospora, **308t**, 708
neutrons, 851
Newton, Sir Isaac, 128
N-formyl methionine (fMet), 212, **213**

nickel (Ni/Ni²⁺), as essential element, 727, **730t**
Nicolson, Garth, 74
Nicotiana (see also tobacco)
glauca, **696**
silvestris, 715, **716**
tabacum, 715, **716**
nicotinamide adenine dinucleotide (see NAD⁺/NADH)
nicotine, 33, **33**, 549
nif genes, 740
nightshade family, 695, 831
Nilson-Ehle, H., 195
Nirenberg, Marshall, 209
Nitella, **47**, **76**
nitrate ions (NO₃⁻), 738
nitrates, 6, 26, 738
assimilation of, 742
nitrification, 736, **736**, 738
nitrifying bacteria, 738
nitrite ions (NO₂⁻), 738
nitrites, 26, 738
Nitrobacter, 738
nitrogen (N), 18
amino acids and, 26
assimilation of, 742, **742**
in atmosphere, 3, 736, **736**
as essential element, 727, 729, **730t**
in prokaryote metabolism, 287, 290, 295, 737–41
nitrogenase, 290, 738
nitrogen cycle, 736–42
human impact on, 743
stages of, 736, **736**
nitrogen dioxide, in photochemical smog, 616
nitrogen fertilizers, 738, 741–42, 745
nitrogen fixation
biological, 290, **736**, 738–42
genetic manipulation of, 747
industrial, 741
manipulation of, 747
nitrogen-fixing bacteria, 287, 290, 738–41, 747
nonsymbiotic, 741
nitrogenous bases, 30, **31**, 198–199, 200
nitrogen oxides, as air pollutants, 616, **617**
Nitrosococcus nitrosus, **285**
Nitrosomonas, 738
Nobel, Park, 809
Noble, David, 488
Noctiluca scintillans, **363**
nodD genes, 740
nodes, **8**, 445, 611, 613
nod genes, 740
nodules, root, 739, 740, **740**
nodulins, 740
Nomuraea rileyi, **319**
noncyclic electron flow, photophosphorylation, **136**, 138
nondisjunction, 173
nonfunctional phloem, **656**, 659
non-heme iron proteins, 116, **117**
nonpolar covalent bonds, 870
nonrandom mating, 242
Norfolk Island pine, 483, **484**
nori, 380
northern spotted owl, 815, **815**
Northwestern coniferous forest, **798–99**
Nostoc, 335, 412
commune, **289**
Nothofagus, 342
menziesii, **523**
no-tillage cropping system, **838**
nuclear division (see meiosis, mitosis)
nuclear envelope, 41, **44**, **45**, 47, 48
endoplasmic reticulum and, 56
of euglenoids, 352
of fungi, 311
nuclear genome, 53
nuclear pores, 47, 48, **48**
nuclear-type endosperm formation, 510
nucleic acid probe, 230, **230**
nucleic acids, 18, 30–32, **31** (see also DNA;

specific compounds)
sequencing of, **229**, 229–30
molecular systematics and, 270–71
nucleoids, 41, **43**, **44**, 52, **52**, 53
nucleolar organizer regions, 48
nucleoli, **45**, 48, 160, **160**
nucleomorphs, 357
nucleoplasm, 47
nucleosides, 102
nucleosomes, **219**, 219–20
nucleotides, 30, **30**, **38t**, 102 (see also specific compounds)
in ATP, 31
in DNA, 209
mutations and, 192–93
replication of DNA and, 203–4
structure of, 198, **198**, 199
genetic code and, 209
protein synthesis and, 209, 210, **211**
in RNA, 208, 209
sequencing of, **229–30**, 270–71
in viral genomes, 297–98
nucleus (atomic), 863–67
nucleus (of cell), 41, 45, **45**, 47–48, **70t**
diploid, haploid, 47, 170–73, 175, 177, 179, 331, 332
division of, 156, **157**, 158, **159**, 160, **160**, 161, 163 (see also meiosis; mitosis)
evolutionary origin of, **273**, 274
functions of, 47
generative, 504
polar, 506, **507**, 509
tube cell, **505**
Nudaria, **380**
nudibranch, 395
numbers, pyramid of, 786, **786**
nutcrackers, 482
nutmeg, 832, **832**
nutrient cycles, 735–36
carbon, 150–51, **151**, 350, 415
human impact on, 743–44
nitrogen, 736, 736–43
phosphorus, 742, **743**
nutrient deficiencies, 727, 729, **730t**, 731
nutrients, inorganic (see also essential elements; macronutrients; micronutrients; nutrition, plant; soils; specific nutrients)
concentration of, in plants, 727, **728t**
exchanged between transpiration and assimilate streams, **763**, 763–64
functions of, 729, **730t**, 731
uptake by roots, 762–64, **763**
nutrition, plant, 726–49 (see also nutrient cycles)
agriculture and, 744–45
of angiosperms, 498
essential elements, 727–31
general requirements, 727–31
mycorrhizae and, 340–41, 762, 774
research on, 745–47
soils and, 731–35
nuts, 545
nyctinastic movements, 719–21 (see also tropisms)
nyctinasty, **723t**
Nymphaeaceae, 521
Nymphaea odorata, **575**, **629**, 787

O

oaks, **577** (see also *Quercus*)
black, 256, **257**, **658**
cladogram of, 268, **268**
cork, 658, **659**
red, 65, 577
scarlet, 256, **257**
white, **657**, **666**, **667**, **669t**
oat, 146, 675
bran, 563
loose smut of, 330
root system, **589**
oca, **830**
ocean currents
dispersal of fruits by, 546
global warming and, 845
oceans

carbon dioxide in, 150
evolution of photosynthetic organisms in, 6–7
primitive, 4
Ochroma lagopus, 659
Ocimum vulgare, 833
odors, of flowers, 531, 539
Oedemeridae, **531**
Oedogonium, 392
  cardiacum, **392**
  foveolatum, **392**
Oenothera, **536**
  gigas, 252
  glazioviana, 192, **192**
  lamarckiana, 252
O horizon of soil, **732**
oil bodies, **23**, **51**, 55, **55**, **71t** (*see also* plastoglobuli)
oil cells, 519
oil palms, 834
oils, 22, **23**, **38t**
  essential, 34, 549
Okazaki fragments, 202, **203**
okra, 827
Olea capensis, 670
oleander, **629**
oleic acid, **23**
Oligocene epoch, 877
oligosaccharides, 74
oligosaccharins, 22, 62
Oliver, F. W., 457
olives, 825
O'Mara, J. G., 840
Onagraceae, **514**
Oncidium sphacelatum, **606**
onion, 162, **562**, **593**
  germination in, **567**, 567–68
Onoclea sensibilis, **458**
oogamy, 371–72, **372**, 430
oogonia, 372, **374**, **385**
oomycetes (*Oomycota*), **171**, 348, 371–74, **372t**, 884
  life cycle of, **374**
  water molds, **171**, 275, 373, 884
oospores, 372, **374**
opaline silica, 376
Oparin, A. I., 3
open system, the Earth as, 98, **98**
operator sequence, 216
operculum, 414, **414**
operons, 216, **217**, 218, **218**
Ophioglossales, 454, 455, **463t**
Ophioglossum, 47, **454**, 454–55
  reticulatum, 455
Ophiostoma ulmi, 317
Ophrys, **535**
  speculum, **534**
opines, 696
opine-synthesizing enzyme, 696
opium poppy, **32**, 550
opposite phyllotaxy, 624
Opuntia, **547**
  inermis, 779, **780**
orange snow, **386**
orbitals, 855
orchard grass, 244
Orchidaceae, **260**, 343, 527, **528**, 528–29
orchids, **528**, 528–29
  commercial production of, 529
  Cymbidium, **278–79**
  epiphytic, 804
  lady slipper, **260**
  mycorrhizae in, 343
  pollination of, **534**, 534–35, **535**
  roots of, 605, **606**
  saprophytic, 529
  underground, 529, **529**
  vanilla, 529, **529**, 833
orders, taxonomic, 264
oregano, 833
organelles, 41, **44** (*see also* by specific type)
organic molecules, 3–4, 18, 869–70 (*see also* by name)
organismal theory, 65
organisms (*see also* by name)

classification of, 262–71, 881–88
  interactions between, 774–82
organ-pipe cactus, **539**
organs, 86 (*see also* leaves; roots; stems)
organ transplants, 308
Origanum vulgare, 833
origin of replication, 201, 202
Origin of Species, On the, 248, 257
Orthotrichum, 421
Oryza sativa, 146, **497**, 563, 684, 826, **827**
Oscillatoria, **289**
osmoregulatory mechanisms, 744
osmosis, 78–81, **79**
  in living organisms, 80–81
osmotic potential (solute potential), 79, **79**
  of guard cells, 693
osmotic pressure, 79
ostrich fern, **458**
outcrossing (outbreeding), 256, 477
  in angiosperms, 510–11
  in gymnosperms, 510
  mechanisms that promote, 242–43
outgroups, 267
ovaries, **499**, 501, 503, **513**
  beetle pollination and, 531
  development into fruit, 509–10, 686
  of early angiosperms, **524**, 527
  inferior, superior, 502
  of orchids, 528
ovary walls, 501, **501**, 506, 510
ovulate cones
  of balsam fir, **483**
  of pines, 476–77, **477**, 478, 480, **480**, 481, 481, **483**
ovules, 184, **468**, 469
  of angiosperms, 499, 501, 506, **507**, 525
  evolution of, 468, 525
  of *Ginkgo*, 490
  of gymnosperms, 472
  insect pollination and, 530
  of pines, **477–479**, 480–82
  of wind-pollinated flowers, 530, 540
ovuliferous scales (cone scales), 480, 481
owl, northern spotted, 815, **815**
Oxalis, **707**
  tuberosa, **830**
oxaloacetate (oxaloacetic acid), 115, 144, **145**, **149**
oxidation, 98, **99**
  beta, 122
  of glucose, **99**, 109, **110**, **121** (*see also* glycolysis)
  of pyruvate, **114**, 122
oxidation-reduction (redox) reactions, 98–99, **99**
oxidative phosphorylation, 108, 109, **110**, 118–20, **121** (*see also* phosphorylation)
Oxydendrum, **813**
oxygen (O), 18 (*see also* respiration)
  in atmosphere, 5–6, 9, 11, 844
  diffusion of, 78
  as essential element, 727, **730t**
  germination and, 565
  nitrogen fixation and, 740
  photosynthesis and, 5–6, 9, 11
  prokaryotes and, 287
Oxyria digyna, 247, 249
oyster mushroom, 333, **333**
ozone (O₃), 616, **617**, 844
  in atmosphere, 6
ozone layer, 844

P

P₆₈₀, P₇₀₀ chlorophyll molecules, 135, **136**, 138, 139, **139**, 152
palea, **541**
paleobotany, 11
paleoherbs, 521, **521**, 526
Paleozoic era, **469**, 470
Palhinhaea, 438
palisade parenchyma, **628**, 629, 630–31, 636
palmately compound leaves, 626, **627**
Palmer's grass, 744
palms, 648
  coconut, **497**, **544**

fan, 518
  oil, 834
  sago, 486
PAN (peroxyacetyl nitrate), 616
Panama pine, 483
Pandorina, **390**
panicle, **499**
Panicum, 827
Papaveraceae, 544
Papaver somniferum, **32**, 544, 550
paper birch, **10**, 541, 658
pappus, 528, **545**, 546
paradermal section, **628**
paraheliotropism, 723
parallel venation, **625**, 631
paramylon, 351, 352
paraphyletic taxa, 266
paraphyses, **417**
parapodia, 396
Para rubber trees, 842
parasexuality, in deuteromycetes, 333–34
parasitic organisms
  angiosperms, 498, **498**
  bacteria, 293
  fungi, 310, 317, 334, 339
  oomycetes, 373
parasitic relationships, 334
parenchyma (parenchyma cells), 48, **571**, 572–73, **573**, 574, 582, 583, 602
  in algae, 381, 392
  auxin and, 676
  of bryophytes, **417**
  interfascicular, 614, 619
  of *Isoetes*, 443
  of leaves, 628
    palisade and spongy, **628**, **629**, 630–31, 636
    pine leaves (needles), 474, 475
  of phloem, 579–80, 582–83
  of pulvinus, 720, **720**
  ray, 573, 648
  of stems, 618, 619, **619**, 620
  storage, 606, **607**
  of xylem, 579, **579**
parenthesomes, 322, **322**
parietal placentation, 501, **501**
Parmelia perforata, **336**
parsimony, principle of, 268
parsley, **752**, 833
  family, 264, **502**, 530, 531, 833
  schizocarp, 545
Parthenium argentatum, 842, **842**
parthenocarpic fruits, 543
Parthenocissus
  quinquefolia, 641
  tricuspidata, 641
particle bombardment, 699
particle model of light, 130
passage cells, 600
passion flower, 537
passive transport, 82
Pasternak, D., **843**
patch-clamp technique, for studying ion channels, 83
pathogen(s) (*see* diseases; plant diseases)
Pauling, Linus, 198, 199
PCR (polymerase chain reaction), 226
pea(s), **545**, 563, **566**, 567, 738, 780, 825, **825**
  ethylene and, 682, **683**
  family (*Fabaceae*), 264, 544, **545**, 636
  garden, **545**
  leaves of, 624
  Mendel's experiments with, 184–90, **185**, **185t**, 242 (*see also* Medelian genetics)
  seed germination in, 566
  tendrils of, 641, 706
peach, 543, 719
peanut(s), 828
pear, **62**, 576, 655
pear decline, 293
peat mosses (*Sphagnidae*), 412, **414**, 414–15
pectic acid, **22**
pectins, 22, **22**, 63, 572
  in cell walls, **62**
  growth of the cell wall and, 66

pedicel, 499
peduncle
  of dinoflagellates, 365, **366**
  of flowers, 499
pelargonidin, **543**
pellicle, 351, **351**
penicillin, 334, 779
penicillinases, 779
Penicillium, **332**, 333, 334
  camemberti, 334
  chrysogenum, 779
  notatum, **332**
  roqueforti, 334
pennate diatoms, **375**, 376, 378
Pennisetum, 827
Pennsylvanian period, 877
3-pentadecanedienyl catechol, 550
pentoses, 18
PEP (phosphoenolpyruvate), 112, **112**, 144, **144**, 146
PEP carboxylase, 144, **144**, 146
Peperomia, 644
peppers, 521, **521**, 832, 833
peptide bond, 28
peptidoglycans, 283, **283**
peptidyl transferase, 212
perennials, 648
perforation plate, 576, **577**
perforations of vessel elements, 576, 577, **577**, 578
perianth, **409**, 501, 502
  of early angiosperms, 524
  of liverworts, 412
pericarp, 510, 547, 564, **557**
  in algae, 362
periclinal divisions, 557, 612, **612**, 649
pericycle, 600, 601, **601**, 602, 603
periderm, 427, 476, **572**, 586–87, **587**, 600, **601**, 602, **653**, 653–54
  gas exchange through, 654
peridinin, 365
peridium, 325, **326**
perigynous flowers, 502, **502**
perigyny, **503**
peripheral proteins, 74, **75**
peripheral zones, 613
periphyses, 327
perisperm, 510, 563
peristome, 421, **421**
perithecium, 319, **319**
periwinkle, 846
permafrost, 819, 820
permanent wilting percentage (point) of soils, 735, **735**
permeability, 78, **79**
Permian period, 429, 431, 457, 474
Peronosporales, 373
peroxisomes, **53**, 53–54, **71t**
  in photorespiration, 143
peroxyacetyl nitrate (PAN), 616
Persea americana, 683
persimmon, **67**, **669t**
pesticides, 781, **837**
pest management systems, integrated, 781
petal primordia, **638**
petals, **499**, 501
  development of, **638**, 639
  of early angiosperms, 524, **524**, 527
  fusion of, 524
  of orchids, **528**, 528–29
petiole, **8**, 624, **625**, 626
petiolule, 626
Petroselinum crispum, 833
petunia, **698**
peyote cactus, **551**
Pfiesteria piscicida, 364, 366, **366**
PGA (3-phosphoglyceric acid), 140, **140**, 142, **143**
PGAL (glyceraldehyde 3-phosphate; 3-phosphoglyceraldehyde), **111**, 111–13, 140–42
pH, 875
  of acid rain, 616–17
  scale, 875, **875t**
  of soils, 735
Phaeocystis, 367

pouchetii, 367
Phaeophyta (*see* brown algae)
phages (*see* bacteriophages)
phagocytes, evolution of, 272, **273**, 274
phagocytosis, 85, **85**
Phalaris canariensis, 675
Phanerochaete chrysosporium, 308
Phaseolus, 58, 738
   vulgaris, **562, 566**, 567, 678, 684, **685, 705**
phellem, 587 (*see also* cork)
phelloderm, 587, **587**, 602, 653, 654, **654**
phellogen (*see* cork cambium)
phenolics, 34–37
phenotypes, 186, 194–96
   environment and, 197
phenotypic ratio(s), 185–89
phenylalanine, 26, **27, 28**
pheophytin, 137
Phlegmariurus, 435, 438
phloem, 8, **8**, 426, **427**, 572, **576**, 579–83, **587t**
   assimilate transport in, 764–69
      aphids and, 766, **767**
      phloem loading, unloading, 767–69, **769**
      pressure-flow hypothesis of, 766–69, **768**
      radioactive tracer experiments, 765–66
   in brown algae, 382
   cell types in, **579t**, 580–83
   development of, 580–582, **583, 594**, 600, 614–21
   florigen transport in, 715–16
   functional, nonfunctional, **656**, 659
   hormones and, 676
   primary, **571, 573**, 600
      of roots, 601, **601**, 602
      of stems, 618, **619, 620**
   secondary, 427, 470
      in pines and other conifers, 476, **476**
      in seedless vascular plants, **435, 442**, **444–45, 448, 455**
   transfer cells and, 574, **574**
   of veins, 631
   viral movement in, 300, **300**
   xylem interchange and, 767, **768**
phloem-immobile ions, 729, 764
phloem loading and unloading, 767–68, **769**
phloem-mobile ions, 729, 764
phloem unloading, 768–69
phlox, 525
Phoenix dactylifera, 825
phosphate ($H_2PO_4^-$; HPO_4^{2-})
   absorption of, 742
   in ATP, **105**, 106
   in nucleic acids, 30, **30**, 31, **31**
   inorganic (P_i), 106, 112, 120, 138
   in phospholipids, 24, **24**
   mycorrhizae and, 340
phosphoenolpyruvate (PEP), 112, **112**, 144, **144**, 146
phosphofructokinase, 111
phosphoglucoisomerase, 111, **111**
3-phosphoglyceraldehyde (PGAL), **111**, 111–13, **140**, 140–42
1,3-bisphosphoglycerate, 112, **112**
3-phosphoglycerate (3-phosphoglyceric acid; PGA), 140, **140**, 142, **143**
3-phosphoglycerate kinase, 140
phosphoglycolate, 142, **142**, 143, **143**
phospholipid bilayer, 24, **24**, 74, **75**, 76
phospholipids, **24**, 24–25, **38t**, 74
   in cellular membranes, 74
phosphorus (P), 18
   as essential element, 727, **730t**
   in fertilizers, 745
   mycorrhizae and, 340
phosphorus cycle, 742, **743**
phosphorylation, 106 (*see also* photophosphorylation)
   oxidative, 108, 109, **110**, 118–20, **121**
   substrate-level, 112, **112**
Photinus pyralis, 698
photobionts, 334–35, **338**, 339
photochemical smog, 616
photoconversion reactions, 712
photoelectric effect, 129
photolysis, 137

photomorphogenic responses (photomorphogenesis), 714, **714**
photons, 130, 135
photoperiodism, 709–14
   in animals, 709
   chemical basis of, 711–14, **712, 713**
   dormancy and, 718–19
   flowering and, 709–11
photophosphorylation, 137–38, **138**
   cyclic, 139, **139**
   noncyclic, **136**, 138
photoreceptors (photoreceptor proteins), 703, **704** (*see also* phytochrome)
photorespiration, 54, **142**, 142–44
photosynthates, 631, 632
photosynthesis, 2, 126–51
   action spectrum for, 131, **131**
   in algae, 350, 395
   CAM, 147, **148**, 149, 443
   carbon cycle and, 150–51, **151**
   carbon-fixation reactions of (*see* carbon-fixation reactions)
   chloroplasts and, 49–50
   competition and, 776
   cotyledons and, 567
   in C_3 plants, 140–42, 144, 146–47
   in C_4 plants, 144–49
   efficiency of, 138, 142–44, 146, 149
   evolution of plants and, **5**, 5–6
   historical perspective on, 127–28
   in lichens, 334, 338–39
   light reactions of, 133–39
   mesophyll and, 630–31
   overall reaction (equation) for, 127
   overview of, **134**
   photochemical smog and, 616
   pigments in (*see* accessory pigments; carotenoids; chlorophylls; phycobilins)
   products of, 142
   in prokaryotes, 127–28, 133, 283, 291
   redox reactions during, 99
   stomatal movements and, 751–54
   transpiration and, 751
photosynthetic autotrophs, 287
photosynthetic bacteria, 127–28, 133, 291 (*see also* cyanobacteria; green bacteria; prochlorophytes; purple bacteria)
photosynthetic organisms (*see also* autotrophs)
   earliest, 5
   ecosystems and, 9, 11
   seashore environment and, 6–7
Photosystems I and II, 135, **136**, 138, **138**, 139
photosystems, 135, **135**, **136**
   in purple and green bacteria, 291
phototropism, 675, **702**, 703–4, **723t**
   in fungi, 315, **315**
phragmoplasts, 159, **159**, 165, **165**, 387, **387**
phragmosome, 158, **159**
phycobilins, 133, 288, 291, 357, 358
phycocyanin, 288, 357
phycoerythrin, 288, 357
phycology, 12
phycoplasts, 386
phyllotaxy, 624
phylogenetic analysis (cladistics), 267
phylogenetic trees, **264**, 265
   traditional, 267, **267**
phylogeny (phylogenetic relationships), 265
   selected characters used in analyzing, **268t**
phylum (phyla), taxonomic, 264
Physarum, **276, 352**
Physoderma
   alfalfae, 313
   maydis, 313
phytoalexins, 32, 62, 780, 782
phytochrome, **712, 713**
   in *Chlorophyta*, 383
   dark reversion of, **713**
   photoconversion of, **713**
   photoperiodism and, **712**, 712–14, **713**
phytohormones (*see* hormones)
phytomeres, 611, **611**
Phytophthora, 373
   cinnamomi, 373
   infestans, 373, **375**, 841

phytoplankton, **785**, 786
   concentration of, **800**
   marine, 348, 349–50, **350**
phytoplasmas, 292–293
Picea, 483
   engelmannii, 777, 791
   glauca, **788, 818**
   rubens, **669t**
pickles, 287
pickleweed, 744
picoplankton, 294
Pierinae, 550
pigments, 130–33 (*see also by name*)
   absorption spectrum of, 131, **131**
   accessory, 133, 288, 291, **349t, 372t**
   action spectrum, 131, **131**
   antenna, 135
   anthocyanins, 35, **35**, 54–55, 542, 543, **543**
   in bacteria, 282, 288, 291, **291**, 292
   carotenoids, 49–51, 133, **134**, 135, 542, 543, **543**
   colors of angiosperm flowers and, 542–43
   P_{680}, P_{700}, 135, **136**, 138, **139**
   photoperiodism and, 712–714, **713**
   photosynthetic (*see* accessory pigments; carotenoids; chlorophylls; phycobilins)
   P_r, P_{fr}, 712–714
   stomatal movements and, 753
   vacuoles and, 54
pigweed, 126
pileus (cap), 324
pili, 284, **284**, 285
Pilobolus, 315, **315**
Pimenta officinalis, 833
Pinaceae, 342, **483** (*see also* pines)
pineapple, **148**, 149, 674
   family, 337
   fruit, 543
pine nuts, 480
pine-oak chapparal, **10**
pines, 474–82, 559 (*see also* conifers; *Pinus*)
   bristlecone, 475, **475, 664**, 664–65
   cladogram of, 268, **268**
   digger, **480**
   eastern white, **480**
   family, 342, **483**
   leaves (needles) of, 474, **475, 476**
   life cycle of, 476–77, **478–79**, 482
   limber, 482
   loblolly, **669t**
   lodgepole, **342, 481, 482**, 791
   longleaf, **474, 475**
   Monterey, 475, **477**
   Norfolk Island, 483, **484**
   Panama, 483
   pinyon, **278, 480**, 482
   ponderosa, **669t**
   red, **480**
   slash, **669t**
   sugar, **278, 480, 669t**, 789
   western white, **669t**
   white, 340, **480**, 660, **660–662, 668, 669t**
   whitebark, 482
   yellow, **480**
Pinguicula grandiflora, 737
pinnae, 455
pinnately compound leaves, 626
pinocytosis, 85, 85–86
Pinus, 659
   albicaulis, 482
   contorta, **342, 481, 482**, 791
   edulis, **475, 480**
   elliottii, 659, **669t, 807**
   flexilis, 482
   lambertiana, **278, 480, 669t, 788, 789**
   life cycle of, **478–479**
   longaeva, 475, **475, 664**, 664–65
   monticola, **669t**
   palustris, **474, 475**
   ponderosa, **480, 669t**
   radiata, **475, 477**
   resinosa, **480**
   sabiniana, **480**
   strobus, **340, 480, 660–662, 668**

taeda, **669t**
pinyon-juniper savannas and woodlands, **810**
pinyon pine, 475, **480**, 482
Piperaceae, 521, **521**
Piper nigrum, 832
Pirozynski, K. A., 343
pisatin, 780
pistil, 501
Pisum sativum (*see* pea(s))
pitcher plant, 642
pit connections, in red algae, 360, **360**
pit-fields, primary, 64, **64**, 573
pith, 428, 573, **573**
   hollow, 445
   of roots, 595, **597**
   of stems, 613, 614, **615**, 618
pith meristems, 613
pith rays, 618, 651
pit-pairs, 64, 66, 577, 659–60, **661, 662**
pit(s), **64**, 66, 576, **576, 577**
   apertures, 660
   bordered, **64**, 66, **661, 662**
   cavity, 66
   coated, 86, **86**
   membranes, **64**, 66, 577, **662**
   ramiform, 576
   simple, 66, **576**
placenta, 501
   of bryophytes, 404–6
placentation, 501, **501**
Placobranchus ocellatus, 395
Plagiogyria, **453**
plainsawn boards, 667, **667**
plankton, 289, 348, **785** (*see also* phytoplankton; zooplankton)
Plantae (plant kingdom), **265t**, 277, **278**, 279, 886 (*see also* bryophytes; vascular plants)
plantains, 827
plant body, 8, 425–30
   internal organization of, 572–73
   primary, 427, 571
   secondary, 427
plant breeding, 837–43, 846
plant morphology, 11
plant physiology, 11
plant(s)
   and adaptations to life on land, 797, 800–2
   ancestors of, 401–3
   biotechnology (*see* biotechnology)
   C_3, C_4, 144–47, **148**, 149
   CAM, 147–49
   classification of, **275**
   domestication of, 824–32
   nutrition and soils, 726–49
   people and, 823–49
   tissues, 573–587
plant diseases, 287, 292–94, 296–303
plant-herbivore interactions, 779–80, 782
plant-pathogen interactions, 779–80, 782
plant pathogens, 292–94, **292–94**
plasma membrane (plasmalemma), 41, **44, 45**, **70t**, **75**, 599 (*see also* endomembrane system; membranes, cellular)
   in algae, 386
   of bacteria, 282–83
   cell-to-cell communication and, 87
   central control hypothesis, 705
   diffusion across, 78–79, **79**
   endomembrane system's evolution from, 272, **273**, 274
   functions of, 46, 74
   growth of cell walls and, 66
   plasmolysis and, 81, **81**
   of prokaryotes, 282–83
   solute transport across, 82–84, 574
   three-layered appearance of, 46, **46**
   transfer cells and, 574
   turgor pressure and, 80–81
   vesicle-mediated transport across, 85
plasmids
   mutations and, 193
   recombinant DNA (DNA cloning) and, 224–26, **225, 226**
   Ti (tumor-inducing), 696, **696**, 697

plasmodesmata, 56, 64, **64**, 65, 66, **67**, 68, **68**, **88**, **164**, 166, 398, 404, 582, **583**
  cell-to-cell communication and, 87–89
  movement of viruses via, 300, **300**
  size exclusion limits of, 88
plasmodia, 352, **352**, **353**, 354
plasmodial slime molds (*Myxomycota*; myxomyocetes), 275, **275t**, **276**, 348, **349t**, 352–54, 884
  life cycle of, 352, **353**
plasmodiocarp, 352, **353**
plasmogamy, 312, **318**
  in *Chlamydomonas*, 389, **389**
  in mushrooms, **321**
  in plasmodial slime molds, **353**
  in wheat rust, 329
plasmolysis, 81, **81**
Plasmopara viticola, 373
plasticity, developmental, 246
plastids, 48–52, **70t**, 581, **581** (*see also* amyloplasts; chloroplasts; chromoplasts; leucoplasts)
  of bryophytes, 404
  cell cycle and, 158
  cytoplasmic inheritance involving genes in, 196
  developmental cycle of, **51**
  of euglenoids, 352
plastocyanin (PC), 137, **138**, **139**
plastoglobuli, 49
plastoquinones (PQ, Qa, Q_b), **136**, **138**
Platanaceae, **518**
Platanus
  occidentalis, 248, **248**, 658, **669t**
  orientalis, 248, **248**
  x *hybrida*, 248, **248**
Plectranthus, **585**
pleiotropy, 195–96
Pleopeltis polypodioides, **453**
Pleurotus ostreatus, 333, **333**
Pliny the Elder, 327, 772
Pliocene epoch, inside front cover
plum, 543
plumule, 562, **562**, 563, **564**, 567, **568**
plurilocular gametangia, 383
plurilocular sporangia, **382**, 383
pneumatophores (air roots), 605, **605**
Pneumocystis carinii, 308
pneumonia, bacterial, 287
Poa
  annua, **633**
  pratensis, 146, 244
Poaceae, 256, **265t**, 511 (*see also* grasses)
pod (fruit), 513 (*see also* legume)
Podandrogyne formosa, **263**
Podophyllum peltatum, **762**
poi, 827
poinsettias, **538**
point mutations (gene mutations), 192
poison ivy, **550**
polar covalent bonds, 869–70, **870**
polar desert, 820
polarity, 874
  of embryos, 556
  of water molecules, 872–73, **872**, **873**
polar microtubules, 163
polar molecules, **24**, 869, **870**
polar nuclei, 506, **507**, 509
polar rings, 358
polio, 296
pollen (pollen grains), **184**, 431, 504–6 (*see also* pollination)
  of angiosperms, 504, 506, **506**, 508
    single-aperture, 519
    walls of, **505**
  bats and, 539
  of cycads, 487
  of eudicots, **506**
  germination of, 473, 481, **508**
  of gymnosperms, 472
  hay fever and, 508
  of monocots, **505**, **540**
  of pines, 477, **477**, 478–**79**, 481
  spores compared to, 506

of wind-pollinated flowers, 540
  of *Wollemia nobilis*, 489
pollen basket, 532
pollen sacs, 501, 504, **504**
pollen tube, 467
pollen tubes, **184**, 431
  of angiosperms, 501, 503, 504, **505**, 508, **508**, 509, **512**, **513**
  of cycads, 487, 488
  of *Ginkgo*, **473**, 490
  of gymnosperms, 472, 473, **473**, 474
  of pines, 477, **479**, 481
pollination, **184**, 431, 508–9
  in angiosperms, 503
  by bats, 538–39, **539**
  by birds, 537–38, **538**
  cross-, **184**, 530 (*see also* outcrossing)
  of cycads, 487
  evolution of angiosperms and, 525
  in gnetophytes, **492**
  in gymnosperms, 472
  by insects, **530**, 530–36
  in pines, 481
  self-, **184**, 514
    homozygotes and, 242
    inhibition of, 243, **243**
  wind, 530, 539–40, **540**, **541**
pollinium, 528
pollution, 616–17
  bryophytes and, 401
  Ginkgo and, 490
  lichens and, 340, 401
poly-A tail, 223
poly-β-hydroxybutyric acid, 284
polyembryony, 472, 482
polygenic inheritance, 195
Polygonaceae, 545
Polygonatum, **508**
polymerase chain reaction (PCR), 226
polymerization, 18
polymers, 18
polynomial names, 262
polypeptides (polypeptide chains)
  of proteins, 28, **28**, **29**, 30
  protein synthesis and, 210, 212, **212**, **213**, 214
  targeting and sorting of, 214, **214**, **215**
Polyphagus euglenae, 309, **313**
polyphyletic taxa, 266
polyploid cells, 170
polyploidy (polyploids), 193 (*see also* allopolyploidy; autopolyploidy)
  speciation and, 249, 252–54
Polypodium, life cycle of, **460–61**
polypores, 323, 324
Polyporus arcularis, **323**
polyribosomes (polysomes), 212, **212**
polysaccharides, 18, **38t** (*see also by name*)
  in bacteria, 283, **283**, 284, 288
  in cell walls, 62–3, **349t**, **372t**
  in fungi, 310
  as storage forms of sugar, 20
  structural role of, 20, 21
Polysiphonia, life cycle of, **362**
polysomes (polyribosomes), 48, **55**, 55–56
polysporangiates, **432**
Polytrichum
  juniperinum, **422**
  life cycle of, **418–19**
  piliferum, **420**
pomegranates, 825
pomes, 544
pondweed, **128**
popcorn, **828**
poplar, 342, 546
poppy
  California, **497**, **534**
  capsules, 544
  family, **544**
  opium, 32, 550
population genetics, 239
populations, 774
  allopatric speciation and geographic separation of, 249

defined, 239
  Hardy-Weinberg law and, 239–40
  human, 835–37
  natural selection within, 245
  rapid evolutionary changes in, 244–45
Populus
  deltoides, **68**, **583**, 631, **669t**
  tremuloides, **788**
pores
  of bryophytes, 403, **403**
  epidermal, **761**
  fungal, 310, 319, **322**, **323**, 324
  in pollen grains (*see* pollen)
  in sieve elements, 579–80, **581–83**
  stomatal (*see* stomata)
  of wood, 665
pore space of soils, 732, 734–35
Porolithon craspedium, **358**
Porphyra, 359, 361, 380
  nereocystis, **359**
porphyrin ring, **117**
portulaca, 543
position effects, 193
Postelsia palmiformis, **276**
post-sieve-tube transport, 769
posttranslational import, 214, **214**, **215**
postzygotic isolating mechanisms, 256
potassium (K), 76
  cycling, 783
  as essential element, 727, **730t**
  in fertilizers, 745
  leaf movements and, 753–54
potassium ions (K+), stomatal movement and, **692**, **693**
potato(es), **21**, 550, 698, **830**, 830–31
  cultivation of, **830**, 834, **840**
  dormancy of, 719
  eye (bud), 642, 719
  genetic diversity of, 841
  late blight of, 373, **375**
  ring rot, 294
  scab, 294
  seed bank for, **840**
  soft rot, 293
  spindle tuber disease of, 302
  sweet, 831, 834
  white (Irish), 642–43, 831
  wilt, **285**, 294
potato famine in Ireland (1846–1847), 373, 841
potato spindle tuber viroid (PSTVd), 302
potential (potential energy)
  of electrons, 95, 865
  osmotic, **79**
  osmotic (solute), 79
  of water, 76–77
Potentilla glandulosa, 246, **247**, 249
Poterioochromonas, 378
P-protein (slime), **580**, 581, 581–82, **583**
P-protein bodies (slime bodies), 582, **582**, **583**
PQ (plastoquinones), **138**
prairies (*see* grasslands)
Precambrian era, inside front cover
precipitation (*see also* rainfall)
  global warming and, 845
  mountains and, 801, **801**, 802
predaceous fungi, 333, **333**
preovules, 468
preprophase bands, 158, 159, 165, 166
  of bryophytes, 404
preprophase spindle, 160
pressure, osmotic, 79
pressure chamber, cohesion-tension theory of water transport and, 757
pressure-flow hypothesis, 767–69, **768**
prezygotic isolating mechanisms, 256
prickles, 642
prickly pear, dispersal of, 547
Priestley, Joseph, 127
primary active transport, 84, **84**
primary consumers, 783 (*see also* herbivores)
primary endosperm nucleus, 509, **513**
primary growth, 8, 426–27, 570–88
  in roots, 591–95
  in stems, 611–14
primary meristems, 557, 559, 570, **570**, **571** (*see*

also ground meristem; procambium; protoderm)
  of roots, 591–92, **593**, **594**
  of shoots, 611, **614**
primary metabolites, definition of, 32
primary phloem, **571**, **573**, 579–83
  of roots, 601, **601**, 602
  of stems, 618, **619**, 620
primary phloem fibers, 618, 650, 651
primary pit-fields, 64, **64**, **573**
primary plant body, 427, 570–87
primary producers, 783
primary roots, 566, 590 (*see also* radicle; taproot)
primary structure
  comparison in root and stem, 595
  of roots, 595–600
  of stems, 610–46
primary structure of proteins, 28, **28**, **29**
primary tissues, 427, 573–87
primary vascular tissues, 576–83
primary xylem, **571**, **573**, 576–79
  of roots, 600, 601, **601**, 602
  of stems, 618, **619**, 620
  vascular cylinder and, 600
primordia
  bud, 611, **611**, 634
  flower, **638**, **639**
  leaf, 611, **611**, **614**, 622, 634, **634**, 635, **635**
    phyllotaxy and, 624
  root, **603**
primrose, **243**
  evening, 192, **192**, 252, 536
  family, **514**
  giant, 252
principle of independent assortment, 188–89
principle of segregation, 185–87
probability, laws of, 188
procambial strands, 614, 618, **618**, 619, **619**, 621, **621**, 622, 634
procambium, 557, **557**, 570, **571**, 576 (*see also* primary meristems)
  of leaves, 634
  of shoots, **611**, 613, 614, **614**
  of stems, 618
Prochlorococcus, 292
Prochloron, 292, **292**
prochlorophytes, 292, **292**
Prochlorothrix, 292
product rule of probability, 188
proembryo, 556, **557**, **558**
progesterone, 846
progymnosperms (*Progymnospermophyta*), 429, 431, 443, **470**, 470–72
  phylogenetic relationships with other embryophytes, **472**
prokaryotes, 6, 11, **43**, **44**, 281–305 (*see also* archaea; bacteria)
  autotrophic, 286, 287, 783
    chemosynthetic, 287
    photosynthetic, 5, 287
  cell division in, **156**
  cell wall of, 283–84
  characteristics of, 282–84
  chromosomes of, 216–18
  commercial uses of, 287
  disease-causing, 287
  DNA replication in, 202, 203
  eukaryotes compared and, 41, 43, 44, **45t**, 272
  fimbriae and pili of, 284, **284**
  flagella of, 284
  forms (shapes) of, 284–85, **285**
  heterotrophic, 5, 286, 783
  inclusion bodies (storage granules) of, 284
  metabolic diversity of, 286–87
  overview of, 282
  oxygen need or tolerance of, 287
  plasma membrane of, 282–83
  regulation of gene expression in, 216
  reproduction of, 285
  world ecosystem and, 287
prolamellar bodies, 51
proline, **27**
promoters, 210

propagation (see asexual reproduction; sexual reproduction)
prophase, 158, 160, **160**, 163
　meiotic, **174**, 174–76, **175**, 177
　mitotic, 160
proplastids, 51, 51–52, **70t**
prop roots, 603, **603**
Prosopis juliflora, 590
prosthetic groups, 103
protandrous plants, 510, **514**
protective layer, 636, **637**
proteinase inhibitors, 550
protein kinases, 691
proteinoid microspheres, 4
proteins, 18, **38t** (see also amino acids; specific types of proteins)
　carrier, **82**, 83
　channel, **82**, 83
　denaturation of, 30
　fibrous, 29
　in food, 825, 838
　globular, 29
　hay fever and, 508
　integral, 74
　iron-sulfur, 116, **117**
　molecular chaperones, 29
　molecular structure of, 26–30
　　primary structure, 28, **28**, 29
　　quaternary structure, **29**, 30
　　secondary structure, 28, **29**
　　tertiary structure, 29, **29**, 30
　movement, 300, 302
　peripheral, 74, **75**
　respiration and, 122
　synthesis of, 208, 210–14
　　DNA strands used as templates, 210
　　RNA synthesis from DNA templates, 210
　　rRNA (ribosomal RNA), 211–12
　　stages in, 212, **212**
　　targeting and sorting of polypeptides, 214, **214**, **215**
　　transfer RNA (tRNA), 210–11
　　translation of mRNA into proteins, 212–14
　transmembrane, 74, 75, **75**
　　multipass, 75, **75**, 82
　transport, 82
　viral, 297–98
prothallial cells, 477, **477**, 479
prothallus, 458–59
Protista (protists; kingdom), 270, 271, 275, **275t**, 275–76, **276**, 347–99, 883
　comparative summary of characteristics of, **349t**, **372t**
　endosymbionts in, 274
　in evolution, 348
　heterotrophic, 351, 356, 365, 367, 378
　photosynthetic, 351, 356, 357, 365, 367, 378, 381, 383
protoderm, 557, **557**, 558, 570, 571, 593 (see also primary meristems)
　of shoots, **611**, 613, 614, **618**
protofilaments, 58, **59**
Protogonyaulax tamarensis, 364
protogynous plants, 510
Protolepidodendron, **426**
proton acceptors, 862
proton donors, 862
protonemata, 407, **416**, **418**
　of *Sphagnidae*, 414
proton gradient, **84**, 110, 118, **119**, 120, **136**, 137, 138, 152
proton pump, 83–84
protons, 863, 874
　in photosynthesis, 135, **136**, 137
protophloem, 579, 595, 614, 621, **622**, **623** (see also primary phloem)
protoplasm, 45
protoplast fusion, 693, 695
protoplasts, 45
　of mature sieve elements, 580, 581
Protosphagnales, 414
protosteles, **427**, **428**, **429**, 435, 439, **442**, **444**, **463t**
protoxylem, 577, 595, **596**, 614, 618, **619**, 621,

622, 623 (see also primary xylem)
protoxylem lacuna, **622**
protoxylem poles, 600, 601, **601**
protozoa, classification of, 275
protracheophytes, 416, 434, **434**
Prunella vulgaris, 245
Prunus, **503**
　persica var. nectarina, 264
　persica var. persica, 264
　serotina, **669t**
Prymnesium, 367
　farvum, **367**
Psaronius, 457
Pseudolycopodiella, 438
Pseudomonas, 293, **294**
　marginalis, **284**
　solanacearum, **285**, 294
　syringa, 699
Pseudomyrmex ferruginea, 775, **775**
pseudoplasmodia (slugs), **354**, 355
pseudopodium, 414
Pseudotrebouxia, 335
Pseudotsuga, 483
　menziesii, **669t**
Psilocybe mexicana, 325
psilocybin, 325
Psilophyton, **426**, **433**
　princeps, **433**
Psilotophyta (psilotophytes), **275t**, 430, 431, 443–45, **463t**, 886 (see also *Psilotum*; *Tmesipteris*)
Psilotum, 443–45, **463t**
　life cycle of, **446–47**
　nudum, **444**
P (peptidyl) site, 212, **212**, **213**
psychedelic chemicals, 550, **551**
psychrophiles, 287
Pteridium aquilinum, 458
pteridophytes, phylogenetic relationships with other embryophytes, **472**
Pteridospermales, 456
Pteridospermophyta, 472
pteridosperms (seed ferns), **456**, 457, **457**, 472
Pterophyta, 431, 449, 452–63, 457–63, **463t**, 887 (see also ferns)
Ptichodiscus brevis, 364
Puccinia graminis, life cycle of, 327, **328–29**
puffballs, 322, 323, 326, **326**
pulvini, 720, **720**
pumpkins, 828
pumps, **82**, 83 (see also by name)
punctuated equilibrium model of evolution, 257–58
Punnett square, 187, 188, **189**
purines, 198, **198**, 199, **199**
purple bacteria, 282, 283
　nonsulfur, **291**
　photosynthesis in, 291
　sulfur, **127**, 128, 291
purple blazing star, 793
purple sage, 779
Puya raimondii, 648
pyramids of biomass, energy, numbers, 785, **786**
pyrenoids, 44, 352, 358, 388, **388**
pyrethrum, 780
pyrimidines, 198, **198**, 199
pyrophosphate bridge, 102
Pyrus communis, 62, **576**, 655
pyruvate (pyruvic acid), 113–15
　in alcohol fermentation, 123, **123**
　in glycolysis, 109, **110**, 113
　in Krebs cycle, 115
　oxidation and decarboxylation of, 114, **114**
　in photosynthesis, 144, **145**, 149
Pythium, 374

Q

Q_a, Q_b (plastoquinones), **136**, 138
quaking aspen, **788**
quantum, of light, 130
quartersawed boards, 667, **667**
quartz (SiO_2), 731
quartzite, 731
Quaternary period, inside front cover

quaternary structure of proteins, **29**, 30
Quercus, 502, 577, 624
　alba, 657, 666, 667, **669t**
　coccinea, 256, **257**
　rubra, 65, 577, 625, 655, 662, 662, 663, **669t**
　suber, 658, **659**
　velutina, 256, **257**, 658
quiescent center, 594, **594**, 604
quillworts (*Isoetes*), 23, 55, 149, **442–43**, **443**
　muricata, 23, 55
　storkii, **442**
quinine, 551, 833
quinoa, 830, **830**
quinones, 118, 137 (see also coenzyme Q)

R

rabies, 296
raceme, **499**
rachilla, **541**
rachis, 453, 455, 626, 825
radial files, 648, 662
radial micellation, 753, **753**
radial pattern, 556, **556**
radial section (surface), 655, 660
radial system, 648
radiant energy (see solar energy)
radicle, 557, 562, **562**, 563, **564**, 566, 568
radioactive decay, 864
radioactive isotopes (radioisotopes), 864
radioactive tracers, 755, 765, 766
radiocarbon dating, 864
radio waves, 129
radish, 570, 595, **759t**
Rafflesia, 498
　arnoldii, **498**
ragweed, western, 505
Raillardiopsis, 251
rainfall, 797 (see also precipitation)
　in deserts, 807
　dispersal of fruits and seeds by, 547
rainforests, 489, **798–800**, **803–5**
　biome map of, **798–99**
　temperate, 804
　tropical, 10, 803–5
ramiform pits, 576
Ranunculaceae, 502, 545
Ranunculus, 596, 598, 614, 619, 620
　peltatus, **197**
rapeseed, 508
Raphanus sativus, 570, 595, **759t**
raphides, 54, 549
Ratibida pinnata, 793
rattlesnake plantain, 778, **778**
ray flowers, 527, 528
ray initials, 648, 649, 650
ray(s), 573, 648, **650**, 655, 661, 663
　of angiosperm woods, 662, **662**
　of conifers, 476
　dilated, **652**
　parenchyma, **661**
　phloem, 476, **652**, 655, **656**
　pith, 651
　vascular, 648, 649, **649**, 650, **650**
　xylem, 476, **652**, 655
ray tracheid, 661
*rbc*L gene, 271, 526
reaction center, 135
reaction wood, 666, **666**
Reboulia hemisphaerica, **406**
Recent epoch, inside front cover
receptacle (algal), **385**
receptacle (flower), 499, **499**, 501
reception of signals, 87, **87**
receptor-mediated endocytosis, 85, **86**
receptors, 87
　hormones and, 690–91, **691**
recessive characteristics, 185, 186
recognition sequences, 224
recombinant DNA, 224–31 (see also genetic engineering)
　Arabidopsis thaliana as model experimental organism, 228, **228**
　DNA libraries and, 226
　DNA sequencing techniques and, **229**, 229–30

nucleic acid hybridization and, 230
plasmids and, 224–26, **225**, **226**
polymerase chain reaction (PCR) and, 226
restriction enzymes and, 224–25, **225**
screening processes in, 225–26
techniques for locating genes of interest, 231
recombination, genetic, 172
recombination nodules, 176, **176**
recombination speciation, 255
red alder, 577, **669t**
red algae (*Rhodophyta*), **275t**, 348, **349t**, 357–61, **358**, 884 (see also algae)
　commercial uses of, 380
　edible, 380
　life cycle of, 360–61, **362**
　unique cellular features of, 358–60
red blood cells, 219
red buckeye, 627
redbud, 54
red mangrove, 603
red maple, 10
red oak, 65, 577, 625, 655, 662, 662, 663, **669t**
redox (oxidation-reduction) reactions, 98–99, **99**
red peppers, 833
red snow, 386
red tides, 350, **363**, 364
reduction, 98, **99**, 109
reduction division (see meiosis; mitosis)
redwoods, 474, **669t**
　coast, 483
　dawn, **485**, 486, **486**
　geographic distribution of, **485**
region, nodal and internodal, **621**
region of cell division, 594, **594**
region of elongation, 594, **595**
region of maturation, **594**, 595, **596**
Regnellidium diphyllum, **55**, 108
regulator genes, 216, **217**
regulatory enzymes, 104
regulatory sequences, 688, **688**
regulatory transcription factors (see transcription factors)
reindeer moss, 337
release factors, 214
Reoviridae, 298
replacement fossils, **239**
replication, 208–9
　bubbles, 201, **202**
　forks, 201, **202**
　origin of, 201
replication of DNA (see DNA (deoxyribonucleic acid), replication of)
reporter genes, 698–99
repressors, 216, 217, **217**, 218
reproduction (reproductive systems) (see also cell division)
　asexual (see asexual reproduction)
　in cellular slime molds, 356
　in conifers other than pines, 482
　in fungi, 311–12
　in lichens, 338
　mitosis and, 166
　in mushrooms, 324
　of prokaryotes, 285–86
　in red algae, 360–61
　sexual (see sexual reproduction)
　in vascular plants, 430–31
　vegetative (see asexual reproduction)
　in yeasts, 330
reproductive apex, 639
reproductive isolation (see genetic isolation)
reservoir, in euglenoids, 351, **351**
resin ducts, 474, 659
resonance energy transfer, 131–32, 135
respiration, 6, **94**, 108–25
　ADP "recharged" to ATP in, 32
　aerobic pathway, 113–22
　anaerobic pathways, 122–23
　fats and proteins in, 122
　germination and, 565
　glucose oxidation and, 109
　mitochondria as sites of, 52
　ripening of fruits and, 683

stages of, 109 (*see also* electron transport chain; glycolysis; Krebs cycle; oxidative phosphorylation)
resting cysts, dinoflagellate, 365
restoration ecology, 792–93
restriction enzymes (restriction endonucleases), 224–25, **225**, 696
DNA sequencing techniques and, 229
restriction fragments, 224
resurrection plant (*Selaginella lepidophylla*), **439**
retinal, **134**
retinol (vitamin A), **134**
reverse transcriptase, 226
R groups, in amino acids, 26, 27, **28**, 29, **29**
rhabdovirus, **299**
Rheum rhabarbarum, **575**, 643
Rhizanthella, **529**
rhizobia, 739, **739**
Rhizobium, **294**, 738–40
rhizoids, 310
of algae, 392, **392**, 395
of bryophytes, 403, **403**, 407, 411, 416, 419, **422t**
of *Equisetum*, **449**
of ferns, 455, 458, 459, **460**, **461**
of fungi, **313**, 316
of *Psilotum*, 443
rhizomes, 180 (*see also* stems)
of *Equisetum*, 445
of ferns, 455, **455**, 458
of Lycopodiaceae, 435
of *Psilotum*, 443, 444
of *Trillium*, 5
tubers attached to, 642, **643**
Rhizophora mangle, 603
Rhizopus, **308t**, 317
life cycle of, **316**
stolonifer, 315, 316, **316**
rhizosphere, 597
Rhodococcus, **294**
Rhodophyta (red algae), 357–61, **362**, 884 (*see also* algae; red algae)
Rhodospirillum rubrum, **291**
Rhopalotria, 487
rhubarb, **575**, 643
Rhynia, **426**
gwynne-vaughanii, **426**, 432–33, **433**
major, 434, **434**
Rhyniophyta (rhyniophytes), 431, **432**, 432–34, **463t**
Rhytidoponera metallica, **549**
riboflavin (vitamin B$_2$), **116**
rib, of leaf, (midrib), **628**, 631
ribonucleic acid (*see* RNA)
ribose, **19**, 30, **30**, **31**
ribosomal RNA (*see* rRNA)
ribosomes, 41, 45, 48, **48**, 52, **52**, 55–56, **71t** (*see also* polysomes)
mitochondrial, **45t**, 52
plastid, **45t**, 52
of prokaryotes, 52, 282
protein synthesis and, 208, 211, 212, **212**, **213**, 214, **215**
ribozymes, 31
ribulose 1,5-bisphosphate (RuBP), 139, 140, **140**, **141**, 141, 142, **142**, 143, 146
ribulose bisphosphate carboxylase/oxygenase (Rubisco), 140, 142–44, 146, **149**
Riccia, 407–08
Ricciocarpus, 407–08
rice, 146, **497**, 826, 827, **827**, 835
amino acids in, 26
Azolla-Anabaena symbiosis and, 741
bran, 563
ethylene and, 682
foolish seedling disease of, 684
Ricinus communis, **562**, 566, 567, **578**
Rick, Charles, 841
Rickettsiae, 882
Rieseberg, Loren H., 255
ring bark, 658
ring-porous woods, **662**, **663**, 665
ring spots, 301
ringworm, 334
ripening of fruits

ethylene and, 683, 697, **698**
genetic engineering and, 697, **698**
RNA (ribonucleic acid), **30**, 31, **31**, 41
DNA compared to, 208, **208**
gene expression and, 208
messenger (*see* mRNA)
in nucleolus, 48
protein synthesis and, 208, 210–14
ribosomal (*see* rRNA)
transfer (*see* tRNA)
of viroids, 302
of viruses, 297, 298, 300
RNA polymerase, 208, 210, **210**, 216, **218**, **688**
RNA primase, **203**
Robert, Karl-Henrik, 845
Robertson, J. David, 74
Robichaux, Robert, 251
Robinia pseudo-acacia, **627**, 656, **657**, 659, 669t
rocks
nutrients derived from, 731
weathering of, 786, **787**, 788
rock spikemoss (*Selaginella rupestris*), **439**
rockweeds, 379, **379**, 383 (*see also Fucus*)
Rohn, Megan, 520
rootcap, 591, 592, **592**, 593, 604
gravitropism and, 704, **705**
root hairs, 342, **585**, **592**, **595**, 595–97, **597**, 604, **604**
density of, **759t**
nitrogen-fixing symbiosis and, 739, **739**
water absorption by roots and, 759–62
root-hair zone, 595
root pressure, 755, **760**, 760–61, **761**
root primordia, 603, **603**
root(s), 7, **8**, 426, 566, 589–609
adaptations of, 606, **607**
adventitious, **439**, 445, 566, **568**, 678
aerial, 603
air, 605, **605**
apical meristems of, **8**, 559, 562, **570**, 591–92, 604, 605
quiescent center, 594
types of organization of, 592, **593**
balance between shoot systems and, 590–91
cortex of, 592, **593**, 597, 598–600, 602
cylindrical form of, 571
cytokinin/auxin ratio and, 681, **681**
embryonic (*see* radicle)
epidermis of, 595–97, **596**, **597**, 602
evolution of, **429**
extent (depth and lateral spread) of, 590
feeder, 590, 596
of ferns, 455
fibrous, 590
fleshy, 606, **607**
functions of, 590
gravitropism in, 704–5, **705**
of gymnosperms, 344
hydrotropism in, 706
hypocotyl-root axis, 482, **482**, **513**, 562, 563, **637**
lateral (branch), 566, 590, **592**, 603
mycorrhizae and (*see* mycorrhizae)
nitrogen fixation and, 738, 739, **739**, 740, **740**
nutrient uptake by, 762–64, **763**
periderm of, 602
primary (taproot), 566, 590, **591**
primary growth of, 591–95, **601**, **602**
near the root tip, **594**, 594–95
primary structure of, 595–600
prop, 603, **603**
secondary growth of, 600, **602**
secondary structure of, 600–2
of seedless vascular plants, 435, 438, **439**, **442**, **443**, 445
soil formation and, 731, **731**
systems, 426, 590–91
tips of, 591–94
transition to shoots, 636–39, **637**
vascular cylinder of, 573, 592, **593**, **596**, **597**, 600, **602**, **603**, 604, **604**
woody, 602
Rosaceae, 256, **502**

rose
family, 256
hips, **772**
Rosemarinus officinalis, **532**
rosemary, **532**
rosettes, 686
rots, soft, 293, **294**, 317
Roundup, 697
roundworm (*see* nematodes)
rRNA (ribosomal RNA), 208
protein synthesis and, 211–12
sequencing of, molecular systematics and, 270, **270**
rubber, 34, 834
guayule as source of, 842, **842**
rubber tree, 34
Rubiaceae, 833
Rubisco (RuBP carboxylase/oxygenase), 140, **140**, **142**, 142–44, **143**, 146, **149**
RuBP (ribulose 1,5-bisphosphate), 139, 140, **140**, **141**, 141, 142, **142**, 143, 146
ruminants, 295
runners (stolons), 180
rusts (fungi) (*see Teliomycetes*)
rutabaga, 763
rye, 146, 335, 590, 595, 719, 762

S

Sabalites montana, **518**
Saccharomyces
carlsbergensis, 331
cerevisiae, 230, 308, 330, 331, **331**, 332, 334
Saccharum, 88, 747
officinale, 834
officinarum, 632
Sachs, Julius von, 65, 693
sacred mushroom, 325
S-adenosylmethionine (SAM), 682, **682**
saffron, 833
sage, purple, 779, **779**
Sagittaria, 558
saguaro cactus, **10**, **497**, **539**, 808
St. Helens, Mount, 789, **792**, **792**
St. Anthony's fire, 335
sake, 334
Salicaceae, 342, 546
Salicornia, 744
salicylic acid, 36, 37, 675, **675t**
saline environments, **744**
Salix, **602**, 603 (*see also* willows)
Salsola, 546, **546**
salt, extreme halophilic archaea and, 295
saltbush, 585, 744, **744**
salt-marsh grass (*Spartina*), 254, **254**
salt secretion, 744
salt-tolerant crops, 843, **843**
salvage pathway, in photorespiration, 142–43, **143**
Salvia leucophylla, 779, **779**
Salvinia, 462, **462**, 463
Salviniales, 454, 462, **462**, **463t**
SAM (S-adenosylmethionine), 682
samaras, 545, **545**
Sambucus canadensis, **573**, 614, 618–19, **618–19**, 650, **651–653**
sandstone, 731
sandy soils, 732, **735**
Sanguinaria canadensis, 549
Sansevieria, **54**
sap, 77
cell (vacuolar), 54
sieve-tube, 766, **767**
velocity of flow of, 758, **758**
wind pollination and, 530
xylem, 757, **757**, **758**, 758
saponins, 549
Saprolegnia, 373, **374**
life cycle of, **374**
saprophytes, 286
angiosperms, 498, **498**
fungi, 310, 312, 313
oomycetes, 373
orchids, 529, **529**
sapwood, **655**, 665
SAR (systemic acquired resistance), 37

Sargassum, 381, **381**, 383
muticum, 381
saturated fats, 24
sauerkraut, 287
Sauromatum guttatum, 37
savannas, **10**, 798–99, 805
biome map of, **798–99**
juniper, 798–99, **810**
Saxifragaceae, **522**
Saxifraga lingulata, **761**
saxifrage, **761**
scale barks, 658, **658**
scalelike outgrowths, 439, 443 (*see also* ligules)
scale mosses, 412
scales
of algae, 349t, 367, 372t, 387
of ferns, 458
of liverworts, **403**
scanning electron microscope (SEM), 43
scarification, 565
breaking of dormancy by, 718
scarlet monkey-flower, **537**
scars
bundle, **657**
leaf, 636, **657**
terminal bud-scale, **657**
Schaller, George, 844
schizocarps, 545, 546
Schizosaccharomyces octosporus, **331**
Schleiden, Matthias, 41
Schwann, Theodor, 41
scientific names, 262–63
sclereids (stone cells), **62**, 575, 575–76, **576**
in xylem, 579
sclerenchyma (cells or tissue), **571**, 575, 619, **620**, 621, **622**, **623**, 632 (*see also* fibers; sclereids)
Scleroderma aurantium, **277**
sclerotia, 335, 354
Scott, D. H., 457
scouring rushes, 445
screening processes
recombinant DNA and, 225–26
scrub communities, 816–17
scutellum, **562**, 563, **564**, 686, **686**
sea lettuce (*Ulva*), **393**, **394**, 394
sea moss, 412
sea-nymph, **542**
sea otters, 371
sea palm, **276**
seashore environment, photosynthetic organisms and, 6–7
sea slugs (nudibranchs), 395, **396**
seasons (*see also* dormancy)
biological clock and, 708–709
seaweeds, 350, 371, 380 (*see also* brown *and* red algae)
edible, 380
Sebdenia polydactyla, **276**
Secale cereale, 146, 335, 590, 719, 762
secondary active transport, 84, **84**
secondary cell walls, 64–65, **65**
pits in, 66
secondary consumers, 783
secondary endosymbiosis, cryptomonads and, 357
secondary growth, 8, 427
in cycads, 486
diffuse, 648
in pines and other conifers, 476, **476**
in roots, 600–602
in stems, 614, 619, 647–71
bark, 654–59
lenticels, **653**, 654
periderm, 653–54
primary body of the stem, effect on, 650–59
stages of development, **651**
in veins, 631
secondary metabolites, 32–37, 54, 334, **549**, 549–51, **551** (*see also* alkaloids; phenolics; terpenoids)
secondary phloem, 427, 470, **576**, 579, 580, 600–2, 648, 649, **649**, 650, **650**, 651, **651**, **652**, 654, 655, **656**, 659, 664

in pines and other conifers, 476, **476**
secondary plant body, 427
secondary plant products (*see* secondary metabolites)
secondary structure of proteins, 28, **28**, 29, **29**
secondary vascular tissues (*see* secondary phloem; secondary xylem; *and* vascular cambium)
secondary xylem, 427, 457, 470, **470**, 476, **476**, **576**, **577**, 578, 600, 601, **601**, 602, 647, 648, 649, **649**, 650, **650–652**, 654, **655**, **656**, 659–70 (*see also* wood)
second law of thermodynamics, 95–97, **96**
second messengers, 87, 691, **691**, 692
secretion, **57**, 58, 591, **592**
secretory vesicles from the trans-Golgi network, **57**, 58, 59, 66
sedge family, 340
sedimentary rocks, 731
Sedum, 644
seed banks, 718, 840, **840**
seed coat, **8**, 470, 522, 562, 563, **564**
  of angiosperms, 509, **513**
  dormancy of seeds and, 565, **565**
  germination and, 565
  of pines, 479, 482, **482**
  of willows and poplars, 546
seed ferns (pteridosperms), 456, 457, **457**, 472
seed leaves, 482, 562–63
seedless vascular plants, 424–66, **463t** (*see also* specific phyla)
  of Carboniferous period, 456–57, **456–57**
  evolution of, 425–31 (*see also* vascular plants, evolution of)
  extinct, 425
  Lycophyta (lycophytes), 435–43, **463t**
  main features of, **463t**
  phyla of, 431
  Psilotophyta, 443–45, **463t**
  Pterophyta (ferns), 449, **452**, **453**, 453–55, 457–63, **463t**
  reproductive systems of, 430–31
  Rhyniophyta (rhyniophytes), 432–34, **463t**
  Sphenophyta, 445–49, **463t**
  Trimerophytophyta (trimerophytes), 443, **463t**
  Zosterophyllophyta (zosterophyllophytes), 434–35, **463t**
seedlings, 555 (*see also* germination)
  Arabidopsis, **560**, 561
    dark-grown, 712–714, **713**
    effects of light on, 714
    establishment of, 566–68
    etiolated, 713, **713**, 714
    gravitropism and, 704
seed pieces, 180
seed plants, **8–9**
  evolution of, 431, 471–72
  phyla of, 470
seeds, 9, 510
  cold treatment of, 717
  of cycads, **487**
  development of, 556
  dispersal of, 546–49
  dormancy of, 565, 717–18
    abscisic acid and, 684
    gibberellin and, 685
  of *Ephedra*, 481
  evolution of, 425, 468–70
  food storage in, 470, 482, 509–10, 563, 567
  germination of (*see* germination)
  of *Ginkgo*, 490, **490**
  of *Gnetum*, **491**
  of gymnosperms, 467
  longevity of, 718
  of pines, 482, **482**
  proteins in, 26
  viability, 718
  of *Wollemia nobilis*, **489**
  of yews, 484
seed-scale complexes, 480, 481
segregation, **186**
segregation, principle of, 185–87, **187**, 190
seismonastic (thigmonastic) movements, 720–21, **720**, **721**

Selaginella, **428**, 559
  kraussiana, 48, **439**
  lepidophylla (resurrection plant), **439**, 439–42
  life cycle of, 440–41
  protostele, **442**
  rupestris (rock spikemoss), **439**
  willdenovii, **439**
Selaginellaceae, 439–42
self-fertilization, 438
self-pollination, **184**, 514
  homozygotes and, 242
  inhibition of, 243, **243**
self-sterility, **194**, 514
semideserts, **798–99**
Senecio, **574**
senescence
  of flowers, genetic engineering and, 698
  of leaves, cytokinins and, 681–82, 698
sepal primordia, **638**
sepals, **499**, 501
  development of, **638**, 639
  of early angiosperms, 524, **524**
  of orchids, 529
separation (abscission) layer, 636, **637**
septa, fungal, **310**, 313, 317, 318, **322**, 323, 333
Sequoia, **485**
  sempervirens, 474, **483**, **669t**
Sequoiadendron giganteum, **483**, **485**
serial endosymbiotic theory, 272–74, 283
serine, **27**, **28**, 143, **143**
sesquiterpenoids, 33, 34
sessile leaves, 624, **625**
setae (stalks), **405**, 406, 414, **417–419**, 420, **422**
Setcreasea purpurea, 89
sex chromosomes, in bryophytes, 404
sex expression and ethylene, 683
sex organs (*see* antheridia; archegonia; gametangia; oogonia)
sexual reproduction, 169, **169**, 170 (*see also* fertilization; meiosis)
  advantages of, 180–81
  in algae, 360–361, **362**, 365, 367, 376, **377**, 379, 382, 383, **384**, **385**, 388, **389**, 390–93, **394**, 396, 397
  in angiosperms, **512–13**
  in ascomycetes, 318, **318**, 320
  in *Basidiomycota*, **321**, 328–29
  in bryophytes, 404, **405**, 406, 410–11, 418–19
  in chytrids, **314**
  in cycads and *Ginkgo*, **488**
  in eukaryotes (*see by group*)
  in fungi, 312
  genetic variability and, 242
  in hymenomycetes, **321**
  in oomycetes, 371, **372**
  in *Pinus*, 476–82
  in seedless vascular plants, 436–37, 440–41, 446–47, 450–51, 460–61
  in slime molds, 352, **353**, 354, 356
  in *Teliomycetes*, 328–29
  in zygomycetes, **316**, 317
shade leaves, 636, **636**
shagbark hickory, **554**, **627**, **647**, **658**
shale, 731
shaman, **325**
shape of plants and plant parts, 571–72 (*see also* morphogenesis)
sheaths
  bundle, **628**, **629**, 632
    in grasses, 632, **632**, **633**
  of hyphae, 342, **342**
  of leaves, 624
  mucigel, 591, **592**
  mucilaginous, 288, 397
  of sclerenchyma cells, 621, **622**
sheep, karakul, **826**
shelf fungi, 322, 323, **323**, 324
shepherd's purse, **557**
shiitake, **325**
shooting star, **500**
shoot(s), **8**, 426, 610–46 (*see also* leaves; plumule; stems)
  apex, **436**, **440**, **446**, **450**, **460**, **479**, 566, 610–12, **612**, 614, 635, **635**
    transition to flowering, 639

apical meristem of, **8**, 556, **557**, 557, **558**, 559, **561**, 562, **562**, 568, **570**, 611–14
  balance between root systems and, 590–91
  cytokinin/auxin ratio and, 681, **681**
  determinate, indeterminate, 474, 498
  emergence during germination, 567, 568
  of *Equisetum*, **448**
  of gymnosperms, 474, **475**
  gravitropism in, 704, **704**
  reproductive (floral), 639–41
  root compared and, 590–91, 636, **637**, 639
  system, 426, 590–91
  transition to root, 636–39, **637**
shoot-tip culture, 695
short-day plants, 709, **710**, 711
short shoots, 474, **483**, 488
shotgun cloning, 226
shrubs, 648
Shull, G. H., 243
shuttle vesicles, **57**, 58
sieve areas, 579, 579–80, **580**
sieve cells, 476, **579**, 579–80, **587t**
sieve elements, 427, **574**, 579–81, 618
sieve plates, 382, **382**, 580, **580**, **581**, 582, **582**, **583**
  in algae, 382, **382**
sieve-plate pores, 580, **581–82**
sieve-tube elements, 579, 580, **580**, **581**, 582, **582**, **583**, **587t**, 651
sieve tubes, **292**, 293, 300, **300**, **382**, 522, 580–82
  assimilate transport in, 765–67, **766**
  sap, 766, 767
Sigillaria, **466**
signals, chemical, 86 (*see also* hormones)
signal transduction, 86–87, **87**
siliceous compounds, 376, 445
silicon, **728**
Silphium, **506**
Silurian period, **424**, **428**, 431, 432, 434, inside front cover
silver maple, **625**
silversword (*Argyroxiphium sandwicense*), 250, **250**, 251
Silvianthemum suecicum, **522**
Simmondsia chinensis, **841**, 841–42
simple diffusion, **82**, 82–83
simple-sequence repeated DNA, 222
simple tissues, 573
Singer, S. Jonathan, 74
single bonds, 869
single-copy DNA, 222
sinigrin, 549
siphonosteles, **428**, **428**, 429, **429**, 455, **455**, **463t**
siphonous algae, 394–95, **395**
sister cells, 157
sister chromatids (*see also* daughter chromosomes)
  in meiosis, 172, **173**
  in mitosis, 160, **160–162**, 163
sitosterol, 25, **25**
size exclusion limits, 88
Skeletonema, **375**
Skoog, Folke, 680
Slack, C. R., 144
slash pine, 659, **807**
slate, 731
sleep movements in leaves, 707, 719
slime (*see* P-protein)
slime bodies (*see* P-protein (slime) bodies)
slime molds (*see* cellular slime molds; plasmodial slime molds)
slime plugs, 582, **582**
slime sheath (*see* mucigel)
slugs
  pseudoplasmodia, **354**, 355
  sea (nudibranchs), 395, **396**
Smilacina racemosa, **762**
smog, 616
smuts, 326 (*see also* Ustomycetes)
snake plant, **54**
snapdragons, 194, **194**, 639
sneezeweed, 780

snow algae, **386**
sodium (Na), **76**
  chloride (NaCl), 873, **873**
  as essential element, 729
  halophytes and, 744
  hydroxide (NaOH), 874, **874**
sodium ions (Na⁺), 744
sodium-potassium pump
  in halophytes, 744
soft rots, 293, **294**, 317
softwoods, 659
soil base, 732, **732**
soil erosion, 743
soils, 726, 731–35
  agriculture and, 744–45
  cation exchange and, 735
  classification of, 732
  deficiencies and toxicities, 729, **730t**
  erosion of, 736, 838, **838**
  formation of, 731–32
  grassland (prairie), 810, **812**
  horizons (layers) of, 731–32, **732**
  inorganic components in, 731
  loam, 732, 735
  mycorrhizae and, 738
  organic components of, 732, **733**
  organisms, **733**
  particles in, 732
  pH of, 735
  plant nutrition and, 735
  pore space of, 732, 734–35
  temperate deciduous forest, 813
  tropical, 805
  water in, 734–35
  weathering of rocks and, 731
Solanaceae, **830**, 831
Solanum, 550
  aviculare, 846
  cheesmaniae, 843
  lycopersicum, **60**, **191**, 675, **704**, 729
  melongana, 843
  tuberosum, **642**, 643, 698, 830, **830**
solar energy (radiant energy), 2, 5, 94
solar tracking, 722, 722–23, **723t**
Solidago
  flexicaulis, **762**
  virgaurea, 246
Solomon's seal, 508
solutes
  definition of, 873
  diffusion of, 78
  osmosis of, 79
  transport across membranes, 82–84
  water potential and, **76**, 77
solutions
  acidic, 874
  basic (alkaline), 874
  definition of, 873
  hypertonic, 79
  hypotonic, 79
  isotonic, 79
solvent, definition of, 873
somatic hybrids, 693, 695
Sonoran Desert, 10, **808**, 809
soredia, 338, **338**
sorghum (*Sorghum vulgare*), 146
sori
  of ferns, **458**, **459**
  of fungi, 322, 326
sorrel, wood, **707**
sorting, polypeptide (or protein), 214, **214**, **215**
source-to-sink pattern of assimilate movement, 764
sourwoods, eastern, **813**
South America, 523
Southern leaf blight of corn, 840, **840**
Southern woodland and scrub, **798–99**
soybean, **51**, 698, 715, 738, 739, **739**, 740, **740**, 826, **826**
  Biloxi, 709, 715
  life cycle of, **512–13**
soy paste (miso), 334
soy sauce (shoyu), 334
Spanish moss, 337, **485**
Spartina, 254, **254**

alterniflora, 254, **254**
anglica, 254, **254**
maritima, 254, **254**
townsendii, 254, **254**
special creation theory, 236
speciation, 248–56
    process of, 249–56
        allopatric speciation, 249
        allopolyploidy, 252, 252–53
        autopolyploidy, 252, **252**
        recombination speciation, 255
        sympatric speciation, 249, 252, 253, **255**
species, taxonomic
        definition of, 239, 248
species (*see also* speciation),
        names of, 262–63
        origin of, 248–49
        reproductive isolation of, 256
Species Plantarum (Linnaeus), 262
specific epithet, 262, 263
specific gravity of wood, 668, 670
spectrophotometer, 712
spectrum
        electromagnetic, **129**
        of visible (white) light, 129
sperm, **170** (*see also* gametes)
        of algae, 377, 383, **384**, 385, 387, **387**, 392, **392**, 397
        of angiosperms, 474, 503, 504, **506**, 508, **508**, 509, **512**
        of bryophytes, 400, 402, 404, **405**, 410, **418**, 420
        of conifers, 474
        of cycads, 487, **488**
        of *Equisetum*, 449, **451**
        of ferns, 459, **461**
        of *Ginkgo*, 488, **488**
        of gnetophytes, 492
        of gymnosperms, 472, 473, **473**, 474
        inheritance and, 184, **186**, **187**, 190, **190**
        of *Lycopodiaceae*, **437**, 438
        of *Marchantia*, **405**, **411**
        of mosses, **419**, 420
        of pine, 479, 481
        of *Psilotum*, 445, **447**
        of seedless vascular plants, 425, 472
        of *Selaginella*, **441**, 442
        of *Zamia*, **488**
spermatangia, 360, **362**
spermatia, 327, 360
spermatogenous (cells or tissue), 404, **405**
spermatogenous cell (body cell), 481
spermogonia, 327, **328–29**
sperm packet, 385
Sphacelia typhina, 335
Sphagnidae (peat mosses), 412, **414**, 414–15
Sphagnum, 278, **414**
S phase (synthesis phase), 157, **157**, 158
Sphenophyta (sphenophytes), 430–31, 445, 448–51, **463t**, 887 (*see also* horsetails; *Equisetum*)
spicebush, 762
        family, 519
spices, 832–33
spike (flower), **499**
spikelet, **541**
spinach, **52**, 710, **710**, 719
Spinacia oleracea, **710**
spindle
        colchicine and, 193
        meiotic, 174, 176, 177, 311
        mitotic, 158–60, **160**, 161, **162**, 163, **165**, 311
        nonpersistent, 386, **387**
        persistent, 386, **387**, 393, 396
spindle pole bodies, 311, **311**
spindle tuber disease of potato, 302
spines, 642, **643**
spiral grain, 667, **667**
Spiranthes, **535**
spirilla, 284, **285**
Spirillum volutans, **285**
Spirogyra, 396, **396**
Spiroplasma citri, 292, **292**
spiroplasmas, 292, **292**
splash cups, 420, **420**

spleenwort, **800**
spongy parenchyma, **628**, **629**, 630, 631
sporangia, **6**, 311, **314**, 315, **315**, 353 (*see also* capsules; eusporangia; leptosporangia; megasporangia; microsporangia; tetrasporangia; zoosporangia; zygosporangia)
        of algae, 382, 383, **384**, 394
        of bryophytes, 407, **408**, 410, 412, **413**, 414, **415**
        of early vascular plants, 432, **432**, 433, **433**, 434, 435
        of *Equisetum*, 448
        evolution of plants and, 425, 429, 431, **432**
        of ferns, 453, 453–54, 458, 462
        of fungi, 311, **314**, **315**, 316
        of hornworts, 412, **413**
        of liverworts, 407, **408**, 410
        of *Lycopodiaceae*, **436**, **437**, 438, **438**
        of mosses, 414, **415**
        of oomycetes, 373, **374**
        of plasmodial slime molds, 352, **353**
        plurilocular, **382**, 383
        of progymnosperms, **471**
        of *Psilotum*, 444, **444**, **446**, **447**
        of *Selaginella*, **440**, **441**, 442
        unilocular, **382**, 383
sporangiophores, 315, **315**, 448
spore dispersal
        in bryophytes, **405**, 408, **413**, 414, 415, 421
        in fungi, 316, 318, 319, 324
        in seedless vascular plants, 448–49, 454
spore mother cells, **446**, 453
spores, 2, 8 (*see also* aeciospores; ascospores; basidiospores; carpospores; conidia; megaspores; microspores; monospores; oospores; teliospores; tetraspores; urediniospores; zoospores; zygospores)
        of algae, **394**
        asexual, 356
        of bacteria, 285, **285**, 286, **286**
        of bryophytes, 402, 406–7, **410**, **418**
        of cellular slime molds, **354**, **355**
        dormant, 316
        of euglenoids, 352, **353**, 354
        of eusporangia, **453**
        of ferns, 458
        of fungi, 308, 311–13, 315, **315**, **316**, **326**
        of heterosporous plants, 431, **441**, **451**, 461
        of homosporous plants, 430, **437**, **447**, 461
        of hornworts, 412, **413**
        of leptosporangia, **453**
        of liverworts, 407, 408, **409–411**
        meiospores, 172
        of mosses, 414, **414**, **415**, 416, **418**, **419**, 421, **421**
        of oomycetes, **374**
        of plasmodial slime molds, 352, **353**
        pollen grains compared to, 506
        of red algae, **359**, 360, 361
        resting, 316
        of seedless vascular plants, 438, 444, 448, **449**
sporic meiosis, **171**, 172
        in brown algae, 383, **384**
sporidia, 330
sporocarps 462, **462**
sporogenous (cells or tissue), 402, **410**, **418**, **436**, **440**, **446**, 450, **460**, 504
sporophylls, **437**, 438, **438** (*see also* megasporophylls; microsporophylls)
        of early vascular plants, **456–57**
        of gymnosperms, **484**
        of *Lycopodiaceae*, **438**
        of *Selaginella*, **439**, 442
        of *Isoetes*, 442
sporophytes, 172, 360
        of brown algae, 383, **384**
        of bryophytes, 402, **402**, 403, 404, **405**, 406
        of *Lycopodiaceae*, 435, **436**, **437**, 438, 459
        of *Equisetum*, **451**
        evolution of vascular plants and, 425, 431
        of ferns, 430, 455, **459–461**, 462
        of gymnosperms, **474–75**, **483–87**, **489–92**
        of homosporous plants, 430

of hornworts, 412, **413**
        of *Isoetes*, 442, **442**
        of liverworts, **405**, 407, 408, **408**, 410, 411
        of *Marchantia*, **405**, 410
        of mosses, 414, 416, **417–419**, 420, 421, **422**
        of *Psilotum*, 443–45, **446**, **447**
        of *Selaginella*, **439**, 442
        of vascular plants, 425, 430, **430**
sporophytic self-incompatibility, 511
sporopollenin, 396, 398, 402, 406–7, 504, **505**
sporozoa, 361
sporulation, 286
spring beauty, 548
spring ephemerals, 812
spruces, 483
        Engelmann, 777, 791
        red, **669t**
        white, **788**, 818
spurge family, 266, **266**
spur shoots, 488, **490**
squash, 567, **576**, 581, 582, **600**
        ethylene and sex expression in, 683
stalk cell (*see* sterile cell)
stalk (*see also* rachis)
        of algae, **379**, **381**, 382, **384**
        of bryophytes, **405**, 406
        of fungus, 324
        of ovule, 506
        of slime molds, **354–55**
stamen primordia, 638
stamens, **499**, 501, 502
        of apple flowers, **503**
        development of, **638**, 639
        of early angiosperms, 524–25
        insect pollination and, 531
        of orchids, 528
        outcrossing and, 510, **511**
        sterilization of, 524, 525
        whorls of, **638**
        of wind-pollinated flowers, 540, **541**
Stamnostoma huttonense, **469**
Stanley, Wendell, 296
Stapelia schinzii, **531**
Staphylococcus, 334
starch, 20, **21**, 22, **38t**
        in algae, 44, **349t**, 365, **372t**, 383
        floridean, **349t**, 358
        in plastids, 44, 45, 49–50, **51**, 126, 142, 149
        in seeds, **51**
starch grains, 45, 49–50, **51**, 126
starch-statolith hypothesis, 705
stearic acid, **23**
Steinbeck, John, 812
steles, 428, **428**, 573 (*see also* eusteles; protosteles; siphonosteles; vascular cylinder)
        of *Psaronius*, 457
stem bundles, 622
Stemonitis splendens, 353
stems, **7** (*see also* rhizomes; shoots)
        apical meristems of, 611–14
        arrangement of leaves on (phyllotaxy), 624
        of calamites (giant horsetails), 456
        of Carboniferous seed plants, 457
        development of, 611–14, **670**
        of eudicots, 614–15, 618–21
        food storage in, 642–43
        functions of, 611
        herbaceous, 619–21
        modifications of, 641–44, **644**
        of monocots, **621–23**
        of pines and other conifers, 474, 476, **476**
        primary growth of, 611–14, **614**
        primary structure of, 614–21, **615**
                types of organization, 614
        primary tissues of, 611–14
        of *Psaronius*, 457
        roots compared and, 595
        secondary growth in, 614, 619, **647–71** (*see also* secondary xylem)
        secondary structure of, 648–56, **658–59**
        of seedless vascular plants, 435, 439, 442, 444, 445, **448**, **455**
        vascular cambium of, 648–50
        vascular tissues of leaves and, 622–24
        water storage in, 644

woody, 618, 650, 653 (*see also* wood)
        external features of, **657**
Stemonitis splendens, 353
Stenocereus thurberi, **539**
stereoisomers, 870
sterigmata, **321**, 323, 324
sterile bract, 480
sterile cell (stalk cell), 481
sterile jacket layer, 402
sterility, male, cytoplasmic, 196
steroids, 25–26, 846, **846**
sterols, 25, **33**, **38t**
        in cellular membranes, 74
Steward, F. C., 695
sticky ends, 224, **225**
Stigeoclonium, 392, **392**
stigmas (eyespot), **184**, 388, **499**, 501, 508
        in euglenoids, 351, **351**
stigmas (flower)
        evolution of angiosperms and, **520**, **524**, **525**
        genetic self-incompatibility and, 511
        outcrossing and, 510
        of wind-pollinated flowers, 540, **540**, **541**
stigmasterol, 74
stigmatic tissue, 508
stinkhorns, 323, 326, **326**
stinking smut of wheat, 330
stipe
        of algae, **379**, **381**, 382, **384**
        of mushrooms, 324
stipules, 624, **624**
stolons
        in fungi, 315, **316**
        in plants (runners), 180, 642
stomata, 7, **7**, **8**
        bryophyte pores as analogous to, **402**, 403
        of C_3, C_4 plants, 146
        of CAM plants, 147, **149**
        of hornworts, 406, 412, **413**
        in leaf epidermis, 584, 584–85, **585**, 626, 627, **628**, **629**, 630, **630**
        of mosses, 406, 420, **420**
        movement (opening and closing) of
                abscisic acid (ABA) and, 684, **692**, 692–93
                cellulose microfibrils in guard cell walls and, 752–53, **753**
                environmental factors and, 753–54
                guard cells and, **692**, 752
                transpiration and, 752
                water loss and, 753
        in photosynthesis, **132**, 139, **140**, 144, **148**, 149, **149**
        in stem epidermis, 618
stomatal crypts, 629
stone cells (sclereids), **62**
stonecrops, 147
stoneworts, 398
stop codons, 214
straight grain, 667
strands
        antiparallel, 199
        complementary, 199
        lagging, 202
        leading, 202
stratification, breaking seed dormancy by, 717
strawberries, 180, 679, **679**
        dispersal of, 547
        wild, 762
streaking, by viral infection, 302
Strelitzia reginae, **537**
Streptococcus lactis, **285**
Streptomyces, **294**
        scabies, **281**
streptomycin, 287
striate venation (parallel venation), **625**, 631
strobili (cones)
        of balsam fir (*Abies*), 483
        of cycads (*Cycas*), 487
        of cypresses (*Cypressus*), 483
        of *Equisetum*, 448, **448**, **450**
        of gnetophytes, 491, 492
        of gymnosperms, 473, 489
        of *Lycopodiaceae*, **438**
        megasporangiate (ovulate)
                of balsam fir, 483

of cycads, 487, **487**
of juniper, **484**
of pines, 476–77, **477**, **478**, 480, **480**, **481**, 481, **483**
of *Wollemia nobilis*, **489**
of yews, **484**
microsporangiate
of cycads, 487
Ephedra, **491**
of pines, 476, 477, **477**, **478**, 481, **481**
of pines, 476–77, **477**, **480**, 480, 481, **481**, 482
of seedless vascular plants, **436**, **437**, 438, **438**, 442, **448**
of *Selaginella*, 442
of yews, **484**
stroma, 48
stroma thylakoids, 49, **50**, **126**, **132**
stromatolites, 289, **289**
structural formulas, 19, A-
structural genes, 216, **217**
introns and exons in, 222, **222**
structural isomers, A-
stubborn disease, 292, **292**
Sturtevant, A. H., 190
style, 499, 501, 508, **524**
Stylites, 443
suberin, 24, 25, **25**, 63, 597
lamellae, 598, **599**, 600, **600**, 654
subsidiary cells, **584**, 585, **585**
subsoil, 732, **732**
subspecies, 264
substomatal chamber, **585**
substrate-level phosphorylation, 112, **112**
substrates, 100–101
subtropical mixed forests, **798–99**, 806, **807**
succession, 786–89, **787**, **788**, 792–93
succinate (succinic acid), **115**
succulent plants, 644, 807, 809
sucrase, **100**
sucrose, **20**
ATP and formation of, 106
photosynthesis and, 142
primary and secondary active transport of, **84**
as transport form of sugar, 20
sucrose-proton cotransport (symport), 84, **84**, 768
sugarbeet, **272**, 606, **607**, 834, 835, **835**
sugarcane, **88**, 146, **632**, 747, 834, 835
sugar maple, 625, **669t**, **762**
sugar pine, **278**, **480**, 788, **789**
sugars, 18 (*see also by name*)
in fruits, 683
in leaves, 636
in nucleic acids, 30–31
from photosynthesis, 142
in seeds, 685, **686**
transport of, 750
in phloem, **764**, 764–66
sulfur (S), 18, **728**
in bacterial metabolism, 128
as essential element, 727, 729, **729**, **730t**
sulfur bacteria (*see* green bacteria; purple bacteria)
sulfur dioxide
as air pollutant, 616, **617**
sulfuric acid, as pollutant, **616–17**, 744
sulfur oxides, as air pollutants, 616, **617**
Sumiki, Y., 684
sum rule of probability, 188
sun (*see* energy)
sunbird, collared, **537**
sundew, 642, **737**
sunflower, 255, **255**, **527**, 563, **722**, 831, **831**, 833–34 (*see also Helianthus*)
family, 250, 264, 511
sun leaves, 636, **636**
sunlight (*see* light)
supernumerary cambia, 606, **607**
supporting cell, 362
suspensors
of angiosperms, 556, **557**, **558**, **559**, 560
in *Lycopodiaceae*, 442
in pines, **479**, 482, 559

of *Selaginella*, 442, 559
sustainability (sustainable agriculture), 699, 844–45
Sutton, Walter, 190, **190**
Svedberg units, 45
swamps (swamp forests), Carboniferous, **433**, 456, 457, **466**
Swedish ivy, **585**
sweet clover, **301**
sweet corn, **828**
sweet gale, 741
sweet peas, 190–91
sweet potato, 606, **607**, 831
Swiss chard, **835**
sycamores, 248, **518**, **658**
American, **669t**
symbiosis (symbiotic relationships) (*see also* endosymbiosis; mutualism; mycorrhizae; serial endosymbiotic theory)
of bacteria, 290
of chloroplasts, 304
in cyanobacteria, 290
of dinoflagellates, 365
of eukaryotes, 53
of fungi, 308, 334–44 (*see also* lichens; mycorrhizae)
mutualistic relationships, 334
parasitic relationships, 334
of green algae, **274**, 335
nitrogen-fixing, 738–41
symmetry (*see* flower, symmetry in)
sympatric speciation, 249, 252, **255** (*see also* polyploidy)
symplast, 87
symplastic domains, 88
symplastic pathway (symplastic transport), 87, 598, 600, 759, **760**
sympodia, 622, **623**
symporters, 83, **84**
synapsis, 175
synaptonemal complex, 175, 176, **176**
Synchytrium endobioticum, 313
synergids, 506, 509
syngamy, 312 (*see also* fertilization)
Synura, **378**, 379
petersenii, **60**
syphilis, 287
Syringa vulgaris, **570**, **628**
Syrphus, 309
system, 98
systematics (systematic classification), 11, 262 (*see also* taxonomy)
analogous features and, 267
homologous features and, 266
methods of, 267–68
molecular, 268–71
traditional method of, 267
systemic acquired resistance (SAR), 37
systemin, 675, **675t**
Szent-Györgyi, Albert, 2

T

2,4,5-T (2,4,5-trichlorophenoxyacetic acid), 679
taiga, **798–99**, 818, 818–19 (*see also* coniferous forests)
biome map of, **798–99**
Takakia, 407, **407**, 416
Tamarix, 590
tandemly repeated DNA, 221–22
tangential section (surface), **655**, **660**
tannins, 35, **36**, 782
tapetum, 454, 504
tapicoa, **831**
taproot, 566, 590, **591**
Taq polymerase, 226
Taraxacum, **545**
officinale (dandelion), **132**, 180, **278**, 528, **545**, **546**, **777**
targeting, polypeptide (or protein), 214, **214**, **215**
taro, 827
tarragon, 833
tarweed, 251
taxa, 264
artificial, 266

monophyletic, 265
natural, 266
paraphyletic, 266
polyphyletic, 266
Taxaceae, 483, **484**
Taxodiaceae, 483, **485**
Taxodium, 483
distichum, **485**
taxol, 34
taxonomy, 11, 262–64, 869–76 (*see also* systematics)
Taxus, **484**
canadensis, **579**
2,3,7,8-TCDD, 679
tea, 549, 833
teeth (*see* peristome)
Tegeticula yucasella, **536**
telegraph plant, **506**
telia, 327
Teliomycetes (rusts), **275t**, 309, 322, 326–28, 871 (*see also* fungi)
life cycle of, 327, **328–29**
teliospores, 327, **328**, **329**, 330
telomerase, 203
telomeres, 203
telophase, 159, **160**, **161**, 163, 165
meiotic, **174**, **175**, 177
mitotic, 163
Teloschistes chrysophthalmus, **337**
temperate deciduous forests, **798–99**, 812–14, **813**, **814**
biome map of, **798–99**
temperate mixed forests, **798–99**, 815–16, **816**
biome map of, **798–99**
temperate rainforests, 804
temperature (*see also* cold)
absolute, 876
after-ripening and, 565
breaking dormancy and, 719
chemical reactions and, 875–76
circadian rhythms and, 708
conversion scale, 880
distribution of organisms and, 800–801
Earth's average, 150–51
enzymatic reactions and, 103–4, **104**
gene expression and, 197
germination and, 565
global, 845 (*see also* global warming)
photosynthesis and, 128, 146, **147**, 149
stomatal movements and, 753, 754
transpiration and, 754
tendrils, **624**, 641, **642**
thigmotropism and, 706, **706**
tension (negative pressure), 755–56
tension wood, 666
teosinte, 829
tepals, 499
terminal buds, 657
terminal bud-scale scars, 657
termination stage of translation, **212**, **213**, 214
terminator, 210
terpenes, **779**
terpenoids, 33, 549
tertiary mycelium, 322
Tertiary period, 474, 531, inside front cover
tertiary structure of proteins, 29, **29**, 30
testcross, 187, **187**
tetanus, 287
tetrads (bivalents), 172, **173**, **175**
tetrads, of spores, **413**, **453**
tetracycline, 287
tetrahydrocannabinol (THC), **551**
Tetrahymena furgasoni, **85**
tetrasporangia, tetraspores, tetrasporophytes, 361, **362**
texture of wood, 667
Thalassiosira pseudonana, 375
Thalictrum dioicum, **762**
thalli
brown algae, 381
of bryophytes, 403
lichen, 338
of liverworts, 407–8, **408**
Thallophyta, 888
Thaxter, Roland, **336**

Thea sinensis, **549**
theca, 363
Theobroma cacao, **549**, **834**
theobromine, 549
Theophrastus, 738
thermal proteinoids, 4
thermodynamics, 875–76
first law of, 95
second law of, 95–97, **96**
thermophilic archaea, extreme, 296, **296**
thermophilic bacteria, **282**
thermophilic prokaryotes (thermophiles), 287
Thermoplasma, 296
Thermopsis montana, **636**
thigmomorphogenesis, 721–22, **723t**
thigmonastic movements (thigmonasty), 720–21, **723t**
thigmotropism, 706, **706**, **723t**
Thiothrix, **286**, 291
thistle, **527**
Thlaspi caerulescens, 747
thorns, 642, **643**
threonine, 26, **27**
Thuidium abietinum, **422**
Thuja plicata, **669t**
thylakoids, 48, 49, **126**, **132** (*see also* grana)
in cyanobacteria, 288, **288**
grana, 49
internal surface of, 135
stroma, 49, **50**
thyme, 833
thymine, 198, **198**, 199, **200**
Tidestromia oblongifolia, 810
Tilia
americana, 575, 577, 580, 614, **655**, **669t** (*see also* basswood; linden)
heterophylla, **762**
Tillandsia usneoides, **485**
Tilletia tritici, 330
tinsel flagella, 60, **60**, 371, **372t**
Ti (tumor-inducing) plasmids, 696, **696**, 697
tissue culture, 693, 694
tissues, 86 (*see also* specific types of tissues)
definition of, 573
tissue systems, 572–86
Tmesipteris, 443, 445, **463t**
lanceolata, 445
parva, **445**
toadstools, 323–24
tobacco, **7**, **33**, 48, **52**, **53**, **140**, 536, 549, 681, **695**, 715 (*see also Nicotiana*)
genetic engineering and, **698**, 699
leaf development in, **635**
Maryland Mammoth, 715, 716
ozone injury in, **617**
tree (*Nicotiana glauca*), 695
tobacco mosaic virus (TMV), 296, 298, **298**, **299**, 300–301, **301**
Tolypocladium inflatum, 308
tomatoes, **191**, 550, 675, **704**, 729
actin filaments in, **60**
breeding improvements, 841
diseases of, **285**, 293, 294, **294**
ripening of, 683
salt-resistant, 843
tonoplast (vascular membrane), 46, **53**, 54, 580–81, **583**
tooth fungi, 323, **323**, 324
topoisomerases, **203**
topsoil, 732, **732**
tori, 659–60, **662**
torpedo stage, **557**, 559, 560
tortoises, 237
Tortula obtusissima, **401**
totipotency, 220, 688, 693, **694**, 695
touch-me-not, 546
touch responses (*see* thigmonastic movements)
toxic blooms (*see* blooms)
Toxicodendron radicans, **550**
toxins (toxic compounds), 34, 334, 340, 486, 745 (*see also* specific toxins and types of toxins)
dinoflagellate, 366
lichens' sensitivity to, 340
trabeculae, **442**

trace elements (micronutrients), 727, 728t
tracers, 864
tracheary elements (water-conducting cells), 427, 428, **428**, 576, 577–78, **578**, 618, **619**
    embolized, **757**
    evolution of, 427–28
tracheids, 428, **432**, 434, 476, 576, **577**, 578, **587t**, 659, **661**
Trachelomonas, 351, **351**
tracheophytes, 416
Tragopogon (goat's beard), 253, **253**
    dubius, 253, **253**
    mirus, 253, **253**
    miscellus, 253, **253**
    porrifolius, 253, **253**
    pratensis, 253, **253**
transcellular pathway, 759, **760**
transcription, 208, 209, **209**, 210
    chromosome condensation and, 220–21
    in eukaryotes, 223, **223**
        specific binding proteins and, 221
    gene regulation in prokaryotes at the level of, 216, **216**
transcription factors, 221, 688, **688**
transduction, 87
    signal, 86–87, **87**
transfer cells, **398**, **406**, **574**
transfer RNA (*see* tRNA)
transformation, in prokaryotes, 286
transformed bacteria, in recombinant DNA, 225
transfusion tissue, **474–75**
transgenic plants, **697**
trans-Golgi network, **57**, 57–58
transition region, 637, **637**
transition vesicles, **57**, 58
translation, 208, 209, **209**, 212–14
translocation, in phloem, 764–69
translocations, chromosomal, 193
transmembrane proteins, 74, 75, **75**
    multipass, 75, **75**, 82
transmission electron microscope (TEM), 42–43, **43**
transmitting tissue, **508**
transpiration, 734, 750, **751t**, 751–64
    abscisic acid and, 684
    cohesion-tension theory and, 754–58
    environmental factors and rate of, 754
    stomatal, **752**, **752**
transpiration stream, 36
    nutrient uptake and, 761, **763**, 763–64, **764**
transport
    active, **82**, 83
        primary, 84, **84**
        secondary, 84, **84**
    apoplastic, 87
    passive, 82
    of solutes across cellular membranes, 82–84
    of sugars, 750
        in phloem, **764**, 764–66
    symplastic, 87
    vesicle-mediated, 85–86
    of water (*see* water transport)
transport proteins, 82
transposons (jumping genes), 193
transverse/cross section (surface), **655**, **660**
Trebouxia, 335, **339**
tree ferns, 453, **455**, 457
trees, 648 (*see also* angiosperms; gymnosperms; *and names of specific trees*)
    dormancy in, 719
    lycophyte, 456, **456**, 457
    mycorrhizae and, 340, **340**
    ringing (girdling), **765**
    water movement in, 754–58
T-region (T-DNA), 696
Trentepohlia, 335, 383
triacylglycerols (*see* triglycerides)
Triassic period, 483, 518, inside front cover
tricarboxylic acid (TCA) cycle (*see* Krebs cycle)
2,4,5–trichlorophenoxyacetic acid (2,4,5–T), 679
Trichoderma, 334

Trichodesmium, 290
trichogyne, 320, 360
Trichomanes, **459**
    speciosum, 462
trichomes (hairs), 34, **140**, 250, 583, **585**, 585–86, **586**, 627, **628**, 634
trichomycetes, 317
Tridachnidae, 365
Trifolium, 738, 740
    repens, 244, **534**
    subterraneum, 719
triglycerides, 22
Trillium (trilliums), **5**, 549, **762**, **813**
    erectum (wake-robin), 219, **219**
    grandiflorum, **762**, **813**
Triloboxylon ashlandicum, **471**
Trimerophyton, **433**
Trimerophytophyta (trimerophytes), 431, **433**, 443, **463t**
triose phosphate isomerase, 111, **111**
triple fusion, 509, **509**
triple response in pea seedlings, 682, **683**
triticale, **839**, 839–40
Triticosecale, 839, **839**
Triticum, 824 (*see also* wheat)
    aestivum, 146, 563, **564**, 835
    vulgare, 47, 156
tritium (³H), **594**
tRNA (transfer RNA), 208 (*see also* RNA)
    protein synthesis and, 210–11, **211**
    structure of, **211**
Tropaeolum tuberosum, **830**
trophic levels, 783–86
tropical forests, 150
tropical mixed forests, **798–99**, 806
tropical plants, cambial activity in, 650
tropical rainforests, 650, **803**, 803–5, **805**
    areas of, 804
    destruction of, 804–5
tropisms, 703–6, **723t** (*see also* nastic movements; phototropism)
truffles, 317, **317**, 342
tryptophan, 26, **27**, 218, 676
tryptophan (trp) operon, 218
Tsang Wang, 486
Tschermak, Erich von, 189
Tsuga, 483
    canadensis, **669t**
    heterophylla, **342**
TTG (*TRANSPARENT TESTA GLABRA*) gene, **586**, 604
tube cells
    of angiosperms, 504, **505**, **512**
    of pines, 477, **477**, **479**
tube nucleus, **479**, **508**
tuberculosis, 287
Tuber, **308t**, 342
    melanosporum, **317**
tubers, 180, 642–43
D-tubocurarine chloride, 846
tubulin, 58, **59**
Tulare apple mosaic virus, **299**
Tulbaghia violacea (family *Liliaceae*), 196
tulips, Rembrandt, **302**
tulip tree, 519, **662**, **669t**
tumbleweeds, 546, **546**
tumor-inducing (Ti) plasmids, 696, **696**, 697
tumors, wound, 301, **301**
tuna, 371
tundra, **10**
    alpine, **798–99**, 817
    Arctic, **798–99**, 819, 819–21
tunica-corpus type of organization, 612, **612**, 613
turgor pressure, 80–81, 425
    cell expansion and, 689, **689**
    leaf dimorphism in aquatic plants and, 626
    leaf movements and, 720, **720**, 721, **721**
    in sieve tubes, 767
    stomatal movements and, 692, 752
twigs, **657**
tyloses, 636, 665, **666**
type specimen, 263, **263**
typhoid fever, 287
tyrosine, **27**

U

ubiquinol, ubiquinone (*see* coenzyme Q)
UDP-glucose, 67
ulcers, stomach, 287
ullucu, **830**
Ullucus tuberosus, **830**
Ulmus, **545**, 623, **669t**
    americana, **657**, **663**
Ulothrix, 392 (*see also* Ulvophyceae)
ultraviolet light (ultraviolet radiation), 6, 12, 129, 192, 203
    bees' perception of, 532, **533**, 543
    carotenoids and, 543
    flavonoids and, 35, 542
Ulva (sea lettuce), 393, **394**, 394
    life cycle of, **394**
Ulvophyceae (ulvophytes), **387**, 393, 393–96, **395**, 886 (*see also* algae; green algae)
Umbelliferae, 264
Umbellularia californica, 833
umbels, **499**
unilocular sporangia, **382**, 383
uniporters, 83, **84**
unit membrane model, 74
University of Wisconsin Arboretum, 792–93
unsaturated fats, 24
uracil, 208, **208**
urea, 738
uredinia, urediniospores, 327, **328**
ureides, 740
Urey, Harold, 3
Uroglena americana, 379
Ustilago
    avenae, 330
    maydis, 330, **330**
Ustomycetes (smuts), **275t**, 309, 322, 330, 883 (*see also* fungi)
uterine contractions, 335
Utricularia vulgaris, **737**

V

Vaccinium vacillans, **762**
vacuolar membrane (*see* tonoplast)
vacuoles, **45**, 46, **54**, 54–55, **71t**
    contractile, 80, 351
    osmosis and, 80
    perialgal, 274
    turgor pressure and, 81
valine, 26, **27**, **28**
Vallisneria, 541
Valonia (see *Ventricaria*)
V/A mycorrhizae, 341
van Helmont, Jan Baptista, 127
Vanilla, 529, **529**, 833
    planifolia, **529**, 833
van Niel, C. B., 127, **128**
van Overbeek, Johannes, 679
variability, genetic (*see* genetic variability)
variations, in Darwin's theory of evolution, 238
varieties, 264
vascular bundles, **132**, 142, 144, **145**, 576
    closed, 619–20, **620**, 621
    of leaves, 631–32
    open, 621
    of pine leaves (needles), 474, 475
    of stems, 614, 615, **615**, 619, **620**, 621, **621**, 622, **622**, **623**, 637
vascular cambium, 8, 427, 576, **576**, 648–50, **649**, **650**
    bifacial, 470
    of *Botrychium*, 453, **454**
    dormancy in winter, 649
    growth rings and, 664
    of *Isoetes*, 443
    of pines and other conifers, 476
    reactivation in spring, 649, 650
    relationship to its derivative tissues, 650
    of roots, 600, 601, **601**, 602, 606, **607**
    of stems, 618–19, 648–50, **651**, **652**
    use of term, 648, 649
vascular cylinder, 603 (*see also* steles)
    of leaves, 622, **623**
    of roots, 573, 592, **593**, 596, 597, 600, **602**, **603**, 604, **604**

of stems, 614, **615**
vascular plants, 8, **8**, 9, 424–553
    bryophytes compared to, 401–4, **402**, 425
    evolution of, 425–31, **432**
        reproductive systems, 430–31
        roots and leaves, 429–30
        tracheary elements, 427–28
        vascular tissues, 428–29
    life cycle of, **430**
    organization of the body of, 425–30
    reproduction in, 430–31
    seedless (*see* seedless vascular plants)
vascular rays, 648, **650**
vascular system, 7–8, 426, **427**, 572
vascular tissues, 428–29, 572–73, **573**, 576–83 (*see also* phloem; xylem)
    auxin and, 676
    of roots, 595
    of stems and leaves, 622–24
    transition from roots to shoots, 636–39, **637**
Vasil, V., 695
vectors, insect, 298, 510, 530–36
vegetables, 834–35
vegetation types (*see* biomes)
vegetative reproduction (*see* asexual reproduction)
veins (venation), 573, 631, 632, **633**, 634, 635, **635**, **636**
    major, 631
    minor, 631
    netted, 631, **631**
    parallel (striate), **625**, 631
velamen, 605, **606**
velvet grass, **762**
venation (*see* veins)
venter (*see also* calyptra), 404, **405**, 406
Ventricaria (also known as *Valonia*), 394
Venus flytrap, 642, 721, **721**, 737
Verbascum thapsus, **585**
vernalization, 719
vernation, circinate, 458
Veronicastrum virginicum, 624, **625**
Verrucaria serpuloides, 336
vesicle-mediated transport, 85–86
vesicles
    of algae, 363, **363**
    coated, 58, **58**, 86, **86**
    mycorrhizal, 341
    transition, **57**, 58
vesicular-arbuscular (V/A) mycorrhizae, 341
vessel elements, 428, 522, 576–77, **577**, 578, **587t**, 662
vessels, 576
Vibrio harveyi, 698
viceroy butterfly, 550
Vicia faba (broad bean), **8**, 738, **766**
vinblastine, **846**
vincristine, **846**
vines, 648
Viola, 549
    quercetorum, **263**
    rostrata, **263**
    tricolor var. *hortensis*, **263**
violets, **179**, 262, **263**, 549
Virchow, Rudolf, 41
Virginia creeper, 641
virions, 297, **297**
viroids, 302–3
virology, 12
viruses, 11, 282, 296–302
    bacterial (*see* bacteriophages)
    capsids of, 298, 299
    diseases caused by, 301–2
    DNA of, 297, 300
    evolutionary origin of, 303
    genetic machinery of the host cell and, 300
    genomes, 297
    movement of, within the plant, 300, **300**
    plant, 297, 300, 301–2
    proteins in, 297, 298
    replication of, 298, 300
    RNA of, 297, 298, 300
    shapes of, 298
    structure of, 297–99

systemic, 300
transmission of, 298, 300
wound tumor, **301**
Vitaceae, **498**
vitamin A (retinol), **134**
vitamins, 102
Vitis, 641
vinifera, 687, **687**, 825
Vittaria, **459**
viviparous mutants, 684, **684**
volcanic eruptions, 789, 793
volva, 324
Volvox, **276**, 389, 390, **390**
carteri, 390, **390**
von Sachs, Julius, 693
voodoo lily, 37
Vorticella, endosymbiosis in, 274, **274**
Voyage of the Beagle, The (Darwin), 236

W

wake-robin (*Trillium erectum*), 219, **219**
Wald, George, 130
wall pressure, 81
walnuts, 563
black, **669t**, 779
wandering Jew, **51**
Wareing, Paul F., 683
wasps, 535
water, 18
absorption by roots, 759–62
hydraulic lift, 761, **762**
passive, 761–62
pathways, 759–60, **760**
positive pressure (root pressure), **760**, 760–61, **761**
dispersal of fruits and seeds by, 546–47
in early Earth, 3
evolutionary transition of plants to land and, 7, 8
germination and, 564–65
ionized, 874, **874**
photolysis of, 137
in photosynthesis, 5–7, 127, 128, **132**, 134, **134**, 135, **136**, 137–39
pollination via, 541–42
in soils, 734, 735, **735**
as a solvent, 873, **873**, 874
structure and properties of, 872–75
transport (movement) of (*see* water transport)
water buttercup (*Ranunculus peltatus*), 197, **197**
water-conducting tissues and cells, 36, 576 (*see also* hydroids; tracheary elements; tracheids; vessel elements; xylem)
of bryophytes, 402
evolution of, 425, 428
water cycle, 734, **734**
water ferns, **462**, 462–63
water-form leaves, 626, **626**
water hemlock, **500**
water hyacinths, **787**
water lily, **575**, **629**, **787**
family, 521
water loss
cuticle as barrier to, 752
prevention of, 24
in bryophytes, 403, **403**
stomatal closing and, 753
by transpiration, **751t**, 751–52
watermelon, 294
water molds, **171**, 275, 373, 884 (*see also* oomycetes)
life cycle of, **374**
waternet, **391**
water potential, **76**, 76–77
osmosis and, 79, **79**, 80
of soils, 735, **735**

water-soluble fibers, **563**
water storage, stems and leaves specialized for, 644
water table, 734
water transport (movement), 751–64 (*see also* transpiration; water loss)
across cellular membranes, 76–79
bulk flow, 77
diffusion, **77**, 77–79, **79**
imbibition, 80
osmosis, 80–81
water potential gradient, **76**, 76–78
cohesion-tension theory of, 754–58, **756**, **757**
lignin and, 36, 37
in osmosis, **77**, 77–81, **79**
in roots, 759–62
stomatal movement and, 750
though xylem, 754, **755**
water vesicles, 644
Watson-Crick model, **199**
Watson, James, 197–99, **199**
wavelengths, 129
wave model of light, 129, 130
waxes, 24–25, **25**, 63
cuticular, 25
epicuticular, 25, 584, **584**
jojoba, **841**, 842
weathering of rocks, 786, **787**, 788
inorganic nutrients derived from, 731
weeds, control of, 679 (*see also* herbicides)
weevils, 487
Weinberg, G., 239, 240
Welwitschia, 149, 490, 492, 518, **519**
mirabilis, 149, **492**
Went, Frits W., 676, **702**, 703
wheat, 327, 825, 835 (see also *Triticum*)
awns of, 196
bran and germ, 563
bread, 47, 563, **835**
collenchyma cells from filament of, **574**
endosperm and embryo, 563, **564**
genetic control of color in, **195t**
hybridization of, 254
kernel of, **564**
leaves of, 633
new strains of, **839**
stinking smut of, 330
wheat germ, 563
wheat rust (*see* black stem rust of wheat)
life cycle of, 328–29
whiplash flagella, 60, **60**
whisk fern, 443
white ash, **669t**
whitebark pine, 482
white fir, 788, **789**
whiteflies, 298
white flowering spurge, 793
white oak, **657**, **666**, 667, **669t**
white pine, 340, **480**, 660, **660–662**, 668, **669t**
white spruce, 788, **818**
whooping cough, 287
whorled phyllotaxy, 624
whorls, 501, 502, 527
Wielandiella, **473**
Wilde, S. A., 774
Wilkesia gymnoxiphium, **251**
Wilkins, Maurice, 198, 199
Williamsoniella coronata, **473**
willows, **36**, 37, 342, 546, **602**, **603**
wilting, 81
wilts (disease), 293–94
wind pollination, 487, 530, 539–40, **540**, **541**
winds, prevailing, **797**
wine-making, **123**
wintercress, **728**
wintergreen, **524**

wire grass, **591**
Wirth, Tim, 845
witch hazel, 546
Witch's hair, **337**
Wolffia, **496**
borealis, **496**
Wollemia nobilis, 488–89, **489**
wood, 21, **647**, 649, 652, 655–56, 659–70 (*see also* secondary xylem)
of angiosperms, **662**, 663
commercial uses of, **669t**
compression, **666**
of conifers, **476**, 659–61
density of, 668
diffuse-porous, **662**, 665
early, late, 665
features and identification of, 667–70
grain of, 667
growth ring (layers; increments), 662, 664–66
knots, 668
methods of sawing, 667
reaction, **666**
ring-porous, 665
of roots, 600–2
specific gravity of, 668
strength of, 668
tension, 666
wood sorrel, **707**
woody magnoliids, 519, **520**, 521, 525
woody plants, 648 (*see also* trees)
woody stems, 618, 650, 653
external features of, **657**
wound callose, 580
wound healing and repair, 574
wound lignin, 37
wound tumor virus, 301, **301**

X

Xanthidium armatum, **397**
Xanthium, **548**
strumarium, 681, 709, **710**
Xanthomonas, 293, **294**
campestris, 294
xanthophylls, 133, **134**
Xanthosoma, 827
X-disease of peach, 293
xerophytes, leaves of, 626, 627, **629**
xylans, 63
xylem, 8, **8**, 426, **427**, 572, 576–79, **587t**
cell types in, **579t**
of early vascular plants, of gymnosperms, 476
primary, **571**, **573**, 577, 578, **578**
of roots, 600, 601, **601**, 602
of seedless vascular plants, **435**, **442**, **444**, 445, **448**, **455**
of stems, 618, **619**, 620
vascular cylinder and, 600
secondary (*see* secondary xylem *and* wood)
tracheary elements of, **427**, 428, **428**, 576, 577–78, **578**
embolized, **757**
of stems, 618, **619**
transfer cells and, 574
of veins, 631
water tranport through, 754, **755**
air bubbles and, **756**, 756–58, **757**
xyloglucans, **22**, 63
xylose, **22**

Y

Yabuta, T., 684
yams, 834
wild, 550, 846, **846**
yeasts, 317, 330–32, 883
classification of, 330

commercial uses of, 331, **331**
yellow pine, **480**
yellow poplar (tulip tree), 519, **662**, **669t**
yellow prairie-cornflower, 793
Yellowstone National Park
1988 fire in, **790**, 790–91, **791**
yews, 579
arils of, 548
European, 34
family, 483, **484**
Pacific, 34
yogurt, 287
yucca, 831
Yucca bevifolia, **808**
yucca moth, 536

Z

Zamia, 487
pumila, 486, **487**, 488
Zea
diploperennis (perennial teosinte), 829
mays ssp. *mays* (maize), **829**, 829
mays ssp. *parviglumis* (annual teosinte), 829, **829**
Zea mays (maize), **567**
grain, **562**, 568
leaf, **145**, 584, 625, 630
root, 593, **594**, 603
stem, 581, 597, 621, 622–23
zeatin, 680, **680**
zeaxanthin, 134
zebra, 10
Zigadenus fremontii, **535**
zinc (Z), as essential element, 727, **729**, **730t**
Zingiber officinale, 832
Zinnia elegans, 676
zooplankton, 348, **785**, 786
zoosporangia, 374
zoospores
of algae, 383, **384**, 389, 391
of chytrids, 314
of oomycetes, 371, **373**, 374
Zooxanthellae, 365, **365**
Zosterophyllophyta (zosterophyllophytes), 431, **432**, 433, 434–35, **463t**
Zosterophyllum, **426**, 433
Zygocactus truncatus, 299
zygomycetes (*Zygomycota*), 275, **308t**, 309, **309**, 312, 313, 315–17
life cycle of, 316
Zygomycota, 313, 315–17, 882
zygosporangia, 316, **316**, 317
zygospores
of algae, 389, **389**
of fungi, 316, 316, 317
zygotes, 170, **170**, **171** (*see also* auxospores; macrocysts; oospores; zygospores)
of angiosperms, 509, 510, **513**
of brown algae, **384**, 385
of bryophytes, 402, 404, 406, 407, **410**, 418, **419**
of cellular slime molds, 356
of diatoms, 376, **376**
of dinoflagellates, 365
embryogenesis and, 556, **557**, 558
of fungi, 312, **314**
of green algae, **386**, 389, **389**, **390**, 391, **392**, 394, 396, **396**, 397, **397**, 398, 402
of gymnosperms, **479**
of oomycetes, 372, **374**
of plasmodial slime molds, 353
of red algae, 360, 361, **362**
of seedless vascular plants, **437**, 438, **441**, **447**, 450, 461
of yeasts, 330
zygotic meiosis, 171, **171**, 379, 389, 391, 392, 397